Fundamental Papers in Wavelet Theory

S0-BOK-533

Fundamental Papers in Wavelet Theory

Edited by

Christopher Heil

and

David F. Walnut

PRINCETON UNIVERSITY PRESS
PRINCETON AND OXFORD

Copyright © 2006 by Princeton University Press

Published by Princeton University Press, 41 William Street,
Princeton, New Jersey 08540
In the United Kingdom: Princeton University Press, 3 Market Place,
Woodstock, Oxfordshire OX20 1SY

All Rights Reserved

Library of Congress: 2005934805

ISBN-13: 978-0-691-11453-8
ISBN-10: 0-691-11453-6

British Library Cataloging-in-Publication Data is available

Printed on acid-free paper.

pup.princeton.edu

Printed in the United States of America

10 9 8 7 6 5 4 3 2 1

Contents

Section II. Precursors in Physics: Affine Coherent States

Section III. Precursors in Mathematics: Early Wavelet Bases

Section IV. Precursors and Development in Mathematics: Atom and Frame Decompositions

Section VII. Selected Applications

Contributor Affiliations

Edward H. Adelson
Department of Brain and Cognitive Sciences
Massachusetts Institute of Technology
Cambridge, MA 02139
adelson@ai.mit.edu

Jean-Pierre Antoine
Institut de Physique Théorique
Université Catholique de Louvain
2, chemin du Cyclotron
B-1348 - Louvain-la-Neuve
Belgium
Antoine@fyma.ucl.ac.be

Erik W. Aslaksen
Senior Principal, Sinclair Knight Merz
100 Christie Street, St. Leonards 2065
Australia
caslaksen@skm.com.au

Thomas P. Barnwell III
School of Electrical and Computer Engineering
Georgia Institute of Technology
Atlanta, GA 30332
tom.barnwell@ee.gatech.edu

Guy Battle
Department of Mathematics
Texas A&M University
College Station, TX 77843
battle@math.tamu.edu

John J. Benedetto
Norbert Wiener Center for Harmonic Analysis and Applications
Department of Mathematics
University of Maryland
College Park, MD 20742
jjb@math.umd.edu

Gregory Beylkin
Department of Applied Mathematics
University of Colorado
Boulder, CO 80309
beylkin@boulder.colorado.edu

Peter J. Burt
Sarnoff Corporation
Princeton, NJ 08543
pburt@sarnoff.com

Albert Cohen
Laboratoire Jacques-Louis Lions
Université Pierre et Marie Curie
4, Place Jussieu
75005 Paris
France
cohen@ann.jussieu.fr

Ronald R. Coifman
Department of Mathematics
Yale University
New Haven, CT 06520
coifman@math.yale.edu

Ronald E. Crochiere
Systems Development Division
SAIC
10260 Campus Point Drive
San Diego, CA 92121
crochiere@trg.saic.com

Ingrid Daubechies
Department of Mathematics
Princeton University
Princeton, NJ 08544
icd@princeton.edu

Ronald A. DeVore
Department of Mathematics
University of South Carolina
Columbia, SC 29208
devore@math.sc.edu

David L. Donoho
Department of Statistics
Stanford University
Stanford, CA 94305
donoho@stat.stanford.edu

Richard J. Duffin
Deceased

Daniel Esteban
Unknown

Hans G. Feichtinger
Department of Mathematics
University of Vienna
Nordbergstrasse 15
A-1090 Vienna,
Austria
hans.feichtinger@univie.ac.at

James L. Flanagan
Vice President for Research
Rutgers University
Piscataway, NJ 08855
jlf@caip.rutgers.edu

Philip Franklin
Deceased

Michael Frazier
Department of Mathematics
Michigan State University
East Lansing, MI 48824
frazier@math.msu.edu

Claude R. Galand
Unknown

Karlheinz Gröchenig
Faculty of Mathematics
University of Vienna
Nordbergstrasse 15
A–1090 Wien
Austria
karlheinz.groechenig@univie.ac.at

Alex Grossmann
Laboratoire Génome
et Informatique
523 Place des Terraces
91034 Evry Cedex
France
grossman@genopole.cnrs.fr

Björn Jawerth
5 Examples, Inc.
1228 Langstonshire Lane
Morrisville, NC 27560
bjawerth@nc.rr.com

Alfred Haar
Deceased

John Horváth
Department of Mathematics
University of Maryland
College Park, MD 20742
jhorvath@wam.umd.edu

CONTRIBUTOR AFFILIATIONS

Stéphane Jaffard
Department of Mathematics
Université Paris 12
61 Avenue du General de Gaulle
94010 Creteil Cedex
France
jaffard@univ-paris12.fr

Iain M. Johnstone
Department of Statistics
Stanford University
Stanford, CA 94305
imj@stanford.edu

John R. Klauder
Departments of Physics and Mathematics
University of Florida
Gainesville, FL 32611
klauder@phys.ufl.edu

Jelena Kovačević
Departments of Biomedical Engineering and Electrical and Computer Engineering
Carnegie Mellon University
Pittsburgh, PA 15213
jelenak@cmu.edu

Wayne Lawton
Department of Mathematics
National University of Singapore
2 Science Drive 2
Singapore 117543
matwml@nus.edu.sg

Pierre-Gilles Lemarié-Rieusset
Université d'Evry Val d'Essonne
Département de Mathématiques
Boulevard F. Mitterrand
F-91025 Evry Cedex
France
pierre-gilles.lemarie@maths.univ-evry.fr

Wolodymyr Madych
Department of Mathematics
University of Connecticut
Storrs, CT 06269
madych@math.uconn.edu

Stéphane Mallat
Centre de Mathématiques Appliquées (CMAP)
Ecole Polytechnique
91128 Palaiseau Cedex
France
stephane.mallat@polytechnique.fr

Yves Meyer
Centre de Mathématiques et de Leurs Applications (CMLA)
Ecole Normale Supérieure de Cachan
94235 Cachan Cedex
France
yves.meyer@cmla.ens-cachan.fr

Fred Mintzer
IBM T. J. Watson Research Center
P.O. Box 704
Yorktown Heights, NY 10598
mintzer@us.ibm.com

Jean Morlet
Society of Exploration Geophysicists

Thierry Paul
C.N.R.S., D.M.A. École Normale Supérieure
45 rue d'Ulm
75230 Paris Cedex 05
France
paul@dma.ens.fr

Vasil A. Popov
Deceased

Vladimir Rokhlin
Departments of Computer Science and Mathematics
Yale University
New Haven, CT 06520
rokhlin-vladimir@yale.edu

Robert D. Ryan
Paris, France
robertdryan@compuserve.com

A. C. Schaeffer
Deceased

Jerome M. Shapiro
The MITRE Corporation
202 Burlington Road M/S E050
Bedford MA 01730-1420
jshapiro@mitre.org

Mark J. T. Smith
Department of Electrical and Computer Engineering
Purdue University
465 Northwestern Avenue
West Lafayette, IN 47907
mjts@purdue.edu

Jan-Olov Strömberg
Department of Mathematics
Royal Institute of Technology
Lindstedtsvägen 25
SE-10044 Stockholm
Sweden
jostromb@math.kth.se

P. P. Vaidyanathan
Department of Electrical Engineering, 136-93
California Institute of Technology
Pasadena, CA 91125
ppvnath@systems.caltech.edu

Martin Vetterli
School of Computer and Communication Sciences
Swiss Federal Institute of Technology (EPFL)
CH-1015 Lausanne
Switzerland
and
Department of Electrical Engineering and Computer Sciences
University of California Berkeley
CA 94720
martin.vetterli@epfl.ch

Susan A. Webber Christensen
Highland Mills, New York

Guido Weiss
Department of Mathematics
Washington University in St. Louis
St. Louis, MO 63130
guido@math.wustl.edu

Mladen Victor Wickerhauser
Department of Mathematics
Washington University in St. Louis
St. Louis, MO 63130
victor@math.wustl.edu

Georg Zimmermann
Institut für Angewandte Mathematik
Universität Hohenheim
D-70593 Stuttgart
Germany
Georg.Zimmermann@uni-hohenheim.de

Preface

This volume traces the development of modern wavelet theory by collecting into one place many of the fundamental original papers in signal processing, physics, and mathematics that stimulated the rise of wavelet theory, along with many major papers in the early development of the subject. The volume is a sourcebook for some of the most significant research papers in the subject and provides a way for a researcher who understands wavelets in his or her own milieu to get a glimpse of the development of the subject from another perspective.

A key feature of this volume is the appearance for the first time of a translation from the German of the original 1910 paper of Alfred Haar in which he described the orthonormal basis that now bears his name, along with translations of six papers heretofore available only in French. We are greatly indebted to John Horváth, Robert Ryan, and Georg Zimmermann for these translations.

The articles in the volume are divided by subject matter into seven sections, and each section is introduced by a prominent researcher in that field. These introductory essays provide a unifying perspective on the papers in each section. We are greatly indebted to Jelena Kovačević, Jean-Pierre Antoine, Hans Feichtinger, Yves Meyer, Guido Weiss, and Victor Wickerhauser for their contributions. We are also indebted to John Benedetto for his essay introducing the entire volume, which provides a personal account of the intellectual precursors to and the historical development of wavelet theory.

The most difficult part of editing this volume was choosing which papers to include. The process of making these choices served to emphasize for us the amazing breadth and impact of wavelets in many disciplines. We tried to choose papers that represented a significant leap forward or a breakthrough that led to significant development in the theory, and did not necessarily choose papers in which a certain concept first appeared. Limitations of space prevented us from including papers on, for example, general time-frequency analysis (including Gabor analysis, which does make a cameo appearance in the paper of Feichtinger and Gröchenig), subdivision schemes, approximation theory, splines, numerical analysis, or differential equations. We are pleased that the introductory essays provide a larger context and describe many of the important developments in wavelet theory.

Two people have been essential in bringing this volume into existence, and we give them our warm thanks. Ingrid Daubechies initially suggested that we create this volume and encouraged our efforts throughout. The unflagging enthusiasm and assistance of our editor at Princeton University Press, Vickie Kearn, was invaluable in bringing this project to completion. The support of National Science Foundation grants DMS-9970524, DMS-9971697, and DMS-0139261 is also gratefully acknowledged.

The development of the modern theory of wavelets provides a powerful example of the way in which the mutual support of mathematics and applications can spur development in many diverse subject areas. We hope that this volume will continue to encourage and enable such interplay as the third decade of the modern wavelet era commences.

Christopher Heil, Atlanta, Georgia
David Walnut, Fairfax, Virginia
July 2004

Acknowledgments

Princeton University Press gratefully acknowledges the support of the following publishers for giving permission to reprint the papers appearing in this volume: IEEE, Bell Laboratories, American Institute of Physics, SIAM, Chapman & Hall, SMF, Universidad Autónoma de Madrid, Springer-Verlag, American Mathematical Society, Indiana University Press, Elsevier Inc., University of Chicago Press, L'Institut Henri Poincaré, John Wiley & Sons, Politecnico di Torino, Johns Hopkins University Press, and American Statistical Association.

Foreword

This volume collects early wavelet papers and precursors. It is a delight to find these diverse papers in one volume, instead of having to comb libraries of the different disciplines to find them. I am sure that many researchers working on, using, or just generally interested in wavelets will welcome this collection.

In the late 1970s and a good part of the 1980s, "wavelet theory" (as I will call it, for want of a better name) emerged as the synthesis of ideas from many sources. In this book, one can trace how mathematics (Littlewood-Paley theory and its developments in harmonic analysis), physics (coherent states in quantum mechanics), electrical engineering (subband coding in signal analysis), and computer science (multiscale descriptions in vision theory) all contributed. Researchers were excited to realize, as they learned from each other, that strands of thought from entirely different disciplines turned out to be quite similar, despite the difference in context and language. Despite these similarities, the interdisciplinary synthesis contributed an essential ingredient: I remain convinced that the full wavelet theory could not have come forth from any single one of these disciplines, without input from the others. For instance, although harmonic analysts had understood most of the fundamental properties of wavelet expansions that we know today, and although electrical engineers had developed the filter banks that are now used to carry out numerically a decomposition into wavelets, only the combination of these two ideas could yield wavelet theory as a tool that is *both* computationally fast and extremely versatile, inheriting its speed from its subband coding parent and its versatility in applications from the powerful mathematical properties of its analysis parent.

Because of the diversity of the roots as well as the applications of wavelets, a course on wavelet theory can be organized in a variety of ways. One can restrict oneself to just one discipline in selecting the material to be covered, or one can boldly decide to teach pieces often viewed as pertaining to different subjects. Even in this latter case, a slant toward the instructor's own specialty and background is probably unavoidable. Whatever the organizing principle for the course, the present collection of papers is bound to be a valuable resource, illustrating the historical depth as well as the disciplinary breadth of wavelet theory. Some of these papers are hard to find elsewhere, even though they were extraordinarily influential (the most extreme case being the Zygmund lecture notes of Yves Meyer, which first circulated as photocopies of handwritten notes); others, like Alfred Haar's paper, are here translated into English for the first time. Most important, students from every one of the different disciplines will find here papers that their own department would typically not stock in its library. I therefore expect that this book will turn up on the reading list of many wavelet courses.

It is always difficult to select the papers to include for publication in a reprint collection; wrenching decisions have to be made, with the result that beautiful papers, definitely worth including, nevertheless are taken off the list because of space constraints. This is the case even more so for the present collection, which seeks to bring together papers from so many different backgrounds. As they planned this book, the editors polled many of the "early" wavelet researchers, asking them what *they* would include if editing a collection like this one. I am sure that most people polled will find that one or more of their suggestions did not make the final list — otherwise this book would have been at least twice as fat.

In particular, there are no papers here about the connection between wavelets and multiresolution analysis, on the one hand, and subdivision algorithms and refinable functions as used in computer graphics, on the other hand. One could easily imagine yet another chapter, for which possible contributions could include early subdivision schemes by Beziér and de Casteljau, papers proving convergence of subdivision schemes, globally by Dyn, Gregory, and Levin, and Dahmen and Micchelli, locally by Daubechies and Lagarias, the comprehensive treatment of shift-invariant spaces by de Boor, DeVore, and Ron, the construction by Schröder and Sweldens of wavelets associated with subdivision schemes for surfaces, together with the lifting scheme that made this construction feasible, as well as some of the first application papers of these wavelets to computer graphics. Such a chapter would interweave nicely with some of the work that is included (such as the Burt and Adelson paper), and introduce yet other important applications of wavelets, as well as the factorization into lifting steps that is crucially important for certain fast implementations of the wavelet transform. It would also merrily add another 150 or 200 pages.

In a different world, where never is heard a discouraging word, the skies are not cloudy all day, and mundane space restrictions don't play a role in designing a book like this, my "missing chapter" on the link with computer graphics would be included, as well as other extra material that (other) nit-picking readers may wish for. As it stands, the book is already substantially heftier than the publisher expected at first. Wavelets have, in twenty years or so, led to such a large volume of papers that even a *list* of all of them requires a book by itself; this observation makes clear what a formidable task was taken on by the editors of this volume. In the inevitable process of deciding which topics or papers made the very final cut, two guiding principles were used: chronology (topics that played an earlier role in the development of wavelet theory were more likely to be included) and accessibility (fields on which review volumes had already appeared, or where papers were already more widely distributed, were less likcly).

By collecting the papers published together in this volume, the editors and the publisher have put together a wonderful gift for the whole community. Whether you are a wavelet expert, a student starting to study wavelet theory, or just curious about the way the wavelet synthesis came about, I am sure you will find this collection most interesting and useful. I wish you joy!

Ingrid Daubechies
Princeton, August 2004

Fundamental Papers in Wavelet Theory

Introduction

John J. Benedetto

1. Background

If ever there was a collection of articles that needed no introduction, this is it. Undaunted, I shall fulfill my charge as introducer by describing some of the intellectual background of wavelet theory and relating this background to the articles in this volume and to their expert introductions by Jelena Kovačević, Jean-Pierre Antoine, Hans Feichtinger, Yves Meyer, Guido Weiss, and Victor Wickerhauser.

I was not a contributor to wavelet theory, but was close enough in the mid-1980s to hear the commotion. I was in the enviable position of having talented graduate students (including the editors of this volume), and so I felt obliged to make a serious judgement about whether to pursue wavelet theory. Timewise, this was before the wavelet stampede and sporadic tarantisms of hype and eventual catharsis. I recall saying to Heil and Walnut late in 1986 that I hoped I was not leading them down a primrose path by studying Meyer's Séminaire Bourbaki article from 1985/1986 (which is translated into English in this volume) with them. I assured them that Meyer was brilliant and deep, and hoped that he was on target since the material seemed so compelling.

Early in 1987 we also read the articles of Grossmann and Morlet and Grossmann, Morlet, and Paul (both reprinted in this volume). In April 1987 I attended the Zygmund lectures by Meyer (this volume) with Ray Johnson. Zygmund appeared and Meyer was dazzling. He gave an alternative proof of (the recently proved and not yet published) Daubechies' theorem constructing smooth compactly supported wavelets; see the article by Daubechies reprinted in this volume. It was during the Zygmund lectures that Salem prize winner Dahlberg reminded me from high in the Hancock Center that the view was better than that in College Park.

My wavelet excitement in the spring of 1987 was juxtaposed with my existing research interests, some of which seemed too inbred, a veritable "Glass Bead Game". These interests included several significant problems which seemed out of reach, for instance, fathoming the arithmetic structure of some of the spectral synthesis problems originally formulated by Wiener and Beurling. On the other hand, and perhaps naively, many of us believed in the regenerative and centralizing power of harmonic analysis, and, then, voila! — les ondelettes arrived.

I became hooked during those exhilarating nascent days of wavelets and made an effort to study wavelet theory: where it came from and where it was going. To the extent that

The author gratefully acknowledges support from NSF DMS Grant 0139759 and ONR Grant N000140210398.

I understand it, this is a beautiful and intricate story relating the signal (sic) contributions of this volume with significant paths through twentieth-century engineering, mathematics, and science. We shall tell parts of the story, in a mix of vignettes and perspectives, as appetizers for the seven course feast that follows. To fix ideas we describe Figure 1 in Section 2. The figure itself is meant as a mise-en-scène to get started. The technical definitions in Section 2 associated with the figure can be omitted by the reader without substantial damage to our storyline. They can also be examined more carefully while reading the descriptive parts of the Introduction related to them.

2. Definitions

The problem of *signal representation* at the top of Figure 1 is to provide effective decompositions of given signals f in terms of harmonics. The terms "effective" and "harmonics" are problem-specific notions.

To define the next level of Figure 1, recall that Fourier series $S(f)$ of 1-periodic functions f have the form

$$S(f)(x) = \sum_{n\in\mathbb{Z}} c_n(f)\, e_n(-x), \tag{2.1}$$

where $e_n(x) = e^{2\pi i x n}$. Equation (2.1) is an example of a *discrete representation*, with harmonics $\{e_n\}_{n\in\mathbb{Z}}$, since we are representing the signal f as a sum (over the integers $\mathbb{Z}$). In a *continuous integral representation* the right side of (2.1) is replaced by an integral such as the Calderón reproducing formula (4.2) below.

These continuous integral representations frequently depend on an underlying locally compact group. The *affine group* in Figure 1 is the underlying group associated with the Calderón reproducing formula or continuous wavelet transform. Section II of the volume is devoted to this topic, and the articles by Grossmann, Morlet, and Paul and Feichtinger and Gröchenig (both reprinted in this volume) are particularly important, cf., our preliminary remarks in Section 4.2 of this introduction.

Formally, the Fourier transform $\hat{f}\colon \mathbb{R}^d \to \mathbb{C}$ of $f\colon \mathbb{R}^d \to \mathbb{C}$ is defined as $\hat{f}(\gamma) = \int_{\mathbb{R}^d} f(x)\, e^{-2\pi i x\cdot\gamma}\, dx$. τ_y denotes translation defined by $(\tau_y f)(x) = f(x-y)$. e_y denotes modulation defined by $(e_y f)(x) = e^{2\pi i x\cdot y} f(x)$. The next part of Figure 1 is described in the following definitions.

Definition 2.1 (Gabor and wavelet systems)

1. Let $\psi \in L^2(\mathbb{R})$. The associated *wavelet* or *affine system* is the sequence $\{\psi_{m,n} : (m,n) \in \mathbb{Z}\times\mathbb{Z}\}$, where $\psi_{m,n}$ is defined by

$$\psi_{m,n}(t) = 2^{m/2}\psi\left(2^m t - n\right). \tag{2.2}$$

   Clearly,

$$\widehat{\psi_{m,n}}(\gamma) = 2^{-m/2}\,(e_{-n}\widehat{\psi})(\gamma/2^m).$$

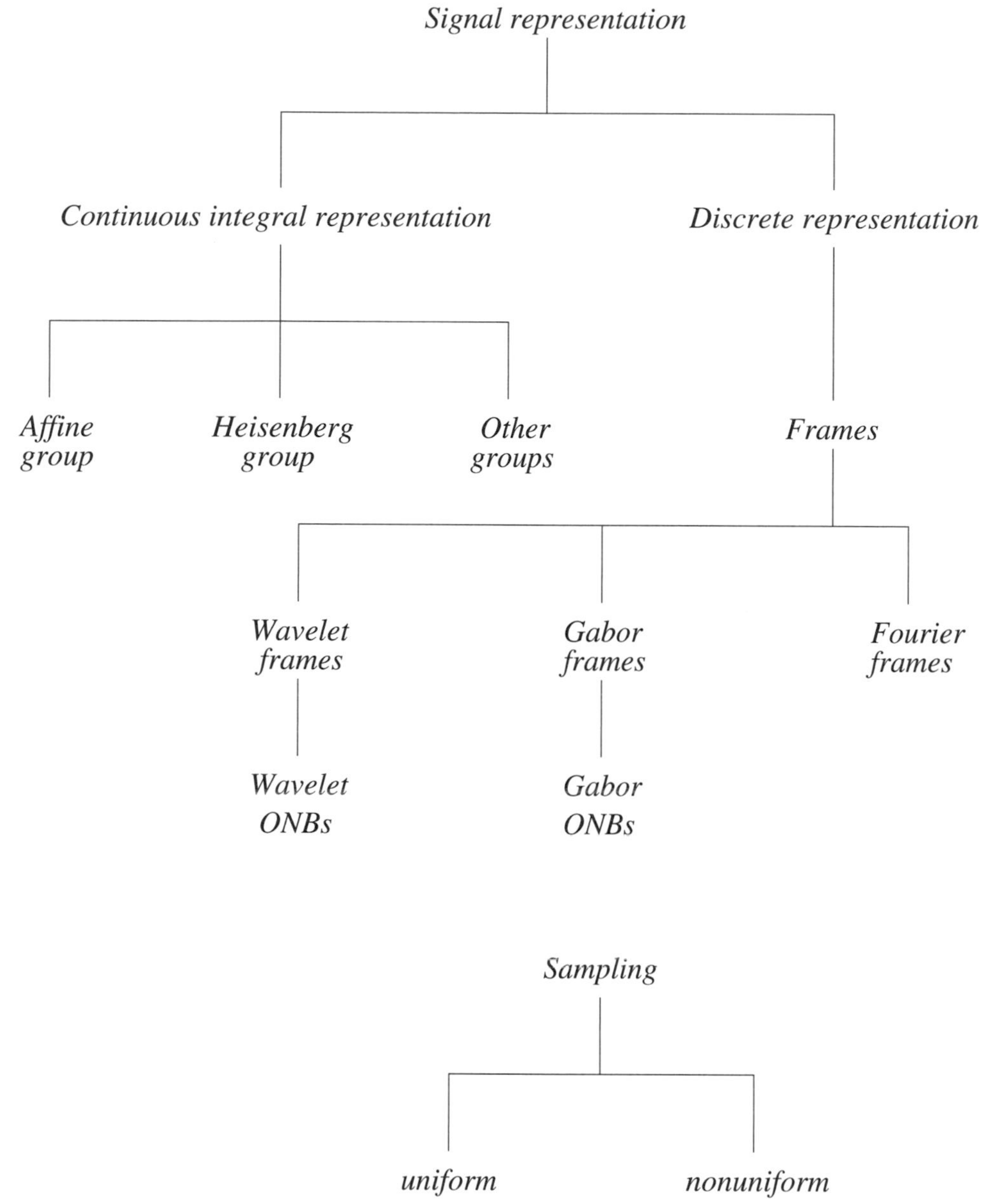

Figure 1. Signal representation

2. Let $g \in L^2(\mathbb{R})$ and let $a, b > 0$. The associated *Gabor* or *Weyl-Heisenberg system* is the sequence $\{g_{m,n} : (m,n) \in \mathbb{Z} \times \mathbb{Z}\}$, where $g_{m,n}$ is defined by

$$g_{m,n}(t) = e^{2\pi i t m b} g(t - na).$$

Clearly,

$$\widehat{g_{m,n}}(\gamma) = \tau_{mb}(e_{-na}\widehat{g})(\gamma).$$

3. Let $\Lambda \subseteq \mathbb{R}$ be countable and let $R > 0$. The associated *Fourier system* is $\{e_\lambda : \lambda \in \Lambda\}$ considered as a subset of $L^2[-R, R]$.

Definition 2.2 (Bases and frames)

Let H be a separable Hilbert space and let $\{x_n : n \in \mathbb{Z}\} \subseteq H$ be a sequence in H.

1. The sequence $\{x_n\}$ is a *basis* or *Schauder basis* for H if each $x \in H$ has a unique decomposition
$$x = \sum_{n \in \mathbb{Z}} c_n(x)\, x_n \quad \text{in } H.$$
A basis $\{x_n\}$ for H is an *orthonormal basis* (ONB) for H if it is *orthonormal.*

2. A basis $\{x_n\}$ for H is an *unconditional basis* for H if

   $\exists\, C > 0$ such that $\forall\, F \subseteq \mathbb{Z}$, where $\operatorname{card} F < \infty$, and $\forall\, b_n, c_n \in \mathbb{C}$, where $n \in F$ and $|b_n| < |c_n|$,
$$\left\| \sum_{n \in F} b_n x_n \right\| \leq C \left\| \sum_{n \in F} c_n x_n \right\|.$$
An unconditional basis is a *bounded unconditional basis* for H if
$$\exists\, A, B > 0 \quad \text{such that} \quad \forall\, n \in \mathbb{Z}, \quad A \leq \|x_n\| \leq B.$$

3. A basis for H is a *Riesz basis* if there is a bounded invertible operator on H mapping $\{x_n\}$ onto an ONB for H.

4. The sequence $\{x_n\}$ is a *frame* for H if there are $A, B > 0$ such that
$$\forall\, x \in H, \quad A\,\|x\|^2 \leq \sum_{n \in \mathbb{Z}} |\langle x, x_n \rangle|^2 \leq B\,\|x\|^2.$$
The constants A and B are *frame bounds*, and a frame is *tight* if $A = B$. A frame is an *exact frame* if it is no longer a frame whenever any of its elements is removed.

Remark 2.3. Frames give rise to discrete representations; see, for example, the articles by Duffin and Schaeffer, Daubechies, Grossmann, and Meyer, and Daubechies (all in this volume). It is natural to analyze wavelet, Gabor, and Fourier frames. The theme of this volume is the wavelet case in both the discrete and continuous setting. However, the Gabor and Fourier cases play a role even when the theme is wavelets.

Finally, the bottom of Figure 1 is meant to indicate that *sampling formulas* are discrete representations associated with various frame decompositions. For example, the Classical Sampling Formula (going back to Cauchy, see [BF01, Chapter 1]),
$$f = T \sum_{n \in \mathbb{Z}} f(nT)\, \tau_{nT} s, \tag{2.3}$$

is a discrete representation of functions f in the Paley-Wiener space of Ω-bandlimited functions, where $2T\Omega \leq 1$, and where the sampling function s is a $1/(2T)$-bandlimited function satisfying some natural properties. In the special case that $2T\Omega = 1$ and $\hat{s} = 1$ on $[-\Omega, \Omega]$, equation (2.3) gives rise to the so-called *Shannon wavelet* ONB for $L^2(\mathbb{R})$; and, in this case, the wavelet decomposition of functions $f \in L^2(\mathbb{R})$ that are not Ω-bandlimited provides an interpretation of aliasing error. The sequence $\{nT\}$ in (2.3) indicates *uniform sampling*; see Figure 1.

3. Frames

3.1 General frames

From the point of view of harmonic analysis, many of us learned about frames from the 1952 article of Duffin and Schaeffer reprinted in this volume, and then from the influential book by Young [You80], now deservedly enjoying a revised first edition. Duffin and Schaeffer defined frames in the Hilbert space setting, but their basic examples were Fourier frames; see Section 3.2.

From a functional analytic point of view, in 1921 Vitali (1875–1932) [Vit21] proved that, if $\{x_n\}$ is a tight frame with $A = B = 1$ and with $\|x_n\| = 1$ for all n, then $\{x_n\}$ is an ONB. Actually, Vitali's result is stronger for the setting $H = L^2[a, b]$ in which he dealt.

In 1936 Köthe [Köt36] proved that bounded unconditional bases are exact frames, and the converse is straightforward. Also, the category of Riesz bases is precisely that of exact frames. Thus, the following three notions are equivalent: Riesz bases, exact frames, and bounded unconditional bases. Besides the article by Duffin and Schaeffer, Bari's characterization of Riesz bases [Bar51] is fundamental in this realm of ideas. From my point of view, her work has all the more impact because it was motivated in part by her early research, with others in the Russian school, in analyzing Riemann's sets of uniqueness for trigonometric series.

Frames have also been studied in terms of the celebrated Naimark dilation theorem (1943), a special case of which asserts that any frame can be obtained by "compression" from a basis. The rank 1 case of Naimark's theorem is the previous assertion for tight frames. The finite decomposition rank 1 case of Naimark's theorem antedates Naimark's paper, and it is due to Hadwiger [Had40] and Gaston Julia [Jul42]. This is particularly interesting in light of modern applications of finite normalized tight frames in communications theory. Because they will arise later, we mention Chandler Davis' use of Walsh functions to give explicit constructions of dilations [Dav77]. Davis [Dav79] also provides an in-depth perspective on the results referred to in this paragraph.

Other applications of Naimark's theorem in the context of frames include feasibility issues for von Neumann measurements in quantum signal processing.

3.2 Fourier frames

Fourier frames go back to Dini (1880) and his book on Fourier series [Din80, pages 190 ff]. There he gives Fourier expansions in terms of the set $\{e_\lambda\}$ of harmonics, where each λ is a solution of the equation

$$x \cos \pi x + a \sin \pi x = 0. \tag{3.1}$$

Equation (3.1) was chosen because of a problem in mathematical physics from Riemann's (1826–1866) and later Riemann-Weber's classical treatise [Rie76, 158–167]. Dini (1845–1918) returned to this topic in 1917, just before his death, with a significant generalization including Fourier frames that are not ONBs [Din54].

The inequalities defining a Fourier frame were explicitly written by Paley and Wiener [PW34, 115, inequalities (30.56)]. The book by Paley and Wiener [PW34] (and to a lesser extent a stability theorem by G. D. Birkhoff [Bir17]) had tremendous influence on mid-twentieth century harmonic analysis. Although nonharmonic Fourier series expansions were developed, the major effort in the study of Fourier systems emanating from [PW34] addressed completeness problems of sequences $\{e_\lambda\} \subseteq L^2[-R, R]$, that is, determining when the closed linear span of $\{e_\lambda\}$ is all of $L^2[-R, R]$. This culminated in the profound work of Beurling and Malliavin in 1962 and 1966 ([BM62], [BM67], [Koo96]; see [BF01, Chapter 1] for a technical overview).

A landmark in this intellectual journey to the heights of Beurling-Malliavin is the article by Duffin and Schaeffer. In retrospect, their paper was underappreciated when it appeared in 1952. The authors defined Fourier frames as well as the general notion of a frame for a Hilbert space H. They emphasized that frames $\{x_n\} \subseteq H$ provide discrete representations $x = \sum a_n x_n$ in norm, as opposed to the previous emphasis on completeness. They understood that the Paley-Wiener theory for Fourier systems is equivalent to the theory of exact Fourier frames. (We noted above that Paley and Wiener used precisely the inequalities defining Fourier frames.) Duffin and Schaeffer also knew that generally they were dealing with overcomplete systems, a useful feature in noise reduction problems.

The next step on this path created by Duffin and Schaeffer is the article by Daubechies, Grossmann, and Meyer reprinted in this volume. From the point of view of the affine and Heisenberg groups (see Figure 1), and inspired by Duffin and Schaeffer, the article by Daubechies, Grossmann, and Meyer establishes the basic theory of wavelet and Gabor frames. Given the nature of this volume, I shall say nothing about Gabor systems except to the extent that they have an impact on wavelet theory — which they do. On the other hand, the wavelet frame results of Daubechies, Grossmann, and Meyer allow us to segue into a broader discussion of wavelets in Section 4.

Before closing Section 3, I'd like to make a brief personal reminiscence about Richard Duffin. In 1990 I spoke about frames and some of their applications at the University of Pittsburgh. Pesi Masani (1919–1999) and Duffin (1909–1996) sat in the front row, both magisterial in their own ways. Masani, an expert on Hilbert spaces, understood the relevance of Naimark's theorem.

Duffin was amused and surprised that frames had reemerged as a tool and theory in time-scale and time-frequency analysis. We talked mathematics through dinner and much later, spirited on by a salubrious beverage or two. I was proud that we "hit it off." There were napkins on which to write (I wish I had kept his calculations), and he told me about networks and his student Raoul Bott, from whom I had taken topology in 1961. Mostly, he was very interested in discussing mathematics and at a genuinely technical level. He was 81 years old! Amazing. Duffin published his last paper with Hans Weinberger in the *Journal of Fourier Analysis and Applications* [DW97] in a lengthy issue on frames that was dedicated to his memory.

4. Wavelet Theory

4.1 A broad and selective outline

What is a *wavelet*? We have already answered this question by defining wavelet systems, frames, and ONBs, and we have commented on the continuous wavelet transform. For example, in the case of wavelet ONBs, $\psi \in L^2(\mathbb{R})$ is a *wavelet* if the sequence $\{\psi_{m,n} : m, n \in \mathbb{Z}\}$, where $\psi_{m,n}$ was defined in (2.2), is orthonormal and if

$$\forall f \in L^2(\mathbb{R}), \quad f = \sum_{m,n\in\mathbb{Z}} \langle f, \psi_{m,n}\rangle \psi_{m,n} \quad \text{in } L^2\text{-norm},$$

where the inner products $\langle f, \psi_{m,n}\rangle$ are the *wavelet coefficients*.

What is *wavelet theory*? This is a much bigger question, one that was first addressed in [Mey90] and [Dau92]. It continues to be answered in diverse ways, as existing methods interact with other branches of mathematics, as the first rush of wavelet results has had a chance to regroup and evolve and mature, and as new applications have tested and created wavelet-based algorithms. Wavelet theory has developed into an imposing mathematical edifice with vitality and depth, as well as with emerging limitations and baroque tendencies. The scope of its applicability exhibits a similar effectiveness and limitation. Furthermore, it is a relief to assert that all that glitters is not a wavelet! Not that anyone ever said that wavelets were a panacea, but, as indicated earlier, there was definitely a period of overprescription of them.

The first wavelet ONB was constructed by Haar in his 1909 dissertation (translated in this volume). The *Haar wavelet* h for the setting of $L^2(\mathbb{R})$ is defined as

$$h(t) = \begin{cases} 1 & \text{if } t \in [0, 1/2), \\ -1 & \text{if } t \in [1/2, 1), \\ 0 & \text{otherwise.} \end{cases} \tag{4.1}$$

Haar's work was followed by Walsh's 1923 construction [Wal23] of ONBs in terms of so-called *Walsh functions*. Actually, about 1900 and without any interest in ONBs, engineers (especially, J. A. Barrett) designed transposition schemes in open-wired lines based on Walsh functions, and they used these schemes to minimize channel "crosstalk." The sequence of Walsh functions on $\mathbb{R}$ is the prototype of *wavelet packets*, just as the Haar wavelet system $\{h_{m,n}\}$ is the prototype of wavelet ONBs on $\mathbb{R}$. The theory of wavelet packets is due to Coifman, Meyer, and Wickerhauser, for example, [Wic94].

It would be nice to give a sequential litany of waveleteers, beginning with Haar and Walsh and taking us to Meyer and Daubechies. Unfortunately, it didn't happen linearly. It is true that the construction of wavelet ONBs is part of a program to construct unconditional bases for many of the important function spaces in analysis. Hence there is a certain lineage after Walsh from the article of Franklin (reprinted in this volume) in 1928, to Ciesielski [Cie81], to Carleson [Car80], and to the article by Strömberg (reprinted in this volume).

However, there were many other paths to the establishment of wavelet theory, not all of which were fully appreciated by the late 1980s. Briefly, and before the appearance of [Mey90] and [Dau92], there were wavelet-oriented traditions and/or developments in spline and approximation theory, in speech and image processing, and in atomic decompositions and the Calderón reproducing formula.

In any case, Carleson's construction of an unconditional basis for the Hardy space H^1 in 1980 led to the spline wavelet ONBs in 1981 in the article by Strömberg. Besides Carleson's construction, there was also the construction of Billard [Bil72]. Using cardinal B-splines, Battle (reprinted in this volume) and Lemarié independently constructed wavelet ONBs in the late 1980s *in the context of* wavelet theory. Research on spline wavelets continues to the present day. Further, there are natural relationships between other aspects of approximation theory and wavelet theory; see, for example, the article by DeVore, Jawerth, and Popov reprinted in this volume.

In harmonic analysis, Coifman's striking decomposition theorem [Coi74] provided a basic theme in the definitive essay by Coifman and Weiss (reprinted in this volume) for the Hardy spaces H^p, $0 < p \leq 1$. This theory had an influence on the development of wavelet theory at the level of expansions in terms of "atoms" (harmonics) having vanishing moments. It is natural, but not necessary, that the harmonics $\psi_{m,n}$ of a wavelet expansion have vanishing moments.

We shall now give a little more detail about the topics in wavelet theory that we have just sketched.

4.2 The Calderón reproducing formula

As mentioned earlier, the Calderón reproducing formula [Cal64, Section 34] is now synonymous with the so-called *continuous wavelet transform.* Calderón's formula is

$$f(t) = \int_0^\infty (\psi_{1/u} * \psi_{1/u} * f)(t)\,\frac{du}{u}, \tag{4.2}$$

where the dilation ψ_x is defined as $\psi_x(t) = x\psi(tx)$ for $x > 0$. Equation (4.2) has been a major influence from harmonic analysis on wavelet theory, and it can be considered as a continuous ("overcomplete") wavelet decomposition of $f \in L^2(\mathbb{R})$. We have used the term "wavelet" in the sense of describing f as an integral ("sum") whose harmonics are dilates and translates of a fixed function ψ. In fact, equation (4.2) is

$$f(t) = \int_0^\infty \int_{\mathbb{R}} \left(\int_{\mathbb{R}} \psi_{1/u}(v - w) f(w) dw \right) \psi_{1/u}(v - t)\,\frac{dv\,du}{u},$$

and so the "sum" we have alluded to is the double integral

$$\int_0^\infty \int_{\mathbb{R}} \cdots \frac{dv\,du}{u}$$

taken over the temporal or spatial variable v and the dilation u. In this case the "wavelet coefficients" of f are

$$\int_{\mathbb{R}} \psi_{1/u}(v - w)\, f(w)\, dw, \quad v \in \mathbb{R}, \quad u > 0.$$

The formal verification of Calderón's formula is elementary, as long as ψ satisfies certain properties. In fact, the Fourier transform of the right side of (4.2) is

$$\widehat{f}(\gamma) \int_0^\infty \widehat{\psi}(u\gamma)^2 \frac{du}{u};$$

and there are even, real-valued, compactly supported, infinitely differentiable functions ψ for which

$$\forall \gamma \in \mathbb{R}\backslash\{0\}, \quad \int_0^\infty \widehat{\psi}(u\gamma)^2 \frac{du}{u} = 1.$$

See the masterpiece on this topic by Frazier, Jawerth, and Weiss [FJW91].

4.3 Haar ONB

In Haar's 1909 dissertation (translated in this volume), where he constructs the Haar ONB, his historical perspective includes contributors such as Poisson, Riemann, Cantor, du Bois-Reymond, Fejér (another Weisz in the field!), and Hilbert. We shall trace his result through the twentieth century, where a host of new names and ideas emerges.

Hoping against hope, many nineteenth-century harmonic analysts desired that Fourier series of continuous functions f converge everywhere or even uniformly to f. Of course, du Bois-Reymond's 1872 example [dBR76] dashed these dreams. Haar's dissertation provides a positive solution to Hilbert's question of finding an ONB $\{h_{m,n}\}$, viz., the Haar ONB, for $L^2[0,1]$ for which $f = \sum \langle f, h_{m,n}\rangle \, h_{m,n}$ uniformly for every continuous 1-periodic function on $\mathbb{R}$. It is not unexpected that the Haar system is unbounded in supremum norm.

From the point of view of Fourier series, Lusin's conjecture [Lus13] directed the reaction of the du Bois-Reymond example to the problem of determining if $S(f) = f$ a.e. for $f \in L^2[0,1]$. This problem was solved affirmatively by Carleson [Car66] and then by Charles Fefferman [Fef73]. We mention this because of the influence of these results on a recent and seemingly preordained phase of wavelet theory, viz., the study of wave packets, not to be confused with wavelet packets. On the road to the solution of Lusin's conjecture, there were two significant results by Kolmogorov in the 1920s. Kolmogorov (1922) proved that there exist $f \in L^1[0,1]$ for which $S(f)$ diverges everywhere [Kol26]. Also, with the Lusin conjecture in mind, Kolmogorov proved that if $f \in L^2[0,1]$, then $S(f) = f$ a.e. when the sums are taken over dyadic blocks [Kol24]. The proof is elementary but ingenious. The extension to $L^p[0,1]$, $p > 1$ is deep and is an integral part of the original Littlewood-Paley theory [LP31], [LP37]. Furthermore, such convergence over dyadic blocks has a natural dyadic wavelet interpretation; see [Mey90].

Besides the aforementioned (Sections 3.1 and 4.1) Walsh functions, to which we shall return in Section 4.4, there were two other Haar-related sequels published in 1928, both of which are inspired exclusively by Haar's article. The first is the article by Philip Franklin reprinted in this volume and the second is the article by Juljusz Schauder [Sch28].

Philip Franklin (1898–1965) addressed and solved the problem of constructing an orthonormal basis $\{f_n\}$ of continuous 1-periodic functions on $\mathbb{R}$ such that $f = \sum c_n(f) \, f_n$ uniformly for every continuous 1-periodic function on $\mathbb{R}$. He did this by orthogonalizing the integrals of the Haar functions. At a personal level, Franklin met and became friendly with Norbert Wiener in 1918 at the U.S. Army Proving Ground in Aberdeen, Maryland.

Both were computers, working on noisy hand-computing machines known as "crashers" — a time-invariant scientific bottom line! Wiener became Franklin's colleague at MIT in 1919, and the two later became brothers-in-law.

Franklin's story for this volume serves as background for Strömberg's article, where "Franklin wavelets" ψ are constructed with the property that $\{\psi_{m,n}\}$ is an ONB for $L^2(\mathbb{R}^d)$. Besides the article by Franklin, Strömberg's work had other influences (some mentioned in Section 4.1), including a body of work by Bočkarev, Carleson, Ciesielski, Domsta, Maurey, Pelcynski, Simon, Sjölin, and Wojtaszczyk from the 1960s to 1980s. One of their themes was to prove the equivalence or not of various bases. For example, Ciesielski, Simon, and Sjölin [CSS77] proved that the Haar and Franklin systems are equivalent in $L^p[0,1]$, $1 < p < \infty$. This brings us back to the 1928 article by Schauder, where he proved that the Haar ONB for $L^2[0,1]$ is a basis for $L^p[0,1]$, $p \geq 1$. Schauder's view helped to set the stage long ago for the wonderful wavelet characterization of Besov spaces and, along with Calderón's profound influence, for the wavelet relationship with Littlewood-Paley-Stein theory; for example, see Section 4.6, Peetre's classic [Pee76], the 1987 article by Meyer translated in this volume, and Meyer's treatise [Mey90]. For perspective, it is well to recall that the Besov spaces $B_p^{s;q}$ are generalizations of the Sobolev spaces (e.g., $W^{s,2} = B_1^{s,2}$) and the Hölder spaces $C^s = B_\infty^{s,\infty}$.

It is interesting to note the following inherent property, one might say limitation, of the Haar or any multiresolution analysis ONB in the setting of $L^2(\mathbb{R})$. If we have a discrete wavelet representation $f = \sum \langle f, h_{m,n} \rangle \, h_{m,n}$ in $L^2(\mathbb{R})$, where h is the Haar wavelet defined by (4.1), then there is "leakage" to infinity of the supports of the $h_{m,n}$ with nonvanishing coefficients $\langle f, h_{m,n} \rangle$, even in the case that f is compactly supported (e.g., see [Dau92]).

4.4 Walsh ONB

Concerning the influence and importance of historical perspective, as we mentioned with Haar, Joseph L. Walsh (1895–1973) could wax rhapsodic. For example, in discussing the Riesz-Fischer theorem in the context of the Walsh ONB, he wrote that its "beauty and simplicity ... was, and still is, almost overwhelming" [Bas70]. Walsh was of course aware of Haar's work, and later he advertised Franklin's theorem. On the other hand, he was probably sensitive to priority vis-à-vis the space of orthonormal Rademacher functions, even though the latter was not an ONB. Rademacher [Rad22] published his results in 1922 and Walsh in 1923, but the discoveries were independent. Rademacher's manuscript was received by the editors of *Mathematische Annalen* on October 8, 1921, and Walsh announced his results to the American Mathematical Society at a meeting on February 25, 1922 (his paper [Wal23] is dated May 1922).

Walsh understood some of the essential differences between the Haar and Walsh ONBs, especially concerning oscillation properties analogous to the trigonometric functions. In his recursive definition, Walsh ordered the Walsh functions according to the average number of zero-crossings of these functions on $[0,1]$; this ordering was also used by Kaczmarz [Kac29], and it is referred to as *sequency ordering*. Another natural ordering of the Walsh functions is due to Paley (1932); it is based on the *binary ordering* of indices. To be specific, the sequence $\{r_n : n = 1, \ldots\}$ of *Rademacher functions* on $[0,1)$ is defined by the 1-periodic functions

$$\forall n = 1, \ldots, \qquad r_n(t) = \text{sgn}\,(\sin 2^n \pi t) \quad \text{on } \mathbb{R}.$$

Paley's binary ordering is based on his theorem asserting that Walsh's original functions w_n can be written as $w_0 = 1$ and

$$\forall n = 1, \ldots, \qquad w_n(t) = \prod_{\epsilon_j = 1} r_j(t) \quad \text{on } \mathbb{R},$$

where $n = \sum_{j=1}^{\infty} \epsilon_j 2^{j-1}$ is the binary expansion of n. A third ordering is based on the orthogonal ± 1 matrices of Sylvester (1867) and Hadamard (1893). It is sometimes called the *Kronecker ordering* and is essentially a binary bit inversion of the binary ordering.

Walsh knew that the Haar and Walsh systems were Hadamard transforms of each other, and that $\{w_n\}$ was uniformly bounded, as opposed to the Haar system. He was not aware that $\{w_n\}$ is the discrete dual group of the compact dyadic group $\mathbb{Z}_2 \times \mathbb{Z}_2 \times \cdots$. ($\mathbb{Z}_2$ is the discrete cyclic group of order 2, not the 2-adic integers.) This duality theorem was proved by Vilenkin [Vil47] and Fine [Fin49]. Paley and Wiener, in their foray into the duality theory of locally compact abelian groups, had announced a similar result at the ICM in Zurich (1932).

The peroration for this review of properties of Walsh functions, especially as compared with the Haar system, is the ultimate distinction (alluded to in Subsection 4.1) between these two systems. This distinction turns out to be a consequence of wavelet theory: the Haar system on $\mathbb{R}$ is a multiresolution analysis wavelet ONB, and the Walsh system on $\mathbb{R}$ is its corresponding family of wavelet packets!

As a personal postscript to this subsection, I was a student of Walsh in potential theory in 1960, and we met again as colleagues at the University of Maryland when we both arrived there in 1965. In the early 1970s he suggested that I study Walsh functions since there were many problems he thought I'd be interested in. Those were the heady days of spectral synthesis, and I remained blissfully ignorant of Walsh functions, especially their applications, until the 1990s. To those less recalcitrant than I, the *Proceedings* from 1970–1974 of *Applications of Walsh Functions*, for example, [SS74], may provide an archetype of things to come and are certainly a quantitative and fascinating portrait of things past.

4.5 Filters and an early patent

On July 29, 1983, Goupillaud, Grossmann, and Morlet filed for a patent on signal representation generators. The patent, based on work of Morlet [Mor81], [MAFG82], was awarded in 1986.

Morlet's idea [Mor81] was to analyze seismic traces by means of sequences of harmonics each having a fixed shape. The trace s can be considered as the real part of a signal f whose Fourier transform is causal, that is, $f \in H^2$. He designed these harmonics to be translates and dilates of a single function ψ. In particular, there are the same number of cycles for high, medium, and low frequencies. The reconstruction of the trace s is then effected by its sequence of "sampled values" $\langle s, \psi_{m,n} \rangle$ (wavelet coefficients) in terms of a wavelet representation. Morlet originally used modulated Gaussians ψ.

As Goupillaud reminded us in 1997, Morlet's set $\{\psi_{m,n}\}$ of harmonics necessarily cannot be orthonormal or linearly independent, because it was vital to achieve noise reduction as well as stable representation in the physical problems being addressed.

Furthermore, Morlet's beautiful idea was first quantified with the proper mathematical tools by Alex Grossmann. This was a significant scientific contribution, fortuitously juxtaposed with the fact that Daubechies was working with Grossmann. Apparently, Roger Balian, who formulated the ONB version of the Balian-Low uncertainty principle for exact Gabor frames, had advised for the Morlet-Grossmann connection, which led to a large body of results including [GGM84] as well as their article reprinted in this volume. Despite my proclivity to extol the many virtues of Gabor systems, these systems arise in the Goupillaud, Grossmann, and Morlet patent as a technological device with shortcomings (inherent undersampling problems in analyzing high frequencies), which are overcome by the Morlet approach.

The first smooth wavelet ONBs for $L^2(\mathbb{R}^d)$ were constructed by Strömberg (reprinted in this volume), for ψ m-times continuously differentiable with exponential decay, and by Meyer (1985), for bandlimited ψ in the Schwartz class on $\mathbb{R}$ and by Lemarié on $\mathbb{R}^d$; see the article by Lemarié and Meyer (translated in this volume). We have already mentioned Daubechies' construction of a compactly supported m times continuously differentiable orthonormal wavelet in 1987. In his Zygmund lectures, Meyer used Mallat's newly packaged (1986) concept of multiresolution analysis (MRA) to prove the Daubechies theorem, see the articles by Mallat reprinted in this volume.

We have introduced MRAs in the same breath while discussing patents because of the signal processing origins of MRAs. These origins are represented by the articles in Section I. At the risk of oversimplification, the article by Burt and Adelson reprinted in this volume and their article [BA83] are a substantial precursor for the structural nature of MRAs, and the remaining articles in Section I provide the decidedly nontrivial nuts and bolts for constructing a wavelet ψ by means of conjugate mirror filters (CMFs) arising in the Fourier analysis of an MRA. Nowadays, the details to clarify and verify the claims in the previous teutonic sentence can be found in many places, but one can do no better than reading the triune treatises by Meyer [Mey90], Daubechies [Dau92], and Mallat [Mal98]. A beautiful component in establishing the relation between MRAs and CMFs is Albert Cohen's equivalence theorem (translated in this volume) for these notions in the case that the Fourier transform of the scaling function is in each of the Sobolev spaces $W^{m,2}(\mathbb{R})$, $m \in \mathbb{N}$. Another gem in this area is the article by Lawton (reprinted in this volume), which constructs compactly supported tight frames for a given trigonometric polynomial CMF.

I had the good fortune to consult for The MITRE Corporation in Washington, D.C., for many years. Consultation for me included an enlightening and respectful exposure to engineering excellence. MITRE's Signal Processing Group was actively involved in designing algorithms for digitizing voice. By 1981, it was not only implementing CMFs from the work of Esteban and Galand and Crochiere, Webber, and Flanagan (both in this volume), but also from a host of related work, for example, Croisier, Esteban, and Galand [CEG76] and Barnwell [BI81]. By 1987, we had discovered (along with many others in industrial or government laboratories) the groundbreaking article by Smith and Barnwell reprinted in this volume, along with [SBI86] and [SBI87], as well as the article by Vetterli in this volume and his earlier work [Vet84]. Furthermore, besides the article by Vaidyanathan in this volume, his book [Vai93] became a staple in our group.

4.6 Harmonic analysis and wavelets in the 1980s

At this point, and dealing with wavelet theoretic harmonic analysis in the mid-1980s, we have commented on the work of Cohen, Daubechies, Lemarié, Mallat, and Meyer. There was contemporary, as well as comparably creative and fundamental, wavelet theoretic harmonic analysis produced by Frazier and Jawerth and Feichtinger and Gröchenig (both in this volume).

An impetus for the work of Frazier and Jawerth was Uchiyama's smooth atomic decomposition of $L^2(\mathbb{R}^d)$ [Uch82], which is a consequence of the Calderón reproducing formula. Bidisc and parabolic versions appeared before the Euclidean version! Another impetus was Michael Wilson's use of the Calderón formula to obtain a smooth atomic decomposition of $B_1^{0,1}$. This led to the Frazier-Jawerth smooth atomic decomposition of $B_p^{s,q}$, and then to Triebel-Lizorkin spaces. They also replaced these smooth atomic decomposition methods with the ϕ-transform, which allows representation independent of the function being decomposed. They had therefore obtained a discrete reformulation of the Littlewood-Paley-Stein theory! Their work was influenced by variations on Hardy space decompositions due to Coifman-Rochberg [CR80] and Ricci-Taibleson [RT83]. The Frazier-Jawerth article in this volume on wavelet frame representations of Besov spaces was submitted on September 17, 1984. The subsequent results on Triebel-Lizorkin spaces were obtained shortly thereafter, but their papers containing them "grew to maturity" before appearing in 1988 and 1990.

The harmonic analysis background for the article by Frazier and Jawerth was firmly in the realm of the Zygmund-Calderón "school." The harmonic analysis background for the article by Feichtinger and Gröchenig is more abstract and equally important.

When I first tried to understand Feichtinger and Gröchenig's atomic decomposition theory (actually, it was one of their infamous preliminary versions), I was unprepared to plumb its depths. It has aged lucidly — a quality personally sought by said plumber. It remains a creative tour d'horizon, extending its tentacles to Banach frames and flexing its formidable technology into a well-developed methodology to address new problems in wavelet theory and its applications. Feichtinger and Gröchenig use representation theory in a fundamental way. They deal with algebraic and structural aspects of classical topics such as Wiener's Tauberian theory. For example, let $S_0(\mathbb{R}^d)$ be the smallest Segal algebra isometrically invariant under translation and modulation. Then, $S_0(\mathbb{R}^d)$ can be identified with the Wiener amalgam space whose global norm is determined by $\ell^1(\mathbb{Z}^d)$ and whose local norm is determined by the space of absolutely convergent Fourier transforms. Their approach to wavelet theory and the construction of unconditional bases uses their methodology for analyzing coorbit spaces. In particular, if one considers the Schrödinger representation of the Heisenberg group, then $S_0(\mathbb{R}^d)$ is obtained as the coorbit of $L^1(\mathbb{R}^{2d})$. With this approach, and independent of the Calderón-Zygmund theory, they proved that sufficiently structured orthonormal wavelets for $L^2(\mathbb{R}^d)$ give rise to unconditional bases for all of the corresponding coorbit spaces including Besov and Triebel-Lizorkin spaces [FG88].

5. Conclusion

At the beginning of section 4, we asked: What is wavelet theory?

For me the answer is both theoretical and concrete. On the one hand, in harmonic analysis, wavelet theory is a natural continuation in the history of *some* of the ideas that define our subject. On the other hand, wavelet theory has become an effective tool to address *some* problems in engineering, mathematics, and the sciences. As such, it also provides a unifying methodology allowing for the possibility of genuine communication between diverse groups. The articles herein indicate a protean body of knowledge so that my answer to the above question is assuredly only one of many. Similarly, this introduction may seem idiosyncratic to others who have also thought about wavelet theory. I hope I have not crafted a procrustean bed with questionable resemblance to the spirit of the articles that follow.

Finally, just as I began by saying that the articles in this volume needed no introduction, so too the introducers. I shall not resist listing two among my favorite publications by each of them: Jelena Kovačević [VK95] and [GKK01]; Jean-Pierre Antoine [AAG00] and [AKLT00]; Hans Feichtinger [FS03] and [Fei02]; Yves Meyer [Mey72] and [JM96]; Guido Weiss [HW96] and [CCMW02]; and Victor Wickerhauser [Wic94] and [Wic03].

The denouement of wavelet theory may lie in the future, or may have already been integrated in the present, but this volume represents its virtuosic roots.

References

[AAG00] S. T. Ali, J.-P. Antoine, and J.-P. Gazeau, *Coherent States, Wavelets, and Their Generalizations*, Springer-Verlag, New York, 2000.

[AKLT00] J.-P. Antoine, Y. B. Kouagou, D. Lambert, and B. Torrésani, *An algebraic approach to discrete dilations: Application to discrete wavelet transforms*, J. Fourier Anal. Appl. **6** (2000), 113–141.

[Bar51] N. K. Bari, *Biorthogonal systems and bases in Hilbert space*, Učen Zap. Mosk. Gos. Univ. 148, Mat. **4** (1951), 69–107.

[BI81] T. P. Barnwell, III, *An experimental study of subband coder design incorporating recessive quadrature filters and optimum ADPCM*, IEEE Int. Conf. Acoustics, Speech, and Signal Processing, 1981.

[Bas70] C. A. Bass (ed.), *Applications of Walsh Functions*, Naval Research Laboratory, Washington, D.C., 1970.

[BF01] J. J. Benedetto and P. J. S. G. Ferreira (eds.), *Modern Sampling Theory: Mathematics and Applications*, Birkhäuser, Boston, 2001.

[BM62] A. Beurling and P. Malliavin, *On Fourier transforms of measures with compact support*, Acta Math. **107** (1962), 291–309.

[BM67] A. Beurling and P. Malliavin, *On the closure of characters and the zeros of entire functions*, Acta Math. **118** (1967), 79–93.

[Bil72] P. Billard, *Bases dans H et bases de sous-espaces de dimension finie dans A*, Linear operators and approximation (Proc. Conf. Oberwolfach, 1971), Internat. Ser. Numer. Math. **20**, Birkhäuser, Basel, 1972, 310–324.

[Bir17] G. D. Birkhoff, *A theorem on series of orthogonal functions with an application to Sturm-Liouville series*, Proc. Nat. Acad. Sci. **3** (1917), 656–659.

[BA83] P. Burt and T. Adelson, *A multiresolution spline with applications to image mosaics*, ACM Trans. Graphics **2** (1983), 217–236.

[Cal64] A. Calderón, *Intermediate spaces and interpolation, the complex method*, Studia Math. **24** (1964), 113–190.

[Car66] L. Carleson, *On convergence and growth of partial sums of Fourier series*, Acta Math. **116** 1966, 135–157.

[Car80] ———, *An explicit unconditional basis in H^1*, Bull. Sci. Math. (2), **104** (1980), 405–416.

[CCMW02] C. K. Chui, W. Czaja, M. Maggioni, and G. Weiss, *Characterization of general tight wavelet frames with matrix dilations and tightness preserving oversampling*, J. Fourier Anal. Appl. **8** (2002), 173–200.

[Cie81] Z. Ciesielski, *The Franklin orthogonal system as unconditional basis in* $\operatorname{Re} H^1$ *and VMO*, Functional analysis and approximation (Oberwolfach, 1980), Internat. Ser. Numer. Math. **60**, Birkhäuser, Basel-Boston, Mass., 1981, 117–125.

[CSS77] Z. Ciesielski, P. Simon, and P. Sjölin, *Equivalence of Haar and Franklin bases in L^p spaces*, Studia Math. **60** (1977), 195–210.

[Coi74] R. R. Coifman, *A real variable characterization of H^p*, Studia Math. **51** (1974), 269–274.

[CR80] R. R. Coifman and R. Rochberg, *Representation theorems for holomorphic and harmonic functions in L^p*, Astérisque **77** (1980), 11–66.

[CEG76] A. Croisier, D. Esteban, and C. Galand, *Perfect channel splitting by use of interpolation, decimation, tree decomposition techniques*, Proc. Int. Conf. Information Sciences/Systems, Patras, 1976, 443–446.

[Dau92] I. Daubechies, *Ten Lectures on Wavelets*, CBMS-NSF Ser. Appl. Math., SIAM, Philadelphia, 1992.

[Dav77] C. Davis, *Geometric approach to a dilation theorem*, Linear Algebra and Appl. **18** (1977), 33–43.

[Dav79] C. Davis, *Some dilation and representation theorems*, Proc. Second International Symposium in West Africa on Functional Analysis and its Applications, Kumasi, 1979, 159–182.

[Din80] U. Dini, *Serie di Fourier e altre rappresentazioni analitiche delle funzioni di una variabile reale*, Tip. T. Nistri e C., Pisa, 1880.

[Din54] U. Dini, *Sugli sviluppi in serie* $\frac{1}{2}a_o + \sum_1^\infty (a_n \cos \lambda_n x + b_n \operatorname{sen} \lambda_n x)$ *dove le* λ_n *sono radici dell'equazione trascendente* $f(z) \cos \pi z + f_1(z) \operatorname{sen} \pi z = 0$, Opere, vol. II, Edizioni Cremonese, Roma, 1954, 183–207.

[dBR76] P. du Bois-Reymond, *Untersuchungen über die Convergenz und Divergenz der Fourierschen Darstellungsformeln*, Bayerischen Abhandlungen **12** (1876), 1–103.

[DW97] R. J. Duffin and H. F. Weinberger, *On dualizing a multivariable Poisson summation formula*, J. Fourier Anal. Appl. **3** (1997), 487–497.

[Fef73] C. Fefferman, *Pointwise convergence of Fourier series*, Ann. of Math. (2) **98** (1973), 551–571.

[Fei02] H. G. Feichtinger, *Modulation spaces of locally compact Abelian groups*, Proc. International Conference on Wavelets and Applications (R. Ramakrishnan and S. Thangavelu, eds.), Chennai, 2002, 1–56.

[FG88] H. G. Feichtinger and K. Gröchenig, *A unified approach to atomic decompositions via integrable group representations*, Function Spaces and Applications (Lund, 1986), Lect. Notes Math. **1302**, Springer, Berlin, 1988, 52–73.

[FS03] H. G. Feichtinger and T. Strohmer, *Advances in Gabor Analysis*, Birkhäuser, Boston, 2003.

[Fin49] N. J. Fine, *On the Walsh functions*, Trans. Amer. Math. Soc. **65** (1949), 372–414.

[FJW91] M. Frazier, B. Jawerth, and G. Weiss, *Littlewood-Paley Theory and the Study of Function Spaces*, CBMS Conf. Lect. Notes, **79**, Amer. Math. Soc., Providence, 1991.

[GGM84] P. Goupillaud, A. Grossmann, and J. Morlet, *Cycle-octave and related transforms in seismic signal analysis*, Geoexploration **23** (1984), 85–102.

[GKK01] V. K. Goyal, J. Kovačević, and J. A. Kelner, *Quantized frame expansions with erasures*, J. Appl. Comput. Harmon. Anal. **10** (2001), 203–233.

[Had40] H. Hadwiger, *Über ausgezeichnete Vektorsterne und reguläre Polytope*, Comment. Math. Helv. **13** (1940), 90–107.

[HW96] E. Hernández and G. Weiss, *A First Course on Wavelets*, CRC Press, Boca Raton, Fl, 1996.

[JM96] S. Jaffard and Y. Meyer, *Wavelet Methods for Pointwise Regularity and Local Oscillations of Functions*, Mem. Amer. Math. Soc. **123**, 1996.

[Jul42] G. Julia, *Sur la représentation analytiques des opérateurs linéaires dans l'espace hilbertien*, C. R. Acad. Sci. Paris **214** (1942), 591–593.

[Kac29] S. Kaczmarz, *Über ein Orthogonal System*, Comptes Rendus Congres Math., Warsaw, 1929.

[Kol24] A. N. Kolmogorov, *Une contribution à l'étude de la convergence des séries de Fourier*, Fundamenta Math. **5** (1924), 96–97.

[Kol26] ———, *Une séries de Fourier-Lebesgue divergente partout*, C. R. Acad. Sci. Paris **183** (1926), 1327–1328.

[Koo96] P. J. Koosis, *Leçons sur le théorème de Beurling and Malliavin*, Université de Montréal, Publications CRM, Montréal, 1996.

[Köt36] G. Köthe, *Das trägheitsgesetz der quadratischen Formen im Hilbertschen Raum*, Math. Zeit. **41** (1936), 137–152.

[LP31] J. E. Littlewood and R. E. A. C. Paley, *Theorems on Fourier series and power series*, J. London Math. Soc. **6** (1931), 230–233.

[LP37] ———, *Theorems on Fourier series and power series III*, Proc. London Math. Soc. **43** (1937), 105–126.

[Lus13] N. Lusin, *Sur la convergence des séries trigonométriques de Fourier*, C. R. Acad. Sci. Paris **156** (1913), 1655–1658.

[Mal98] S. Mallat, *A Wavelet Tour of Signal Processing*, Academic Press, Boston, 1998.

[Mey72] Y. Meyer, *Algebraic Numbers and Harmonic Analysis*, North-Holland, Amsterdam, 1972.

[Mey90] Y. Meyer, *Ondelettes et Opérateurs*, Hermann, Paris, 1990.

[Mor81] J. Morlet, *Sampling theory and wave propogation*, Proc. 51st Annual International Meeting of the Society of Exploration Geophysicists, 1981.

[MAFG82] J. Morlet, G. Arens, E. Fourgeau, and D. Giard, *Wave propogation and sampling theory: Part I. Complex signal and scattering in multilayered media, and Part II. Sampling theory and complex waves*, Geophysics **47** (1982), 203–221, 222–236.

[PW34] R.E.A.C. Paley and N. Wiener, *Fourier Transforms in the Complex Domain*, Amer. Math. Soc. Colloquium Publications **19**, American Mathematical Society Providence 1987 (reprint of 1934 original).

[Pee76] J. Peetre, *New Thoughts on Besov Spaces*, Duke University Math. Series **1**, Mathematics Department, Duke University, Durham, 1976.

[Rad22] H. A. Rademacher, *Einige Sätze über Reihen von allgemeinen Orthogonalenfunktionen*, Math. Annalen **87** (1922), 112–138.

[RT83] F. Ricci and M. Taibleson, *Boundary values of harmonic functions in mixed norm spaces and their atomic structure*, Ann. Scuola Norm. Sup. Pisa Cl. Sci. (4) **10** (1983), 1–54.

[Rie76] B. Riemann, *Partielle Differentialgleichungen*, 2nd edition, Braunschweig, 1876.

[SS74] G. F. Sandy and H. Schreiber (eds.), *Applications of Walsh Functions and Sequency Theory*, IEEE, New York, 1974.

[Sch28] J. Schauder, *Eine Eigenschaft des Haarschen Orthogonalsystems*, Math. Zeit. **28** (1928), 317–320.

[SBI86] M. J. T. Smith and T. P. Barnwell III, *Exact reconstruction for tree-structured subband coders*, IEEE Trans. Acoustics, Speech, and Signal Processing **34** (1986), 434–441.

[SBI87] ———, *A new filter bank theory for time-frequency representation*, IEEE Trans. Acoustics, Speech, and Signal Processing **35** (1987), 314–327.

[SF67] B. Sz.-Nagy and C. Foiaş, *Analyse harmonique des opérateurs de l'espace de Hilbert*, Masson et Cie, Paris, Akadémiai Kiadó, Budapest, 1967.

[Uch82] A. Uchiyama, *A constructive proof of the Fefferman-Stein decomposition of* BMO($\mathbb{R}^n$), Acta Math. **148** (1982), 215–241.

[Vai93] P. P. Vaidyanathan, *Multirate Systems and Filter Banks*, Prentice-Hall, Englewood Cliffs, NJ, 1993.

[Vet84] M. Vetterli, *Multidimensional subband coding: some theory and algorithms*, Signal Processing **6** (1984), 97–112.

[VK95] M. Vetterli and J. Kovačević, *Wavelets and Subband Coding*, Signal Processing Series, Prentice-Hall, Englewood Cliffs, NJ, 1995.

[Vil47] N. Ya. Vilenkin, *A class of complete orthonormal series*, Izv. Akad. Nauk SSSR, Ser. Mat., 11 (1947), 363–400. *On a class of complete orthonormal systems*, Amer. Math. Soc. Transl. (2), **28** (1963) 1–35; translation from Izv. Akad. Nauk SSSR, Ser. Mat. 11 (1947), 363–400.

[Vit21] G. Vitali, *Sulla condizione di chiusura di un sistema di funzioni ortoganali*, Atti R. Acad. Naz. Lincei, Rend. Cl. Sci. Fis. Mat. Nat. **30** (1921), 498–501.

[Wal23] J. L. Walsh, *A closed set of normal orthogonal functions*, Amer. J. Math. **45** (1923), 5–24.

[Wic94] M. V. Wickerhauser, *Adapted Wavelet Analysis from Theory to Software*, A. K. Peters, Wellesley, MA, 1994, With a separately available computer disk (IBM-PC or Macintosh).

[Wic03] M. V. Wickerhauser, *Mathematics for Multimedia*, Academic Press, San Diego, 2003.

[Wil85] J. M. Wilson, *On the atomic decomposition for Hardy spaces*, Pacific J. Math. **116** (1985), 201–207.

[You80] R. M. Young, *An Introduction to Nonharmonic Fourier Series*, Academic Press, New York, 1980.

SECTION I

Precursors in Signal Processing

Introduction

Jelena Kovačević

What a treat to look back and observe what happened within the field of wavelets over the past twenty years. We witnessed tremendous advances, wavelets became a commonplace technique, several standards are wavelet-based, wavelet software packages such as Matlab are used regularly. As icing on the cake, wavelets brought people from many disparate areas together. It is commonplace today to attend a wavelet meeting and sit in a room with mathematicians, engineers, statisticians, and physicists.

Where did it all start? This volume gives the answer: many places. One of those many is signal processing. If you are not familiar with signal processing, do not immediately start thinking of DSP chips. Signal processing is the mathematical framework for acquisition, representation, and analysis of signals (and many other tasks). It is heavily motivated by applications and is (mostly) discrete-time oriented; your signal is a sequence of numbers representing a speech or audio signal, an EEG signal, a heart rate signal. It can also be a matrix of numbers representing illumination of individual picture elements (pixels) or even a three-dimensional volume representing moving images.

The beautiful collection of papers that follows is a glimpse into the history of the birth of wavelets. The beauty of these solutions comes from trying to solve real problems and ingenious ways of doing it. Looking back, you will see that multiresolution ideas kept on coming up, showing that multiresolution is inherent within a certain class of problems.

So sit down, relax, and enjoy the trip to yesteryear ...

The selection of papers in this section could be split into two uneven parts. The first is the work of Burt and Adelson, while the other contains the rest. The reason for this is that, historically, the second set developed more or less in a sequence, and the papers were influenced by the previous ones. The work of Burt and Adelson was picked up later by the signal processing people but developed initially on its own and did not influence the early signal processing papers.

A small aside about notation. Discrete-time sequences are denoted by h_n, with $n \in \mathcal{Z}$. Their discrete-time Fourier transform is denoted by $H(e^{j\omega})$, a 2π-periodic function. Finally, for the non-DSP people: the z-transform is defined as $H(z) = \sum_{n \in \mathcal{Z}} h_n z^{-n}$ and reduces to the discrete-time Fourier transform on the unit circle, that is, for $z = e^{j\omega}$. If you are wondering where the z-transform came from, you can think of it as a counterpart of the Laplace transform in continuous time.

The Laplacian pyramid as a compact image code

The paper by Burt and Adelson designs an efficient image coding system based on the novel idea of pyramid coding. The key idea is to produce a prediction of the image by lowpass filtering and downsampling the original image and then calculate the difference between the original and the prediction based on that lowpass image. This produces an instant compression system since the difference image is of low energy that can be appropriately quantized. Further compression can be achieved by iterating the process on the prediction, resulting in a "pyramid" of difference images and a final lowpass signal.

If you are not completely wavelet-immune, you will, in the preceding paragraph, immediately recognize elements of wavelet theory. What is more amazing, though, is that Burt and Adelson, without the benefit of our twenty years of wavelet knowledge base, did the same — albeit not calling them wavelets. To illustrate this point, consider the following few quotations from their paper:

"It [pyramid] represents an image as a series of quasi-bandpassed images, each sampled at successively sparser densities. The resulting code elements, which form a self-similar structure, are localized in both space and spatial frequency." Moreover, "The scale of the Laplacian operator doubles from level to level of the pyramid, while the center frequency of the passband is reduced by an octave."

It is worth noting that, since the difference signals are not sampled, the scheme is oversampled (overcomplete) and thus corresponds to a wavelet frame rather than a wavelet basis.

Digital coding of speech in subbands

This work by Crochiere, Webber, and Flanagan truly starts the area of subband coding that later led to connections with wavelets. What is also worth noting is that the reason subbands are used is to allow perception to play a part in coding — a concept much used in compression starting with the late 1980s.

This is a precursor of a precursor. It does divide a speech signal into subbands, by bandpass filtering the original speech signal. These bandpass filters are narrow to allow for efficient quantization and to eliminate the effect that noise in one subband might have on another. So sampled bands are then quantized according to perceptual criteria.

In wavelet terms, this work would amount to doing a discrete wavelet transform with one level but not quite using the correct mother wavelet to allow the scheme to be inverted. Moreover, one might consider it undersampled, since guard bands are allowed, amounting to parts of the spectrum not being reconstructed. In signal processing terms, aliasing — overlapping of the spectrum due to downsampling — is still present.

The quest for a perfectly reconstructed signal starts.

Application of quadrature mirror filters to split-band voice coding schemes

This work by Esteban and Galand continues seamlessly along the lines of the previous article. It, quite naturally, addresses the problem of how to eliminate aliasing. They consider the two-channel filter bank (see Figure 1), where the channels (subbands) are obtained by lowpass (H_0) and highpass (H_1) filtering the original signal followed by downsampling by two. The authors come up with an ingenious solution to remove aliasing, which became famous in the community as QMF (quadrature mirror filters). Many later solutions that would eventually construct bases (perfect reconstruction in signal processing terms) would still be called QMF. The deceptively simple solution is to form the

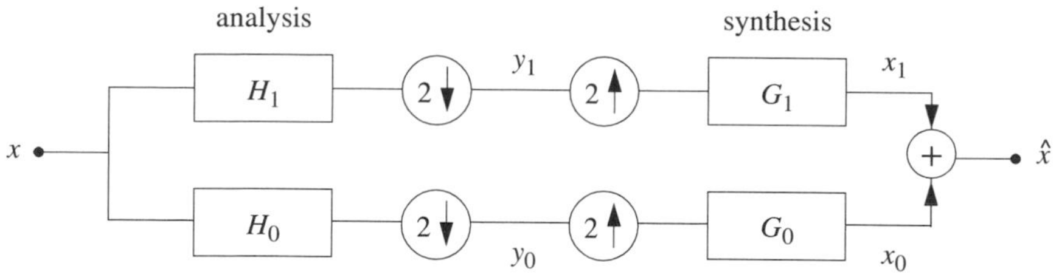

Figure 1. Two-channel filter bank

highpass filter $H_1(z)$ as

$$H_1(z) = -H_0(-z).$$

The quest for a perfectly reconstructed signal continue ...

A procedure for designing exact reconstruction filter banks for tree-structured subband coders and Filters for distortion-free two-band multirate filter banks

Why the titles of two papers in this subsection? The first is by Smith and Barnwell and the second by Mintzer. The reason is that these two papers solved the same problem independently. Namely, they both found the solution to having perfect reconstruction, two-channel filter banks. Moreover, their solution is orthogonal (corresponds to a unitary transform) and the filter design method is by spectral factorization. For you waveleters, spectral factorization is exactly what Ingrid Daubechies used in her famous paper to design "Orthonormal bases of compactly supported wavelets" (reprinted later in this volume), except that the factors she chose were slightly different.

In short, the solution is to form the filter bank with the following filters: $H_0(z)$, $H_1(z) = H_0(-z^{-1}), G_0(z) = H_0(z^{-1}), G_1(z) = H_0(-z)$. The condition for perfect reconstruction then boils down to

$$H_0(z)H_0(z^{-1}) + H_0(-z^{-1})H_0(-z) = P(z) + P(-z) = 1.$$

Thus, to solve for the filter $H(z)$, one has to find an appropriate $P(z)$ under very mild constraints, and then factor it into its spectral factors. Given the above, $P(z)$ can be factored into a product of reciprocally paired zeros. The distribution of zeros between $H(z)$ and $H(z^{-1})$ can be performed many ways. Smith and Barnwell choose one solution, while Daubechies is governed by the smoothness of the corresponding wavelets and chooses to put the maximum number of zeros at $z = -1$ into $H(z)$ and the rest inside the unit circle.

The quest for a perfectly reconstructed signal is over. Or, is it?

Filter banks allowing perfect reconstruction

As you might have guessed, humans have a insatiable intellect, and as soon as one solved how to obtain perfect reconstruction filter banks for two channels (dyadic case), the quest for doing it with the least harsh constraints on the filters (biorthogonal solution), and for the same solutions when there are more than two channels, started. In his work, Vetterli solves both problems. Moreover, he establishes, for many years to come, the polyphase analysis as a standard for analyzing filter banks, thus transforming a linear shift-variant system (because of downsampling) into a multi-input, multi-output linear shift-invariant system. The polyphase domain — the domain of the cosets of the regular lattice — becomes the preferred tool. It brings the convenience and elegance of matrix notation and analysis into play. Finally, Vetterli also solved the dual problem — transmultiplexers, a communications application.

Theory and design of M-channel maximally decimated quadrature mirror filters with arbitrary M, having the perfect reconstruction property

The final paper in this section introduces an important concept into the filter bank community — that of losslessness. The idea, coming from circuit theory, deals with systems in which there is no loss of energy. In the z-domain, these systems become paraunitary on the unit circle and thus correspond to orthonormal wavelet bases.

Moreover, Vaidyanathan proposes efficient schemes for factoring and implementing perfect reconstruction filter banks. These are based on lossless building blocks — lattice structures.

In conclusion

So how does this all tie in? Why do we say these papers are precursors to wavelets in signal processing?

Suppose we take the two-channel filter bank from figure 1, and iterate on the lowpass filter. That is, at the output of the lowpass branch we add another two-channel filter bank. We continue the process J times. What we get is the *discrete-time wavelet transform* with J levels. Unless you are involved in wavelet research, this is the only wavelet transform you will actually use. This is what people use in Matlab to implement the wavelet transform.

The discrete-time wavelet transform illustrates many of the concepts we have mentioned so far. It is critically sampled; that is, the number of samples at any point in the system is the same. If the polyphase matrix of the initial analysis filter bank is invertible, we do have a transform — and we call it a biorthogonal transform (it is the most general solution to the problem). The filters used can have compact support (FIR filters) or infinite support (IIR) filters. If the polyphase matrix is paraunitary, our transform is orthonormal. In practical terms, this means that the norm (energy) is preserved throughout the system and the synthesis filters are the same as the analysis ones (within reversal). The outputs of the last lowpass branch are scaling (coarse) coefficients, while the outputs of all the bandpass/highpass branches are wavelet (detail) coefficients. If we draw the equivalent filters through each branch of the system, we get wavelet filters, which are (discretely) streched versions of the same basic mother wavelet (highpass filter). If we increase the number of channels to N and iterate on the lowpass branch as before, we obtain a wavelet transform with a stretching (dilation) factor of N. The wavelet filters are now N-times stretched versions of the same mother wavelet. If we relax the constraint of critical sampling and sample with a number smaller then N, we get an overcomplete system (frame).

This was all discrete. What about the real — continuous-time — wavelet bases? Well, if the lowpass filter is smooth enough and we iterate to infinity, we obtain the continuous-time wavelet bases. This construction is the one Daubechies has in her paper. The other way around, getting the discrete-time version from the continuous-time one, is always possible: just assign the coefficients in the two-scale equations for the scaling function and the wavelet to the lowpass and highpass filters, respectively.

I hope you will enjoy reading the historic papers in this section and the rest of the volume.

532 IEEE TRANSACTIONS ON COMMUNICATIONS, VOL. COM-31, NO. 4, APRIL 1983

The Laplacian Pyramid as a Compact Image Code

PETER J. BURT, MEMBER, IEEE, AND EDWARD H. ADELSON

***Abstract*—We describe a technique for image encoding in which local operators of many scales but identical shape serve as the basis functions. The representation differs from established techniques in that the code elements are localized in spatial frequency as well as in space.**

Pixel-to-pixel correlations are first removed by subtracting a low-pass filtered copy of the image from the image itself. The result is a net data compression since the difference, or error, image has low variance and entropy, and the low-pass filtered image may represented at reduced sample density. Further data compression is achieved by quantizing the difference image. These steps are then repeated to compress the low-pass image. Iteration of the process at appropriately expanded scales generates a pyramid data structure.

The encoding process is equivalent to sampling the image with Laplacian operators of many scales. Thus, the code tends to enhance salient image features. A further advantage of the present code is that it is well suited for many image analysis tasks as well as for image compression. Fast algorithms are described for coding and decoding.

INTRODUCTION

A COMMON characteristic of images is that neighboring pixels are highly correlated. To represent the image directly in terms of the pixel values is therefore inefficient: most of the encoded information is redundant. The first task in designing an efficient, compressed code is to find a representation which, in effect, decorrelates the image pixels. This has been achieved through predictive and through transform techniques (cf. [9], [10] for recent reviews).

In predictive coding, pixels are encoded sequentially in a raster format. However, prior to encoding each pixel, its value is predicted from previously coded pixels in the same and preceding raster lines. The predicted pixel value, which represents redundant information, is subtracted from the actual pixel value, and only the difference, or prediction error, is encoded. Since only previously encoded pixels are used in predicting each pixel's value, this process is said to be causal. Restriction to causal prediction facilitates decoding: to decode a given pixel, its predicted value is recomputed from already decoded neighboring pixels, and added to the stored prediction error.

Noncausal prediction, based on a symmetric neighborhood centered at each pixel, should yield more accurate prediction and, hence, greater data compression. However, this approach does not permit simple sequential coding. Noncausal approaches to image coding typically involve image transforms, or the solution to large sets of simultaneous equations. Rather than encoding pixels sequentially, such techniques encode them all at once, or by blocks.

Both predictive and transform techniques have advantages. The former is relatively simple to implement and is readily adapted to local image characteristics. The latter generally provides greater data compression, but at the expense of considerably greater computation.

Here we shall describe a new technique for removing image correlation which combines features of predictive and transform methods. The technique is noncausal, yet computations are relatively simple and local.

The predicted value for each pixel is computed as a local weighted average, using a unimodal Gaussian-like (or related trimodal) weighting function centered on the pixel itself. The predicted values for all pixels are first obtained by convolving this weighting function with the image. The result is a low-pass filtered image which is then subtracted from the original.

Let $g_0(ij)$ be the original image, and $g_1(ij)$ be the result of applying an appropriate low-pass filter to g_0. The prediction error $L_0(ij)$ is then given by

$$L_0(ij) = g_0(ij) - g_1(ij).$$

Rather than encode g_0, we encode L_0 and g_1. This results in a net data compression because a) L_0 is largely decorrelated, and so may be represented pixel by pixel with many fewer bits than g_0, and b) g_1 is low-pass filtered, and so may be encoded at a reduced sample rate.

Further data compression is achieved by iterating this process. The reduced image g_1 is itself low-pass filtered to yield g_2 and a second error image is obtained: $L_2(ij) = g_1(ij) - g_2(ij)$. By repeating these steps several times we obtain a sequence of two-dimensional arrays $L_0, L_1, L_2, \cdots, L_n$. In our implementation each is smaller than its predecessor by a scale factor of 1/2 due to reduced sample density. If we now imagine these arrays stacked one above another, the result is a tapering pyramid data structure. The value at each node in the pyramid represents the difference between two Gaussian-like or related functions convolved with the original image. The difference between these two functions is similar to the "Laplacian" operators commonly used in image enhancement [13]. Thus, we refer to the proposed compressed image representation as the Laplacian-pyramid code.

The coding scheme outlined above will be practical only if required filtering computations can be performed with an efficient algorithm. A suitable fast algorithm has recently been developed [2] and will be described in the next section.

Paper approved by the Editor for Signal Processing and Communication Electronics of the IEEE Communications Society for publication after presentation in part at the Conference on Pattern Recognition and Image Processing, Dallas, TX, 1981. Manuscript received April 12, 1982; revised July 21, 1982. This work was supported in part by the National Science Foundation under Grant MCS-79-23422 and by the National Institutes of Health under Postdoctoral Training Grant EY07003.

P. J. Burt is with the Department of Electrical, Computer, and Systems Engineering, Rensselaer Polytechnic Institute, Troy, NY 12181.

E. H. Adelson is with the RCA David Sarnoff Research Center, Princeton, NJ 08540.

0090-6778/83/0400-0532$01.00 © 1983 IEEE

THE GAUSSIAN PYRAMID

The first step in Laplacian pyramid coding is to low-pass filter the original image g_0 to obtain image g_1. We say that g_1 is a "reduced" version of g_0 in that both resolution and sample density are decreased. In a similar way we form g_2 as a reduced version of g_1, and so on. Filtering is performed by a procedure equivalent to convolution with one of a family of local, symmetric weighting functions. An important member of this family resembles the Gaussian probability distribution, so the sequence of images $g_0, g_1, \cdots, g_n$ is called the Gaussian pyramid.[1]

A fast algorithm for generating the Gaussian pyramid is given in the next subsection. In the following subsection we show how the same algorithm can be used to "expand" an image array by interpolating values between sample points. This device is used here to help visualize the contents of levels in the Gaussian pyramid, and in the next section to define the Laplacian pyramid.

Gaussian Pyramid Generation

Suppose the image is represented initially by the array g_0 which contains C columns and R rows of pixels. Each pixel represents the light intensity at the corresponding image point by an integer I between 0 and $K-1$. This image becomes the bottom or zero level of the Gaussian pyramid. Pyramid level 1 contains image g_1, which is a reduced or low-pass filtered version of g_0. Each value within level 1 is computed as a weighted average of values in level 0 within a 5-by-5 window. Each value within level 2, representing g_2, is then obtained from values within level 1 by applying the same pattern of weights. A graphical representation of this process in one dimension is given in Fig. 1. The size of the weighting function is not critical [2]. We have selected the 5-by-5 pattern because it provides adequate filtering at low computational cost.

The level-to-level averaging process is performed by the function REDUCE.

$$g_k = \text{REDUCE}\,(g_{k-1}) \tag{1}$$

which means, for levels $0 < l < N$ and nodes i, j, $0 \leqslant i < C_l$, $0 \leqslant j < R_l$,

$$g_l(i,j) = \sum_{m=-2}^{2} \sum_{n=-2}^{2} w(m,n) g_{l-1}(2i+m, 2j+n).$$

Here N refers to the number of levels in the pyramid, while C_l and R_l are the dimensions of the lth level. Note in Fig. 1 that the density of nodes is reduced by half in one dimension, or by a fourth in two dimensions from level to level. The dimensions of the original image are appropriate for pyramid construction if integers M_C, M_R, and N exist such that $C = M_C 2^N + 1$ and $R = M_R 2^N + 1$. (For example, if M_C and M_R are both 3 and N is 5, then images measure 97 by 97 pixels.) The dimensions of g_l are $C_l = M_C 2^{N-l} + 1$ and $R_l = M_R 2^{N-l} + 1$.

[1] We will refer to this set of low-pass filtered images as the Gaussian pyramid, even though in some cases it will be generated with a trimodal rather than unimodal weighting function.

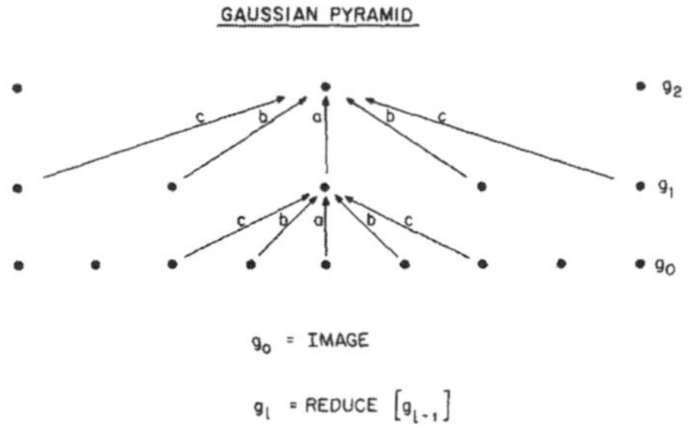

Fig. 1. A one-dimensional graphic representation of the process which generates a Gaussian pyramid. Each row of dots represents nodes within a level of the pyramid. The value of each node in the zero level is just the gray level of a corresponding image pixel. The value of each node in a high level is the weighted average of node values in the next lower level. Note that node spacing doubles from level to level, while the same weighting pattern or "generating kernel" is used to generate all levels.

The Generating Kernel

Note that the same 5-by-5 pattern of weights w is used to generate each pyramid array from its predecessor. This weighting pattern, called the generating kernel, is chosen subject to certain constraints [2]. For simplicity we make w separable:

$$w(m,n) = \hat{w}(m)\hat{w}(n).$$

The one-dimensional, length 5, function $\hat{w}$ is normalized

$$\sum_{m=-2}^{2} \hat{w}(m) = 1$$

and symmetric

$$\hat{w}(i) = \hat{w}(-i) \qquad \text{for } i = 0,\ 1, 2.$$

An additional constraint is called equal contribution. This stipulates that all nodes at a given level must contribute the same total weight (=1/4) to nodes at the next higher level. Let $\hat{w}(0) = a$, $\hat{w}(-1) = \hat{w}(1) = b$, and $\hat{w}(-2) = \hat{w}(2) = c$. In this case equal contribution requires that $a + 2c = 2b$. These three constraints are satisfied when

$$\hat{w}(0) = a$$

$$\hat{w}(-1) = \hat{w}(1) = 1/4$$

$$\hat{w}(-2) = \hat{w}(2) = 1/4 - a/2.$$

Equivalent Weighting Functions

Iterative pyramid generation is equivalent to convolving the image g_0 with a set of "equivalent weighting functions" h_l:

$$g_l = h_l \oplus g_0$$

or

$$g_l(i,j) = \sum_{m=-M_l}^{M_l} \sum_{n=-M_l}^{M_l} h_l(m,n) g_0(i2^l + m \cdot j2^l + n).$$

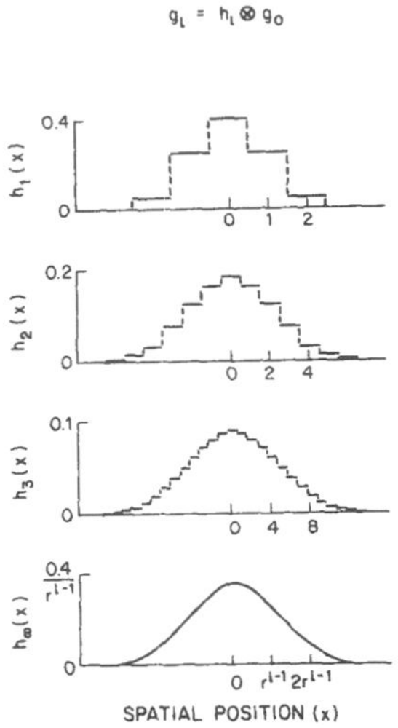

Fig. 2. The equivalent weighting functions $h_l(x)$ for nodes in levels 1, 2, 3, and infinity of the Gaussian pyramid. Note that axis scales have been adjusted by factors of 2 to aid comparison. Here the parameter a of the generating kernel is 0.4, and the resulting equivalent weighting functions closely resemble the Gaussian probability density functions.

The size M_l of the equivalent weighting function doubles from one level to the next, as does the distance between samples.

Equivalent weighting functions for Gaussian-pyramid levels 1, 2, and 3 are shown in Fig. 2. In this case $a = 0.4$. The shape of the equivalent function converges rapidly to a characteristic form with successively higher levels of the pyramid, so that only its scale changes. However, this shape does depend on the choice of a in the generating kernel. Characteristic shapes for four choices of a are shown in Fig. 3. Note that the equivalent weighting functions are particularly Gaussian-like when $a = 0.4$. When $a = 0.5$ the shape is triangular; when $a = 0.3$ it is flatter and broader than a Gaussian. With $a = 0.6$ the central positive mode is sharply peaked, and is flanked by small negative lobes.

Fast Filter

The effect of convolving an image with one of the equivalent weighting functions h_l is to blur, or low-pass filter, the image. The pyramid algorithm reduces the filter band limit by an octave from level to level, and reduces the sample interval by the same factor. This is a very fast algorithm, requiring fewer computational steps to compute a set of filtered images than are required by the fast Fourier transform to compute a single filtered image [2].

Example: Fig. 4 illustrates the contents of a Gaussian pyramid generated with $a = 0.4$. The original image, on the far left, measures 257 by 257. This becomes level 0 on the pyramid. Each higher level array is roughly half as large in each dimension as its predecessor, due to reduced sample density.

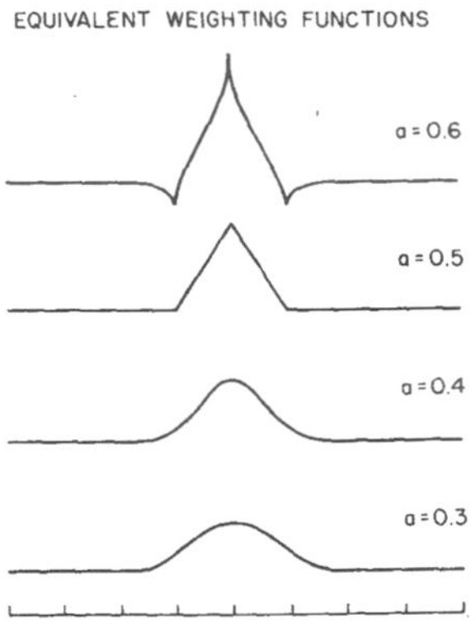

Fig. 3. The shape of the equivalent weighting function depends on the choice of parameter a. For $a = 0.5$, the function is triangular; for $a = 0.4$ it is Gaussian-like, and for $a = 0.3$ it is broader than Gaussian. For $a = 0.6$ the function is trimodal.

Gaussian Pyramid Interpolation

We now define a function EXPAND as the reverse of REDUCE. Its effect is to expand an $(M + 1)$-by-$(N + 1)$ array into a $(2M + 1)$-by-$(2N + 1)$ array by interpolating new node values between the given values. Thus, EXPAND applied to array g_l of the Gaussian pyramid would yield an array $g_{l,1}$ which is the same size as g_{l-1}.

Let $g_{l,n}$ be the result of expanding g_l n times. Then

$$g_{l,0} = g_l$$

and

$$g_{l,n} = \text{EXPAND}\,(g_{l,n-1}).$$

By EXPAND we mean, for levels $0 < l \leqslant N$ and $0 \leqslant n$ and nodes i, j, $0 \leqslant i < C_{l-n}$, $0 \leqslant j < R_{l-n}$,

$$g_{l,n}(ij) = 4 \sum_{m=-2}^{2} \sum_{n=-2}^{2} w(m, n) \cdot g_{l,n-1}\left(\frac{i-m}{2}, \frac{j-n}{2}\right). \tag{2}$$

Only terms for which $(i - m)/2$ and $(j - n)/2$ are integers are included in this sum.

If we apply EXPAND l times to image g_l, we obtain $g_{l,l}$, which is the same size as the original image g_0. Although full expansion will not be used in image coding, we will use it to help visualize the contents of various arrays within pyramid structures. The top row of Fig. 5 shows image $g_{0,0}$, $g_{1,1}$, $g_{2,2}$, $\cdots$ obtained by expanding levels of the pyramid in Fig. 4. The low-pass filter effect of the Gaussian pyramid is now shown clearly.

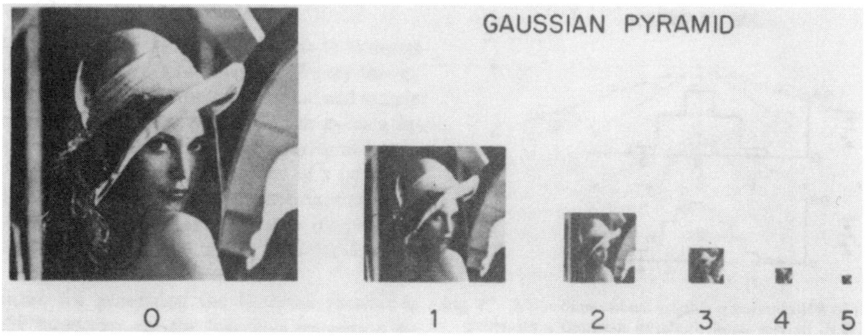

Fig. 4. First six levels of the Gaussian pyramid for the "Lady" image. The original image, level 0, measures 257 by 257 pixels, and each higher level array is roughly half the dimensions of its predecessor. Thus, level 5 measures just 9 by 9 pixels.

THE LAPLACIAN PYRAMID

Recall that our purpose for constructing the reduced image g_1 is that it may serve as a prediction for pixel values in the original image g_0. To obtain a compressed representation, we encode the error image which remains when an expanded g_1 is subtracted from g_0. This image becomes the bottom level of the Laplacian pyramid. The next level is generated by encoding g_1 in the same way. We now give a formal definition for the Laplacian pyramid, and examine its properties.

Laplacian Pyramid Generation

The Laplacian pyramid is a sequence of error images L_0, L_1, $\cdots$, L_N. Each is the difference between two levels of the Gaussian pyramid. Thus, for $0 \leqslant l < N$,

$$\begin{aligned} L_l &= g_l - \text{EXPAND}\,(g_{l+1}) \\ &= g_l - g_{l+1,1}. \end{aligned} \tag{3}$$

Since there is no image g_{N+1} to serve as the prediction image for g_N, we say $L_N = g_N$.

Equivalent Weighting Functions

The value at each node in the Laplacian pyramid is the difference between the convolutions of two equivalent weighting functions h_l, h_{l+1} with the original image. Again, this is similar to convolving an appropriately scaled Laplacian weighting function with the image. The node value could have been obtained directly by applying this operator, although at considerably greater computational cost.

Just as we may view the Gaussian pyramid as a set of low-pass filtered copies of the original image, we may view the Laplacian pyramid as a set of bandpass filtered copies of the image. The scale of the Laplacian operator doubles from level to level of the pyramid, while the center frequency of the pass-band is reduced by an octave.

In order to illustrate the contents of the Laplacian pyramid, it is helpful to interpolate between sample points. This may be done within the pyramid structure by Gaussian interpolation. Let $L_{l,n}$ be the result of expanding L_l n times using (2). Then, $L_{l,l}$ is the size of the original image.

The expanded Laplacian pyramid levels for the "Lady" image of Fig. 4 are shown in the bottom row of Fig. 5. Note that image features such as edges and bars appear enhanced in the Laplacian pyramid. Enhanced features are segregated by size: fine details are prominent in $L_{0,0}$, while progressively coarser features are prominent in the higher level images.

Decoding

It can be shown that the original image can be recovered exactly by expanding, then summing all the levels of the Laplacian pyramid:

$$g_0 = \sum_{l=0}^{N} L_{l,l}.$$

A more efficient procedure is to expand L_N once and add it to L_{N-1}, then expand this image once and add it to L_{N-2}, and so on until level 0 is reached and g_0 is recovered. This procedure simply reverses the steps in Laplacian pyramid generation. From (3) we see that

$$g_N = L_N \tag{4}$$

and for $l = N-1, N-2, \cdots, 0$,

$$g_l = L_l + \text{EXPAND}\,(g_{l+1}).$$

Entropy

If we assume that the pixel values of an image representation are statistically independent, then the minimum number of bits per pixel required to exactly encode the image is given by the entropy of the pixel value distribution. This optimum may be approached in practice through techniques such as variable length coding.

The histogram of pixel values for the "Lady" image is shown in Fig. 6(a). If we let the observed frequency of occurrence $f(i)$ of each gray level i be an estimate of its probability of occurrence in this and other similar images, then the entropy

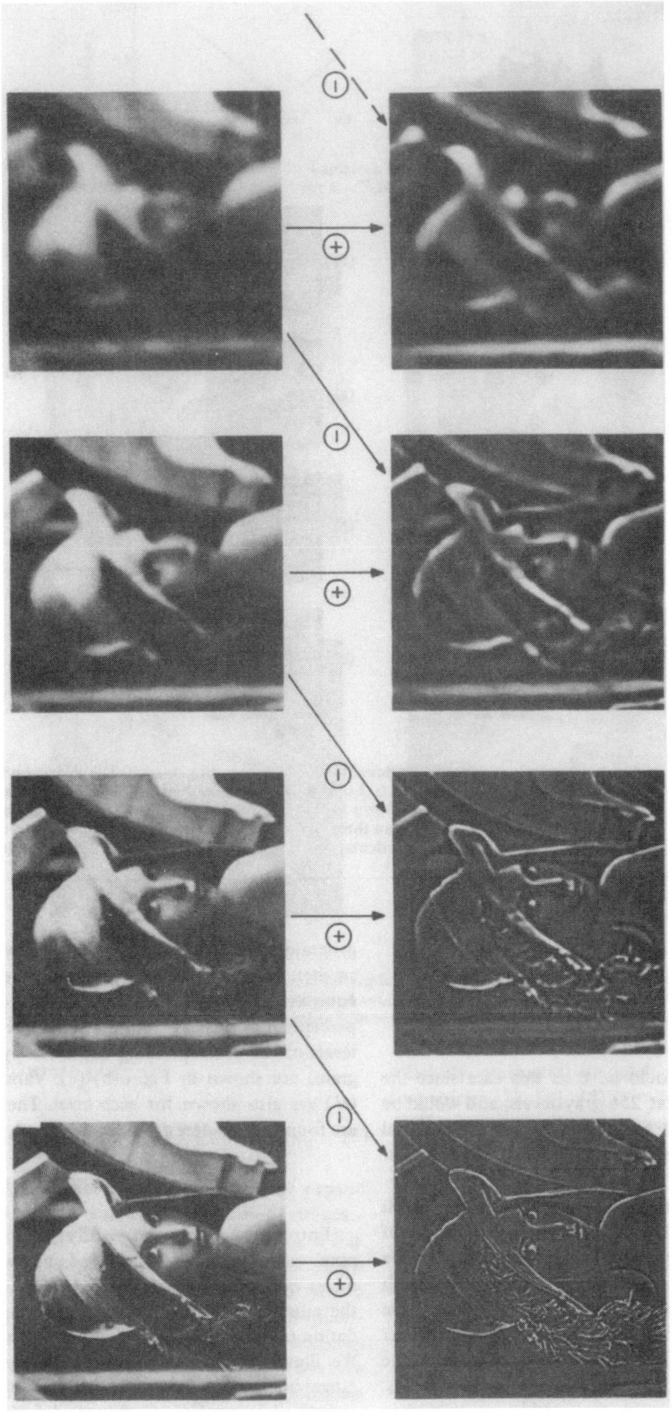

Fig. 5. First four levels of the Gaussian and Laplacian pyramids. Gaussian images, upper row, were obtained by expanding pyramid arrays (Fig. 4) through Gaussian interpolation. Each level of the Laplacian pyramid is the difference between the corresponding and next higher levels of the Gaussian pyramid.

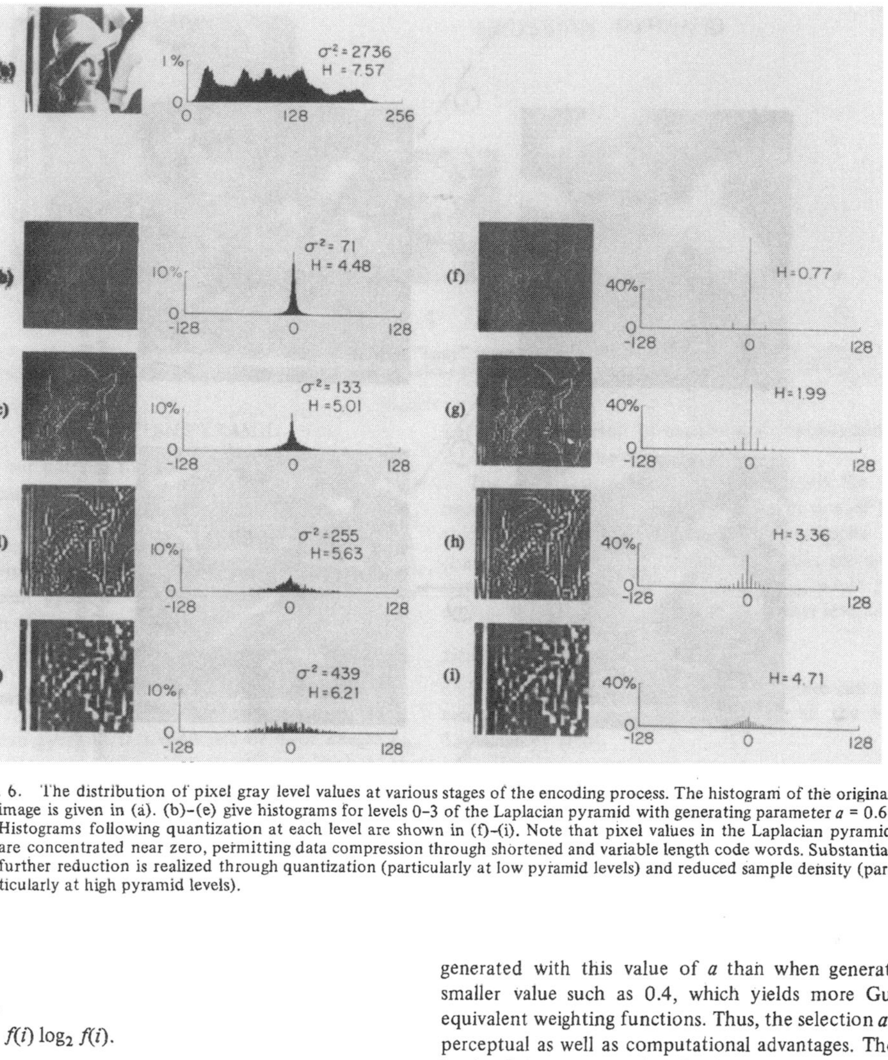

Fig. 6. The distribution of pixel gray level values at various stages of the encoding process. The histogram of the original image is given in (a). (b)-(e) give histograms for levels 0-3 of the Laplacian pyramid with generating parameter $a = 0.6$. Histograms following quantization at each level are shown in (f)-(i). Note that pixel values in the Laplacian pyramid are concentrated near zero, permitting data compression through shortened and variable length code words. Substantial further reduction is realized through quantization (particularly at low pyramid levels) and reduced sample density (particularly at high pyramid levels).

is given by

$$H = -\sum_{i=0}^{255} f(i) \log_2 f(i).$$

The maximum entropy would be 8 in this case since the image is initially represented at 256 gray levels, and would be obtained when all gray levels were equally likely. The actual entropy estimate for "Lady" is slightly less than this, at 7.57.

The technique of subtracting a predicted value from each image pixel, as in the Laplacian pyramid, removes much of the pixel-to-pixel correlation. Decorrelation also results in a concentration of pixel values around zero, and, therefore, in reduced variance and entropy. The degree to which these measures are reduced depends on the value of the parameter "a" used in pyramid generation (see Fig. 7). We found that the greatest reduction was obtained for $a = 0.6$ in our examples. Levels of the Gaussian pyramid appeared "crisper" when generated with this value of a than when generated with a smaller value such as 0.4, which yields more Guassian-like equivalent weighting functions. Thus, the selection $a = 0.6$ had perceptual as well as computational advantages. The first four levels of the corresponding Laplacian pyramid and their histograms are shown in Fig. 6(b)-(e). Variance (σ^2) and entropy (H) are also shown for each level. These quantities generally are found to increase from level to level, as in this example.

QUANTIZATION

Entropy can be substantially reduced by quantizing the pixel values in each level of the Laplacian pyramid. This introduces quantization errors, but through the proper choice of the number and distribution of quantization levels, the degradation may be made almost imperceptible to human observers. We illustrate this procedure with uniform quantization. The range of pixel values is divided into bins of size n, and the quantized value $C_l(i, j)$ for pixel $L_l(i, j)$ is just the middle

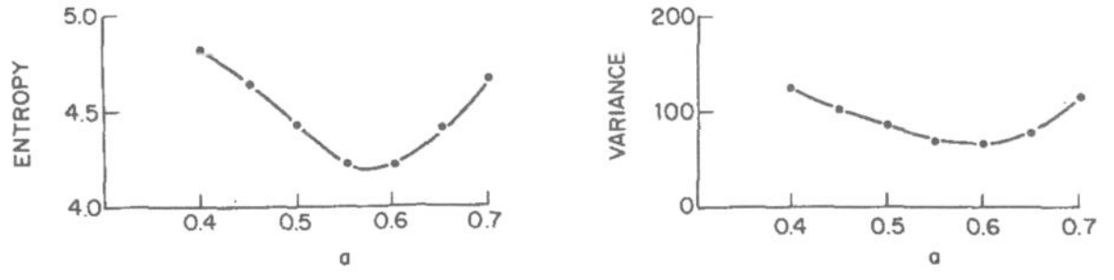

Fig. 7. Entropy and variance of pixel values in Laplacian pyramid level 0 as a function of the parameter "a" for the "Lady" image. Greatest reduction is obtained for $a \cong 0.6$. This estimate of the optimal "a" was also obtained at other pyramid levels and for other images.

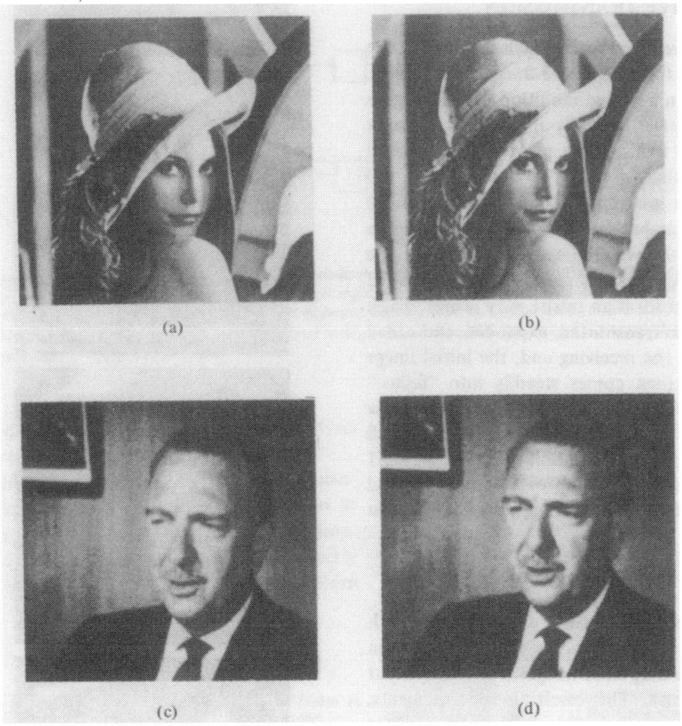

Fig. 8. Examples of image data compression using the Laplacian pyramid code. (a) and (c) give the original "Lady" and "Walter" images, while (b) and (d) give their encoded versions. The data rates are 1.58 and 0.73 bits/pixel for "Lady" and "Walter," respectively. The corresponding mean square errors were 0.88 percent and 0.43 percent, respectively.

value of the bin which contains $L_l(i, j)$:

$$C_l(i, j) = mn \quad \text{if } (m - 1/2)n < L_l(i, j) \leqslant (m + 1/2)n. \quad (5)$$

The quantized image is reconstructed through the expand and sum procedure (4) using C values in the place of L values.

Results of quantizing the "Lady" image are shown in Fig. 6(f)-(i). The bin size for each level was chosen by increasing n until degradation was just perceptible when viewed from a distance of approximately five times the image width (pixel-pixel separation $\cong$ 3 min arc). Note that bin size becomes smaller at higher levels (lower spatial frequencies). Bin size at a given pyramid level reflects the sensitivity of the human observer to contrast errors within the spatial frequency bands represented at that level. Humans are fairly sensitive to contrast perturbations at low and medium spatial frequencies, but relatively insensitive to such perturbations at high spatial frequencies [3], [4], [7].

This increased observer sensitivity along with the increased data variance noted above means that more quantization levels must be used at high pyramid levels than at low levels. Fortunately, these pixels contribute little to the overall bit rate for the image, due to their low sample density. The low-level (high-frequency) pixels, which are densely sampled, can be coarsely quantized (cf. [6], [11], [12]).

RESULTS

The final result of encoding, quantization, and reconstruction are shown in Fig. 8. The original "Lady" image is shown in Fig. 8(a); the encoded version, at 1.58 bits/pixel, is shown in Fig. 8(b). We assume that variable-length code words are used to take advantage of the nonuniform distribution of node values, so the bit rate for a given pyramid level is its estimated entropy times its sample density, and the bit rate for the image is the sum of that for all levels. The same procedure was performed on the "Walter" image; the original is shown in Fig. 8(c). while the version encoded at 0.73 bits/pixel is shown in Fig. 8(d). In both cases, the encoded images are almost indistinguishable from the originals under viewing conditions as stated above.

PROGRESSIVE TRANSMISSION

It should also be observed that the Laplacian pyramid code is particularly well suited for progressive image transmission. In this type of transmission a coarse rendition of the image is sent first to give the receiver an early impression of image content, then subsequent transmission provides image detail of progressively finer resolution [5]. The observer may terminate transmission of an image as soon as its contents are recognized, or as soon as it becomes evident that the image will not be of interest. To achieve progressive transmission, the topmost level of the pyramid code is sent first, and expanded in the receiving pyramid to form an initial, very coarse image. The next lower level is then transmitted, expanded, and added to the first, and so on. At the receiving end, the initial image appears very blurry, but then comes steadily into "focus." This progression is illustrated in Fig. 9, from left to right. Note that while 1.58 bits are required for each pixel of the full transmission (rightmost image), about half of these, or 0.81 bits, are needed for each pixel for the previous image (second from right, Fig. 9), and 0.31 for the image previous to that (third from right).

SUMMARY AND CONCLUSION

The Laplacian pyramid is a versatile data structure with many attractive features for image processing. It represents an image as a series of quasi-bandpassed images, each sampled at successively sparser densities. The resulting code elements, which form a self-similar structure, are localized in both space and spatial frequency. By appropriately choosing the parameters of the encoding and quantizing scheme, one can substantially reduce the entropy in the representation, and simultaneously stay within the distortion limits imposed by the sensitivity of the human visual system.

Fig. 10 summarizes the steps in Laplacian pyramid coding. The first step, shown on the far left, is bottom-up construction of the Gaussian pyramid images $g_0, g_1, \cdots, g_N$ [see (1)]. The Laplacian pyramid images $L_0, L_1, \cdots, L_N$ are then obtained as the difference between successive Gaussian levels [see (3)]. These are quantized to yield the compressed code represented by the pyramid of values $C_l(ij)$ [see (5)]. Finally, image reconstruction follows an expand-and-sum procedure [see (4)] using C values in the place of L values. Here we designate the reconstructed image by r_0.

It should also be observed that the Laplacian pyramid encoding scheme requires relatively simple computations. The computations are local and may be performed in parallel, and the same computations are iterated to build each pyramid level from its predecessors. We may envision performing Lapla-

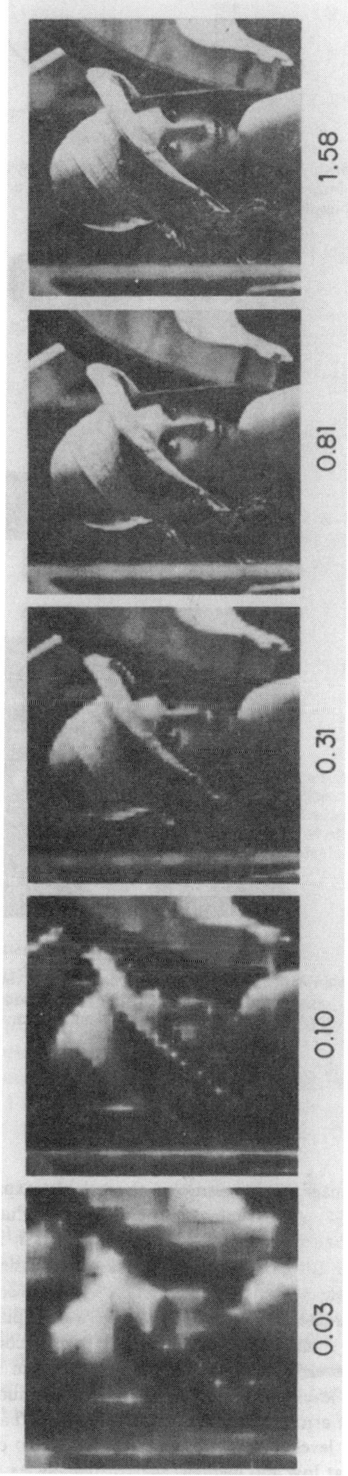

Fig. 9. Laplacian pyramid code applied to progressive image transmission. High levels of the pyramid are transmitted first to give the receiver a quick but very coarse rendition of the image. The receiver's image is then progressively refined by adding successively lower pyramid levels as these are transmitted. In the example shown here, the leftmost figure shows reconstruction using pyramid levels 4–8, or just 0.03 bits/pixel. The following four figures show the reconstruction after pyramid levels 3, 2, 1, and 0 have been added. The cumulative data rates are shown under each figure in bits per pixel.

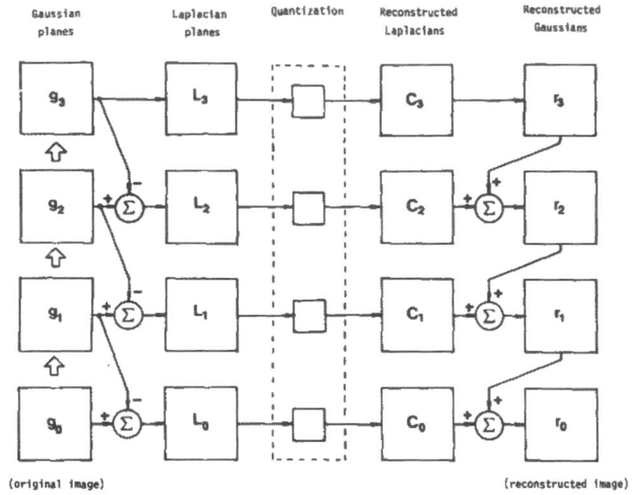

Fig. 10. A summary of the steps in Laplacian pyramid coding and decoding. First, the original image g_0 (lower left) is used to generate Gaussian pyramid levels $g_1, g_2, \cdots$ through repeated local averaging. Levels of the Laplacian pyramid $L_0, L_1, \cdots$ are then computed as the differences between adjacent Gaussian levels. Laplacian pyramid elements are quantized to yield the Laplacian pyramid code $C_0, C_1, C_2, \cdots$. Finally, a reconstructed image r_0 is generated by summing levels of the code pyramid.

cian coding and decoding in real time using array processors and a pipeline architecture.

An additional benefit, previously noted, is that in computing the Laplacian pyramid, one automatically has access to quasi-bandpass copies of the image. In this representation, image features of various sizes are enhanced and are directly available for various image processing (e.g., [1]) and pattern recognition tasks.

REFERENCES

[1] K. D. Baker and G. D. Sullivan, "Multiple bandpass filters in image processing," *Proc. IEE*, vol. 127, pp. 173–184, 1980.

[2] P. J. Burt, "Fast filter transforms for image processing," *Comput. Graphics, Image Processing*, vol. 16, pp. 20–51, 1981.

[3] C. R. Carlson and R. W. Cohen, "Visibility of displayed information," Off. Naval Res., Tech. Rep., Contr. N00014-74-C-0184, 1978.

[4] ——, "A simple psychophysical model for predicting the visibility of displayed information," *Proc. Soc. Inform. Display*, pp. 229–246, 1980.

[5] K. Knowlton, "Progressive transmission of grayscale and binary pictures by simple, efficient, and lossless encoding schemes," *Proc. IEEE*, vol. 68, pp. 885–896, 1980.

[6] E. R. Kretzmer, "Reduced-alphabet representation of television signals," in *IRE Nat. Conv. Rec.*, 1956, pp. 140–147.

[7] J. J. Kulikowski and A. Gorea, "Complete adaptation to patterned stimuli: A necessary and sufficient condition for Weber's law for contrast," *Vision Res.*, vol. 18, pp. 1223–1227, 1978.

[8] A. N. Netravali and B. Prasada, "Adaptive quantization of picture signals using spatial masking," *Proc. IEEE*, vol. 65, pp. 536–548, 1977.

[9] A. N. Netravali and J. O. Limb, "Picture coding: A review," *Proc. IEEE*, vol. 68, pp. 336–406, 1980.

[10] W. K. Pratt, Ed., *Image Transmission Tecniques*. New York: Academic, 1979.

[11] W. F. Schreiber, C. F. Knapp, and N. D. Key, "Synthetic highs, an experimental TV bandwidth reduction system," *J. Soc. Motion Pict. Telev. Eng.*, vol. 68, pp. 525–537, 1959.

[12] W. F. Schreiber and D. E. Troxel, U.S. Patent 4 268 861, 1981.

[13] A. Rosenfeld and A. Kak, *Digital Picture Processing*. New York: Academic, 1976.

★

Peter J. Burt (M'80) received the B.A. degree in physics from Harvard University, Cambridge, MA, in 1968, and the M.S. and Ph.D. degrees in computer science from the University of Massachusetts, Amherst, in 1974 and 1976, respectively.

From 1968 to 1972 he conducted research in sonar, particularly in acoustic imaging devices, at the USN Underwater Sound Laboratory, New London, CT, and in London, England. As a Postdoctoral Fellow, he has studied both natural vision and computer image understanding at New York University, New York, NY (1976–1978), Bell Laboratories (1978–1979), and the University of Maryland, College Park (1979–1980). He has been a member of the faculty at Rensselaer Polytechnic Institute, Troy, NY, since 1980.

★

Edward H. Adelson received the B.A. degree in physics and philosophy from Yale University, New Haven, CT, in 1974, and the Ph.D. degree in experimental psychology from the University of Michigan, Ann Arbor, in 1979.

From 1979 to 1981 he was Postdoctoral Fellow at New York University, New York, NY. Since 1981, he has been at RCA David Sarnoff Research Center, Princeton, NJ, as a member of the Technical Staff in the Image Quality and Human Perception Research Group. His research interests center on visual processes in both human and machine visual systems, and include psychophysics, image processing, and artificial intelligence.

Dr. Adelson is a member of the Optical Society of America, the Association for Research in Vision and Ophthalmology, and Phi Beta Kappa.

Copyright © 1976 American Telephone and Telegraph Company
THE BELL SYSTEM TECHNICAL JOURNAL
Vol. 55, No. 8, October 1976
Printed in U.S.A.

Digital Coding of Speech in Sub-bands

By R. E. CROCHIERE, S. A. WEBBER, and J. L. FLANAGAN

(Manuscript received March 26, 1976)

A rationale is advanced for digitally coding speech signals in terms of sub-bands of the total spectrum. The approach provides a means for controlling and reducing quantizing noise in the coding. Each sub-band is quantized with an accuracy (bit allocation) based upon perceptual criteria. As a result, the quality of the coded signal is improved over that obtained from a single full-band coding of the total spectrum. In one implementation, the individual sub-bands are low-pass translated before coding. In another, "integer-band" sampling is employed to alias the signal in an advantageous way before coding. Other possibilities extend to complex demodulation of the sub-bands, and to representing the sub-band signals in terms of envelopes and phase-derivatives. In all techniques, adaptive quantization is used for the coding, and a parsimonious allocation of bits is made across the bands. Computer simulations are made to demonstrate the signal qualities obtained for codings at 16 and 9.6 kb/s.

I. DIVISION OF SPEECH SPECTRUM INTO SUB-BANDS

For digital transmission a signal must be sampled and quantized. Quantization is a nonlinear operation and produces distortion products that are typically broad in spectrum. Because of the characteristics of the speech spectrum, quantizing distortion is not equally detectable at all frequencies. Coding the signal in narrower sub-bands offers one possibility for controlling the distribution of quantizing noise across the signal spectrum and, hence, for realizing an improvement in signal quality. In earlier work, splitting of the spectrum by high-pass and low-pass filtering has been used advantageously for video and speech transmission.[1,2]

A question, then, is what design of sub-bands makes sense for speech coding? A choice based upon perceptual criteria is suggested, namely, band-partitioning such that each sub-band contributes equally to the so-called articulation index (AI).[3] The AI concept is based upon a nonuniform division of the frequency scale for the speech spectrum. Twenty nonuniform contiguous bands are derived in which each elemental band contributes 5 percent to the total AI.

Appealing to this notion, one partitioning of the frequency range 200 to 3200 Hz into four "equal-contribution" bands is given below and shown in Fig. 1.

Sub-band Number	Frequency Range (Hz)
1	200–700
2	700–1310
3	1310–2020
4	2020–3200

Each sub-band in its original analog form contributes 20 percent to AI. The total AI, therefore, is 80 percent, which corresponds to a word intelligibility of approximately 93 percent.[4]

II. LOW-PASS TRANSLATION OF SUB-BANDS

A straightforward approach to processing the sub-bands is to make a low-pass translation before coding. This facilitates sampling-rate reduction and realizes any benefits which might accrue from coding the low-pass signal.

The low-pass translation can be accomplished in a variety of ways. One method is shown in Fig. 2. The input speech signal is filtered with a bandpass filter of width W_n for the nth band. W_{1n} is the lower edge of the band and W_{2n} is the upper edge of the band. The resulting signal $s_n(t)$ is modulated by a cosine wave, $\cos(W_{1n}t)$, and filtered

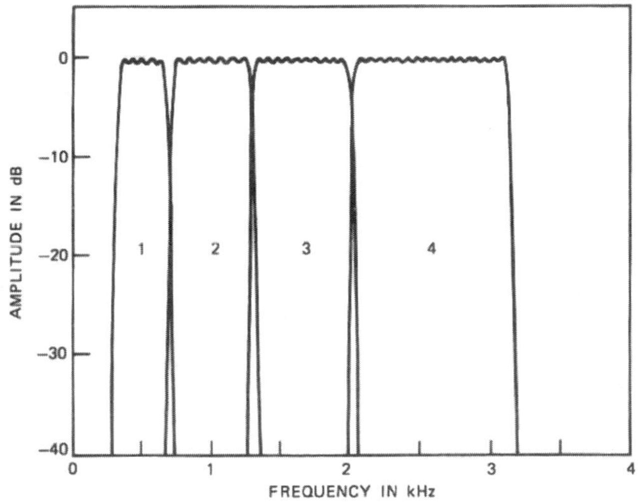

Fig. 1—Partitioning of the speech spectrum into four contiguous bands that contribute equally to articulation index. The frequency range is 200 to 3200 Hz.

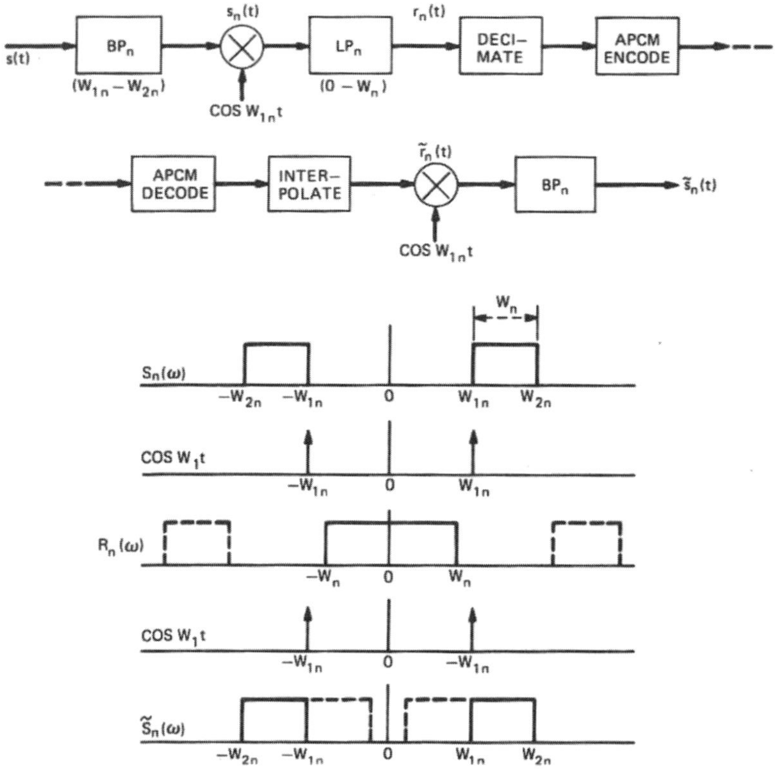

Fig. 2—Sequence of operations for low-pass translation of speech sub-bands, adaptive PCM encoding, transmission, decoding, and band restoration.

by a low-pass filter $h_n(t)$ with bandwidth $(0 - W_n)$. This filter is necessary to remove the unwanted signal images above $2W_{1n}$, as shown in Fig. 2. The resulting signal $r_n(t)$ corresponds to the low-pass translated version of $s_n(t)$ and can be expressed in the form:

$$r_n(t) = [s_n(t) \cos (W_{1n}t)]^* h_n(t). \tag{1}$$

Notice, in this instance, that a constraint is implied by the convolution, namely, that the passband width $W_n \leqq 2W_{1n}$, or that $W_{2n} \leqq 3W_{1n}$. Practically this poses no problem.*

The signal $r_n(t)$ is sampled at rate $2W_n$. If it is already in digital form, the sampling rate is decimated (reduced) to the rate $2W_n$. This signal is digitally encoded and multiplexed with encoded signals from other channels as shown in Fig. 3. At the receiver the data is demulti-

* For example, this constraint requires that W be increased slightly, from 200 to 233 Hz, for $n = 1$ in Fig. 1.

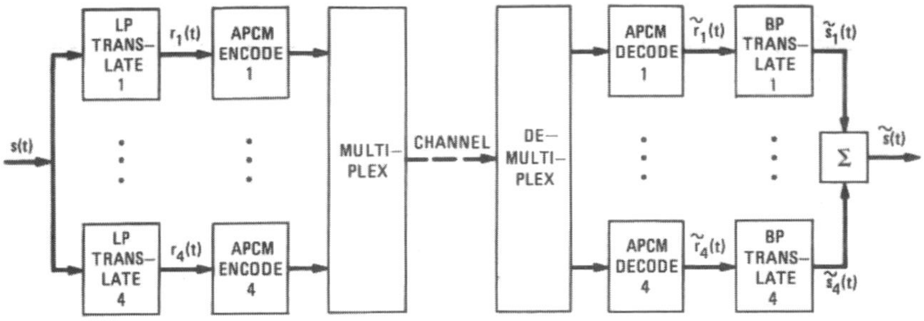

Fig. 3—Four-band encoder using low-pass translation and APCM encoding in each band.

plexed into separate channels, decoded, and interpolated to give the estimate $\tilde{r}_n(t)$ for the nth channel. Reconstruction of the detected signal is simply done by the reverse band translation. That is, it is modulated by cos $(W_{1n}t)$ and bandpass filtered to the original passband, as shown in Fig. 2. The sub-band signal $\tilde{s}_n(t)$ is then summed with the other bands to give the full-band signal $\tilde{s}(t)$.

An alternate implementation of the low-pass translation method, which avoids the above-mentioned restriction on W_n, follows from a modification of the complex demodulation process. In this approach, $s(t)$ is complex modulated by $e^{j\omega_n t}[\omega_n = (W_{1n} + W_{2n})/2 =$ center frequency of band $n]$ and filtered by a low-pass filter $h_n'(t)$ with bandwidth $(0 - W_n/2)$. The resulting complex signal $a_n(t) + jb_n(t)$,

$$a_n(t) = [s(t) \cos \omega_n t]^* h_n'(t) \tag{2a}$$

$$b_n(t) = [s(t) \sin \omega_n t]^* h_n'(t) \tag{2b}$$

corresponds exactly to the output of the phase vocoder.[5] The conjugate of this signal $a_n(t) - jb_n(t)$ corresponds to a modulation of $s(t)$ by $e^{-j\omega_n t}$. If the complex signal $a_n(t) + jb_n(t)$ is complex modulated by $e^{-j(W_n/2)t}$ and its conjugate complex modulated by $e^{j(W_n/2)t}$, the two resulting complex signals correspond to the negative and positive frequency components of the low-pass translated signal $r_n(t)$, as shown in Fig. 4. The sum of these two signals gives a real signal corresponding to the desired low-pass translated signal $r_n(t)$; i.e.,

$$r_n(t) = [a_n(t) + jb_n(t)]e^{-j(W_n/2)t} + [a_n(t) - jb_n(t)]e^{+j(W_n/2)t}, \tag{3}$$

or

$$r_n(t) = 2\left[a_n(t) \cos\left(\frac{W_n}{2}t\right) + b_n(t) \sin\left(\frac{W_n}{2}t\right)\right]. \tag{4}$$

For reconstruction, it can be shown that $a_n(t)$ and $b_n(t)$ can be recovered from the low-pass translated signal $r_n(t)$ by the following relations

$$a_n(t) = [r_n(t) \cos (W_n t/2)]*h_n'(t) \tag{5a}$$

$$b_n(t) = [r_n(t) \sin (W_n t/2)]*h_n'(t). \tag{5b}$$

Equations (4) and (5) suggest a method of implementation of the low-pass translation and reconstruction with a phase vocoder. For a digital implementation of the low-pass translation, this approach is particularly appealing. For example, at the sampling rate $f_s = 2W_n/2\pi$, the sequences corresponding to $\cos (W_n t/2)$ and $\sin (W_n t/2)$ are 1, 0, −1, 0, 1, ···, and 0, 1, 0, −1, 0, ···, respectively. Therefore, an efficient way to generate $r_n(t)$ is to sample a_n and b_n (or decimate if they are in digital form) to one half of this sampling rate (i.e., $W_n/2\pi$) and form $r_n(t)$ by interleaving samples of a_n and b_n (with appropriate sign changes). A similar approach can be used in the reconstruction process by recognizing that alternate samples of $r_n(t) \cos (W_n t/2)$ and

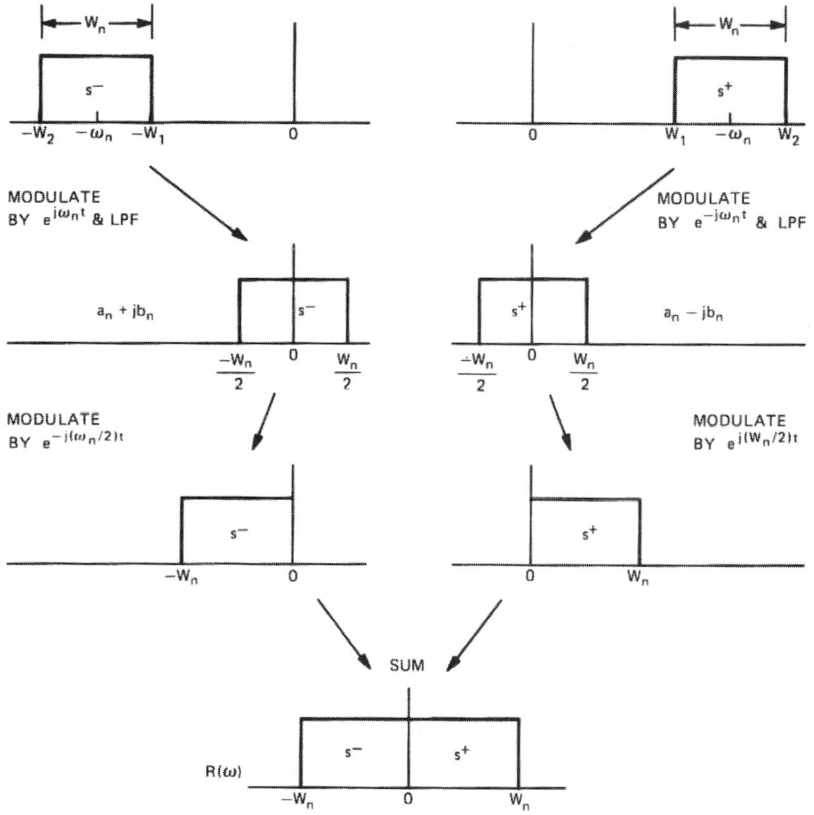

Fig. 4—Frequency-domain interpretation of complex demodulation method for low-pass translation.

$r_n(t)$ sin $(W_n t/2)$ (at sampling rate $2W_n/2\pi$) are zero valued. Thus, the two input sequences to the interpolators (which can be sampled at half of this rate or $W_n/2\pi$) can be generated by selecting alternate samples of $r_n(t)$ (with appropriate sign changes).

A further modification on this approach can be made by noting that, since adaptive coding is used to encode $r_n(t)$, the sign changes in the construction and separation of $r_n(t)$ are not necessary. That is, an alternate sequence $r_n'(t)$ can be generated by interleaving samples of a_n and b_n without sign changes. This sequence can be encoded and decoded and inputs to the interpolators can be formed from alternate samples of $\tilde{r}_n'(t)$ (without sign changes). Figure 5 shows an implementation of this method. The signal $s(t)$ is modulated by $\cos \omega_n t$ and $\sin \omega_n t$, where ω_n is the center frequency of band n. These signals are filtered with low-pass filters $h_n'(t)$ with bandwidth $(0 - W_n/2)$. The outputs are decimated (if they are in digital form) or sampled (if analog) at a sampling rate W_n. The low-pass translated signal $r_n'(t)$ is obtained (at sampling rate $2W_n$) by interleaving samples of a_n and b_n. $r_n'(t)$ is encoded, transmitted, and decoded as in Fig. 3. On reconstruction a_n and b_n are recovered by selecting alternate samples of $\tilde{r}_n'(t)$. These signals are then interpolated, filtered, modulated, and

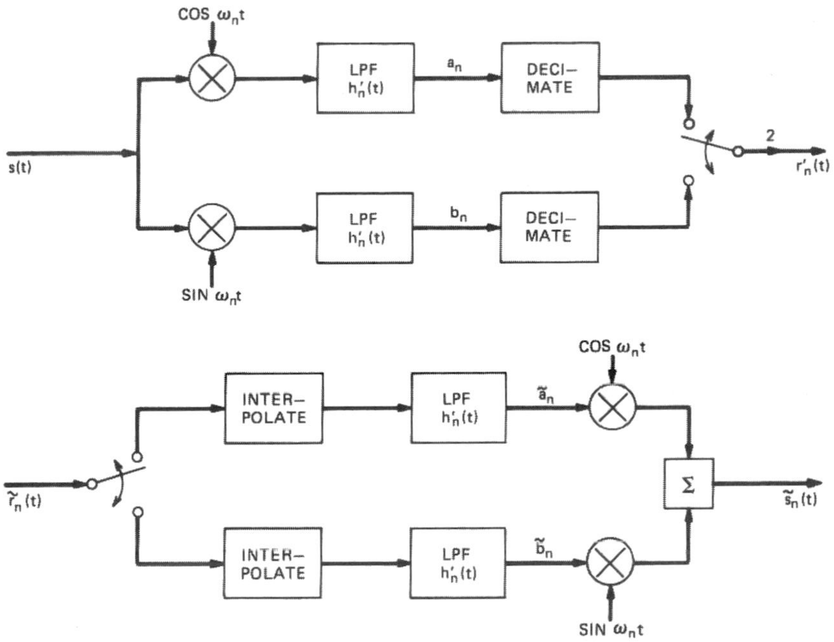

Fig. 5—Implementation of complex demodulation for low-pass translation with interleaving of samples of a_n and b_n.

summed as shown in Fig. 5 to give the reconstructed sub-band signal $\tilde{s}_n(t)$.

For digital implementation h_n' can be realized with a digital filter. Decimation, or sampling-rate reduction by an integer factor M, can be achieved by retaining only one out of every M samples of the output of the filter. The filter is necessary to avoid aliasing. Interpolation by an integer factor M is accomplished by increasing the sampling rate by filling in $M - 1$ zero-valued samples between each pair of input samples. The filter h_n' then removes the unwanted harmonic images of the base-band signal and smooths (i.e., interpolates) these samples to appropriate values of the base-band waveform. Efficient methods for implementing digital decimators and interpolators are discussed in Ref. (6).

III. ENCODING OF THE SUB-BAND SIGNALS

Digital encoding of the low-pass translated signal $r_n(t)$ is best accomplished using adaptive-PCM (APCM).[7,8] APCM encoding is preferred over adaptive-differential PCM (ADPCM) methods in this case due to the low sample-to-sample correlation of the low-pass-translated, Nyquist-rated, sampled signals.

For computer simulations, APCM coders based on a one-word step-size memory were used according to methods proposed by Jayant, Flanagan, and Cummiskey.[7–9] Step-size adaption is achieved according to the relation

$$\Delta_r = \Delta_{r-1} \times M, \qquad (6)$$

where Δ_r is the quantizer step-size used for the rth sample and Δ_{r-1} is the step-size of the $(r - 1)$th sample. M is a multiplication factor whose value depends on the quantizer level at the $(r - 1)$th sample. For example, in a two-bit quantizer, two magnitude levels and the sign can be represented. If the smaller magnitude level is used at time $r - 1$, M is chosen to have a value $M = M_1 < 1$, and if the larger magnitude level is chosen, $M = M_2 > 1$ is used. For a three-bit quantizer, four magnitude levels and the sign can be represented. In this case, there are four choices for M. Through simulations, appropriate values of M for a two-bit quantizer were found to be $M_1 = 0.845$ and $M_2 = 1.96$. For a three-bit quantizer, they are $M_1 = 0.845$, $M_2 = 1.0$, $M_3 = 1.0$, and $M_4 = 1.4$. Note that the three-bit quantizer does not change its step-size at time r unless the largest or smallest quantizer level is encountered at time $r - 1$. The above values of M are in approximate agreement with values proposed by Jayant[7] for full-band APCM encoding.

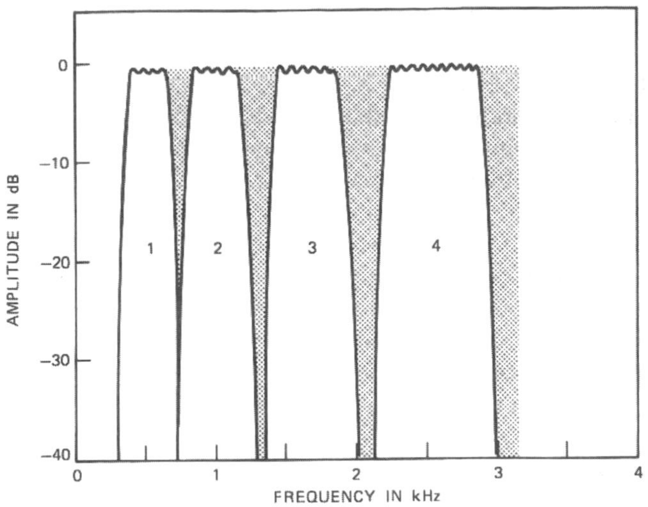

Fig. 6—Partitioning of the speech spectrum into four noncontiguous bands to achieve reduced bit-rate coding.

IV. SUB-BAND CODING FOR TRANSMISSION AT DATA RATES

The transmission bit rate of the sub-band coder can be reduced into the range of conventional data speeds by further limiting the sub-bands in width and tolerating some spectral gaps as shown in Fig. 6. Carried to excess, the noncontiguous bands produce a reverberant quality in the signal, such as one gets from comb filtering. In moderation, however, some highly useful compromises can be achieved between transmission bit rate and quality. The coded bands still cover a respectable range of the speech spectrum, and provide a quality considerably better than coding a single full-band signal.

V. INTEGER-BAND SAMPLING AND HARDWARE CONSIDERATIONS

Another attractive alternate implementation of these ideas is to use "integer-band" sampling to code a signal that is aliased in an advantageous way. The technique is illustrated in Fig. 7.

The signal sub-bands $s_n(t)$ are chosen to have a lower cutoff frequency of mf_n and an upper cutoff frequency of $(m + 1)f_n$, where m is an integer and f_n is the bandwidth of the nth band. This bandpassed signal is sampled at $2f_n$ to produce the sampled spectrum shown in Fig. 7 (for $m = 2$). The received signal is recovered by decoding and bandpassing to the original signal band. Typically, values of m from 1 to 3 are most useful for coder applications with lower bands using values of $m = 1$ and upper bands using $m = 2$ or $m = 3$. This integer-band sampling technique achieves the theoretical maximum efficiency in sampling.[10]

A very attractive advantage of the integer-band sampling approach is that it does not require the use of modulators. A slight disadvantage is that the above restrictions prevent the choice of bands strictly on the basis of equal contribution to AI. However, little loss in performance is observed if this equal contribution to AI condition is only approximate (within a factor of 2). This implementation was used for perceptual comparisons, which will be discussed later.

This approach is especially attractive for implementing the bandpass filters as charge-coupled-device (CCD) transversal filters. The analog to discrete-time conversion is inherently accomplished by the CCD filter with little or no analog prefiltering or post filtering required for the prevention of aliasing. The initial signal sampling can be conveniently high, say 15 kHz, to realize the CCD filter, and the filter output can be decimated to the $2f_n$ rate for coding. After transmission and decoding, the $2f_n$ rate can be interpolated to the 15-kHz rate for the final bandpass filtering, again by the analog CCD filter.

Another advantage of CCD filters (and also digital filters) is that the filter cutoff frequencies are inherently normalized to the initial sampling frequency. Therefore, the sampling frequency and, consequently, the bit rate of the coder, can be varied over a limited range by

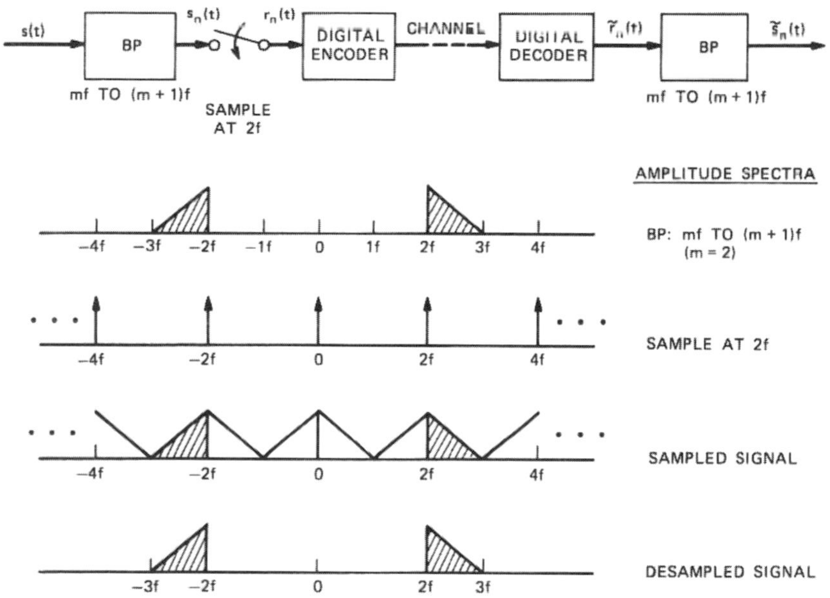

Fig. 7—Integer-band sampling technique for digital encoding of speech sub-bands.

Table I — Frequencies and sampling rates for the 16-kb/s coder

Sub-band No.	ω_n Center Freq (Hz)	$r_n'(t)$ Sampling Rate (s^{-1})	Decimation (From 10 kHz)	Quantization (Bits)
1	448	1250	16	3
2	967	1429	14	3
3	1591	1667	12	2
4	2482	2500	8	2

varying the master clock frequency. This cannot be achieved with analog filters.

Present technology is able to provide four 100-tap CCD transversal filters on a single integrated-circuit chip or one 200-tap filter on a chip with all necessary drivers and control logic.

VI. COMPUTER SIMULATIONS OF SUB-BAND ENCODERS

The sub-band coder has been implemented by computer simulation for transmission bit rates of approximately 16 kb/s and 9.6 kb/s. The complex demodulation approach in Fig. 5 was used for low-pass translation of the bands. An initial sampling rate of 10 kHz was employed in both cases.

The 16-kb/s coder was implemented with the band center frequencies and sub-band sampling rates shown in Table I. Bandwidths are equal to one half of the sampling rates and correspond to those shown in Fig. 1. Three-bit coders were used in the two lower bands, and two-bit coders were used for the upper bands. The filters were 125-tap FIR filters. As can be observed in Fig. 1, the filters overlap in their transition bands and give an overall flat frequency response from 200 Hz to 3100 Hz.

The 9.6-kb/s coder was implemented with the bands given in Table II and illustrated in Fig. 6. In this case gaps were allowed between bands. Larger filter orders, 175-tap (FIR), were used to reduce transition bands and conserve bandwidth. Only the lower band used a three-bit coder. Upper bands used 2-bit coders.

Table II — Frequencies and sampling rates for the 9.6-kb/s coder

Sub-band No.	ω_n Center Freq (Hz)	$r_n'(t)$ Sampling Rate (s^{-1})	Decimation (From 10 kHz)	Quantization (Bits)
1	448	800	25	3
2	967	952	21	2
3	1591	1111	18	2
4	2482	1538	13	2

Illustrations of the signal coded for 16 kb/s and 9.6 kb/s by the above-band-translation technique are given by the spectrograms of Figs. 8 and 9, respectively. In each figure, the upper spectrogram corresponds to the original sentence. The middle spectrogram corresponds to the signal played through the filters, decimators, and interpolators—but without coders. The bottom spectrogram illustrates the sub-band encoded speech at the designated bit rate.

Other simulations have also been made for encoding the signals $a_n(t)$ and $b_n(t)$ directly and also for encoding the magnitude and phase derivative (as in the phase vocoder). Similar quality results were found in these simulations.

VII. SUBJECTIVE COMPARISONS WITH OTHER ENCODING METHODS

Informal listening tests were made to compare the quality of the sub-band coder simulations with that of full-band encoding. For the 16-kb/s coder, comparisons were made with 2- and 3-bit ADPCM. For the 9.6-kb/s coder, comparisons were made with adaptive delta modulation (ADM) (i.e., 1-bit ADPCM). Results for the 16-kb/s coder comparisons are given in Table III.

Twelve listeners were asked to compare pairs of sentences for signal quality and indicate which was better. Two speakers were used in the experiments and sentence pairs were played in a randomly selected order. Each listener made a total of 16 comparisons in each of the experiments.

In comparing 16-kb/s sub-band encoding to 16-kb/s (2 bits/sample) ADPCM, listeners rated the sub-band encoded sentence as having higher quality in 94 percent of the sentence pairs. When the bit rate of the ADPCM coder was increased to 24 kb/s (3 bits/sample), they rated the sub-band encoded sentence as having higher quality in 34 percent of the sentence pairs. Experiment I demonstrates that the quality of the 16-kb/s sub-band coder is clearly preferred over that of ADPCM at the same bit rate. In Experiment II listeners exhibited much greater indecision, indicating that the quality of the 16-kb/s sub-band coder is close to that of 24-kb/s ADPCM, but that preference leans slightly in favor of the ADPCM.

Also included in Table III are signal-to-quantizing-noise ratios (s/n) measured on the speech signals, averaged for the two speakers for each of the coding methods. s/n data is not found to be a reliable indicator of listener preference. This observation is not surprising and has been previously recognized in the speech coding literature.[7,8]

A second series of listening experiments compared 9.6-kb/s sub-band coding with ADM. The sub-band encoder in this case is implemented with the integer-band method described earlier. The ADM coder is a

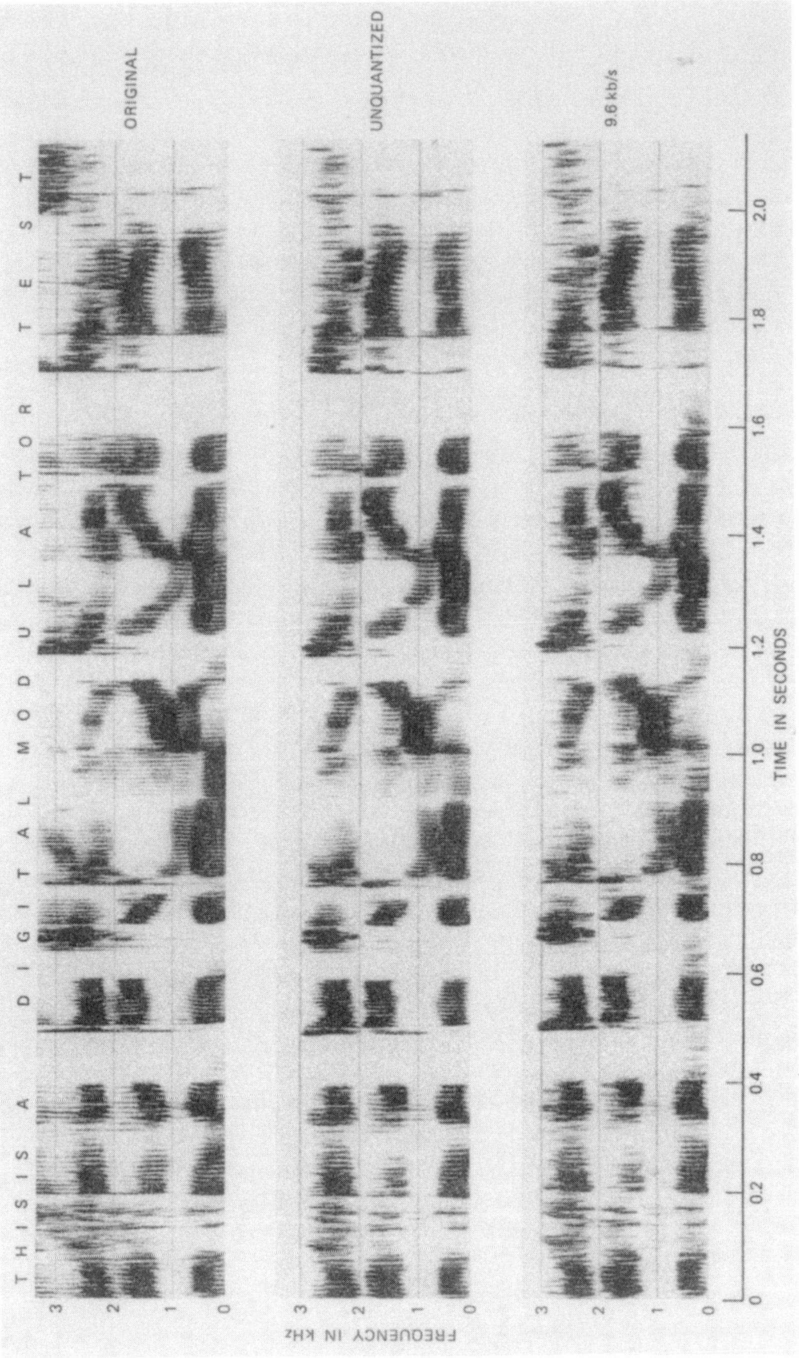

Fig. 8—Sound spectrograms comparing original, unquantized speech output and the 16-kb/s output from the sub-band coder.

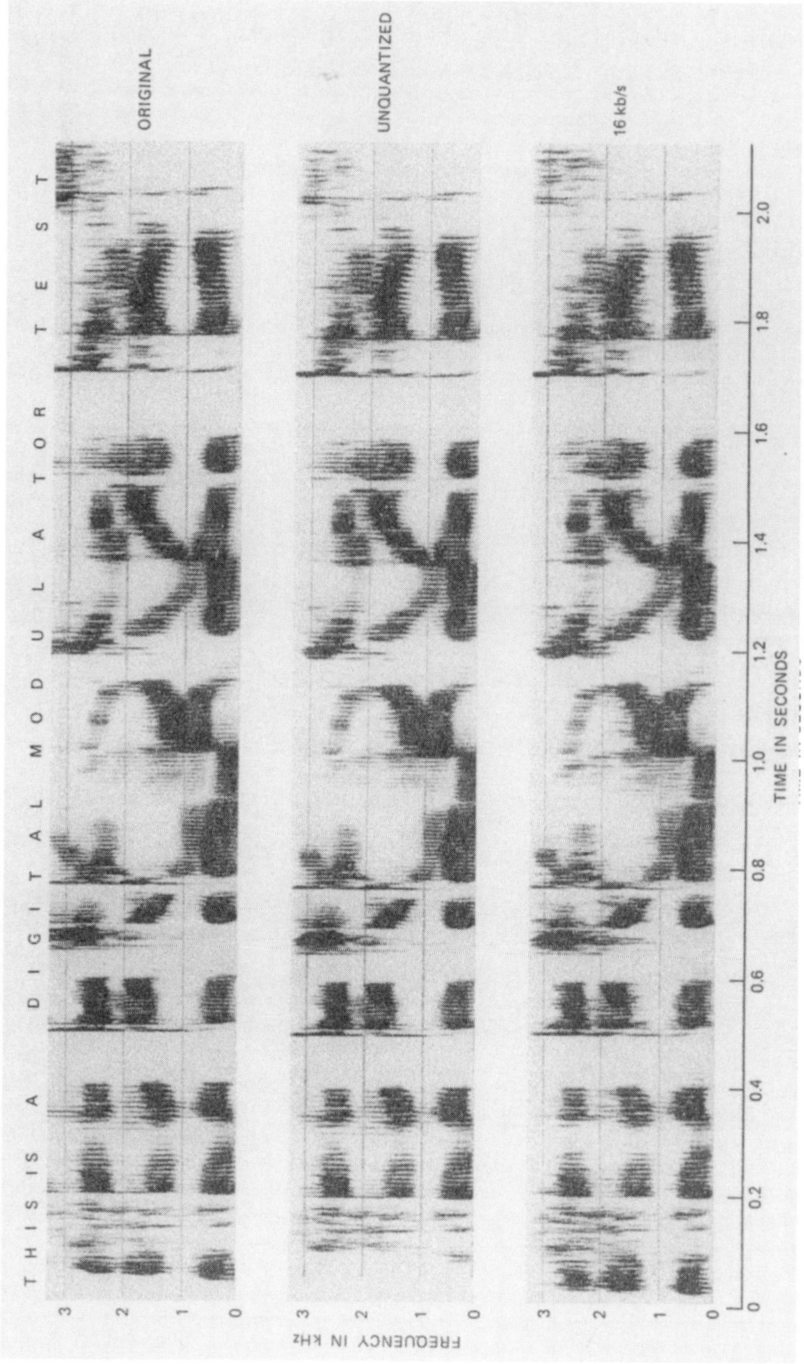

Fig. 9—Sound spectrograms comparing original, unquantized speech output and the 9.6-kb/s output from the sub-band coder.

Table III — Comparison of 16-kb/s sub-band coder with ADPCM

Experiment 1: 16-kb/s Sub-band vs 16-kb/s ADPCM

	Listener Preference (%)	S/N (dB)
16-kb/s Sub-band	94	11.1
16-kb/s ADPCM (2 Bits)	6	10.9

Experiment 2: 16-kb/s Sub-band vs 24-kb/s ADPCM

	Listener Preference (%)	S/N (dB)
16-kb/s Sub-band	34	11.1
24-kb/s ADPCM	66	14.5

forward step-size transmitting coder shown by Jayant[11] to have improved performance over conventional ADM. Table IV shows the results of these experiments. Three different bit rates, 10.3, 12.9, and 17.2 kb/s, were used for the ADM coder. In the first two experiments, the 9.6-kb/s sub-band coder was clearly preferred. In the third experiment, there was greater indecision with preference leaning slightly in favor of the sub-band coder. Note that this is true despite the opposite ordering of the s/n values! In other words, the perceptual palatability is not well reflected in the s/ns as has been observed previously.[8]

Table IV — Comparison of 9.6-kb/s sub-band coder with ADM

Experiment 1: 9.6-kb/s Sub-band vs 10.2-kb/s ADM

	Listener Preference (%)	S/N (dB)
9.6-kb/s Sub-band	96	9.9
10.3-kb/s ADM	4	8.2

Experiment 2: 9.6-kb/s Sub-band vs 12.9-kb/s ADM

	Listener Preference (%)	S/N (dB)
9.6-kb/s Sub-band	82	9.9
12.9-kb/s ADM	18	9.7

Experiment 3: 9.6-kb/s Sub-band vs 17.2-kb/s ADM

	Listener Preference (%)	S/N (dB)
9.6-kb/s Sub-band	61	9.9
17.2-kb/s ADM	39	11

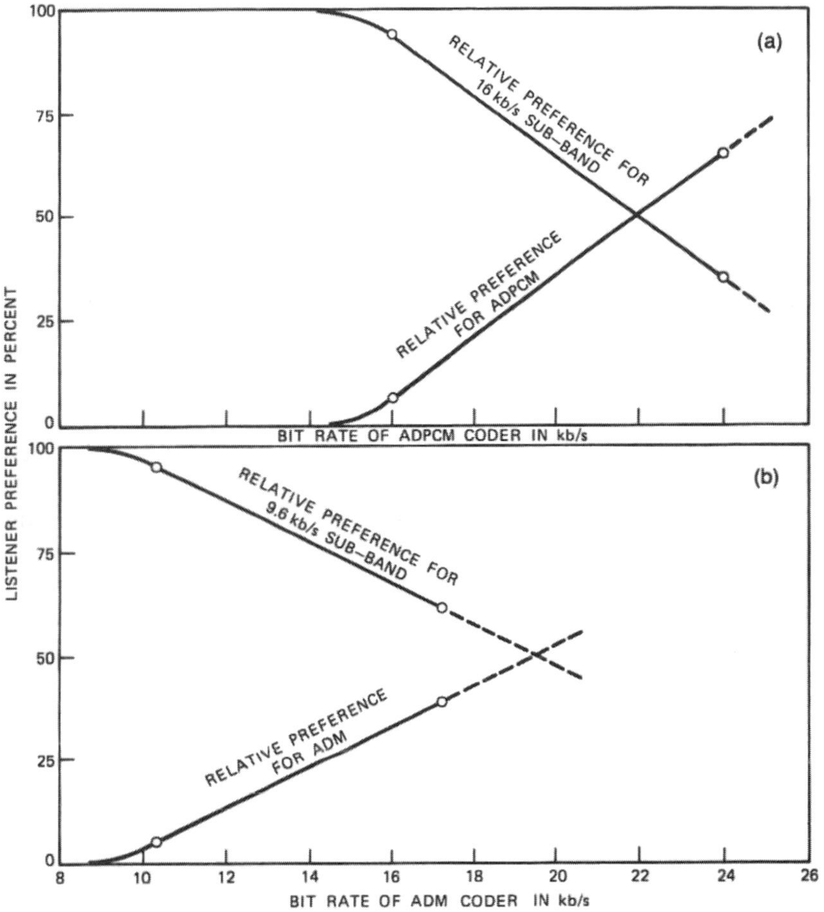

Fig. 10—(a) Relative comparison of quality of 16-kb/s sub-band coding against ADPCM coding (based on listener preference) for different ADPCM coder bit rates. (b) Relative comparison of quality of 9.6-kb/s sub-band coding against ADM coding for different ADM coder bit rates.

Figure 10 summarizes the results of the listener preference tests in Tables III and IV. Listener preference is plotted against the ADPCM and ADM coder bit rates. The crossover points of the curves in the two comparisons determine the point at which the two types of coders have approximately equal subjective quality. In the first comparison, the quality of the 16-kb/s sub-band coder is seen to be comparable to that of 22-kb/s ADPCM; i.e., it has a 6-kb/s advantage over the ADPCM coder. In the second comparison, the 9.6-kb/s coder has a subjective quality that is comparable to the 19-kb/s ADM and, therefore, has a 9.4-kb/s advantage over ADM.

It is clear from the listener preference tests that the sub-band coding technique is considerably better in quality than full-band ADPCM or ADM coding methods. We have carried this coding down to 7.2 kb/s and find that the quality is only slightly poorer than that at 9.6 kb/s. We have also pressed the coding rate down to 4.8 kb/s and find that the quality becomes considerably poorer owing to the increased band limiting and gaps between bands.

VIII. CONCLUSION

We have described a method for digitally coding speech in sub-bands of the total signal spectrum. Partitioning into sub-bands has several distinct advantages. Bit allocations for quantization of each band can be made on a perceptually palatable basis. Quantization products in a given band are confined to that band and do not "spill over" into adjacent frequency ranges. Selection of sub-band widths can also be made according to perceptual criteria, namely, for equal contributions to AI (and hence to signal intelligibility). As a result, the sub-band coding produces a quality signal that is better than a single full-band coding at the same total bit rate. The price paid is the band-filtering and the individual coding.

"Integer-band" sampling is demonstrated to be an economical and effective method for implementing the sub-band coder. Emerging technologies in device fabrication (such as CCDs) suggest economical implementations of the band filtering in terms of analog transversal filters.

The sub-band coder, implemented by integer band sampling, is demonstrated for speech transmission at rates of 16, 9.6, and 7.2 kb/s. The latter two transmission rates push down into the data range and are attractive for "voice-coordination" over data channels.

Informal perceptual experiments demonstrate that the signal quality of speech coded at 9.6 kb/s by the sub-band method is approximately equivalent to a 19-kb/s coding of the full-band signal. For a given transmission bit rate, therefore, the sub-band technique provides a significant improvement in signal quality. Or alternatively, for a given signal quality, the sub-band system can provide the transmission at a significantly reduced bit rate.

REFERENCES

1. E. R. Kretzmer, "Reduced-Alphabet Representation of Television Signals," IRE Convention Record, *4* (1956), pp. 140–147.
2. E. E. David and H. S. McDonald, "A Bit-Squeezing Technique Applied to Speech Signals," IRE Convention Record, *4* (1956), pp. 148–153.
3. L. L. Beranek, "The Design of Communications Systems," Proc. IRE, *35* (September 1947), pp. 880–890.

4. K. D. Kryter, "Methods for the Calculation and Use of the Articulation Index," J. Acoust. Soc. Amer., *34* (1962), pp. 1689–1697.
5. J. L. Flanagan and R. M. Golden, "Phase Vocoder," B.S.T.J., *45* (November 1966), pp. 1493–1509.
6. R. E. Crochiere and L. R. Rabiner, "Optimum FIR Digital Filter Implementations for Decimation, Interpolation, and Narrowband Filtering," IEEE Trans. Acoust., Speech, and Signal Proc., *ASSP-23*, No. 5, October 1975, pp. 444–456.
7. N. S. Jayant, "Digital Coding of Speech Waveforms: PCM, DPCM, and DM Quantizers," Proc. IEEE, *62* (May 1974), pp. 611–632.
8. P. Cummiskey, N. S. Jayant, and J. L. Flanagan, "Adaptive Quantization in Differential PCM Coding of Speech," B.S.T.J., *52* (September 1973), pp. 1105–1118.
9. N. S. Jayant, "Adaptive Quantization With a One-Word Memory," B.S.T.J., *52* (September 1973), pp. 1119–1144.
10. C. B. Feldman and W. R. Bennett, "Band Width and Transmission Performance," B.S.T.J., *28* (July 1949), pp. 490–595.
11. N. S. Jayant, "Step-Size Transmitting Differential Coders for Mobile Telephony," B.S.T.J., *54* (November 1975), pp. 1557–1581.

SECTION I

APPLICATION OF QUADRATURE MIRROR FILTERS TO SPLIT BAND VOICE CODING SCHEMES

D. Esteban and C. Galand

IBM Laboratory
06610 La Gaude, France

Abstract

This paper deals with applications of Quadrature Mirror Filters (QMF) to coding of voice signal in sub-bands. Use of QMF's enables to avoid the aliasing effects due to samples decimation when signal is split into sub-bands. Each sub-band is then coded independently with use of Block Companded PCM (BCPCM) quantizers. Then a variable number of bits is allocated to each sub-band quantizer in order to take advantage of the relative perceptual effect of the quantizing error.

The paper is organized as follows :

- First, splitting in two sub-bands with QMF's is analysed.
- Then, a general description of a splitband voice coding scheme using QMF's is made.
- Finally, two coding schemes are considered, operating respectively at 16 KBps and 32 KBps. Averaged values of S/N performances are given when encoding both male and female voices. Comparisons are made with conventional BCPCM and CCITT A-Law.

Taped results will be played at the conference.

1) Introduction

Decomposition of the voice spectrum in sub-bands has been proposed by R. Crochiere et al. /1/ as a means to reduce the effect of quantizing noise due to coding. The main advantages of this approach are the following :

- first, to localize the quantizing noise in narrow frequency sub-bands, thus preventing noise interference between these sub-bands,
- second, to enable the attribution of bit resources to the various frequency bands according to perceptual criteria.

As a result, the quantizing noise is perceptually more acceptable, and the signal to noise ratio is improved.

The implementation proposed in /1/ is straightforward and takes advantage of a bank of non-overlapping band-pass filters. Unfortunately, for a non perception of aliasing effects due to decimation, this approach needs sophisticated band-pass filters. The split-band coding scheme we propose here avoids these inconveniences. Quasi perfect sub-band splitting can be achieved by use of Quadrature Mirror Filters (QMF) /2/ associated with decimation/interpolation techniques.

2) QMF band splitting

Principle

Let us consider for explanation purposes Fig. 1 in which we describe the decomposition of a sampled signal in two contiguous sub-bands, where :

- . H_1 is a sampled half band low pass filter with an impulse response $h_1(n)$,
- . H_2 is the corresponding half band mirror filter, i.e. which satisfies the following magnitude relation :

$$|H_1(e^{j\omega T})| = |H_2(e^{j(\frac{\omega s}{2} - \omega)T})| \qquad (1)$$

  where $\omega s = 2\pi f_s = 2\pi/T$ denotes the sampling rate and $H_1(e^{j\omega T})$ denotes the Fourier Transform of $h_1(n)$.
- . K_1 is a half band low pass filter with an impulse response $k_1(n)$ and K_2 is the corresponding mirror filter of K_1.

After frequency limiting to $f_s/2$, the signal $x(t)$ is sampled at f_s and filtered by H_1 and H_2. The obtained signals $x_1(n)$ and $x_2(n)$ represent respectively the low and high half-bands of $x(n)$. As their spectra occupy half the Nyquist bandwidth of the original signal, the sampling rate in each band can be halved by ignoring every second sample. For reconstruction, the signals $y_1(n)$ and $y_2(n)$ are interpolated by inserting one zero valued sample between each sample and filtered by K_1 and K_2 before being added to give the signal $s(n)$.

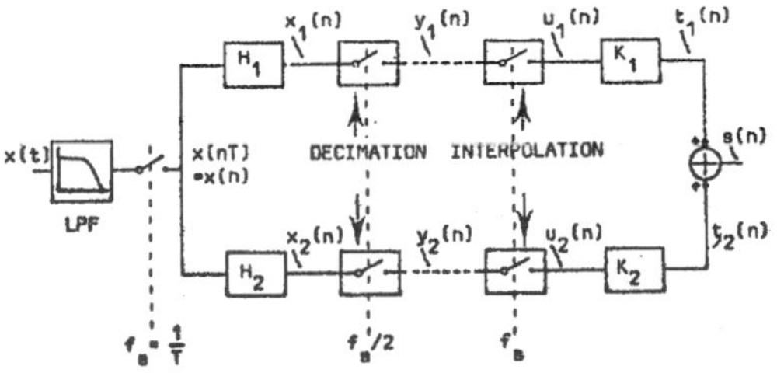

Fig 1 Principle of 2 sub-bands splitting by use of half-band mirror filters

Let us analyse the structure of Fig. 1. If $X(z)$, $H(z)$ and $X_1(z)$ represent respectively the z transforms of $x(n)$, $h_1(n)$ and $x_1(n)$, then :

$$X_1(z) = H_1(z)X(z) \qquad (2)$$

The z transform $Y_1(z)$ of the decimated signal $y_1(n)$ and the z transform $U_1(z)$ of the interpolated signal $u_1(n)$ are given by /3/ :

$$Y_1(z) = \frac{1}{2}\{X_1(z^{\frac{1}{2}})+X_1(-z^{\frac{1}{2}})\} \qquad (3)$$

$$U_1(z) = Y_1(z^2) \qquad (4)$$

After final filtering, the z transform of $t_1(n)$ is :

$$T_1(z) = K_1(z)U_1(z) \qquad (5)$$

where $K_1(z)$ represents the z transform of $k_1(n)$.

Combining relations (2)-(5) gives :

$$T_1(z) = \frac{1}{2}\{H_1(z)X(z)+H_1(-z)X(-z)\}K_1(z) \qquad (6)$$

The z transform $T_2(z)$ is derived in a similar manner :

$$T_2(z) = \frac{1}{2}\{H_2(z)X(z)+H_2(-z)X(-z)\}K_2(z) \qquad (7)$$

The z transform $S(z)$ of the signal $s(n)$ is obtained by adding relations (6) and (7) :

$$\begin{aligned} S(z) &= \frac{1}{2}\{H_1(z)K_1(z)+H_2(z)K_2(z)\}X(z) \\ &+\frac{1}{2}\{H_1(-z)K_1(z)+H_2(-z)K_2(z)\}X(-z) \end{aligned} \qquad (8)$$

The second term of this sum represents aliasing effects due to decimation and can be eliminated if we choose K_1 and K_2 appropriately. First, we must satisfy the symmetry relation (1). This is elegantly solved if H_1 is a finite impulse response (FIR) filter :

$$H_1(z) = \sum_{n=0}^{N-1} h_1(n)z^{-n} \qquad (9)$$

It can be seen that the impulse response $h_2(n)$ of the mirror filter H_2 is obtained by inverting every second sample of $h_1(n)$.

$$H_2(z) = \sum_{n=0}^{N-1} h_1(n)(-1)^n z^{-n} = H_1(-z) \qquad (10)$$

We can now cancel the second term of (8) by choosing :

$$K_1(z) = H_1(z) \qquad (11)$$

$$K_2(z) = -H_2(z) = -H_1(-z) \qquad (12)$$

Equation (8) now becomes :

$$S(z) = \frac{1}{2}\{H_1^2(z)-H_1^2(-z)\}X(z) \qquad (13)$$

Let us evaluate this relation on the unit circle

$$S(e^{j\omega T}) = \frac{1}{2}\{H_1^2(e^{j\omega T}) - H_1^2(e^{j(\omega+\frac{\omega_s}{2})T})\}X(e^{j\omega T}) \qquad (14)$$

If we choose for H_1 a symmetrical FIR filter, its Fourier transform $H_1(e^{j\omega T})$ can be expressed in term of its magnitude $H_1(\omega)$:

$$H_1(e^{j\omega T}) = H_1(\omega)e^{-j(N-1)\pi\frac{\omega}{\omega_s}} \qquad (15)$$

Substituting in (14) gives :

$$\begin{aligned} S(e^{j\omega T}) &= \frac{1}{2}\{H_1^2(\omega)-H_1^2(\omega+\frac{\omega_s}{2})e^{-j(N-1)\pi}\} \\ &\times e^{-j(N-1)2\pi\frac{\omega}{\omega_s}} \cdot X(e^{j\omega T}) \end{aligned} \qquad (16)$$

Two cases are to be considered, depending on the parity of N .

. First case, N even

$$S(e^{j\omega T}) = \frac{1}{2}\{H_1^2(\omega)+H_1^2(\omega+\frac{\omega_s}{2})\}e^{-j(N-1)\omega T}.X(e^{j\omega T}) \qquad (17)$$

Considering the case of perfect filters,

$$H_1^2(\omega) + H_1^2(\omega+\frac{\omega_s}{2}) = 1 \qquad (18)$$

we get :

$$S(e^{j\omega T}) = \frac{1}{2}\, e^{-j(N-1)\omega T}.X(e^{j\omega T}) \qquad (19)$$

or

$$s(n) = \frac{1}{2}\, x(n-N+1) \qquad (20)$$

The signal is perfectly reconstructed (neglecting the gain factor 1/2) with a delay of (N-1) samples.

. Second case, N odd

In this case, the original signal cannot be perfectly reconstructed, it can be seen from (16) that the amplitude at $\omega = \omega_s/4$ is always zero.

To summarize, we have defined a set of conditions for perfect reconstruction :

H_1 = Symmetrical FIR filter of even order ;
$H_2(z) = H_1(-z)$;
$K_1(z) = H_1(z)$; $H_1^2(\omega) + H_1^2(\omega+\omega_s/2) = 1$;
$K_2(z) = -H_2(z)$;

Implementation

Fig. 2a gives an efficient implementation of the QMF band splitting, using a symmetrical FIR half band filter with an even number of coefficients. The input signal $x(t)$ is sampled at f_s and filtered by H_1 and H_2, giving the low-band channel $x_1(n)$ and the high-band channel $x_2(n)$. Then the sampling rate is decreased to $f_s/2$ by decimating every second sample, giving the signals $y_1(n)$ and $y_2(n)$.

Fig. 2b shows the reconstruction of the initial signal with the same filter. First, the sampling rate is increased to f_s by inserting one zero valued sample between each sample of $y_1(n)$ and $y_2(n)$, giving two signals $u_1(n)$ and $u_2(n)$. Then these signals are filtered by H_1 and H_2, and the signal s(n) is obtained by subtracting the filtered signals $t_1(n)$ and $t_2(n)$.

The total number of multiplications to perform per initial sampling interval (splitting and reconstruction) is equal to the filter length N, the number of additions if of the order of N.

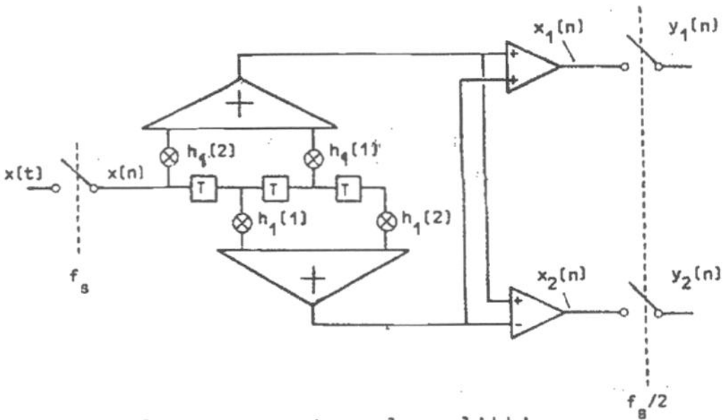

Fig 2a Quadrature channels splitting

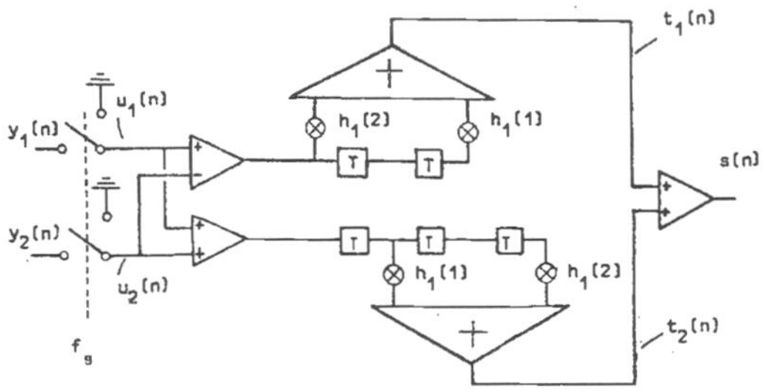

Fig 2b Quadrature channels reconstruction

3) Split-band voice coding scheme based on QMF 2^p sub-bands tree decomposition

In the previously described implementation, a signal x(t) was sampled at f_s to give a signal x(n), and split into two signals $y_1(n)$ and $y_2(n)$ with reduction of the sampling rate to $f_s/2$. This decomposition can be extanded to more than two sub-bands by applying to $y_1(n)$ and $y_2(n)$, which represent respectively the low sub-band and the high sub-band of x(n), the same decomposition process as to the initial signal x(n) (see Fig. 4). Four signals are thus obtained with reduction of the sampling rate to $f_s/4$. The spectrum of each of these signals represents the spectrum of x(n) in the corresponding sub-band.

This decomposition can be generalized by repeating the processus p times. The initial signal is thus split into 2^p signals sampled at $f_s/2^p$ by a p-stage tree arrangement of decimation filters of the type shown on Fig. 2a. As the ith stage includes 2^{i-1} filters, the total number of filters is 2^{p-1}. The resulting information rate after p stages is the same as the one of the original signal.

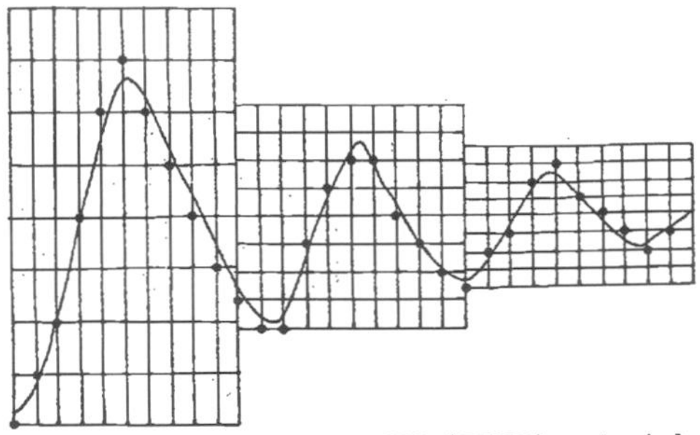

Fig 3 Block Companded PCM (BCPCM) principle

Quantization of the sub-band signals

As mentioned in /1/ and due to the fact that the sub-band signals are narrow band and Nyquist sampled, the sample-to-sample correlation of these signals is low. Consequently, straight PCM encoding techniques are preferred to differential methods.

An efficient and simple approach to code the sub-bands signals is obtained by means of Block Companded PCM (BCPCM) coding scheme /4/. This type of companding has been initially proposed for full band coding of speech waveforms, but can be straightforwardly applied to sub-band encoding. The principle of BCPCM coding can be summarized as follows :

The samples are encoded on a block basis. For each block of M samples, a scale factor is chosen in such a way that the larger sample in the block will not fall out of the coded range.

- Then, the M samples of the block are quantized with respect to the obtained scale factor and both the coded values and the scale factor are transmitted.

The overhead bit rate necessary to the transmission of the scale factor is inversely proportional to the length of the blocks, but this length must be chosen so as to take in account the formant evolution. For a full band coding, a length of 8 to 16 ms has been found satisfying.

The main advantages of BCPCM are a low overhead information rate, a very large dynamic range, and no transient clipping. Fig. 3 shows the adaptation of the scale factor to the signal, considering three consecutive blocks, and assuming 3 bits quantization.

The BCPCM coding scheme has been used with success in conjunction with the QMF band splitting,

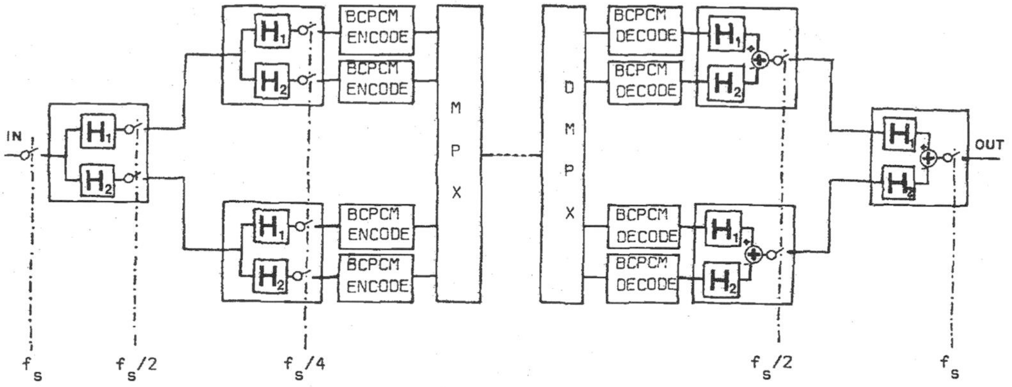

Fig 4 Four sub-bands SVCS with QMF and BCPCM

assuming different number of bits to code each frequency sub-band so as to weight the perceptual effect of the quantizing noise in the voice spectrum. Examples of bit allocation will be discussed in section 4. After quantization (see Fig. 4), the signals and scale factors from all channels are time multiplexed and transmitted.

2^p sub-bands reconstruction

At the receiving end, the data is demultiplexed and decoded. The reconstruction of the speech signal is made by a p-stage tree arrangement of filters of the type of the one shown in Fig. 2b.

If a same filter of N taps is used for each stage, the number of multiplies per input sample for the whole 2^p sub-bands decomposition/reconstruction is Np. In fact, filter constraint can be reduced from stage to stage with respect to the bandwidth so as to optimize the total processing. It has been shown in section 2, that there is a delay of (N-1) samples between the original and reconstructed signals in case of two sub-bands splitting. Consequently, the number of delayed samples is $(2^p-1)(N-1)$ for the 2^p sub-bands splitting.

4) Simulation of Split-band Voice Coding Scheme

In this section, two Split-band Voice Coding Schemes (SVCS) are considered. The first one operates at a bit rate of 16 KBps and provides a quality sufficient for telephony applications, the second operates at a bit rate of 32 KBps and gives a quality comparable to that provided by standard companded laws. The characteristics of these two coders are given hereafter.

- 16 KBps SVCS

  . input signal band limited to 0-4000 Hz
  . sampling rate : 8 KHz
  . number of sub-bands : 8
  . bit allocation : 3 3 3 1 1 1 1 1
  . block duration : 20 ms (160 samples)
  . number of overhead bits : 40

- 32 KBps SVCS

The characteristics of this coder are the same as the previous one, excepted the bit allocation that has been increased to :

5 5 5 4 4 3 3 1

- Performance

The performance of the two considered SVCS has been evaluated by comparison with conventional BCPCM coders operating at the same bit rate. For convenience, two types of BCPCM coders have been considered, the first one operating in PCM mode, the second one being able to take a PCM/DPCM decision /4/, so as to encode the high-correlated blocks of samples in differential mode.

The experimentations were made on a set of utterances pronounced by 7 speakers (4 female voices and 3 male voices) representing a total duration of 3.5 minutes of continuous speech. The averaged signal to noise ratios are given in table 1.

Table 1 Comparative performances (dB) of BCPCM and SVCS coders.

Coder \ Bit Rate	16 KBPS	32 KBPS
BCPCM (PCM Mode)	8	21
BCPCM (PCM/DPCM Mode)	11	24
SVCS	14	25

It must be noted that, for BCPCM coders, the PCM/DPCM decision enables a signal-to-noise improvement (SNRI) of 3dB. This improvement is not surprising and is in accordance with the well-known results of conventional PCM /5/. Moreover, it can be seen that split-band coding techniques provide SNRI over full-band techniques. This improvement is 3dB in case of 16 KBps, and only 1dB in case of 32 KBps. However, as noticed in /1/, it has been observed that for SVCS, the subjective level of the quantizing noise is less than for BCPCM, resulting in a more pleasant voice quality.

The previously described 16 KBps SVCS provides a speech quality which is sufficient for telephony applications. Furthermore, listening tests have shown that it is not possible to tell the difference between the 32 KBps SVCS and the CCITT 64 KBps A-Law, although the measured signal to noise ratios are respectively 25dB and 37dB.

5) Conclusions

The application of Quadrature Mirror Filters to Split-band Voice Coding Schemes has been discussed. As noticed in /1/, sub-band coding results in a signal to noise improvement over full-band coding. Moreover, the subjective effects of quantizing noise are less, resulting in a more pleasant coding quality.

Use of QMF enables to avoid aliasing effects due to decimation. Consequently, band splitting can be performed up to a large number of sub-bands without using sophisticated filters.

Two SVCS have been described, using BCPCM techniques and operating at 16 KBps and 32 KBps. The first one gives a speech quality which is sufficient for telephony applications. The second allows a quality comparable to that provided by the standard 64 KBps PCM code, thus achieving a halving of the bit rate for speech encoding.

References

/1/ R.E. Crochiere, S.A. Webber, J.L. Flanagan, "Digital coding of speech in sub-bands", 1976 Int'l IEEE Conf. on ASSP, Philadelphia.

/2/ A. Croisier, D. Esteban, C. Galand, "Perfect channel splitting by use of interpolation/decimation/tree decomposition techniques" 1976 Int'l Conf. on Information Sciences and Systems, Patras.

/3/ R. Schaffer, L. Rabiner, "A digital signal processing approach to interpolation", Proc. IEEE, Vol. 61, pp. 692-702, June 1973.

/4/ A. Croisier, "Progress in PCM and delta modulation : block companded coding of speech signals", 1974 Int'l Zürich seminar.

/5/ K.W. Cattermole, "Principles of pulse code modulation", London Iliffe Books Ltd.

A PROCEDURE FOR DESIGNING EXACT RECONSTRUCTION FILTER BANKS FOR TREE-STRUCTURED SUBBAND CODERS

M. J. T. Smith and T. P. Barnwell, III

School of Electrical Engineering
Georgia Institute of Technology
Atlanta, GA 30332

ABSTRACT

In recent years, tree-structured analysis/reconstruction systems have been extensively studied for use in subband coders for speech. In such systems, it is important that the individual channel signals be decimated in such a way that the number of samples coded and transmitted does not exceed the number of samples in the original speech signal. Under this constraint, the systems presented in the past have sought to remove the aliasing distortion while minimizing the overall analysis/reconstruction distortion. In this paper, it is shown that it is possible to design tree-structured analysis/reconstruction systems which meet the sampling rate condition and which also result in exact reconstruction of the input signal. This paper develops the conditions for exact reconstruction and presents a general method for designing the corresponding high quality analysis and reconstruction filters.

INTRODUCTION

In a recent paper by Barnwell [1], it was shown that alias-free reconstruction using recursive and non-recursive filters was possible where the analysis/reconstruction section had no frequency distortion or no phase distortion, but not both. In the development that follows, the coefficient symmetry condition on the analysis filters is lifted and exact reconstruction free of aliasing, phase distortion and frequency distortion is shown to be possible using FIR filters. In addition, the filter constraints that enable perfect reconstruction are discussed and an easily implementable design procedure providing high quality filters is presented.

Exact Reconstruction

The frequency division of the subband coder is performed in the analysis stage as shown in Figure 1, where $H_o(e^{j\omega})$ and $H_1(e^{j\omega})$ are lowpass and highpass filters respectively. To preserve the system sampling rate, both channels are decimated resulting in the two down-sampled signals, $Y_o(e^{j\omega})$ and $Y_1(e^{j\omega})$. In the reconstruction section, the bands are recombined by up-sampling and filtering to give the reconstructed signal

$$\hat{X}(e^{j\omega})=\frac{1}{2}[H_o(e^{j\omega})G_o(e^{j\omega})+H_1(e^{j\omega})G_1(e^{j\omega})]X(e^{j\omega}) \quad (1)$$
$$+\frac{1}{2}[H_o(-e^{j\omega})G_o(e^{j\omega})-H_1(-e^{j\omega})G_1(e^{j\omega})]X(-e^{j\omega})$$

The frequency response of the 2-band system is represented by the first term in equation (1), while the second term is the aliasing.

To obtain exact reconstruction, consider the case in which the analysis/reconstruction filters are designed using the assignments

$$G_o(e^{j\omega})=H_o(e^{-j\omega}) \quad (2a)$$

$$G_1(e^{j\omega})=H_o(-e^{j\omega}) \quad (2b)$$

$$H_1(e^{j\omega})=H_o(-e^{-j\omega}) \quad (2c)$$

where $H_o(e^{j\omega})$ is not required to be linear phase.

This assignment eliminates the aliasing term in equation (1) and the resulting overall system function is given by

$$C(e^{j\omega})=\frac{1}{2}H_o(e^{j\omega})H_o(e^{-j\omega})+\frac{1}{2}H_o(-e^{j\omega})H_o(-e^{-j\omega})$$
$$=\frac{1}{2}F_o(e^{j\omega})+\frac{1}{2}F_1(e^{j\omega}) \quad (3)$$

where $F_o(e^{j\omega})$ and $F_1(e^{j\omega})$ will be called the "product filters." It is clear from equation (3) that

$$F_1(e^{j\omega})=F_o(-e^{j\omega})$$

Now assume that the product filters are zero phase FIR filters. Then the exact reconstruction condition is given by

$$c(n) = \mathscr{F}^{-1}\{C(e^{j\omega})\} = \delta(n) \quad (4)$$

where $\mathscr{F}^{-1}\{\cdot\}$ denotes the inverse Fourier transform. The corresponding condition on the product filter is given by

$$f_o(n)\frac{[1+(-1)^n]}{2} = \delta(n) \quad (5)$$

27.1.1

CH1945-5/84/0000-0285 $1.00 © 1984 IEEE

So in order for exact reconstruction to be obtained, there are two conditions on the product filters which must be met. First, they must meet the condition of equation (5). Second, they must be decomposable into analysis and reconstruction filters in such a way that equation (3) is valid.

In order to see how these conditions can be met, it is appropriate to decompose the product filters into their zero time and nonzero time components, giving

$$F_o(e^{j\omega})=V(e^{j\omega})+A \qquad (6)$$

in the frequency domain or, equivalently,

$$f_o(n)=v(n)+A\delta(n) \qquad (7)$$

in the time domain. In these expressions, $v(n)$ is constrained to have no zero time component, i.e., $v(0) = 0$, and the exact reconstruction condition of equation (10) is met so long as $A = 1$ and

$$v(2n)=0 \qquad n = 0,\pm,1,\pm,2,... \qquad (8)$$

or, equivalently

$$V(e^{j\omega})=-V(-e^{j\omega}) \qquad (9)$$

This is the class of all filters which are anti-symmetric about the Nyquist frequency, as illustrated in Figure 3. This constraint can be met in several ways and a number of powerful filter design tools are available to design the required filters.

Filter Design

In the design procedure, the approach is to first design the product filter, $F_o(z)$, and then to decompose $F_o(z)$ into $H_o(z)$ and $H_o(z^{-1})$. There are a number of techniques which can be used to design $F_o(z)$.

For example, any filter of the form

$$f(n)=w(n)\frac{\sin\frac{\pi n}{2}}{\pi n} \qquad (10)$$

where $w(n)$ is a window function, will satisfy the exact reconstruction condition of equation (5). Similarly, optimal equiripple product filters can be constructed using a Remez exchange algorithm. In such designs (in which the Chebyshev error is minimized with equal weighting over the entire frequency band), only those solutions with the largest number of extrema, i.e. the extraripple filters, will satisfy equations (8) and (9). Both the Parks-McClellan and the Hofstetter algorithms for equiripple filter design may be used to design the product filters.

The attenuation of the analysis filter and product filter are related in the following way

$$A_p = 2A_a + 20\log_{10}(2-10^{-A_a/10}) \qquad (11)$$

where A_p is the product filter attenuation and A_a is the analysis filter attenuation. The second term in this expression is due to the difference in the transition width of $F_o(z)$ and $H_o(z)$. This tends to make the Hofstetter algorithm [2] more attractive for the equal ripple designs since the attenuation is an explicit input to the design procedure. It is important to notice that the transition width of the analysis filter will be larger than that of the product filter and therefore care should be exercised in using the Parks-McClellan algorithm.

The second condition which must be met is that the product filters must be factorable in such a way that equation (2c) is valid. The additional constraint that equation (2c) places on $F_o(z)$ can be stated as follows: for every zero of $F_o(z)$ at $re^{j\phi}$, there must be another corresponding zero at $(1/r)e^{j\phi}$. If this condition is met, then $F_o(z)$ can always be written as the product of reciprocally paired zeros

$$F_o(z)=G\prod_{m=1}^{M}(z-z_m)(z^{-1}-z_m) \qquad (12)$$

where G is a real constant. An important point to note is that any FIR filter which is both real and whose coefficients are symmetric in the time domain (zero phase) comes close to meeting this condition. In particular, the zeros of any such real, symmetric FIR filter must either meet the above condition, lie in complex zero pairs on the unit circle or occur at $z = \pm 1$. These two conditions become identical if there is a further requirement that any zero on the unit circle must be a double zero. The important point here is that any product filter which meets the exact reconstruction condition of equation (10) can be easily transformed into a new product filter which also meets the factorization condition required to satisfy equation (2c). The required transformation is simply

$$f_o(n) \rightarrow af_o(n)+b \qquad (13)$$

where "a" and "b" are real constants.

This is exactly the strategy outlined by Herrmann and Schussler [3] for designing minimum phase FIR filters. The starting point is a symmetric, zero phase half-band filter with an odd number of coefficients. Figure 3 shows a real and symmetric product filter (equation 13) with "b" equal to zero. Picture now, if you will, the same filter with "b" equal to a half. The stopband ripples now occur above and below the real axis. These zero crossings mark the location of roots on the unit circle. As "b" is increased, the symmetry conditions are not changed, but the zeros on the unit circle migrate toward one another and leave the unit circle in pairs. When "b" is large enough, all of the zeros will have either been driven off the unit circle or will have become double zeros on the unit circle (Figure 2). For equiripple filters, it is possible to have many double zeros on the unit circle, while for window designs there will generally only be one double zero pair. The effect of the multiplier constant "a" is to rescale the impulse response to have the correct passband gain. It can be easily shown that

$$a=1+10^{-A_p/20} \tag{14a}$$

$$b=^{-A_p/20} \tag{14b}$$

Once the appropriate product filters have been designed, it is straightforward to extract the analysis and reconstruction filters. For this case, the product filter can be characterized as in equation (12) and, therefore

$$F_1(z) = F_o(-z) = G \prod_{m=1}^{M} (z+z_m)(z^{-1}+z_m) \tag{15}$$

where G is a constant. The corresponding analysis and reconstruction filters are given by

$$H_o(z)=\sqrt{G} \prod_{m=1}^{M} (z-z_m) \tag{16a}$$

$$H_1(z)=\sqrt{G} \prod_{m=1}^{M} (z^{-1}+z_m) \tag{16b}$$

$$G_o(z)=\sqrt{G} \prod_{m=1}^{M} (z^{-1}-z_m) \tag{16c}$$

$$G_1(z)=\sqrt{G} \prod_{m=1}^{M} (z+z_m) \tag{16d}$$

Filters of this type give the desired exact reconstruction.

Discussion

A close examination of equations (16a-16d) reveals some interesting points. First, for each zero pair in $F_o(z)$ (one at $re^{j\phi}$ and one at $(1/r)e^{j\phi}$) one of the two zeros is always included in the analysis filter while the other is included in the reconstruction filter. Hence, the analysis filters and the reconstruction filters always have identical magnitude responses, and this magnitude response is exactly the square root of $F_o(e^{j\omega})$. Therefore, $F_o(e^{j\omega})$ must be designed to be the square of the desired analysis filter magnitude.

Second, note that for a given product filter, there are many choices of $H_o(z)$ that have a magnitude response equal to the square root of the product filter magnitude. Among these is the minimum phase analysis filter consisting of roots strictly inside the unit circle and one root from each double zero pair on the unit circle. The remaining roots comprise the corresponding reconstruction filter which, in this case, would have maximum phase. Figs. 4 and 5 show the normalized magnitude response and group delay respectively of a 32 coefficient minimum phase analysis filter. Also shown (Fig. 6) is an approximate linear phase analysis filter obtained by selecting conjugate pairs of product filter roots that occur alternately inside and outside of the unit circle as the frequency increases.

Third, note that the analysis filters are related in that $h_1(n)$ is formed by reversing $h_o(n)$ in time and multiplying by $(-1)^n$. The implication is that the filter coefficient symmetry present in QMFs is absent here and therefore the polyphase structure [1,4] cannot be used in implementation.

The major point is that exact reconstruction analysis and reconstruction filters are not difficult to design. In addition, the exact reconstruction filters generated by this technique generally have better characteristics in terms of their transition widths and attenuation than the most published QMFs of the same length, and, of course, they also give exact reconstruction.

REFERENCES

[1] T. P. Barnwell, "Subband Coder Design Incorporating Recursive Quadrature Filters and Optimum ADPCM Coders," IEEE Trans. on Acoustics, Speech and Signal Processing, Vol. ASSP-30, No. 5, Oct. 1982, pp. 751-765.

[2] E. Hofstetter, A. Oppenheim, J. Siegel, "A New Technique for the Design of Nonrecursive Digital Filters," Proc. Fifth Annual Princeton Conf. Inform. Sci. Systems, 1971, pp. 64-72.

[3] O. Herrmann and W. Schussler, "Design of Nonrecursive Digital Filters with Minimum Phase," Electronics Letters, Vol. 6, No. 11, April 27, 1970, pp. 329-330.

[4] R. E. Crochiere, "On the Design of Sub-Band Coders for Low-Bit-Rate Speech Communications," Bell System Tele. Journal, 56, (May-June, 1977), pp. 747-770.

[5] A. Croisier, D. Esteban and C. Galand, "Perfect Channel Splitting by Use of Interpolation/Decimation/Tree Decomposition Techniques," Proc. of the 1976 IEEE Int. Conf. on Information Science and Systems, Patras, Greece.

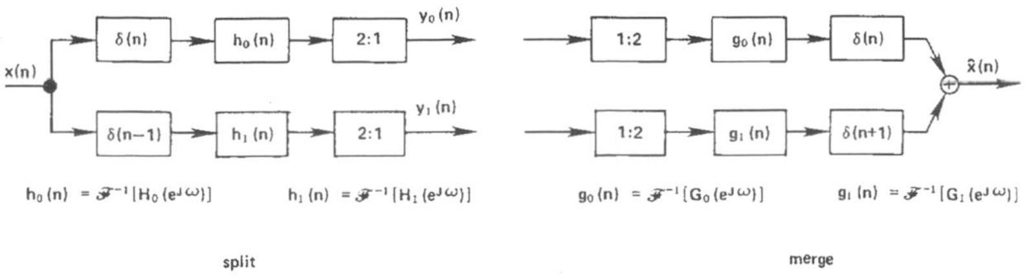

FIGURE 1. 2-BAND SUBBAND CODER.

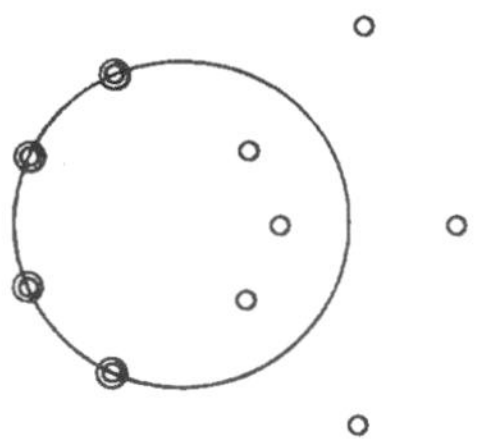

FIGURE 2. POLE/ZERO PLOT OF PRODUCT FILTER.

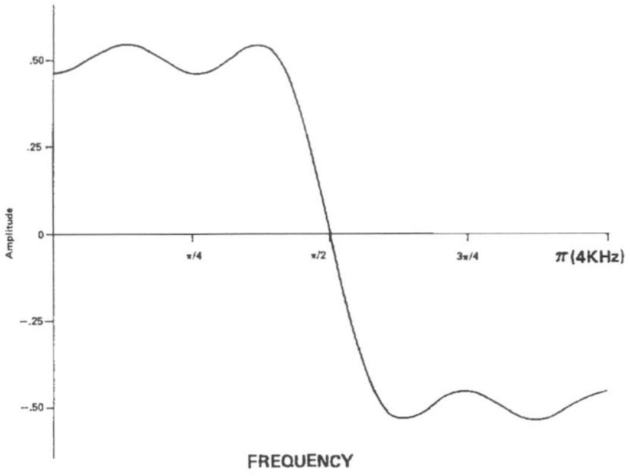

FIGURE 3. NONZERO TIME COMPONENT OF PRODUCT FILTER. $V(e^{j\omega})$.

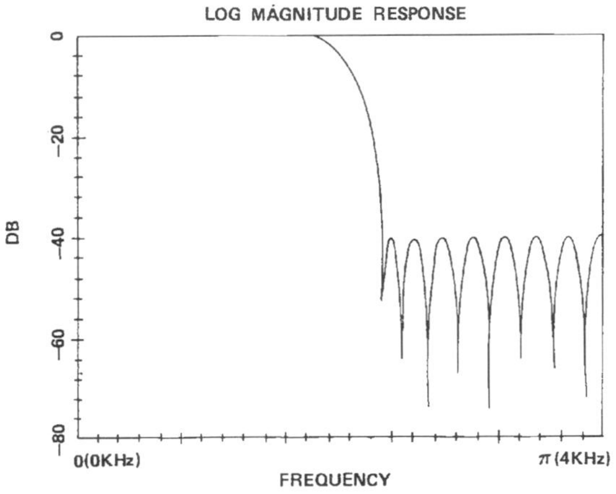

FIGURE 4. 32-TAP MINIMUM PHASE FILTER.

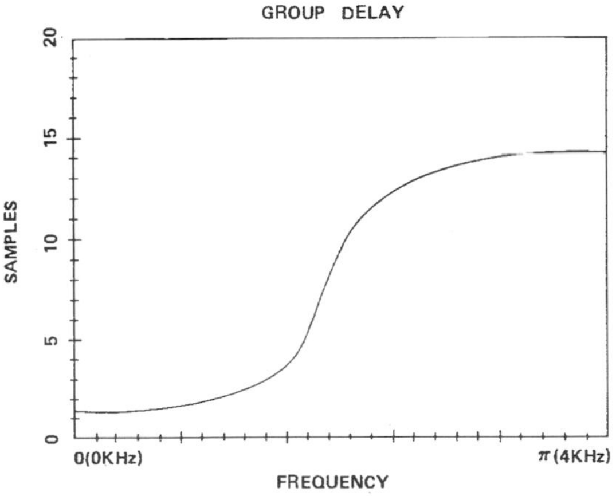

FIGURE 5. 32-TAP MINIMUM PHASE FILTER.

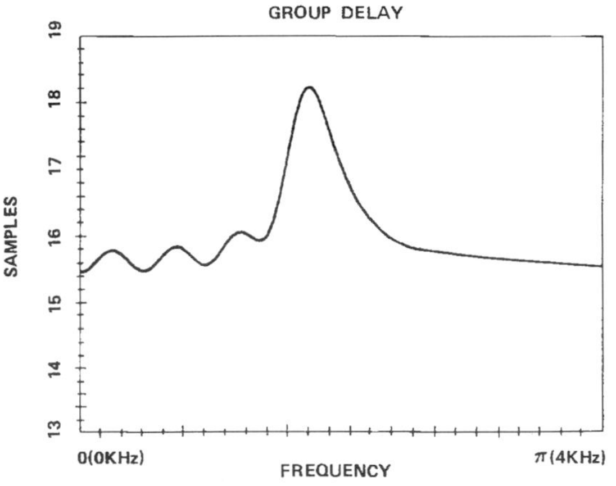

FIGURE 6. 32-TAP APPROXIMATELY LINEAR PHASE FILTER.

27.1.4

626 IEEE TRANSACTIONS ON ACOUSTICS, SPEECH, AND SIGNAL PROCESSING, VOL. ASSP-33, NO. 3, JUNE 1985

Filters for Distortion-Free Two-Band Multirate Filter Banks

FRED MINTZER, MEMBER, IEEE

Abstract—In this paper, conditions are given for a two-band multirate filter bank to be alias free and to have a unity frequency response. It is shown that the class of quadrature mirror filters (QMF's) that satisfies these conditions is quite limited. A class of filters which does satisfy these conditions is given, and a simple procedure for designing filters from this class is presented with an example.

Introduction

MULTIRATE filter banks are used in a number of digital signal processing applications. In speech applications, for example, they are found in subband coding, adaptive transform coding, and some noise reduction techniques. With a multirate filter bank, the incoming signal is filtered into several narrow-band components. Each narrow-band component is decimated, modified, and interpolated to the original sampling frequency. Then the narrow-band components are summed to form a modified version of the input signal. The presence of this operation is not always apparent since it often appears in a different form: the reconstruction of a signal from its modified short-time Fourier transform (STFT).

In addition to the errors caused by the signal modifications, aliasing and frequency response distortions may also be contributed by the multirate filter bank. Controlling these distortions can be very important to the quality of the reconstructed signal. The general problem of designing filters which provide a distortion-free N-band multirate filter bank is very difficult. It will not be addressed in this paper. Instead, the scope will be restricted to two-band multirate filter banks.

A significant breakthrough in controlling distortions in two-band filter banks was the proposal of quadrature mirror filters (QMF's) by Esteban *et al.* [1]. They have achieved great popularity in speech applications because of their control of aliasing. However, as is shown in this paper, the class of QMF's which produces zero distortion is very limited. In common usage, QMF's are designed to permit limited channel response distortion, and the filter design procedure, a nonlinear optimization, is not straightforward.

Conditions for Distortion-Free Two-Band Multirate Filter Banks

Let us consider the two-band multirate filter bank illustrated in Fig. 1, which has input signal u_n with Fourier transform $U(e^{j\omega})$ and output signal z_n with Fourier transform $Z(e^{j\omega})$. Note that in the figure, the blocks labeled "Increase Sampling Rate by 2" entail only zero padding. In their book, Crochiere and Rabiner [2] calculate $Z(e^{j\omega})$ in the absence of any signal modification. $Z(e^{j\omega})$ can be then expressed as

$$2\,Z(e^{j\omega}) = C(e^{j\omega})\,U(e^{j\omega}) + A(e^{j\omega})\,U(e^{j(\omega+\pi)}) \quad (1)$$

where $C(e^{j\omega})$, the channel filter, is given by

$$C(e^{j\omega}) = G^0(e^{j\omega})\,H^0(e^{j\omega}) + G^1(e^{j\omega})\,H^1(e^{j\omega}) \quad (2)$$

and $A(e^{j\omega})$, the alias filter, is given by

$$A(e^{j\omega}) = G^0(e^{j\omega})\,H^0(e^{j(\omega+\pi)}) + G^1(e^{j\omega})\,H^1(e^{j(\omega+\pi)}). \quad (3)$$

It is clear from (1) that alias-free reconstruction is achieved when

$$A(e^{j\omega}) = 0, \quad (4)$$

in which case the output simply is

$$2\,Z(e^{j\omega}) = C(e^{j\omega})\,U(e^{j\omega}). \quad (5)$$

Distortion-free reconstruction is achieved when it is also true that

$$C(e^{j\omega}) = e^{j\omega p}, \quad (6)$$

in which case $2Z_n$ is merely u_{n-p} and

$$2\,Z(e^{j\omega}) = U(e^{j\omega})\,e^{-j\omega p}. \quad (7)$$

In the remainder of this paper, we will examine those combinations of the filters $H^0(e^{j\omega})$, $H^1(e^{j\omega})$, $G^0(e^{j\omega})$, and $G^1(e^{j\omega})$ which satisfy (4) and (6).

Quadrature Mirror Filters

With the quardrature mirror filter scheme, a single even-length symmetric low-pass filter f_n of length N, with frequency response $F_q(e^{j\omega})$, is designed and represented as the product of its magnitude term $F(e^{j\omega})$ and its delay term $e^{-j(N-1)\omega/2}$. The filters $H^0(e^{j\omega})$, $H^1(e^{j\omega})$, $G^0(e^{j\omega})$, and $G^1(e^{j\omega})$ are constructed from $F_q(e^{j\omega})$ according to

$$H^0(e^{j\omega}) = F_q(e^{j\omega}) = F(e^{j\omega})\,e^{-j(N-1)\omega/2}$$

$$H^1(e^{j\omega}) = F_q(e^{j(\omega+\pi)}) = F(e^{j(\omega+\pi)}) \cdot e^{-j(N-1)\omega/2}\,e^{-j(N-1)\pi/2}$$

$$G^0(e^{j\omega}) = F_q(e^{j\omega}) = F(e^{j\omega})\,e^{-j(N-1)\omega/2}$$

Manuscript received November 29, 1983; revised October 22, 1984.

The author is with the IBM Thomas J. Watson Research Center, Yorktown Heights, NY 10598.

0096-3518/85/0600-0626$01.00 © 1985 IEEE

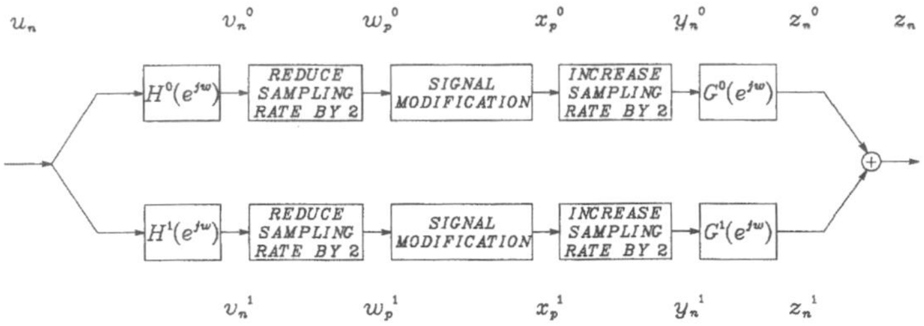

Fig. 1. Two-band multirate filter bank with signal modification.

$$G^1(e^{j\omega}) = -F_q(e^{j(\omega+\pi)}) = -F(e^{j(\omega+\pi)}) \cdot e^{-j(N-1)\omega/2}\, e^{-j(N-1)\pi/2}. \tag{8}$$

By substituting these values into (2) and (3), it is found that

$$A(e^{j\omega}) = 0 \tag{9}$$

and

$$C(e^{j\omega}) = [F^2(e^{j\omega}) - F^2(e^{j(\omega+\pi)})\, e^{-j(N-1)\pi}]\, e^{-j(N-1)\omega} \tag{10}$$

which, for N even, reduces to

$$C(e^{j\omega}) = [F^2(e^{j\omega}) + F^2(e^{j(\omega+\pi)})]\, e^{-j(N-1)\omega}. \tag{11}$$

If we define $\Theta(e^{j\omega})$ as $F^2(e^{j\omega})$, then

$$C(e^{j\omega}) = [\Theta(e^{j\omega}) + \Theta(e^{j(\omega+\pi)})]\, e^{-j(N-1)\omega}. \tag{12}$$

For distortion-free output, it is required that $C(e^{j\omega}) = e^{-j\omega p}$, which in this case requires that $\Theta(e^{j\omega}) + \Theta(e^{j(\omega+\pi)}) = 1$. It is shown in [4] that the condition $\Theta(e^{j\omega}) + \Theta(e^{j(\omega+\pi)}) = 1$ requires that $\Theta(e^{j\omega})$ be a half-band filter, one whose coefficients θ_n satisfy $\theta_{2p} = 0$ for $p \neq 0$ and $\theta_0 = 0.5$. One drawback of the QMF scheme is that the class of half-band filters $\Theta(e^{j\omega})$ which satisfy (11) is limited to filters of the form $0.5 + 0.5 \cos[(2N-1)\omega]$, which is shown in the Appendix. Since this class does not contain any low-pass filters with sharp transitions, it is of limited practical interest. In practice, $F(e^{j\omega})$ is constrained in the design procedure so that $C(e^{j\omega})$ closely approximates unity, as was done in the filter design procedure described in [3].

Filters for Distortion-Free Output

Another choice of filters which produce no aliasing in the two-band filter bank is the filter set

$$\begin{aligned} H^0(e^{j\omega}) &= m(e^{j\omega}) \\ H^1(e^{j\omega}) &= M(e^{j(\omega+\pi)})\, e^{j\omega} \\ G^0(e^{j\omega}) &= M(e^{j\omega}) \\ G^1(e^{j\omega}) &= m(e^{j(\omega+\pi)})\, e^{-j\omega}. \end{aligned} \tag{13}$$

By substituting these values into (2) and (3), it is found that

$$A(e^{j\omega}) = 0 \tag{14}$$

and

$$C(e^{j\omega}) = m(e^{j\omega})\, M(e^{j\omega}) + m(e^{j(\omega+\pi)})\, M(e^{j(\omega+\pi)}). \tag{15}$$

If we define $\chi(e^{j\omega})$ as $m(e^{j\omega})\, M(e^{j\omega})$, then

$$C(e^{j\omega}) = \chi(e^{j\omega}) + \chi(e^{j(\omega+\pi)}). \tag{16}$$

For distortion-free channel response, it is sufficient that $C(e^{j\omega}) = 1$, a condition that is satisfied if $\chi(e^{j\omega})$ is a half-band filter. With the choice of filters of (13), there are many satisfactory solutions to (2) and (3) since all that is required is that $m(e^{j\omega})$ and $M(e^{j\omega})$ are the two factors of a half-band filter. In the following section, a design procedure is given which produces one of the satisfactory solutions.

A Design Procedure for Filters for Distortion-Free Output

Of all the many combinations of filters which produce distortion-free output for the two-band multirate filter bank, let us focus on the one with the following characteristics.

1) $\chi(e^{j\omega})$ is a symmetric odd-length half-band filter with nonnegative frequency response.

2) $m(e^{j\omega})$ is the minimum phase part of $\chi(e^{j\omega})$.

3) $M(e^{j\omega})$ is the maximum phase part of $\chi(e^{j\omega})$.

The following design procedure produces filters with these characteristics.

1) Design a symmetric half-band filter χ_p^0 using the filter design program of McClellan *et al.* [5]. Detailed instructions on how to do this are found in [4]. Without loss of generality, assume that the center top is χ_0^0. This filter is zero phase, with a frequency response minimum at $-\delta_s$.

2) Construct the filter χ_n^1, with frequency response $\chi^1(e^{j\omega}) + \delta_s$, via the transformation $\chi_n^1 = \chi_n^0$ for $n \neq 0$, $\chi_0^1 = \chi_0^0 + \delta_s$. The filter $\chi^1(e^{j\omega})$ has a nonnegative frequency response, as required by the design procedure in step 5), but it is no longer a half-band filter since $\chi_0^1 \neq 0.5$.

3) Rescale the coefficients of χ_n^1 to produce the half-band filter $\chi(e^{j\omega})$ via $\chi_n = \chi_n^1/(2 * \chi_0^1)$. This filter is a half-band filter with a nonnegative frequency response.

4) Find the minimum-phase factor of $\chi(e^{j\omega})$. Several procedures have been proposed to perform this task, including those in [6]–[8]. This is $m(e^{j\omega})$.

5) The maximum-phase factor $M(e^{j\omega})$ is constructed by time-reversing the coefficients of $m(e^{j\omega})$.

This choice of filters has several advantages, in addition to producing distortion-free output. The magnitudes of the frequency responses of the two low-pass filters $H^0(e^{j\omega})$ and $G^0(e^{j\omega})$ are equal, and are equal to the square root of $\chi(e^{j\omega})$. Also, the magnitudes of the frequency responses of the two high-pass filters $H^1(e^{j\omega})$ and $G^1(e^{j\omega})$ are equal, and are equal to the square root of $\chi(e^{j(\omega+\pi)})$. The frequency responses of the decimators and interpolators are thus related in a simple way to $\chi(e^{j\omega})$, and are easily controlled in the design of $\chi(e^{j\omega})$. Furthermore, $M(e^{j\omega})$ and $m(e^{j\omega})$ have the same coefficients, which can lead to both storage and computational advantages in many implementations.

Example

Let us consider the design of filters for a simple two-band filter bank. The signal was sampled at 8 kHz. It is desired that the low-pass filters in the filter bank $H^0(e^{j\omega})$ and $G^0(e^{j\omega})$ have stop-band cutoff frequencies of 2.1 kHz. Our processing budget requires that the low-pass and high-pass filters have at most 50 taps.

The first stage of the procedure is to design $\chi^0(e^{j\omega})$. From the requirements above, the normalized stopband cutoff frequency for $\chi^0(e^{j\omega})$ is $2.1/8 = 0.2625$. The rest of the parameters needed to design $\chi^0(e^{j\omega})$ are computed according to the rules given for designing half-band filters in [4]. The passband cutoff frequency for $\chi^0(e^{j\omega})$ is given by $0.5000 - 0.2625 = 0.2375$. The passband and stopband ripples of $\chi^0(e^{j\omega})$ are equal. The order of $\chi^0(e^{j\omega})$, required to produce length 50 minimum phase and maximum phase factors, is $2 * 50 - 1 = 99$. The computer program of [5] was then used to design a filter with these input parameters. Setting the appropriate coefficients of the filter designed to zero produces the half-band filter $\chi^0(e^{j\omega})$. The frequency response of this filter is given in Fig. 2(a).

Steps 2) and 3) of the design procedure of the preceding section are then applied to produce $\chi(e^{j\omega})$. The frequency response of this filter is given in Fig. 2(b). The filter design program of [6] was then used to find the minimum phase part of $\chi(e^{j\omega})$. The magnitude frequency response of the filters $m(e^{j\omega})$, $M(e^{j\omega})$, $H^0(e^{j\omega})$, and $G^0(e^{j\omega})$ is the same, and is given in Fig. 2(c). Finally, the high-pass filters $H^1(e^{j\omega}) = G^1(e^{j\omega})$ are constructed from $m(e^{j\omega})$ and $M(e^{j\omega})$ according to (13). The magnitude frequency response of these filters is given in Fig. 2(d). By its construction, the channel response $C(e^{j\omega})$ is symmetric, and hence has linear phase. The magnitude of $C(e^{j\omega})$, which ideally should be unity, is given in Fig. 2(e).

Conclusions

In this paper, conditions for a two-band multirate filter bank to be both alias free and to have a unity frequency response were reviewed. A class of filters that satisfies these conditions was given, and a straightforward way of designing filters of this class was presented with an example.

Appendix

Let us assume that θ_n is a half-band filter generated from a symmetric filter f_n convolved with itself.

$$\theta_n = \sum_p f_n f_{n-p}. \tag{A1}$$

For f_n symmetric and of length M, θ_n will also be symmetric and of length $2M - 1$. However, because of the zero coefficients, not all lengths are possible for symmetric half-band filters. A symmetric half-band filter must have a length of the form $4N - 1$. Thus, $M = 2N$, and f_n must have length $2N$.

Assume without loss of generality that the center tap of θ_n occurs at $n = 2N - 1$ and that f_n is nonzero only in the range $0 \le n \le 2N - 1$. (This differs from the rest of the paper, which assumes that the center tap of θ_n occurs at $n = 0$.)

The half-band filter constraints for θ_n, so positioned, require that

$$\theta_k = 0, \quad \text{for } 1 \le k \le 2N - 3,\ k \text{ odd} \tag{A2}$$

and

$$\theta_{2N-1} = 0.5. \tag{A3}$$

It will now be shown that (A2) requires that

$$f_k = 0, \quad \text{for } 1 \le k \le 2N - 3,\ k \text{ odd}. \tag{A4}$$

Although a formal proof will not be given in the interest of brevity, a proof could be constructed along these lines.

First, θ_0 is evaluated using (A1) as $f_0 f_0$, which for $\theta_0 \ne 0$ requires that $f_0 \ne 0$. Next, θ_1 is evaluated as $2[f_0 f_1]$, which for $f_0 \ne 0$ and $\theta_1 = 0$ requires that $f_1 = 0$. θ_3 is evaluated as $2[f_0 f_3 + f_1 f_2]$. This reduces to $f_0 f_3$ since $f_1 = 0$. Thus, for $\theta_3 = 0$, it is required that $f_3 = 0$. θ_5 is evaluated as $2[f_0 f_5 + f_1 f_4 + f_2 f_3]$. Since $f_1 = 0$ and $f_3 = 0$, θ_5 reduces to $2[f_0 f_5]$, which for $\theta_5 = 0$ requires that $f_5 = 0$. Proceeding similarly, each θ_k, k odd, $1 \le k \le 2N - 3$ reduces to $2 f_0 f_k$ once the zero terms have been eliminated, and for $\theta_k = 0$, $f_k = 0$, as stated in (A4).

θ_{2N-1} is evaluated as $2[\Sigma_{k=0}^{2N-1} f_k f_{2N-k-1}]$ which, after elimination of the zero terms, becomes $2 f_0 f_{2N-1}$. Then (A3) requires

$$\theta_{2N-1} = 2 f_0 f_{2N-1} = 0.5. \tag{A5}$$

Finally, the symmetry of f_n is invoked. Equation (A4) then requires

$$f_k = 0, \quad \text{for } 1 \le k \le 2N - 3. \tag{A6}$$

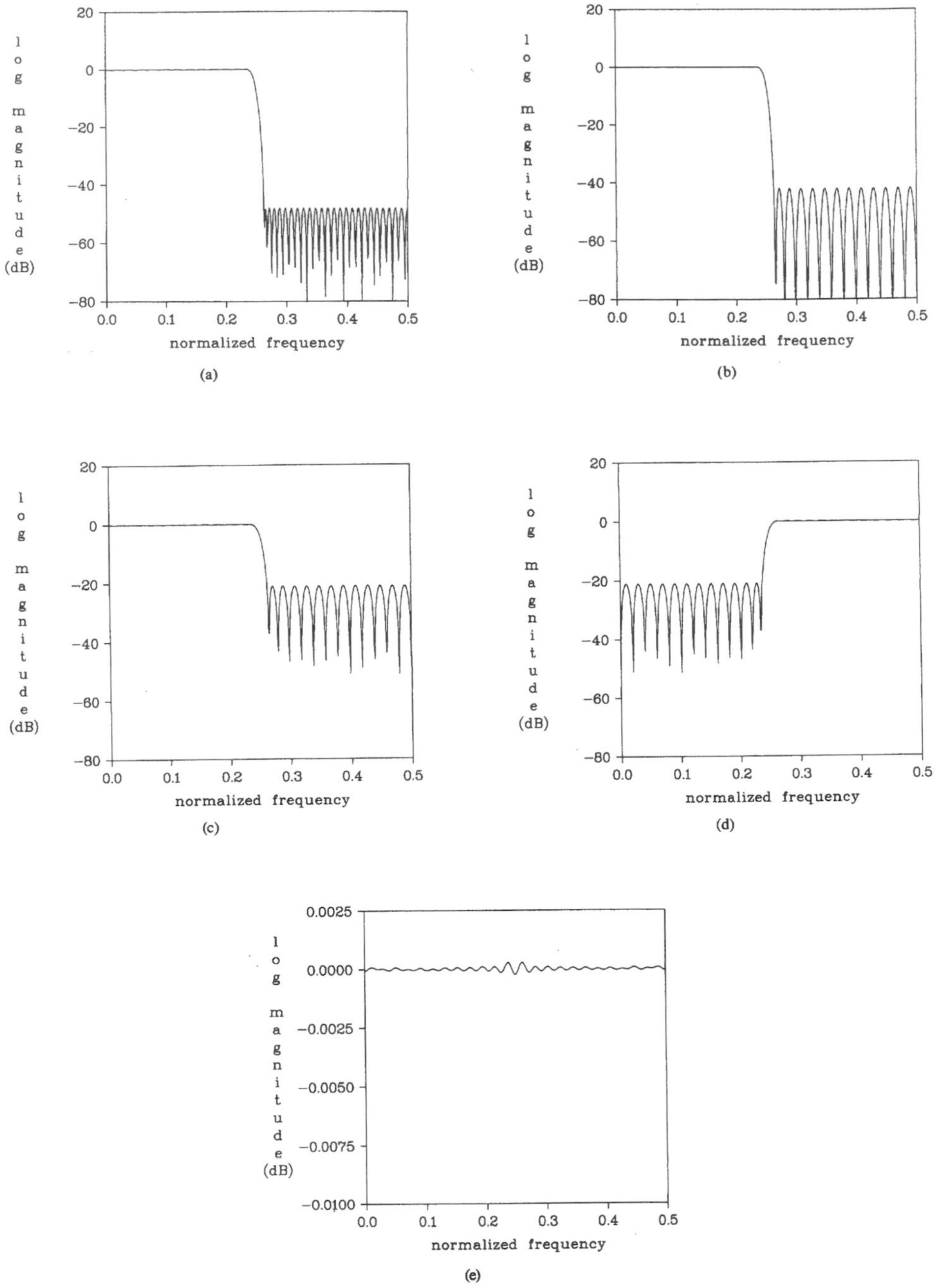

Fig. 2. (a) Frequency response of $\chi(e^{j\omega})$. (b) Frequency response of $\chi^0(e^{j\omega})$. (c) Frequency response of $H^0(e^{j\omega})$. (d) Frequency response of $H^1(e^{j\omega})$. (e) Frequency response of $C(e^{j\omega})$.

Equation (A5) then requires

$$f_0 = f_{2N-1} = \sqrt{0.5}. \tag{A7}$$

Using (A6) and (A7), θ_n is evaluated as

$$\theta_n = \begin{cases} 0.5 & n = 2N - 1 \\ 0.25 & n = 0, 4N - 2 \\ 0.0 & \text{otherwise.} \end{cases} \tag{A8}$$

$\Theta(e^{j\omega})$ is evaluated from (A8) as

$$\Theta(e^{j\omega}) = e^{j\omega(2N-1)}\{0.5 + 0.5 \cos [(2N - 1)\omega]\}. \tag{A9}$$

References

[1] D. Esteban and C. Galand, "Application of quadrature mirror filters to split band voice coding schemes." in *Proc. ICASSP'77*, May 1977, pp. 191-195.

[2] R. E. Crochiere and L. R. Rabiner, *Multirate Digital Signal Processing*. Englewood Cliffs, NJ: Prentice Hall, 1983, p. 380, eq. (7.244).

[3] J. D. Johnston, "A filter family designed for use in quadrature mirror filter banks," in *Proc. ICASSP'80*, Apr. 1980, pp. 291-294.

[4] F. Mintzer, "On half-band, third-band, and *N*th-band FIR filters and their design," *IEEE Trans. Acoust., Speech, Signal Processing*, vol. ASSP-30, pp. 734-738, Oct. 1982.

[5] J. H. McClellan, T. W. Parks, and L. R. Rabiner, "A computer program for designing optimum FIR linear phase digital filters," *IEEE Trans. Audio Electroacoust.*, vol. AU-21, pp. 506-526, Dec. 1973.

[6] G. A. Mian and A. P. Nainer, "A fast procedure to design equiripple minimum-phase FIR filters," *IEEE Trans. Circuits Syst.*, vol. CAS-29, pp. 327-331, May 1982.

[7] R. Boite and H. Leich "A new procedure for the design of high order minimum phase FIR digital or CCD filters," *Signal Processing*, vol. 3, pp. 101-108, Apr. 1981.

[8] S. Ebert and U. Heute, "Accelerated design of linear or minimum phase FIR digital filters with a Chebyshev magnitude response," *Proc. IEE*, vol. 130, part G, pp. 267-270.

Fred Mintzer (S'75-M'79) was born in Wilkes-Barre, PA, on June 23, 1948. He received the B.S. degree in electrical engineering from Rutgers University, New Brunswick, NJ, in 1970, and the Ph.D. degree, also in electrical engineering, from Princeton University, Princeton, NJ, in 1978.

From 1971 to 1974 he was employed as an Applications Engineer by the Commonwealth Telephone Company, Dallas, PA. He joined the IBM Thomas J. Watson Research Center, Yorktown Heights, NY, in 1978 as a member of the project that developed the RSP architecture, and engaged in research on distributed digital signal processing and data communications. In 1980 he became the Manager of the Signal Processing Applications Project, and continued in research on signal processing applications, architectures, and algorithms. In 1983 he joined the Image Technologies Department as Manager of the NCI Architectures Project. His current research interests are in image display and print algorithms, signal processing algorithms, and specialized architectures.

Dr. Mintzer is a member of Tau Beta Pi and Eta Kappa Nu.

Signal Processing 10 (1986) 219–244
North-Holland

FILTER BANKS ALLOWING PERFECT RECONSTRUCTION

Martin VETTERLI (Member EURASIP)

Département d'Électricité, Laboratoire d'Informatique Technique, École Polytechnique Fédérale de Lausanne, 16 Chemin de Bellerive, CH-1007 Lausanne, Switzerland

Received 29 May 1985
Revised 23 August 1985 and 31 October 1985

Abstract. Splitting a signal into N filtered channels subsampled by N is an important problem in digital signal processing. A fundamental property of such a system is that the original signal can be perfectly recovered from the subsampled channels. It is shown that this can always be done, and that FIR solutions exist. This is done by mapping the NM-dimensional nonlinear problem (where N is the number of channels and M the length of the FIR filters) into an M-dimensional linear problem. For $N = 2$, a general class of FIR solutions is derived, together with methods to find filters. The dual problem of mixing N signals into one channel upsampled by N is also addressed. Several applications are proposed. All results are obtained by looking at the N filter bank as a true N channel system, rather than N separate channels.

Zusammenfassung. Die Aufspaltung eines Signals in N Kanäle, die jeweils N-fach unterabgetastet werden, ist ein wichtiges Problem der Signalverarbeitung. Eine grundlegende Eigenschaft solch eines Systems ist es, daß das Originalsignal unverzerrt aus den unterabgetasteten Kanalsignalen rekonstruiert werden kann. Es wird gezeigt, daß dies stets gelingt und daß es FIR-Lösungen gibt. Dazu dient eine Abbildung des NM-dimensionalen, nicht-linearen Entwurfsproblems (wobei N die Kanalzahl und M die Filterlänge ist) auf ein M-dimensionales, lineares Problem. Für den Fall $N = 2$ wird eine allgemeingültige Klasse von FIR-Lösungen hergeleitet. Zudem werden Verfahren zum Auffinden der Koeffizienten angegeben. Das duale Problem wird ebenfalls behandelt: Es besteht darin, N Einzelsignale nach N-facher Erhöhung der Datenrate zu einem Gesamtsignal zu vermischen. Verschiedene Anwendungen werden vorgeschlagen. Alle Ergebnisse beruhen darauf, daß nicht jeder einzelne der N Kanäle, sondern die Filterbank wirklich als N-Kanal-System betrachtet wird.

Résumé. La séparation d'un signal en N canaux sous-échantillonnés d'un facteur N est un problème important en traitement numérique des signaux. Une propriété fondamentale d'un tel système est que le signal original puisse être reconstruit parfaitement à partir des canaux sous-échantillonnés. Il est montré dans la suite que ceci est toujours possible, et que des solutions RIF existent. Celles-ci sont obtenues en transformant un problème non linéaire de dimension NM (où N est le nombre de canaux et M la longueur des filtres RIF) en un problème linéaire de dimension M. Pour $N = 2$, on dérive une classe générale de solutions RIF, ainsi que des méthodes pour trouver des filtres. Le problème dual du multiplexage de N signaux en un signal de fréquence d'échantillonnage N fois plus élevée est également considéré. Plusieurs applications sont proposées. Tous les résultats sont obtenus en considérant le problème des bancs de N filtres comme un problème à N dimensions plutôt que N problèmes séparés.

Keywords. Filter banks, multirate systems, decimation, interpolation, quadrature mirror filters.

1. Introduction

Let us first briefly state the basic problem we want to solve. Suppose an infinite sequence of samples $x(n)$. This sequence is filtered into N sequences $y_0(n) \dots y_{N-1}(n)$ (with linear, time invariant filters). The sequences $y_i(n)$ are subsampled by a factor N, that is only every Nth sample is kept, or $y_i'(n) = y_i(Nn)$. Now, the problem is to recover $x(n)$ from the subsampled sequences $y_i'(n)$ (see Fig. 1).

Obviously, there is the same number of samples per unit of time in $x(n)$ and in all the $y_i'(n)$ together, thus a solution should exist. Nevertheless, there were not many practical solutions to the problem up to

0165-1684/86/$3.50 © 1986, Elsevier Science Publishers B.V. (North-Holland)

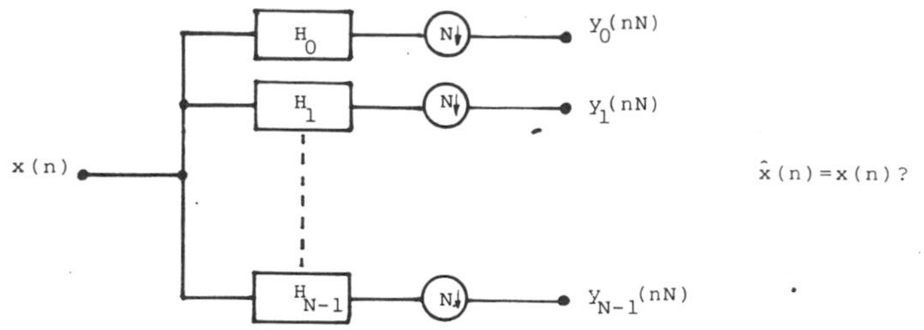

Fig. 1. General problem statement. Can the initial signal be recovered from the subsampled channel signals?

recently. On the one hand, the downsampling/upsampling process creates aliased versions of the original signal (unless perfectly sharp bandpass filters are used before subsampling, leading to infinitely long filters) which have to be cancelled in the reconstruction process, but, on the other hand, the original signal appears filtered at the output (and this filtering has to be cancelled, which might be impossible for stability reasons).

The work on subsampled filter banks was initiated by the introduction of the quadrature mirror filter concept [5, 6, 7, 8]. The two channel QMF filter bank, while not solving the problem perfectly, annihilates the aliasing perfectly, a very useful feature in speech processing [3]. The work on efficient implementation of filter banks started with the computation of the transmultiplexers by polyphase networks and FFTs [1]. The merging of the two approaches was first proposed by Nussbaumer [13] and was further investigated by many authors [12, 15, 16, 20, 30]. The first perfect FIR solution for the two channel subsampled filter bank was proposed by Smith and Barnwell [22] and Wackersreuther [29]. The matrix notation we developed to address the general case [28] was also independently introduced by Ramstad [19] and Smith and Barnwell [23]. Thorough treatments of the filter bank problems were done by Vary and co-authors [9, 25, 26].

The main results appearing below are briefly stated hereafter:

- emphasis is put on FIR analysis and FIR synthesis filters, because IIR solutions lead to implicit pole/zero cancellations,
- a general class of FIR solutions for $N=2$ is derived,
- linear phase solutions are shown to exist,
- two new methods to generate FIR filters that will satisfy the perfect reconstruction requirement for $N=2$ are developed,
- for $N>2$, the NM-dimensional nonlinear problem (with M being the filter length) that has to be solved to find FIR solutions is shown to reduce to an M-dimensional linear problem,
- solutions are shown to exist and a method to find them is given,
- it is shown that aliasing can always be cancelled,
- the case where the N filters are derived from a single prototype filter by frequency shifting is shown to only have an IIR solution,
- the dual problem of mixing N signals onto a single channel upsampled by N is solved.

All these results (except the aliasing cancellation property and one design procedure [29]) are, to our knowledge, original. Most results are obtained by using a general polyphase representation of the filters appearing in a subsampled filter bank and by reducing the general problem to the analysis of the determinant of the polyphase filter matrix. While a matrix notation was also used in [19, 23], this generalized polyphase representation seems to be original. As applications, a subband coder incorporating linear prediction, a scrambling scheme for analog signals, noncritically subsampled filter banks and filter banks on finite fields are proposed.

The outline of the paper is the following: Section 2 states the general problem and two known but unpractical solutions. Section 3 thoroughly investigates the two channel case (which is simple yet important in practice). Section 4 looks at the general case, and shows that FIR solutions for both the analysis and synthesis can exist. Section 5 looks at the dual problem which is closely related to the initial one. Section 6, finally, proposes a couple of applications. Appendix A shows that only FIR analysis and synthesis does not produce implicit pole/zero cancellation and Appendix B gives some more results on the two channel case.

In the following, all signals and filters are assumed to be complex unless specified otherwise, for simplicity and generality reasons only. Note as well that both filters and matrices (or vectors) use the numbering starting from 0 to $N-1$ (N being the dimension), which is unusual for linear algebra, but makes notations more coherent and simple.

2. The problem and two obvious solutions

In Fig. 1, the problem addressed below is stated pictorially: the signal $x(n)$ with z-transform $X(z)$ [17, 18] is filtered into N channels, which are then subsampled by N. Can the original signal be recovered from the N subsampled channels? In order to solve the problem stated in Fig. 1, we consider the system depicted in Fig. 2. There, the operations in the analysis part have been matched by equivalent operations

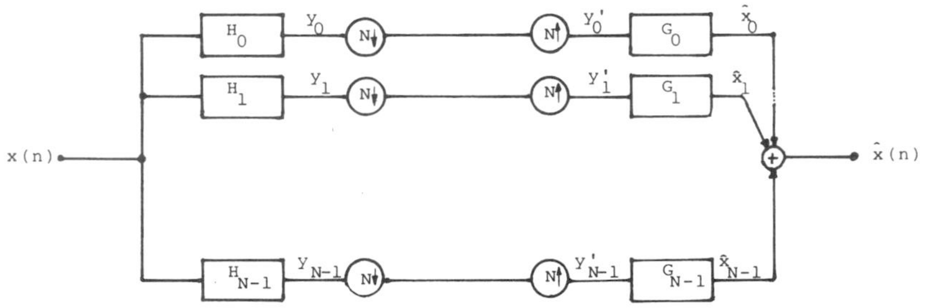

Fig. 2. Solution to the general problem involving upsampling and interpolation.

in the synthesis part in order to recover the initial signal. The function $N\downarrow$ means subsampling by N, that is replacing the sequence $x(n)$ by the sequence $x'(n) = x(nN)$. In the z-domain, this can be shown to be equal to [4, 21]

$$X'(z) = (1/N) \sum_{k=0}^{N-1} X(W^k z^{1/N}), \quad W = e^{-j2\pi/N}. \tag{1}$$

The function $N\uparrow$ means upsampling by N, which corresponds to replacing $x'(n)$ by $x''(n) = x'(n)$ for $n = lN$, and zero otherwise. This leads to the following z-transform:

$$X''(z) = X'(z^N). \tag{2}$$

Actually, the cascading of subsampling and upsampling by the factor N corresponds to the modulation of the sequence $x(n)$ by the function $l(n)$ given by

$$l(n) = (1/N) \sum_{k=0}^{N-1} W^{-nk}, \tag{3}$$

which gives the following z-transform for $x''(n) = x(n)l(n)$:

$$X''(z) = (1/N) \sum_{k=0}^{N-1} X(W^k z). \tag{4}$$

Thus, downsampling and subsequent upsampling of a signal produces an output containing the signal itself as well as $N-1$ aliased versions (which are undesired).

In order to reproduce the input signal exactly at the output in the system of Fig. 2, one can use perfect bandpass filters (infinitely sharp and nonoverlapping) with a bandwidth of $1/N$ (assume a normalized sampling frequency of 1). Then, subsampling by N is allowed (since no spectral overlapping occurs). After upsampling, interpolation with the perfect bandpass and subsequent addition of the bands reproduce the input signal perfectly as can readily be verified. The only problem is that the required filters have to be infinitely long, otherwise the reconstructed signal is an approximated version of the original signal only, and, in particular, aliased versions of the original signal will appear in the output signal. In the above approach, one tried to verify the sampling theorem before subsampling (thus requiring perfect bandpass filters), and thus the problem was approached on one channel at a time basis. No use was made of the fact that all channels are computed simultaneously.

Another solution [27], where the simultaneity of the process in the N channels is used, appears when the analysis filters are of length N (equal to the number of filters). Then, the vector of the subsampled signals at time n can be seen to be equal to the product of a matrix $\boldsymbol{H}$ with a vector $\boldsymbol{x}$ containing the N last samples of the input signal ($[x(n), x(n-1), \ldots, x(n-N+1)]$). The rows of the matrix $\boldsymbol{H}$ are obtained from the coefficients of the input filters. Similarly, the N reconstructed outputs ($[\hat{x}(n), \hat{x}(n-1), \ldots, \hat{x}(n-N+1)]$) are equal to the product of a matrix $\boldsymbol{G}$ with the vector of the subsampled signals. The matrix $\boldsymbol{G}$ has its columns equal to the coefficients of the synthesis filters (which are also of length N). Then, if $\boldsymbol{H}$ is invertible (that is, the N analysis filters are linearly independent) and $\boldsymbol{G} = \boldsymbol{H}^{-1}$, it can be shown that the reconstruction is perfect. The problem here is that length N filters in an N channel filter bank are in general too short for practical applications.

Thus, the question that will be addressed next is: are there length M FIR filters, $N < M < \infty$, that will allow perfect reconstruction?

At this point, a remark is already appropriate: the first solution above divided the N channel problem into N separate problems of down-upsampling by N and required therefore infinitely long filters. The second solution simply solves the problem by looking at all the channels at the same time, and while not satisfactory because of the short filter length, it is nevertheless perfect. Thus, the N channel filter bank problem should always be considered as a whole, 'N-dimensional problem', which has to be solved as such.

3. The general two channel case

The two channel case is depicted in Fig. 3. For channel 0, the following holds (using (4) and the convolution property of the z-transform):

$$Y_0(z) = H_0(z)X(z), \tag{5}$$

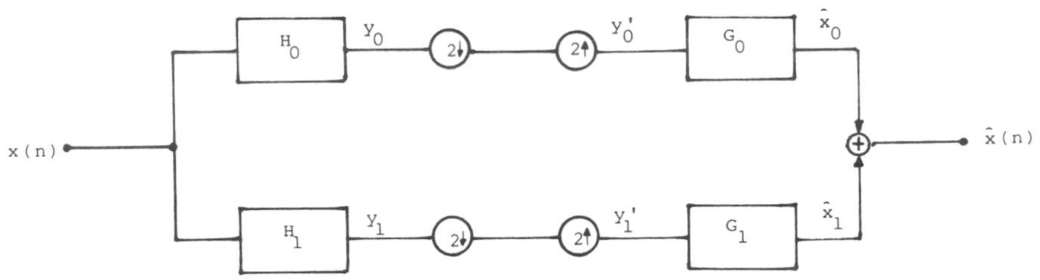

Fig. 3. Two channel case.

$$Y_0'(z)=\tfrac{1}{2}[H_0(z)X(z)+H_0(-z)X(-z)], \tag{6}$$

$$\hat{X}_0(z)=\tfrac{1}{2}[H_0(z)X(z)+H_0(-z)X(-z)]G_0(z). \tag{7}$$

Similar relations hold for channel 1. Thus, $\hat{X}(z)$ can be written as

$$\hat{X}(z)=\tfrac{1}{2}([H_0(z)G_0(z)+H_1(z)G_1(z)]X(z)+[H_0(-z)G_0(z)+H_1(-z)G_1(z)]X(-z)). \tag{8}$$

The reconstructed signal is therefore a function of the original signal $X(z)$, plus a function of the modulated signal $X(-z)$, as shown in Fig. 4 and in equation (9):

$$\hat{X}(z)=F_0(z)X(z)+F_1(z)X(-z), \tag{9}$$

where

$$F_0(z)=\tfrac{1}{2}[H_0(z)G_0(z)+H_1(z)G_1(z)], \tag{10}$$

$$F_1(z)=\tfrac{1}{2}[H_0(-z)G_0(z)+H_1(-z)G_1(z)]. \tag{11}$$

A necessary and sufficient condition for perfect reconstruction (since $F_0(z)$ and $F_1(z)$ are linear and time invariant) is that $F_0(z)$ is a pure delay and $F_1(z)$ is equal to zero.

In matrix notation, this is equivalent to

$$\begin{bmatrix} F_0(z) \\ F_1(z) \end{bmatrix}=\frac{1}{2}\begin{bmatrix} H_0(z) & H_1(z) \\ H_0(-z) & H_1(-z) \end{bmatrix}\cdot\begin{bmatrix} G_0(z) \\ G_1(z) \end{bmatrix}=\begin{bmatrix} z^{-k} \\ 0 \end{bmatrix} \tag{12}$$

or

$$\boldsymbol{f}(z)=\boldsymbol{M}(z)\boldsymbol{g}(z)=[z^{-k}\ \ 0]^{\mathrm{T}}, \tag{13}$$

where the meanings of $\boldsymbol{f}$, $\boldsymbol{M}$, and $\boldsymbol{g}$ in (13) are obvious from (12). Now, setting

$$\boldsymbol{g}(z)=\boldsymbol{M}^{-1}(z)[z^{-k}\ \ 0]^{\mathrm{T}} \tag{14}$$

will solve the given problem of exact reconstruction. This is shown in Fig. 5 and the resulting output is

$$\hat{X}(z)=[\boldsymbol{f}(z)]^{\mathrm{T}}\boldsymbol{x}(z)=z^{-k}X(z), \tag{15}$$

where

$$\boldsymbol{x}(z)=[X(z)\ \ X(-z)]^{\mathrm{T}}. \tag{16}$$

While (14) gives a perfect solution, the problem of existence (causality, type of synthesis filters) has yet

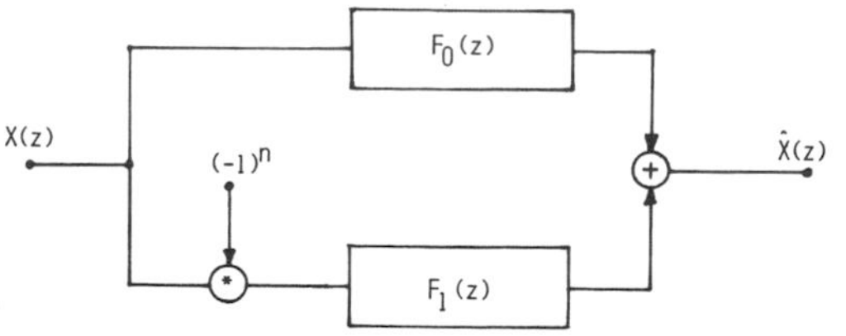

Fig. 4. Equivalent two channel system, where the output is a linear function of the original and the modulated input.

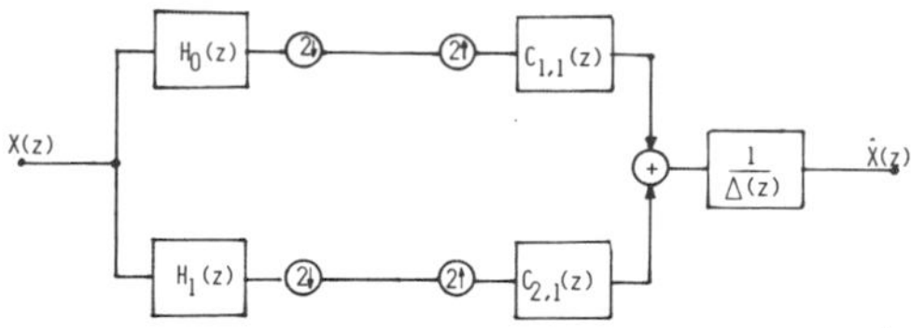

Fig. 5. Two channel system with perfect reconstruction.

to be addressed. Consider now the inverse of the filter matrix $\boldsymbol{M}(z)$:

$$\boldsymbol{M}^{-1}(z)=(1/\Delta(z))\boldsymbol{C}(z), \tag{17}$$

where

$$\Delta(z)=\tfrac{1}{4}[H_0(z)H_1(-z)-H_0(-z)H_1(z)] \tag{18}$$

is the determinant of $\boldsymbol{M}(z)$ and

$$\boldsymbol{C}(z)=\frac{1}{2}\begin{bmatrix} H_1(-z) & -H_1(z) \\ -H_0(-z) & H_0(z) \end{bmatrix} \tag{19}$$

is the cofactor matrix of $\boldsymbol{M}(z)$.

We will consider separately $1/\Delta(z)$ and $\boldsymbol{C}(z)$, because $1/\Delta(z)$ can be seen as a common post-filter for all synthesis filters (and can thus be applied after summation) but also because $1/\Delta(z)$ and $\boldsymbol{C}(z)$ lead in general to different types of filters.

Thus, choose the synthesis filters as

$$\boldsymbol{g}(z)=\boldsymbol{C}(z)[1\ \ 0]^{\mathrm{T}} \tag{20}$$

or $G_0(z)=\frac{1}{2}H_1(-z)$ and $G_1(z)=-\frac{1}{2}H_0(-z)$. Then, we have the following transmission vector:

$$\boldsymbol{f}(z)=\boldsymbol{M}(z)\boldsymbol{C}(z)[1\ \ 0]^{\mathrm{T}}=[\Delta(z)\ \ 0]^{\mathrm{T}}, \tag{21}$$

and the following reconstructed signal:

$$\hat{X}(z)=\Delta(z)X(z). \tag{22}$$

The following general remark can be made.

Remark. Choosing the synthesis filters as the first column elements from the cofactor matrix $\boldsymbol{C}(z)$ of the analysis filter matrix $\boldsymbol{M}(z)$ leads to the following properties:

(a) The aliasing is perfectly cancelled.

(b) The input/ouput transfer function is equal to the determinant $\Delta(z)$ of the analysis filter matrix.

Several possibilities are now open in order to achieve a perfect input/output transfer function.

(i) Use a post-filter equal to $1/\Delta(z)$. As shown in Appendix A, this means implicit pole/zero cancellation, which can lead to numerical problems (besides the fact that care has to be taken when choosing the analysis filters so that $1/\Delta(z)$ is a stable filter).

(ii) Use a post-filter $1/\Delta'(z)$ such that $\Delta(z)/\Delta'(z)$ is an all-pass filter. This leads to perfect amplitude reconstruction but to phase distortion.

(iii) Choose the analysis filters in such a way that $\Delta(z)$ is a pure delay, that is, a monomial in z or z^{-1} (a monomial is a polynomial having a single nonzero coefficient). This achieves perfect reconstruction within a delay, and does not have the problems of (i) or (ii).

In the following, we will look for solutions of the third kind, since they meet all our requirements. As shown in Appendix A, only FIR analysis and synthesis filters lead to perfect reconstruction without implicit pole/zero cancellation. Thus, only FIR analysis filters are considered below.

The power of the method shown so far is that the whole problem of filtering/decimation allowing perfect reconstruction has been reduced to investigate properties of the determinant of the filter matrix $\boldsymbol{M}(z)$ from (13).

Assume that $H_0(z)$ and $H_1(z)$ are FIR filters of length M_0 and M_1, respectively. We define $P(z)$ by

$$P(z) = H_0(z)H_1(-z) = \sum_{i=0}^{M_0+M_1-2} p_i z^{-i}. \tag{23}$$

Then, the determinant $\Delta(z)$ is simply given by

$$\Delta(z) = \tfrac{1}{4}[P(z) - P(-z)] = 2 \sum_{i=0}^{M_s-2} p_{2i+1} z^{-2i-1}, \tag{24}$$

where $M_s = \frac{1}{2}(M_0 + M_1)$ when $M_0 + M_1$ is even and $M_s = \frac{1}{2}(M_0 + M_1 + 1)$ when $M_0 + M_1$ is odd.

Now, if the p_{2i+1} are all zero but one (equal to 2), and the p_{2i} are arbitrary, then the reconstruction will be perfect using the FIR synthesis filters given by (20).

Three methods are now possible in order to derive FIR analysis filters that will allow perfect synthesis with FIR filters. They are simply different ways to meet the requirement that equation (24) should reduce to a monomial.

Method 1. The first method is outlined below:

(a) Take a polynomial $P(z)$ satisfying the following conditions:

- degree $= M_0 + M_1 - 2 = M_s - 2$.
- P_{2i} arbitrary,
- $P_{2i+1} = \begin{cases} 0, & i \neq k, \\ 2, & i = k. \end{cases}$

(b) Factorize $P(z)$ into its $M_s - 2$ factors containing one zero each:

$$P(z) = d_0 d_1 \prod_{i=0}^{M_s-3} (z^{-1} + \alpha_i), \tag{25}$$

where d_0 and d_1 are scalar normalizing factors.

(c) Divide the set of zeros into two sets, and this arbitrarily:

$$P(z) = d_0 \prod_{i=0}^{M_0-2} (z^{-1} + \alpha_i)\, d_1 \prod_{l=M_0-1}^{M_s-3} (z^{-1} + \alpha_l). \tag{26}$$

(d) Set $H_0(z)$ and $H_1(z)$ equal to

$$H_0(z) = d_0 \prod_{i=0}^{M_0-2} (z^{-1} + \alpha_i), \qquad H_1(z) = d_1 \prod_{l=M_0-1}^{M_s-3} (-z^{-1} + \alpha_l). \tag{27), (28}$$

From this construction, we have

$$\Delta(z)=\tfrac{1}{4}[H_0(z)H_1(-z)-H_0(-z)H_1(z)]=\tfrac{1}{4}[P(z)-P(-z)]=z^{-2k-1}, \tag{29}$$

and thus, using $G_0(z)=\frac{1}{2}H_1(-z)$ and $G_1(z)=-\frac{1}{2}H_0(-z)$ as in (20), we find $\hat{X}(z)=X(z)z^{-2k-1}$ at the output.

That this construction can produce reasonable filters is shown in a simple example. Take a perfect halfband filter [18] whose z-transform is given by

$$H(z)=2\sum_{i=-\infty}^{\infty}\sin(\tfrac{1}{2}i\pi)/(\tfrac{1}{2}i\pi)z^{-i}. \tag{30}$$

Truncate it below $-2k-1$ and above $2k+1$, apply a window if desired and delay it by z^{-2k-1}. The resulting filter $H'(z)$, given as

$$H'(z)=2\sum_{i=0}^{4k+2}\sin(\tfrac{1}{2}(i-2k-1)\pi)/(\tfrac{1}{2}(i-2k-1)\pi)z^{-i}, \tag{31}$$

has the required property described in (a), namely

$$H'(z)-H'(-z)=4z^{-2k-1}. \tag{32}$$

Factorizing $H'(z)$ into its $4k+2$ zeros, grouping conjugate pairs, and distributing the pairs and zeros among $H_0(z)$ and $H_1(-z)$ will produce a low-pass (H_0) and a high-pass (H_1). This produces a perfect system in the sense described above.

Method 2. Another simple method is to choose $H_0(z)$ and then derive $H_1(z)$ by solving the following equation:

$$\Delta(z)=\tfrac{1}{4}[H_0(z)H_1(-z)-H_0(-z)H_1(z)]=z^{-2k-1}, \tag{33}$$

where k can be arbitrarily chosen within the range $0,\ldots,M-2$ (if we assume that both $H_0(z)$ and $H_1(z)$ are length M filters). Then, (33) leads to a set of $M-1$ equations of the following form:

$$M-1\left\{\overbrace{\begin{bmatrix} h_{0,1} & h_{0,0} & 0 & 0 & 0 & \cdots & & 0 \\ h_{0,3} & h_{0,2} & h_{0,1} & h_{0,0} & 0 & \cdots & & 0 \\ 0 & & & & & & & \\ \vdots & & & & & & & \vdots \\ & & & & & & & 0 \\ 0 & & & & h_{0,M-2} & & h_{0,M-3} & h_{0,M-4} \\ 0 & & \cdots & & 0 & & h_{0,M-1} & h_{0,M-2} \end{bmatrix}}^{M}\right. \cdot \begin{bmatrix} h_{1,0} \\ -h_{1,1} \\ \vdots \\ h_{1,k} \\ \vdots \\ h_{1,M-2} \\ -h_{1,M-1} \end{bmatrix} = \begin{bmatrix} 0 \\ 0 \\ \vdots \\ 0 \\ 2 \\ 0 \\ \vdots \\ 0 \\ 0 \end{bmatrix}, \tag{34}$$

where $h_{i,j}$ means the jth coefficient of the ith filter (here, we assume both k and M to be even). Choosing one of the coefficients of H_1, for example $h_{1,0}$ equal to 1, leads to an equivalent set of $M-1$ equations for the $M-1$ unknowns $h_{1,1}, \ldots, h_{1,M-1}$:

$$
M-1\left\{\overbrace{\begin{bmatrix}
h_{0,0} & 0 & 0 & 0 & & \cdots & 0 \\
h_{0,2} & h_{0,1} & h_{0,0} & 0 & & \cdots & 0 \\
0 & & & & & & \\
\vdots & & & & & & \vdots \\
& & & & & & \\
0 & \cdots & & h_{0,M-2} & & h_{0,M-3} & h_{0,M-4} \\
0 & \cdots & 0 & 0 & & h_{0,M-1} & h_{0,M-2}
\end{bmatrix}}^{M-1}\right.
\cdot
\begin{bmatrix}
-h_{1,1} \\ h_{1,2} \\ \vdots \\ h_{1,k} \\ \vdots \\ h_{1,M-2} \\ -h_{1,M-1}
\end{bmatrix}
=
\begin{bmatrix}
0 \\ 0 \\ \vdots \\ 0 \\ 2 \\ 0 \\ \vdots \\ 0 \\ 0
\end{bmatrix}
-
\begin{bmatrix}
h_{0,1} \\ h_{0,3} \\ 0 \\ \vdots \\ 0 \\ 0
\end{bmatrix}. \tag{35}
$$

This system is in general solvable. If not, one has to change one or more coefficients of the filter $H_0(z)$ until the determinant of the matrix in (35) differs from zero.

Since the choice of the arbitrary values in (34) to get (35) can be delicate, it is possible to use a different method to solve the system in (34). If one has an a priori knowledge of what $H_1(z)$ should be, one can add an equation and thus get M equations for the M unknowns of $H_1(z)$. For example, if $H_1(z)$ is a high-pass filter, one can place a zero at $z=1$ by adding an equation stating that the sum of the $h_{1,i}$'s should be zero. Of course, this supplementary equation should be linearly independent of the previous ones, otherwise the system becomes unsolvable.

Method 3. Yet another method is to choose the filters $H_0(z)$ and $H_1(z)$ in a class of fiters that will automatically satisfy the determinant constraint. One such class, proposed in [22, 29], consists in choosing $H_0(z)$ such that its autocorrelation function has zero even-index samples (except the zeroth term of course). Assuming $H_0(z)$ of length M and setting $H_1(z)=z^{-M+1}H_0(-z^{-1})$ will satisfy the determinant constraint automatically as can be verified.

To conclude this section on two channel FIR filter banks, we show in Appendix B the case where one of the channels is delayed by z^{-1}. Then, the determinant of $\boldsymbol{M}(z)$ is equal to

$$\Delta(z)=\tfrac{1}{4}[H_0(z)H_1(-z)+H_0(-z)H_1(z)]. \tag{36}$$

Thus, all results can be carried over by simply noting that $\Delta(z)$ is a polynomial with even-index samples only, which corresponds to the previously analysed determinant delayed by one sample. Still, by using the powerful determinant analysis, we show also in Appendix B that there exists a linear phase solution for the two channel case, but only for even-length filters, where $H_0(z)$ and $H_1(z)$ have different symmetry.

4. The N channel case

While the two channel case described above could be treated without using matrix notation, the N channel case calls for a matrix approach. As will be seen, the representation form of the filter bank problem is of crucial importance, and this for the following three reasons:

(a) understanding of the system and physical interpretation,

(b) analysis of the computational complexity,

(c) numerical properties of the system.

Thus, we will introduce different representations of the problem, namely:

(a) the modulated filter matrix $\boldsymbol{M}(z)$, which is the initial representation of the problem of perfect reconstruction with aliasing cancellation,

(b) the polyphase filter matrix $\boldsymbol{P}(z)$, which gives physical insight into the problem and allows a simpler mathematical treatment,

(c) the diagonal filter matrix $\boldsymbol{D}(z)$, which appears when the filter bank is obtained from a single prototype filter by modulation and permits a treatment of the computational complexity.

The various representations are obtained with multiplication by Fourier matrices and can be seen as basis changes.

Assume a system as depicted in Fig. 2. Similarly to equations (5)-(7), we get, for the ith channel,

$$\hat{X}_i = (1/N)\left[\sum_{k=0}^{N-1} H_i(W^k z)X(W^k z)\right]G_i(z), \quad W = \mathrm{e}^{-\mathrm{j}2\pi/N}. \tag{37}$$

Now, the reconstructed signal $\hat{X}(z)$ is a linear combination of $X(z)$ and its $N-1$ aliased components $X(W^k z)$, or, similarly to (9),

$$\hat{X}(z) = \sum_{k=0}^{N-1} F_k(z)X(W^k z), \tag{38}$$

where

$$F_i(z) = (1/N)\sum_{l=0}^{N-1} H_l(W^i z)G_l(z). \tag{39}$$

Again, we want $F_0(z)$ to be equal to a perfect delay and $F_i(z)$, $i \neq 0$, to be equal to zero. In matrix notation, this leads to the following system:

$$\begin{bmatrix} F_0(z) \\ F_1(z) \\ \vdots \\ F_{N-1}(z) \end{bmatrix} = (1/N)\begin{bmatrix} H_0(z) & H_1(z) & \dots & H_{N-1}(z) \\ H_0(Wz) & H_1(Wz) & \dots & H_{N-1}(Wz) \\ \vdots & \vdots & & \vdots \\ H_0(W^{N-1}z) & \dots & & H_{N-1}(W^{N-1}z) \end{bmatrix} \cdot \begin{bmatrix} G_0(z) \\ G_1(z) \\ \vdots \\ G_{N-1}(z) \end{bmatrix} = \begin{bmatrix} z^{-k} \\ 0 \\ \vdots \\ 0 \end{bmatrix}, \tag{40}$$

or

$$\boldsymbol{F}(z) = \boldsymbol{M}(z)\boldsymbol{g}(z) = \boldsymbol{u}(z), \tag{41}$$

with

$$f_i(z) = F_i(z), \quad \boldsymbol{M}_{i,j}(z) = (1/N)H_j(W^i z), \quad g_i(z) = G_i(z), \quad \boldsymbol{u}(z) = [z^{-k} \;\; 0 \dots 0].$$

Again, we assume the $H_i(z)$ to be FIR filters, and we choose $\boldsymbol{g}(z)$ to be

$$\boldsymbol{g}(z) = \boldsymbol{C}(z)[1 \;\; 0 \dots 0]^{\mathrm{T}}, \tag{42}$$

where $\boldsymbol{C}(z)$ is the cofactor matrix of $\boldsymbol{M}(z)$. Then, the reconstructed signal is equal to, similarly to (22),

$$\hat{X}(z) = \Delta(z)X(z), \tag{43}$$

where Δ is the determinant of the filter matrix $\boldsymbol{M}(z)$.

Similarly to (22) and Fig. 5, the determinant can be cancelled at the output, thus yielding perfect reconstruction. Note that this implies implicit pole/zero cancellation. If this is undesired (for the reasons explained in Appendix A), we require that both the analysis and synthesis filters be FIR, and, therefore, we want the determinant to reduce to a pure delay.

In order to analyse the determinant of the modulated filter matrix $\boldsymbol{M}(z)$, we premultiply $\boldsymbol{M}(z)$ by the Fourier matrix $\boldsymbol{F}$ and we obtain the polyphase filter matrix representation

$$\boldsymbol{P}(z)=\boldsymbol{F}\boldsymbol{M}(z), \tag{44}$$

where:

$$\boldsymbol{P}_{ij}(z)=z^{-j}H_{ij}(z^N), \qquad \boldsymbol{F}_{n,m}=W^{nm}, \tag{45}$$

and $H_{ij}(z^N)$ is the jth polyphase component of the ith filter, that is, the filter

$$H_{ij}(z^N)=h_{i,j}+h_{i,j+N}z^{-N}+h_{i,j+2N}z^{-2N}+\cdots. \tag{46}$$

The ith filter $H_i(z)$ is related to its polyphase components in the following way:

$$H_i(z)=\sum_{j=0}^{N-1} z^{-j}H_{i,j}(z^N) \tag{47}$$

and, reciprocally,

$$H_{i,j}(z^N)=z^j(1/N)\sum_{k=0}^{N-1} H_i(W^{jk}z). \tag{48}$$

Fig. 6 shows a pictorial representation of the polyphase filter interpretation of the filtering/subsampling process, and an equivalent figure could be drawn for the upsampling/filtering process.

Now, the determinants $\Delta_\mathrm{p}(z)$ of $\boldsymbol{P}(z)$ and $\Delta(z)$ of $\boldsymbol{M}(z)$ are equal, except for a constant factor given by the determinant Δ_f of $\boldsymbol{F}$:

$$\Delta_\mathrm{p}(z)=\Delta_\mathrm{f}\Delta(z). \tag{49}$$

Similarly, the inverses and cofactor matrices are related:

$$[\boldsymbol{P}(z)]^{-1}=[\boldsymbol{M}(z)]^{-1}\boldsymbol{F}^{-1}, \qquad \boldsymbol{C}_\mathrm{p}(z)=\frac{\Delta_\mathrm{p}(z)}{\Delta(z)}\boldsymbol{C}(z)\boldsymbol{F}^{-1}. \tag{50), (51}$$

Now, we choose the reconstruction filters as

$$\boldsymbol{g}(z)=\boldsymbol{C}_\mathrm{p}(z)[1\ 1\ 1\ldots 1]^\mathrm{T}. \tag{52}$$

Then, the transmission vector $\boldsymbol{F}(z)$ becomes

$$\begin{aligned}\boldsymbol{f}(z)&=\boldsymbol{M}(z)\boldsymbol{C}_\mathrm{p}(z)[1\ 1\ 1\ldots 1]^\mathrm{T}=\Delta_\mathrm{p}(z)\boldsymbol{M}(z)(1/\Delta(z))\boldsymbol{C}(z)\boldsymbol{F}^{-1}[1\ 1\ 1\ldots 1]^\mathrm{T}\\&=[\Delta_\mathrm{p}(z)\ 0\ 0\ldots 0]^\mathrm{T},\end{aligned} \tag{53}$$

and the reconstructed signal is equal to

$$\hat{X}(z)=\Delta_\mathrm{p}(z)X(z). \tag{54}$$

Therefore, we can consider the polyphase filter matrix alone, thus avoiding the matrix $\boldsymbol{M}(z)$ which is redundant (each filter coefficient appearing N times) and in general complex.

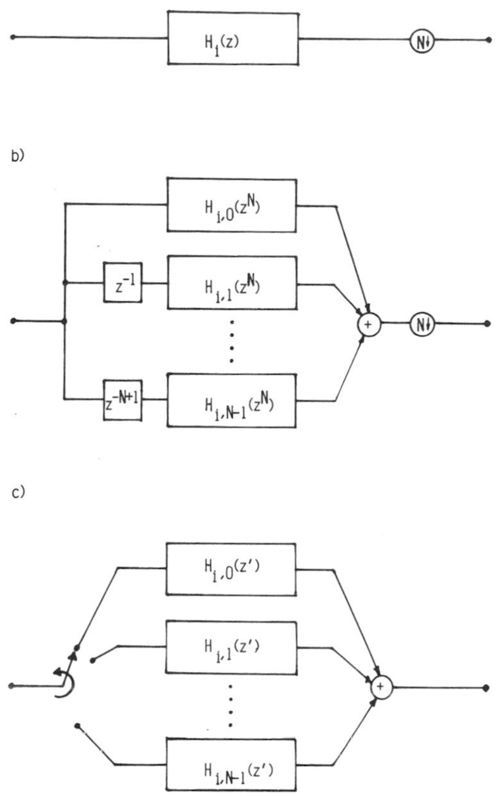

Fig. 6. Equivalence between subsampled FIR filtering and polyphase representation. (a) Initial FIR filter. (b) Equivalent FIR filter. (c) Equivalent polyphase filters.

Consider the determinant of the polyphase filter matrix $\boldsymbol{P}(z)$. Instead of the usual recursive determinant formula, we use the one which consists of the sum of all possible products where not two terms are taken from the same row or column [24]. The determinant of an $N\times N$ matrix $\boldsymbol{A}$ becomes the sum of $N!$ terms:

$$\det[\boldsymbol{A}]=\sum_{\theta}(a_{0,\alpha}a_{1,\beta}\cdots a_{N-1,\gamma})\det[\boldsymbol{K}_\theta], \tag{55}$$

where $\boldsymbol{K}_\theta$ is a permutation matrix indicating which terms appear in the θth product. The determinant of $\boldsymbol{K}_\theta$ is either 1 or -1 and $\alpha, \beta, \ldots, \gamma$ are all different (because there are no repetitions). Using this formula for the determinant, we can write $\Delta_\mathrm{p}(z)$ as follows:

$$\Delta_\mathrm{p}(z)=z^{-N(N-1)/2}\sum_{\theta}(H_{\alpha,0}(z^N)H_{\beta,1}(z^N)\cdots H_{\gamma,N-1}(z^N))\det[\boldsymbol{K}_\theta]. \tag{56}$$

The factor $z^{-N(N-1)/2}$ is obtained because we take an element from each row of the matrix $\boldsymbol{P}(z)$: an element from the ith row has an associated delay of z^{-i} (from (45)) and the product of N elements from different rows produces a delay equal to $1 \times z^{-1} \cdots z^{-N+1}$, that is, $z^{-N(N-1)/2}$. The product of the N polyphase filters in (56) is a polynomial in z^{-N}, since each polyphase filter is already a polynomial in z^{-N}. As can be verified, the lowest power of this product is zero, and the highest is $N(M-N)$ when all N filters are of length M. Thus, the determinant of the polyphase filter matrix $\boldsymbol{P}(z)$ has at most $M-N+1$ nonzero coefficients, because only coefficients with indexes multiple of N can be different from zero. Therefore, we can rewrite $\Delta_p(z)$ in the following manner:

$$\Delta_p(z) = d_0 z^{-N(N-1)/2} + d_1 z^{-N(N-1)/2-N} + \cdots + d_{M-N} z^{-NM+N(N+1)/2}. \tag{57}$$

Now, the requirement that the overall transmission should be a delay means that the $M-N+1$ coefficients of $\Delta_p(z)$ should all be zero but one. Therefore, $M-N+1$ equations have to be satisfied, but we have N filters with M coefficients each, that is, NM unknowns. Note that the z-transforms of the filters are multiplied with one another, that is, the equations to be solved are nonlinear. In order to reduce these nonlinear equations to a linear system of equations of dimension $M-N+1$ with $M-N+1$ unknowns, one can perform the following steps:

(a) Choose the coefficients appearing in $N-1$ columns of $\boldsymbol{P}(z)$, that is, choose $N-1$ filters (actually, one could also choose $N-1$ rows, that is $N-1$ polyphase components of the N filters).

(b) Choose $N-1$ coefficients of the last filter (or of the remaining polyphase components).

In this manner, there are only $M-N+1$ unknowns left, and it can be seen that the corresponding equations are now linear. Note that the resulting system is not always solvable (depending on the a priori coefficients), but that these singularities are rather scarce.

As an example, we derive a filter bank for $N=3$ and $M=7$. For simplicity, we take the following analysis filters:

$$H_0(z) = 1 + z^{-1} + z^{-2} + z^{-3} + z^{-4} + z^{-5} + z^{-6}, \tag{58}$$

$$H_1(z) = 1 - z^{-1} + z^{-2} - z^{-3} + z^{-4} - z^{-5} + z^{-6}, \tag{59}$$

$$H_2(z) = 1 + h_1 z^{-1} + h_2 z^{-2} + h_3 z^{-3} + h_4 z^{-4} + h_5 z^{-5} + z^{-6}. \tag{60}$$

One can verify that the determinant of the equivalent polyphase filter matrix $\boldsymbol{P}(z)$, given as

$$\boldsymbol{P}(z) = \begin{bmatrix} H_{0,0}(z^3) & H_{1,0}(z^3) & H_{2,0}(z^3) \\ z^{-1}H_{0,1}(z^3) & z^{-1}H_{1,1}(z^3) & z^{-1}H_{2,1}(z) \\ z^{-2}H_{0,2}(z^3) & z^{-2}H_{1,2}(z^3) & z^{-2}H_{2,2}(z^3) \end{bmatrix}, \tag{61}$$

is equal to

$$\Delta_p(z) = \tfrac{2}{3}[(h_4-1)z^{-15} + (h_1-h_3)z^{-12} + (h_3-h_5)z^{-6} + (1-h_2)z^{-3}]. \tag{62}$$

Setting $h_1 = 1$, $h_2 = -1$, $h_3 = 1$, $h_4 = 1$, and $h_5 = 1$, reduces (62) to

$$\Delta_p(z) = \tfrac{4}{3} z^{-3}. \tag{63}$$

From the cofactor matrix $C_p(z)$ we obtain the following synthesis filters (where unnecessary delays were cancelled):

$$G_0(z) = \tfrac{1}{6}[1 + z^{-1} - z^{-4} - z^{-5} + z^{-7} + z^{-8} - z^{-10}], \tag{64}$$

$$G_1(z) = \tfrac{1}{6}[-z^{-1} + z^{-2} - z^{-4} + z^{-5} - z^{-7}], \tag{65}$$

$$G_2(z) = \tfrac{1}{6}[-1 + z^{-2} - z^{-8} + z^{-10}]. \tag{66}$$

Using these filters in a system like the one depicted in Fig. 2 leads to zero aliasing transmission, and the reconstructed signal is equal to

$$\hat{X}(z) = z^{-2}X(z). \tag{67}$$

Note that, in this example, $H_0(z)$ and $H_1(z)$ are 'reasonable', since we could choose them, but that $H_3(z)$ is dictated by solving (62) to a monomial, and can be 'strange'. Also, there is a problem with the size of the synthesis filters. Since they are obtained from the cofactor matrix, their length is upperbounded by $(M-1)(N-1)+1$ in the worst case, that is, in general much longer than the analysis filters. Only in the case $N=2$ they are guaranteed to be of the same length as the analysis filters.

In practice, both for the computational ease and for application reasons, one would like to have filters which are modulated versions of a single prototype filter $H_p(z)$:

$$h_{i,n} = W^{in}h_{p,n}, \qquad H_i(z) = H_p(W^i z). \tag{68), (69}$$

Thus, if $H(z)$ is a lowpass filter (with a bandwidth of the order of $1/N$), then the $H_i(z)$ are bandpass filters displaced by i/N.

Unfortunately, when the filters are chosen as in (68) and (69), the determinant cannot be a pure delay unless $N=M$ (note that in the solution from [22, 29] for $N=2$, the second filter is modulated but also time reversed, and thus the method cannot readily be generalized to $N>2$).

Consider the polyphase filter matrix in the case when the filters are modulated. Then, $\boldsymbol{P}(z)$ has the following form:

$$\boldsymbol{P}_{n,m}(z) = W^{-nm}z^{-n}H_{p,n}(z^N), \tag{70}$$

where $H_{p,n}(z^N)$ is the nth polyphase component of the prototype filter $H_p(z)$. Postmultiplying $\boldsymbol{P}(z)$ by the Fourier matrix and dividing by N yields the diagonal filter matrix $D(z)$, or, expressed in terms of $\boldsymbol{M}(z)$,

$$\boldsymbol{D}(z) = (1/N)\boldsymbol{F}\boldsymbol{M}(z)\boldsymbol{F} = \begin{bmatrix} H_{p,0}(z^N) & & 0 \\ & z^{-1}H_{p,1}(z^N) & \\ 0 & & z^{-N+1}H_{p,N-1}(z^N) \end{bmatrix}. \tag{71}$$

Now, the determinants of $\boldsymbol{D}(z)$ and $\boldsymbol{M}(z)$ are equal except for a sign factor:

$$\Delta_d(z) = \begin{cases} \Delta(z), & \lfloor\tfrac{1}{2}(N+1)\rfloor \text{ odd}, \\ -\Delta(z), & \lfloor\tfrac{1}{2}(N+1)\rfloor \text{ even}, \end{cases} \tag{72}$$

as can be verified by evaluating the determinant of the Fourier matrix. Now, $\Delta_d(z)$ is equal to

$$\Delta_d(z) = z^{-N(N-1)/2}\prod_{i=0}^{N-1} H_{p,i}(z^N), \tag{73}$$

and, in order for $\Delta_d(z)$ to be a monomial, all the factors of the product in (73) have to be monomials. Thus, $H_p(z)$ cannot have more than N nonzero coefficients (actually, it requires exactly one nonzero coefficient in each polyphase filter, otherwise $\Delta_d(z)$ is either zero or not a monomial).

If we allow infinite impulse response (IIR) synthesis filters, we can invert $\boldsymbol{D}(z)$ and use the following synthesis filters:

$$\begin{aligned} g(z) &= z^{-N+1}[\boldsymbol{M}(z)]^{-1}[1 \; 0 \ldots 0]^T \\ &= z^{-N+1}\boldsymbol{F}[[H_{p,0}(z^N)]^{-1}, [z^{-1}H_{p,1}(z^N)]^{-1}, \ldots, [z^{-N+1}H_{p,N-1}(z^N)]^{-1}]^T, \end{aligned} \tag{74}$$

from where the reconstructed signal $\hat{X}(z)$ follows to

$$\hat{X}(z) = z^{-N+1}X(z). \tag{75}$$

Note that, in (74), it is required that $[\boldsymbol{D}(z)]^{-1}$ is stable, or that each polyphase filter has its zeroes strictly within the unit circle.

Of course, in the case of modulated filters, both the analysis and the synthesis filter bank can be computed with reduced computational complexity, by using the polyphase/FFT approach first introduced in [1]. In the analysis filter bank, one has to compute the following result:

$$\begin{bmatrix} Y_0(z) \\ Y_1(z) \\ \vdots \\ Y_{N-1}(z) \end{bmatrix} = \begin{bmatrix} H_{p,0}(z^N) & z^{-1}H_{p,1}(z^N) & z^{-N+1}H_{p,N-1}(z^N) \\ H_{p,0}(z^N) & W^{-1}z^{-1}H_{p,1}(z^N) & W^{-N+1}z^{-N+1}H_{p,N-1}(z^N) \\ \vdots & \vdots & \vdots \\ H_{p,0}(z^N) & \cdots & \end{bmatrix} \cdot \begin{bmatrix} X(z) \\ X(z) \\ \vdots \\ X(z) \end{bmatrix}, \tag{76}$$

which is equal to

$$\begin{bmatrix} Y_0(z) \\ Y_1(z) \\ \vdots \\ Y_{N-1}(z) \end{bmatrix} = \begin{bmatrix} 1 & 1 & \cdots & 1 \\ 1 & W & \cdots & W^{N-1} \\ \vdots & \vdots & & \vdots \\ 1 & W^{N-1} & \cdots & \end{bmatrix} \cdot \begin{bmatrix} H_{p,0}(z^N)X(z) \\ z^{-1}H_{p,1}(z^N)X(z) \\ \vdots \\ z^{-N+1}H_{p,N-1}(z^N)X(z) \end{bmatrix}, \tag{77}$$

or, assuming that the prototype filter length is a multiple of N such that $M = kN$, the computational load per set of subsampled output samples is equivalent to the evaluation of N polyphase filters of length k as well as an FFT of size N.

In the case of the synthesis filter bank, the output $\hat{X}(z)$ is obtained from the channel signals $Y_i'(z)$ in the following way:

$$\begin{aligned} \hat{X}(z) &= [\boldsymbol{g}(z)]^{\mathrm{T}}[Y_0'(z) \;\; Y_1'(z) \;\; \ldots \;\; Y_{N-1}'(z)]^{\mathrm{T}} \\ &= z^{-N+1}[1 \;\; 1 \ldots 1][\boldsymbol{D}(z)]^{-1}\boldsymbol{F}[Y_0(z) \;\; \ldots \;\; Y_{N-1}(z)]^{\mathrm{T}}, \end{aligned} \tag{78}$$

that is, with a Fourier transform and N all-pole filters. Note that the $Y_i'(z)$ have only every Nth sample different from zero, thus the synthesis requires the same order of complexity as the analysis. Fig. 7 shows an N channel system with modulated filters computed by Fourier transforms.

If one does not want IIR solutions, one can evaluate the cofactor matrix in order to get FIR synthesis filters and neglect the determinant. Note that the synthesis filters are also modulated in that case, thus allowing efficient implementations (even if they are longer than the analysis filters). Of course, approximations can be made, and especially products of filters which are nonadjacent can be regarded as being zero. This approach is known as 'pseudo-QMF' filter banks [12, 13, 15, 20, 30] and allows both efficient implementation and quasi-QMF behaviour. While that kind of filters requires further investigation, the above framework is certainly helpful.

Concluding this section, we recall that the proposed approach always allows the cancelling of the aliasing in a subsampled filter bank (by choosing the synthesis filters accordingly to the cofactor matrix) and that FIR analysis/synthesis systems can be found for $N > 2$ and $M > N$. Finally, the modulated filter bank case does not have an FIR solution, but has an efficient analysis and synthesis implementation.

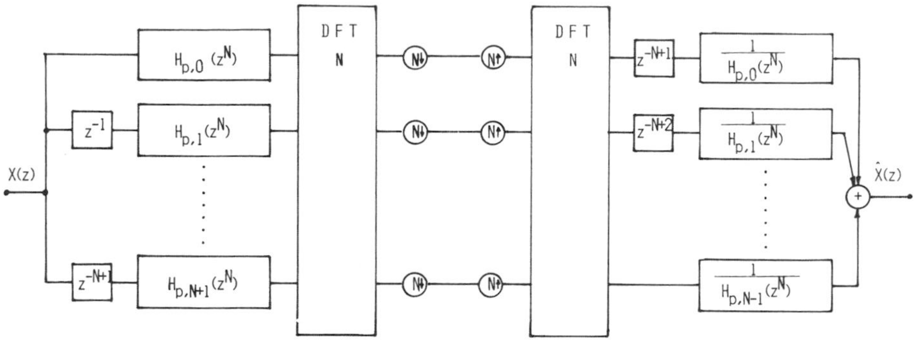

Fig. 7. Modulated filter bank implemented with polyphase filters and DFTs.

5. The dual problem

In the previous sections, we have considered the problem of separating a signal (at sampling frequency f) into N channel signals (at sampling frequency f/N) by filtering, in such a way that the original can be perfectly reconstructed.

In the following, we look at the dual problem which consists of multiplexing N signals (with initial sampling frequency f) into one signal (at sampling frequency Nf) by filtering, in a way that permits the perfect reconstruction of the N original signals (Fig. 8). Note that the two obvious solutions from Section 2 hold as well here. Both when the filters are perfect bandpass or length $M = N$ FIR filters, the signal can be reconstructed perfectly. Again, the interesting case appears for $M > N$, but not perfectly bandpass.

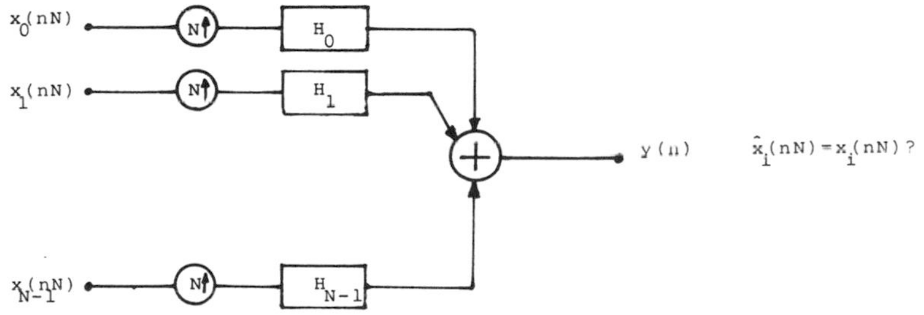

Fig. 8. General dual problem statement. Can the initial signals be recovered from the upsampled channel?

The case $N = 2$ is investigated below. Assume a system as depicted in Fig. 9. Then, using (2) and then (1), the signals in Fig. 9 equal

$$X_i'(z) = X_i(z^2), \quad i = 0, 1, \tag{79}$$

$$Y(z) = H_0(z)X_0(z^2) + H_1(z)X_1(z^2), \tag{80}$$

$$\hat{X}_i'(z) = G_i(z)Y(z), \quad i = 0, 1, \tag{81}$$

$$\hat{X}_i(z) = \tfrac{1}{2}[\hat{X}_i'(z^{1/2}) + \hat{X}_i'(-z^{1/2})], \tag{82}$$

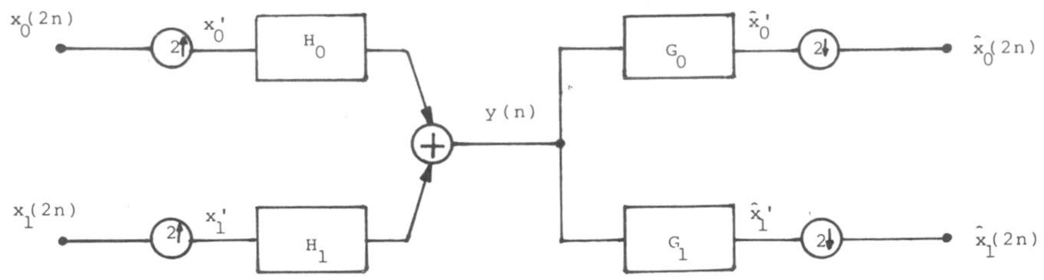

Fig. 9. Solution to the two channel dual problem.

or, explicitly,

$$\hat{X}_i(z)=\tfrac{1}{2}(G_i(z^{1/2})[H_0(z^{1/2})X_0(z)+H_1(z^{1/2})X_1(z)] + G_i(-z^{1/2})[H_0(-z^{1/2})X_0(z)+H_1(-z^{1/2})X_1(z)]),\quad i=0,1. \tag{83}$$

Thus, a necessary and sufficient condition in order to recover $\hat{X}_i(z)$ equal to $X_i(z)$ is the following:

$$\tfrac{1}{2}[G_i(z^{1/2})H_j(z^{1/2})+G_i(-z^{1/2})H_j(-z^{1/2})]=\begin{cases} z^{-k}, & i=j,\\ 0, & i\neq j.\end{cases} \tag{84}$$

Below, for reasons similar to those mentioned in Section 3 and Appendix A, we consider the case where all involved filters are FIR. Introducing the following notation:

$$T_{i,j}(z)=H_i(z)G_j(z),\quad i,j=0,1, \tag{85}$$

it is easy to see that (84) is equivalent to say,

$$\tfrac{1}{2}[T_{i,j}(z)+T_{i,j}(-z)]=\begin{cases} z^{-2k}, & i=j,\\ 0, & i\neq j,\end{cases} \tag{86}$$

or simply that:

- $T_{i,i}(z)$ has a single even-index coefficient different from zero, while having arbitrary odd-index ones,
- $T_{i,j}(z)$, $i\neq j$, has all even-index coefficients equal to zero, while having arbitrary odd-index ones.

This has an obvious interpretation: $X_i'(z)$ has only information on the even-index terms of its z-transform (being zero otherwise because of the upsampling) and all odd-index terms of $\hat{X}_i'(z)$ are disregarded due to the subsampling. Thus, if the transmission from input i to output i has a single even coefficient and the transmission from i to j has no even coefficient different from zero, then an impulse appearing (on even time) at the input of filter i will be transmitted to output i only, and this without distortion. Thus, the signal at input i is perfectly reconstructed at output i, disregarding a delay.

Calling $H_{ie}(z)$ and $H_{io}(z)$ the polynomials incorporating the even and odd parts of $H_i(z)$, that is,

$$H_{ie}(z)=\tfrac{1}{2}[H_i(z)+H_i(-z)],\qquad H_{io}(z)=\tfrac{1}{2}[H_i(z)-H_i(-z)], \tag{87),(88}$$

and noting that (86) includes only the even part of $T_{i,j}(z)$, we rewrite (86) as follows:

$$\tfrac{1}{2}[H_{ie}(z)G_{je}(z)+H_{io}(z)G_{jo}(z)]=\begin{cases} z^{-2k}, & i=j,\\ 0, & i\neq j.\end{cases} \tag{89}$$

This leads to the following matrix equation:

$$\frac{1}{2}\begin{bmatrix} H_{0e}(z) & H_{0o}(z) & 0 & 0 \\ H_{1e}(z) & H_{1o}(z) & 0 & 0 \\ 0 & 0 & H_{0e}(z) & H_{0o}(z) \\ 0 & 0 & H_{1e}(z) & H_{1o}(z) \end{bmatrix} \cdot \begin{bmatrix} G_{0e}(z) \\ G_{0e}(z) \\ G_{1e}(z) \\ G_{1o}(z) \end{bmatrix} = \begin{bmatrix} z^{-2k} \\ 0 \\ 0 \\ z^{-2k} \end{bmatrix}. \tag{90}$$

Then, it is sufficient to invert the 2×2 nonzero blocks in (90). For example, $G_0(z)$ is obtained from

$$\begin{bmatrix} G_{0e}(z) \\ G_{0o}(z) \end{bmatrix} = [1/(2\Delta(z))]\begin{bmatrix} H_{1o}(z) & -H_{0o}(z) \\ -H_{1e}(z) & H_{0e}(z) \end{bmatrix} \cdot \begin{bmatrix} z^{-2k} \\ 0 \end{bmatrix}, \tag{91}$$

where

$$\Delta(z) = \tfrac{1}{4}[H_{0e}(z)H_{1o}(z) - H_{0o}(z)H_{1e}(z)]. \tag{92}$$

Now, if Δ is a delay (actually an odd delay since Δ is an odd function of z) and the output filters are chosen accordingly to (90) and (91), then both $\hat{X}_0(z)$ and $\hat{X}_1(z)$ are equal to $X_0(z)$ and $X_1(z)$ within a delay.

Note that the matrix in (91) is the cofactor matrix of the 2×2 block in (90). A remark is again appropriate here. Choosing

$$[G_{0e}(z)\ \ G_{0o}(z)]^{\mathrm{T}} = [H_{1o}(z)\ \ -H_{1e}(z)]^{\mathrm{T}} z^{-1}, \tag{93}$$

$$[G_{1e}(z)\ \ G_{1o}(z)]^{\mathrm{T}} = [-H_{0o}(z)\ \ H_{0e}(z)]^{\mathrm{T}} z^{-1} \tag{94}$$

(or respectively as the first and second column of the cofactor matrix) leads to the following properties of the system in Fig. 9 (note the z^{-1} factor in order to be in phase with the subsampling):

- there is no crosstalk from one input channel to another output channel,
- the transmission of a channel to its output is equal to the determinant of the submatrix in (90) times z^{-1}.

The above method can be extended to the case $N>2$. Due to the upsampling and subsampling by N, one is only interested in the transmission at every multiple of N. This leads, through relations similar to (86)-(90), to the inversion of a matrix of the following form:

$$\boldsymbol{M}_{\mathrm{d}}(z) = (1/N)\begin{bmatrix} H_{0,0}(z) & H_{0,1}(z) & \cdots & H_{0,N-1}(z) \\ H_{1,0}(z) & & & \\ \vdots & \vdots & & \vdots \\ H_{N-1,0}(z) & & \cdots & \end{bmatrix}, \tag{95}$$

where

$$H_{i,j}(z) = h_{i,j}z^{-j} + h_{i,j+N}z^{-j-N} + \cdots. \tag{96}$$

Again, one can choose the synthesis filters as elements of the cofactor matrix (thus eliminating the cross over), and the determinant can be reduced to a delay, thus allowing perfect reconstruction.

Obviously, the dual and the initial problem are closely related, since both require the inversion of a polyphase filter matrix. Therefore, the same analysis filters can be used for the initial and the dual problem, and the synthesis filters in the dual problem can be deduced from the ones of the initial problem by some scaling and shifting.

6. Applications

The purpose of this section is not to show impletation results, but rather to point to potential applications.

6.1. Subband coder incorporating an LPC filter

It was shown in Section 3 that, in the two channel case, one could choose one filter, $H_0(z)$, and derive the other, $H_1(z)$, by solving a linear system of equations given in (35). An interesting choice for $H_0(z)$ is the following:

$$H_0(z) = 1 - L(z), \tag{97}$$

where

$$L(z) = l_1 z^{-1} + l_2 z^{-2} + \cdots + l_{N-1} z^{-N+1}. \tag{98}$$

$L(z)$ is the filter whose current output is the best linear prediction of the current sample value from the previous $N-1$ samples [11]. The output of $H_0(z)$ is the so-called residual, that is, loosely stated, the 'unpredictable' part of the signal.

Now, $H_1(z)$ is evaluated by solving (35). In some sense, the output of $H_1(z)$ is the complementary signal with regard to the residual. Both signals are then subsampled by 2 and the synthesis filters are chosen according to (20). The system is depicted in Fig. 10. This approach is currently under investigation, but sufficient results are not yet available to judge the power of the method. The idea is simply to split a signal into its residual and complementary parts, in a way similar to the splitting into low and high pass components with a conventional QMF bank. It is hoped that the matching of the filters to the signal (rather than having fixed filters as in a conventional system) will improve the coding gain without destroying the quality associated with subband coders. Of course, the process can be iterated, especially on the complementary signal, since the subsampled residual might be crudely quantized.

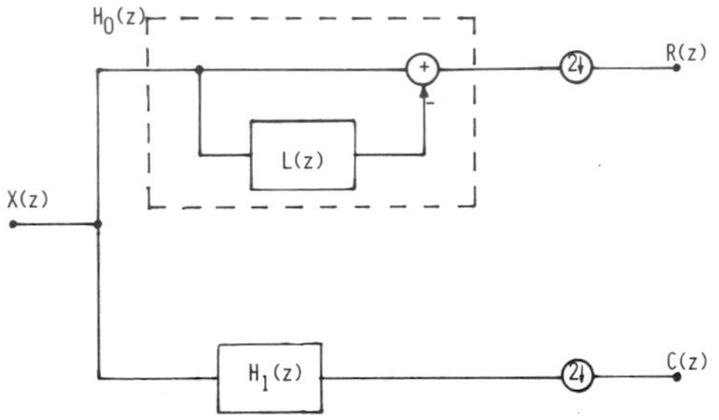

Fig. 10. Example of a subband coder using an LPC filter. $R(z)$ is the subsampled residual and $C(z)$ is the complementary signal.

6.2. Scrambling of analog signals

When scrambling analog signals, the bandwidth should not grow. Therefore, methods have been proposed where the analog signal is filtered into bands which are then interchanged. This can be done in the analog or in the digital domain. Using the filterbanks developed so far, which allow separation or mixing of signals permitting exact recovery, one can devise such a scrambling system. Since the filters are not restricted to bandpass filters, complex functions can be used, which can make decryption even more difficult. Of course, since the transmitted signal is analog and will be distorted, the reconstruction will not be perfect.

The system, depicted in Fig. 11, works as follows. The signal is first split into N subsampled channels using filters as developed in Section 3. Then, the channels are permuted and mixed together as described in Section 5. Note that the splitting and the mixing filters could be similar, but are different in general. At the receiver, the mixing is undone first, then the channels are backpermuted, and, finally, the initial splitting is undone as well, thus reproducing the original. In Fig. 11, the case for $N=4$ is shown (for practical purposes, N should be bigger, of course), using two stages of two filters each. The number of possible permutations is

$$N_p = N! \tag{99}$$

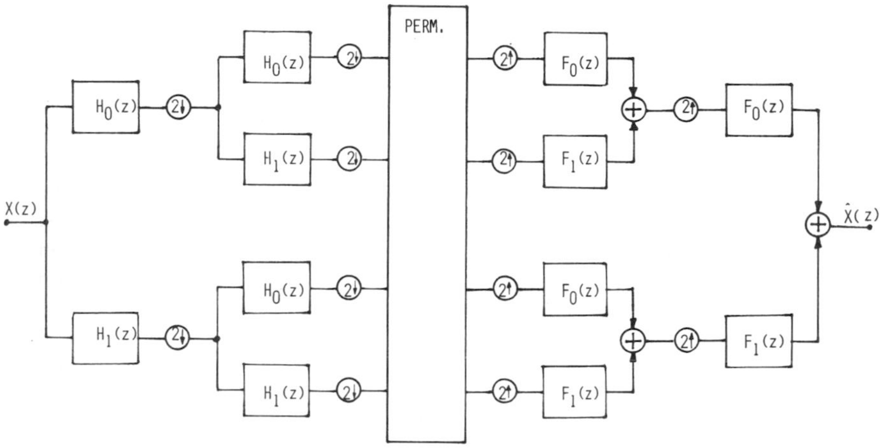

Fig. 11. Example of scrambling. After downsampling and permutation, the signals are recombined to form the channel signal.

6.3. *Filter banks with noncritical subsampling*

A filter bank is critically subsampled when the output of an N *channel* bank is subsampled by N. Noncritical subsampling is the case when the subsampling factor N' is smaller than N, the number of channels. In the latter case, we obtain a rectangular system of N' equations involving N filters instead of the square system of (40). Obviously, since there are less contraints, solutions will exist as well (ad absurdum, one could take $N-N'$ filters equal to zero, and solve (40) of dimension N'). The new degrees of freedom can be used to meet new, self-imposed constraints. Take as a simple example a two channel filter bank without subsampling. Then, the only equation to be met is

$$H_0(z)G_0(z)+H_1(z)G_1(z)=z^{-k}. \tag{100}$$

Choosing, for example, $H_0(z)=H(z)$ and $H_1(z)=H(-z)$, where $H(z)$ is a length M FIR filter, as well as $G_0(z)=G(z)$ and $G_1(z)=G(-z)$, $G(z)$ also being FIR, leads to the following condition:

$$H(z)G(z)+H(-z)G(-z)=z^{-2k}. \tag{101}$$

This condition can be met using Method 1 of Section 3, for example. Note that the two analysis filters used here would not work in the subsampled case, but are possible here because there is no aliasing cancellation to be met. The point here is to simply show that the approach from the previous sections is quite general and useful for N' going from 1 to N.

6.4. Fractional sampling rate change

An interesting application of the initial together with the dual system is the fractional sampling rate change. Assume, for example, five signals with sampling frequency f that should be multiplexed onto three channels with sampling frequency $\frac{5}{3}f$. Again, the number of samples per unit of time is the same in the five input channels and in the three output channels. One solution is to first multiplex the five channels onto one channel with sampling frequency $f' = 5f$, and then to demultiplex this channel into three channels, which yields the desired sampling frequency $f'' = \frac{5}{3}f$. Such a system is depicted in Fig. 12. If the filters $H_i(z)$ and $F_i(z)$ meet the requirement of determinants being monomials, a perfect FIR reconstruction of the original signals can be achieved.

6.5. Filter banks on finite fields

If one really wants 'perfect' reconstruction, for example in error detection applications, one can resort to arithmetic on finite fields [2]. It is not difficult to generalize the methods shown so far to filter banks on finite fields. To keep things simple, we look at arithmetic modulo a prime number p, that is, to arithmetic over GF(p). Then, all filter coefficients and signal values belong to GF(p). The z-transform is defined as usual. For simplicity, we only look at the two channel case. In order to express the down-upsampling by N, we require an Nth root of unity in GF(p), which means that p must be strictly greater than N. Calling α in GF(p) the element such that $\alpha^2 = 1$ and β such that $\beta + \beta = 1$, we express the down/upsampling by 2 as

$$X'(z) = \beta[X(z) + X(\alpha z)]. \tag{102}$$

Thus, similarly to (12), we have to invert a matrix $\boldsymbol{M}(z)$ equal to

$$\boldsymbol{M}(z) = \begin{bmatrix} H_0(z) & H_1(z) \\ H_0(\alpha z) & H_1(\alpha z) \end{bmatrix}. \tag{103}$$

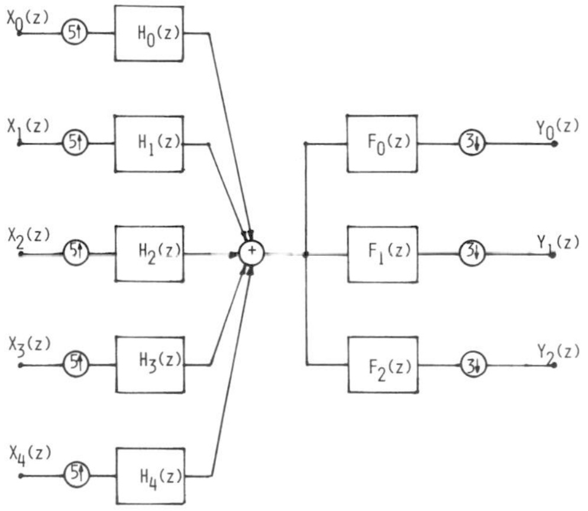

Fig. 12. Example of a fractional sampling rate change. The inputs have sampling frequency f and the outputs a sampling frequency $\frac{5}{3}f$.

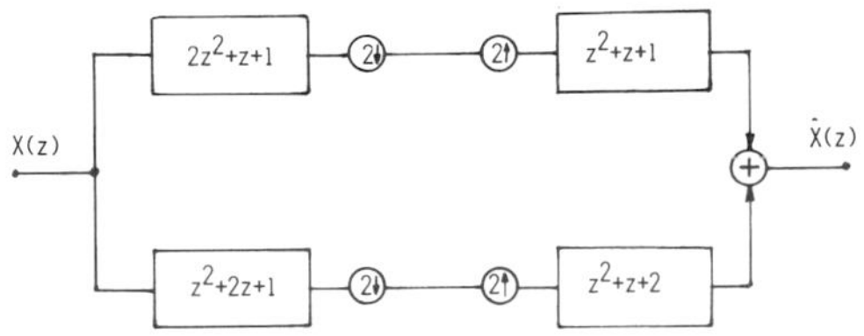

Fig. 13. Two channel system over GF(3).

Its determinant has odd terms only. Thus, if we want it to be a monomial, we start with a polynomial having arbitrary even terms but only a single nonzero odd term. Then, we factor it into two polynomials which can be used as filters $H_0(z)$ and $H_1(z)$. Of course, the factorization is done in GF(p). Such a simple system is given in Fig. 13, where all computations are done in GF(3).

Note here that implicit pole/zero cancellation (Appendix A) does not cause problems as in the case of systems over the field of real or complex numbers. Thus, one could cancel the determinant with an all-pole filter as well. The generalization for N greater than 2 (when an Nth root of unity exists in GF(p)) follows the same lines as in Section 4, where W is now the Nth root of unity in GF(p).

7. Conclusion

The problem of subsampled filter banks has been addressed by developing an analysis framework. Using a powerful matrix notation, several new results were obtained, especially for FIR analysis and synthesis filters. In that case, the $N \times M$ nonlinear design problem (where N is the number of channels, and M the filter length) was shown to reduce to an M-dimensional linear problem. This is obtained by analysing the matrix inversion (where the matrix elements are polynomials) and, in particular, the determinant of the matrix.

For $N=2$, the class of FIR solutions allowing perfect reconstruction has been demonstrated, which includes the solutions known so far [22, 29], but also new solutions (for example, a solution with two linear phase analysis filters). Two new methods are given in order to find actual filters. The original QMF filters [5, 6, 7, 8] are shown to be the solution when the synthesis filters are obtained from the cofactor matrix (but neglecting the determinant), and this method can be generalized to arbitrary N, thus allowing perfect aliasing cancellation. Furthermore, for $N>2$, it is shown that the determinant can be reduced to a pure delay (by solving a linear system of equations of size M), thus allowing perfect reconstruction with FIR analysis and synthesis filters. Actual solutions have been shown, but note that while there is no reason to doubt the existence of a solution for arbitrary N and M, the conjecture has yet to be proven.

The case where the filters are obtained by modulation from a single prototype filter is shown to only have an IIR perfect synthesis solution, but the efficient polyphase/FFT analysis [1] is extended to a perfect and efficient synthesis as well.

The dual problem of multiplexing N signals onto a single channel upsampled by N was addressed as well and shown to be equivalent to the initial problem. Finally, several applications are proposed, among others a new speech coding scheme and filter banks on finite fields.

In conclusion, looking at the N channel filter bank problem as a global, N channel system has proven to be fruitful. The design of N 'simultaneous' filters is therefore quite different from conventional, single filter design.

Acknowledgment

Many thanks go to Prof. Nussbaumer for fruitful discussions, for first suggesting to seek a general solution to the problem and for then giving the polyphase interpretation; to R. Leonardi for carefully rereading the manuscript as well as for giving many suggestions; and to Prof. Zwahlen, Prof. Wohlhauser and C. El Hayek of the Département de Mathématiques, EPFL, for useful discussions. The use of the symbolic manipulation system MACSYMA is acknowledged (being very helpful for deriving examples and doing exploratory work) and thanks go to Dr. Calinon for giving access to this system. This work was supported by the "Fonds National Suisse de la Recherche Scientifique", whose help is greatly appreciated.

The author is indebted to one of the reviewers for many constructive suggestions as well as for a very careful rereading of the manuscript.

Appendix A

Below, we prove that a perfect analysis/synthesis system without implicit pole/zero cancellation requires both FIR analysis and FIR synthesis filters.

First we define what we mean by implicit or explicit pole zero cancellation. By explicit pole/zero cancellation, we mean numerator/denominator simplications done on the transfer function of a single filter. This is done before implementing a given filter.

By implicit pole/zero cancellation, we mean numerator/denominator simplifications between two cascaded or two parallel filters, but where the two filters are physically separated. A cascade example is given below, where $H_0(z)$ and $H_1(z)$ are two separate filters (appearing, for example, in the analysis and the synthesis part respectively):

$$H_0(z)H_1(z) = [(A(z)B(z))/C(z)][D(z)/(E(z)B(z))] = (A(z)D(z))/(C(z)E(z)). \tag{A.1}$$

Therefore, $B(z)$ was implicitly cancelled between the two filters. A parallel example would be the following:

$$\begin{aligned} H_0(z)+H_1(z) &= (A(z)B(z))/(C(z)[A(z)+D(z)]) + (D(z)B(z))/(C(z)[A(z)+D(z)]) \\ &= (A(z)B(z)+D(z)B(z))/(C(z)[A(z)+D(z)]) = B(z)/C(z). \end{aligned} \tag{A.2}$$

Here, the factor $[A(z)+D(z)]$ was implicitly cancelled between the two filters. While explicit pole/zero cancellation is obviously always permitted, implicit cancellation has two problems associated with it.

The first one is internal stability, that is, a transfer function that is externally stable (from input to output) may have an unstable part inside [10]. This problem can be avoided by careful analysis.

The second and more fundamental problem for practical realizations is the precision problem. While, theoretically, the pole/zero cancellation is realized, in a physical system with finite precision arithmetic the cancellation will in general not be done perfectly. Note that the effect is nonlinear and thus difficult to track and quantify. While our purpose is not to solve this particular problem, we will try to avoid the implicit cancellations in the following.

It will be shown below that perfect reconstruction without implicit pole/zero cancellation can only be achieved when both the analysis and the synthesis filters are FIR.

Consider the input/output transmission $F_0(z)$ from (13):

$$F_0(z) = \tfrac{1}{4}\{[H_1(-z)/\Delta(z)]H_0(z) - [H_0(-z)/\Delta(z)]H_1(z)\}. \tag{A.3}$$

Explicit pole/zero cancellation can be done, for example, between $H_1(-z)$ and $\Delta(z)$ in (A.3), but cancellation between the two summands in (A.3) are implicit ones since they are physically separated filters. Similarly, cancellations between $H_0(z)$ and $\Delta(z)$ in the first summand are implicit as well.

Assume now that all explicit cancellations have been made in (A.3). Then we do not want implicit cancellations between the two summands and we want the transmission $F_0(z)$ to have no poles (except at infinity or zero, which means time shifts only). Thus, the two summands have to be FIR. Now, we do not want implicit cancellations between the factors of the summands. Therefore, both factors need to be

FIR, which means that $H_0(z)$ has to be FIR (because it is a factor of the first summand) and that $H_1(z)$ has to be FIR (because it is a factor of the second summand).

Now, the fact of analysis filters being FIR leads to the requirement that the synthesis filters have to be FIR as well (otherwise, there would be implicit cancellations between the analysis and the synthesis part of the system). This is achieved when the determinant constraint is fulfilled ($\Delta(z)$ being a monomial) because then perfect reconstruction is guaranteed with FIR synthesis filters.

Appendix B

Further properties of two channel systems are explored below. First we look at the delayed channel case, and then we prove that linear phase systems exist only for even length filters where the two filters have different symmetry.

B.1. Delayed channel case

Assume that channel 1 is delayed by z^{-1} at the input and that channel 0 is delayed by z^{-1} at the output. Then, $\hat{X}_0(z)$ and $\hat{X}_1(z)$ satisfy the following equalities:

$$\hat{X}_0(z)=\tfrac{1}{2}[H_0(z)X(z)+H_0(-z)X(-z)]G_0(z)z^{-1}, \tag{B.1}$$

$$\hat{X}_1(z)=\tfrac{1}{2}[z^{-1}H_1(z)X(z)-z^{-1}H_1(-z)X(-z)]G_1(z). \tag{B.2}$$

Thus, $F_0(z)$ and $F_1(z)$ are equal to

$$\begin{bmatrix} F_0(z) \\ F_1(z) \end{bmatrix} = \frac{1}{2}\begin{bmatrix} H_0(z) & z^{-1}H_1(z) \\ H_0(-z) & -z^{-1}H_1(-z) \end{bmatrix} \cdot \begin{bmatrix} z^{-1}G_0(z) \\ G_1(z) \end{bmatrix}, \tag{B.3}$$

and the determinant equals

$$\Delta(z)=\tfrac{1}{4}[H_0(z)H_1(-z)+H_0(-z)H_1(z)]z^{-1}. \tag{B.4}$$

$\Delta(z)$ is a function with odd powers of z^{-1} when both $H_0(z)$ and $H_1(z)$ are FIR filters.

Consider now the term $H_0(z)H_1(-z)$. If this polynomial has arbitrary odd-index terms but only one nonzero even-index term, then $\Delta(z)$ is a pure delay and the signal can be reconstructed with FIR synthesis filters as well. This condition is equivalent to the one in Section 3, since it simply means that $H_0(z)H_1(-z)$ is shifted by one sample.

B.2. Linear phase solutions

The perfect FIR solution for the two channel case given in [22, 29] leads to minimum phase filters, but the question remains: are there linear phase solutions as well? This question is interesting because one often wants channel signals which are in phase, typically in subband coding applications. Below, we look at the nondelayed case with two filters of the same length.

A linear phase FIR filter has either a symmetric or an antisymmetric impulse response [18]. We introduce the function sym$[H(z)]$ which is defined as follows:

$$\text{sym}[H(z)] = \begin{cases} 1 & \text{if } H(z) \text{ has a symmetric impulse response,} \\ -1 & \text{if } H(z) \text{ has an antisymmetric impulse response,} \\ 0 & \text{if } H(z) \text{ has no symmetry in its impulse response.} \end{cases}$$

This function has the following properties:

$$\text{sym}[H(z)G(z)] = \text{sym}[H(z)]\,\text{sym}[G(z)], \tag{B.5}$$

$$\text{sym}[H(-z)] = \begin{cases} \text{sym}[H(z)] & \text{if } H(z) \text{ is of odd length,} \\ -\text{sym}[H(z)] & \text{if } H(z) \text{ is of even length.} \end{cases} \tag{B.6}$$

When two filters have the same length and the same symmetry, then this symmetry is preserved by addition of the filters; in other words:

$$\text{sym}[H(z)+G(z)] = \text{sym}[H(z)]\text{sym}[G(z)]\tfrac{1}{2}(\text{sym}[H(z)+\text{sym}[G(z)]). \tag{B.7}$$

When the lengths are different, the result has no symmetry.

Using this function, we analysis the determinant $\Delta(z)$ given by (17) when both $H_0(z)$ and $H_1(z)$ are length M linear phase filters. Look at $H_0(z)H_1(-z)$. When M is odd, because of (B.5) and (B.6) we have

$$\text{sym}[H_0(z)H_1(-z)] = \text{sym}[H_0(z)]\,\text{sym}[H_1(z)]. \tag{B.8}$$

Since the product filter has odd length as well, we have

$$\text{sym}[H_0(z)H_1(-z)] = \text{sym}[H_0(-z)H_1(z)], \tag{B.9}$$

and therefore,

$$\text{sym}[\Delta(z)] = \text{sym}[H_0(z)]\,\text{sym}[H_1(z)]. \tag{B.10}$$

Now, the center of symmetry of Δ is the coefficient with index $M-1$, that is, an even number. Since $\Delta(z)$ is an odd function of z^{-1}, the coefficient of z^{-M+1} is zero. All the nonzero coefficients of $\Delta(z)$ appear therefore twice, that is, $\Delta(z)$ can never be a monomial.

When M is even,

$$\text{sym}[H_0(z)H_1(-z)] = (-1)\,\text{sym}[H_0(z)]\,\text{sym}[H_1(z)]. \tag{B.11}$$

since the product filter is of odd length,

$$\text{sym}[H_0(z)H_1(-z)] = \text{sym}[H_0(-z)H_1(z)] \tag{B.12}$$

or

$$\text{sym}[\Delta(z)] = \text{sym}[H_0(z)H_1(-z)] = (-1)\,\text{sym}[H_0(z)]\,\text{sym}[H_1(z)]. \tag{B.13}$$

The center of symmetry is the coefficient with index $M-1$, an odd number. All other coefficients of $\Delta(z)$ except the center one appear twice. Thus, in order for $\Delta(z)$ to be a monomial, the $(M-1)$st coefficient is the only one that can and must be different from zero. Thus, $\Delta(z)$ should have even symmetry, that is, $H_0(z)$ and $H_1(z)$ should have different symmetry.

In conclusion, one can see that minimum phase filters are not the only solution to the two channel perfect decomposition scheme. It is possible to derive a two channel system with perfect reconstruction and using linear phase filters. This is achieved when:

- M is even,
- $H_0(z)$ and $H_1(z)$ are of different symmetry.

Obviously, the same type of analysis can be applied to other cases as well (delayed channels, filters of different lengths, etc.) in order to find the set of possible solutions.

References

[1] M.G. Bellanger and J.L. Daguet, "TDM-FDM transmultiplexer: Digital polyphase and FFT", *IEEE Trans. on Comm.*, Vol. COM-22, No. 9, September 1974, pp. 1199-1204.

[2] R.E. Blahut, *Theory and Practice of Error Control Codes*, Addison-Wesley, Reading, MA, 1983.

[3] R.E. Crochiere, S.A. Webber and J.L. Flanagan, "Digital coding of speech in sub-bands", *Bell System Tech. J.*, Vol. 55, No. 8, October 1976, pp. 1069-1085.

[4] R.E. Crochiere and L.R. Rabiner, *Multirate Digital Signal Processing*, Prentice-Hall, Englewood Cliffs, NJ, 1983.

[5] A. Croisier, D. Esteban and C. Galand, "Perfect channel splitting by use of interpolation, decimation, tree decomposition techniques", *Internat. Conf. on Information Sciences/Systems*, Patras, August 1976, pp. 443-446.

[6] D. Esteban and C. Galand, "Application of quadrature mirror filters to split-band coding", *Internat. conf. on ASSP, ICASSP 77,* Hartford, May 1977, pp. 191-195.

[7] C. Galand, "Codage en sous-bandes: Théorie et application à la compression numérique du signal de parole", Thèse d'Etat, Université de Nice, 1983.

[8] C.R. Galand and H.J. Nussbaumer, "New quadrature mirror filter structures", *IEEE Trans. Acoust. Speech Signal Process.*, Vol. ASSP-32, No. 3, June 1984, pp. 552-531.

[9] U. Heute and P. Vary, "A digital filter bank with polyphase network and FFT hardware: Measurements and applications", *Signal Processing*, Vol. 3, No. 4, October 1981, pp. 307-319.

[10] T. Kailath, *Linear Systems*, Prentice-Hall, Englewood Cliffs, NJ. 1980.

[11] J. Makhoul, "Linear prediction: A tutorial review", *Proc. IEEE*, Vol. 63, No. 4, April 1975, pp. 561-580.

[12] J. Masson and Z. Picel, "Flexible design of computationally efficient nearly perfect QMF filter banks", *Proc. 1985 IEEE Conf. on ASSP*, Tampa, March 1985, pp. 541-544.

[13] H.J. Nussbaumer, "Pseudo QMF filter bank", *IBM Tech. Disclosure Bull.*, Vol. 24, No. 6, November 1981, pp. 3081-3087.

[14] H.J. Nussbaumer, *Fast Fourier Transform and Convolution algorithms*, Springer, Berlin, 1982.

[15] H.J. Nussbaumer and M. Vetterli, "Computationally efficient QMF filter banks", *Proc. 1984 Internat. IEEE Conf. on ASSP*, San Diego, March 1984, pp. 11.3.1-4.

[16] H.J. Nussbaumer and M. Vetterli, "Pseudo quadrature mirror filters", *Proc. Internat. Conf. on Digital Signal Processing*, Florence, September 1984, pp. 8-12.

[17] A.V. Oppenheim and R.W. Schafer, *Digital Signal Processing*, Prentice-Hall, Englewood Cliffs, NJ, 1975.

[18] L.R. Rabiner and B. Gold, *Theory and Application of Digital Signal Processing*, Prentice-Hall, Englewood Cliffs, NJ, 1975.

[19] T.A. Ramstad, "Analysis/synthesis filterbanks with critical sampling", *Internat. Conf. on DSP*, Florence, September 1984, pp. 130-134.

[20] J.H. Rothweiler, "Polyphase quadrature filters—a new subband coding technique", *Proc. 1983 Internat. IEEE Conf. on ASSP*, Boston, MA, March 1983. pp. 1280-1283.

[21] R.W. Shafter and L.R. Rabiner, "A digital signal processing approach to interpolation", *Proc. IEEE*, Vol. 61, No. 6, June 1973, pp. 692-702.

[22] M.J.T. Smith and T.P. Barnwell, "A procedure for designing exact reconstruction filterbanks for tree structured sub-band coders", *Proc. IEEE ICASSP-84,* San Diego, March 1984, pp. 27.1.1-4.

[23] M.J.T. and T.P. Barnwell, "A unifying framework for analysis/synthesis systems based on maximally decimated filter banks", *Proc. IEEE ICASSP-85,* Tampa, March 1985, pp. 521-521.

[24] G. Strang, *Linear Algebra and Its Applications*, Academic Press, New York, 1980.

[25] P. Vary and U. Heute, "A short-time spectrum analyzer with polyphase-network and DFT", *Signal Processing*, Vol. 2, No. 1, January 1980, pp. 55-65.

[26] P. Vary and G. Wackersreuther, "A unified approach to digital polyphase filter banks", *AEU*, Vol. 37, No. 1/2, January 1983, pp. 29-34.

[27] M. Vetterli, "Tree structures for orthogonal transforms and application to the Hadamard transform", *Signal Processing*, Vol. 5, No. 6, November 1983, pp. 473-484.

[28] M. Vetterli, "Splitting a signal into subsampled channels allowing perfect reconstruction", *Proc. IASTED Conf. on Applied Signal Processing and Digital Filtering*, Paris, June 1985.

[29] G. Wackersreuther, "On the design of filters for ideal QMF and polyphase filter banks", *AEU*, Vol. 39, No. 2, February 1985, pp. 123-130.

[30] P. L. Chu, "Quadrature mirror filter design for an arbitrary number of equal bandwidth channels", *IEEE Trans. Acoust. Speech Signal Process*, Vol. ASSP-33, No. 1, February 1985, pp. 203-218.

476 IEEE TRANSACTIONS ON ACOUSTICS, SPEECH, AND SIGNAL PROCESSING, VOL. ASSP-35, NO. 4, APRIL 1987

Theory and Design of *M*-Channel Maximally Decimated Quadrature Mirror Filters with Arbitrary *M*, Having the Perfect-Reconstruction Property

P. P. VAIDYANATHAN, MEMBER, IEEE

Abstract—**Based on the concept of losslessness in digital filter structures, this paper derives a general class of maximally decimated *M*-channel quadrature mirror filter banks that lead to perfect reconstruction. The perfect-reconstruction property guarantees that the reconstructed signal $\hat{x}(n)$ is a delayed version of the input signal $x(n)$, i.e., $\hat{x}(n) = x(n - n_0)$. It is shown that such a property can be satisfied if the alias component matrix (AC matrix for short) is unitary on the unit circle of the z plane. The number of channels M is arbitrary, and when M is two, the results reduce to certain recently reported 2-channel perfect-reconstruction QMF structures. A procedure, based on recently reported FIR cascaded-lattice structures, is presented for optimal design of such FIR *M*-channel filter banks. Design examples are included.**

I. Introduction

QUADRATURE mirror filter (QMF) banks have received considerable attention during the past several years because of a wide variety of engineering applications [1]–[13]. An M-channel QMF bank is shown in Fig. 1, where $H_0(z)$, $H_1(z)$, $\cdots$, $H_{M-1}(z)$ are the transfer functions of analysis bank filters, and $F_0(z)$, $F_1(z)$, $\cdots$, $F_{M-1}(z)$ represent the synthesis filters. In the analysis bank, the incoming signal $x(n)$ is split into M frequency bands by filtering, and each subband signal is maximally decimated, i.e., decimated by a factor of M. The M decimated signals are then processed in the synthesis bank by interpolating each signal, filtering, and then adding the M filtered signals. In a typical application of such a system, the M decimated signals in the analysis bank are coded and transmitted. The motivation for such signal splitting and coding before transmission is a well-understood topic, and is covered well in the literature [1]–[5].

A common requirement in most applications is that the reconstructed signal $\hat{x}(n)$ should be "as close" to $x(n)$ as possible, in a certain well-defined sense. The reconstructed signal in general suffers from aliasing error because the analysis bank filters $H_k(z)$ that precede the decimators are not ideal. In practice, for a given set of analysis filters $H_k(z)$, the synthesis filters $F_k(z)$ can be

Manuscript received May 19, 1986; revised September 8, 1986. This work was supported in part by the National Science Foundation under Grant ECS 84-04245 and in part by Caltech's Program in Advanced Technology sponsored by Aerojet General, General Motors, GTE, and TRW.

The author is with the Department of Electrical Engineering, California Institute of Technology, Pasadena, CA 91125.

IEEE Log Number 8612887.

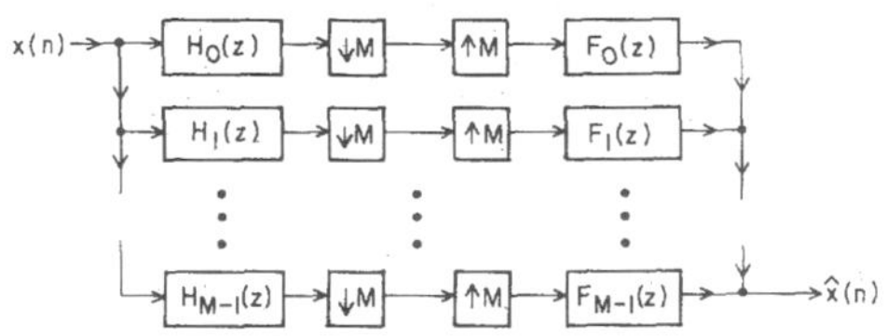

Fig. 1. The M-channel maximally decimated parallel QMF bank.

chosen so as to reduce the effect of this aliasing caused by the decimation operation.

It is well known that in two-channel QMF banks (i.e., $M = 2$) perfect cancellation of aliasing can be accomplished by a simple choice of the functions $F_0(z)$ and $F_1(z)$ [1]–[5]. For the case of M channels, if M is a power of two, tree structures with two-channel QMF banks can be built that are free from aliasing. For arbitrary M, approximate cancellation of aliasing can be accomplished in an elegant manner [8]–[10], whereas perfect cancellation of aliasing can be accomplished with somewhat more complicated synthesis bank filters [11]–[15].

The most general expression for $\hat{X}(z)$ is of the form [3], [7]

$$\hat{X}(z) = \frac{1}{M} \sum_{l=0}^{M-1} X(zW^{-l}) \sum_{k=0}^{M-1} H_k(zW^{-l})\, F_k(z) \quad \text{(1a)}$$

where $W = e^{-j2\pi/M}$. In (1a), $X(zW^{-l})$, $l \neq 0$ represents the aliasing terms. Aliasing is cancelled if and only if the following set of relations holds:

$$\begin{bmatrix} H_0(z) & H_1(z) & \cdots & H_{M-1}(z) \\ H_0(zW^{-1}) & H_1(zW^{-1}) & \cdots & H_{M-1}(zW^{-1}) \\ \vdots & & & \\ H_0(zW^{-M+1}) & H_1(zW^{-M+1}) & \cdots & H_{M-1}(zW^{-M+1}) \end{bmatrix} \cdot \begin{bmatrix} F_0(z) \\ F_1(z) \\ \vdots \\ F_{M-1}(z) \end{bmatrix} = \begin{bmatrix} T(z) \\ 0 \\ \vdots \\ 0 \end{bmatrix}. \quad \text{(1b)}$$

The above $M \times M$ matrix has been referred to as the alias component matrix (AC matrix) in the literature [7]. Once

0096-3518/87/0400-0476$01.00 © 1987 IEEE

aliasing has been cancelled, the structure of Fig. 1 is time invariant (and of course linear), and $\hat{X}(z)$ is related to $X(z)$ by a transfer function $T(z)$:

$$T(z) \triangleq \frac{\hat{X}(z)}{X(z)} = \frac{1}{M} \sum_{k=0}^{M-1} F_k(z) H_k(z). \qquad (2)$$

Thus, $T(z)$ represents the "distortion" caused by the alias-free analysis-synthesis system. An alias-free system is said to have no amplitude distortion if $T(z)$ is a (stable) all-pass function, whereas if $T(z)$ is a linear-phase FIR function, then the system is free from phase distortion. Depending upon the application in hand, it is always possible to choose the set of filters $\{F_k(z)\}$ (for a given set of analysis filters $\{H_k(z)\}$) such that either the amplitude distortion is zero or the phase distortion is zero [11]–[13].

If an alias-free system is such that the quantity $T(z)$ is a delay, then both amplitude and phase distortions are zero. Such QMF banks are called perfect-reconstruction banks and satisfy

$$\hat{X}(z) = cz^{-n_0}X(z) \qquad (3)$$

for some positive integer n_0. Here c is an arbitrary constant.

In principle, one can always set $T(z) = z^{-n_0}$ in (1b) and invert the AC matrix in order to obtain the synthesis filters $F_k(z)$ that would lead to perfect reconstruction. This approach, however, is of little use in practice as it often leads to synthesis filters of very high order which in addition are typically unstable. This motivates us to look at the perfect-reconstruction problem from other points of view that do not involve the inversion of the AC matrix.

A simple way to obtain perfect reconstruction is to choose the analysis and synthesis filters according to

$$H_k(z) = z^{-k}, \qquad F_k(z) = z^{-(M-1-k)}. \qquad (4)$$

It is easily verified that such a choice satisfies (3) with $n_0 = M - 1$. (See Appendix A.) However, the filtering functions $H_k(z)$ are trivial, and such a structure has little practical value.

A fundamental result has recently been established by Smith and Barnwell [6] who showed that two-channel perfect-reconstruction QMF banks can indeed be constructed, while at the same time accomplishing nontrivial FIR filtering functions $H_0(z)$, $H_1(z)$. The result is based on an important property satisfied by linear-phase FIR half-band filters, and the designs of $H_0(z)$ and $H_1(z)$ are based on the spectral-factorization of an appropriately conditioned half-band filter. This problem has also been recently addressed by Mintzer [16].

One of the main aims of our paper here is the extension of these perfect-reconstruction results for the case of M channels, with *arbitrary* M. Referring to the maximally decimated QMF structure of Fig. 1, we show how the transfer functions $\{H_k(z)\}$ and $\{F_k(z)\}$ can be constructed such that aliasing is perfectly cancelled, and in addition (3) is exactly satisfied. Our solution is such that $H_k(z)$ and $F_k(z)$ are FIR, for all k, $0 \le k \le M - 1$. Moreover, if $N - 1$ is the order of each of the filters $H_k(z)$, then the order of each of $F_k(z)$ is also $N - 1$.

The notion of polyphase networks in signal splitting and reconstruction [3], [30], [31] enables one to address many theoretical and practical issues in a unified manner [11]–[13]. In this paper, we make further use of this tool by defining generalized polyphase structures which can be useful even when the analysis-bank filters are entirely unrelated to each other. The role of all-pass functions and losslessness in signal splitting and reconstruction applications has been noticed and analyzed by some authors in the past [11], [12], [32], [34], thus leading to IIR QMF banks with (no aliasing and) no amplitude distortion. Such IIR QMF banks do lead to phase distortion (because $T(z)$ of (2) is an all-pass function in these examples) which can be compensated by an equalizer. In this paper, we take advantage of the result that multivariable FIR lossless functions can be appropriately employed in a QMF bank in order to reduce all types of distortion to zero, thereby resulting in perfect reconstruction.

Certain simple solutions to the perfect-reconstruction problem, which are of restricted use, have recently been presented [11]–[13] and do not in general have FIR components.[1] The question of *existence* of perfect-reconstruction QMF banks for arbitrary M, with FIR analysis and synthesis filters, need not bother us. A simple example of such a system is obtained in Appendix A. More examples can be found in [15]. However, our purpose here is to provide new solutions based on the observation that the concept of losslessness in digital networks [17], [20] is closely related to the concept of signal reconstruction in maximally decimated QMF banks. Our results are such that the FIR filters in the analysis bank have the same length as those in the synthesis bank. In order to render the paper readable in a self-contained manner, we define the notion of losslessness in Section II. We then review the fundamental two-channel perfect-reconstruction results [6] in order to place them in the context of the lossless structural framework. Section III introduces the general M-channel perfect-reconstruction circuit. This section derives a set of *sufficient* conditions for perfect reconstruction with arbitrary M. A set of *necessary* and *sufficient* conditions is also included in this section.

When M is a power of two, it is well known that tree structures based on the two-channel building blocks in [6] can be used in order to obtain perfect reconstruction. Section IV includes a proof that such Smith–Barnwell tree structures satisfy the set of sufficient conditions developed in Section III. Such a proof is encouraging because it shows that the sufficient conditions we develop are not unduly restrictive, and do not disable us from obtaining good stopband attenuation for the analysis filters.

Since the design of the analysis bank in the two-channel

[1]In a recent conversation with M. Vetterli at the ICASSP'86 Tokyo Conference, we learned that Dr. Vetterli has made similar observations [14], [15].

case [6], [16] is based on factorization of a half-band filter, a natural attempt in the case of M channels is to try designing the analysis bank based on the factorization of FIR Mth band filters [21]. Surprisingly, this attempt does *not* work for all M (but works only for restricted M, viz. $M = 2$), i.e., the resulting analysis filters do not have good frequency response. The theoretical reason for this is explained in Section V. We feel that it is useful to be aware of this result.

In Section VI we develop a class of FIR lattice structures for the analysis and synthesis banks based on some recent work [18]–[20] on FIR lattice filters. The purpose is to enable us to set up an optimization algorithm that designs $H_k(z)$ so as to have good stopband attenuation. The lattice structures are such that they *automatically* satisfy the set of sufficient conditions required for perfect reconstruction. Accordingly, if we optimize the parameters of this lattice, it is equivalent to finding the best set of transfer functions $\{H_k(z)\}$ by searching *only* among the class of transfer functions that already satisfy conditions for alias-free perfect reconstruction. We feel that this is a major advantage of using the lattice structures in the optimization. A design example is included here. Finally, Section VII examines recursive QMF banks in the context of lossless matrices.

A. Notations Used in the Paper

Superscript T stands for matrix (or vector) transposition, whereas superscript dagger (†) stands for transposition followed by complex conjugation. Boldface italic letters indicate matrices and vectors. Superscript asterisk (*) stands for complex conjugation, while subscript asterisk denotes conjugation of coefficients of the function or matrix. The tilde accent on a function $F(z)$ is defined such that, on the unit circle, $\tilde{F}(z) = F^{\dagger}(z)$. Thus, for arbitrary z, $\tilde{F}(z) = F_*^T(z^{-1})$, and for functions with real coefficients, $\tilde{F}(z) = F^T(z^{-1})$.

Since our work was primarily motivated by the original contributions in [6], these results are often referenced in this paper. For ease of reference, the perfect-reconstruction two-channel structure in [6] will be called the Smith–Barnwell structure (and abbreviated as the SB structure).

II. The Smith–Barnwell QMF Bank in the Context of Lossless Structures

A single-input single-output digital transfer function $G(z)$ is said to be lossless [17] if it is stable and satisfies $|G(e^{j\omega})| = 1$ for all ω. Such a function is merely an all-pass function. An m-input p-output transfer function (i.e., a general matrix transfer function) $\boldsymbol{T}(z)$ is said to be lossless if it is stable and satisfies $\boldsymbol{T}^{\dagger}(e^{j\omega})\,\boldsymbol{T}(e^{j\omega}) = \boldsymbol{I}$ for all ω or equivalently, by analytic continuation,

$$\tilde{\boldsymbol{T}}(z)\,\boldsymbol{T}(z) = \boldsymbol{T}(z)\,\tilde{\boldsymbol{T}}(z) = \boldsymbol{I}, \qquad \text{for all } z. \tag{5}$$

In essence, a lossless function or matrix is stable and is unitary on the unit circle. Such a $\boldsymbol{T}(z)$ can be looked upon as a multiinput, multioutput all-pass function. These concepts are discrete-time versions of the notions of lossless-

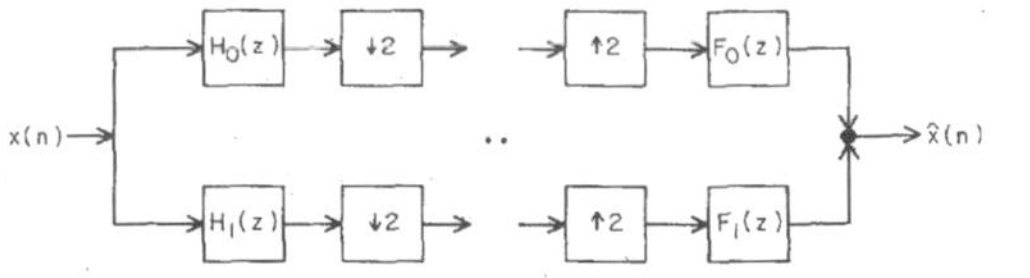

Fig. 2. The two-channel QMF bank.

ness in continuous-time passive network theory [22], [23] and have been used in the past in a completely different context, viz., for low-sensitivity digital filter design [17], [18], [24].

If $G(z)$ is FIR and lossless, then it has to be a pure delay, i.e., $G(z) = z^{-\alpha}$, α = integer. However, it is possible to have matrix-valued FIR functions $\boldsymbol{T}(z)$ that are more complicated than a delay. For example,

$$\boldsymbol{T}(z) = \frac{\begin{bmatrix} 1 + z^{-1} & 1 - z^{-1} \\ 1 - z^{-1} & 1 + z^{-1} \end{bmatrix}}{2}$$

is easily verified to be lossless, as it satisfies (5).

A. Revisiting the Perfect-Reconstruction Structure

Now consider the two-channel QMF structure of Fig. 2. Here $H_0(z)$ and $H_1(z)$ are low-pass and high-pass transfer functions, respectively. The signal $\hat{X}(z)$ is given by

$$\hat{X}(z) = [H_0(z)\,F_0(z) + H_1(z)\,F_1(z)]\,X(z)/2 + [H_0(-z)\,F_0(z) + H_1(-z)\,F_1(z)]\,X(-z)/2. \tag{6}$$

The term involving $X(-z)$ is the aliasing term and is required to be made equal to zero. In the scheme due to Smith and Barnwell [6], the following relation between the transfer functions is enforced:[2]

$$H_1(z) = z^{-(N-1)}\,H_0(-z^{-1}) \tag{7}$$

$$F_0(z) = z^{-(N-1)}\,H_0(z^{-1}) \tag{8}$$

$$F_1(z) = z^{-(N-1)}\,H_1(z^{-1}) \tag{9}$$

where $N - 1$ is the order of $H_0(z)$. Here $H_0(z)$ and $H_1(z)$ are constrained to satisfy

$$H_0(z^{-1})\,H_0(z) + H_1(z^{-1})\,H_1(z) = 1, \qquad \text{for all } z. \tag{10}$$

Without loss of generality, $N - 1$ can be assumed to be odd. (See Section V.) The conditions (7)–(9) are sufficient to enforce the following condition:

$$H_0(-z)\,F_0(z) + H_1(-z)\,F_1(z) = 0 \tag{11}$$

and thereby cancel aliasing. The relation (6) then becomes

[2]The relations (7)–(9) do not appear to be the same as those in [6] because we have displayed a causal version of those equations.

$$\hat{X}(z) = \tfrac{1}{2}z^{-(N-1)}\left[H_0(z^{-1})\,H_0(z) + H_1(z^{-1})\,H_1(z)\right]X(z) \tag{12}$$

which in view of (10) reduces to (3). In order to ensure that condition (10) indeed holds, a design procedure is proposed in [6] based on the spectral factorization of a linear-phase FIR half-band filter with positive amplitude response.

The AC matrix for a two-channel QMF bank is

$$\boldsymbol{H}(z) = \begin{bmatrix} H_0(z) & H_1(z) \\ H_0(-z) & H_1(-z) \end{bmatrix}. \tag{13}$$

It can be verified that condition (7) implies that the columns of (13) are "orthogonal," that is,

$$H_0(z^{-1})\,H_1(z) + H_0(-z^{-1})\,H_1(-z) = 0. \tag{14}$$

Moreover, each column of (13) satisfies

$$H_k(z^{-1})\,H_k(z) + H_k(-z^{-1})\,H_k(-z) = 1, \qquad k = 0, 1 \tag{15}$$

as can be verified by employing (7) and (10). Properties (14) and (15) are precisely equivalent to the statement that $\boldsymbol{H}(z)$ is lossless, i.e., satisfies the condition $\tilde{\boldsymbol{H}}(z)\,\boldsymbol{H}(z) = \boldsymbol{I}$.

In contrast, consider the more standard two-channel QMF bank [1]–[5]. For such a filter bank, the transfer function $H_0(z)$ is restricted to be a linear-phase FIR filter of odd order $N-1$, and the remaining transfer functions are given by

$$H_1(z) = H_0(-z),\; F_0(z) = H_0(z),$$
$$F_1(z) = -H_1(z). \tag{16}$$

With such a choice of transfer functions, aliasing is cancelled and we have [5]

$$\hat{X}(z) = \left[H_0^2(z) - H_0^2(-z)\right]X(z)/2. \tag{17}$$

It can furthermore be shown that $\boldsymbol{H}(z)$ satisfies

$$\tilde{\boldsymbol{H}}(z)\,\boldsymbol{H}(z) = \left[H_0(z^{-1})\,H_0(z) + H_1(z^{-1})\,H_1(z)\right]\cdot \boldsymbol{I} \tag{18}$$

which reduces to

$$\tilde{\boldsymbol{H}}(z)\,\boldsymbol{H}(z) = z^{(N-1)}\left[H_0^2(z) - H_0^2(-z)\right]\cdot \boldsymbol{I}, \tag{19}$$

because of (16) and the fact that $H_0(z)$ is an odd-order filter with symmetric impulse response. In order for $\boldsymbol{H}(z)$ to be lossless, it is again necessary to satisfy (10). But it is well known that two linear-phase FIR transfer functions $H_0(z)$ and $H_1(z)$ *cannot* satisfy (10) unless they are trivial combinations of delays [29].

In summary, the AC matrix in the Smith–Barnwell perfect-reconstruction structure is lossless, whereas the AC matrix in the standard two-channel alias-free structures is not[3] except in trivial situations.

[3]In this paper, the term "standard two-channel QMF bank" stands for a structure as in Fig. 2, with the transfer functions satisfying (16) and with $H_0(z)$ representing an odd order ($= N-1$) linear-phase FIR filter, with symmetric impulse response.

B. Revisiting Polyphase Implementations of Two-Channel Structures

Let the transfer functions $H_0(z)$ and $H_1(z)$ be written in the form

$$H_0(z) = E_{00}(z^2) + z^{-1}E_{01}(z^2) \tag{20}$$

$$H_1(z) = E_{10}(z^2) + z^{-1}E_{11}(z^2). \tag{21}$$

Clearly such a representation is always possible. We therefore have

$$\begin{bmatrix} H_0(z) & H_0(-z) \\ H_1(z) & H_1(-z) \end{bmatrix} = \begin{bmatrix} E_{00}(z^2) & E_{01}(z^2) \\ E_{10}(z^2) & E_{11}(z^2) \end{bmatrix}\begin{bmatrix} 1 & 1 \\ z^{-1} & -z^{-1} \end{bmatrix}. \tag{22}$$

Defining the 2×2 matrix $\boldsymbol{E}(z) = [E_{kl}(z)]$, $0 \le k, l \le 1$, we obtain from (22)

$$\boldsymbol{H}(z) = \begin{bmatrix} 1 & z^{-1} \\ 1 & -z^{-1} \end{bmatrix}\boldsymbol{E}^T(z^2). \tag{23}$$

Thus,

$$\tilde{\boldsymbol{H}}(z)\,\boldsymbol{H}(z) = 2\boldsymbol{E}(z^{-2})\,\boldsymbol{E}^T(z^2) \tag{24}$$

which shows that $\boldsymbol{H}(z)$ is lossless if and only if $\sqrt{2}\boldsymbol{E}(z)$ is lossless. In view of our earlier conclusions, it thus follows that $\sqrt{2}\boldsymbol{E}(z)$ is lossless for the Smith–Barnwell scheme, whereas it is not for the standard QMF scheme. For the standard QMF structure it can be verified that $E_{kl}(z)$ satisfy

$$E_{10}(z) = E_{00}(z), \qquad E_{01}(z) = z^{-m_1}E_{00}(z^{-1}),$$
$$E_{11}(z) = -z^{-m_1}E_{00}(z^{-1}) \tag{25}$$

leading to the result

$$\tilde{\boldsymbol{E}}(z)\,\boldsymbol{E}(z) = \boldsymbol{E}(z)\,\tilde{\boldsymbol{E}}(z) = 2E_{00}(z^{-1})\,E_{00}(z)\cdot \boldsymbol{I}. \tag{26}$$

In order to develop a physical feeling for these properties, let us redraw Fig. 2 in terms of the $\boldsymbol{E}(z^2)$ matrix. For the SB structure,

$$\begin{bmatrix} F_0(z) \\ F_1(z) \end{bmatrix} = z^{-(N-1)}\begin{bmatrix} H_0(z^{-1}) \\ H_1(z^{-1}) \end{bmatrix} = z^{-(N-1)}\boldsymbol{E}(z^{-2})\begin{bmatrix} 1 \\ z \end{bmatrix} \tag{27}$$

whence

$$\left[F_0(z) \;\; F_1(z)\right] = \left[z^{-1} \;\; 1\right] z^{-2m_1}\boldsymbol{E}^T(z^{-2}) \tag{28}$$

where $m_1 = (N-2)/2$. Thus, Fig. 2 can be redrawn as in Fig. 3. For the standard QMF structure also, it can be verified, based on the relations (25), that Fig. 2 can be redrawn as in Fig. 3. Based on standard identities for multirate systems [3], Fig. 3 can in turn be redrawn as in Fig. 4. For the SB structure, since $\sqrt{2}\boldsymbol{E}(z)$ is lossless, the structure of Fig. 2 is therefore eventually equivalent to Fig. 5(a), whereas for the standard QMF structure, it is equivalent to Fig. 5(b), which clearly shows us that the standard structure is alias free (see Appendix A), while the SB structure, in addition to being alias free, has the perfect-reconstruction property. From Fig. 5 it follows

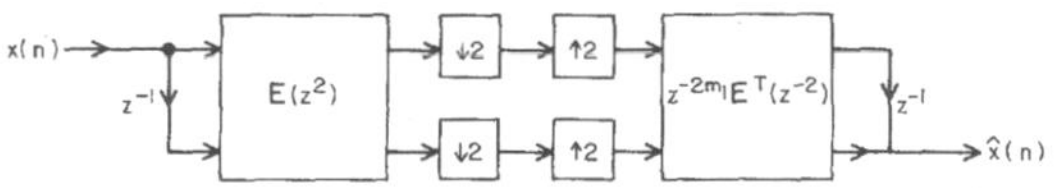

Fig. 3. An equivalent structure for Fig. 2.

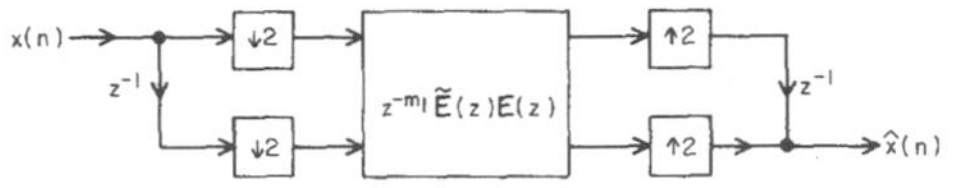

Fig. 4. A redrawing of Fig. 3.

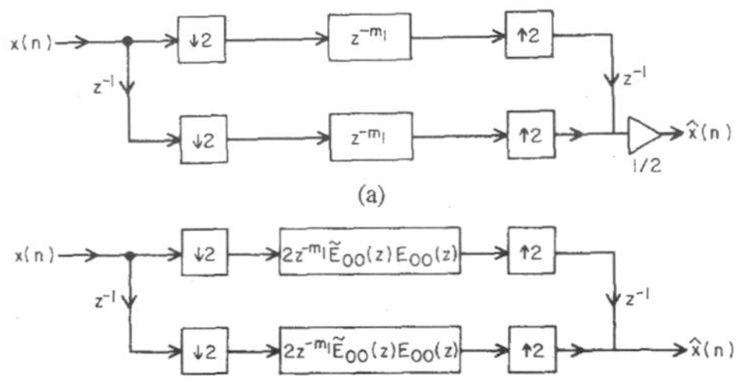

Fig. 5. The simplified equivalent structures (a) for the SB QMF bank, and (b) for the standard QMF bank.

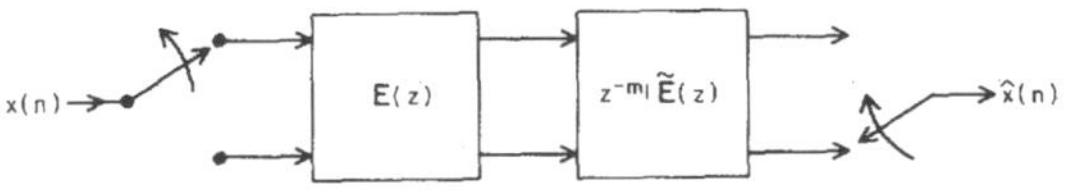

Fig. 6. "Polyphase" implementation of Fig. 3.

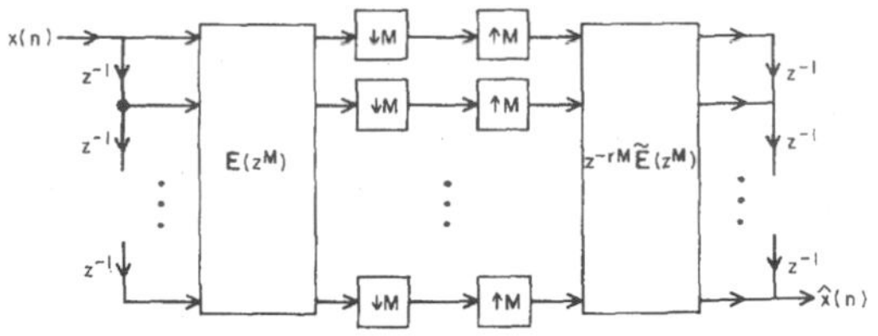

Fig. 7. Extension to M-channel QMF banks.

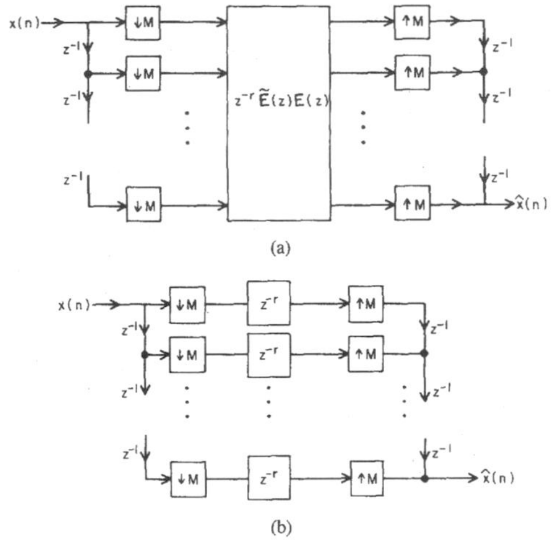

Fig. 8. (a) A structure that is equivalent to Fig. 7. (b) The final equivalent version of Fig. 7.

(Appendix A) that

$$\hat{X}(z) = \begin{cases} \frac{1}{2}z^{-(N-1)}\,X(z) & \text{(SB structure)} \\ 2z^{-(N-1)}\,E_{00}(z^{-2})\,E_{00}(z^{2})\,X(z) = 2z^{-1}E_{00}(z^{2})\,E_{01}(z^{2})\,X(z) & \text{(standard structure)} \end{cases} \tag{29}$$

which are, of course, well-known results.

Notice finally that a "polyphase" implementation results by redrawing Fig. 3 as in Fig. 6. This structure differs from the more well-known polyphase structures [3] (which have diagonal $\boldsymbol{E}(z)$ matrices).

III. Maximally Decimated M-Channel Perfect-Reconstruction Structures

The interpretation of the two-channel SB structure as in Fig. 3 immediately tells us how the idea can be extended for M channels. Thus, consider Fig. 7 where $\tilde{\boldsymbol{E}}(z^M)$ stands for $\boldsymbol{E}^T(z^{-M})$ (and more generally stands for $\boldsymbol{E}_*^T(z^{-M})$ if the coefficients in $\boldsymbol{E}(z)$ can be complex). In Fig. 7, r is an integer large enough so that $z^{-r}\tilde{\boldsymbol{E}}(z)$ has no positive powers of z. Fig. 7 is clearly equivalent to Fig. 8(a). If $\boldsymbol{E}(z)$ is lossless, this in turn is equivalent to Fig. 8(b). For Fig. 8(b) we can easily show that

$$\hat{X}(z) = z^{-(M-1+rM)}X(z) \tag{30}$$

which implies perfect reconstruction. Notice that the analysis filters and synthesis filters are given by

$$\boldsymbol{h}(z) = \begin{bmatrix} H_0(z) \\ H_1(z) \\ \vdots \\ H_{M-1}(z) \end{bmatrix} = \boldsymbol{E}(z^M)\begin{bmatrix} 1 \\ z^{-1} \\ \vdots \\ z^{-(M-1)} \end{bmatrix} \tag{31a}$$

$$\boldsymbol{f}(z) = \begin{bmatrix} F_0(z) \\ F_1(z) \\ \vdots \\ F_{M-1}(z) \end{bmatrix} = z^{-rM}\boldsymbol{E}(z^{-M})\begin{bmatrix} z^{-(M-1)} \\ z^{-(M-2)} \\ \vdots \\ z^{-1} \\ 1 \end{bmatrix} \tag{31b}$$

Thus, we can state the following result.

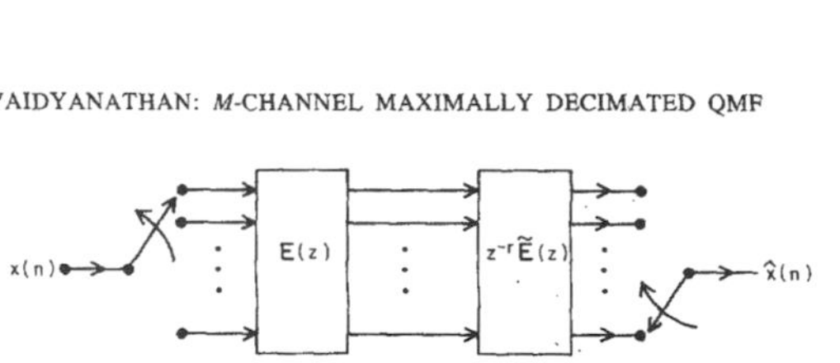

Fig. 9. The polyphase implementation of Fig. 7.

Lemma 3.1: Let $H_0(z), H_1(z), \cdots, H_{M-1}(z)$ be a set of analysis filters in the maximally decimated structure of Fig. 1. Define $E_{kl}(z)$ for $0 \leq k, l \leq M-1$ by

$$H_k(z) = \sum_{l=0}^{M-1} z^{-l} E_{kl}(z^M). \tag{32}$$

If the matrix

$$\boldsymbol{E}(z) \triangleq [E_{kl}(z)] \tag{33}$$

is lossless, then the set of synthesis filters $\{F_k(z)\}$ defined according to (31b) leads to perfect reconstruction.

Notice that the orders of the synthesis-bank filters and analysis-bank filters are the same. Notice also that if $H_k(z)$ are IIR, then there does not exist finite r such that $z^{-r}\tilde{\boldsymbol{E}}(z)$ is causal and stable, hence, the above result is meant to be used for FIR QMF banks alone. Finally, notice that a polyphase implementation of Fig. 7 can immediately be drawn as shown in Fig. 9. Once again, this differs from standard polyphase structures [3] that have diagonal $\boldsymbol{E}(z)$.

The most important practical question is now the following: how do we construct a set of M FIR transfer functions $H_0(z), H_1(z), \cdots, H_{M-1}(z)$ such that the FIR matrix $\boldsymbol{E}(z)$ defined above is lossless? A trivial solution, of course, is to take $H_k(z)$ as in (4) which makes $\boldsymbol{E}(z) = \boldsymbol{I}$ which is clearly lossless. We would like to obtain nontrivial solutions whereby $H_k(z)$ are "good" band-pass filters. Fortunately, such a design scheme is rendered feasible because of certain recently reported FIR lattice structures [18]–[20], and is the topic of Section VI.

For the rest of this section, we study certain properties of QMF banks constructed based on the lossless property. The first one pertains to the AC matrix $\boldsymbol{H}(z)$ given by

$$\boldsymbol{H}(z) = \begin{bmatrix} H_0(z) & H_1(z) & \cdots & H_{M-1}(z) \\ H_0(zW^{-1}) & H_1(zW^{-1}) & \cdots & H_{M-1}(zW^{-1}) \\ \vdots & & & \\ H_0(zW^{-M+1}) & H_1(zW^{-M+1}) & \cdots & H_{M-1}(zW^{-M+1}) \end{bmatrix}. \tag{34}$$

From the definition of $E_{kl}(z)$ as in (32), it follows that

$$[\boldsymbol{H}(z)]_{s,k} = H_k(zW^{-s}) = \sum_{l=0}^{M-1} z^{-l} W^{ls} E_{kl}(z^M), \quad 0 \leq s, k \leq M-1 \tag{35}$$

whence $\boldsymbol{H}(z)$ can be written as

$$\boldsymbol{H}(z) = \boldsymbol{W}\boldsymbol{A}(z)\,\boldsymbol{E}^T(z^M) \tag{36}$$

where $\boldsymbol{W}$ is the $M \times M$ DFT matrix, and $\boldsymbol{A}(z)$ is a diagonal matrix defined as

$$\boldsymbol{A}(z) = \begin{bmatrix} 1 & & & \boldsymbol{0} \\ & z^{-1} & & \\ & & \ddots & \\ \boldsymbol{0} & & & z^{-(M-1)} \end{bmatrix}.$$

This implies

$$\tilde{\boldsymbol{H}}(z)\,\boldsymbol{H}(z) = M\boldsymbol{E}(z^{-M})\,\boldsymbol{E}^T(z^M) \tag{37}$$

establishing the following property.

Property 3.1: $\boldsymbol{H}(z)$ is lossless if and only if $\sqrt{M}\boldsymbol{E}(z)$ is lossless.

The second property pertains to the relation between $\{H_k(z)\}$ and $\{F_k(z)\}$ of *any* perfect-reconstruction QMF bank. Recall that perfect reconstruction implies

$$\begin{bmatrix} H_0(z) & H_1(z) & \cdots & H_{M-1}(z) \\ H_0(zW^{-1}) & H_1(zW^{-1}) & \cdots & H_{M-1}(zW^{-1}) \\ \vdots & & & \\ H_0(zW^{-M+1}) & H_1(zW^{-M+1}) & \cdots & H_{M-1}(zW^{-M+1}) \end{bmatrix} \cdot \begin{bmatrix} F_0(z) \\ F_1(z) \\ \vdots \\ F_{M-1}(z) \end{bmatrix} = \begin{bmatrix} cz^{-n_0} \\ 0 \\ 0 \\ \vdots \\ 0 \end{bmatrix} \tag{38}$$

for all z. Thus, if z is replaced with zW^{-1} in (38), it continues to hold:

$$\begin{bmatrix} H_0(zW^{-1}) & H_1(zW^{-1}) & \cdots & H_{M-1}(zW^{-1}) \\ H_0(zW^{-2}) & H_1(zW^{-2}) & \cdots & H_{M-1}(zW^{-2}) \\ \vdots & & & \\ H_0(z) & H_1(z) & \cdots & H_{M-1}(z) \end{bmatrix} \cdot \begin{bmatrix} F_0(zW^{-1}) \\ F_1(zW^{-1}) \\ \vdots \\ F_{M-1}(zW^{-1}) \end{bmatrix} = \begin{bmatrix} cz^{-n_0}W^{n_0} \\ 0 \\ 0 \\ \vdots \\ 0 \end{bmatrix} \tag{39}$$

which can be rewritten as

$$\begin{bmatrix} H_0(z) & H_1(z) & \cdots & H_{M-1}(z) \\ H_0(zW^{-1}) & H_1(zW^{-1}) & \cdots & H_{M-1}(zW^{-1}) \\ \vdots & & & \\ H_0(zW^{-M+1}) & H_1(zW^{-M+1}) & \cdots & H_{M-1}(zW^{-M+1}) \end{bmatrix} \cdot \begin{bmatrix} F_0(zW^{-1}) \\ F_1(zW^{-1}) \\ \vdots \\ F_{M-1}(zW^{-1}) \end{bmatrix} = \begin{bmatrix} 0 \\ cz^{-n_0}W^{n_0} \\ 0 \\ \vdots \\ 0 \end{bmatrix}. \tag{40}$$

In general, by replacing z with zW^{-k}, $0 \le k \le M-1$ in (38), we arrive at a set of M matrix–vector equations of the form in (40). These can be put together into a compact equation:

$$H(z)\,F(z) = cz^{-n_0}\begin{bmatrix} 1 & & & O \\ & W^{n_0} & & \\ & & W^{2n_0} & \\ O & & \ddots & \\ & & & W^{(M-1)n_0}\end{bmatrix} \tag{41}$$

where $F(z)$ is defined by

$$F(z) = \begin{bmatrix} F_0(z) & F_0(zW^{-1}) & \cdots & F_0(zW^{-M+1}) \\ F_1(z) & F_1(zW^{-1}) & \cdots & F_1(zW^{-M+1}) \\ \vdots & \vdots & \vdots & \vdots \\ F_{M-1}(z) & F_{M-1}(zW^{-1}) & \vdots & F_{M-1}(zW^{-M+1})\end{bmatrix}. \tag{42}$$

The above argument therefore establishes the following result.

Property 3.2: The maximally decimated M-channel QMF structure of Fig. 1 gives rise to perfect reconstruction if and only if (41) holds, where $H(z)$ is the AC matrix and $F(z)$ is defined as in (42).

This property gives rise to an important corollary. Thus, let $H(z)$ be lossless. Then

$$H(z)\,H_*^T(z^{-1}) = I. \tag{43}$$

If, in addition, $F_k(z)$ are such that there is perfect reconstruction, then (41) holds; hence,

$$F(z) = cz^{-n_0}H_*^T(z^{-1})\begin{bmatrix} 1 & & O \\ & W^{n_0} & \\ & & \ddots \\ O & & W^{(M-1)n_0}\end{bmatrix} \tag{44}$$

which shows that $F_k(z)$ *must* be of the form

$$F_k(z) = cz^{-n_0}H_k(z^{-1}). \tag{45}$$

In fact, we can state the following.

Property 3.3: Let the AC matrix $H(z)$ be lossless. Then the structure of Fig. 1 gives rise to perfect reconstruction if and only if $F_k(z)$ are related to $H_k(z)$ as in (45) where c is an arbitrary constant.

Notice, as a verification, that for the case of $M = 2$, the SB structure satisfies property 3.3. The importance of this property is that, if we know how to choose $H_k(z)$ such that $H(z)$ is lossless, then there exists a unique way to choose $F_k(z)$ so that the reconstruction is perfect. Notice also that if $H(z)$ is lossless but $H_k(z)$ not FIR, then $F_k(z)$ in (45) are unstable (assuming $H_k(z)$ are stable, of course), and hence, there does not exist a stable perfect-reconstruction technique. Thus, for IIR filters, the strategy would be to avoid forcing $H(z)$ to be lossless. Finally, it should be emphasized that (45) is a necessary *condition for perfect reconstruction only if* $H(z)$ *is* lossless.

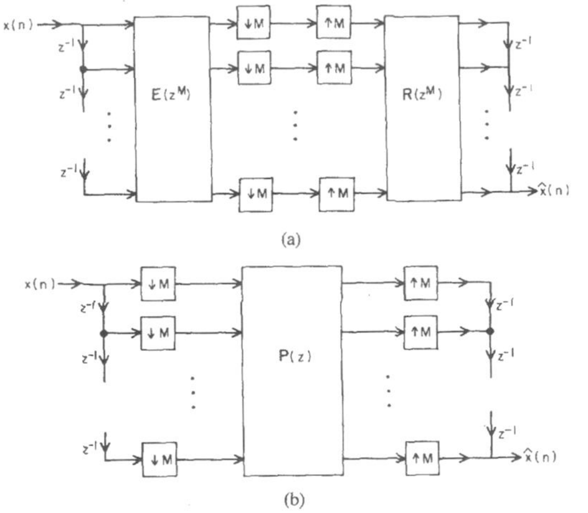

Fig. 10. Pertaining to lemma 3.2.

Next, let us assume that Fig. 1 performs perfect reconstruction so that (41) holds. If in addition (45) holds, then by direct substitution into (41) we can verify that

$$H(z)\,H_*^T(z^{-1}) = I. \tag{46}$$

In other words, we have the following.

Property 3.4: Let the structure of Fig. 1 be such that it performs perfect reconstruction. Furthermore, let $F_k(z)$ be related to $H_k(z)$ as in (45). Then $H(z)$ is necessarily lossless.

The above properties are essentially various useful manifestations of certain sets of sufficient conditions for perfect reconstruction. A natural question that arises in this context is: what is a set of necessary *and sufficient* conditions? We now turn our attention to this issue.

Given an arbitrary set of filters $\{H_k(z)\}$ and $\{F_k(z)\}$, we can also express $H_k(z)$ in the form (32) and $F_k(z)$ in the form

$$F_k(z) = \sum_{l=0}^{M-1} z^{-(M-1-l)}R_{lk}(z^M). \tag{47}$$

Letting

$$R(z) \triangleq [R_{lk}(z)], \tag{48}$$

Fig. 1 can therefore be redrawn as in Fig. 10(a). Letting $P(z) = R(z)\,E(z)$, Fig. 10(a) can be redrawn as in Fig. 10(b). What is a set of necessary and sufficient conditions on $P(z)$ so that (3) holds? In order to answer this question, first note that $\hat{X}(z)$ can be expressed as

$$\hat{X}(z) = \frac{z^{-(M-1)}}{M}\sum_{k=0}^{M-1} X(zW^k)\sum_{l=0}^{M-1} W^{-kl} \cdot \sum_{s=0}^{M-1} z^{-(l-s)}P_{s,l}(z^M) \tag{49}$$

where $P_{s,l}(z)$ is the (s, l)th element of $P(z)$. Perfect reconstruction for arbitrary $X(z)$ occurs if and only if

$$\sum_{l=0}^{M-1} W^{-kl} \sum_{s=0}^{M-1} z^{-(l-s)} P_{s,l}(z^M) = \alpha z^{-k_0}\delta(k) \quad (50)$$

for some constant α and some nonnegative integer k_0. Defining

$$q(l) = \sum_{s=0}^{M-1} z^{-(l-s)} P_{s,l}(z^M), \qquad 0 \le l \le M-1, \quad (51)$$

equation (50) says that the IDFT of the M-point sequence $q(l)$ must be an impulse. Accordingly, (50) is equivalent to the requirement

$$\sum_{s=0}^{M-1} z^{-(l-s)} P_{s,l}(z^M) = \frac{\alpha}{M} z^{-k_0}. \quad (52)$$

Let the coefficients of $P_{s,l}(z)$ be denoted $p_{s,l}(n)$, i.e.,

$$P_{s,l}(z) = \sum_{n=0}^{\infty} p_{s,l}(n) z^{-n}. \quad (53)$$

Equation (52) implies

$$p_{s,l}(n) = 0 \quad (54)$$

for all n except when n satisfies

$$l - s + nM = k_0. \quad (55)$$

For every s in $0 \le s \le M-1$, there is a unique l in $0 \le l \le M-1$ satisfying (55). This l is given by

$$l = \begin{cases} s + k_{00}, & s < M - k_{00} \\ s + k_{00} - M, & s \ge M - k_{00} \end{cases} \quad (56)$$

where k_{00}, which satisfies $0 \le k_{00} \le M-1$, is defined by

$$k_{00} = k_0 \bmod M. \quad (57)$$

The value of n for which l and s satisfy (56) is given by

$$n = \begin{cases} k_{01}, & s < M - k_{00} \\ k_{01} + 1, & s \ge M - k_{00} \end{cases} \quad (58)$$

where k_{01} is such that

$$k_0 = k_{00} + k_{01}M. \quad (59)$$

In summary, $\boldsymbol{P}(z)$ takes on the form

$$\boldsymbol{P}(z) = \frac{\alpha}{M}\begin{bmatrix} \boldsymbol{O} & z^{-k_{01}}\boldsymbol{I}_1 \\ z^{-(k_{01}+1)}\boldsymbol{I}_2 & \boldsymbol{O} \end{bmatrix} \quad (60)$$

where $\boldsymbol{I}_1$ is the $(M - k_{00}) \times (M - k_{00})$ identity matrix and $\boldsymbol{I}_2$ is the $k_{00} \times k_{00}$ identity matrix. If $k_{00} = 0$, then $\boldsymbol{P}(z)$ is proportional to the identity matrix, but more generally, a nondiagonal $\boldsymbol{P}(z)$ is permissible. We summarize these results as follows.

Lemma 3.2: Consider a maximally decimated M-channel structure as in Fig. 1 with arbitrary M. Let $\boldsymbol{E}(z)$ and $\boldsymbol{R}(z)$ be defined in terms of $\{H_k(z)\}$ and $\{F_k(z)\}$, according to (32) and (47), respectively. Then the structure

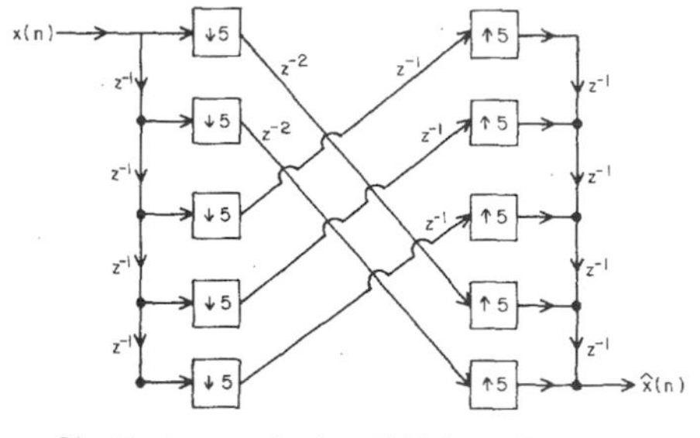

Fig. 11. An example where $P(z)$ is nondiagonal.

gives rise to perfect reconstruction, if and only if the matrix $\boldsymbol{P}(z)$ defined as $\boldsymbol{P}(z) = \boldsymbol{R}(z)\,\boldsymbol{E}(z)$ is of the form (60). Under such a condition we get

$$\hat{X}(z) = \frac{\alpha}{M} z^{-(k_0+M-1)} X(z). \quad (61)$$

Fig. 11 shows a example where $k_0 = 7$, and $M = 5$ so that $k_{00} = 2$ and $k_{01} = 1$. In this example, $\hat{X}(z) = \alpha z^{-11} X(z)/5$.

Given an FIR analysis bank (for which $\boldsymbol{E}(z)$ has polynomial entries), there exists an FIR synthesis bank giving rise to perfect reconstruction if $\boldsymbol{R}(z)$ given by

$$\boldsymbol{R}(z) = \boldsymbol{P}(z)\,\boldsymbol{E}^{-1}(z) \quad (62)$$

has polynominal entries. This in turn happens if $\boldsymbol{E}(z)$ has a determinant equal to a power of z.[4] This is equivalent to saying that det $\boldsymbol{H}(z)$ is a power of z. Notice that if $\boldsymbol{E}(z)$ is lossless, then an obvious choice for $\boldsymbol{R}(z)$ in order to obtain perfect reconstruction is

$$\boldsymbol{R}(z) = z^{-\beta}\boldsymbol{E}^T(z^{-1}) \quad (63)$$

where β is a large enough positive integer such that the RHS of (63) has no positive powers of z.

Finally, consider the special case when the analysis filters constitute a uniform DFT bank:

$$H_k(z) = H_0(zW^k). \quad (64)$$

Under this condition, with

$$H_0(z) = \sum_{l=0}^{M-1} E_{0l}(z^M) z^{-l}, \quad (65)$$

we have

$$H_k(z) = \sum_{l=0}^{M-1} E_{0l}(z^M) W^{-kl} z^{-l}, \quad (66)$$

whence

$$E_{kl}(z) = W^{-kl} E_{0l}(z). \quad (67)$$

[4]Such matrix polynomials are commonly referred to as "unimodular" in the literature [25]. More strictly, a unimodular matrix-polynomial has a *constant* nonzero determinant. The use of such matrices in perfect-reconstruction banks for arbitrary M has also been independently noted in [15].

Thus,

$$E(z) = W^{\dagger} \begin{bmatrix} E_{00}(z) & & & 0 \\ & E_{01}(z) & & \\ & & \ddots & \\ 0 & & & E_{0,M-1}(z) \end{bmatrix}. \quad (68)$$

An obvious choice of $\boldsymbol{R}(z)$ in order to get perfect reconstruction is

$$\boldsymbol{R}(z) = \begin{bmatrix} E_{00}^{-1}(z) & & & 0 \\ & E_{01}^{-1}(z) & & \\ & & \ddots & \\ 0 & & & E_{0,M-1}^{-1}(z) \end{bmatrix} W \quad (69)$$

so that

$$\boldsymbol{P}(z) = \boldsymbol{R}(z)\,\boldsymbol{E}(z) = M \cdot \boldsymbol{I}. \quad (70)$$

This choice is indeed practicable, provided the numerators of $E_{0l}(z)$ have minimum phase. This observation has also been made in some earlier publications [11]–[15]. However, in this special case, the synthesis bank in general is IIR, even with an FIR analysis bank.

IV. Losslessness of AC Matrix in Tree-Structured SB Filter Banks

The two-channel SB circuit can be used in a tree structure [6] in order to generate perfect-reconstruction QMF banks with M channels when M is a power of two. Fig. 12 is such a demonstration for $M = 2^2 = 4$. The quantities $A_0(z)$, $A_1(z)$ are the analysis filters, and $B_0(z)$, $B_1(z)$ are the synthesis filters for the basic two-channel prototype, and are related by

$$A_1(z) = z^{-(N-1)} A_0(-z^{-1}) \quad (71)$$

$$B_0(z) = z^{-(N-1)} A_0(z^{-1}) \quad (72)$$

$$B_1(z) = z^{-(N-1)} A_1(z^{-1}) \quad (73)$$

where $N - 1$ (which must be odd) is the order of the lowpass filter $A_0(z)$. Since the two-channel prototype has perfect-reconstruction property, repeated application of this property shows that the tree structure also has this property. It can be verified that

$$\hat{X}(z) = z^{-(N-1)(M-1)} X(z) \quad (74)$$

for such tree structures.

In Section II we saw that the two-channel SB structure has a lossless AC matrix. In Section III we found that for arbitrary M, if $\boldsymbol{H}(z)$ is lossless, then there exists a unique way to choose $F_k(z)$ so as to obtain perfect reconstruction. In this context, it is natural to raise the following question: does the tree structured SB QMF bank also exhibit a lossless AC matrix? The answer is in the affirmative as one might intuitively expect (but is not entirely obvious) and can be established as follows.

The tree structure can be drawn in an equivalent parallel form as in Fig. 1. Such a redrawing is demonstrated

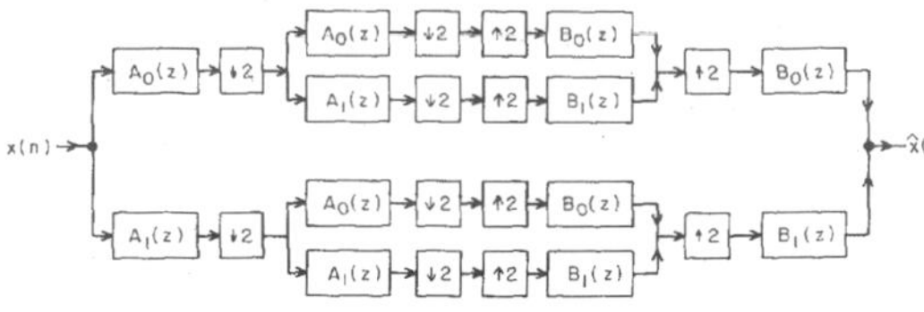

Fig. 12. The tree structure for $M = 4$.

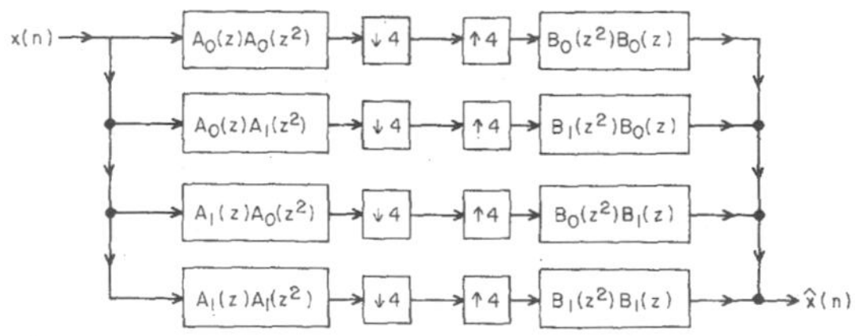

Fig. 13. The parallel equivalent of Fig. 12.

in Fig. 13 for $M = 4$, and is justified because of well-known equivalence relations for multirate systems [3, ch. 3]. It can be seen, in general, that the kth analysis filter $H_k(z)$ of the equivalent parallel structure for $M = 2^L$ is given by

$$H_k(z) = A_{i0(k)}(z)\, A_{i1(k)}(z^2) \cdots A_{i_{L-1}(k)}(z^{M/2}) \quad (75)$$

where

$$i_0(k)\, i_1(k) \cdots i_{L-1}(k) \quad (76)$$

is the "binary representation" of the integer k, i.e.,

$$k = \sum_{l=0}^{L-1} i_l(k)\, 2^{L-1-l}. \quad (77)$$

For example, in Fig. 13, $H_2(z) = A_1(z)\, A_0(z^2)$. Similarly, the kth synthesis filter $F_k(z)$ of the equivalent parallel structure is given by

$$F_k(z) = B_{i0(k)}(z)\, B_{i1(k)}(z^2) \cdots B_{i_{L-1}(k)}(z^{M/2}). \quad (78)$$

In view of the relations (72) and (73), we obtain from (75) and (78) the following relation:

$$F_k(z) = z^{-\beta} H_k(z^{-1}) \quad (79)$$

where

$$\begin{aligned} \beta &= (N-1)(1 + 2 + 4 + \cdots + 2^{L-1}) \\ &= (N-1)(M-1). \end{aligned} \quad (80)$$

Thus, the equivalent parallel structure satisfies the relation (45). Moreover, we *know* that it has perfect-reconstruction property because (74) holds. So, by invoking property 3.4 of Section III, it is immediately clear that the AC matrix $\boldsymbol{H}(z)$ is indeed lossless.

V. Relation Between Mth Band FIR Filters and M-Channel Perfect-Reconstruction QMF Banks

Smith and Barnwell [6] and Mintzer [16] have outlined elegant techniques for the design of the transfer function

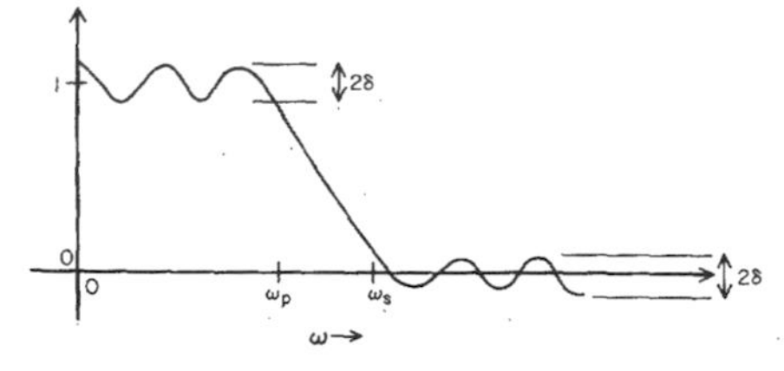

Fig. 14. The amplitude response of a linear-phase half-band filter.

$H_0(z)$ in the two-channel SB structure, such that $H_0(z)$ has prescribed stopband attenuation, while at the same time satisfying (10) [where $H_1(z)$ in (10) is related to $H_0(z)$ as in (7)]. Their procedure is based on the design of a linear-phase half-band FIR (low-pass) transfer function $G(z)$ of order $2(N-1)$ [where $N-1$ is the order of $H_0(z)$]. The frequency response of $G(z)$ has the form

$$G(e^{j\omega}) = e^{-j\omega(N-1)} G_0(e^{j\omega}) \quad (81)$$

where $G_0(e^{j\omega})$ is the amplitude response and is as shown in Fig. 14. This response exhibits symmetry with respect to $\pi/2$; in particular, $\omega_s = \pi - \omega_p$, and $\delta_1 = \delta_2 = \delta$. Accordingly, $G(z)$ satisfies

$$G(z) + (-1)^{N-1} G(-z) = z^{-(N-1)}. \quad (82)$$

If we now construct $G_+(z) \triangleq G(z) + \delta z^{-(N-1)}$, then $G_+(z)$ has a positive-valued amplitude response and satisfies

$$G_+(z) + (-1)^{N-1} G_+(-z) = (1 + 2\delta) z^{-(N-1)}. \quad (83)$$

Let $H_0(z)$ be a spectral factor of $G_+(z)$, and define $H_1(z)$ as in (7). [Clearly, $H_1(z)$ is then a spectral factor of $G_+(-z)$.] Then (10) is satisfied automatically (up to a scale factor) because of (83). The order of $G(z)$ can be estimated depending upon the required stopband attenuation of $H_0(z)$. This then is the essence of the design procedures described in [6] and [16]. The impulse response $g(n)$ of $G(z)$ satisfies the condition $g(n) = 0$ if $[n - (N - 1)]$ is a (nonzero) even number. This follows from (82). Accordingly, if $N-1$ is even, then $g(0) = g(2(N-1)) = 0$ anyway, and hence, $N-1$ can be assumed to be odd without loss of generality.

In designing the analysis bank filters of an M-channel perfect-reconstruction QMF structure, the above approach based on half-band filters can be extended. A digital Mth band filter [21]

$$G(z) = \sum_{n=0}^{L-1} g(n) z^{-n}$$

is a linear-phase (low-pass) FIR filter with cutoff frequency nearly equal to π/M, and satisfies

$$g(n) = 0, \qquad n - \frac{L-1}{2} = \text{nonzero multiple of } M. \quad (84)$$

We assume $g(n)$ to be real for all n. For convenience of discussion, define a zero-phase FIR filter

$$G_1(z) = z^{(L-1)/2} G(z). \quad (85)$$

Because of (84), $G_1(z)$ satisfies the condition

$$\sum_{k=0}^{M-1} G_1(zW^{-k}) = Mg\left(\frac{L-1}{2}\right) \quad (86)$$

where $W = e^{-j2\pi/M}$. Assume that $G_1(e^{j\omega})$ (which is real) is also nonnegative for all ω (if this is not true, it can be ensured simply by adding $\epsilon z^{-(L-1)/2}$ to $G(z)$ where ϵ is sufficiently large). We can now define (causal) spectral factors $H_k(z)$ of $G_1(zW^{-k})$, for each k:

$$H_k(z) H_{k,*}(z^{-1}) = G_1(zW^{-k}), \qquad 0 \le k \le M-1. \quad (87)$$

In view of (86), the spectral factors therefore satisfy the condition

$$\sum_{k=0}^{M-1} H_{k,*}(z^{-1}) H_k(z) = Mg\left(\frac{L-1}{2}\right). \quad (88)$$

This condition is an extension of (10) which was satisfied by the two-channel SB structure. Accordingly, if we neglect the aliasing effects in an M-channel QMF bank, and design $H_k(z)$ in the above manner, then the choice of synthesis filters as in (45) ensures perfect reconstruction. Perfect-reconstruction results, under the assumption that aliasing is negligible, can be found in [35].

Let us now explore the aliasing problem when the filters are chosen in this particular manner. Replacing z with zW^{-l}, (88) implies

$$\sum_{l=0}^{M-1} H_{k,*}(z^{-1}W^{l}) H_k(zW^{-l}) = Mg\left(\frac{L-1}{2}\right). \quad (89)$$

Consider an M-channel QMF bank as in Fig. 1 with $H_k(z)$ defined in the above manner. If the corresponding AC matrix $\boldsymbol{H}(z)$ (34) can be forced to be lossless, then we can immediately obtain the synthesis bank that would enable perfect reconstruction. Recall that losslessness of $\boldsymbol{H}(z)$ is equivalent to

$$\boldsymbol{H}_*^T(z^{-1}) \boldsymbol{H}(z) = \boldsymbol{I}, \quad (90)$$

i.e.,

$$\sum_{s=0}^{M-1} H_{k,*}(z^{-1}W^{s}) H_l(zW^{-s}) = \delta(k-l). \quad (91)$$

Since $H_k(z)$ have been constructed by spectral factorization of an Mth band filter, they satisfy (89). Hence, (91) is automatically satisfied for $k = l$ (except for a scale factor of no consequence). Now, if (91) holds for all k, l with $0 \le k, l \le M-1$, perfect reconstruction is possible. Let us next examine the side effects of satisfying (91) for $k \ne l$. For example, let $k = 1$, $l = 0$, then (91) implies

$$\sum_{s=0}^{M-1} H_{1,*}(z^{-1}W^{s}) H_0(zW^{-s}) = 0. \quad (92)$$

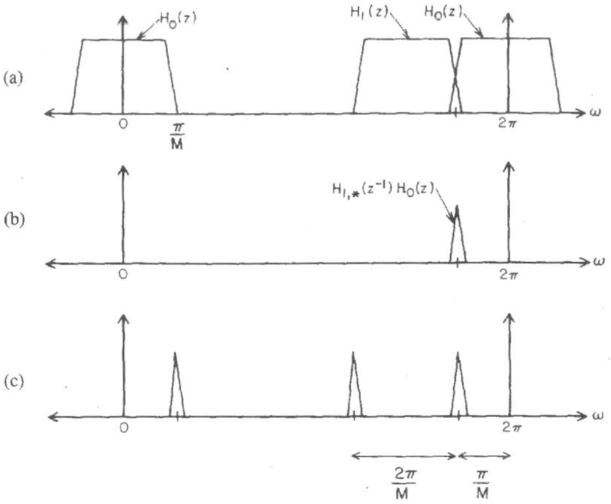

Fig. 15. Pertaining to design based on Mth-band filters.

The magnitudes of $H_0(z)$, $H_1(z)$, and $H_{1,*}(z^{-1})\,H_0(z)$ on the unit circle are sketched in Fig. 15. In Fig. 15(c) is shown the typical appearance of the magnitudes of the quantities appearing in the summation of (92). It is clear that if $H_0(z)$ is a good low-pass filter with cutoff π/M, then there is hardly an overlap between the successive waveforms in Fig. 15(c), unless $M = 2$. Accordingly, for $M > 2$, it is not possible to satisfy (92), unless $H_0(z)$ is a "poor" low-pass function which would permit overlap between adjacent spectral shapes in Fig. 15(c). But for subband coding applications, the stopband rejection of $H_k(z)$ is of crucial importance [3].

In conclusion, if we design $H_k(z)$ by spectral factorizing an Mth band filter, and if $H_0(z)$ is a sharp-cutoff filter with reasonably large stopband attenuation, then $\boldsymbol{H}(z)$ cannot be made lossless unless $M = 2$. Now, property 3.4 says that if (45) holds, then losslessness of $\boldsymbol{H}(z)$ is a necessary condition for perfect reconstruction. Accordingly, if (45) holds, and if $H_k(z)$ are "good" filters obtained as above, then since $\boldsymbol{H}(z)$ cannot be lossless, we cannot have perfect reconstruction either. In conclusion, if $H_k(z)$ are designed based on Mth band filters, and if $F_k(z)$ are chosen as in (45), we cannot have perfect reconstruction unless $H_k(z)$ are poor filters (or unless we choose to ignore aliasing).

Finally, notice that the spectral factors $H_k(z)$ defined in (87) will have complex coefficients except when $k = 0$, and hence, the subband signals for such a scheme would be complex, even for real signals. This would not always be desirable.

In the next section we turn our attention to better design procedures which ensure that $\boldsymbol{H}(z)$ is lossless (hence permitting perfect reconstruction), at the same time permitting $H_0(z)$ to have arbitrarily sharp cutoff and large stopband attenuation. In addition, all transfer functions $H_k(z)$ have real coefficients.

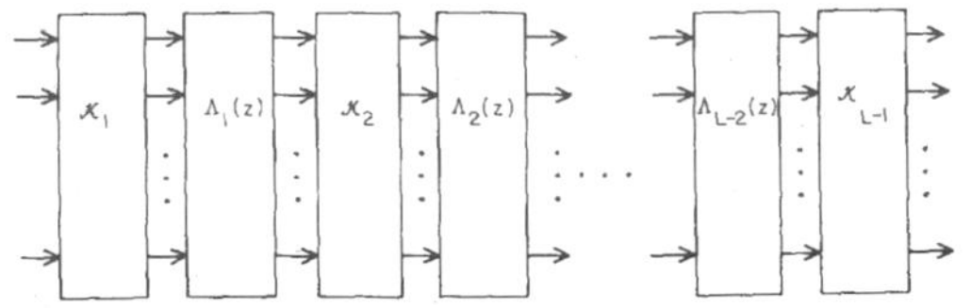

Fig. 16. A simple means of obtaining an $M \times M$ lossless matrix $\boldsymbol{E}(z)$.

VI. Design Based on Numerical Optimization of Lossless FIR Lattice Structures

A technique has recently been proposed [18]–[20] for the implementation of FIR filters and filter banks based on lossless building blocks. The basic ideas of such a technique can be exploited in order to design the analysis (and synthesis) bank filters of the perfect-reconstruction QMF structure. We include here a self-contained but brief description of the procedure for doing this.

Consider again the QMF structure of Fig. 7. According to the results of Section III, if the M-input M-output system $\boldsymbol{E}(z)$ is FIR and lossless,[5] then Fig. 7 represents a perfect-reconstruction QMF bank. From the definition of losslessness, it is clear that a cascade of lossless systems is lossless. For example, if $\boldsymbol{E}(z) = \boldsymbol{E}_1(z)\,\boldsymbol{E}_2(z)$, and if $\boldsymbol{E}_1(z)$ and $\boldsymbol{E}_2(z)$ are lossless, then

$$\tilde{\boldsymbol{E}}\boldsymbol{E}(z) = \tilde{\boldsymbol{E}}_2(z)\,\tilde{\boldsymbol{E}}_1(z)\,\boldsymbol{E}_1(z)\,\boldsymbol{E}_2(z) = \tilde{\boldsymbol{E}}_2(z)\,\boldsymbol{E}_2(z) = \boldsymbol{I}.$$

A simple means, therefore, of obtaining an $M \times M$ lossless system $\boldsymbol{E}(z)$ is indicated in Fig. 16, which is a cascade of two kinds of lossless building blocks. The building blocks K_i are constant unitary matrices, i.e.,

$$\mathrm{K}_i^{\dagger}\mathrm{K}_i = \boldsymbol{I}, \qquad 1 \le i \le L - 1. \tag{93}$$

In this section we consider QMF banks with real coefficients, hence, K_i have real coefficients (i.e., they are *orthogonal* matrices).

The second type of building blocks $\Lambda_i(z)$ which separate successive K_i are diagonal matrices with delay elements. The diagonal nature ensures that $\Lambda_i(z)$ are lossless. For example, a typical $\Lambda_i(z)$ for $M = 3$ could be

$$\Lambda_i(z) = \begin{bmatrix} 1 & 0 & 0 \\ 0 & 1 & 0 \\ 0 & 0 & z^{-1} \end{bmatrix}. \tag{94}$$

This choice of $\Lambda_i(z)$ is clearly not the only possible diagonal matrix of delay elements, and is meant to serve as an example. Our numerical example will be based on (94) because the form (94) has been used in an earlier paper [20] under a different context. The "best" choice of the diagonal matrix $\Lambda_i(z)$ that maximizes attenuation provided by $H_k(z)$ is not clear at this point in time.

[5] In the case of FIR filters, the term "lossless" is synonymous with "unitary on the unit circle of z-plane" because stability is automatically guaranteed.

If we employ the M-input M-output system of Fig. 16 in place of $E(z)$ in Fig. 7, then perfect reconstruction is guaranteed! It only remains to adjust the parameters of the matrices K_i such that $H_k(z)$ will have good attenuation characteristics. This can be accomplished by nonlinear optimization techniques, and some details are included in this section. It should be noticed that once $H_k(z)$ are computed in this fashion, the designer has the choice of either using the structure of Fig. 7 (with $E(z)$ as in Fig. 16), or simply implementing $H_k(z)$ in *direct form* and obtaining the structure of Fig. 1 [the filters $F_k(z)$ can be obtained from $H_k(z)$ as in (45)]. In either case, the reconstruction is perfect (as long as signal rounding effects and coefficient quantization are small enough to be ignored).

A. An Optimization Procedure

A simple way to generate an $M \times M$ orthogonal matrix K_i is as a sequence of planar rotations. For example,

$$K_i = \begin{bmatrix} \cos\theta_{1,i} & \sin\theta_{1,i} & \mathbf{0} \\ \sin\theta_{1,i} & -\cos\theta_{1,i} & \mathbf{0} \\ \mathbf{0} & \mathbf{0} & I_1 \end{bmatrix} \begin{bmatrix} 1 & 0 & 0 & \mathbf{0} \\ 0 & \cos\theta_{2,i} & \sin\theta_{2,i} & \mathbf{0} \\ 0 & \sin\theta_{2,i} & -\cos\theta_{2,i} & \mathbf{0} \\ \mathbf{0} & \mathbf{0} & \mathbf{0} & I_2 \end{bmatrix} \cdots \begin{bmatrix} I_{M-1} & \mathbf{0} & \mathbf{0} \\ \mathbf{0} & \cos\theta_{M-1,i} & \sin\theta_{M-1,i} \\ \mathbf{0} & \sin\theta_{M-1,i} & -\cos\theta_{M-1,i} \end{bmatrix} \tag{95}$$

where I_k are identity matrices of appropriate dimensions (not necessarily equal to the subscript k). It should be realized that (95) does not represent a most general $M \times M$ orthogonal matrix (which would involve more than $M - 1$ rotational angles [26]), and is merely intended as a possibility. Assuming that each K_i in Fig. 16 has the above form, an objective function ϕ can be formulated which measures the filtering accuracy of $H_k(z)$ (such as stopband attenuation). Such a function depends upon $\theta_{k,i}$, $1 \le k \le M - 1$, $1 \le i \le L - 1$ in a nonlinear fashion, and can be optimized by employing standard gradient algorithms.

We now demonstrate this procedure with an example for the case of three channels.

A Design Example: Let $M = 3$. In this example, we constrain $E(z)$ to be as in Fig. 17, so that $\Lambda_i(z)$ are as in (94). We restrict K_i to be as in (95), i.e.,

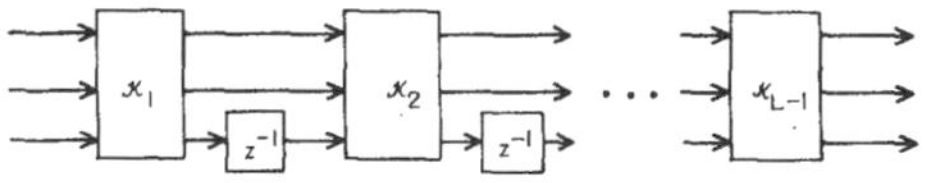

Fig. 17. The 3×3 lossless $E(z)$ for the design example.

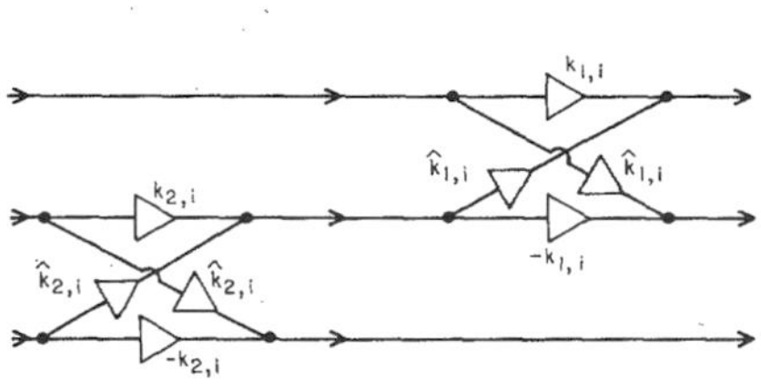

Fig. 18. The lattice structure implementing K_i.

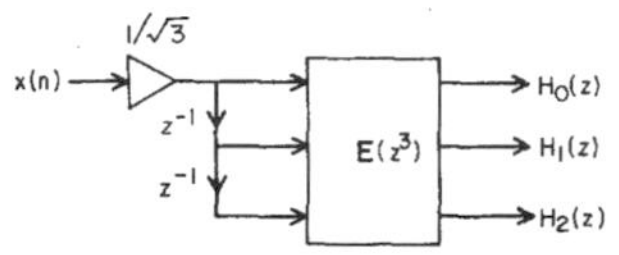

Fig. 19. The analysis bank for the 3-channel case.

$$K_i = \begin{bmatrix} \cos\theta_{1,i} & \sin\theta_{1,i} & 0 \\ \sin\theta_{1,i} & -\cos\theta_{1,i} & 0 \\ 0 & 0 & 1 \end{bmatrix} \begin{bmatrix} 1 & 0 & 0 \\ 0 & \cos\theta_{2,i} & \sin\theta_{2,i} \\ 0 & \sin\theta_{2,i} & -\cos\theta_{2,i} \end{bmatrix}. \tag{96}$$

The lattice structure implementing K_i is shown in Fig. 18, where $k_{1,i} = \cos\theta_{1,i}$, $\hat{k}_{1,i} = \sin\theta_{1,i}$, $k_{2,i} = \cos\theta_{2,i}$, $\hat{k}_{2,i} = \sin\theta_{2,i}$. The three transfer functions $H_0(z)$, $H_1(z)$, and $H_2(z)$ of the analysis bank (Fig. 19) are automatically guaranteed to satisfy the condition

$$\left|H_0(e^{j\omega})\right|^2 + \left|H_1(e^{j\omega})\right|^2 + \left|H_2(e^{j\omega})\right|^2 = 1 \tag{97}$$

because the losslessness of $E(z)$ [induced by the orthogonality of (96)] ensures that the AC matrix $H(z)$ is lossless. We wish the frequency responses to be of the form in Fig. 20 and accordingly formulate an objective function

$$\phi = \int_{(\pi/3)+\epsilon}^{\pi} \left|H_0(e^{j\omega})\right|^2 d\omega + \int_0^{(2\pi/3)-\epsilon} \left|H_2(e^{j\omega})\right|^2 d\omega + \int_0^{(\pi/3)-\epsilon} \left|H_1(e^{j\omega})\right|^2 d\omega + \int_{(2\pi/3)+\epsilon}^{\pi} \left|H_1(e^{j\omega})\right|^2 d\omega. \tag{98}$$

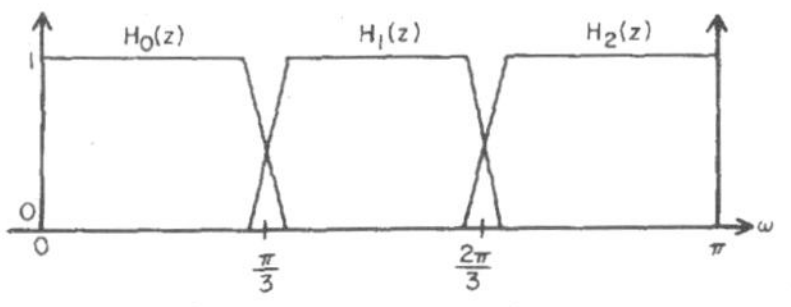

Fig. 20. The magnitude responses desired in the analysis bank.

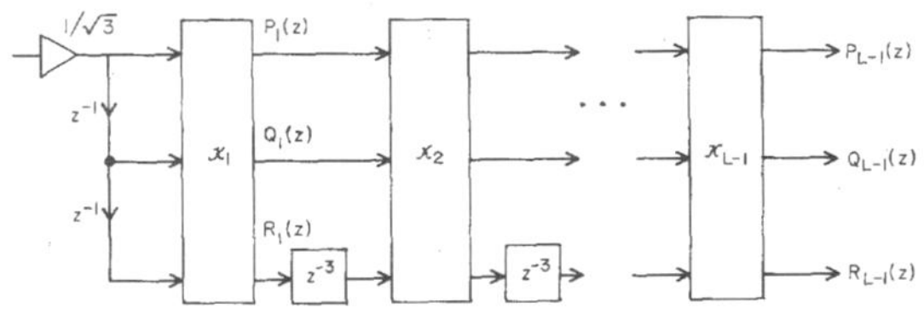

Fig. 21. The structure for the optimization of analysis filters.

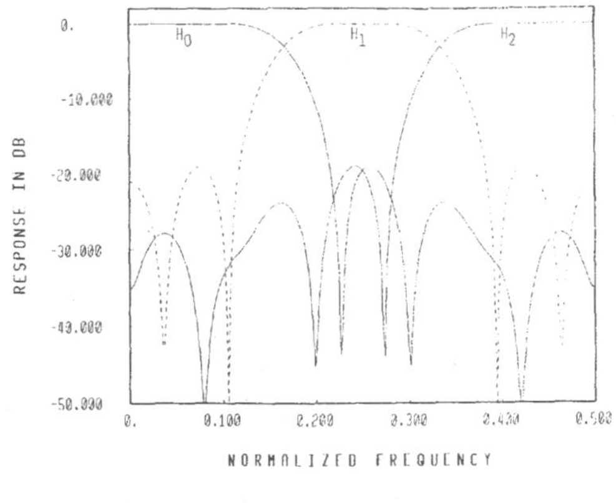

Fig. 22. Magnitude response plots for the optimized analysis filters.

The quantity ϵ depends on the desired stopband edges. The function ϕ involves only the stopband energies of the various transfer functions. However, if we minimize ϕ, it automatically ensures good passband responses, since the constraint (97) is enforced by the structure of Fig. 17. For example, in the frequency region $(0, \pi/3 - \epsilon)$, $|H_1(e^{j\omega})|^2$ and $|H_2(e^{j\omega})|^2$ are "small," since ϕ is minimized. Accordingly, $|H_0(e^{j\omega})|$ is close to unity, because of (97).

The minimization of ϕ can be accomplished by invoking a nonlinear optimization software package. We employed a routine (called ZXMWD) of the IMSL software package [27], which is based on a modified quasi-Newton technique [28]. In order to invoke this routine, the user supplies a subroutine that computes ϕ for a given set of values of $\theta_{k,i}$. (The user is not required to supply the gradient.) Such a subroutine should first compute the coefficients of the polynomials $H_0(z)$, $H_1(z)$, $H_2(z)$, and then evaluate ϕ. The computation of the polynomial coefficients can be done recursively as follows (see Fig. 21).

Recursion:

$$P_m(z) = k_{1,m}P_{m-1}(z) + k_{2,m}\hat{k}_{1,m}Q_{m-1}(z) + z^{-3}\hat{k}_{2,m}\hat{k}_{1,m}R_{m-1}(z) \tag{99a}$$

$$Q_m(z) = \hat{k}_{1,m}P_{m-1}(z) - k_{2,m}k_{1,m}Q_{m-1}(z) - z^{-3}\hat{k}_{2,m}k_{1,m}R_{m-1}(z) \tag{99b}$$

$$R_m(z) = \hat{k}_{2,m}Q_{m-1}(z) - z^{-3}k_{2,m}R_{m-1}(z). \tag{99c}$$

Initialization:

$$P_1(z) = (k_{1,1} + z^{-1}k_{2,1}\hat{k}_{1,1} + z^{-2}\hat{k}_{2,1}\hat{k}_{1,1})/\sqrt{3} \tag{100a}$$

$$Q_1(z) = (\hat{k}_{1,1} - z^{-1}k_{2,1}\,k_{1,1} - z^{-2}\,\hat{k}_{2,1}k_{1,1})/\sqrt{3} \tag{100b}$$

$$R_1(z) = (z^{-1}\hat{k}_{2,1} - z^{-2}k_{2,1})/\sqrt{3}. \tag{100c}$$

In the above equations, $k_{i,m} = \cos\theta_{i,m}$ and $\hat{k}_{i,m} = \sin\theta_{i,m}$. At the end of the recursion we obtain the polynomials

$$H_0(z) = P_{L-1}(z), \quad H_1(z) = Q_{L-1}(z), \quad H_2(z) = R_{L-1}(z) \tag{101}$$

which enables us to compute ϕ in (98). The order of $H_k(z)$ is $N - 1 = 3(L-2) + 2$. Fig. 22 shows the magnitude response plots of the three optimized analysis filters $H_k(z)$, for an example where $L - 1 = 5$. The number of theta parameters is 10, and the orders of $H_k(z)$ are $N - 1 = 3 \times (L-2) + 2 = 14$. Table I shows the parameters $k_{i,m}$ and $\hat{k}_{i,m}$, whereas Table II shows the impulse response coefficients of $H_k(z)$. The synthesis filters can be taken to be $F_k(z) = z^{-14}H_k(z^{-1})$, in order to obtain perfect reconstruction. (Note that $F_k(z)$ have the same order as the analysis filters.) Even though the attenuation in Fig. 22 is not sufficient for standard QMF applications, this can be improved by increasing the number of stages $L - 1$ in Fig. 17 so that the analysis bank has a higher order.

The way in which the objective function is defined in (98) is such that, upon convergence of the optimization algorithm, $|H_2(e^{j\omega})|$ is the image of $|H_0(e^{j\omega})|$ with respect to $\pi/2$, and moreover, $|H_1(e^{j\omega})|$ is symmetric with respect to $\pi/2$. This can also be seen from Fig. 22. Accordingly, once the optimization converges, we will have $|H_2(e^{j\omega})| = |H_0(-e^{j\omega})|$. This does not necessarily imply $H_2(z) = \pm H_0(-z)$, and the fact that this is (approximately) the case in Table II is only a coincidence. In fact, the lattice structure itself does not even impose the restriction $|H_2(e^{j\omega})| = |H_0(-e^{j\omega})|$. Next, from Table II we see that $|H_2(e^{j\omega})|$ is only approximately equal to $|H_0(-e^{j\omega})|$, but this relation can be made more exact simply by requesting a more stringent convergence criterion for optimization. We wish to emphasize, however, that perfect-reconstruction property will be *exactly* satisfied even if the optimization is inaccurate, simply by virtue of the fact that the transfer functions are derived from the lattice structure. Thus, perfect reconstruction is *structurally induced*. The values of $\theta_{i,j}$ determine only the shapes of the individual $H_k(z)$'s.

TABLE I
THE VALUE OF THE LATTICE COEFFICIENTS IN THE OPTIMIZED ANALYSIS BANK. NUMBER OF SECTIONS $L - 1 = 5$; FILTER ORDER $N - 1 = 14$

m	$k_{1,m}$	$\hat{k}_{1,m}$	$k_{2,m}$	$\hat{k}_{2,m}$
1	-0.70922170d+00	0.70498552d+00	-0.93600000d-04	-0.10000000d+01
2	0.95631350d+00	0.29234310d+00	-0.35865230d+00	-0.93347122d+00
3	0.15709290d+00	0.98758383d+00	-0.99999990d+00	0.44721358d-03
4	0.32816120d+00	-0.94462174d+00	-0.15896640d+00	-0.98728399d+00
5	-0.11370000d-03	0.99999999d+00	-0.70708230d+00	0.70713126d+00

TABLE II
IMPULSE RESPONSE COEFFICIENTS OF THE OPTIMIZED ANALYSIS BANK FILTERS

n	$h_0(n)$	$h_1(n)$	$h_2(n)$
0	-0.42975335986199d-01	-0.92770412381267d-01	0.42988860238211d-01
1	0.13938012584211d-04	0.82043979687156d-06	-0.13939070914603d-04
2	0.14891039020466d+00	0.87653824068124d-02	-0.14892169715324d+00
3	0.29711060779911d+00	-0.89793582734887d-05	0.29732024389502d+00
4	0.35375400443686d+00	0.18640257224072d+00	-0.35374966894441d+00
5	0.26733022920753d+00	-0.43884653241662d-04	0.26709728401356d+00
6	0.87062827411073d-01	-0.35433035625168d+00	-0.87063835222800d-01
7	-0.52122033341321d-01	-0.21534684870289d-03	-0.52084419980068d-01
8	-0.87593136078151d-01	0.35645950964714d+00	0.87579833045549d-01
9	-0.42709632397534d-01	-0.48560852349886d-05	-0.42706675473313d-01
10	0.47426369238237d-01	-0.19310826169468d+00	-0.47471783144698d-01
11	0.42961831070467d-01	0.22960198324688d-04	0.42967738845495d-01
12	0. d+00	0. d+00	0. d+00
13	-0.23276534497968d-01	-0.26465419895258d-05	-0.23274922989163d-01
14	0.21786836385535d-05	0.24771633130473d-09	0.21785328013287d-05

TABLE III
IMPULSE RESPONSE OF THE COMPOSITE SYSTEM

	The sequence $\hat{x}(n)$		
n	for x(n)=δ(n)	for x(n)=δ(n-1)	for x(n)=δ(n-2)
0	0.000000000	0.	0.
1	-0.000000000	0.000000000	0.
2	-0.000000000	-0.000000000	0.000000000
3	0.000000000	-0.000000000	-0.000000000
4	-0.000000000	0.000000000	-0.000000000
5	-0.000000000	-0.000000000	0.000000000
6	0.000000000	-0.000000000	-0.000000000
7	-0.000000000	0.000000000	-0.000000000
8	0.000000000	-0.000000000	0.000000000
9	0.000000000	0.000000000	-0.000000000
10	-0.000000000	0.000000000	0.000000000
11	-0.000000000	-0.000000000	0.000000000
12	0.000000000	-0.000000000	-0.000000000
13	-0.000000000	0.000000000	-0.000000000
14	1.000000000	-0.000000000	0.000000000
15	-0.000000000	1.000000000	-0.000000000
16	0.000000000	-0.000000000	1.000000000
17	0.000000000	0.000000000	-0.000000000
18	-0.000000000	0.000000000	0.000000000
19	0.000000000	-0.000000000	0.000000000
20	-0.000000000	0.000000000	-0.000000000
21	-0.000000000	-0.000000000	0.000000000
22	0.000000000	-0.000000000	-0.000000000
23	-0.000000000	0.000000000	-0.000000000
24	-0.000000000	-0.000000000	0.000000000
25	0.000000000	-0.000000000	-0.000000000
26	0.000000000	0.000000000	-0.000000000
27	-0.000000000	0.000000000	0.000000000
28	0.000000000	-0.000000000	0.000000000
29	0.	0.000000000	-0.000000000
30	0.	0.	0.000000000

In order to demonstrate the perfect-reconstruction property of the QMF bank characterized by the parameters in Tables I and II, the complete system of Fig. 7 has been

TABLE IV
AN ARBITRARY INPUT SEQUENCE $x(n)$ AND THE RECONSTRUCTED SEQUENCE $\hat{x}(n)$ FOR THE DESIGN EXAMPLE. HERE $\hat{x}(n + N - 1)$ IS SHOWN, IN ORDER TO ALIGN THE SAMPLES

n	x(n)	$\hat{x}$(n+N-1)
1	1.000000000	1.000000000
2	0.500000000	0.500000000
3	0.600000000	0.600000000
4	0.300000000	0.300000000
5	-0.300000000	-0.300000000
6	0.400000000	0.400000000
7	0.350000000	0.350000000
8	3.120000000	3.120000000
9	1.003000000	1.003000000
10	-0.450000000	-0.450000000

simulated on a computer. With the input $x(n)$ taken as an impulse function $\delta(n - k)$ with $k = 0$, 1, and 2, respectively, the response $\hat{x}(n)$ is shown in the three columns of Table III. This response is clearly an impulse (up to an accuracy of at least 10 significant decimal digits). Since the columns of Table III are successively shifted versions, the time invariance of the complete system is demonstrated (thereby demonstrating alias cancellation). Table IV shows an arbitrary input $x(n)$ and the reconstructed signal $\hat{x}(n)$. It is clear that $\hat{x}(n)$ is a delayed version of $x(n)$. These two tables thus demonstrate the perfect-reconstruction property. Finally, Fig. 23 shows a plot of $|T(e^{j\omega})| = |H_0(e^{j\omega})|^2 + |H_1(e^{j\omega})|^2 + |H_2(e^{j\omega})|^2$, where $T(z)$ is the composite transfer function

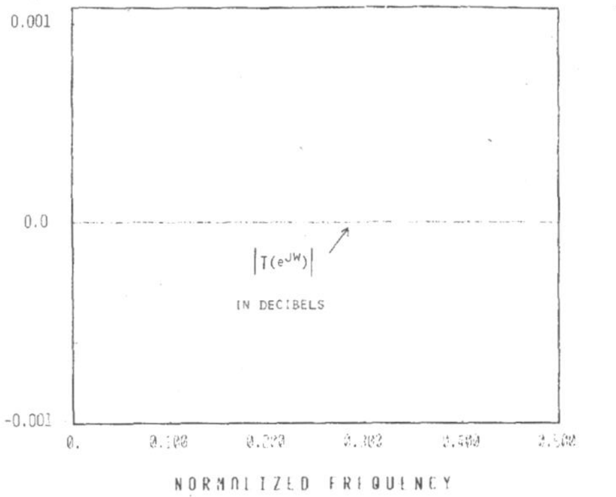

Fig. 23. The composite response in decibels for the design example.

$\hat{X}(z)/X(z)$. Evidently, $|T(e^{j\omega})| = 1$ for all ω, in accordance with perfect-reconstruction property.

B. Comments and Further Generalizations

In a practical implementation of a perfect-reconstruction system, there are three additional sources of error. First, the elements characterizing the matrices K_i have to be quantized, which means they will not be exactly orthogonal. Second, there is computational roundoff noise when the analysis and synthesis banks are digitally implemented. Finally, since the subband signals themselves are encoded before transmission (and subsequently decoded at the synthesis end), the coding error gets reflected in the reconstructed signal. These three errors are ignored in this sequel, as they require further study.

A considerable amount of work remains to be done in connection with improved optimal designs of $H_k(z)$. First, the matrices in (95) and (96) are not the most general orthogonal forms. (For example, a general 3×3 orthogonal matrix requires three planar angles in order to be completely characterized.) More general orthogonal forms can offer improved stopband attenuations for the same order $N - 1$. Second, there is no reason in practice to restrict oneself to the simplified form of $\Lambda_i(z)$ as in (94). Finally, losslessness of $\boldsymbol{H}(z)$ (which is the basis of obtaining the above optimization procedure) is itself only a sufficient (rather than necessary) condition for perfection of reconstruction, and the question is whether we can obtain more efficient designs in other ways.

As in most nonlinear optimization problems, it is possible to have local extrema, and it is judicious to try out several initial parameter estimates in order to obtain a solution that is (reasonably close to) the global minimum.

Explicit computation of the gradient of ϕ (rather than gradient-approximation using differences) is often preferred [28] in the procedure for optimization of ϕ. Gradient computation is rendered easy in the case of our objective function because of the way in which the $\theta_{k,i}$ parameters enter (95) and (96). For example, $\partial\phi/\partial\theta_{1,i}$ can be found simply by replacing the matrix

$$\begin{bmatrix} \cos\theta_{1,i} & \sin\theta_{1,i} & 0 \\ \sin\theta_{1,i} & -\cos\theta_{1,i} & 0 \\ 0 & 0 & 1 \end{bmatrix} \tag{102}$$

with the derivative

$$\begin{bmatrix} -\sin\theta_{1,i} & \cos\theta_{1,i} & 0 \\ \cos\theta_{1,i} & \sin\theta_{1,i} & 0 \\ 0 & 0 & 0 \end{bmatrix}. \tag{103}$$

The objective function obtained by replacing (102) with (103) is precisely the gradient $\partial\phi/\partial\theta_{1,i}$. In the design example reported above, this gradient computation has not been utilized.

Summarizing, the design results presented in this subsection are by no means the best in the sense that more general orthogonal matrices can be used during optimization, and $\Lambda_i(z)$ can be more general than in (94). The example is meant only to demonstrate the basic result on perfect reconstruction in maximally decimated QMF banks for arbitrary M. The scope for improvement is wide.

VII. Signal Reconstruction in QMF Banks with Recursive Filters

In Section III it was indicated that, under certain (rather stringent) restrictions, it is possible to obtain perfect-reconstruction IIR QMF banks. More useful solutions in the IIR case can be obtained if phase distortion in the reconstruction process can be tolerated. This can be accomplished by forcing $T(z)$ in (2) to be (stable) and all-pass. Examples can be found in [11]-[13], [32], and [34]. In this section we wish to indicate a general means of accomplishing this, based on the losslessness concept.

The inverse of the AC matrix is

$$\boldsymbol{H}^{-1}(z) = \frac{Adj\ \boldsymbol{H}(z)}{A(z)} \tag{104}$$

where $A(z) = \det \boldsymbol{H}(z)$. Clearly, $A(z)$ is stable (assuming $H_k(z)$ are stable) and the entries in $Adj\ \boldsymbol{H}(z)$ are stable, but $\boldsymbol{H}^{-1}(z)$ is not necessarily stable unless the zeros of $A(z)$ are restricted to be strictly inside the unit circle. Let us now consider the consequence of choosing the synthesis filters $F_k(z)$ to be

$$\boldsymbol{f}(z) \triangleq \begin{bmatrix} F_0(z) \\ F_1(z) \\ \vdots \\ F_{M-1}(z) \end{bmatrix} = Adj\ \boldsymbol{H}(z) \begin{bmatrix} 1 \\ 0 \\ \vdots \\ 0 \end{bmatrix}. \tag{105}$$

Such a choice is clearly stable and satisfies (1b) with $T(z) = A(z) = \det \boldsymbol{H}(z)$. It can be shown (Appendix B) that $\det \boldsymbol{H}(z)$ is all-pass if $\boldsymbol{H}(z)$ is lossless. Hence, the choice (105) ensures that (2) holds with $T(z)$ equal to an all-pass function. In other words, aliasing and amplitude distortion have been completely eliminated.

This observation can be reinterpreted in terms of Fig.

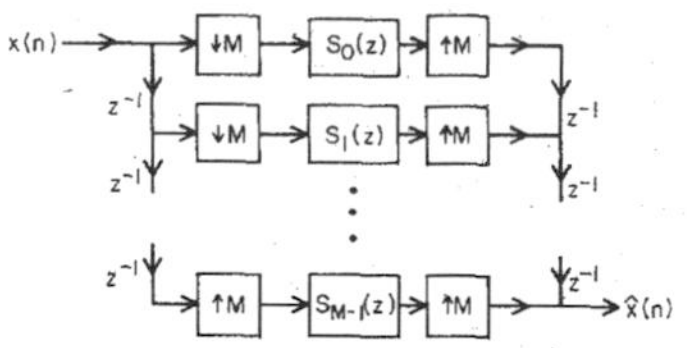

Fig. 24. Pertaining to Appendix A.

10 also. Thus, by property 3.1, $\sqrt{M}\,\boldsymbol{E}(z)$ is lossless when $\boldsymbol{H}(z)$ is lossless. Let us choose $\boldsymbol{R}(z)$ in Fig. 10 according to

$$\boldsymbol{R}(z) = Adj\,\boldsymbol{E}(z) \tag{106}$$

where

$$\boldsymbol{E}^{-1}(z) = \frac{Adj\,\boldsymbol{E}(z)}{B(z)} \tag{107}$$

and $B(z) = \det \boldsymbol{E}(z)$. Then

$$\begin{aligned}\boldsymbol{P}(z) &= \boldsymbol{R}(z)\,\boldsymbol{E}(z)\\ &= B(z)\,\boldsymbol{E}^{-1}(z)\,\boldsymbol{E}(z) = B(z)\cdot \boldsymbol{I}.\end{aligned} \tag{108}$$

With this choice of $\boldsymbol{R}(z)$, Fig. 10 reduces to Fig. 24, where $S_k(z) = B(z)$ for all k. Thus, by Appendix A, aliasing is absent, and $T(z) = \hat{X}(z)/X(z) = z^{-(M-1)} B(z^M)$. Since $\sqrt{M}\boldsymbol{E}(z)$ is lossless, $B(z)$ is again all-pass (except for a scale factor), and thus, amplitude distortion has been eliminated.

In summary, because of the properties of the determinants of lossless matrices, it is possible to eliminate (aliasing and) amplitude distortion in IIR QMF banks, when $\boldsymbol{H}(z)$ [or equivalently $\sqrt{M}\boldsymbol{E}(z)$] is lossless. Specific instances of such reconstruction can be found in recent publications [32].

VIII. Summary and Conclusions

In this paper we have considered maximally decimated M-channel parallel QMF structures of the form shown in Fig. 1. We showed in Section III how perfect reconstruction can be accomplished by restricting $\boldsymbol{E}(z)$ to be FIR and lossless (i.e., unitary on the unit circle). This is equivalent to forcing the AC matrix $\boldsymbol{H}(z)$ to be FIR and lossless. A lossless AC matrix $\boldsymbol{H}(z)$ implies that the analysis filters $\{H_k(z)\}$ satisfy the "power-complementary" property, i.e.,

$$|H_0(e^{j\omega})|^2 + |H_1(e^{j\omega})|^2 + \cdots + |H_{M-1}(e^{j\omega})|^2 = 1. \tag{109}$$

Conversely, (109) does *not* imply that $\boldsymbol{H}(z)$ is lossless and is by no means a sufficient condition for perfect reconstructibility. In Section III several properties pertaining to perfect reconstruction were presented and, based on some of these, it was also verified in Section IV that the tree structured SB QMF bank has a lossless AC matrix.

We feel that for $M > 2$, the problem of designing $H_k(z)$ should not be approached from the point of view of spectral factorization of Mth band filters, because such an approach cannot lead to $H_k(z)$ with good stopband attenuation if $\boldsymbol{H}(z)$ is at the same time forced to be lossless. The results of Section VI place in evidence one method for obtaining a set of analysis filters which satisfy the sufficient conditions for perfect reconstruction. A procedure has also been presented whereby the coefficients of such an FIR analysis bank can be optimized in order to provide a good stopband attenuation for each of $H_k(z)$. The optimization problem in Section VI opens up several possibilities for improving the design algorithm, and some of these are currently under study.

Appendix A

Consider the M-channel structure shown in Fig. 24. It is easily seen that $\hat{X}(z)$ in this figure is given by

$$\hat{X}(z) = \frac{z^{-(M-1)}}{M} \sum_{l=0}^{M-1} X(zW^l) \sum_{k=0}^{M-1} S_k(z^M) W^{-kl}. \tag{A.1}$$

From here we can deduce that the structure is free from aliasing if and only if $S_k(z)$ is independent of k, i.e.,

$$S_k(z) = S(z) \quad \text{for all} \quad k. \tag{A.2}$$

Under this condition, we have $\hat{X}(z) = z^{-(M-1)} S(z^M) X(z)$. In particular, if $S(z) = 1$ then $\hat{X}(z) = z^{-(M-1)} X(z)$, which in turn is also clear by inspection of Fig. 24. This also provides a simple example of a perfect-reconstruction system.

Appendix B

Let $\boldsymbol{H}(z)$ be lossless. Then

$$\boldsymbol{H}_*^T(z^{-1})\,\boldsymbol{H}(z) = I, \qquad \text{for all } z. \tag{A.3}$$

Taking the determinant of (A.3),

$$\det \boldsymbol{H}_*^T(z^{-1}) \det \boldsymbol{H}(z) = 1, \qquad \text{for all } z, \tag{A.4}$$

i.e.,

$$A_*(z^{-1})\,A(z) = 1, \qquad \text{for all } z \tag{A.5}$$

where $A(z) \triangleq \det \boldsymbol{H}(z)$. This shows that $A(z)$ is all-pass.

Acknowledgment

We wish to acknowledge encouraging and useful discussions we have had during the ICASSP'86 with R. E. Crochiere of the AT&T Bell Labs, T. Barnwell, III, of the Georgia Institute of Technology, T. E. Ramstad of the Norwegian Institute of Technology, and M. Vetterli of the Ecole Polytechnique Federale de Lausanne.

References

[1] A. Croisier, D. Esteban, and C. Galand, "Perfect channel splitting by use of interpolation/decimation/tree decomposition techniques," presented at the Int. Conf. Inform. Sci. Syst., Patras, Greece, 1976.
[2] D. Esteban and C. Galand, "Application of quadrature mirror filters to split-band voice coding schemes," in *Proc. IEEE Int. Conf. Ac-*

coust., Speech, Signal Processing, Hartford, CT, May 1977, pp. 191–195.

[3] R. E. Crochiere and L. R. Rabiner, *Multirate Digital Signal Processing*. Englewood Cliffs, NJ: Prentice-Hall, 1983.

[4] V. K. Jain and R. E. Crochiere, "Quadrature mirror filter design in the time domain," *IEEE Trans. Acoust., Speech, Signal Processing*, vol. ASSP-32, pp. 353–361, Apr. 1984.

[5] C. R. Galand and H. J. Nussbaumer, "New quadrature mirror filter structures," *IEEE Trans. Acoust., Speech, Signal Processing*, vol. ASSP-32, pp. 522–531, June 1984.

[6] M. J. T. Smith and T. P. Barnwell, III, "A procedure for designing exact reconstruction filter banks for tree structured subband coders," in *Proc. IEEE Int. Conf. Acoust., Speech, Signal Processing*, San Diego, CA, Mar. 1984, pp. 27.1.1–27.1.4.

[7] — "A unifying framework for analysis/synthesis systems based on maximally decimated filter banks," in *Proc. IEEE Int. Conf. Acoust., Speech, Signal Processing*, Tampa, FL, Mar. 1985, pp. 521–524.

[8] H. J. Nussbaumer, "Pseudo QMF filter bank," *IBM Tech. Disc. Bull.*, vol. 24, no. 6, pp. 3081–3087, Nov. 1981.

[9] P. L. Chu, "Quadrature mirror filter design for an arbitrary number of equal bandwidth channels," *IEEE Trans. Acoust., Speech, Signal Processing*, vol. ASSP-33, pp. 203–218, Feb. 1985.

[10] J. H. Rothweiler, "Polyphase quadrature filters—A new subband coding technique," in *Proc. 1983 IEEE Int. Conf. Acoust., Speech, Signal Processing*, pp. 1280–1283, Boston, MA, Mar. 1983.

[11] K. Swaminathan and P. P. Vaidyanathan, "Theory of uniform DFT, parallel quadrature mirror filter banks," presented at the IEEE Int. Conf. Acoust., Speech, Signal Processing, Tokyo, Japan, Apr. 1986.

[12] —, "Design of uniform DFT, parallel FIR quadrature mirror filters," presented at the IEEE Int. Symp. Circuits Syst., San Jose, CA, May 1986.

[13] —, "Theory and design of uniform DFT, parallel quadrature mirror filter banks," *IEEE Trans. Circuits Syst.*, vol. CAS-33, pp. 1170–1191, Dec. 1986.

[14] M. Vetterli, "Splitting a signal into subsampled channels allowing perfect reconstruction," in *Proc. IASTED Conf. Appl. Signal Processing Digital Filtering*, Paris, France, June 1985.

[15] —, "Filter banks allowing for perfect reconstruction," *Signal Processing*, vol. 10, no. 3, pp. 219–244, Apr. 1986.

[16] F. Mintzer, "Filters for distortion-free two-band multirate filter banks," *IEEE Trans. Acoust., Speech, Signal Processing*, vol. ASSP-33, pp. 626–630, June 1985.

[17] P. P. Vaidyanathan and S. K. Mitra, "Low passband sensitivity digital filters: A generalized viewpoint and synthesis procedures," *Proc. IEEE*, pp. 404–423, Apr. 1984.

[18] P. P. Vaidyanathan, "New cascaded lattice structures for FIR filters having extremely low coefficient sensitivity," in *Proc. IEEE Int. Conf. Acoust., Speech, Signal Processing*, Tokyo, Japan, Apr. 1986, pp. 497–500.

[19] —, "Implementation of arbitrary FIR transfer functions as passive cascaded lattice structures," presented at the IEEE Int. Symp. Circuits Syst., San Jose, CA, May 1986.

[20] —, "Passive cascaded lattice structures for low sensitivity FIR filter design, with applications to filter banks," *IEEE Trans. Circuits Syst.*, vol. CAS-33, pp. 1045–1064, Nov. 1986.

[21] F. Mintzer, "On half-hand, third-band, and *N*th-band FIR filters and their design," *IEEE Trans. Acoust., Speech, Signal Processing*, vol. ASSP-30, pp. 734–738, Oct. 1982.

[22] V. Belevitch, *Classical Network Synthesis*. San Francisco, CA: Holden-Day, 1968.

[23] B. D. O. Anderson and S. Vongpanitlerd, *Network Analysis and Synthesis*. Englewood Cliffs, NJ: Prentice-Hall, 1973.

[24] A. Fettweis, "Pseudopassivity, sensitivity, and stability of wave digital filters," *IEEE Trans. Circuit Theory*, vol. CT-19, pp. 668–673, Nov. 1972.

[25] T. Kailath, *Linear Systems*. Englewood Cliffs, NJ: Prentice-Hall, 1980.

[26] F. D. Murnaghan, *The Unitary and Rotation Groups*. Washington, DC: Spartan Books, 1962.

[27] The IMSL Library, A set of Fortran subroutines for mathematics and statistics.

[28] P. E. Gill, W. Murray, and M. H. Wright, *Practical Optimization*. New York: Academic, 1981.

[29] P. P. Vaidyanathan, "On power-complementary FIR filters," *IEEE Trans. Circuits Syst.*, vol. CAS-32, pp. 1308–1310, Dec. 1985.

[30] M. G. Bellanger, G. Bonnerot, and M. Coudreuse, "Digital filtering by polyphase network: Application to sample-rate alteration and filter banks," *IEEE Trans. Acoust., Speech, Signal Processing*, vol. ASSP-24, pp. 109–114, Apr. 1976.

[31] A. G. Constantinides and R. A. Valenzuela, "An efficient and modular transmultiplexer design," *IEEE Trans. Commun.*, vol. COM-30, pp. 1629–1641, July 1982.

[32] A. Fettweis, J. A. Nossek, and K. Meerkotter, "Reconstruction of signals after filtering and sampling rate reduction," *IEEE Trans. Acoust., Speech, Signal Processing*, vol. ASSP-33, pp. 893–902, Aug. 1985.

[33] T. H. Ramstad, "Analysis/synthesis filter banks with critical sampling," presented at the Int. Conf. Digital Signal Processing, Florence, Italy, Sept. 1984.

[34] T. P. Barnwell, III, "Subband coder design incorporating recursive quadrature filters and optimum ADPCM coders," *IEEE Trans. Acoust., Speech, Signal Processing.*, vol. ASSP-30, pp. 751–765, Oct. 1982.

[35] G. Wackersreuther, "Some new aspects of filters for filter banks," *IEEE Trans. Acoust., Speech, Signal Processing*, vol. 34, pp. 1182–1200, Oct. 1986.

P. P. Vaidyanathan (S'80–M'83) was born in Calcutta, India, on October 16, 1954. He received the B.Sc. (Hons.) degree in physics, and the B. Tech. and M. Tech. degrees in radiophysics and electronics from the University of Calcutta, India, in 1974, 1977, and 1979, respectively, and the Ph.D. degree in electrical and computer engineering from the University of California, Santa Barbara, in 1982.

He was a Postdoctoral Fellow at the University of California, Santa Barbara, from September 1982 to February 1983. Since March 1983 he has been with the California Institute of Technology, Pasadena, as an Assistant Professor of Electrical Engineering. His main research interests are in digital signal processing, linear systems, and filter design.

Dr. Vaidyanathan served as the Vice Chairman of the Technical Program Committee for the 1983 IEEE International Symposium on Circuits and Systems. He currently serves as an Associate Editor for the IEEE TRANSACTONS ON CIRCUITS AND SYSTEMS. He was the recipient of the Award for Excellence in Teaching at the California Institute of Technology for 1983–1984. He was also a recipient of the National Science Foundation's Presidential Young Investigator Award in 1986.

SECTION II

Precursors in Physics: Affine Coherent States

Introduction

Jean-Pierre Antoine

The story of the origin of wavelet analysis has been told many times (see the vivid account of B. B. Hubbard [Hub98]). The geophysicist Jean Morlet was analyzing microseismic data in the context of oil exploration. The technique consists in sending short impulsions (called wavelets by geophysicists) into the ground and analyzing the signals which have been reflected on density discontinuities. The result is usually a mess, a very noisy and confusing signal. Fourier methods are normally used for unraveling it, more precisely the windowed Fourier transform (WFT), but with mixed results. Then Morlet had the idea of exchanging roles in the WFT: instead of a fixed window containing a variable number of oscillations, he chose a fixed number of oscillations in a window of variable length. The resulting signal he called "wavelet of constant shape." It turned out that much better results were obtained in this way, but Morlet could not understand why. He turned for advice to a physicist colleague, Roger Balian, who suggested he consult a quantum physicist from Marseille named Alex Grossmann. Morlet and Grossmann worked together for a year and rapidly discovered that the key was to be found in group theory — and the result was the several pages in this section.

The idea is simple. Morlet's wavelets of constant shape were obtained by translating and dilating a fixed oscillating function ψ and the wavelet transform was simply the projection of the signal (i.e., scalar product) onto the resulting wavelet family. Now dilations and translations on the line together constitute the so-called *affine group* G_{aff} of the line (also called the $ax + b$ group):

$$\mathbb{R} \ni t \mapsto at + b, \ a > 0, \ b \in \mathbb{R}. \tag{1}$$

Since it is natural to consider finite-energy signals, one restricts signals to functions $s \in L^2(\mathbb{R}, dt)$. Then one needs to determine how the transformations (1) act on such signals. In technical terms, one needs a representation of G_{aff} in the space of signals $L^2(\mathbb{R}, dt)$. Fortunately, the representations of G_{aff} had been studied earlier by Aslaksen and Klauder, the first paper in this section. It turns out that there are only two unitary irreducible representations, acting in Fourier (i.e., frequency) space on the subspace of L^2-functions with support on $\mathbb{R}^+$ and $\mathbb{R}^-$, respectively. Clearly the former is the one to use, since it corresponds to signals with positive frequencies only (such signals are called *progressive*). It reads

$$[U(b,a)\psi](t) = \frac{1}{\sqrt{a}}\,\psi\Big(\frac{t-b}{a}\Big), \quad a > 0, \ b \in \mathbb{R}, \ \psi \in L^2(\mathbb{R}^+, dt), \tag{2}$$

where the normalization factor $1/\sqrt{a}$ ensures the unitarity of the representation. Then the wavelet transform of the signal s, with respect to the progressive wavelet ψ (i.e., $\widehat{\psi}(\omega) = 0$, $\omega < 0$), reads, in time and frequency domains, respectively,

$$S(b,a) = \frac{1}{\sqrt{a}} \int_{-\infty}^{\infty} dt\, \overline{\psi\left(a^{-1}(t-b)\right)}\, s(t) \tag{3}$$

$$= \sqrt{a} \int_{0}^{\infty} d\omega\, e^{i\omega b}\, \overline{\widehat{\psi}(a\omega)}\, \widehat{s}(\omega)\,, \tag{4}$$

the second formula resulting from the first by Plancherel's theorem.

The next question is whether this transform is invertible, that is, whether the signal s can be retrieved from its transform S. The outcome depends on the choice of the wavelet: the answer is positive if the progressive wavelet ψ satisfies the so-called *admissibility condition*:

$$\int_{0}^{\infty} d\omega\, \frac{|\widehat{\psi}(\omega)|^2}{\omega} < \infty. \tag{5}$$

In practice, this condition essentially reduces to the (only slightly weaker) requirement that ψ has zero mean:

$$\widehat{\psi}(0) = 0 \iff \int_{-\infty}^{\infty} dt\, \psi(t) = 0. \tag{6}$$

It turns out that the admissibility condition, too, has a group-theoretical origin: it expresses the fact that the representation U of G_{aff} has the property called *square integrability*, that is, the matrix element $\langle U(b,a)\psi|\psi\rangle$ is a square integrable function of $(b,a) \in G_{\mathrm{aff}}$ (for the appropriate invariant (Haar) measure).

All this is demonstrated in Morlet and Grossmann's paper, which is thus the genuine founding paper of wavelet analysis. Amusingly enough, the name "wavelets of constant shape" did not survive this first paper; it was immediately abbreviated to "wavelets." As a result, the name is often thought as meaning "little wave," because (6) implies that ψ must be an oscillating function, but, as we have seen, the reality is quite different!

As expected from the various remarks on group-theoretical aspects, the theory developed so far is not limited to the affine group G_{aff}; it extends to all groups that possess a square integrable representation. A prime example is the Weyl-Heisenberg group underlying the canonical coherent states of quantum mechanics, a theory extensively studied by John Klauder, for instance, in his paper here and its predecessors. By analogy, wavelets may be called the (generalized) coherent states associated to the affine group. The full scope of this extension did not escape the founding fathers. Again the mathematics was available and Grossmann knew it, as did his then Ph.D. student Thierry Paul (it is highly suggestive that the title of his Ph.D. thesis was "Wavelets and Quantum Mechanics"!). So, quite immediately, Grossmann, Morlet, and Paul wrote the theory in full generality, and this was the third paper in this section. Here the full connection is made between square integrable group representations and the associated (generalized) coherent states. In a companion paper [GMP86], the same authors applied the theory in full detail to the two most obvious examples, namely, the Weyl-Heisenberg group and the affine group of the line.

At this stage, the theory started to diversify. New actors entered the game, notably Yves Meyer and Ingrid Daubechies (who had been a de facto Ph.D. student of Grossmann), and soon the whole Centre de Physique Théorique in Marseille was abuzz with wavelets. Actually it sounded like a fairy tale, with rumors of mysterious facts, miraculous cancellations, and so on. The notion of frame was rediscovered, and soon Meyer and Mallat would invent the technique of multiresolution analysis, leading to the discrete wavelet transform (DWT). This brought wavelets in contact with the world of signal processing and quickly the theory exploded. But this is another story, which will be told in other sections.

Actually the real power of our three papers lies elsewhere. The generality of the formalism established in the third paper allows its application to a large number of situations, for which the DWT offers only poor solutions, such as wavelets in higher dimensions, wavelets on manifolds, time-dependent wavelets, or wavelets associated with groups that do not have square integrable representations in the standard sense. Thus it is fitting to say a few words about these developments as well.

At the starting point, we find Alex Grossmann again. Sometime in spring 1987, he was visiting us at the Université Catholique de Louvain, and we were discussing a possible Ph.D. topic for a young African student, Romain Murenzi. The idea came up, why not try to do in two dimensions what had been so successful in one, namely, wavelet analysis? The topic seemed tractable, involving moderate amounts of mathematics and some simple computing technology, and if it worked out, there could be very interesting practical applications. The problem was that nobody knew how to do it! The next summer, Murenzi went down to Marseille and started to work with Grossmann and Daubechies, who happened to be there too. And when he came back three months later, the solution was clear. The key is to start from the operations that one wants to apply to an image, namely, translations in the image plane, rotations for choosing a direction of sight, and global magnification (zooming in and out). The problem is to combine these three elements in such a way that the wavelet machine could start rolling. The result of Murenzi was that the so-called similitude group yields a solution, actually, the only one, since this group has, up to equivalence, only one unitary irreducible representation, in the space $L^2(\mathbb{R}^2, d^2x)$ — that is, the space of finite-energy images. And this representation *is* square integrable! Thus it remained only to put it all together and to turn the mathematical crank described in Grossmann, Morlet, and Paul's paper, and the two-dimensional continuous wavelet transform was born [ACMP93]. Since then, it has been applied in many domains of physics, engineering, and applied mathematics [AMVA04].

It is instructive to write down the explicit form of the two-dimensional continuous wavelet transform. Given an image $s \in L^2(\mathbb{R}^2, d^2\vec{x})$, its transform S with respect to the wavelet ψ is given by

$$S(\vec{b}, a, \theta) = a^{-1} \int_{\mathbb{R}^2} d^2\vec{x}\; \overline{\psi(a^{-1}\, r_{-\theta}(\vec{x} - \vec{b}))}\; s(\vec{x}) \tag{7}$$

$$= a \int_{\mathbb{R}^2} d^2\vec{k}\; e^{i\vec{b}\cdot\vec{k}}\; \overline{\widehat{\psi}(a r_{-\theta}(\vec{k}))}\; \widehat{s}(\vec{k}). \tag{8}$$

Here $\vec{b} \in \mathbb{R}^2$ is a translation, $a > 0$ a dilation, and r_θ the familiar 2×2 rotation matrix of angle $\theta \in [0, 2\pi)$. One should notice the exact parallelism between these formulas and the corresponding ones in one dimension, (3) and (4); the only difference is the occurrence of the rotation parameter θ. Similar considerations apply to the inversion formula, the covariance with respect to the group operations, and so on. We really see here in action the general formalism of the paper.

Actually, what is true in 2-D applies almost verbatim to 3-D, and in fact to any dimension. More general situations can also be treated in much the same way. The prime example is that of wavelets on the 2-sphere, although this case is a bit more complicated. Here the set of relevant transformations (local dilations and movements on the surface of the sphere) is not a group but only a subset (not a subgroup!) of the Lorentz group (the common conformal group of the sphere and of the (tangent) plane) [AV99], and one faces the situation where the representation is not square integrable on the group itself, but only modulo a subgroup. Time-dependent wavelets can also be treated along the same lines, which leads to applications in motion estimation (video sequences, TV). A survey of all these extensions may be found in the recent monographs [AAG00] and [AMVA04].

Altogether, the conclusion is that the formalism initiated in Grossmann, Morlet, and Paul has indeed been extraordinarily fruitful. Treating wavelets as coherent states associated to square-integrable group representations has led to a very fertile cross-breeding between mathematics, quantum mechanics, and signal processing and has opened the way to an incredible wealth of applications.

References

[AAG00] S. T. Ali, J.-P. Antoine, and J.-P. Gazeau, *Coherent States, Wavelets, and Their Generalizations*, Springer-Verlag, New York, 2000.

[ACMP93] J.-P. Antoine, P. Carrette, R. Murenzi, and B. Piette, *Image analysis with two-dimensional continuous wavelet transform*, Signal Process **31** (1993), 241–272.

[AV99] J.-P. Antoine and P. Vandergheynst, *Wavelets on the 2-sphere: A group-theoretical approach*, Appl. Comput. Harmon. Anal. **7** (1999), 262–291.

[AMVA04] J.-P. Antoine, R. Murenzi, P. Vandergheynst, and S. T. Ali, *Two-Dimensional Wavelets and Their Relatives*, Cambridge University Press, Cambridge, 2004.

[GMP86] A. Grossmann, J. Morlet, and T. Paul, *Integral transforms associated to square integrable representations. II. Examples*, Ann. Inst. H. Poincaré **45** (1986), 293–309.

[Hub98] B. B. Hubbard, *The World According to Wavelets*, 2nd ed., A. K. Peters, Wellesley, MA, 1998.

JOURNAL OF MATHEMATICAL PHYSICS VOLUME 10, NUMBER 12 DECEMBER 1969

Continuous Representation Theory Using the Affine Group*

ERIK W. ASLAKSEN
Bell Telephone Laboratories, Inc., Holmdel, New Jersey
AND
JOHN R. KLAUDER†
Bell Telephone Laboratories, Inc., Murray Hill, New Jersey

(Received 9 May 1969)

We present a continuous representation theory based on the affine group. This theory is applicable to a mechanical system which has one or more of its classical canonical coordinates restricted to a smaller range than $-\infty$ to ∞. Such systems are especially troublesome in the usual quantization approach since, as is well known from von Neumann's work, the relation $[P, Q] = -iI$ implies that P and Q must have a spectrum from $-\infty$ to ∞ if they are to be self-adjoint. Consequently, if the spectrum of either P or Q is restricted, at least one of the operators, say Q, is not self-adjoint and does not have a spectral resolution. Thus Q cannot generate a coordinate representation. This leads us to consider a different pair of operators, P and B, both of which are self-adjoint and which obey $[P, B] = -iP$. The Lie group corresponding to this latter algebra is the affine group, which has two unitarily inequivalent, irreducible representations, one in which the spectrum of P is positive. Using the affine group as our kinematical group, we have developed continuous representations analogous to those Klauder and McKenna developed for the canonical group, and have shown that the former representations have almost all the desirable properties of the latter.

1. INTRODUCTION

The general concepts and properties of continuous representation theory (CRT) have been developed by Klauder[1]; for convenience we briefly recall them here: Let $\mathcal{H}$ denote abstract Hilbert space and let $U[l]$ be a family of unitary operators on $\mathcal{H}$. If we now choose an arbitrary but fixed unit vector $\Phi_0 \in \mathcal{H}$, called the *fiducial vector*, then we can generate a subset of $\mathcal{H}$ by operating on Φ_0 with $U[l]$. Denote this subset by $\mathfrak{S}$; then

$$\mathfrak{S} = \{U[l]\Phi_0 : l \in \mathfrak{L}\},$$

where $\mathfrak{L}$ is some label space. With any vector $\Psi \in \mathcal{H}$ we can now associate the complex, bounded, continuous function

$$\psi(l) = (U[l]\Phi_0, \Psi),$$

and the set $\mathfrak{C} \equiv \{\psi(l) : \Psi \in \mathcal{H}\}$ is called a *continuous representation* of $\mathcal{H}$.

For the further development it is convenient to let the $U[l]$ be the elements of a kinematical group, and to interpret the labels l as the classical canonical coordinates p and q for a system with one degree of freedom, as we shall be considering here. Without going into any details at this point, we just mention that use of the classical canonical coordinates as the labels leads to a particularly simple physical interpretation of the theory.[2] For the common case, which we shall refer to as the *canonical case*, when the classical Cartesian coordinates p and q can take on any value on the real line, the CRT has been developed in detail by Klauder and McKenna.[3] In this development, the unitary operators of interest are the Weyl operators $U[p, q] = \exp[i(pQ - qP)]$, where Q and P are the familiar self-adjoint operators satisfying $[Q, P] = iI$.

In this paper we develop a CRT appropriate to a different group and suitable for different dynamical systems. Suppose, for example, that the range of the classical variable p is restricted to be positive, $p > 0$. Such restricted coordinates are not unknown; in particular, we were motivated to undertake the present investigation by the case of the gravitational field. There the metric has to satisfy certain positivity requirements,[4] which lead to restrictions on the range of the components $g_{\mu\nu}$. Such restrictions must be reflected in the quantum theory; in our example this requires that the spectrum of the operator P be positive, i.e., $P > 0$. According to a theorem of von Neumann,[5] such a restriction is not compatible with having Q and P both be self-adjoint, and thus the appropriate unitary operators cannot be the familiar Weyl operators of the usual canonical theory.

Elsewhere[6] we have argued that the affine group is

* This paper is based on a thesis submitted by E. W. Aslaksen to Lehigh University in partial fulfillment of the requirement for the Ph.D. degree.
† Part of this work was done while the author was at Syracuse University, with National Science Foundation support.
[1] J. R. Klauder, J. Math. Phys. **4**, 1055 (1963); **5**, 177 (1964).

[2] J. R. Klauder, Talk given at Seminar on Unified Theories of Elementary Particles, Munich, 1965 (unpublished); J. R. Klauder, J. Math. Phys. **8**, 2392 (1967).
[3] J. R. Klauder and J. McKenna, J. Math. Phys. **5**, 878 (1964); **6**, 68 (1965).
[4] C. Moller, *The Theory of Relativity* (Oxford University Press, London, 1962), p. 235.
[5] J. von Neumann, Math. Ann. **104**, 570 (1931).
[6] I. M. Gel'fand and M. A. Naimark, Dokl. Akad. Nauk SSSR, **55**, 570 (1947); E. W. Aslaksen and J. R. Klauder, J. Math. Phys. **9**, 206 (1968).

Downloaded 14 Oct 2004 to 128.112.156.90. Redistribution subject to AIP license or copyright, see http://jmp.aip.org/jmp/copyright.jsp

pertinent for this problem. This group is abstractly defined as the group of linear transformations without reflections on the real line: $x \to (P_0/p)x - q$. With this parameterization the unitary group elements may be given as

$$U[p, q] = e^{-iqP}e^{i\ln(p/p_0)B}, \qquad (1.1)$$

where $p_0 > 0$, and where P and B are self-adjoint generators which fulfill

$$[B, P] = iP. \qquad (1.2)$$

Although this group is formally "close" to the canonical group, as demonstrated by multiplying both sides of the canonical commutation relation

$$[Q, P] = iI$$

by P and making the identification

$$B = \tfrac{1}{2}(PQ + QP),$$

the actual unitary representations of the affine group are sufficiently different from those in the canonical case to necessitate a reexamination of the associated continuous representations. It is the purpose of this paper to carry out that reexamination for a finite number of degrees of freedom.

2. THE OVERCOMPLETE FAMILY OF STATES

The unitary representations of the affine group have been studied,[6] and it is known that there exist two and only two unitarily inequivalent, irreducible representations, one for which P is positive and one for which it is negative. In particular, if we take our representation space $\mathfrak{R}$ to be $L^2(R)$ and denote by R^+ and R^- the positive and negative half of the real line, respectively, then $\mathfrak{R}$ can be written as the direct sum of two subspaces which are invariant under $U[p, q]$:

$$\mathfrak{R} = \mathfrak{R}_+ \oplus \mathfrak{R}_-,$$

where

$$\mathfrak{R}_+ \equiv \{\phi(k): \phi \in L^2(R),\ \phi(k) = 0 \text{ if } k < 0\},$$

$$\mathfrak{R}_- \equiv \{\phi(k): \phi \in L^2(R),\ \phi(k) = 0 \text{ if } k > 0\}.$$

If $\phi(k) \in \mathfrak{R}$ and we choose the particular representation where P is just multiplication by k, then

$$U[p, q]\phi(k) = \left(\frac{p_0}{p}\right)^{\frac{1}{2}} e^{-iqk}\phi\left(\frac{p_0}{p}k\right). \qquad (2.1)$$

We can evidently treat both inequivalent representations in $L^2(R^+)$ by writing

$$U[p, q]\phi(k) = \left(\frac{p_0}{p}\right)^{\frac{1}{2}} e^{\mp iqk}\phi\left(\frac{p_0}{p}k\right), \qquad (2.2)$$

where $\phi(k) \in L^2(R^+)$, and the minus sign corresponds to the representation where P has positive spectrum, the plus sign to the representation where P has negative spectrum. Of course, since we have chosen $P > 0$, we shall always use the corresponding irreducible representation.

From the commutation relation (1.2) we can immediately deduce the following relations, which are frequently used in this paper:

$$U[p, q]U[p', q'] = U[pp'/p_0, q + (p_0/p)q'], \qquad (2.3)$$

$$U^\dagger[p, q] = U[p_0^2/p, -qp/p_0], \qquad (2.4)$$

$$PU[p, q] = \frac{p}{p_0}U[p, q]P, \qquad (2.5)$$

$$BU[p, q] = U[p, q]\left(B + \frac{pq}{p_0}P\right), \qquad (2.6)$$

$$QU[p, q] = U[p, q]\left(\frac{p_0}{p}Q + qI\right). \qquad (2.7)$$

In order for the set $\mathfrak{S}$ to be suitable for constructing a continuous representation, we shall demand that it have the following three properties:

(1) For each $\Phi \in \mathfrak{S}$ and every $\delta > 0$, there exists a vector $\Phi' \in \mathfrak{S}$, $\Phi' \neq \Phi$, such that $\|\Phi - \Phi'\| < \delta$.

(2) The mapping $l \to \Phi[l]$ is a many-one continuous map of a separated topological space $\mathfrak{L}$ onto $\mathfrak{S}$. By continuity in $\mathfrak{S}$, we mean the usual weak continuity in $\mathcal{H}$. Thus, if $l_n \to l$, then $(\Phi[l_n], \Psi) \to (\Phi[l], \Psi)$ for all $\Psi \in \mathcal{H}$.

(3) The span of $\mathfrak{S}$ is dense in $\mathcal{H}$.

Such a subset $\mathfrak{S}$ is called an overcomplete family of states (OFS). Following Klauder and McKenna,[3] we show that the subset $\mathfrak{S}$ generated by an irreducible representation of the $U[p, q]$ in Eq. (1.1) is indeed an OFS.

Lemma 2.1: The function $V[q]$, defined by $V[q] \equiv e^{-iqP}$, is a strongly continuous function of q, i.e., $q_\alpha \to q$ implies that $\|(V[q_\alpha] - V[q])\Psi\| \to 0$ for each $\Psi \in \mathcal{H}$. The same is true for $W[p] = \exp[i \ln(p/p_0)B]$.

Proof: Let $\epsilon > 0$ be given, and $\delta = |q_\alpha - q| > 0$. Then

$$\|(V[q_\alpha] - V[q])\Psi\| = \|e^{-iqP}(e^{\pm i\delta P} - I)\Psi\| \le \|(e^{\pm i\delta P} - I)\Psi\|.$$

By assumption $e^{\pm i\delta P}$ is weakly continuous, hence strongly continuous, so there exists a δ_0 such that $\delta < \delta_0$ implies that the last expression is less than ϵ.

Downloaded 14 Oct 2004 to 128.112.156.90. Redistribution subject to AIP license or copyright, see http://jmp.aip.org/jmp/copyright.jsp

Lemma 2.2: The mapping, $R^+ \times R \rightarrow \mathfrak{S}$, defined by $\Phi[p, q]$, is continuous when $R^+ \times R$ has the product topology and $\mathcal{H}$ the strong topology.

Proof: Let (p_0, q_0) and $\epsilon > 0$ be given. Then

$$\begin{aligned}\|\Phi[p_0, q_0] - \Phi[p, q]\| &= \|V[q_0]W[p_0]\Phi_0 - V[q]W[p]\Phi_0\| \\ &\leq \|V[q_0]W[p_0]\Phi_0 - V[p]W[p_0]\Phi_0\| \\ &\quad + \|V[q]W[p_0]\Phi_0 - V[q]W[p]\Phi_0\| \\ &= \|V[q_0]W[p_0]\Phi_0 - V[q]W[p_0]\Phi_0\| \\ &\quad + \|W[p_0]\Phi_0 - W[p]\Phi_0\| \equiv A + D.\end{aligned}$$

Because of Lemma 2.1, there exists a $\delta > 0$ such that $|p_0 - p| < \delta$, $|q_0 - q| < \delta$ implies that $A < \epsilon/2$, $D < \epsilon/2$.

Consider the functions $\psi(p, q) \equiv (\Phi[p, q], \Psi)$, $\Psi \in \mathcal{H}$. Since $\Phi[p, q]$ is strongly continuous, it is certainly weakly continuous; thus $\psi(p, q)$ is a continuous function. Schwartz's inequality, $|(\Psi, \Phi)| \leq \|\Psi\| \cdot \|\Phi\|$, gives $|\psi(p, q)| \leq \|\Psi\|$. Furthermore, $\psi(p, q)$ is square integrable. To show this, denote the unitary map from $\mathcal{H}$ to $\mathfrak{R}_\pm$ by $T_\pm$, and let $\phi_0(k) \in \mathfrak{R}$ and $\psi(k) \in \mathfrak{R}$ be the functions corresponding to Φ_0 and Ψ, respectively, under $T_\pm$. Then

$$T_\pm\Phi[p, q] = \left(\frac{p_0}{p}\right)^{\frac{1}{2}} e^{\mp iqk}\phi_0\left(\frac{p_0}{p}k\right)$$

and

$$\begin{aligned}\psi(p, q) &= (\Phi[p, q], \Psi) \\ &= (T_\pm\Phi[p, q], T_\pm\Psi) \\ &= \int e^{\pm iqk}\left(\frac{p_0}{p}\right)^{\frac{1}{2}} \phi_0^*\left(\frac{p_0}{p}k\right)\psi(k)\, dk \\ &= (2\pi)^{-\frac{1}{2}}\int e^{\pm iqk}\left[\left(\frac{2\pi p_0}{p}\right)^{\frac{1}{2}}\phi_0^*\left(\frac{p_0}{p}k\right)\psi(k)\right] dk \\ &= (2\pi)^{-\frac{1}{2}}\int e^{\pm iqk}h(p, k)\, dk,\end{aligned}$$

where

$$h(p, k) = \left(\frac{2\pi p_0}{p}\right)^{\frac{1}{2}}\phi_0^*\left(\frac{p_0}{p}k\right)\psi(k).$$

The functions $\phi_0(k)$ and $\psi(k)$ are measurable on R^+, the positive half of the real line. Since $p > 0$, p^{-1} is a measurable function; therefore $\phi_0(p_0/pk)$ is also measurable[7] as a function of p, so $h(p, k)$ is measurable on $R^+ \times R$. However,

$$\int |h(p, k)|^2 \frac{dp}{p} = 2\pi\, |\psi(k)|^2 \int \left|\phi_0\left(\frac{p_0}{p}k\right)\right|^2 \frac{dp}{p},$$

with $p_0/pk = a$, $dp = -p^2/p_0k\, da$, and

$$\int \left|\phi_0\left(\frac{p_0}{p}k\right)\right|^2 \frac{dp}{p} = \int |\phi_0(a)|^2 \frac{da}{a} \equiv \langle P^{-1}\rangle.$$

The last integral is not finite for all $\phi_0 \in \mathfrak{R}$, so let $\mathfrak{R}_0$ be the subset of $\mathfrak{R}$ for which the integral is finite, and let the value of the integral be M. Then

$$\int |h(p, k)|^2\, dp = 2\pi M\, |\psi(k)|^2$$

and

$$\int dk \int |h(p, k)|^2\, dp = 2\pi M\, \|\Psi\|^2.$$

By the theorem of Tonelli,[8] $|h(p, k)|^2$ is integrable over $R^+ \times R^+$, and then Fubini's theorem[8] shows that the integration can be performed in any order; in particular, $\int |h(p, k)|^2\, dk$ exists for all p except possibly for a set of measure zero.

Further, $h(p, k)$ is integrable in k since it is the product of two functions which are both square integrable in k. So, for almost all fixed p, $\psi(p, q)$ is the Fourier transform of a function which is both integrable and square integrable on R^+ and, as a function of q, square integrable for almost all p. By Parseval's theorem,

$$\int |\psi(p, q)|^2\, dq = \int |h(p, k)|^2\, dk,$$

and since we have shown that the right-hand side is an integrable function of p, Tonelli's theorem finally gives that $\psi(p, q) \in L^2(R^+ \times R)$.

If we set $\lambda(p, q) = (\Phi[p, q], \Lambda)$, then the same arguments give

$$\iint \psi^*(p, q)\lambda(p, q)\, dp\, dq = 2\pi M(\Psi, \Lambda).$$

In the foregoing we chose the measure with respect to which the integrals over $\mathfrak{S}$ exists as simply $dp\, dq$ times an arbitrary constant. Klauder[1] has shown that, so far as the result of the integration being proportional to the inner product goes, there is no loss of generality in taking the measure to be the left-invariant group measure. For the affine group the left-invariant group measure is (in the usual notation for Lie groups[9])

$$\mu_\sigma^\nu(a) = \left[\frac{\partial\phi^\nu(t, a)}{\partial t^\sigma}\right]_{t=e},$$

[7] P. R. Halmos, *Measure Theory* (D. Van Nostrand Co., Princeton, N.J., 1950), p. 81.

[8] E. J. McShane, *Integration* (Princeton University Press, Princeton, N.J., 1944), pp. 137, 145.

[9] P. M. Cohn, *Lie Groups* (Cambridge University Press, London, 1965).

Downloaded 14 Oct 2004 to 128.112.156.90. Redistribution subject to AIP license or copyright, see http://jmp.aip.org/jmp/copyright.jsp

with $a = (p', q')$, $t = (p, q)$, and $e = (p_0, 0)$. Then

$$\mu_l = \begin{bmatrix} \frac{p}{p_0} & 0 \\ 0 & \frac{p_0}{p} \end{bmatrix},$$

and $|\mu_e| = \Delta_l = 1$. Thus, the measure is just a constant times $dp\,dq$ and, in view of the previous calculations, we choose

$$d\mu(p, q) \equiv (2\pi M)^{-1}\, dp\, dq.$$

We mention that the right-invariant group measure is not equal to the left-invariant group measure; we find

$$\mu_r = \begin{vmatrix} p & 0 \\ q & 1 \end{vmatrix}$$

and $\Delta_r = p$; thus the right-invariant measure is proportional to $p^{-1}\, dp\, dq$.

Having obtained these results, we can simply refer to the arguments of Klauder and McKenna[3]—in particular, those arguments leading to their Theorem 3.2 and Lemmas 3.3 and 3.4, from which our first theorem immediately follows:

Theorem 2.1:
(a) Let $\Psi \in \mathcal{H}$; then

$$\Psi = \iint (\Phi[p, q], \Psi)\Phi[p, q]\, d\mu(p, q).$$

(b) The span of $\mathfrak{S}$ is dense in $\mathcal{H}$.
(c) The identity operator may be written as

$$I = \iint \Phi[p, q]\Phi^\dagger[p, q]\, d\mu(p, q).$$

Consequently, $\mathfrak{S}$ is an overcomplete family of states.

3. THE CONTINUOUS REPRESENTATION

Let the map $C: \mathcal{H} \to \mathfrak{C}$ be defined by $C\Psi = \psi(p, q) \equiv (\Phi[p, q], \Psi)$ for each $\Psi \in \mathcal{H}$. Thus $\mathfrak{C} \equiv \{\psi(p, q)\}$. Since $\Phi[p, q] = U[p, q]\Phi_0$, it is clear that $\mathfrak{C}$ will depend on Φ_0; this dependence will be discussed later. Also, depending upon which of the two irreducible representations of the affine group we use, we get two spaces, $\mathfrak{C}_+$ and $\mathfrak{C}_-$. To see what these spaces consist of, consider

$$\psi(p, q) = \int \left(\frac{p_0}{p}\right)^{\frac{1}{2}} e^{iqk} \phi_0^*\left(\frac{p_0}{p}\, k\right) \psi(k)\, dk.$$

Let $q = q' + iq''$; then the space $\mathfrak{C}_+$ arising from $\mathfrak{R}_+$ will consist of those functions which are the limit as $q'' \to 0$ of functions analytic in the lower half of the complex q plane, whereas $\mathfrak{C}_-$ will consist of those functions which are the limit as $q'' \to 0$ of functions analytic in the upper half plane. The two spaces have, except for the function which is identically zero, no elements in common, i.e., $\mathfrak{C}_+ \cap \mathfrak{C}_- = \{0\}$. By $\mathfrak{C}$ we mean either $\mathfrak{C}_+$ or $\mathfrak{C}_-$.

We define the inner product in $\mathfrak{C}$ by

$$(\psi', \psi) = \iint \psi'^*(p, q)\psi(p, q)\, d\mu(p, q). \quad (3.1)$$

Using the results of the previous section, we can show that the following theorem is true:

Theorem 3.1: The set $\mathfrak{C}$, given by $\psi(p, q) = (\Phi[p, q], \Psi)$ for all $\Psi \in \mathcal{H}$, is a family of bounded, continuous, and square-integrable functions. When supplied with the inner product displayed in Eq. (3.1), the set $\mathfrak{C}$ is a Hilbert space which is unitarily equivalent to the original space $\mathcal{H}$ under the unitary mapping

$$C\Psi = (\Phi[p, q], \Psi) = \psi(p, q),$$

$$C^{-1}\psi(p, q) = \iint \psi(p, q)\Phi[p, q]\, d\mu(p, q) = \Psi.$$

The set $\mathfrak{C}$ is called a continuous representation of $\mathcal{H}$.

We now investigate the existence of the derivatives of $\psi(p, q)$. Let $\mathfrak{D}_P$ and $\mathfrak{D}_B$ be the domains of P and B on $\mathcal{H}$. We want to see if $U[p, q]\mathfrak{D}_P \subset \mathfrak{D}_P$ and $U[p, q]\mathfrak{D}_B \subset \mathfrak{D}_B$. Let $\Phi \in \mathfrak{D}_P$. Using Stone's theorem,[8] we write P as

$$\lim_{q' \to 0} \frac{i}{q'} (V[q'] - I).$$

Then

$$\frac{i}{q'} (V[q'] - I)U[p, q]\Phi$$

$$= \frac{i}{q'}\Big(\exp\left(-iq'\left(1 - \frac{p_0}{p}\right)P\right) \times U[p, q]V[q'] - U[p, q]\Big)\Phi$$

$$= \frac{i}{q'} \exp\left(-iq'\left(1 - \frac{p_0}{p}\right)\right) U[p, q](V[q'] - I)\Phi + \frac{i}{q'}\, U[p, q]\left(\exp\left(-iq'\left(\frac{p}{p_0} - 1\right)P\right) - I\right)\Phi,$$

and in the limit $q' \to 0$, $PU[p, q]\Phi = U[p, q](p/p_0)P\Phi$; thus $U[p, q]\mathfrak{D}_P \subset \mathfrak{D}_P$.

Downloaded 14 Oct 2004 to 128.112.156.90. Redistribution subject to AIP license or copyright, see http://jmp.aip.org/jmp/copyright.jsp

Let $\Phi \in \mathfrak{D}_B$. Then

$$\frac{i}{\ln(p'/p_0)}(W[p'] - I)U[p,q]\Phi$$
$$= \frac{i}{\ln(p'/p_0)}\left(\exp\left(iq\left(1 - \frac{p_0}{p}\right)P\right) \times U[p,q]W[p'] - U[p,q]\right)\Phi$$
$$= \frac{i}{\ln(p'/p_0)}\exp\left(iq\left(1 - \frac{p_0}{p}\right)P\right)U[p,q](W[p'] - I)\Phi + \frac{i}{\ln(p'/p_0)}U[p,q]\left(\exp\left(iqp\left(1 - \frac{p_0}{p}\right)P\right) - I\right)\Phi,$$

and in the limit $p'/p_0 \to 1$,

$$BU[p,q]\Phi = U[p,q](B + pqp_0^{-1}P)\Phi.$$

So $U[p,q]\Phi \in \mathfrak{D}_B$ if and only if $\Phi \in \mathfrak{D}_B \cap \mathfrak{D}_P$, or $U[p,q]\mathfrak{D}_B \cap \mathfrak{D}_P \subset \mathfrak{D}_B \cap \mathfrak{D}_P$. We define $\mathcal{R}_B$ and $\mathcal{R}_P$ by $T:\mathfrak{D}_B \to \mathcal{R}_B$, $T:\mathfrak{D}_P \to \mathcal{R}_P$, where $T:\mathcal{H} \to \mathcal{R}$.

Assume $\phi_0 \in \mathcal{R}_B \cap \mathcal{R}_P \cap \mathcal{R}_0$, and let $\ln(p/p_0) \equiv \alpha$; then

$$\frac{\partial\psi}{\partial\alpha} = \lim_{\Delta\alpha\to 0}\frac{\psi(\alpha + \Delta\alpha, q) - \psi(\alpha, q)}{\Delta\alpha} = \lim_{\Delta\alpha\to 0}\left(U[\alpha,q]\left\{\frac{W[\Delta\alpha] - I}{\Delta\alpha}\right\}\Phi_0, \Psi\right).$$

From Stone's theorem

$$\lim_{\Delta\alpha\to 0}\frac{W[\Delta\alpha] - I}{\Delta\alpha} = iB;$$

thus

$$\frac{\partial\psi}{\partial\alpha} = -i(U[\alpha,q]B\Phi_0, \Psi).$$

Further, $d\alpha/dp = p^{-1}$, so $\partial\psi/\partial\alpha = p\partial\psi/\partial p$, and

$$\frac{\partial\psi}{\partial p} = -ip^{-1}(U[p,q]B\Phi_0, \Psi),$$

$$\left|\frac{\partial\psi}{\partial p}\right| \le p^{-1}\|B\Phi_0\| \cdot \|\Psi\|.$$

Similarly,

$$\frac{\partial\psi}{\partial q} = \lim_{\Delta q\to 0}\left(\left\{\frac{V[\Delta q] - I}{\Delta q}\right\}U[p,q]\Phi_0, \Psi\right) = i(PU[p,q]\Phi_0, \Psi) = ipp_0^{-1}(U[p,q]P\Phi_0, \Psi).$$

Thus,

$$\left|\frac{\partial\psi}{\partial q}\right| \le pp_0^{-1}\|P\Phi_0\| \cdot \|\Psi\|.$$

The continuity of the derivatives follows directly from the strong continuity of the family of operators $U[p,q]$.

4. DIAGONAL EXPECTATION VALUES

As we have seen, p and q cannot be the eigenvalues of P and Q. But we can introduce a connection between the operator formalism and c numbers by requiring that p and q be the expectation values of P and Q, respectively, and also that the expectation value of B be pq, with respect to the states $\Phi[p,q]$. We have

$$(\Phi[p,q], P\Phi[p,q]) = (U[p,q]\Phi_0, PU[p,q]\Phi_0) = (\Phi_0, U^\dagger[p,q]PU[p,q]\Phi_0) = pp_0^{-1}(\Phi_0, P\Phi_0) \equiv p.$$

Thus, the fiducial vector Φ_0 must correspond to a state of the system in which the expectation value of P is p_0. In the representation space $\mathcal{R}$, this means

$$\int \phi_0^*(k)k\phi_0(k)\,dk = p_0. \tag{4.1}$$

In addition, we always have the normalization

$$\int |\phi_0(k)|^2\,dk = 1.$$

This shows the reason for introducing the constant p_0; its value is determined by Eq. (4.1).

The expectation value of B is

$$(\Phi[p,q], B\Phi[p,q]) = \left(\Phi_0, U^\dagger[p,q]U[p,q]\left(B + \frac{pq}{p_0}P\right)\Phi_0\right) = \left(\Phi_0, \left(B + \frac{qp}{p_0}P\right)\Phi_0\right) = (\Phi_0, B\Phi_0) + qp.$$

We can secure that the expectation value of B equals pq by demanding that $(\Phi_0, B\Phi_0)$ vanish, for which it is sufficient that Φ_0 be real.

The expectation value of Q is determined as follows:

$$QU = Q\exp[-iqP]\exp\left[i\ln\left(\frac{p}{p_0}\right)B\right] = \exp[-iqP](Q + qI)\exp\left[i\ln\left(\frac{p}{p_0}\right)B\right] = \exp[-iqP]\exp\left[i\ln\left(\frac{p}{p_0}\right)B\right]\left(\frac{p_0}{p}Q + qI\right) = U\left(\frac{p_0}{p}Q + qI\right),$$

so

$$U^\dagger QU = U^\dagger U\left(\frac{p_0}{p}Q + qI\right) = \frac{p_0}{p}Q + qI.$$

Then

$$(\Phi[p,q], Q\Phi[p,q]) = \frac{p_0}{p}(\Phi_0, Q\Phi_0) + q.$$

Downloaded 14 Oct 2004 to 128.112.156.90. Redistribution subject to AIP license or copyright, see http://jmp.aip.org/jmp/copyright.jsp

We restrict Φ_0 to satisfy

$$(\Phi_0, Q\Phi_0) = 0; \tag{4.2}$$

this is always satisfied if $\phi_0(k)$ is real; thus $(\Phi[p,q], Q\Phi[p,q]) = q$.

In summary, we list the requirements on the fiducial vector Φ_0:

(a) The unit vector Φ_0 lies in the domain of P, P^{-1}, Q, and B.

(b) The expectation values of these operators are respectively chosen as

$$(\Phi_0, P\Phi_0) = \int |\phi_0(k)|^2\, k\, dk \equiv p_0,$$

$$(\Phi_0, P^{-1}\Phi_0) = \int |\phi_0(k)|^2\, k^{-1}\, dk \equiv M,$$

$$(\Phi_0, Q\Phi_0) = (\Phi_0, B\Phi_0) = 0.$$

These restrictions are not severe and there is a large class of allowed vectors Φ_0. For example, it suffices that $\phi_0(k)$ be real, several times differentiable, and vanish sufficiently fast at 0 and ∞.

5. A WEYL-LIKE REPRESENTATION OF GENERAL OPERATORS

Throughout CRT we make use of the operators $V[q]$ and $W[p]$. Analogous operators were introduced in the canonical case by Weyl[10] as a means of going from classical to quantum mechanics; in particular, he asserted that the quantum operator corresponding to the classical quantity

$$f(p, q) = \iint e^{i(\sigma p + \tau q)} \xi(\sigma, \tau)\, d\sigma\, d\tau$$

should be

$$F(P, Q) = \iint e^{i(\sigma P + \tau Q)} \xi(\sigma, \tau)\, d\sigma\, d\tau.$$

We now investigate the use of the analog of these operators in the affine case, and spell out more clearly under what conditions Weyl's representation holds. We are always assuming an irreducible representation of $U[p,q]$ and that the spectrum of P is positive, such that P^{-1} exists and $P^{\frac{1}{2}}$ is uniquely defined.

First of all, consider the following calculation. We know from the development in Sec. 2, particularly Theorem 2.1, that

$$(\Phi_0, P^{-1}\Phi_0)(\Lambda, \Psi) = \iint (\Lambda, \Phi[p,q])(\Phi[p,q], \Psi) \frac{dp\, dq}{2\pi}.$$

Then

$$(X, P^{-1}\Phi)(\Lambda, \Psi) = \iint (\Lambda, U[p,q]\Phi)(U[p,q]X, \Psi) \frac{dp\, dq}{2\pi}, \tag{5.1}$$

so that

$$|\Psi\rangle\langle X|\, P^{-1} = \iint (X, U^\dagger[p,q]\Psi) U[p,q] \frac{dp\, dq}{2\pi}$$

and

$$|\Psi\rangle\langle X| = \iint (X, PU^\dagger[p,q]\Psi) U[p,q] \frac{dp\, dq}{2\pi}.$$

The latter equation can be generalized to

$$A = \iint \mathrm{Tr}\,\{APU^\dagger[p,q]\} U[p,q] \frac{dp\, dq}{2\pi}, \tag{5.2}$$

and we define the *kernel* of A to be

$$\tilde{a}(p, q) \equiv \mathrm{Tr}\,\{APU^\dagger[p,q]\}.$$

In the canonical case the kernel does not contain the operator P, and it is straightforward to show that the appearance of this P is connected with the fact that the right- and left-invariant measures are not the same for the affine group. In Eq. (5.1), the left-hand side remains invariant if we operate on both Λ and Ψ with $U[r,s]$, the right-hand side must be invariant under left multiplication, i.e., we must use the left-invariant measure. However, we can rewrite the equation by letting $\Phi = P\Theta$; using the relation

$$U[p,q]P = p_0 p^{-1} P U[p,q],$$

we obtain

$$(X, \Theta)(\Lambda, \Psi) = \iint (\Lambda, PU[p,q]\Theta)(U[p,q]X, \Psi) \frac{p_0}{p} \frac{dp\, dq}{2\pi},$$

and if we now operate on X and Θ with $U[r,s]$, the left-hand side remains invariant. But this corresponds to multiplication on the right under the integral, so that we must now have the right-invariant measure, which indeed we do. If we commute P through $U^\dagger[p,q]$, the measure becomes left invariant again, and we are back to the first case. If the P were not present, the right- and left-invariant measures would have to be identical.

Leaving aside the question of the relation between the type of operator and type of kernel for the moment, we first show that the closure of the set $U[p,q]$ (in the weak topology) equals the set of all linear operators on $\mathcal{H}$. Since the representation of the Weyl-like operators is irreducible, Schur's lemma says that if $[B, U[p,q]] = 0$ for all p, q, then $B \sim I$. Let a prime on an operator algebra denote the set of all bounded operators which commute with all the

[10] H. Weyl, *The Theory of Groups and Quantum Mechanics* (Dover Publications, Inc., New York, 1931).

Downloaded 14 Oct 2004 to 128.112.156.90. Redistribution subject to AIP license or copyright, see http://jmp.aip.org/jmp/copyright.jsp

operators in the algebra. Then we have

$$\{U[p, q]\}' = \{I\}.$$

But further we clearly have

$$\{U[p, q]\}'' = \{I\}' = \mathcal{B}(\mathcal{H}),$$

where $\mathcal{B}(\mathcal{H})$ is the set of all bounded operators on Hilbert space.

We need the following theorem, due to von Neumann:

Theorem 5.1: Let $\mathcal{A}$ be a $*$-algebra of operators on $\mathcal{H}$ such that $I \in \mathcal{A}$. Then $\mathcal{A}''$ is the weak closure of $\mathcal{A}$. A proof of this theorem can be found in Ref. 11, p. 44.

Applied to our case, this theorem says that any bounded operator B is given by the limit of sequence like

$$B_N = \sum_{n=1}^{N} c_n U[p_n, q_n].$$

But such a limit is given by the expression in Eq. (5.2) if we interpret the kernel as a "distribution." In particular, a matrix element has the form

$$\langle\Lambda| B |\Phi\rangle = \iint \tilde{b}(p, q) \langle\Lambda| U[p, q] |\Phi\rangle \frac{dp\, dq}{2\pi}.$$

Here $\langle\Lambda| U[p, q] |\Phi\rangle$ is the "test function" on which the "distribution" $\tilde{b}(p, q)$ operates. Since the set of all bounded linear operators is dense (in the weak topology) in the set of all linear operators, we have arrived at the desired result.

If we now try to find more general relationships between an operator and its kernel, e.g., $\tilde{a} \in L^2$ if A is bounded, we immediately run into difficulties due to the appearance of the operator P in the definition of the kernel, and it has not been possible to find any such general relationships for the expansion displayed in Eq. (5.2). But if we study the effect of P in the expansion, it becomes evident that we can define a slightly different expansion which has some nice properties. Going back to Eq. (5.1), we let the P^{-1} on the left-hand side operate on X instead of Φ,

$$(P^{-1}X, \Phi)(\Lambda, \Psi) = \iint (\Lambda, U[p, q]P\Phi)(U[p, q]P^{-1}X, \Psi) \frac{dp\, dq}{2\pi},$$

so that Eq. (5.2) now reads

$$A = \iint \mathrm{Tr}\, \{AU^\dagger[p, q]\} U[p, q]P \frac{dp\, dq}{2\pi}.$$

By writing $(X, P^{-1}\Phi) = (P^{-\frac{1}{2}}X, P^{-\frac{1}{2}}\Phi)$, we find that

$$A = \iint \mathrm{Tr}\, \{AP^{\frac{1}{2}}U^\dagger[p, q]\} U[p, q]P^{\frac{1}{2}} \frac{dp\, dq}{2\pi}. \quad (5.3)$$

It then follows immediately that

$$\mathrm{Tr}\,(B^\dagger A) = \iint \mathrm{Tr}\, \{BP^{\frac{1}{2}}U^\dagger[p, q]\}^* \times \mathrm{Tr}\, \{AP^{\frac{1}{2}}U^\dagger[p, q]\} \frac{dp\, dq}{2\pi}$$
$$\equiv \iint b^*(p, q) a(p, q) \frac{dp\, dq}{2\pi}. \quad (5.4)$$

Since Hilbert–Schmidt operators are those for which $\mathrm{Tr}\,(A^\dagger A) < \infty$, we evidently have the following result:

Theorem 5.2: Every Hilbert–Schmidt operator A can be written in the form

$$A = \iint a(p, q) U[p, q] P^{\frac{1}{2}} \frac{dp\, dq}{2\pi},$$

with

$$a(p, q) \equiv \mathrm{Tr}\, \{AP^{\frac{1}{2}}U^\dagger[p, q]\} \in L^2(R^+ \times R).$$

To proceed further, we introduce a space of test functions similar to the space $\mathcal{S}$ consisting of infinitely differentiable functions of fast decrease, introduced by Schwartz.[12] We define $\mathcal{E}$ to be the linear space of all real-valued functions of two variables, $a(p, q)$, which are infinitely differentiable in p and q, fall off faster than any power of q for $q \to \pm\infty$, and faster than any power of p for $p \to 0$ and $p \to \infty$. It would be straightforward to introduce a topology which would make $\mathcal{E}$ a locally convex Frechet space, but since we shall not go into questions of continuity or any others which require a knowledge of the topology, we content ourselves with letting $\mathcal{E}$ be a linear space.

We now want to show that $a(p, q) \in \mathcal{E}$ implies that A is trace class, and to this end we represent the operator A by a sequence, in the following manner:

$$A = \int a(p, q) \sum_{m,n} |m\rangle\langle m| U[p, q]P^{\frac{1}{2}} |n\rangle\langle n| \frac{dp\, dq}{2\pi}$$
$$= \sum_{m,n} A_{mn} |m\rangle\langle n|, \quad (5.5)$$

with

$$A_{mn} = \int a(p, q) \langle m| U[p, q]P^{\frac{1}{2}} |n\rangle \frac{dp\, dq}{2\pi}, \quad (5.6)$$

and where $\{|n\rangle, n = 0, 1, \cdots\}$ is a complete orthonormal sequence of vectors.

[11] J. Dixmier, *Les algebres d'operateurs dans l'espace hilbertien* (Gauthier-Villars, Paris, 1957).

[12] L. Schwartz, *Théorie des distributions* (Hermann & Cie., Paris, 1957), Vol. 2.

Downloaded 14 Oct 2004 to 128.112.156.90. Redistribution subject to AIP license or copyright, see http://jmp.aip.org/jmp/copyright.jsp

For an arbitrary vector $|\psi\rangle \in \mathcal{H}$, we certainly have

$$\langle\psi| A^\dagger A |\psi\rangle = \sum_{n,q} \langle\psi \mid n\rangle\langle q \mid \psi\rangle \sum_m A^*_{qm} A_{nm}$$
$$\leq \tfrac{1}{2} \sum_{n,q} [\langle\psi \mid n\rangle\langle n \mid \psi\rangle + \langle\psi \mid q\rangle\langle q \mid \psi\rangle] \sum_m A^*_{qm} A_{nm}.$$

Since $A^\dagger A$ is symmetric and positive, we have

$$\langle\psi| A^\dagger A |\psi\rangle \leq \tfrac{1}{2} \sum_{n,q} [\langle\psi \mid n\rangle\langle n \mid \psi\rangle + \langle\psi \mid q\rangle\langle q \mid \psi\rangle] \times \sum_m |A^*_{qm} A_{nm}|$$
$$= \langle\psi| \sum_n E_n |n\rangle\langle n \mid \psi\rangle$$
$$\equiv \langle\psi| E |\psi\rangle,$$

where

$$E_n \equiv \sum_{q,m} |A^*_{qm} A_{nm}|.$$

The operator A is trace class if and only if $\mathrm{Tr}\,(A^\dagger A)^{\frac{1}{2}}$ is finite, but we just showed that $(A^\dagger A)^{\frac{1}{2}} \leq E^{\frac{1}{2}}$, so A is trace class if

$$\sum_n \Big\{\sum_{q,m} |A^*_{qm} A_{nm}|\Big\}^{\frac{1}{2}} < \infty. \tag{5.7}$$

It is not difficult to show that the inequality in Eq. (5.7) will be satisfied if there exist a constant C, two positive integers n_0 and m_0, and an $\epsilon > 0$ such that, for $m > m_0$ and $n > n_0$,

$$|A_{mn}| \leq C[m + n]^{-4-\epsilon}. \tag{5.8}$$

The sequence $\{A_{mn}\}$ is related to the kernel $a(p, q)$ by Eq. (5.6), and let us in particular assume that the sequence $\{|n\rangle\}$ is generated by an operator H,

$$H |n\rangle = n |n\rangle.$$

From Eq. (5.6) we then obtain

$$n^\beta m^\alpha A_{mn} = \int a(p, q) \langle m| H^\alpha U[p, q] P^{\frac{1}{2}} H^\beta |n\rangle \frac{dp\, dq}{2\pi}.$$

But the operator H to the left of $U[p, q]$ can be written in terms of a differential operator which operates on $U[p, q]$, say

$$R\left[p, q, \frac{\partial}{\partial p}, \frac{\partial}{\partial q}, p^{-1}\right],$$

i.e.,

$$RU[p, q] = HU[p, q].$$

Similarly, the operator H to the right of $P^{\frac{1}{2}}$ can also be written in terms of a differential operator on $U[p, q]$, say

$$S\left[p, q, \frac{\partial}{\partial p}, \frac{\partial}{\partial q}, p^{-1}\right],$$

i.e.,

$$SU[p, q]P^{\frac{1}{2}} = U[p, q]P^{\frac{1}{2}}H. \tag{5.9}$$

We thus find

$$n^\beta m^\alpha A_{mn} = \int a(p, q) \langle m| S^\beta R^\alpha U[p, q] P^{\frac{1}{2}} |n\rangle \frac{dp\, dq}{2\pi}. \tag{5.10}$$

Let us now further assume that H is a polynomial in P and B. It follows that R is a polynomial in p, q, $\partial/\partial p$, $\partial/\partial q$, and p^{-1}. Also, commuting H through $U[p, q]P^{\frac{1}{2}}$ in Eq. (5.9) will yield a polynomial in P, Q, p, q, and p^{-1} [see Eqs. (2.5)–(2.7)], so that S is also a polynomial in $p, q, \partial/\partial p, \partial/\partial q$, and p^{-1}. The operator $S^\beta R^\alpha$ is still a polynomial in p, q, $\partial/\partial p$, $\partial/\partial q$, and p^{-1}, and we can then, by repeated integrations by parts on each individual term in the polynomial, bring the operator over to operate on $a(p, q)$ in Eq. (5.10). The resulting operator, say $E_{\alpha\beta}$, must also be a polynomial in p, q, $\partial/\partial p$, $\partial/\partial q$, and p^{-1}, and we finally obtain

$$n^\beta m^\alpha A_{mn} \leq \left\{\int |E_{\alpha\beta} a(p, q)|^2 \frac{dp\, dq}{2\pi}\right\}^{\frac{1}{2}} \tag{5.11}$$

by use of Schwarz's inequality.

Since $a(p, q) \in \mathcal{E}$, the right-hand side of Eq. (5.11) is finite, and so by choosing $\alpha > 4$, $\beta > 4$, we have shown that $a(p, q) \in \mathcal{E}$ implies that A is trace class.

Let A be a trace class operator such that its kernel $a(p, q)$ is an element of $\mathcal{E}$, and let B be an arbitrary bounded operator with kernel $b(p, q)$. Then

$$\mathrm{Tr}\,(AB) = \int a(p, q) b(p, q) \frac{dp\, dq}{2\pi} < \infty,$$

since AB is also a trace class operator. This relation defines a valid linear functional on the space $\mathcal{E}$ and leads to our next theorem.

Theorem 5.3: Every bounded operator B can be written in the form

$$B = \iint b(p, q) U[p, q] P^{\frac{1}{2}} \frac{dp\, dq}{2\pi},$$

with

$$b(p, q) = \mathrm{Tr}\,\{BP^{\frac{1}{2}} U^\dagger[p, q]\} \in \mathcal{E}'.$$

The space $\mathcal{E}'$, which is the dual of $\mathcal{E}$, is the analog of the tempered distributions in this case where $p > 0$. In the canonical case it is known that the Weyl kernel of an arbitrary bounded operator is a tempered distribution.[13]

We conclude this section by remarking that in the expansion

$$A = \iint a(p, q) U[p, q] P^n \frac{dp\, dq}{2\pi}, \tag{5.12}$$

[13] G. Loupias, Compt. Rend., Ser. A **262**, 799 (1966); S. Miracle-Solé, *ibid.*, 1478 (1966).

Downloaded 14 Oct 2004 to 128.112.156.90. Redistribution subject to AIP license or copyright, see http://jmp.aip.org/jmp/copyright.jsp

with

$$a(p, q) = \mathrm{Tr}\,\{AP^{1-n}U^\dagger[p, q]\},$$

any value of n is admissible so far as forming a legitimate expansion is concerned. It is only necessary to treat the kernel $a(p, q)$ as a "distribution," which can always be found by the following limiting operation:

$$a(p, q) = \lim_{M} \sum_{n=1}^{M} \langle m|\, AP^{1-n}U^\dagger[p, q]\,|m\rangle,$$

where $\{|m\rangle, m = 1, 2, \cdots\}$ is a complete sequence of orthonormal vectors. However, the properties of the expansion, i.e., the particular relationships between classes of operators and their corresponding classes of kernels, do depend on the value of n in Eq. (5.12), and we have seen that the choice $n = 0$, which is the closest analog to an expansion in Weyl operators, is not the best choice from this viewpoint.

6. UNIQUENESS OF THE DIAGONAL MATRIX ELEMENTS

To what extent do the diagonal matrix elements $(\Phi[p, q]\, A\Phi[p, q])$ of an operator A determine the complete matrix $(\Phi[p, q], A\Phi[p', q'])$? Assume that two operators A_1 and A_2 lead to the same diagonal elements, and let $D = A_1 - A_2$. Then

$$(\Phi[p, q], D\Phi[p, q]) = 0 \tag{6.1}$$

for all p and q. If, in the fashion of (5.2), we write D as

$$D = \iint d(p', q')U[p', q'] \frac{dp'\, dq'}{2\pi},$$

then Eq. (6.1) becomes

$$\iint (\Phi_0, U^\dagger[p, q]U[p', q']U[p, q]\Phi_0) \times d(p', q') \frac{dp'\, dq'}{2\pi} = 0.$$

Now,

$$U^\dagger[p, q]U[p', q']U[p, q] = U\left[p', \frac{p}{p_0}(q' - q) + \frac{p}{p'}q\right],$$

so Eq. (6.1) finally can be written

$$\iint \mathcal{K}\left(p_0, 0; p', \frac{p}{p_0}(q' - q) + \frac{p}{p'}q\right) \times d(p', q') \frac{dp'\, dq'}{2\pi} = 0, \tag{6.2}$$

where

$$\mathcal{K}(p, q; p', q') \equiv (\Phi[p, q], \Phi[p', q']). \tag{6.3}$$

The function $\mathcal{K}(p, q; p', q')$ is, of course, dependent on our choice of the fiducial vector Φ_0. For each choice of Φ_0 there will be a class of operators which are uniquely determined by their diagonal matrix elements; for some choices of Φ_0 it may even be that all operators are uniquely defined by their diagonal matrix elements. A complete analysis of this uniqueness problem has not been carried out, but we demonstrate that operators of the general form

$$A = \sum_{m=-M}^{M} \sum_{n=0}^{N} a_{mn} P^m B^n, \tag{6.4}$$

i.e., polynomials in P, P^{-1}, and B, are uniquely determined by their diagonal matrix elements. Let

$$\begin{aligned} J(p, q) &\equiv \sum_{m,n} a_{mn}(\Phi[p, q], P^m B^n \Phi[p, q]) \\ &= \sum_{m,n} a_{mn}\left(\Phi_0, \left(\frac{p}{p_0}P\right)^m \left(B + \frac{pq}{p_0}P\right)^n \Phi_0\right); \end{aligned}$$

then, if $b \equiv pq/p_0$, we have

$$J\left(p, \frac{p_0 b}{p}\right) = \sum_{m,n} a_{mn}\left(\frac{p}{p_0}\right)^m (\Phi_0, P^m(B + bP)^n\Phi_0). \tag{6.5}$$

Fix n at its maximum value, $n = N$, and choose b large enough so that we only have to consider the leading term $(bP)^N$ in $(B + bP)^N$. Then, if the left-hand side of Eq. (6.5) is identically zero, we have

$$\sum_{m=-M}^{M} a_{mN}\left(\frac{p}{p_0}\right)^m b^N(\Phi_0, P^{m+N}\Phi_0) = 0.$$

But P^{m+N} is a positive operator, so this implies that $a_{mN} = 0$ for all m. Then we set $n = N - 1$ and go through the same argument, and so by induction we see that $a_{mn} = 0$ for all m and n. Since the operators of the form displayed in Eq. (6.4) form a linear space and the mapping $A \to (\Phi[p, q], A\Phi[p, q])$ is linear, we have proved the following theorem:

Theorem 6.1: Any polynomial in P, P^{-1}, and B of the form

$$A = \sum_{m=-M}^{M} \sum_{n=0}^{N} a_{mn} P^m B^n$$

is uniquely determined by its diagonal matrix elements in any affine phase-space continuous representation.

Although we have treated only one degree of freedom in this paper, the extension to finitely many degrees of freedom is relatively straightforward and consists mainly of introducing a sufficient number of indices and a convenient notation for handling these. We shall omit going into this matter here, but in a forthcoming paper where we extend the present results to field theories, the necessary notation will be displayed.

Downloaded 14 Oct 2004 to 128.112.156.90. Redistribution subject to AIP license or copyright, see http://jmp.aip.org/jmp/copyright.jsp

SIAM J. MATH. ANAL.
Vol. 15, No. 4, July 1984

© 1984 Society for Industrial and Applied Mathematics
009

DECOMPOSITION OF HARDY FUNCTIONS INTO SQUARE INTEGRABLE WAVELETS OF CONSTANT SHAPE*

A. GROSSMANN† AND J. MORLET‡

Abstract. An arbitrary square integrable real-valued function (or, equivalently, the associated Hardy function) can be conveniently analyzed into a suitable family of square integrable wavelets of constant shape, (i.e. obtained by shifts and dilations from any one of them.) The resulting integral transform is isometric and self-reciprocal if the wavelets satisfy an "admissibility condition" given here. Explicit expressions are obtained in the case of a particular analyzing family that plays a role analogous to that of coherent states (Gabor wavelets) in the usual L_2-theory. They are written in terms of a modified Γ-function that is introduced and studied. From the point of view of group theory, this paper is concerned with square integrable coefficients of an irreducible representation of the nonunimodular $ax+b$-group.

1. Introduction.

1.1. It is well known that an arbitrary complex-valued square integrable function $\psi(t)$ admits a representation by Gaussians, shifted in direct and Fourier transformed space. If $g(t)=2^{-1/2}\pi^{-3/4}e^{-t^2/2}$ and t_0, ω_0 are arbitrary real, consider

$$g^{(t_0,\omega_0)}(t)=e^{-i\omega_0 t_0/2}e^{i\omega_0 t}g(t-t_0) \tag{1.1}$$

and form the inner product

$$\Psi(t_0,\omega_0)=\int \bar{g}^{(t_0,\omega_0)}(t)\psi(t)\,dt. \tag{1.2}$$

Then

$$\iint |\Psi(t_0,\omega_0)|^2 dt_0\,d\omega_0=\int |\psi(t)|^2 dt. \tag{1.3}$$

The function $\psi(t)$ can be recovered from the function $\Psi(t_0,\omega_0)$ through

$$\psi(t)=\iint g^{(t_0,\omega_0)}(t)\Psi(t_0,\omega_0)\,dt_0\,d\omega_0. \tag{1.4}$$

The above statements remain true if the Gaussian g is replaced by an arbitrary square integrable function. The advantages of the Gaussian are (i) maximal concentration in direct and Fourier transformed space and (ii) the possibility of a simple intrinsic characterization of the space of functions $\Psi(t_0,\omega_0)$.

This representation of functions has been used in quantum mechanics, quantum optics and signal theory. (See e.g. [1], [4], [5], [6].)

1.2. Consider now the case where the object of interest is not a complex-valued function $\psi(t)$, but a square integrable real-valued function $s(t)$, say the wiggle of a seismograph. It has been known for a long time that it is very useful to consider $s(t)$ as the real part of a complex-valued square integrable function $h(t)$ which has the special property that its Fourier transform vanishes on a half-line (say $\tilde{h}(\omega)=0$ for $\omega<0$). The

*Received by the editors September 21, 1982.

†Centre de Physique Théorique, Section II, Centre National de la Recherche Scientifique, Marseille, France.

‡Elf Aquitaine Company, O.R.I.C. Lab., 370 bis Av. Napoléon Bonaparte, 92500 Rueil-Malmaison, France.

723

space of such functions $h(t)$ is denoted by $\mathbf{H}^2$ and called the Hardy space on the line. It is a closed subspace of the space L_2 $(\mathbb{R}, dt)$ of all square integrable functions. The functions s and h are in a natural one-to-one correspondence, and special properties of the function $h(t)$ (in particular its phase) make it a valuable tool.

1.3. This paper is concerned with the decomposition of functions $h \in \mathbf{H}^2$ into square integrable "elementary wavelets", and with the corresponding reconstruction problem. One can of course analyze the function $h(t)$ by applying to it the general results described in §1.1, applicable to any function in L_2. This is indeed what is done traditionally (see e.g. the famous paper [4]). It is however clear that, when we follow this procedure, we are not taking advantage of the special features of the function $h(t)$ which led us to introduce it in the first place; we are analyzing a function that belongs to the subspace $\mathbf{H}^2 \subset L_2$ in terms of wavelets that do not belong to this subspace (the Fourier transform of a Gaussian does not vanish on a half-line). It will not help (at least in principle) to replace the Gaussian by an elementary wavelet that belongs to $\mathbf{H}^2$, since we have to consider all of its shifts in Fourier transformed space, and these are sure to bring it out of $\mathbf{H}^2$.

1.4. In several papers devoted to the study of seismic traces [7], [8], one of us has suggested analyzing them in terms of wavelets of *fixed shape*, and has produced strong numerical evidence for the soundness of such analysis. The aim of the present paper is to give mathematical underpinnings for this procedure, which also avoids the objections that were raised in §1.3. The main idea is to analyze functions in terms of wavelets obtained by shifts (only in direct space, not in Fourier transformed space) and *dilations* from a suitable basic wavelet.

1.5. The group G_2 of shifts and dilations (which is the only two-parameter Lie group and thus the "smallest" noncommutative Lie group), acts on $\mathbf{H}^2$ through a natural irreducible unitary representation $U(\gamma)$ $(\gamma \in G_2)$. If we fix a function $g \in \mathbf{H}^2$ ("the analyzing wavelet"), we obtain a correspondence between an arbitrary $h \in \mathbf{H}^2$, and the matrix element $m_h^{(g)}(\gamma) = (U(\gamma)g, h)$ considered as a function on the group G_2. The main question, both from a conceptual and practical point of view, is whether the correspondence $h \to m_h^{(g)}$ has a well-behaved inverse, allowing a "stable" reconstruction of h from $m_h^{(g)}$. Stated somewhat differently, the question is whether, for a suitable invariant measure $d\gamma$ on G_2, one has

$$\int |m_h^{(g)}(\gamma)|^2 d\gamma = \int |h(t)|^2 dt \tag{1.5}$$

in analogy to (1.3).

1.6. It turns out that the answer depends on the choice of the analyzing wavelet g. For (1.5) to hold, the wavelet g, in addition to being in $\mathbf{H}^2$, has to satisfy an "admissibility" condition.

The main general result, proved in §3, can be stated without reference to group theory:

Let $h(t)$ (the function to be analyzed) satisfy

$$\int |h(t)|^2 dt < \infty \tag{i}$$

and

(ii) $$\tilde{h}(\omega)=0 \quad \text{for } \omega\leq 0.$$

(Conditions (i) and (ii) say that $h\in\mathbf{H}^2$.)

Let $g(t)$ (the analyzing wavelet) satisfy (i), (ii), and also the "admissibility condition"

(iii) $\int du\, e^u \int_0^\infty |\tilde{g}(\omega)\tilde{g}(e^u\omega)|^2\, d\omega<\infty$. Associate to h the function $(\mathcal{C}h)(u,v)$ of two variables, defined by

$$(\mathcal{C}h)(u,v)=\frac{1}{\sqrt{c_g}}e^{u/2}\int \bar{g}(e^u t-v)h(t)\,dt \tag{1.6}$$

where

$$c_g=\frac{2\pi}{\|g\|^2}\int du\, e^u\int |\tilde{g}(\omega)\tilde{g}(e^u\omega)|^2 d\omega=2\pi\int_0^\infty \frac{|\tilde{g}(\omega)|^2}{\omega}d\omega \tag{1.7}$$

and

$$\|g\|^2=\int |g(t)|^2 dt.$$

Then

(a)

$$\iint |(\mathcal{C}h)(u,v)|^2 du\, dv=\int |h(t)|^2 dt \tag{1.8}$$

and

(b) $h(t)$ can be recovered from $(\mathcal{C}h)(u,v)$ through

$$h(t)=\frac{1}{\sqrt{c_g}}\iint e^{u/2}g(e^u t-v)(\mathcal{C}h)(u,v)\,du\,dv.$$

If g is not admissible, then $c_g=\infty$, and the transformation (1.6) is not defined.

The need for an admissibility condition may seem surprizing. It stems from the fact that G_2, in contrast to all other "everyday" groups, is nonunimodular (i.e. has no right-and-left-invariant measure) (compare [2], [3]).

Section 4 is devoted to a more detailed study of the transformation $\mathcal{C}$ in the special case where g is a "particularly good" wavelet which plays a role analogous to that of the Gaussian in the conventional theory. We find it convenient to introduce a special function $\Gamma_\alpha(z)$ which may be of independent interest. In §4 we gather the results necessary for an intrinsic characterization of the range of $\mathcal{C}$, which will be given in a forthcoming paper.

This paper can also be viewed as the description of a natural quantum-mechanical representation for particles that "only know how to move in one direction".

This interpretation and further developments will also be found in forthcoming papers.

2. Notation and preliminaries.

2.1. The inner product of square integrable functions is written as

$$(f,g)=\int \bar{f}(t)g(t)\,dt$$

where $\bar{f}$ is the complex conjugate of f.

The Fourier transform of $f(t)$ is

$$\tilde{f}(\omega)=(2\pi)^{-1/2}\int e^{-i\omega t}f(t)\,dt$$

inverted by

$$f(t)=(2\pi)^{-1/2}\int e^{i\omega t}\tilde{f}(\omega)\,d\omega.$$

We also write

$$\tilde{f}=\mathfrak{F}f, \qquad f=\mathfrak{F}^{-1}\tilde{f}.$$

The shift operator T^v is defined by

$$(T^v f)(t)=f(t-v) \qquad (v\in\mathbb{R}).$$

The corresponding multiplication operator is E^v;

$$(E^v\tilde{f})(\omega)=e^{iv\omega}\tilde{f}(\omega).$$

The dilation operator Z^u is defined by

$$(Z^u f)(t)=e^{-u/2}f(e^{-u}t).$$

The relations

$$(2.1)\qquad \begin{aligned} T^vZ^u&=Z^uT^{v/\exp u}, & E^vZ^u&=Z^uE^{v\exp u},\\ Z^uT^v&=T^{v\exp u}Z^u, & Z^uE^v&=E^{v/\exp u}Z^u \end{aligned}$$

will be basic for all that follows. They correspond to

$$(T^vZ^u f)(t)=e^{-u/2}f(e^{-u t}-e^{-u}v),$$
$$(Z^uT^v f)(t)=e^{-u/2}f(e^{-u}t-v).$$

The commutation properties with $\mathfrak{F}$ are

$$(2.2)\qquad \mathfrak{F}T^v=E^{-v}\mathfrak{F},$$
$$(2.3)\qquad \mathfrak{F}Z^u=Z^{-u}\mathfrak{F}.$$

We have $T^{v_1}T^{v_2}=T^{v_1+v_2}$, and $Z^{u_1}Z^{u_2}=Z^{u_1+u_2}$. The operators $\mathfrak{F}$, T^v, E^v, Z^u are all unitary in $L_2(\mathbb{R})$.

2.2. We say that a function $h\in L_2(\mathbb{R},dt)$ belongs to the Hardy space $\mathbf{H}^2\subset L_2$ if $\tilde{h}(\omega)=0$ for $\omega<0$.

$\mathbf{H}^2$ is a closed subspace of L_2.

A real-valued function cannot belong to $\mathbf{H}^2$.

If $h\in\mathbf{H}^2$ then $\bar{h}$ and $\check{h}$ (where $\check{h}(t)=h(-t)$) are orthogonal to all of $\mathbf{H}^2$. However, $h^*\in\mathbf{H}^2$, where $h^*(t)=\bar{h}(-t)$.

The Hilbert transform in $L_2(\mathbb{R},dt)$ can be defined as $H=-\mathcal{F}^{-1}\varepsilon\mathcal{F}$, where ε is the operator of multiplication by $\operatorname{sgn}\omega$ (the sign of ω). We have $H^2=-1$; H is unitary, anti-Hermitian and real (commutes with complex conjugation). If $h\in\mathbf{H}^2$, then $h=iHh$, giving $\operatorname{Im}h=H\operatorname{Re}h$ and $\operatorname{Re}h=-H\operatorname{Im}h$.

If $s(t)$ is any real-valued, square integrable function then

$$h_s=s+iHs$$

belongs to $\mathbf{H}^2$. We have

$$(h_s,g)=2(s,g) \tag{2.4}$$

for every $g\in\mathbf{H}^2$. Also

$$\|h_s\|^2=2\|s\|^2 \tag{2.5}$$

where $\|f\|^2=(f,f)$

2.3. Shifts and dilations in $\mathbf{H}^2$. If $h\in\mathbf{H}^2$, then $T^vh\in\mathbf{H}^2$ and $Z^uh\in\mathbf{H}^2$ for all u,v.

For our purposes it is crucial to remark that the family T^v, Z^u acts *irreducibly* in $\mathbf{H}^2$. That is: If V is a closed subspace of $\mathbf{H}^2$, containing at least one nonzero vector; if V is stable under all Z^u, T^v (which means $T^uZ^vh\in V$ whenever $h\in V$), then V is all of $\mathbf{H}^2$. For a proof see e.g. [9].

Another important, if obvious, remark is that shifts and dilations are "real", in the sense that

$$\operatorname{Re}(T^vh)=T^v(\operatorname{Re}h) \tag{2.6}$$

and

$$\operatorname{Re}(Z^uh)=Z^u(\operatorname{Re}h). \tag{2.7}$$

3. The $\mathcal{C}$-transform: arbitrary admissible wavelet.

3.1. Admissible analyzing wavelet. We shall say that a function g, not identically zero, is an *admissible analyzing wavelet*, if

(i) g belongs to $\mathbf{H}^2$

and

(ii) g satisfies the condition

$$\iint|(Z^{-u}T^vg,g)|^2\,du\,dv<\infty. \tag{3.1}$$

By (2.2) and (2.3), the condition (3.1) can also be written as

$$\iint|(Z^uE^{-v}\tilde{g},\tilde{g})|^2\,du\,dv=2\pi\int du\,e^u\int|\tilde{g}(w)\tilde{g}(e^u\omega)|^2\,d\omega<\infty \tag{3.2}$$

$$=2\pi\|g\|^2\int_0^\infty\frac{|\tilde{g}(\omega)|^2}{\omega}\,d\omega.$$

Examples. 1) Let $0<a<b<\infty$. Define $g(t)$ through its Fourier transform: $\tilde{g}(\omega)=1$ if $a<\omega<b$, and 0 otherwise. Then $g(t)$ is an admissible analyzing wavelet, as can be shown by a simple calculation.

2) Let $\alpha>0$. Define $g_\alpha(t)$ through its Fourier transform $\tilde{g}_\alpha(\omega)=\exp(-(\alpha/2)\ln^2\omega)$ $=\omega^{-\alpha\ln\omega/2}$ for $\omega>0$, and $g_\alpha(\omega)=0$ for $\omega<0$. Then $g_\alpha(t)$ is an admissible analyzing wavelet which will be studied in §4.

Remarks. 1) There exist functions in $\mathbf{H}^2$ that are not admissible analyzing wavelets; this is the case e.g. if the Fourier transform of g is defined by

$$\tilde{g}(\omega)=\begin{cases}\omega^{-1/2+\varepsilon} & (0<\omega\leq 1)\\ \omega^{-1/2-\varepsilon} & (1<\omega<\infty)\end{cases}\qquad (\varepsilon>0)$$

and $\tilde{g}(\omega)=0$ for $\omega\leq 0$.

2) A Gaussian cannot be an admissible analyzing wavelet since it does not belong to $\mathbf{H}^2$. However, if ω_0 is positive and sufficiently large, the function $e^{i\omega_0 t}\exp(-t^2/2)$ is very close to an admissible analyzing wavelet.

3) From (3.2) we see: If $\tilde{g}(\omega)$ is the Fourier transform of an admissible wavelet then, for any real-valued $\varphi(\omega)$, the function $e^{i\varphi(\omega)}\tilde{g}(\omega)$ is also the Fourier transform of an admissible wavelet.

3.2. The number c_g. If g is an admissible analyzing wavelet, we denote by c_g the number

$$c_g=\frac{1}{\|g\|^2}\iint\left|(Z^{-u}T^{v}g,g)\right|^2 du\,dv.$$

By (2.2), c_g can also be written as

$$c_g=\frac{1}{\|g\|^2}\iint\left|(Z^{u}E^{-v}\tilde{g},\tilde{g})\right|^2 du\,dv$$

which gives

$$c_g=\frac{2\pi}{\|g\|^2}\int_0^\infty d\omega\int du\,e^u\left|\tilde{g}(\omega)\tilde{g}(e^u\omega)\right|^2=2\pi\int_0^\infty\frac{|\tilde{g}(\omega)|^2}{\omega}d\omega. \tag{3.3}$$

3.3. The $\mathcal{C}$-transform. Let g be a fixed admissible analyzing wavelet. For arbitrary real u, v, define

$$g^{(u,v)}=Z^{-u}T^{v}g. \tag{3.4}$$

For every $h\in\mathbf{H}^2$, define the function $\mathcal{C}h$ of variables u, v by

$$(\mathcal{C}h)(u,v)=\frac{1}{\sqrt{c_g}}\left(g^{(u,v)},h\right) \tag{3.5}$$

i.e.

$$(\mathcal{C}h)(u,v)=\frac{1}{\sqrt{c_g}}e^{u/2}\int\bar{g}(e^u t-v)h(t)\,dt. \tag{3.6}$$

In words: $(\mathcal{C}h)(u,v)$ is obtained by "testing" the function h with the help of dilated and shifted analyzing wavelet. The dilation parameter is u, and the shift parameter is v. The result of testing is multiplied by a normalization factor which depends on the choice of the admissible analyzing wavelet.

We call $\mathcal{C}h$ the $\mathcal{C}$-transform of h (with respect to g). By (2.4) we have, with $s(t)=\mathrm{Re}\,h(t)$,

$$(\mathcal{C}h)(u,v)=\frac{2}{\sqrt{c_g}}e^{u/2}\int\bar{g}(e^u t-v)s(t)\,dt.$$

Alternative ways of writing $(\mathcal{C}h)(u,v)$ are, by (2.2), (2.3),

$$(3.7)\qquad (\mathcal{C}h)(u,v)=\frac{1}{\sqrt{c_g}}(Z^{-u}T^{v}g,h)=\frac{1}{\sqrt{c_g}}(Z^{u}E^{-v}\tilde{g},\tilde{h})$$

$$=\frac{1}{\sqrt{c_g}}e^{-u/2}\int_0^{\infty}\bar{\tilde{g}}(e^{-u}\omega)\tilde{h}(\omega)\exp[ive^{-u}\omega]\,d\omega.$$

From (3.7) one sees that

$$|(\mathcal{C}h)(u,v)|\leq\frac{1}{\sqrt{c_g}}\|h\|\,\|g\|$$

for all u,v.

The correspondence $h\to\mathcal{C}h$ is linear.

3.4. Isometry of $\mathcal{C}$. We claim: *For every $h\in\mathbf{H}^2$, the function $(\mathcal{C}h)(u,v)$ is square integrable, and*

$$(3.8)\qquad \iint|(\mathcal{C}h)(u,v)|^2\,du\,dv=\|h\|^2.$$

Proof. (i) The equality (3.8) holds for $h=g$, since, by (3.5),

$$\iint|(\mathcal{C}g)(u,v)|^2\,du\,dv=\frac{1}{c_g}\iint|(Z^{-u}T^{v}g,g)|^2\,du\,dv=\frac{c_g}{c_g}\|g\|^2.$$

(ii) Equation (3.8) also holds for every h of the form $h=Z^{-u_0}T^{v_0}g$. Indeed,

$$(\mathcal{C}h)(u,v)=\frac{1}{\sqrt{c_g}}(Z^{-u}T^{v}g,Z^{-u_0}T^{v_0}g)=\frac{1}{\sqrt{c_g}}(T^{-v_0}Z^{u_0-u}T^{v}g,g)$$

$$=\frac{1}{\sqrt{c_g}}(Z^{u_0-u}T^{v-v_0\exp(u-u_0)}g,g)$$

$$=(\mathcal{C}g)(u-u_0,v-v_0e^{u-u_0}),$$

by (2.1), (3.5) and (3.7).

Now

$$\iint|(\mathcal{C}g)(u-u_0,v-v_0e^{u-u_0})|^2\,du\,dv=\iint|(\mathcal{C}g)(u',v-v_0e^{u'})|^2\,du'\,dv$$

$$=\iint|(\mathcal{C}g)(u',v')|^2\,du'\,dv'=\|g\|^2$$

with $u'=u-u_0$, $v'=v-v_0e^{u-u_0}$.

(iii) By standard arguments, (3.8) is extended to all finite linear combinations of vectors of the form $Z^{u}T^{v}g$. By irreducibility (§2.3.), these vectors are dense in $\mathbf{H}^2$, and so (3.8) is extended by continuity to all of $\mathbf{H}^2$. This completes the proof.

By the polarization identity, one has

$$(3.9)\qquad (\mathcal{C}h_1,\mathcal{C}h_2)_{L_2(\mathbf{R}^2,du\,dv)}=(h_1,h_2)$$

for all $h_1\in\mathbf{H}^2$, $h_2\in\mathbf{H}^2$.

3.5. Inversion of $\mathcal{C}$. We now sketch a verification of the fact that $\mathcal{C}$, considered as an integral transform, is self-reciprocal. In other words: If $\mathcal{C}h$ is given by (3.6), then h can be recovered from $\mathcal{C}h$ through the formula

$$h(t)=\frac{1}{\sqrt{c_g}}\iint e^{u/2}g(e^u t-v)(\mathcal{C}h)(u,v)\,du\,dv. \tag{3.10}$$

Since the integral (3.10) cannot converge for every h and every t, the formula (3.10) has a sense that is familiar from the L^2-theory of Fourier transforms or from [1].

In order to obtain (3.10) we write, with a slight stretching of notations, and using (2.4), (3.9),

$$h(t_0)=(\delta_{t_0},h)_{L_2(\mathbf{R},dt)}=\frac{1}{2}(\delta_{t_0}^{(+)},h)_{\mathbf{H}^2}=\frac{1}{2}(\mathcal{C}\delta_{t_0}^{(+)},\mathcal{C}h)_{L_2(\mathbf{R}^2,du\,dv)} \tag{3.11}$$

where $\delta_{t_0}(t)=\delta(t-t_0)$ (Dirac measure) and $\delta^{(+)}(t)=\delta(t-t_0)+(i/\pi)P/(t-t_0)$ (principal part). The function $(\mathcal{C}\delta_{t_0}^{(+)})(u,v)$ can be found by using (2.4):

$$(\mathcal{C}\delta_{t_0}^{(+)})(u,v)=\frac{1}{\sqrt{c_g}}e^{u/2}\int\bar{g}(e^u t-v)\delta_{t_0}^{(+)}(t)\,dt \tag{3.12}$$

$$=2\frac{1}{\sqrt{c_g}}e^{u/2}\int\bar{g}(e^u t-v)\delta(t-t_0)\,dt=2\frac{1}{\sqrt{c_g}}e^{u/2}\bar{g}(e^u t_0-v).$$

Inserting (3.12) into (3.11) gives (3.10).

Remark on redundancy. Equation (3.10) is a way of recovering the function $h(t)$ (and $s(t)=\operatorname{Re}h(t)$) from the function $(\mathcal{C}h)(u,v)$. The function $h(t)$ can also be recovered from values of $(\mathcal{C}h)(u,v)$ on suitable subsets of the plane, e.g. from the function

$$(\mathcal{C}h)(0,v)=\frac{1}{\sqrt{c_g}}\int_0^\infty\bar{\tilde{g}}(\omega)\tilde{h}(\omega)e^{iv\omega}\,d\omega.$$

We see that $\tilde{h}(\omega)$ can be obtained from $(\mathcal{C}h)(0,v)$ through Fourier transformation and division by $\bar{\tilde{g}}(\omega)$. The last step, however, corresponds—at best—to an unbounded operator. This makes the recovery of $h(t)$ from $(\mathcal{C}h)(0,v)$ an impractical proposition in general, and shows the advantage of working with the isometric transformation (3.10) or with suitable discrete approximations to it.

Covariance of $\mathcal{C}$. By the construction of $\mathcal{C}$, we have: If $h_1=Z^{u_1}h$, then

$$(\mathcal{C}h_1)(u,v)=(\mathcal{C}h)(u+u_1,v).$$

If $h_2=T^{v_2}h$, then

$$(\mathcal{C}h_2)(u,v)=(\mathcal{C}h)(u,v-v_2e^u).$$

3.6. Reproducing equation. The range of $\mathcal{C}$ is not all of $L_2(\mathbb{R}^2,du\,dv)$. In this section we derive a condition that has to be satisfied by all functions of the form $\mathcal{C}h$, with $h\in\mathbf{H}^2$. More specific results are given in §4, for a particular analyzing wavelet.

Define a kernel $G(u,v;u',v')$ by

$$G(u,v;u',v')=(g^{(u,v)},g^{(u',v')})=(Z^{-u}T^{v}g,Z^{-u'}T^{v'}g) \tag{3.13}$$

$$=(Z^uE^{-v}\tilde{g},Z^{u'}E^{-v'}\tilde{g})=(Z^{u_1}E^{-v_1}\tilde{g},\tilde{g})$$

with

$$u_1 = u - u' \quad \text{and} \quad v_1 = v - v'e^{u-u'}. \tag{3.14}$$

Then a function $f(u,v)$ that belongs to the range of $\mathcal{C}$ must satisfy

$$f(u,v) = \frac{1}{c_g} \iint G(u,v;u',v') f(u',v')\, du'\, dv'. \tag{3.15}$$

Indeed, by the definition and isometry of $\mathcal{C}$,

$$(\mathcal{C}h)(u,v) = \frac{1}{\sqrt{c_g}} (g^{(u,v)}, h) = \frac{1}{\sqrt{c_g}} (\mathcal{C}g^{(u,v)}, \mathcal{C}h)$$

$$= \frac{1}{\sqrt{c_g}} \iint \left(\overline{\mathcal{C}g^{(u,v)}}\right)(u',v')(\mathcal{C}h)(u',v')\, du'\, dv'.$$

Now

$$(\mathcal{C}g^{(u,v)})(u',v') = \frac{1}{\sqrt{c_g}} (g^{(u',v')}, g^{(u,v)})$$

giving (3.15).

3.7. Cycle-octave representations.

THEOREM. *Let $s(t)$ be any real-valued square integrable function, and g an admissible analyzing wavelet. Associate to s the function $S(u,\tau)$ defined by*

$$S(u,\tau) = \frac{2}{\sqrt{c_g}} \int \bar{g}(e^u t - e^{-u}\tau) s(t)\, dt. \tag{3.16}$$

Then $s(t)$ can be recovered from $S(u,\tau)$ through

$$s(t) = \operatorname{Re} h(t) \tag{3.17}$$

where

$$h(t) = \frac{1}{\sqrt{c_g}} \iint g(e^u t - e^{-u}\tau) S(u,\tau)\, du\, d\tau \tag{3.18}$$

One has

$$\iint |S(u,\tau)|^2 du\, d\tau = 2 \int s(t)^2 dt. \tag{3.19}$$

The function $h(t)$ defined by (3.18) *belongs to* $\mathbf{H}^2$.

An approximate discrete version of (3.16), (3.18), was discovered by one of us [7].

The statements of this theorem are an immediate consequence of the results proved so far, if we introduce the variable

$$\tau = e^u v.$$

3.8. Group-theoretical comments. The objects that we study, namely

$$\sqrt{c_g}(\mathcal{C}h)(u,-v)=(g,T^v Z^u h),$$

are matrix elements (coefficients, in another terminology) of the irreducible representation, in $\mathbf{H}^2$, of the two-parameter group of shifts and dilations. We have shown that these coefficients, considered as functions on the group, are square integrable with respect to the right Haar measure $dx_R = du\,dv$, if the vector g is suitably chosen.

If the standard theory of square integrable representations were applicable here (see e.g. [2]), it would follow that all coefficients of this representation are square integrable i.e. that all wavelets are admissible. However, the standard theory holds only for unimodular groups (i.e. groups possessing a right- and left-invariant Haar measure), while the group here is the prime example of nonunimodularity. (The left-invariant Haar measure is $dx_L = e^{-u}du\,dv$). Our results fit into the general theory of square integrable representations of nonunimodular groups, developed by Duflo and Moore [3].

4. The $\mathcal{C}$-transform: wavelet g_α.

4.1. The function $\tilde{g}_\alpha$. Among all admissible wavelets there is one that plays—in the $\mathbf{H}^2$-theory that we are concerned with—the same privileged role that the Gaussian plays in L^2-theory. The Fourier transform of this wavelet is just the image of a Gaussian under a natural map.

Let $\alpha > 0$. Consider the function $\tilde{g}_\alpha(\omega)$ defined by

$$\tilde{g}_\alpha(\omega)=\begin{cases}\exp\left(-\frac{\alpha}{2}\ln^2\omega\right) & \text{for } \omega>0,\\ 0 & \text{for } \omega\le 0.\end{cases} \tag{4.1}$$

Notice that $\tilde{g}_\alpha(\omega)$ is infinitely differentiable everywhere, in particular at $\omega=0$. Furthermore, $\tilde{g}_\alpha(\omega)$ tends to zero at infinity faster than any inverse polynomial.

We shall first verify that g_α is admissible, by using the criterion (3.2): one has

$$\int_0^\infty |\tilde{g}_\alpha(\omega)\tilde{g}_\alpha(e^u\omega)|^2 d\omega = \sqrt{\frac{\pi}{2\alpha}}\, e^{1/8\alpha} e^{-\alpha u^2/2-u/2}$$

and consequently

$$\int du\, e^u \int_0^\infty |\tilde{g}_\alpha(\omega)\tilde{g}_\alpha(e^u\omega)|^2 d\omega = \frac{\pi}{\alpha} e^{1/(4\alpha)}$$

which shows admissibility, and also gives

$$c_g = 2\pi^{3/2}\alpha^{-1/2}. \tag{4.2}$$

The basic functional equation satisfied by $\tilde{g}_\alpha(\omega)$ is

$$(Z^u\tilde{g}_\alpha)(\omega)=e^{-\alpha u^2/2-u/2}\omega^{\alpha u}\tilde{g}_\alpha(\omega). \tag{4.3}$$

It corresponds to the equation relating a shifted Gaussian to the Gaussian multiplied by an exponential.

4.2. Condition satisfied by functions in the range of $\mathcal{C}$. We can use (4.3) to derive conditions satisfied by all functions in the range of $\mathcal{C}$. Writing

$$(\mathcal{C}h)(u,v)=\frac{1}{\sqrt{c_g}}(Z^uE^{-v}\tilde{g}_\alpha,\tilde{h})=(E^{-v\exp(-u)}Z^u\tilde{g}_\alpha,\tilde{h})\frac{1}{\sqrt{c_g}}$$

and evaluating $Z^u\tilde{g}$ by (4.3), we obtain

$$(\mathcal{C}h)(u,v)=\frac{1}{\sqrt{c_g}}e^{-\alpha u^2/2-u/2}\int_0^\infty \exp(ive^{-u}\omega)\omega^{\alpha u}\bar{\tilde{g}}(\omega)\tilde{h}(\omega)\,d\omega.$$

Introducing variables

$$z=\alpha u-1 \quad \text{and} \quad q=ve^{-u}$$

and writing

$$\Psi(z,q)=(\mathcal{C}h)(u,v),$$

we see that

$$\Psi(z,q)=\frac{1}{\sqrt{c_g}}e^{-(z^2-z)/2\alpha}\int_0^\infty e^{iq\omega}\omega^{z-1}\bar{\tilde{g}}_\alpha(\omega)\tilde{h}(\omega)\,d\omega. \tag{4.4}$$

It follows from (4.4) that $\Psi(z,q)$ satisfies

$$\frac{\partial\Psi}{\partial q}=ie^{z/\alpha}\Psi(z+1,q),$$

and, more generally

$$\Psi(z+n,q)=(-i)^n e^{-(nz+(n-1)n)/2\alpha}\frac{\partial^n\Psi}{\partial q^n}.$$

4.3. The function $\Gamma_\alpha(z)$. The wavelet $g_\alpha(t)$ is given by the inverse Fourier transform

$$g_\alpha(t)=(2\pi)^{-1/2}\int_0^\infty e^{i\omega t}\tilde{g}_\alpha(\omega)\,d\omega,$$

which does not seem expressible in closed form through special functions known to us. In order to evaluate it and related quantities we have found it convenient to introduce the function $\Gamma_\alpha(z)$ which will now be discussed and which, we believe, is also intrinsically interesting.

If $\alpha>0$ and if $z=x+iy$ is arbitrary complex, define $\Gamma_\alpha(z)$ by

$$\Gamma_\alpha(z)=\int_0^\infty \omega^{z-1}e^{-\omega}\exp\left(-\frac{\alpha}{2}\ln^2\omega\right)d\omega. \tag{4.5}$$

This definition is modeled on the definition

$$\Gamma(z)=\int_0^\infty \omega^{z-1}e^{-\omega}\,d\omega \qquad (\mathrm{Re}\,z>0)$$

of Euler's gamma function. Because of the factor $\exp(-\alpha\ln^2\omega/2)$, the function $\Gamma_\alpha(z)$ is entire analytic in z, in contrast to $\Gamma(z)$. If the factor $\exp(-\alpha\ln^2\omega/2)$ were replaced by a step function, the resulting integral would be an incomplete gamma function.

The substitution $\omega = e^s$ brings $\Gamma_\alpha(z)$ to the form

$$\Gamma_\alpha(z) = \int_{-\infty}^{\infty} e^{zs} \exp\left(-e^s - \frac{\alpha}{2}s^2\right) ds. \tag{4.6}$$

We may think of $\Gamma_\alpha(z)$ as being a hybrid between a Gaussian and the Γ-function. This is made precise e.g. through the following statement:

If $\operatorname{Re} z > 0$, then

$$\Gamma_\alpha(z) = \frac{1}{\sqrt{2\pi\alpha}} \int_{-\infty}^{\infty} e^{-u^2/2\alpha}\Gamma(z - iu)\, du. \tag{4.7}$$

The function $\Gamma_\alpha(z)$ satisfies a functional equation that goes over into the classical $\Gamma(z) = (z-1)\Gamma(z-1)$ in the limit $\alpha \to 0$. Denote by $\Gamma_\alpha^{(n)}$ the nth derivative of Γ_α with respect to z. We have then

$$\Gamma_\alpha(z) = (z-1)\Gamma_\alpha(z-1) - \alpha\Gamma_\alpha^{(1)}(z-1) \tag{4.8}$$

and, more generally,

(4.9)

$$\Gamma_\alpha^{(n)}(z) = (z-1)\Gamma_\alpha^{(n)}(z-1) - \alpha\Gamma_\alpha^{(n+1)}(z-1) + n\Gamma_\alpha^{(n-1)}(z-1) \qquad (n = 1, 2, \cdots).$$

The function $\Gamma_\alpha(z)$ has asymptotic expansions for large $|z|$. For instance, we shall write the analogue of the formula $\Gamma(z) \simeq (2\pi)^{1/2} e^{-z+(z-1/2)\ln z}(1 + 1/12z)$:

Denote by z_1 the solution of

$$z_1 = z - \alpha \ln z_1,$$

which is positive when z is large and positive. Then

$$\Gamma_\alpha(z) \simeq \sqrt{\frac{2\pi}{z_1 + \alpha}}\, e^{-z_1 + (z^2 - z_1^2)/2\alpha}\left[1 - \frac{z_1}{8(z_1+\alpha)^2} + \frac{5z_1}{24(z_1+\alpha)^3}\right]. \tag{4.10}$$

The expression (4.10) is numerically quite accurate even for small values of $|z|$.

If for fixed y, x is let go to $-\infty$, one has

$$\Gamma_\alpha(x + iy) \sim \sqrt{\frac{2\pi}{\alpha}}\, e^{(x+iy)^2/2\alpha}.$$

We consider next the variation of $\Gamma_\alpha(z)$ with α. From (4.6) we see that $\Gamma_\alpha(z)$ satisfies

$$\frac{\partial \Gamma_\alpha}{\partial \alpha} = -\frac{1}{2}\frac{\partial^2 \Gamma_\alpha}{\partial z^2}. \tag{4.11}$$

Notice the minus sign in (4.11). We obtain the usual heat equation if we consider $\Gamma_\alpha(x+iy)$ as function of y for fixed x:

$$\frac{\partial \Gamma_\alpha(x+iy)}{\partial \alpha} = \frac{1}{2}\frac{\partial^2 \Gamma_\alpha(x+iy)}{\partial y^2}. \tag{4.12}$$

The function $\Gamma_\alpha(x+iy)$ is bounded by its values on the real axis: $|\Gamma_\alpha(x+iy)| < \Gamma_\alpha(x)$. Along parallels to the imaginary axis, it decreases faster than any inverse polynomial. Considered as a function of α, $\Gamma_\alpha(x)$ is monotonically decreasing. On the positive real axis, $\Gamma_\alpha(x)$ is bounded by Euler's Γ-function to which it tends as $\alpha\to 0$. For every x,y, we have $|\Gamma_\alpha(x+iy)|<\sqrt{2\pi/\alpha}\, e^{x^2/2\alpha}$. If we let α tend to $+\infty$ while keeping z fixed, then $\Gamma_\alpha(z)$ behaves as $\sqrt{2\pi/\alpha}\, e^{\alpha z^2/2}$.

Details and further results will be given elsewhere.

4.4. A class of integrals. With the help of the function $\Gamma_\alpha(z)$ one can evaluate integrals of the form

$$h_{\alpha,\beta}(q)=\int_0^\infty \omega^{\beta-1}\exp\left(-\frac{\alpha}{2}\ln^2\omega\right)e^{i\omega q}\,d\omega \tag{4.13}$$

which will be needed below. Here β is complex and arbitrary, $\alpha>0$, $q\neq 0$, and $\operatorname{Im} q\geq 0$.

It is convenient to introduce the variable

$$\kappa=\ln\left(\frac{1}{i}q\right)=\ln|q|+i\arg q-i\frac{\pi}{2}. \tag{4.14}$$

If q is real, then $\operatorname{Im}\kappa=-(\pi/2)\operatorname{sgn} q$.

The integral (4.13) is first transformed into

$$h_{\alpha,\beta}(q)=\int_{-\infty}^{\infty}\exp\left(-e^{\kappa+s}-\frac{\alpha}{2}s^2+\beta s\right)ds$$

by the substitution $\omega=e^s$. Then the substitution $s'=s+\kappa$ and a shift of the path of integration back to the real axis bring it to the form

$$h_{\alpha,\beta}(q)=\exp\left(-\frac{\alpha}{2}\kappa^2-\beta\kappa\right)\int\exp\left[-e^s+(\alpha\kappa+\beta)s-\frac{\alpha}{2}s^2\right]ds,$$

giving the result

$$h_{\alpha,\beta}(q)=\exp\left(-\frac{\alpha}{2}\kappa^2-\beta\kappa\right)\Gamma_\alpha(\alpha\kappa+\beta). \tag{4.15}$$

Remarks. The value of the integral (4.13) for $q=0$ can be computed directly. It is

$$h_{\alpha,\beta}(0)=\sqrt{\frac{2\pi}{\alpha}}\exp\left(\frac{\beta^2}{2\alpha}\right).$$

The function $h_{\alpha,\beta}(q)$ defined by (4.15) and (4.16) is infinitely differentiable on the real axis, and it decreases at infinity faster than any inverse polynomial. Furthermore, $h_{\alpha,\beta}$ belongs to $\mathbf{H}^2$.

As a function of $\beta, h_{\alpha,\beta}$ is entire analytic, and square integrable on every parallel to the imaginary axis. We have, from (4.13)

$$\frac{d^n}{dq^n}h_{\alpha,\beta}=i^n h_{\alpha,\beta+n}.$$

4.5. Explicit expressions. We can now write down expressions for various quantities of interest:

1) The wavelet $g_\alpha(t)$ is given by

$$g_\alpha(t)=(2\pi)^{-1/2}e^{-\alpha\theta^2/2-\theta}\Gamma_\alpha(\alpha\theta+1)\qquad\left(\theta=\ln\left(\frac{1}{i}t\right)\right).$$

2) The $\mathcal{C}$-transform of g_α is given by

$$(\mathcal{C}g_\alpha)(u,v)=2^{-1/2}\pi^{-3/4}\alpha^{1/4}e^{-u^2\alpha/4}e^{-\alpha\eta^2-\eta}\Gamma_{2\alpha}(2\alpha\eta+1)$$

where $\eta=\ln((1/i)ve^{-u/2})$.

3) The kernel of the integral equation satisfied by the functions in the range of $\mathcal{C}$ is (compare (3.15))

$$G(u,v;u',v')=e^{-\alpha u_1^2/4-\alpha\eta_1^2-\eta_1}\Gamma_{2\alpha}(2\alpha\eta_1+1)$$

where $u_1=u-u'$ and

$$\eta_1=\ln\left(\frac{1}{i}ve^{(u-u')/2}-\frac{1}{i}v'e^{-(u-u')/2}\right).$$

Acknowledgments. One of us (A. G.) would like to thank H. Bacry for stimulating conversations.

REFERENCES

[1] V. BARGMANN, *On a Hilbert space of analytic functions and an associated integral transform*, *Part* I, Comm. Pure Appl. Math, 14 (1961), pp. 187–214; *Part* II, ibid. 20 (1967), pp. 1–101.
[2] J. DIXMIER, *Les C*-algèbres et leurs représentations*, Gauthier-Villars, Paris, 1969.
[3] M. DUFLO AND C. C. MOORE, *On the regular representation of a nonunimodular locally compact group*, J. Funct. Anal., 21 (1976), pp. 209–243.
[4] D. GABOR, *Theory of Communication*, J. Inst. Electr. Engin. (London) 93 (III), 1946, pp. 429–457.
[5] C. W. HELSTROM, *An expansion of a signal in Gaussian elementary signals*, IEEE Trans. Infor. Theory, IT 12 (1966), pp. 81–82.
[6] J. R. KLAUDER AND E. C. SUDARSHAN, *Fundamentals of Quantum Optics*, Benjamin, New York, 1968.
[7] J. MORLET, G. ARENS, E. FOURGEAU AND D. GIARD, *Wave propagation and sampling theory*, *Part* II, Geophys. 47 (1982), pp. 222–236.
[8] J. MORLET, *Sampling theory and wave propagation*, Proc. 51st Annual International Meeting of the Society of Exploration Geophysicists, Los Angeles, 1981.
[9] N. YA. VILENKIN, *Special Functions and the Theory of Group Representations*, American Mathematical Society, Providence, RI, 1968.

Transforms associated to square integrable group representations. I. General results

A. Grossmann, J. Morlet,[a] and T. Paul[b]
Centre de Physique Théorique,[c] Section II, Centre National de la Recherche Scientifique, Luminy, Case 907, F-13288 Marseille Cedex 9, France

(Received 20 December 1984; accepted for publication 12 April 1985)

Let G be a locally compact group, which need not be unimodular. Let $x \rightarrow U(x)$ ($x \in G$) be an irreducible unitary representation of G in a Hilbert space $\mathscr{H}(U)$. Assume that U is square integrable, i.e., that there exists in $\mathscr{H}(U)$ at least one nonzero vector g such that $\int |(U(x)g,g)|^2 \, dx < \infty$. We give here a reasonably self-contained analysis of the correspondence associating to every vector $f \in \mathscr{H}(U)$ the function $(U(x)g,f)$ on G, discussing its isometry, characterization of the range, inversion, and simplest interpolation properties. This correspondence underlies many properties of generalized coherent states.

I. INTRODUCTION

This paper is the first of a series concerned with applications of various families of "generalized coherent states" to quantum mechanics, wave propagation, and signal analysis.

Many properties of the classical (canonical) coherent states[1,2] are closely tied to the Weyl–Heisenberg group. In particular, the fundamental formula

$$\mathbf{1} = \int |z\rangle d^2 z \langle z| \tag{1.1}$$

is a way of writing the orthogonality relations[3,4] for the irreducible representation of that group.

Aslaksen and Klauder[5] have considered the analogous states for the two-parameter group of shifts and dilations and found that the "fiducial vector" ("analyzing wavelet" in our terminology) cannot be arbitrary, in contrast to the Weyl–Heisenberg case.

The same two-parameter group appeared in Ref. 6 in the study of decomposition of signals into "wavelets of constant shape"; the restriction on the analyzing wavelet was there called an "admissibility condition."

Another (equivalent) representation of the same group, together with an appropriate choice of the analyzing wavelet, has given rise to a realization of quantum mechanics on a Hilbert space of function analytic on a half-plane.[7,8] Other groups were used to define coherent states: SU(2) for spin coherent states,[9] SU(1,1) in Ref. 10, and a general definition was proposed by Perelomov.[11]

In this paper we shall be concerned with "coherent states" associated with certain representations of arbitrary (in particular not necessarily unimodular) locally compact groups.

Let G be a locally compact group, U a continuous irreducible representation of G in a Hilbert space $\mathscr{H}(U)$, and g a vector in $\mathscr{H}(U)$.

We consider the family of vectors

$$|x\rangle = U(x)g \quad (x \in G) \tag{1.2}$$

in $\mathscr{H}(U)$. This family depends on the choice of g.

[a] Permanent address: ELF Aquitaine Company, ORIC Lab. 370 Bis Av. Napoleon Bonaparte, Rueil-Malmaison, France.
[b] Allocataire D.G.R.S.T.
[c] Laboratoire propre.

Since U is irreducible, the linear span of the vectors $|x>$ is dense in $\mathscr{H}(U)$.

One can then ask the question whether there exists a (suitably normalized) invariant measure $d\mu(x)$ on G, such that

$$\int |x\rangle d\mu(x) \langle x| = \mathbf{1}, \tag{1.3}$$

where $|x\rangle\langle x|$ is defined by $|x\rangle\langle x|f\rangle = (U(x)g,f)U(x)g$, and $\mathbf{1}$ is the identity operator in $\mathscr{H}(U)$.

The answer, in general, is no; this can be seen by taking $G = \mathbf{R}$ (additive), and by considering the one-dimensional irreducible representation space $\mathbf{C}$, with $U(x)$ the operator of multiplication by e^{ix}.

However, if the representation U is "square integrable" in a sense that will be defined below, then there exists a dense set of vectors in $\mathscr{H}(U)$, which give rise to (1.3). If g is such a vector (called "admissible") then the correspondence $f \rightarrow \psi$, with

$$\psi(x) = \langle x|f\rangle = (U(x)g,f) \quad (x \in G, \quad f \in \mathscr{H}(U)), \tag{1.4}$$

can be shown to be a multiple of an isometry between $\mathscr{H}(U)$ and $L^2(G,d\mu(x))$ and so (1.3) holds. The range of this transform is a closed subspace of $L^2(G,d\mu(x))$, and can be characterized by a reproducing kernel. If the group is unimodular, the set of admissible vectors is the whole space $\mathscr{H}(U)$, but this is not the case if the group is not unimodular (e.g., the affine group).

The purpose of this first paper is to give general results about transformations defined by (1.4). All the results derived here can be found in the mathematical literature and are part of the study of orthogonality relations for generalized square integrable representations. See, in particular, Refs. 12 and 13. We write them here in a form that is covenient for the applications we have in mind, using tools familiar to mathematical physicists (e.g., we give another proof of orthogonality relations with the help of quadratic forms).

The second paper of the series will be devoted to the particular case of the "$ax + b$" group, which has given rise to applications in applied mathematics.[6]

Further papers will be concerned with discrete versions of (1.2) and (1.3), with analyticity properties, special cases, and applications.

0022-2488/85/102473-07$02.50
© 1985 American Institute of Physics

Downloaded 14 Oct 2004 to 128.112.156.90. Redistribution subject to AIP license or copyright, see http://jmp.aip.org/jmp/copyright.jsp

II. SQUARE INTEGRABLE REPRESENTATIONS; ADMISSIBLE VECTORS

In this section we present basic notions that will be used in the construction of transforms defined in Sec. IV.

A. Notations

G will denote a locally compact group, with identity e, and $x, y, \ldots$ elements of G. It is well known that there exists in such a group a left-invariant and a right-invariant (Haar) measure.[3,4] The left-invariant measure, with a fixed normalization, will be written dx. So

$$d(yx) = dx \quad (y \in G). \tag{2.1}$$

The right-invariant measure will be denoted by $d_R x$:

$$d_R(xy) = d_R(x) \quad (y \in G). \tag{2.2}$$

One has

$$d_R x = \Delta^{-1}(x) dx, \quad d(xy) = \Delta(y) dx, \tag{2.3}$$

where $\Delta(x)$ (the modular function) is a positive-valued character

$$\Delta(e) = 1, \quad \Delta(x) > 0, \quad \Delta(xy) = \Delta(x)\Delta(y). \tag{2.4}$$

If $\Delta(x) \equiv 1$, the group G is said to be unimodular. Notice a collision of terminologies: the group $GL(n,\mathbf{R})$ of nonsingular $n \times n$ matrices is unimodular, even though it contains nonunimodular matrices.

The inhomogeneous group $IGL(n,\mathbf{R})$, a semidirect product of translations in $\mathbf{R}^n$ and of $GL(n;\mathbf{R})$, is not unimodular.

When one deals with semidirect products, it is convenient to use the right-invariant measure $d_R x$, since it is the product of the right-invariant measures of the factors (see Ref. 14, p. 210).

We have

$$d(x^{-1}) = \Delta^{-1}(x) dx = d_R x. \tag{2.5}$$

If $x \to \Phi(x)$ is any function on the group, define

$$\check{\Phi}: \check{\Phi}(x) = \Phi(x^{-1}). \tag{2.6}$$

Notice that $\check{\check{\Phi}} = \Phi$.

Then

(Φ is left integrable)$\Leftrightarrow$($\check{\Phi}$ is right integrable)

and the corresponding integrals are equal, since

$$\int \check{\Phi}(x) d_R x = \int \Phi(x^{-1}) d(x^{-1}) = \int \Phi(x') dx'.$$

We shall need the following statement: Let $\Phi(x)$ be a complex-valued function on the group G, such that

$$\bar{\Phi}(x) = \Phi(x^{-1}) \quad (x \in G) \tag{2.7}$$

[here $\bar{\Phi}(x)$ is the complex conjugate of $\Phi(x)$]. If, for some p, with $1 < p < +\infty$, we have $\Phi \in L^p(G,dx)$ and if Φ satisfies (2.7), then also $\bar{\Phi} \in L^p(G,dx)$.

Here $\Phi \in L^p(G,dx)$ means

$$\int |\Phi(x)|^p \, dx < \infty,$$

the integral being taken over G.

B. Left and right regular representation

Left regular: If $\Phi \in L^2(G,dx)$ and $a \in G$, we define $\lambda(a)\Phi \in L^2(G,dx)$ by

$$(\lambda(a)\Phi)(x) = \Phi(a^{-1}x) \quad (x \in G). \tag{2.8}$$

Right regular: If $\Psi \in L^2(G,d_R x)$ and $a \in G$, we define $\rho(a)\Psi \in L^2(G,d_R x)$ by

$$(\rho(a)\Psi)(x) = \Psi(xa) \quad (x \in G). \tag{2.9}$$

The two representations λ and ρ act unitarily—in different spaces in general, namely $L^2(G,dx)$ and $L^2(G,d_R x)$.

C. Definition of square integrable representations and of admissible vectors

Let $x \to U(x)$ be a strongly continuous unitary representation of the locally compact group G in a complex Hilbert space $\mathscr{H}(U)$.

A vector $g \in \mathscr{H}(U)$ is said to be *admissible* if

$$\int |(U(x)g,g)|^2 \, dx < +\infty. \tag{2.10}$$

In (2.10) the left-invariant measure dx can be replaced by the right-invariant measure $d_R x$, since

$$(U(x)g,g) = (g, U(x^{-1})g);$$

so

$$\int |(U(x)g,g)|^2 \, dx = \int |(g, U(x^{-1})g)|^2 \, dx = \int |(U(x^{-1})g,g)|^2 \, dx = \int |(U(x)g,g)|^2 \, d_R x.$$

Definition (2.1): U will be called square integrable if (i) U is irreducible and, (ii) there exists in $\mathscr{H}(U)$ at least one nonzero admissible vector.

Remarks: (a) Any representation unitarily equivalent to a square-integrable representation is also square-integrable.

(b) If G is compact, any irreducible representation U of G is square integrable. We shall see below examples of square-integrable representations of groups that are not compact.

(c) As an example of an irreducible representation that is *not* square-integrable, consider the one-dimensional representation $x \to e^{iax}$ of $\mathbf{R}$.

(d) If G is unimodular and if U is a square-integrable representation of G, then every vector in $\mathscr{H}(U)$ is admissible. (See, e.g., Ref. 3.) We shall see that the situation is different if G is not unimodular.

III. ORTHOGONALITY RELATIONS

A. Historical comments

Orthogonality relations were derived by Schur (beginning of this century) for finite groups, Weyl (in the 1920's) for compact groups, and Bargmann and Godement for (square-integrable representations of) unimodular groups in the 1950's; nonunimodular groups have been investigated more recently.[12,13]

Downloaded 14 Oct 2004 to 128.112.156.90. Redistribution subject to AIP license or copyright, see http://jmp.aip.org/jmp/copyright.jsp

Orthogonality relations in the unimodular case are expressed by the following equality [3,4]:

$$\int_G \overline{(U(x)g_1,f_1)}(U(x)g_2,f_2)dx = \lambda\,(g_2,g_1)(f_1,f_2)$$

for every g_1, g_2, f_1, f_2 in $\mathscr{H}(U)$, where λ depends only on the square-integrable representation U.

B. Statement of orthogonality relations

We have the following[12,13] theorem.

Theorem 3.1: Let U be a square integrable representation of G, acting on the Hilbert space $\mathscr{H}(U)$. Then there exists in $\mathscr{H}(U)$ a unique self-adjoint positive operator C such that the following hold.

(i) The set of admissible vectors coincides with the domain of C.

(ii) Let g_1 and g_2 be any two admissible vectors. Let f_1 and f_2 be any two vectors in $\mathscr{H}(U)$. Then

$$\int_G \overline{(U(x)g_1,f_1)}(U(x)g_2,f_2)dx = (Cg_2,Cg_1)(f_1,f_2). \quad (3.1)$$

(iii) If the group G is unimodular, then C is a multiple of the identity.

The proof in the general case uses an extension of the Schur lemma and is given in the Appendix.

C. A special case

If $g_1 = g_2 = f_1 = f_2$, then (3.1) gives that for any admissible vector g, one has

$$(Cg,Cg) = \frac{1}{\|g\|^2}\int |(U(x)g,g)|^2\,dx, \quad (3.2)$$

where $\|\cdot\|$ denotes the norm in $\mathscr{H}(U)$.

If $g_1 = g_2 = g$, one has

$$\int \overline{(U(x)g,f_1)}(U(x)g,f_2)dx = \frac{\int |(U(x)g,g)|^2\,dx}{\|g\|^2}(f_1,f_2). \quad (3.3)$$

IV. L_g TRANSFORM AND R_g TRANSFORM

In this section, we define the transforms described in the Introduction and prove their isometry.

A. Definitions

Let U be a square-integrable representation of G acting on $\mathscr{H}(U)$, and let g be a nonzero admissible vector [see (2.10)].

Associate to g the positive number

$$c_g = \frac{1}{\|g\|^2}\int |(U(x)g,g)|^2\,dx, \quad (4.1)$$

which, by Sec. II A, is also

$$c_g = \frac{1}{\|g\|^2}\int |(U(x)g,g)|^2\,d_R x.$$

Notice that, by the results of Sec. II A one has

$$c_{U(x_0)g} = \Delta(x_0)^{-1}c_g, \quad (4.2)$$

where $\Delta(x_0)$ is the modular function.

For any $f\in\mathscr{H}(U)$ consider the complex-valued functions $L_g f$ and $R_g f$ on G, defined by

$$(L_g f)(x) = (1/\sqrt{c_g})(U(x)g,f) \quad (x\in G), \quad (4.3)$$

$$(R_g f)(x) = (1/\sqrt{c_g})(g,U(x)f) \quad (x\in G). \quad (4.4)$$

One has

$$(R_g f)(x) = (L_g f)(x^{-1}). \quad (4.5)$$

The function $(L_g f)(x)$ will be called the L_g *transform* of f. It depends on the representation U and on the choice of the admissible vector g. For reasons which will become clear, we shall sometimes call g the *analyzing wavelet*.

Similarly, $(R_g f)(x)$ will be called the R_g transform of f.

Remark: By (4.2) and (4.3) we have

$$(L_{U(x_0)g}f)(x) = \Delta(x_0)^{1/2}(L_g f)(xx_0). \quad (4.6)$$

By the same argument we have also

$$(R_{U(x_0)g}f)(x) = \Delta(x_0)^{1/2}(R_g f)(x_0^{-1}x). \quad (4.7)$$

B. Continuity and boundedness

By the continuity of $U(x)$ and continuity of scalar product, the functions $(L_g f)(x)$ and $(R_g f)(x)$ are continuous on G.

By the Schwarz inequality, $L_g f$ and $R_g f$ are bounded on G:

$$|(L_g f)(x)| \leqslant (1/\sqrt{c_g})\|f\|\,\|g\|, \quad (4.8)$$

$$|(R_g f)(x)| \leqslant (1/\sqrt{c_g})\|f\|\,\|g\|, \quad (4.9)$$

for every $x\in G$.

C. Intertwining (covariance)

By the definition of L_g, we have, for every $f\in\mathscr{H}(U)$, $a\in G$, $x\in G$,

$$(L_g f)(a^{-1}x) = (U(a^{-1}x)g,f) = (U(x)g,U(a)f),$$

which can be written as

$$\lambda(a)L_g = L_g U(a). \quad (4.10)$$

Similarly we have

$$(R_g f)(xa) = (g,U(x)U(a)f),$$

giving

$$\rho(a)R_g = R_g U(a). \quad (4.11)$$

D. Isometry of L_g and of R_g

We have the following proposition.

Proposition 4.1: (i) The correspondence $f\to L_g f$ is isometric from $\mathscr{H}(U)$ into $L^2(G,dx)$; that is, for every $f_1\in\mathscr{H}(U)$, $f_2\in\mathscr{H}(U)$, we have

$$\int \overline{(L_g f_1)(x)}(L_g f_2)(x)dx = (f_1,f_2). \quad (4.12)$$

(ii) The correspondence $f\to R_g f$ is isometric from $\mathscr{H}(U)$ into $L^2(G,d_R x)$: for every $f_1\in\mathscr{H}(U)$, $f_2\in\mathscr{H}(U)$, we have

$$\int \overline{(R_g f_1)(x)}\,(R_g f_2)(x)d_R x = (f_1,f_2). \quad (4.13)$$

Downloaded 14 Oct 2004 to 128.112.156.90. Redistribution subject to AIP license or copyright, see http://jmp.aip.org/jmp/copyright.jsp

Proof: This proposition is a corollary of Theorem 3.1; by (3.3), we have

$$(f_1,f_2)=\frac{\|g\|^2}{\int|(U(x')g,g)|^2\,dx'}\int\overline{(U(x)g,f_1)}\,(U(x)g,f_2)dx$$
$$=\frac{1}{c_g}\int\overline{(U(x)g,f_1)}\,(U(x)g,f_2)dx$$
$$=\int\overline{(L_gf_1)(x)}\,(L_gf_2)(x)dx.$$

The change of variable $x\to x^{-1}$, $d(x^{-1})=d_R(x)$ gives

$$(f_1,f_2)=\frac{1}{c_g}\int\overline{(U(x^{-1})g,f_1)}\,(U(x^{-1})g,f_2)d(x^{-1})$$
$$=\frac{1}{c_g}\int\overline{(g,U(x)f_1)}(g,U(x)f_2)d_R(x)$$
$$=\int\overline{(R_gf_1)(x)}\,(R_gf_2)(x)d_R(x),$$

and the proposition is proved.

Remark: Since L_g is isometric from $\mathscr{H}(U)$ into $L^2(G,dx)$, L_g is unitary from $\mathscr{H}(U)$ to $L_g\mathscr{H}(U)\subset L^2(G,dx)$. By (4.10) we see that $L_g\mathscr{H}(U)$ is invariant under the left regular representation. So L_g is an intertwining operator between U and the restriction of the left regular representation of G.

This construction allows us to consider the representation U as a subrepresentation of the left regular representation of G. The following section will give a characterization of the range of L_g.

The same remark is also valid for R_g.

V. CHARACTERIZATION OF THE RANGES OF R_g and L_g

Let g be an admissible vector for U. Consider, on G, the complex-valued function $p_g(x)$, defined by

$$p_g(x)=(1/c_g)(U(x)g,g)=(1/\sqrt{c_g})(L_gg)(x)$$
$$=(1/\sqrt{c_g})(R_gg)(x^{-1}). \tag{5.1}$$

The function p_g satisfies $\bar{p}_g(x)=p_g(x^{-1})$ and belongs, by Sec. II A, to $L^2(G;dx)\cap L^2(G,d_Rx)$.

Proposition 5.1: (i) Let Φ belong to $L^2(G,dx)$. Then Φ belongs to $L_g\mathscr{H}(U)\subset L^2(G,dx)$ if and only if the equation

$$\Phi(x)=\int p_g(y^{-1}x)\Phi(y)dy \tag{5.2}$$

holds for every $x\in G$.

(ii) Let Ψ belong to $L^2(G,d_Rx)$. Then Ψ belongs to $R_g\mathscr{H}(U)\subset L^2(G,d_Rx)$ if and only if the equation

$$\Psi(x)=\int p_g(yx^{-1})\Psi(y)d_R(y) \tag{5.3}$$

holds for every $x\in G$.

Proof: Notice first that the integrals (5.2) and (5.3) converge for every x, since the integrand is the product of two square-integrable functions.

(i) Suppose Φ belongs to $L_g\mathscr{H}(U)$. This means that $\Phi(y)=(L_gf)(y)$ for some $f\in\mathscr{H}(U)$. By the definition of p_g [(5.1)] and of L_g [(4.3)], we have

$$\int p_g(y^{-1}x)\Phi(y)dy=\frac{1}{\sqrt{c_g}}\int(L_gg)(y^{-1}x)\Phi(y)dy$$
$$=\frac{1}{\sqrt{c_g}}\int\overline{L_gg(x^{-1}y)}\Phi(y)dy$$
$$=\frac{1}{\sqrt{c_g}}\int\overline{(\lambda(x)L_g)(y)}\Phi(y)dy.$$

From (4.10) we obtain

$$\int p_g(y^{-1}x)\Phi(y)dy=\frac{1}{\sqrt{c_g}}\int\overline{L_g(U(x)g)(y)}\Phi(y)dy$$
$$=(1/\sqrt{c_g})(U(x)g,f)$$
$$=(L_gf)(x)=\Phi(x)$$

by isometry and definition of L_g.

An analogous proof holds for R_g.

We must now prove the converse part of the proposition. To do this we must have an explicit expression of the inverse of L_g.

Lemma 5.2: For every $\Phi\in L^2(G,dx)$ the expression

$$\varphi=\frac{1}{\sqrt{c_g}}\int_G\Phi(x)U(x)g\,dx \tag{5.4}$$

defines a vector in $\mathscr{H}(U)$.

Proof: The integral (5.4) is weakly convergent. Indeed, for any $\psi\in\mathscr{H}(U)$, the function $x\to(\psi,U(x)g)$ is in $L^2(G,dx)$ [it is up to a constant $\overline{(L_g\Psi)(x)}$]. Since $\Phi(x)\in L^2(G,dx)$, the integral

$$\frac{1}{\sqrt{c_g}}\int(\psi,U(x)g)\Phi(x)dx=\int\overline{(L_g\psi)(x)}\Phi(x)dx$$

exists. Furthermore, by the Schwarz inequality in $L^2(G,dx)$ and isometry of L_g we have

$$\left|\frac{1}{\sqrt{c_g}}\int(\Psi,U(x)g)\Phi(x)dx\right|\leqslant\|\Phi\|_{L^2(G,dx)}\|\psi\|_{\mathscr{H}(U)}.$$

Then, by the Riesz theorem, the integral $(1/\sqrt{c_g})\times\int_G\Phi(x)U(x)g\,dx$ defines a vector in $\mathscr{H}(U)$.

Lemma 5.3: If Φ satisfies (5.2) then

$$L_g\left(\frac{1}{\sqrt{c_g}}\int\Phi(x)U(x)g\,dx\right)(y)=\Phi(y). \tag{5.5}$$

Proof: The computation of the left-hand side of (5.5) gives

$$\frac{1}{\sqrt{c_g}\sqrt{c_g}}\int\Phi(x)(U(y)g,U(x)g)dx$$
$$=\frac{1}{c_g}\int\Phi(x)(U(x^{-1}y)g,g)dx$$
$$=\int\Phi(x)p_g(x^{-1}y)dx=\Phi(y)\ \text{by hypothesis.}$$

Lemma 5.3 implies that if Φ satisfies (5.2), then Φ belongs to $L_g\mathscr{H}(U)$.

The same proof holds for R_g.

During the proof of the proposition we have proved the following.

Proposition 5.4: The inverses of L_g and R_g, on their respective domains $L_g\mathscr{H}(U)$ and $R_g\mathscr{H}(U)$ are given by

Downloaded 14 Oct 2004 to 128.112.156.90. Redistribution subject to AIP license or copyright, see http://jmp.aip.org/jmp/copyright.jsp

$$L_g^{-1}\Phi = \frac{1}{\sqrt{c_g}} \int \Phi(x) U(x) g \, dx, \tag{5.6}$$

$$R_g^{-1}\psi = \frac{1}{\sqrt{c_g}} \int \psi(x) U(x^{-1}) g \, d_R x, \tag{5.7}$$

where the integrals are taken in the weak sense.

Remarks: (1) The formulas (5.6) and (5.7) express the fact that, on the range of an isometric operator, the adjoint coincides with the inverse.

(2) The condition (5.2) can be rephrased by saying that the space $L_g \mathscr{H}(U) \subset L^2(G,dx)$ is a Hilbert space with reproducing kernel[15,16] (functional Hilbert space in the terminology of Ref. 16). The evaluation functional in $L_g \mathscr{H}(U)$ is

$$\Phi \to \Phi(x) = (e_x, \Phi)_{L^2(G,dx)}, \tag{5.8}$$

with

$$e_x(y) = \bar{p}_g(y^{-1}x) = p_g(x^{-1}y). \tag{5.9}$$

Similarly, (5.3) says that $R_g \mathscr{H}(U) \subset L^2(G, d_R x)$ is a Hilbert space with reproducing kernel in which the evaluation functional is

$$\Psi \to \Psi(x) = (h_x, \Psi)_{L^2(G,d_R x)}, \tag{5.10}$$

with

$$h_x(y) = \bar{p}_g(yx^{-1}) = p_g(xy^{-1}). \tag{5.11}$$

(3) In terms of the dyadic notation used in the Introduction [(1.2)], the isometry property (4.12) can be written as

$$\int \frac{dx}{c_g} |x\rangle \langle x| = 1, \tag{5.12}$$

where the integral is taken in the weak sense.

Indeed, from (5.12) we get

$$\int \frac{\langle f_1 | x \rangle}{\sqrt{c_g}} \frac{\langle x | f_2 \rangle}{\sqrt{c_g}} dx = (f_1, f_2),$$

which is (4.12).

The reproducing property (5.2) can be also deduced from (5.12), since (5.12) implies

$$\int \frac{\langle y | x \rangle}{c_g} \frac{\langle x | f \rangle}{\sqrt{c_g}} dx = \frac{\langle y | f \rangle}{\sqrt{c_g}},$$

which is (5.2).

The same remark holds for R_g.

VI. COVARIANT INTERPOLATION

A. Interpolation for the left transform

Proposition: Let $x_1,\dots,x_n$ be n points in G. For $1 \leqslant i,j \leqslant n$ consider the number

$$M_{ij} = (U(x_j)g, U(x_i)g) = (U(x_i^{-1}x_j)g, g) = c_g p_g(x_i^{-1}x_j). \tag{6.1}$$

Let M be the $n \times n$ positive definite Hermitian matrix with entries M_{ij}. Assume that

$$\det M \neq 0. \tag{6.2}$$

For $1 \leqslant j \leqslant n$, define a function $\Phi_j(x)$ on G, by

$$\Phi_j(x) = (U(x_j^{-1}x)g, g) = c_g p_g(x_j^{-1}x). \tag{6.3}$$

Let $\zeta_1,\dots,\zeta_n$ be any n complex numbers.

Define on G the function $\Phi(x) = \Phi(x_1,\dots,x_n;\zeta_1,\dots,\zeta_n;x)$ as

$$\Phi(x) = -\frac{1}{\det M} \times \det \begin{vmatrix} 0 & \Phi_1(x) & \cdots & \Phi_n(x) \\ \zeta_1 & M_{11} & \cdots & M_{1n} \\ \vdots & & & \vdots \\ \zeta_n & M_{n1} & \cdots & M_{nn} \end{vmatrix}.$$

Then we have the following.

(i) $\Phi(x)$ belongs to the range of L_g,

$\Phi \in L_g \mathscr{H}(U)$.

(ii) $\Phi(x)$ satisfies

$\Phi(x_i) = \zeta_i \quad (i = 1,\dots,n)$,

i.e., is a solution of the interpolation problem.

(iii) $\Phi(x)$ is of minimal norm, in the following sense: If Φ^{other} is any other function on G satisfying (i) and (ii), then

$\|\Phi^{\text{other}}\| > \|\Phi\|$,

the norm being taken in $L^2(G;dx)$.

(iv) The interpolation procedure is invariant under left multiplication in G, in the following sense: Let a be any element of G. Then

$$M(ax_1,\dots,ax_n) = M(x_1,\dots,x_n) \tag{6.4}$$

and

$$\Phi(ax_1,\dots,ax_n;\zeta_1,\dots,\zeta_n;ax) = \Phi(x_1,\dots,x_n;\zeta_1,\dots,\zeta_n;x), \tag{6.5}$$

so that the left-displaced interpolation problem

$$\Phi_1(ax_i) = \zeta_i \tag{6.6}$$

is solved by the function

$$\Phi_1(x) = \Phi(a^{-1}x). \tag{6.7}$$

B. Interpolation for the right transform

Proposition: Let $x_1,\dots,x_n$ be n points in G and $\zeta_1,\dots,\zeta_n$ n complex numbers. Consider the $n \times n$ Hermitian matrix $N = N(x_1,\dots,x_n)$,

$$N_{ij} = c_g p_g(x_i x_j^{-1}) = (U(x_j^{-1})g, U(x_i^{-1})g).$$

Assume that N_{ij} is invertible, and define on G the function $\Psi(x) = \Psi(x_1,\dots,x_n;\zeta_1,\dots,\zeta_n;x)$ as

$$\psi(x) = -\frac{1}{\det N} \times \det \begin{vmatrix} 0 & \Psi_1(x) & \cdots & \Psi_n(x) \\ \zeta_1 & & & \\ \vdots & & N & \\ \zeta_n & & & \end{vmatrix}, \tag{6.8}$$

where $\Psi_j(x) = (U(xx_j^{-1})g, g) \quad (j = 1,\dots,n)$.

Then, (i) Ψ belongs to $R_g \mathscr{H}(U) \subset L^2(G, d_R x)$; (ii) Ψ takes the prescribed values $\zeta_1,\dots,\zeta_n$ at the prescribed points $x_1,\dots,x_n$,

$$\Psi(x_j) = \zeta_j \quad (j = 1,\dots,n); \tag{6.9}$$

(iii) Ψ is of minimal norm, subject to (i) and (ii); and (iv) for any $a \in G$, the function

Downloaded 14 Oct 2004 to 128.112.156.90. Redistribution subject to AIP license or copyright, see http://jmp.aip.org/jmp/copyright.jsp

$$\Psi_1(x) = \Psi(xa^{-1}) \qquad (6.10)$$

takes the values $\zeta_1,...,\zeta_n$ at the points $x_1a,...,x_na$,

$$\Psi_1(x_ja) = \zeta_j \quad (j = 1,...,n). \qquad (6.11)$$

An alternative way of writing the interpolating functions $\Phi(x)$ and $\Psi(x)$ is

$$\Phi(x) = \sum_{i=1}^{n}\sum_{j=1}^{n}\zeta_i(M^{-1})_{ij}p_g(x_j^{-1}x), \qquad (6.12)$$

$$\Psi(x) = \sum_{i=1}^{n}\sum_{j=1}^{n}\zeta_i(N^{-1})_{ij}p_g(xx_j^{-1}), \qquad (6.13)$$

where M^{-1} is the matrix inverse of M, and N^{-1} is the matrix inverse of N.

The interpolations are typical of Hilbert spaces with reproducing kernels and the proof can be adapted from Meschkowsky.[15]

APPENDIX: PROOF OF THEOREM 3.1

In the proof of Theorem 3.1, we shall use the following extension of Schur's lemma.

Proposition A.1: Suppose that (i) G is a group; (ii) U is a unitary irreducible representation of G in a Hilbert space $\mathscr{H}$; (iii) π is a unitary (not necessarily irreducible) representation of G in a Hilbert space $\mathscr{H}'$; (iv) T is a closed operator from $\mathscr{H}$ to $\mathscr{H}'$ with domain $\mathscr{D} \subset \mathscr{H}$ dense in $\mathscr{H}$ and stable under U; and (v) $TU(x) = \pi(x)T$ on $\mathscr{D}$ for every x in G; then T is a multiple of an isometry, and $\mathscr{D} = \mathscr{H}$.

Proof: Let us denote by $(\cdot,\cdot)$ and $(\cdot,\cdot)'$ the scalar products in $\mathscr{H}$ and $\mathscr{H}'$, and by $\|\cdot\|$ and $\|\cdot\|'$ the associated norms.

Consider on $\mathscr{D}$ the scalar product

$$(g,f)_T = (g,f) + (Tg,Tf)'$$

and the associated norm

$$\|g\|_T^2 = \|g\|^2 + \|Tg\|'^2.$$

Then $\mathscr{D}$, equipped with the scalar product $(\cdot,\cdot)_T$, is a Hilbert space which we call $\mathscr{D}_T$.

Since

$$\frac{(\|Tg\|')^2}{\|g\|_T^2} = \frac{(\|Tg\|')^2}{(\|Tg\|')^2 + \|g\|^2} \leqslant 1,$$

T is bounded from $\mathscr{D}_T$ to $\mathscr{H}'$.

Moreover $U(x)$ is unitary in $\mathscr{D}_T$

$$\begin{aligned}\|U(x)g\|_T^2 &= \|U(x)g\|^2 + (\|TU(x)g\|')^2\\ &= \|g\|^2 + (\|\pi(x)Tg\|')^2\\ &= \|g\|^2 + (\|Tg\|')^2 = \|g\|_T^2,\end{aligned}$$

for every x in G and g in $\mathscr{D}$, and $U(x)|_{\mathscr{D}}$ is surjective, since for every $g \in \mathscr{D}$, $g = U(x)U(x^{-1})g$, and $\mathscr{D}$ is stable under $U(x^{-1})$.

By the usual Schur's lemma[17] we have that T is a multiple of an isometry from $\mathscr{D}_T$ to $\mathscr{H}'$, that is

$$(\|Tg\|')^2 = \lambda\, \|g\|_T^2,$$

for every g in $\mathscr{D}$, which gives

$$(\|Tg\|')^2 = \lambda\, \|g\|^2 + \lambda\, (\|Tg\|')^2.$$

From this equality we see that $\lambda \neq 1$ and that

$$(\|Tg\|')^2 = [\lambda/(1-\lambda)]\,\|g\|^2,$$

so T is a multiple of an isometry from $\mathscr{D}$ to $\mathscr{H}'$ and consequently extends to a multiple of an isometry from $\mathscr{H}$ to $\mathscr{H}'$. Since T was assumed closed, one has $\mathscr{D} = \mathscr{H}$, and the proposition is proved.

If we take $\mathscr{H}' = \mathscr{H}$ and $\pi = U$, the same argument together with the classical lemma of Schur shows that T is a multiple of the identity.

In order to prove Theorem 3.1, we first compute the integral (3.1).

For g admissible, consider the following operator T_g from $\mathscr{H}(U)$ to $L^2(G,dx)$: the domain $\mathscr{D}$ of T_g is the set of vectors f in $\mathscr{H}(U)$ such that

$$\int |(U(x)g,f)|^2 dx < +\infty;$$

for f in $\mathscr{D}$, T_gf is defined by

$$(T_gf)(x) = (U(x)g,f).$$

For f in $\mathscr{D}$ and y in G we have

$$\begin{aligned}\int |(U(x)g,U(y)f)|^2\, dx &= \int |(U(y^{-1}x)g,f)|^2\, dx\\ &= \int |(U(x)g,f)|^2\, dx < +\infty,\end{aligned}$$

by the left invariance of dx. So $U(y)f$ belongs to $\mathscr{D}$ for every y in G, i.e., $\mathscr{D}$ is stable under U and

$$T_g U(y) = L(y)T_g \quad \text{on } \mathscr{D}.$$

We see that $\mathscr{D}$ contains the linear span of the set of vectors $U(x)g$, $x \in G$. This linear span is dense in $\mathscr{H}(U)$ by irreducibility of U, so $\mathscr{D}$ is dense in $\mathscr{H}(U)$.

We prove now that T_g is closed: Take a sequence $\{f_n\}$, with $f_n \in \mathscr{D}$ for every n, converging to f in $\mathscr{H}(U)$, and such that T_gf_n converges in $L^2(G,dx)$ to $\varphi \in L^2(G,dx)$. Then T_gf_n converges to φ weakly in $L^2(G,dx)$ and the sequence of $L^2(G,dx)$ norms $\|T_gf_n\|$ is bounded.

By the continuity of the scalar product in $\mathscr{H}(U)$, the sequence of numbers $(U(x)g,f_n)$ converges to $(U(x)g,f)$ for every $x \in G$. Then we have (see Ref. 18, p. 207) that

$$(U(x)g,f) = \varphi(x)$$

so

$$\int |(U(x)g,f)|^2\, dx = \int |\varphi(x)|^2 dx < +\infty,$$

which implies that f belongs to $\mathscr{D}$ and $T_gf = \varphi$. So T_g is closed.

By the extended Schur's lemma (A.1), T_g is a multiple of an isometry: so $\mathscr{D} = \mathscr{H}(U)$ and T_g is bounded.

Now take g_1 and g_2 admissible; then T_{g_1} and T_{g_2} are bounded and $T_{g_1}^* T_{g_2}$ is a bounded operator in $\mathscr{H}(U)$. Since

$$L(x)T_{g_i} = T_{g_i}U(x) \quad (i = 1,2),$$

for every x in G, we have

$$U(x)T_{g_1}^* T_{g_2} = T_{g_1}^* T_{g_2} U(x),$$

for every x in G.

So, by Schur's lemma, $T_{g_1}^* T_{g_2}$ is a multiple of the identity:

$$T_{g_1}^* T_{g_2} = C_{g_1g_2}\mathbf{1}.$$

Downloaded 14 Oct 2004 to 128.112.156.90. Redistribution subject to AIP license or copyright, see http://jmp.aip.org/jmp/copyright.jsp

This means that, for every f_1, f_2 in $\mathscr{H}$,

$$\int \overline{(U(x)g_1,f_1)}(U(x)g_2,f_2)dx = (T_{g_1} f_1, T_{g_2} f_2)_{L^2(G,dx)}$$
$$= (f_1, T^*_{g_1} T_{g_2} f_2)_{\mathscr{H}(U)}$$
$$= C_{g_1 g_2}(f_1,f_2). \qquad \text{(A1)}$$

Let us now consider the number $C_{g_1 g_2}$ defined by

$$C_{g_1 g_2} = \frac{\int \overline{(U(x)g_1,f)}(U(x)g_2,f)dx}{\|f\|^2}. \qquad \text{(A2)}$$

Here, $C_{g_1 g_2}$ is, by (A1), independent of $f \neq 0$.

Let us denote by $\mathscr{A}$ the set of admissible vectors.

The correspondence q: $\mathscr{A} \times \mathscr{A} \to \mathbb{C}$ defined by $q(g_1,g_2) = C_{g_1 g_2}$ for g_1, g_2 in $\mathscr{A}$ is by (A2) a positive, symmetric quadratic form with form domain $\mathscr{A}$. Moreover q is closed: indeed consider on $\mathscr{A}$ the norm $\|\cdot\|_q$ defined by

$$\|g\|_q^2 = \|g\|^2 + q(g,g).$$

Take a $\|\cdot\|_q$-Cauchy sequence of vectors g_n in $\mathscr{A}$. This implies that (i) $\{g_n\}$ is a Cauchy sequence with respect to the $\mathscr{H}(U)$ norm, [so $\{g_n\}$ converges to g in $\mathscr{H}(U)$], and (ii) that

$$\lim_{n,m\to\infty} q(g_n - g_m, g_n - g_m) = 0,$$

which implies that the sequence of functions φ_n of $L^2(G,dx)$ defined by $\varphi_n(x) = (U(x)g_n, f)$ is a Cauchy sequence in $L^2(G,dx)$ and so converges strongly to $\varphi \in L^2(G,dx)$. Consequently φ_n converges weakly to φ and the sequence of norms $\|\varphi_n\|$ is bounded.

Moreover the sequence $(U(x)g_n f)$ converges for each x in G to $(U(x)g,f)$. Then (see Ref. 18, p. 207), $(U(x)g,f) = \varphi(x)$ which means that, for every f in $\mathscr{H}$, $\int |(U(x)g,f)|^2\, dx < +\infty$. In particular $\int |(U(x)g,g)|^2\, dx < +\infty$ and so g is admissible. Furthermore

$$\lim_{n\to\infty} \|g_n - g\|_q^2 = \lim_{n\to\infty} \|g_n - g\|^2 + \lim_{n\to\infty} q(g_n - g, g_n - g)$$
$$= 0 + \lim_{n\to\infty} \frac{\|\varphi_n - \varphi\|^2_{L^2(G,dx)}}{\|f\|^2} = 0,$$

which shows that g_n converges in $\|\cdot\|_q$ norm to $g \in \mathscr{A}$.

So $\mathscr{A}$ is complete and then,[19] q is closed. Being as q is a densely defined closed symmetric positive form, by the second representation theorem,[20] there exists a unique positive operator C with domain $\mathscr{A}$ such that

$$q(g_1,g_2) = C_{g_1 g_2} = (Cg_1, Cg_2).$$

This proves parts (i) and (ii) of Theorem 3.1.

Suppose now G is unimodular; then we can see that

$$q(U(y)g_1, U(y)g_2) = \frac{1}{\|f\|^2}\int \overline{(U(xy)g_1,f)}(U(xy)g_2,f)dx$$
$$= \frac{1}{\|f\|^2}\int \overline{(U(x)g_1,f)}(U(x)g_2,f)d(xy^{-1})$$
$$= q(g_1,g_2).$$

This implies

$$(U(y)^{-1}CU(y)g_1, U(y)^{-1}CU(y)g_2) = (Cg_1, Cg_2),$$

and then $U(y)^{-1}CU(y) = C$ on $\mathscr{D}$.

By the remark at the end of the proof of Proposition A.1, it follows that C is a multiple of the identity since C is closed, $\mathscr{D}$ dense in $\mathscr{H}(U)$ and stable under U.

This proves point (iii) of Theorem 3.1.

In particular, if G is unimodular, then $\mathscr{A} = \mathscr{H}(U)$; that is, if one vector is admissible, then all vectors are admissible.

[1] J. R. Klauder and E. Sudarshan, *Fundamentals of Quantum Optics* (Benjamin, New York, 1968).
[2] J. R. Klauder and B. S. Skagerstam, *Coherent States. Applications in Physics and Mathematical Physics* (World Scientific, Singapore, 1984).
[3] G. Warner, *Harmonic Analysis on Semi-simple Lie Groups. I and II* (Springer, Berlin, 1972).
[4] S. A. Gaal, *Linear Analysis and Representation Theory* (Springer, Berlin, 1973).
[5] E. W. Aslaksen, and J. R. Klauder, J. Math. Phys. **10**, 2267 (1969).
[6] A. Grossmann and J. Morlet, SIAM J. Math. Anal. **15**, 723 (1984).
[7] A. Grossmann and J. Morlet, "Decomposition of functions into wavelets of constant shape and related transforms," to appear in *Mathematics + Physics, Lectures on Recent Results* (World Scientific, Singapore).
[8] T. Paul, J. Math. Phys. **25**, 3252 (1984).
[9] T. Paul, "Affine coherent states for the radial Schrödinger equation," in preparation.
[10] A. O. Barut and L. Girardello, Commun. Math. Phys. **21**, 41 (1972).
[11] A. M. Perelomov, Commun. Math. Phys. **26**, 222 (1972).
[12] M. Duflo and C. C. Moore, J. Funct. Anal. **21**, 209 (1976).
[13] A. L. Carey, Bull. Austral. Math. Soc. **15**, 1 (1976).
[14] E. Hewitt and K. A. Ross, *Abstract Harmonic Analysis. I.* (Springer, Berlin, 1963).
[15] H. Meschkowsky, *Hilbertsche Räume mit Kernfunktion* (Springer, Berlin, 1962).
[16] R. Young, *An Introduction to Non-harmonic Fourier Series* (Academic, New York, 1980).
[17] A. A. Kirillov, *Elements of the Theory of Representations* (Springer, Berlin, 1976).
[18] E. Hewitt and K. Stromberg, *Real and Abstract Analysis* (Springer, Berlin, 1969).
[19] M. Reed and B. Simon, *Methods of Modern Mathematical Physics. I. Functional Analysis* (Academic, New York, 1980).
[20] T. Kato, *Perturbation Theory for Linear Operators* (Springer, Berlin, 1976).

Downloaded 14 Oct 2004 to 128.112.156.90. Redistribution subject to AIP license or copyright, see http://jmp.aip.org/jmp/copyright.jsp

SECTION III

Precursors in Mathematics: Early Wavelet Bases

Introduction

Hans G. Feichtinger

The plain fact that wavelet families are very interesting orthonormal systems for $L^2(\mathbf{R})$ makes it natural to view them as an important contribution to the field of orthogonal expansions of functions. This classical field of mathematical analysis was particularly flourishing in the first thirty years of the twentieth century, when detailed discussions of the convergence of orthogonal series, in particular of trigonometric series, were undertaken.

Alfred Haar describes the situation in his 1910 paper in *Math. Annalen* appropriately as follows: for any given (family of) orthonormal system(s) of functions on the unit interval $[0, 1]$ one has to ask the following questions:

- *convergence theory* (sufficient conditions that a series is convergent);
- *divergence theory* (in contrast to convergence theory, it exhibits examples of relatively "decent" functions for which nevertheless no good convergence, e.g., at that time mostly in the pointwise or uniform sense, takes place);
- *summability theory* (to what extent can summation methods help to overcome the problems of divergence);
- *uniqueness theory* (under what circumstances can one be assured that in case of convergence of the series of partial sums of an orthogonal expansion of a function, its limit equals the original function).

Alfred Haar's 31-page paper is mainly concerned with the properties of the partial sum operators, by studying the integral kernel of the corresponding projection operators and deriving corresponding properties from it, not only for the case of the classical trigonometric functions, but also for orthonormal systems related to Sturm-Liouville differential equations. Recall that only a few years earlier Fejér [Fej04] (also published in *Math. Annalen*) had shown that Cesàro summability (also denoted as C1-summability) was a way to overcome the problem of divergence of a classical Fourier series for the case of continuous functions, while in the years before obviously inherent problems with the questions of pointwise convergence of the partial sums of Fourier series had been revealed. In fact, at the time of the writing of Haar's paper it was not at all clear whether the divergence problem even for continuous functions was shared by general orthonormal systems of functions or only by those considered up to that time.

In this sense the construction of what is nowadays called Haar's orthonormal basis for $L^2([0,1])$ in chapter 3 of his paper provides an answer to this very question and had probably no additional relevance to the author at that time. Indeed, only the last 9 pages of the paper are concerned with the description of the Haar system, verifying the orthogonality and completeness relations, and above all the uniform convergence of the partial sums of the Haar series to f, for any continuous function f on $[0,1]$. More specifically, he shows that one has pointwise convergence of the Haar partial sums at all the Lebesgue points of f, that is, at those points where the pointwise derivative of the antiderivative of f exists. It is well known that this is true for almost all points in $[0,1]$ for a given Lebesgue integrable function on the interval. Moreover, the limit of the Haar partial sums is exactly this real number.

It is certainly of great value to the community that the effort of compiling this volume also brought the translation of Haar's article from German into English, so that readers can now check what Haar really did in his paper. Obviously he did not see the structure of a wavelet basis (at least, not as something worth commenting on), as he was working over the interval $[0,1]$ only. So it was left to Yves Meyer to point out (what is nowadays almost the usual way of explaining the basics of wavelet orthonormal systems) that one can see the Haar system (naturally extended to the whole real line) as the first orthonormal system of this kind, with the disadvantage that it consists of discontinuous functions only.

Obviously Haar's example did not exclude the possibility that uniform convergence for continuous functions could only arise for orthonormal systems of functions that are discontinuous. Hence it was left as an open problem whether one could also have an orthonormal system of *continuous* functions on $[0,1]$ with the same properties. The answer was provided eighteen years later in volume 100 of *Math. Annalen*, in the year 1928, by Philip Franklin, a professor at MIT, in a concise 8-page paper.[1]

Franklin's paper provides the description of an orthonormal basis (again over $[0,1]$), which is nowadays called *Franklin's system*, consisting of appropriately defined continuous, piecewise linear functions, with a node sequence at points of the form $k/2^n$, successively inserted within $[0,1]$. He proves that this series expansion is uniformly convergent for continuous functions on the interval, and in the quadratic mean for square-integrable, measurable functions in the Lebesgue sense. In a final remark Franklin also mentions that one may obtain the Haar basis by starting from his basis, taking derivatives, and then orthonormalizing the resulting system. He also indicates that several of the properties of the Haar system could be derived easily in this way.

The next paper — in chronological order — in this section is due to J.-O. Strömberg, entitled: "A Modified Franklin System and Higher-Order Spline Systems on $\mathbf{R}^n$ as Unconditional Bases for Hardy Spaces." It appeared in Vol. II of the *Conference on Harmonic Analysis in honor of A. Zygmund* (on the latter's eightieth birthday), which took place more than fifty years after the publication of Franklin's paper, in March 1981 (the volume was published in 1983), long before Yves Meyer's construction was carried out (this fact was recognized by him on several occasions later on). Peter Jones, one of the editors of that volume, described the situation at a wavelet conference in Hong Kong with these

[1] For the historically interested reader we note that at this time, along with David Hilbert, Otto Blumenthal, and Constantin Carathéodory, Albert Einstein was an editor of this journal.

words: "here we were all sitting, listening to his talk, without realizing the relevance of the construction he had given, or putting it into the appropriate context."

Looking at the paper nowadays, at a time when the standard facts about wavelets are well known to a wide community, one easily recognizes that Strömberg is really describing an orthonormal wavelet system in the "classical sense" for the Hilbert space $L^2(\mathbf{R})$ or even $L^2(\mathbf{R}^n)$, $n \geq 1$, inspired by Franklin's basis and the concept of higher-order spline bases. Again, the motivation for the paper was a very specific question: can one have unconditional bases for the (real) Hardy spaces $H^p(\mathbf{R}^n)$ (even for $p < 1$)? As a matter of fact, Strömberg's approach did overcome some difficulties in an earlier construction given by L. Carleson [Car80], whose result in turn followed the nonconstructive existence proof given by B. Maurey [Mau79], [Mau80].

Strömberg's paper itself proceeds from a description of his new wavelet system to results stating that it is not only an orthonormal basis for $L^2(\mathbf{R})$, but also an unconditional basis for $H^p(\mathbf{R})$ (for $p > 1/(m+5/2)$, where m is the degree of the splines used). Moreover, it is shown that the H^p-norm is equivalent to a certain solid sequence space norm (i.e., a norm on the space of coefficients which has the property that coefficients which are smaller in absolute value yield functions with smaller H^p-norm).

When Y. Meyer got into contact with J. Morlet and his system of functions (generated by the "Mexican hat function," the second derivative of a Gauss function), which according to numerical experiments worked almost like an orthonormal system, he was aware of the fact that according to Balian's result of 1981 [Bal81] one could not have an orthonormal basis of Weyl-Heisenberg form with good joint time-frequency concentration, and therefore, according to his own words, he was going to verify that it was also impossible to have a wavelet orthonormal system, as we know it today. To his surprise his first conjecture turned out to be false, because he himself was able to produce a system of bandlimited Schwartz functions (in other words, the Fourier transforms of what are nowadays called *Meyer wavelets* have compact support and are infinitely smooth), which is an orthonormal basis for $L^2(\mathbf{R})$. In fact, from the group-theoretical perspective (which had been already brought into view by Alex Grossmann, who was the intermediator between Morlet and Meyer) the analogy between the two cases was quite tempting. In each case a certain irreducible (square) integrable group representation providing a continuous — abstract — wavelet transform played a key role, also providing a continuous inversion formula due to an abstract version of Calderón's reproducing formula, which was then also the basis for other reconstruction principles (such as the atomic decomposition methods for coorbit spaces developed by Feichtinger and Gröchenig [FG86]). Meyer's wavelet construction shows that there are indeed decisive differences between the case of the nonunimodular $(ax+b)$-group and the Heisenberg group, which among other properties is nilpotent and even possesses compact neighborhoods of the identity which are invariant under inner automorphisms.

Yves Meyer's first paper — although certainly widely circulated — was actually not published in one of the standard mathematical journals, but rather in the prestigious "Séminaire Bourbaki," 38eme année, 1985/86, in the issue of February 1986 (published in 1987 in *Astérisque* 145–146). Nevertheless it is fair to say that it marks the beginning of "modern wavelet theory" and that due to his enthusiasm for the possible far-reaching consequences of this discovery wavelets became a hot topic within a short time. Reportedly the construction was carried out in the summer of 1985, and presented in the seminar in October of the same year. When I met Yves Meyer in Paris in February 1986 he told me that he himself, with P. G. Lemarié, had found two constructions of orthonormal wavelet systems, but that they did not yet have a general scheme for constructing wavelets, so it was not yet clear at that time how one could systematically construct wavelets with certain additional properties.

Aside from the construction itself it is quite remarkable to observe, reading this paper more than twenty years after its appearance, how clearly the relevance of the new wavelet system is already explained, based on the fact that it is a universal (in some sense) family of Schwartz functions which provides an unconditional basis for quite a wide range of Banach spaces of functions or tempered distributions, including Sobolev spaces, but also H^1 and BMO (in the w*-sense). Also a bit surprisingly, no direct reference is made to the pioneering work of J. Peetre or H. Triebel, two mathematicians who had developed many details concerning the relevant family of Banach spaces of functions (resp., distributions) (nowadays called Besov-Triebel-Lizorkin spaces), using a Fourier approach that in turn makes essential use of Littlewood-Paley theory.

The second construction, due to Lemarié and Meyer, was published in 1986 (submitted in December 1985) in *Revista Matematica Iberoamericana*, vol. 2. It refers to Wojtaszczyk's paper [Woj82] on the Franklin system as an unconditional basis for H^1, but does not mention Strömberg's paper. It provides a new systems of functions, obtained by dyadic dilations and integer translations of $2^n - 1$ functions ψ_k, such that the overall family is an orthonormal basis for $L^2(\mathbf{R}^n)$. These functions are Schwartz functions, in fact their Fourier transforms are C^∞-functions with compact support, hence they decay faster than the inverse of any polynomial. Furthermore they have vanishing moments of all orders, or, equivalently, the Taylor series of $\hat{\psi}_k$ at zero is trivial. Obviously this also implies that the functions $\hat{\psi}_k$ cannot be analytic, and hence one cannot have exponential decay for the members of the family ψ_k. A recent construction of bandlimited wavelets with subexponential decay can be found in [DH98], which in turn is based on a systematic description of Lemarié-Meyer (bandlimited) wavelets as given in [BSW93].

One should mention here an alternative construction, due to Lemarié ([Lem88], cited in the paper of Guy Battle as a preprint), which has not been included in the present volume. It gives the first construction of exponentially decaying wavelet functions of a given smoothness and any given (finite) number of vanishing moments. The main part of his paper uses spline functions in order to construct for any given natural number m a suitable function ψ of class C^{m-2} for which ψ and its first $m-2$ derivatives are $O(e^{-\varepsilon|t|})$ for a certain $\varepsilon > 0$, and $\int_{\mathbf{R}} t^k \psi(t)\, dt = 0$ for $0 \leq k \leq m-1$. He also generalizes the result to the multidimensional case there.

Recall that it was Ingrid Daubechies who was able to construct compactly supported wavelet systems with similar properties. Her paper [Dau88] (reprinted in this volume) marks one of the cornerstones of the field, for at least two reasons: compactly supported wavelets are important for applications, and moreover the so-called cascade algorithm made the calculation of wavelet coefficients numerically efficient.

The practical importance of wavelet systems actually depends very much on the fact that already appears in the included paper by Lemarié and Meyer from 1986: wavelet orthonormal bases of "nice functions" are also unconditional bases for a variety of Banach spaces of functions or distributions on $\mathbf{R}^n$. In fact, once one started to consider compression and denoising methods based on the wavelet expansion, such as the methods of hard or soft thresholding (which simply consists of the idea of discarding the small wavelet coefficients, interpreting them as noise), the fact that they provide unconditional bases for the different spaces of smooth functions automatically implied that the (nonlinear) thresholding operators are in fact uniformly bounded on all of those spaces. In fact, if the convergence of the wavelet expansion was conditional (based on a specific order), one could not expect that such a property is granted a priori. That the wavelet system has to consist of "nice functions" is obvious, since it is clear that such a property cannot be valid for the simple Haar system (except for the L^2 context), because a finite approximating partial sum of a wavelet series of a smooth function would of course be just a step function, having no smoothness at all. In contrast, the partial sums of the wavelet expansion for a function in some Sobolev space (to give an example) are convergent in that very space, whenever the function expanded belongs to that space and the wavelet system satisfies sufficiently high moment conditions.

The last paper in this series is due to Guy Battle, and is entitled "A Block Spin Construction of Ondelettes, I: Lemarié Functions." It appeared in *Comm. Math. Phys.* in 1987, and was motivated by potential applications in physics. In fact, the author refers to an earlier construction (a polynomial generalization of Haar's system to $\mathbf{R}^n$), which he had published together with G. Federbush in 1982 and 1983 under the name of "phase cell cluster expansions." Their early wavelet-like systems (using a finite number of generators) did have good concentration and vanishing moments, but poor regularity. Stimulated by the constructions of Lemarié and Meyer, he provides a cell-cluster approach to the construction of orthonormal wavelet systems. The two main steps are first an orthogonalization at the level of dyadic scales, which is then complemented by the procedure (which has now become standard) of "symmetric orthonormalization" (preserving the group structure of the system: starting from a Riesz basis of functions, obtained by shifts along a lattice, one obtains an equivalent orthonormal basis, with the same structure, with identical closed linear span within the Hilbert space L^2).

In summary, this section provides access to a selection of papers which play a historical role within mathematical analysis, and which belong certainly to the most cited papers over an extended period of time. Since some of those papers have been available so far only in their original language (German in the case of Haar's paper, translated by Georg Zimmermann, and French in the case of Yves Meyer's papers, translated by John Horváth) it is a valuable service to the community that we are seeing carefully translated English versions in this volume. Recall that, for example, Haar's basis (extended to all of $\mathbf{R}$), is quite often taken as the most natural and simple example of a wavelet orthonormal system, with all the algebraic properties easily verified, which have played an important role, for example, in Daubechies' construction of compactly supported wavelets of a prescribed smoothness.

It is almost characteristic of the constructions provided here that they arose from very specific questions in a functional analytic context, which by themselves would have been considered as interesting mathematical problems by a small group of experts only. However, the resulting constructions turned out to be of high relevance for many applications, and in a number of contexts completely different from the original ones. Hence they provide good examples of "how mathematical research may work," and that the investigation of fundamental mathematical questions often yields surprising developments that benefit the entire scientific community.

At the end, let us mention that additional historical information on the early times of wavelet theory can also be found in [Mey93].

References

[Bal81] R. Balian, *Un principe d'incertitude fort en théorie du signal ou en mécanique quantique*, C. R. Acad. Sci. Paris **292** (1981), 1357–1362.

[BSW93] A. Bonami, F. Soria, and G. Weiss, *Band-limited wavelets*, J. Geom. Anal. **3** (1993), 543–578.

[Car80] L. Carleson, *An explicit unconditional basis in* H^1, Bull. Sci. Math. (2) **104** (1980), 405–416.

[Dau88] I. Daubechies, *Orthonormal bases of compactly supported wavelets*, Comm. Pure Appl. Math. **41** (1988), 909–996.

[DH98] J. Dziubański and E. Hernández, *Band-limited wavelets with subexponential decay*, Canad. Math. Bull. **41** (1998), 398–403.

[FG86] H. G. Feichtinger and K. Gröchenig, *Banach spaces related to integrable group representations and their atomic decompositions*, I, J. Funct. Anal. **86** (1989), 307–340.

[Fej04] L. Fejér, *Untersuchungen über Fouriersche Reihen*, Math Ann. **58** (1904), 51–69.

[Lem88] P. G. Lemarié, *Ondelettes à localisation exponentielle*, J. Math. Pures Appl. **67** (1988), 227–236.

[Mau79] B. Maurey, *Isomorphismes entre espaces* H_1, C. R. Acad. Sci., Paris, Sér A **288** (1979), 271–273.

[Mau80] ———, *Isomorphismes entre espaces* H_1, Acta Math. **145** (1980), 79–120.

[Mey93] Y. Meyer, *Wavelets: Algorithms and Applications*, SIAM, Philadelphia, 1993.

[Woj82] P. Wojtaszczyk, *The Franklin system is an unconditional basis in* H_1, Ark. Mat. **20** (1982), 293–300.

On the Theory of Orthogonal Function Systems

Alfred Haar
Translated by Georg Zimmermann

Introduction

In the theory of series expansions of real functions, the so-called *orthogonal function systems* play a major rôle. By this, we mean a system of infinitely many functions $\varphi_1(s), \varphi_2(s), \ldots$ which have, with respect to an arbitrary, measurable set M of points, the *orthogonality property*

$$\int_M \varphi_p(s)\, \varphi_q(s)\, ds = 0 \qquad (p \neq q,\ p, q = 1, 2, \ldots),$$
$$\int_M (\varphi_p(s))^2\, ds = 1 \qquad (p = 1, 2, \ldots),$$

where the integrals are taken in the Lebesgue sense; if they furthermore satisfy the so-called *completeness relation*

$$\int_M (u(s))^2\, ds = \left\{ \int_M u(s)\, \varphi_1(s)\, ds \right\}^2 + \left\{ \int_M u(s)\, \varphi_2(s)\, ds \right\}^2 + \cdots$$

for all functions $u(s)$ which together with their squares are integrable over the set M, then, following Hilbert, we denote the system a *complete orthogonal function system*, or, for short, a *complete orthogonal system* for the measure space M.

The formal infinite series

$$\varphi_1(s) \int_M f(t)\, \varphi_1(t)\, dt + \varphi_2(s) \int_M f(t)\, \varphi_2(t)\, dt + \cdots$$

is denoted the *Fourier series* of $f(s)$ with respect to the orthogonal function system $\varphi_1(s)$, $\varphi_2(s), \ldots$.

The most simple orthogonal system is the system of trigonometric functions (for the interval $0 \leq s \leq 2\pi$)

$$\frac{1}{\sqrt{2\pi}},\ \frac{1}{\sqrt{\pi}} \cos s,\ \frac{1}{\sqrt{\pi}} \sin s,\ \ldots,\ \frac{1}{\sqrt{\pi}} \cos ns,\ \frac{1}{\sqrt{\pi}} \sin ns,\ \ldots.$$

Author's Note: This paper is, except for insignificant modifications, a reprint of my inaugural dissertation from Göttingen, which appeared in July 1909.

Translator's Note: As nearly a century has elapsed since its publication, the original text sounds somewhat archaic to the modern ear. The translator has attempted to preserve the author's sound and style as much as possible. Some minor typographical errors have been corrected, and some notation has been modernized.

A large and interesting class of orthogonal function systems stems from the so-called *eigenvalue problem* for self-adjoint differential equations. This problem consists of determining those values for the parameter λ, for which the differential equation

$$\frac{d}{dx}\Big(p(x)\,\frac{du}{dx}\Big) + q(x)\,u + \lambda\,u = 0$$

has a solution which at two points $x = \alpha$ and $x = \beta$, say, satisfies homogeneous boundary conditions, like, for example,

$$u(\alpha) = 0 \quad \text{and} \quad u(\beta) = 0\,,$$

or

$$\frac{du}{dx} - h\,u = 0 \quad \text{for} \quad x = \alpha \quad \text{and} \quad \frac{du}{dx} + H\,u = 0 \quad \text{for} \quad x = \beta\,.$$

It can be shown that, if the functions $p(x)$ and $q(x)$ satisfy certain continuity conditions, then there always exist countably many such parameter values, and the associated solutions form a complete orthogonal function system for the interval under consideration. Of particular importance is the so-called *regular* case, where the function $p(x)$ does not vanish on the interval (including the endpoints). The functions obtained in this manner are denoted a *Sturm-Liouville function system.* The case where the function $p(x)$ vanishes at one or both ends of the interval $[\alpha, \beta]$ is no less important; the *spherical harmonics* and the *Bessel functions* satisfy such a differential equation, as we know.

If we consider the by now classical theory of trigonometric series, we find that the results in this theory can be classified in four groups. First, we should name the

theory of convergence, whose duty it is to determine sufficient conditions on a function ensuring convergence of its trigonometric series. Right next to these studies, there is the

theory of divergence, which complements the former in many ways; it draws the lines showing how far the theory of convergence can reach at most. The most important result of this theory is the Du Bois-Reymond theorem, predicting the existence of a continuous function whose trigonometric series does not converge. This makes necessary a

theory of summation, which is called upon to help out in the cases of divergence. Indeed, various summation methods are known with the aid of which it is possible to "sum" the trigonometric series of all continuous functions. The modern theory of summation of trigonometric series was founded by L. Fejér; later on, various results by Poisson and Riemann were interpreted as summation methods by various authors. The last and most difficult problems are encountered in the

theory of uniqueness, which by its main problem — under which circumstances a convergent trigonometric series is the Fourier series of the represented function — forms the keystone of the whole theory. By the famous papers of Riemann, Cantor, and Du Bois-Reymond we are already able to answer also these questions.

As to the theory of the orthogonal functions originating from second order differential equations, which are closely related to the trigonometric functions, only the theory of convergence has been studied up to now. By a series of papers,[1] it has been proven that the conditions stated in the theory of trigonometric series, here also are sufficient to ensure

[1] Of the many papers addressing this question I only name the more recent studies by Stekloff, Zaremba, Kneser, Hilbert, and Hobson.

convergence of the series. Only the theory of spherical harmonics has been pursued beyond these results in a recently published paper by Mr. Fejér,[2] in which the author discusses the summation theory of this function system.

In the paper at hand, we are dealing with the *theory of divergence and the theory of summation of orthogonal function systems.*

In Chapter I, the theory of divergence is discussed; § 1 presents a general sufficient condition which in many cases enables us to construct for a given orthogonal function system a continuous function whose Fourier series with respect to this orthogonal system does not converge. In § 2 and § 3 this theorem is applied to the theory of Sturm-Liouville functions and to spherical harmonics in order to construct a continuous function that cannot be expanded with respect to these function systems.

Chapter II is dedicated to the theory of summation; in § 1, we prove a general lemma, which enables us to establish the converse of the theorem proved in § 1 of the first chapter. § 2 and § 3 present the application of this lemma to Sturm-Liouville series. This yields the result (which is a generalization of a theorem proved for trigonometric series by L. Fejér) that if a continuous function — which possibly has to satisfy certain boundary conditions — is expanded into a Sturm-Liouville series, and from the partial sums s_n of this series, the arithmetic means

$$s_1, \frac{s_1+s_2}{2}, \frac{s_1+s_2+s_3}{3}, \ldots, \frac{s_1+s_2+\cdots+s_n}{n}, \ldots$$

are formed, then the sequence of functions thus defined converges uniformly to the given function. § 4 presents a general criterion which allows us to decide whether a given summation method has the property that by its means, the Fourier series with respect to a given orthogonal system of all functions in the "range" of this system are summable.

The investigations in Chapter I suggest the question: does there exist at all an orthogonal function system with the property that *every continuous function can be expanded in the Fourier manner into a uniformly convergent series*, according to the functions of this system? In Chapter III, we shall encounter a *whole class* of orthogonal systems having this property. But these function systems are also of interest from a different point of view, namely, because of a number of properties distinguishing this class. These properties point to the fact that in certain problems, where the orthogonal systems are used as an auxiliary means only, it will be advisable to employ just these special systems, whereby in many cases we gain a simpler presentation of the proof. In many cases still the nature of the very problem requires the application of such a special function system, without which the solution of the problem does not seem possible.

Chapter I. Divergent Series

If the functions

$$\varphi_1(s), \varphi_2(s), \ldots, \varphi_n(s), \ldots$$

defined on the interval $[\alpha, \beta]$ form a complete orthogonal function system, then the formal series

$$\varphi_1(s)\int_\alpha^\beta f(t)\,\varphi_1(t)\,dt + \varphi_2(s)\int_\alpha^\beta f(t)\,\varphi_2(t)\,dt + \cdots$$

[2] *Math. Annalen*, Vol. 67 (1909), p. 76.

shall be denoted the *Fourier series* of the function $f(s)$ with respect to this orthogonal system. Terminating this infinite series at the nth term, we obtain the finite sum

$$\varphi_1(s) \int_\alpha^\beta f(t)\, \varphi_1(t)\, dt + \cdots + \varphi_n(s) \int_\alpha^\beta f(t)\, \varphi_n(t)\, dt,$$

which we will refer to as $[f(s)]_n$ from now on. Writing

$$K_n(s,t) = \varphi_1(s)\, \varphi_1(t) + \varphi_2(s)\, \varphi_2(t) + \cdots + \varphi_n(s)\, \varphi_n(t)$$

for short yields

$$[f(s)]_n = \int_\alpha^\beta K_n(s,t)\, f(t)\, dt.$$

§ 1. A General Criterion

We base our investigations on an arbitrary orthogonal function system for the interval $[\alpha, \beta]$:

$$\varphi_1(s),\ \varphi_2(s),\ \ldots.$$

We denote by a an arbitrary point of this interval and consider the infinitely many numbers

$$\omega_n = \int_\alpha^\beta |K_n(a,t)|\, dt;$$

if the numbers ω_n thus defined do not all lie below a finite bound, that is, if from the sequence

$$\omega_1,\ \omega_2,\ \omega_3,\ \ldots,$$

we can take a subsequence

$$\omega_{\nu_1} \leq \omega_{\nu_2} \leq \omega_{\nu_3} \leq \cdots$$

whose elements grow beyond all bounds, then it is always possible to construct a continuous function whose Fourier series with respect to the orthogonal system at hand diverges at the point $s = a$.

The construction of this function $F(s)$ takes place in three steps.

1) *First*, we construct the both integrable and square integrable functions

$$v_{\nu_1}(s),\ v_{\nu_2}(s),\ v_{\nu_3}(s),\ \ldots,$$

defined by the equation

$$v_{\nu_p}(s) = \text{sign of } K_{\nu_p}(a,s);$$

that is,

$$\begin{aligned} v_{\nu_p}(s) &= 1 \quad \text{if } K_{\nu_p}(a,s) > 0, \\ &= -1 \quad \text{if} \qquad\qquad < 0, \\ &= 0 \quad \text{if} \qquad\qquad = 0; \end{aligned}$$

thus it always holds that

$$v_{\nu_p}(t)\, K_{\nu_p}(a,t) = |K_{\nu_p}(a,t)|,$$

and consequently we have, in our notation,

$$[v_{\nu_p}(a)]_{\nu_p} = \int_\alpha^\beta |K_{\nu_p}(a,t)|\,dt = \omega_{\nu_p}.$$

The functions $v_{\nu_p}(s)$, which have absolute value ≤ 1 everywhere, thus have the property that the ν_pth partial sum of their Fourier series at the point $s = a$ has value ω_{ν_p}.

2) Next, we construct a sequence of continuous functions

$$f_{\nu_1}(s),\ f_{\nu_2}(s),\ f_{\nu_3}(s),\ \dots$$

of absolute value less than 1 and having the property that

$$\int_\alpha^\beta (v_{\nu_p}(s) - f_{\nu_p}(s))^2\,ds < \delta_p \quad (p = 1, 2, 3, \dots),$$

where δ_p stands for an arbitrarily small positive quantity.[3]

Forming the ν_pth partial sum of the expansion of $f_{\nu_p}(s)$ yields

$$\begin{aligned}
\left|[f_{\nu_p}(a)]_{\nu_p}\right| &= \left|\int_\alpha^\beta K_{\nu_p}(a,t)\, f_{\nu_p}(t)\,dt\right| \\
&= \left|\int_\alpha^\beta K_{\nu_p}(a,t)\, v_{\nu_p}(t)\,dt - \int_\alpha^\beta K_{\nu_p}(a,t)\,\big(v_{\nu_p}(t) - f_{\nu_p}(t)\big)\,dt\right| \\
&\geq \omega_{\nu_p} - \left|\int_\alpha^\beta K_{\nu_p}(a,t)\,\big(v_{\nu_p}(t) - f_{\nu_p}(t)\big)\,dt\right| \\
&\geq \omega_{\nu_p} - \sqrt{\int_\alpha^\beta \big(K_{\nu_p}(a,t)\big)^2\,dt \int_\alpha^\beta \big(v_{\nu_p}(t) - f_{\nu_p}(t)\big)^2\,dt}.
\end{aligned}$$

If we now choose the quantity δ_p such that

$$\sqrt{\delta_p \int_\alpha^\beta \big(K_{\nu_p}(a,t)\big)^2\,dt} < \frac{\omega_{\nu_p}}{2},$$

then

$$\left|[f_{\nu_p}(a)]_{\nu_p}\right| > \frac{\omega_{\nu_p}}{2}.$$

[3] The construction of these functions does not pose any difficulties. A possible procedure is the following: let us assume for simplicity that $[0, 2\pi]$ is the interval under consideration, which after all is no substantial restriction; we let

$$f_{\nu_p}(r,s) = \int_0^{2\pi} \frac{1 - r^2}{1 - 2r\,\cos(s-t) + r^2}\, v_{\nu_p}(t)\,dt \quad (0 < r < 1).$$

In the theory of trigonometric series it is shown that the continuous functions f_{ν_p} remain smaller than the maximum of $|v_{\nu_p}(s)|$ in absolute value, and that

$$\lim_{r\to 1} \int_0^{2\pi} [f_{\nu_p}(r,s) - v_{\nu_p}(s)]^2\,ds = 0.$$

Thus it is possible to determine r such that $f_{\nu_p}(r,s)$ satisfies all conditions posed.

In other words, the continuous functions $f_{\nu_p}(s)$, which remain less than 1 in absolute value, have the property that the ν_pth partial sum of their Fourier series at the point $s = a$ turns out to be larger than $\omega_{\nu_p}/2$.

3) We now reach the sought-after function $F(s)$ by the following consideration: the ν_1th partial sum of the Fourier expansion of the continuous function

$$F'(s) = f_{\nu_1}(s)$$

at the point $s = a$ is larger than $\omega_{\nu_1}/2$ in absolute value. If this series does not diverge at this point, we can determine a number G' such that all partial sums of the Fourier series of $F'(s)$ for $s = a$ are less than G', that is, that

$$|[F'(a)]_n| < G' \quad (n = 1, 2, 3, \dots).$$

We now pick out from the sequence of indices

$$\nu' = \nu_1,\ \nu_2,\ \nu_3,\ \dots \tag{1}$$

an index which we will call ν'', say, in such a manner that

$$\omega_{\nu''} > 6 \cdot 4(G' + 1),$$

and then form with the associated function $f_{\nu''}(s)$ the continuous function

$$F''(s) = f_{\nu'}(s) + \frac{1}{4} f_{\nu''}(s).$$

If the Fourier series of this continuous function $F''(s)$ is not divergent, we can determine a number G'' such that for each n, we have

$$|[F''(a)]_n| < G''.$$

Then we determine in the index sequence (1) an index ν''' in such a manner that the associated $\omega_{\nu'''}$ satisfies

$$\omega_{\nu'''} > 6 \cdot 4^2(G'' + 2),$$

and form the function

$$F'''(s) = f_{\nu'}(s) + \frac{1}{4} f_{\nu''}(s) + \frac{1}{4^2} f_{\nu'''}(s).$$

We keep proceeding in this manner: if the Fourier series of the continuous function

$$F^{(q-1)}(s) = f_{\nu'}(s) + \frac{1}{4} f_{\nu''}(s) + \dots + \frac{1}{4^{q-2}} f_{\nu^{(q-1)}}(s)$$

does not diverge at the point $s = a$, then we determine $G^{(q-1)}$ such that, for each n,

$$|[F^{(q-1)}(a)]_n| < G^{(q-1)} \tag{2}$$

and then pick out from the sequence (1) an index $\nu^{(q)}$ in such a manner that

$$\omega_{\nu^{(q)}} > 6 \cdot 4^{q-1}(G^{(q-1)} + q - 1); \tag{3}$$

the possibility of this choice is guaranteed by the assumption that the ω_{ν_p} grow beyond all bounds.

Now I claim that the *infinite series*

$$F(s) = f_{\nu'}(s) + \frac{1}{4} f_{\nu''}(s) + \cdots + \frac{1}{4^{q-1}} f_{\nu^{(q)}}(s) + \cdots$$

represents a continuous function whose Fourier series with respect to the orthogonal system at hand diverges at the point $s = a$.

The uniform convergence of the series $F(s)$ follows immediately from the fact that all $f_{\nu^{(q)}}$ remain less than 1 in absolute value. To prove the divergence of the Fourier series of $F(s)$ at $s = a$, we show that the *number sequence*

$$[F(a)]_{\nu'},\ [F(a)]_{\nu''},\ [F(a)]_{\nu'''},\ \ldots$$

grows beyond all bounds. To estimate $[F(a)]_{\nu^{(q)}}$, say, we decompose the function $F(s)$ in three summands, as indicated in the formula

$$F(s) = \left(f_{\nu'}(s) + \cdots + \frac{1}{4^{q-2}} f_{\nu^{(q-1)}}(s) \right) + \frac{1}{4^{q-1}} f_{\nu^{(q)}}(s) + \left(\frac{1}{4^q} f_{\nu^{(q+1)}}(s) + \cdots \right)$$

by the inserted parentheses, and consider the $\nu^{(q)}$th partial sum of the Fourier series of each individual summand at the point $s = a$. The first summand — which in our notation is $F^{(q-1)}(s)$ — contributes in the expression for $[F(a)]_{\nu^{(q)}}$ an amount that by inequality (2) is smaller than $G^{(q-1)}$. The last summand is smaller than $1/(3 \cdot 4^{q-1})$ in absolute value, and its contributed amount is thus less than $\omega_{\nu^{(q)}}/(3 \cdot 4^{q-1})$.[4] Since finally

$$\left|[f_{\nu^{(q)}}(a)]_{\nu^{(q)}}\right| > \frac{\omega_{\nu^{(q)}}}{2},$$

this implies

$$\left|[F(a)]_{\nu^{(q)}}\right| > \frac{\omega_{\nu^{(q)}}}{2 \cdot 4^{q-1}} - G^{(q-1)} - \frac{\omega_{\nu^{(q)}}}{3 \cdot 4^{q-1}} = \frac{\omega_{\nu^{(q)}}}{6 \cdot 4^{q-1}} - G^{(q-1)}.$$

Hence, according to inequality (3),

$$\left|[F(a)]_{\nu^{(q)}}\right| > q - 1.$$

Thus our claim is proved.

The condition that the ω_n do not remain below a bound independent of n thus turns out to be *sufficient* for a continuous function to exist whose Fourier series with respect to the considered orthogonal system does not converge. We shall see in the next section that for a very extensive class of orthogonal systems, this condition is also *necessary*.

[4]Indeed, we have that, if $\varphi(s)$ denotes an arbitrary function which on the entire interval $[\alpha, \beta]$ is less than M in absolute value,

$$|[\varphi(a)]_{\nu^{(q)}}| = \left| \int_\alpha^\beta K_{\nu^{(q)}}(a,t)\,\varphi(t)\,dt \right| \le M \int_\alpha^\beta |K_{\nu^{(q)}}(a,t)|\,dt = M\,\omega_{\nu^{(q)}}.$$

§ 2. Application to Sturm-Liouville Series[5]

The theorem just derived has an immediate application to the theory of Sturm-Liouville series.

If the coefficients $p(x)$ and $q(x)$ of the self-adjoint differential equation

$$L(u) \equiv \frac{d}{dx}\left(p(x)\,\frac{du}{dx}\right) + q\,u + \lambda\,u = 0 \tag{4}$$

are different from zero on the entire interval $[\alpha, \beta]$ (including the boundaries), then the parameter λ can — in infinitely many ways — be determined such that the present equation possesses a solution satisfying the boundary conditions

$$\frac{du}{dx} - h\,u = 0 \quad \text{for} \quad x = \alpha, \qquad \frac{du}{dx} + H\,u = 0 \quad \text{for} \quad x = \beta. \tag{5}$$

The infinitely many functions

$$u_1(x),\ u_2(x),\ u_3(x),\ \cdots$$

thus obtained form a complete orthogonal function system; we will call it a *Sturm-Liouville orthogonal system* for short and remark immediately that instead of the boundary conditions (5), an arbitrary pair of homogeneous boundary conditions may be chosen.

To study the orthogonal system $u_n(x)$, we apply to the present differential equation a transformation common to this theory, stemming from Liouville. We put

$$z = \int_\alpha^x (p(x'))^{-1/2}\,dx' \qquad v(z) = (p(x))^{-1/4}\,u(x).$$

Our differential equation then passes into the new differential equation

$$\frac{d^2v}{dz^2} + Q\,u + \lambda\,v = 0, \tag{4'}$$

where $Q(z)$ stands for a function easily expressible in terms of the functions $p(x)$, $q(x)$.

The boundary conditions become

$$\frac{dv}{dz} - h'\,v = 0 \quad \text{for} \quad z = 0, \qquad \frac{dv}{dz} - H'\,v = 0 \quad \text{for} \quad z = \pi, \tag{5'}$$

where for simplicity we have assumed

$$\int_\alpha^\beta (p(x))^{-1/2}\,dx = \pi$$

— which after all can always be obtained by multiplying the independent variable by a constant; h' and H' are two constants which can be expressed easily in terms of h, H. We denote the Sturm-Liouville functions arising from the differential equation (4′) by

$$v_1(z),\ v_2(z),\ v_3(z),\ \ldots$$

[5]With a similar method, Mr. Lebesgue has constructed a continuous function whose trigonometric series is divergent, resp., not uniformly convergent (cf. Lebesgue, Séries trigonométriques, p. 87). In a treatise that recently appeared in the *Annales de Toulouse* (3e série, t. I, p. 25), Mr. Lebesgue has generalized his results. Quite some common ground with the paper at hand can be found therein.

and, to begin with, show that there *exists a continuous function whose Fourier series with respect to this orthogonal system does not converge.*

To this purpose, we employ an asymptotic representation of the nth term of this function system, due to Liouville and improved by Hobson.[6] We assume it to be normalized such that

$$\int_0^\pi (v_n(z))^2 \, dz = 1.$$

Then we have for every point of the interval $[0, \pi]$ that

$$v_n(z) = \sqrt{\frac{2}{\pi}} \cos nz \left\{ 1 + \frac{\alpha_n(z)}{n^2} \right\} + \sin nz \left\{ \frac{\beta(z)}{n} + \frac{\gamma_n(z)}{n^2} \right\},$$

where the functions $\alpha_n(z)$, $\gamma_n(z)$, and $\beta(z)$ remain below a bound A, independent of n and z. To prove the existence of a continuous function whose Fourier series diverges at the point $z = a$, it suffices — according to the theorem derived in § 1 — to show that the quantities

$$\int_0^\pi |K_n(a,t)| \, dt = \int_0^\pi |v_1(a)\, v_1(t) + \cdots + v_n(a)\, v_n(t)| \, dt$$

grow beyond all bounds. To this end, we set

$$K_n(a,t) = \frac{2}{\pi} \sum_{p=1,\ldots,n} \cos pa \, \cos pt + \Phi_n(a,t)$$

and prove that $|\Phi_n(a,t)|$ remains below a bound independent from n, a, and t, but

$$\int_0^\pi \left| \sum_{p=1,\ldots,n} \cos pa \, \cos pt \right| dt$$

grows beyond all bounds. Namely, if we form $\Phi_n(a,t)$, we obtain first the series

$$\sqrt{\frac{2}{\pi}} \left\{ \beta(t) \sum_{p=1,\ldots,n} \frac{\cos pa \, \sin pt}{p} + \beta(a) \sum_{p=1,\ldots,n} \frac{\cos pa \, \sin pt}{p} \right\} \tag{6}$$

and second three finite trigonometric series whose pth terms have denominator p^2, p^3, p^4, respectively. Since in each term of these latter series, the absolute values of the numerators are less than A^2, these series are certainly less than $A^2 \sum_{p=1,2,\ldots} \frac{1}{p^2}$ in absolute value. To show now also that the series (6) or — which is equivalent — the series

$$\sum_{p=1,\ldots,n} \frac{\cos pa \, \sin pt}{p} \quad \text{and} \quad \sum_{p=1,\ldots,n} \frac{\cos pt \, \sin pa}{p},$$

[6] Cf. Hobson, *Proceedings of the London Mathematical Society*, Ser. 2, Vol. 6 (1908), p. 349.

respectively, remain below a bound independent from n, a, and t, we decompose

$$\sum_{p=1,\ldots,n} \frac{\cos pa \,\sin pt}{p} = \frac{1}{2}\Big\{ \sum_{p=1,\ldots,n} \frac{\sin p(t+a)}{p} + \sum_{p=1,\ldots,n} \frac{\sin p(t-a)}{p} \Big\},$$

$$\sum_{p=1,\ldots,n} \frac{\cos pt \,\sin pa}{p} = \frac{1}{2}\Big\{ \sum_{p=1,\ldots,n} \frac{\sin p(t+a)}{p} + \sum_{p=1,\ldots,n} \frac{\sin p(a-t)}{p} \Big\}.$$

Since, however, the series $\sum_{p=1,2,\ldots} \frac{\sin pt}{p}$ is the Fourier series of the function $\frac{\pi-t}{2}$, the sums $|\sum_{p=1,\ldots,n} \frac{\sin pt}{p}|$ — as taught in the theory of trigonometric series[7] — remain below an upper bound independent from t and n, and thus it is shown that $|\Phi_n(a,t)|$ remains finite.

It remains to prove that the quantities

$$\omega_n = \int_0^\beta \Big| \sum_{p=1,\ldots,n} \cos pt \,\cos pa \Big| \,dt$$

become infinitely large as n grows. To this end, we proceed similarly as Mr. Lebesgue does at the place mentioned above.

To shorten the calculations, we assume that the arbitrarily chosen point $z = a$ lies between 0 and $\frac{\pi}{2}$, that is,

$$0 < \delta < a < \frac{\pi}{2} - \delta.^{8}$$

Now

$$2 \sum_{p=1,\ldots,n} \cos pt \,\cos pa = \frac{\sin(2n+1)(\frac{t+a}{2})}{2\sin(\frac{t+a}{2})} + \frac{\sin(2n+1)(\frac{t-a}{2})}{2\sin(\frac{t-a}{2})} - 1;$$

since, however, for each value of n and t under consideration, the first summand in this formula in absolute value remains less than the smaller of the two quantities

$$\left| \frac{1}{2\sin(\frac{\delta}{2})} \right| \quad \text{and} \quad \left| \frac{1}{2\sin(\frac{3\pi}{4} - \frac{\delta}{2})} \right|,$$

it obviously suffices to show that

$$\omega_n' = \int_0^\pi \left| \frac{\sin \frac{(2n+1)(t-a)}{2}}{\sin(\frac{t-a}{2})} \right| dt = \int_{-a/2}^{(\pi-a)/2} \left| \frac{\sin(2n+1)\vartheta}{\sin\vartheta} \right| d\vartheta$$

becomes infinitely large as n grows. We obviously have

$$\int_{-a/2}^{(\pi-a)/2} \left| \frac{\sin(2n+1)\vartheta}{\sin\vartheta} \right| d\vartheta > \int_{-a/2}^{(\pi-a)/2} \left| \frac{\sin(2n+1)\vartheta}{\vartheta} \right| d\vartheta.$$

---

[7] Compare, e.g., Kneser, *Math. Ann.*, Vol. 60 (1905), p. 402.

[8] Should this not be the case, we first would have to modify the integral ω_ν slightly in order to be able to apply our further considerations; however, since we only care to show that there exist continuous functions whose Sturm-Liouville series does not converge, this restriction is irrelevant.

Let us consider the intervals where

$$\left|\sin(2n+1)\vartheta\right| > \sin\left(\frac{\pi}{8}\right) = \mu$$

holds;

$$i_p = \left[\pi\,\frac{\frac{1}{8}+p}{2n+1},\ \pi\,\frac{\frac{7}{8}+p}{2n+1}\right]$$

is such an interval; since

$$\int_{i_p} \frac{\mu}{\vartheta}\,d\vartheta = \mu\,\log\left(1+\frac{6}{8p+1}\right),$$

we certainly have

$$\omega_n' \geq \mu \sum_{p=1,\ldots,\nu} \log\left(1+\frac{6}{8p+1}\right),$$

where ν stands for the smallest integer of the property that

$$\pi\,\frac{\frac{7}{8}+\nu}{2n+1} \leq \frac{\pi-a}{2}.$$

Now, however, this number ν — who still depends on n — grows beyond all bounds as n grows; since furthermore the infinite series

$$\sum_{p=1,2,\ldots} \log\left(1+\frac{6}{8p+1}\right)$$

diverges,[9] n can be chosen so large that ω_n becomes larger than an arbitrary number.

This allows us — on account of our general theorem in § 1 — to conclude that there exists a continuous function $F(z)$, whose Fourier series with respect to the $v_n(z)$ diverges at the point $z = a$.

Having proved the existence of this continuous function $F(z)$, it is now easy to show that *the Fourier series of the continuous function*

$$\overline{F}(x) = \big(p(x)\big)^{-1/4}\,F\left(\int_\alpha^x \big(p(x')\big)^{-1/2}\,dx'\right)$$

with respect to the $u_n(x)$ is divergent. Namely, since by virtue of our substitution

$$v_n(z) = \big(p(x)\big)^{-1/4}\,u_n(x),$$

we have

$$dz = (p(x))^{-1/2}\,dx,$$

[9]The easiest way to prove the divergence of this series is to show that the limit of the product $p\log\left(1+\frac{6}{8p+1}\right)$ for $p\to\infty$ equals $\frac{3}{4}$.

we thus find

$$\int_{\alpha}^{\beta} \overline{F}(x)\, u_n(x)\, dx = \int_{\alpha}^{\beta} (p(x))^{-1/4} F\Big(\int_{\alpha}^{x} (p(x'))^{-1/2}\, dx'\Big) (p(x))^{-1/4}\, v_n(z)\, dx$$
$$= \int_0^{\pi} F(z)\, v_n(z)\, dz.$$

Since, however, the series

$$\sum_{n=1,2,\ldots} v_n(z) \int_0^{\pi} F(z)\, v_n(z)\, dz$$

diverges at the point $z = a$, the same holds for the series

$$\sum_{n=1,2,\ldots} u_n(x) \int_{\alpha}^{\beta} \overline{F}(x)\, u_n(x)\, dx = (p(x))^{-1/4} \sum_{n=1,2,\ldots} v_n(z) \int_0^{\pi} F(z)\, v_n(z)\, dz$$

at the point $x = b$, which by virtue of the transformation

$$z = \int_{\alpha}^{x} (p(x'))^{-1/4}\, dx'$$

corresponds to the point $z = a$.[10] Thus our claim is proved.

§ 3. Application to Spherical Harmonics

As nth spherical harmonic or nth Legendre polynomial $P_n(x)$ we denote that solution of the differential equation

$$\frac{d}{dx}\left((1-x^2)\frac{dy}{dx}\right) + n(n+1)\, y = 0, \tag{7}$$

which remains finite at the points $x = -1$ and $x = +1$. If n and m differ, we have

$$\int_{-1}^{+1} P_n(x)\, P_m(x)\, dx = 0,$$

and it is common to normalize the $P_n(x)$ in such a way that

$$\int_{-1}^{+1} (P_n(x))^2\, dx = \frac{2}{2n+1}.$$

The system of Legendre polynomials is not a Sturm-Liouville orthogonal system, since in the differential equation (7), the coefficient of $\frac{d^2y}{dx^2}$ vanishes at the points $x = 1$ and $x = -1$. We will show that there exist continuous functions whose spherical harmonic series diverges.

If we put, as before,

$$K_n(x,t) = \sum_{p=0,1,2,\ldots,n} \frac{2p+1}{2} P_p(x)\, P_p(t),$$

[10] The existence of such a point b for which $a = \int_{\alpha}^{b} p(x)^{-1/4}\, dx$, if only $0 < a < \pi$, follows immediately from the fact that the continuous function $z = \int_{\alpha}^{x} p(x')^{-1/4}\, dx'$ at the points $x = \alpha$ and $x = \beta$ assumes the values 0 and π, respectively; therefore there has to exist a point b between α and β where it assumes the value a.

then the nth partial sum of the spherical harmonic series of a function $f(x)$ is given by

$$[f(x)]_n = \int_{-1}^{+1} K_n(x,t)\, f(t)\, dt,$$

and thus we have to show that the infinitely many quantities $\int_{-1}^{+1} |K_n(x,t)|\, dt$ do not remain below a bound independent of n. We show this for the point $x = 0$.

A well-known formula states[11] that

$$K_n(x,t) = \frac{n+1}{2} \frac{P_{n+1}(x)\, P_n(t) - P_n(x)\, P_{n+1}(t)}{x-t};$$

and since $P_n(0)$ vanishes for each odd index, it suffices to show that the quantities

$$\omega_{2n} = \frac{2n+1}{2} |P_{2n}(0)| \int_{-1}^{+1} \left| \frac{P_{2n+1}(t)}{t} \right| dt$$

grow arbitrarily large with increasing n. To this end, we apply the frequently used approximation formula

$$P_n(\cos\theta) = \sqrt{\frac{2}{n\pi\, \sin\theta}} \left[\cos\left(\left(n + \frac{1}{2}\right)\theta - \frac{\pi}{4} \right) + \frac{\alpha_n(\theta)}{n} \right],$$

which within the interval $[-1+\varepsilon, 1-\varepsilon]$ represents the spherical harmonics for each value of the index n, where ε stands for a positive number different from zero; the functions $\alpha_n(\theta)$ remain below a bound independent of n and θ, as $\cos\theta$ varies within the interval $[-1+\varepsilon, 1-\varepsilon]$. This formula shows immediately that

$$\lim_{n\to\infty} \sqrt{n\pi}|P_{2n}(0)| = 1,$$

and from this it follows that the ω_{2n} will certainly grow beyond all bounds, if the quantities

$$\sqrt{2n+1} \int_{-1}^{+1} \left| \frac{P_{2n+1}(t)}{t} \right| dt > \sqrt{2n+1} \int_{-1/\sqrt{2}}^{0} \left| \frac{P_{2n+1}(t)}{t} \right| dt = \omega'_{2n}$$

become infinitely large with growing n. To show this, we put

$$t = \cos\left(\vartheta + \frac{\pi}{2}\right) = -\sin\vartheta.$$

We obtain

$$\omega'_{2n} = \sqrt{2n+1} \int_0^{\pi/4} \left| \frac{P_{2n+1}(\cos(\vartheta + \frac{\pi}{2}))}{\sin\vartheta} \cos\vartheta \right| d\vartheta;$$

since, however, in the entire interval $[0, \pi/4]$, we have

$$\cos\vartheta \geq \frac{1}{\sqrt{2}} \quad \text{and} \quad \sin\vartheta \leq \vartheta,$$

we thus find

$$\omega'_{2n} > \sqrt[4]{2} \int_0^{\pi/4} \left| \frac{\sqrt{\frac{2n+1}{2}} \cos\vartheta\, P_{2n+1}(\cos(\vartheta + \frac{\pi}{2}))}{\vartheta} \right| d\vartheta.$$

[11]Cf. Christoffel, *Journal für Mathematik*, Vol. 55, p. 73.

However, according to our approximation formula, we have

$$\sqrt{\frac{2n+1}{2}\cos\vartheta}\; P_{2n+1}\Big(\cos\Big(\vartheta+\frac{\pi}{2}\Big)\Big)$$
$$=\frac{1}{\sqrt{\pi}}\left(\cos\Big(\Big(2n+\frac{3}{2}\Big)\Big(\vartheta+\frac{\pi}{2}\Big)-\frac{\pi}{4}\Big)+\frac{\alpha_{2n+1}(\vartheta+\frac{\pi}{2})}{2n+1}\right)$$
$$=\frac{1}{\sqrt{\pi}}\left((-1)^{n+1}\sin\Big(2n+\frac{3}{2}\Big)\vartheta+\frac{\alpha_{2n+1}(\vartheta+\frac{\pi}{2})}{2n+1}\right).$$

Let us write $\sin(\pi/8)=\mu$ for short, and choose from now on n so large that we have

$$\left|\frac{\alpha_{2n+1}(\vartheta+\frac{\pi}{2})}{2n+1}\right|<\frac{\mu}{2}$$

for each value of ϑ in the interval $[0,\pi/4]$. If ϑ is enclosed between the bounds

$$\pi\,\frac{\frac{1}{8}+p}{2n+\frac{3}{2}}\qquad\text{and}\qquad\pi\,\frac{\frac{7}{8}+p}{2n+\frac{3}{2}},$$

where p denotes an integer, we certainly have

$$\left|\sin\left(2n+\frac{3}{2}\right)\vartheta\right|>\mu,$$

and since

$$\left|\frac{\alpha_{2n+1}(\vartheta+\frac{\pi}{2})}{2n+1}\right|<\frac{\mu}{2},$$

we thus find

$$\left|\sqrt{\frac{2n+1}{2}\cos\vartheta}\,P_{2n+1}\Big(\cos\Big(\vartheta+\frac{\pi}{2}\Big)\Big)\right|>\frac{\mu}{2\sqrt{\pi}}.$$

Consequently, we have for the integral taken between these bounds

$$\int_{i_p}\left|\frac{\sqrt{\frac{2n+1}{2}\cos\vartheta}\,P_{2n+1}(\cos(\vartheta+\frac{\pi}{2}))}{\vartheta}\right|d\vartheta>\frac{\mu}{2\sqrt{\pi}}\log\Big(1+\frac{6}{8p+1}\Big).$$

Now, however, the intervals

$$\left[\pi\,\frac{\frac{1}{8}+p}{2n+\frac{3}{2}},\ \pi\,\frac{\frac{7}{8}+p}{2n+\frac{3}{2}}\right]$$

lie within the interval $\big[0,\frac{\pi}{4}\big]$ as soon as $0\le p\le\frac{n-1}{2}$, and thus we obtain

$$\omega'_{2n}>\frac{\sqrt[4]{2}\,\mu}{2\sqrt{\pi}}\sum_{p=1,2,\ldots,\frac{n-1}{2}}\log\Big(1+\frac{6}{8p+1}\Big).$$

From the divergence of the infinite series $\sum_{p=1,\ldots}\log\Big(1+\frac{6}{8p+1}\Big)$, however, it follows immediately that the ω'_{2n} grow beyond all bounds, and thus our claim is proved.

Our general criterion is applicable without any difficulties also in such cases where the orthogonal series are supposed to be "*summed up*" with some *summation method.* By this we understand the following: the infinitely many functions

$$a_1(n),\ a_2(n),\ a_3(n),\ \ldots$$

defined on the point set M have the property that for a certain value $n = n_0$, say, which is an accumulation point of the point set M, we have

$$\lim_{n\to n_0} a_p(n) = 1 \quad (p = 1, 2, 3, \ldots).$$

Let us now consider an arbitrary infinite series

$$u_1 + u_2 + u_3 + \cdots,$$

which has the property that the series

$$a_1(n)u_1 + a_2(n)u_2 + a_3(n)u_3 + \cdots$$

converges for each value of n belonging to the point set M. If now the limit

$$S = \lim_{n\to n_0} \left(a_1(n)u_1 + a_2(n)u_2 + a_3(n)u_3 + \cdots\right)$$

exists, we say that *the present series is summable with the aid of the summation method given by the functions $a_p(n)$, and assign S to it as its "sum."*

If the infinite series

$$K(n;\, a, t) = a_1(n)\,\varphi_1(a)\,\varphi_1(t) + a_2(n)\,\varphi_2(a)\,\varphi_2(t) + \cdots$$

converges for each value of n under consideration, then the investigations of § 1 yield the following theorem: *if we have*

$$\limsup_{n\to\infty} \int_\alpha^\beta |K(n;\, a, t)|\, dt = \infty,$$

then it is possible to specify a continuous function, whose Fourier series (with respect to the $\varphi_p(s)$) at the point $s = a$ is not summable with the aid of the summation method given by the functions $a_p(n)$.

Chapter II. Theory of Summation

If

$$\varphi_1(s),\ \varphi_2(s),\ \ldots,\ \varphi_n(s),\ \ldots$$

is an arbitrary orthogonal function system defined on the interval $[\alpha, \beta]$, then we say that a function $f(s)$ defined on the interval $[\alpha, \beta]$ belongs to the "range" of this function system, if for each arbitrarily small number δ, there can be determined n constants $c_1, \ldots, c_n$ such that in the entire interval, we have

$$|f(s) - c_1\varphi_1(s) - c_2\varphi_2(s) - \cdots - c_n\varphi_n(s)| < \delta.$$

The set of these functions $f(s)$ form the *range* of the present orthogonal system. This notion plays an exceedingly important rôle in summation theory, since only for functions in the range can the Fourier series be summed up in such a way that the sequence of functions arising from the summation will be uniformly convergent. Incidentally, this notion of range is a very comprehensive one in the known examples; for instance, for the trigonometric functions, for the Legendre polynomials, or even for an arbitrary orthogonal system arising from a differential equation, it consists of all continuous functions, which if need be satisfy certain boundary conditions. In general, we can say: if all *analytical* functions can be expanded with respect to the functions of an orthogonal system, then all *continuous* functions of the interval — by virtue of the well-known Weierstraß theorem — belong to the range of this function system.

§ 1. A Lemma

We have proved in Chapter I that if the present orthogonal function system satisfies a certain condition, it is always possible to specify a continuous function whose Fourier series with respect to this system diverges at a given point. We shall show now that for those orthogonal systems whose range comprises all continuous functions, this condition is also *necessary* for such a function to exist. To this end, we prove the following simple lemma:[12]

To each function $f(s)$ of a certain class of functions, let there be assigned a sequence of real functions $f_1(s), f_2(s), \ldots$; in symbols

$$f(s) \sim f_1(s),\ f_2(s),\ \ldots.$$

This assignment shall have the following properties:

A) *If*

$$f(s) \sim f_1(s),\ f_2(s),\ \ldots$$

and

$$g(s) \sim g_1(s),\ g_2(s),\ \ldots,$$

then

$$f(s)+g(s) \sim f_1(s)+g_1(s),\ f_2(s)+g_2(s),\ \ldots.$$

B) *For each s, let $\left|f_p(s)\right|$ always be less than the upper bound of $\left|f(s)\right|$ multiplied by a quantity M, which for all functions of the class is the same:*

$$|f_p(s)| < M \cdot \mathrm{Max}\ |f(s)|.$$

If now $f'(s), f''(s), \ldots$ are a sequence of functions which converge to the function $f(s)$ uniformly in s, and if the function sequences assigned to $f^{(n)}(s)$ by virtue of our assignment converge to $F^{(n)}(s)$, respectively, uniformly in s, that is, there hold the limit equations

$$\lim_{p\to\infty} f_p^{(n)}(s) = F^{(n)}(s) \tag{8}$$

uniformly in s, then the function sequence assigned to $f(s)$ converges uniformly to a function $F(s)$, and we have

$$\lim_{n\to\infty} F^{(n)}(s) = F(s).$$

[12] As a special case of this lemma, there arises a theorem pronounced by Mr. H. Lebesgue, *Rendiconti del Circolo Matematico di Palermo*, Vol. 26 (1908), p. 325.

Namely, the convergence of the sequence $f'(s)$, $f''(s)$, ... to the function $f(s)$ implies that for sufficiently large q and q', we have

$$\left|f^{(q)}(s) - f^{(q')}(s)\right| < \varepsilon,$$

however small ε be chosen; since, furthermore, by our first assumption, the sequence assigned to the function $f^{(q)}(s) - f^{(q')}(s)$ consists of the differences $f_p^{(q)}(s) - f_p^{(q')}(s)$, it follows from the second assumption that

$$\left|f_p^{(q)}(s) - f_p^{(q')}(s)\right| < \varepsilon M \tag{9}$$

for sufficiently large q and q' and arbitrary p. Since, however, the limit equations (8) hold, we can, after fixing the indices q, q', still choose the index p so large that

$$\left|F^{(q')}(s) - f_p^{(q')}(s)\right| < \varepsilon \quad \text{and} \quad \left|F^{(q)}(s) - f_p^{(q)}(s)\right| < \varepsilon.$$

From the last three inequalities, however, it follows by addition that

$$\left|F^{(q)}(s) - F^{(q')}(s)\right| < (M+2)\,\varepsilon,$$

and this inequality states that *the functions $F^{(n)}(s)$ converge uniformly to a function $F(s)$.*

In order to show that this function $F(s)$ is the uniform limit of the function sequence $f_1(s)$, $f_2(s)$, ... assigned to $f(s)$, we note that for sufficiently large n and arbitrary p, it turns out that

$$\left|f_p^{(n)}(s) - f_p(s)\right| < \varepsilon M.$$

This follows from our assumptions A) and B) just like the inequality (9) derived above. Moreover, we choose n so large that

$$|F(s) - F^{(n)}(s)| < \varepsilon,$$

and then determine for this fixed index n a quantity P such that

$$\left|F^{(n)}(s) - f_p^{(n)}(s)\right| < \varepsilon$$

whenever $p > P$. By adding the last three inequalities we realize that

$$|F(s) - f_p(s)| < (M+2)\,\varepsilon$$

for any sufficiently large p; and so our theorem is proved.[13]

We will immediately draw an important conclusion from this theorem.

We assign to the function $f(s)$ the functions $f_p(s)$ which in the entire interval $[\alpha, \beta]$ have the value that the pth partial sum of the Fourier series of $f(s)$, formed with respect to the orthogonal system $\varphi_1(s)$, $\varphi_2(s)$, ..., assumes at an arbitrary point $s = a$, say:

$$f_p(s) = [f(a)]_p = \int_\alpha^\beta K_p(a,t)\,f(t)\,dt.$$

[13]It is worth noting that in the proof of this theorem, the circumstance that the occurring functions depend on only one variable was not made use of at all. The theorem thus remains correct if all functions depend on several independent variables.

This assignment obviously satisfies condition A). If we furthermore have

$$\int_\alpha^\beta |K_p(a,t)|\,dt < M,$$

where M denotes a number independent of p, then also our second assumption B) is satisfied. If we understand by $\varphi(s)$ any finite aggregate

$$\varphi(s) = a_1\,\varphi_1(s) + \cdots + a_n\,\varphi_n(s)$$

of our orthogonal functions, then obviously the function sequence assigned to $\varphi(s)$ converges uniformly to the value of this function at the point $s = a$. If now $f(s)$ is an arbitrary function in the range of our orthogonal system, we can pick out a sequence $\varphi'(s)$, $\varphi''(s)$, ... of the functions $\varphi(s)$ considered just now, converging uniformly to $f(s)$. According to the lemma just proved, the sequence

$$[f(a)]_1,\ [f(a)]_2,\ \ldots$$

assigned to $f(s)$ then also has to converge to $f(a)$. With that, though, it is shown that *if we have*

$$\int_\alpha^\beta |K_p(a,t)|\,dt < M \quad (p = 1, 2, 3, \ldots),$$

then the Fourier series of each function belonging to the range of this orthogonal system converges at the point $s = a$. This theorem tells us that in the quite general case that the range of our orthogonal system contains all continuous functions, the sufficient condition given in § 1 of Chapter 1 for the existence of a continuous function which cannot be expanded with respect to this orthogonal system is also *necessary.*

§ 2. Application to the Theory of Trigonometric and Sturm-Liouville Series

Out of consideration for the subsequent explanations, we will derive in advance two theorems from the classical theory of trigonometrics series, based on our lemma in § 1 of Chapter 1:

1) The so-called Poisson integral

$$f_r(s) = \frac{1}{2\pi}\int_0^{2\pi} \frac{1-r^2}{1-2r\cos(s-t)+r^2}\,f(t)\,dt$$

assigns to each function $f(s)$ a function set $f_r(s)$. This assignment obviously satisfies condition A) of our lemma. Since, however,

$$\frac{1-r^2}{1-2r\cos(s-t)+r^2}$$

is always positive whenever $r < 1$, and since consequently we have

$$\frac{1}{2\pi}\int_0^{2\pi}\left|\frac{1-r^2}{1-2r\cos(s-t)+r^2}\right| dt$$
$$= \frac{1}{2\pi}\int_0^{2\pi}\Big\{1+2\sum_{n=1,2,\ldots} r^n \cos n(s-t)\Big\}dt = 1$$

for any value of r and s taken into consideration, it follows that $f_r(s)$ taken in absolute value is less than the maximum of $|f(s)|$, no matter how r and s are chosen. In other words, the assignment given by the Poisson integral also satisfies assumption B). As is well known, we have for every value $r < 1$

$$f_r(s) = \frac{1}{2\pi}\int_0^{2\pi} f(t)\,dt$$
$$+\frac{1}{\pi}\sum_{n=1,2,\ldots} r^n\Big\{\cos ns\int_0^{2\pi} f(t)\,\cos nt\,dt + \sin ns\int_0^{2\pi} f(t)\,\sin nt\,dt\Big\},$$

where for $r < 1$ the series on the right converges absolutely and uniformly. From this we see that the function sets assigned to the functions $\cos ns$ and $\sin ns$, respectively, converge uniformly to these functions, as the parameter r converges to 1. An immediate consequence thereof is the fact that if $\Phi(s)$ stands for an arbitrary trigonometric polynomial:

$$\Phi(s) = a_0 + a_1\,\cos s + a_1'\,\sin s + \cdots + a_n\,\cos ns + a_n'\,\sin ns,$$

then the function set assigned to it converges uniformly to $\Phi(s)$. If $F(s)$ now denotes any continuous function with period 2π, we can pick out from the trigonometric polynomials a sequence $\Phi'(s)$, $\Phi''(s)$, ... which converges uniformly to $F(s)$. Our lemma states, however,[14] that then also *the functions*

$$F_r(s) = \frac{1}{2\pi}\int_0^{2\pi}\frac{1-r^2}{1-2r\cos(s-t)+r^2}F(t)\,dt$$

assigned to $F(s)$ *converge uniformly to* $F(s)$ *as* $r \to 1$, *if* $F(s)$ *denotes a continuous periodic function.*

2) *The Fejér summation method of trigonometric series.* If $F(s)$ is a periodic function, then, as is well known, the series

$$[F(s)]_n = \frac{1}{2\pi}\int_0^{2\pi}\frac{\sin(\frac{2n+1}{2})(s-t)}{\sin(\frac{s-t}{2})}F(t)\,dt$$

need not converge. If we put

$$[F^*(s)]_n = \frac{[F(s)]_0 + [F(s)]_1 + \cdots + [F(s)]_{n-1}}{n},$$

[14] The fact that the assigned function set is not countable, but of the cardinality of the continuum, is obviously irrelevant.

however, then the sequence of these $[F^*(s)]_n$, as Mr. Fejér has shown, converges uniformly to $F(s)$. Indeed, we find by a simple calculation

$$[F^*(s)]_n = \frac{1}{2n\pi}\int_0^{2\pi} \frac{\sin^2 n(\frac{s-t}{2})}{\sin^2(\frac{s-t}{2})}\, F(t)\, dt.$$

By this formula, each function $F(s)$ is assigned the function sequence $[F^*(s)]_n$; this assignment satisfies assumption A), and since we have

$$\frac{1}{2n\pi}\int_0^{2\pi} \frac{\sin^2 n(\frac{s-t}{2})}{\sin^2(\frac{s-t}{2})}\, dt = 1,$$

assumption B) is satisfied also. It is also easy to see that the function sequences assigned to the trigonometric polynomials converge uniformly to these functions. From this, there follows, by our lemma and the Weierstraß theorem mentioned above, the theorem by Mr. Fejér.

3) Finally, we make one more application of our lemma to the *convergence theory of Sturm-Liouville series.*

Keeping to the terminology of the previous chapter, we consider the Sturm-Liouville orthogonal system $v_1(z)$, $v_2(z)$, ... treated there and again put

$$K_n(z,t) = v_1(z)\, v_1(t) + \cdots + v_n(z)\, v_n(t).$$

We have proved in the previous chapter that the difference

$$\Phi_n(z,t) = K_n(z,t) - \frac{2}{\pi}\sum_{p=1,\ldots,n} \cos pz\, \cos pt$$

in absolute value remains below an upper bound Φ independent of n, z, and t; from this we conclude that the assignment given by the integral

$$f_n(z) = \int_0^{\pi}\left\{K_n(z,t) - \frac{2}{\pi}\sum_{p=1,\ldots,n} \cos pz\, \cos pt\right\} f(t)\, dt = \int_0^{\pi} \Phi_n(z,t)\, f(t)\, dt \qquad (10)$$

satisfies the assumptions of our lemma. If $\varphi(z)$ now denotes some *analytic* function, thus belonging to the range of *both* orthogonal systems

$$1,\ \cos z,\ \cos 2z,\ \ldots, \quad \text{and} \quad v_1(z),\ v_2(z),\ v_3(z),\ \ldots,$$

then, as we know, both integrals

$$\frac{2}{\pi}\int_0^{\pi} \sum_{p=1,\ldots,n} \cos pz\, \cos pt\, \varphi(t)\, dt \quad \text{and} \quad \int_0^{\pi} K_n(z,t)\, \varphi(t)\, dt$$

converge with increasing n uniformly to $\varphi(z)$, since the Fourier series of these functions with respect to both orthogonal systems converge uniformly. Consequently, the integral

$$\int_0^{\pi} \Phi_n(z,t)\, \varphi(t)\, dt$$

has limit zero. Our lemma now states that the integral (10) also converges to zero, if $f(z)$ is an arbitrary function that can be approximated uniformly by the functions $\varphi(z)$. In other words, our integral converges to zero if $f(z)$ is an arbitrary continuous function.[15] This immediately implies the following theorem:

Let $f(z)$ be an arbitrary continuous function; *the Fourier series of this function with respect to the orthogonal system* $v_1(z)$, $v_2(z)$, ... *converges or diverges, respectively, according as the cosine series of this function is convergent or divergent.*

If now $F(z)$ is an *arbitrary* function, integrable in the Lebesgue sense, then we construct a sequence of *continuous* functions $f'(z)$, $f''(z)$, ... such that we have

$$\lim_{p\to\infty}\int_0^\pi |F(z)-f^{(p)}(z)|\,dz = 0.$$

Then we obviously have for each p

$$\left|\int_0^\pi \Phi_n(z,t)\,F(t)\,dt\right| \le \Phi\int_0^\pi \left|F(t)-f^{(p)}(t)\right|dt + \left|\int_0^\pi \Phi_n(z,t)\,f^{(p)}(t)\,dt\right|.$$

From this inequality, we deduce by means of the theorem just proved that we have

$$\lim_{n\to\infty}\int_0^\pi \Phi_n(z,t)\,F(t)\,dt = 0;$$

that is, *the Sturm-Liouville expansion of an integrable function is convergent or divergent at some point, respectively, according as the cosine series of this function at this point is convergent or divergent.*[16]

Also, it is immediate how these theorems are to be modified in the case that instead of the boundary conditions (5), some other pair of homogeneous boundary conditions is assumed.

§ 3. Summation of Sturm-Liouville series

We will now show that the summation method applied to the trigonometric series by Mr. Fejér can be applied to the Sturm-Liouville series with equal success.

We keep to the terminology from § 2 of the first chapter, and consider on the interval $[\alpha,\beta]$ the eigenfunctions $u_1(x)$, $u_2(x)$, ... of the differential equation

$$L(u) \equiv \frac{d}{dx}\left(p\,\frac{du}{dx}\right) + q\,u + \lambda\,u = 0, \quad (p(x)>0 \text{ in the interval } [\alpha,\beta]), \tag{4}$$

with a pair of homogeneous boundary conditions

$$\frac{du}{dx} - h\,u = 0 \quad \text{for} \quad x=\alpha \quad \text{and} \quad \frac{du}{dx} + H\,u = 0 \quad \text{for} \quad x=\beta. \tag{5}$$

The eigenfunctions of the differential equation arising from (4) by virtue of the Liouville transformation shall again be denoted $v_1(z)$, $v_2(z)$, ... Because of the relations derived in Chapter 1, § 2 between the two orthogonal systems $u_n(x)$ and $v_n(z)$, it obviously suffices

[15] Namely, if we assume the boundary conditions (5) or (5′), respectively, the range of the function system $v_1(z)$, $v_2(z)$, ... comprises *all* continuous functions.

[16] This remark allows for a new proof of the theorem that there exist continuous functions whose Sturm-Liouville series diverge.

to restrict oneself to this latter system, since the results obtained for this are transferable to the function system $u_n(x)$ without any difficulty.

Let $f(z)$ be an arbitrary function with Fourier series

$$\sum_{n=1,2,\ldots} v_n(z) \int_0^{\pi} f(t)\, v_n(t)\, dt;$$

we consider the *arithmetic means formed from the partial sums of this series*

$$\begin{aligned}
[f^*(z)]_1 &= [f(z)]_1, \\
[f^*(z)]_2 &= \frac{[f(z)]_1 + [f(z)]_2}{2}, \\
[f^*(z)]_3 &= \frac{[f(z)]_1 + [f(z)]_2 + [f(z)]_3}{3}, \ldots .
\end{aligned}$$

If we now put in the same manner as before

$$K_n^*(z,t) = \frac{K_1(z,t) + K_2(z,t) + \cdots + K_n(z,t)}{n}, \tag{11}$$

then we have

$$[f^*(z)]_n = \int_0^{\pi} K_n^*(z,t)\, f(t)\, dt\,.$$

If we now assign to each function $f(z)$ the functions

$$[f^*(z)]_1,\ [f^*(z)]_2,\ [f^*(z)]_3,\ \ldots$$

so defined, this assignment obviously satisfies assumption A) of our lemma in § 1 of this chapter. If we furthermore have $f(z) = v_n(z)$, then the function sequence assigned to $v_n(z)$ is given by

$$0,\ 0,\ \ldots,\ 0,\ \frac{v_n(z)}{n},\ \frac{2\,v_n(z)}{n+1},\ \ldots,\ \frac{p\,v_n(z)}{n+p},\ \ldots$$

and we see that this function sequence converges uniformly to $v_n(z)$ as $p \to \infty$. From this, it also follows immediately that, if $v(z)$ denotes a finite aggregate of the form

$$v(z) = a_1\, v_1(z) + \cdots + a_n\, v_n(z)$$

(with constant coefficients a), then the function sequence assigned to this function converges uniformly to $v(z)$. If this assignment now satisfies assumption B) of our lemma also, we may conclude therefrom that *the Fourier series of each function lying in the range of the orthogonal function system $v_n(z)$ is summable by the method of the arithmetic mean.*

To show this, it obviously suffices to prove that the integral

$$\int_0^{\pi} \left|K_n^*(z,t)\right| dt$$

remains below a bound M independent of n and z. Now we have (cf. § 2 of Chapter 1)

$$K_n(z,t) = \frac{2}{\pi} \sum_{p=1,\ldots,n} \cos pz\, \cos pt + \Phi_n(z,t)\,,$$

and we have shown at that place that $\Phi_n(z,t)$ remains below a bound independent of n, z, and t. Since we have, though,

$$K_n^*(z,t) = \frac{1}{n}\Big(\sum_{p=1} \cos pz \cos pt + \sum_{p=1,2} \cos pz \cos pt + \cdots + \sum_{p=1,\ldots,n} \cos pz \cos pt\Big)$$
$$+ \frac{1}{n}\Big(\Phi_1(z,t) + \cdots + \Phi_n(z,t)\Big),$$

this implies that $\int_0^\pi |K_n^*(z,t)|\,dt$ definitely lies below a finite bound, since we have

$$\int_0^\pi \left|\frac{1}{n}\Big(\sum_{p=1} \cos pz \cos pt + \cdots + \sum_{p=1,\ldots,n} \cos pz \cos pt\Big)\right| dt$$
$$= \frac{1}{2n}\int_0^\pi \left| -(n+1) + \frac{1}{2}\left\{\frac{\sin^2(n+1)(\frac{z+t}{2})}{\sin^2(\frac{z+t}{2})} + \frac{\sin^2(n+1)(\frac{z-t}{2})}{\sin^2(\frac{z-t}{2})}\right\}\right| dt$$
$$\le \frac{n+1}{2n}\pi + \frac{1}{4n}\int_0^\pi \left|\frac{\sin^2(n+1)(\frac{z+t}{2})}{\sin^2(\frac{z+t}{2})} + \frac{\sin^2(n+1)(\frac{z-t}{2})}{\sin^2(\frac{z-t}{2})}\right| dt$$
$$= \frac{n+1}{2n}\pi + \frac{n+1}{4n}\pi = \frac{3(n+1)}{4n}\pi\,,$$

and $\frac{1}{n}(\Phi_1(z,t) + \cdots + \Phi_n(z,t))$ remains less than a number independent of n, z, t.

Thus it is shown that the second assumption of our lemma is fulfilled also, and we obtain the result:[17]

The Sturm-Liouville expansion of a function in the range of the Sturm-Liouville orthogonal system under consideration is always summable by the method of the arithmetic mean.

If we assume the boundary conditions (5), then the range of the orthogonal system under consideration consists of all continuous functions (since all twice differentiable functions satisfying the boundary condition (5) are expandable); this implies the following theorem:

If a continuous function is expanded in the Fourier manner into a series that progresses according to the eigenfunctions of the differential equation (4), *satisfying the boundary conditions* (5), *then the sequence* $[f^*(x)]_n$ *of arithmetic means converges uniformly to the function* $f(x)$.[18]

If instead of the boundary conditions (5), a different pair of conditions is assumed,

$$u(\alpha) = 0 \quad \text{and} \quad u(\beta) = 0, \tag{12}$$

say, then in the very same way the theorem can be proved that the sequence of arithmetic means $[f^*(x)]_n$ converges uniformly to $f(x)$, if $f(x)$ lies in the range of the $u_n(x)$. In this case, though, we have to bear in mind that the range is now made up from the continuous functions vanishing at the points $x = \alpha$ and $x = \beta$, that is, *if a continuous function*

[17] Of course, this theorem can also be deduced from the theorem in § 2 of this chapter, but it seems expedient to take this very generalizable course.

[18] An immediate consequence of this theorem is the fact that if the Sturm-Liouville series of a continuous function $f(s)$ converges at the point $s = a$, then its sum equals $f(a)$.

satisfying the boundary condition (12) *is expanded with respect to the eigenfunctions of the differential equation* (4) *which satisfy the boundary condition* (12), *then this series is according to the usual terminology "Cesàro summable," that is, the sequence of arithmetic means* $\frac{1}{n}([f(z)]_1 + \cdots + [f(z)]_n)$ *formed from the partial sums* $[f(z)]_n$ *converges uniformly to the function* $f(z)$. The theorem is to be modified accordingly, if some other pair of homogeneous boundary conditions is assumed.

§ 4. Generalizations

The investigations of this chapter are applicable immediately to the expansion of functions with several variables, since our lemma — which constitutes the sole basis for the proofs of this section — remains true also in these more general cases (cf. § 1 of this chapter).

If we terminate the formal Fourier series of a function $f(s,\sigma)$ with respect to the orthogonal system $\varphi_1(s)$, $\varphi_2(s)$, ...

$$\sum_{(p)}\sum_{(q)} \varphi_p(s)\,\varphi_q(\sigma) \int_\alpha^\beta \int_\alpha^\beta f(t,\tau)\,\varphi_p(t)\,\varphi_q(\tau)\,dt\,d\tau \tag{13}$$

at the term with indices n, m, we denote the finite sum thus obtained by $[f(s,\sigma)]_{n,m}$:

$$[f(s,\sigma)]_{n,m} = \sum_{p=1,\ldots,n}\ \sum_{q=1,\ldots,m} \varphi_p(s)\,\varphi_q(\sigma) \int_\alpha^\beta \int_\alpha^\beta f(t,\tau)\,\varphi_p(t)\,\varphi_q(\tau)\,dt\,d\tau.$$

In our terminology (introduced in Chapter 1), we have

$$[f(s,\sigma)]_{n,m} = \int_\alpha^\beta \int_\alpha^\beta K_n(s,t)\,K_m(\sigma,\tau)\,f(t,\tau)\,dt\,d\tau. \tag{13'}$$

If this assignment satisfies the conditions of our lemma, we can conclude that the expansion of each function of two variables lying in the range of our orthogonal system converges uniformly to this function.

Let us return now to the Sturm-Liouville function systems. It is easily shown that for this system, equation (13′) does not satisfy assumption B) of our lemma. The assignment

$$[f^*(s,\sigma)]_{n,n} = \int_0^\pi \int_0^\pi K_n^*(s,t)\,K_n^*(\sigma,\tau)\,f(t,\tau)\,dt\,d\tau, \tag{14}$$

where $K_n^*(s,t)$ denotes the function defined by equation (11) does satisfy both assumptions of our lemma, though, since we have shown that

$$\int_0^\pi \left|K_n^*(s,t)\right| dt$$

lies below an upper bound independent of n, s, and t. This immediately implies that the function sequence defined by (14) *converges uniformly to* $f(s,\sigma)$, *if this function lies in the range of the Sturm-Liouville system under consideration.*

To this fact there corresponds the following summation method: *from the partial sums* $[f]_{n,m}$ *of the doubly infinite series* (13), *form the simple sequence:*

$$[f^*]_{n,n} = \frac{1}{n^2} \sum_{p=1,\dots,n} \sum_{q=1,\dots,n} [f]_{p,q} \quad (n = 1, 2, \dots). \tag{15}$$

If the function $f(s,\sigma)$ *lies in the range of the Sturm-Liouville function system under consideration, then the sequence* (15) *converges uniformly to* $f(s,\sigma)$.[19]

Finally, we make one more application of our lemma to the general theory of the summation of orthogonal series. Let there be given a summation method by the infinitely many functions

$$a_1(n),\ a_2(n),\ a_3(n),\ \dots.$$

If the sum

$$K(n;a,t) = \sum_{p=1,2,\dots} a_p(n)\,\varphi_p(a)\,\varphi_p(t)$$

converges, then the necessary and sufficient condition for the Fourier series of each function belonging to the range of the present orthogonal system to be "summable" at the point a with the aid of this summation method, is that the integral

$$\int_\alpha^\beta |K(n;a,t)|\,dt$$

remains below an upper bound independent of n.

Chapter III. On a Class of Orthogonal Function Systems

The purpose of this section is to treat a class of orthogonal function systems which, besides a series of other noteworthy properties, are particularly distinguished by the fact that the *Fourier series* with respect to these systems *of each continuous function converge and represent the function.* In §§ 1–3 we consider the simplest representative of this class; § 4 will then present the generalization of the theorems we gained to further systems.

§ 1. The Orthogonal Function System χ

The complete orthogonal function system χ, the simplest representative of that class of orthogonal systems, we define as follows:

Let $\chi_0(s) = 1$ on the entire interval $[0,1]$ including the boundaries; thereupon let

$$\begin{aligned} \chi_1(s) &= \ \ 1 \quad \text{for} \quad 0 \le s < \tfrac{1}{2}, \\ &= -1 \quad \text{for} \quad \tfrac{1}{2} < s \le 1. \end{aligned}$$

[19]Instead of the simple sequence $[f^*]_{n,n}$, we could also define a double sequence $[f^*]_{n,m}$ of the same property, but this "simple" summation is to be preferred to the other one.

We furthermore put

$$\begin{aligned}
\chi_2^{(1)}(s) &= \sqrt{2} \quad \text{and} \quad \chi_2^{(2)}(s) = 0 \quad \text{for} \quad 0 \le s < \tfrac{1}{4}\,,\\
&= -\sqrt{2} \qquad\qquad\qquad\quad\; = 0 \quad \text{for} \quad \tfrac{1}{4} < s < \tfrac{1}{2}\,,\\
&= 0 \qquad\qquad\qquad\qquad\;\;\; = \sqrt{2} \quad \text{for} \quad \tfrac{1}{2} < s < \tfrac{3}{4}\,,\\
&= 0 \qquad\qquad\qquad\qquad\;\;\; = -\sqrt{2} \quad \text{for} \quad \tfrac{3}{4} < s \le 0\,.
\end{aligned}$$

In this manner we proceed; in general, we define the functions of our system in the following way: we divide the interval $[0, 1]$ into 2^n equal parts and denote these subintervals in turn by $i_n^{(1)}, i_n^{(2)}, \ldots, i_n^{(2^n)}$. Now we put

$$\begin{aligned}
\chi_n^{(k)} &= 0 \quad \text{within the intervals} \quad i_n^{(1)}, i_n^{(2)}, \ldots, i_n^{(2k-2)};\\
&= \sqrt{2^{n-1}} \quad \text{within the interval} \quad i_n^{(2k-1)};\\
&= -\sqrt{2^{n-1}} \quad \text{within the interval} \quad i_n^{(2k)};\\
&= 0 \quad \text{within the intervals} \quad i_n^{(2k+1)}, \ldots, i_n^{(2^n)}\\
&\qquad (k = 1, 2, \ldots, 2^{n-1}).
\end{aligned}$$

At the points 0 and 1 we assign to each function $\chi_n^{(k)}(s)$ being constant on the interval $\left[0, \frac{1}{2^n}\right]$ or $\left[1-\frac{1}{2^n}, 1\right]$, respectively, the value it assumes in these respective intervals. Thus $\chi_n^{(k)}(s)$ is a piecewise constant function which is continuous with the exception of the points $\frac{2k-2}{2^n}$, $\frac{2k-1}{2^n}$, $\frac{2k}{2^n}$, where it suffers a finite jump. We now make the agreement that at these points, $\chi_n^{(k)}$ be equal to the arithmetic means of the values it assumes in the intervals adjoining at that exact point.

We now claim that the countably infinitely many functions

$$\chi: \qquad \chi_0(s),\ \chi_1(s),\ \chi_2^{(1)}(s), \chi_2^{(2)}(s),\ \chi_3^{(1)}(s),\ \chi_3^{(2)}(s),\ \ldots$$

so defined *form a complete orthogonal function system.* Indeed, if $\chi_n^{(k)}(s)$ and $\chi_\nu^{(\varkappa)}(s)$ are two different functions of the system χ and we have $n > \nu$, then $\chi_\nu^{(\varkappa)}(s)$ has a constant value on the entire interval where $\chi_n^{(k)}(s)$ is different from zero. Therefore, we have

$$\int_0^1 \chi_n^{(k)}(s)\, \chi_\nu^{(\varkappa)}(s)\, ds = \text{const.} \int_0^1 \chi_n^{(k)}(s)\, ds = 0,$$

from which we conclude immediately that the function system χ possesses the *orthogonality condition.* In order to show that also the *completeness relation* is satisfied, it obviously suffices to prove that any function $f(s)$ which is integrable in the Lebesgue sense and which for all pairs n, k coming into question satisfies the relation

$$\int_0^1 f(s)\, \chi_n^{(k)}(s)\, ds = 0, \tag{16}$$

vanishes identically up to a null set. For this purpose, let us consider the function

$$F(s) = \int_0^s f(s')\, ds';$$

because of the equation

$$\int_0^1 f(s)\, \chi_0(s)\, ds = 0,$$

we have

$$F(1) = 0.$$

The last equation yields in connection with the equation

$$\int_0^1 f(s)\, \chi_1(s)\, ds = \int_0^{1/2} f(s)\, ds - \int_{1/2}^1 f(s)\, ds = 0$$

the statement that $F\left(\frac{1}{2}\right) = 0$. From this and from the equations

$$\int_0^1 f(s)\, \chi_2^{(1)}(s)\, ds = \sqrt{2}\left\{\int_0^{1/4} f(s)\, ds - \int_{1/4}^{1/2} f(s)\, ds\right\} = 0,$$

$$\int_0^1 f(s)\, \chi_2^{(2)}(s)\, ds = \sqrt{2}\left\{\int_{1/2}^{3/4} f(s)\, ds - \int_{3/4}^{1} f(s)\, ds\right\} = 0,$$

we deduce that we have

$$F\left(\tfrac{1}{4}\right) = F\left(\tfrac{3}{4}\right) = 0,$$

and so on. We can conclude in this way that the function $F(s)$ always equals zero if s is a finite binary fraction of the form $\frac{1}{2^{p_1}} + \cdots + \frac{1}{2^{p_n}}$, and these points form an everywhere dense point set. As we know, $F(s)$ is a continuous function, however, and we have with the exception of a set of measure zero that

$$f(s) = \frac{d}{ds}(F(s)).$$

We deduce from this that $F(s)$ vanishes identically on the entire interval, and that $f(s)$ — apart from a point set of measure zero — also is zero everywhere. Thus the completeness of the considered orthogonal system is also proven and it is shown that *each integrable function satisfying the relations* (16) *vanishes except for a null set.*

§ 2. Expansions with Respect to the Orthogonal Function System χ

We now get to the most important point of this investigation by showing that the *Fourier series of each continuous function on the interval* $[0, 1]$ *taken with respect to the orthogonal system* χ *defined just now converges uniformly to this function.*

Let us cut off the infinite series

$$\chi_0(s)\int_0^1 f(t)\,\chi_0(t)\,dt + \chi_1(s)\int_0^1 f(t)\,\chi_1(t)\,dt + \cdots$$
$$+\,\chi_n^{(1)}(s)\int_0^1 f(t)\,\chi_n^{(1)}(t)\,dt + \cdots + \chi_n^{(p)}(s)\int_0^1 f(t)\,\chi_n^{(p)}(t)\,dt + \cdots$$

at some term, at $\chi_n^{(p)}(s)\int_0^1 f(t)\,\chi_n^{(p)}(t)\,dt$, say; we obtain a finite sum, which we from now on denote by $[f(s)]_n^{(p)}$:

$$[f(s)]_n^{(p)} = \chi_0(s)\int_0^1 f(t)\,\chi_0(t)\,dt + \cdots + \chi_n^{(p)}(s)\int_0^1 f(t)\,\chi_n^{(p)}(t)\,dt.$$

If we set analogously as before

$$K_n^{(p)}(s,t) = \chi_0(s)\,\chi_0(t) + \cdots + \chi_n^{(1)}(s)\,\chi_n^{(1)}(t) + \cdots + \chi_n^{(p)}(s)\,\chi_n^{(p)}(t),$$

then we have

$$[f(s)]_n^{(p)} = \int_0^1 K_n^{(p)}(s,t)\,f(t)\,dt.$$

The last equation defines infinitely many functions of two variables

$$K_0(s,t),\ K_1(s,t),\ K_2^{(1)}(s,t),\ K_2^{(2)}(s,t),\ \ldots,$$

and we now turn to the investigation of the properties of these functions.

The function $K_0(s,t)$ defined in the square $0 \le s \le 1$, $0 \le t \le 1$ equals 1 everywhere. The function $\chi_1(s)\,\chi_1(t)$ equals 1 within the squares

$$Q_{11} : 0 \le s \le \tfrac{1}{2},\quad 0 \le t \le \tfrac{1}{2} \qquad \text{and} \qquad Q_{22} : \tfrac{1}{2} \le s \le 1,\quad \tfrac{1}{2} \le t \le 1;$$

it equals -1 in the other two squares

$$Q_{12} : \tfrac{1}{2} \le s \le 1,\quad 0 \le t \le \tfrac{1}{2} \qquad \text{and} \qquad Q_{21} : 0 \le s \le \tfrac{1}{2},\quad \tfrac{1}{2} \le t \le 1;$$

therefore, $K_1(s,t)$ equals 2 in Q_{11} and Q_{22}, but equals 0 in Q_{12}, Q_{21}.

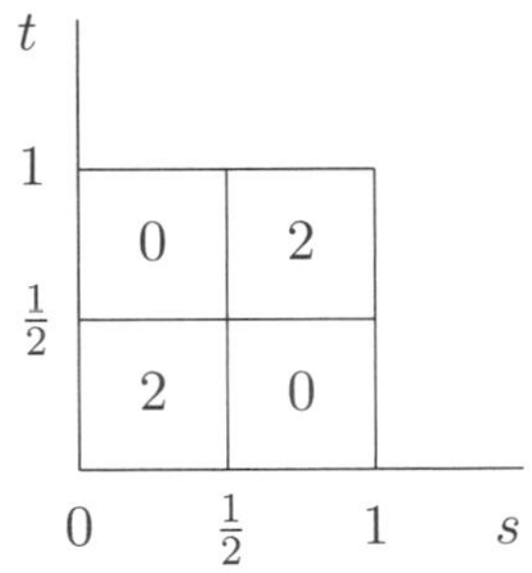

$K_1(s,t)$

At the lines $s = \frac{1}{2}$ and $t = \frac{1}{2}$, respectively, the function $K_1(s,t)$ of course equals the arithmetic means of the values it assumes in the squares adjoining at that exact point. In order to obtain furthermore $K_2^{(1)}(s,t)$ and $K_2^{(2)}(s,t)$, we indicate in the figure those values the functions $\chi_2^{(1)}(s)\,\chi_2^{(1)}(t)$ and $\chi_2^{(2)}(s)\,\chi_2^{(2)}(t)$ assume, respectively.

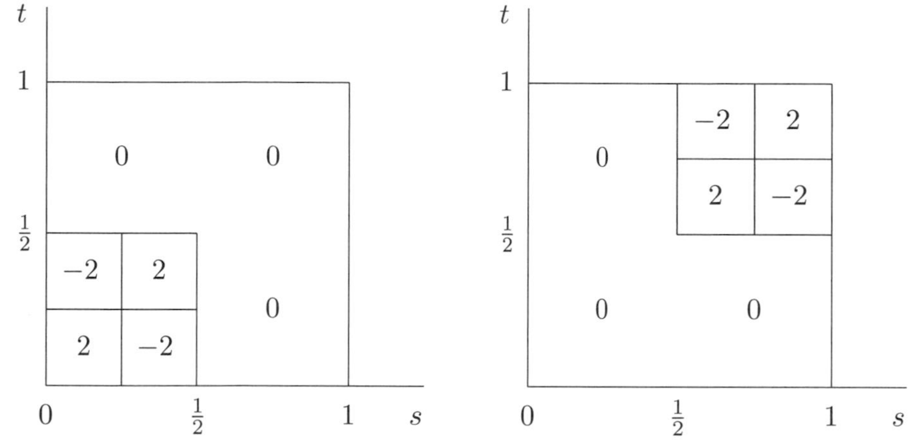

The values of $K_2^{(1)}(s,t)$ and $K_2^{(2)}(s,t)$ are thus represented graphically by the following figures.

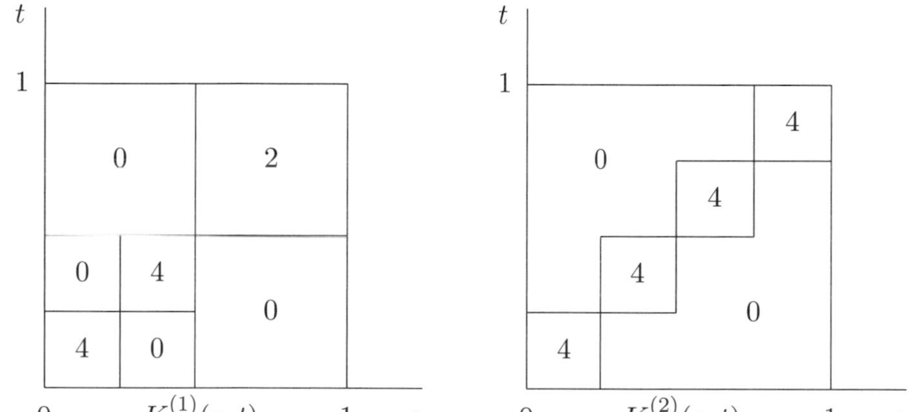

$K_2^{(1)}(s,t)$ $K_2^{(2)}(s,t)$

From this, the formation priciple of the functions $K_n^{(p)}(s,t)$ is already apparent: To obtain the range of the function $K_n^{(2^n-1)}(s,t)$, we divide the unit square Q into 2^{2^n} equal subsquares; in the subsquares $q_1, \ldots, q_{2^n}$ lying along the diagonal $s = t$ of the square Q, we have

$$K_n^{(2^n-1)}(s,t) = 2^n;$$

within the other squares, $K_n^{(2^n-1)}(s,t)$ equals zero. At the points where this function suffers a jump, it assumes the arithmetic means of the values it has in the squares adjoining at that exact point. To obtain now $K_{n+1}^{(p)}(s,t)$, say, we subdivide each of the first p subsquares $q_1, q_2, \ldots, q_p$ of the former division which lie along the diagonal into four equal squares, as the figure suggests:

$$q_\varkappa = q_\varkappa^{(1,1)} + q_\varkappa^{(1,2)} + q_\varkappa^{(2,1)} + q_\varkappa^{(2,2)} \quad (\varkappa = 1, \ldots, p).$$

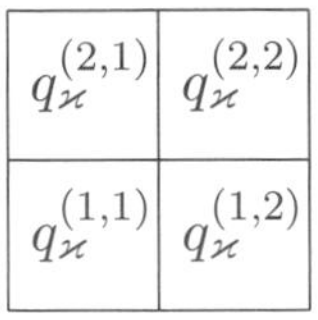

$q_\varkappa$

Then $K_{n+1}^{(p)}(s,t)$ is given by the following rule: In the subsquares $q_\varkappa^{(1,1)}$, $q_\varkappa^{(2,2)}$, we have $K_{n+1}^{(p)}(s,t) = 2^{n+1}$; in the subsquares $q_\varkappa^{(1,2)}$, $q_\varkappa^{(2,1)}$, though, it equals zero. Within all the other squares, we have $K_{n+1}^{(p)}(s,t) = K_n^{(2^n-1)}(s,t)$.[20] At the points where $K_{n+1}^{(p)}(s,t)$ becomes discontinuous (that is, at those points s, t where one of the quantities s and t is a finite binary fraction of the form $\frac{1}{2^{p_1}} + \cdots + \frac{1}{2^{n+1}}$), we determine it according to the rule mentioned above.

In order to prove the correctness of this rule, we assume it to be correct for $K_n^{(2^n-1)}(s,t)$; now we have

$$K_{n+1}^{(1)}(s,t) = K_n^{(2^n-1)}(s,t) + \chi_{n+1}^{(1)}(s)\,\chi_{n+1}^{(1)}(t).$$

Since $\chi_{n+1}^{(1)}(s)$ differs from zero only in the interval $0 \le s \le \frac{1}{2^n}$, $K_{n+1}^{(1)}(s,t)$ can only differ from $K_n^{(2^n-1)}(s,t)$ in the square

$$0 \le s \le \frac{1}{2^n}, \quad 0 \le t \le \frac{1}{2^n},$$

that is, in q_1. Since we have

$$\begin{aligned} \chi_{n+1}^{(1)}(s)\,\chi_{n+1}^{(1)}(t) &= 2^n \quad \text{in the subsquares } q_1^{(1,1)}, q_1^{(2,2)}, \\ &= -2^n \quad \text{in the subsquares } q_1^{(1,2)}, q_1^{(2,1)}, \end{aligned}$$

it follows that $K_{n+1}^{(1)}(s,t)$ has the value just given:

$$\begin{aligned} K_{n+1}^{(1)}(s,t) &= 2^{n+1} \quad \text{in } q_1^{(1,1)}, q_1^{(2,2)}, \\ &= 0 \qquad\;\; \text{in } q_1^{(1,2)}, q_1^{(2,1)}. \end{aligned}$$

Let us denote by $f(s)$ an arbitary function, integrable in the Lebesgue sense and defined in the interval $[0,1]$, and by $s = a$ an arbitrary point of the interval. Then we have

$$[f(a)]_n^{(p)} = \int_0^1 K_n^{(p)}(a,t)\,f(t)\,dt;$$

[20] That is, it equals 2^n in the subsquares $q_{p+1}, \ldots, q_{2^n}$, and equals 0 if the point s, t does not lie within one of these squares.

if we assume for the moment that a not be a finite binary fraction of the above form, then the function $K_n^{(p)}(a,t)$ of t equals zero everywhere except for an interval $i_n^{(p)}$ whose length $l_n^{(p)}$ equals $1/2^{n-1}$ or $1/2^n$. In this interval $i_n^{(p)}$, though, we have $K_n^{(p)}(a,t) = 1/l_n^{(p)}$ and we thus find that

$$[f(a)]_n^{(p)} = \frac{1}{l_n^{(p)}} \int_{i_n^{(p)}} f(t)\,dt.$$

If, however, a is a finite binary fraction, then $K_n^{(p)}(a,t)$ is different from zero in an interval $\bar{i}_n^{(p)}$ whose length is

$$\bar{l}_n^{(p)} = \frac{1}{2^{n-2}} \quad \text{or} \quad = \frac{1}{2^{n-1}};$$

the value of $K_n^{(p)}(a,t)$ in this interval equals $1/\bar{l}_n^{(p)}$, however, and thus we obtain in this case also

$$[f(a)]_n^{(p)} = \frac{1}{\bar{l}_n^{(p)}} \int_{\bar{i}_n^{(p)}} f(t)\,dt.$$

In both cases the length of the integration interval — which contains the point $t = a$ — thus equals the reciprocal value of the factor standing in front of the interval.

But now $l_n^{(p)}$ and $\bar{l}_n^{(p)}$ converge to zero with increasing n, and therefore the partial sums $[f(a)]_n^{(p)}$ converge to

$$\lim_{n\to\infty} \frac{1}{l_n^{(p)}} \int_{i_n^{(p)}} f(t)\,dt.$$

Since the intervals $i_n^{(p)}$ shrink to the point $t = a$ with increasing n, this limit is nothing else but the value of the differential quotient of $\int_0^s f(t)\,dt$ with respect to s at the point $s = a$:

$$\lim_{n\to\infty} \frac{1}{l_n^{(p)}} \int_{i_n^{(p)}} f(t)\,dt = \left[\frac{d}{ds}\left(\int_0^s f(t)\,dt\right)\right]_{s=a},$$

and we have the result: *If $f(s)$ is an arbitrary function, then the partial sums of its expansion converge at every point $s = a$ where the differential quotient $\frac{d}{ds}\left(\int_0^s f(t)\,dt\right)$ exists, and represent this value.*

Since, however, — by a theorem of Lebesgue — everywhere except for a set of measure zero,

$$\frac{d}{ds}\left(\int_0^s f(t)\,dt\right)$$

exists and agrees with $f(s)$, this implies: *the expansion of an arbitrary function with respect to the functions of our orthogonal system converges at every point with the exception of a point set of measure zero.*

Is $f(s)$ continuous at every point of the interval, however, then we have, as is well known,

$$f(s) = \frac{d}{ds}\left(\int_0^s f(t)\,dt\right)$$

for every point without exception, that is, *the Fourier expansion (with respect to our orthogonal system χ) of an arbitrary continuous function converges at every point of the interval* $[0,1]$.[21]

§ 3. Further Properties of the Orthogonal Function System χ

Now we will derive some further properties of our function system, corresponding to those theorems in the theory of trigonometric series obtained there from various summation methods.

From the circumstance that the functions $K_n^{(p)}(s,t)$ are always positive we conclude the following theorem:

If the function $f(s)$, integrable in the Lebesgue sense in the interval $[0,1]$, remains between the bounds m and M, that is,

$$m \leq f(s) \leq M,$$

then all partial sums of the Fourier series of $f(s)$ with respect to χ also remain between these bounds, that is,

$$m \leq [f(s)]_n^{(p)} \leq M.$$

Indeed, we have, for example,

$$[f(s)]_n^{(p)} = \int_0^1 K_n^{(p)}(s,t)\, f(t)\, dt \leq M \int_0^1 K_n^{(p)}(s,t)\, dt = M.$$

Let now $s = a$ be an arbitrary point of the interval $[0,1]$. Since for sufficiently large n, the functions $K_n^{(p)}(a,t)$ will certainly vanish if

$$0 \leq t \leq a - \varepsilon$$

or if

$$a + \varepsilon \leq t \leq 1,$$

however small the positive number ε has been chosen, we have

$$[f(a)]_n^{(p)} = \int_{a-\varepsilon}^{a+\varepsilon} K_n^{(p)}(a,t)\, f(t)\, dt,$$

whenever n exceeds a certain bound. Since in this formula, the behavior of the function $f(s)$ in the intervals $0 \leq s \leq a - \varepsilon$ and $a + \varepsilon \leq s \leq 1$ is not being expressed at all, we conclude from this that *the convergence of the Fourier series of an arbitrary function taken with respect to χ at the point $s = a$ only depends on the behavior of this function in the neighborhood of this point.*

If the functions $f(s)$ and $g(s)$ agree in however small an interval, then we can specify an index N such that whenever we have $n > N$, all the $[f(s)]_n^{(p)}$ and $[g(s)]_n^{(p)}$ also agree in this interval.

[21] In a note that appeared in *Jahresberichte der deutschen Mathematiker-Vereinigung* (1910), Mr. Faber has made my orthogonal system the object of his investigations and derives my theorems afresh. His method of proof is not essentially different from the one given here, though.

In connection with the main theorem of the paragraph before, these latter results yield the following theorem: *the Fourier series with respect to* χ *of a function* $f(s)$ *converges to this function at every point of continuity of* $f(s)$.

§ 4. Various Generalizations

We can generalize the orthogonal system just defined in various directions without destroying its essential properties. We will now outline some of these generalizations.

To begin with, we can use a more general method in dividing up the s-axis for the construction of a similar orthogonal system in the following way: the first function of the system again be equal to 1; then we choose an arbitrary point in the interval $[0,1]$, say, α_1, and construct a function $\overline{\chi}_1(s)$ which is constant on the intervals $[0,\alpha_1]$ and $[\alpha_1,1]$, respectively, and furthermore satisfies the two conditions

$$\int_0^1 \overline{\chi}_1(s)\,ds = 0\,, \qquad \int_0^1 \left(\overline{\chi}_1(s)\right)^2 ds = 1.$$

Then we choose two points $\alpha_2^{(1)}$ and $\alpha_2^{(2)}$, respectively, arbitrarily in the intervals $[0,\alpha_1]$ and $[\alpha_1,1]$, and determine the functions $\overline{\chi}_2^{(1)}(s)$ and $\overline{\chi}_2^{(2)}(s)$ according to the following rule: $\overline{\chi}_2^{(1)}(s)$ shall vanish on $[\alpha_1,1]$ and assume on the intervals $[0,\alpha_2^{(1)}]$ and $[\alpha_2^{(1)},\alpha_1]$ a constant value each which shall be chosen in such a way that we have

$$\int_0^1 \overline{\chi}_2^{(1)}(s)\,ds = 0, \qquad \int_0^1 \left(\overline{\chi}_2^{(1)}(s)\right)^2 ds = 1.$$

$\overline{\chi}_2^{(2)}(s)$, though, shall vanish on $[0,\alpha_1]$ and assume on the intervals $[\alpha_1,\alpha_2^{(2)}]$ and $[\alpha_2^{(2)},0]$ a constant value each which shall be chosen in such a way that the relations

$$\int_0^1 \overline{\chi}_2^{(2)}(s)\,ds = 0, \qquad \int_0^1 \left(\overline{\chi}_2^{(2)}(s)\right)^2 ds = 1$$

are satisfied. Now we choose four points $\alpha_3^{(1)}$, $\alpha_3^{(2)}$, $\alpha_3^{(3)}$, $\alpha_3^{(4)}$, which shall lie in the intervals $[0,\alpha_2^{(1)}],\ldots,[\alpha_2^{(2)},1]$, respectively, and construct in corresponding manner the functions $\overline{\chi}_3^{(1)}(s)$, $\ldots$, $\overline{\chi}_3^{(4)}(s)$; and we proceed in this way. Again we make the agreement that at the points of discontinuity, the functions be equal to the arithmetic means of the values they assume in the intervals adjoining at that exact point.

If the points $\alpha_n^{(p)}$ chosen in this manner form an everywhere dense point set, then we can conclude in the exactly same way as earlier in § 1 of this chapter that the *orthogonal system so defined is complete*. In order to show that this function system also has the property that all continuous functions can be expanded in the Fourier manner into a series progressing according to the functions of this system, we note that the functions

$$\overline{K}_n^{(p)}(s,t) = \overline{\chi}_0(s)\,\overline{\chi}_0(t) + \cdots + \overline{\chi}_n^{(1)}(s)\,\overline{\chi}_n^{(1)}(t) + \cdots + \overline{\chi}_n^{(p)}(s)\,\overline{\chi}_n^{(p)}(t)$$

remain positive for every pair of values n, p and that we have

$$\int_0^1 \overline{K}_n^{(p)}(s,t)\,dt = 1\,.$$

If we now by $[f(s)]_n^{(p)}$ again denote the finite sum we obtain by terminating the Fourier series of the function $f(s)$ at the term $\overline{\chi}_n^{(p)}(s) \int_0^1 f(t)\,\overline{\chi}_n^{(p)}(t)\,dt$, then we have

$$[f(s)]_n^{(p)} = \int_0^1 \overline{K}_n^{(p)}(s,t)\,f(t)\,dt\,; \tag{17}$$

and since the functions $\overline{K}_n^{(p)}(s,t)$ are always positive, we may conclude that $[f(s)]_n^{(p)}$ always remains between the maximum and the minimum of $f(s)$. In other words, the assignment given by dint of equation (17) satisfies all conditions of our lemma in § 1 of Chapter 2, which implies that the *Fourier series with respect to the orthogonal system under consideration of an arbitrary function $f(s)$ converges uniformly to this function, if $f(s)$ lies in the range of the orthogonal system.*

Now, however, it is immediately clear that every finite aggregate of our orthogonal functions is a piecewise constant function; conversely, every piecewise constant function which suffers a jump only at finitely many points $\alpha_n^{(p)}$ and at such a point equals the arithmetic means of the values it assumes in the intervals adjoining at that exact point, is a finite aggregate of our orthogonal functions. Since, however, the points of discontinuity $\alpha_n^{(p)}$ are distributed everywhere dense in the interval $[0,1]$, we see immediately that every continuous function can be approximated arbitrarily by such a piecewise constant function. Thus it is shown that *the range of our orthogonal system comprises all continuous functions, and thus the Fourier series of any continuous function converges uniformly in the whole interval.* It also does not cause any further difficulties to furnish the proof that the series of any integrable function is convergent everywhere with the exception of a null set.

A further generalization of our orthogonal system could be obtained by performing, instead of the bisection of the intervals, a trisection, or quadrusection, and so on; then we define, in the same manner as before, a complete orthogonal function system by constructing at each subdivision of the intervals a finite system of piecewise constant functions which are orthogonal to each preceding function, and which in the divided intervals assume a constant value each. If we define — which after all is always possible — at *each* subdivision such a number of functions that there exists no piecewise constant function not vanishing identically, which is orthogonal to all functions defined *up to now*, and which only suffers a finite jump at the division points of the *current* subdivision; then — if the subdivision points form an everywhere dense point set — the function systems thus obtained are complete in the sense that every Lebesgue integrable function which is orthogonal to all functions of the system vanishes except for a set of measure zero. They all own the convergence property specified above, and finally, the theorem also holds that the convergence of the Fourier series (with respect to these function systems) of a function at a point only depends on the behavior of the function in the neighborhood of this point.

A Set of Continuous Orthogonal Functions.

Von

Philip Franklin in Cambridge (Mass.) (U. S. A.).

1. Introduction.

There exist continuous functions for which, at some points of the interval of orthogonality the classical Fourier series fails to converge. The analogous expansions in orthogonal functions arising from the simpler boundary value problems seem to share this property with the Fourier expansion[1]). This led A. Haar to ask if the property was common to all sets of orthogonal functions. He showed that it was not by exhibiting a set of orthogonal functions giving, as the expansion of any continuous function, a series converging uniformly to the function throughout the fundamental interval. The individual functions of his set, however, are *discontinuous*, so that his example does not exclude the possibility of the property being common to all sets of *continuous* orthogonal functions. In this paper we construct a set of continuous orthogonal functions similar to Haar's set in that the expansion of any continuous function represents the function everywhere.

2. Definition of the functions.

Consider the set of functions defined for $0 \leqq x \leqq 1$ by

$$
\begin{aligned}
&v_0 = 1, \qquad v_1 = x, \\
(1)\qquad &v_2 = 0, \quad x \leqq \tfrac{1}{2} \quad \text{and} \quad v_2 = x - \tfrac{1}{2}, \quad x \geqq \tfrac{1}{2}, \\
&\cdots\cdots\cdots\cdots\cdots\cdots \\
&v_n = 0, \quad x \leqq a_n \quad \text{and} \quad v_n = x - a_n, \quad x \geqq a_n
\end{aligned}
$$

where $a_n = (2n - 1 - 2^k)/2^k$, k integral and such that the highest power of 2 contained in $2n - 1$ is 2^k.

[1]) A. Haar, Math. Annalen **69** (1910), pp. 331—371.

Thus a_n is the nth term of the series

$$0,\ \tfrac{1}{2},\ \tfrac{1}{4},\ \tfrac{3}{4},\ \tfrac{1}{8},\ \tfrac{3}{8},\ \tfrac{5}{8},\ \tfrac{7}{8},\ \tfrac{1}{16},\ \ldots$$

and all the a_n are distinct.

Each of these functions v_i is continuous for $0 \leqq x \leqq 1$, which we take as our fundamental interval. Furthermore, these functions are linearly independent, since the first two obviously are so, and if any v_n $(n \geqq 2)$ were linearly dependent on a finite number of functions of the set not including v_n, it could not have a discontinuity in its derivative at a_n. Thus we may apply the process of orthogonalization[2]) to this set and so obtain the normal and orthogonal set f_n, where

$$f_0 = 1,$$

$$(2)\qquad f_{n+1} = \left[v_{n+1} - \sum_{h=1}^{n} f_h \int_0^1 f_h(t)\, v_{n+1}(t)\, dt\right] \Big/ \pm \sqrt{\int_0^1 v_{n+1}(t)^2\, dt}\,.$$

We find, for example

$$\begin{aligned}
&f_0 = 1, \qquad f_1 = \sqrt{3}(1-2x),\\
&f_2 = \sqrt{3}(1-4x), \qquad x \leqq \tfrac{1}{2} \text{ and } \sqrt{3}(4x-3), \qquad x \geqq \tfrac{1}{2},\\
(3)\quad &f_3 = \sqrt{3}(10-76x)/19, \qquad x \leqq \tfrac{1}{4},\\
&\sqrt{3}(52x-22)/19, \qquad \tfrac{1}{4} \leqq x \leqq \tfrac{1}{2},\\
&\sqrt{3}(10-12x)/19, \qquad x \geqq \tfrac{1}{2}.
\end{aligned}$$

The sign of these functions is not determined by (2), we take that sign before the radical which makes the first sign change of each f_n one from plus to minus. This defines in a unique way a set of continuous functions normal and orthogonal on the unit interval.

3. Convergence Properties.

The series expansion of a given function $F(x)$ in terms of the functions f_n is

$$(4)\qquad \sum_{i=0}^{\infty} A_i f_i(x), \quad \text{where} \quad A_i = \int_0^1 F(t)\, f_i(t)\, dt.$$

As we shall derive the convergence properties from approximation considerations, we recall that, for any linear combination of the first $(n+1)\, f_i$, $T_n = \sum_{i=0}^{n} C_i f_i$, we have:

[2]) cf. e. g. Courant-Hilbert, Methoden der math. Physik, Berlin 1924, p. 34.

$$(5)\qquad E(T_n) = \int_0^1 (F(x) - T_n(x))^2\,dx$$
$$= \int_0^1 F(x)^2\,dx - \sum_{i=0}^{n} A_i^2 + \sum_{i=0}^{n} (C_i - A_i)^2.$$

This shows that, for n fixed, the mean square error $E(T_n)$ is least when $C_i = A_i$, and T_n is a partial sum of the series (4), $S_n = \sum_{i=1}^{n} A_i f_i$, and also that, as n increases, $E(S_n)$ decreases.

We shall now show that, if $F(x)$ is continuous, for any positive ε there exists an $N(\varepsilon)$ such that, if $n > N$, $E(S_n) < \varepsilon$. We first define a broken line function of type n, B_n, to be a continuous function, linear in each of the intervals bonded by consecutive points of the set (b_j),

$$(6)\qquad 0, \frac{1}{2^k}, \frac{2}{2^k}, \ldots a_n, a_n + \frac{1}{2^k}, a_n + \frac{3}{2^k}, \ldots 1.$$

These (except for 1) are simply the a_i, $i \leq n$, arranged in order of magnitude. Every function B_n is clearly a linear combination of the first $(n+1)v_i$, for we may reduce it to zero by subtracting the constant times v_0 which makes it zero at $x = 0$, then the constant times v_1 which coincides with the remainder in the first interval, then the constant times $v_i \left(i = 2^{k-1} + 1 \text{ so that } a_i = \frac{1}{2^k}\right)$ which coincides with the new remainder in the first two intervals, and so on. As it is evident from (2) that each v_i is linearly dependent on the first $(i+1)f_j$, we see that each B_n is a linear combination of the first $(n+1)f_j$, i. e. it is a T_n.

From the continuity of $F(x)$ in the closed interval $0 \leq x \leq 1$, we infer the existence of a $\delta(\varepsilon)$ such that $|F(x_1) - F(x_2)| < \varepsilon$, whenever $|x_1 - x_2| < \delta$. If K is an integer for which

$$(7)\qquad \frac{1}{2^K} < \delta(\varepsilon), \qquad N = 2^K$$

may be taken as $N(\varepsilon)$. For, as all the points $m/2^K$ $(m = 1, 2, 3 \ldots 2^K - 1)$ are points a_i with $i \leq N$, the broken line function which agrees with $F(x)$ at 0, 1 and these points, and is linear in the intervals determined by them is a B_n and hence a T_n for any $n > N$. But, for this function, from (7) we have

$$(8)\qquad |F(x) - T_n(x)| < \varepsilon,$$

so that $E(T_n)$ and hence $E(S_n) < \varepsilon$.

We shall next prove that when n exceeds the N of (7), $S_n(x)$ uniformly approximates $F(x)$. A characteristic property of $S_n(x)$ is the fact that it is the T_n which minimizes $E(T_n)$. To apply this property, we need a lower limit for $E_{12}(T_n)$, the contribution to $E(T_n)$ from an interval $x_1 \leqq x \leqq x_2$, in which T_n is a linear function of x. We assume that

$$(9) \qquad x_2 - x_1 = D < \delta(\varepsilon).$$

Consequently, if $L(x)$ is the linear function for which

$$(10) \qquad L(x_1) = F(x_1) \quad \text{and} \quad L(x_2) = F(x_2),$$

we will have throughout the interval $x_1 x_2$

$$(11) \qquad |F(x) - L(x)| < \varepsilon \qquad (x_1 \leqq x \leqq x_2).$$

It follows from this that, in this interval

$$(12) \qquad |F(x) - T_n(x)| > |L(x) - T_n(x)| - \varepsilon.$$

Recalling that $L(x)$ und $T_n(x)$ are both linear in the interval $x_1 x_2$ and putting

$$(13) \qquad H_1 = |F(x_1) - T_n(x_1)|, \qquad H_2 = |F(x_2) - T_n(x_2)|,$$

we find for the right member of (12):

$$(14) \quad |L(x) - T_n(x)| - \varepsilon = \frac{1}{D}|(H_2 \pm H_1)x - H_2 x_1 \mp H_1 x_2| - \varepsilon,$$

the upper and lower signs corresponding to the cases in which $T_n(x_1)$ and $T_n(x_2)$ are on opposite sides of, or on the same side of $L(x)$. If we take the upper signs, and integrate the square of this expression over the parts of the interval $x_1 x_2$ where the expression is positive, we find

$$(15) \qquad E_{12}(T_n) > \frac{D[(H_1-\varepsilon)^3 + (H_2-\varepsilon)^3]}{3(H_1+H_2)},$$

when H_1 and H_2 are both $\geqq \varepsilon$. If either of these is $< \varepsilon$ the corresponding parenthesis does not appear in (15), but as it is negative, it may be inserted without destroying the inequality. If at least one of the H's is $> 4\varepsilon$, we shall have

$$(16) \qquad \frac{1}{H_1 + H_2} > \frac{1}{2[(H_1-\varepsilon)+(H_2-\varepsilon)]},$$

and in view of this, (15) becomes:

$$(17) \quad E_{12}(T_n) > \frac{D}{12}[(H_1 - \varepsilon)^2 + (H_2 - \varepsilon)^2], \qquad \text{Max}(H_1, H_2) > 4\varepsilon.$$

When the lower signs in (14) are used, we have in place of (15):

$$(18) \qquad E_{12}(T_n) > \frac{D[(H_1-\varepsilon)^3 - (H_2-\varepsilon)^3 - \varepsilon^2(H_1 - H_2)]}{3(H_1 - H_2)}.$$

When H_1 and H_2 are both $\geqq \varepsilon$, the last term in the numerator does not appear. As after division by the denominator it is negative, its presence merely strengthens the inequality. We have inserted it so that (18) will apply to the case in which one of the H's, say H_1 is $< \varepsilon$. Here the first term does not appear, and may not be supplied by itself since it is a positive multiple of the denominator. However, in view of our second condition, when $H_1 < \varepsilon$, $H_2 > 4\varepsilon$, and

$$\frac{(H_1 - \varepsilon)^3}{H_1 - H_2} - \varepsilon^2 < 0, \tag{19}$$

so that these two terms may be inserted together. As it is easily seen that (18) implies (17), this last inequality gives a lower limit for $E_{12}(T_n)$ which holds in all cases.

Let us now consider S_n, which is linear in the intervals bounded by the b_j of (6). We put

$$H_j = |F(b_j) - S_n(b_j)|. \tag{20}$$

Let M be the greatest of the quantities H_j, and b_g one of the points at which $H_g = M$. Suppose that $M > 4\varepsilon$, and let b_p be the first point to the left of b_g at which $H_p \leqq 4\varepsilon$ (or 0, if no such point exists) and b_q (or 1) the first such point to the right. We will compare S_n with a particular T_n, obtained from it by the following process. For values of x outside the interval $b_p \leqq x \leqq b_q$, $T_n(x)$ coincides with $S_n(x)$. Inside this interval, we put

$$\begin{aligned}(21)\quad T_n(b_p) = S_n(b_p) \ (\text{or } F(0)), \quad & T_n(b_q) = S_n(b_q) \ (\text{or } F(1)),\\ T_n(b_i) = F(b_i) \qquad\qquad & (p < i < q).\end{aligned}$$

In view of the linear character of T_n in the intervals between the b_j, it is defined by these values.

To compare the mean square errors $E(S_n)$ and $E(T_n)$, we need merely compare the contributions from the interval $b_p b_q$. Applying (17), we find

$$E_{pq}(S_n) > \frac{3}{2}(b_q - b_p)\varepsilon^2 + \frac{D}{12}(M - \varepsilon)^2 - 4D\varepsilon^2. \tag{22}$$

Here $D = \frac{1}{2^k}$, so that the elementary b_j intervals are at least D, and at most $2D$ in width. For T_n, we obviously have

$$E_{pq}(T_n) < (b_q - b_p)\varepsilon^2 + 100\, D\varepsilon^2, \tag{23}$$

since the numerical error is at most ε, cf. (8), except in the end intervals, where it is at most 5ε. From (22) and (23) we find:

(24) $E_{pq}(S_n) - E_{pq}(T_n) > \frac{D}{12}(M - \varepsilon)^2 - 104\,D\,\varepsilon^2 + \frac{1}{2}(b_q - b_p)\,\varepsilon^2,$

and since

(25) $$E_{pq}(S_n) - E_{pq}(T_n) = E(S_n) - E(T_n) < 0,$$

we must have[3])

(26) $$M < 37\,\varepsilon.$$

Thus all the $H_j < 37\,\varepsilon$, and in consequence

(27) $$|F(x) - S_n(x)| < 38\,\varepsilon \qquad (n > N(\varepsilon)),$$

which proves our contention about uniform approximation. We may express this result as

Theorem I. *If any function $F(x)$, continuous in the interval $0 \leqq x \leqq 1$, be expanded in a series of constant multiples of the functions f_j* (3), *the resulting series* (4) *will converge to $F(x)$ at all points of the unit interval, and, in fact, uniformly.*

Let us next treat the case in which $F(x)$ is a measurable function with summable square in the unit interval. Such a function can be approximated to an arbitrary degree of exactness, in the sense of least squares, by a continuous function[4]), $C(x)$, so that

(28) $$\int_0^1 [F(x) - C(x)]^2\,dx < \eta^2.$$

As we have shown above, (8), that there exists a T_n for which

(29) $$\int_0^1 [C(x) - T_n(x)]^2\,dx < \eta^2,$$

for this function $E(T_n) < 4\,\eta^2$. This implies the same relation for $E(S_n)$, and shows that for the functions now treated the expansion will converge in the mean.

If, at any point of the unit interval, $F(x)$ is continuous, we may show that the expansion converges there in the ordinary sense. Let, then, x_c be a point of continuity of $F(x)$, and $x_1 x_2$ an interval such that

(30) $$|F(x) - F(x_c)| < \varepsilon, \quad x_1 \leqq x \leqq x_2; \quad x_2 - x = x - x_1.$$

Let us take n so large that $\frac{1}{2^k}$ is small compared with $x_2 - x_1$, and

---

[3]) By a longer calculation, which treats the cases more in detail, we may show that, in fact, $M < 17\,\varepsilon$.

[4]) cf. e. g. E. W. Hobson; Theory of functions of a Real Variable, 2nd ed., Cambridge 1921, vol. 1, p. 584.

for S_n consider the H_j for the b_j of the interval $x_1 x_2$ here defined by

$$H_j = |F(x_c) - S_n(b_j)| \qquad (x_1 \leqq b_j \leqq x_2). \tag{31}$$

If more than $1/P$ of these H_j were $\geqq 4\varepsilon$, from (17) we would have

$$E(S_n) \geqq E_{12}(S_n) > (x_2 - x_1)\varepsilon^2/P. \tag{32}$$

As $E(S_n)$ approaches zero as n increases, we may take n so large that $1/P < \frac{1}{4}$. When this is done, $\frac{3}{4}$ of the $H_j < 4\varepsilon$, and in particular b_j for which this holds exist on both sides of x_c in the interval $x_1 x_2$. Calling b_p the first such point to the left of x_c, and b_q the first one to the right, we may use the argument given above to establish (27), to show that each of the H_j for b_j adjacent to x_c (or for x_c itself if that is a b_j) $< 37\varepsilon$, and hence

$$|F(x_c) - S_n(x_c)| < 38\varepsilon \qquad (n > N_1(\varepsilon)). \tag{33}$$

This proves the convergence of S_n to $F(x)$ at any point of continuity, and an obvious modification of the argument shows that the convergence is uniform for any closed interval in which $F(x)$ is continuous. We have thus established:

Theorem II. *If any measurable function $F(x)$, with summable square in the interval $0 \leqq x \leqq 1$, be expanded in a series of constant multiples of the functions f_j (3), the resulting series (4) will converge to $F(x)$ at all points of continuity. Further, in any closed interval of continuity the convergence will be uniform. Over the whole interval, the series converges in the mean to $F(x)$.*

The consideration of simple examples shows that if the function $F(x)$ is continuous in each of the intervals $x_1 x_2$ and $x_2 x_3$, but discontinuous at x_2, the series will in general oscillate between two finite values at x_2.

4. Other Functions.

We have based the set of function used, f_i (3), on the functions v_i (1) with breaks at a_n, the proper fractions with denominators integral powers of 2. Similarly, we could obtain an orthogonal set of broken line functions from any other set of points p_n. For such a set to have the convergence properties of theorems I and II, it is sufficient that the points p_n be everywhere dense on the unit interval and enumerated in such a way that, of the intervals marked off at any stage by these points, the ratio of the greatest interval to the least remains uniformly bounded for the entire set. With this restriction on the p_n, the theorems can be proved essentially as above.

5. Application to the Haar set.

The set of discontinuous functions used by Haar may be obtained by applying the process of orthogonalization (2) to the derivatives of our functions v_i (1). As the linear combinations of them are step functions, the deductions made from (5) show that the set is complete. As, in a step function, any one step may be altered without disturbing the rest of the function, it is obvious from the minimum property that in any interval between adjacent points of discontinuity b_j, $S_n(x)$ must lie between the greatest and least values of $F(x)$ in the interval. From this remark most of the convergence theorems given by Haar may be deduced immediately without further calculation.

Massachusetts Institute of Technology.

(Eingegangen am 20. 12. 1927.)

A MODIFIED FRANKLIN SYSTEM AND HIGHER-ORDER SPLINE SYSTEMS ON $\mathbf{R}^n$ AS UNCONDITIONAL BASES FOR HARDY SPACES

Jan-Olov Strömberg
University of Stockholm
Princeton University

1. INTRODUCTION

Let $H^p(I)$ be the subspace of distributions in the Hardy space $H^p(\mathbf{R})$ that are supported in $I = [0, 1]$. The Franklin system $(m = 0)$ and higher-order spline system $(m > 0)$ have been studied and used as unconditional basis for $H^p(I)$, $p > 1/(m + 2)$ in [1, 4, 7, and 9-11]. The existence of an unconditional basis for H^1 was first shown by B. Maurey [8] and an explicit construction was made by L. Carleson [3]. The spline system as unconditional basis for the bi-Hardy space $H^1(I \times I)$ has been studied by S. Y. A. Chang.

The purpose of this paper is to define m-order spline systems on $\mathbf{R}$ and especially get a modified Franklin system $(m = 0)$. We will also get spline systems on $\mathbf{R}^n$, $n > 1$. All these systems can also be made periodical to become spline systems on $\mathbf{T}^n$ by a simple summation. (The spline systems on I do not extend periodically to the circle $\mathbf{T} = \mathbf{R}/\mathbf{Z}$ and cannot be used as unconditional basis for the real Hardy spaces $H^p(\mathbf{T})$, $0 < p \leq 1$, without modification.)

We are especially interested to see how these spline systems on $\mathbf{R}^n$ can be used as unconditional basis for $H^p(\mathbf{R}^n)$, $n \geq 1$.

2. CONSTRUCTION OF TWO FUNCTIONS

Let $m \geq 0$. We will construct one function on $\mathbf{R}$ from which we get the m-order spline system on $\mathbf{R}$ and another function on $\mathbf{R}$ that we will use together with the first one to get the m-order spline system on $\mathbf{R}^n$.

Let $A_0 = \mathbf{Z}_+ \cup \{0\} \cup \frac{1}{2}\mathbf{Z}_-$ and $A_1 = A_0 \cup \left\{\frac{1}{2}\right\}$. A_0 splits $\mathbf{R}$ into intervals $\{I_\sigma\}_{\sigma \in A_0}$ (σ = left endpoint of I_σ). Let S_0^m be the subspace of functions f in $L^2(\mathbf{R})$ such that $f \in C^m(\mathbf{R})$ and is a real polynomial of degree $\leq m + 1$ on each I_σ, $\sigma \in A_0$. Let S_1^m be the corresponding subspace of $L^2(\mathbf{R})$ with the set A_0 replaced by A_1. Note that $S_1^m = \{f;\ f(\cdot + 1) \in S_0^m\}$ and also that the set A_1 splits $\mathbf{R}$ into the same collection of intervals as A_0 did except for the interval $I_0 = [0, 1]$, which is split into two intervals $\left[0, \frac{1}{2}\right]$ and $\left[\frac{1}{2}, 1\right]$. It follows that S_0^m has codimension 1 in S_1^m, since the $m + 1$ order derivative of a function f in S_1^m may have a jump at $\frac{1}{2}$ and if not, it is contained in S_0^m. Thus there is a function τ in $L^2(\mathbf{R})$ that is a uniquely defined modulo sign by

(i) $\tau \in S_1^m$,

(ii) $\tau \perp S_0^m$, that is, $\int_{\mathbf{R}} \tau(x)f(x)\,dx = 0$ for all $f \in S_0^m$,

(iii) $\|\tau\|_{L^2} = 1$.

Next let $\tilde{S}_0^m$ be the subspace of functions f in S_0^m that are supported in $[0, \infty)$ and let $\tilde{S}_1^m$ be the subspace of functions f in S_1^m that are supported in $[1, \infty)$. We observe that $\tilde{S}_1^m$ is a subspace of $\tilde{S}_0^m$ with codimension 1. Thus there is a function ρ in $L^2(\mathbf{R})$ that is a uniquely defined modulo sign by

(iv) $\rho \in \tilde{S}_0^m$,

(v) $\rho \perp \tilde{S}_1^m$,

(vi) $\|\rho\|_{L^2} = 1$.

By definition, both τ and ρ are piecewise polynomials of degree $\leq m + 1$ and are in $C^m(\mathbf{R})$; furthermore, the $(m+1)$-order derivatives of them are piecewise constant with discontinuities in A_1 resp. $\mathbf{Z}_+ \cup \{0\}$ and the following estimates hold with some constant r, $0 < r < 1$ (which may change from line to line):

$$|D^k\tau(x)| \leq Cr^{|x|}, \; k = 0, \ldots, m + 1, \tag{1}$$

$$|D^k\rho(x)| \leq \begin{matrix} Cr^x, & x \geq 0 \\ 0, & x < 0 \end{matrix}, \; k = 0, \ldots, m + 1. \tag{2}$$

The function τ also satisfies the moment conditions

$$\int_{\mathbf{R}} \tau(x)x^\alpha \, dx = 0, \; \alpha = 0, \ldots, m + 1. \tag{3}$$

3. MAIN RESULTS

First we consider the one-dimensional case. Let $\nu = (j, k) \in \mathbf{Z} \times \mathbf{Z}$ and set

$$f_\nu(x) = 2^{j/2}\tau(2^j x - k).$$

Then we have:

Theorem 1

(a) $\{f_\nu\}_{\nu \in \mathbf{Z}\times\mathbf{Z}}$ *is a complete orthonormal system in* $L^2(\mathbf{R})$.

(b) *Let* $p > 1/(m + 2)$. *Then* f_ν *is in the dual space of* $H^p(\mathbf{R})$, $\nu \in \mathbf{Z} \times \mathbf{Z}$, *and if* $c_\nu = (f_\nu, f)$, *then the sum* $\sum_{\nu \in \mathbf{Z}\times\mathbf{Z}} c_\nu f_\nu$ *converges unconditionally to* f *in the* $H^p(\mathbf{R})$ *norm. Furthermore, the coefficients in a converging sum are uniquely defined; that is, if* $\sum c_\nu f_\nu$ *converges to a function* $f \in H^p(\mathbf{R})$ *in the norm, then* $c_\nu = (f_\nu, f)$.

(c) *Let* J_ν *be the interval* $[2^{-j}k, 2^{-j}(k + 1)]$ *and let* χ_ν *be its characteristic function. Then for any collections of coefficients* $\{c_\nu\}_\nu$ *such that only finitely many* c_ν *are non-vanishing, we have*

$$\left\| \sum_\nu c_\nu f_\nu \right\|_{H^p(\mathbf{R})} \le C \left\| \left\{ \sum_\nu |c_\nu|^2 2^j \chi_\nu \right\}^{1/2} \right\|_{L^p(\mathbf{R})}, \quad p > 1/(m + 5/2). \quad (4)$$

On the other hand, if $f \in H^p(\mathbf{R})$, $p > 1/(m+2)$ *and* $c_\nu = (f_\nu, f)$, *then*

$$\left\| \left\{ \sum_\nu |c_\nu|^2 2^j \chi_\nu \right\}^{1/2} \right\|_{L^p(\mathbf{R})} \le C \| f \|_{H^p(\mathbf{R})}. \quad (5)$$

In higher dimensions we will use both the functions τ and ρ. We want to define a system of functions on $\mathbf{R}^n$, $n > 1$. Let Ω be the set of n-tuples of functions $\omega = (\omega_1, \ldots, \omega_n)$, where ω_i is either τ or ρ but $\omega_i = \tau$ for at least one i, $1 \le i \le n$. Thus Ω consists of $2^n - 1$ n-tuples. Now we define the function τ^ω on $\mathbf{R}^n$ by

$$\tau^\omega(x) = \tau^\omega(x_1, \ldots, x_n) = \prod_{i=1}^{n} \omega_i(x_i), \quad x \in \mathbf{R}^n, \ \omega \in \Omega.$$

For example, when $n = 2$, we get the three functions $\tau(x_1)\tau(x_2)$, $\tau(x_1)\rho(x_2)$, and $\rho(x_1)\tau(x_2)$. Let $\nu = (j, k_1, \ldots, k_n) \in \mathbf{Z} \times \mathbf{Z}^n$ and define the function f by

$$f_\nu^\omega(x) = 2^{nj/2}\tau^\omega(2^j x - (k_1, \ldots, k_n)), \quad x \in \mathbf{R}^n.$$

Then:

Theorem 2

(a) $\{f_\nu^\omega\}_{\Omega \times \mathbf{Z}^{n+1}}$ *is a complete orthonormal system in* $L^2(\mathbf{R}^n)$.

(b) $\{f_\nu^\omega\}_{\Omega \times \mathbf{Z}^{n+1}}$ *is an unconditional basis for* $H^p(\mathbf{R}^n)$ *when* $p > n/(n + m + 1)$; *that is, the corresponding statements of* Theorem 1(b) *hold with the sum taken over both* ν *and* ω.

(c) *The* n*-dimensional version of* (4) *holds when* $p > p(n, m)$, *and the* n*-dimensional version of* (5) *holds when* $p > n/(n + m + 1)$. (*The sum is taken over both* ν *and* ω *and* $\mathbf{R}$ *is replaced by* $\mathbf{R}^n$ *and* J_ν *is a cube.*)

Remark 1. With the methods that we give in detail when $n = 1$, one can get $p(n, m) = 1/\big(1/2 + (m + 2)/n\big)$.

Remark 2. $2^{(n)j}\chi_\nu$ may as well be replaced by $|f_\nu^{(\omega)}|^2$ in (4) and (5).

Remark 3. If we set $f_\nu^{\omega,T}(x) = \sum_{t \in \mathbf{Z}^n} f_\nu^{\omega}(x - t)$ when $\nu = (j, k_1, \dots, k_n)$, $j \geq 0$, we get functions on $\mathbf{T}^n$, which together with the constant function 1 will be a complete orthonormal system. This system is an unconditional basis for the real Hardy space $H^p(\mathbf{T}^n)$ (defined with one dilation) when $p > n/(n + m + 1)$.

4. EXPLICIT DESCRIPTION OF τ AND ρ, m = 0

Since τ and ρ are linear on the intervals between the points $\sigma \in A_1$, it is enough to find their values at these points. We will in this section use the notation $\tau_\sigma = \tau(\sigma)$ and $\rho_\sigma = \rho(\sigma)$, $\sigma \in A_1$.

Definition. The *simple tent* Λ_σ in S_0^0, $\sigma \in A_0$, is the function in S_0^0 that is 1 at $x = \sigma$ but is 0 at all $\sigma_1 \in A_0$, $\sigma_1 \neq \sigma$.

Since τ is orthogonal to all simple tents in S_0^0 and ρ is orthogonal to all simple tents in S_0^0 with support in $[1, \infty)$, we will get the following equations after computing the integrals:

$$\tau_{\sigma-1} + 4\tau_\sigma + \tau_{\sigma+1} = 0, \qquad = 2, 3, 4, \dots,$$
$$\tau_{\sigma-1/2} + 4\tau_\sigma + \tau_{\sigma+1/2} = 0, \qquad = -1/2, -1, -3/2, \dots,$$
$$\tau_0 + 6\tau_{1/2} + 13\tau_1 + 4\tau_2 = 0,$$
$$2\tau_{-1/2} + 9\tau_0 + 6\tau_{1/2} + \tau_1 = 0,$$

and

$$\rho_{\sigma-1} + 4\rho_\sigma + \rho_{\sigma+1} = 0.$$

The characteristic equation $r^2 + 4r + 1 = 0$ has the two roots $r_1 = \sqrt{3} - 2$ and $r_2 = r_1^{-1} = -(2 + \sqrt{3})$. Since the functions must go to zero at infinity, we get

$$\tau_\sigma = \tau_1(\sqrt{3} - 2)^{\sigma-1}, \ \sigma = 1, 2, 3, \dots,$$
$$\tau_\sigma = \tau_0(-\sqrt{3} - 2)^{2\sigma}, \ \sigma = 0, -1/2, -1, \dots,$$

$$\rho_\sigma = \rho_1(\sqrt{3} - 2)^{\sigma - 1}, \qquad = 1, 2, 3, \ldots,$$

and since ρ is linear on $[0, 1]$ and vanishes on the negative axis, we have $\rho_{1/2} = \frac{1}{2}\rho_1$ and $\rho_\sigma = 0$ for all $= 0, -1/2, -1, \ldots$. From the third and fourth rows of the equations above, we now get $\tau_0 = 2(\sqrt{3} - 1)\tau_1$ and $\tau_{1/2} = -(\sqrt{3} + 1/2)\tau_1$. Finally, by the normalization (iii) and (vi), one can get the values of $|\tau_1|$ and $|\rho_1|$. We leave the computation of these values to the interested readers.

It is now easy to check the estimates (1) and (2) when $k = 0, 1$ and $m = 0$. To see that the inequality (3) holds ($m = 0$), we can truncate the functions 1 and x outside a large compact set so that the truncated functions are in S_0^0. The corresponding part of the integrals will then be zero by orthogonality and the remainder of the integral can be made arbitrarily small if the compact set is chosen large enough, since τ is rapidly decreasing at infinity.

5. IDEAS FOR (1)–(3) WHEN m > 0

The derivatives $D^{m+2}\tau$ and $D^{m+2}\rho$ (taken in the sense of distributions) are measures supported in A_1, and in this section we use the notations τ_σ and ρ_σ, $\sigma \in A_1$ for the coefficients given by

$$D^{m+2}\tau = \sum_{\sigma \in A_1} \tau_\sigma \delta_\sigma,$$

$$D^{m+2}\rho = \sum_{\sigma \in A_1} \rho_\sigma \delta_\sigma$$

(δ_σ is the delta function at σ).

Let h be the function $h(x) = \Delta^{2m+4}(x_+^{2m+3})$, where $x_+^i = \max(x^i, 0)$ and Δ^i is defined by $\Delta^1 f(x) = f(x) - f(x - 1)$, $\Delta^i = \Delta^1\Delta^{i-1}$, $i = 2, 3, \ldots$, and set $h_k(x) = h(x - k)$, $k = 1, 2, \ldots$. Then $D^{m+2}h_k$ are functions in S_0^m, $k = 1, 2, \ldots$ (and also in $\widetilde{S}_1^m$) and by orthogonality we have

$$\int D^{m+2} h_k \tau \, dx = 0,$$

which after integration by parts, gives us the equations

$$\sum_{\sigma \in A_1} h_k(\sigma)\tau_\sigma = 0, \; k = 1, 2, \ldots .$$

Observe that $h_k(\sigma)$ is zero for all $\sigma \in A_1$ except $\sigma = k + 1, \ldots,$ $k + 2m + 3$. We get the characteristic equation

$$\sum_j h(j) r^{j-1} \equiv Q_{2m+2}(r),$$

where Q_{2m+2} is a polynomial of degree $2m + 2$. By differentiation of geometric series, we can write

$$Q_j(r) = \frac{c_j (1-r)^{j+2}}{r} \left(r \frac{d}{dr} \right)^{j+1} \frac{1}{1-r}, \; j = 0, 1, \ldots .$$

Using a simple change of variable ($t = 1/r$), we see that if r_1 is a root to the equation $Q_j(r) = 0$, then $1/r_1$ is a root to the same equation. By this and an elementary mean value theorem, one can conclude by induction over j that the characteristic equation $Q_{2m+2}(r) = 0$ has $2m + 2$ distinct roots of which $m + 1$ roots $r_1, \ldots, r_{m+1}$ are in the open interval $(0, -1)$ and the remaining $m + 1$ roots are in the interval $(-\infty, -1)$. We conclude that

$$\tau_\sigma = C_1 r_1^\sigma + \cdots + C_{m+1} r_{m+1}^\sigma + C_{m+2} r_1^{-\sigma} + \cdots + C_{2m+2} r_{m+1}^{-\sigma}, \; \sigma = 2, 3, \ldots,$$

for some constants $C_1, \ldots, C_{2m+2}$, and since τ is in $L^2(\mathbf{R})$, $C_{m+2}, \ldots, C_{2m+2}$ are all equal to zero. We now get the estimate (1) for $x \geq 0$ by repeated integration. In the same way we also get (1) for $x < 0$ and (2). With the same truncation argument as we used in the case $m = 0$, we will get (3) also when $m > 0$.

6. ORTHOGONALITY

First we will see that $\{f_\nu\}_{\nu \in \mathbf{Z}^2}$ is an orthogonal system in $L^2(\mathbf{R})$. Let $A_\nu = \{x;\; 2^j x - k \in A_0\}$. The basic observation is that the collection of sets $\{A_\nu\}_{\mathbf{Z}^2}$ is totally ordered by inclusion; that is, if $\nu \neq \nu'$, then either $A_\nu \nsubseteq A_{\nu'}$ (we write then $\nu \prec \nu'$) or $A_{\nu'} \subsetneq A_\nu$ (we write then $\nu' \prec \nu$). Thus if $\nu = (j, k)$ and $\nu' = (j', k')$, then

$$\nu \prec \nu' \quad \text{is equivalent to} \quad \begin{cases} j < j' \\ j = j',\ k < k'. \end{cases}$$

Let $\nu \prec \nu'$ and consider the functions f_ν and $f_{\nu'}$. By a linear transformation, $f_{\nu'}$ can be transformed to (a multiple of) τ and, by the same linear transformation, f_ν will be transformed to a function in S_0^m, and since $\tau \perp S_0^m$, we conclude that $f_{\nu'} \perp f_\nu$. This proves that $\{f_\nu\}_{\mathbf{Z}^2}$ is an orthogonal system in $L^2(\mathbf{R})$.

Next we want to see that $\{f_\nu^\omega\}_{\Omega \times \mathbf{Z}^{n+1}}$ is an orthogonal system in $L^2(\mathbf{R}^n)$, $n > 1$. Let $\omega = (\omega_1, \ldots, \omega_n)$ and $\nu = (j, k_1, \ldots, k_n)$. Set

$$\omega_{i;j,k_i}(x_i) = 2^{j/2}\omega_i(2^j x_i - k_i), \quad i = 1, \ldots, n.$$

Then

$$f_\nu^\omega(x) = \prod_{i=1}^n \omega_{i;j,k_i}(x_i),$$

and $\omega_{i;j,k_i}$ is of the form $2^{j/2}\tau(2^j x_i - k_i) = f_{j,k_i}(x_i)$ or $2^{j/2}\rho(2^j x_i - k_i)$. Let $\omega' = (\omega_1', \ldots, \omega_n')$ and $\nu' = (j', k_1', \ldots, k_n')$. Then we have

$$\int_{\mathbf{R}^n} f_\nu^\omega(x) f_{\nu'}^{\omega'}(x)\,dx = \prod_{i=1}^n \int_{\mathbf{R}} \omega_{i;j,k_i}(x_i)\omega'_{i;j',k_i'}(x_i)\,dx_i.$$

If $(\omega, \nu) \neq (\omega', \nu')$, we will see that there is an $i = i_0$ such that the integral in the x_i-direction vanishes in the product on the right-hand side. There are the following four cases:

Case 1. $j < j'$ (or similarly $j' < j$). Then we use that $\omega_i' = \tau$ for some $i = i_0$.

Case 2. $j = j'$, $\omega \neq \omega'$; then $\omega_i \neq \omega_i'$ for some $i = i_0$, say, $\omega_{i_0}' = \tau$, $\omega_{i_0} = \rho$ (or similarly, if $\omega_{i_0} = \tau$, $\omega_{i_0}' = \rho$).

Case 3. $j = j'$, $\omega = \omega'$, $(k_1, \ldots, k_n) \neq (k_1', \ldots, k_n')$; that is, $k_i \neq k_i'$ for some $i = i_0$, and $\underline{\omega_{i_0}' = \omega_{i_0} = \tau}$, say, $k_{i_0}' > k_{i_0}$ (or similarly $k_{i_0} > k_{i_0}'$).

Case 4. $j = j'$, $\omega = \omega'$, $(k_1, \ldots, k_n) \neq (k_1', \ldots, k_n')$; that is, $k_i \neq k_i'$ for some $i = i_0$ and $\underline{\omega_{i_0}' = \omega_{i_0} = \rho}$, say, $k_{i_0}' < k_{i_0}$ (or similarly $k_{i_0} < k_{i_0}'$).

By a linear transformation we can now transform the function $\omega_{i_0; j', k_{i_0}'}'$ to get the function τ (in cases 1-3) or the function ρ (in case 4), and by the same linear transformation $\omega_{i_0; j, k_{i_0}}$ will be transformed to a function in S_0^m (cases 1-3) resp. $\tilde{S}_1^m$ (case 4). Since τ is orthogonal to S_0^m and ρ is orthogonal to $\tilde{S}_1^m$, we conclude that the integral in the x_{i_0}-direction vanishes. This proves the orthogonality of the system $\{f_\nu^\omega\}_{\Omega \times \mathbf{Z}^{n+1}}$.

7. COMPLETENESS IN L^2

We will first show that $\{f_\nu\}_{\mathbf{Z}^2}$ is a complete system in $L^2(\mathbf{R})$.

Let $S_\nu = \{f;\ f(x) = g(2^j x - k) \text{ for some } g \in S_0^m\}$, $\nu = (j, k) \in \mathbf{Z}^2$, and let P_ν be the projection operator $P_\nu : L^2 \to S_\nu$. If $\nu' < \nu$, then $S_{\nu'} \not\subset S_\nu$ and $P_\nu P_{\nu'} f = P_{\nu'} P_\nu f = P_{\nu'} f$, $f \in L^2(\mathbf{R})$.

Let $f \in L^2(\mathbf{R})$ and let $c_\nu = \int_{\mathbf{R}} f\, f_\nu\, dx$; then

$$c_\nu f_\nu = P_{j,k+1} f - P_{j,k} f,$$
$$\nu = (j, k),$$

and

$$f - \sum_{|j| \le j_0} \sum_{|k| \le k_0} c_\nu f_\nu = f - \sum_{|j| \le j_0} (P_{j,k_0+1} f - P_{j,-k_0} f)$$
$$= f - P_{j_0, k_0+1} f - \sum_{j=-j_0}^{j_0-1} (P_{j+1,-k_0} f - P_{j,k_0+1} f) + P_{-j_0,-k_0} f.$$

The completeness of the system $\{f_\nu\}_{\mathbf{Z}^2}$ in $L^2(\mathbf{R})$ now follows easily if we can show that if $f \in L^2(\mathbf{R})$, then

(i) $\| P_{j,k} f - f \|_{L^2} \to 0$ as $j \to +\infty$ independent of k,

(ii) $\| P_{j,k} f \|_{L^2} \to 0$ as $j \to -\infty$ independent of k,

(iii) $\| P_{j+1,-k} f - P_{j,k+1} f \|_{L^2} \to 0$ as $k \to +\infty$ for each fixed j.

(i) follows by approximation in $L^2(\mathbf{R})$ by spline functions, which we leave to the reader.

For (ii) we use that $P_{j,k} f$ is piecewise a polynomial of degree $\leq m + 1$ to get

$$\| P_{j,k} f \|_\infty \leq C 2^{j/2} \| P_{j,k} f \|_{L^2} \leq C 2^{j/2} \| f \|_{L^2} .$$

Thus $f - P_{j,k} f$ converges uniformly to f as j tends to $-\infty$. We conclude that

$$\| f \|_{L^2} \geq \| f - P_{j,k} f \|_{L^2} \to \| f \|_{L^2}$$

and

$$\| P_{j,k} f \|_{L^2}^2 = \| f \|_{L^2}^2 - \| f - P_{j,k} f \|_{L^2}^2 \to 0 \text{ as } j \to -\infty.$$

To show (iii), we may assume that $f \in S_{j+1,0}$ and has compact support. Then

$$P_{j+1,-k} f = f + P_{j+1,-k} f - P_{j+1,0} f = f - \sum_{i=-k}^{-1} c_{j+1,i} f_{j+1,i}$$

and

$$\sum_{i=-k}^{-1} c_{j+1,i} f_{j+1,i}(x) = \int \left\{ \sum_{i=-k}^{-1} f_{j+1,i}(y) f_{j+1,i}(x) \right\} f(y)\, dy$$

and by (1) we get

$$\left| \sum_{i=-k}^{-1} f_{j+1,i}(y) f_{j+1,i}(x) \right| \leq C 2^{j/2} r^{2^{j+1}|x-y|}.$$

We conclude from this that $|P_{j+1,-k} f(x)| \leq C_f r^{2^j |x|}$ with the constant C_f independent of k. For any $\varepsilon > 0$, we can now find a large compact set K_ε independent of k such that we can write $P_{j+1,-k} f = g_k + g_k'$ with $g_k, g_k' \in S_{j+1,-k}$, $\| g_k \|_{L^2} < \varepsilon$, and g_k' supported in K_ε. We observe that $g_k' \in S_{j,k+1}$ if k is chosen large enough. Thus $P_{j,k+1} g_k' = g_k'$ and we get

$$\begin{aligned} P_{j+1,-k}f - P_{j,k+1}f &= P_{j+1,-k}f - P_{j,k+1}P_{j+1,-k}f \\ &= g_k + g_k' - P_{j,k+1}g_k - P_{j,k+1}g_k' \\ &= g_k - P_{j,k+1}g_k \end{aligned}$$

and

$$\| P_{j+1,-k}f - P_{j,k+1}f \|_{L^2} = \| g_k - P_{j,k+1}g_k \|_{L^2} \leq \| g_k \|_{L^2} \leq \varepsilon,$$

provided k is chosen large enough. This proves (iii) and finishes the proof of the completeness of the system $\{f_\nu\}_{\mathbf{Z}^2}$ in $L^2(\mathbf{R})$.

Next we will see that $\{f_\nu^\omega\}_{\Omega \times \mathbf{Z}^{n+1}}$ is a complete system in $L^2(\mathbf{R}^n)$. For simplicity we will only look at the case when $n = 2$. From the one-dimensional case, we conclude that the collection of products

$$\{f_{j_1,k_1}(x_1)f_{j_2,k_2}(x_2)\}_{\mathbf{Z}^2 \times \mathbf{Z}^2}$$

is a complete orthonormal system in $L^2(\mathbf{R}^2)$. However, it involves two independent dilations and that is the reason we use the system $\{f_\nu^\omega\}_{\Omega \times \mathbf{Z}^2}$, which involves only one dilation. We split the products into three cases:

Case 1. $j_1 = j_2$. These products are still in our system as the functions f_ν^ω with $\omega = (\tau, \tau)$.

Case 2. $j_1 > j_2$. We will see that these products are in the closure of the span of the functions f_ν^ω with $\omega = (\tau, \rho)$.

Case 3. $j_1 < j_2$. These products are in the closure of the span of the functions f_ν^ω with $\omega = (\rho, \tau)$.

We will only look at case 2, as case 3 can be treated in the same way. We claim that the product $f_{j_1,k_1}(x_1)f_{j_2,k_2}(x_2)$ is in the closure of the span of functions

$$\{f_{j_1,k_1}(x_1)2^{j/2}\rho(2^{j_1}x_2 - k)\}_{k \in \mathbf{Z}}$$

provided $j_1 > j_2$. This will again be a one-dimensional problem. By dilation argument we may assume that $j_1 = 0$ and $j_2 < 0$. It is enough to show that the function f_{j_2,k_2} that is in S_{j_2,k_2+1} can be approximated in $L^2(\mathbf{R})$ by linear combinations of the functions $\{\rho(\cdot - k)\}_{k\in\mathbf{Z}}$.

First we approximate f_{j_2,k_2} by a function f in S_{j_2,k_2+1} with compact support. Let k_0 be so large that the interval $[-k_0, k_0]$ contains the support of f. Set $\widetilde{S}_k = \{f : f(\cdot + k) \in \widetilde{S}_0^m\}$ and let $\widetilde{P}_k$ be the projection operator $\widetilde{P}_k : L^2(\mathbf{R}) \to \widetilde{S}_k$. Then $\widetilde{P}_{-k_0}f = f$ and $\widetilde{P}_{k_0}f = 0$, since $f \in \widetilde{S}_{-k_0}$ and since

$$\|f\|_{L^2}^2 - \|\widetilde{P}_{k_0}f\|_{L^2}^2 = \|f - \widetilde{P}_{k_0}f\|_{L^2}^2 = \|f\|_{L^2}^2 + \|\widetilde{P}_{k_0}f\|_{L^2}^2 .$$

The last inequality holds, since f and $\widetilde{P}_{k_0}f$ have disjoint supports. Thus

$$f = \widetilde{P}_{-k_0}f - \widetilde{P}_{k_0}f = \sum_{k=-k_0}^{k_0-1} \widetilde{P}_k f - \widetilde{P}_{k+1}f,$$

and $\widetilde{P}_k f - \widetilde{P}_{k+1}f$ is a multiple of $\rho(\cdot - k)$. This shows that f can be written as a linear combination of the functions $\{\rho(\cdot - k)\}_{k=-k_0}^{k_0-1}$ and hence f_{j_2,k_2} is in the closure of the span of the functions $\{\rho(\cdot - k)\}_{k\in\mathbf{Z}}$. This completes the proof of the completeness of the system $\{f_\nu^\omega\}_{\Omega\times\mathbf{Z}^3}$ in $L^2(\mathbf{R}^2)$.

8. PROOF OF THEOREM 1(a) AND (b)

First we observe that f is in $\mathrm{Lip}(m + 1)$, that is, $D^m f$ is in Lip 1 and hence is in the dual space of $H^p(\mathbf{R})$ when $1/p \leq 2 + m$. We will first prove Theorem 1(c) by an area integral approach. The inequalities (4) and (5) will then be the basic ingredients in the proof of Theorem 1(b). At the end of this paper, we will show how (b) can be proved directly with atoms.

Proof of (4) and (5). It is enough to show (5) for f in a dense subspace of $H^p(\mathbf{R})$, since if f_j converges to f, so will the corresponding coefficients $c_{j;\nu}$ converge to c_ν by the duality, and hence

(5) will hold for a general f if the sum over ν is restricted to be finite. The estimate for an infinite sum then follows by monotone convergence. Thus we may assume that the Fourier transform of f is C^∞, compactly supported and vanishing near the origin.

Let $\tilde{\nu} = (2^{-j}k, 2^{-j}) \in \mathbf{R}^2_+$, $\nu = (j, k) \in \mathbf{Z}^2$, and let μ be the measure on $\mathbf{R}^2_+$ defined by $\mu = \sum_\nu \delta_{\tilde{\nu}}$, where $\delta_{\tilde{\nu}}$ is the Dirac measure at $\tilde{\nu}$. Set $\tau_t(x) = t^{-1}\tau(x/t)$, $\tilde{\tau}_t(x) = \tau_t(-x)$, and define the function F on $\mathbf{R}^2_+$ by

$$F(x, t) = \tilde{\tau}_t * f(x),$$

where f is the finite sum $\sum_\nu c_\nu f_\nu$ [in showing (4)] or the function given in (5). Observe that in both cases we have $F(\tilde{\nu}) = 2^{j/2}c_\nu$, $\nu \in \mathbf{Z}^2$. We define the functions

$$AF(z) = \left\{ \iint_{0 \le z - x \le t} |F(x, t)|^2 d\mu(x, t) \right\}^{1/2}$$

and

$$g^*_\lambda F(z) = \left\{ \iint_{\mathbf{R}^2_+} (1 + |x - z|/t)^{-2\lambda} |F(x, t)|^2 d_\mu(x, t) \right\}^{1/2}, \quad \lambda > 0.$$

Then AF is identical to the expression

$$\left\{ \sum_\nu |c_\nu|^2 2^j \chi_\nu \right\}^{1/2}.$$

Now we let ψ be a C^∞ function supported in $[-1, 1]$ such that

$$\int \psi(x)x^\alpha\, dx = 0, \quad \alpha = 0, 1, \ldots, m + 1 \tag{6}$$

and

$$\int_0^\infty |\hat{\psi}(s\xi)|^2 \frac{ds}{s} = 1 \text{ for } \xi \neq 0. \tag{7}$$

Let $\psi_s(x) = s^{-1}\psi(x/s)$ and define the function G on $\mathbf{R}^2_+$ by $G(y, s) = \psi_s * f(y)$ and set

$$AG(z) = \left\{ \iint_{|y - z| < s} |G(y, s)|^2 \frac{dy\, ds}{s^2} \right\}^{1/2}$$

and

$$g^*_\lambda G(z) = \left\{ \iint_{\mathbf{R}^2_+} (1 + |y - z|/s)^{-2\lambda} |G(y, x)|^2 \frac{dy\, ds}{s^2} \right\}^{1/2}, \quad \lambda > 0.$$

SECTION III

By A. P. Calderón and A. Torchinsky [2, Theorem 6.9, p. 56, and Theorem 3.5, p. 20], we get

$$\| AG \|_{L^p} \sim \| f \|_{H^p}, \ 0 < p < \infty$$

and

$$\| g_\lambda^*(\cdot) \|_{L^p} \leq C \| A(\cdot) \|_{L^p}, \ 1/\lambda < \min(p, 2).$$

To obtain the inequalities (4) and (5), we have only to show the pointwise estimates

$$AG(z) \leq Cg_\lambda^* F(z), \ \lambda < m + 5/2, \tag{8}$$

$$AF(z) \leq Cg_\lambda^* G(z), \ \lambda < m + 2. \tag{9}$$

By (5) we have the identity

$$F(x, t) = \iint_{\mathbf{R}_+^2} \tau_t * \psi_s(y - x) G(y, s) \frac{dy\, ds}{s},$$

and since $F(\tilde{\nu}) = c_\nu 2^{j/2}$, we also have the identity

$$G(y, s) = \iint \tau_t * \psi_s(y - x) F(x, t) t \, d\mu(x, t).$$

We have to make estimates on the convolution $\tau_t * \psi_s$. Let $D^{-1}g$ denote the primitive function of g, that is,

$$D^{-1}g(x) = \int_0^x g(y) dy.$$

By (3), $D^k\tau$ will satisfy (1) even for $k = -1, \ldots, -m - 2$; and similarly, by (6), $D^k\psi$ will be supported in $[-1, 1]$ even for $k = -1, \ldots, -m - 2$. By repeated integration by parts we get

$$|\tau_t * \psi_s(y)| = |D^{m+1}\tau_t * D^{-m-1}\psi_s(y)| \leq C(s/t)^{m+1}(1 + |y|/t)^{-N} t^{-1},$$

for any $N > 0$ when $s \leq t$, and

$$|\tau_t * \psi_s(y)| = |D^{-m-2}\tau_t * D^{m+2}\psi_s(y)| \leq C(t/s)^{m+2}(1 + |y|/s)^{-N} s^{-1},$$

for any $N > 0$ when $s \geq t$. If we use that $D^{m+2}\tau$ is a sum of point-measures, rapidly decreasing at infinity, we can also get the estimate

$$\| (1 + |\cdot|/t)^N(\tau_t * \psi_s)\|_{L^2} \leq C(s/t)^{m+3/2}t^{-1/2},$$

for any $N > 0$ when $s \leq t$.

To estimate $AG(z)$, we let $|y - z| < s$ and get from these inequalities

$$|\tau_t * \psi_s(y - x)| \leq C(t/s)^\varepsilon (1 + |x - z|/t)^{-3-m+\varepsilon}t^{-1}, \text{ when } t \leq s,$$

and

$$|\tau_t * \psi_s(y - x)| \leq C(s/t)^\varepsilon (1 + |x - z|/t)^{-N}t^{-1}, \text{ when } t \geq s,$$

for any $N > 0$ and $\varepsilon > 0$. Plugging these estimates into the identity for $G(y, s)$ above, we get, after using the Cauchy-Schwarz inequality,

$$|G(y, s)|^2 \leq C\iint_{\mathbf{R}^2_+} (\min(s/t, t/s))^\varepsilon$$

$$\times (1+|x - z|/t)^{-5-2m+3\varepsilon}|F(x, t)|^2 d\mu(x, t).$$

Integrating over the cone $|y - z| < s$, we get, after changing the order of integration, the inequality (8) with $\lambda = m + 5/2 - 3\varepsilon/2$.

To estimate $AF(z)$, we let $|x - z| < t$ and get from the estimates of $\tau_t * \psi_s$

$$|\tau_t * \psi_s(y - x)| \leq C(t/s)^\varepsilon (1 + |y - z|/s)^{-N}s^{-1}, \text{ when } s \geq t$$

and

$$\| (1 + |(\cdot - z)|/s)^{m+2-\varepsilon}(\tau_t * \psi_s(\cdot - x))\|_{L^2} \leq C(s/t)^\varepsilon s^{-1/2}, \text{ when } s \leq t,$$

for any $N > 0$ and $\varepsilon > 0$. Using these estimates in the identity for $F(x, t)$ together with the Cauchy-Schwarz inequality, we get

$$|F(x, t)|^2 \leq C\iint_{\mathbf{R}^2_+} (\min(s/t, t/s))^\varepsilon$$

$$\times (1 + |y - z|/s)^{-4-2m+2\varepsilon}|G(y, s)|^2 \frac{dy\, ds}{s^2}.$$

Integrating over the cone $|x - z| < t$, we get, after changing the order of integration, the inequality (9) with $\lambda = m + 2 - \varepsilon$. This finishes the proof of Theorem 1(c).

Proof of Theorem 1(b). As a consequence of Theorem 1(c), the partial sums $\Sigma_\nu (f_\nu, f)f_\nu$ are uniformly bounded in the H^p norm by the H^p norm of f, $f \in H^p(\mathbf{R})$, $p > 1/(m+2)$. We will also get the following lemma.

Lemma. The set of finite linear combinations of the functions $\{f_\nu\}$ *are dense* $H^p(\mathbf{R})$, $p > 1/(m+2)$.

Proof of the Lemma. Define $S_N f$ by

$$S_N f = \sum_{\substack{|j|<N \\ |k|<N}} (f_\nu, f)f_\nu, \ (\nu = (j, k)), \ f \in H^p(\mathbf{R}).$$

Since $L^2(\mathbf{R}) \cap H^p(\mathbf{R})$ is dense in $H^p(\mathbf{R})$, it is enough to show that

$$\| f - S_N f \|_{H^p} \to 0 \text{ as } N \to 0, \ f \in L^2(\mathbf{R}) \cap H^p(\mathbf{R}).$$

Using dominated convergence, we see that the left side of (5) goes to zero as N tends to infinity if the sum is only taken over $\nu = (j, k)$ with $|j|$ or $|k|$ larger than N. It follows that for each $\varepsilon > 0$ there is an N_ε such that $\| S_N f - S_{N_\varepsilon} f \|^p_{H^p} < \varepsilon$ for all $N \geq N_\varepsilon$. Let $M(\ )$ be the nontangential maximal function. Then we have

$$\| f - S_N f \|^p_{H^p} \leq \int_{|x|<R} |M(f - S_N f)|^p dx + C\int_{|x|>R} |M(f - S_{N_\varepsilon} f)|^p dx$$
$$+ C\int_{|x|>R} |M(S_N f - S_{N_\varepsilon} f)|^p dx.$$

The third integral is less than $C\varepsilon$ independent of R, and since $f - S_{N_\varepsilon}$ is in $H^p(\mathbf{R})$, the second integral will be less than ε if R is large enough; and finally, using Hölder's inequality, the first integral is bounded by $C_R \| M(f - S_N f) \|^p_{L^2} \leq C_R \| f - S_N f \|^p_{L^2}$, which goes to zero as N tends to infinity by the completeness of $\{f_\nu\}$ in $L^2(\mathbf{R})$. This finishes the proof of the lemma. (We have assumed that $p < 2$, which is the only case of interest.)

Now let $\{\nu_i\}_{i=1}^{\infty}$ be any enumeration of $\{\nu\}_{\mathbf{Z}^2}$ and let $S_K^P f$ denote the partial sum

$$\sum_{i=1}^{K} (f_{\nu_i}, f) f_{\nu_i}, \quad f \in H^p(\mathbf{R}).$$

Let $f \in H^p(\mathbf{R})$. For each $\varepsilon > 0$ we can find a finite linear combination f_ε of the functions $\{f_\nu\}$ such that $\| f - f_\varepsilon \|_{H^p} < \varepsilon$. Then $S_K^p f_\varepsilon = f_\varepsilon$ if K is large enough. Since the partial sums are uniformly bounded in the H^p norm, we get

$$\| f - S_K^p f \|_{H^p} \leq C \| f - f_\varepsilon \|_{H^p} + C \| S_K^p (f - f_\varepsilon) \|_{H^p} \leq C \| f - f \|_{H^p} < C\varepsilon,$$

for any $\varepsilon > 0$ provided K is large enough. This proves the unconditional convergence. The uniqueness of the coefficients is now easy. If $\sum_\nu c_\nu f_\nu$ converges to f in the H^p norm, we apply the partial sums to f_{ν_0}, which is in the dual space. On the one hand, we get the limit c_{ν_0}; on the other hand, this limit must be (f_{ν_0}, f), which shows that $c_\nu = (f_\nu, f)$. With this we have finished the proof of Theorem 1.

9. PROOFS OF THEOREM 2(b) AND (c)

The area integral approach will work in higher dimension in exactly the same way as in the one-dimensional case. It involves a summation over ω that is harmless, since Ω is a finite set. To estimate the convolutions $\tau_t^{(\omega)} * \psi_s$, we can no longer use integration by parts, but instead we can, for instance, make estimates on the Fourier transform side. With this method, we get (4) in Theorem 2(c) when $p > 1/\left(1/2 + (m + 3/2)/n\right)$ and (5), and consequently also Theorem 2(b) when $p > 1/\left(1/2 + (m + 2)/n\right)$.

Using atoms we obtain inequality (5) and Theorem 2(b) when $p > n/(n + m + 1)$. Since atoms have been used on Franklin systems on [0, 1] (see, for example, [5 and 10]), we will only sketch the proof briefly.

Let f be an atom supported in a cube Q with side s and center x_0 such that

$$\int f x^{\alpha} dx = 0, \ |\alpha| \leq m, \ \| f \|_{\infty} \leq s^{-n/p}$$

and let x_{ν} denote the point $(2^{-j}k_1, \ldots, 2^{-j}k_n) \in \mathbf{R}^n$ and $|\nu| = 2^{-j}$ for $\nu = (j, k_1, \ldots, k_n) \in \mathbf{Z}^{n+1}$. Then

$$\begin{aligned} |(f_{\nu}^{\omega}, f)| &\leq \| f_{\nu}^{\omega} \|_{L^1(Q)} \| f \|_{\infty} \\ &\leq C s^{-n/p} |\nu|^{n/2} \min\left\{1, r^{(|x_0 - x_{\nu}| - ns)/|\nu|}\right\}, \text{ when } |\nu| \leq s, \end{aligned}$$

and using the mth-order Taylor expansion of f_{ν}^{ω} at x_0 and the moment conditions of f, we get

$$|(f_{\nu}^{\omega}, f)| \leq C s^{1+m+n-n/p} |\nu|^{-n/2-m-1} r^{|x_{\nu} - x_0| |\nu|}, \text{ when } |\nu| \geq s.$$

Let $M(\ )$ be the nontangential maximal function and let Sf be *any* partial sum

$$\sum_{\nu, \omega} (f_{\nu}^{\omega}, f) f_{\nu}^{\omega}.$$

Looking at the moment conditions of τ^{ω}, we find that $|M(\tau^{\omega})(x)| \leq C(1 + |x|)^{-m-n-2}$ and consequently

$$|M(f_{\nu}^{\omega})(x)| \leq C |\nu|^{-n/2} (1 + |x - x_{\nu}| / |\nu|)^{-m-n-2}.$$

By summation we now get for $\varepsilon > 0$ and $|x - x_0| \geq 2ns$

$$|M(Sf)(x)| \leq \sum_{\nu, \omega} |(f_{\nu}^{\omega}, f)| |M(f_{\nu}^{\omega})(x)| \leq C(|x - x_0|/s)^{-m-n-1+\varepsilon} s^{-n/p}$$

and

$$\left\{ \sum_{\nu, \omega} |(f_{\nu}^{\omega}, f)|^2 |\nu|^{-n} \chi_{\nu}(x) \right\}^{1/2} \leq C(|x - x_0|/s)^{-m-n-1+\varepsilon} s^{-n/p}.$$

By these estimates and L^2 estimates when $|x - x_0| < 2ns$, we get

$$\| Sf \|_{H^p} \leq \| M(Sf) \|_{L^p} \leq C, \ p > n/(n + m + 1),$$

and

$$\left\| \left\{ \sum_{\nu, \omega} |(f_{\nu}^{\omega}, f)|^2 |\nu|^n \chi_{\nu} \right\}^{1/2} \right\|_{L^p} \leq C, \ p > n/(n + m + 1).$$

Taking limits with finite sums of atoms, we get inequality (5) and $\| Sf \|_{H^p} \leq C \| f \|_{H^p}$ for all $f \in H^p(\mathbf{R}^n)$, $p > n/(n + m + 1)$.

To prove that finite linear combinations of $\{f_{\nu}^{\omega}\}$ are dense in $H^p(\mathbf{R}^n)$, $p > n/(n + m + 1)$, is now easy if we use the

L^2-completeness and that

$$|M(Sf - f)(x)| \leq C_f |x|^{-n-m-1+\varepsilon} \text{ for } |x| > c_f$$

uniformly for all partial sums Sf when f is a finite sum of atoms. The rest of the proof of Theorem 2(b) is the same as in the one-dimensional case.

An open problem is to find a better range of p for the n-dimensional version of inequality (4) when $n \geq 2$. (Added in proof:) A better range of p for the n-dimensional version of (4) has been found by Peter Sjögren, and the author of this paper is now aware of methods that will give the sharp range.

REFERENCES

1. Bockariev, S. V. Existence of basis in the space of analytic functions in the disc and some properties of the Franklin system. *Mat. Sbornik* 95 (1974):3-18 (in Russian).

2. Calderón, A. P., and Torchinsky, A. Parabolic maximal functions associated with a distribution. *Adv. Math.* 16 (1975): 1-64.

3. Carleson, L. An explicit unconditional basis in H^1. Institut Mittag-Leffler, Report No. 2, 1980.

4. Ciesielski, Z. Equivalence, unconditionality and convergence a.e. of the spline bases in L_p spaces. Approximation theory, Banach Center Publications, vol. 4, pp. 55-68.

5. Ciesielski, Z. The Franklin orthogonal system as unconditional basis in Re H^1 and VMO. Preprint, 1980.

6. Ciesielski, Z., and Domsta, J. Estimates for the spline orthonormal functions and for their derivatives. *Studia Math.* 44 (1972):315-320.

7. Ciesielski, Z.; Simon, P.; and Sjölin, P. Equivalence of Haar and Franklin bases in L^p spaces. *Studia Math.* 60 (1976):195-211.

8. Maurey, B. Isomorphismes entre espace H^1. *Acta Math.* 145 (1980):79-120.

9. Schipp, F., and Simon, P. Investigation of Haar and Franklin series in the Hardy spaces. Preprint, 1980.

10. Sjölin, P., and Strömberg, J.-O. Basis properties of Hardy spaces. University of Stockholm, Report No. 19, 1981.

11. Wojtaszczyk, P. The Franklin system is an unconditional basis in H^1. To appear in *Arkiv för Matematik*.

Uncertainty Principle, Hilbert Bases and Algebras of Operators

Yves Meyer
Translated by John Horváth

1. Introduction

The search for Hilbert bases connected to the uncertainly principle is motivated in the following way by R. Balian [1].

> One can be interested, in the theory of communications, to represent an oscillating signal as a superposition of elementary wavelets, each of which possesses at the same time a sufficiently well-defined frequency and a localization in time. The useful information is, in fact, often carried at the same time by the emitted frequencies and by the structure of the signal in time (the example of music is characteristic). The representation of a signal as a function of time exhibits badly the spectrum of frequencies at play, while to the contrary its Fourier analysis masks the instant of its emission and the duration of each of the elements of the signal. An adequate representation should combine the advantages of these two complementary descriptions, while presenting a discrete character better adapted to the theory of communications.
>
> An analogous problem arises in quantum mechanics, but this time for probability waves.
>
> The uncertainty principle prohibits making precise at the same time the position and the impulsion of the particle. But it might be comfortable, for pedagogical reasons or to understand better certain phenomena with the help of concepts from classical mechanics, to work in an intermediary representation in which every basis function would be sufficiently well-defined in both position and impulsion.

Translator's Note: The references have been updated, and a few minor corrections made with the author's concurrence. A new version of Figure 1 was generated by Norbert Kaiblinger.

Author's Note: This paper and the paper "Wavelets and Hilbert bases," which is reprinted next in this volume, were written in 1985. The construction of a smooth orthonormal wavelet basis in these papers has awkward proofs. Simpler proofs use tools such as multiresolution analysis, which was introduced by S. Mallat and the author during the fall of 1986 (see the papers on multiresolution analysis reprinted elsewhere in this volume). Furthermore, both papers omit any reference to Jan-Olov Strömberg's fundamental achievement of constructing spline orthonormal wavelet bases (Strömberg's paper is reprinted preceding this one). The author rediscovered Strömberg's paper in 1988 and immediately acknowledged his priority. Strömberg's construction does not provide us with wavelets in the Schwartz class, as do the constructions in this paper.

Let us denote by $x \in \mathbf{R}^n$ the position variable (denoted q in quantum mechanics) and by $\xi \in \mathbf{R}^n$ the impulsion variable (which corresponds to frequency in the case of music). Let us recall the statement of the uncertainty principle. One denotes by $\psi(x)$ a function in $L^2(\mathbf{R}^n; dx)$ normalized by $(\int_{\mathbf{R}^n} |\psi(x)|^2 \, dx)^{1/2} = \|\psi\|_2 = 1$. One assumes furthermore that $\int_{\mathbf{R}^n} |x|^2 \, |\psi(x)|^2 \, dx$ converges, as well as $\int_{\mathbf{R}^n} |\xi|^2 \, |\widehat{\psi}(\xi)|^2 \, d\xi$, where $\widehat{\psi}(\xi) = \int e^{-ix\cdot\xi} \, \psi(x) \, dx$ is the Fourier transform of ψ. Under these conditions one defines the mean values x_0 and ξ_0 of x and ξ with respect to the probability measures $|\psi(x)|^2 \, dx$ and $|\widehat{\psi}(\xi)|^2 \, \frac{d\xi}{(2\pi)^n}$ by $x_0 = \int x \, |\psi(x)|^2 \, dx$ and $\xi_0 = \int \xi \, |\widehat{\psi}(\xi)|^2 \, \frac{d\xi}{(2\pi)^n}$. Then one computes the variances and standard deviations by

$$(\Delta x)_\psi = \left(\int |x - x_0|^2 \, |\psi(x)|^2 \, dx \right)^{1/2} \tag{1.1}$$

and

$$(\Delta \xi)_\psi = \left(\int |\xi - \xi_0|^2 \, |\widehat{\psi}(\xi)|^2 \, \frac{d\xi}{(2\pi)^n} \right)^{1/2}. \tag{1.2}$$

The uncertainty principle gives a numerical bound to the possibility of localizing simultaneously ψ around x_0 and $\widehat{\psi}$ around ξ_0. One has $(\Delta x)_\psi \, (\Delta \xi)_\psi \geq n/2$ and the minimum is attained when ψ is a Gaussian density. The problem which we will solve is the following. Can one find a Hilbert basis ψ_i, $i \in I$, of $L^2(\mathbf{R}^n)$ and a constant C such that for all $i \in I$ one has

$$(\Delta x)_{\psi_i} \, (\Delta \xi)_{\psi_i} \leq C \; ? \tag{1.3}$$

The coherent states of quantum mechanics (whose definition we will recall) furnish a continuous version of the basis we seek to construct. These coherent states will be obtained from the unitary action of a locally compact group G on a vector $\varphi_0 \in L^2(\mathbf{R}^n)$ chosen in such a way that $(\Delta x)_{\varphi_0} \, (\Delta \xi)_{\varphi_0}$ shall be finite. This group action will leave the product invariant.

The passage to the construction of a Hilbert basis is more delicate. One replaces G by a suitable lattice $\Lambda \subset G$ and φ_0 by a finite set of $2^n - 1$ functions $\psi^{(\varepsilon)}$, $\varepsilon \in E$.

So let G be a locally compact group (whose elements will be noted g), dg a left-invariant Haar measure on G and U a unitary representation of G on the Hilbert space H. The representation U is said to be *square integrable* if it is irreducible and if there exists a nonzero vector $\varphi_0 \in H$ such that $\int_G |\langle U(g)\varphi_0, \varphi_0 \rangle|^2 \, dg$ converges. One denotes by $c(\varphi_0)$ the value of this integral and then one has the remarkable identity

$$f = \frac{1}{c(\varphi_0)} \int_G \langle f, U(g)\varphi_0 \rangle \, U(g)\varphi_0 \, dg, \tag{1.4}$$

allowing one to represent every vector $f \in H$ as a combination of vectors of the orbit of φ_0 under the unitary action of G. Consequently one has the "Parseval formula"

$$\langle f_1, f_2 \rangle = \frac{1}{c(\varphi_0)} \int_G \langle f_1, U(g)\varphi_0 \rangle \, \overline{\langle f_2, U(g)\varphi_0 \rangle} \, dg. \tag{1.5}$$

Let G be the group of affine transformations $g(y) = ty + x$ of $\mathbf{R}^n$ ($t > 0$, $x \in \mathbf{R}^n$) and $\widetilde{G}$ the group obtained by adjoining to G the orthogonal transformations $\rho \in O_n$. The left-invariant

Haar measure on $\widetilde{G}$ is then $dg = dx\, t^{-n-1}\, dt\, d\rho$ when $g(y) = t\,\rho(y) + x$, $x \in \mathbf{R}^n$, $t > 0$, $\rho \in O_n$. The Haar measure $d\rho$ on O_n is normalized by $\rho(O_n) = 1$.

One considers the unitary representation U of $\widetilde{G}$ defined by

$$U(g)f(y) = t^{-n/2}\, f(g^{-1}(y)) \quad \text{if} \quad g(y) = t\,\rho(y) + x. \tag{1.6}$$

If $\varphi_0 \in L^2(\mathbf{R}^n)$ has sufficient decay at infinity and regularity so that $(\Delta x)_{\varphi_0}$ and $(\Delta\xi)_{\varphi_0}$ are both finite, then with the preceding notation $(\Delta x)_{U(g)\varphi_0} = t\,(\Delta x)_{\varphi_0}$ while $(\Delta\xi)_{U(g)\varphi_0} = t^{-1}\,(\Delta\xi)_{\varphi_0}$.

Thus the product which figures in the uncertainty principle of Heisenberg does not change under the unitary action of $\widetilde{G}$. The representation U is irreducible and (1.4) holds as soon as $\int_{\widetilde{G}} |\langle U(g)\varphi_0, \varphi_0\rangle|^2\, dg$ converges. Let us suppose, to simplify, that φ_0 is radial and set $\psi_t(y) = t^{-n}\,\varphi_0(t^{-1}y)$. The condition on φ_0 is equivalent to the convergence of $\int_0^\infty (\widehat{\varphi}_0(t\xi))^2\, \frac{dt}{t}$ for $\xi \neq 0$. Then this integral is independent of $\xi \in \mathbf{R}^n \setminus \{0\}$ and will be denoted $c(\varphi_0)$. Finally (1.4) can be written

$$f = \frac{1}{c(\varphi_0)} \int_0^\infty f * \psi_t * \psi_t\, \frac{dt}{t}. \tag{1.7}$$

Identities of the type (1.7) were introduced and used by A. P. Calderón in the theory of interpolation since the 1960s.

If φ_0 is radial and satisfies $|\varphi_0(x)| \leq C\,(1 + |x|)^{-n-1}$, then (1.7) is equivalent to $\int_{\mathbf{R}^n} \varphi_0(x)\, dx = 0$. The example we have in sight is given by

$$\varphi_0(x) = c_n\, \frac{|x|^2 - n}{(|x|^2 + 1)^{(n+3)/2}},$$

where c_n is chosen so that $\widehat{\varphi}_0(\xi) = |\xi| \exp(-|\xi|)$. If $g(y) = ty + x$, then $\langle f, U(g)\varphi_0\rangle = t^{n/2}\,(f * \psi_t)(x) = t^{n/2}\, v(x,t)$ where $v(x,t) = -t\,\frac{\partial}{\partial t} u(x,t)$ and where $u(x,t)$ is the solution of the Dirichlet problem in the half-space $\mathbf{R}^n \times]0, +\infty[$ with $f(x) = u(x,0)$.

The Littlewood-Paley-Stein theory taught us to describe the classical function spaces by conditions concerning the norm of the gradient of the harmonic extension (or, simply, on $|v(x,t)|$). The representation of functions by series of wavelets will allow us to obtain a discrete analogue of the Littlewood-Paley-Stein theory: the classical function spaces will be characterized by the absolute values of the wavelet coefficients (Theorem 3 below).

As one can guess, we will obtain a Hilbert basis compatible with the uncertainty principle replacing (1.4) by a discrete version, where the integral over G becomes a sum over the dyadic lattice $\Lambda \subset G$ which we shall now define.

The dyadic lattice is the set of all affine transformations of $\mathbf{R}^n$ of the form $D_j R_k$ where $j \in \mathbf{Z}$, $k \in \mathbf{Z}^n$, $R_k(y) = y + k$ and $D_j(y) = 2^j y$. Let us denote by $K \subset G$ the compact set of affine transformations $\alpha x + \beta$, where $1 \leq \alpha \leq 2$ and $0 \leq \beta_j \leq 1$ $(1 \leq j \leq n)$. Then it is easy to see that the compact subsets λK, $\lambda \in \Lambda$ form a partition of G up to sets with measure zero. This is why Λ is called a lattice.

2. Wavelets and Hilbert Bases

We intend to establish a discrete version of (1.4), namely

$$f = \sum_{\lambda\in\Lambda}\sum_{\varepsilon\in E}\langle f, U(\lambda)\psi^{(\varepsilon)}\rangle\, U(\lambda)\psi^{(\varepsilon)}, \tag{2.1}$$

where E is a finite set containing $2^n - 1$ elements, Λ is the dyadic lattice, and where the vectors $U(\lambda)\psi^{(\varepsilon)}$, $\lambda \in \Lambda$, $\varepsilon \in E$, form a Hilbert basis of $L^2(\mathbf{R}^n)$. Since furthermore we shall apply (2.1) to tempered distributions f, we require that the functions $\psi^{(\varepsilon)}$ belong to the class $\mathcal{S}(\mathbf{R}^n)$ of Schwartz. It follows that the $\psi^{(\varepsilon)}$ will belong to the subspace $\mathcal{S}_0$ of $\mathcal{S}$ of functions whose moments are all zero.

Contrary to what happens in the continuous case, we do not have at present any general method to construct the functions $\psi^{(\varepsilon)}$. The existence of algorithms having the properties we have just described seems to be an accident.

The various functions $U(\lambda)\psi^{(\varepsilon)}$ satisfy condition (1.3) uniformly in $\lambda \in \Lambda$. These functions will be called "wavelets." The reason for this terminology is that in dimension one the Fourier transform of ψ is supported by $2\pi/3 \le |\xi| \le 8\pi/3$; it is therefore rather well localized in frequencies and the same applies to the functions $U(\lambda)\psi$ provided that one uses an exponential scale (the octaves in music, ...). The localization of ψ itself is the best possible taking into account that of $\widehat{\psi}$. As G. David has shown, one can in the definition of the lattice Λ replace 2 by $\theta = 1 + 1/m$; $m \in \mathbf{N}$, $m \ge 1$, and ψ has to be changed as a consequence. In the choice we shall make below, the frequencies which figure in each term $U(\lambda)\psi$ cover exactly two octaves, which is not very precise. In the choice of David these two octaves will become (in the exponential scale of frequencies) an arbitrarily small interval.

Let us denote by $Q = Q(j,k)$ the dyadic cube defined by $2^j x - k \in [0,1[^n$, by $\mathcal{Q}_j$ the set of all the cubes $Q(j,k)$, $k \in \mathbf{Z}^n$, and by $\mathcal{Q}$ the union of all the $\mathcal{Q}_j$, $j \in \mathbf{Z}$. If $\lambda(x) = 2^{-j}(x+k)$, then $U(\lambda)\psi^{(\varepsilon)}$ has the same position relative to $Q(j,k)$ as $\psi^{(\varepsilon)}$ relative to $[0,1[^n$. For this reason we write $U(\lambda)\psi^{(\varepsilon)} = \psi_Q^{(\varepsilon)}$.

Now let us enter into the details of the construction of our Hilbert bases. We begin by defining the remarkable functions $\theta(t)$, $\alpha(t)$, $\varphi(t)$, and $\psi(t)$ of one real variable, the last three belonging to the class $\mathcal{S}(\mathbf{R})$ of Schwartz. We require $\theta(t)$ to be odd, infinitely differentiable, equal to $\pi/4$ if $t \ge \pi/3$ (thus to $-\pi/4$ if $t \le -\pi/3$). Then we construct $\alpha(t)$ which is infinitely differentiable, even, and has compact support. Furthermore this function $\alpha(t)$ is zero if $0 \le t \le 2\pi/3$ or if $t \ge 8\pi/3$. Lastly $\alpha(t) = \pi/4 + \theta(t-\pi)$ if $2\pi/3 \le t \le 4\pi/3$ and $\alpha(2t) = \pi/2 - \alpha(t)$ for the same values of t. Finally $\psi \in \mathcal{S}(\mathbf{R})$ is defined by its Fourier transform $\widehat{\psi}(t) = e^{-it/2}\sin\alpha(t)$, while $\varphi \in \mathcal{S}(\mathbf{R})$ is given by $\widehat{\varphi}(t) = \cos\alpha(t)$ if $|t| \le 4\pi/3$ and $\widehat{\varphi}(t) = 0$ if $|t| \ge 4\pi/3$. Therefore one has

$$\psi(x) = \frac{1}{\pi}\int_0^\infty \cos\Big(\Big(x - \frac{1}{2}\Big)t\Big)\, \sin\alpha(t)\, dt. \tag{2.4}$$

The function $\psi(x)$ is real, rapidly decreasing and satisfies $\psi(1-x) = \psi(x)$. We denote by $\mathcal{I}$ the collection of all dyadic intervals $I(j,k) = [2^{-j}k, 2^{-j}(k+1)[$, and then we set $\psi_I(x) = 2^{j/2}\,\psi(2^j x - k)$. With this we have

Theorem 1 *The collection ψ_I, $I \in \mathcal{I}$, of wavelets is a Hilbert basis of $L^2(\mathbf{R})$.*

To show it, one defines a remarkable sequence E_j of regularizing operators imitating the conditional expectation operators $\mathcal{E}_j$ with respect to the σ-field $\mathcal{F}_j$ generated by the intervals $[k\,2^{-j}, (k+1)\,2^{-j}[$, $k \in \mathbf{Z}$. One defines, for every tempered distribution $f \in \mathcal{S}'(\mathbf{R})$, the "softened" mean of f on the interval $I = [2^{-j}\,k, 2^{-j}\,(k+1)[$ by $\lambda_{(j,k)} = 2^j \int \varphi(2^j x - k)\, f(x)\, dx$.

One then constructs the regularizing approximation operator E_j, $j \in \mathbf{Z}$, by

$$E_j(f)(x) = \sum_{-\infty}^{\infty} \lambda_{(j,k)}(f)\, \varphi(2^j x - k), \tag{2.5}$$

the summation being taken in k. One shows without difficulty the convergence of $E_j(f)$ to f, for all norms of homogeneous function spaces, when j tends to $+\infty$.

Let us call $D_{(j,k)}\colon L^2(\mathbf{R}) \to L^2(\mathbf{R})$ the orthogonal projection operator onto the wavelet ψ_I, $I = I(j,k)$; that is, $D_{(j,k)}(f) = \langle f, \psi_I\rangle\, \psi_I$. Then one has the following remarkable identity (where the summation is with respect to k):

$$\sum_{-\infty}^{\infty} D_{(j,k)} = E_{j+1} - E_j. \tag{2.6}$$

If f belongs to $L^2(\mathbf{R})$, then $\lim_{j\downarrow-\infty} \|E_j(f)\|_2 = 0$ and $\lim_{j\uparrow\infty} \|f - E_j(f)\|_2 = 0$. So one has

$$\sum_{-\infty}^{\infty}\sum_{-\infty}^{\infty} D_{(j,k)} = 1. \tag{2.7}$$

Let us then pass to the proof of Theorem 1. The orthogonality between the ψ_I, $I \in \mathcal{I}$, is verified by hand, while the completeness follows from (2.7).

To situate better the basis which we will construct in dimension n, it is preferable to recall the construction of the Haar system for $L^2(\mathbf{R}^n)$. One calls $h^{(0)}(x)$ the characteristic function of the interval $[0,1[$, and $h^{(1)}(x)$ the function equal to 1 if $0 \le x < 1/2$, to -1 if $1/2 \le x < 1$, and to 0 elsewhere.

One denotes by E the set of the $2^n - 1$ sequences $(\varepsilon_1, \varepsilon_2, \ldots, \varepsilon_n)$ of zeros or ones, with the exception of the sequence consisting only of zeros. One defines successively $h^{(\varepsilon)}(x) = h^{(\varepsilon_1)}(x_1) \cdots h^{(\varepsilon_n)}(x_n)$, and then $h_Q^{(\varepsilon)}(x) = 2^{nj/2}\, h^{(\varepsilon)}(2^j x - k)$, when the dyadic cube Q is defined by $2^j x - k \in [0,1[^n$. Then the collection of the $h_Q^{(\varepsilon)}$, $\varepsilon \in E$, $Q \in \mathcal{Q}$, is a Hilbert basis of $L^2(\mathbf{R}^n)$; it is the Haar system. It is appropriate to observe that $(\Delta\xi)_{h_Q^{(\varepsilon)}} = +\infty$.

Let us denote by D_j the operator defined by

$$D_j(f) = \sum_{\varepsilon\in E}\sum_{Q\in\mathcal{Q}_j} \langle f, \psi_Q^{(\varepsilon)}\rangle\, \psi_Q^{(\varepsilon)}.$$

Then $D_j = \mathcal{E}_{j+1} - \mathcal{E}_j$, where $\mathcal{E}_j$ is the conditional expectation operator with respect to the σ-field $\mathcal{F}_j$ generated by the cubes $Q \in \mathcal{Q}_j$.

We shall imitate this construction introducing a regularity of class C^∞. One sets $\psi^{(0)}(x) = \varphi(x)$, $\psi^{(1)}(x) = \psi(x)$, then, this time on $\mathbf{R}^n$, $\psi^{(\varepsilon)}(x) = \psi^{(\varepsilon_1)}(x_1) \cdots \psi^{(\varepsilon_n)}(x_n)$,

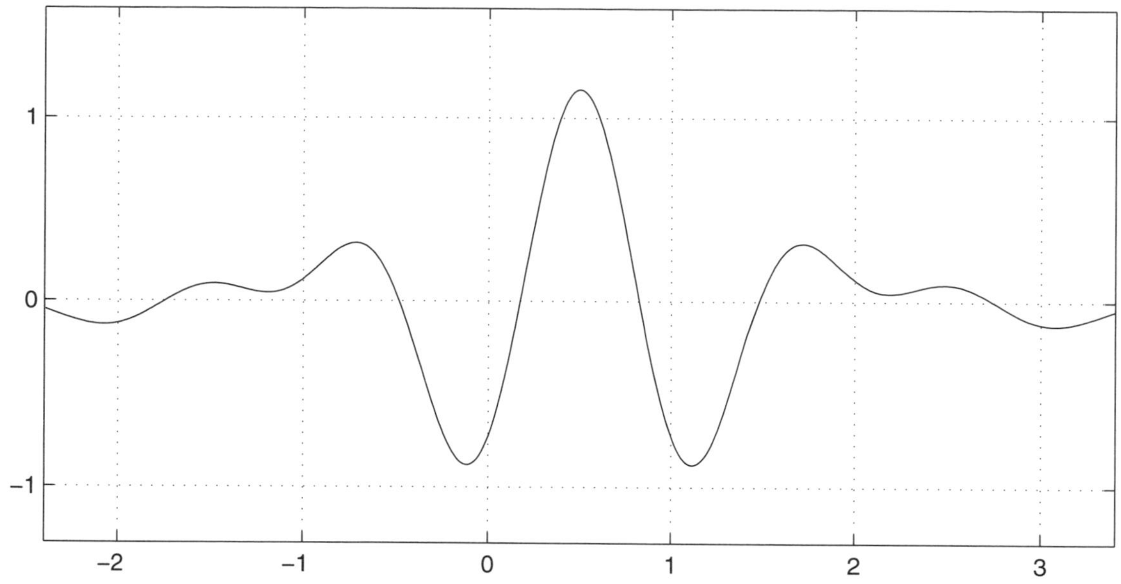

Figure 1. Graph of the function $\psi(t)$

where $\varepsilon \in E$. Finally, for every dyadic cube Q defined by $2^j x - k \in [0,1[^n$ one sets $\psi_Q^{(\varepsilon)}(x) = 2^{nj/2}\,\psi^{(\varepsilon)}(2^j x - k)$.

Theorem 2 *The collection of the $\psi_Q^{(\varepsilon)}$, $\varepsilon \in E$, $Q \in \mathcal{Q}$, is a Hilbert basis of $L^2(\mathbf{R}^n)$.*

Let us show how this result follows from Theorem 1. For the convenience of the notation we shall limit ourselves to $n = 2$. One writes, taking up again the identity (2.6),

$$E_{j+1} \otimes E_{j+1} - E_j \otimes E_j = \sum_k \sum_{k'} D_{j,k} \otimes D_{j,k'} + \sum_k \sum_{k'} E_{j,k} \otimes D_{j,k'} + \sum_k \sum_{k'} D_{j,k} \otimes E_{j,k'}. \qquad (2.8)$$

We denoted by $E_{j,k}$ (resp. $D_{j,k}$) the orthogonal projection operator onto $2^{j/2}\varphi(2^j x - k)$ (resp. $2^{j/2}\psi(2^j x - k)$). The three terms on the right-hand side of (2.8) correspond to $\varepsilon = (1,1)$, $\varepsilon = (0,1)$ and $\varepsilon = (1,0)$. The orthogonality between the different functions $\psi_Q^{(\varepsilon)}$ is easy and left to the reader.

G. David has remarked that the preceding construction can be generalized. One replaces the dyadic scale 2^j by θ^j, where $\theta = 1 + 1/m$, $m \geq 1$, $m \in \mathbf{N}$. The interval $[-\pi/3, \pi/3]$ which figures in the construction of the function $\theta(t)$ is then replaced by $[-m\pi/(2m+1), m\pi/(2m+1)]$, and similarly the numbers $2\pi/3$, $4\pi/3$, $8\pi/3$ which figure in the construction of $\widehat{\psi}$ become $m\pi - m\pi/(2m+1)$, $m\pi + m\pi/(2m+1)$, $(m+1)\pi - (m+1)\pi/(2m+1)$. Finally $\widehat{\psi}(t) = |\widehat{\psi}(t)|\, e^{-it/2}$.

The function φ is defined by its Fourier transform which is equal to 1 on the interval $[-m\pi + m\pi/(2m+1), m\pi - m\pi/(2m+1)]$ and to 0 outside of $[-m\pi - m\pi/(2m+1), m\pi + m\pi)/2m + 1]$. The details are left to the reader.

3. Wavelets and Unconditional Bases

We have just established the fundamental identity

$$f = \sum_{\varepsilon \in E} \sum_{Q \in \mathcal{Q}} \langle f, \psi_Q^{(\varepsilon)} \rangle \, \psi_Q^{(\varepsilon)} \tag{3.1}$$

giving the decomposition of an arbitrary function $f \in L^2(\mathbf{R}^n)$ in a Hilbert basis of wavelets. This algorithm has a remarkable property which distinguishes it from every other decomposition in a Hilbert basis: The regularity of f is characterized by the decay of the absolute values of the coefficients of the wavelets, and if f belongs to one of the classical function spaces which serve to measure this regularity, then the series (3.1) converges to f for the corresponding norm. To make precise what we have just said, we shall introduce a scale of homogeneous function spaces. Let us begin by describing the "bottom of the ladder," to wit, the space $\mathcal{S}_0(\mathbf{R}^n)$ of functions of the class $\mathcal{S}(\mathbf{R}^n)$ whose moments are all zero. We will denote this subspace, equipped with the topology induced from $\mathcal{S}(\mathbf{R}^n)$, by V. The dual space V' is the quotient space of the tempered distributions modulo the polynomials.

The functions $f \in V$ are characterized by the rapid decay of their wavelet coefficients: For every $m \in \mathbf{N}$,

$$\alpha_f(\varepsilon, Q) = \langle f, \psi_Q^{(\varepsilon)} \rangle = O((1 + |k|)^{-m} \, 2^{-m|j|})$$

as $|k| + |j|$ tends to infinity. The series of wavelets converges then to f in the topology of V.

The convergence of the corresponding series of wavelets to $S \in V'$ takes place in the weak topology $\sigma(V', V)$. In other words, there is numerical convergence after having integrated against $f \in V$. If S is the constant 1, all its wavelet coefficients are zero and one has indeed $1 = 0$ after integrating against a function with integral zero.

We consider now an intermediary situation where $V \subset B \subset V'$ and B is a Banach space such that the two inclusions are continuous and have dense image. We shall restrict ourselves to spaces B such that the series of wavelets converges in the norm of B independently of the order of terms. This leads us to the following definitions.

The Banach spaces which are most similar to Hilbert spaces are those which have an *unconditional basis.* This means that there exists a sequence $(e_k)_{k \in \mathbf{N}}$ of vectors of the Banach space B in question having the following two properties:

(3.2) *Every vector $x \in B$ can be written $x = \sum_0^\infty \alpha_k \, e_k$, where the α_k are scalars such that $\lim_{m \uparrow \infty} \| x - \sum_0^m \alpha_k \, e_k \| = 0$, the α_k are then unique.*

(3.3) *There exists a constant C such that for every $m \geq 1$ and every sequence α_k, $0 \leq k \leq m$, one has $\| \sum_0^m \lambda_k \, \alpha_k \, e_k \|_B \leq C \, \| \sum_0^m \alpha_k \, e_k \|_B$ as soon as $\sup |\lambda_k| \leq 1$.*

Then the criterion of a convergent series $\sum_0^\infty \alpha_k \, e_k$ belonging to B depends only on the sequence $|\alpha_k|$, $k \in \mathbf{N}$. Furthermore, the series in question are unconditionally convergent.

We remark that in any situation where a family of vectors is not labeled by an ordered set, the natural concept of convergence of a series implies unconditional convergence. Specifically, given vectors $f_\lambda \in B$, $\lambda \in \Lambda$, we say that the series $\sum_{\lambda \in \Lambda} f_\lambda$ is a *summable family* if for every $\epsilon > 0$ there exists a finite subset $F \subset \Lambda$ such that for any finite set G disjoint from F, $\| \sum_{\lambda \in G} f_\lambda \| < \varepsilon$. This statement, that $\sum_{\lambda \in \Lambda} f_\lambda$ is a summable family, is equivalent to the statement that the series $\sum_{\lambda \in \Lambda} f_\lambda$ is unconditionally convergent.

Once we have excluded the spaces $L^1(\mathbf{R}^n)$, $L^\infty(\mathbf{R}^n)$ and those which are tied to them (they are excluded because they have no unconditional bases), the basis of the wavelets $\psi_Q^{(\varepsilon)}$, $\varepsilon \in E$, $Q \in \mathcal{Q}$, is an unconditional basis for all the classical function spaces (L^p, $1 < p < +\infty$, $W^{s,p}$, Sobolev spaces, Besov spaces, ...).

Many usual spaces are duals B^* of Banach spaces B having an unconditional basis, but do not have an unconditional basis because they do not satisfy (3.2).

Let $(e_k)_{k\in\mathbf{N}}$ be an unconditional basis of B. Then there exists a sequence E_k^*, $k \in \mathbf{N}$ of elements of B^* biorthogonal to the sequence e_k, $k \in \mathbf{N}$. Every $y \in B^*$ can be written uniquely as $y = \sum_0^\infty \beta_k\, e_k^*$, where the series converges for the topology $\sigma(B^*, B)$. In general the convergence is not in the sense of the norm of B^*. However, one still has property (3.3). The membership of y in B^* is characterized by the order of magnitude of the absolute values $|\beta_k|$ of the coefficients.

The examples of pairs (B, B^*) we have in view are (H^1, BMO) or $(B_1^{-r,1}, C^r)$ for $r > 0$. We have denoted by $B_q^{s,p}$ the homogeneous Besov space defined by $2^{sj}\, \|\Delta_j(f)\|_p \in \ell^q(\mathbf{Z})$ with the notation of [15]. The dual of $B_1^{-r,1}$ is the usual homogeneous Hölder space C^r.

The limiting case $r = 0$ is also interesting. The Besov space $B_1^{0,1}$ has been studied systematically by G. de Souza and O'Neil. Its dual is, in its complex version, the Bloch space, in its real version, the vector space generated by the first derivatives of the functions of the Zygmund class (defined by $|f(x+y) + f(x-y) - 2f(x)| \le C|y|$ for all $x \in \mathbf{R}^n$ and all $y \in \mathbf{R}^n$).

Theorem 3 *A distribution $f \in \mathcal{S}'(\mathbf{R}^n)$ belongs to $L^p(\mathbf{R}^n)$, $1 < p < +\infty$, if and only if its wavelet coefficients $\alpha_Q^{(\varepsilon)}(f)$ satisfy*

$$\left(\sum_{Q\ni x} |\alpha_Q^{(\varepsilon)}(f)|^2\, |Q|^{-1}\right)^{1/2} \in L^p(\mathbf{R}^n; dx).$$

If $p = 1$ then this condition characterizes the space H^1 of Stein and Weiss. For $s \in \mathbf{R}$ one has $f \in W^{s,p}$ $(1 < p < +\infty)$ if and only if its wavelet coefficients $\alpha_Q^{(\varepsilon)}(f)$ satisfy

$$\left(\sum_{Q\ni x} |\alpha_Q^{(\varepsilon)}(f)|^2\, |Q|^{-1}\, (1 + 4^j)^s\right)^{1/2} \in L^p(\mathbf{R}^n; dx).$$

One has $f \in B_q^{s,p}$ (homogeneous Besov space) if and only if $\sup_{\varepsilon\in E} |\langle f, \psi_Q^{(\varepsilon)}\rangle| = \alpha(k,j)$ satisfies

$$\left(\sum_{-\infty}^{\infty}\left\{\left(\sum_{k\in\mathbf{Z}^n} (\alpha(k,j))^p\right)^{1/p} 2^{j(s+n(1/2-1/p))}\right\}^q\right)^{1/q} < +\infty,$$

with the usual changes if $p = +\infty$ or $q = +\infty$.

One has $f \in \mathrm{BMO}(\mathbf{R}^n)$, the space of John and Nirenberg, if and only if its wavelet coefficients $\alpha_Q^{(\varepsilon)}(f)$ satisfy the condition of Carleson: there exists a constant C such that for every dyadic cube R one has $\sum_{Q\subset R} |\alpha_Q^{(\varepsilon)}(f)|^2 < C\,|R|$ where the sum is over the dyadic subcubes $Q \subset R$.

One can make Theorem 3 more precise by defining a unitary operator $U\colon L^2(\mathbf{R}^n) \to L^2(\mathbf{R}^n)$ by $U(\psi_Q^{(\varepsilon)}) = h_Q^{(\varepsilon)}$, $\varepsilon \in E$, $Q \in \mathcal{Q}$. Then U restricted to $\mathcal{S}_0(\mathbf{R}^n)$ can be extended to an isomorphism between the Hardy space $H^1(\mathbf{R}^n)$ and its dyadic version.

Similarly, U is an isomorphism of $L^p(\mathbf{R}^n)$ onto itself for $1 < p < +\infty$, and U defines (by transposition) an isomorphism between the space $\mathrm{BMO}(\mathbf{R}^n)$ and its dyadic version [13]. So one finds again a famous theorem of B. Maurey.

4. The Local Version of the Wavelets

In the decomposition of a function into a series of wavelets there are two sums. One is indexed by $k \in \mathbf{Z}^n$ and corresponds to a change of origin and amounts to displacing our instrument of analysis ("the analyzing wavelet"). The other is a sum in $j \in \mathbf{Z}$. The interpretation of 2^{-j} is the "resolution": if one wants to see details of size 2^{-m} $(m \geq 1)$ it is appropriate to push the summation in j up to $j = m$. In fact, one sets $\varphi(x) = \varphi(x_1) \cdots \varphi(x_n)$, and one calls $\alpha^{(\varepsilon)}(k, j)$ the wavelet coefficients of f and $\beta(k, m)$ the scalar products of f with $2^{mn}\,\varphi(2^m x - k)$. One then has the identity

$$\sum_{j<m} \sum_{k \in \mathbf{Z}^n} \sum_{\varepsilon \in E} \alpha^{(\varepsilon)}(k, j)\, \psi_{j,k}^{(\varepsilon)}(x) = \sum_{k \in \mathbf{Z}^n} \beta(k, m)\, \varphi(2^m x - k).$$

This implies that as long as j does not go beyond m, the sums obtained are too smooth to be able to see small and complicated details of size smaller than 2^{-m}.

Let us push this point of view to the extreme and assume that the singular support of f is contained in a compact subset K. Then to reconstruct f modulo an infinitely differentiable error it suffices, for every $\varepsilon > 0$, to restrict the series of wavelets to the cubes Q whose sides are not larger than ε and whose distance from K is at most ε.

A second local version of wavelets is obtained by defining the periodized wavelets. Those periodized wavelets will serve to analyze the function spaces consisting of functions f defined on $\mathbf{R}^n$, which are periodic with period 1 in each variable.

If $j \geq 0$ and if $r = (r_1, \ldots, r_n)$ is a sequence of integers such that $0 \leq r_1 < 2^j, \ldots, 0 \leq r_n < 2^j$, one sets

$$\psi_{j,r}^{(\varepsilon)}(x) = 2^{nj/2} \sum_{k \in Z^n} \psi^{(\varepsilon)}(2^j x + 2^j k - r). \tag{4.1}$$

We recall that ε is a sequence $(\varepsilon_1, \ldots, \varepsilon_n)$ of zeros and ones with the exception of the sequence $(0, 0, \ldots, 0)$, and that $\psi^{(\varepsilon)}(x) = \psi^{(\varepsilon_1)}(x_1) \cdots \psi^{(\varepsilon_n)}(x_n)$ with $\psi^{(0)}(x) = \varphi(x)$ and $\psi^{(1)}(x) = \psi(x)$. Each function $\psi_{j,r}^{(\varepsilon)}$ is periodic with period 1 in each variable. One has $\psi_{j,r}^{(\varepsilon)}(x) = \theta_j^{(\varepsilon)}(x - 2^{-j} r)$ and, restricted to $[-\ell, \ell]^n$, $0 < \ell < 1$, each function $\theta_j^{(\varepsilon)}(x)$ is written

$$\theta_j^{(\varepsilon)}(x) = 2^{nj/2}\, \psi^{(\varepsilon)}(2^j x) + R_j^{(\varepsilon)}(x), \tag{4.2}$$

where, for every $\alpha \in \mathbf{N}^n$ and every integer N, the function $2^{jN}\, \partial^\alpha R_j^{(\varepsilon)}(x)$ converges uniformly to 0 on $[-\ell, \ell]^n$. In other words, the periodized wavelets are asymptotically truncated wavelets.

To finish, we will reorder the sequence $\psi_{j,r}^{(\varepsilon)}$. One sets $\gamma_0(x) = 1$; then $\gamma_1(x), \gamma_2(x), \ldots, \gamma_{2^n-1}(x)$ are, up to order, the functions $\psi_{0,0}^{(\varepsilon)}(x)$, where $\varepsilon \in E$. Next, if $2^{nj} \leq m < 2^{n(j+1)}$, the functions $\gamma_m(x)$ are, up to order, the $(2^n-1)\,2^{nj}$ functions $\psi_{j,r}^{(\varepsilon)}$ (where $\varepsilon \in E$ and $0 \le r_1 < 2^j$ etc.). The properties of this sequence are even better than those of the sequence of wavelets. Indeed, this sequence $\gamma_m(x)$ is a basis for the limit spaces as the following theorem indicates.

Theorem 4 *The sequence of functions $\gamma_m(x)$, $m \in \mathbf{N}$, which we have defined is a Hilbert basis of $L^2([0,1]^n)$.*

It is a basis for the space of periodic continuous functions with period 1 in each variable. It is also a basis for the space of periodic functions of class C^m, $m \in \mathbf{N}$, with period 1 in each variable. Furthermore it is a basis for the space $L^1([0,1]^n)$.

It is an unconditional basis for the periodic versions of the spaces described by Theorem 3 (L^p, $W^{s,p}$, Sobolev spaces, Besov spaces, H^1, BMO, C^r for $r \notin \mathbf{N}$, etc.).

It is appropriate to observe that the periodic-BMO space which figures here is not the space of restrictions to $[0,1]^n$ of the functions of $\mathrm{BMO}(\mathbf{R}^n)$, but the smaller space of restrictions to $[0,1]^n$ of periodic BMO functions with period 1 in each variable (e.g., the function $\log(x)$, restricted to $[0,1]$, does not belong to the periodic-BMO space).

The periodic-H^1 space is the H^1 space of the torus $(\mathbf{R}/\mathbf{Z})^n$ and not the space of restrictions to $[0,1]^n$ of functions of $H^1(\mathbf{R}^n)$.

We cannot resist the pleasure of writing the characterization of the Hölder spaces C^r (periodic with period 1 in each variable) in the basis $\gamma_m(x)$. The function f belongs to C^r if and only if $\langle f, \gamma_m\rangle = O(m^{-(1/2+r/n)})$. In particular, every series $\sum_0^\infty \alpha_m\, \gamma_m(x)$ such that $|\alpha_m| = m^{-(1/2+r/n)}$ for $m \ge 1$ defines a function $f(x)$ of class C^r, periodic with period 1 in each variable, which does not satisfy $f(x) - f(x_0) = o(|x - x_0|^r)$ at any point $x_0 \in \mathbf{R}^n$. This construction is much more supple and natural than that of Weierstrass. The space C^r, $r > 0$, is defined in [15]. In particular if $r = 1$ we have to substitute for the usual space C^1 the Zygmund class defined in [15].

5. Wavelets and Operators

Let V be the space $\mathcal{S}_0(\mathbf{R}^n)$ of test functions whose moments are all zero, and let V' be the (topological) dual of V. Denote by B, $V \subset B \subset V'$, one of those function spaces (described by Theorem 3) for which the wavelets form an unconditional basis. One can also choose for B the dual of such a space.

Call φ the function associated with ψ by Theorem 1. Set $\varphi_Q(x) = 2^{nj}\,\varphi(2^j x - k)$ when Q is defined by $2^j x - k \in [0,1[^n$. Then there exists a remarkable action of $L^\infty(\mathbf{R}^n)$ on B given by

$$\pi(a, f) = \sum_{\varepsilon \in E} \sum_{Q \in \mathcal{Q}} \langle a, \varphi_Q\rangle\, \langle f, \psi_Q^{(\varepsilon)}\rangle\, \psi_Q^{(\varepsilon)}. \tag{5.1}$$

We assumed $a \in L^\infty(\mathbf{R}^n)$ and $f \in B$. In other words, the operator which with f associates $\pi(a, f)$ is diagonal in the basis of wavelets; the corresponding eigenvalues $\langle a, \varphi_Q\rangle$ are in fact bounded. The continuity of this action follows from (3.3).

Next we propose to show that one has

$$\|\pi(f, g)\|_2 \le C\, \|f\|_2\, \|g\|_{\mathrm{BMO}}. \tag{5.2}$$

According to an idea of J. L. Journé, one fixes g in BMO and one studies the operator which with f associates $\pi(f,g)$. Its distributional kernel satisfies $|\frac{\partial}{\partial y_j} K(x,y)| \leq C\,|x-y|^{-n-1}$ ($x \neq y$, $1 \leq j \leq n$). This operator sends L^∞ into BMO. This implies (owing to the atomic decomposition of H^1) that it sends H^1 into L^1. By interpolation this operator is bounded on $L^2(\mathbf{R}^n)$ and one obtains (5.2).

We shall now write the paraproduct operators of J. M. Bony ([2]) in the basis of wavelets. If $0 < r < 1$ and if $a(x)$ belongs to $C^r(\mathbf{R}^n)$ and to $L^\infty(\mathbf{R}^n)$, then the paraproduct $\pi(a,f)$ is given, modulo an r-regularizing operator, by the identity (5.1). An operator is r-regularizing if it is continuous from the Sobolev space H^s into H^{s+r} for every $s \in \mathbf{R}$.

One can even, modulo a second error which is r-regularizing, replace the softened mean $\langle a, \varphi_Q\rangle$ by the value taken by a at $2^{-j}k$. In this realization of the paraproduct one has thus $\pi_0(a, \psi_Q^{(\varepsilon)}) = a(2^{-j}k)\,\psi_Q^{(\varepsilon)}$.

If $m < r < m+1$, the algebra of paradifferential operators is not commutative modulo the r-regularizing operators anymore. The realization $\pi_m(a, \psi_Q^{(\varepsilon)})$ is then given by

$$\pi_m(a, \psi_Q^{(\varepsilon)}) = \left(\sum_{|\alpha|\leq m} \frac{1}{\alpha!}\,(x - 2^{-j}k)^\alpha\, \partial^\alpha a(2^{-j}k)\right) \psi_Q^{(\varepsilon)}. \tag{5.3}$$

The wavelet basis allows to obtain unexpected results concerning certain operator algebras of Calderón-Zygmund. Let us first recall the definition of the algebra $\mathcal{A}$ of P. G. Lemarié ([10], [11], [14]). An operator T belongs to $\mathcal{A}$ if, on the one hand, it is continuous on $L^2(\mathbf{R}^n)$ and if furthermore the following two supplementary conditions are satisfied:

(5.4) *the restriction to the open subset $y \neq x$ of $\mathbf{R}^n \times \mathbf{R}^n$ of the distributional kernel $K(x,y)$ of T is a C^∞-function satisfying*

$$|\partial_x^\alpha\, \partial_y^\beta\, K(x,y)| \leq C_{\alpha,\beta}\,|x-y|^{-n-|\alpha|-|\beta|}.$$

(5.5) *T and its adjoint T^* are endomorphisms of $\mathcal{S}_0(\mathbf{R}^n)$.*

The operators $T \in \mathcal{A}$ are characterized by their matrices with respect to the wavelet basis $\psi_Q^{(\varepsilon)}$, $\varepsilon \in E$, $Q \in \mathcal{Q}$. The conditions on the coefficients $\gamma^{(\varepsilon,\varepsilon')}(Q,R) = \langle T(\psi_Q^{(\varepsilon)}), \psi_R^{(\varepsilon')}\rangle$, $Q \in \mathcal{Q}$, $R \in \mathcal{Q}$, are symmetric in (Q,R). Call $[Q,R]$ the smallest integer m such that $mQ \supset R$ and $mR \supset Q$, where mQ denotes the cube whose center is that of Q, and whose side is m times that of Q. Then the conditions are simply the existence of a sequence C_q, $q \in \mathbf{N}$, such that one has

$$|\gamma^{(\varepsilon,\varepsilon')}(Q,R)| \leq C_q\,[Q,R]^{-q} \tag{5.6}$$

(for all $q \geq 1$, all Q, all R, etc.).

Let us call $\mathcal{G}$ the group of all the permutations of the product set $E \times \mathcal{Q}$ such that for a certain constant C one has $[Q,Q'] \leq C$ as soon as $g(\varepsilon,Q) = (\varepsilon',Q')$. Let $U_g\colon L^2(\mathbf{R}^n) \to L^2(\mathbf{R}^n)$ be the unitary operator defined, with the same notation, by $U_g(\psi_Q^{(\varepsilon)}) = \psi_{Q'}^{(\varepsilon')}$.

We have thus constructed a representation of $\mathcal{G}$ by unitary operators $U_g \in \mathcal{A}$. These operators U_g are bounded on all the spaces H^s. Let us denote by $\mathcal{I} \subset \mathcal{A}$ the two-sided ideal formed by the operators T which, for a certain $\varepsilon > 0$, send H^s into $H^{s+\varepsilon}$ for every $s \in \mathbf{R}$.

The study of the symbolic calculus for the algebra $\mathcal{A}$ amounts to that of the quotient algebra $\mathcal{A}/\mathcal{I}$. But one can easily construct two permutations g_1 and g_2 of $\mathcal{G}$ such that the corresponding unitary operators U_1 and U_2 generate, modulo $\mathcal{I}$, the free group of two elements, which contradicts the existence of a symbolic calculus in the usual sense.

6. History of the Subject

The transformation into wavelets ("wavelet transform") was discovered and made explicit in its continuous version by J. Morlet. The motivation of the works of J. Morlet and of his collaborators was the numerical treatment of the seismic signal by reflection (in the framework of petroleum research of the group Elf-Aquitaine).

One finds again the identity (1.7) in the works of A. P. Calderón in the theory of interpolation.

In parallel, and with a different motivation, various authors wrote discrete algorithms using functions of the form $2^{nj/2}\psi(2^j x - k)$, $j \in \mathbf{Z}$, $k \in \mathbf{Z}^n$. For instance L. Carleson smoothed the Haar system to obtain an unconditional basis of the space H^1, while Frazier and Jawerth, faithful to the program of Guido Weiss, wrote atomic decomposition algorithms for Besov spaces. It is needless to say that the paths followed by these mathematicians cannot lead to the above described orthonormal basis. The discovery of this basis is accidental.

Theorems 1 and 2 owe much to R. R. Coifman and P. G. Lemarié. Theorem 3 is essentially contained in the already quoted works of L. Carleson, M. Frazier, and B. Jawerth, and Theorem 4 was discovered by G. David.

A. Grossmann was our bandleader. It is thanks to him that the members of the R. C. P. "Ondelettes" were able to collaborate efficiently, and that the present work was realized.

References

[1] R. Balian, *Un principe d'incertitude fort en théorie du signal ou en mécanique quantique*, C. R. Acad. Sci. Paris Sér. II Méc. Phys. Chim. Sci. Univers Sci. Terre **292** (1981), 1357–1362.

[2] J. M. Bony, *Calcul symbolique et propagation des singularités pour les équations aux dérivées partielles non linéaires*, Ann. Sci. École Norm. Sup. (4) **14** (1981), 209–246.

[3] A. P. Calderón and A. Torchinsky, *Parabolic maximal functions associated with a distribution*, Advances in Math. **16** (1975), 1–64; *Parabolic maximal functions associated with a distribution. II*, Advances in Math. **24** (1977), 101–171.

[4] L. Carleson, *An explicit unconditional basis in H^1*, Bull. Sci. Math. (2) **104** (1980), 405–416.

[5] I. Daubechies, A. Grossmann, and Y. Meyer, *Painless nonorthogonal expansions*, J. Math. Phys. **27** (1986), 1271–1283.

[6] G. David and J. L. Journé, *A boundedness criterion for generalized Calderón-Zygmund operators*, Ann. Math. (2) **120** (1984), 371–397.

[7] M. Frazier and B. Jawerth, *Decomposition of Besov spaces*, Indiana Univ. Math. J. **34** (1985), 777–799.

[8] P. Goupillaud, A. Grossmann, and J. Morlet, *Cycle octave and related transforms in seismic signal analysis*, Geoexploration **23** (1984/1985), 85–102.

[9] A. Grossmann and J. Morlet, *Decomposition of Hardy functions into square integrable wavelets of constant shape*, SIAM J. Math. Anal. **15** (1984), 723–736.

[10] P. G. Lemarié, *Continuité sur les espaces de Besov des opérateurs définis par des intégrales singulières*, Ann. Inst. Fourier (Grenoble) **35** (1985), 175–187.

[11] P. G. Lemarié, Thèse (c/o Mme Dumas, Bâitment 425, Faculté des Sciences, 91405 Orsay).

[12] P. G. Lemarié and Y. Meyer, *Ondelettes et bases hilbertiennes*, Rev. Mat. Iberoamericana **2** (1986), 1–18.

[13] B. Maurey, *Isomorphisms entre espaces* H^1, Acta Math. **145** (1980), 79–120.

[14] Y. Meyer, *Real analysis and operator theory*, Pseudodifferential operators and applications (Notre Dame, Ind., 1984), Proc. Sympos. Pure Math. **43**, Amer. Math. Soc., Providence, 1985, 219–235.

[15] Y. Meyer, *Régularité des solutions des équations aux dériveés partielles non linéaires (d'aprés J.-M. Bony)*, Séminaire Bourbaki, 1979/80, Lecture Notes in Math. **842**, Springer, Berlin, 1981, 293–302,

[16] Th. Paul, Thése (de Doctorat d'Etat), Centre de Physique Théorique, Marseille Luminy, 1985.

[17] E. Stein, Singular Integrals and Differentiability Properties of Functions, Princeton University Press, Princeton, 1970.

[18] P. Wojtaszczyk, *The Franklin system is an unconditional basis in* H^1, Ark. Mat. **20** (1982), 293–300.

Wavelets and Hilbert Bases

P. G. Lemarié and Y. Meyer
In homage to A. P. Calderón
Translated by John Horváth

1. Introduction

Wavelet transforms appear implicitly in a famous work of A. P. Calderón [2]. It was rediscovered and made explicit by J. Morlet and his collaborators [4], [7], [8] as an efficient technique of numerical analysis allowing signal processing in connection with oil prospecting.

Wavelet transforms are similar to the Fourier transformation but the imaginary exponentials $\exp(ix \cdot \xi)$ indexed by the frequencies $\xi \in \mathbf{R}^n$ are replaced by the "wavelets" ψ_Q indexed by the collection of all the cubes $Q \subset \mathbf{R}^n$. These "wavelets" ψ_Q are all copies (by translation and change of scale) of the same regular function ψ, decreasing at infinity as strongly as possible. To be more precise, we choose ψ in the class $\mathcal{S}(\mathbf{R}^n)$ of L. Schwartz and if the cube Q is defined by $a_j \leq x_j \leq a_j + d$, $1 \leq j \leq n$, where $a = (a_1, \ldots, a_n)$, then $\psi_Q(x) = d^{-n/2}\,\psi((x-a)/d)$. This means that ψ_Q is *localized* and *adjusted to the cube* Q.

The goal is to be able to write, by a judicious choice of ψ, any function f as a sum or as an integral of the $c_Q\,\psi_Q(x)$, where the coefficients c_Q are calculated like Fourier coefficients:

$$c_Q = \int_{\mathbf{R}^n} f(x)\,\overline{\psi}_Q(x)\,dx.$$

The authors mentioned above have obtained conditions on ψ allowing us to reconstitute f with the help of the *redundant information* given by the knowledge of all the "wavelet coefficients" c_Q.

Following a procedure made explicit by L. Carleson in [3], we propose to make this redundancy disappear by replacing the collection of all the cubes $Q \subset \mathbf{R}^n$ by that, denoted $\mathcal{Q}$, of the *dyadic cubes*. A dyadic cube Q is defined by two indices $j \in \mathbf{Z}$, $k \in \mathbf{Z}^n$, and one has

$$Q = \{x \in \mathbf{R}^n; 2^j x - k \in [0,1]^n\}. \tag{1.1}$$

The construction that will follow is, as in [3], an imitation of the Haar system of which we recall the definition.

We call $h^{(1)}(t)$ the function equal to 1 if $0 \leq t < 1/2$, to -1 if $1/2 \leq t < 1$, and to 0 elsewhere. On the other hand, $h^{(0)}(t)$ is the characteristic function of the interval

Translator's Note: The references have been updated, and a new version of Figure 1 was generated by Norbert Kaiblinger.

Authors' Note: As is the case for the paper "Uncertainty Principle, Hilbert Bases, and Algebras of Operators" reprinted preceding this one, the notion of multiresolution analysis is still missing from this paper, and so is any reference to Strömberg's work.

[0, 1]. We denote by $E \subset \{0, 1\}^n$ the set of the $2^n - 1$ sequences $(\epsilon_1, \ldots, \epsilon_n)$ of zeros and ones, excluding the sequence $(0, 0, \ldots, 0)$. Then the Haar system for $L^2(\mathbf{R}^n)$ consists of the functions $h_Q^{(\epsilon)}$, $\epsilon \in E$, $Q \in \mathcal{Q}$, defined by

$$h_Q^{(\epsilon)}(x) = 2^{nj/2}\, h^{(\epsilon_1)}(2^j x_1 - k_1) \cdots h^{(\epsilon_n)}(2^j x_n - k_n).$$

The Haar system is well adapted to the function spaces $L^p(\mathbf{R}^n)$, $1 < p < +\infty$, but does not permit the study of Sobolev and Besov spaces, or of the Hardy spaces $H^1(\mathbf{R}^n)$ of Stein and Weiss.

The generally admitted idea is that "to smooth the Haar system makes the orthogonality disappear." This is what appears reading [3] or [6]. We shall see, however, that there exists an exceptional choice of two functions $\psi^{(0)}$ and $\psi^{(1)}$ of $\mathcal{S}(\mathbf{R})$ such that the sequence of functions

$$\psi_Q^{(\epsilon)}(x) = 2^{nj/2}\, \psi^{(\epsilon_1)}(2^j x_1 - k_1) \cdots \psi^{(\epsilon_n)}(2^j x_n - k_n) \tag{1.2}$$

is a Hilbert basis of $L^2(\mathbf{R}^n)$ ($\epsilon \in E$, $Q \in \mathcal{Q}$).

This basis suits all the classical function spaces: Sobolev, Besov, Hardy, ..., spaces that translate isomorphically into sequence spaces. In particular, the coefficients of a function $f \in H^1(\mathbf{R}^n)$ with respect to the basis $\psi_Q^{(\epsilon)}$, $\epsilon \in E$, $Q \in \mathcal{Q}$, and those of a function g in the dyadic version of H^1 with respect to the basis $h_Q^{(\epsilon)}$ are characterized by the same condition. This remark yields a very simple proof of Maurey's theorem (see Theorem 4 below).

In the fundamental identity

$$f(x) = \sum_{\epsilon \in E} \sum_{Q \in \mathcal{Q}} \langle f, \psi_Q^{(\epsilon)} \rangle\, \psi_Q^{(\epsilon)}(x) \tag{1.3}$$

we dispose of several levels of reading.

If $f \in L^2(\mathbf{R}^n)$, then we deal with the usual decomposition with respect to an orthonormal basis. The convergence is in the quadratic mean and almost everywhere.

If $f(x)$ and its derivatives of order 1 belong to $L^2(\mathbf{R}^n)$, then we have also convergence in $L^2(\mathbf{R}^n)$ of the termwise differentiated series. If more generally f and its derivatives up to order s ($s \in \mathbf{N}$) belong to $L^2(\mathbf{R}^n)$, then the series (1.3) will converge automatically with respect to the corresponding norm, that is, that of the Sobolev space H^s. In other words, the regularity of f accelerates the convergence (as this is the case for Fourier series but certainly not for the Haar system). Finally, we will show that this new algorithm permits a very simple writing of the paraproduct $\pi(a, f)$ of J. M. Bony [1].

2. Statement of the Fundamental Theorem

We begin by constructing special functions of one real variable. We denote by S the interval $[2\pi/3, 4\pi/3]$. Thus we have $2S = [4\pi/3, 8\pi/3]$, $S - 2\pi = -S = [-4\pi/3, -2\pi/3]$ and hence also $2\pi - S = S$. We denote by $\theta(t)$ an *odd* function of the real variable t, of class C^∞, with values in $[-\pi/4, \pi/4]$, equal to $\pi/4$ if $t \geq \pi/3$ (hence to $-\pi/4$ if $t \leq -\pi/3$). One defines the even function $\omega(t)$ by $\omega(t) = 0$ if $0 \leq t \leq 2\pi/3$ or if $t \geq 8\pi/3$, $\omega(t) = \pi/4 + \theta(t - \pi)$ if $2\pi/3 \leq t \leq 4\pi/3$, and $\omega(t) = \pi/4 - \theta(t/2 - \pi)$ if $4\pi/3 \leq t \leq 8\pi/3$.

The fundamental identities satisfied by $\omega(t)$ are $\omega(-t) = \omega(t)$, $\omega(2t) = \pi/2 - \omega(t)$, $t \in S$, and

$$\omega(t - 2\pi) = \omega(2\pi - t) = \frac{\pi}{2} - \omega(t), \quad t \in S.$$

By construction $\omega(t)$ is infinitely differentiable.

With the help of $\omega(t)$ one defines $\psi \in \mathcal{S}(\mathbf{R})$ by

$$\widehat{\psi}(t) = e^{-it/2} \sin \omega(t) = e^{-it/2}\, \widetilde{\omega}(t). \tag{2.1}$$

Thus we have

$$\psi(t) = \frac{1}{\pi} \int_0^\infty \cos \left[\left(t - \frac{1}{2} \right) s \right] \widetilde{\omega}(s)\, ds. \tag{2.2}$$

As J. Morlet made us observe, one can also start from an *odd* function $\omega_1(t)$ defined by $\omega_1(t) = -\omega(t)$ if $t \geq 0$ (and hence $\omega_1(t) = \omega(t)$ if $t \leq 0$). We associate with it $\psi_1 \in \mathcal{S}(\mathbf{R})$ defined by

$$\widehat{\psi}_1(t) = e^{-it/2} \sin \omega_1(t), \tag{2.3}$$

which implies

$$\psi_1(t) = \frac{1}{\pi} \int_0^\infty \sin \left[\left(t - \frac{1}{2} \right) s \right] \sin \omega(s)\, ds. \tag{2.4}$$

In the theorems that follow ψ can be replaced systematically by ψ_1. If $I \subset \mathbf{R}$ is the dyadic interval defined by $2^j x - k \in [0, 1]$, we define ψ_I by $\psi_I(x) = 2^{j/2}\, \psi(2^j x - k)$.

Then we have

Theorem 1 *The collection of wavelets $\psi_I(x)$ is a Hilbert basis of $L^2(\mathbf{R})$.*

We shall give the corresponding statement in $\mathbf{R}^n$ resuming the notation of the introduction and imitating the construction of the Haar system. The function ψ stays the same as above, and we associate with it a function $\varphi \in \mathcal{S}(\mathbf{R})$ having integral 1, defined by $\widehat{\varphi}(t) = \cos \omega(t)$ if $|t| \leq 4\pi/3$ and by $\widehat{\varphi}(t) = 0$ otherwise.

One sets $\psi^{(0)}(t) = \varphi(t)$, $\psi^{(1)}(t) = \psi(t)$ and defines $\psi^{(\epsilon)} \in \mathcal{S}(\mathbf{R}^n)$ by

$$\psi^{(\epsilon)}(x) = \psi^{(\epsilon_1)}(x_1)\, \psi^{(\epsilon_2)}(x_2) \cdots \psi^{(\epsilon_n)}(x_n),$$

where $\epsilon = (\epsilon_1, \ldots, \epsilon_n) \in E$; this means that $\epsilon_j = 0$ or 1 but the sequence consisting only of zeros is excluded.

Let us recall that a dyadic cube Q is defined by $j \in \mathbf{Z}$, $k \in \mathbf{Z}^n$ and is the set of the $x \in \mathbf{R}^n$ such that $2^j x - k \in [0, 1]^n$. We then set

$$\begin{aligned} \psi_Q^{(\epsilon)}(x) &= 2^{nj/2}\, \psi^{(\epsilon)}(2^j x - k) \\ &= 2^{nj/2}\, \psi^{(\epsilon_1)}(2^j x_1 - k_1) \cdots \psi^{(\epsilon_n)}(2^j x_n - k_n). \end{aligned} \tag{2.5}$$

With this notation, we intend to prove the following result.

Theorem 2 *The collection of functions $\psi_Q^{(\epsilon)}$, $\epsilon \in E$, $Q \in \mathcal{Q}$, is a Hilbert basis of $L^2(\mathbf{R}^n)$.*

Before proving this result, let us introduce some notation. In spite of the fact that the choice of ψ is not unique, the functions $\psi_Q^{(\epsilon)}$, $\epsilon \in E$, $Q \in \mathcal{Q}$, will be called the *wavelets*. Any series $\sum_{\epsilon \in E} \sum_{Q \in \mathcal{Q}} \alpha(\epsilon, Q)\, \psi_Q^{(\epsilon)}(x)$ is called a series of wavelets.

If $f(x)$ belongs to $L^2(\mathbf{R}^n)$ or, more generally, is a tempered distribution, the coefficients

$$\alpha(\epsilon, Q) = \int f(x)\, \psi_Q^{(\epsilon)}(x)\, dx \tag{2.6}$$

will be called the *wavelet coefficients*. An easy consequence of Theorem 2 is that one will always have the inversion formula

$$f(x) = \sum_{\epsilon \in E} \sum_{Q \in \mathcal{Q}} \alpha(\epsilon, Q)\, \psi_Q^{(\epsilon)}(x) \tag{2.7}$$

in the following sense: for every test function $u(x) \in \mathcal{S}(\mathbf{R}^n)$ whose moments are all zero,

$$\langle f, u\rangle = \sum_{\epsilon \in E} \sum_{Q \in \mathcal{Q}} \alpha(\epsilon, Q) \int_{\mathbf{R}^n} \psi_Q^{(\epsilon)}(x)\, u(x)\, dx.$$

Finally, the transformation that associates with the tempered distribution f the numerical sequence $\alpha(\epsilon, Q)$ will be called the *wavelet transform*. It plays the role of a local Fourier transformation (in fact the integration is extended over the whole $\mathbf{R}^n$, but there is a strong decay due to the rapid decrease of ψ). We split the proof of Theorem 2 into two parts.

First we shall prove that $\psi_Q^{(\epsilon)}$ is orthogonal to $\psi_{Q'}^{(\epsilon')}$ if $(Q, \epsilon) \neq (Q', \epsilon')$. Then we shall show that any function $f \in L^2(\mathbf{R}^n; dx)$ orthogonal to each of the $\psi_Q^{(\epsilon)}$, $\epsilon \in E$, $Q \in \mathcal{Q}$, is necessarily zero.

The orthogonality relations follow from the corresponding properties in one real variable (which we will describe).

Lemma 1 *One has*

$$\int_{-\infty}^{\infty} \widehat{\varphi}(t)\, \overline{\widehat{\psi}}(2^{-j}t)\, e^{ik2^{-j}t}\, dt = 0$$

for all $j \in \mathbf{N}$ and all $k \in \mathbf{Z}$. Similarly one has

$$\int_{-\infty}^{\infty} \widehat{\psi}(t)\, \overline{\widehat{\psi}}(2^{-j}t)\, e^{ik2^{-j}t}\, dt = 0$$

for $j \geq 1$ and $k \in \mathbf{Z}$.

Let us begin by proving the first relation. If $j \geq 1$, the supports of $\widehat{\varphi}(t)$ and of $\widehat{\psi}(2^{-j}t)$ are disjoint.

If $j = 0$, one has $\widehat{\varphi}(t)\, \overline{\widehat{\psi}}(t) = e^{it/2} \sin\omega(t) \cos\omega(t)$ if $|t| \leq 4\pi/3$, and $= 0$ otherwise. The support of $\widehat{\varphi}\overline{\widehat{\psi}}$ is thus composed of the two intervals $S = [2\pi/3, 4\pi/3]$ and $S - 2\pi$. The identity $\omega(t - 2\pi) = \pi/2 - \omega(t)$ and the relation $e^{i\pi} = -1$ conclude the verification.

The proof of the second assertion is the same. If $j \geq 2$, the supports of $\widehat{\psi}(t)$ and of $\widehat{\psi}(2^{-j}t)$ are disjoint. It remains to consider the case $j = 1$ and one makes in the integral the change of variables $t = 2s$. Returning to the variable t, one observes that $\widehat{\psi}(2t)\widehat{\psi}(t) = e^{-it/2} \sin\omega(t) \cos\omega(t)$ if $|t| \leq 4\pi/3$, and $= 0$ elsewhere. We are brought back to the preceding case.

We shall now elucidate the case $j = 0$.

Lemma 2 *For every $k \in \mathbf{Z}$, $k \neq 0$ one has*

$$\int_{-\infty}^{\infty} |\widehat{\varphi}(t)|^2 e^{ikt}\, dt = \int_{-\infty}^{\infty} |\widehat{\psi}(t)|^2 e^{ikt}\, dt = 0.$$

To see this, we will prove that

$$\sum_{-\infty}^{\infty} |\widehat{\psi}(t+2\pi j)|^2 = \sum_{-\infty}^{\infty} |\widehat{\varphi}(t+2\pi j)|^2 = 1.$$

We shall limit ourselves to the first identity. The sum to be calculated defines a 2π-periodic function which it suffices to know on an interval of length 2π. We choose the interval $2\pi/3 \leq t \leq 8\pi/3$. If $2\pi/3 \leq t \leq 4\pi/3$, the only values of j that arise are $j = 0$ and $j = -1$. Because $\omega(t - 2\pi) = \pi/2 - \omega(t)$, the identity follows from $\sin^2 \omega(t) + \cos^2 \omega(t) = 1$. If $4\pi/3 \leq t \leq 8\pi/3$, one has to take $j = 0$ and $j = -2$, and one concludes in the same way.

Let us return to the orthogonality between the $\psi_Q^{(\epsilon)}$. One is led to calculate the integral

$$I(j, j', k, k', \epsilon, \epsilon') = \int_{\mathbf{R}^n} \exp[i(k2^{-j} - k'2^{-j'})\xi]\, \widehat{\psi}^{(\epsilon_1)}(2^{-j}\xi_1)\, \overline{\widehat{\psi}^{(\epsilon_1')}}(2^{-j'}\xi_1) \cdots \widehat{\psi}^{(\epsilon_n)}(2^{-j}\xi_n)\, \overline{\widehat{\psi}^{(\epsilon_n')}}(2^{-j'}\xi_n)\, d\xi_1 \cdots d\xi_n.$$

By symmetry we may assume that $j \leq j'$. One makes immediately the change of variables $2^{-j}\xi = u$ and is thereby reduced to the case $j = 0$, which we will suppose henceforth. If $j' \geq 1$, we call m a subscript such that $\epsilon_m = 1$ (such a subscript exists since $\epsilon \in E$). One integrates first with respect to the variable x_m, and Lemma 1 ensures that I is zero. If $j' = 0$ and $k \neq k'$, there exists a subscript m such that $k_m \neq k'_m$. One integrates first with respect to the variable x_m and I is zero, either because of the first assertion of Lemma 1 or because of Lemma 2.

Finally, the last case to consider is $j' = 0$, $k = k'$, and $\epsilon \neq \epsilon'$. Then there is a subscript m such that $\epsilon_m = 1$ and $\epsilon'_m = 0$ and the first assertion of Lemma 1 implies the nullity of the integral.

One verifies without effort that $\|\psi_Q^{(\epsilon)}\|_2 = 1$ for all ϵ and all Q.

3. The Sequence of the $\psi_Q^{(\epsilon)}$ is a Total Subset in $L^2(\mathbf{R}^n)$

We arrive at the most technical part of the proof of Theorem 2.

We will verify that $\langle f, \psi_Q^{(\epsilon)} \rangle = 0$ for all $\epsilon \in E$ and all $Q \in \mathcal{Q}$ implies $f = 0$.

For this we will write these relations using the Fourier transformation. Changing the notation as convenient, we have for a certain function $f \in L^2(\mathbf{R}^n)$

$$\int_{\mathbf{R}^n} f(\xi)\, \exp(ik2^{-j}\xi)\, \widehat{\psi}^{(\epsilon_1)}(2^{-j}\xi_1) \cdots \widehat{\psi}^{(\epsilon_n)}(2^{-j}\xi_n)\, d\xi_1 \cdots d\xi_n = 0 \tag{3.1}$$

for every $j \in \mathbf{Z}$, every $k \in \mathbf{Z}^n$, and every sequence $(\epsilon_1, \ldots, \epsilon_n) \neq (0, \ldots, 0)$ of zeros and ones.

We must deduce from this that $f = 0$. We proceed to the change of variables $2^{-j}\xi = u$. Then (3.1) becomes

$$\int_{\mathbf{R}^n} f_j(\xi)\, \exp(ik\xi)\, \widehat{\psi}^{(\epsilon_1)}(\xi_1) \cdots \widehat{\psi}^{(\epsilon_n)}(\xi_n)\, d\xi = 0 \text{ where } f_j(\xi) = f(2^j\xi). \tag{3.2}$$

It is classical that for a fixed j and all $k \in \mathbf{Z}^n$ (3.2) is equivalent to the identity

$$\sum_{k\in\mathbf{Z}^n} f_j(\xi + 2k\pi)\, \widehat{\psi}^{(\epsilon_1)}(\xi_1 + 2k_1\pi) \cdots \widehat{\psi}^{(\epsilon_n)}(\xi_n + 2k_n\pi) = 0. \tag{3.3}$$

This identity must be satisfied for all $j \in \mathbf{Z}$, all $\epsilon \in E$, and all $\xi \in \mathbf{R}^n$. The ideal situation would be if there existed a subset $\Omega \subset \mathbf{R}^n$ having the following properties: the translates $\Omega + 2k\pi$, $k \in \mathbf{Z}^n$ are pairwise disjoint; for every $\xi \in \mathbf{R}^n$ there exists a $j \in \mathbf{Z}^n$ such that $2^j\xi \in \Omega$; and finally, for every $\xi \in \Omega$ there exists $\epsilon \in E$ such that $\widehat{\psi}^{(\epsilon_1)}(\xi_1) \neq 0, \ldots, \widehat{\psi}^{(\epsilon_n)}(\xi_n) \neq 0$. Then (3.3) would imply that $f_j(\xi) = 0$ on Ω for all $j \in \mathbf{Z}$ and thus $f = 0$. Naturally this is not the case, but we will try to come as close to this ideal situation as possible.

To simplify the notation which follows, we shall restrict ourselves to dimension $n = 2$. Let us write again (3.3) in detail:

$$\sum_k \sum_\ell f_j(u + 2k\pi, v + 2\ell\pi)\, \widehat{\psi}(u + 2k\pi)\, \widehat{\psi}(v + 2\ell\pi) = 0, \tag{3.4}$$

$$\sum_k \sum_\ell f_j(u + 2k\pi, v + 2\ell\pi)\, \widehat{\varphi}(u + 2k\pi)\, \widehat{\psi}(v + 2\ell\pi) = 0, \tag{3.5}$$

$$\sum_k \sum_\ell f_j(u + 2k\pi, v + 2\ell\pi)\, \widehat{\psi}(u + 2k\pi)\, \widehat{\varphi}(v + 2\ell\pi) = 0. \tag{3.6}$$

The sums (3.4) to (3.6) define obviously $(2\pi\mathbf{Z})^2$-periodic functions, which we will analyze on a "fundamental domain." First we shall suppose that

$$(u, v) \in \left[\frac{2\pi}{3}, \frac{4\pi}{3}\right]^2 = \Omega_1$$

and consider that (3.4), (3.5), and (3.6) are coupling $f_j(u, v)$ (which we try to calculate) to the undesirable values $f_j(u + 2k\pi, v + 2\ell\pi)$, which we will try to eliminate.

For $(u, v) \in \Omega_1$ we have necessarily $k = 0$ or $k = -1$ and $\ell = 0$ or $\ell = -1$. Setting $f_j = f_j(u, v)$, $g_j = f_j(u - 2\pi, v)$, $\widetilde{f}_j = f_j(u, v - 2\pi)$, and $\widetilde{g}_j = f_j(u - 2\pi, v - 2\pi)$, we have

$$\begin{aligned} &\sin\omega(u)\, \sin\omega(v)\, f_j - \cos\omega(u)\, \sin\omega(v)\, g_j \\ &\quad - \sin\omega(u)\, \cos\omega(v)\, \widetilde{f}_j + \cos\omega(u)\, \cos\omega(v)\, \widetilde{g}_j = 0, \end{aligned} \tag{3.7}$$

$$\begin{aligned} &\cos\omega(u)\, \cos\omega(v)\, f_j + \sin\omega(u)\, \sin\omega(v)\, g_j \\ &\quad - \cos\omega(u)\, \cos\omega(v)\, \widetilde{f}_j - \sin\omega(u)\, \cos\omega(v)\, \widetilde{g}_j = 0, \end{aligned} \tag{3.8}$$

$$\begin{aligned} &\sin\omega(u)\, \cos\omega(v)\, f_j - \cos\omega(u)\, \cos\omega(v)\, g_j \\ &\quad + \sin\omega(u)\, \sin\omega(v)\, \widetilde{f}_j - \cos\omega(u)\, \sin\omega(v)\, \widetilde{g}_j = 0. \end{aligned} \tag{3.9}$$

These three relations obviously do not allow us to calculate the four unknowns, which are f_j, g_j, $\widetilde{f}_j$, and $\widetilde{g}_j$.

We consider (3.4) for the point $(u', v') = (2u, 2v)$.

Then the sole values of k and ℓ that need be considered are $k = 0$ or $k = -2$, $\ell = 0$ or $\ell = -2$.

We observe that $f_j(u', v') = f_{j+1}(u, v)$, $f_j(u' - 4\pi, v') = f_{j+1}(u - 2\pi, v)$, etc, $\ldots$. Shifting the subscript j by a unit, we therefore obtain the missing relation in the form

$$\sin\omega(u)\,\sin\omega(v)\,f_j + \cos\omega(u)\,\sin\omega(v)\,g_j + \sin\omega(u)\,\cos\omega(v)\,\widetilde{f}_j + \cos\omega(u)\,\cos\omega(v)\,\widetilde{g}_j = 0. \tag{3.10}$$

The determinant of the four equations written is equal to 1, and we obtain thus $f_j = g_j = \widetilde{f}_j = \widetilde{g}_j = 0$ for every $j \in \mathbf{Z}$.

Let us denote now by Ω_2 the rectangle $\pi/3 \le u \le 2\pi/3$, $2\pi/3 \le v \le 4\pi/3$. We must observe that $\widetilde{\psi}(u) = 0$ on Ω_2 while $\widetilde{\varphi}(u) = 1$. The only relation pertinent for calculating $f_j(u, v)$ if $(u, v) \in \Omega_2$ is therefore (3.5). The only values of k and ℓ that really occur are $k = 0$ and $\ell = 0$ or $\ell = -1$.

We still set $f_j = f_j(u, v)$ and $\widetilde{f}_j = f_j(u, v - 2\pi)$ and have

$$f_j\,\sin\omega(v) - \widetilde{f}_j\,\cos\omega(v) = 0. \tag{3.11}$$

Obviously this sole relation does not permit us to calculate f_j and $\widetilde{f}_j$. But here again we consider the point $(2u, 2v)$ by which one writes (3.4) and (3.5). The only values of k or ℓ that occur are $k = 0$ and $k = -1$, $\ell = 0$ or $\ell = -2$. Let us set $\sigma_j = f_j(u, v)\,\cos\omega(v) + f_j(u, v - 2\pi)\,\sin\omega(v)$ and $\tau_j = f_j(u - \pi, v)\,\cos\omega(v) + f_j(u - \pi, v - 2\pi)\,\sin\omega(v)$ (having taken care to shift the subscripts by a unit).

Thus we have $\sigma_j \cos\omega(2u) + \tau_j \sin\omega(2u) = 0$ and $\sigma_j \sin\omega(2u) - \tau_j \cos\omega(2u) = 0$. These two relations imply $\sigma_j = 0$ or

$$f_j\,\cos\omega(v) - \widetilde{f}_j\,\sin\omega(v) = 0. \tag{3.12}$$

The relations (3.11) and (3.12) imply $f_j = \widetilde{f}_j = 0$ for every $j \in \mathbf{Z}$. Naturally, the same approach suits the other seven rectangles obtained by symmetry with respect to the coordinate axes or their bisectors, starting with Ω_2. Finally, one studies the rectangle Ω_3 defined by $0 \le u \le \pi/3$ and $2\pi/3 \le v \le 4\pi/3$. The only significant relation is again $f_j\,\sin\omega(v) - \widetilde{f}_j\,\cos\omega(v) = 0$. We use the same strategy as above, writing (after the necessary shift of j) the corresponding relation on the rectangle $2\Omega_3$ for the point $(2u, 2v)$. It follows that $f_j\,\cos\omega(v) - \widetilde{f}_j\,\sin\omega(v) = 0$, which yields $f_j = \widetilde{f}_j = 0$ on Ω_3. Here again we obtain the same conclusion for all the rectangles deduced from Ω_3 by all the symmetries of the problem.

Due to the examination of these three "fundamental domains," the whole annulus $2\pi/3 \le \sup(|u|, |v|) \le 8\pi/3$ is filled and $f(2^j u, 2^j v) = 0$ in this annulus. It follows immediately that $f = 0$ almost everywhere on $\mathbf{R}^2$.

4. Convergence of the Series of Wavelets in the Spaces of Test Functions or Distributions

Let $\mathcal{S}(\mathbf{R}^n)$ be the space of test functions equipped with the usual topology and $\mathcal{S}_0 \subset \mathcal{S}$ the subspace formed by the functions $f \in \mathcal{S}$ that verify $\int_{\mathbf{R}^n} x^\alpha f(x)\,dx = 0$ for all $\alpha \in \mathbf{N}^n$ (i.e., all its moments are zero). Then all the wavelets $\psi_Q^{(\epsilon)}$ belong to $\mathcal{S}_0$. If a series of wavelets converges in $\mathcal{S}$, it also converges in $\mathcal{S}_0$ to a function of $\mathcal{S}_0$. One proves easily that the converse is true.

Lemma 3 *If $f \in \mathcal{S}_0$, then the series of wavelets converges to f in the sense of the topology of $\mathcal{S}$.*

What happens, then, when f belongs to $\mathcal{S}$? It is easy to show that the series of wavelets converges to f uniformly on $\mathbf{R}^n$ and that the same is true for the differentiated series.

Let us now start with the function 1 (identically equal to 1 on the whole $\mathbf{R}^n$). Every coefficient (in wavelets) of 1 is 0. Why does one obtain an absurd numerical equality?

Let us consider the topological dual space of $\mathcal{S}_0(\mathbf{R}^n)$. This is the quotient space $\mathcal{S}'(\mathbf{R}^n)/\mathbf{C}[x]$ of the tempered distributions modulo the polynomials. If S is such a tempered distribution, then the corresponding series of wavelets $\sum_{Q\in\mathcal{Q}} \lambda_Q\, \psi_Q^{(\epsilon)}(x)$ converges in the following sense: for every function $f \in \mathcal{S}_0$ there is numerical convergence of the series obtained by integration against f. This is the reason why one obtains $1 = 0$ (modulo the constant functions).

5. Characterization of the Spaces $L^p(\mathbf{R}^n)$, $1 < p \leq +\infty$, BMO($\mathbf{R}^n$) and of its Predual $H^1(\mathbf{R}^n)$ by Transformation into Wavelets

Let E be a separable Banach space. A sequence $(e_k)_{k\in\mathbf{N}}$ of elements of E is a basis of E if every vector $x \in E$ can be written uniquely as $x = \sum_0^\infty \xi_k\, e_k$, where the ξ_k are scalars and where the partial sums of the written series converge to x in the norm of E. The basis is said to be unconditional if there exists a constant $C \geq 1$ such that for every $m \geq 1$ and every sequence ξ_k, $k \in \mathbf{N}$, of scalars one has $\|\sum_0^m \lambda_k\, \xi_k\, e_k\|_E \leq C\,\|\sum_0^m \xi_k\, e_k\|_E$ as soon as $\sup|\lambda_k| \leq 1$.

This means that the condition for a series $\sum_0^\infty \xi_k\, e_k$ to belong to E concerns only the sequence $|\xi_k|$, $k \in \mathbf{N}$, and that if a sequence satisfies this condition then automatically every other sequence satisfies it whose absolute values are dominated termwise by those of the first one.

The space BMO($\mathbf{R}^n$) is not separable and therefore does not possess an unconditional basis. We will replace it by the separable version VMO($\mathbf{R}^n$), which is the closure in the norm of BMO($\mathbf{R}^n$) of the space $\mathcal{S}(\mathbf{R}^n)$ of test functions.

Theorem 3 *The sequence $\psi_Q^{(\epsilon)}$, $\epsilon \in E$, $Q \in \mathcal{Q}$ is an unconditional basis for $L^p(\mathbf{R}^n; dx)$, $1 < p \leq +\infty$, VMO($\mathbf{R}^n$), and $H^1(\mathbf{R}^n)$.*

To prove it one constructs explicitly the operator T that transforms $\psi_Q^{(\epsilon)}$ into $\lambda(Q,\epsilon)\, \psi_Q^{(\epsilon)}$ if $|\lambda(Q,\epsilon)| \leq 1$. The distributional kernel of T is $K(x,y) = \sum_{\epsilon\in E} \sum_{Q\in\mathcal{Q}} \lambda(Q,\epsilon)$

$\psi_Q^{(\epsilon)}(x)\,\psi_Q^{(\epsilon)}(y)$. One obtains easily the Calderón-Zygmund estimates

$$|\partial_x^\alpha\,\partial_y^\beta K(x,y)| \le C(\alpha,\beta)\,|x-y|^{-n-|\alpha|-|\beta|} \tag{5.1}$$

for all $x \in \mathbf{R}^n$, all $y \neq x$, $\alpha \in \mathbf{N}^n$, and $\beta \in \mathbf{N}^n$.

On the other hand, T is obviously bounded on $L^2(\mathbf{R}^n)$ since it is diagonalized in an orthonormal basis. Finally T verifies in the sense of the theorem of David and Journé $T(1) = T^*(1) = 0$.

It follows that T is bounded on $L^p(\mathbf{R}^n)$, $1 < p < \infty$, and on the "boundary spaces" $H^1(\mathbf{R}^n)$ and $\mathrm{VMO}(\mathbf{R}^n)$.

It remains to write down explicit criteria of belonging to these function spaces. To this effect we recall the construction of the Haar basis of $L^2(\mathbf{R}^n; dx)$. One denotes by $\widetilde{\psi}(x)$ the function of a real variable equal to 1 on $[0, 1/2[$, to -1 on $[1/2, 1[$, and to 0 elsewhere. One calls $\widetilde{\varphi}$ the characteristic function of the interval $[0, 1]$ (equal to 1 on this interval and to 0 elsewhere). One then constructs for all $\epsilon \in E$ and all $Q \in \mathcal{Q}$

$$\widetilde{\psi}_Q^{(\epsilon)} = 2^{nj/2}\,\widetilde{\psi}^{(\epsilon_1)}(2^j x_1 - k_1)\cdots\widetilde{\psi}^{(\epsilon_n)}(2^j x_n - k_n),$$

where $\widetilde{\psi}^{(0)} = \widetilde{\varphi}$, $\widetilde{\psi}^{(1)} = \widetilde{\psi}$.

Theorem 4 *The isometry $U\colon L^2(\mathbf{R}^n; dx) \to L^2(\mathbf{R}^n; dx)$ which associates $\widetilde{\psi}_Q^{(\epsilon)}$ with $\psi_Q^{(\epsilon)}$ extends to an isomorphism of $L^p(\mathbf{R}^n)$ onto itself if $1 < p < +\infty$, to an isomorphism of the Hardy space $H^1(\mathbf{R}^n; dx)$ of Stein and Weiss onto its dyadic version H_d^1, and to an isomorphism of the space $\mathrm{BMO}(\mathbf{R}^n; dx)$ of John and Nirenberg onto its dyadic version.*

This means that the (wavelet) coefficients $\lambda(Q,\epsilon)$ of $f \in L^p(\mathbf{R}^n)$ are characterized by the condition

$$\left(\sum_\epsilon \sum_Q |\lambda(Q,\epsilon)|^2 |Q|^{-1} \chi_Q(x)\right)^{1/2} \in L^p(\mathbf{R}^n)$$

where $1 < p < \infty$. One denotes by $|Q|$ the volume of Q and by $\chi_Q(x)$ the characteristic ("indicator") function of Q.

If $p = 1$, then this condition characterizes the coefficients of $f \in H^1(\mathbf{R}^n)$. Finally we will show, and this will be the beginning of the proof, that the wavelet coefficients $\lambda(Q,\epsilon)$ of $f \in \mathrm{BMO}(\mathbf{R}^n)$ are characterized by Carleson's condition

$$\sum_\epsilon \sum_{Q \subset R} |\lambda(Q,\epsilon)|^2 \le C|R| \tag{5.2}$$

(for every dyadic cube R, the summation being extended over all dyadic subcubes $Q \subset R$).

Let us first prove that (5.2) implies the convergence for the topology $\sigma(\mathrm{BMO}, H^1)$ of the series $\sum_\epsilon \sum_Q \lambda(Q,\epsilon)\,\psi_Q^{(\epsilon)}$ to a function $f \in \mathrm{BMO}$. To do this we disregard the index ϵ and reason on finite sums. Passing to the limit is routine once the inequality $\|\sum_Q \lambda(Q)\,\psi_Q(x)\|_{\mathrm{BMO}} \le C_0 \;|\!|\!|\; \lambda(Q) \;|\!|\!|$ is established for these finite sums, C_0 depends only on n, and $|\!|\!| \cdot |\!|\!|$ is the lower bound of the $\sqrt{C}$ in (5.2).

Let $\varphi \in \mathcal{D}(\mathbf{R}^n)$ be a function having integral 1 and supported by the unit cube $0 \le x_j \le 1$ $(1 \le j \le n)$.

We set $\varphi_Q(x) = 2^{nj}\,\varphi(2^j x - k)$ so that $\varphi_Q(x)$ is supported by the dyadic cube Q and has integral equal to 1. Then one considers the distribution

$$K(x, y) = \sum_{Q \in \mathcal{Q}} \lambda(Q)\, \psi_Q(x)\varphi_Q(y).$$

One has $|\lambda(Q)| \leq (C|Q|)^{1/2}$, from which it follows immediately that $K(x, y)$ satisfies (5.1). Let us show that the operator T, whose distributional kernel is $K(x, y)$, is bounded on $L^2(\mathbf{R}^n)$. One has

$$\begin{aligned} \|T(f)\|_2 &= \left\| \sum_{Q \in \mathcal{Q}} \lambda(Q) \left(\int f\, \varphi_Q \right) \psi_Q(x) \right\|_2 \\ &= \left(\sum_{Q \in \mathcal{Q}} |\lambda(Q)|^2 \left| \int f\, \varphi_Q \right|^2 \right)^{1/2} \\ &\leq C_0\ \|\!|\ \lambda(Q)\ |\!\|\ \ \|f\|_2 \end{aligned}$$

due to the dyadic version of Carleson's inequality stated in the following lemma.

Lemma 4 *Let p_Q, $Q \in \mathcal{Q}$, x_Q, $Q \in \mathcal{Q}$, be two arbitrary positive sequences, and $\omega(x)$, $x \in \mathbf{R}^n$, the maximal function defined by $\omega(x) = \sup_{Q \ni x} x_Q$. Let us assume that for every dyadic cube $R \in \mathcal{Q}$ we have $\sum_{Q \subset R} p_Q \leq |R|$. Then one has $\sum_{Q \in \mathcal{Q}} p_Q\, x_Q \leq \int_{R^n} \omega(x)\, dx$.*

Therefore one has continuity of $T\colon L^2(\mathbf{R}^n; dx) \to L^2(\mathbf{R}^n; dx)$.

This, combined with (5.1), implies $T(1) \in \mathrm{BMO}$, which is in fact what we tried to prove.

Conversely let us assume that f belongs to BMO.

We denote by R a dyadic cube and by R_j, $j \geq 1$, the (nondyadic) cubes with the same center as R whose sides are 2^j times those of R. We decompose, as usual, f in the series

$$f = c(R) + f_0(x) + f_1(x) + \cdots + f_j(x) + \cdots ,$$

where $c(R) = m_R f$ is the mean of f on R and where $f_0(x) = f(x) - c(R)$ on the cube R, $f_0(x) = 0$ elsewhere, and $f_j(x) = f(x) - c(R)$ on $R_j \setminus R_{j-1}$, $f_j(x) = 0$ elsewhere. The condition $f \in \mathrm{BMO}$ implies $|m_{R_{j-1}} f - m_{R_j} f| \leq C\|f\|_{\mathrm{BMO}}$, hence $|m_{R_j} f - m_R f| \leq C\, j\, \|f\|_{\mathrm{BMO}}$. Finally, $\|f_j\|_2 \leq C(1 + j)\, |R_j|^{1/2}\, \|f\|_{\mathrm{BMO}}$. One then writes $\langle f, \psi_Q\rangle = \sum_0^\infty \langle f_j, \psi_Q\rangle$.

For $j \geq 2$, one has

$$\langle f_j, \psi_Q\rangle \leq \|f_j\|_2 \left(\int_{R_j \setminus R_{j-1}} |\psi_Q(x)|^2\, dx \right)^{1/2} \leq C_m\, \|f_j\|_2 \left(\frac{|Q|}{|R_j|} \right)^m$$

for every $m \geq 1$ (due to the rapid decay of ψ and the geometric condition $Q \subset R$).

For $j = 0$ or $j = 1$, one uses Plancherel's identity and one has

$$\sum_Q |\langle f_j, \psi_Q\rangle|^2 = \|f_j\|_2^2 \leq C\, |R|\, \|f\|_{\mathrm{BMO}}^2.$$

One concludes observing that

$$\sum_{Q \subset R} \sum_{j \geq 2} j^2 \frac{|Q|^2}{|R_j|} = c_n \, |R|.$$

The end of the proof of Theorem 3 is now obvious. In fact it is classical that the dyadic BMO space is also characterized by (5.2) by decomposing the functions in the Haar system. Thus U establishes an isomorphism between these two spaces. By duality, U (which is its own transpose) defines an isomorphism between the usual Hardy space $H^1(\mathbf{R}^n)$ and its dyadic version. By interpolation U is an isomorphism on all the spaces $L^p(\mathbf{R}^n)$, $1 < p < +\infty$.

6. Characterization of the Homogeneous Hölder Spaces C^r, $r > 0$, of the Sobolev Spaces H^s and of the Besov Spaces $B_q^{s,p}$

Let us recall that if $0 < r < 1$, then f belongs to the homogeneous Hölder space C^r when $|f(x) - f(y)| \leq C\,|x-y|^r$ for all $x \in \mathbf{R}^n$ and all $y \in \mathbf{R}^n$. If $r = 1$, then we agree to replace C^r by the Zygmund class defined by $|f(x+y) + f(x-y) - 2f(x)| \leq C|y|$ ($x \in \mathbf{R}^n$, $y \in \mathbf{R}^n$).

If $r = m + s$, $0 < s \leq 1$, $m \in \mathbf{N}$, then one writes $f \in C^r$ when $\partial^\alpha f \in C^s$ for all $\alpha \in \mathbf{N}^n$ such that $|\alpha| = m$.

The Sobolev space H^s, $s \in \mathbf{R}$, is defined by

$$\int_{\mathbf{R}^n} |\widehat{f}(\xi)|^2 \, (1 + |\xi|^2)^s \, d\xi < \infty,$$

and finally the homogeneous Besov spaces $B_q^{s,p}$ are characterized by the Littlewood-Paley decomposition. One starts with a function $\theta \in \mathcal{D}(\mathbf{R}^n)$, zero in a neighborhood of 0 and such that $\sum_{-\infty}^{\infty} |\theta(2^j\xi)| \geq 1$ for every nonzero $\xi \in \mathbf{R}^n$. Then a tempered distribution f, modulo the polynomials, belongs to $B_q^{s,p}$ if and only if $\|\Delta_j(f)\|_p \leq 2^{-js}\,\epsilon_j$, where $\epsilon_j \in \ell^q(\mathbf{Z})$ and where $\mathcal{F}(\Delta_j(f))(\xi) = \theta(2^{-j}\xi)\,(\mathcal{F}f)(\xi)$ denoting by $\mathcal{F}$ the Fourier transform (acting on the tempered distribution f). We denote by $\|\cdot\|_p$ the norm in $L^p(\mathbf{R}^n; dx)$.

This space $B_q^{s,p}$ does not depend on the choice of θ. One has $B_\infty^{s,\infty} = C^s$ (homogeneous Hölder space) and $H^s = L^2 \cap B_2^{s,2}$ for $s \geq 0$.

Theorem 5 *A tempered distribution f (modulo the polynomials) belongs to $B_q^{s,p}$ if and only if $\sup_{\epsilon \in E} |\langle f, \psi_Q^{(\epsilon)} \rangle| = \alpha(k,j)$ satisfies*

$$\left(\sum_{j=-\infty}^{\infty} \left\{ \left(\sum_{k \in \mathbf{Z}^n} (\alpha(k,j))^p \right)^{1/p} 2^{j(s+n(1/2-1/p))} \right\}^q \right)^{1/q} < +\infty \tag{6.1}$$

(with the usual changes if $p = +\infty$ or $q = +\infty$).

Let us first show the direct part. To simplify the writing we restrict ourselves to dimension 1.

Then the function $\overline{\widehat{\psi}}$ which serves to define the wavelets will be our function θ and one has

$$\langle f, \psi_Q \rangle = 2^{-j/2}(2\pi)^{-1} \int \widehat{f}(\xi)\, e^{i2^{-j}k\xi}\, \overline{\widehat{\psi}}(2^{-j}\xi)\, d\xi = 2^{-j/2}\, \Delta_j(f)(2^{-j}k).$$

To conclude one uses the evident sampling lemma.

Lemma 5 *There exists a constant $C = C(n)$ such that for every $R > 0$, every $p \in [1, +\infty]$, and every function $f \in L^p(\mathbf{R}^n; dx)$ whose Fourier transform is carried by the ball $|\xi| \leq 20R$, one has $(\sum_{k\in\mathbf{Z}^n} R^{-n}|f(kR^{-1})|^p)^{1/p} \leq C(n)\, \|f\|_p$.*

Naturally 20 can be replaced (changing, if necessary, the constant $C(n)$) by another constant. For us $20 \geq 8\pi/3$.

Thus one has for every $j \in \mathbf{Z}$

$$\left(2^{-nj}|\Delta_j f(2^{-j}k)|^p\right)^{1/p} \leq C(n)\, \|\Delta_j(f)\|_p,$$

which obviously implies (6.1).

Conversely, assume that $\alpha(k,j)$ satisfies (6.1) and let us show that

$$f(x) = \sum\sum \alpha(k,j)\, 2^{j/2}\, \psi(2^j x - k)$$

belongs to $B_q^{s,p}$. One tests this belonging on the unit ball of the dual $B_{q'}^{-s,p'}$ for which one uses the direct part. The details are left to the reader.

7. Calculation of the Partial Sums of a Wavelet Expansion

Let f be a tempered distribution (modulo the polynomials). What can one say of the difference

$$g(x) = f(x) - \sum_{\epsilon\in E}\sum_{|Q|\leq 1} \langle f, \psi_Q^{(\epsilon)}\rangle\, \psi_Q^{(\epsilon)}\ ?$$

It seems to be intuitive that the small cubes are responsible for the small details of $f(x)$. These small details are very important or significant if $f(x)$ is very irregular. So that our difference $g(x)$ should be infinitely differentiable.

Let us denote by $\mathcal{D}_Q^{(\epsilon)}$, $\epsilon \in E$, $Q \in \mathcal{Q}$, the orthogonal projection operator onto the vector $\psi_Q^{(\epsilon)}$ of the Hilbert space $L^2(\mathbf{R}^n; dx)$. Similarly denote by $\mathcal{E}_Q$, $Q \in \mathcal{Q}$, the orthogonal projection operator onto the vector φ_Q. The functions φ and ψ are defined by Theorem 2.

Let us call $\mathcal{Q}_j$ the collection of dyadic cubes with side 2^{-j} and set

$$\mathcal{D}_j = \sum_{\epsilon\in E}\sum_{Q\in\mathcal{Q}_j} \mathcal{D}_Q^{(\epsilon)} \quad \text{and similarly} \quad \mathcal{E}_j = \sum_{Q\in\mathcal{Q}_j} \mathcal{E}_Q.$$

One has the following remarkable identity:

Theorem 6 *For every $j \in \mathbf{Z}$, $\mathcal{D}_j = \mathcal{E}_{j+1} - \mathcal{E}_j$.*

Let us observe that $\mathcal{E}_j$ is a "regularizing version" of the conditional expectation operator with respect to the σ-algebra $\mathcal{F}_j$ generated by the cubes $Q \in \mathcal{Q}_j$. In fact, this conditional expectation is $E_j(f) = \sum_{Q \in \mathcal{Q}_j} \chi_Q \, 2^{nj} \langle f, \chi_Q \rangle$ and $\mathcal{E}_j$ is given by the same identity with the difference that $2^{nj}\,\chi_Q$ is replaced by the "C^∞ version" φ_Q of the normalized characteristic function of the cube Q.

Let us give the structure of the proof of Theorem 6 for dimension 1. One denotes by Δ_j the operator of convolution with $2^j\,\psi(2^j x)$ and by M_j the operator of pointwise multiplication by $\exp(2\pi i 2^j x)$. Then a harmless application of the Poisson summation formula shows that

$$\mathcal{D}_j = \Delta_j \Delta_j^* + \Delta_j M_j \Delta_j^* + \Delta_j M_{j+1} \Delta_j^* + \Delta_j M_j^* \Delta_j^* + \Delta_j M_{j+1}^* \Delta_j^*.$$

One denotes by S_j the operator of convolution with $2^j \varphi(2^j x)$, and one then has the following remarkable identities:

$$\begin{aligned} \Delta_j \Delta_j^* &= S_{j+1} S_{j+1}^* - S_j S_j^*, \\ \Delta_j M_j \Delta_j^* &= -S_j M_j S_j^*, \\ \Delta_j M_{j+1}^* \Delta_j^* &= S_{j+1} M_{j+1} S_{j+1}^*. \end{aligned}$$

On the other hand, one performs for $\mathcal{E}_j$ the calculation we sketched for $\mathcal{D}_j$ and obtains $\mathcal{E}_j = S_j S_j^* + S_j M_j S_j^* + S_j M_j^* S_j^*$. It follows from all this that

$$\mathcal{D}_j = \mathcal{E}_{j+1} - \mathcal{E}_j.$$

Let us pass to the case of higher dimensions. For instance, if $n = 2$ one has, keeping the notation of dimension 1,

$$\mathcal{E}_{j+1} \otimes \mathcal{E}_{j+1} - \mathcal{E}_j \otimes \mathcal{E}_j = \mathcal{D}_j \otimes \mathcal{D}_j + \mathcal{D}_j \otimes \mathcal{E}_j + \mathcal{E}_j \otimes \mathcal{D}_j,$$

and these three terms correspond to the three wavelets $\psi_0^{(\epsilon)}$, $\epsilon \in E$, $Q \in \mathcal{Q}_j$, necessary to obtain our basis in dimension 2.

Thus Theorem 6 is proved, and these remarks yield a new proof of Theorem 2.

Returning to $g(x)$, we then have

$$\begin{aligned} g(x) &= f(x) - \sum_{j \geq 0} \mathcal{D}_j(f) \\ &= f(x) - \sum_{j \geq 0} (\mathcal{E}_{j+1}(f) - \mathcal{E}_j(f)) \\ &= \mathcal{E}_0(f) \\ &= \sum_{k \in \mathbf{Z}^n} c_k \, \varphi(x-k), \end{aligned}$$

where the coefficients c_k are given by $c_k = \int f(u)\,\varphi(u-k)\,du$.

This means that $f(x)$ and $g(x)$ differ by a trivial error term.

8. Paraproducts and Wavelet Transforms

The function φ is the same as in Theorem 2. The operator S_j is the convolution with $2^{nj}\varphi(2^j x)$. One sets $\Delta_j = S_{j+1} - S_j$ and finally

$$\pi(a, f) = \sum_{-\infty}^{\infty} S_{j-3}(a)\, \Delta_j(f).$$

One assumes $a(x) \in L^\infty \cap C^r$ and one reasons modulo the r-regularizing operators, that is, those which transform H^s into H^{s+r} for all $s \in \mathbf{R}$. The error terms are analyzed by the following lemma.

Lemma 6 *Let $f_j(x)$, $j \in \mathbf{Z}$, be functions in $L^2(\mathbf{R}^n)$ whose Fourier transforms $\widehat{f_j}$ satisfy, for two constants $R_2 > R_1 > 0$,*

$$\widehat{f_j}(\xi) = 0 \quad \textit{if} \quad |\xi| \le R_1 2^j \quad \textit{or if} \quad |\xi| \ge R_2 2^j. \tag{8.1}$$

Then for every $s \in \mathbf{R}$ there exists a constant $C = C(R_1, R_2, s, n)$ such that

$$\left\| \sum_{-\infty}^{\infty} f_j(x) \right\|_{H^s} \le C \left(\sum_{-\infty}^{\infty} \|f_j\|_2^2 \,(1 + 4^j)^s \right)^{1/2}. \tag{8.2}$$

Using this lemma one begins by replacing $S_{j-3}(a)$ by $S_{j-10}(a)$. Following that, one calculates $\pi(a, \psi_Q^{(\epsilon)})$ when Q is defined by $2^m - k \in [0,1]^n$. Then $\Delta_j(\psi_Q^{(\epsilon)}) = 0$ unless $|m-j| \le 2$.

One defines (by linearity) the operator R_a by

$$R_a(\psi_Q^{(\epsilon)}) = \pi(a, \psi_Q^{(\epsilon)}) - S_{m-10}(a)\, \psi_Q^{(\epsilon)}$$

and, using our lemma and the characterization by series of wavelets of the Sobolev spaces, one shows that R_a is r-regularizing.

One can then state

Theorem 7 *If $r > 0$ and $a(x) \in C^r \cap L^\infty$, the paraproduct $\pi(a, f)$ is defined by linearity by*

$$\pi(a, \psi_Q^{(\epsilon)}) = S_Q(a)\, \psi_Q^{(\epsilon)}$$

where, by an abuse of language, $S_Q(a) = S_{j-10}(a)$ when

$$Q = \{x \in \mathbf{R}^n; 2^j x - k \in [0,1]^n\};$$

if furthermore $0 < r < 1$, one has, modulo an r-regularizing operator,

$$\pi(a, \psi_Q^{(\epsilon)}) = a(2^{-j}k)\, \psi_Q^{(\epsilon)}.$$

In other words, the paraproduct is diagonalized in the Hilbert basis of the wavelets.

9. Analysis of the Earlier Works on the Subject

The "continuous version" of the transformation into wavelets has a long history going back to the works of A. Calderón and his collaborators. The discrete version appears in 1980 when L. Carleson proves that the Haar system, smoothed correctly, is an unconditional basis of the space $H^1(\mathbf{T})$. Naturally Carleson's wavelets do not form a Hilbert basis, but there exists an associated biorthogonal system.

Another process was introduced by Frazier and Jawerth [6] and, independently, in [4] and [9].

The question is to exhibit ψ in $\mathcal{S}(\mathbf{R}^n)$ so that one has, for everyfunction$f \in L^2(\mathbf{R}^n; dx)$,

$$f(x) = \sum_{Q \in \mathcal{Q}} \langle f, \psi_Q \rangle \, \psi_Q(x) \tag{9.1}$$

without uniqueness.

It is sufficient for this that the Fourier transform of ψ be supported by the cube $-\pi \leq x_j \leq \pi$, $1 \leq j \leq n$, and that one has, for all $\xi \neq 0$,

$$1 = \sum_{-\infty}^{\infty} |\widehat{\psi}(2^j \xi)|^2. \tag{9.2}$$

If one has uniqueness in (9.1), then the functions $\psi_Q(x)$ are automatically orthogonal but this is excluded by the condition on the support of $\widehat{\psi}$.

References

[1] J. M. Bony, *Calcul symbolique et propagation des singularités pour les équations aux dérivées partielles non linéaires*, Ann. Sci. École Norm. Sup. (4) **14** (1981), 209–246.

[2] A. P. Calderón, *Intermediate spaces and interpolation, the complex method*, Studia Math. **24** (1964), 113–190.

[3] L. Carleson, *An explicit unconditional basis in H^1*, Bull. Sci. Math. (2) **104** (1980), 405–416.

[4] I. Daubechies, A. Grossmann, and Y. Meyer, *Painless nonorthogonal expansions*, J. Math. Phys. **27** (1986), 1271–1283.

[5] G. David and J. L. Journé, *A boundedness criterion for generalized Calderón-Zygmund operators*, Ann. of Math. (2) **120** (1984), 371–397.

[6] M. Frazier and B. Jawerth, *Decomposition of Besov spaces*, Indiana Univ. Math. J. **34** (1985), 777–799.

[7] P. Goupillaud, A. Grossmann, and J. Morlet, *Cycle octave and related transforms in seismic signal analysis*, Geoexploration **23** (1984/1985), 85–102.

[8] A. Grossmann and J. Morlet, *Decomposition of Hardy functions into square integrable wavelets of constant shape*, SIAM J. Math. Anal. **15** (1984), 723–736.

[9] Y. Meyer, *La transformation en ondelettes et les nouveaux paraproduits*, (à paraître aus Actes du Colloque d'Analyse non linéaire du CEREMADE, Univ. de Paris-Dauphine).

[10] P. Wojtaszczyk, *The Franklin system is an unconditional basis in H^1*, Ark. Mat. **20** (1982), 293–300.

This work was executed in the framework of "R.C.P. ondelettes" of the French National Scientific Research Center M727.1285, December 1985.

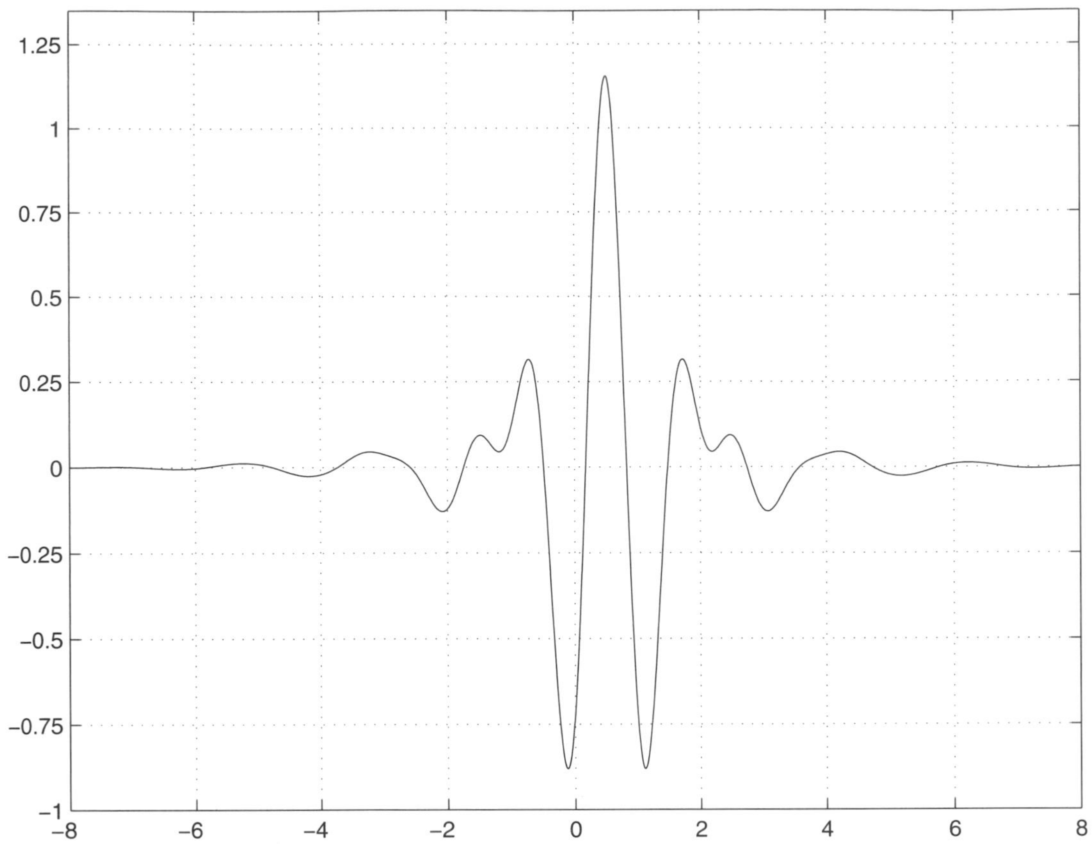

Figure 1. Wavelet

Commun. Math. Phys. 110, 601–615 (1987)

Communications in Mathematical Physics
© Springer-Verlag 1987

A Block Spin Construction of Ondelettes[1]. Part I: Lemarié Functions

Guy Battle[2]

Mathematics Department, Cornell University, Ithaca, New York 14853, USA

Abstract. Using block spin assignments, we construct an L^2-orthonormal basis consisting of dyadic scalings and translates of just a finite number of functions. These functions have exponential localization, and for even values of a construction parameter M one can make them class C^{M-1} with vanishing moments up to order M inclusive. Such a basis has an important application to phase cell cluster expansions in quantum field theory.

1. Introduction

Quite recently Y. Meyer et al. [1, 2, 3] have constructed very useful bases of ondelettes (wavelets) to solve certain problems in functional analysis. These new functions are now expected to have applications to several areas of physics. They have already had an impact on constructive quantum field theory [4, 5, 6].

A basis of ondelettes is defined to be an orthonormal basis—say for $L^2(\mathbb{R}^d)$—whose functions are dyadic scalings (from $2^{-\infty}$ to 2^{∞}) and translates of just a finite number of them. The most familiar example is the standard basis of Haar functions on $\mathbb{R}^d$. Indeed, Battle and Federbush [4, 5] used a polynomial generalization of the Haar basis to develop a phase cell cluster expansion a few years ago. This basis has the following useful properties:

(a) The basis consists of all dyadic scalings and translates of a finite collection $\psi_1, \ldots, \psi_n$ of functions.

(b) ψ_i is a piecewise polynomial supported on the cube associated with it. Thus we have sharp localization but poor regularity.

(c) For all multi-indices α for which $|\alpha|$ is less than or equal to a certain construction parameter,

$$\int \psi_i(x)\, x^\alpha\, dx = 0. \tag{1.1}$$

Equivalently, $\hat{\psi}_i(p)$ vanishes to some finite order at $p = 0$.

---

1 Supported in part by the National Science Foundation under Grant No. DMS 8603795 and by the Mathematical Sciences Institute

2 On leave from the Mathematics Department, Texas A & M University, College Station, Texas 77843 USA

Remark. The construction parameter is the maximum degree of the polynomials used, and it determines the size of the collection $\psi_1, \ldots, \psi_n$. For the standard Haar basis the construction parameter is zero and $n = 2^d - 1$.

Although these functions were instrumental in introducing important new ideas in constructive quantum field theory, their lack of regularity created serious technical inconvenience. Battle and Federbush developed a very natural expansion for the hierarchical version [5] of the celebrated ϕ_3^4 model, but it took the more complicated expansion of Williamson [7] to control the real model with these ondelettes.

This past summer Meyer and his co-workers announced the existence of a basis of ondelettes with the following remarkable properties:

(a) The basis consists of all dyadic scalings and translates of functions $\psi_1, \ldots, \psi_{2^d-1}$.

(b) ψ_i is a Schwartz function. (In fact, $\hat{\psi}_i$ is a compactly supported C^∞ function!)

(c) (1.1) holds for *all* multi-indices α. Equivalently, $\hat{\psi}_i(p)$ vanishes to infinite order at $p = 0$.

In addition, a co-worker of Meyer, P. Lemarié, found a basis [3] of ondelettes whose properties complement the preceding properties in the following interesting way:

(a) The basis consists of all dyadic scalings and translates of functions $\psi_1, \ldots, \psi_{2^d-1}$.

(b) ψ_i is class C^N (where N can be made arbitrarily large) and ψ_i has *exponential* localization. (Although the preceding ondelettes fall off faster than any negative power of distance, they cannot have exponential decay, because the vanishing of *all* moments implies that their Fourier transforms cannot be analytic at $p = 0$.)

(c) (1.1) holds for $|\alpha| \leqq N + 1$.

The functions of Lemarié are better suited for constructive field theory because exponential localization is a more useful property than smoothness. In the Battle–Federbush expansion one needs no more regularity for the ondelettes than class C^1. This will be discussed in greater detail in Part II.

In this paper we introduce a machine that actually *finds* the ondelettes of Lemarié. It is a "block spin" construction consisting of two natural stages. The first stage orthogonalizes levels and is carried out in Sects. 2 & 3; the second stage is an orthogonalization (on each scale) preserving translation properties and is carried out in Sect. 4. In Sect. 5 we verify the properties of the functions, and in Sect. 6 we verify completeness.

The orthogonalization of levels is based on a very familiar idea. Let $f_1, \ldots, f_n$ be arbitrary L^2 functions that are not necessarily orthogonal and minimize $\| \varphi \|_2$ with respect to the constraints

$$(\varphi, f_i) = 0, \quad i = 1, \ldots, n. \tag{1.2}$$

Then the solution φ_0 is orthogonal to any f for which some (f, f_i) has a non-zero value. This is why our "block spin assignment" rules schematically look like

$$oooo + - oooooo$$

The solution to this constraint is guaranteed to be orthogonal to the solution of any non-zero block spin assignment we could possibly make at any larger scale, because *this* assignment determines zero block spin values for all larger scales. The essence of this idea goes back to Kupiainen and Gawedzki [8], but we have to introduce a clever modification to obtain class C^N solutions.

The translation invariant orthogonalization is classical. There is a very natural candidate for an orthonormal basis for each subspace, and our task is to show that it makes sense and has all of the desired properties in our case.

The Lemarié functions themselves are given in Sect. 4 by Eq. (4.7). They are quite explicit, and the fact that our construction reproduces them is an indication of how natural they are. The construction guarantees orthogonality of levels, but since the functions *are* explicitly given, one may wish to see a *direct* verification, so we give here such a calculation. We concentrate on the case $d=1$, where at the unit scale level the functions are even translates of the function ϕ given by

$$\hat{\phi}(p) = c\,w(p)^{-1/2}(1-e^{ip})^{M+1}\frac{\hat{\chi}(p)^{M+1}}{\sum_{n=-\infty}^{\infty}|\hat{\chi}(p+2\pi n)|^{2M+2}}, \tag{1.2}$$

where the function w is given by (4.6) in Sect. 4, χ is the characteristic function of $[0,1]$, and M is an even integer. If we let ϕ_m denote the $2m$-translate of ϕ, then

$$\int \phi_m(p)\overline{\hat{\phi}_{m'}(2^r p)}dp = c\int \frac{e^{i2mp - i2^{r+1}m'p}}{\sqrt{w(p)}\sqrt{w(2^r p)}}(1-e^{ip})^{M+1}(1-e^{-i2^r p})^{M+1} \cdot \frac{\hat{\chi}(p)^{M+1}\overline{\hat{\chi}(2^r p)^{M+1}}}{[\sum_n|\hat{\chi}(p+2\pi n)|^{2M+2}][\sum_n|\hat{\chi}(2^r p+2\pi n)|^{2M+2}]}dp \tag{1.3}$$

for a given positive integer r. Using the identities

$$\hat{\chi}(2^r p) = (1+e^{i2^{r-1}p})\cdots(1+e^{i2p})(1+e^{ip})\hat{\chi}(p), \tag{1.4}$$

$$(1+e^{i2^{r-1}p})\cdots(1+e^{i2p})(1+e^{ip})(1-e^{ip}) = e^{-i(2^r-1)p}(1-e^{i2^r p}), \tag{1.5}$$

we get

$$\sum_n \int_0^{2\pi} \frac{e^{i2mp - i2^{r+1}m'p}e^{-i(2^r-1)(M+1)p}|1-e^{i2^r p}|^{2M+2}}{\sqrt{w(p)}\sqrt{w(2^r p)}\sum_n|\hat{\chi}(2^r p+2\pi n)|^{2M+2}}dp \tag{1.6}$$

because $\sum_n|\hat{\chi}(p+2\pi n)|^{2M+2}$ cancels out. On the other hand, $w(p)$ *has period* π. So does everything else remaining in the integrand, except

$$e^{-i(2^r-1)(M+1)p}. \tag{1.7}$$

Since M is even, the integral is zero.

We close the Introduction with the claim that we have a basis of exponentially localized ondelettes that are orthogonal with respect to the massless Sobelev norm $\|\vec{\nabla}\varphi\|_2$. This result is non-trivial, because the $|\vec{\nabla}|^{-1}$ potential of the L^2 ondelettes constructed here cannot have exponential decay. We describe the new basis in Part II.

2. Block-Spin Constraints

Consider the set $\mathscr{R}_s$ of mutually disjoint integer-valued translates Γ of the rectangular solid

$$R_s = \{x \in \mathbb{R}^d | 0 \leqq x_\mu \leqq 2 \text{ for } 1 \leqq \mu \leqq s-1 \text{ and } 0 \leqq x_\mu \leqq 1 \text{ for } s \leqq \mu \leqq d\}.$$

obviously this involves even-integer translates in each μ-direction for which $1 \leqq \mu \leqq s-1$. Let R_s' be the unit translate of R_s in the positive s-direction and define the block spin assignment σ_s as follows:

$$\sigma_s(R_s) = 1, \quad \sigma_s(R_s') = -1, \quad \sigma_s(\Gamma) = 0, \quad \Gamma \neq R_s, R_s'.$$

The first step in our construction is to find the function φ_s that minimizes $\|\varphi\|_2^2$ with respect to constraints that are easiest to understand if we initially write them in an ill-posed form:

$$\prod_{\mu=1}^{d} \left(\int dp_\mu\right)\left(\prod_{\mu=1}^{d} p_\mu^{-1}\right)^M \hat{\varphi}(p)\overline{\hat{\chi}_\Gamma(p)} = \sigma_s(\Gamma), \tag{2.1}$$

where χ_Γ is the characteristic function of Γ. First, it is clear that if $s' > s$, then (2.1) implies

$$\prod_{\mu=1}^{d} \left(\int dp_\mu\right)\left(\prod_{\mu=1}^{d} p_\mu^{-1}\right)^M \hat{\varphi}(p)\overline{\hat{\chi}_\Gamma(p)} = 0, \quad \Gamma \in \mathscr{R}_{s'}. \tag{2.2}$$

Second, it is equally obvious that *if* $r > 0$, *then (2.1) implies*

$$\prod_{\mu=1}^{d} \left(\int dp_\mu\right)\left(\prod_{\mu=1}^{d} p_\mu^{-1}\right)^M \hat{\varphi}(p)\overline{\hat{\chi}_\Gamma(p)} = 0, \quad \Gamma \in 2^r\mathscr{R}_{s'}, \quad 1 \leqq s' \leqq d. \tag{2.3}$$

Third, it is easy to see that (2.1) is $(2m_1, \ldots, 2m_s, m_{s+1}, \ldots, m_d)$-translation-covariant—i.e., (2.1) is equivalent to

$$\prod_{\mu=1}^{d} \left(\int dp_\mu\right)\left(\prod_{\mu=1}^{d} p_\mu^{-1}\right)^M \hat{\varphi}_m(p)\overline{\hat{\chi}_\Gamma(p)} = \sigma_{s,m}(\Gamma), \tag{2.4}$$

where φ_m is the $(2m_1, \ldots, 2m_s, m_{s+1}, \ldots, m_d)$-translate of φ and $\sigma_{s,m}$ is the same translate of the block spin assignment σ_s. Finally, note that the scaling of the block spin assignment can be chosen such that (2.1) is scale-covariant. If

$$\sigma_s^{(r)}(\Gamma) \equiv 2^{r(M+1/2)d}\sigma_s(2^{-r}\Gamma), \quad \Gamma \in 2^r\mathscr{R}_s, \tag{2.5}$$

then (2.1) is equivalent to

$$\prod_{\mu=1}^{d} \left(\int dp_\mu\right)\left(\prod_{\mu=1}^{d} p_\mu^{-1}\right)^M \hat{\varphi}^{(r)}(p)\overline{\hat{\chi}_\Gamma(p)} = \sigma_s^{(r)}(\Gamma), \quad \Gamma \in 2^r\mathscr{R}_s, \tag{2.6}$$

where

$$\varphi^{(r)}(x) = 2^{-rd/2}\varphi(2^{-r}x). \tag{2.7}$$

Now the functions χ_Γ generate a hierarchy of L^2-subspaces that is ordered by containment. In particular, the block-spin constraints at a given level determine the block-spin constraints at all higher levels. It is obvious from (2.2) and (2.3) that the ill-posed constraint (2.1) yields zero constraints for all higher levels. This is a familiar kind of set-up, which has been used by Federbush [9] in the gauge field setting. The idea goes back to Gawedzki and Kupiainen [8], and the point is that the minimization of the Hilbert space norm with respect to such constraints creates a complete set of functions where any two that live on different levels are orthogonal with respect to that norm.

This constrained minimization of the L^2 norm is in ill-posed form because *the linear functionals defining the constraints are unbounded with respect to the L^2 norm.* Our motivation for introducing negative powers of the p_μ is to create orthogonal levels of *smooth* functions for our scale hierarchy, but for our scaling property we also need the homogeneity in the p_μ, so we have an infrared difficulty to cure. The orthogonality of scalings cannot be guaranteed by our constrained minimization unless the constraints can be expressed in terms of *bounded* linear functionals.

However, we can actually form bounded linear functionals from *finite linear combinations* of the unbounded ones. Since multiplication by e^{ip_μ} in momentum space corresponds to translation by a unit in the μ-direction, (2.1) implies

$$\prod_{\mu=1}^{d} \left(\int dp_\mu\right)\left(\prod_{\mu=1}^{d} p_\mu^{-1}\right)^M \hat{\phi}(p) \overline{\prod_{\mu=1}^{s-1} (1-e^{i2p_\mu})^M \prod_{\mu=s}^{d} (1-e^{ip_\mu})^M \hat{\chi}_\Gamma(p)}$$

$$= \Big(\prod_{\mu \neq s} P_M^-(-m_\mu)\Big)\big(P_M^-(-m_s) - P_M^-(1-m_s)\big), \tag{2.8}$$

where Γ is the $(2m_1, \ldots, 2m_{s-1}, m_s, \ldots, m_d)$-translate of R_s and

$$P_M^\pm(n) = (\pm 1)^n \binom{M}{n}, \quad 0 \leqq n \leqq M,$$

$P_M^\pm(n) = 0$, otherwise.

The point is that

$$(1-e^{i\nu p_\mu})^M = \sum_n P_M^-(n) e^{in\nu p_\mu}, \tag{2.9}$$

and that our linear functionals are now bounded because $1-e^{i\nu p_\mu}$ cancels the infrared singularity of p_μ^{-1}.

Now (2.8) alone implies that all such integrals are zero for higher levels. The algebra is no longer a triviality, but it is interesting to check. The relevant identities are

$$1-e^{i2p_\mu} = (1+e^{ip_\mu})(1-e^{ip_\mu}),$$

$$2(1-e^{i2p_\mu})^M \hat{\chi}(2p_\mu) = (1+e^{ip_\mu})^{M+1}(1-e^{ip_\mu})^M \hat{\chi}(p_\mu),$$

$$\sum_n P_{M+1}^+(n)\big(P_M^-(-n) - P_M^-(1-n)\big) = 0,$$

where χ denotes the characteristic function of $[0,1]$.

Since

$$\hat{\chi}(p_\mu) = i\frac{1 - e^{ip_\mu}}{p_\mu}, \tag{2.10}$$

we will make the replacement

$$p_\mu^{-M}(1 - e^{ip_\mu})^M \hat{\chi}(p_\mu) = i^{-M}\hat{\chi}(p_\mu)^{M+1} \tag{2.11}$$

from this point on.

It is clear from our scaling properties that we may restrict our attention to the sth sub-level of the unit scale level for the remainder of our discussion.

3. Construction of the sth Sub-Level

The next phase in our construction is to derive the formula for φ_s—i.e., to solve the constrained minimization problem posed in the last section. We first minimize

$$\|\varphi\|_2^2 + \alpha^2 \sum_m \Bigg[\prod_{\mu=1}^{d} (\textstyle\int dp_\mu)\hat{\varphi}(p) \prod_{\mu=1}^{s-1} \overline{\hat{\chi}(2p_\mu)^{M+1}} \prod_{\mu=s}^{d} \overline{\hat{\chi}(p_\mu)^{M+1}}$$
$$\cdot \exp\left(- i2 \sum_{\mu=1}^{s-1} m_\mu p_\mu - i \sum_{\mu=s}^{d} m_\mu p_\mu \right) - \Big(\prod_{\mu \neq s} P_M^-(-m_\mu)\Big)\big(P_M^-(-m_s) - P_M^-(1 - m_s)\big) \Bigg]^2, \tag{3.1}$$

and then take the limit of our α-dependent solution as $\alpha \to \infty$. Since (3.1) is quadratic in φ, we need only to collect terms and complete the square for the inner product to see that (3.1) has the form

$$\|(1 + \alpha^2 \sum_m F_m^s)^{1/2}\varphi - \alpha^2 \sum_l \beta_l^s (1 + \alpha^2 \sum_m F_m^s)^{-1/2} f_l^s\|_2^2, \tag{3.2}$$

plus constant terms, where f_l^s is given by

$$\widehat{f_l^s}(p) = \prod_{\mu=1}^{s-1} \hat{\chi}(2p_\mu)^{M+1} \prod_{\mu=s}^{d} \hat{\chi}(p_\mu)^{M+1} \exp\left(i2 \sum_{\mu=1}^{s-1} l_\mu p_\mu + i \sum_{\mu=s}^{d} l_\mu p_\mu \right). \tag{3.3}$$

F_m^s is the un-normalized orthogonal projection $(\cdot, f_m^s) f_m^s$, and

$$\beta_l^s = \Big(\prod_{\mu \neq s} P_M^-(-l_\mu)\Big)\big(P_M^-(-l_s) - P_M^-(1 - l_s)\big). \tag{3.4}$$

The operator calculus makes sense because $\sum_m F_m^s \geqq 0$, but of course the F_m^s are not mutually orthogonal. Thus our problem reduces to solving the linear equation

$$\varphi + \alpha^2 \sum_m F_m^s \varphi = \alpha^2 \sum_l \beta_l^s f_l^s. \tag{3.5}$$

Now for *small* α we have the Neumann series solution

$$\varphi = \alpha^2 \sum_l \beta_l^s \sum_{n=0}^{\infty} (-\alpha^2)^n \Big(\sum_m F_m^s\Big)^n f_l^s, \tag{3.6}$$

which can be transformed in much the same manner that the corresponding Neumann series was transformed in [9]. The point is that (with $\vec{\lambda}_0 = l$)

$$\begin{aligned}(\sum_m F_m)^n f_l^s &= \sum_{\vec{\lambda}_1,\dots,\vec{\lambda}_n}\left[\prod_{i=1}^{n}(f^s_{\vec{\lambda}_{i-1}}, f^s_{\vec{\lambda}_i})\right] f^s_{\vec{\lambda}_n} \\ &= \sum_{\vec{\lambda}_1,\dots,\vec{\lambda}_n}\left[\prod_{i=1}^{n}(f^s_{\vec{\lambda}_{i-1}-\vec{\lambda}_i}, f^s_0)\right] f^s_{\vec{\lambda}_n}, \\ &= \prod_{\mu=1}^{d}\left(\int_0^{2\pi} dk_\mu\right) e^{-il\cdot k} g(k)^n \sum_m e^{im\cdot k} f^s_m, \end{aligned} \tag{3.7}$$

where

$$g(k) = \sum_m (f^s_m, f^s_0) e^{im\cdot k}. \tag{3.8}$$

Remark. $\sum_m e^{im\cdot k} f^s_m$ is a finite sum on any compact set and $g(k)$ is just a trigonometric polynomial, so there are no convergence problems here. Also, $g(k)$ does not depend on s as we see below.

(3.6) becomes

$$\varphi = \alpha^2 \sum_l \beta^s_l \prod_{\mu=1}^{d}\left(\int_0^{2\pi} dk_\mu\right) \frac{e^{-il\cdot k}}{1+\alpha^2 g(k)} \sum_m e^{im\cdot k} f^s_m, \tag{3.9}$$

and by analytic continuation in α, this formula holds for large real α as well. ($g(k) \geqq 0$ because (f^s_m, f^s_0) is of positive type.) Thus our solution φ_s of the constrained minimization problem is given by the limit of this expression as $\alpha \to \infty$. Now the Poisson summation formula enables us to write $g(k)$ in a more useful form:

$$g(k) = \prod_{\mu=1}^{d}\left(\sum_{n=-\infty}^{\infty} |\hat{\chi}(k_\mu + 2\pi n)|^{2M+2}\right). \tag{3.10}$$

Thus $g(k)$ does not vanish anywhere, and so our solution

$$\varphi_s = \sum_l \beta^s_l \prod_{\mu=1}^{d}\left(\int_0^{2\pi} dk_\mu\right) e^{-il\cdot k} g(k)^{-1} \sum_m e^{im\cdot k} f^s_m \tag{3.11}$$

is well-defined and well-behaved. Indeed, φ_s is an L^2 function because $g(k)^{-1}$ is bounded.

The expression for $\hat{\varphi}_s$ is quite explicit. If we set

$$a_m = \prod_{\mu=1}^{d}\left(\int_0^{2\pi} dk_\mu\right) e^{-im\cdot k} g(k)^{-1}, \tag{3.12}$$

then $g(k)^{-1} = \sum_m a_m e^{im\cdot k}$ and

$$\varphi_s = \sum_m \beta^s_l \sum_m a_{l-m} f^s_m. \tag{3.13}$$

Hence

$$\hat{\varphi}_s(p) = \sum_l \beta_l^s \sum_m a_{l-m} \exp\left(i2 \sum_{\mu=1}^{s-1} m_\mu p_\mu + i \sum_{\mu=s}^{d} m_\mu p_\mu \right)$$

$$\cdot \prod_{\mu=1}^{s-1} \hat{\chi}(2p_\mu)^{M+1} \prod_{\mu=s}^{d} \hat{\chi}(p_\mu)^{M+1}$$

$$= \sum_l \beta_l^s \exp\left(i2 \sum_{\mu=1}^{s-1} l_\mu p_\mu + i \sum_{\mu=s}^{d} l_\mu p_\mu \right) g(-2p_1, \ldots,$$

$$-2p_{s-1}, -p_s, \ldots, -p_d)^{-1} \prod_{\mu=1}^{s-1} \hat{\chi}(2p_\mu)^{M+1} \prod_{\mu=s}^{d} \hat{\chi}(p_\mu)^{M+1}$$

$$= \sum_l \beta_l^s \exp\left(i2 \sum_{\mu=1}^{s-1} l_\mu p_\mu + i \sum_{\mu=s}^{d} l_\mu p_\mu \right)$$

$$\cdot \prod_{\mu=1}^{s-1} \frac{\hat{\chi}(2p_\mu)^{M+1}}{\sum_{n=-\infty}^{\infty} |\hat{\chi}(2p_\mu + 2\pi n)|^{2M+2}} \prod_{\mu=s}^{d} \frac{\hat{\chi}(p_\mu)^{M+1}}{\sum_{n=-\infty}^{\infty} |\hat{\chi}(p_\mu + 2\pi n)|^{2M+2}}. \tag{3.14}$$

Combining this with (3.4), we obtain

$$\hat{\varphi}_s(p) = \prod_{\mu=1}^{s-1} \frac{(1 - e^{-i2p_\mu})^M \hat{\chi}(2p_\mu)^{M+1}}{\sum_{n=-\infty}^{\infty} |\hat{\chi}(2p_\mu + 2\pi n)|^{2M+2}} \cdot \frac{(1 - e^{-ip_s})^{M+1} \hat{\chi}(p_s)^{M+1}}{\sum_{n=-\infty}^{\infty} |\hat{\chi}(p_s + 2\pi n)|^{2M+2}}$$

$$\cdot \prod_{u=s+1}^{d} \frac{(1 - e^{-ip_\mu})^M \hat{\chi}(p_\mu)^{M+1}}{\sum_{n=-\infty}^{\infty} |\hat{\chi}(p_\mu + 2\pi n)|^{2M+2}}. \tag{3.15}$$

Remark. The manipulation above is justified by the fact that $g(k)^{-1}$ is smooth—i.e., that $\{a_m\}$ has rapid fall-off.

4. Translation-Invariant Orthogonalization

Having decomposed $L^2(\mathbb{R}^d)$ into subspaces corresponding to sublevels of scales, we now seek a function in each subspace with the property that all translates associated with the subspace yield an orthonormal basis for the subspace. This is the final step in our construction. As before, we consider the sth sublevel for the unit scale without loss. Let $\mathcal{H}_s$ be the subspace that we have constructed for this sublevel. $\mathcal{H}_s$ is spanned by the set of functions $\varphi_{s,m}$ given by

$$\hat{\varphi}_{s,m}(p) = \exp\left(i2 \sum_{\mu=1}^{s} m_\mu p_\mu + i \sum_{\mu=s+1}^{d} m_\mu p_\mu \right) \hat{\varphi}_s(p). \tag{4.1}$$

We should first point out that there is a canonical solution of such a problem—under the right conditions. The natural candidate ϕ_s is given by

$$\hat{\phi}_s(p) = h_s(p)^{-1/2} \hat{\varphi}_s(p), \tag{4.2}$$

$$h_s(p) = \sum_l |\hat{\varphi}_s(p_1 + l_1\pi, \ldots, p_s + l_s\pi, p_{s+1} + 2\pi l_{s+1}, \ldots, p_d + 2\pi l_d)|^2 \tag{4.3}$$

and the translates $\phi_{s,m}$ are given by

$$\begin{aligned}\hat{\phi}_{s,m}(p) &= \exp\left(i2 \sum_{\mu=1}^{s} m_\mu p_\mu + i \sum_{\mu=s+1}^{d} m_\mu p_\mu \right) \hat{\phi}_s(p) \\ &= h_s(p)^{-1/2} \phi_{s,m}(p). \end{aligned} \tag{4.4}$$

There are a number of things to be verified, however.

First we insert (3.15) in (4.3) to obtain a more explicit expression for h_s. We have

$$\begin{aligned} h_s(p) = &\prod_{\mu=1}^{s-1} \frac{|1 - e^{i2p_\mu}|^{2M}}{\sum_{n=-\infty}^{\infty} |\hat{\chi}(2p_\mu + 2\pi n)|^{2M+2}} w(p_s) \\ &\cdot \prod_{\mu=s+1}^{d} \frac{|1 - e^{ip_\mu}|^{2M}}{\sum_{n=-\infty}^{\infty} |\hat{\chi}(p_\mu + 2\pi n)|^{2M+2}}, \end{aligned} \tag{4.5}$$

$$w(t) = \frac{|1 - e^{it}|^{2M+2}}{\sum_{n=-\infty}^{\infty} |\hat{\chi}(t + 2\pi n)|^{2M+2}} + \frac{|1 + e^{it}|^{M+2}}{\sum_{n=-\infty}^{\infty} |\hat{\chi}(t + 2\pi n + \pi)|^{2M+2}}. \tag{4.6}$$

Thus (since we are assuming M to be even)

$$\begin{aligned} \hat{\phi}_s(p) = &\prod_{\mu=1}^{s-1} \frac{(-1)^{(1/2)M} e^{iMp_\mu} \hat{\chi}(2p_\mu)^{M+1}}{\left(\sum_{n=-\infty}^{\infty} |\hat{\chi}(2p_\mu + 2\pi n)|^{2M+2} \right)^{1/2}} \\ &\cdot \frac{(1 - e^{ip_s})^{M+1} \hat{\chi}(p_s)^{M+1}}{w(p_s)^{1/2} \sum_{n=-\infty}^{\infty} |\hat{\chi}(p_s + 2\pi n)|^{2M+2}} \prod_{\mu=s+1}^{d} \frac{(-1)^{(1/2)M} e^{i(1/2)Mp_\mu} \hat{\chi}(p_\mu)^{M+1}}{\left(\sum_{n=-\infty}^{\infty} |\hat{\chi}(p_\mu + 2\pi n)|^{2M+2} \right)^{1/2}}. \end{aligned} \tag{4.7}$$

Now the denominators in (4.6) are periodic functions that do not vanish anywhere, so $w(t)$ is a periodic C^∞ function. Furthermore $w(t)$ is a positive function that cannot vanish anywhere, so all of the denominators in (4.7) are well-defined and do not vanish anywhere. Indeed, by continuity and periodicity they are bounded below by positive constants, and so we have established:

Theorem 4.1. *ϕ_s is an L^2 function.*

Now consider the inner product $(\phi_{s,m}, \phi_{s,m'})$ with $m' \neq m$. Returning to (4.4) we see that

$$\begin{aligned} (\phi_{s,m}, \phi_{s,m'}) &= \prod_{\mu=1}^{d} (\textstyle\int dp_\mu) h_s(p)^{-1} \phi_{s,m}(p) \overline{\phi_{s,m'}(p)} \\ &= \prod_{\mu=1}^{d} (\textstyle\int dp_\mu) h_s(p)^{-1} |\hat{\phi}_s(p)|^2 \\ &\quad \cdot \exp\left(i2 \sum_{\mu=1}^{s} (m_\mu - m'_\mu) p_\mu + i \sum_{\mu=s+1}^{d} (m_\mu - m'_\mu) p_\mu \right). \end{aligned} \tag{4.8}$$

We decompose the integration as follows:

$$\int dp_\mu = \sum_{n=-\infty}^{\infty} \int_{\pi n}^{\pi n+\pi} dp_\mu, \quad 1 \leqq \mu \leqq s,$$

$$\int dp_\mu = \sum_{n=-\infty}^{\infty} \int_{2\pi n}^{2\pi n+2\pi} dp_\mu, \quad s+1 \leqq \mu \leqq d.$$

Then for the nth term we make the change of variable $p'_\mu = p_\mu - \pi n$ or $p'_\mu = p_\mu - 2\pi n$ as the case may be. Both h_s and the exponential are unaffected by these changes, so by (4.3) our inner product reduces to

$$\prod_{\mu=1}^{s}\left(\int_0^{\pi} dp'_\mu\right)\prod_{\mu=s+1}^{d}\left(\int_0^{2\pi} dp'_\mu\right)\exp\left(i2\sum_{\mu=1}^{s}(m_\mu - m'_\mu)p'_\mu + i\sum_{\mu=s+1}^{d}(m_\mu - m'_\mu)p'_\mu\right)$$

which is obviously zero for $m' \neq m$. Thus we have proven:

Theorem 4.2. $\{\phi_{s,m}\}$ *is an orthogonal set.*

We shall not bother to normalize $\phi_{s,m}$. We observe that the normalization constant is independent of m, so we need say no more about it.

Our next concern is to verify:

Theorem 4.3. $\phi_{s,m} \in \mathscr{H}_s$.

Proof. Let $\varphi^{(N)}_{s,m}$ be defined inductively by

$$\varphi^{(N)}_{s,m'} = \sum_m \left[\prod_{\mu=1}^{s}\left(\int_0^{\pi} dp'_\mu\right)\prod_{\mu=s+1}^{d}\left(\int_0^{2\pi} dp'_\mu\right)\exp\left(i2\sum_{\mu=1}^{s}(m_\mu - m'_\mu)p'_\mu \right.\right.$$
$$\left.\left. + i\sum_{\mu=s+1}^{d}(m_\mu - m'_\mu)p'_\mu\right)h_s(p')^{(-1/6)M}\right]\varphi^{(N-1)}_{s,m} \tag{4.9}$$

with $\varphi^{(0)}_{s,m} = \varphi_{s,m}$. The point of this definition is that we have to worry about the zeros of h_s. Let $b_{m'-m}$ be the coefficient defined by this negative fractional power of h_s. Then

$$\varphi^{(N)}_{s,m'} = \sum_m b_{m'-m}\varphi^{(N-1)}_{s,m}, \tag{4.10}$$

$$\widehat{\varphi^{(N-1)}_{s,m}}(p) = \exp\left(i2\sum_{\mu=1}^{s} m_\mu p_\mu + i\sum_{\mu=s+1}^{d} m_\mu p_\mu\right)\widehat{\varphi^{(N-1)}_s}(p), \tag{4.11}$$

$$h_s(p)^{(-1/6)M} = \sum_m b_m \exp\left(i2\sum_{\mu=1}^{s} m_\mu p_\mu + i\sum_{\mu=s+1}^{d} m_\mu p_\mu\right), \tag{4.12}$$

so we have

$$\widehat{\varphi^{(N)}_{s,m'}}(p) = h_s(p)^{-(1/6)M}\,\widehat{\varphi^{(N-1)}_{s,m'}}(p). \tag{4.13}$$

Thus (with $\vec{\lambda}_0 = m$)

$$\phi_{s,m} = \varphi^{(3M)}_{s,m} = \sum_{\vec{\lambda}_1,\ldots,\vec{\lambda}_{3M}} \prod_{i=1}^{3M} b_{\vec{\lambda}_{i-1}-\vec{\lambda}_i}\varphi_{s,\vec{\lambda}_{3M}}, \tag{4.14}$$

so $\phi_{s,m} \in \mathscr{H}_s$ provided this multiple sum converges in L^2. Now, in particular, $\{\varphi^{(N)}_{s,m}(x)\}$ is uniformly summable on compact sets for $N \leqq 3M$. On the other hand

$\{b_m\}$ *is square summable because* $h(p)^{-(1/6)M}$ *is square integrable on* $[0,\pi]^s \times [0,2\pi]^{d-s}$. Hence

$$\sum_m b_{m'-m}\varphi_{s,m}^{(N-1)}(x) \tag{4.15}$$

converges uniformly on compact sets in the m'-square-summable sense, and so by translation decomposition of integrals it converges in the L^2 sense. Thus $\varphi_{s,m}^{(N)} \in \mathscr{H}_s$ for all m provided $\varphi_{s,m}^{(N-1)} \in \mathscr{H}_s$ for all m. The induction carries us up to $N = 3M$, so $\phi_{s,m} \in \mathscr{H}_s$ for all m. $\square$

Remark. The uniform summability of $\{\varphi_{s,m}^{(N)}(x)\}$ on compact sets follows from

$$\widehat{\varphi_{s,m}^{(N)}}(p) = h_s(p)^{-(N/6)M}\widehat{\varphi_{s,m}}(p) \tag{4.16}$$

together with the observation that $h_s(p)^{-(N/6)M}\hat{\phi}_s(p)$ is an integrable C^∞ function whose derivatives are integrable, provided $N \leqq 3M$.

Finally, we need to establish:

Theorem 4.4. $\{\phi_{s,m}\}$ *spans* $\mathscr{H}_s$.

Proof. Let

$$c_m = \prod_{\mu=1}^{s}\left(\int_0^{\pi} dp'_\mu\right)\prod_{\mu=s+1}^{d}\left(\int_0^{2\pi} dp'_\mu\right)\exp\left(-i2\sum_{\mu=1}^{s} m_\mu p'_\mu - i\sum_{\mu=s+1}^{d} m_\mu p'_\mu\right)\sqrt{h_s(p)}. \tag{4.17}$$

Thus

$$\sqrt{h_s(p)} = \sum_m c_m \exp\left(i2\sum_{\mu=1}^{s} m_\mu p_\mu + i\sum_{\mu=s+1}^{d} m_\mu p_\mu\right), \tag{4.18}$$

and since

$$\hat{\varphi}_{s,m'}(p) = \sqrt{h_s(p)}\,\hat{\phi}_{s,m'}(p), \tag{4.19}$$

it follows that

$$\hat{\varphi}_{s,m'}(p) = \sum_m c_{m'-m}\hat{\phi}_{s,m}(p). \tag{4.20}$$

Since $\{c_m\}$ is square summable and $\{\phi_{s,m}(x)\}$ is summable uniformly on compact sets, we know that

$$\sum_m c_{m'-m}\phi_{s,m}(x) \tag{4.21}$$

converges in the L^2 sense. Having expressed the $\varphi_{s,m}$ as linear combinations of the $\phi_{s,m}$ we have shown that the latter span $\mathscr{H}_s$. $\square$

5. Moment Properties and Exponential Decay

In this section we examine the properties of ϕ_s. By (4.7) we have

$$\hat{\phi}_s(p) = O\left(\prod_\mu |p_\mu|^{-M-1}\right) \tag{5.1}$$

for large p, and

$$\hat{\phi}_s(p) = O(|p_s|^{M+1}) \tag{5.2}$$

for small p_s. The former property implies that ϕ_s is class C^{M-1} smooth, while the latter property means that

$$\int_{-\infty}^{\infty} \phi_s(x) x_s^N \, dx_s = 0 \tag{5.3}$$

for arbitrary $\hat{x}_s \equiv (x_1, \ldots, x_{s-1}, x_{s+1}, \ldots, x_d)$ and $N \leqq M$. Equation (5.3) is a strong condition; it implies that

$$\int \phi_s(x) x^\alpha \, dx = 0 \tag{5.4}$$

for all multi-indices α for which $\alpha_s \leqq M$. There are more than enough vanishing moments to give us the desired power law for the long-distance fall-off of a massless potential of ϕ_s—provided that ϕ_s itself has enough decay. The fact that $\hat{\phi}_s$ is an integrable C^∞ function with integrable derivatives gives us as much as we want for that purpose.

Indeed, our final task is to show that ϕ_s has *exponential* fall-off. This property is a consequence of the following theorem.

Theorem 5.1. *$\hat{\phi}_s$ extends to an analytic function bounded by $c \prod_\mu (1 + |\mathrm{Re}\, z_\mu|)^{-M-1}$ on the product of strips $|\mathrm{Im}\, z_\mu| < \delta$ for some $\delta > 0$ ($p_\mu = \mathrm{Re}\, z_\mu$).*

Proof. Returning to (4.7), we first note that the numerators extend to entire functions. Moreover, they are bounded on arbitrary strips centered about the real axis because

$$|e^{i\xi}| \leqq e^{|\mathrm{Im}\,\xi|},$$

$$|\hat{\chi}(\xi)| = \frac{|1 - e^{i\xi}|}{|\xi|} = \begin{cases} O(1), & |\xi| \leqq 1, \\ O\left(\dfrac{e^{|\mathrm{Im}\,\xi|}}{(1 + |\mathrm{Re}\,\xi|)}\right), & |\xi| \geqq 1. \end{cases}$$

To control the denominators we must appeal to:

Lemma 5.2. $\sum_{n=-\infty}^{\infty} |\hat{\chi}(t + 2\pi n)|^{2M+2}$ *extends to an entire function whose real part is positive and bounded away from zero on the strip $|\mathrm{Im}\,\xi| < \delta$ for some $\delta > 0$.*

Proof. The entire function is

$$4^{M+1} \sum_{n=-\infty}^{\infty} \frac{\sin^{2M+2}(\frac{1}{2}\xi)}{(\xi + 2\pi n)^{2M+2}}, \tag{5.5}$$

and it is periodic. Hence, its derivative is periodic and therefore bounded on any strip centered about the real axis. On the other hand, the function is strictly positive on the real axis, so the periodicity also tells us that it is greater than some positive constant on the whole real axis. The boundedness of the derivative certainly implies boundedness of the partial derivative of the real part of (5.5) with respect to Im ξ, so we have the desired conclusion. □

Proof of Theorem 5.1. (continued) An immediate consequence of the lemma is that

$$\left(\sum_{n=-\infty}^{\infty} |\hat{\chi}(t+2\pi n)|^{2M+2}\right)^{-\alpha}, \quad \alpha>0$$

extends to a function that is both bounded *and analytic* on the strip $|\text{Im}\,\xi|<\delta$. Our argument is complete if we can show that $w(t)^{-\alpha}$ can also be extended in this way. Now Lemma 5.2 applies to the function

$$\sum_{n=-\infty}^{\infty} |\hat{\chi}(t+2\pi n+\pi)|^{2M+2}$$

as well, so $w(t)$ itself extends to a bounded analytic function on some strip centered about the real axis. But $w(t)$ is also periodic and strictly positive on the real axis, so we may use the same reasoning as before. □

6. Completeness

It is easy to convince oneself that our construction guarantees completeness of the orthogonal set of Lemarié functions, provided that our original set of block-spin constraints is complete. Now recall that $\varphi_{s,m}$ was constructed with the set of constraints

$$\prod_{\mu=1}^{d}\left(\int dp_\mu\right)\hat{\varphi}(p)\prod_{\mu=1}^{s-1}\overline{\hat{\chi}(2p_\mu)^{M+1}}\prod_{\mu=s}^{d}\overline{\hat{\chi}(p_\mu)^{M+1}}\exp\left(-i2\sum_{\mu=1}^{s-1}l_\mu p_\mu - i\sum_{\mu=s}^{d}l_\mu p_\mu\right)$$
$$=\left(\prod_{\mu\neq s}P_M^-(m_\mu-l_\mu)\right)\left(P_M^-(m_s-l_s)-P_M^-(1+m_s-l_s)\right) \tag{6.1}$$

with l running over all d-tuples of integers. These constraints follow from (2.8), (2.10), and (4.1), and the first observation to make is that the functions on $\mathbb{Z}^d$ given by

$$l\mapsto\left(\prod_{\mu\neq s}P_M^-(m_\mu-l_\mu)\right)\left(P_M^-(m_s-l_s)-P_M^-(1+m_s-l_s)\right) \tag{6.2}$$

span the Hilbert space $l^2(\mathbb{Z}^d)$. This is an obvious consequence of the fact that the functions

$$(1-e^{ik_s})^{M+1}\prod_{\mu\neq s}(1-e^{ik_\mu})^M\prod_\mu e^{im_\mu k_\mu} \tag{6.3}$$

span $L^2(T^d)$. (Linear combinations form an algebra that separates points.)

On the other hand, we know that $\hat{\varphi}_{s,m}$ is a square-summable combination of the functions

$$\widehat{f_l^s}(p)=\prod_{\mu=1}^{s-1}\hat{\chi}(2p_\mu)^{M+1}\prod_{\mu=s}^{d}\hat{\chi}(p_\mu)^{M+1}\exp\left(i2\sum_{\mu=1}^{s-1}l_\mu p_\mu+i\sum_{\mu=s}^{d}l_\mu p_\mu\right) \tag{6.4}$$

because $\varphi_{s,m}$ minimizes $\|\varphi\|_2$ with respect to (6.1). The implication of our remark above is that the $\hat{\phi}_{s,m}$ span the same subspace of $L^2(\mathbb{R}^d)$ as the $\widehat{f_l^s}$. Our proof of completeness therefore reduces to establishing the following theorem:

Theorem 6.1. *Let* $\zeta \in L^2(\mathbb{R}^d)$ *such that*

$$\int \hat{\zeta}(p)\widehat{f_l^s}(2^r p)dp = 0 \tag{6.5}$$

for all integers r, *all* $l \in \mathbb{Z}^d$, *and* $1 \leqq s \leqq d$. *Then* $\zeta = 0$.

Proof. Pick an arbitrary *negative* value for r. We know that

$$\hat{\chi}(2^v p_\mu) = (1 + e^{i2^{v-1}p_\mu})(1 + e^{i2^{v-2}p_\mu})\cdots (1 + e^{i2^{v+r+1}p_\mu})(1 + e^{i2^{v+r}p_\mu})\hat{\chi}(2^{v+r}p_\mu), \quad v = 0, 1, \tag{6.6}$$

so by taking appropriate linear combinations of (6.5) we can infer that

$$\int \hat{\xi}(p) \prod_{\mu=1}^{s-1} \hat{\chi}(2^{r+1}p_\mu)^M \prod_{\mu=s}^{d} \hat{\chi}(2^r p_\mu)^M \exp\left(i2^{r+1} \sum_{\mu=1}^{s-1} l_\mu p_\mu + i2^r \sum_{\mu=s}^{d} l_\mu p_\mu \right) dp = 0 \tag{6.7}$$

for all $l \in \mathbb{Z}^d$, where $\xi \in L^2(\mathbb{R}^d)$ is given by

$$\hat{\xi}(p) = \hat{\zeta}(p) \prod_{\mu=1}^{s-1} \hat{\chi}(2p_\mu) \prod_{\mu=s}^{d} \hat{\chi}(p_\mu). \tag{6.8}$$

Since $\hat{\zeta} \in L^2(\mathbb{R}^d)$, we see that $\hat{\xi} \in L^1(\mathbb{R}^d)$, and so ξ is continuous. Now

$$\lim_{r \to -\infty} \hat{\chi}(2^{v+r}p_\mu) = 1, \tag{6.9}$$

so if we consider $l_\mu = 2^{r'+1}l'_\mu$, $r' > r$ and then take the limit of (6.7) for fixed l'_μ as $r \to -\infty$, it follows from dominated convergence that

$$\int \hat{\xi}(p) \exp\left(i2^{r'+1} \sum_{\mu=1}^{s-1} l'_\mu p_\mu + i2^{r'} \sum_{\mu=s}^{d} l'_\mu p_\mu \right) = 0, \tag{6.10}$$

i.e.,

$$\xi(2^{r'+1}l'_1, \ldots, 2^{r'+1}l'_{s-1}, 2^{r'}l'_s, \ldots, 2^{r'}l'_d) = 0. \tag{6.11}$$

Since this equation holds for arbitrarily negative r', ξ vanishes at all dyadic points, and so by continuity ξ vanishes everywhere. Hence $\zeta = 0$. □

Acknowledgements. This work was motivated by a talk given at the 1986 IAMP Congress by Ingrid Daubechies and also by conversations with Paul Federbush. I also thank Yves Meyer, who urged me to make this construction known.

References

1. Lemarié, P., Meyer, Y.: Ondelettes et bases Hilbertiannes. Rev. Mat. Ibero-Am. (to appear)
2. Meyer, Y.: Sémin. Bourbaki **38**, 662 (1985–86)
3. Lemarié, P.: Une nouvelle base d' ondelettes de $L^2(\mathbb{R}^n)$. Ecole Normale Supérieure. Preprint
4. Battle, G., Federbush, P.: A phase cell cluster expansion for euclidean field theories. Ann. Phys. **142**, (1), 95 (1982)
5. Battle, G., Federbush, P.: A phase cell cluster expansion for a hierarchical ϕ_3^4 model. Commun. Math. Phys. **88**, 263 (1983)

6. Battle, G., Federbush, P.: Ondelettes and phase cell cluster expansions, a vindication. Commun. Math. Phys. (in press)
7. Williamson, C.: A phase cell cluster expansion of ϕ_3^4. Ann. Phys. (to appear)
8. Gawedzki, K., Kupiainen, A.: A rigorous block spin approach to massless lattice theories. Commun. Math. Phys. **77**, 31 (1980)
9. Federbush, P.: A phase cell approach to Yang–Mills theory. I. Modes, lattice-continuum duality. Commun. Math. Phys. **107**, 319–329 (1986)

Communicated by A. Jaffe

Received November 7, 1986

SECTION IV

Precursors and Development in Mathematics: Atom and Frame Decompositions

Introduction

Yves Meyer

There are no doubts. We owe *wavelet analysis* to Jean Morlet. How did Morlet discover wavelets? Here is the story. During the late seventies Morlet was working as a geophysicist for the Elf-Aquitaine company. He had to process the backscattered seismic signals which carry the information related to the geological layers. These seismic signals present transient patterns. Processing such signals with conventional tools such as *windowed Fourier analysis* or *Gabor wavelets* creates numerical artifacts. Morlet elaborated wavelet analysis to overcome this problem. Quoting the geophysicist Pierre Goupillaud,

> A product of the renowned Ecole Polytechnique, Morlet performed the exceptional feat of discovering a novel mathematical tool which has made the Fourier transform obsolete after 200 years of uses and abuses, particularly in its fast version.

Goupillaud is exaggerating. Around 1945, Léon Brillouin, Dennis Gabor, Claude Shannon, and Jean Ville were already aware of the limitations of the Fourier transformation. They were looking at *time-frequency* representations that would play the role of an audiogram or a sonogram for audio signals, see [Bri56].

Gabor conjectured that his famous wavelets

$$w_{(k,l)}(x) = \exp(2\pi ilx)\,\exp(-(x-k)^2/2),\ \ (k,l) \in \mathbf{Z}^2 \tag{1}$$

might be the building blocks (or *time-frequency atoms*) of speech signals. That is why he used the name *logons*. More precisely, Gabor suggested that any signal $f(x)$ could be written as a linear combination of logons:

$$f(x) = \sum_{(k,l)\in\mathbf{Z}^2} c(k,l)\,w_{(k,l)}(x). \tag{2}$$

For proving this identity, Gabor made three observations: (a) the *time-frequency plane* can be paved with congruent squares $Q_{(k,l)} = [k, k+1) \times [2\pi l, 2\pi(l+1))$, $(k,l) \in \mathbf{Z}^2$; (b) the *phase-space localization* of $w_{(k,l)}(x)$ is precisely given by $Q_{(k,l)}$; and (c) the dimensionality of the space of functions that are localized in a large domain Ω of the time-frequency plane is $|\Omega|/2\pi$. Let us quote Gabor. In *La théorie des communications et la physique* (1951) Dennis Gabor writes:

> On obtient donc une représentation complète d'un signal quelconque en inscrivant dans chaque cellule un nombre complexe, l'amplitude de la fonction élémentaire associée à la cellule. Si l'on multiplie ces fonctions par leurs

coefficients complexes, leur somme verticale donne le signal à l'instant choisi... Les signaux élémentaires ne sont pas orthogonaux, et il faut déterminer leurs coefficients $C_{j,k}$ par une méthode d'approximations successives, qui, pourtant, converge très rapidement.

Today we know that Gabor was both wrong and right. The iterative algorithm Gabor had in mind does not converge. However, Gabor was right, since his claim presaged the theory of frames, as developed by Ingrid Daubechies in her two papers in this section.

Morlet decided to replace *Gabor wavelets* which have a fixed duration, by new *building blocks* $\psi_{(a,b)}(x) = a^{-1}\psi(\frac{x-b}{a})$, $(b,a) \in P = \mathbf{R} \times (0,\infty)$, which have an arbitrarily small duration a. Morlet's wavelet is $\psi(x) = \exp(5ix)\exp(-x^2/2)$. If negligible errors are disregarded, the Fourier transform of ψ vanishes on $(-\infty, 0]$ and ψ belongs to the Hardy space $\mathcal{H}^2(\mathbf{R})$. These *building blocks* $\psi_{(a,b)}(x)$ are the *time-scale wavelets* and the small duration a of $\psi_{(a,b)}(x)$ matches rapid changes in the transient components of a signal.

Morlet knew that the type of analysis he was promoting was not limited to seismic signals. In 1981 he contacted the physicist Roger Balian at the Ecole Polytechnique in Paris. Balian directed him to Alexander Grossmann. Grossmann interpreted Morlet's wavelets as *coherent states* and gave an elegant proof of Morlet's reconstruction algorithm. This algorithm says that any $f \in \mathcal{H}^2(\mathbf{R})$ can be written as a linear combination of the time-scale wavelets $\psi_{(a,b)}(x)$. In other words, we have

$$f(t) = \iint_P W(a,b)\,\psi_{(a,b)}(t)\,db\,\frac{da}{a}, \tag{3}$$

where the wavelet coefficients of f are $W(a,b) = \langle f, \psi_{(a,b)}\rangle$.

Grossmann used representation theory for proving (3), and this approach imposed some limitations. In fact the unitary representation of the affine group $ax+b$, $a>0, b \in \mathbf{R}$, is not irreducible on $L^2(\mathbf{R})$, since the Hardy space $\mathcal{H}^2(\mathbf{R})$ is invariant. Everything is repaired if $L^2(\mathbf{R})$ is replaced by $\mathcal{H}^2(\mathbf{R})$. This explains why Grossmann and Morlet assumed that the wavelet ψ and the function f are analytic. These restrictions limited wavelet analysis to one-dimensional signals. Finally the only requirement imposed upon the analyzing wavelet ψ is to belong to the Hardy space $\mathcal{H}^2(\mathbf{R})$ and to be localized. The *Morlet wavelet* $\psi(x) = \exp(5ix)\exp(-x^2/2)$ does not satisfy these requirements unless some small error terms are disregarded. This group representation approach to wavelet analysis was successfully generalized in the fifth paper in this section.

The computational load of calculating the double integral that appears in the right-hand side of (3) is prohibitive. Therefore Morlet approximated this integral by suitable Riemann sums. The domain of integration P is then paved with disjoint Heisenberg boxes $R_{(j,k)} = [k2^{-\alpha j}, (k+1)2^{-\alpha j}) \times [2^{\alpha j}, 2^{\alpha(j+1)})$, $j,k \in \mathbf{Z}$, where $\alpha \in (0,1)$ is small. This paving of the *time-frequency plane* replaces the congruent squares used by Brillouin, Gabor, and Shannon in the 1940s. Finally Morlet was led to the following alternative to (3):

$$f(x) = \sum_j \sum_k c(j,k) 2^{\alpha j/2} \psi(2^{\alpha j}x - k) \tag{4}$$

Morlet viewed (4) as an expansion in a basis when he wrote:

> It is possible to decompose a signal on a discrete set of elementary complex valued wavelets used as a basis of the corresponding expansion, as first proposed by D. Gabor in 1946. ... It is suitable to propose constant shape wavelets as the basic wavelets.

Morlet believed that a dictionary could exist between wavelet bases and partitions of the *time-frequency plane.* He attributed this idea to Gabor. We will later return to this crucial point.

Morlet's scientific vision was decisive for the success of wavelet analysis. But as Pierre Goupillaud tells us, this had no effect on Morlet's career at the Elf-Aquitaine company. In fact *Morlet's only reward for years of perseverance and creativity in producing this extraordinary tool was an early retirement.*

Wavelet analysis could not survive without being reshaped. Let me indicate how the theory evolved from 1984 to 1987. I became acquainted with Morlet's work in 1984. I related the Grossmann and Morlet reconstruction algorithm (3) to *Calderón's identity.* This identity was discovered by Alberto Calderón in the late 1960s. This observation provided us with the generalization of Morlet's work to several dimensions.

The next problem concerned the validity of (4). For attacking this problem Ingrid Daubechies and Alexander Grossmann borrowed the definition of a *frame* from R. J. Duffin and A. C. Schaeffer in the first paper in this section. A sequence x_j, $j \in J$, of vectors in a Hilbert space $\mathcal{H}$ is a *frame* if two positive constants C_1, C_2 exist such that

$$C_1 \|x\|^2 \le \sum_{j \in J} |\langle x, x_j \rangle|^2 \le C_2 \|x\|^2, \quad x \in H. \tag{5}$$

Then the operator T defined by $T(x) = \sum_{j \in J} \langle x, x_j \rangle \, x_j$ is invertible. In other words, we have a reconstruction scheme which reads

$$x = \sum_{j \in J} \langle x, y_j \rangle \, x_j, \ \ y_j = (T^*)^{-1}(x_j). \tag{6}$$

The sequence y_j, $j \in J$, is the *dual frame.* Roughly speaking, a frame is an overcomplete basis. Using a frame, we pay more but we gain flexibility and stability.

If ψ is a function in $L^2(\mathbf{R})$ and $\alpha \in (0, 1)$ is a small exponent, we write $\psi_{(j,k)}(x) = 2^{\alpha j/2} \psi(2^{\alpha j} x - k)$. Daubechies asked herself whether $\psi_{(j,k)}$, $(j,k) \in \mathbf{Z}^2$, is a frame when α is small enough. She found some sufficient conditions on ψ and α. Then (6) yields

$$f(x) = \sum_j \sum_k c(j,k) \, 2^{\alpha j/2} \, \psi(2^{\alpha j} x - k), \ \ c(j,k) = \int f(x) \, \tilde{\psi}_{(j,k)} \, dx, \tag{7}$$

where $\tilde{\psi}_{(j,k)}$, $(j,k) \in \mathbf{Z}^2$, is the dual frame.

Daubechies' approach is extremely appealing since it is quite flexible and most of Morlet's claims can be given some firm ground. Daubechies' book or her second paper in this section are the best references.

In my opinion this approach suffers from several drawbacks. The first and the main one concerns the status of the dual frame $\tilde{\psi}_{(j,k)}$. We do not have access to these functions $\tilde{\psi}_{(j,k)}$.

That is why Daubechies uses an iteration scheme for obtaining (7). The second drawback concerns the convergence of (7). Daubechies focused on L^2 estimates. The reason for this choice is not that this norm is the most relevant in signal or image processing, but that the L^2-norm is the only one for which estimates are available. In fact I proved that (7) might diverge if any other functional norm is used. That is why I tried to prove some versions of (7) in which the dual frame was explicit.

I observed that Grossmann and Morlet were using the same dyadic partition of the frequency axis as Littlewood and Paley did in the 1930s. The outputs of this filter-bank are the famous dyadic blocks $\Delta_j f(x)$, $j \in \mathbf{Z}$. Sampling each $\Delta_j f(x)$ at the Nyquist rate yields the wavelet coefficients $c(j,k) = \int f(x)\,\psi_{(j,k)}(x)\,dx$. This remark and Shannon's sampling theorem yield a stronger version of (7), where $\psi_{(j,k)} = \tilde{\psi}_{(j,k)}$ and ψ belongs to the Schwartz class. This construction, which appeared in the third paper, was discovered independently by Frazier and Jawerth and named the ϕ-transform in the fourth paper.

A few months later I constructed the first orthonormal wavelet basis of $L^2(\mathbf{R})$ of the form

$$\psi_{j,k}(x) = 2^{j/2}\psi(2^j x - k), \quad j \in \mathbf{Z},\ k \in \mathbf{Z}, \tag{8}$$

where ψ belongs to the Schwartz class $\mathcal{S}(\mathbf{R})$. I observed that this basis was a universal unconditional basis for all classical functional Banach spaces $\mathcal{B}$ that admit an unconditional basis. It implies numerical stability of the expansion of a function into my wavelet basis. If the coefficients are slightly perturbed nothing bad happens. This is impossible with Fourier expansions unless $\mathcal{B} = L^2$ or a Sobolev space based on L^2. The fact that such wavelet bases are unconditional bases for most of the functional spaces is an easy corollary of the Calderón-Zygmund theory.

These discoveries sprang out as a revolution. But I soon found out that orthonormal wavelet bases already existed. In 1938, in the last issue of *Studia Mathematica* published before the Second World War, Marcinkiewicz proved that the Haar system is an unconditional basis of L^p when $1 < p < \infty$. The Haar system is the prototype of an orthonormal wavelet basis. Marcinkiewicz's theorem is obviously false if $p = 1$. That is why the existence of an unconditional basis for the Hardy space $\mathcal{H}^1$ became a hot issue. This Hardy space should be viewed as the "limit" of L^p as p tends to 1. Bernard Maurey solved the problem in 1980. Then Lennart Carleson simplified Maurey's approach and constructed a basis whose structure is close to our orthonormal wavelet basis. Finally, Wojtaszczyk proved in 1981 that the usual Franklin orthonormal basis is an unconditional basis of $\mathcal{H}^1$ and J. O. Strömberg constructed an unconditional basis of $\mathcal{H}^1$ that has the same structure as our basis. In Strömberg's basis ψ is a spline and cannot be $\mathcal{C}^\infty$.

In 1986, I had the chance of having some discusssions with Stéphane Mallat at the University of Chicago. We were sharing Antoni Zygmund's office. We emphasized the role played by the *scaling function* in the construction of orthonormal wavelet bases. Then orthonormal wavelet bases were given a new route of access. The localization of the wavelets in the time-frequency plane became a byproduct of the orthogonality. Furthermore, wavelets could be incorporated into the larger family of *finite elements* and this remark opened the road to many applications in numerical analysis. These observations were incorporated inside a new scheme that is now called *multiresolution analysis. Multiresolution analysis* generalizes refinement of meshes. Any *multiresolution analysis* yields an orthonormal wavelet basis and a partial converse was later proved by Pierre-Gilles Lemarié-Rieusset.

Mallat made another remarkable discovery. He elaborated a dictionary between orthonormal wavelet bases and quadrature mirror filters. These filters were discovered in the late 1970s by a team of scientists working at the IBM research center of La Gaude. In fact every pair (ϕ, ψ) of a scaling function and a fellow wavelet proceeds from a pair $(\mathcal{F}_0, \mathcal{F}_1)$ of two quadrature mirror filters.

Multiresolution analysis was used by Ingrid Daubechies (1987) to construct a family of bases of type (8), where $\psi = \psi_m$, $m \in \mathbf{N}$, belongs to $\mathcal{C}^m$ and is compactly supported. This outstanding achievement can be compared to the design of the Fast Fourier transform. After Daubechies' discovery, wavelets became an efficient tool in signal and image processing.

We focused on wavelets and mathematics. This can be blamed since wavelets are present in many other scientific fields that range from quantum field theory to acoustics. But these isolated findings needed a unified scientific treatment. In other words, the strings were waiting for the conductor of the symphony. Morlet was this conductor. Before Jean Morlet and Alexander Grossmann, no one would have related signal processing or image processing to some mathematical problems concerning the geometry of Banach spaces or to quantum field theory. Even inside mathematics, the wavelet spirit brought some fresh air. "Wild" Calderón-Zygmund operators that lie beyond pseudodifferential operators and cannot be studied with Fourier tools have a sparse representation in a wavelet basis. This remark was decisive in extending and generalizing the Fast Multipole algorithm of V. Rokhlin. Here is another example. In the 1970s, R. Coifman and Guido Weiss in the second paper in this section proved that any f belonging to the Hardy space $\mathcal{H}^1$ can be written as a convex combination of *atoms*. These atoms are the simplest building blocks in $\mathcal{H}^1$. This nonlinear decomposition needed to be related to the decomposition of f into a wavelet basis. Today we know that the former is obtained from the latter by a thresholding that is similar to Donoho's denoising algorithms. Taking into account the magnitude of the wavelet coefficients of $f \in \mathcal{H}^1$, one peels off its wavelet expansion and obtains the Coifman and Weiss atomic decomposition of f. Details can be found in my book [Mey92].

As often happens in science, a new discovery brings a new collection of problems. In the case of Morlet's findings, let us raise two questions. The first one reads as follows:

> Given a partition of the time-frequency plane by a family of sets Ω_j, $j \in J$, is it possible to construct an orthonormal basis w_j, $j \in J$, such that for each $j \in J$, the Wigner transform of w_j is almost supported by Ω_j?

We do not impose any limitations on the positions or orientations of these Ω_j, $j \in J$. The converse is true: any orthonormal basis yields a partition of the time-frequency plane. The proof runs as follows: (a) The Wigner transform W_j of w_j is the Weyl symbol of the projection P_j from L^2 onto w_j; (b) Moyal's identity implies that these Wigner transforms W_j form an orthonormal sequence in the time-frequency plane; and (c) $I = \sum P_j$ implies $1 = \sum W_j$.

The second problem addresses an even more fundamental issue:

> Given a class $\mathcal{C}$ of signals, what is the optimal unfolding of a signal $f \in \mathcal{C}$ in the time-frequency plane?

In other words, what *time-frequency atoms* should be used to decompose signals belonging to $\mathcal{C}$? Optimality is defined by a criterion which is averaged over $\mathcal{C}$. This line of research started with Wigner and goes far beyond ordinary wavelets. Among the most promising results, let me mention the *curvelets* constructed by Emmanuel Candès and David Donoho. *Curvelets* surpass ordinary wavelets in image processing when images are modeled by regions with smooth boundaries. A competing algorithm was designed by Mallat and Erwan Le Pennec. Let me also mention the *best basis* approach by Ronald Coifman and Victor Wickerhauser where the search for the optimal unfolding in the time-frequency plane is incorporated inside the processing. We should not forget the fascinating *independent component analysis* where the signal is decomposed into a series of statistically independent time-frequency atoms. Let us acknowledge our debt again:

> Jean Morlet launched a scientific program which already offered fruitful alternatives to Fourier analysis and is now moving beyond wavelets.

References

[Bri56] L. Brillouin, *Science and Information Theory*, Academic Press, New York, 1956.

[Gab51] D. Gabor, *La théorie des communications et la physique*, in *La cybernétique. Théorie du signal et de l'information.* Réunions d'études et de mises au point tenues sous la présidence de Louis de Broglie, Editions de la Revue d'Optique, Paris, 1951, 115–149.

[Mey92] Y. Meyer, *Wavelets and Operators*, Cambridge University Press, Cambridge, 1992.

A CLASS OF NONHARMONIC FOURIER SERIES([1])

BY

R. J. DUFFIN AND A. C. SCHAEFFER

1. **Introduction.** A sequence $\{\lambda_n\}$, $n=0, \pm 1, \pm 2, \cdots$, of real or complex numbers we shall say has *uniform density* 1 if there are constants L and δ such that $|\lambda_n - n| \leqq L$ and $|\lambda_n - \lambda_m| \geqq \delta > 0$ for $n \neq m$. This is a more restrictive notion than density, for, considering only those λ_n for which $n>0$, it is clear that a sequence of uniform density 1 has a density as defined by Pólya equal to 1, but the converse is not true. Sequences of uniform density d are defined in a later part of the present paper for any $d>0$. If $f(z)$ is an entire function of exponential type γ, $0 \leqq \gamma < \pi$, that is,

$$f(z) = O(e^{\gamma|z|})$$

uniformly in all directions as $|z| \to \infty$, then $f(z)$ is completely determined by its values at any sequence of uniform density 1. Some properties of entire functions of exponential type extend in a natural way from a sequence of uniform density to all points of the real axis or of a strip parallel to the real axis. For example, the authors have shown [6] that if an entire function of exponential type γ, $0 \leqq \gamma < \pi$, is uniformly bounded at a sequence of uniform density 1, then it is uniformly bounded on the entire real axis. It also has a bound in every strip parallel to the real axis. This result was applied to questions concerning the coefficients of power series.

In the present paper a further property of sequences of uniform density is proved. It is shown that if $f(z)$ is an entire function of exponential type γ, $0 \leqq \gamma < \pi$, belonging to $L_2(-\infty, \infty)$ on the real axis and $\{\lambda_n\}$ is a sequence of uniform density 1, then the ratio $\{\sum_n |f(\lambda_n)|^2\}/\int_{-\infty}^{\infty}|f(x)|^2 dx$ has positive upper and lower bounds independent of the function. An essentially equivalent statement is that if $g(t) \in L_2(-\gamma, \gamma)$ where $0<\gamma<\pi$, and $\{\lambda_n\}$ is a sequence of uniform density 1, then there are positive constants A and B independent of the function $g(t)$ such that

$$A \leqq \frac{\dfrac{1}{2\pi}\displaystyle\sum_n \left| \int_{-\gamma}^{\gamma} g(t) e^{i\lambda_n t} dt \right|^2}{\displaystyle\int_{-\gamma}^{\gamma} |g(t)|^2 dt} \leqq B. \tag{1}$$

Presented to the International Congress of Mathematicians, August 31, 1950; received by the editors August 1, 1951.

([1]) Part of the work on this paper was done while the authors were at Purdue University, and another part while they were under contracts with the Office of Naval Research, the Office of Ordnance Research, and the Flight Research Laboratory.

There are other sequences for which this inequality is true; we shall say that a sequence of functions $\{\exp(i\lambda_n t)\}$ is a *frame* over the interval $(-\gamma, \gamma)$ if there are positive constants A and B such that (1) is true for all $g(t) \in L_2(-\gamma, \gamma)$. If $\gamma=\pi$ and $\lambda_n=n$, then $A=B=1$ is Parseval's relation. The proof that a constant B exists is quite direct, the proof of the existence of A in case $\{\lambda_n\}$ is a sequence of uniform density 1 and $0<\gamma<\pi$ is one of the central results of the present paper.

It was shown by Paley and Wiener [9], who initiated much of the work in nonharmonic Fourier series, that if λ_n is real and $|\lambda_n - n| < \pi^{-2}$, in which case the condition $|\lambda_n - \lambda_m| \geqq \delta > 0$ for $n \neq m$ is automatically satisfied, then $\{\exp(i\lambda_n x)\}$ is closed over $(-\pi, \pi)$. Boas [1] pointed out that their results imply that (1) is true in this case with $\gamma=\pi$. Duffin and Eachus [5] showed that this inequality is true in the case $\gamma=\pi$ if λ_n are real or complex numbers satisfying $|\lambda_n - n| \leqq 0.22 \cdots$. In the present paper it is shown that (1) is true in the case $\gamma=\pi$ if $\{\lambda_n\}$ is a sequence such that $|\operatorname{Re}(\lambda_n) - n| \leqq 0.22 \cdots$ and the imaginary part of λ_n is uniformly bounded. Boas [3] has considered problems analogous to some of these for L_p spaces.

If relation (1) is true, then the sequence of functions $\{\exp(i\lambda_n t)\}$ is clearly complete over $(-\gamma, \gamma)$; that is, if $g(t) \in L_2(-\gamma, \gamma)$, then the set of relations

$$\int_{-\gamma}^{\gamma} g(t) e^{i\lambda_n t} dt = 0$$

imply that $g(t)$ vanishes almost everywhere in $(-\gamma, \gamma)$. A proof of completeness in a more general case than any mentioned above was given by Levinson [8]. Levinson's result, without stating the most general form, shows that if the points lie in a strip parallel to the real axis and if $|\operatorname{Re}(\lambda_n) - n| \leqq \alpha < 1/4$, then the set of functions $\{\exp(i\lambda_n t)\}$ are complete in $(-\pi, \pi)$. On the other hand, completeness is a less strong conclusion than the frame condition. We shall show, for example, that if $\{\lambda_n\}$ is a sequence of uniform density 1, then the sequence of functions $\{\exp(i\lambda_n t)\}$ for which $n>0$ form a complete set in any interval $(-\gamma, \gamma)$ where $0<\gamma<\pi$, but they do not constitute a frame in any such interval.

Relation (1) gives the set of functions $\{\exp(i\lambda_n t)\}$ properties quite similar to an orthonormal set such as $\{\exp(int)\}$ in Hilbert space. However, the situation is more complicated because the set $\{\exp(i\lambda_n t)\}$ is highly overcomplete on an interval of length less than 2π. Most of the previous study of nonharmonic Fourier series has been for the exactly complete case; that is, the case in which the sequence of functions is complete but becomes incomplete by the omission of any one of them. It has therefore seemed worthwhile to give in detail some of the elementary relationships between moment sequences, expansion coefficients, etc. It is shown that these relations are consequences of well known properties of positive definite transformations in Hilbert space.

These considerations give information only about mean convergence of the series. However, combining properties of mean convergence with properties of the Dirichlet kernel gives conditions for pointwise convergence. It results that nonharmonic Fourier series have to a large extent the same convergence and summability properties as ordinary Fourier series.

2. **Fourier frames.** We begin the proofs by making the following more precise definitions.

DEFINITION. A sequence $\{\lambda_n\}$ of real or complex numbers has *uniform density* d, $d>0$, if there are constants L and δ such that

$$\left|\lambda_n - \frac{n}{d}\right| \leqq L, \qquad n = 0, \pm 1, \pm 2, \cdots, \tag{2}$$

$$|\lambda_n - \lambda_m| \geqq \delta > 0, \qquad n \neq m. \tag{3}$$

DEFINITION. A set of functions $\{\exp(i\lambda_n t)\}$ is a *frame* over an interval $(-\gamma, \gamma)$ if there exist positive constants A and B which depend exclusively on γ and the set of functions $\{\exp(i\lambda_n t)\}$ such that

$$A \leqq \frac{\frac{1}{2\pi}\sum_n \left|\int_{-\gamma}^{\gamma} g(t) e^{i\lambda_n t} dt\right|^2}{\int_{-\gamma}^{\gamma} |g(t)|^2 dt} \leqq B \tag{4}$$

for every function $g(t) \in L_2(-\gamma, \gamma)$.

In the case of a frame, we shall suppose in part 2, except where the contrary is stated, that the index n runs through all positive and negative integers and zero; however, we do not suppose that the λ_n are distinct. The following theorem is to be proved.

THEOREM I. *If $\{\lambda_n\}$ is a sequence of uniform density d, then the set of functions $\{\exp(i\lambda_n t)\}$ is a frame over the interval $(-\gamma, \gamma)$ where $0<\gamma<\pi d$.*

The following equivalent theorem is also to be proved.

THEOREM I′. *Let $\{\lambda_n\}$ be a sequence of uniform density d and let $0<\gamma<\pi d$. If $f(z)$ is an entire function of exponential type γ such that $f(x) \in L_2(-\infty, \infty)$, then*

$$A \leqq \frac{\sum_n |f(\lambda_n)|^2}{\int_{-\infty}^{\infty} |f(x)|^2 dx} \leqq B. \tag{5}$$

Here A and B are positive constants which depend exclusively on γ and $\{\lambda_n\}$.

It will be shown that either of these theorems implies the other, and that any set of positive constants A and B which suffices in (4) or (5) also suffices in the other inequality. It will be clear from the sequel that the constant B exists under conditions much milder than are necessary to infer the existence of A. The proof of these theorems depends on several lemmas. The following lemma is a now classical result of Paley and Wiener [9].

LEMMA I. *If $f(z)$ is an entire function of exponential type γ and if $f(x) \in L_2(-\infty, \infty)$, then there is a function $g(t) \in L_2(-\gamma, \gamma)$ such that*

$$f(z) = \frac{1}{(2\pi)^{1/2}} \int_{-\gamma}^{\gamma} g(t) e^{izt} dt. \tag{6}$$

It is clear in this lemma that $g(t)$ is the Fourier transform of $f(x)$, that is, the Fourier transform of $f(x)$ vanishes almost everywhere outside $(-\gamma, \gamma)$, and so Plancherel's theorem states that

$$\int_{-\infty}^{\infty} |f(x)|^2 dx = \int_{-\gamma}^{\gamma} |g(t)|^2 dt. \tag{7}$$

From these relations it follows that if $f(z)$ is an entire function of exponential type γ such that $f(x) \in L_2(-\infty, \infty)$, then

$$|f(x+iy)| \leqq \left(\frac{\gamma}{\pi}\right)^{1/2} e^{\gamma|y|} \left\{ \int_{-\infty}^{\infty} |f(x)|^2 dx \right\}^{1/2}. \tag{8}$$

Differentiating (6) k times we also have

$$f^{(k)}(z) = \frac{1}{(2\pi)^{1/2}} \int_{-\gamma}^{\gamma} g(t)(it)^k e^{izt} dt,$$

so, by Plancherel's theorem,

$$\int_{-\infty}^{\infty} |f^{(k)}(x)|^2 dx = \int_{-\gamma}^{\gamma} |g(t)|^2 t^{2k} dt$$

$$\leqq \gamma^{2k} \int_{-\gamma}^{\gamma} |g(t)|^2 dt.$$

Thus,

$$\int_{-\infty}^{\infty} |f^{(k)}(x)|^2 dx \leqq \gamma^{2k} \int_{-\infty}^{\infty} |f(x)|^2 dx \tag{9}$$

for every entire function $f(z)$ of exponential type γ.

From Lemma I it clearly follows that Theorem I and Theorem I′ are equivalent. More generally, a set of functions $\{\exp(i\lambda_n t)\}$ is a frame over $(-\gamma, \gamma)$ if and only if there exist positive constants A and B such that

$$A \leqq \frac{\sum_n |f(\lambda_n)|^2}{\int_{-\infty}^{\infty} |f(x)|^2 dx} \leqq B \tag{10}$$

for every entire function $f(z)$ of exponential type γ satisfying $f(x) \in L_2(-\infty, \infty)$.

Results similar to Lemma II were obtained by Plancherel and Pólya [10] and by Boas [2] under different conditions.

LEMMA II. *Let $\{\exp(i\lambda_n t)\}$ be a frame over the interval $(-\gamma, \gamma)$. If M is any constant and $\{\mu_n\}$ is a sequence satisfying $|\mu_n - \lambda_n| \leqq M$, then there is a number $C = C(M, \gamma, \{\lambda_n\})$ such that*

$$\frac{\sum_n |f(\mu_n)|^2}{\sum_n |f(\lambda_n)|^2} \leqq C$$

for every entire function $f(z)$ of exponential type γ.

Proof. It is clearly sufficient to prove this lemma under the additional hypothesis that $f(x) \in L_2(-\infty, \infty)$. Let $\{\exp(i\lambda_n t)\}$ be a frame over $(-\gamma, \gamma)$ such that inequality (10) is satisfied, and, if M is a given positive number, let $\{\mu_n\}$ be a sequence such that $|\lambda_n - \mu_n| \leqq M$. It is to be shown that if $f(z)$ is an entire function of exponential type γ belonging to $L_2(-\infty, \infty)$ on the real axis and ρ is a given positive number, then

$$\sum_n |f(\mu_n) - f(\lambda_n)|^2 \leqq T \sum_n |f(\lambda_n)|^2 \tag{11}$$

where

$$T = \frac{B}{A}(e^{\gamma^2/\rho^2} - 1)(e^{M^2\rho^2} - 1). \tag{12}$$

If $f(z)$ satisfies these conditions, then Taylor's theorem shows that

$$f(\mu_n) - f(\lambda_n) = \sum_{k=1}^{\infty} \frac{f^{(k)}(\lambda_n)}{k!}(\mu_n - \lambda_n)^k.$$

Multiplying and dividing the last series termwise by ρ^k, we have from Cauchy's inequality

$$|f(\mu_n) - f(\lambda_n)|^2 \leqq \left\{\sum_{k=1}^{\infty} \frac{|f^{(k)}(\lambda_n)|^2}{\rho^{2k} k!}\right\}\left\{\sum_{k=1}^{\infty} \frac{(M\rho)^{2k}}{k!}\right\}.$$

The second sum on the right side of this inequality is $\exp(M^2\rho^2) - 1$. The function $f^{(k)}(z)$ is an entire function of exponential type γ, and, according to inequality (9), it belongs to class $L_2(-\infty, \infty)$ on the real axis. Then the function $f^{(k)}(z)$ satisfies inequality (10), so

$$\sum_n |f^{(k)}(\lambda_n)|^2 \leqq B\int_{-\infty}^{\infty} |f^{(k)}(x)|^2 dx \leqq B\gamma^{2k}\int_{-\infty}^{\infty} |f(x)|^2 dx.$$

Since the function $f(z)$ also satisfies (10), it is clear that the last expression is equal to or less than $\{B\gamma^{2k}/A\}\sum |f(\lambda_n)|^2$, and inequality (11) follows. Then Minkowski's inequality shows that

$$\left(\sum_n |f(\mu_n)|^2\right)^{1/2} \leqq \left(\sum_n |f(\lambda_n)|^2\right)^{1/2} + \left(T\sum_n |f(\lambda_n)|^2\right)^{1/2},$$

so the lemma follows with $C=(1+T^{1/2})^2$.

Lemma II shows that the constant B of Theorems I, I′ exists. For in the case $d=1$ it is well known that the sequence of functions e^{int} is a frame over $(-\gamma, \gamma)$ where $0<\gamma\leqq\pi$. Since $|\lambda_n-n|\leqq L$, the existence of B follows from Lemma II. The case $d\neq 1$ may be reduced to the case $d=1$ by a change of variables.

The following result shows that the set of points $\{\lambda_n\}$ such that $\{\exp(i\lambda_n t)\}$ is a frame over a fixed interval is in a sense an open set.

Lemma III. *Let $\{\exp(i\lambda_n t)\}$ be a frame over $(-\gamma, \gamma)$. There is a $\delta_1>0$ such that $\{\exp(i\mu_n t)\}$ is a frame over the same interval whenever $|\mu_n-\lambda_n|\leqq\delta_1$.*

Proof. Let $f(z)$ be an entire function of exponential type γ such that $f(x) \in L_2(-\infty, \infty)$, and let $\{\exp(i\lambda_n t)\}$ be a frame over $(-\gamma, \gamma)$. Then $f(z)$ satisfies inequality (11) where $|\mu_n-\lambda_n|\leqq M$, and T is defined by (12) with ρ and M any positive numbers. From (11) and Minkowski's inequality we see that

$$\left(\sum_n |f(\lambda_n)|^2\right)^{1/2} \leqq \left(\sum_n |f(\mu_n)|^2\right)^{1/2} + \left(T\sum_n |f(\lambda_n)|^2\right)^{1/2}. \tag{13}$$

Now let $\delta_1=M=1/\rho$ and choose ρ so large that $T<1/4$. Then inequalities (10) and (13) show that

$$\frac{A}{4}\int_{-\infty}^{\infty} |f(x)|^2 dx \leqq \sum_n |f(\mu_n)|^2.$$

Lemma II completes the proof.

The above method is similar to that used by Duffin and Eachus [5] to show that $\{\exp(i\mu_n t)\}$ is a frame over $(-\pi, \pi)$ if $|\mu_n-n|\leqq M<(\log 2)/\pi$. To obtain this result let $\lambda_n=n$, $\gamma=\pi$, $\rho=(\gamma/M)^{1/2}$. Then since in this case $B=A$, we see from (12) that $T=(e^{\gamma M}-1)^2$, so $T<1$ if $M<(\log 2)/\pi$. From (10) and (13) it is seen that $\{\exp(i\mu_n t)\}$ is a frame over $(-\pi, \pi)$ if $|\mu_n-n| \leqq M<(\log 2)/\pi$. Theorem II will serve to strengthen this result by showing that $\{\exp(i\mu_n t)\}$ is a frame over $(-\pi, \pi)$ if there are constants β and M such that $|I(\mu_n)| \leqq \beta$, $|\text{Re}(\mu_n) - n| \leqq M < (\log 2)/\pi$.

The proof of the existence of the positive constant A in Theorem I depends on several lemmas. The following result has previously been proved by the authors [6].

LEMMA IV. *If $f(z)$ is an entire function of exponential type γ and $\{\lambda_n\}$ is a sequence of uniform density d, $d > \gamma/\pi$, then*

$$|f(x + iy)| \leqq Me^{\gamma|y|} \sup |f(\lambda_n)|$$

where the constant M is independent of $f(z)$.

The following lemma is closely related to results of Bourgin [4] and Ibragimov [7]; however, these authors were not concerned with sequences of uniform density. It is stated in a more general form than needed because of its intrinsic interest.

LEMMA V. *Let A_2 be the closed subspace of $L_2(-\pi, \pi)$ generated by the set of functions $1, e^{i\theta}, e^{2i\theta}, \cdots$. If $f(z)$ is an entire function of exponential type γ such that $f^{(n)}(0) \neq 0$, $n = 0, 1, 2, \cdots$, and the sequence $\{\lambda_n\}$ has uniform density d, $d > \gamma/\pi$, then the set of functions $f(\lambda_n e^{i\theta})$, $n = 0, \pm 1, \pm 2, \cdots$, is complete in A_2.*

Proof. Let

$$F(z) = \int_{-\pi}^{\pi} f(ze^{i\theta}) g^*(\theta) d\theta \tag{14}$$

where $g(\theta) \in A_2$, that is,

$$g(\theta) = \underset{N \to \infty}{\text{l.i.m.}} \sum_0^N c_\nu e^{i\nu\theta}$$

with $\sum |c_\nu|^2 < \infty$. Here and elsewhere the * represents the complex conjugate. Since $f(z)$ is an entire function of exponential type γ, it is clear that $F(z)$ is also. Under the hypothesis

$$F(\lambda_n) = \int_{-\pi}^{\pi} f(\lambda_n e^{i\theta}) g^*(\theta) d\theta = 0, \qquad n = 0, \pm 1, \pm 2, \cdots,$$

it is to be shown that $g(\theta) = 0$ almost everywhere. Since $F(\lambda_n) = 0$, Lemma IV shows that $F(z)$ vanishes identically. Then differentiating (14) we have

$$F^{(k)}(0) = \int_{-\pi}^{\pi} f^{(k)}(0) e^{ik\theta} g^*(\theta) d\theta$$

for $k = 0, 1, 2, \cdots$. Since $f^{(k)}(0) \neq 0$, the lemma follows.

LEMMA VI. *Given R satisfying $0 < R < \pi$, let $p(z)$ be regular in the circle*

$|z| \leqq R$ and let $\{\lambda_n\}$ be a sequence of uniform density 1. For each positive number h there is an integer N and a finite set of numbers $a_{-N}, a_{-N+1}, \cdots, a_0, \cdots, a_N$ such that

$$p(z) - \sum_{-N}^{N} a_n e^{i\lambda_n z} = \sum_{k=0}^{\infty} b_k z^k \tag{15}$$

and

$$|b_k| \leqq \frac{h}{R^k}, \qquad |a_n| \leqq N. \tag{16}$$

Moreover, given h, R, $p(z)$, L, δ, the same N suffices for all sequences satisfying

$$|\lambda_n - n| \leqq L, \qquad |\lambda_n - \lambda_m| \geqq \delta > 0 \quad \text{for} \quad n \neq m. \tag{17}$$

Proof. the function $f(z) = e^{iRz}$ satisfies the conditions of Lemma V with $\gamma = R < \pi$, so the set of functions $\{\exp(i\lambda_n Re^{i\theta})\}$ is complete in A_2. It is well known that closure and completeness are equivalent in A_2 so there is a finite set of numbers $a_{-M}, a_{-M+1}, \cdots, a_M$ such that if

$$\tau(Re^{i\theta}) = p(Re^{i\theta}) - \sum_{-M}^{M} a_n e^{i\lambda_n Re^{i\theta}}$$

then

$$\left\{\frac{1}{2\pi}\int_{-\pi}^{\pi} |\tau(Re^{i\theta})|^2 d\theta\right\}^{1/2} \leqq \frac{h}{2}. \tag{18}$$

The Taylor's series of $\tau(z)$ about the origin

$$\tau(z) = p(z) - \sum_{-M}^{M} a_n e^{i\lambda_n z} = \sum_{k=0}^{\infty} b_k z^k$$

converges in $|z| < R$. Then making use of Cauchy's integral representation of b_k we have from (18) and Schwarz's inequality

$$|b_k| = \left|\frac{1}{2\pi i}\int_{|z|=R} \tau(z)\frac{dz}{z^{k+1}}\right| \leqq \frac{h}{2R^k}.$$

Let $a_n = 0$ for $|n| > M$. Then choosing a sufficiently large integer N, the first part of the lemma follows.

The proof of the second part of the lemma is by contradiction. Let h, R, $p(z)$, L, δ be fixed and suppose there is no N which suffices for all sequences $\{\lambda_n\}$ satisfying (17). Then there are sequences $\{\lambda_n^{(1)}\}, \{\lambda_n^{(2)}\}, \cdots, \{\lambda_n^{(j)}\}, \cdots$ such that the least integer $N = N(j)$ for which (15) and (16) are true for some $a_n^{(j)}$, $b_k^{(j)}$ satisfies

$$N(j) > j.$$

From relation (17) it is clear that there is a subset of the sequences $\{\lambda_n^{(j)}\}$, which by a renumbering we suppose is the entire set, which converges to a limit sequence $\{\lambda_n^{(0)}\}$,

$$\lim_{j\to\infty} \lambda_n^{(j)} = \lambda_n^{(0)}, \qquad n = 0 \pm 1, \pm 2, \cdots.$$

The sequence $\{\lambda_n^{(0)}\}$ satisfies (17) and so has uniform density 1. Then, by what has been shown, there is an N_0 and a set of numbers $a_n^{(0)}$ such that

$$\left\{\frac{1}{2\pi}\int_{-\pi}^{\pi}\left|p(Re^{i\theta}) - \sum_{-N_0}^{N_0} a_n^{(0)} \exp\left(i\lambda_n^{(0)} Re^{i\theta}\right)\right|^2 d\theta\right\}^{1/2} \leqq \frac{h}{2},$$

$$\left|a_n^{(0)}\right| \leqq N_0, \qquad n = 0, \pm 1, \pm 2, \cdots, \pm N_0.$$

The finite sum

$$\sum_{-N_0}^{N_0} a_n^{(0)} \exp\left(i\lambda_n Re^{i\theta}\right)$$

is a continuous function of the $2N_0+1$ variables λ_n, so for sufficiently large j

$$\left\{\frac{1}{2\pi}\int_{-\pi}^{\pi}\left|p(Re^{i\theta}) - \sum_{-N_0}^{N_0} a_n^{(0)} \exp\left(i\lambda_n^{(j)} Re^{i\theta}\right)\right|^2 d\theta\right\}^{1/2} \leqq h.$$

This gives a contradiction since $N = N_0$ suffices in (15) and (16) for all large j.

Proof of Theorem I. It has been shown that the constant B of inequality (4) exists. To show the existence of the positive constant A, it is sufficient to consider the case $d=1$, $0<\gamma<\pi$. Given the sequence $\{\lambda_n\}$ of uniform density 1, let

$$\lambda_n^{(\nu)} = \lambda_{n+\nu} - \nu. \tag{19}$$

Then for each positive and negative integer ν the sequence $\{\lambda_n^{(\nu)}\}$ is of uniform density 1 with the same bounds L, δ as the given sequence,

$$\left|\lambda_n^{(\nu)} - n\right| \leqq L, \qquad \left|\lambda_n^{(\nu)} - \lambda_m^{(\nu)}\right| \geqq \delta > 0 \qquad \text{for } n \neq m.$$

In Lemma VI define $p(z)$, R, h by

$$p(z) = 1, \qquad R = \frac{1}{2}(\gamma + \pi), \qquad h = \frac{1}{2R}(R - \gamma).$$

Then the lemma asserts that for suitable $a_n^{(\nu)}$, $b_k^{(\nu)}$, N we have

$$1 - \sum_{n=-N}^{N} a_n^{(\nu)} \exp\left(i\lambda_n^{(\nu)} x\right) = \sum_{k=0}^{\infty} b_k^{(\nu)} x^k \tag{20}$$

where

$$|b_k^{(\nu)}| \leqq \frac{h}{R^k}, \qquad |a_n^{(\nu)}| \leqq N; \tag{21}$$

and the same constant N suffices for $\nu=0, \pm 1, \pm 2, \cdots$. If we define

$$\psi_\nu(x) = e^{i\nu x} - e^{i\nu x}\sum_{k=0}^{\infty} b_k^{(\nu)} x^k, \tag{22}$$

it follows from (19) and (20) that

$$\psi_\nu(x) = \sum_{n=-N}^{N} a_n^{(\nu)} \exp(i\lambda_{n+\nu}x). \tag{23}$$

Writing

$$\zeta^{(\nu)}(x) = e^{i\nu x}\sum_{k=0}^{\infty} b_k^{(\nu)} x^k, \tag{24}$$

we have

$$\psi_\nu(x) = e^{i\nu x} - \zeta^{(\nu)}(x).$$

It is to be shown that $\zeta^{(\nu)}(x)$ has in some sense a small average value, and it will follow that $\psi_\nu(x)$ has some average behavior similar to $e^{i\nu x}$. In the remainder of the proof of Theorem I we shall use the notation

$$(\phi, g) = \frac{1}{2\pi}\int_{-\gamma}^{\gamma}\phi(x)g(x)dx, \qquad \|g\| = \left\{\frac{1}{2\pi}\int_{-\gamma}^{\gamma}|g(x)|^2dx\right\}^{1/2}.$$

Given a function $g(x)\in L_2(-\gamma, \gamma)$, define $g(x)=0$ in the part of $(-\pi, \pi)$ that lies outside $(-\gamma, \gamma)$. Then Parseval's relation takes the form

$$\|g\|^2 = \sum_{-\infty}^{\infty}|(e^{i\nu x}, g)|^2.$$

Since $R>\gamma$, inequality (21) shows that the series (24) converges uniformly in $-\gamma\leqq x\leqq\gamma$; hence after multiplying (24) by $g(x)$ and integrating termwise we obtain

$$(\zeta^{(\nu)}, g) = \sum_{k=0}^{\infty} b_k^{(\nu)}(e^{i\nu x}, x^k g).$$

If this series is multiplied and divided termwise by $(R\gamma)^{k/2}$, then Cauchy's inequality and (21) show that

$$|(\zeta^{(\nu)}, g)|^2 \leqq h^2\left\{\sum_{k=0}^{\infty}\left(\frac{\gamma}{R}\right)^k\right\}\sum_{k=0}^{\infty}|(e^{i\nu x}, x^k g)|^2 R^{-k}\gamma^{-k}.$$

Then making use of Parseval's relation, we have

$$\sum_{\nu=-\infty}^{\infty} |(\zeta^{(\nu)}, g)|^2 \leqq \frac{h^2R}{R-\gamma} \sum_{k=0}^{\infty} \sum_{\nu=-\infty}^{\infty} |(e^{i\nu x}, x^k g)|^2 R^{-k}\gamma^{-k}$$
$$= \frac{h^2R}{R-\gamma} \sum_{k=0}^{\infty} \frac{\|x^k g\|^2}{R^k\gamma^k}.$$

In the last series the integration defining $\|x^k g\|$ need only be carried over $(-\gamma, \gamma)$, so $\|x^k g\| \leqq \gamma^k \|g\|$. Thus, recalling the definition of h, we have

$$\sum_{\nu=-\infty}^{\infty} |(\zeta^{(\nu)}, g)|^2 \leqq \|g\|^2/4. \tag{25}$$

Since $e^{i\nu x} = \psi_\nu(x) + \zeta^{(\nu)}(x)$, we have $(e^{i\nu x}, g) = (\psi_\nu, g) + (\zeta^{(\nu)}, g)$, and then Minkowski's inequality shows that

$$\left(\sum_{-\infty}^{\infty} |(e^{i\nu x}, g)|^2\right)^{1/2} \leqq \left(\sum_{-\infty}^{\infty} |(\psi_\nu, g)|^2\right)^{1/2} + \left(\sum_{-\infty}^{\infty} |(\zeta^{(\nu)}, g)|^2\right)^{1/2}.$$

The left side of this inequality is equal to $\|g\|$ by Parseval's theorem, and the second of the two terms on the right we have shown in (25) is dominated by $\|g\|/2$. Thus

$$\sum_{-\infty}^{\infty} |(\psi_\nu, g)|^2 \geqq \|g\|^2/4. \tag{26}$$

Now substituting in (23), we have

$$(\psi_\nu, g) = \sum_{n=-N}^{N} a_n^{(\nu)} (e^{\lambda_{n+\nu} x}, g),$$

so Cauchy's inequality and the estimate of $a_n^{(\nu)}$ given in (21) shows that

$$|(\psi_\nu, g)|^2 \leqq (2N+1)N^2 \sum_{n=-N}^{N} |(e^{i\lambda_{n+\nu} x}, g)|^2.$$

If the last inequality is summed on ν from $-\infty$ to ∞, then in the double summation the sum $n+\nu$ runs through each integer precisely $2N+1$ times. Thus

$$\sum_{-\infty}^{\infty} |(\psi_\nu, g)|^2 \leqq (2N+1)^2 N^2 \sum_{k=-\infty}^{\infty} |(e^{i\lambda_k x}, g)|^2. \tag{27}$$

If we combine (26) and (27, Theorem I follows; and with $d=1$ we have $A \geqq 1/\{4N^2(2N+1)^2\}$. The magnitude of N is not determined by the previous argument so this is not an estimate of A.

As previously remarked, Theorem I is equivalent to Theorem I′. However, the conditions of Theorem I′ are unnecessarily restrictive because it is not necessary to suppose that $f(x)\in L_2(-\infty, \infty)$. If $\{\lambda_n\}$, γ, d satisfy the conditions of Theorem I′, then every entire function $f(z)$ of exponential type γ satisfies inequality (5). If either of $\int_{-\infty}^{\infty}|f(x)|^2dx$, $\sum|f(\lambda_n)|^2$ is finite, then the other is also and (5) holds. For if $\int_{-\infty}^{\infty}|f(x)|^2dx<\infty$, then inequality (5) follows from Theorem I′, so suppose that $\sum|f(\lambda_n)|^2<\infty$. Then $f(z)$ is bounded at the points $z=\lambda_n$, so Lemma IV shows that $f(z)$ is bounded on the entire real axis. The function

$$F_\epsilon(z) = f(z)\,\frac{\sin \epsilon z}{\epsilon z}$$

where $0<\epsilon<(\pi d-\gamma)/2$ is an entire function of exponential type $(\pi d+\gamma)/2$ and $F_\epsilon(x)\in L_2(-\infty, \infty)$. By Theorem I′ it follows that there are positive constants $A^\Delta=A((\gamma+\pi d)/2)$, $B^\Delta=B((\gamma+\pi d)/2)$ such that

$$A^\Delta \leqq \frac{\sum |F_\epsilon(\lambda_n)|^2}{\int_{-\infty}^{\infty} |F_\epsilon(x)|^2dx} \leqq B^\Delta.$$

Letting ϵ approach zero, the constants A^Δ, B^Δ may be supposed fixed since $F_\epsilon(z)$ remains of type $(\pi d+\gamma)/2$. It then follows that the limit $f(z)$ of $F_\epsilon(z)$ also satisfies this inequality. Thus $f(x)\in L_2(-\infty, \infty)$, so Theorem I′ shows that $f(z)$ satisfies inequality (5).

This remark leads to a strengthening of Theorem I. Without attempting a complete analogy to the stronger form of Theorem I′, let $\{\lambda_n\}$, γ satisfy the conditions of Theorem I with $d=1$. Thus $0<\gamma<\pi$ and $\{\lambda_n\}$ has uniform density 1. Let $G(t)$ be a function of bounded variation over $(-\gamma, \gamma)$. If

$$\frac{1}{2\pi}\sum_{-\infty}^{\infty}\left|\int_{-\gamma}^{\gamma} e^{i\lambda_n t}dG(t)\right|^2 < \infty,$$

then $G(t)$ is essentially the indefinite integral of a function of class $L_2(-\infty, \infty)$, and

$$A \leqq \frac{\frac{1}{2\pi}\sum_{-\infty}^{\infty}\left|\int_{-\gamma}^{\gamma} e^{i\lambda_n t}dG(t)\right|^2}{\int_{-\gamma}^{\gamma} |G'(t)|^2dt} \leqq B.$$

This follows from Lemma I and the stronger form of Theorem I′ since

$$f(z) = \frac{1}{(2\pi)^{1/2}}\int_{-\gamma}^{\gamma} e^{izt}dG(t)$$

is an entire function of exponential type γ.

Not every frame over an interval $(-\gamma, \gamma)$ can be strengthened in the same manner to Stieltjes integrals. This is shown by the example in which $\gamma=\pi$, $\lambda_n=n$, and $G(t)=0$ for $-\pi<t<\pi$, $G(-\pi)=1$, $G(\pi)=1$. For in this case, although $\{e^{int}\}$ is a frame over $(-\pi, \pi)$, we see that $G(t)$ is not absolutely continuous but

$$\int_{-\pi}^{\pi} e^{int}dG(t) = 0, \qquad n = 0, \pm 1, \pm 2, \cdots .$$

Let $\{\lambda_n\}$ be a sequence of uniform density 1, and let $0<\gamma<\pi$. Over the interval $(-\gamma, \gamma)$, the set of functions $\{\exp (i\lambda_n t)\}$, $n=0, \pm 1, \pm 2, \cdots$, is a frame, and is therefore complete. The subset of functions $\{\exp (i\lambda_n t)\}$ for which $n>0$ is complete over $(-\gamma, \gamma)$ according to a result of Levinson [8, p. 3], but it is not a frame over this interval. For let $g(t)=e^{i\alpha t}$ where $\alpha>0$. Then

$$f(z) = \frac{1}{(2\pi)^{1/2}}\int_{-\gamma}^{\gamma} e^{i\alpha t}e^{izt}dt = \left(\frac{2}{\pi}\right)^{1/2} \frac{\sin \gamma(\alpha + z)}{\alpha + z}. \tag{28}$$

Now $\int_{-\gamma}^{\gamma}|g(t)|^2dt$ is a positive constant independent of α, but

$$\sum_{n=1}^{\infty}| f(\lambda_n) |^2 = \frac{2}{\pi}\sum_{n=1}^{\infty}\left|\frac{\sin \gamma(\alpha + \lambda_n)}{\alpha + \lambda_n}\right|^2 \leqq \frac{2}{\pi} e^{2\gamma L} \sum_{n=1}^{\infty} \frac{1}{|\alpha + \lambda_n|^2}$$

and this approaches zero as $\alpha\to\infty$. Thus there is no positive constant A such that (4) is satisfied.

We shall say that a set of functions $\{\exp (i\lambda_n t)\}$ is an exact frame over an interval I if it is a frame over I but fails to be a frame over I by the removal of any function of the set. This use of exact is analogous to that of Paley and Wiener in the case of exactly complete sets. If $\{\exp (i\lambda_n t)\}$ is a frame over I but is not an exact frame over I, then we say it is an overcomplete frame over I.

If $\{\lambda_n\}$ is a set of uniform density 1 and $0<\gamma<\pi$, then, according to Theorem I, the set of functions $\{\exp (i\lambda_n t)\}$ is a frame over $(-\gamma, \gamma)$. This set of functions is an overcomplete frame over $(-\gamma, \gamma)$, indeed if any finite number of the functions are deleted the remaining functions are a frame over $(-\gamma, \gamma)$. This follows from the fact that if a finite number of the λ_n are deleted, the remaining λ_n may be reindexed so as to satisfy inequalities (2), (3) with the bound L replaced by a larger number.

LEMMA VII. *If $\{\exp (i\lambda_n t)\}$ is a frame over $(-\gamma, \gamma)$ but fails to be a frame over this interval by the removal of some function of the set, then it fails to be a frame over $(-\gamma, \gamma)$ by the removal of any function of the set.*

Proof. Suppose that there are positive constants A, B such that for all $g(t) \in L_2(-\gamma, \gamma)$,

$$A \leqq \frac{\frac{1}{2\pi}\sum_n \left| \int_{-\gamma}^{\gamma} e^{i\lambda_n t} g(t)dt \right|^2}{\int_{-\gamma}^{\gamma} | g(t) |^2 dt} \leqq B$$

but that if λ_m is omitted there are no such A, B. The failure must be in A rather than B. Then there is a sequence of functions $g_k(t)$, $k=1, 2, 3, \cdots$, whose Fourier transforms we write $f_k(z)$,

$$f_k(z) = \frac{1}{(2\pi)^{1/2}} \int_{-\gamma}^{\gamma} g_k(t) e^{izt} dt, \tag{29}$$

which are normalized by the relation

$$\int_{-\infty}^{\infty} | f_k(x) |^2 dx = \int_{-\gamma}^{\gamma} | g_k(t) |^2 dt = 1$$

and such that

$$\sum_n | f_k(\lambda_n) |^2 \geqq A, \qquad \sum_{n\neq m} | f_k(\lambda_n) |^2 \leqq \frac{1}{k} \cdot$$

Clearly $f_k(\lambda_n)$ tends to zero as $k\to\infty$ for each $n\neq m$, but $|f_k(\lambda_m)|$ has a lower bound equal to or greater than $A^{1/2}>0$ as $k\to\infty$. Now $f_k(z)$ is an entire function of exponential type γ, and, according to (8), it satisfies

$$| f_k(x + iy) | \leqq (\gamma/\pi)^{1/2} e^{\gamma |y|}, \qquad k = 1, 2, 3, \cdots.$$

The sequence of entire functions being uniformly dominated, there is a subsequence which converges to a limit $f(z)$, and uniformly in every bounded domain. Thus $f(z)$ is an entire function of exponential type γ and

$$| f(x + iy) | \leqq \left(\frac{\gamma}{\pi}\right)^{1/2} e^{\gamma |y|}, \qquad \int_{-\infty}^{\infty} | f(x) |^2 dx \leqq 1.$$

Now $f(\lambda_n)=0$ for $n\neq m$, and $f(\lambda_m)\neq 0$, in particular, $f(z)$ does not vanish identically. Therefore, if j is any integer and

$$F(z) = \frac{z - \lambda_m}{z - \lambda_j} f(z),$$

then $F(z)$ is a not identically vanishing entire function of exponential type γ. It is readily shown that $F(x)\in L_2(-\infty, \infty)$ and $F(\lambda_n)=0$ for $n\neq j$. According to Lemma I there is a function $G(t)\in L_2(-\gamma, \gamma)$ such that

$$F(z) = \frac{1}{(2\pi)^{1/2}} \int_{-\gamma}^{\gamma} G(t) e^{izt} dt.$$

$G(t)$ is not equivalent to zero in $(-\gamma, \gamma)$, but

$$\frac{1}{(2\pi)^{1/2}}\int_{-\gamma}^{\gamma} G(t)e^{i\lambda_n t}dt = 0, \qquad n \neq j,$$

so the lemma follows.

THEOREM II. *Let* $\lambda_n = \alpha_n + i\beta_n$ *where* α_n *and* β_n *are real and* $|\beta_n| \leqq \beta$ *for some constant* β. *If the set of functions* $\{\exp(i\alpha_n t)\}$ *is a frame over an interval* $(-\gamma, \gamma)$, *then* $\{\exp(i\lambda_n t)\}$ *is a frame over the same interval.*

Proof. Let $f(z)$ be an entire function of exponential type γ such that $f(x) \in L_2(-\infty, \infty)$. According to Hadamard's factorization theorem it may be written in the form

$$f(z) = z^k e^{az+b}\prod_{\nu=1}^{\infty}\left(1 - \frac{z}{z_\nu}\right)e^{z/z_\nu} \tag{30}$$

where $z_1, z_2, z_3, \cdots$ are the zeros of $f(z)$ other than at the origin. Now $f(z)$ satisfies inequality (8), so Carleman's formula [11] written for the upper half-plane and for the lower half-plane in turn shows that $\sum \operatorname{Im}(1/z_\nu)$ is an absolutely convergent series. If

$$1/z_\nu = p_\nu + iq_\nu$$

where p_ν and q_ν are real, Hadamard's formula may be written in the form

$$f(z) = z^k e^{(c+id)z+b}\prod_{\nu=1}^{\infty}\left(1 - \frac{z}{z_\nu}\right)e^{p_\nu z} \tag{31}$$

where c and d are real,

$$c = \operatorname{Re}(a), \qquad d = \operatorname{Im}(a) + \sum_{\nu=1}^{\infty} q_\nu.$$

This product converges uniformly in every bounded domain since (30) does and $\sum q_\nu$ is an absolutely convergent series.

We are going to define a sequence $\{\lambda_n^{(1)}\}$ and an entire function $f_1(z)$ of exponential type γ belonging to $L_2(-\infty, \infty)$ on the real axis such that

$$\lambda_n^{(1)} = \alpha_n + i\beta_n^{(1)}, \qquad |\beta_n^{(1)}| \leqq \beta/2, \tag{32}$$

and

$$\frac{\sum_n |f_1(\lambda_n^{(1)})|^2}{\int_{-\infty}^{\infty}|f_1(x)|^2dx} \leqq e^{\beta\gamma}\frac{\sum_n |f(\lambda_n)|^2}{\int_{-\infty}^{\infty}|f(x)|^2dx}. \tag{33}$$

There are two cases to consider, $d \geqq 0$ and $d<0$. Suppose first that $d \geqq 0$. Now form from the given function $f(z)$ another entire function whose zeros are obtained by reflecting in the real axis those zeros of $f(z)$ that lie in the lower half-plane. Let

$$\overset{\Delta}{z}_\nu = \begin{cases} z_\nu & \text{if} \quad \text{Im}\,(z_\nu) \geqq 0, \\ \overset{*}{z}_\nu & \text{if} \quad \text{Im}\,(z_\nu) < 0 \end{cases}$$

and define

$$f^\Delta(z) = z^k e^{(c+id)z+b} \prod_{\nu=1}^{\infty}\left(1 - \frac{z}{z_\nu^\Delta}\right) e^{p_\nu z}. \tag{34}$$

Then $f^\Delta(z)$ is an entire function of exponential type satisfying

$$\left| f^\Delta(x) \right| = \left| f(x) \right|. \tag{35}$$

To show this we note that the zeros of $f(z)$ that lie in the lower half-plane may be reflected in the real axis one at a time. Thus, if m is any positive integer, the function

$$\overset{\Delta}{f}_m(z) = z^k e^{(c+id)z+b} \prod_{\nu=1}^{m}\left(1 - \frac{z}{z_\nu^\Delta}\right) e^{p_\nu z} \prod_{\nu=m+1}^{\infty}\left(1 - \frac{z}{z_\nu}\right) e^{p_\nu z}$$

has the same modulus on the real axis as $f(x)$, and it is an entire function of exponential type γ since the ratio $f_m^\Delta(z)/f(z)$ tends to 1 as z tends to infinity in any direction. Thus, $f_m^\Delta(z)$ is dominated by the right side of (8) for $m=1, 2, 3, \cdots$, and hence its limit $f^\Delta(z)$ is dominated by the right side of (8). The product (34) defining $f^\Delta(z)$ converges uniformly in every bounded domain. If we compare (31) and (34), it is clear that

$$\left| f^\Delta(x+iy) \right| \leqq \left| f(x+iy) \right|, \qquad y \geqq 0.$$

Since $d \geqq 0$, we see also that

$$\left| f^\Delta(x+iy) \right| \leqq \left| f(x-iy) \right|, \qquad y \geqq 0.$$

Now define a new sequence $\{\lambda_n^\Delta\}$ by reflecting in the real axis those points of $\{\lambda_n\}$ that lie in the lower half-plane. Let

$$\overset{\Delta}{\lambda}_n = \begin{cases} \lambda_n & \text{if} \quad \text{Im}\,(\lambda_n) \geqq 0, \\ \overset{*}{\lambda}_n & \text{if} \quad \text{Im}\,(\lambda_n) < 0. \end{cases}$$

Then, according to the inequalities stated in the preceding paragraph,

$$\left| \overset{\Delta}{f}(\overset{\Delta}{\lambda}_n) \right| \leqq \left| f(\lambda_n) \right|.$$

The points λ_n were assumed to lie in a strip of width 2β which is symmetric about the real axis. The points λ_n^Δ lie in a strip of width β which lies in the upper half-plane. We now translate the points λ_n^Δ to obtain a new sequence which lies in a strip of width β symmetric about the real axis. Let

$$\lambda_n^{(1)} = \lambda_n^\Delta - i\beta/2, \qquad f_1(z) = f^\Delta(z + i\beta/2).$$

Then the points $\lambda_n^{(1)}$ lie in the strip $|y| \leqq \beta/2$ and

$$|f_1(\lambda_n^{(1)})| = |f^\Delta(\lambda_n^\Delta)| \leqq |f(\lambda_n)|.$$

Also, according to Lemma I, there is a function $g^\Delta(t) \in L_2(-\gamma, \gamma)$ such that

$$f^\Delta(z) = \frac{1}{(2\pi)^{1/2}} \int_{-\gamma}^{\gamma} g^\Delta(t) e^{izt} dt$$

and, by (35),

$$\int_{-\infty}^{\infty} |f(x)|^2 dx = \int_{-\infty}^{\infty} |f^\Delta(x)|^2 dx = \int_{-\gamma}^{\gamma} |g^\Delta(t)|^2 dt.$$

Thus, writing $g_1(t) = g^\Delta(t) e^{-\beta t/2}$, we see that

$$f_1(z) = \frac{1}{(2\pi)^{1/2}} \int_{-\gamma}^{\gamma} g_1(t) e^{izt} dt.$$

Hence

$$\int_{-\infty}^{\infty} |f_1(x)|^2 dx = \int_{-\gamma}^{\gamma} |g^\Delta(t)|^2 e^{-\beta t} dt \geqq e^{-\beta\gamma} \int_{-\infty}^{\infty} |f(x)|^2 dx.$$

The sequence $\lambda_n^{(1)}$ and the function $f_1(z)$ thus defined satisfy (32) and (33).

If $d < 0$, we reflect in the real axis the zeros of $f(z)$ and the points that lie in the upper half-plane. Then after a translation we obtain a sequence $\{\lambda_n^{(1)}\}$ and an entire function $f_1(z)$ satisfying (32) and (33).

The above process may be iterated. If $f(z)$ is an entire function of exponential type γ which belongs to $L_2(-\infty, \infty)$ on the real axis, then for $k = 1, 2, 3, \cdots$ there is a sequence $\{\lambda_n^{(k)}\}$ and an entire function $f_k(z)$ of exponential type γ which belongs to $L_2(-\infty, \infty)$ on the real axis such that

$$(36) \qquad \lambda_n^{(k)} = \alpha_n + \beta_n^{(k)}, \qquad |\beta_n^{(k)}| \leqq \beta/2^k,$$

and

$$(37) \qquad \frac{\sum_n |f_k(\lambda_n^{(k)})|^2}{\int_{-\infty}^{\infty} |f_k(x)|^2 dx} \leqq e^{2\beta\gamma} \frac{\sum_n |f(\lambda_n)|^2}{\int_{-\infty}^{\infty} |f(x)|^2 dx}.$$

The constant $\exp(2\beta\gamma)$ is here written in place of $\exp(\gamma\beta+\gamma\beta/2+\cdots+\gamma\beta/2^{k-1})$.

Suppose now that $\{\exp(i\alpha_n t)\}$ is a frame over $(-\gamma, \gamma)$. According to Lemma III there is a $\delta_1>0$ such that the set of functions $\{\exp(i\mu_n t)\}$ is a frame whenever $|\mu_n-\alpha_n|\leqq\delta_1$. Let k be so large that $\beta/2^k<\delta_1$. Then $\{\exp(i\lambda_n^{(k)}t)\}$ is a frame over $(-\gamma, \gamma)$, and the left side of (37) therefore has a positive lower bound A'. Thus

$$A'e^{-2\beta\gamma} \leqq \frac{\sum_n |f(\lambda_n)|^2}{\int_{-\infty}^{\infty} |f(x)|^2 dx}.$$

Since $|\lambda_n-\alpha_n|\leqq\beta$, Lemma II asserts that the right side of (37) has a finite upper bound. It follows that the set of functions $\{\exp(i\lambda_n t)\}$ is a frame over $(-\gamma, \gamma)$.

3. **Abstract frames.** For two vectors u and v of Hilbert space, let $(u, v)=(v, u)^*$ be the complex scalar product which defines the norm $\|u\|=(u, u)^{1/2}$. We define a *frame* to be an infinite sequence of nonzero vectors $\phi_1, \phi_2, \phi_3, \cdots$ such that for an arbitrary vector v,

$$A\|v\|^2 \leqq \sum_n |(v, \phi_n)|^2 \leqq B\|v\|^2. \tag{38}$$

Here A and B are positive constants called *bounds* of the frame. The numbers $\alpha_n=(v, \phi_n)$, $n=1, 2, \cdots$, are called the moment sequence of the vector v relative to the frame. Since a frame is clearly a complete set, the finite linear combinations of the ϕ_n are everywhere dense. For the space $L_2(0, 1)$ the scalar product is defined as $(u, v)=\int_0^1 u(x)v^*(x)dx$.

If $\{\phi_n\}$ is a frame and $\{c_n\}$ is a sequence of numbers such that $\sum|c_n|^2<\infty$, then $\sum c_n\phi_n$ converges and

$$\left\|\sum_1^{\infty} c_n\phi_n\right\|^2 \leqq B\sum_1^{\infty}|c_n|^2. \tag{39}$$

For if

$$v_k = \sum_{n=1}^{k} c_n\phi_n, \qquad k \geqq 1,\ v_0 = 0,$$

then for $k\geqq j$ we have from Schwarz' inequality and the frame condition,

$$\|v_k - v_j\|^2 = \sum_{n=j+1}^{k} c_n(\phi_n, v_k - v_j)$$

$$\leqq \left\{\sum_{n=j+1}^{k} |c_n|^2\right\}^{1/2} \{B\|v_k - v_j\|^2\}^{1/2}.$$

Hence

$$\|v_k - v_j\|^2 \leqq B \sum_{n=j+1}^{k} |c_n|^2,$$

and (39) follows.

This shows that a linear transformation S is defined by the relation

$$Sv = \sum_{n=1}^{\infty} (v, \phi_n)\phi_n. \tag{40}$$

The transformation S is self-adjoint, and, if we make use of (38), it follows that

$$A\|v\|^2 \leqq (Sv, v) \leqq B\|v\|^2. \tag{41}$$

This states that S is positive definite with positive upper and lower bounds. Hence the inverse S^{-1} exists as a self-adjoint transformation, and

$$B^{-1}\|v\|^2 \leqq (S^{-1}v, v) \leqq A^{-1}\|v\|^2. \tag{42}$$

LEMMA VIII. *Let* $\{\phi_n\}$ *be a frame and let* v *be an arbitrary vector. Then there exists a moment sequence* $\{\beta_n\}$ *such that*

$$v = \sum_{1}^{\infty} \beta_n\phi_n \tag{43}$$

and

$$B^{-1}\|v\|^2 \leqq \sum_{1}^{\infty} |\beta_n|^2 \leqq A^{-1}\|v\|^2. \tag{44}$$

If $\{b_n\}$ *is any other sequence such that* $v = \sum_1^\infty b_n\phi_n$, *then* $\{b_n\}$ *is not the moment sequence of any vector, and*

$$\sum_{1}^{\infty} |b_n|^2 = \sum_{1}^{\infty} |\beta_n|^2 + \sum_{1}^{\infty} |b_n - \beta_n|^2. \tag{45}$$

Proof. The first part of the lemma follows from the preceding discussion when we write $v = Su$, $\beta_n = (u, \phi_n)$. To prove (45) there is no loss of generality in supposing $\sum |b_n|^2 < \infty$. Then $0 = v - v = \sum (b_n - \beta_n)\phi_n$, so, with $v = Su$, $\beta_n = (u, \phi_n)$, we have $0 = \sum (b_n - \beta_n)\beta_n^*$ and (45) follows. The uniqueness of the transformation S^{-1} shows that $\{b_n\}$ is not a moment sequence.

Now define a new sequence $\{\phi_n'\}$ by $\phi_n' = S^{-1}\phi_n$. Then $\{\phi_n'\}$ is a frame with the positive bounds A^{-1}, B^{-1}. For if v is any vector and $u = S^{-1}v$, then

$$\sum_n |(v, \phi_n')|^2 = \sum_n |(Su, \phi_n')|^2 = \sum_n |(u, \phi_n)|^2 = (Su, u) = (S^{-1}v, v)$$

so the result follows from (42). If v is any vector, then it may be expanded by *conjugate frames* in the form

$$v = \sum_n (v, \phi_n')\phi_n = \sum_n (v, \phi_n)\phi_n'. \tag{46}$$

The following result also gives a method of finding the expansion coefficients β_n by a rapidly converging process of successive approximations.

THEOREM III. *Let* $\{\phi_n\}$ *be a frame, and let* $\rho = 2/(A+B)$. *If* v *is an arbitrary vector, define*

$$v^{(1)} = v - \rho \sum_n (v, \phi_n)\phi_n,$$

$$v^{(k+1)} = v^{(k)} - \rho \sum_n (v^{(k)}, \phi_n)\phi_n, \qquad k \geqq 1.$$

Let

$$\beta_n^{(k)} = \rho(v + v^{(1)} + v^{(2)} + \cdots + v^{(k-1)}, \phi_n),$$

$$v_k = \sum_{n=1}^{\infty} \beta_n^{(k)}\phi_n.$$

Then

$$\|v - v_k\| \leqq \left(\frac{B - A}{B + A}\right)^k \|v\|. \tag{47}$$

Proof. The transformation $T = I - \rho S$ satisfies

$$|(Tv, v)| \leqq \theta\|v\|^2, \qquad \theta = \frac{B - A}{B + A}.$$

Since T is self-adjoint, $\|Tv\| \leqq \theta\|v\|$. Thus $\|v^{(1)}\| \leqq \theta\|v\|$, and in general, $\|v^{(k)}\| \leqq \theta^k\|v\|$. Adding the relations $v^{(k+1)} - v^{(k)} = -\rho S v^{(k)}$ for $k = 0, 1, \cdots, m-1$, we see that $v^{(m)} - v = -v_m$. Thus $\|v_m - v\| = \|v^{(m)}\| \leqq \theta^m\|v\|$.

A frame which fails to be a frame on the removal of any one of its vectors is termed an *exact frame*. It is not difficult to show by an example that the abstract analogue of Lemma VII is false.

LEMMA IX. *The removal of a vector from a frame leaves either a frame or an incomplete set.*

Proof. Suppose that ϕ_m is removed from the frame $\{\phi_n\}$. As a special case of (43) we may write

$$\phi_m = \sum \beta_n \phi_n \tag{48}$$

where $\phi_m' = S^{-1}\phi_m$ and $\beta_n = (\phi_m', \phi_n)$.

If $\beta_m \neq 1$, then

$$\phi_m = (1 - \beta_m)^{-1} \sum{}' \beta_n \phi_n$$

where $\sum'$ indicates the omission of the mth term. Thus if v is an arbitrary vector, then

$$|(v, \phi_m)|^2 \leqq |1 - \beta_m|^{-2} \{ \sum{}' |\beta_n|^2 \} \{ \sum{}' |(\phi_n, v)|^2 \}$$

so

$$\sum |(v, \phi_n)|^2 \leqq \{1 + |1 - \beta_m|^{-2} \sum{}' |\beta_n|^2\} \sum{}' |(v, \phi_n)|^2.$$

It follows that the subset of $\{\phi_n\}$ with $n \neq m$ is a frame, for in place of (38) we have $A'\|v\|^2 \leqq \sum' |(v, \phi_n)|^2 \leqq B\|v\|^2$ where

$$A' = \frac{A}{1 + |1 - \beta_m|^{-2} \sum' |\beta_n|^2}.$$

Now suppose that in (48) we have $\beta_m = 1$. We show in this case that $\{\phi_n\}$, $n \neq m$, is incomplete. Since $\beta_m = 1$, we obtain

$$0 = \sum{}' \beta_n \phi_n$$

and this may be written

$$0 = \sum \beta_n' \phi_n$$

where $\beta_n' = \beta_n$ if $n \neq m$, $\beta_m' = 0$. But $0 = \sum 0 \cdot \phi_n$ so relation (45) with $b_n = 0$ shows that

$$2 \sum |\beta_n'|^2 = 0.$$

Thus $\beta_n = 0$ for $n \neq m$. Now $\beta_n = (\phi_m', \phi_n)$ where $\phi_m' = S^{-1}\phi_m$, and the vector ϕ_m' is therefore orthogonal to all vectors of the set $\{\phi_n\}$, $n \neq m$. Thus

(49) $$(\phi_m', \phi_n) = \delta_{mn}.$$

LEMMA X. *If $\{\phi_n\}$ is an exact frame, then $\{\phi_n\}$ and $\{\phi_n'\}$, where $\phi_n' = S^{-1}\phi_n$, are biorthogonal. Any sequence of numbers $\{c_n\}$ such that $\sum |c_n|^2 < \infty$ is the moment sequence of some vector with respect to $\{\phi_n\}$, and*

(50) $$A \sum |c_n|^2 \leqq \|\sum c_n \phi_n\| \leqq B \sum |c_n|^2.$$

Proof. If $\{\phi_n\}$ is an exact frame, then (49) is true for all m as well as for all n so $\{\phi_n\}$ and $\{\phi_n'\}$ are biorthogonal. Given the sequence $\{c_n\}$, according to (39) the vector v, $v = \sum c_n \phi_n$, has finite norm. Then

$$c_n = (v, \phi_n') = (u, \phi_n)$$

where $u = S^{-1}v$, so $\{c_n\}$ is the moment sequence of the vector u. Then (50) follows from (44).

Paley and Wiener have also given a Hilbert space development of their theory. It is of interest to show the precise relationship with the present theory. Their theory concerns an infinite sequence of vectors $f_1, f_2, \cdots$ which is close to a complete orthonormal sequence $\psi_1, \psi_2, \cdots$. By "close" is meant that for any sequence of complex numbers $\{c_n\}$,

$$\|\sum c_n(f_n - \psi_n)\|^2 \leqq \theta^2 \sum |c_n|^2 \tag{51}$$

where θ, $0<\theta<1$, is a constant independent of $\{c_n\}$. An arbitrary vector v may be represented as $v=\sum c_n\psi_n$ and it may be shown that (51) implies that the sequence $\{f_n\}$ satisfies the frame condition (38). This was first pointed out by Boas [1]. Applying the triangle inequality to (51) gives

$$(1-\theta)^2\sum |c_n|^2 \leqq \|\sum c_n f_n\|^2 \leqq (1+\theta)^2 \sum |c_n|^2. \tag{52}$$

Hence $\{f_n\}$ fails to be complete on the removal of any one of the f_n. Thus $\{f_n\}$ is an exact frame.

Conversely suppose that $\{\phi_n\}$ is an exact frame. Let $f_n = 2(B^{1/2}+A^{1/2})^{-1}\phi_n$ where A and B are the upper and lower bounds of the sequence $\{\phi_n\}$. Then (50) may be expressed in the form (52) with $\theta=(B^{1/2}-A^{1/2})/(B^{1/2}+A^{1/2})$. It was shown by Duffin and Eachus [5] that relation (52) for a complete set $\{f_n\}$ implies the existence of a complete orthonormal set $\{\psi_n\}$ satisfying (51). *Thus the theory of Paley and Wiener and the theory of exact frames are equivalent.*

The following theorem gives a new example of an exact frame. The proof is omitted. *Let $\{\lambda_n\}$ be a sequence of uniform density d such that for some positive constant τ, $\{\lambda_n\}$ and $\{\lambda_n+\tau\}$ are the same set of points. Then the set of functions $\{\exp(i\lambda_n t)\}$ is an exact frame over $(-\pi d, \pi d)$.*

4. **Pointwise convergence.** Let $\{\lambda_n\}$ be a sequence of uniform density 1 and suppose that $g(x)\in L_2(-\pi, \pi)$. Then Theorem I and Lemma VIII together imply that corresponding to each positive constant γ, $0<\gamma<\pi$, there exist expansion coefficients c_n such that $\sum|c_n|^2<\infty$ and

$$g(x) = \underset{N\to\infty}{\text{l.i.m.}} \sum_{-N}^{N} c_n e^{i\lambda_n x}, \qquad -\gamma \leqq x \leqq \gamma.$$

We shall show that indeed the limit in the mean $\sum_{-N}^{N} c_n \exp(i\lambda_n x)$ exists over the larger interval $(-\pi, \pi)$, defining a function $g(x)$ in the latter interval. It is also to be shown that the nonharmonic Fourier series actually converges to $g(x)$ at a given point of $(-\pi, \pi)$ if and only if the ordinary Fourier series of $g(x)$ converges to $g(x)$ at this point. A similar statement could be made for summability. Among all sequences $\{c_n\}$ corresponding to the same $g(x)$ in $(-\gamma, \gamma)$ there exists a unique sequence which minimizes $\sum|c_n|^2$, according to Lemma VIII. Whether or not $\{c_n\}$ is this minimizing sequence does not effect the convergence of the nonharmonic Fourier series.

Paley and Wiener have also obtained convergence properties for their class of nonharmonic Fourier series. Theorem IV is a generalization of their result and has a sharper conclusion. The proof is along different lines than that of these authors. The convergence theory of Paley and Wiener has been generalized along different lines by Levinson [8].

In order to investigate the convergence of the nonharmonic Fourier series, we write

$$e^{i\lambda_n x} = e^{inx}e^{i(\lambda_n - n)x} = e^{inx}\sum_{k=0}^{\infty}\frac{i^k(\lambda_n - n)^k x^k}{k!}$$

and we recall that since $\{\lambda_n\}$ is a sequence of uniform density 1, $|\lambda_n - n|$ is bounded. Thus, in the notation of Theorem IV below,

$$g_N(x) = \sum_{-N}^{N} c_n e^{\lambda_n x} = \sum_{-N}^{N} c_n \zeta_n(x)$$

are the partial sums of the nonharmonic Fourier series of $g(x)$.

THEOREM IV. *Let $\{b_k\}$ be a sequence of positive constants such that $\sum_0^\infty b_k \pi^k < \infty$, and write*

$$\zeta_n(x) = e^{inx}\sum_{k=0}^{\infty} b_{nk}x^k \tag{53}$$

where $|b_{nk}| \leqq b_k$, $n = 0, \pm 1, \pm 2, \cdots$. If $\{c_n\}$ is any set of complex numbers such that $\sum |c_n|^2 < \infty$, then

$$g_N(x) = \sum_{n=-N}^{N} c_n \zeta_n(x) \tag{54}$$

converges in mean to a function $g(x)$ of class $L_2(-\pi, \pi)$. If $\{a_n\}$ are the Fourier coefficients of $g(x)$ over $(-\pi, \pi)$, then

$$\phi_N(x) = (\pi^2 - x^2)\sum_{-N}^{N}(c_n\zeta_n(x) - a_n e^{inx}) \tag{55}$$

converges uniformly to zero for $-\pi \leqq x \leqq \pi$.

Proof. We have

$$g_N(x) = \sum_{n=-N}^{N} c_n \zeta_n(x) = \sum_{k=0}^{\infty} \psi_{Nk}(x)x^k$$

where

$$\psi_{Nk}(x) = \sum_{n=-N}^{N} c_n b_{nk} e^{inx}. \tag{56}$$

Then

$$\|\psi_{Nk}(x)\|^2 = \sum_{n=-N}^{N} |c_n b_{nk}|^2 \leqq c^2 b_k^2$$

where $c^2 = \sum_{-\infty}^{\infty} |c_n|^2$. It is clear that

(57) $$\psi_k(x) = \underset{N\to\infty}{\text{l.i.m.}}\ \psi_{Nk}(x)$$

exists and that

$$\|\psi_k(x)\| \leqq cb_k, \qquad \psi_k(x) \sim \sum_{-\infty}^{\infty} c_n b_{nk} e^{inx}.$$

Then

$$\left\| \sum_{k=\mu}^{\nu} x^k \psi_k(x) \right\| \leqq \sum_{k=\mu}^{\nu} \|x^k \psi_k(x)\| \leqq \sum_{k=\mu}^{\nu} \pi^k c b_k$$

so it follows that $\sum_0^{\nu} x^k \psi_k(x)$ converges in mean to some function of class $L_2(-\pi, \pi)$. We define this function as $g(x)$ and show that $g(x)$ has the properties given in the statement of the theorem. Thus

$$g(x) = \underset{\nu\to\infty}{\text{l.i.m.}} \sum_{k=0}^{\nu} x^k \psi_k(x).$$

Then

$$\|g(x) - g_N(x)\| \leqq \sum_{k=0}^{\infty} \|x^k(\psi_{Nk}(x) - \psi_k(x))\|$$

$$\leqq \sum_{k=0}^{\infty} \pi^k \|\psi_{Nk}(x) - \psi_k(x)\| \leqq \sum_{k=0}^{\infty} \pi^k c b_k.$$

But $\|\psi_{Nk}(x) - \psi_k(x)\| \to 0$ as $N \to \infty$ for each fixed k so it follows that

$$g(x) = \underset{N\to\infty}{\text{l.i.m.}}\ g_N(x).$$

To investigate the question of convergence we write

$$D_N(u) = \frac{\sin (N + 1/2)u}{\sin (u/2)}$$

for the Dirichlet kernel. Now (56), (57) show that

$$g_N(x) = \sum_{k=0}^{\infty} x^k(\psi_k(t), D_N(x - t)).$$

Now define $f_N(x)$ by the relation

$$f_N(x) = (g(t), D_N(x - t)) = \sum_{k=0}^{\infty} (t^k \psi_k(t), D_N(x - t)).$$

Then $f_N(x)$ is the Nth partial sum of the Fourier series of $g(x)$ over $(-\pi, \pi)$ so

$$
\begin{aligned}
\phi_N(x) &= (\pi^2 - x^2)(g_N(x) - f_N(x)) \\
&= \frac{1}{2\pi}\sum_{k=0}^{\infty}\int_{-\pi}^{\pi}(\pi^2 - x^2)(x^k - t^k)\psi_k(t)D_N(x-t)dt \qquad (58)\\
&= (\pi^2 - x^2)\sum_{k=0}^{\infty}(\psi_k(t), (x^k - t^k)D_N(x-t)).
\end{aligned}
$$

We now show that there is a constant A independent of k such that

$$
\left|\frac{(\pi^2 - x^2)(x^k - t^k)}{\sin((x-t)/2)}\right| \leqq A\pi^k, \qquad |x| \leqq \pi, \ |t| \leqq \pi. \tag{59}
$$

By a change of variables, this is equivalent to the relation

$$
\left|\frac{(1 - x^2)(x^k - t^k)}{\sin(\pi(x-t)/2)}\right| \leqq A\pi^{-2}, \qquad |x| \leqq 1, \ |t| \leqq 1.
$$

Clearly it is sufficient to take $x \geqq 0$, in which case $(x-t)/2$ lies in the interval $(-1/2, 1)$. But $u(u-1)/\sin \pi u$ is bounded when u lies in this interval, so it is sufficient to show that

$$
h = \left|\frac{(1-x)(x^k - t^k)}{(x-t)(2+t-x)}\right|
$$

is bounded by 4 for $-1 \leqq t \leqq 1$, $0 \leqq x \leqq 1$. If $-1 \leqq t \leqq -1/2$, then $h \leqq 4(1-x)/(2+t-x) \leqq 4$. If $-1/2 < t \leqq 1$, then $h \leqq 2(1-x)(x^k - t^k)/(x-t) = 2(1-x)(x^{k-1} + tx^{k-2} + \cdots + t^{k-1}) \leqq 2$.

From (59) we see that

$$
|(\psi_k(t), (\pi^2 - x^2)(x^k - t^k)D_N(x-t))| \leqq A\pi^k \|\psi_k(t)\|
$$

so

$$
\sum_{k=M}^{\infty}|(\psi_k(t), (\pi^2 - x^2)(x^k - t^k)D_N(x-t))| \leqq Ac\sum_{k=M}^{\infty}\pi^k b_k. \tag{60}
$$

Given ϵ, $\epsilon > 0$, choose M so large that

$$
Ac\sum_{k=M}^{\infty}\pi^k b_k < \frac{\epsilon}{3}\cdot
$$

It is clear that

$$
(\pi^2 - x^2)\sum_{k=0}^{M-1}(\psi_k(t), (x^k - t^k)D_N(x-t))
$$

tends to zero as $x \to \pm\pi$, so there is a positive number $\delta = \delta(\epsilon)$ such that this sum is bounded by $\epsilon/3$ in the intervals $(-\pi, -\pi+\delta)$ and $(\pi-\delta, \pi)$. If x lies in the interval $-\pi+\delta \leqq x \leqq \pi-\delta$ and $k=0, 1, 2, \cdots, M-1$, then

$$H_k(x, t) = \frac{x^k - t^k}{\sin((x-t)/2)}$$

is a continuous function of x and t, $-\pi \leqq t \leqq \pi$. Then

$$(\psi_k(t), (x^k - t^k)D_N(x-t)) = \frac{1}{2\pi}\int_{-\pi}^{\pi} \psi_k(t) H_k(x, t) \sin\left(N + \frac{1}{2}\right)(x-t)dt,$$

and a simple extension of the Riemann-Lebesgue lemma shows that this tends to zero uniformly in x, $|x| \leqq x-\delta$, as N tends toward infinity. Thus choose N so large that

$$\left|(\pi^2 - x^2)\sum_{k=0}^{M-1}(\psi_k(t), (x^k - t^k)D_N(x-t))\right| < \frac{\epsilon}{3}$$

for $|x| \leqq \pi - \delta$. Then

$$|\phi_N(x)| < \epsilon, \qquad -\pi \leqq x \leqq \pi,$$

and the theorem follows.

References

1. R. P. Boas, Jr., *A trigonometric moment problem*, J. London Math. Soc. vol. 14 (1939) pp. 242–244.

2. ———, *Entire functions bounded on a line*, Duke Math J. vol. 6 (1940) pp. 148–149.

3. ———, *A general moment problem*, Amer. J. Math. vol. 63 (1941) pp. 361–370.

4. D. G. Bourgin, *A class of sequences of functions*, Trans. Amer. Math. Soc. vol. 60 (1946) pp. 478–518.

5. R. J. Duffin and J. J. Eachus, *Some notes on an expansion theorem of Paley and Wiener*, Bull. Amer. Math. Soc. vol. 48 (1942) pp. 850–855.

6. R. J. Duffin and A. C. Schaeffer, *Power series with bounded coefficients*, Amer. J. Math. vol. 67 (1945) pp. 141–154.

7. I. Ibragimov, *On the completeness of systems of analytic functions*, Bull. Acad. Sci. URSS. Sér. Math. vol. 13 (1949) pp. 45–54.

8. N. Levinson, *Gap and density theorems*, Amer. Math. Soc. Colloquium Publications, vol. 26, 1940, chaps. I and IV.

9. R. E. A. C. Paley and N. Wiener, *Fourier transforms in the complex domain*, Amer. Math. Soc. Colloquium Publications, vol. 19, 1934, chaps. I and VII.

10. M. Plancherel and G. Pólya, *Functions entières et intégrales de Fourier multiples*, Comment. Math. Helv. vol. 10 (1937–1938) pp. 110–163.

11. E. C. Titchmarsh, *The theory of functions*, Oxford, 1932, chap. III.

Carnegie Institute of Technology,
Pittsburgh, Pa.
University of Wisconsin,
Madison, Wis.

BULLETIN OF THE
AMERICAN MATHEMATICAL SOCIETY
Volume 83, Number 4, July 1977

EXTENSIONS OF HARDY SPACES AND THEIR USE IN ANALYSIS

BY RONALD R. COIFMAN AND GUIDO WEISS[1]

1. Introduction. It is well known that the theory of functions plays an important role in the classical theory of Fourier series. Because of this certain function spaces, the H^p spaces, have been studied extensively in harmonic analysis. When $p > 1$, L^p and H^p are essentially the same; however, when $p \leqslant 1$ the space H^p is much better adapted to problems arising in the theory of Fourier series. We shall examine some of the properties of H^p for $p \leqslant 1$ and describe ways in which these spaces have been characterized recently. These characterizations enable us to extend their definition to a very general setting that will allow us to unify the study of many extensions of classical harmonic analysis.

The theory of H^p spaces on $\mathbf{R}^n$ has recently received an important impetus from the work of C. Fefferman and E. M. Stein **[29]**. Their work resulted in many applications involving sharp estimates for convolution operators. It is not immediately apparent how much of a role the differential structure of $\mathbf{R}^n$ plays in obtaining these results. Our purpose is to isolate from this theory some of the measure theoretic and geometric properties that enable us to obtain in a unified form many of these applications as well as other results in harmonic analysis. We shall not deal with those questions involving H^p spaces that are not relevant to our purpose. Some general references involving harmonic analysis and H^p spaces are **[23]**, **[64]**, **[27]**, **[62]**, **[57]** and **[55]**.

The main tool in our development is an extension and a refinement of the Calderón-Zygmund decomposition of a function into a "good" and "bad" part. This tool is presented in the proof of Theorem A and is of a somewhat technical nature. It is included here in order to make the presentation of the theory we develop essentially self-contained. In some examples we give applications of this theory that require material not presented here. We do, however, give the necessary references. In this sense, we hope that this exposition is accessible to a general audience.

Before beginning our presentation we would like to thank our colleagues A. Baernstein, Y. Meyer, R. Rochberg and E. M. Stein who read a large part of this manuscript and made many useful suggestions.

Suppose f is a *real-valued* integrable function on T, the perimeter of the unit disc in the plane (which we identify in the usual way with $[-\pi, \pi)$). Suppose f

[1] This paper is based on the material presented by the last-named author in an Invited Address to the American Mathematical Society meeting in Saint Louis on April 11–12, 1975. This lecture, in turn, was based on results obtained in a collaboration by the two authors of this paper; received by the editors January 20, 1976.

AMS (MOS) subject classifications (1970). Primary 30A78, 42A18, 42A40; Secondary 32A07, 42A56, 43A85.

© American Mathematical Society 1977

has the Fourier series $\sum_{-\infty}^{\infty} c_k e^{ik\theta}$. Let F be the analytic function defined by the power series $c_0 + 2\sum_1^{\infty} c_k z^k$, $|z| < 1$. This association of analytic functions in the disc $D = \{z \in \mathbf{C}: |z| < 1\}$ with real functions on T is the basis of a general technique, known as the use of *complex methods*, for obtaining results in the theory of Fourier series. The essential ingredients of this procedure can be described briefly in the following way. Certain basic operations on functions f on T (often realized as multiplication of the Fourier coefficients of f by specific sequences) can be interpreted as operations on the associated analytic functions F. Powerful tools from the theory of functions (such as factorization of zeroes, conformal mappings, etc.) can be used to study the properties of such operations on the functions F. The properties can then be translated as results about functions on T since f can be recovered from F by taking the almost everywhere existing radial limits

$$f(\theta) = \lim_{\substack{r\to 1\\ r<1}} \Re e\{F(re^{i\theta})\}. \tag{1.1}$$[2]

If $f \in L^p(T)$, $1 < p < \infty$, the M. Riesz inequality (see **[64]** for a proof) is satisfied by the analytic function F and the original function f:

$$\left(\int_{-\pi}^{\pi} |F(re^{i\theta})|^p\, d\theta\right)^{1/p} \leqslant A_p \|f\|_p, \tag{1.2}$$

where A_p is independent of f and r, $0 \leqslant r < 1$. Moreover, the radial limits (1.1) can be taken in the L^p norm as well.

In 1915 **[33]** Hardy considered the means

$$\mu_p(F; r) = \left(\int_{-\pi}^{\pi} |F(re^{i\theta})|^p\, d\theta\right)^{1/p},$$

$0 < p < \infty$, as functions of r and showed that they behave similarly to the "maximum modulus" $\mu_\infty(F; r) = \max\{|F(re^{i\theta})|;\ \theta \in [-\pi, \pi)\}$ (among other things, he showed that $\mu_p(F; r)$ increases with r and satisfies the conclusion of the "three circle theorem"). In 1923 F. Riesz (see **[48]**) introduced the class of functions F, analytic in D, for which

$$\|F\|_{H^p} = \sup_{0 \leqslant r < 1} \mu_p(F; r) < \infty, \tag{1.3}$$

where $0 < p \leqslant \infty$, and denoted this class by the symbol H^p in obvious homage to Hardy. Since that time these spaces have become known as *Hardy spaces*.[3]

[2] Many examples of this technique can be found in the books of Duren **[23]**, Hoffman **[35]** and Zygmund **[64]**.

[3] In his 1923 paper F. Riesz established many of the important properties of these spaces (e.g. the Blaschke product factorization, the existence of boundary values and other results). His brother M. Riesz also made important contributions in this field. Thus, it could be argued fairly that the name "Riesz" should also be attached to these spaces. We shall not enter into such an argument but would also like to point out that subsequent contributions by Hardy and Littlewood were most basic to the development of the theory of H^p spaces.

For the next few pages we shall assume that $L^p(T)$ consists only of real-valued functions and, thus, is a vector space over the real numbers. Though not essential, this assumption will simplify some of the assertions we are going to make.

An immediate consequence of the M. Riesz inequality is that $F \in H^p$ whenever $f \in L^p(T)$, $1 < p < \infty$. Conversely, if an analytic function F belongs to H^p then the limit function f in (1.1) exists (in the L^p norm and a.e.); moreover, F can be reconstructed from f (up to an imaginary constant) in the manner described above, provided $1 < p < \infty$ (see [**64**, Chapter VII]). Hence, for these values of p, $L^p(T)$ is the space $\mathfrak{Re}\,H^p$ consisting of the boundary values of the real parts of functions in H^p (in the sense of (1.1)).[4] The situation is completely different when $p = 1$. In fact, for a nonnegative $f \in L^1(T)$, $F \in H^1$ if and only if $\int_{-\pi}^{\pi} f(1 + \log^+ f) < \infty$ (see Chapter VII in [**64**]). Characterizations of $\mathfrak{Re}\,H^1$ have been obtained only recently; in a sense, this is the origin of this exposition. We shall explain these characterizations in great detail later.

It was observed by Hardy and Littlewood, as well as many others, that there are many results in Fourier analysis that hold for $L^p(T)$, $1 < p < \infty$, fail to be true for $L^1(T)$ and, yet, remain true for $\mathfrak{Re}\,H^1$. Since $L^q(T) \subsetneq L^p(T) \subsetneq \mathfrak{Re}\,H^1 \subsetneq L^1(T)$ when $1 < p < q \leqslant \infty$ we see that $\mathfrak{Re}\,H^1$ can be regarded as a good "substitute" for $L^1(T)$ that is endowed with many of the properties enjoyed by the L^p spaces for $1 < p < \infty$ and, moreover, contains all these spaces. Let us examine some specific examples of this situation.

Suppose $f \in L^p(T)$, $1 < p \leqslant 2$, has the Fourier series $\sum_{-\infty}^{\infty} c_k e^{ik\theta}$. A well-known result of Paley (see [**64**, Vol. 2. p. 111]) asserts, in part, that

$$\sum_{-\infty}^{\infty}{}' |c_k|^p |k|^{p-2} < \infty.^5 \tag{1.4}$$

This result is false when $p = 1$. In fact, there exists an integrable function whose Fourier series is $\sum_2^\infty \cos k\theta/\log k$ (see [**64**, Vol. 1, p. 184]). On the other hand, Hardy has shown that if $f \in \mathfrak{Re}\,H^1$ we have

$$\sum_{-\infty}^{\infty}{}' \frac{|c_k|}{|k|} < \infty. \tag{1.5}$$

This last inequality is more striking when we realize that no condition on the rate of convergence to 0 can be given for the Fourier coefficients of a function in $L^1(T)$ (see [**64**, Vol. 1, p. 183]); in this sense, the Riemann-Lebesgue theorem, asserting that these coefficients tend to zero, cannot be improved. Another example of this phenomenon is Paley's result:

$$\sum_{-\infty}^{\infty} |c_{2^k}|^2 < \infty \tag{1.6}$$

[4] It is not hard to show, making use of (1.2), that the space of complex-valued $L^p(T)$ functions and H^p (with norm given by (1.3)) are equivalent Banach spaces when $1 < p < \infty$ (see Boas [**1**]).

[5] We are adopting the standard notation: $\sum'$ means that we are summing over all nonzero integers k.

is satisfied by the Fourier coefficients of a function f in $\mathfrak{Re}H^1$ but does not hold for an arbitrary function in $L^1(T)$.

Perhaps the most basic difference between $L^1(T)$ and $\mathfrak{Re}H^1$ that was discovered by Hardy and Littlewood involves the Poisson integral of f:

$$(P_r * f)(\theta) = \frac{1}{2\pi}\int_{-\pi}^{\pi} \frac{1-r^2}{1-2r\cos(\theta-\psi)+r^2} f(\psi)\,d\psi \equiv u_f(re^{i\theta}),$$

$0 \leqslant r < 1$. Let

$$(P^*f)(\theta) = \sup_{|w-e^{i\theta}|\leqslant 2(1-|w|)} |u_f(w)|$$

denote the (nontangential) maximal operator associated with this integral. Hardy and Littlewood showed that $P^*f \in L^1(T)$ when $f \in \mathfrak{Re}H^1$, while this is not true for a general $f \in L^1(T)$. It has been understood for a long time that this result represents a basic difference between these two function spaces; however, it was only recently shown by Burkholder, Gundy and Silverstein **[2]** that we have here a real-variable characterization of $\mathfrak{Re}H^1$: *for* $f \in L^1(T)$, $P^*f \in L^1(T)$ *if and only if* $f \in \mathfrak{Re}H^1$.[6]

If $f \in \mathfrak{Re}H^1$ then, by definition, $f(\theta) = \mathfrak{Re}F(e^{i\theta})$, where $F \in H^1$. If we make the restriction that the imaginary part of F vanishes at 0, the function F is unique. We can then put a norm on $\mathfrak{Re}H^1$ by letting $\|f\|_{H^1} \equiv \|F\|_{H^1}$ (see (1.3)). With this norm $\mathfrak{Re}H^1$ is a Banach space.

The next important step toward an understanding of $\mathfrak{Re}H^1$ was taken by C. Fefferman who characterized the dual, $\{\mathfrak{Re}H^1\}^*$, of this space as the class of functions of *Bounded Mean Oscillation* (BMO) originally introduced by John and Nirenberg **[38]**: a periodic function l (of period 2π) is said to be in the class BMO provided it belongs to $L^1(T)$ and

$$\int_I |l(\theta) - m_I(l)| \frac{d\theta}{|I|} \leqslant c$$

for all finite intervals I, where c is independent of I and $m_I(l) = (1/|I|)\int_I l(\theta)\,d\theta$. The space BMO is a Banach space if we let

$$\|l\|_{BMO} = \left|\frac{1}{2\pi}\int_0^{2\pi} l\right| + \sup_I \frac{1}{|I|}\int_I |l - m_I(l)|$$

be its norm. C. Fefferman showed that the mappings

$$f \to \int_{-\pi}^{\pi} f(\theta) l(\theta)\,d\theta,$$

induced by the various $l \in BMO$, yield all the bounded linear functionals on $\mathfrak{Re}H^1$ and their norms as linear functionals are equivalent to $\|\cdot\|_{BMO}$. (The

[6] We mentioned before that, for $f \geqslant 0$, being a member of $\mathfrak{Re}H^1$ is equivalent to the integrability of $f \log^+ f$. On the other hand, this last condition is equivalent to the integrability of P^*f (in this connection, and for higher dimensional extensions, see E. M. Stein **[53]**).

integrals $\int_{-\pi}^{\pi} fl$ have to be interpreted appropriately; we shall consider this point later in §3.)[7]

This characterization of $\{\Re e H^1\}^*$ is easily seen to be equivalent to the following description of the functions belonging to $\Re e H^1$. We begin by introducing a basic "building block" for the elements of $\Re e H^1$:

DEFINITION. An *atom* (more precisely, a 1-*atom*) is either the function $a_0(\theta) \equiv 1/2\pi$ or a function $a(\theta)$, $-\pi \leqslant \theta < \pi$, satisfying:

(i) The support of a is contained in an interval $I \subset [-\pi, \pi)$;[8]

(ii) $|a(\theta)| \leqslant |I|^{-1}$ for all $\theta \in [-\pi, \pi)$;

(iii) $\int_0^{2\pi} a(\theta)\, d\theta = 0$.

The above-mentioned characterization is, then, simply: *a function f belongs to $\Re e H^1$ if and only if*

$$f = \sum_{j=0}^{\infty} \lambda_j a_j \tag{1.7}$$

where a_j is an atom, $0 \leqslant j$, and $\sum_{j=0}^{\infty} |\lambda_j| < \infty$. Moreover, if F is the analytic function corresponding to f (in the manner described above) there exist absolute constants c_1 and c_2 such that

$$c_1 \|F\|_{H^1} \leqslant \inf \sum_0^{\infty} |\lambda_j| \leqslant c_2 \|F\|_{H^1}, \tag{1.8}$$

where the infimum is taken over all decompositions (1.7).

This characterization of $\Re e H^1$ leads us to a corresponding real-variable characterization of $\Re e H^p, 0 < p \leqslant 1$. The example

$$F(z) = (1/2\pi)[(1 + z)/(1 - z)]$$

of a function in H^p, $0 < p < 1$, whose real part converges to 0 as $z = re^{i\theta} \to e^{i\theta}$ (as $r \to 1$) for $\theta \in (0, 2\pi)$, indicates that $\Re e H^p$ is not a function space; rather, it should be defined as a space of distributions (in this example, the boundary value is the Dirac measure concentrated at $\theta = 0$). When $1/2 < p \leqslant 1$ these distributions are obtained by a simple renormalization of the atoms we have just introduced: We call a function $a(\theta)$ a *p-atom* if it satisfies properties (i) and (iii) and, instead of (ii), if it satisfies $|a(\theta)| \leqslant |I|^{-1/p}$.[9]

The characterization of $\Re e H^p$ can be stated as follows (see **[10]**):

THEOREM I. *A distribution f belongs to $\Re e H^p$, $0 < p \leqslant 1$, if and only if it can be represented in the form $f = \sum_{j=0}^{\infty} \lambda_j a_j$, where a_j is a p-atom, $0 \leqslant j$, and $\sum_0^{\infty} |\lambda_j|^p < \infty$.*

It is a common practice of modern mathematics to transform a theorem characterizing a mathematical notion into a definition of that notion. Doing

[7] This will become evident later on when we discuss the more general setting for the theory of H^p spaces.

[8] We are really dealing with functions defined on the perimeter of the unit circle and our "intervals" are intervals in this perimeter. Thus, we allow I to have the form $\{\theta: -\pi \leqslant \theta < a$ or $b < \theta \leqslant \pi$, where $a \leqslant b\}$.

[9] For the cases $0 < p < 1/2$ we also have to modify condition (iii). We shall discuss this matter later.

this is particularly fruitful if the characterization has three basic properties:

(a) Known results follow easily from it.

(b) New results follow easily from it.

(c) The notion being characterized can be extended to new situations and this leads to a theory as useful and powerful as the original one.

Theorem I has these features. First of all, it is clear that the defining properties of an atom make sense in any topological measure space where a basis for open sets (such as the intervals used above) is prescribed. This is one of the major points of this presentation. First, however, we shall give some illustrations of properties (a) and (b).

We begin by establishing Hardy's inequality (1.5). Let us first observe that it suffices to prove (1.5) for the Fourier coefficients of an atom. In fact, we shall show that

$$\sum_{k=-\infty}^{\infty}{}' \frac{|a_k|}{|k|} \leqslant 4 \tag{1.9}$$

whenever $a_k = (1/2\pi)\int_{-\pi}^{\pi} a(\theta)e^{-ik\theta}\,d\theta$ are the Fourier coefficients of a 1-atom $a(\theta)$. Hardy's inequality then follows immediately from (1.9) and the representation (1.7) of functions in $\mathfrak{Re}\,H^1$.[10] Thus, let $a(\theta)$ be a 1-atom whose supporting interval, I, in conditions (i) and (ii) is centered at 0 (there is no loss in generality since $|a_k|$ is unchanged by translations). Using first (iii) and then (ii) we have

$$|a_k| = \left|\frac{1}{2\pi}\int_I a(\theta)[e^{-ik\theta} - 1]\,d\theta\right| \leqslant \frac{1}{2\pi}\int_I |a(\theta)|\,|k\theta|\,d\theta$$
$$\leqslant \frac{|k|\,|I|}{2\pi}\int_I |I|^{-1}\,d\theta = \frac{|k|\,|I|}{2\pi}.$$

Moreover, again using (ii) we have

$$\sum_{k=-\infty}^{\infty} |a_k|^2 = \frac{1}{2\pi}\int_{-\pi}^{\pi} |a(\theta)|^2\,d\theta \leqslant \frac{1}{2\pi}\int_I |I|^{-2}\,d\theta = \frac{1}{2\pi}|I|^{-1}.$$

Making use of these two estimates we have

$$\sum_{k=-\infty}^{\infty}{}' \frac{|a_k|}{|k|} = \sum_{|k|\leqslant|I|^{-1}} \frac{|a_k|}{|k|} + \sum_{|k|>|I|^{-1}} \frac{|a_k|}{|k|}$$
$$\leqslant \frac{1}{\pi}\sum_{|k|\leqslant|I|^{-1}} |I| + \left(\sum_{-\infty}^{\infty} |a_k|^2\right)^{1/2}\left(\sum_{|k|>|I|^{-1}} k^{-2}\right)^{1/2} \leqslant 4$$

and (1.9) is established. In view of (1.8) we obtain

[10] This is an important feature of the use of atoms: many problems can be reduced to considerations involving only atoms and, because of this, the resolution of these problems is made simpler.

(1.10) $$\sum_{-\infty}^{\infty}{}' \frac{|a_k|}{|k|} \leqslant 4c_2 \|F\|_{H^1}$$

for any $F \in \Re e H^1$.

Our next observation is that we can easily identify the dual spaces of $\Re e H^p, 0 < p \leqslant 1$, if we know Theorem I. We restrict our attention to the case $1/2 < p \leqslant 1$, since we have defined a p-atom only for this range of p. If $L \in \{\Re e H^p\}^*$ it is easy to show (and we will do so later in a more general context) that the action of this linear functional on atoms $a(\theta)$ is given by an integrable function $l(\theta)$ as follows:

$$La = \int_{-\pi}^{\pi} a(\theta) l(\theta)\, d\theta.$$

Moreover, the continuity of L is equivalent to the "boundedness" $|La| \leqslant \|L\|$. We claim that l satisfies

(1.11) $$\sup_{I \subset [-\pi,\pi)} |I|^{-1/p} \int_I |l(\theta) - m_I(l)|\, d(\theta) \leqslant 2\|L\|.$$

To see this we choose a function f supported in I such that $\|f\|_\infty \leqslant 1$. Obviously, if χ_I is the characteristic function of I,

$$a(\theta) = 2^{-1}|I|^{-1/p}\{f(\theta) - m_I(f)\chi_I(\theta)\}$$

is a p-atom. We thus have

$$\left|\int_I f(\theta)[l(\theta) - m_I(l)]\, d\theta\right| = \left|\int_I [f(\theta) - m_I(f)] l(\theta)\, d\theta\right|$$
$$= 2|I|^{1/p}\left|\int_I a(\theta) l(\theta)\, d\theta\right| \leqslant 2|I|^{1/p}\|L\|.$$

Inequality (1.11) now follows by taking the supremum over all such functions f. It is also clear, by retracing our steps, that if the supremum in (1.11) is finite then the integral above this inequality defines a bounded linear functional on $\Re e H^p$.

When $p = 1$ (1.11) is the defining condition for the space BMO. This condition has been studied extensively for $p < 1$ as well (see [7] and [46]). In particular, as we shall show later, when $1/2 < p < 1$ (1.11) is equivalent to

(1.12) $$|l(\theta) - l(\psi)| \leqslant c|\theta - \psi|^\alpha, \qquad \alpha = 1/p - 1.$$

This yields the characterization of the dual of $\Re e H^p$ given by Duren, Romberg and Shields [24]. At this point it is relevant to point out that results involving H^p spaces are translated by duality into results on Lipschitz spaces (i.e. spaces of functions satisfying inequalities of the type (1.11) or (1.12)). We shall return to this point after we have introduced our extensions of the Hardy spaces.

We have encountered many other situations where Theorem I can be used directly and efficiently to produce known results in the theory of H^p spaces. Let us now illustrate the fact that Theorem I enjoys property (b); that is, new results can be obtained from it. Perhaps the most striking development in this direction is that we can obtain a simple and useful characterization of the Fourier coefficients of elements in $\mathfrak{Re}H^p$. It is well known that such a characterization exists for $L^2(T)$ via the Plancherel theorem; that is, a necessary and sufficient condition for a sequence $\{a_k\}$, $k = 0, \pm 1, \pm 2, \ldots$, to be the sequence of Fourier coefficients of an $f \in L^2(T)$ is $\sum_{k=-\infty}^{\infty} |a_k|^2 < \infty$. This condition enables us to easily characterize those linear operators commuting with translations that are bounded on $L^2(T)$. In fact, it is easy to see that if A is such an operator then its action is determined by an associated sequence $\{\mu_k\}$ in such a way that if f has Fourier coefficients $\{a_k\}$ then Af has Fourier coefficients $\{\mu_k a_k\}$.[11] It is immediate to check that $\{\mu_k a_k\}$ is square summable for all square summable sequences $\{a_k\}$ if and only if $\{\mu_k\}$ is bounded. The study of convolution operators on $\mathfrak{Re}H^p$ is not quite so simple; however, we shall see that the characterization we will give of the Fourier coefficients of the elements of these spaces is very useful for such a study.

We first observe that, if $a(\theta)$ is a real-valued (nonconstant) 1-atom whose supporting interval is centered at 0, then, making use of (i), (ii) and the estimate $|e^{i\theta} - 1|^2 \leq |\theta|^2$,

$$\left(\int_{-\pi}^{\pi} |a(\theta)|^2\, d\theta\right)^{1/2}\left(\int_{-\pi}^{\pi} |a(\theta)|^2 |e^{i\theta} - 1|^2\, d\theta\right)^{1/2} \leq \frac{1}{\sqrt{2}}. \tag{1.13}$$

Let $\sum_{-\infty}^{\infty} a_k e^{ik\theta}$ be the Fourier series of $a(\theta)$ and $A(z) = 2\sum_1^{\infty} a_k z^k$ the analytic function associated with $a(\theta)$ described in the beginning of this exposition. It is a well-known fact (see Zygmund's book **[64]**) that the boundary values $A(e^{i\theta}) = \lim_{r\to 1} A(re^{i\theta})$ cannot vanish on a set of positive measure. Thus, $A(e^{i\theta})$ could be an atom, or a multiple of one, only if it were bounded by $1/2\pi$; but simple examples (we can choose $a(\theta) = (2\pi)^{-1} \operatorname{sgn} \theta$, $-\pi \leq \theta < \pi$) show that this cannot be expected. Nevertheless, $A(e^{i\theta})$ does inherit the very basic property (1.13). In fact, using Plancherel's theorem and the fact that $a_{-k} = \bar{a}_k$ we have

$$\begin{aligned}
&\left(\int_{-\pi}^{\pi} |A(e^{i\theta})|^2\, d\theta\right)^{1/2}\left(\int_{-\pi}^{\pi} |A(e^{i\theta})|^2 |1 - e^{i\theta}|^2\, d\theta\right)^{1/2} \\
&\qquad = 8\pi\left(\sum_{k=1}^{\infty} |a_k|^2\right)^{1/2}\left(\sum_{k=1}^{\infty} |a_{k+1} - a_k|^2 + |a_1|^2\right)^{1/2} \\
&\qquad = 4\pi\left(\sum_{k=-\infty}^{\infty} |a_k|^2\right)^{1/2}\left(\sum_{k=-\infty}^{\infty} |a_{k+1} - a_k|^2\right)^{1/2} \\
&\qquad = \left(\int_{-\pi}^{\pi} |a(\theta)|^2\, d\theta\right)^{1/2}\left(\int_{-\pi}^{\pi} |a(\theta)|^2 |1 - e^{i\theta}|^2\, d\theta\right)^{1/2}.
\end{aligned} \tag{1.14}$$

---

[11] The property of commuting with translations easily implies that the functions $e^{ik\theta}$, $k = 0, \pm 1, \pm 2, \ldots$, are proper vectors for the operator A. The sequence $\{\mu_k\}$ is the corresponding sequence of proper values of A. Such operators are called *multiplier operators* or (more appropriately) *convolution operators*; $\{\mu_k\}$ is referred to as the sequence of *multipliers*.

More generally, let us consider (possibly complex-valued) functions $M \in L^2(T)$ satisfying

$$(1.15)\qquad \left(\int_{-\pi}^{\pi} |M(\theta)|^2\,d\theta\right)^{1/4}\left(\int_{-\pi}^{\pi} |M(\theta)|^2|e^{i\theta}-1|^2\,d\theta\right)^{1/4} \leqslant 1,\qquad \int_{-\pi}^{\pi} M(\theta)\,d\theta = 0.$$

Such functions can be shown to be sums of atoms centered at 0 that are "neatly stacked one on top of the other". We shall return to this point later; we mention this now, however, to justify calling such functions *molecules*.[12] It turns out that molecules are practically as basic as atoms in the study of H^p-space theory.

We first observe for any M in $L^2(T)$

$$(1.16)\qquad \int_{-\pi}^{\pi} |M(\theta)|\,d\theta \leqslant 3\pi\left(\int_{-\pi}^{\pi} |M(\theta)|^2\,d\theta\right)^{1/4}\left(\int_{-\pi}^{\pi} |M(\theta)|^2|e^{i\theta}-1|^2\,d\theta\right)^{1/4}$$

In fact, let $M_0(\theta) = |1-e^{i\theta}|M(\theta)$; then

$$\begin{aligned}\int_{-\pi}^{\pi} |M(\theta)|\,d\theta &= \left(\int_{|\theta|\leqslant A^2/2} + \int_{|\theta|>A^2/2}\right)|M(\theta)|\,d\theta\\ &\leqslant \|M\|_2\left(2\int_0^{A^2/2} d\theta\right)^{1/2}\\ &\quad+\left(\int_{-\pi}^{\pi} |M(\theta)|^2|e^{i\theta}-1|^2\,d\theta\right)^{1/2}\left(2\int_{A^2/2}^{\pi}|e^{i\theta}-1|^{-2}\right)^{1/2}\\ &\leqslant \|M\|_2A + \|M_0\|_2 2\pi\left(2\int_{A^2/2}^{\infty}\theta^{-2}\,d\theta\right)^{1/2} = \|M\|_2A + 2\pi\|M_0\|_2A^{-1}.\end{aligned}$$

Letting $A = (2\pi\|M_0\|_2\|M\|_2^{-1})^{1/2}$ we obtain (1.16). Thus, any molecule satisfies $\|M\|_1 \leqslant 3\pi$. Incidentally, (1.16) is a version of an inequality of Carlson [**9**].

It now follows easily that any f satisfying (1.7) does, indeed, belong to $\mathfrak{Re}\,H^1$. In order to see this it suffices to consider an atom $a(\theta)$ and observe that the associated analytic function $A(z)$ satisfies $\|A\|_{H^1} \leqslant \sqrt{2}(1+\pi)$. But this is an immediate consequence of (1.13), (1.14), (1.16) and the well-known fact that $\mu_1(F; r) = \int_{-\pi}^{\pi} |F(re^{i\theta})|\,d\theta$ is a nondecreasing function of r. Thus, one half of Theorem I is established in case $p = 1$.

Similar considerations apply when $p < 1$. Note that a p-atom is obtained from a 1-atom $a(\theta)$, when $1/2 < p \leqslant 1$, by multiplying the latter by $|I|^{1-(1/p)} \leqslant \|a\|_2^{2(1-p)/p}$. Consequently, we define a p-molecule (centered at 0), to be a 1-molecule multiplied by its L^2-norm to the power $2(1-p)/p$. The reader can then check that if (1.15) is replaced by

$$(1.17)\qquad \|M\|_2^{3p-2}\|M(1-e^{i\theta})\|_2^{2-p} \leqslant 1$$

---

[12] More precisely, M is a 1-*molecule centered at the origin*; for a molecule centered at $\theta_0 \in (-\pi, \pi]$ the defining condition is obtained by replacing $|e^{i\theta}-1|^2$ in (1.15) by $|e^{i\theta}-e^{i\theta_0}|^2$.

the above arguments can be easily modified to provide a proof of one of the implications in Theorem I when $2/3 < p \leqslant 1$.

Let us now return to the study of multiplier operators and the Fourier coefficients of H^1 functions. We first make the general observation that if B is a linear operator mapping 1-atoms into molecules satisfying (1.15) then, by (1.16), B clearly has a bounded extension as an operator from $\mathfrak{Re}\, H^1$ to $L^1(T)$. We can say more, however, by observing that the argument involving the identity (1.14) can be applied to $M = Ba$ ($a(\theta)$ being a 1-atom) to show that $M \in \mathfrak{Re}\, H^1$ and its associated analytic function belongs to H^1 with H^1 norm not exceeding 2π (because of (1.16)). Hence, B maps $\mathfrak{Re}\, H^1$ boundedly into itself. Thus, we see that any linear operator mapping atoms into molecules has an extension that is a bounded operator on $\mathfrak{Re}\, H^1$. In §3 we shall show that a molecule belongs to H^1 in a general setting in which the notion of "analytic H^1" is completely absent.

We are thus led to examine the question: Which operators map atoms into molecules? If the operator is a multiplier two basic properties of the Fourier coefficients $\{a_k\}$ of atoms centered at 0 are particularly useful in trying to resolve this question. The first of these comes from (1.13) and Plancherel's theorem:

$$(1.18) \qquad \left(\sum_{k=-\infty}^{\infty} |a_k|^2\right)^{1/2} \left(\sum_{k=-\infty}^{\infty} |a_{k+1} - a_k|^2\right) \leqslant \frac{\sqrt{2}}{2\pi}.$$

The second one was established in the proof of Hardy's inequality (see the argument following (1.9)) and can be restated as follows:

$$(1.19) \qquad |a_k| \leqslant \min\left\{1, \frac{k\|a\|_2^{-2}}{\pi}\right\}, \qquad k = 0, \pm 1, \pm 2, \ldots.$$

We shall illustrate the usefulness of these two properties of Fourier coefficients of atoms by establishing the following result:

THEROEM (1.20). *Suppose $\{\mu_k\}$ is a bounded sequence satisfying $|\mu_{k+1} - \mu_k| \leqslant c_0/k$, $k = 0, \pm 1, \pm 2, \ldots$, then the multiplier operator associated with this sequence maps $\mathfrak{Re}\, H^1$ boundedly into itself.*[13]

All we have to show is that if $a(\theta) = \sum_{-\infty}^{\infty} a_k e^{ik\theta}$ is an atom then $M(\theta) = \sum_{-\infty}^{\infty} \mu_k a_k e^{ik\theta}$ satisfies (1.15) independently of a. We have:

$$\|M\|_2 \|M(e^{i\theta} - 1)\|_2 = 2\pi \left(\sum_{-\infty}^{\infty} |\mu_k a_k|^2\right)^{1/2} \left(\sum_{-\infty}^{\infty} |\mu_{k+1} a_{k+1} - \mu_k a_k|^2\right)^{1/2}.$$

Let $\|\mu\|_\infty$ be the $\sup_k\{|\mu_k|\}$ and observe that

$$\mu_{k+1} a_{k+1} - \mu_k a_k = (\mu_{k+1} - \mu_k) a_{k+1} + \mu_k (a_{k+1} - a_k).$$

[13] The condition $|\mu_{k+1} - \mu_k| \leqslant c_0/k$ implies $2^N \sum_{2^{N-1}}^{2^N} \{|\mu_{k+1} - \mu_k|^2 + |\mu_{-k+1} - \mu_{-k}|^2\} \leqslant c_0'$ for all $N > 0$. Such a condition is often called a *Hörmander condition* (see [36]). The proof we give can easily be changed to show that the Hörmander condition yields the conclusion of (1.20). Similar theorems are also announced in [29, p. 150].

Thus the product involving the two series above is dominated by

$$\left(\sum_{-\infty}^{\infty} |\mu_k a_k|^2\right)^{1/2}\left(\sum_{-\infty}^{\infty} |\mu_k|^2 |a_{k+1} - a_k|^2\right)^{1/2}$$
$$+\left(\sum_{-\infty}^{\infty} |\mu_k a_k|^2\right)^{1/2}\left(\sum_{-\infty}^{\infty} |\mu_{k+1} - \mu_k|^2 |a_{k+1}|^2\right)^{1/2}.$$

By (1.18) we see that the first summand does not exceed $\sqrt{2}\,\|\mu\|_\infty^2/2\pi$. Using (1.19), on the other hand, we have

$$\left(\sum_{-\infty}^{\infty} |\mu_k a_k|^2\right)\left(\sum_{-\infty}^{\infty} |\mu_{k+1} - \mu_k|^2 |a_{k+1}|^2\right)$$
$$\leqslant \|\mu\|_\infty^2\left(\sum_{-\infty}^{\infty} |a_k|^2\right)\left(\frac{c_0^2}{\pi^2}\sum_{|k|\leqslant \|a\|_2^2} \frac{1}{k^2}k^2\|a\|_2^{-4} + \sum_{|k|\geqslant \|a\|_2^2}\frac{c_0^2}{k^2}\right).$$

But this last expression is clearly less than a constant depending only on $\|\mu\|_\infty$ and c_0 and Theorem (1.20) is proved.

When $p < 1$, $\Re e H^p$ is not naturally converted into a Banach space. Nevertheless, we can introduce a "norm" by letting $\|f\|_{H^p} \equiv \|F\|_{H^p}$, where F is the analytic function associated with the distribution f and $\operatorname{Im} F(0) = 0$ (see (1.13)). This is not a norm; however, $d_p(f,g) = \|f - g\|_p^p$ defines a metric. Thus, we can discuss continuity questions concerning operators on $\Re e H^p$. In particular, a linear operator mapping p-atoms into p-molecules is continuous on $\Re e H^p$, $1/2 < p \leqslant 1$. We can translate (1.17) into a condition on the Fourier coefficients of a p-atom that is analogous to (1.18). Condition (1.19) also has a simple analog. All this can be stated by saying that the Fourier coefficients $\{a_k\}$ of a p-atom $a(\theta)$, $1/2 < p \leqslant 1$, are such that $\{\|a\|_2^{p/(2-p)-1} a_k\}$ are the Fourier coefficients of a 1-atom. Using these facts one can show that the assumptions in Theorem (1.20) imply that the multiplier operator is bounded on $\Re e H^p$ as long as $2/3 < p \leqslant 1$.

Several examples of multipliers satisfying the hypotheses of Theorem (1.20) can be given. The case where $\mu_k = |k|^{i\gamma}$, γ real, has been studied by several people. The operator associating to $f \in \Re e H^1$ the boundary value F of the analytic function in H^1 such that $\Re e F = f$ is a multiplier operator given by $\mu_k = 0$ if $k < 0$, $\mu_0 = 1$ and $\mu_k = 2$ if $k > 0$. The fact that H^1 is stable under the action of certain convolution operators was first pointed out by E. M. Stein. (See [54] and [55]; there this situation is discussed in higher dimensional settings. We shall do the same later on.)

Several considerations are in order once the continuity of such operators on the spaces $\Re e H^p, p \leqslant 1$, is established. First, we remark that a duality argument shows that the corresponding Lipschitz spaces (see (1.12)) are preserved.

Secondly, the question of interpolating the boundedness of these operators becomes a natural one. We have already remarked that any such multiplier operator is bounded on $L^2(T)$. At this point we would like to point out that here, again, the atomic characterization of $\Re e H^p$ is particularly well suited for obtaining a variety of results on interpolation of operators. We shall develop this topic later in a more general setting. For the moment, however, we shall illustrate the simplicity of some of the methods involved by examining some aspects of the following theorem:

THEOREM (1.21). *Fix $p_0 \leqslant 1$. Let B be a bounded linear operator on $L^2(T)$ satisfying the condition*

$$\|Ba\|_{p_0} \leqslant c \tag{1.22}$$

for all p_0-atoms. Then, $\|Bf\|_p \leqslant c_p\|f\|_{H^p}$ for $p_0 \leqslant p \leqslant 1$, and $\|Bf\|_p \leqslant a_p\|f\|_p$ for $1 < p \leqslant 2$, where c_p and a_p are independent of $f \in H^p$ or $f \in L^p(T)$ (though they do depend on c in (1.22) and p_0).

To illustrate our point we shall only consider the case $1/2 < p_0 < 1$ and we shall deduce the conclusion for $p = 1$. As we have already seen on several occasions it suffices to obtain the result for $f(\theta) = a(\theta)$ a 1-atom. Then $|I|^{1-(1/p_0)}a$ is a p_0-atom (I being the supporting interval of a). Thus, by (1.22) and the L^2-boundedness of B we obtain

$$\|Ba\|_{p_0} \leqslant c|I|^{(1/p_0)-1} \quad \text{and} \quad \|Ba\|_2 \leqslant c'|I|^{-1/2}. \tag{1.23}$$

Using Hölder's inequality and (1.23) we obtain the desired conclusion:

$$\begin{aligned}\int |Ba| &= \int |Ba|^{p_0/(2-p_0)}|Ba|^{(2-2p_0)/(2-p_0)} \\ &\leqslant \left(\int |Ba|^{p_0}\right)^{1/(2-p_0)}\left(\int |Ba|^2\right)^{(1-p_0)/(2-p_0)} \\ &\leqslant cc'|I|^{(1-p_0)/(2-p_0)}|I|^{(p_0-1)/(2-p_0)} = cc'.\end{aligned}$$

We thus see that the boundedness of many operators on H^p, or *merely on atoms*, for some $p < 1$, implies its continuity as an operator on L^q for certain $q > 1$. In particular, the familiar results of Calderón-Zygmund (see [5]) have a natural setting in this atomic approach to H^p spaces. In fact, let $k(\theta)$ be integrable for $0 < \varepsilon \leqslant |\theta| \leqslant \pi$ for all $\varepsilon > 0$ and satisfy the two conditions

(a) the sequence $\hat{k}(j) = \lim_{\varepsilon\to 0} \int_{\varepsilon<|\theta|\leqslant\pi} e^{-ij\theta}k(\theta)\,d\theta$ exists and is bounded by c_1;

(b) $\int_{2|\theta|<|\psi|\leqslant\pi} |k(\psi - \theta) - k(\psi)|\,d\psi \leqslant c_2$.

Let us also assume that the convolution $\int_{-\pi}^{\pi} k(\theta - \varphi)f(\varphi)\,d\varphi$ is suitably defined for $f \in L^2(T)$. We shall show that

$$\|k * a\|_1 = \int_{-\pi}^{\pi}\left|\int_{-\pi}^{\pi} k(\theta - \varphi)a(\varphi)\,d\varphi\right|d\theta \leqslant c_1 + c_2 \tag{1.24}$$

for all 1-atoms a. From this it will follow that there exists a constant c_p such that

$$\|k * a\|_p \leqslant c_p\|f\|_p \tag{1.25}$$

whenever $1 < p < \infty$.

Before doing this we would like to point out that these considerations provide us yet another method for showing that the atomic $\mathcal{R}e H^1$ space is, indeed, included in the real part of the classical H^1 space. This follows from the fact that the kernel $k(\theta) = (1 + e^{i\theta})/(1 - e^{i\theta}) \sim 1 + 2\sum_{j=1}^{\infty} e^{ij\theta}$ obviously maps (as a convolution operator) f into the boundary value of the analytic function having real part f (for any reasonably "nice" f) and, moreover, the mean value theorem yields the inequality

$$|k(\psi - \theta) - k(\psi)| \leqslant c|\theta|/|\psi|^2, \quad \text{if } |\psi| > 2|\theta|,$$

which certainly implies (b).

To establish (1.24) we argue as follows: First observe that condition (a) tells us that $f \to k * f$ is a bounded operator on $L^2(T)$. Now let $a(\theta)$ be a 1-atom supported in an interval I centered at 0. Since the result is trivial for the constant atom, we can assume $\int_{-\pi}^{\pi} a = 0$. Then,

$$\begin{aligned} \|k * a\|_1 &= \int_{-\pi}^{\pi} |(k * a)(\psi)|\, d\psi \\ &= \int_{2|I|<|\psi|\leqslant\pi} |(k * a)(\psi)|\, d\psi + \int_{2|I|\geqslant|\psi|} |(k * a)(\psi)|\, d\psi. \end{aligned}$$

To estimate the first of these integrals we make use of the fact that a has mean 0 and, thus, the integral equals

$$\begin{aligned} &\int_{2|I|<|\psi|\leqslant\pi} \left| \int_{-\pi}^{\pi} [k(\psi - \theta) - k(\psi)]a(\theta)\, d\theta \right| d\psi \\ &\qquad \leqslant \int_{|I|} |a(\theta)| \left(\int_{2|I|<|\psi|\leqslant\pi} |k(\psi - \theta) - k(\psi)|\, d\psi \right) d\theta. \end{aligned}$$

But, by condition (b), $\int_{2|I|<|\psi|\leqslant\pi} |k(\psi - \theta) - k(\psi)|\, d\psi \leqslant c_2$ and we obtain the inequality

$$\int_{2|I|<|\psi|\leqslant\pi} |(k * a)(\psi)|\, d\psi \leqslant c_2 \int_I |a(\theta)|\, d\theta \leqslant c_2 .$$

To estimate the second integral we use Schwarz's inequality and the L^2-boundedness of the convolution operator

$$\int_{|\psi|\leqslant 2|I|} |k * a| \leqslant \|k * a\|_2 2|I|^{1/2} \leqslant c_1 \|a\|_2 |I|^{1/2} \leqslant c_1$$

and (1.24) is established. Inequality (1.25) then follows by interpolation and duality (since convolution operators are the negative of their adjoints).

Let us summarize very briefly some of the points we have made favoring the study of H^p spaces, $p \leqslant 1$, and, in particular, the atomic approach to this study:

(1) Many results that are valid for $L^q(T)$, $1 < q < \infty$, fail for $L^1(T)$ but remain valid for $\mathcal{R}e H^1$ (or $\mathcal{R}e H^p$).

(2) Several estimates on operators (in particular, convolution operators) are easily made on atoms and imply corresponding estimates for these operators applied to general functions.

(3) We can characterize the Fourier coefficients of the distributions $f \in \Re e H^p$ and from this obtain results on multiplier operators.[14]

(4) The boundedness of some operator on $\Re e H^p$, $p \leqslant 1$, implies (via interpolation theorems) the boundedness of this operator on certain $L^q(T)$ spaces, $1 < q$. Moreover, by duality we see that such an operator must preserve certain smoothness properties. Atoms are very useful for obtaining such results easily.

Let us now describe briefly the Hardy spaces associated with the real line **R** and their atomic characterization. The classical definition of these spaces is the following one: $F \in H^p(\mathbf{R})$, $0 < p$, if and only if F is an analytic function defined on the upper half-plane $\mathbf{R}^2_+ = \{z = x + iy \in \mathbf{R}^2; x \in \mathbf{R}, y > 0\}$ satisfying

$$\sup_{y>0}\left(\int_{-\infty}^{\infty} |F(x+iy)|^p\, dx\right)^{1/p} \equiv \|F\|_{H^p} < \infty. \tag{1.26}$$

Their theory was developed by Krylov [**44**]. In particular, it is known that the boundary values

$$\lim_{y\to 0+} F(x+iy) = F(x) \tag{1.27}$$

exist almost everywhere and in the L^p norm ($p > 0$). As was the case in the circle, these boundary values should be considered as distributions when $p < 1$ and, in this case, their real parts can be characterized in terms of atoms that can be defined in a manner that is completely analogous to the one we gave above. This characterization applies to the case $p = 1$ as well. We shall now, however, introduce a variant of the notion of an atom that has the advantage of having a completely characterizable Fourier transform:

DEFINITION. A function $a \in L^2(\mathbf{R})$ is a *$(p, 2)$-atom, $p \leqslant 1$*, if

(i) it is supported in a finite interval I;

(ii) $(\int_I |a(x)|^2 dx/|I|)^{1/2} \leqslant |I|^{-1/p}$;

(iii) $\int_{-\infty}^{\infty} x^k a(x)\, dx = 0$ for $k = 0, 1, \ldots, [1/p] - 1$, where $[\alpha]$ denotes the largest integer less than or equal to α.[15]

With this definition the statement of Theorem I on **R** remains unchanged and the $H^p(\mathbf{R})$ analog of (1.8) holds.

Recall that we had made the assumption that $L^p(T)$ consisted of real-valued functions and $\Re e H^p$ consisted of the boundary values of the real parts of H^p functions on the unit disc. We shall abandon this convention and we allow our atoms to be complex-valued. $H^p(\mathbf{R})$ will denote the complexification of the

---

[14] We are aware of the fact that despite announcing a characterization of the Fourier coefficients of functions $f \in \Re e H^1$ we have not yet given a characterization. Actually, for our purposes, we needed only to know properties (1.18) and (1.19) for Fourier coefficients of atoms. The reader can easily verify that $\sum c_k e^{ik\theta}$ is a Fourier series of $f \in \Re e H^1$ if and only if $c_k = \sum \lambda_n e^{i\alpha_n k} a_k^{(n)}$ where $\sum |\lambda_n| < \infty$; each sequence $a^{(n)} = \{\ldots, a_{-2}^{(n)}, a_{-1}^{(n)}, a_0^{(n)}, a_1^{(n)}, a_2^{(n)}, \ldots\}$ satisfies (1.18) and (1.19); the numbers α_n reflect the fact that we translated the atoms in the decomposition (1.7) so that their supporting interval is centered at 0.

[15] That is, as p tends to zero we impose the condition that more moments of a are 0. Similar conditions on the p-atoms on T are the ones that are needed for Theorem I when $p \leqslant 1/2$ (see [**10**]).

space $\mathfrak{Re} H^p(\mathbf{R})$. This space (of distributions) should not be confused with the space of boundary values of analytic functions satisfying (1.26).

Let us now suppose $a(t)$ is a $(p, 2)$-atom supported on an interval I which is centered about the origin. There is no loss in generality if we assume $\|a\|_2^2 = |I|^{(p-2)/p}$. Then the function $A(z) = (1/2\pi)\int_{-\infty}^{\infty} e^{-izt}a(t)\,dt$ is an entire function in the complex plane $\mathbf{C}$. Moreover, $|A(z)| \leqslant e^{|z||I|}\|a\|_1$; that is, A is of exponential type. The Plancherel theorem and (ii) yield $\int_{-\infty}^{\infty}|A(x)|^2\,dx \leqslant |I|^{1-(2/p)}$; from (iii) we also have the fact that the derivatives $A^{(k)}(0) = 0$ for $k = 0, 1, \ldots, [1/p] - 1$. It follows from the Paley-Wiener theorem (see (7.2) in Vol. II of [64]) that the converse of all this is also true: If $A \in L^2(\mathbf{R})$ is the restriction of an entire function of exponential type σ, is normalized by $\|A\|_2^2 = \sigma^{1-(2/p)}$ and satisfies $A^{(k)}(0) = 0$ for $k = 0, 1, \ldots, [1/p] - 1$, then A is the Fourier transform of a $(p, 2)$-atom. This gives us the following characterization of the Fourier transforms of the elements of $H^p(\mathbf{R})$:

THEOREM II. *$\hat{f}$ is the Fourier transform of a member of $H^p(\mathbf{R})$, $p \leqslant 1$, if and only if*

$$\hat{f} = \sum_{j=0}^{\infty} \lambda_j e^{i\alpha_j x} A_j(x),$$

where $\sum |\lambda_j|^p < \infty$, α_j are real numbers and $A_j(x)$ are the restrictions of entire functions of exponential type σ_j such that $\|A_j\|_2^2 = \sigma_j^{1-(2/p)}$; moreover, $A_j^{(k)}(0) = 0$ for $0 \leqslant k \leqslant [1/p] - 1$.

In complete analogy with (1.13) any $(1, 2)$-atom $a(t)$ satisfies $\|a\|_2\|ta(t)\|_2 \leqslant 1$. Since the Fourier transform of $ta(t)$ is $-iA'(x)$ it follows from Plancherel's theorem that $\|A\|_2\|A'\|_2 \leqslant 1$. We are led, as before, to introduce the notions of a molecule and of a *Fourier-molecule* centered at 0. We define the latter to be a function $\mathfrak{M}$ on $\mathbf{R}$ satisfying

$$\text{(1.28)} \qquad \text{(i)}\ \ \|\mathfrak{M}\|_2\|\mathfrak{M}'\|_2 \leqslant 1, \qquad \text{(ii)}\ \ |\mathfrak{M}(x)| \leqslant |x|/\|\mathfrak{M}\|_2^2.$$

We leave it to the reader to verify that $A = \hat{a}$ satisfies (ii) when a is a $(2, 1)$-atom centered at 0, that $\check{\mathfrak{M}} + i(\operatorname{sgn} \mathfrak{M})^{\vee} \in L^1$ with uniformly bounded L^1 norm (thus, $\check{\mathfrak{M}}$ is in $H^1(\mathbf{R})$) and, finally, that Theorem II holds if we replace the functions A_j by Fourier-molecules.

For $2/3 < p \leqslant 1$ we say that $\mathfrak{M}$ is a *Fourier-molecule for $H^p(\mathbf{R})$* (centered at 0) if $\sigma^{1-(1/p)}\mathfrak{M}$ is a Fourier-molecule for $H^1(\mathbf{R})$, where $\|\mathfrak{M}\|_2^2 = \sigma^{(2/p)-1}$. Again, it can be easily verified that Theorem II can be restated in terms of Fourier-molecules (similar definitions can be given for $p < 2/3$).

Theorem (1.20) has the following analog:

THEOREM (1.29). *The multiplier operator mapping $f \in H^p(\mathbf{R})$ into $(\mu\hat{f})^{\vee}$ is bounded on $H^p(\mathbf{R})$ when $2/3 < p$, provided $\mu \in L^\infty(\mathbf{R})$ satisfies $\sup_{R>0} R\int_{R\leqslant|x|<2R}|\mu'(x)|^2\,dx < \infty$.*

This result follows immediately from the observation that the conditions we just imposed on μ are necessary and sufficient for the multiplier to preserve the defining properties (1.28) of the Fourier-molecules in question (the slight alteration of the inequalities (i) and (ii) by fixed constants depending only on μ are of minor importance).

Another application of Theorem II stated in terms of Fourier-molecules follows from the simple general inequality $||\mathfrak{M}|'| \leqslant |\mathfrak{M}'|$ valid for any differentiable function $\mathfrak{M}$. Since this inequality implies that $|\mathfrak{M}|$ is a Fourier-molecule whenever $\mathfrak{M}$ is one we have

THEOREM (1.30). *Suppose* $f \in H^1(\mathbf{R})$. *Then there exists* $g \in H^1(\mathbf{R})$ *such that* $|\hat{f}| \leqslant \hat{g}$ *and* $\|g\|_{H^1} \leqslant c\|f\|_{H^1}$, *where c is independent of f.*

(See p. 287 of Vol. I of [64] for the version of this result on T.) We shall restrict ourselves to these remarks concerning $H^p(\mathbf{R})$. It will become clear in the sequel that the atomic approach plays an important role here as well: duality, the study of other operators commuting with translations, maximal operators and other subjects can be studied from this point of view.

We shall now extend these notions so that we can see how they apply to the study of convolution operators on n-dimensional Euclidean space $\mathbf{R}^n$. Let us begin by describing an extension of the theory of Hardy spaces to $\mathbf{R}^n$ developed in [59]. The first observation that motivated this extension is that the Cauchy-Riemann equations admit the following natural generalization to higher dimensions: A vector-valued function $F(x,y) = (u_0(x,y), u_1(x,y), \ldots, u_n(x,y))$ defined on the upper half-space $\mathbf{R}^{n+1}_+ = \{(x,y) \in \mathbf{R}^{n+1}: x \in \mathbf{R}^n, y > 0\}$ is called a *generalized analytic function* or a *Riesz system* if and only if $u_j, j = 0, 1, \ldots, n$, are harmonic functions satisfying:

$$\sum_{j=0}^{n} \frac{\partial u_j}{\partial x_j} = 0 \quad \text{and} \quad \frac{\partial u_j}{\partial x_k} = \frac{\partial u_k}{\partial x_j}, \tag{1.31}$$

where $\partial/\partial x_0$ is defined to be $\partial/\partial y$.

Such a generalized analytic function F is said to belong to $H^p(\mathbf{R}^n)$, $p > 0$, provided

$$\begin{aligned} \|F\|_{H^p} &\equiv \sup_{y>0}\left(\int_{\mathbf{R}^n} |F(x,y)|^p\,dx\right)^{1/p} \\ &\equiv \sup_{y>0}\left(\int_{\mathbf{R}^n}\left[\sum_{j=0}^{n} |u_j(x,y)|^2\right]^{p/2} dx\right)^{1/p} < \infty. \end{aligned} \tag{1.32}$$

It can be shown that if $p > (n-1)/n$ then $\lim_{y\to 0} F(x,y) = (f_0(x), f_1(x), \ldots, f_n(x)) = F(x)$ exists a.e. and in the L^p-norm. As is the case in one dimension,when $p < 1$ one should consider these boundary values in the sense of distributions. When $p = 1$, however, $F(x)$ is a bona fide vector-valued function in $L^1(\mathbf{R}^n)$ and its components are related by the Riesz transforms. In terms of the Fourier transform these relations can be expressed in the following way:

$$\hat{f}_j(x) = c_n(x_j/|x|)\hat{f}_0(x), \qquad j = 1, 2, \ldots, n.$$

Moreover,

$$(P^* F)(x) = \sup_{|t-x|<y} |F(t,y)|$$

belongs to $L^p(\mathbf{R}^n)$ whenever $F \in H^p(\mathbf{R}^n)$, $(n-1)/n < p$ (see [59] for all these results).

C. Fefferman and E. M. Stein [29] recently proved that this property actually characterizes the distributions that are boundary values of functions in $H^p(\mathbf{R}^n)$. More precisely, a distribution f is the boundary value of the first component of a Riesz system $F \in H^p$ if and only if

$$(P^*f)(x) = \sup_{|t-x|<y} |(P_y^*f)(t)| \tag{1.33}$$

belongs to $L^p(\mathbf{R}^n)$, where P_y^*f is the Poisson integral of f:

$$c_n \int_{\mathbf{R}^n} f(s) \frac{y}{[|s-t|^2 + y^2]^{(n+1)/2}} ds.$$

We shall call such an f a boundary distribution of $H^p(\mathbf{R}^n)$. Other maximal operators, besides the maximal Poisson integral (1.33) can be used for this characterization. For example, if $\varphi \geqslant 0$ is a C^∞ function with compact support and $\varphi_\varepsilon(x) = \varepsilon^{-n}\varphi(\varepsilon^{-1}x)$ then $\sup_{\varepsilon>0}|(f * \varphi_\varepsilon)(x)|$ belongs to L^p if and only if f is a boundary distribution of $H^p(\mathbf{R}^n)$. The discovery of this "maximal function" characterization of $H^p(\mathbf{R}^n)$ was an important breakthrough in the real-variable understanding of these spaces as well as in the study of operators on them.

An interesting application of the Fefferman-Stein results is the observation made by L. Carleson [8] that $H^p(\mathbf{R}^n)$ can also be defined in terms of holomorphic functions of several complex variables. In fact, let $\Gamma_j, j = 1, 2, \ldots, n+1$, be open convex cones with vertex at the origin such that their union covers $\mathbf{R}^n$, except for the origin. Assume that none of the Γ_j contains an entire straight line. Then f is a boundary distribution for $H^p(\mathbf{R}^n)$ if and only if

$$f = f_1 + f_2 + \cdots + f_{n+1} \tag{1.34}$$

where each f_j is a boundary distribution of a holomorphic function F_j in the H^p space associated with the tube domain $T_{\Gamma_j^*}$.[16]

Briefly, we obtain this representation in the following way: Suppose f is a boundary distribution of $H^p(\mathbf{R}^n)$. We can construct $n+1$ functions φ_j that are homogeneous of degree 0 such that $\sum_{j=1}^n \varphi_j(x) \equiv 1$ $(x \neq 0)$, the support of φ_j lies in Γ_j^* and each φ_j belongs to the class $C^\infty(\mathbf{R}^n - \{0\})$. It follows from the Fefferman-Stein theory that $f_j = (\varphi_j\hat{f})\check{}$ is also a boundary distribution of $H^p(\mathbf{R}^n)$. The integral representation of F_j in terms of $\hat{f}_j$ (see footnote 16) is then in $H^p(T_{\Gamma_j^*})$. Conversely, if f_j is the boundary value of a function in $H^p(T_{\Gamma_j^*})$ it follows from the theory of Hardy spaces on tubes that a Poisson-type maximal function of f_j belongs to $L^p(\mathbf{R}^n)$. This, in turn, forces f_j to be a boundary distribution of $H^p(\mathbf{R}^n)$ because of the Fefferman-Stein theory.

There are several other extensions of Hardy spaces in terms of generalized Cauchy-Riemann equations (see [6] and [60]).[17] Fefferman and Stein have

[16] These are yet another extension of Hardy spaces to the n-dimensional case. We shall not consider them further in this paper. For their definition and properties see Chapter III of [57]. In this case Γ_j^* is the cone dual to Γ_j and $T_{\Gamma_j^*} = \{z = x + iy \in \mathbf{C}^n : x \in \mathbf{R}^n \text{ and } y \in \Gamma_j^*\}$. f_j and F_j are related as follows: the support of $\hat{f}$ lies in Γ_j,

$$F_j(z) = \int_{\Gamma_j} e^{iz\cdot t}\hat{f}_j(t)\,dt \text{ and } \sup_{y\in\Gamma_j^*} \int_{\mathbf{R}^n} |F(x+iy)|^p\,dx < \infty.$$

shown that their maximal function characterization remains valid for these extensions. Moreover, they show that the dual of $H^1(\mathbf{R}^n)$ is the space $BMO(\mathbf{R}^n)$ of functions of bounded mean oscillation on $\mathbf{R}^n$ (the definition of this space is the same as in the one dimensional case except that intervals are replaced by solid spheres). Atoms on $\mathbf{R}^n$ can also be introduced as before (using solid spheres instead of intervals). Thus, the *atomic* $\mathfrak{H}^1(\mathbf{R}^n)$ space is well defined on $\mathbf{R}^n$: it consists of all functions $f = \sum_{j=1}^{\infty} \lambda_j a_j$, where $\sum |\lambda_j| < \infty$ and each a_j is an atom (either a 1-atom, defined similarly to those introduced when discussing T, or a (1, 2)-atom. We shall show later that this makes no difference). If we define the norm of f, $\|f\|_{\mathfrak{H}^1(\mathbf{R}^n)} \equiv \|f\|_{\mathfrak{H}^1}$, to be the infimum of all the sums $\sum |\lambda_j|$ for which $f = \sum \lambda_j a_j$ we obtain a Banach space. Similarly the *atomic* $\mathfrak{H}^p(\mathbf{R}^n)$ space can be introduced by making use of p-atoms. We shall do all this carefully in a much more general setting. We end this section by making some observations concerning these spaces. The first one is that *the elements of* $\mathfrak{H}^p(\mathbf{R}^n)$ *are always boundary distributions of* $H^p(\mathbf{R}^n)$ *when* $n/(n+1) < p \leqslant 1$. This follows easily from the general theory of $H^p(\mathbf{R}^n)$ spaces developed in [59] once the following inequality is established:

(1.35) *If a is a p-atom, $n/(n+1) < p \leqslant 1$, then* $\|R_j a\|_p \leqslant c_{p,n}$,

where R_j is the jth Riesz transform of a and $c_{p,n}$ is a constant independent of the atom a. This is as easy to show as was (1.24): R_j is the integral operator with kernel $k_j(x) = x_j/|x|^{n+1}$, $j = 1, 2, \ldots, n$ (up to a multiplicative constant). $(R_j a)(x)$ is, then, the principal value integral

$$\lim_{\varepsilon \to 0+} \int_{|x-y| \geqslant \varepsilon} k_j(x-y) a(y)\, dy \equiv \text{P.V.} \int_{\mathbf{R}^n} k_j(x-y) a(y)\, dy.$$

An immediate calculation using the mean value theorem yields

$$|k_j(x-y) - k_j(x)| \leqslant (n+2)2^n \frac{|y|}{|x|^{n+1}} \quad \text{if } 2|y| < |x|. \tag{1.36}$$

Because of the translation invariance of the convolution operator R_j there is no loss in generality if we assume a is supported in the sphere S_r about the origin of radius $r > 0$ and $|a(x)| \leqslant |S_r|^{-1/p} = cr^{-n/p}$ (from here on let us agree that c denotes a constant, not always the same one, that is independent of the atom a). We have

$$\begin{aligned}\|R_j a\|_p^p &= \int_{\mathbf{R}^n} \left| \text{P.V.} \int_{|y|<r} k_j(x-y) a(y)\, dy \right|^p dx \\ &= \left(\int_{|x| \leqslant 2r} + \int_{|x|>2r} \right) \left| \text{P. V.} \int_{|y|<r} k_j(x-y) a(y)\, dy \right|^p dx \equiv \text{I} + \text{II}.\end{aligned}$$

We estimate I by using Hölder's inequality and the L^2-boundedness of R_j:

$$\begin{aligned}\text{I} &= \int_{|x| \leqslant 2r} 1 \cdot |(R_j a)(x)|^p\, dx \leqslant c(r^n)^{(2-p)/2} \left(\int_{\mathbf{R}^n} |(R_j a)(x)|^2\, dx \right)^{p/2} \\ &\leqslant cr^{n(2-p)/2} \|a\|_2^p \leqslant cr^{n(2-p)/2} r^{n(p-2)/2} = c.\end{aligned}$$

[17] In fact, this approach can be carried out even in the setting of compact Lie groups [16].

We estimate II by making use of the fact that a has mean 0 and property (1.36):

$$\begin{aligned} \mathrm{II} &= \int_{|x|>2r} \left| \int_{|y|<r} [k(x-y) - k(x)] a(y)\, dy \right|^p dx \\ &\leqslant c_n^p \int_{|x|>2r} |x|^{-p(n+1)} \left\{ \int_{|y|<r} |y|\, |a(y)|\, dy \right\}^p dx \\ &\leqslant c \int_{2r}^{\infty} s^{n-1-p(n+1)}\, ds \left(\int_{|y|<r} |y| r^{-n/p}\, dy \right)^p \leqslant c r^{n-p(n+1)} r^{-n+p(n+1)} = c. \end{aligned}$$

This establishes (1.35).

Let us examine some of the consequences of (1.35) when $p = 1$. By a simple extension of the reasoning used to establish (1.11) it follows that $BMO(\mathbf{R}^n)$ represents the dual of $\mathfrak{H}^1(\mathbf{R}^n)$. In particular, $\|f\|_{\mathfrak{H}^1(\mathbf{R}^n)}$ is equivalent to the norm obtained by considering the maximum of all values Lf where L is a bounded linear functional on $\mathfrak{H}^1(\mathbf{R}^n)$ represented by a BMO function having BMO norm 1. But, as we stated above, Fefferman and Stein showed that this last norm is equivalent to that of $H^1(\mathbf{R}^n)$. Thus, (1.35) shows that $\mathfrak{H}^1(\mathbf{R}^n)$ is not only contained in $H^1(\mathbf{R}^n)$ but the two norms are equivalent. Moreover, it follows from the work of Fefferman and Stein that BMO represents $(H^1(\mathbf{R}^n))^*$ in the same manner it represents $(\mathfrak{H}^1(\mathbf{R}^n))^*$ (see §3). The Hahn-Banach theorem can be invoked to argue that $\mathfrak{H}^1(\mathbf{R}^n) = H^1(\mathbf{R}^n)$.

For $p < 1$ the identification of $\mathfrak{H}^p$ with H^p cannot be done using duality. The decomposition theorem obtained in [10] has recently been extended by R. Latter to $\mathbf{R}^n$ and provides a direct proof that any $f \in H^p$ admits an atomic decomposition; that is, $\mathfrak{H}^p = H^p$.[18]

2. Hardy spaces associated with spaces of homogeneous type. In [17] we introduced certain topological measure spaces having properties that permitted us to extend the Calderón-Zygmund theory of singular integrals to these spaces. It turns out that these properties can also be used to obtain a very useful theory if we employ Theorem I in order to define Hardy spaces. As can be seen from §1, by doing this we completely avoid the problem of finding an appropriate system of partial differential equations generalizing the Cauchy-Riemann equations. Nevertheless, we shall see that in certain examples, where the structure of the space allows us to introduce natural extensions of the Cauchy-Riemann equations, the spaces introduced here are closely related to Hardy spaces that could be defined in terms of such partial differential equations.

We begin by introducing the notion of a *space of homogeneous type*: this is a topological space X endowed with a Borel measure μ and a *quasi-metric* (*or quasi-distance*) d. The latter is a mapping $d\colon X \times X \to \mathbf{R}^+ = \{t \in \mathbf{R}\colon t \geqslant 0\}$ satisfying (a) $d(x,y) = d(y,x)$, (b) $d(x,y) > 0$ if and only if $x \neq y$ and (c) there exists a constant K such that $d(x,y) \leqslant K[d(x,z) + d(z,y)]$ for all x, y, z in X. We postulate that the *spheres* $B_r(x) = \{y \in X\colon d(x,y) < r\}$ centered at

[18] Latter's proof was communicated to us by a letter from J. Garnett while this manuscript was still being written.

x and of radius $r > 0$ form a basis of open neighborhoods of the point x and, also, $\mu(B_r(x)) > 0$ whenever $r > 0$. Our basic assumption relating the measure and the quasi-distance is the existence of a constant A such that

$$\mu(B_r(x)) \leqslant A\mu(B_{r/2}(x)). \tag{2.1}$$

Before developing the theory of Hardy spaces in this setting we shall give several examples of spaces of homogeneous type.

(1) $X = \mathbf{R}^n$, $d(x,y) = |x-y| = (\sum_{j=1}^n (x_j - y_j)^2)^{1/2}$ and μ equals Lebesgue measure;

(2) $X = \mathbf{R}^n$, $d(x,y) = \sum_{j=1}^n |x_j - y_j|^{\alpha_j}$, where $\alpha_1, \alpha_2, \ldots, \alpha_n$ are positive numbers, not all equal, and μ is Lebesgue measure (we shall refer to such distances as *nonisotropic*);

(3) Example (1) can be further generalized by letting X and d be as in (2) and μ a measure for which (2.1) holds. An important class of such measures is obtained by letting $d\mu(x) = [d(x,0)]^\beta dx$, $\beta > 0$, where dx is the element of Lebesgue measure;

(4) $X = [0,1)$, μ is Lebesgue measure and $d(x,y)$ is the length of the smallest dyadic interval containing x and y (we consider the intervals to be closed on the left and open on the right);

(5) $X = [-1,1]$, μ is Lebesgue measure and $d\mu(x) = (1-x)^\alpha(1+x)^\beta dx$, where $\alpha, \beta > -1$ (this space is associated with the study of Jacobi polynomials);

(6) $X = \mathbf{R}_+ = \{r \in \mathbf{R}; r \geqslant 0\}$, $d\mu(r) = r^{n-1} dr$, d is the usual distance (this space is important for the study of radial functions in $\mathbf{R}^n$);

(7) Compact Riemannian manifolds with natural distances and measures furnish us with a wide variety of these spaces. Particular examples are compact Lie groups G and the homogeneous spaces G/K obtained from the closed subgroups of G. We shall study in detail the case of the surface Σ_{n-1} of the unit sphere in $\mathbf{R}^n$ (which can be identified with $SO(n)/SO(n-1)$). The special case when $n = 4$ is especially interesting since Σ_3 can be identified with $SU(2)$ (see [**17**]). For Σ_{n-1}, $d(x,y) = |x-y|$ and μ is the Lebesgue measure;

(8) X is the boundary of a Lipschitz domain in $\mathbf{R}^n$, μ is harmonic measure and d is Euclidean distance;

(9) When X is the boundary of a smooth and bounded pseudo-convex domain in $\mathbf{C}^n$ one can introduce a nonisotropic quasi-distance that is related to the complex structure (see [**42**], [**56**] and example (13)) in such a way that we obtain a space of homogeneous type by using the Lebesgue surface measure. A concrete example that will be studied here is the surface of the unit sphere $\Sigma_{2n-1} = \{z \in \mathbf{C}^n: z \cdot \bar{z} = \sum_{j=1}^n z_j \bar{z}_j = 1\}$. The nonisotropic distance is given by $d(z,w) = |1 - z \cdot \bar{w}|^{1/2} = |1 - \sum_{j=1}^n z_j \bar{w}_j|^{1/2}$ (this is *not* equivalent to the distance given in the seventh example);

(10) Similarly, a quasi-distance can be introduced on the closure of a bounded, smooth pseudo-convex domain in order to obtain an example of one of our spaces (again, see [**56**] and example (13)). In the solid unit ball $B_{2n} = \{z \in \mathbf{C}^n: |z|^2 = z \cdot \bar{z} \leqslant 1\}$ the quasi-distance can be expressed in the form

$$d(z, w) = \big||z|-|w|\big| + \frac{\big||z|\,|w|-\bar{z}\cdot w\big|}{|z|\,|w|}$$

when z and w are close (in the Euclidean sense) to the boundary. We shall consider this situation in greater detail later when we discuss the Bergman kernel associated with B_{2n};

(11) Other nonisotropic distances, besides those given in the second example, arise naturally on Euclidean spaces. For each $x = (x_1, x_2, x_3) \in \mathbf{R}^3$ consider the matrix

$$\begin{pmatrix} 1 & x_1 & x_3 + (x_1 x_2/2) \\ 0 & 1 & x_2 \\ 0 & 0 & 1 \end{pmatrix}$$

which we shall also denote by x. Matrix multiplication gives us a group structure on $\mathbf{R}^3$ and Haar measure coincides with Lebesgue measure. The reader can verify that the product of $x = (x_1, x_2, x_3)$ and $y = (y_1, y_2, y_3)$ is the point $(x_1 + y_1, x_2 + y_2, x_3 + y_3 + (x_1 y_2 - x_2 y_1)/2)$. The dilations $T_\lambda x = (\lambda x_1, \lambda x_2, \lambda^2 x_3)$, $\lambda \in \mathbf{R}_+$, obviously preserve this operation; in fact they are automorphisms of this group of 3×3 matrices. In studying operators that commute with these automorphisms it is useful to employ the quasi-distance induced by the "norm" $\|x\| = |x_1| + |x_2| + |x_3|^{1/2}$. That is,

$$d(x,y) = \|xy^{-1}\| = |x_1 - y_1| + |x_2 - y_2| + |x_3 - y_3 + (y_1 x_2 - x_1 y_2)/2|^{1/2}.$$

This usefulness stems from the fact that d is homogeneous with respect to T_λ; that is, $d(T_\lambda x, T_\lambda y) = \lambda d(x,y)$ for all $\mathbf{R}_+$. This is just a special case of a class of groups having such dilations (see [**40**]). This group is an example of a Heisenberg group. It can also be identified with the boundary of a "generalized half-space" that is the image of the ball $B_4 \subset \mathbf{C}^2$ under an appropriate Cayley transform (see [**42**]).

(12) Another class of examples is obtained if X is a locally compact group for which there exists a countable base $\{U_j\}$ of open neighborhoods of the identity $\mathbf{1}, j = 0, \pm 1, \pm 2, \ldots$ (if G is compact we consider only the indices $j \leqslant 0$) satisfying (a) $U_j = U_j^{-1}$, (b) $U_j U_j \subset U_{j+1}$, (c) $0 < \mu(U_{j+1}) < c\mu(U_j)$, where μ is Haar measure and c is a constant independent of j, (d) $\bigcup_j U_j = X$. We can then introduce a quasi-distance having the property that the solid spheres of radius $r > 0$ have a measure that is essentially r. This is done in the following way: We first define a distance m from $\mathbf{1}$ by letting $m(x) = \inf\{\mu(U_j)\colon x \in U_j\}$; the desired quasi-distance is then $d_m(x,y) = m(xy^{-1})$ (see [**17**, p. 78], for a proof of this fact). If X is the group of 3×3 matrices we considered in example (11) we obtain a countable base of open neighborhoods of the identity by letting $U_j = T_{2^j} B^0, j = 0, \pm 1, \pm 2, \ldots$, where B^0 is the interior of the ordinary solid sphere of radius 1 in $\mathbf{R}^3$. The distance d_m is then equivalent to the distance d introduced in the previous example. Another interesting special case is furnished by the integers (or, more generally, by the group $\mathbf{Z}^n$ of lattice points in $\mathbf{R}^n$). Observe that in this case we have no differentiable structure. A similar situation arises when X is the field of p-adic numbers. For example, we discuss briefly the 3-adic field Q_3. Formally the

nonzero elements of Q_3 are represented by the sums $x = \sum_{l=k}^{\infty} a_l 3^l$, where $a_l = 0$, 1 or 2, with $a_k \neq 0$. In this case the norm of x is denoted by $\|x\|$ and is equal to 3^{-k}. A neighborhood system for the origin is given by the "fractional ideals" $P^k = \{x \in Q_3 : \|x\| \leqslant 3^{-k}\}$. $P^0 \equiv R$ is the ring of integers in Q_3. If Haar measure (with respect to addition) μ is normalized so that $\mu(R) = 1$ then it can be shown that $\mu(P^k) = 3^{-k}$. It follows that $U_j = P^{-j}$ satisfy properties (a), (b), (c) and (d).

(13) In most of the examples we have given X was either a group or had associated with it a natural group acting on it. A general method for obtaining spaces of homogeneous type that does not involve a group structure is the following. (Actually we can define the concept of a space of homogeneous type by this method.) Let X be a topological space on which there is defined a Borel measure μ. Suppose that at each point $x \in X$ there exists a basis of open neighborhoods $\{U_x^\alpha\}$, $\alpha \in \mathfrak{A}$, such that (a) $\bigcup_\alpha U_x^\alpha = X$, (b) $\bigcap_\alpha U_x^\alpha = \{x\}$, (c) either: $U_x^\alpha \subset U_x^\beta$ or $U_x^\beta \subset U_x^\alpha$, (d) there exists a constant c such that if $y \in U_x^\alpha$ then there exists $\beta \in \mathfrak{A}$ such that $U_x^\alpha \subset U_y^\beta$ and $\mu(U_y^\beta) \leqslant c\mu(U_x^\alpha)$. It is easy to see that the spheres of a space of homogeneous type satisfy these properties. An appropriate quasi-distance d is obtained by letting

$$d(x,y) = \inf\{\mu(U_x^\alpha): y \in U_x^\alpha\} + \inf\{\mu(U_y^\beta): x \in U_y^\beta\}$$

for all $x, y \in X$. If, for $r > 0$, there exist neighborhoods $U_x^\alpha \subset U_x^\beta$ such that $c_1 r \leqslant \mu(U_x^\alpha) \leqslant \mu(U_x^\beta) \leqslant c_2 r$, where c_1, c_2 are independent of x and r, then the spheres $B_r(x)$ defined by the quasi-distance d have measure comparable to r. This situation occurs often and provides a technically convenient fact enabling one to obtain certain normalizations independent of the dimension.

When X is the boundary of a smooth and bounded strongly pseudo-convex domain in $\mathbf{C}^n$ one can construct a quasi-distance by this method. A system of open neighborhoods at a point x can be obtained by considering the surface of the osculating sphere Σ at x: The distance introduced in example (9) gives us a natural basis of open neighborhoods of x as a point of Σ. These neighborhoods can then be projected onto X and they, in turn, can be used to define a quasi-distance by the method we have just described.

If $\mathfrak{D}$ is a bounded domain in $\mathbf{C}^n$ with smooth boundary $\partial\mathfrak{D}$ define a smooth mapping $x \to \bar{x}$ of $\mathfrak{D}$ to $\partial\mathfrak{D}$ with the property that whenever x is close to $\partial\mathfrak{D}$ then $\bar{x}$ is the normal projection of x to $\partial\mathfrak{D}$. Let $\nu_{\bar{x}}$ be the unit vector in the direction of the outward normal to $\partial\mathfrak{D}$ at $\bar{x}$. Denote by T_x^2 the 1-dimensional complex vector space generated by $\nu_{\bar{x}}$. Let T_x^1 be the orthogonal complement of T_x^2. The sets

$$U_x^r = \{z \in \mathfrak{D}: z = x + z_1 + z_2, z_1 \in T_x^1, z_2 \in T_x^2, |z_1|^2 + |z_2| < r\},$$

$r > 0$, form a system of neighborhoods at x satisfying (a), (b), (c) and (d).[19]

[19] The pseudo-convexity assumption is not necessary when we consider boundaries of smooth domains in $\mathbf{C}^n$. In fact, if we restrict the quasi-metric just defined on $\mathfrak{D}$ to $\partial\mathfrak{D}$ we obtain a distance that is equivalent to the one defined in the previous paragraph. We can make use of pseudo-convexity if we wish to avail ourselves of certain estimates for Szegõ and Bergman kernels in order to study the operators they define on H^p spaces and L^p spaces associated with these domains (see **[31]**, **[56]**).

Before discussing the H^p-space theory associated with a space of homogeneous type we make some additional bibliographical comments. As we mentioned at the beginning of this section, spaces of homogeneous type were introduced in [12], [17] as a geometric setting for the basic real variable theory of singular integrals; at the same time Korányi and Vági [42] and N. Rivière [49] independently introduced this theory in the setting of example (12). The basic motivating examples were (9) and (11). The quasi-distances for pseudo-convex domains in $\mathbf{C}^n$ were given by Stein [56] who also proved that they give a space of homogeneous type. Various generalizations of these distances are in Rothschild and Stein [50].

The example discussed in the last part of (12) is a particular case of the martingale setup for H^p. It is in this setting that the atomic characterization of the H^p space was obtained first by C. Herz [34]. For a discussion of BMO and H^1 for martingales see Garsia's book *Martingale inequalities*, Benjamin, 1973.

We are now ready to introduce the Hardy spaces associated with a space of homogeneous type X. If $0 < p < q, p \leqslant 1 \leqslant q \leqslant \infty$, we say that a function $a(x)$ is a (p, q)-*atom* if

(i) the support of a is contained in a sphere $B_r(x_0)$;

(ii) $\{[1/\mu(B_r(x_0))] \int |a(x)|^q \, d\mu(x)\}^{1/q} \leqslant \{\mu(B_r(x_0))\}^{-1/p}$;

(iii) $\int a(x) \, d\mu(x) = 0$.

In case $\mu(X) < \infty$ the constant function having value $[\mu(X)]^{-1/p}$ is also considered to be an atom. (In order to simplify certain calculations we shall henceforth assume $\mu(X) = 1$ whenever X has finite measure.) Observe that a (p, q)-atom is in $L^1(X, \mu)$ and is normalized in such a way that its $L^p(X, \mu)$ norm does not exceed 1.

Recall that when we discussed H^p spaces associated with T or $\mathbf{R}$, when $p < 1$, we pointed out that these were *not* spaces of *functions* defined on T or $\mathbf{R}$. In fact we considered the elements of H^p to be distributions. In the present setting we are, again, forced to introduce an appropriate space of linear functionals in order to define the Hardy spaces. In order to do this we introduce the *Lipschitz* (or *Hölder*) spaces $\mathfrak{L}_\alpha$, $\alpha > 0$, consisting of those functions l on X for which

$$|l(x) - l(y)| \leqslant C[\mu(S)]^\alpha, \tag{2.2}$$

where S is any sphere containing both x and y and C depends only on l.[20] Let $\mathfrak{N}^{(\alpha)}(l)$ be the infimum of all C for which (2.2) holds. If we define

[20] Unless α is sufficiently small it can happen that the only functions satisfying (2.2) are the constants. This is true for T and $\mathbf{R}$ if $\alpha > 1$. On the other hand if X is the space described in example (4) (or is a p-adic field) any finite sum $\sum a_i \chi_{S_i}$, where the χ_{S_i}'s are characteristic functions of disjoint spheres, belongs to $\mathfrak{L}_\alpha$ for all $\alpha > 0$. In any case, for each of our examples, there is an interval of positive α's where $\mathfrak{L}_\alpha$ is nontrivial. Condition (2.2) can also be written in the more familiar form $|l(x) - l(y)| \leqslant Cm(x,y)^\alpha$ if m is a quasi-distance such that the spheres of radius r have measure comparable to r (this can be done quite generally; see p. 78 of [17]). When $\mu(X) = \infty$, $\mathfrak{L}_\alpha$ really consists of equivalence classes of such functions: l_1 and l_2 are equivalent if $l_1 - l_2$ is constant. We will consider this point in greater detail in §3.

$$\|l\|^{(\alpha)} = \begin{cases} \mathfrak{N}^{(\alpha)}(l), & \text{if } \mu(X) = \infty, \\ \mathfrak{N}^{(\alpha)}(l) + \left|\int_X l(x)\,d\mu(x)\right|, & \text{if } \mu(X) = 1, \end{cases}$$

then $\|\cdot\|^{(\alpha)}$ is a norm. A straightforward argument shows that $\mathfrak{L}_\alpha$, so normed, is a Banach space.

If a is a (p, q)-atom, $0 < p < 1$ and $\alpha = 1/p - 1$, then using (i), (iii), (2.2) and (ii), in this order, we have

$$\left|\int_X a l\,d\mu\right| = \left|\int_{B_r(x_0)} a l\,d\mu\right| = \left|\int_{B_r(x_0)} a(x)[l(x) - l(x_0)]\,d\mu(x)\right|$$
$$\leqslant \|l\|^{(\alpha)}[\mu(B_r(x_0))]^{1/p-1}\int_{B_r(x_0)} |a(x)|\,d\mu(x) \leqslant \|l\|^{(\alpha)}.$$

That is, the mapping $l \to \int_X a l\,d\mu$ is a bounded linear functional on $\mathfrak{L}_\alpha$ with norm not exceeding 1.

We now define the space $H^{p,q}$, for $0 < p < 1 \leqslant q$, to be the subspace of the dual $\mathfrak{L}_\alpha^*$ of $\mathfrak{L}_\alpha$, where $\alpha = 1/p - 1$, consisting of those linear functionals admitting an *atomic decomposition*

$$h = \sum_{j=0}^{\infty} \lambda_j a_j, \tag{2.3}$$

where the a_j's are (p, q)-atoms, $\sum_0^\infty |\lambda_j|^p < \infty$. (This last inequality implies that the series in (2.3) converges in the norms of $\mathfrak{L}_\alpha^*$ and we consider h to be the limit, in this norm, of the partial sums of this series.) The infimum of the numbers $\sum_0^\infty |\lambda_j|^p$ taken over all such representations of h will be denoted by the symbol $\|h\|_{p,q}$. The mapping $h \to \|h\|_{p,q}$ is clearly not a norm (it is homogeneous of order $p < 1$); But $d_{p,q}(h, g) = \|h - g\|_{p,q}$ is a metric. When $p = 1 < q \leqslant \infty$ the sums (2.3) are convergent in the L^1 norm whenever $\sum_0^\infty |\lambda_j| < \infty$ and the a_j's are $(1, q)$-atoms; hence, they define a subspace $H^{1,q}$ of $L^1(X)$. In this case the mapping $h \to \|h\|_{1,q}$ is a norm.

A straightforward argument shows that $H^{p,q}$ is complete. In particular $H^{1,q}$ is a Banach space. It also follows easily from the above definitions that

$$H^{p,\infty} \subset H^{p,q_2} \subset H^{p,q_1} \tag{2.4}$$

whenever $1 < q_1 < q_2 < \infty$ and $0 < p \leqslant 1$ (q_1 may equal 1 when $p < 1$). A basic result concerning these spaces is that the converses of the set-theoretic inclusions (2.4) hold:

THEOREM A. *$H^{p,q} = H^{p,\infty}$ whenever $p < q < \infty$. Moreover, the metrics $d_{p,q}$ and $d_{p,\infty}$ are equivalent; that is, there exists a constant c (depending on p and q) such that*

$$d_{p,q} \leqslant d_{p,\infty} \leqslant c d_{p,q}.$$

The proof of this theorem is rather technical. In order not to disrupt the flow of this presentation we shall give it later.

Theorem A enables us to define the space $H^p = H^p(X)$ for $p \leqslant 1$ to be any one of the spaces $H^{p,q}$ for $p < q \leqslant \infty$, $1 \leqslant q$. Any one of the metrics $d_{p,q}$ can be used in order to turn H^p into a topological space. It is often useful to choose the space $H^{p,2}$ and the metric $d_{p,2}$ to work with when the space X has a "Fourier transform" associated with it, since the $(p, 2)$-atoms are L^2-functions.

In order to characterize the dual of $H^1(X)$ we introduce the function spaces BMO_q[21]: a function l on X is said to belong to BMO_q provided it is locally in $L^q(X)$ and

$$\left[\frac{1}{\mu(S)}\int_S |l(x) - m_S(l)|^q \, d\mu(x)\right]^{1/q} \leqslant C \tag{2.5}$$

for all spheres S, where

$$m_S(l) = \frac{1}{\mu(S)}\int_S l(y)\, d\mu(y)$$

and C is independent of S. Let $\mathfrak{N}_q(l)$ denote the infimum of all constants C for which (2.5) holds. We can then introduce a "norm" on BMO_q by defining

$$\|l\|^{(q)} = \|l\|^{(q)}_{BMO} = \begin{cases} \mathfrak{N}_q(l), & \text{if } \mu(X) = \infty, \\ \mathfrak{N}_q(l) + \left|\int_X l(x)\, d\mu(x)\right|, & \text{if } \mu(X) = 1. \end{cases}$$

When the measure of X is finite, BMO_q, together with this norm, is a Banach space. If $\mu(X) = \infty$ then we consider the set of equivalence classes of functions defined by the relation "l_1 and l_2 in BMO_q are equivalent if and only if $l_1 - l_2$ is constant". If l_1 and l_2 are equivalent then, clearly, $\|l_1\|^{(q)} = \|l_2\|^{(q)}$. Thus, we can define the norm of each equivalence class to be the norm of any of its members and we obtain a Banach space which we also denote by BMO_q (as is usually done when dealing with L^p-spaces we call the members of BMO_q "functions" even when we really are talking about equivalence classes).

We shall show (in §3)

THEOREM B. *If $p < 1$ and $\alpha = (1/p) - 1$ then $\mathfrak{L}_\alpha$ is the dual of $H^p(X)$. That is, each continuous linear functional on H^p is a mapping of the form $h \to \sum \lambda_j \int a_j l \, d\mu$, where $l \in \mathfrak{L}_\alpha$ and $h = \sum \lambda_j a_j \in H^p$ (see (2.3)). If $p = 1$ then BMO is the dual of $H^1(X)$. More precisely, if $h = \sum \lambda_j a_j \in H^1$ then*

$$\lim_{n\to\infty} \sum_1^n \lambda_j \int l a_j \, d\mu \tag{2.6}$$

is a well-defined continuous linear functional $h \to \langle h, l\rangle$, for each $l \in BMO$, whose norm is equivalent to $\|l\|_{BMO}$; moreover, each continuous linear functional on H^1 has this form.

In the proof of Theorem B it will be shown that BMO_{q_1} and BMO_{q_2},

---

[21] "BMO" stands for *bounded mean oscillation.* These functions were introduced by John and Nirenberg [38]. We shall often write BMO instead of BMO_1.

$1 \leqslant q_1, q_2 < \infty$, are equal as vector spaces and the norms $\| \; \|^{(q_1)}$ and $\| \; \|^{(q_2)}$ are equivalent.[22] This will follow from the fact that $BMO_{q'}$ characterizes $(H^{1,q})^*$.

We introduced the notion of a molecule in the first section and showed how molecules can be used in order to obtain multiplier theorems. This notion extends to spaces of homogeneous type and, as we shall see, is quite useful for studying operators associated with some of the particular spaces we have described above. Let $\varepsilon > 0$. We say that M is an *ε-molecule for* $H^1 = H^1(X)$ centered at x_0 if and only if

$$\text{(2.7)} \qquad \left\{ \int_X |M(x)|^2 \, d\mu(x) \right\} \left\{ \int_X |M(x)|^2 [m(x, x_0)]^{1+\varepsilon} \, d\mu(x) \right\}^{1/\varepsilon} \leqslant 1,$$

where $m(x, x_0)$ is the infimum of the measures of the spheres containing both x and x_0, and

$$\text{(2.7')} \qquad \int_X M(x) \, d\mu(x) = 0.$$

The function $m(x, x_0)$, defined on $X \times X$, will be called the *measure distance*. It can be shown quite generally that it is, indeed, a quasi-metric such that, if we consider it together with X and μ, we have a space of homogeneous type that is "equivalent" to (X, μ, d) in particular, the spaces $H^p(X)$ defined in terms of d are the same as those defined in terms of m. Assuming this to be the case we shall now show

THEOREM C. *If M is a molecule for H^1 centered at x_0 then $M \in H^1$. Moreover, $\|M\|_{H^1}$ depends only on the constant ε in* (2.7).

The proof of this result illustrates again the importance of Theorem A since we shall show that M is decomposable into a sum of $(1, 2)$- and $(1, \infty)$-atoms. To fix our ideas let us assume X has infinite measure and that if $B_{r,m}(x_0) = \{x : m(x, x_0) < r\}$ then $\mu(B_{r,m}(x_0))$ is between $C_1 r$ and $C_2 r$, where $0 < C_1 < C_2$ are constants independent of x_0 and r. We write $\mu(B_{r,m}(x_0)) \sim r$ to denote this relation between the radius of a sphere and its measure.[23]

Set $\sigma = \|M\|_2^{-2}$, let χ_0 denote the characteristic function of $B_\sigma(x_0) \equiv B_{\sigma,m}(x_0)$ and, for $i \geqslant 1$, let χ_i be the characteristic function of $\{x : \sigma 2^{i-1} \leqslant m(x, x_0) < \sigma 2^i\}$. Define M_i to be the function $M\chi_i - ((1/\int \chi_i) \int M\chi_i \, d\mu)$ for $i = 0, 1, 2, \ldots$. We shall show that (up to an unimportant multiplicative constant) $2^{i(\varepsilon/2)} M_i$ is a $(1, 2)$-atom. Clearly, $\int M_i \, d\mu = 0$ for $i \geqslant 0$. Moreover,

$$\left(\frac{1}{\sigma} \int |M_0|^2 \, d\mu \right)^{1/2} \leqslant \left(\frac{1}{\sigma} \int_{B_\sigma(x_0)} |M|^2 \, d\mu \right)^{1/2} + \left(\frac{1}{\sigma} \int_{B_\sigma(x_0)} |M| \, d\mu \right) \leqslant \frac{2}{\sigma}.$$

---

22 This last assertion extends only part of a result of John and Nirenberg [38]. They show that BMO is equivalent to "exponential BMO" as well. Their proof can be adapted to spaces of homogeneous type.

23 It can be shown that, in general, $\mu(B_{r,m}(x_0))$ is of the same order of magnitude as r provided we take into account certain obvious nonessential difficulties (if $\mu(X) < \infty$, we cannot allow r to be too large; if $\mu(\{x_0\}) > 0$ we cannot allow r to be too small). These are technical points that we do not wish to elaborate on at the present time.

Thus, $(1/2)M_0$ is a (1, 2)-atom. If $i \geqslant 1$ we make the following estimate in which (2.7) is used:

$$\left(\int |M_i|^2 d\mu\right)^{1/2} \leqslant 2\left(\int |M|^2 \chi_i\right)^{1/2}$$

$$= 2\left(\int |M(x)|^2 [m(x, x_0)]^{1+\varepsilon} [m(x, x_0)]^{-1-\varepsilon} \chi_i(x)\, d\mu(x)\right)^{1/2}$$

$$\leqslant (\sigma 2^{i-1})^{-(1+\varepsilon)/2} C^{\varepsilon/2} \sigma^{\varepsilon/2} = C'(2^i\sigma)^{-1/2}(2^{i\varepsilon})^{-1/2}.$$

It follows that

$$\left(\int |M_i|^2 \frac{d\mu}{\mu(B_{2^i\sigma}(x_0))}\right)^{1/2} \leqslant (\text{const}) 2^{-i}\sigma 2^{-i\varepsilon/2}$$

and, thus, $2^{i\varepsilon/2} M_i$ is a constant multiple of an (1, 2)-atom. We have $M = \sum_{i=0}^{\infty} M_i + \sum_{i=0}^{\infty} m_i \chi_i'$, where $m_i = \int M\chi_i\, d\mu$ and $\chi_i' = \chi_i / \int \chi_i$. The argument we just gave shows that $\sum_{i=0}^{\infty} M_i \in H^1$ and the H^1 norm of this function is, up to a constant, less than or equal to $2^{\varepsilon/2}/(2^{\varepsilon/2} - 1)$. Let $N_j = \sum_{i=j}^{\infty} m_i$. Then, summing by parts and using $\sum_0^{\infty} m_i = \int M\, d\mu = 0$ we have

$$\sum_{i=0}^{\infty} m_i \chi_i' = \sum_{j=0}^{\infty} (N_j - N_{j+1})\chi_j' = \sum_{j=0}^{\infty} N_{j+1}(\chi_{j+1}' - \chi_j').$$

Since $\int \chi_j'\, d\mu = 1$ for all j we have $\int (\chi_{j+1}' - \chi_j')\, d\mu = 0$. Moreover, the support of $\chi_{j+1}' - \chi_j'$ lies within $B_{2^{j+1}\sigma}$ which has measure $\sim 2^{j+1}\sigma$. From the definition of χ_j' we also have $|\chi_{j+1}' - \chi_j'| \leqslant 1/\int \chi_{j+1} + 1/\int \chi_j \leqslant 2/\int \chi_j$. Since $\mu(B_{2^j\sigma}(x_0)) \sim \sigma 2^j$ it follows that $\int \chi_j \sim \mu(B_{2^j\sigma}(x_0)) - \mu(B_{2^{j-1}\sigma}(x_0)) \sim 2^j\sigma - 2^{j-1}\sigma = 2^{j-1}\sigma = 2^{j\sigma}/2$.[24] This shows that, up to a multiplicative constant $\chi_{j+1}' - \chi_j'$ is a $(1, \infty)$-atom. Furthermore, making use of (2.7) again we have

$$|N_j| \leqslant \sum_{i=j}^{\infty} \int |M\chi_i|\, d\mu \lesssim \sum_{i=j}^{\infty} \left(\int |M|^2 \chi_i\right)^{1/2} (\sigma 2^i)^{1/2}$$

$$\leqslant C \sum_{i=j}^{\infty} (\sigma 2^i)^{-1/2} 2^{-i\varepsilon/2} (\sigma 2^i)^{1/2} = C_\varepsilon (2^{-\varepsilon/2})^j.$$

This shows that $\sum_{i=0}^{\infty} m_i \chi_i'$ belongs to H^1 $(= H^{1,\infty})$ and, up to a multiplicative constant, its norm does not exceed $2^{\varepsilon/2}/(2^{\varepsilon/2} - 1)$. This establishes Theorem C.

Before developing the general theory further let us make three observations. First, recall that immediately following (1.17) we pointed out that any linear operator mapping atoms into molecules satisfying (1.15) has a bounded extension mapping H^1 into itself. The argument we gave, however, used the fact that the conjugate function operator maps H^1 boundedly into itself.

[24] Recall that the notation $\mu(B_r(x_0)) \sim r$ meant that for appropriate constants $0 < C_1 < C_2$ we have $C_1 r \leqslant \mu(B_r(x_0)) \leqslant C_2 r$. We are tacitly assuming that $2C_1 > C_2$. If this were not the case then $qC_1 > C_2$ would hold for a sufficiently large q. The argument we are giving would still be valid if q^j is used when 2^j occurs.

Theorem C shows that this was not necessary and that, in general, a linear operator mapping atoms into molecules satisfying (2.7) and (2.7′) has a bounded extension mapping H^1 into itself.

Secondly, it should be remarked that any $(1, \infty)$- (or (1, 2)-) atom is a molecule for H^1. If we define such an atom in terms of the measure distance this follows from a straightforward computation.

Thirdly, when we introduced inequality (1.17) we did so in order to indicate the role played by molecules for H^p for some values of p less than 1. This can be done in general. We leave it for the reader to verify that appropriate molecules for H^p can be defined and that the values of p for which this can be done depend on ε.

We shall now discuss two interpolation theorems for operators acting on $H^p(X)$ and $L^q(X)$. The first result is a Marcinkiewicz-type interpolation theorem (see [**64**]). In order to state it we need to introduce the notion of *weak-type* operators. Suppose B is a map from a vector space into measurable functions on a measure space (Y,ν); we say that B is *sublinear* provided $|B(f+g)| \leqslant |Bf| + |Bg|$ and $|Baf| = |a|\,|Bf|$ a.e. whenever f and g belong to the domain of B and a is a scalar. If B is defined on $H^p(X)$, for some $p \leqslant 1$, we say that it is of *weak-type* (H^p,p) provided $\nu(\{y \in Y: |(Bf)(y)| > \lambda\}) \leqslant (M/\lambda)^p \|f\|_{H^p}$ for all $f \in H^p$, where $\|f\|_{H^p}$ denotes any one of the "norms" $\|f\|_{p,q}$.[25] If B is defined on $L^p(X)$, $0 < p < \infty$, we say that it is of *weak-type* (p, p) provided $\nu(\{y \in Y: |(Bf)(y)| > \lambda\}) \leqslant (M\|f\|_p/\lambda)^p$ for all $f \in L^p$; we define weak-type (∞, ∞) to be the same as boundedness: $\|Bf\|_\infty \leqslant M\|f\|_\infty$ for all $f \in L^\infty(X)$. For p fixed, the infimum of all constants M for which any one of these inequalities hold (independently of f) is called the *weak-type norm* of B.

We can now announce the first interpolation theorem:

THEOREM D. *Suppose* $0 < p_1 \leqslant 1 \leqslant p_2 \leqslant \infty$, $p_1 < p_2$, *and B is a sublinear operator of weak-types* (H^{p_1}, p_1) *and* (p_2, p_2) *having weak-type norms* M_1 *and* M_2. *If* $1 < p < p_2$ *then B is defined on* $L^p(X)$ *and*

$$\|Bf\|_p \leqslant M\|f\|_p,$$

where M depends on M_1, M_2, p_1 *and* p_2 *but is independent of* $f \in L^p$. *If* $p_1 < p \leqslant 1$ *then B is defined on* $H^p(X)$ *and*

$$\|Bf\|_p \leqslant M\|f\|_{H^p}^{1/p},$$

where M depends on M_1, M_2, p_1 *and* p_2 *but is independent of* $f \in H^p$.

This result is proved in the Ph.D. thesis of one of our students, R. Macias [**45**]. It was obtained in the classical situation by Igari [**37**] (for Riesz-systems when $p_1, p_2 \geqslant 1$) and C. Fefferman, Rivière and Sagher [**30**].

The original theorem of Marcinkiewicz dealt only with L^p-spaces and was proved by truncating the function f at heights $\lambda \in (0, \infty)$ and making

[25] Of course, when we vary q the constant M changes accordingly. The important fact is that $\|\ \|_{p,q_1}$ and $\|\ \|_{p,q_2}$ are equivalent. We could have defined the notion of weak type (H^{p_1}, p_2) with $p_1 \neq p_2$ as well. For simplicity we restrict ourselves to the case $p_1 = p_2$.

appropriate estimates. In our case a substitute for these truncations is obtained by a variant of the Calderón-Zygmund decomposition. We shall give more details about this proof in the next section.

The second interpolation theorem involves an adaptation of what is known as the "complex method" and extends a result on interpolation of *analytic families of operators* on classical H^p spaces (see **[58]**). Before stating this theorem we need to make some definitions. Let $D = \{z = x + iy \in \mathbf{C}: 0 < x < 1\}$. A complex-valued function F defined on $\overline{D}$ is said to be of *admissible growth* if there exists a constant $a < \pi$ such that $e^{-a|y|} \log |F(z)|$, $z = x + iy$, is uniformly bounded from above in the strip D. Let us first consider a space of homogeneous type X such that $\mu(X) = \infty$. Suppose that, for each $z \in \overline{D}$, B_z is a linear operator defined on $L^\infty(X)$-functions having bounded support and mean 0 (we shall denote this class of functions by $L_0^\infty(X)$); we assume that the range of B_z consists of measurable functions on a measure space (Y, ν) and that $(B_z f)g \in L^1(Y)$ whenever $f \in L_0^\infty(X)$ and g is a simple function on Y. The family $\{T_z\}$ is called an *analytic family of operators on* $\overline{D}$ provided $F(z) = \int_Y (T_z f) g\, d\nu$ is continuous on $\overline{D}$, analytic in D and of admissible growth whenever $f \in L_0^\infty(X)$ and g is a simple function on Y. If X has finite measure we replace $L_0^\infty(X)$ by $L^\infty(X)$ in the definitions above.

THEOREM E. *Suppose* $\{B_z\}$ *is an analytic family of operators satisfying*

$$\|B_{iy} f\|_{p_0} \leqslant A_0(y) \|f\|_{H^{p_0}}^{1/p_0} \quad \textit{and} \quad \|B_{1+iy} f\|_{p_1} \leqslant A_1(y) \|f\|_{p_1}$$

for all $y \in (-\infty, \infty)$ *and* $f \in L_0^\infty(X)$ $(f \in L^\infty(X)$ *if* $\mu(X) < \infty)$, *where* $\log A_j(y) \leqslant C_j e^{d_j|y|}$, $C_j > 0$, $\pi > d_j > 0$, $j = 1, 2$, *and* $p_0 \leqslant 1 \leqslant p_1$. *If* $1/p = t/p_0 + (1 - t)/p_1 < 1$ $(0 < t < 1)$ *then*

$$\|B_t f\|_p \leqslant M \|f\|_p, \tag{2.8}$$

while, if $1/p = t/p_0 + (1 - t)/p_1 \geqslant 1$ $(0 < t < 1)$,

$$\|B_t f\|_p \leqslant M \|f\|_{H^p}^{1/p}, \tag{2.9}$$

where M *depends on* $C_0, C_1, d_0, d_1, p_0, p_1$ *but is independent of* $f \in L_0^\infty(X)$ $(L^\infty(X)$ *if* $\mu(X) < \infty)$.

The proof of this theorem (to be found in the thesis of R. Macias **[45]**) uses techniques developed by Stein and Fefferman (see **[29]**) in order to obtain inequality (2.8), and a modification of an argument found in **[58]** in order to obtain inequality (2.9). We shall give details of this proof in §3.

We shall now illustrate how these results involving the atomic H^p spaces we have introduced can be applied to problems in harmonic analysis associated with the spaces of homogeneous type we described above.

Let us begin with example (1) where $X = \mathbf{R}^n$, μ is Lebesgue measure and d is Euclidean distance. In §1 we discussed the Riesz transforms and some of their properties. These operators are particular examples of the convolution operator

(2.10) $$\tilde{f}(x) = \lim_{\varepsilon \to 0+} \int_{\varepsilon < |y| < 1/\varepsilon} k(y) f(x-y)\, dy \equiv (k * f)(x),$$

where

$$k(y) = \Omega(y')/|y|^n, \quad y \neq 0, \quad y' = y/|y|, \quad \Omega \in L^1(\Sigma_{n-1})$$

and

$$\int_{\Sigma_{n-1}} \Omega(y')\, dy' = 0$$

(here dy' denotes the element of Lebesgue surface measure). If Ω satisfies the Lipschitz condition $|\Omega(x') - \Omega(y')| \leqslant C'|x' - y'|$ then the kernel k satisfies inequality (1.36)

(2.11) $$|k(x-y) - k(x)| \leqslant C|y|/|x|^{n+1}$$

for an appropriate constant C whenever $2|y| < |x|$. But we showed above that this property and L^2-boundedness was all that we need in order to conclude that the operator defined by (2.10) is bounded from H^1 into L^1 (or, for that matter, from H^p for $n/(n+1) < p$ to L^p–see (1.35)). Theorem D can now be used to conclude that this operator is bounded on $L^p(\mathbf{R}^n)$ for $1 < p \leqslant 2$ (since the adjoint of this operator is its negative, this operator is bounded on L^p for $1 < p < \infty$).[26]

It is not hard to show $f \to k * f$ maps H^1 boundedly into itself. Because of Theorem C it suffices to show that $\tilde{a} = k * a$ is a constant multiple of a molecule whenever a is an atom. To see this let $a(x)$ be a (1, 2)-atom supported in a sphere of radius r about $x_0 \in \mathbf{R}^n$. It is clear that $\tilde{a}$ has mean 0. Since a has mean 0 we have

$$\tilde{a}(x) \equiv \lim_{\varepsilon \to 0} \int_{\varepsilon < |x-y| < 1/\varepsilon} k(x-y) a(y)\, dy = \int_{\mathbf{R}^n} [k(x-y) - k(x-x_0)] a(y)\, dy.$$

Thus, using (2.11), we have, for $|x - x_0| > 2r$,

$$|\tilde{a}(x)| \leqslant \int_{|y-x_0|<r} \frac{|y-x_0|}{|x-x_0|^{n+1}} |a(y)|\, dy$$
$$\leqslant \frac{r}{|x-x_0|^{n+1}} \int |a(y)|\, dy \leqslant \frac{r}{|x-x_0|^{n+1}}.$$

It follows immediately from property (ii) of a (1, 2)-atom that $\int |a(x)|^2 dx \leqslant (\text{const}) r^{-n}$. Making use of these estimates on $\tilde{a}(x)$ and $\|a\|_2^2$ as well as the L^2-boundedness of the operator $a \to \tilde{a}$ we have that a constant multiple of

$$\left(\int |\tilde{a}(x)|^2 dx\right)\left(\int |\tilde{a}(x)|^2 |x - x_0|^{n+1} dx\right)^n$$

[26] The L^2-boundedness of such operators can be established easily by calculating the Fourier transform of $k * f$ (see [57]). In other situations where the Fourier-transform is not available one can use a general method developed by Cotlar, Knapp and Stein that is adaptable to spaces of homogeneous type (see [40]).

is less than or equal to

$$r^{-n}\left[\left(\int_{|x-x_0|<2r} |\tilde{a}(x)|^2 |x-x_0|^{n+1}\,dx\right)^n + \left(\int_{|x-x_0|\geqslant 2r} |\tilde{a}(x)|^2 |x-x_0|^{n+1}\,dx\right)^n\right]$$

$$\leqslant r^{-n}(r^{n+1}\|\tilde{a}\|_2^2)^n + r^{-n}\left(\int_{|x-x_0|\geqslant 2r}\left(\frac{r}{|x-x_0|^{n+1}}\right)^2 |x-x_0|^{n+1}\,dx\right)^n$$

$$\leqslant (\text{const})r^{-n}(r^{n+1}\|a\|_2^2)^n + r^{-n}r^{2n}\left(\int_{|x-x_0|\geqslant 2r}\frac{dx}{|x-x_0|^{n+1}}\right)^n$$

$$\leqslant (\text{const})[r^{-n}r^n + r^n r^{-n}].$$

This shows that, up to a multiplicative constant independent of a,

$$\left(\int |\tilde{a}(x)|^2\,dx\right)\left(\int |\tilde{a}(x)|^2(|x-x_0|^n)^{(1+(1/n))}\,dx\right)^n$$

satisfies condition (2.7) with $\varepsilon = 1/n$ (observe that $m(x,x_0) = |x-x_0|^n$ is a measure distance in this case).

Inequality (2.11) implies the weaker condition

$$\int_{|x|>2|y|} |k(x-y)-k(x)|\,dx \leqslant C. \tag{2.12}$$

In establishing (1.24) we gave an argument that is easily modified to show that this condition, together with the boundedness of $\hat{k}$, implies that the operator $f \to k * f$ maps H^1 boundedly into L^1. Both (2.11) and (2.12) have natural extensions to spaces of homogeneous type. In fact, it is easy to see that the argument we have just given shows that if an integral operator

$$(Bf)(x) = \int_X f(y)k(x,y)\,d\mu(y)$$

is bounded on $L^2(X)$, Bf has mean 0 when f is an atom and has a kernel satisfying, for $m(x,y_0) > Cm(y,y_0)$,

$$|k(x,y)-k(x,y_0)| \leqslant C\left[\frac{m(y,y_0)}{m(x,y_0)}\right]^{\varepsilon}\frac{1}{m(x,y_0)} \tag{2.13}$$

then it maps $H^1(X)$ boundedly into itself. (In fact it maps H^p into itself for p close enough to 1.) If the kernel satisfies

$$\int_{d(x,x_0)>C_1 d(y,x_0)} |k(x,y)-k(x,x_0)|\,d\mu(x) \leqslant C_2 \tag{2.14}$$

and B is bounded on $L^2(X)$ then B maps H^1 boundedly into $L^1(X)$ with a norm that depends on the constants C_1, C_2 and the L^2-operator norm of B. This follows from a simple modification of the argument used to establish (1.24).

We remark that molecules can be defined in terms of other norms besides the L^2 norm. For example, a function $M \in L^{1+\delta}$, $\delta > 0$, satisfying

$$\int M\,d\mu = 0 \quad \text{and} \quad \left(\int |M|^{1+\delta}\,d\mu\right)^{1/\delta}\left(\int |M(x)|^{1+\delta}[m(x,x_0)]^{\delta+\varepsilon}\,d\mu(x)\right)^{1/\varepsilon} \leqslant 1$$

can easily be shown to belong to H^1 (or, with a change in normalization, to H^p, for some $p < 1$). Suppose

$$(Bf)(x) = \int k(x,y)f(y)\,d\mu(y)$$

is a bounded operator on L^2, then a sufficient condition on $k(x,y)$ that implies that atoms are mapped by B onto these (more general) molecules (and, thus, B is a bounded transformation of H^p into itself for some $p < 1$) is

$$\int_{m(x,y_0)>Cm(y,y_0)} |k(x,y) - k(x,y_0)|^{1+\delta}[m(x,y_0)]^{\delta+\varepsilon}\,d\mu(x) \leqslant C[m(y,y_0)]^{\varepsilon}. \tag{2.15}$$

This condition holds whenever (2.13) is valid and reduces to (2.14) if $\delta = \varepsilon = 0$.

These observations illustrate how the atomic theory of H^p-spaces can be used to study certain Calderón-Zygmund singular integrals. In order to obtain the boundedness of these operators on $L^p(\mathbf{R}^n)$, $1 < p < \infty$, much less stringent conditions are needed. In fact, if Ω is an odd function in $L^1(\Sigma_{n-1})$ one obtains this boundedness (see [**57**]). When Ω is even we need to know more; Calderón-Zygmund [**4**] show that if $\Omega \in L \log L(\Sigma_{n-1})$ then B is bounded on $L^p(\mathbf{R}^n)$.

If $n = 2$ then this boundedness follows from the assumption that $\Omega \in H^1(\Sigma_1) = H^1(T)$ (see p. 160 of [**21**]). We shall now show that this class of operators together with the identity forms an algebra. Let us begin by characterizing the Fourier transforms of the operators having a kernel $k(y) = \Omega(y')/|y|^2$ with $\Omega \in H^1(T)$. In terms of polar coordinates any such operator has the form

$$(Bf)(x) = \lim_{\varepsilon\to 0+}\int_\varepsilon^{1/\varepsilon}\int_0^{2\pi}\frac{\Omega(\theta)}{r}f(x - re^{i\theta})\,dr\,d\theta.$$

If we write $y = re^{i\psi}$ then it can be shown (see p. 162 of [**57**]) that if $k_j(re^{i\theta}) = \Omega_j(\theta)/r^2 = e^{ij\theta}/r^2$ then

$$\hat{k}_j(y) = \lim_{\varepsilon\to 0}\int_\varepsilon^{1/\varepsilon}\int_0^{2\pi}\frac{e^{ij\theta}}{r}e^{-iy\cdot re^{i\theta}}\,dr\,d\theta = \frac{2\pi}{|j|}e^{ij\psi}i^{-|j|}.$$

It follows from Hardy's inequality (1.5) that if

$$\Omega(\theta) \sim \sum_{j\neq 0} a_j e^{ij\theta}$$

belongs to $H^1(T)$ then

$$\hat{k}(y) = 2\pi \sum_{j \neq 0} \frac{i^{-|j|} a_j}{|j|} e^{ij\psi}$$

and this last series converges absolutely. Let

$$\Omega = \Omega_o + \Omega_e \quad \text{where } [\Omega(\theta + \pi) + \Omega(\theta)]/2 = \Omega_e(\theta)$$

(Ω_e and Ω_o are the even and odd parts of Ω as a function on $\mathbf{R}^2$). Then the function $\hat{k}(e^{i\psi}) = \hat{k}(y)$ has the property

$$((d/d\psi)\hat{k})(e^{i\psi}) = 2\pi[\Omega_o(\psi - \pi/2) + \tilde{\Omega}_e(\psi - \pi/2)],$$

where $\tilde{\Omega}_e$ is the conjugate function of Ω_e (if we consider the imaginary part $Q(\theta)$ of the kernel $(1 + e^{i\theta})/(1 - e^{i\theta})$ discussed in the paragraph following (1.25), then $\tilde{\Omega}_e$ is the generalized convolution $\Omega_e * Q$). It follows that $(d/d\psi)\hat{k} \in H^1(T)$. Conversely, if $\hat{k}(y)$ is a function on $\mathbf{R}^2$ that is homogeneous of degree 0 and satisfies

$$\frac{d}{d\psi}\hat{k} \in H^1 \quad \text{and} \quad \int_T \hat{k}(e^{i\psi})\, d\psi = 0$$

then there exists a function k on $\mathbf{R}^2$ of the form $\Omega(\theta)/r^2$ with $\Omega \in H^1$ such that $\hat{k}$ is the Fourier transform of k.

We shall now show that the vector space generated by these operators together with the identity form an algebra by showing that $\hat{k}\hat{h}$ has a derivative (with respect to ψ) in H^1 whenever $\hat{h}$ and $\hat{k}$ are homogeneous of degree 0 and have derivatives in H^1. Suppose, then, that $\hat{h}(e^{i\psi}) = \sum \lambda_j A_j(\psi)$ and $\hat{k}(e^{i\psi}) = \sum \mu_l B_l(\psi)$ where A_j', B_l' are atoms, $\sum |\lambda_j| < \infty$ and $\sum |\mu_l| < \infty$. Obviously, $\sum_{j,l} |\lambda_j \mu_l| < \infty$ and we only have to show that $(1/2)(AB)'$ is an atom whenever A' and B' are nonconstant atoms. Suppose A has a supporting interval I and B has a supporting interval J (since the atoms have mean zero A and A' have the same supporting interval). The support of AB is contained in $I \cap J$. Moreover, $|(AB)'| = |A'B + AB'| \leqslant 1/|I| + 1/|J|$ (the absolute value of the integral of an atom is always less than or equal to 1). Thus, $|(AB)'| \leqslant 2/|I \cap J|$ and our claim is established.

It is not hard to show that in dimension $n > 2$ the condition $\Omega \in H^1(\Sigma_{n-1})$ implies the boundedness of the associated singular integral operator on $L^p(\mathbf{R}^n)$, $1 < p < \infty$. It is natural to raise the question whether such operators, together with the identity, generate an algebra.

Let us now turn to example (2) in which we introduced nonisotropic distances on $\mathbf{R}^n$. We shall show how the theory we have described, applied to this situation, can be used in order to obtain a priori estimates for certain partial differential equations. Let us consider the special case of the heat equation

$$\partial u/\partial x_2 - \partial^2 u/\partial x_1^2 = g(x_1, x_2),$$

where $x = (x_1, x_2) \in \mathbf{R}^2$. It will become clear that the arguments presented here apply to a wide variety of equations. Suppose $g \in L^p(\mathbf{R}^2)$, for an appropriate p, can we find a solution u, such that $\partial u/\partial x_2$ and $\partial^2 u/\partial x_1^2$ also belong to $L^p(\mathbf{R}^2)$? By applying (formally) the Fourier transform we see that our equation is equivalent to

$$(2.16) \qquad (y_1^2 + iy_2)\hat{u}(y_1, y_2) = \hat{g}(y_1, y_2);$$

thus,

$$\left(\frac{\partial u}{\partial x_2}\right)^{\wedge}(y_1, y_2) = \frac{iy_2}{y_1^2 + iy_2}\hat{g}(y_1, y_2) \quad \text{and}$$

$$\left(\frac{\partial^2 u}{\partial x_1^2}\right)^{\wedge}(y_1, y_2) = \frac{-y_1^2}{y_1^2 + iy_2}\hat{g}(y_1, y_2).$$

We are, therefore, led to consider the operator assigning to g the function

$$(Mg)(x) = \left[\frac{iy_2}{y_1^2 + iy_2}\hat{g}(y_1, y_2)\right]^{\vee}(x)$$

(as well as the operator induced by the multiplier $y_1^2/(y_1^2 + iy_2)$). M is obviously a bounded operator on $L^2(\mathbf{R}^2)$. We claim that by choosing an appropriate distance on $\mathbf{R}^2$ we will find it easy to show that M preserves the H^p spaces associated with this distance, for certain values of p; consequently, M is bounded on the corresponding Lipschitz spaces and the L^q spaces for $1 < q < \infty$.

A natural way for introducing this distance is obtained by considering the family of "dilations" on $\mathbf{R}^2$ mapping x into $T_\lambda x = (\lambda^{1/3}x_1, \lambda^{2/3}x_2)$, $\lambda > 0$. Let us also write $(T_\lambda g)(x) \equiv g(T_\lambda x)$. It is then clear that $M(T_\lambda g) = T_\lambda(Mg)$ for all $\lambda > 0$. That is, the operators M and T_λ commute.

We now introduce the "norm" $\|x\| = |x_1|^3 + |x_2|^{3/2}$ in order to obtain families of spheres in $\mathbf{R}^2$ that are invariant under the action of these dilations. Observe that

$$(2.17) \qquad \|T_\lambda x\| = \lambda\|x\|$$

for all $\lambda > 0$. By letting $B_\lambda = \{x \in \mathbf{R}^2 \colon \|x\| < \lambda\}$, $\lambda > 0$, we obtain the family of spheres about the origin associated with this norm. Since the Jacobian of the transformation T_λ is λ we see that $B_\lambda = T_\lambda B_1$ has measure $c\lambda$, where c is an absolute constant.

Suppose that M is an operator given (formally) by $g \to Mg = (m\hat{g})^{\vee}$, where the multiplier m is bounded and satisfies

$$(2.18) \qquad \begin{aligned} m(T_\lambda y) = m(y), \qquad (\Delta_h^2 m)(y) &\equiv m(y+h) - 2m(y) + m(y-h) \\ &= O\left[\left(\frac{\|h\|}{\|y\|}\right)^{2/3}\right] \end{aligned}$$

for $\lambda > 0$ and $y, h \in \mathbf{R}^2$. We claim that M is a bounded operator on H^1 as well as on H^p if p is close to but less than 1. (Using the fact that $|h|^2 \leqslant c\|h\|^{2/3}$ for $|h| \leqslant 1$ we see that condition (2.18) is satisfied by the two multipliers

$iy_2/(y_1^2 + iy_2)$ and $y_1^2/(y_1^2 + iy_2)$ we are considering here.) This claim follows from

LEMMA (2.19). *Suppose a is a 1-atom supported in the sphere B_1, then $Ma = (m\hat{a})^{\vee}$ is a molecule centered at 0. More precisely, $\int Ma = 0$ and*

$$\left(\int |(Ma)(x)|^2\,dx\right)\left(\int |(Ma)(x)|^2\|x\|^{1+\eta}\,dx\right)^{1/\eta} \leqslant c_\eta,$$

where $0 < \eta < 1/3$ and c_η is independent of a.

Once this lemma is established we see from Theorem C that Ma belongs to H^1. The fact that the operator M commutes with the translations of $\mathbf{R}^2$ and the dilations T_λ shows that it is well defined and is bounded on H^1.

In order to establish (2.19) we shall need the following result:

LEMMA (2.20). *Suppose $f \in L^2(\mathbf{R}^2)$ satisfies*

$$\int_{\mathbf{R}^2} |\hat{f}(x+h) - 2\hat{f}(x) + \hat{f}(x-h)|^2\,dx = \int_{\mathbf{R}^2} |(\Delta_h^2 \hat{f})(x)|^2\,dx \leqslant C\|h\|^{2\alpha}$$

for $2\alpha > 1$ then $\int_{\mathbf{R}^2} |f(x)|^2 \|x\|^\beta\,dx < \infty$ when $0 < \beta < 2\alpha$.

PROOF. Suppose $R > 1$ and $h = T_{1/R}h'$ where $|h'| = \sqrt{(h_1')^2 + (h_2')^2} = 1$. If $R/2 < \|x\| \leqslant R$ then $1/2 < \|T_{1/R}x\| \leqslant 1$. It follows that $|e^{i\theta} - 2 + e^{-i\theta}| \geqslant C_0\theta^2$ for an appropriate constant C_0, where $\theta = (T_{1/R}h')\cdot x = h\cdot x = (T_{1/R}x)\cdot h'$. Thus, by Plancherel's theorem and the assumption made in this lemma

$$\begin{aligned} C_0\int_{R/2<\|x\|\leqslant R} |x\cdot h|^4 |f(x)|^2\,dx &\leqslant \int_{R/2<\|x\|\leqslant R} |e^{ix\cdot h} - 2 + e^{-ix\cdot h}|^2 |f(x)|^2\,dx \\ &\leqslant C\|h\|^{2\alpha} = CR^{-2\alpha}\|h'\|. \end{aligned}$$

We can now integrate both sides of this inequality with respect to h' over the unit circle in $\mathbf{R}^2$. Keeping in mind that $x\cdot h = (T_{1/R}x)\cdot h'$ is (essentially) the inner product of two unit vectors we obtain

$$\int_{R/2<\|x\|\leqslant R} |f(x)|^2\,dx \leqslant (\text{const})R^{-2\alpha}.$$

Thus,

$$\begin{aligned} \int_{1<\|x\|} |f(x)|^2 \|x\|^\beta\,dx &\leqslant \sum_1^\infty 2^{k\beta} \int_{2^{k-1}<\|x\|\leqslant 2^k} |f(x)|^2\,dx \\ &\leqslant (\text{const}) \sum_1^\infty 2^{k(\beta-2\alpha)} < \infty. \end{aligned}$$

Since $\int_{\|x\|\leqslant 1} |f(x)|^2 \|x\|^\beta\,dx$ is finite (since $f \in L^2(\mathbf{R}^2)$) the lemma is established.

Lemma (2.19) is a consequence of the result we have just obtained if we can establish that $\hat{f} = m\hat{a}$ satisfies the hypothesis of Lemma (2.20) with $\alpha = 2/3$. In fact,

$$(\Delta_h^2 m\hat{a})(y) = m(y+h)(\Delta_h^2 \hat{a})(y) + \hat{a}(y)(\Delta_h^2 m)(y)$$
$$+[m(y+h) - m(y-h)][\hat{a}(y) - \hat{a}(y-h)].$$

To estimate the first term we use the fact that a has support in B_1 and m is bounded. Thus,

$$\int |m(y+h)(\Delta_h^2 \hat{a})(y)|^2\, dy \leqslant \|m\|_\infty \int |e^{ix\cdot h} - 2 + e^{-ix\cdot h}|^2 |a(x)|^2\, dx$$
$$\leqslant (\text{const})\,|h|^4 \leqslant (\text{const})\,\|h\|^{4/3}$$

when $|h| \leqslant 1$ (this is clearly the only case we need to consider since $\|a\|_2 \leqslant 1$).

In order to estimate the second term we first observe that $|\hat{a}(y)| \leqslant (\text{const}) \cdot \min\{1, |y|\}$. Thus,

$$(\text{const}) \int |\hat{a}(y)|^2 |(\Delta_h^2 m)(y)|^2\, dy$$
$$\leqslant \|h\|^{4/3} \left\{ \int_{\|y\| \leqslant 1} \frac{|y|^2}{\|y\|^{4/3}}\, dy + \int_{\|y\|>1} \|y\|^{-4/3}\, dy \right\}$$
$$\leqslant (\text{const})\,\|h\|^{4/3}.$$

Finally, in order to estimate the last term we observe that the assumption on the second difference $\Delta_h^2 m$ in (2.18) implies

$$|m(y+h) - m(y)| = O[(\|h\|/\|y\|)^{1/3}].$$

This last estimate follows from the argument preceding (3.5) on p. 44 of **[64]** applied to each variable separately. Secondly, we observe that

$$|\hat{a}(y+h) - \hat{a}(y)| = O(|h|) \quad \text{and} \quad \int |\hat{a}(y+h) - \hat{a}(y)|^2\, dy = O(|h|^2).$$

Thus,

$$\int |m(y+h) - m(y-h)|^2 |\hat{a}(y) - \hat{a}(y-h)|^2\, dy$$
$$\leqslant C\|h\|^{2/3} \left(\int_{\|y\|<1} \frac{|h|^2}{\|y\|^{2/3}}\, dy + \int_{\|y\| \geqslant 1} |\hat{a}(y) - \hat{a}(y-h)|^2\, dy \right).$$

Using, again, the fact that $|h|$ is dominated by a constant times $\|h\|^{1/3}$ when $|h| \leqslant 1$, we see that the last expression is $O(\|h\|^{4/3})$ when $\|h\| \leqslant 1$. Lemma (2.19) is thus established.

Recall the question we raised originally when we introduced the heat equation: given $g \in L^p(\mathbf{R}^2)$ can we find a solution u such that $\partial u/\partial x_2$ and $\partial^2 u/\partial x_1^2$ belong to $L^p(\mathbf{R}^2)$? We then proceeded formally and exhibited a solution by means of equality (2.16). It is not clear, however, that such a solution is a function belonging to some natural function space associated with $L^p(\mathbf{R}^2)$, nor is it clear that the multiplier operators we considered act naturally on $L^p(\mathbf{R}^2)$. Another useful feature of the atomic theory of H^p spaces is that if $g \in H^1$ ($H^1 = H^1(\mathbf{R}^2, \| \|)$ being defined with respect to the quasi-distance induced by $\| \|$) then the solution

$$u = [(1/(y_1^2 + iy_2))\hat{g}(y_1, y_2)]^{\check{}} \tag{2.21}$$

belongs to $L^3(\mathbf{R}^2)$. This is easily seen in the following way: Suppose a is an atom supported in B_1. Then from the estimates $|\hat{a}(x)| \leqslant (\text{const})\min\{1, |x|\} \leqslant (\text{const})\min\{1, \|x\|^{1/3}\}$ we obtain

$$(\text{const})\int \frac{|\hat{a}(x)|}{\|x\|}dx \leqslant \int_{\|x\|\leqslant 1} \|x\|^{-2/3}dx + \|a\|_2\left(\int_{\|x\|>1} \|x\|^{-2}dx\right)^{1/2} \leqslant C,$$

where C depends only on the dimension. Since

$$\int |\hat{a}(T_\lambda x)|\frac{dx}{\|x\|} = \int |\hat{a}(x)|\frac{dx}{\|x\|} \leqslant C \quad \text{for all } \lambda > 0$$

it follows that we have a nonisotropic version of Hardy's inequality

$$\int_{\mathbf{R}^2} \frac{|\hat{g}(x)|}{\|x\|}dx < \infty \tag{2.22}$$

whenever $g \in H^1(\mathbf{R}^2, \| \|)$. Since $|\hat{a}(x)| \leqslant 1$ whenever a is an atom in $H^1(\mathbf{R}^2, \| \|)$ we must have, if $1 \leqslant p$,

$$\int \frac{|\hat{a}(x)|^p}{\|x\|}dx \leqslant \int \frac{|\hat{a}(x)|}{\|x\|}dx \leqslant C.$$

Thus, if $g = \sum \alpha_i a_i \in H^1(\mathbf{R}^2, \| \|)$ then

$$\left(\int_{\mathbf{R}^2} |\hat{g}(x)|^p \frac{dx}{\|x\|}\right)^{1/p} \leqslant \sum |\alpha_i|\left(\int_{\mathbf{R}^2} |\hat{a}_i(x)|^p \frac{dx}{\|x\|}\right)^{1/p} \leqslant C\sum |\alpha_i|.$$

Consequently,

$$\left(\int_{\mathbf{R}^2} |\hat{g}(x)|^p \frac{dx}{\|x\|}\right)^{1/p} \leqslant C\|g\|_{H^1} \tag{2.23}$$

whenever $g \in H^1(\mathbf{R}^2, \| \|)$.

Equation (2.16) gives us a solution u satisfying

$$|\hat{u}(y)| = \left|\frac{\hat{g}(y)}{y_1^2 + iy_2}\right| \leqslant (\text{const})\frac{|\hat{g}(y)|}{\|y\|^{2/3}}.$$

Applying (2.23) with $p = 3/2$ we see that $\hat{u} \in L^{3/2}(\mathbf{R}^2)$. It now follows from the Hausdorff-Young theorem that $u \in L^3(\mathbf{R}^2)$. Furthermore, we have shown that the derivatives $\partial u/\partial x_2$ and $\partial^2 u/\partial x_1^2$ (which certainly exist in the sense of distributions) belong to H^1. The fact that the operator defined by (2.21) maps H^1 into $L^3(\mathbf{R}^2)$ implies, by duality, that it maps $L^{3/2}(\mathbf{R}^2)$ into $BMO(\mathbf{R}^2, \| \|)$. By applying an appropriate interpolation argument it can be shown that this operator maps L^p into L^q, where $1/p - 1/q = 2/3$ and $1 < p < 3/2$.

Singular integrals associated with the heat equation were studied first by Jones [39] and later by Fabes and Rivière [26]. Perhaps some of the results described here for H^p spaces are new. Multipliers that are homogeneous relative to generalized dilations were studied by Rivière [49] and Krée [43].

In example (3) we considered other measures on $\mathbf{R}^n$, besides Lebesgue measure, which, together with the Euclidean distance, satisfied the basic inequality (2.1). Such measures arise naturally in several ways. For example radial functions in $L^1(\mathbf{R}^n)$ give rise to the study of the integrable functions on $(0, \infty)$ with respect to the measure $r^{n-1}\,dr$. The study of central functions on certain Lie groups also gives rise to measures $\omega(x)\,dx$ satisfying (2.1), where ω is an appropriate weight function. The case $\omega(x) = e^{b(x)}$, where b is a BMO function, arises in factorization problems we shall discuss later. In all these cases we are dealing with weights belonging to the class A^∞ of Muckenhoupt (see [13]) characterized by the inequality

$$\frac{\omega(E \cap Q)}{\omega(Q)} \leqslant C\left(\frac{|E \cap Q|}{|Q|}\right)^{\delta}, \tag{2.24}$$

where C and δ are constants depending only on ω, $|E|$ denotes the Lebesgue measure of $E \subset \mathbf{R}^n$, $\omega(E) = \int_E \omega(x)\,dx$ and Q is a cube with sides parallel to the axes.

It is an immediate consequence of (2.24) that $\omega(\overline{Q}) \leqslant C\omega(Q)$ whenever $\overline{Q}$ is the cube concentric with Q having sides twice the length of the latter (we shall use the letter C to denote a constant, not always the same, that depends only on the dimension n and the weight ω). Thus, $(\mathbf{R}^n, \omega)$, together with the Euclidean distance, is an example of a space of homogeneous type. Let $H^1(\mathbf{R}^n, \omega\,dx)$ denote the corresponding atomic Hardy space.

A basic general result relating these spaces to the more classical space $H^1(\mathbf{R}^n, dx)$ is the following:

THEOREM (2.25). $H^1(\mathbf{R}^n, \omega\,dx) = \{f\omega^{-1} : f \in H^1(\mathbf{R}^n, dx)\}$.

In other words, the map $f \to f\omega$ is an isomorphism between the "weighted" Hardy space $H^1(\mathbf{R}^n, \omega\,dx)$ and the "ordinary" Hardy space $H^1(\mathbf{R}^n, dx)$.

In order to prove this theorem we make use of the following properties of weights satisfying the A^∞ condition (2.24) (see [13]):

(2.26)

(a) there exists η, $C > 0$ such that

$$\left(\frac{1}{|Q|}\int_Q \omega^{1+\eta}\,dx\right)^{1/(1+\eta)} \leqslant C\frac{\omega(Q)}{|Q|},$$

(b) there exists $p > 1$ and $C > 0$ such that

$$\frac{\omega(Q)}{|Q|}\left(\frac{1}{|Q|}\int_Q \omega^{-1/(p-1)}\,dx\right)^{p-1} \leqslant C$$

for all cubes Q. Suppose, now, that a is a $(1, \infty)$-atom in $H^1(\mathbf{R}^n, \omega\,dx)$; that is, a is supported in a cube Q, $|a(x)| \leqslant 1/\omega(Q)$ and $\int a(x)\omega(x)\,dx = 0$. We claim that $A = a\omega$ is (up to a multiplicative constant independent of a) a $(1, 1 + \eta)$-atom in $H^1(\mathbf{R}^n, dx)$. The fact that A has mean zero is obvious; moreover, from (2.26) (a) we have

$$\left(\frac{1}{|Q|}\int |A(x)|^{1+\eta}\,dx\right)^{1/(1+\eta)} \leqslant \frac{1}{\omega(Q)}\left(\int [\omega(x)]^{1+\eta}\frac{dx}{|Q|}\right)^{1/(1+\eta)} \leqslant \frac{C}{|Q|}.$$

Conversely, (2.26) (b) shows that if A is a $(1,\infty)$-atom in $H^1(\mathbf{R}^n, dx)$ then $a = A/\omega$ is a $(1,p)$-atom in $H^1(\mathbf{R}^n, \omega\,dx)$. Theorem (2.25) now follows from Theorem A.

A consequence of (2.25) is the following characterization of radial functions belonging to $H^1(\mathbf{R}^n, dx)$:

COROLLARY (2.27). *Suppose f is defined on $[0,\infty)$ then $x \to f(|x|)$ is a function in $H^1(\mathbf{R}^n, dx)$ if and only if $r \to f(|r|)|r|^{n-1}$ is a function in $H^1(\mathbf{R}^1, dr)$.*

In order to show this result let us first assume that $f(|x|) = \sum \alpha_i a_i(x)$, where a_i is a $(1,\infty)$-atom on $\mathbf{R}^n$ (not necessarily radial) and $\sum |\alpha_i| < \infty$. We claim that

$$A_i(|r|) = \int_{|x|=1} a_i(|r|x)\,d\sigma(x),$$

where $d\sigma(x)$ is the element of surface measure on the unit sphere in $\mathbf{R}^n$, is an atom in $H^1(\mathbf{R}^1, |r|^{n-1}\,dr)$ times a constant that depends only on the dimension. If the support of $a = a_i$ contains the origin this is obviously true.[27] If the support of a is at a distance $R > 0$ from the origin and d is the diameter of this support, then there are $N \geqslant C[R/d]^{n-1}$ rotations $\rho_1, \rho_2, \ldots, \rho_N$ such that $a(\rho_j x), j = 1, \ldots, N$, have disjoint supports. Hence,

$$A(|r|) = \frac{1}{N}\int_{|x|=1}\left(\sum_{j=1}^{N} a(r\rho_j x)\right)d\sigma(x)$$

is bounded by $(1/N)(1/d^n) \leqslant C(1/dR^{n-1})$ (since the supports of the functions appearing in the sum are disjoint). But the last expression is dominated by a constant times

$$1\Big/\left(\int_R^{R+d} r^{n-1}\,dr\right).$$

In order to obtain the converse we start out with a $(1,\infty)$-atom A in $H^1(\mathbf{R}^1, |r|^{n-1}\,dr)$ and observe that if its support contains 0 then the radial function it defines is (essentially) an atom in $H^1(\mathbf{R}^n, dx)$. If the support of A is the interval $(R, R+d)$, $R > 0$, we can "reverse" the construction we just gave in order to obtain an appropriate atomic decomposition in $\mathbf{R}^n$ of the radial function defined by A. More generally, many results concerning weighted H^p spaces on $\mathbf{R}^n$ were developed in the thesis of one of our Ph.D. students, J. Garcia-Cuerva [32]. The class of weights considered here on $\mathbf{R}^n$ admits a useful straightforward extension to spaces of homogeneous type.

The notion of weighted H^1 spaces has interesting applications in the compact case as well. The following result, as we shall see, can be obtained by using this notion:

---

27 We remark, once and for all, that when we talk about the support of an atom we really mean the smallest sphere containing the support of an atom.

THEOREM (2.28). *Suppose $f \in L^1(\mathbf{R})$ has mean 0 and is supported in the interval $[-1,1]$. Then $f = \sum \alpha_j a_j$, where $\sum |\alpha_j| < \infty$ and the a_j's are $(1,\infty)$-atoms supported in $[-1,1]$, if and only if $\int_{-1}^{1} |\tilde{f}(x)|\,dx < \infty$, where*

$$\tilde{f}(x) = \text{P. V.} \frac{1}{\pi} \int \frac{f(t)}{x-t}\,dt.$$

Since the Hilbert transform $\tilde{a}$ of a $(1,\infty)$-atom satisfies $\int_{-\infty}^{\infty} |\tilde{a}(x)|\,dx = \|\tilde{a}\|_1 \leqslant C$, C independent of a (see the argument following (2.12)), we have

COROLLARY (2.29). *If $f \in L^1(\mathbf{R})$ is supported in an interval I and has mean zero then $\tilde{f} \in L^1(I)$ implies $\tilde{f} \in L^1(\mathbf{R})$.*

Using Theorem (2.28) and the same argument we gave to establish Theorem (2.25) we obtain a characterization of an H^1 space associated with the space of homogeneous type we introduced in example (5). In order to state this result let us write $\omega_{\alpha,\beta}(x) = (1-x)^\alpha(1+x)^\beta$, $-1 \leqslant x \leqslant 1$, $-1 < \alpha, \beta$,

$$(H_{\alpha,\beta} f)(x) = \frac{1}{\omega_{\alpha,\beta}(x)} \text{P.V.} \int_{-1}^{1} \frac{f(t)\omega_{\alpha,\beta}(t)}{x-t}\,dt$$

and let $H^1([-1,1], \omega_{\alpha,\beta}(t)\,dt)$ be the atomic Hardy space generated by the $(1,\infty)$-atoms defined in terms of the measure $\omega_{\alpha,\beta}(t)\,dt$. We then have

COROLLARY (2.30). *$f \in H^1([-1,1], \omega_{\alpha,\beta}(t)\,dt)$ if and only if f and $H_{\alpha,\beta} f$ belong to $L^1(\omega_{\alpha,\beta}(t)\,dt)$.*

We now pass to the proof of (2.28). Let f satisfy the hypotheses of this theorem; then, $f_e(\theta) = f(\cos\theta)|\sin\theta|$ is a function in $L^1([-\pi,\pi], d\theta)$. If $x = \cos\psi$ then

$$\tilde{f}(x) = \frac{1}{\sin\psi} \text{P.V.} \frac{1}{\pi} \int_0^{\pi} f_e(\theta) \frac{\sin\psi}{\cos\psi - \cos\theta}\,d\theta$$

$$= -\frac{1}{2\sin\psi} \text{P.V.} \frac{1}{\pi} \int_{-\pi}^{\pi} f_e(\theta) \cot\frac{\psi-\theta}{2}\,d\theta.$$

But

$$f_e^c(\psi) = \frac{1}{\pi} \text{P.V.} \int_{-\pi}^{\pi} f_e(\theta) \cot\frac{\psi-\theta}{2}\,d\theta$$

is the conjugate function of f_e. That is, $f_e + i f_e^c$ is the boundary value of the analytic function associated with f_o we described at the beginning of §1. Moreover, it follows from the change of variables we made above (and the fact that f_e^c is an odd function) that $\int_{-1}^{1} |\tilde{f}(x)|\,dx = C \int_{-\pi}^{\pi} |f_e^c(\psi)|\,d\psi$. Consequently, f_e belongs to $H^1(T)$ $(= \Re e\, H^1(T))$ and, therefore, has an atomic decomposition

$$f_e(\theta) = \sum \alpha_j a_j(\theta), \tag{2.31}$$

where $\sum |\alpha_j| < \infty$ and a_j is a $(1,\infty)$-atom of mean 0 supported in the interval I_j. We now claim that it can be assumed that if $0 \leqslant \theta \leqslant \pi$ the sum in (2.31) involves only atoms a_j whose support I_j lies in $[0,\pi]$. To see this we first observe that since f_e is even

$$f_e(\theta) = \sum \alpha_j \frac{a_j(\theta) + a_j(-\theta)}{2}.$$

If $0 \in I_j$ then $a_j'(\theta) = (a_j(\theta) + a_j(-\theta))/2$ is supported in $I_j' = \{\theta\colon \theta \in I_j$ or $-\theta \in I_j\}$. We also have $\int_{-\pi}^{\pi} a_j' = 0$ and, since a_j' is even, $\int_0^{\pi} a_j' = 0$. Moreover, $|a_j'(\theta)| \leqslant 1/|I_j| \leqslant 2/|[0,\pi) \cap I_j'|$. Thus, $a_j'/2$, restricted to $[0,\pi]$ is a $(1,\infty)$-atom. A similar argument applies to the case $\pi \in I_j$ (see footnote 8). If neither 0 nor π belong to I_j then either $a_j(\theta)$ or $a_j(-\theta)$ is 0 for $0 \leqslant \theta \leqslant \pi$ and there is nothing to prove.

Having made this restriction of the representation (2.31) to $[0,\pi]$ we shall now show that A_j, defined on $[-1,1]$ by

$$A_j(\cos\theta)\sin\theta = a_j(\theta)$$

is a $(1,p)$-atom in the space $H^1([-1,1],dt)$ for certain values of $p > 1$ (up to an inessential multiplicative constant). In fact, let J_j be the image of I_j under the map $\theta \to \cos\theta = t$. Then A_j is supported by J_j, $|J_j| = \int_{I_j} \sin\theta\, d\theta$ and $\int_{J_j} A_j(t)\,dt = \int_{I_j} a_j(\theta)\,d\theta = 0$. Thus, we need only show that

$$\left(\int_{J_j} |A_j(t)|^p \frac{dt}{|J_j|}\right)^{1/p} \leqslant \frac{C}{|J_j|}.$$

We can rewrite this inequality in the form:

$$\left(\int_{I_j} |A_j(\cos\theta)\sin\theta|^p \frac{(\sin\theta)^{1-p}}{|J_j|}\,d\theta\right)^{1/p} \leqslant \frac{C}{|J_j|}.$$

Since $|A_j(\cos\theta)\sin\theta| = |a_j(\theta)| \leqslant 1/|I_j|$ the desired inequality follows from the easily derived condition

$$\left(\frac{1}{|I_j|}\int_{I_j} \sin\theta\,d\theta\right)^{1-(1/p)} \left(\frac{1}{|I_j|}\int_{I_j} (\sin\theta)^{1-p}\,d\theta\right)^{1/p} \leqslant C,$$

for $1 < p < 2$. (In fact, this is the A^q condition of Muckenhoupt [13] for the weight $\sin\theta$ valid for $q > 2$, where $1/p + 1/q = 1$.) Theorem (2.28) now follows from Theorem A.

It would be of interest to obtain higher dimensional extensions of Theorem (2.28) and Corollary (2.29).

Corollary (2.30) is relevant to the study of Jacobi polynomial expansions. These are polynomials, $P_n^{\alpha,\beta}$, of order n that are eigenfunctions of the differential operator

$$D_{\alpha,\beta} = (1-x^2)\frac{d^2}{dx^2} + (\beta - \alpha - (\alpha+\beta+2)x)\frac{d}{dx}$$

corresponding to the eigenvalue $-n(n+\alpha+\beta+1)$, α, $\beta > -1$. The Jacobi functions, $Q_n^{\alpha,\beta}$, of the second kind are also eigenfunctions of this differential operator corresponding to the same eigenvalue; moreover (see p. 171 of [25]), they are related to the Jacobi polynomials by the formula

$$Q_n^{\alpha,\beta}(x) = (H_{\alpha,\beta} P_n^{\alpha,\beta})(x).$$

Thus, expansions in terms of Jacobi polynomials are transformed into expansions in terms of Jacobi functions of the second kind. Corollary (2.30) gives us some information about these expansions when the former represents $H^1([-1,1], \omega_{\alpha,\beta}(t)\,dt)$.

When $\alpha = \beta$ the $P_n^{\alpha,\alpha}$ are the ultraspherical polynomials. Muckenhoupt and Stein [47] have developed a theory of H^p spaces involving expansions in terms of these polynomials. When $\alpha = (n-2)/2$, n an integer, it is well known that these expansions arise from the study of spherical harmonics. Using this geometrical interpretation and the techniques used to establish Theorem (2.25) one can see that the space $H^1([-1,1], \omega_{\alpha,\alpha}(t)\,dt)$ introduced here and the H^1 space studied by Muckenhoupt and Stein are identical when $\alpha = (n-2)/2$. It would be of interest to see what is the relation between these two spaces for other indices α. The analogous problem involving Hankel transforms has been studied by J. Garcia-Cuerva [32].

An important topic involving expansions in terms of Jacobi polynomials is the study of the multiplier operators associated with these expansions. These operators can be defined in a manner completely analogous to the way they are defined in the theory of Fourier series. We recall that the polynomials $P_n^{\alpha,\beta}$ can be obtained by the Gram-Schmidt orthonormalization process from the monomials (relative to the natural inner product defined by the weight $\omega_{\alpha,\beta}$). Thus, if $f \in L^2([-1,1], \omega_{\alpha,\beta}(x)\,dx)$ then

$$f = \sum_{n=0}^{\infty} \hat{f}_n P_n^{(\alpha,\beta)},$$

where $\hat{f}_n = \int_{-1}^{1} f(x) P_n^{\alpha,\beta}(x) \omega_{\alpha,\beta}(x)\,dx$. Hence, if $\{m_n\}$ is a bounded sequence the operator M mapping f into the function having expansion $\sum_{n=0}^{\infty} m_n \hat{f}_n P_n^{\alpha,\beta}$ is well defined and maps into $L^2([-1,1], \omega_{\alpha,\beta}(x)\,dx)$. These are the operators commuting with the differential operator $D_{\alpha,\beta}$. As is the case in the study of Fourier series and integrals it is natural to inquire what properties of f are preserved under the action of M. Since $[-1,1]$, together with the measure induced by the weight $\omega_{\alpha,\beta}$, is a space of homogeneous type, the theory of singular integrals connected with these spaces can be applied to obtain sufficient conditions under which M is a bounded operator on

$$L^p([-1,1], \omega_{\alpha,\beta}(x)dx), \qquad 1 < p < \infty$$

(see [17] and [20]). A careful reading of these results and the methods used will show that these conditions also imply the boundedness of M on the corresponding atomic H^p spaces when $p\ (\leqslant 1)$ is close to 1 (compare with the discussion preceding and following conditions (2.14) and (2.15)). Because of Corollary B it then follows that M preserves the "natural" Lipschitz condition associated with this space of homogeneous type:

$$|f(x) - f(y)| \leq C\left|\int_x^y (1-t)^\alpha (1+t)^\beta \, dt\right|^{1/p-1}$$

Specifically, one can establish the following extension of a multiplier theorem of Connett and Schwartz [**20**]:

THEOREM (2.32). *Suppose $\alpha \geq \beta > -1/2$ and k is the largest integer not exceeding $\alpha + 2$, then if $\{m_n\}$ is a bounded sequence satisfying*

$$\sum_{R \leq n < 2R} |\Delta^k m_n|^2 \leq cR^{1-2k},$$

where Δ^k is the kth difference operator, the multiplier operator M associated with $\{m_n\}$ is bounded on $L^p([-1,1], \omega_{\alpha,\beta}(x)\,dx)$ for $1 < p < \infty$ and on $H^1([-1,1], \omega_{\alpha,\beta}(x)\,dx)$.

Let us now pass to the situation encountered when $X = [0, 1)$, μ is Lebesgue measure and the quasi-distance is given in terms of dyadic intervals as described in example (4). We shall refer to this situation as "the dyadic case". Many aspects of harmonic analysis become particularly simple and elegant in this dyadic case. In particular, the atomic H^1 space can easily be seen to coincide with the space of all L^1 functions whose Hardy-Littlewood maximal function f^* is integrable. That is, if we let $f^*(x) = \sup_I (1/|I|)|\int_I f|$, where the supremum is taken over all dyadic intervals I containing x, then f is in atomic H^1 if and only if $f^* \in L^1$.[28]

In order to establish this equivalence we first observe that it is easy to see that $\|a^*\|_1 \leq 1$ whenever a is a $(1, \infty)$-atom. In fact, if I and J are two nondisjoint dyadic intervals in $[0, 1)$ then either $I \subset J$ or $I \supset J$. From this it follows immediately that if a is supported in I then a^* is also supported in I and satisfies $a^*(x) \leq 1/(|I|)$; therefore, $\|a^*\|_1 \leq 1$ and we see that

$$\|f^*\|_1 \leq \|f\|_{H^1}$$

whenever f is in atomic H^1. Conversely, if f and f^* are in L^1 we shall show that f has an atomic decomposition $\sum \alpha_j a_j$ with $\sum |\alpha_j| \leq 8\|f^*\|_1$. Without loss of generality we can assume $\int_0^1 f = 0$. We shall now obtain the desired representation of f by means of the Calderón-Zygmund decomposition of f. We first consider $U'_k = \{x \in [0,1): f^*(x) > 2^k\}$ for k an integer and we let $U'_k = \bigcup_i I_k^i$ where the I_k^i's are maximal dyadic open subintervals of U'_k (and, thus, for k fixed, they are disjoint). Now, each I_k^i is contained in a (unique) dyadic interval, J_k^i, of twice its length. Since each I_k^i is maximal, J_k^i contains a point outside U'_k and, thus, $m_{J_k^i}(f) = (1/|J_k^i|)\int_{J_k^i} f$ has absolute value less than or equal to 2^k. Let $U_k = \bigcup_i J_k^i$; we may assume that J_k^i, for k fixed, are mutally disjoint (recall that two intersecting dyadic intervals have the property that one must contain the other). The Calderón-Zygmund decomposition is then, $f = g_k + b_k$, where $g_k = (1 - \chi_{U_k})f + \sum_i m_{J_k^i}(f)\chi_{J_k^i}$. Clearly, $|g_k| \leq 2^k$ and the b_k's are supported in U_k and satisfy $\int_{J_k^i} b_k = 0$. As $k \to \infty$ we have $g_k \to f$ (trivially in the a.e. sense and, also, in the L^1-sense since the convergence is

[28] f^* is the natural analog of the nontangential maximal operator P^* we introduced in the beginning of §1 (see [2]).

dominated by $2f^*$). Also, $\lim_{k\to-\infty}\|g_k\|_1 = 0$. Thus, $f = \sum_{-\infty}^{\infty}(g_{k+1} - g_k) = \sum_{-\infty}^{\infty}(b_k - b_{k+1})$. Moreover, $|b_k - b_{k+1}| = |g_{k+1} - g_k| \leqslant 2^{k+1}$. Since $U_{k+1} \subset U_k$ and each interval J_{k+1}^j must be a subinterval of some J_k^i we have $\int (b_k - b_{k+1})\chi_{J_k^i} = 0$. Therefore, $a_k^i = (1/2^{k+1}|J_k^i|)[b_k - b_{k+1}]\chi_{J_k^i}$ is a $(1,\infty)$-atom and $f = \sum_{k=-\infty}^{\infty} 2^{k+1}\sum_i |J_k^i| a_k^i$. Furthermore,

$$\|f^*\|_1 = \int_0^\infty |\{x: [0,1): f^*(x) > \lambda\}|\, d\lambda \geqslant \sum_{k=-\infty}^{\infty} |U'_{k+1}| 2^k$$

$$\geqslant \sum_{k=-\infty}^{\infty} |U_k| 2^k 2^{-2} = \sum_{k=-\infty}^{\infty} 2^{k+1} 2^{-3} \sum_i |J_k^i|.$$

But this shows that $\|f\|_{H^1} \leqslant 8\|f^*\|_1$ and the claim is established.

Let us now pass to some illustrations of how these methods apply to the study of harmonic analysis on homogeneous spaces (example (7)). We begin with an extension of Hardy's inequality associated with the surface Σ_{n-1} of the unit sphere in $\mathbf{R}^n$, $n \geqslant 3$. Let t', x', y', ... denote points of Σ_{n-1}, dt', dx', dy', ... the elements of Lebesgue surface measure on Σ_{n-1} (which is normalized so that Σ_{n-1} has measure 1). H_k denotes the space of homogeneous harmonic polynomials of degree k restricted to Σ_{n-1} and $Z_{x'}^{(k)}$ the zonal harmonic of degree k with pole x'. The defining property of $Z_{x'}^{(k)}$ is that it is the (unique) element of H_k such that

$$Y(x') = \int_{\Sigma_{n-1}} Y(t') Z_{x'}^{(k)}(t')\, dt' \tag{2.33}$$

whenever $Y \in H_k$. $Z_{x'}^{(k)}$ is also determined (up to a multiplicative constant) by the property that it is a member of H_k that is invariant under the action of all rotations ρ of $\mathbf{R}^n$ that leave x' fixed (see pp. 146–147 of [**57**]). From this and the fact that $Z_{x'}^{(k)}(x') = d_k$ = dimension of H_k we obtain the following simple expression for zonal harmonics:

$$Z_{x'}^{(k)}(t') = d_k \int_{\Sigma_{n-2}} [t' \cdot (iy' + x')]^k\, dy',$$

where Σ_{n-2} is being identified with all those vectors $y' \in \Sigma_{n-1}$ orthogonal to x'. From this we see that

$$\left|\frac{\partial^2}{\partial t_i \partial t_j} Z_{x'}^{(k)}(t')\right| \leqslant 2d_k k(k-1). \tag{2.34}$$

Now let us consider

$$Z_{x'}^{(k)}(t') - Z_{x'}^{(k)}(x') = (\nabla Z_{x'}^{(k)})(x') \cdot (t' - x') + E^{(k)}(x', t').$$

The error term involves the second derivatives of $Z_{x'}^{(k)}$ (evaluated at some point on the segment joining x' and t') and products of the type $(x'_l - t'_l) \cdot (x'_j - t'_j)$. Because of (2.34), therefore, we have $|E^{(k)}(x', t')| \leqslant c d_k k^2 \cdot |x' - t'|^2$. Moreover, it follows from the integral representation of $Z_{x'}^{(k)}$ that

$$(\nabla Z_{x'}^{(k)})(x') = k d_k x'.$$

Consequently, $\nabla Z_{x'}^{(k)}(x') \cdot (t' - x') = kd_k(x' \cdot t' - 1) = -(kd_k/2)|x' - t'|^2$. It follows that

$$(2.35) \qquad |Z_{x'}^{(k)}(t') - d_k| = |Z_{x'}^{(k)}(t') - Z_{x'}^{(k)}(x')| \leqslant cd_k k^2 |x' - t'|^2.$$

If $f \in L^1(\Sigma_{n-1})$ its projection onto H_k is

$$f_k(x') = \int_{\Sigma_{n-1}} f(t') Z_{x'}^{(k)}(t')\, dt'$$

(this follows from the defining property of the zonal harmonics). We thus have the orthogonal development of f in terms of spherical harmonics (see [**57**, Chapter 4, §2]): $f \sim \sum f_k$. We shall now derive the following extension of Hardy's inequality:

THEOREM (2.36). *If $f \in H^1(\Sigma_{n-1})$ then*

$$\sum_{k=1}^{\infty} \frac{\|f_k\|_2}{k^{n/2}} = \sum_{k=1}^{\infty} k^{-n/2} \left(\int_{\Sigma_{n-1}} |f_k(x')|^2\, dx' \right)^{1/2} \leqslant c\|f\|_{H^1}.$$

In order to establish (2.36) we choose a $(1, \infty)$-atom supported in a "sphere" of radius R about x'. Then, by the "mean 0" property of a:

$$a_k(y') = \int_{\Sigma_{n-1}} a(t') Z_{y'}^{(k)}(t')\, dt' = \int_{\Sigma_{n-1}} a(t')[Z_{y'}^{(k)}(t') - Z_{y'}^{(k)}(x')]\, dt'.$$

Thus, by Minkowski's integral inequality

$$\|a_k\|_2 \leqslant \int_{\Sigma_{n-1}} |a(t')| \left\{ \int_{\Sigma_{n-1}} |Z_{y'}^{(k)}(t') - Z_{y'}^{(k)}(x')|^2\, dy' \right\}^{1/2} dt'.$$

Since $Z_{v'}^{(k)}(u') = Z_{u'}^{(k)}(v')$ (see p. 143 of [**57**]) we have

$$\begin{aligned} &\int_{\Sigma_{n-1}} |Z_{y'}^{(k)}(t') - Z_{y'}^{(k)}(x')|^2\, dy' \\ &\quad = \int_{\Sigma_{n-1}} \{[Z_{t'}^{(k)}(y')]^2 + [Z_{x'}^{(k)}(y')]^2 - 2Z_{t'}^{(k)}(y')Z_{x'}^{(k)}(y')\}\, dy' \\ &\quad = Z_{t'}^{(k)}(t') + Z_{x'}^{(k)}(x') - 2Z_{x'}^{(k)}(t') = 2[d_k - Z_{x'}^{(k)}(t')]. \end{aligned}$$

Hence, we can use (2.35) to obtain the estimate

$$\|a_k\|_2 \leqslant c \int_{\Sigma_{n-1}} |a(t')| k \sqrt{d_k}\, |x' - t'|\, dt'.$$

Since $d_k \leqslant ck^{n-2}$ (see [**57**, p. 145]) and $|x' - t'| \leqslant R$ for t' in the support of a we have

$$\|a_k\|_2 \leqslant cRk \cdot k^{(n-2)/2} = cRk^{n/2}.$$

Making use of this inequality and the obvious estimate for the L^2 norm of an atom we finally obtain

$$\sum_{k=1}^{\infty} \frac{\|a_k\|_2}{k^{n/2}} = \sum_{k\leqslant 1/R} \frac{\|a_k\|_2}{k^{n/2}} + \sum_{k>1/R} \frac{\|a_k\|_2}{k^{n/2}}$$

$$\leqslant \sum_{k\leqslant 1/R} cR + \left(\sum_{k=1}^{\infty} \|a_k\|_2^2\right)^{1/2} \left(\sum_{k>1/R} k^{-n}\right)^{1/2}$$

$$\leqslant c'(1 + \|a\|_2 R^{(n-1)/2}) \leqslant c''.$$

Theorem (2.36) is an immediate consequence of this inequality.

The version of Hardy's inequality we have just discussed assumes a special form on the sphere Σ_3. This sphere can be identified in a natural way with the group $SU(2)$ of unitary 2×2 matrices with determinant 1. This group is the simplest noncommutative semisimple compact Lie group to analyze and is particularly useful for obtaining insight into problems in harmonic analysis associated with more general Lie groups. We shall need to introduce some notation in order to illustrate some of these features.

We write $z = (z_1, z_2) = (x_1 + ix_2, x_3 + ix_4)$ ($\in \mathbf{C}^2$) for $x = (x_1, x_2, x_3, x_4)$ $\in \mathbf{R}^4$. To x (or z) we can associate the 2×2 matrix

$$u = u(x) = -i\begin{pmatrix} -\bar{z}_2 & z_1 \\ \bar{z}_1 & z_2 \end{pmatrix}.$$

Clearly, $\det u = |z_1|^2 + |z_2|^2 = x_1^2 + x_2^2 + x_3^2 + x_4^2$ and, hence, Σ_3 corresponds to $SU(2)$ and normalized Lebesgue measure on Σ_3 corresponds to Haar measure on $SU(2)$. The irreducible representations of $SU(2)$ can be realized explicitly on the spaces $\mathcal{P}^l$ of homogeneous polynomials of degree $2l$, $l = 0, 1/2, 1, 3/2, 2, \ldots$, in $(\omega_1, \omega_2) = \omega \in \mathbf{C}^2$. That is, $\mathcal{P}^l$ consists of all the polynomials of the form $p(\omega) = \sum_{j=-l}^{l} a_j \omega_1^{l-j} \omega_2^{l+j}$ and the mapping of $u \in SU(2)$ into the transformation T_u^l defined by

$$(T_u^l p)(\omega) = p(u'\omega),$$

where u' is the transpose of u, is a representation. Moreover, T_u^l is unitary with respect to an appropriate inner product for which

$$p_j(\omega) = \frac{1}{\sqrt{(l-j)!}\sqrt{(l+j)!}} \omega_1^{l-j} \omega_2^{l+j},$$

$j = -l, -l+1, \ldots, l$, form an orthonormal basis (see **[17]** for further details). The matrix entries of T^l relative to this basis are usually denoted by $t_{k,j}^l(u)$, $-l \leqslant k, j \leqslant l$, and are defined by the relation

$$(-i)^{2l} \frac{(-\bar{z}_2\omega_1 + \bar{z}_1\omega_2)^{l-j}(z_1\omega_1 + z_2\omega_2)^{l+j}}{\sqrt{(l-j)!\,(l+j)!}}$$

$$= \sum_{k=-l}^{l} t_{k,j}^l(u(z)) \frac{\omega_1^{l-k}\omega_2^{l+k}}{\sqrt{(l-k)!\,(l+k)!}}.$$

It follows that $t_{k,j}^l(u(z)) = t_{k,j}^l(u(x))$ are homogeneous polynomials in x of degree $2l$. Moreover, their restrictions to Σ_3 form an orthogonal basis for the space H^{2l} of spherical harmonics of degree $2l$. The expansion of functions on Σ_3 in terms of spherical harmonics turns out to be equivalent to the Peter-Weyl expansion of functions on $SU(2)$. The latter can be expressed in the following way: if $f \in L^2(SU(2))$ then

$$f(u) = \sum_{2l=0}^{\infty} (2l+1)\,\mathrm{tr}\,(\hat{f}(l)T_u^l),$$

where

$$\hat{f}(l) = \int_{SU(2)} f(u)T_{u^{-1}}^l\,du,$$

and the convergence of this series to $f(u)$ is valid in the L^2-norm. Moreover,

$$\int |f(u)|^2\,du = \sum_{2l=0}^{\infty} (2l+1)|||\hat{f}(l)|||^2, \tag{2.37}$$

where $|||\hat{f}(l)|||$ denotes the Hilbert-Schmidt norm of $\hat{f}(l)$. It follows from these considerations and, in particular, (2.37), that the projection, f_{2l}, of f onto H^{2l} satisfies $\|f_{2l}\|_2^2 = (2l+1)|||\hat{f}(l)|||^2$. Thus, Hardy's inequality (2.36) when $n = 4$ is equivalent to

$$\sum_{2l=0}^{\infty} \frac{|||\hat{f}(l)|||}{(2l+1)^{3/2}} < c\|f\|_{H^1} \tag{2.38}$$

whenever $f \in H^1(SU(2))$. By using the interpolation Theorem D and the two inequalities (2.37) and (2.38) we obtain the following generalization of Paley's inequality:

$$\sum_{2l=0}^{\infty} (2l+1)^{(5p/2)-4}|||\hat{f}(l)||| \leqslant c_p\|f\|_p \tag{2.39}$$

whenever $f \in L^p(SU(2))$, $1 < p \leqslant 2$.

As is the case in $\mathbf{R}^n$ and T, the study of the behavior of multiplier operators is enhanced by appropriate use of properties of H^p spaces. We first remark that bounded operators M on $L^2(SU(2))$ that commute with *left* translations correspond to matrix multiplier operators; that is, if $f \in L^2(SU(2))$ then Mf has the Peter-Weyl expansion

$$(Mf)(u) = \sum_{2l=0}^{\infty} (2l+1)\,\mathrm{trace}\,[\hat{M}(l)\hat{f}(l)T_u^l],$$

where $\hat{M}(l)$ is a $(2l+1)\times(2l+1)$ matrix whose operator norm $\|\hat{M}(l)\|$ does not exceed a constant C independently of l. It turns out that the space $H^1(\Sigma_3) = H^1(SU(2))$, defined in terms of atoms associated with the usual Euclidean distance on Σ_3,[29] can be characterized by certain "Riesz" transforms that correspond to the following matrix multiplier operators:

---

[29] On $SU(2)$ this distance is more naturally written in terms of the Hilbert-Schmidt norm: $d(u,v) = |||u - v|||$ when $u, v \in SU(2)$.

$$[\hat{M}_{\pm}(l)]_{k,j} = \frac{\sqrt{(l \pm j)(l \mp j + 1)}}{2l+1}\delta_{j\mp 1,k}, \qquad [\hat{H}_{\pm}(l)]_{k,j} = \frac{l \pm k}{2l+1}\delta_{k,j},$$

where $\delta_{k,j}$ is the Kronecker delta. These operators are related to certain generalized Cauchy-Riemann equations in $\mathbf{R}^4$ in a manner similar to the way the conjugate function operator on the circle T is related to the Cauchy-Riemann equations in $\mathbf{R}^2$ (again, see **[17]** for further details and motivation).

The use of the atomic Hardy spaces in the study of multiplier operators is somewhat complicated by the fact that the multipliers consist of sequences of matrices. As in the previous cases, one can characterize the matricial Fourier transform of a molecule centered at the identity. In order to show that a multiplier preserves H^1 it is enough to show that it maps atoms into molecules (this follows from Theorem C). Let us discuss this situation in greater detail. Let us write $\rho(u) = |||u - \mathbf{1}|||^3$ where $\mathbf{1}$ is the identity element of $SU(2)$. Then a molecule centered at $\mathbf{1}$ is a function α (u) satisfying

$$\text{(2.40)} \quad \int \alpha(u)\,du = 0 \quad \text{and} \quad \left(\int |\alpha(u)|^2\,du\right)\left(\int |\alpha(u)|^2[\rho(u)]^{1+\varepsilon}\right)^{1/\varepsilon} \leqslant 1.$$

If we let $\varepsilon = 1/3$ we obtain a weighted Plancherel theorem (see p. 114 of **[17]**) which, applied to α, takes the form

$$\int_{SU(2)} |\alpha(u)|^2\, |||u - \mathbf{1}|||^4\,du = \sum_{2l=0}^{\infty} (2l+1)\,|||\Delta^2\hat{\alpha}(l)|||^2,$$

where

$$\Delta^2\hat{\alpha}(l) = \int \alpha(u)\,|||u - \mathbf{1}|||^2\, T^l_{u^{-1}}\,du$$

is a certain second order "difference" operator defined on sequences of matrices $\hat{\alpha}(l)$. It can be shown that

$$[\Delta^2\hat{\alpha}(l)]_{k,j} = 2[\hat{\alpha}(l)]_{k,j} - \sum_{\gamma,\delta=\pm 1/2} A^l_k(\gamma,\delta)A^l_j(\gamma,\delta)[\hat{\alpha}(l+\gamma)]_{k+\delta,j+\delta}$$

where

$$A^l_k(\gamma,\delta) = \sqrt{\frac{1}{2} + \frac{\gamma(1+4\delta k)}{2l+1}}.$$

Thus, with $\varepsilon = 1/3$ (2.40) can be expressed by

$$\text{(2.41)} \quad \begin{gathered} \hat{\alpha}(0) = 0 \quad \text{and} \\ \left(\sum_{2l=0}^{\infty} (2l+1)\,|||\hat{\alpha}(l)|||^2\right)\left(\sum_{2l=0}^{\infty} (2l+1)\,|||\Delta^2\hat{\alpha}(l)|||^2\right)^{1/3} \leqslant 1. \end{gathered}$$

The direct method for checking whether the matrices $\hat{\alpha}(l) = \hat{M}(l)\hat{a}(l)$ are the coefficients of a molecule when a is an atom centered at 0 can be used here. This procedure, however, is technically complicated since it involves estimates on $\Delta^2(\hat{M}(l)\hat{a}(l))$. Fortunately, it can be shown that, if the inequality

$$\sum_{2l=0}^{\infty} (2l+1)|||\Delta^2(\hat{M}(l)\hat{a}(l))|||^2 \leqslant c \sum_{2l=0}^{\infty} (2l+1)|||\Delta^2 \hat{a}(l)|||^2$$

holds for some special central atoms a, then M maps H^1 boundedly into itself (a careful reading of the proof of Theorem (3.1) on p. 81 of [17] shows that M is a convolution operator with a kernel satisfying condition (2.15)). Conditions on $\hat{M}(l)$ that insure the validity of the last inequality can be stated in a relatively simple manner; we thus obtain the following result:

THEOREM (2.42). *Let* $\{\hat{M}(l)\}$ *be a sequence of* $(2l+1)\times(2l+1)$ *matrices satisfying* $\|\Delta^j \hat{M}(l)\| \leqslant cl^{-j+1/2}, j = 0, 1, 2,$ *where*

$$\Delta^0 \hat{M}(l) = \hat{M}(l) \quad \text{and} \quad \Delta^1 \hat{M}(l) = \sum_{\varepsilon,\delta=\pm 1/2} \varepsilon A_k^l(\varepsilon,\delta) A_j^l(\varepsilon,\delta)[\hat{M}(l+\varepsilon)]_{k+\delta,j+\delta},$$

then M is a bounded multiplier operator on $L^p(SU(2))$, $1 < p < \infty$, and on H^p for p sufficiently close to, but not exceeding, 1.

Two remarks are in order: (a) the conditions on $\Delta^j \hat{M}(l)$ are quite easy to check for the Riesz transforms $M_\pm$ and $H_\pm$ (see pp. 134 and 135 of [17]); (b) If the matrices $\hat{M}(l)$ are diagonal, Theorem (2.42) assumes a more familiar form: *Let* μ_m^l, $-l \leqslant m \leqslant l$, *be the entries of the diagonal matrix* $\hat{M}(l)$; *if* μ_m^l *and* $l^2(2\mu_m^l - \mu_{m+\delta}^{l+1/2} - \mu_{m-\delta}^{l-1/2})$ *are* $O(1)$ *for* $\delta = -1/2, 1/2$ *then M is bounded on $L^p(SU(2))$ for $1 < p < \infty$ and on H^1.*

The analysis of the Fourier transform described here for $SU(2)$ can, in principle, be extended to other classical compact groups. The main difficulties involve the choice of a special basis in each representation space and the calculation of certain Clebsch-Gordan coefficients that lead to formulae for operators corresponding to Δ^1 and Δ^2. The spaces H^p for compact semisimple Lie groups were introduced in [16] using generalized Cauchy-Riemann equations. The basic subharmonicity result obtained there permits one to establish the duality of H^1 and BMO by the methods of Fefferman and Stein. This duality result, in turn, shows that the H^1 defined in terms of the Cauchy-Riemann equations is the atomic H^1 space.

We now discuss example (9) and its connection with the theory of analytic functions of several complex variables. If

$$B_n = \{z \in \mathbf{C}^n \colon |z|^2 = |z_1|^2 + |z_2|^2 + \cdots + |z_n|^2 < 1\}$$

then $\partial B_n = \Sigma_{2n-1}$.

A function $F(z)$ holomorphic in B_n is said to belong to $\mathcal{H}^p(\partial B_n)$ if

$$\|F\|_{\mathcal{H}^p}^p = \sup_{r<1} \int_{|z'|=1} |F(rz')|^p \, dz' < \infty,$$

where dz' is normalized Lebesgue measure on Σ_{2n-1}. The space $\mathcal{H}^2$ is of particular interest. It can be identified with the closed subspace of $L^2(\partial B_n)$ consisting of boundary functions of holomorphic functions.

The projection operator P from $L^2(\partial B_n)$ onto $\mathcal{H}^2$ is given in terms of the Szegö kernel $(1 - z \cdot \bar{\omega})^{-n}$. We first define the transformation S:

$$(Sf)(z) = \int_{\partial B_n} \frac{f(\omega')}{(1 - z \cdot \bar{\omega}')^n} d\omega' \quad \text{for } |z| < 1.$$

We now observe that $d(z, \omega) = |1 - z \cdot \bar{\omega}|^{1/2}$ is a distance on ∂B_n which is particularly well adapted to the study of this kernel. In fact, $m(z, \omega) = |1 - z \cdot \bar{\omega}|^n$ is a quasi-distance equivalent to the measure of the smallest d-sphere containing z and ω, $|z| = |\omega| = 1$, and the basic inequality (2.14) is verified for the Szegö kernel: for $|z| = |\omega| = |\omega_0| = 1$

$$\left| \frac{1}{(1 - z \cdot \bar{\omega})^n} - \frac{1}{(1 - z \cdot \bar{\omega}_0)^n} \right| \leq \left[\frac{m(\omega, \omega_0)}{m(z, \omega_0)} \right]^{1/2n} \frac{C_1}{m(z, \omega_0)} \tag{2.43}$$

if $m(z, \omega_0) > C_2 m(\omega, \omega_0)$.

It can also be shown (see [42]) that for $z \in \partial B_n$, and $f \in L^2(\partial B_n)$

$$(Pf)(z) = \lim_{r \to 1} (Sf)(rz) = \frac{1}{2} f(z) + \lim_{\varepsilon \to 0} \int_{m(z,\omega) > \varepsilon} \frac{f(\omega)}{(1 - z \cdot \bar{\omega})^n} d\omega.$$

It now follows from the general theory developed previously that the projection operator can be defined as a bounded operator on L^p, $1 < p < \infty$, and H^p, $1 - \eta < p \leq 1$, and it preserves the Lipschitz character of f (with respect to d). Using the methods of Fefferman and Stein [29] it can be shown that the dual of $\mathcal{H}^1(\partial B_n)$ is $P(BMO)$ (see [14]). This result is equivalent to:

THEOREM (2.44). *Let $F \in \mathcal{H}^1(\partial B_n)$. Then there exist atoms a_j such that*

$$F(z) = \sum \lambda_j (Pa_j)(z) \equiv \sum \lambda_j A_j(z)$$

with

$$\sum |\lambda_j| \leq c \|F\|_{\mathcal{H}^1};$$

thus, $P(H^1) = \mathcal{H}^1$.

The functions $A_j(z) = (Pa_j)(z)$, called *holomorphic atoms*, are easily seen to be molecules. A related fact (see [14]) is that any holomorphic atom A can be written in the form

$$A(z) = \sum_{i=1}^{N} B_i(z) C_i(z) \tag{2.45}$$

with B_i, $C_i \in \mathcal{H}^2$ satisfying $\sum_{i=1}^N \|B_i\|_{\mathcal{H}^2} \|C_i\|_{\mathcal{H}^2} \leq c$, where N and c depend only on the dimension n. From this follows the factorization result:

THEOREM (2.46). *Let $F \in \mathcal{H}^1(\partial B_n)$. Then there are functions G_j, $H_j \in \mathcal{H}^2$ such that $F = \sum_j G_j H_j$ and*

$$\sum_j \|G_j\|_{\mathcal{H}^2} \|H_j\|_{\mathcal{H}^2} \leq c \|F\|_{\mathcal{H}^1}.$$

Let us indicate briefly how (2.45) is proved. We may assume without loss of generality that an atom a is supported in a sphere S centered at $\mathbf{1}$

$= (0,0,\ldots,i)$ (the "North pole"). We let $\varphi_S = (1/|S|)\chi_S$ and write

$$1 - \bar{\omega}\cdot z = -\bar{\omega}_0\cdot z_0 + (1+iz_n)(i\bar{\omega}_n) + (1 - i\bar{\omega}_n)$$

for $z = (z_1, z_2, \ldots, z_n)$, $\omega = (\omega_1, \omega_2, \ldots, \omega_n) \in \partial B_n$, where $z_0 = (z_1, z_2, \ldots, z_{n-1}, 0)$ and $\omega_0 = (\omega_1, \omega_2, \ldots, \omega_{n-1}, 0)$. We have

$$(1 - \bar{\omega}\cdot z)^n = \sum_{|J|+k+l=n} d_{J,k,l}\bar{\omega}_0^J(1 - i\bar{\omega}_n)^k \bar{\omega}_n^l z_0^J (1+iz_n)^l,$$

where $J = (j_1, \ldots, j_{n-1})$, $|J| = j_1 + j_2 + \cdots + j_{n-1}$ and $z_0^J = z_1^{j_1} z_2^{j_2}\cdots z_{n-1}^{j_{n-1}}$. We write

$$1 = \int_{\partial B_n} \varphi_S(\omega)\,d\omega = \sum_{|J|+k+l=n} z_0^J(1+iz_n)^l D_{J,k,l}(z),$$

where

$$D_{J,k,l}(z) = d_{J,k,l}\int_{\partial B_n} \frac{\bar{\omega}_0^J(1-i\bar{\omega}_n)^k\bar{\omega}_n^l\varphi_S(\omega)}{(1 - z\cdot\bar{\omega})^n}\,d\omega.$$

Thus,

$$A(z) = \sum_{|J|+k+l=n} z_0^J(1+iz_n)^l D_{J,k,l}(z) A(z).$$

We now claim that each summand can be split into a product of the form $B(z)C(z)$ with $\|B\|_{\mathcal{H}^2}\|C\|_{\mathcal{H}^2} \leqslant C$. This is achieved by using inequality (2.43) in order to estimate $A(z)$ and by making rough estimates for $D_{J,k,l}(z)$ when z is far from S (see **[14]** for details).

Other spaces $\mathcal{H}^p$ of holomorphic functions for which this theory can be used are the spaces of functions F that are holomorphic in B_n and satisfy

$$\int_{|z|<1} |F(z)|^p(1 - |z|^2)^m\,dv(z) < \infty,$$

where $dv(z)$ is the element of Lebesgue measure on $\mathbf{C}^n$. The case $m = 0$ is of particular interest. Here the projection, P, of $L^2(B_n)$ onto $\mathcal{H}^2(B_n)$ is given in terms of the Bergman kernel:

$$(Pf)(z) = \frac{n!}{\pi^n}\int_{B_n} \frac{f(\omega)}{(1 - z\cdot\bar{\omega})^{n+1}}\,dv(\omega).$$

Unlike the preceding case, this projection is not a singular integral transform; however, P is unbounded on $L^1(B_n)$. If we introduce the quasi-distance d of example (10), B_n becomes a space of homogeneous type and P maps (atomic) H^1 into itself. As was the case for ∂B_n, $PH^1 = \mathcal{H}^1(B_n)$ and factorizations analogous to the ones in Theorem (2.46) are obtainable.

In this connection we point out that when $n = 1$ the dual space of $\mathcal{H}^1(B_1)$ can be represented by the class Λ_* of *smooth functions* of Zygmund (see **[14]**). The fact that $PH^1 = \mathcal{H}^1(B_1)$ allows us also to represent this dual space by holomorphic BMO functions on the solid disc B_1. From this it follows easily that smooth functions are characterized by the property that the derivatives of their analytic extensions to the interior of the disc belongs to the class $BMO(B_1)$. It would be of interest to extend this relation to higher dimensions.

The methods we have just described extend to more general strictly pseudo-convex domains (see [56]). It would be interesting to obtain similar factorization results in this more general setting. Another important related study involves the polydisc and its distinguished boundary T^n. In this situation the possibility of obtaining corresponding results is completely open.

We have considered $\Sigma_{2n-1} = \partial B_n$ as a point set that furnished us two different examples of a space of homogeneous type. We did this by first employing the ordinary Euclidean metric and, then, the nonisotropic distance $d(z,\omega) = |1 - z \cdot \bar{\omega}|^{1/2}$ (in both cases Lebesgue measure was used). It is natural to compare the Hardy spaces that arise in these two situations and one does find basic differences. At this point, however, we would like to point out that when dealing with *holomorphic* functions many results are easily reduced to the case of the 1-dimensional torus (where $n = 1$). Let us illustrate this point by considering a version of Hardy's inequality associated with $\mathcal{H}^1(\partial B_n)$.

Let $F(z) = \sum_0^\infty F_k(z)$ be holomorphic in B_n, where $F_k(z) = \sum_{|\alpha|=k} c_\alpha z^\alpha$ is the projection of F onto the polynomials in $z = (z_1, z_2, \ldots, z_n)$ that are homogeneous of degree k. Let $g_z(e^{i\theta}) = F(e^{i\theta}z) = \sum_0^\infty F_k(z)e^{ik\theta}$. By (1.10) we then have

$$\sum_1^\infty \frac{|F_k(z)|}{k} \leqslant \int_{-\pi}^{\pi} |g_z(e^{i\theta})|\, d\theta.$$

Integrating $z \in \partial B_n$ (we now assume $F \in \mathcal{H}^1(\partial B_n)$) we obtain

$$\sum_1^\infty \frac{\|F_k\|_1}{k} \leqslant c\|F\|_1. \tag{2.47}$$

Since $\|F_k\|_2 \leqslant k^{(n-1)/2}\|F_k\|_1$, (2.47) implies the following result involving the Taylor coefficients of F:

$$\sum_{k=1}^\infty \frac{1}{\sqrt{k^{n+1}(n+k-1)!}} \Big(\sum_{|\alpha|=k} \alpha_1!\,\alpha_2! \cdots \alpha_n!\,|c_\alpha|^2\Big)^{1/2} \leqslant c\|F\|_1.$$

Other results on $\mathcal{H}^p(\partial B_n)$ that are reducible to the 1-dimensional case by similar arguments can be found in [51].

Consider a function $B(z)$ on B_n for which $b_z(e^{i\theta}) = B(e^{i\theta}z)$ satisfies

$$\sup_z \|b_z(e^{i\theta})\|_{BMO(\partial B_1)} = c < \infty.$$

From the previous argument we then obtain

$$\left|\int_{\partial B_n} F\bar{B}\, d\sigma\right| = \lim_{r\to 1}\left|\int_{\partial B_n} F(rz)\overline{B(rz)}\, d\sigma(z)\right| \leqslant c\int_{\partial B_n} |F|\, d\sigma.$$

Consequently, $B \in BMO(\partial B_n)$. There are, however, functions B in $BMO(\partial B_n)$ for which $\sup_z \|b_z(e^{i\theta})\|_{BMO(\partial B_1)} = \infty$. For example, $B(z_1, z_2) = \sum_0^\infty z_1^{2^k}$ furnishes us an example when $n = 2$. For such a function the last inequality cannot be obtained by a reduction to the 1-dimensional case.

In these last few examples the spaces of homogeneous type involved were compact. There exists a noncompact space that is closely connected to B_n and ∂B_n. Let us give some details about this situation when $n = 2$. In this case we shall consider the *generalized half-plane*

$$D = D_2 = \{(z_1, z_2) \in \mathbf{C}^2 \colon \operatorname{Im} z_1 > |z_2|^2\}$$

which is the image of B_2 under the *Caley transform*

$$Z = (z_1, z_2) = i\left(\frac{1-\omega_1}{1+\omega_1}, \frac{\omega_2}{1+\omega_1}\right).$$

This mapping of (ω_1, ω_2) onto (z_1, z_2) extends to the boundary ∂B_2 which is mapped onto $\partial D = \{(z_1, z_2)\colon \operatorname{Im} z_1 = |z_2|^2\} = \{(t + i|z|^2, z)\colon t \in \mathbf{R}, z \in \mathbf{C}\}$. There is a group G that acts in a simply transitive fashion on ∂D. The elements of G can be identified with the set $\mathbf{R} \times \mathbf{C}$; the image of $(t + i|z|^2, z) \in \partial D$ under the action on $(\tau, \xi) \in G$ is

$$(t + \tau + 2 \operatorname{Im} z\bar{\xi} + i|z + \xi|^2, z + \xi).$$

If we write $(\tau, \xi) = (x_3, (x_1 + ix_2)/2)$ and make the correspondence of (τ, ξ) with the matrix

$$\begin{pmatrix} 1 & x_1 & x_3 + (x_1 x_2/2) \\ 0 & 1 & x_2 \\ 0 & 0 & 1 \end{pmatrix}$$

(see example (11)) we obtain an isomorphism between the group of transformations G and the group H of such matrices (under matrix multiplication). The latter group is an example of a *Heisenberg group*. This gives us a natural identification of the spaces ∂D, G, and H in each of which we can introduce ordinary Lebesgue measure; it is easy to check that the latter is a Haar measure for G and H.

The Cauchy-Szegö kernel associated with D has the form

$$k(Z, W) = k((z_1, z_2), (\omega_1, \omega_2)) = c(i(\bar{\omega}_1 - z_1) - 2z_2\bar{\omega}_2)^{-2}.$$

Using this kernel we obtain the projection operator of $L^2(\partial D)$ onto the subspace consisting of boundary values of holomorphic functions (the method is the same as that used in the study of B_n). Observe that if $Z = (z_1, z_2) = (t + i|z|^2, z) \in \partial D$ then $k(Z, 0) = c(|z|^2 - it)^{-2}$. We see, thus, that the "measure norm" $m(Z) = (t^2 + |z|^4)$ is particularly well adapted to the study of this kernel. Moreover, $m(Z)$ is homogeneous with respect to the dilations δ_λ, $\lambda > 0$, given by $\delta_\lambda Z = (\lambda^{1/2} t + i|\lambda^{1/4} z|^2, \lambda^{1/4} z)$. We obtain a space of homogeneous type whose point set is ∂D (or G, or H). The kernel $k(Z, W)$ satisfies inequality (2.14) and, thus, preserves the associated space H^1 (or H^p for p near but less than 1).

These H^p spaces are also useful for the study of partial differential operators commuting with the dilations δ_λ (these generalize homogeneous differential operators). An interesting example, studied by Folland and Stein **[31]**, is the operator

$$\mathcal{L}_\alpha = -\frac{\partial^2}{\partial \bar{z}\partial z} - |z|^2\frac{\partial^2}{\partial t^2} + i\frac{\partial}{\partial t}\left(z\frac{\partial}{\partial z} - \bar{z}\frac{\partial}{\partial \bar{z}}\right) + i\alpha\frac{\partial}{\partial t},$$

when α is not an odd integer. The fundamental solutions for $\mathcal{L}_\alpha$ turn out to be

$$c_\alpha\left(\frac{|z|^2 + it}{|z|^2 - it}\right)^{\alpha/2}\frac{1}{(|z|^4 + t^2)^{1/2}}.$$

Folland and Stein were able to use this, together with the theory of singular integrals, in order to obtain sharp estimates for solutions of the equation $\mathcal{L}_\alpha f = g$. Singular integrals, H^p spaces and their maximal characterizations for the Heisenberg group are also treated by D. Geller in his forthcoming thesis at Princeton University under the direction of E. M. Stein. The theory of Hardy spaces developed here, and its uses in the study of singular integrals, can also be used for this purpose and provides corresponding estimates in H^p and BMO.

The Heisenberg group H we have considered is an example of a class of Lie groups considered by Rothschild and Stein **[50]**: the *free nilpotent Lie groups of step r*. Briefly, these groups arise in the following important construction. Let M be a manifold and $X_1, \dots, X_n$ vector fields which, together with their commutators up to order r, span the tangent space at each point of M. Let us assume that these vector fields are *free* up to step r; that is, at each point of M the dimension of the space spanned by $\{X_1, \dots, X_n\}$ and their commutators up to order r is equal to the dimension of the free nilpotent Lie algebra of step r. As was shown by Rothschild and Stein, this is not a handicap in the study of the general situation. Then, there exists, at each point of M, a local coordinate system with respect to which the X_j's are (essentially) the generators of a free nilpotent Lie algebra. The homogeneous norm function on the Lie algebra can be used in order to define a quasi-distance on M that is well adapted for the study of differential operators involving the vector fields X_j. (In the above example, $M = \partial D$, the X_j's can be chosen as a real basis for the holomorphic vector fields on ∂D.) This program is carried out in general in **[50]** for generalizations of the operators $\mathcal{L}_\alpha$. A priori estimates are then obtained by studying singular integrals, first on the corresponding group and then on the space of homogeneous type M. It would be of interest to characterize the spaces $H^1(M)$ in terms of these singular integrals.

Perhaps the simplest example of a noncompact group is the group $\mathbf{Z}$ (or $\mathbf{Z}^n$) of integers. This group, with the usual distance and counting measure, is a space of homogeneous type. The spaces H^p are easily characterizable in terms of the discrete Hilbert transform:

$$\{f_k\} \in H^1 \text{ if and only if } \sum |f_k| < \infty \text{ and } \sum |\tilde{f}_k| < \infty,$$

where $\tilde{f}_k = \sum_{j \neq k} f_j/(k-j)$.

Various characterizations of the Fourier transform

$$\hat{f}(\theta) = \sum f_k e^{ik\theta}$$

can be given as before.

This example shows that the differentiable structure of the group is not strictly necessary in order to obtain a singular integral characterization of the space H^1. A striking illustration of this fact occurs in the p-adic case (see example (12)). There, it turns out that there exists a system of singular integral operators, analogous to the Riesz transforms, which characterize H^1. (See **[61]** where it is also shown that some of the subharmonicity inequalities of $\mathbf{R}^n$ have an appropriate analogue.)

3. Proofs of the principal results. In order to establish Theorem A we shall need to know some general facts about the geometry and analysis of spaces of homogeneous type X. Many of these facts are proved in **[17]**; when such a result is not contained in this monograph we shall give a proof of it here. Throughout this section A denotes the constant in the inequality (2.1) and K is the constant occurring in the "triangle inequality" satisfied by the quasi-distance d.

THEOREM (3.1) (VITALI-WIENER TYPE COVERING LEMMA). *Suppose E is a bounded set (i.e. contained in a sphere) in X such that for each $x \in E$ there exists $r(x) > 0$ (thus, $\{S_{r(x)}(x)\}$ is a covering of E), then there exists a sequence of points $x_j \in E$ such that $\{S_{r(x_j)}(x_j)\}$ is a disjoint family of spheres while $\{S_{4Kr(x_j)}(x_j)\}$ is a covering of E.*

(See Theorem (1.2) on p. 69 of **[17]**.)

THEOREM (3.2) (WHITNEY TYPE COVERING LEMMA). *Suppose $U \subsetneqq X$ is an open bounded set and $C \geqslant 1$. Then there exists a sequence of spheres $\{S_j\} = \{S_{s_j}(x_j)\}$ satisfying*

(i) $U = \bigcup_j S_j$;

(ii) *there exists a constant $M = M(A, C, K)$ such that no point of X belongs to more than M of the spheres $\bar{S}_j \equiv S_{Cs_j}(x_j)$;*

(iii) $\bar{\bar{S}}_j \cap (X - U) \neq \varnothing$ *for each j, where $\bar{\bar{S}}_j = S_{3KCs_j}(x_j)$.*

PROOF. For each x, $r(x) = (1/8CK^2)d(x, U') = (1/8CK^2)\min\{d(x,y): y \notin U\} > 0$ since U is open and $X - U = U' \neq \varnothing$. By (3.1) we can find a mutally disjoint sequence of spheres $\{S_{r(x_j)}(x_j)\}$ such that U is covered by the spheres $S_{4Kr(x_j)}(x_j)$. On the other hand, if $y \in S_{4KCr(x_j)}(x_j)$ then

$$d(x_j,y) < 4KCr(x_j) = (1/2K)d(x_j, U') < d(x_j, U').$$

It follows that if we put $s_j = 4Kr(x_j)$ we have $S_{s_j}(x_j) = S_j \subset S_{Cs_j}(x_j) = \bar{S}_j \subset U$. Thus,

$$U = \bigcup S_j \subseteqq \bigcup \bar{S}_j \subset U. \tag{3.3}$$

In particular, (i) is established. Moreover, since $3KCs_j = 12K^2Cr(x_j) = (3/2)\rho(x_j, U')$ we must have $\bar{\bar{S}}_j \cap U' \neq \varnothing$ and (iii) is established.

In order to show that (ii) is valid we shall make use of the following geometric property of spaces of homogeneous type (see Lemma (1.1) on p. 68 of **[17]**):

(3.4) *There exists a constant* $N = N(A,K)$ *such that* $S_r(x)$ *cannot contain more than* N^n *points* $\{x_i\}$ *satisfying* $d(x_i, x_j) > r2^{-n}$ *when* $i \neq j$ $(n = 1, 2, 3, \ldots)$.

Now suppose $x \in \bar{S}_j$. We claim that $Cs_j < d(x, U') \equiv R$ since

$$2CKs_j = d(x_j, U') \leqslant K[d(x_j, x) + d(x, U')] < K[Cs_j + d(x, U')].$$

Consequently, if $y \in \bar{S}_j$, $d(x,y) \leqslant K[d(x,x_j) + d(x_j,y)] < 2KR$ and we must have $\bar{S}_j \subset S_{2KR}(x)$. Moreover,

$$R = d(x, U') \leqslant K[d(x,x_j) + d(x_j, U')] < K[Cs_j + 2CKs_j]$$

and it follows that $s_j \geqslant R/(1+2K)CK$.

If x also belong to $\bar{S}_i$, $i \neq j$, the same argument shows $\bar{S}_i \subset S_{2KR}(x)$. Thus, all such points $x_i, x_j, \ldots$ belong to $S_{2KR}(x)$. But $S_{r(x_i)}(x_i) \cap S_{r(x_j)}(x_j) = \varnothing$ when $i \neq j$ and this implies

$$d(x_i, x_j) \geqslant \min\{r(x_i), r(x_j)\} = \frac{1}{4K} \min\{s_i, s_j\} \geqslant \frac{R}{4CK^2(1+2K)}.$$

It now follows from (3.4) that there cannot be more than $M \leqslant NN^{\log_2 4CK^2(1+2K)}$ spheres $\bar{S}_j$ containing x and (ii) is established.

Property (ii) will be referred to as the *M-disjointness* of the collection $\{\bar{S}_j\}$.

We shall make repeated use of the following extension to spaces of homogeneous type of the *Hardy-Littlewood maximal function*

$$(\mathfrak{M}f)(x) = \sup\left\{\frac{1}{\mu(S)}\int_S |f|\,d\mu\text{: } S \text{ is a sphere}, x \in S\right\}.$$

One can show, by using (3.1), that $\mathfrak{M}f$ is defined and finite a.e. for $f \in L^1(X)$; moreover, we have the weak-type inequality:

THEOREM (3.5). *If* $f \in L^1(X)$ *there exists a constant* $C_0 = C_0(A,K)$ *such that*

$$\mu(\{x \in X\text{: } (\mathfrak{M}f)(x) > \alpha\}) \leqslant (C_0/\alpha)\|f\|_1.$$

(See Theorem (2.1) on p. 71 of [17].)

From this result and the obvious boundedness of this maximal operator on L^∞ one establishes the existence of a constant A_p such that

(3.6) $$\|\mathfrak{M}f\|_p \leqslant A\|f\|_p$$

when $p > 1$ (one can prove this directly as is done in the second chapter of [57] or one can appeal to the Marcinkiewicz interpolation theorem).

A useful variant of the Hardy-Littlewood maximal operator is $\mathfrak{M}_{p_0}f = (\mathfrak{M}|f|^{p_0})^{1/p_0}$. From (3.5) and (3.6) we obtain the inequalities

(3.7) $$\mu(\{x\text{: } (\mathfrak{M}_{p_0}f)(x) > \alpha\}) \leqslant C_0(\|f\|_{p_0}/\alpha)^{p_0}$$

and, for $p > p_0$,

(3.8) $$\|\mathfrak{M}_{p_0}f\|_p \leqslant A_{(p/p_0)}^{1/p_0}\|f\|_p.$$

It is often useful to use the *centered maximal function* of f:

$$(\mathfrak{M}_c f)(x) = \sup_{r>0} \frac{1}{\mu(S_r(x))} \int_{S_r(x)} |f|\, d\mu.$$

The operators $\mathfrak{M}$ and $\mathfrak{M}_c$ are "equivalent" in the sense that

$$(\mathfrak{M}_c f)(x) \leqslant (\mathfrak{M} f)(x) \leqslant D(\mathfrak{M}_c f)(x), \tag{3.9}$$

where $D = D(A, K)$. This follows from the simple observation that if $x \in S_r(y)$ then $S_r(y) \subset S_{2Kr}(x) \subset S_{K(1+2K)r}(y)$. Thus,

$$\frac{1}{\mu(S_r(y))} \int_{S_r(y)} |f|\, d\mu \leqslant \left\{ \frac{\mu(S_{K(1+2K)r}(y))}{\mu(S_r(y))} \right\} \frac{1}{\mu(S_{2Kr}(x))} \int_{S_{2Kr}(x)} |f|\, d\mu$$

and we obtain (3.9) by making use of inequality (2.1).

The centered maximal function is useful, for example, in the proof the following geometric result:

LEMMA (3.9). *If $f \in L^1(X)$ has support in $S_0 = S_{r_0}(x_0)$ then there exists $C_1 = C_1(A, K)$ such that*

$$U^\alpha = \{x \in X: (\mathfrak{M} f)(x) > \alpha\} \subset S_{2Kr_0}(x_0)$$

whenever $\alpha > C_1 m_{S_0}(|f|) = C_1 (1/\mu(S_0)) \int_{S_0} |f|\, d\mu$.

PROOF. We first observe that if $d(x, x_0) \geqslant 2Kr_0$ and $r \leqslant r_0$ then $S_r(x) \cap S_0 = \varnothing$. To see this, suppose $y \in S_r(x) \cap S_0$. Then

$$2Kr_0 \leqslant d(x, x_0) \leqslant K[d(x,y) + d(y, x_0)] < K[r + r_0] \leqslant 2K_0$$

which shows that such a y cannot exist. On the other hand, if $r = d(x, x_0) \geqslant r_0$ then $S_0 \subset S_{2Kr}(x)$. This is also immediate: if $y \in S_0$

$$d(x,y) \leqslant K[d(x, x_0) + d(x_0, y)] < K[r + r_0] \leqslant 2Kr.$$

From the first observation we see that if $d(x, x_0) \geqslant 2Kr_0$ then $\int_{S_r(x)} |f|\, d\mu = 0$ if $r \leqslant r_0$. Thus, the centered maximal function of f satisfies

$$(\mathfrak{M}_c f)(x) = \sup_{r \geqslant r_0} \frac{1}{\mu(S_r(x))} \int_{S_r(x)} |f|\, d\mu.$$

Since f has support in S_0 we have

$$\frac{1}{\mu(S_r(x))} \int_{S_r(x)} |f|\, d\mu \leqslant \left\{ \frac{\mu(S_0)}{\mu(S_r(x))} \right\} \frac{1}{\mu(S_0)} \int_{S_0} |f|\, d\mu.$$

If $r \geqslant r_0$, the second observation implies (we also use (2.1))

$$\frac{\mu(S_0)}{\mu(S_r(x))} \leqslant \frac{\mu(S_{2Kr}(x))}{\mu(S_r(x))} \leqslant A^{2+\log_2 K} = C.$$

Thus, if $r \geqslant r_0$, $m_{S_r(x)}(|f|) \leqslant C m_{S_0}(|f|)$ and we see that if $\beta > C m_{S_0}(|f|)$

then $\{x: (\mathfrak{M}_c f)(x) > \beta\} \subset S_{2Kr_0}(x)$. By (3.9) it now follows that if $\alpha > DCm_{S_0}(|f|) \equiv C_1 m_{S_0}(|f|)$ then $U^\alpha \subset S_{2Kr_0}(x)$ and the lemma is proved.

We are now ready to prove Theorem A. We have already observed that $H^{p,q} \supset H^{p,\infty}$ for $0 < p \leqslant 1 \leqslant q, p < q$; thus, we need to establish the opposite inclusion relation. We shall do so by showing that a (p, q)-atom $a(x)$ has the representation

$$a = \sum_j \alpha_j a_j \tag{3.10}$$

where each a_j is a (p, ∞)-atom and $\sum |\alpha_j|^p \leqslant B$, B independent of a. More explicitly, we shall show that the linear functionals a and $\sum_j \alpha_j a_j$, acting on $\mathfrak{L}_{(1/p)-1}$, are equal when $p < 1$; when $p = 1$ (3.10) is an equality in $L^1(X)$. We shall give an inductive proof. In order to state the induction hypothesis we need to introduce some notation:

For each positive integer n let N^n denote the n-fold cartesian product of the natural numbers N, $N^0 \equiv \{0\}$. We write j_n to represent the general element of N^n, the support of a is contained in the sphere S_0 (the three defining properties of a (p, q)-atom are satisfied with respect to S_0), and we set $b = [\mu(S_0)]^{1/p} a$. We shall apply Theorem (3.2) with the constant $C = 2K$; thus, the M occurring in part (ii) of (3.2) depends only on A and K. From (2.1) we can, therefore, deduce that the ratio $\mu(\overline{\overline{S}}_j)/\mu(S_j)$ does not exceed some fixed constant $k_0 = k_0(A, K)$. Given a sphere S of radius s, $\overline{S}$ and $\overline{\overline{S}}$ will denote spheres with the same center as S having radii $2Ks$ and $6K^2 s$ respectively.

The induction hypothesis we shall establish is the following one:

There exists a collection of spheres $\{S_{j_l}\}$, $j_l \in N^l$ for $l = 0, 1, \ldots,$ such that for each natural number n

$$b = Mk_0 \alpha \sum_{l=0}^{n-1} \alpha^l \sum_{j_l \in N^l} [\mu(\overline{\overline{S}}_{j_l})]^{1/p} a_{j_l} + \sum_{j_n \in N^n} h_{j_n}, \tag{3.11}$$

where $\alpha = \alpha(p, q, A, K)$ is sufficiently large (we shall be explicit later) and

(I) *a_{j_l} is a (p, ∞)-atom supported in $\overline{\overline{S}}_{j_l}$, $l = 0, 1, \ldots, n-1$;*
(II) $\bigcup_{j_n \in N^n} \overline{S}_{j_n} \subset \{x: (\mathfrak{M}_q b)(x) > \alpha^n/2\}$;
(III) *$\{\overline{S}_{j_l}\}$ is an M^l-disjoint collection;*
(IV) *the functions h_{j_n} are supported in $\overline{\overline{S}}_{j_n}$;*
(V) $\int h_{j_n} d\mu = 0$;
(VI) $|h_{j_n}(x)| \leqslant |b(x)| + 2k_0^{1/q} \alpha^n \chi_{\overline{\overline{S}}_{j_n}}(x)$;
(VII) $[m_{\overline{\overline{S}}_{j_n}}(|h_{j_n}|^q)]^{1/q} \leqslant 2k_0^{1/q} \alpha^n$.

Let us first show that if these properties are valid for each $n \in N$ then (3.10) holds. We begin by proving

$$\frac{(Mk_0 \alpha)^p}{\mu(S_0)} \sum_{n=0}^{\infty} \alpha^{np} \sum_{j_n \in N^n} \mu(\overline{\overline{S}}_{j_n}) \leqslant B, \tag{3.12}$$

where B is independent of a. From (II) and (III) we have

$$\sum_{j_n \in N^n} \mu(\overline{\overline{S}}_{j_n}) \leqslant k_0 \sum_{j_n \in N^n} \mu(\overline{S}_{j_n}) \leqslant k_0 M^n \mu\Big(\bigcup_{j \in N^n} \overline{S}_{j_n}\Big)$$
$$\leqslant k_0 M^n \mu(\{x: (\mathfrak{M}_q b)(x) > \alpha^n/2\}).$$

From (3.7) we thus deduce that

$$\sum_{j_n \in N^n} \mu(\bar{S}_{j_n}) \leqslant C_0 k_0 M^n \left(\frac{2}{\alpha^n}\right)^q \|b\|_q^q.$$

Consequently,

$$\sum_{n=0}^{\infty} \alpha^{np} \sum_{j_n \in N^n} \mu(\bar{S}_{j_n}) \leqslant C_0 k_0 2^q \sum_{n=0}^{\infty} (M\alpha^{p-q})^n \|b\|_q^q.$$

Since $\|b\|_q^q \leqslant \mu(S_0)$ we obtain (3.12) with B depending on p, q, A, K and α (as long as $\alpha^{p-q} M < 1$) but not on a.

From (VII) we have

$$\int |h_{j_n}|\, d\mu \leqslant \mu(S_{j_n}) 2k_0^{1/q} \alpha^n = C\alpha^n \mu(S_{j_n});$$

thus, if we let $H_n = \sum_{j_n \in N^n} h_{j_n}$, using the M^n-disjointness of $\{S_{j_n}\}$ and the above estimate for $\sum \mu(S_{j_n})$ we see that

$$\int |H_n|\, d\mu \leqslant \sum_{j_n \in N^n} \int |h_{j_n}|\, d\mu \leqslant C\alpha^n \sum_{j_n \in N^n} \mu(S_{j_n}) \leqslant C(M\alpha^{1-q})^n \|b\|_q^q.$$

This shows that if $p = 1$ (and, thus, $q > 1$) the first series on the right-hand side of (3.11) converges to b in $L^1(X)$. Equality (3.11) is to be interpreted in the following way when $p < 1$: If $l \in \mathfrak{L}_\beta$, $\beta = 1/p - 1$, then

$$\int bl\, d\mu = Mk_0 \alpha \sum_{k=0}^{n-1} \alpha^k \sum_{j_k \in N^k} [\mu(\bar{S}_{j_k})]^{1/p} \int a_{j_k} l\, d\mu + \sum_{j_n \in N^n} h_{j_n} l\, d\mu.$$

Inequality (3.12) assures us that the double sum on the right is well defined. The fact that $\|H_n\|_1 \leqslant \sum \|h_{j_n}\|_1 < \infty$ and H_n has bounded support (this follows from (II), (IV) and (3.9) if $\alpha^q > C_1$) imply that the last sum equals $\int H_n l\, d\mu$ and is well defined. Theorem A is then proved if we show that $\lim_{n\to\infty} \int H_n l\, d\mu = 0$. To see that this is true let x_{j_n} be the center of S_{j_n} and suppose $\|l\|^{(\beta)} = 1$. Then, using (V) and (VII) and the above estimate for $\sum \mu(S_{j_n})$

$$\left|\int H_n l\, d\mu\right| = \sum_{j_n \in N^n} \left|\int_{S_{j_n}} h_{j_n} l\, d\mu\right| \leqslant \sum \left|\int_{S_{j_n}} h_{j_n}(l - l(x_{j_n}))\, d\mu\right|$$

$$\leqslant \sum [\mu(S_{j_n})]^{-1+1/p} \int_{S_{j_n}} |h_{j_n}|\, d\mu \leqslant C\alpha^n \sum [\mu(S_{j_n})]^{1/p}$$

$$\leqslant C\alpha^n [(M\alpha^{-q})^n \|b\|_q^q]^{1/p} = C\|b\|_q^{q/p} (M\alpha^{p-q})^{n/p}.$$

Since $p < q$ the last term tends to 0 and the theorem is proved if the induction hypothesis is established.

Let us show that the hypothesis is valid for $n = 1$. Let $U^\alpha = \{x \in X: (\mathfrak{M}_q b)(x) > \alpha\}$. By (3.9) $U^\alpha \subset \bar{S}_0$ provided $\alpha^q > C_1$. Assuming this condition on α, we see that U^α is a bounded open set. By (3.7) we have

$$\mu(U^\alpha) \leqslant C_0(\|b\|_q/\alpha)^q \leqslant C_0 \alpha^{-q} \mu(S_0).$$

Hence, if $\alpha^q > C_0$ (which we assume to be the case) $\mu(U^\alpha) < \mu(S_0) \leqslant \mu(X)$ and $X - U^\alpha$ cannot be empty. We can, therefore, apply Theorem (3.2) and obtain a Whitney-type covering of U^α by spheres $\{S_j\}$ satisfying conditions (i), (ii) and (iii) with $C = 2K$. Letting χ_j be the characteristic function of S_j we define

$$\eta_j(x) = \begin{cases} \dfrac{\chi_j(x)}{\Sigma\chi_i(x)} & \text{if } x \in U^\alpha, \\ 0 & \text{if } x \notin U^\alpha \end{cases}$$

(observe that the M-disjointness of $\{S_j\}$ implies that $\Sigma\chi_i(x)$ is a finite sum for each $x \in U^\alpha$). We then put

$$g_0(x) = \begin{cases} b(x) & \text{if } x \notin U^\alpha, \\ \sum m_{S_j}(\eta_j b)\chi_j(x) & \text{if } x \in U^\alpha \end{cases}$$

and $h_j(x) = b(x)\eta_j(x) - m_{S_j}(\eta_j b)\chi_j(x)$ for all $x \in X$. It follows that

$$b(x) = g_0(x) + \sum_{j \in N} h_j(x) \tag{3.13}$$

and we shall show that, by an appropriate normalization, this is the representation (3.11) of b when $n = 1$. This equality is a version of the Calderón-Zygmund decomposition of the function b into the "good" function g_0 and "bad" function $H = \Sigma h_j$. The induction proof we are presenting will show how we can keep refining this decomposition by applying it repeatedly to the bad function.

If $x \notin U^\alpha$ then $|g_0(x)| = |b(x)| \leqslant (\mathfrak{M}_q b)(x) \leqslant \alpha$;[30] while if $x \in U^\alpha$

$$|g_0(x)| \leqslant \sum_{\substack{\text{at most} \\ M \text{ terms}}} \frac{1}{\mu(S_j)} \int_{S_j} |b\eta_j|\, d\mu \leqslant \sum_{\text{ibid}} \frac{\mu(\bar{S}_j)}{\mu(S_j)} \left[\frac{1}{\mu(\bar{S}_j)} \int_{\bar{S}_j} |b|^q\, d\mu\right]^{1/q} \leqslant \sum_{\text{ibid}} k_0 \alpha$$
$$\leqslant (Mk_0)\alpha.$$

This shows that

(1) $|g_0(x)| \leqslant (Mk_0)\alpha$.

We have already observed that $U^\alpha \subset \bar{S}_0$; for $x \notin U^\alpha$, $g_0(x) = b(x)$. These facts, together with the inclusions Supp $b \subset S_0 \subset \bar{S}_0$ imply

(2) *the support of* g_0 *is within* $\bar{S}_0$.

We also clearly have:

(3) *the support of* h_j *is within* S_j;

(4) $\int h_j\, d\mu = 0$.

Since $\|h_j\|_1 \leqslant 2\|b\chi_j\|_1 = 2\int_{S_j} |b|\, d\mu$ and the spheres S_j are M-disjoint we must have

$$\sum_j \|h_j\|_1 \leqslant 2\sum_j \int_{S_j} |b|\, d\mu \leqslant 2M \int_{U^\alpha} |b|\, d\mu \leqslant 2M\|b\|_1$$
$$\leqslant 2M\|b\|_q[\mu(S_0)]^{1/q'} \leqslant 2M\mu(S_0).$$

[30] That $|b(x)| \leqslant (\mathfrak{M}_q b)(x)$ holds for almost every x is true if we assume that μ is a regular measure. This assumption is tacitly made throughout this paper.

In particular, this shows that the sum in (3.13) is convergent in the L^1-norm; from (4) and $\int b\,d\mu = 0$ we thus can conclude that

(5) $\int g_0\,d\mu = 0$.

It follows that $a_0 \equiv g_0/Mk_0\alpha[\mu(\bar{S}_0)]^{1/p}$ is a (p, ∞)-atom. Thus, we can write (3.13) in the form

$$b = Mk_0\alpha[\mu(\bar{S}_0)]^{1/p}a_0 + \sum_{j\in N} h_j$$

which is the special case $n = 1$ of (3.11). Moreover, property (I) has just been established. (3) and (4) are parts (IV) and (V) of the induction hypothesis when $n = 1$. Since

$$\bigcup_j S_j = U^\alpha = \{x: (\mathfrak{M}_q b)(x) > \alpha\} \subset \{x: (\mathfrak{M}_q b)(x) > \alpha/2\}$$

property (II) is also satisfied. Property (III) is a consequence of the Whitney-type covering we are using (property (ii) of (3.2)). Since $\eta_j(x) \leqslant 1$ we have

$$|h_j(x)| \leqslant |b(x)| + |m_{S_j}(\eta_j b)|\chi_j(x) \leqslant |b(x)| + k_0^{1/q}\alpha\chi_j(x)$$

and, thus, property (VI) holds. Similarly, we obtain (VII):

$$\left(\frac{1}{\mu(S_j)}\int_{S_j}|h_j|^q\,d\mu\right)^{1/q} \leqslant \left(\frac{1}{\mu(S_j)}\int_{S_j}|b|^q\,d\mu\right)^{1/q} + \alpha k_0^{1/q}$$

$$\leqslant \left(\frac{\mu(\bar{S}_j)}{\mu(S_j)}\frac{1}{\mu(\bar{S}_j)}\int_{\bar{S}_j}|b|^q\,d\mu\right)^{1/q} + \alpha k_0^{1/q} \leqslant 2\alpha k_0^{1/q}.$$

This shows that the induction hypothesis holds for $n = 1$. We now assume it true for n and will show that this implies its validity for $n + 1$.

Let $U_{j_n} = \{x \in X: (\mathfrak{M}_q h_{j_n})(x) > \alpha^{n+1}\}$. Hypothesis (IV) tells us that the support of h_{j_n} is within S_{j_n}. It follows from (VII), provided $\alpha^q > 2^q k_0 C_1$, that

$$C_1 m_{S_{j_n}}(|h_{j_n}|^q) \leqslant C_1 k_0(2\alpha^n)^q < \alpha^{q(n+1)}.$$

From Lemma (3.9), therefore, we obtain

$$U_{j_n} = \{x \in X: (\mathfrak{M}|h_{j_n}|^q)(x) > \alpha^{q(n+1)}\} \subset \bar{S}_{j_n}.$$

Let $\{S_{j_n,i}\}$ be a Whitney covering for U_{j_n} with constant $C = 2K$ (see Theorem (3.2)). From (3.3) and (3.2)(ii) we have $\bigcup_i \bar{S}_{j_n,i} = U_{j_n} \subset \bar{S}_{j_n}$ and $\{\bar{S}_{j_n,i}\}$ is M-disjoint. Since, from (III), we know that $\{\bar{S}_{j_n}\}$ are M^n-disjoint it follows that the totality of spheres in the family $\{\bar{S}_{j_n,i}\}$ are M^{n+1} disjoint. This establishes hypothesis (III) for $n + 1$.

We now put

$$g_{j_n}(x) = \begin{cases} h_{j_n}(x) & \text{if } x \notin U_{j_n}, \\ \sum_i m_{S_{j_n,i}}(h_{j_n}\eta_{j_n}^i)\chi_{S_{j_n,i}}(x) & \text{if } x \in U_{j_n}, \end{cases}$$

and

$$h_{j_n,i} = h_{j_n}\eta^i_{j_n} - m_{S_{j_n,i}}(h_{j_n}\eta^i_{j_n})\chi_{S_{j_n,i}}(x),$$

where $\eta^i_{j_n}(x) = \chi_{S_{j_n,i}}(x)/\sum_l \chi_{S_{j_n,l}}(x)$ for $x \in U_{j_n}$ and is 0 if $x \notin U_{j_n}$. If $x \in U_{j_n}$, therefore,

$$|g_{j_n}(x)| \leqslant \sum_{\substack{\text{at most} \\ M \text{ terms}}} |m_{S_{j_n,i}}(h_{j_n}\eta^i_{j_n})\chi^i_{j_n}(x)| \leqslant k_0 M\alpha^{n+1}$$

while, if $x \in U_{j_n}$ then $|g_{j_n}(x)| = |h_{j_n}(x)| \leqslant (\mathfrak{M}_q h_{j_n})(x) \leqslant \alpha^{n+1}$. In any case $\|g_{j_n}\|_\infty \leqslant k_0 M\alpha^{n+1}$. Since the support of h_{j_n} is within $S_{j_n} \subset \bar{S}_{j_n}$ and $U_{j_n} \subset \bar{S}_{j_n}$ it follows that the support of g_{j_n} is within $\bar{S}_{j_n}$. Moreover, $\int h_{j_n,i}\,d\mu = 0$ (which shows that property (V) is valid for $n+1$). In the same way we showed that $\sum \|h_j\|_1 \leqslant 2M\|b\|_1$ when the case $n = 1$ was considered, we can show $\sum_i \|h_{j_n,i}\|_1 \leqslant 2M\|h_{j_n}\|_1$. It follows, therefore, that

$$h_{j_n} = g_{j_n} + \sum_i h_{j_n,i}$$

is valid in $L^1(X)$ (it is also true a.e. since, for each x, the sum on the right has at most M terms) and $\int g_{j_n}\,d\mu = 0$. Hence

$$a_{j_n} = g_{j_n}/[k_0 M\alpha^{n+1}(\mu(\bar{S}_{j_n}))^{1/p}]$$

is a (p, ∞)-atom supported in the sphere $\bar{S}_{j_n}$. From this it follows that the representation (3.11) is true for $n+1$ and so is (I). Property (IV) is trivially true. We have, from the definition of $h_{j_n,i}$ and (VI),

$$\begin{aligned} |h_{j_n,i}(x)| &\leqslant \left\{|h_{j_n}(x)| + \left[k_0\frac{1}{\mu(\bar{S}_{j_n,i})}\int_{\bar{S}_{j_n,i}} |h_{j_n}|^q\,d\mu\right]^{1/q}\right\}\chi^i_{j_n}(x) \\ &\leqslant \{|b(x)| + 2k_0^{1/q}\alpha^n + k_0^{1/q}\alpha^{n+1}\}\chi^i_{j_n}(x) \\ &\leqslant \{|b(x)| + 2k_0^{1/q}\alpha^{n+1}\}\chi^i_{j_n}(x) \end{aligned}$$

as long as $\alpha > 2$. This establishes (VI) for $n+1$. Property (VII) has a similar proof: from the definition of $h_{j_n,i}$

$$\begin{aligned} [m_{S_{j_n,i}}(|h_{j_n,i}|^q)]^{1/q} &\leqslant 2[m_{S_{j_n,i}}(|h_{j_n}|^q)]^{1/q} \\ &\leqslant 2[k_0 m_{\bar{S}_{j_n,i}}(|h_{j_n}|^q)]^{1/q} \leqslant 2k_0^{1/q}\alpha^{n+1}. \end{aligned}$$

Finally, we establish (II): From (VI) we have $(\mathfrak{M}_q h_{j_n})(x) \leqslant (\mathfrak{M}_q b)(x) + 2k_0^{1/q}\alpha^n$. Thus, if $x \in U_{j_n}$

$$\alpha^{n+1} < (\mathfrak{M}_q h_{j_n})(x) \leqslant (\mathfrak{M}_q b)(x) + 2k_0^{1/q}\alpha^n.$$

It follows that $\alpha^{n+1}/2 < (\mathfrak{M}_q b)(x)$ if $\alpha > 4k_0^{1/q}$ and, thus,

$$\bigcup_{j_n,i} S_{j_n,i} = \bigcup_{j_n}\Big(\bigcup_i S_{j_n,i}\Big) \subset \bigcup_{j_n} U_{j_n} \subset \Big\{x \in X: (\mathfrak{M}_q b)(x) > \frac{\alpha^{n+1}}{2}\Big\}$$

and (II) is valid for $n + 1$.

Observe that we used the fact that α exceeds each of the following numbers: $4k_0^{1/q}$, 2, $2(k_0 C_1)^{1/q}$, $C_0^{1/q}$, $M^{1/(q-p)}$, $M^{1/(q-1)}$. Each of these numbers depends on p, q, A and K and are independent of n. Theorem A is, therefore, proved.

Let us now pass to the study of the dual space of $H^p(X)$. We begin with the case $p = 1$ and, toward this end, we first make the following observation:

LEMMA (3.14). *If $b \in BMO_q$, $1 \leqslant q < \infty$, then $|b| \in BMO_q$.*

In order to see this we note that if a function b satisfies: for each sphere S there is a constant b_S such that

$$\left\{\frac{1}{\mu(S)}\int_S |b(x) - b_S|^q \, d\mu\right\}^{1/q} \leqslant C,$$

where C is independent of S then $b \in BMO_q$ and $\|b\|_{BMO}^{(q)}$ does not exceed $2C$, if $\mu(X) = \infty$, or $2C + |\int_X b\, d\mu|$ if $\mu(X) = 1$. This follows from

$$\begin{aligned}\left\{\frac{1}{\mu(S)}\int_S |b - m_S(b)|^q \, d\mu\right\}^{1/q} &\leqslant \left\{\frac{1}{\mu(S)}\int_S |b - b_S|^q \, d\mu\right\}^{1/q} \\ &\quad + \left\{\frac{1}{\mu(S)}\int_S |b_S - m_S(b)|^q \, d\mu\right\}^{1/q} \\ &\leqslant C + \left|\int_S (b_S - b(x))\frac{d\mu(x)}{\mu(S)}\right| \\ &\leqslant C + \left\{\int_S |b_S - b(x)|^q \frac{d\mu(x)}{\mu(S)}\right\}^{1/q} \leqslant 2C.\end{aligned}$$

Lemma (3.14) now follows from this and the inequality

$$\big|\,|b(x)| - |m_S(b)|\,\big| \leqslant |b(x) - m_S(b)|.$$

Since the infimum (supremum) of two numbers x and y is

$$(x + y - |x - y|)/2((x + y + |x - y|)/2)$$

it follows immediately from (3.14) that BMO_q is a lattice. In particular, the function

$$b_N(x) = \begin{cases} N & \text{if } b(x) \geqslant N, \\ b(x) & \text{if } -N < b(x) < N, \\ -N & \text{if } b(x) \leqslant -N, \end{cases}$$

belongs to BMO_q if b does. Recall that $\|b\|_{BMO}^{(q)}$ equals $\mathfrak{N}_q(b)$, if $\mu(X) = \infty$, or $\mathfrak{N}_q(b) + |\int b\, d\mu|$, if $\mu(X) = 1$. It follows from the above considerations that

$$\|b_N\|_{BMO}^{(q)} \leqslant \begin{cases} 3\mathfrak{N}_q(b) & \text{if } \mu(X) = \infty, \\ 3\mathfrak{N}_q(b) + \left|\int b_N\,d\mu\right| & \text{if } \mu(X) = 1. \end{cases}$$

Thus, in either case, we have (from the dominated convergence theorem)

$$\|b_N\|_{BMO}^{(q)} \leqslant 4\|b\|_{BMO}^{(q)} \tag{3.15}$$

provided N is large enough.

Let q' be the conjugate index to q: $(1/q') + (1/q) = 1$. We assume $1 \leqslant q < \infty$ and, thus, $1 < q' \leqslant \infty$. Suppose $f = \sum \alpha_j a_j$ belongs to $H^{1,q'}$, where a_j is a $(1, q')$-atom supported in the sphere S_j, $\sum |\alpha_j| < \infty$ and b is a bounded function on X. Then, assuming $\int a_j\,d\mu = 0$ for all j (which is always true if $\mu(X) = \infty$).

$$\begin{aligned} \left|\int fb\,d\mu\right| &\leqslant \sum |\alpha_j| \left|\int a_j b\,d\mu\right| = \sum |\alpha_j| \left|\int a_j(b - m_{S_j}(b))\,d\mu\right| \\ &\leqslant \sum |\alpha_j| \left(\int |a_j|^{q'}\,d\mu\right)^{1/q'} \left(\int_{S_j} |b(x) - m_{S_j}(b)|^q\,d\mu(x)\right)^{1/q} \\ &\leqslant \sum |\alpha_j| \left(\frac{1}{\mu(S_j)}\int_{S_j} |b(x) - m_{S_j}(b)|^q\,d\mu(x)\right)^{1/q} \leqslant \left(\sum |\alpha_j|\right)\mathfrak{N}_q(b). \end{aligned}$$

From this we can easily deduce $|\int fb\,d\mu| \leqslant \|f\|_{(1,q')}\|b\|_{BMO}^{(q)}$ whenever $f \in H^{1,q'}$ and $b \in L^\infty$. From this last inequality and (3.15), therefore, we obtain

$$\left|\int fb_N\,d\mu\right| \leqslant 4\|f\|_{(1,q')}\|b\|_{BMO}^{(q)} \tag{3.16}$$

whenever $f \in H^{1,q'}$, $b \in BMO_{q'}$ and N is sufficiently large.

Let $L_0^{q'} = \{f \in L^{q'}(X)$: Support of f is bounded$\}$ and put $\mathfrak{D} = H^{1,q'} \cap L_0^{q'}$. Then, the linear functional L_b mapping f into $\int fb\,d\mu$ is well defined on $\mathfrak{D}$ whenever $b \in BMO_q$ (since b is locally in L^q). By the dominated convergence theorem,

$$\lim_{N\to\infty} \int fb_N\,d\mu = \int fb\,d\mu$$

for $f \in \mathfrak{D}$. This equality and (3.16) give us $|L_b f| \leqslant 4\|f\|_{(1,q')}\|b\|_{BMO}^{(q)}$ for all $f \in \mathfrak{D}$ and $b \in BMO_q$. But $\mathfrak{D}$ is dense in $H^{(1,q)}$ (a partial sum $\sum_1^n \alpha_j a_j$ of an atomic representation $f = \sum \alpha_j a_j$ with $\sum |\alpha_j| \leqslant (1 + \varepsilon)\|f\|_{1,q'}$ can be used to approximate an $f \in H^{1,q'}$ by elements of $\mathfrak{D}$). Thus, L_b has a unique bounded extension on $H^{1,q'}$ which we also denote by L_b. In this sense we have

$$BMO_q \subset (H^{1,q'})^*$$

and the linear functional norm $\|L_b\|$, $b \in BMO_q$, satisfies

$$\|L_b\| \leqslant 4\|b\|_{BMO}^{(q)}. \tag{3.17}$$

Now suppose L is a bounded linear functional on $H^{1,q'}$. Let $\|L\|$ be the norm of L. If S is a sphere in X let $L_0^{q'}(S) = \{f \in L^{q'}(S): \int_S f\,d\mu = 0\}$. It follows that, if $f \in L_0^{q'}(S)$, then

$$a(x) = f(x)[\mu(S)]^{-1/q}\|f\|_{L^{q'}(S)}^{-1}$$

is a $(1,q')$-atom supported in S. From this we also have $\|f\|_{1,q'} \leqslant [\mu(S)]^{1/q}\|f\|_{L^{q'}(S)}$. Hence, Lf is defined and

$$|Lf| \leqslant \|L\|[\mu(S)]^{1/q}\|f\|_{L^{q'}(S)};$$

that is, L is a bounded linear functional on $L_0^{q'}(S)$. Consequently, using the Hahn-Banach theorem and the Riesz representation theorem we conclude that there exists $l \in L^q(S)$ such that

$$Lf = \int_S fl\,d\mu \tag{3.18}$$

for all $f \in L_0^{q'}(S)$.[31] The function l is uniquely determined up to a constant; or, equivalently, if $\int_S fl\,d\mu = 0$ for all $f \in L_0^{q'}(S)$ it follows that l is constant. To see this choose h in $L^{q'}(S)$; since $h - m_S(h) \in L_0^{q'}(S)$ we have

$$0 = \int_S l(h - m_S(h))\,d\mu = \int_S h(l - m_S(l))\,d\mu.$$

Since this equality holds for all $h \in L^{q'}(S)$ it follows that $l(x) = m_S(l)$ for a.e. $x \in S$.

Let $\{S_j\}_{j=1}^{\infty}$ be an increasing sequence of spheres converging to X. We obtain a function $l_j = l$ satisfying (3.18) for each S_j and, from the above argument, we see that the condition $\int_{S_1} l_j\,d\mu = 0$ gives us a function l for which (3.18) holds for any sphere S. In particular, if a is a $(1,q')$-atom supported in S we have

$$\|L\| \geqslant |La| = \left|\int_S la\,d\mu\right| = \left|\int_S [l - m_S(l)]a\,d\mu\right|.$$

If f is supported in S and has $L^{q'}$ norm 1 then $a = 2^{-1}[\mu(S)]^{-1/q}(f - m_S(f))$ is a $(1,q')$-atom. Using this atom in the last expression and using the fact that $l - m_S(l)$ has mean 0 on S we obtain

$$\left|\int_S f[l - m_S(l)]\frac{d\mu}{[\mu(S)]^{1/q}}\right| \leqslant 2\|L\|.$$

Taking the supremum of the left side over all f supported in S such that $\|f\|_{q'} = 1$ we obtain

$$\left[\int_S |l - m_S(l)|^q \frac{d\mu}{\mu(S)}\right]^{1/q} \leqslant 2\|L\|.$$

[31] In order to apply the Riesz representation theorem we must assume $q' < \infty$. This is not a restriction, however, since, by Theorem A, $H^{1,\infty} = H^{1,r}$ for $1 < r < \infty$.

This shows that $l \in BMO_q$ with $\mathfrak{N}_q(l) \leqslant 2\|L\|$. In case $\mu(X) = 1$ we also must have $|\int_X l\,d\mu| = |L\chi_X| \leqslant 1$. Thus, in general

$$\|l\|_{BMO}^{(q)} \leqslant 3\|L\|. \tag{3.18}$$

The characterization of the dual of H^p for $p < 1$ is simpler to obtain. Using Theorem A we see that we can restrict our attention to $H^{p,1}$. Let $\alpha = 1/p - 1$ and suppose $l \in \mathfrak{L}_\alpha$. Recall that the elements, h, of $H^{p,1}$ are, by definition, continuous linear functionals in $\mathfrak{L}_\alpha^*$. Let us write $\langle h, l\rangle$ to denote the value of the linear functional h at $l \in \mathfrak{L}_\alpha$. Then the mapping L_l: $h \to \langle h, l\rangle$ is a well-defined linear functional on $H^{p,1}$. If $h = \sum \alpha_j a_j$ is an atomic representation of h in terms of $(p, 1)$-atoms we then have (see the estimates preceding (2.3))

$$|\langle h, l\rangle| = \left|\sum \alpha_j \int l a_j\,d\mu\right| \leqslant \|l\|^{(\alpha)} \sum |\alpha_j| \leqslant \|l\|^{(\alpha)}\Big(\sum |\alpha_j|^p\Big)^{1/p}.$$

We see, therefore, that

$$|L_l h| \leqslant \|l\|^{(\alpha)} \|h\|_{p,1}^{1/p}. \tag{3.19}$$

Now suppose L is a bounded[32] linear functional on $H^{p,1}$. As was the case when $p = 1$ we can show the existence of a function l, bounded on each sphere, such that $La = \int la\,d\mu$ for each $(p, 1)$-atom a (we use the same argument as before except that now $q' = 1$ and $q = \infty$). In particular, if f is supported in a sphere S and $\|f\|_1 = 1$ then $(1/2)[\mu(S)]^{1-1/p}(f - m_S(f)) = a$ is a $(p, 1)$-atom and, thus,

$$\|L\| \geqslant |La| = \left|\int_S [l - m_S(l)]a\,d\mu\right| = 2^{-1}[\mu(S)]^{1/p-1}\left|\int_S [l - m_S(l)]f\,d\mu\right|.$$

Taking the supremum over all such f we obtain

$$\|\chi_S(l - m_S(l))\|_\infty \leqslant 2\|L\|\,[\mu(S)]^\alpha.$$

It now follows from a straightforward argument that there exists $l_0 = l$ a.e. for which

$$|l_0(x) - l_0(y)| \leqslant 4\|L\|\,[\mu(S)]^\alpha,$$

where S is any sphere containing x and y. We then have

$$\|l_0\|^\alpha \leqslant 5\|L\| \tag{3.20}$$

whether the measure of X is finite or infinite.

This establishes Theorem B. Inequalities (3.17), (3.18), (3.19) and (3.20) show that the norms of the functions representing the linear functionals are equivalent to the norms of the linear functionals.

In the argument preceding (3.19) we showed that $|\langle h, l\rangle| \leqslant \|l\|^{(\alpha)} \sum |\alpha_j|$, for

[32] That is, there exists $\|L\| < \infty$ such that $|Lh| \leqslant \|L\|\,\|h\|_{p,1}^{1/p}$. The space $H^{p,1}$ is not a normed linear space, but it is obvious (by the usual arguments) that this notion of boundedness is equivalent to the continuity of L.

$h \in H^{p,1}$. In view of this inequality it is not surprising that the space $B_p \subset \mathfrak{L}_\alpha^*$ consisting of linear functionals of the form $h = \sum \alpha_j a_j$, where $\sum |\alpha_j| < \infty$ and a_j are $(p, 1)$- (or (p, ∞)-) atoms, has a dual that is also characterized by $\mathfrak{L}_\alpha$ (with the norm $\|h\|^{(p)}$ taken as the infimum over all sums $\sum |\alpha_j|$, where $h = \sum \alpha_j a_j$, B_p is a Banach space). In the classical case $X = T$, B_p can easily be identified with the space of distributions h whose Poisson integral $u_h(z) = u_h(x + iy)$ satisfies

$$\|h\|_p = \iint_{|z|<1} |u_h(z)|(1 - |z|)^{1/p-2}\, dx\, dy < \infty;$$

moreover, the norm $\|h\|_p$ is equivalent to the infimum of all sums $\sum |\alpha_j|$ arising from all the representations $h = \sum \alpha_j a_j$. From our point of view these spaces are not useful since the intermediate interpolation spaces between L^2 and B_p are not $L^{p_0}(T)$ spaces when $p_0 < 1$ (in fact, they are spaces of harmonic functions in $L^{p_0}[(1 - |z|)^{1/p_0-2}\, dx\, dy]$.

Let us now pass to the proof of Theorem D. We use the notation introduced in §2: in particular, B is a sublinear operator of weak-types (H^{p_1}, p_1) and (p_2, p_2) where $0 < p_1 \leqslant 1 \leqslant p_2 \leqslant \infty$, $p_1 < p_2$. Let us first suppose $1 < p$, where $1/p = t/p_1 + (1 - t)/p_2$ $(0 < t < 1)$. Choose q so that $1 < q < p$ and let $\mathfrak{M}_q$ be the maximal function operator introduced just before inequality (3.7). Let $f \in L^p(X)$ and, for $\alpha > 0$,

$$0^\alpha = \{x \in X: (\mathfrak{M}_q f)(x) > \alpha\}.$$

As was the case in the proof of Theorem A we obtain a Calderón-Zygmund decomposition of f from a Whitney covering of 0^α by balls S_j satisfying (i), (ii) and (iii) in (3.2) with $C = 2K$:

$$f = g_\alpha + h_\alpha \tag{3.21}$$

where $g_\alpha(x) = g(x) = f(x)$ if $x \notin 0^\alpha$, $g_\alpha(x) = g(x) = \sum_j m_{S_j}(\eta_j f)\chi_j(x)$ when $x \in 0^\alpha$, $h_j = f\eta_j - m_{S_j}(\eta_j f)\chi_j$ and $h_\alpha = h = \sum_j h_j$. We then have

$$\left(\frac{1}{\mu(S_j)} \int_{S_j} |h_j|^q\, d\mu\right)^{1/q} \leqslant c\alpha$$

(see the proof of induction hypothesis (VII) when $n = 1$; the constant $c = 2k_0^{1/q}$ depends only on X). It follows, therefore, that

$$a_j = h_j / c\alpha[\mu(S_j)]^{1/p_1}$$

is a (p_1, q)-atom. Thus, $h = c\alpha \sum_j [\mu(S_j)]^{1/p_1} a_j \in H^{p_1}$ $(= H^{p_1,q})$ with norm $\|h\|_{p_1} \sim \|h\|_{p_1,q}$ not exceeding

$$(c\alpha)^{p_1} \sum_j \mu(S_j) \leqslant M(c\alpha)^{p_1} \mu(0^\alpha). \tag{3.22}$$

Let us first suppose $p_2 < \infty$. Then, the weak-type hypotheses imply

$$\begin{aligned} p^{-1}\|Bf\|_p^p &= \int_0^\infty \alpha^{p-1}\nu(\{y \in Y: |(Bf)(y)| > \alpha\})\,d\alpha \\ &\leqslant \int_0^\infty \alpha^{p-1}[\nu(\{y \in Y: |(Bg_\alpha)(y)| > \alpha/2\}) \\ &\qquad + \nu(\{y \in Y: |(Bh_\alpha)(y)| > \alpha/2\})]\,d\alpha \\ &\leqslant \int_0^\infty \alpha^{p-1}\left(\frac{2M_2\|g_\alpha\|_{p_2}}{\alpha}\right)^{p_2} d\alpha + \int_0^\infty \alpha^{p-1}\left(\frac{2M_1}{\alpha}\right)^{p_1}\|h_\alpha\|_{p_1,q}\,d\alpha. \end{aligned}$$

In order to estimate the second integral we make use of (3.22) and (3.8) in order to dominate it by

$$M(2cM_1)^{p_1}\int_0^\infty \alpha^{p-1}\mu(0^\alpha)\,d\alpha = p^{-1}M(2cM_1)^{p_1}\|\mathfrak{M}_q f\|_p^p \leqslant (\text{const})\|f\|_p^p.$$

In order to estimate the first integral we write

$$\begin{aligned} &\int_0^\infty \alpha^{p-p_2-2}\left\{\int_X |g_\alpha(x)|^{p_2}\,d\mu(x)\right\} d\alpha \\ &\qquad = \int_0^\infty \alpha^{p-p_2-2}\left\{\int_{0^\alpha} + \int_{X-0^\alpha} |g(x)|^{p_2}\,d\mu(x)\right\} d\alpha. \end{aligned}$$

But, as in the proof of Theorem A, we can show that $|g(x)| \leqslant c\alpha$ for all $x \in X$; thus,

$$\begin{aligned} \int_0^\infty \alpha^{p-p_2-2}\left\{\int_{0^\alpha} |g(x)|^{p_2}\,d\mu(x)\right\} d\alpha &\leqslant (\text{const})\int_0^\infty \alpha^{p-1}\mu(0^\alpha)\,d\alpha \\ &= (\text{const})\|\mathfrak{M}_q f\|_p^p \leqslant (\text{const})\|f\|_p^p. \end{aligned}$$

Moreover,

$$\begin{aligned} &\int_0^\infty \alpha^{p-p_2-1}\left\{\int_{X-0^\alpha} |g(x)|^{p_2}\,d\mu(x)\right\} d\alpha \\ &\quad = \int_0^\infty \alpha^{p-p_2-1}\,d\alpha \int_{(\mathfrak{M}_q f)(x)\leqslant\alpha} |f(x)|^{p_2}\,d\mu(x) \\ &\quad = \int_X |f(x)|^{p_2}\left\{\int_{(\mathfrak{M}_q f)(x)}^\infty \alpha^{p-p_2-1}\,d\alpha\right\} d\mu(x) \\ &\quad \leqslant \int_X |f(x)|^{p_2}\left\{\int_{|f(x)|} \alpha^{p-p_2-1}\,d\alpha\right\} d\mu(x) = \frac{1}{p_2-p}\int_X |f(x)|^p\,d\mu(x). \end{aligned}$$

Collecting all these estimates we obtain the desired inequality $\|Bf\|_p \leqslant (\text{const}) \cdot \|f\|_p$ and it is clear that the constant in this inequality depends on M_1, M_2, p_1, p_2 and t, but not on $f \in L^p(X)$. If $p_2 = \infty$ the situation is simpler since $\|Bg_\alpha\|_\infty \leqslant M_2\|g_\alpha\|_\infty \leqslant cM_2\alpha$ and, consequently, $\{y \in Y: |(Bg_\alpha)(y)| > cM_2\alpha\}$ is empty. Thus,

$$\begin{aligned}\int_0^\infty \beta^{p-1}\nu(\{y\in Y\colon |(Bf)(y)|>\beta\})\,d\beta &\\ \leqslant (2cM_2)^p\int_0^\infty \alpha^{p-1}[\nu(\{y\colon |(Bg_\alpha)(y)|>cM_2\alpha\})&\\ +\nu(\{y\colon |(Bh_\alpha)(y)|>cM_2\alpha\})]\,d\alpha&\\ =(2cM_2)^p\int_0^\infty \alpha^{p-1}\nu(\{y\colon |(Bh_\alpha)(y)|>cM_2\alpha\})\,d\alpha.&\end{aligned}$$

Now we can use the assumed weak-type (H^{p_1},p_1) assumption and proceed with an argument similar to that used above to obtain the L^p-boundedness of B.

The proof of the theorem for the case $p\leqslant 1$ is simpler. It suffices to estimate $\|Ba\|_p$ when a is a (p,∞)-atom. When this is the case it is clear that $a=[\mu(S)]^{1/p_1-1/p}b$ (S is a supporting sphere for a), where b is a (p_1,p_2)-atom. Thus $\|a\|_{H^{p_1}}\leqslant[\mu(S)]^{1-1/p_1}$. Using this estimate and the weak-type hypotheses we then have, putting $\sigma=[\mu(S)]^{-1/p}$,

$$\begin{aligned}p^{-1}\|Ba\|_p^p &= \int_0^\infty \alpha^{p-1}\nu(\{y\colon |(Ba)(y)|>\alpha\})\,d\alpha\\ &\leqslant \int_0^\sigma \alpha^{p-1}\left(\frac{M_1}{\alpha}\right)^{p_1}\sigma^{p_1-p}\,d\alpha+\int_\sigma^\infty \alpha^{p-1}\left(\frac{M_2}{\alpha}\right)^{p_2}\sigma^{p_2-p}\,d\alpha\\ &= M_1^p/(p-p_1)+M_2^p/(p_2-p).\end{aligned}$$

The conclusion of Theorem D now follows easily from this.

The proof of Theorem E, when $p_0<p\leqslant 1$, is an adaptation to spaces of homogeneous type of the argument given in [58] which dealt with the classical H^p spaces. When $1<p\leqslant p_1$ one can use an extension of the proof given by Fefferman and Stein in [29]. We shall not give a detailed proof of Theorem E (it is written up in [45]); rather, we shall indicate what is needed in order to extend these arguments to our situation.

Let us first consider the case $1<p\leqslant p_1$. Here, as in [29], the theorem is obtained by showing that the adjoint operator B_t^* maps $L^{p'}$ boundedly into $L^{p'}$. In order to achieve this one makes use of the just established duality result for H^1 in order to convert boundedness from H^1 to L^1 into boundedness from L^∞ to BMO. By employing a maximal operator that is particularly well adapted to BMO functions one can then reduce the problem to a classical interpolation theorem involving only L^p-spaces. This maximal operator was introduced for this purpose by Fefferman and Stein [29] when $X=\mathbf{R}^n$. Its definition and basic properties extend naturally to spaces of homogeneous type: Given a locally integrable function f we let

$$f^\#(x)=\sup\left\{\frac{1}{\mu(S)}\int_S|f(y)-m_S(f)|\,d\mu(y)\colon x\in S\right\}$$

(as usual, S denotes a ball in X). Clearly $f^\#\leqslant 2\mathfrak{M}f$ and $\|f^\#\|_\infty\leqslant\|f\|_{BMO}$. A basic property of this maximal operator is that, for $1\leqslant p<\infty$, it dominates the Hardy-Littlewood maximal operator. When $\mu(X)=\infty$, in fact, we have $\|f^*\|_p\leqslant A_p\|f^\#\|_p$. The precise general statement of this result and its use for the interpolation result are found in [45].

When $p_0 < p \leqslant 1$ this interpolation result was obtained independently by Calderón and Torchinski in the case dealing with $\mathbf{R}^n$ endowed with the distance of example (2) of [3].

4. Further observations. We have seen how H^1 can be used effectively as a substitute for L^1. It is natural to suppose that the space BMO plays the same role vis-a-vis L^∞. The proof of Theorem E which we have just described furnishes us an example of this situation. In fact, many results can be obtained efficiently by making use of BMO. This point is fully exploited in the work of Fefferman and Stein [29]. We shall see further evidence of this situation after having made some comments concerning the functional analysis properties of $H^1(X)$ and $BMO(X)$. Before doing this, however, we would like to express, again, our gratitude to Y. Meyer since many of the observations we shall make are either due to him or took place when discussing this work with him.

It follows easily from property (3.4) that if X is complete with respect to the quasi-distance then it is locally compact (this was pointed out to us by R. Levy). Since this is the situation in all the examples we have given of spaces of homogeneous type, we shall assume henceforth that X is locally compact (or complete). It turns out that H^1 is one of the few examples of a separable, nonreflexive Banach space which is a dual space. In fact, let VMO[33] denote the closure in the BMO norm of the space $C_0(X)$ of continuous functions with compact support. One can then prove the following result:

THEOREM (4.1). *H^1 is the dual of VMO. More precisely, each continuous linear functional on VMO has the form $\langle f, v\rangle = \int_X fv\,d\mu$ for all $v \in C_0$, where $f \in H^1$, and $\|f\|_{H^1}$ is equivalent to the linear functional norm.*

This result is a generalization of the theorem of F. and M. Riesz (see Sarason [52] in this connection). It shows that any Borel measure ν satisfying, for $v \in C_0(X)$,

$$\left|\int v\,d\nu\right| \leqslant c\|v\|_{BMO}$$

must be absolutely continuous with Radon-Nikodým derivative in H^1.

Theorem (4.1) is a consequence of a general result in functional analysis and a lemma that will be proved below. The functional analysis result (see Exercise 41 on p. 439 of [22]) is the following: *Let X be a locally convex linear topological space and T a linear subspace of X^*. Then T is X-dense in X^* if and only if T is a total set of functionals on X.* The lemma we need is:

LEMMA (4.2). *Suppose $\|f_k\|_{H^1} \leqslant 1$, $k = 1, 2, \ldots$, then there exists $f \in H^1$ and a subsequence (f_{k_j}) such that*

$$\lim_{j\to\infty} \int f_{k_j} v\,d\mu = \int fv\,d\mu$$

for all $v \in C_0$.

[33] These initials stand for "vanishing mean oscillation". A space VMO was introduced in the classical setting by Sarason [52]. It is somewhat different from ours; we adopted a modified version for which (4.1) holds.

In our case $X = VMO$ and $T = H^1$. The fact that T is a subspace of X^* follows from the case $p = 1$ of Theorem B (since X is a subspace of BMO). In particular, if $\langle f, v\rangle = 0$ for all $f \in H^1 = T$ then v must be the zero element of VMO ($\subset BMO$). Thus, T is total. By the above stated functional analysis result, it follows that T is (weak*) dense in X^*. Since the spaces we are dealing with are separable this is equivalent to the statement: if $x^* \in X^*$ then we can find a sequence $\{f_k\}$ in $T (= H^1)$ such that $\langle f_k, v\rangle \to \langle x^*, v\rangle$ for all $v \in X (= VMO)$. By the Banach-Steinhaus theorem we can conclude that the sequence of norms $\|f_k\|_{H^1}$ is bounded (this requires an elementary, but somewhat technical, argument showing that the supremum of $|\int v\, d\mu|$ over all $v \in C_0$ with $\|v\|_{BMO} \leqslant 1$ defines a norm that is equivalent to $\|f\|_{H^1}$ whenever $f \in H^1$. The principal step in this argument involves showing that $\|b_\varepsilon\|_{BMO} \leqslant c\|b\|_{BMO}$ where $b_\varepsilon(x) = [1/\mu(S_\varepsilon(x))] \int_{S_\varepsilon(x)} b(y)\, d\mu(y)$). We can now invoke Lemma (4.2) to obtain a subsequence $\{f_{k_j}\}$ and an $f \in H^1$ such that

$$\langle x^*, v\rangle = \lim_{j\to\infty} \langle f_{k_j}, v\rangle = \lim_{j\to\infty} \int f_{k_j} v\, d\mu = \int fv\, d\mu$$

for all $v \in C_0$. From this it follows that the linear functional x^* is represented by f and we obtain Theorem (4.1).

In order to establish (4.2) we will need two lemmas. The first follows from a simple diagonalization argument:

LEMMA (4.3). *Suppose* $\lambda_j^k \geqslant 0, j, k = 1, 2, \ldots,$ *satisfies* $\sum_{j=1}^\infty \lambda_j^k \leqslant 1$ *for each* $k = 1, 2, \ldots,$ *then there exists an increasing sequence of natural numbers,* $k_1 < k_2 < \cdots < k_n < \cdots$ *such that* $\lim_{n\to\infty} \lambda_j^{k_n} = \lambda_j$ *for each* j *and* $\sum_{j=1}^\infty \lambda_j \leqslant 1$.

The second lemma is of a geometric nature. For simplicity we consider the case where the spheres of radius r have measure r (and $\mu(X) = \infty$). By an "atom" we shall mean a $(1, \infty)$-atom.

LEMMA (4.4). *Suppose n is an integer. Then X is the union of balls* S_i^n, $i = 1, 2, \ldots,$ *of measure* β^n, *where* $\beta = \beta(X) > 1$, *such that no point belongs to more than N of these balls (N is independent of n) and each* $f \in H^1$ *has the representation*

$$f = \sum_{i=1}^{\infty} \sum_{n=-\infty}^{\infty} \lambda_i^n a_i^n,$$

where a_i^n *is an atom supported in* S_i^n *and* $\sum_{i,n} |\lambda_i^n| \leqslant c\|f\|_{H^1}$.

We shall sketch the proof of this lemma. If B is any sphere and $\bar{B}$ is the smallest sphere containing all those balls intersecting B and having the same radius as B then the ratio $\mu(\bar{B})/\mu(B)$ is bounded independently of B. Let β be such a bound. An argument similar to the one used to establish (3.2) shows that X can be covered with balls $B_1^n, B_2^n, \ldots, B_j^n, \ldots$ of measure β^n that are almost disjoint. Moreover, this can be done in such a way that $\{b_j^n\}$ is a sequence of mutually disjoint balls and, for each j, b_j^n has the same center as B_j^n and a radius a fixed multiple (depending only on A and K) of the radius of B_j^n. Let $f = \sum \lambda_k a_k$. Without loss of generality we can assume each a_k is an atom supported by a sphere of measure β^n for some n (the infimum of $\sum |\lambda_j|$ over all such decompositions differs from $\|f\|_{H^1}$ by a constant multiple that depends on β but not on f). We make our choice of the representation of f so that $\sum |\lambda_j| \leqslant c\|f\|_{H^1}$ (c independent of f). Let $n = 0$ and consider all atoms a_j whose supporting spheres have measure 1 and intersect B_1^0. The sum of terms $\lambda_k a_k$ involving these atoms is supported in a sphere S_1^1 of measure β and can be written in the form $\lambda_1^1 a_1^1$, where $a_{1,1}$ is an atom supported in S_1^1. Now consider the remaining atoms whose supporting sphere has measure 1. Collect those intersecting B_2^0 and, similarly, obtain $\lambda_2^1 a_2^1$. Continue inductively until all atoms supported by spheres of measure 1 are exhausted. Now let $n = -1$ and do the same for atoms whose supporting spheres have measure β^{-1} and intersect B_1^{-1}. We obtain $\lambda_1^0 a_1^0, \lambda_2^0 a_2^0, \ldots$. Next, consider $n = 1$, etc. Observe that, by definition, $\lambda_j^n a_j^n$ is a sum $\sum \lambda_k a_k = \alpha_j^n$ over only indices k such that a_k has a supporting sphere of measure β^{n-1} meeting B_j^{n-1}. Thus, $\|\alpha_j^n\|_\infty \leqslant \sum |\lambda_k| \beta^{1-n}$ (the sum is over this set of indices). Hence, we let $\lambda_j^n = \|\alpha_j^n\|_\infty \mu(S_j^n) \leqslant \beta \sum |\lambda_k|$. By our construction, these indices occur only once. Thus, we obtain the last inequality in (4.4). The almost disjointness of the sequence $S_1^n, S_2^n, \ldots$ follows from the disjointness of the sequence $b_1^{n-1}, b_2^{n-1}, \ldots$ and the fact that the balls in each of the two sequences have the same radius.

We are now ready to prove Lemma (4.2). We write $f_k = \sum_{i,n} \lambda_i^n(k) a_i^n(k)$ as in Lemma (4.4). We can suppose that the coefficients $\lambda_i^n(k)$ are nonnegative and, applying Lemma (4.3) we can assume $\lim_{k\to\infty} \lambda_i^n(k) = \lambda_i^n$ exists for each (i, n) and $\sum_{i,n} \lambda_i^n \leqslant 2$. For each fixed (i, n) the atoms $a_i^n(k)$ are supported by the sphere S_i^n for all k and are uniformly bounded. Thus, there exists a subsequence of $a_i^n(1), a_i^n(2), \ldots, a_i^n(k), \ldots$ converging weakly to a function a_i^n (that is

$$\int_{S_i^n} a_i^n(k_j)\varphi\, d\mu \to \int_{S_i^n} a_i^n \varphi\, d\mu$$

for all $\varphi \in L^1(S_i^n)$). Applying, again, a diagonalization argument we can assume $a_i^n(k) \to a_i^n$ as $k \to \infty$ for all (i, n). We let $f = \sum \lambda_i^n a_i^n$. It is clear that $f \in H^1$. It remains to be shown that $\lim_{k\to\infty} \int f_k v\, d\mu = \int fv\, d\mu$ for all $v \in C_0$. We write

$$\int f_k v\, d\mu = \left(\sum_{-N\leqslant n\leqslant N} + \sum_{n<-N} + \sum_{n>N} \right) \sum_i \lambda_i^n(k) \int a_i^n(k) v\, d\mu.$$

The last sum is dominated by a constant multiple of $\beta^{-N} \sum_{n>N} \sum_i \lambda_i^n(k) \|v\|_1$ which is small provided N is large. Also if N is large we have $|v(x) - v(x_i^n)| < \varepsilon$ for $x \in S_i^n$ if x_i^n is the center of S_i^n and $n < -N$. Thus, the second sum is dominated by

$$\sum_{n<-N} \sum_i \lambda_i^n(k) \left| \int a_i^n(k)(x)\{v(x) - v(x_i^n)\}\, d\mu(x) \right| \leqslant \varepsilon.$$

The first sum involves only a finite number of terms, independently of k (here we use the almost disjointness, for each n, of the spheres S_i^n and the fact that the compact support of v can only meet a finite number of these spheres since their radii range only from β^{-N} to β^{N}). Thus, the first sum tends to $\int (\sum_{-N\leqslant n\leqslant N} \sum_i \lambda_i^n a_i^n) v\, d\mu$ which, for the above reasons, is close to $\int fv\, d\mu$ if N is large. This establishes (4.2) and, consequently, Theorem (4.1) is proved.

This theorem shows, in particular, that whenever $L^1(X)$ is not a dual space[34] $H^1(X)$ is a basically different space. Another way to see that $H^1(X)$ differs from $L^1(X)$ is to exhibit an unbounded function in $BMO(X)$. For a wide class of spaces of homogeneous type we obtain a collection of such functions by letting $g = \log \nu^*$, where ν is a nonnegative locally finite measure such that

$$\nu^*(x) = \sup_{x\in S} \frac{\nu(S)}{\mu(S)} < \infty$$

for almost every x (with respect to μ). In particular, if ν is a point mass at x_0 then $\log m(x, x_0) \in BMO$ (here m denotes the "measure distance" mentioned in footnote 20). These functions cannot be bounded unless x has a finite number of points.

Let us recall, once again, that a most basic result in the theory of H^p spaces is their characterization in terms of maximal averages (in the classical case we are referring to [2]; for the n-dimensional case see [29]). During the course of a conversation with Y. Meyer, the three of us observed that a recent proof of the duality result between H^1 and BMO, due to L. Carleson [8] can be extended to the general setting of spaces of homogeneous type provided a certain geometric assumption is added. From this one can then obtain a "maximal function" characterization of H^1. The result of Carleson, in the general setting, can be formulated in the following way:

Let $\{k_r(x,y)\}$, $r > 0$, be a family of nonnegative valued functions on $X\times X$ satisfying

(1) $k_r(x,y) = 0$ if $d(x,y) \geqslant r$;

(2) there exists constants c and C such that for all $x, y \in X$

$$k_r(x,x) \geqslant \frac{c}{\mu(B_r(x))} \quad \text{and} \quad k_r(x,y) \leqslant \frac{C}{\mu(B_r(x))};$$

(3) $\int_X k_r(x,y)\, d\mu(y) = 1$ for all $x \in X$;

(4) there exists an $\alpha > 0$ such that for all $x_1, x_2, y \in X$

$$|k_r(x_1,y) - k_r(x_2,y)| \leqslant \frac{C}{\mu(B_r(y))}\left[\frac{d(x_1,x_2)}{r}\right]^{\alpha}.$$

When $X = \mathbf{R}$ such a family is obtained by choosing a nonnegative "bell-shaped" Lipschitz function φ supported by the interval $(-1/2, 1/2)$, having integral 1 and putting $k_r(x,y) = (1/r)\varphi((x-y)/r)$. The more general geometric assumptions needed for constructing such a family can be stated in terms of the measure distance m as follows: there exists $\varepsilon > 0$ such that

---

[34] This is, in fact, the usual situation. The space l^1 of absolutely summable sequences is exceptional: it is the dual of c_0, the space of sequences converging to 0. The proof we have just given, among other things, shows that $H^1(X)$, in general, is similar to l^1.

$$\mu(B_x \cap B_y) \leqslant \left\{ \frac{m(x,y)}{\mu(B_x \cup B_y)} \right\}^{\varepsilon} \mu(B_x \cup B_y)$$

whenever $B_x(B_y)$ is a ball containing $x(y)$. If this condition is satisfied Carleson's duality proof is based on a characterization of BMO functions:

THEOREM (4.5). *Suppose $\varphi \in BMO$ has support within a ball $B \subset X$ then there exists a constant $C = C(X) > 0$, a measurable function $r(y) > 0$ and two bounded functions b, ψ such that*
(a) $\varphi(x) = \int k_{r(y)}(x,y)b(y)\,d\mu(y) + \psi(x)$;
(b) $\|b\|_\infty \leqslant C\|\varphi\|_{BMO}$ *and* $\|\psi\|_\infty \leqslant C\|\varphi\|_{BMO}$;
(c) *the supports of b and ψ are within B.*

The maximal function of f, with respect to the family $\{k_r\}$ is defined by letting its value at $x \in X$ be

$$f^*(x) = \sup_{r>0} \int k_r(x,y) f(y)\,d\mu(y).$$

Making use of (4.5) one can show that an integrable f belongs to H^1 if and only if $f^* \in L^1(X)$.

It is not clear to us to what extent the above geometric condition is necessary for the validity of this characterization of H^1. Moreover, the original one-dimensional methods in **[10]** and their extension to n dimensions by Latter are applicable in several of the examples of spaces of homogeneous type we have presented here; they yield a direct atomic decomposition of these "maximal function" H^p spaces, even for some $p < 1$. It would be of interest to clarify this situation further.

Before ending this presentation we would like to say a few words about an important class of Hardy spaces for which an "atomic" theory has not been developed. These are the H^p spaces on the polydiscs, orginally introduced by Zygmund **[64]**, and their extensions to Reinhardt regions and tubes (see **[57]** and **[51]**). It is reasonable to suppose that, in $\mathbf{R}^n$, these spaces are related to an atomic H^1 space generated by atoms whose support lies within rectangles with sides parallel to the axes and the cancellation properties occur along each axis. For example, a *rectangular* $(1, \infty)$-atom on $\mathbf{R}^2$ is a function supported in the rectangle R, satisfying $|a(x,y)| \leqslant 1/|R|$, for all $(x,y) \in \mathbf{R}^2$ and

$$\int_{-\infty}^{\infty} a(x,y)\,dy = 0 = \int_{-\infty}^{\infty} a(x,y)\,dx.$$

These atoms generate a subspace $\mathfrak{H}^1$ of our "cubic" $H^1(\mathbf{R}^2)$. The latter is characterized by the Riesz transforms. It is easy to see that $\mathfrak{H}^1$ is invariant under the iterated Hilbert transforms. Carleson and Fefferman showed, however, that this property does not characterize $\mathfrak{H}^1$. Nevertheless, it may turn out to be another useful substitute for L^1. A natural question is to see if the interpolation theorems we have considered in this exposition are valid here. A positive result along these lines would certainly imply some interesting multiplier theorems. Another question is to see if there are "natural operators" characterizing $\mathfrak{H}^1$.

REFERENCES

1. R. Boas, *Isomorphism between H^p and L^p*, Amer. J. Math. **77** (1955), 655–656. MR **17**, 1080.

2. D. L. Burkholder, R. F. Gundy and M. L. Silverstein, *A maximal function characterization of the class H^p*, Trans. Amer. Math. Soc. **157** (1971), 137–153. MR **43** #527.

3. A. P. Calderón and A. Torchinsky, *Spaces of distributions in $L^p(\mathbf{R}^n)$* (to appear).

4. A. P. Calderón and A. Zygmund, *Singular integral operators and differential equations*, Amer. J. Math. **79** (1959), 901–921. MR **20** #7196.

5. ———, *On the existence of certain singular integrals*, Acta Math. **88** (1952), 85–139. MR **14**, 637.

6. ———, *On higher gradients of harmonic functions*, Studia Math. **24** (1964), 211–226. MR **29** #4903.

7. S. Campanato, *Proprietà di hölderianità di alcune classi di funzioni*, Ann. Scuola Norm. Sup. Pisa (3) **17** (1963), 175–188. MR **27** #6119.

8. L. Carleson, *Two remarks on H^1 and BMO*, Analyse Harmonique d'Orsay **164** (1975), 1–11.

9. F. Carlson, *Une inégalité*, Ark. Mat. Astronom. Fys. 25B, no. 1, (1934), 1–5.

10. R. R. Coifman, *A real variable characterization of H^p*, Studia Math. **51** (1974), 269–274. MR **50** #10784.

11. ———, *Characterizations of Fourier transforms of Hardy spaces*, Proc. Nat. Acad. Sci. U. S. A. **71** (1974), no. 10, 4133–4134.

12. R. R. Coifman and M. de Guzmán, *Singular integrals and multipliers on homogeneous spaces*, Rev. Un. Mat. Argentina **25** (1970/71), 137–143. MR **47** #9180.

13. R. R. Coifman and C. Fefferman, *Weighted norm inequalities for maximal functions and singular integrals*, Studia Math. **51** (1974), 241–250. MR **50** #10670.

14. R. R. Coifman, R. Rochberg and G. Weiss, *Factorization theorems for Hardy spaces in several variables*, Ann. of Math. **103** (1976), 611–635.

15. R. R. Coifman and G. Weiss, *On subharmonicity inequalities involving solutions of generalized Cauchy-Riemann equations*, Studia Math. **36** (1970), 77–83. MR **42** #6265.

16. ———, *Invariant systems of conjugate harmonic functions associated with compact Lie groups*, Studia Math. **44** (1972), 301–308.

17. ———, *Analyse harmonique non-commutative sur certains espaces homogenes*, Lecture Notes in Math., vol. 242, Springer-Verlag, Berlin and New York, 1971.

18. ———, *Théorèmes sur les multiplicateurs de Fourier sur $SU(2)$ et Σ_2*, C. R. Acad. Sci. Paris Sér. A-B **271** (1970), pp. A928–A930. MR **42** #6157.

19. ———, *A multiplier theorem for $SU(2)$ and Σ_2*, Rev. Un. Mat. Argentina **25** (1970), 145–166.

20. W. Connett and A. Schwartz, *A multiplier theorem for ultraspherical series*, Studia Math. **51** (1974), 51–70. MR **50** #10674.

21. M. Cotlar, *A unified theory of Hilbert transforms and ergodic theorems*, Rev. Mat. Cuyana **1** (1955), 105–167 (1956). MR **18**, 893.

22. N. Dunford and J. Schwartz, *Linear operators*. I, Interscience, New York and London, 1964 (1st ed., 1958; MR **22** #8302).

23. P. L. Duren, *Theory of H^p spaces*, Academic Press, New York, 1970. MR **42** #3552.

24. P. L. Duren, B. W. Romberg and A. L. Shields, *Linear functionals on H^p spaces with $0 < p < 1$*, J. Reine Angew. Math. **238** (1969), 32–60. MR **41** #4217.

25. A. Erdélyi et al., *Higher transcendental functions*, Vol. II, McGraw-Hill, New York, 1953. MR **15**, 419; **34** #1565.

26. E. B. Fabes and N. M. Rivière, *Singular integrals with mixed homogeneity*, Studia Math. **27** (1966), 19–38. MR **35** #683.

27. C. Fefferman, *H^p spaces*, Harmonic Analysis (Studies in Math., Vol. 13, M. Ash, Ed.), Math. Assoc. of Amer., 1976, 38–75.

28. ———, *Characterizations of bounded mean oscillation*, Bull. Amer. Math. Soc. **77** (1971), 587–588. MR **43** #6713.

29. C. Fefferman and E. M. Stein, *H^p spaces of several variables*, Acta Math. **129** (1972), 137–193.

30. C. Fefferman, N. M. Rivière and Y. Sagher, *Interpolation between H^p spaces: The real method*, Trans. Amer. Math. Soc. **191** (1974), 75–81.

31. G. B. Folland and E. M. Stein, *Estimates for the $\bar{\partial}_b$-complex and analysis on the Heisenberg group*, Comm. Pure Appl. Math. **27** (1974), 429–522.

32. J. Garcia-Cuerva, *Weighted H^p-spaces*, Ph.D. Thesis, Washington Univ., Missouri, 1975.

33. G. H. Hardy, *The mean value of the modules of an analytic function*, Proc. London Math. Soc. (2) **14** (1915), 269–277.

34. C. Herz, *Bounded mean oscillation and regulated martingales*, Trans. Amer. Math. Soc. **193** (1974), 199–215. MR **50** #5930.

35. K. Hoffman, *Banach spaces of analytic functions*, Prentice-Hall, Englewood Cliffs, N. J., 1962. MR **24** #2844.

36. L. Hörmander, *Estimates for translation invariant operators in L^p spaces*, Acta Math. **104** (1960), 93–140. MR **22** #12389.

37. S. Igari, *An extension of the interpolation theorem of Marcinkiewicz*. II, Tôhoku Math. J. (2) **15** (1963), 343–358. MR **28** #1431.

38. F. John and L. Nirenberg, *On functions of bounded mean oscillation*, Comm. Pure Appl. Math. **14** (1961), 415–426. MR **24** #A1348.

39. B. F. Jones, Jr., *A class of singular integrals*, Amer. J. Math. **86** (1964), 441–462. MR **28** #4308.

40. A. W. Knapp and E. M. Stein, *Intertwining operators for semi-simple Lie groups*, Ann. of Math. (2) **93** (1971), 489–578.

41. ———, *Singular integrals and the principal series*. I, Proc. Nat. Acad. Sci. U.S.A. **63** (1969), 281–284. MR **41** #8588.

42. A. Korányi and S. Vági, *Singular integrals on homogeneous spaces and some problems of classical analysis*, Ann. Scuola Norm. Sup. Pisa **25** (1971), 575–648.

43. P. Krée, *Distributions quasi homogènes. Généralisations des intégrals singulières et du calcul symbolique de Calderón-Zygmund*, C. R. Acad. Sci. Paris **261** (1965), 2560–2563. MR **32** #6211.

44. V. Krylov, *Functions regular in the half-plane*, Sbornik **6** (1939), 95–138. (Russian)

45. R. Macias, *H^p-spaces interpolation theorems*, Ph.D. Thesis, Washington Univ., Missouri, 1975.

46. N. G. Meyers, *Mean oscillation over cubes and Hölder continuity*, Proc. Amer. Math. Soc. **15** (1964), 717–721. MR **29** #5969.

47. B. Muckenhoupt and E. M. Stein, *Classical expansions and their relation to conjugate harmonic functions*, Trans. Amer. Math. Soc. **118** (1965), 17–92. MR **33** #7779.

48. F. Riesz, *Sur les valeurs moyennes du module des fonctions harmoniques et des fonctions analytiques*, Acta Sci. Math. **1** (1922/23), 27–32.

49. N. M. Rivière, *Singular integrals and multiplier operators*, Ark. Mat. **9** (1971), no. 2, 243–278.

50. L. P. Rothschild and E. M. Stein, *Hypoelliptic differential operators and nilpotent groups* (preprint).

51. W. Rudin, *Function theory in polydiscs*, Benjamin, New York, 1969. MR **41** #501.

52. D. Sarason, *Functions of vanishing mean oscillation*, Trans. Amer. Math. Soc. **207** (1975), 391–405.

53. E. M. Stein, *Note on the class L* log *L*, Studia Math. **32** (1969), 305–310. MR **40** #799.

54. ———, *Classes H^p, multiplicateurs et fonctions de Littlewood-Paley*, C. R. Acad. Sci. Paris Sér. A-B **263** (1966), A716–A719, A780–A781; ibid. **264** (1967), A107–A108. MR **37** #695a, b, c.

55. ———, *Singular integrals and differentiability properties of functions*, Princeton Univ. Press, Princeton, N. J., 1970. MR **44** #7280.

56. ———, *Boundary behavior of holomorphic functions of several complex variables*, Math. Notes, Princeton Univ. Press, Princeton, N. J., 1972.

57. E. M. Stein and G. Weiss, *Introduction to Fourier analysis on Euclidean spaces*, Princeton Univ. Press, Pinceton, N. J., 1971. MR **46** #4102.

58. ———, *On the interpolation of analytic families of operators acting on H^p-spaces*, Tôhoku Math. J. (2) **9** (1957), 318–339. MR **20** #1216.

59. ———, *On the theory of harmonic functions of several variables*. I, Acta Math. **103** (1960), 25–62. MR **22** #12315.

60. ———, *Generalization of the Cauchy-Riemann equations and representations of the rotation group*, Amer. J. Math. **90** (1968), no. 1, 163–196. MR **36** #6540.

61. M. H. Taibleson, *Fourier analysis on local fields*, Math. Notes, Princeton Univ. Press, Princeton, N. J., 1975.

62. G. Weiss, *Complex methods in harmonic analysis*, Amer. Math. Monthly **77** (1970), no. 5, 465–474. MR **41** #5886.

63. ———, *Harmonic analysis*, Studies in real and complex analysis (Studies in Math., Vol. III), Math. Assoc. of Amer., Buffalo, N. Y.; Prentice-Hall, Englewood Cliffs, N. J., 1965, pp. 124–178. MR **32** #1080.

64. A. Zygmund, *Trigonometric series*, 2nd. rev. ed., Vols. I, II, Cambridge Univ. Press, New York, 1959. MR **21** #6498.

DEPARTMENT OF MATHEMATICS, WASHINGTON UNIVERSITY, ST. LOUIS, MISSOURI 63130

Painless nonorthogonal expansions[a)]

Ingrid Daubechies[b)]
Theoretische Natuurkunde, Vrije Universiteit Brussel, Pleinlaan 2, B 1050 Brussels, Belgium

A. Grossmann
Centre de Physique Theorique II, CNRS-Luminy, Case 907, F 13288 Marseille Cedex 9, France

Y. Meyer
Centre de Mathématiques, Ecole Polytechnique, 91128 Palaiseau Cedex, France

(Received 18 September 1985; accepted for publication 11 December 1985)

In a Hilbert space $\mathscr{H}$, discrete families of vectors $\{h_j\}$ with the property that $f = \Sigma_j \langle h_j | f \rangle h_j$ for every f in $\mathscr{H}$ are considered. This expansion formula is obviously true if the family is an orthonormal basis of $\mathscr{H}$, but also can hold in situations where the h_j are not mutually orthogonal and are "overcomplete." The two classes of examples studied here are (i) appropriate sets of Weyl–Heisenberg coherent states, based on certain (non-Gaussian) fiducial vectors, and (ii) analogous families of affine coherent states. It is believed, that such "quasiorthogonal expansions" will be a useful tool in many areas of theoretical physics and applied mathematics.

I. INTRODUCTION

A classical procedure of applied mathematics is to store some incoming information, given by a function $f(x)$ (where x is a continuous variable, which may be, e.g., the time) as a discrete table of numbers $\langle g_j | f \rangle = \int dx\, g_j(x) f(x)$ rather than in its original (sampled) form. In order to have a mathematical framework for all this, we shall assume that the possible functions f are elements of a Hilbert space $\mathscr{H}$ [we take here $\mathscr{H} = L^2(\mathbf{R})$]; the functions g_j are also assumed to be elements of this Hilbert space.

One can, of course, choose the functions g_j so that the family $\{g_j\}$ ($j \in J$, J a denumerable set) is an orthonormal basis of $\mathscr{H}$. The decomposition of f into the g_j is then quite straightforward: one has

$$f = \sum_j \langle g_j | f \rangle g_j ,$$

where the series converges strongly. The requirement that the g_j be orthonormal leads, however, to some less desirable features. Let us illustrate these by means of two examples.

Take first $g_j(x) = p_j(x)\, w(x)^{1/2}$, where the p_j are orthonormal polynomials with respect to the weight function w. In this case local changes of the function f will affect the whole table of numbers $\langle g_j | f \rangle$ ($j \in J$), which is a feature we would like to avoid.

An orthonormal basis $\{g_j\}$ ($j \in J$), which would enable us to keep nonlocality under control, is given by our second example. We cut $\mathbf{R}$ (the set of real numbers, which is the range of the continuous variable x) into disjoint intervals of equal length, and we construct the g_j starting from an orthonormal basis for one interval. Schematically, consider h_n, an orthonormal basis of $L^2([0,a))$,

$$J = \{(n,m);\ n,m \in \mathbf{Z},\ \text{the set of integers}\},$$

$$g_{n,m}(x) = \begin{cases} h_n(x - ma), & \text{for } ma < x < (m+1)a, \\ 0, & \text{otherwise.} \end{cases}$$

If now the function f undergoes a local change, confined to the interval $[ka, la]$, only the numbers $\langle g_{n,m} | f \rangle$ with $k < m < l - 1$ will be affected, reflecting the locality of the change. This choice for the g_j also has, however, its drawbacks: some of the functions g_j are likely to be discontinuous at the edges of the intervals, thereby introducing discontinuities in the analysis of f, which need not have been present in f itself. This is particularly noticeable if one takes the following natural choice for the h_n:

$$h_n(x) = a^{-1/2} e^{i2\pi x/a} .$$

In this case even very smooth functions f will give values $g_{n,m}(f)$ significantly different from zero for rather high values of n, reflecting high-frequency components artificially introduced by the cutting of $\mathbf{R}$ into intervals.

We shall now see how these undesirable features can be avoided by taking radically different options for the choice of the g_j. In particular, we shall not restrict ourselves to orthonormal bases. Let us start by asking which properties we want to require for the g_j.

The storage of the function f in the form of a discrete table of numbers $\langle g_j | f \rangle$ ($j \in J$) only makes sense if one is certain that f is completely characterized by the numbers $\langle g_j | f \rangle$ ($j \in J$). In other words, we want

$$\langle g_j | f \rangle = \langle g_j | h \rangle, \quad \text{for all } j \text{ in } J,$$

to imply $f = h$, which is equivalent to saying that the vectors $\{g_j\}$ ($j \in J$) span a dense set, i.e., that the orthonormal complement $\{g_j; j \in J\} = \{0\}$. This will be our first requirement.

In all the cases we shall discuss, the set $\{g_j\}$ ($j \in J$) is such that the map

$$T:\ f \to (\langle g_j | f \rangle)_{j \in J}$$

defines a bounded operator from $\mathscr{H}$ to $l^2(J)$, the Hilbert space of all square integrable sequences labeled by J. In other

[a)] This work was carried out within the framework of the R. C. P. "Ondelettes."

[b)] "Bevoegdverklaard Navorser" at the National Foundation for Scientific Research, Belgium.

 0022-2488/86/051271-13$02.50 © 1986 American Institute of Physics

Downloaded 14 Oct 2004 to 128.112.156.90. Redistribution subject to AIP license or copyright, see http://jmp.aip.org/jmp/copyright.jsp

words, J is countable, and there exists a positive number B such that for all f in $\mathscr{H}$ one has

$$\sum_{j\in J}|\langle g_j|f\rangle|^2 \leqslant B\,\|f\|^2 .$$

This can also be stated in the following, equivalent, form: If $|g_j\rangle\,\langle g_j|$ is defined as the operator associating to every vector h in $\mathscr{H}$ the vector $\langle g_j|h\rangle g_j$, then

$$\sum_{j\in J}|g_j\rangle\,\langle g_j|\in\mathscr{B}(\mathscr{H})$$

(the set of all bounded operators in $\mathscr{H}$), with

$$\Big\|\sum_{j\in J}|g_j\rangle\,\langle g_j|\Big\|\leqslant B .$$

In order to reconstruct f from the discrete table $(\langle g_j|f\rangle)_{j\in J}$, one needs to invert the map T

$$T:\ f\mapsto(\langle g_j|f\rangle)_{j\in J}$$

from $\mathscr{H}$ to $l^2=l^2(J)$.

In general the image $T\mathscr{H}$ is not all of l^2, but only a subspace of l^2; one can see this, for instance, if the g_j constitute what is often called, in the physics literature, an "overcomplete" set, i.e., if each g_j is in the closed linear span of the remaining ones: $\{g_k;\ k\in J,\ k\neq j\}$. Strictly speaking, there is then no inverse map T^{-1}. This is, of course, no real difficulty: One can define a map $\widetilde{T}$ from $l^2(J)$ to $\mathscr{H}$, which is zero on the orthogonal complement of $T\mathscr{H}$ and which inverts T when restricted to $T\mathscr{H}$.

If the spectrum of the positive operator $\Sigma_{j\in J}|g_j\rangle\,\langle g_j|$ reaches down to zero, this inverse is an unbounded operator, and the recovery of f from $(\langle g_j|f\rangle)_{j\in J}$ becomes an ill-posed problem. This is avoided if we require the spectrum of $\Sigma_{j\in J}|g_j\rangle\,\langle g_j|$ to be bounded away from zero, i.e., if we impose that there exist positive constants A,B such that

$$A\mathbf{1}\leqslant\sum_{j\in J}|g_j\rangle\,\langle g_j|\leqslant B\mathbf{1} .$$

(Here $\mathbf{1}$ is the identity operator in $\mathscr{H}$. The inequality sign $\leqslant$ between two operators L and T means that their difference $T-L$ is positive definite.) Equivalently, for all f in $\mathscr{H}$, we require

$$A\,\|f\|^2\leqslant\sum_{j\in J}|\langle g_j|f\rangle|^2\leqslant B\,\|f\|^2 . \tag{1.1}$$

This is the second condition we impose on the set $\{g_j\}$ $(j\in J)$. The only new condition is the lower bound.

A set of vectors $\{g_j\}$ $(j\in J)$ in a Hilbert space $\mathscr{H}$, satisfying condition (1.1) with $A,B>0$, is called a *frame*.[1] Note that in general the vectors $\{g_j\}_{j\in J}$ will not be a basis in the technical sense, even though their closed linear span is all of $\mathscr{H}$. This is so because the vectors g_j need not be "ω independent," even though they will usually be linearly independent. That is, a vector g_j usually cannot be written as a finite linear combination of vectors $g_{j'}$ (with $j'\neq j$) but it may well belong to the closed linear span of the infinitely many remaining members of the family. Frames were introduced in the context of nonharmonic Fourier series, where the functions g_j are exponentials.[1,2] As far as we know, this is the only context in which frames have been put to use. One of the aims of the present paper is to provide examples of frames in other contexts. Notice that the results on frames in connection with nonharmonic Fourier series can be rewritten as estimates for entire functions in the Paley–Wiener space[1,2]; one of the results we shall derive here can be rewritten as an analogous estimate for entire functions of growth less than $(2,\frac{1}{2})$ (see Ref. 3).

Notice that, even for functions g_j satisfying the condition (1.1), the effective inversion of the map T: $f\mapsto(\langle g_j|f\rangle)_j$ may be a complicated matter. The condition (1.1) on the g_j ensures that the operator is $\widetilde{T}$ is bounded ($\|\widetilde{T}\|\leqslant A^{-1/2}$) but does not provide a way of calculating it. We are still left with a problem where we have to invert large matrices, although some convergence questions are under control. Assuming for a moment that $\widetilde{T}$ is given, we may define the family $e_k=\widetilde{T}d_k$, where the d_k $(k\in J)$ form the natural orthonormal basis of $l^2(J)$. For $c=(c_j)_{j\in J}\in l^2$, the image $\widetilde{T}c$ is then given by

$$\widetilde{T}c=\sum_j c_j e_j ,$$

where the series converges strongly, by the boundedness of $\widetilde{T}$. This then implies, for all f in $\mathscr{H}$,

$$f=\sum_j\langle g_j|f\rangle e_j , \tag{1.2}$$

again with strong convergence of the series. While (1.2) looks identical to the familiar expansion of f into biorthogonal bases, it really is very different because the $(g_j)_{j\in J}$ need not be a basis at all, technically speaking.

There exists, however, a particular class of frames for which these computational problems do not arise. These are the frames for which the ratio B/A reaches its "optimal" value, $B/A=1$. One has then, for all f in $\mathscr{H}$,

$$\sum_{j\in J}|\langle g_j|f\rangle|^2=A\,\|f\|^2 \tag{1.3}$$

or, equivalently,

$$\sum_{j\in J}|g_j\rangle\,\langle g_j|=A\mathbf{1} .$$

So the map T is now a multiple of an isometry from $\mathscr{H}$ into l^2; as such, it is inverted, on its range, by a multiple of its adjoint T^*. Moreover TT^* is a multiple of the orthogonal projection operator on the range T, which can be thus easily characterized.

It is evident that (1.3) is satisfied whenever the g_j constitute an g_j constitute an orthonormal basis (with $A=1$ then). We shall see that there are other, more interesting examples of frames satisfying (1.3), in which the vectors g_j are not mutually orthogonal, and where the set $\{g_j\}$ $(j\in J)$ is "overcomplete" in the sense defined above. We shall say that a frame is *tight* if it satisfies condition (1.3) or, equivalently, if the inequalities in (1.1) can be tightened into equalities. The inversion formula allowing one to recover the vector f from $(\langle g_j|f\rangle)_{j\in J}$ is particularly simple for tight frames. For any f in $\mathscr{H}$ one has

$$f=A^{-1}\sum_{j\in J}\langle g_j|f\rangle g_j , \tag{1.4}$$

where the series converges strongly (as in the case of a general frame). The expansion (1.4) is thus entirely analogous

Downloaded 14 Oct 2004 to 128.112.156.90. Redistribution subject to AIP license or copyright, see http://jmp.aip.org/jmp/copyright.jsp

to an expansion with respect to an orthonormal basis, even though the g_j need not be orthogonal. We believe that tight frames and the associated simple (painless!) quasiorthogonal expansions will turn out to be very useful in various questions of signal analysis, and in other domains of applied mathematics. Closely related expansions have already been used in the analysis of seismic signals.[4]

The vectors g_j constituting a tight frame need not be normalized. On the other hand, an orthogonal basis consisting of vectors of different norm, does not constitute a tight frame.

In real life, of course, one will have to deal with finite sets of vectors g_j, i.e., one will have to truncate the infinite set J to a finite subset. The reconstruction problem then becomes ill-posed, and extra conditions, using *a priori* information on f, will be needed to stabilize the reconstruction procedure.[5] We shall not address this question here.

In this paper, we shall discuss two classes of examples of sets $\{g_j\}$ ($j \in J$). In both cases, this discrete set of vectors is obtained as a discrete subset of a continuous family which forms an orbit of a unitary representation of a particular group. Schematically, such families can be described as follows. Consider the following.[6]

(i) $U(\cdot)$ is an irreducible unitary representation, on $\mathscr{H}$, of a locally compact group $\mathscr{G}$.

(ii) $d\mu(\cdot)$ is the left-invariant measure on $\mathscr{G}$.

(iii) Let g be an admissible vector in $\mathscr{H}$ for U (see Ref. 6), i.e., a nonzero vector such that

$$c_g = \|g\|^{-2} \int d\mu(y) \, |\langle g, U(y) g \rangle|^2 < \infty \,, \tag{1.5}$$

the integral being taken over $\mathscr{G}$.

(Notice that there are many irreducible unitary representations for which no admissible vectors exist. However, if there is one admissible vector, there is a dense set of them, and we call the representation square integrable.[6])

(iv) Then

$$\int d\mu(y) \, U(y) \, |g\rangle \, \langle g| \, U(y)^* = c_g \mathbf{1} \,, \tag{1.6}$$

where the integral is to be understood in the weak sense. If the group $\mathscr{G}$ is unimodular [i.e., if $d\mu(\cdot)$ is both left and right invariant], the existence of one admissible vector in $\mathscr{H}$ implies that all vectors in $\mathscr{H}$ are admissible; moreover, one has in this case that $c_g = c\|g\|^2$ for some c independent of g [see Ref. 6(b)].

(v) In order to obtain possible sets $\{g_j\}$ ($j \in J$) we choose (1) an admissible vector g in $\mathscr{H}$ and (2) a "lattice" of discrete values for the group element y: $\{ y_j;\ j \in J\}$.

The vectors g_j are then defined as

$$g_j = U(y_j) g \,.$$

By imposing appropriate restrictions on g and on J, we shall obtain families $\{g_j\}$ that are frames—or tight frames—in $\mathscr{H}$.

With this procedure it is possible to adjust the "spacing" of the "lattice" $\{ y_j;\ j \in J\}$ according to the desired degree of "oversampling." In the two cases that we shall consider, this flexibility can be exploited at little computational cost, since the action of $U(y)$ on g is very simple and the new g_j—obtained after an adjustment of the "lattice"—can be easily and quickly calculated.

In this paper, we shall discuss sets $\{g_j\}$ ($j \in J$) constructed along the lines described above for two different groups; the Weyl–Heisenberg group, and the affine or $ax + b$ group.

In Sec. II we treat the Weyl–Heisenberg case. We start, in Sec. II A, by giving a short review of the definition and main properties of this group and of the associated "overcomplete" set, generally called the set of coherent states. A particular discrete set of coherent states is associated to the so-called von Neumann lattice and to a particular choice of g; it has been discussed and used many times (see, e.g., Refs. 7 and 8). It is well known that the set of coherent states associated to the von Neumann lattice is complete, i.e., that its linear span is dense in $\mathscr{H}$ (see Refs. 8–10). It thus meets the first of the two requirements listed above. We show in Sec. II B that the second requirement is not met: the coherent states associated to the von Neumann lattice do *not* constitute a frame. In Sec. II C we shall see that a similar lattice, with density twice as high, does lead to a frame. In II D we concentrate on analogous families of states based on function g with compact support, as opposed to the most commonly discussed canonical coherent states, where g is a Gaussian. We derive sufficient conditions ensuring that the $g_j = U(y_j) g$ constitute a frame. In Sec. II E we show how g can be chosen in such a way that the frame generated is tight. In Sec. II F we analyze this situation and describe in more detail the necessary and sufficient conditions that g has to satisfy in order to generate a tight frame.

In Sec. III we discuss the $ax + b$ group. Again we start, in Sec. III A, with a short review of definitions and properties, including the so-called affine coherent states. The affine coherent states were first defined in Ref. 11; detailed studies of them can be found, e.g., in Refs. 4 and 12; for applications of these states to signal analysis, see Ref. 4. In Sec. III B we discuss discrete "lattices" of affine coherent states based on "band-limited" functions g, i.e., on functions such that the Fourier transform of g has compact support. We derive sufficient conditions for these discrete sets to be frames. In Sec. III C we show how certain specific choices of g lead to tight frames; in Sec. III D we again analyze the construction, and derive necessary and sufficient conditions on g, ensuring that certain frames will be tight.

As can be readily seen from Sec. II E and Sec. III C, the construction of tight frames associated with the Weyl–Heisenberg group is essentially the same as that of tight frames associated with the $ax + b$ group. Tight frames associated with the $ax + b$ group were first introduced[13] in a different context closer to pure mathematics. In Ref. 13(b) one can find a definition of "quasiorthogonal families" very close to our tight frames, and a short discussion of the similarities between a "quasiorthogonal family" and an orthonormal basis. For the many miraculous properties of this orthonormal basis, see Ref. 14.

Finally, let us note that while we have restricted our discussion to $\mathscr{H} = L^2(\mathbf{R})$, it is possible to extend the discussion to $L^2(\mathbf{R}^n)$, as well for the Weyl–Heisenberg group as for the $ax + b$ group. In the latter case the unitary representation $U(\cdot)$ underlying the construction of frames, is no

Downloaded 14 Oct 2004 to 128.112.156.90. Redistribution subject to AIP license or copyright, see http://jmp.aip.org/jmp/copyright.jsp

longer irreducible. A more detailed analysis shows, however, that the essential feature is cyclicity of the representation rather than its irreducibility.[15]

II. THE WEYL–HEISENBERG CASE

A. Review of definitions and basic properties

The Weyl–Heisenberg group is the set $\mathbf{T}\times\mathbf{R}\times\mathbf{R}$ (where $\mathbf{T}$ is the set of complex numbers of modulus 1), with the group multiplication law

$$(z,q,p)(z',q',p') = (e^{i(pq'-p'q)/2}zz',q+q',p+p') .$$

We shall here be concerned with the irreducible unitary representation of this group acting in the Hilbert space $\mathscr{H} = L^2(\mathbf{R},dx)$, and given by

$$(W(z,q,p)f)(x) = ze^{-ipq/2}e^{ipx}f(x-q) .$$

The *Weyl operators* $W(q,p)$ are defined as

$$W(q,p) = W(1,q,p);$$

they satisfy the relations

$$W(q,p)W(q',p') = \exp[i(pq'-p'q)/2]\,W(q+q',p+p'),$$

an exponentiated form of the Heisenberg commutation relations. By a theorem of von Neumann, the above relations determine the irreducible family W up to unitary equivalence. A well-known property[16] of Weyl operators is the following; for all f_1,f_2,g_1,g_2, in $\mathscr{H}$ one has

$$\int\int dpdq\langle f_1,W(q,p)g_1\rangle\langle W(q,p)g_2,f_2\rangle = 2\pi\langle f_1|f_2\rangle\,\langle g_2|g_1\rangle. \tag{2.1}$$

Comparing this with (1.5) and (1.6), one sees that all elements of $\mathscr{H}$ are admissible and that $c_g = 2\pi\|g\|^2$. These two features are a consequence of the unimodularity of the Weyl–Heisenberg group.

The family of canonical coherent states is defined as a particular orbit under this set of unitary operators. The canonical coherent states can be defined as the family of vectors $W(q,p)\Omega$, where Ω is the ground state of the harmonic oscillator:

$$\Omega(x) = \pi^{-1/4}\exp(-x^2/2) .$$

One readily sees that this is equivalent to the customary definition of a canonical coherent state as the function

$$\pi^{-1/4}e^{-ipq/2}e^{ipx}\exp(-(x-q)^2/2) .$$

We shall often work with orbits of Weyl operators other than the canonical coherent states. We therefore introduce the notation

$$|q,p;g\rangle = W(q,p)g , \tag{2.2}$$

where g is any nonzero element of $\mathscr{H}$. The canonical coherent states are thus $|p,q;\Omega\rangle$. As a consequence of (2.1), one has

$$\int\int dqdp|q,p;g\rangle\,\langle q,p;g| = 2\pi\|g\|^2\mathbf{1} .$$

B. The von Neumann lattice and Zak transform

Take $a,b>0$. For any integer m,n, consider

$$|ma,nb;\Omega\rangle = W(ma,nb)\Omega . \tag{2.3}$$

It is known[8–10] that the linear span of the set $\{|ma,nb;\Omega\rangle;\ m,n \text{ in } \mathbf{Z}\}$ is dense in $\mathscr{H}$ if and only if $ab\leqslant 2\pi$. At the critical density $ab=2\pi$, this set of points $\{(ma,nb)\}$ in phase space is called a von Neumann lattice.[7] In quantum mechanics, the associated set of canonical coherent states has a nice physical interpretation. It corresponds to choosing exactly one state per "semiclassical Gibbs cell," i.e., per cell of area h (Planck's constant).

Notice that the discrete set of Weyl operators $\{W(ma,n2\pi/a);\ m,n\in\mathbf{Z}\}$ is Abelian. This feature is exploited in the construction of the kq transform, or Zak transform,[17] which will turn out to be useful in what follows.

Denote by $\square$ the semiopen rectangle $\square = [-\pi/a,\pi/a)\times[-a/2,a/2)$.

The Zak transform is a unitary map from $L^2(\mathbf{R})$ onto $L^2([-\pi/a,\pi/a)\times[-a/2,a/2)) = L^2(\square)$ and is defined as follows. For a function f in $C_0^\infty(\mathbf{R})$ (infinitely differentiable functions with compact support), one defines its Zak transform Uf by

$$(Uf)(k,q) = \left(\frac{a}{2\pi}\right)^{1/2}\sum_l e^{ikal}f(q-la) , \tag{2.4}$$

where, for any q, only a finite number of terms in the sum contribute, due to the compactness of the support of f. The map U, defined by (2.4), is isometric from $C_0^\infty(\mathbf{R})\subset L^2(\mathbf{R})$ to $L^2(\square)$; there exists therefore an extension, which we shall also denote by U, to all of $L^2(\mathbf{R})$. It turns out that this extension maps $L^2(\mathbf{R})$ onto all of $L^2(\square)$; this is the Zak transform.

We ask now whether the family (2.3) constitutes a frame, i.e., whether the spectrum of the positive operator

$$P = \sum_m\sum_n|ma,n2\pi/a;\Omega\rangle\,\langle ma,n2\pi/a;\Omega|$$

is bounded away from zero. We shall see that the answer to this question is straightforward for the unitarily equivalent operator UPU^{-1}. The same technique was used in Ref. 10(a) to investigate the question whether the linear span of (2.3) and of similar families is dense.

An easy calculation leads to

$$[UW(ma,n2\pi/a)f](k,q) = e^{-ikma}e^{iqn2\pi/a}(Uf)(k,q) . \tag{2.5}$$

Hence, for $fL^2(\square)$, we have

$$\langle f,UPU^{-1}f\rangle = \sum_m\sum_n\left|\int\int_\square dk\,dq\,e^{ikma}e^{iqn2\pi/a}(U\Omega)(k,q)f(k,q)\right|^2 = \int dk\int dq|U\Omega(k,q)f(k,q)|^2 ,$$

where we have used the basic unitary property of Fourier series expansions. This shows that the operator P is unitarily equivalent to multiplication by $|U\Omega(k,q)|^2$ in $L^2(\square)$. The spectrum of P is therefore exactly the numerical range of the function $|U\Omega(k,q)|^2$. The function $U\Omega$ is given by

Downloaded 14 Oct 2004 to 128.112.156.90. Redistribution subject to AIP license or copyright, see http://jmp.aip.org/jmp/copyright.jsp

$$U\Omega(k,q) = [a\pi^{-3/2}/2]^{1/2}\sum_l \exp[ikla - (q - la)^2/2]$$
$$= [a\pi^{-3/2}/2]^{1/2} \exp(-q^2/2)$$
$$\times\theta_3(a(k - iq)/2, \exp(-a^2/2)),$$

where θ_3 is one of Jacobi's theta functions[18]

$$\theta_3(z,u) = 1 + 2\sum_l u^{l^2} \cos(2lz).$$

The zeros of $U\Omega$ are therefore completely determined by the zeros of θ_3; one finds that the function $U\Omega$ has zeros at the corner of the semiopen rectangle $[-\pi/a,\pi/a) \times [-a/2,a/2)$, and nowhere else. (The fact that $U\Omega$ is zero at the corner can also be seen easily from its series expansion.) This is enough, however, to ensure that the spectrum of the multiplication operator by $|U\Omega(k,q)|^2$, and therefore also of P, contains zero. Therefore the family (2.3) associated to the von Neumann lattice is *not* a frame.

C. A frame of canonical coherent states

Since the family (2.3) with $ab = 2\pi$ is not a frame, it is clear that we have to look at lattices with higher density, i.e., with $ab < 2\pi$. [If $ab > 2\pi$, the linear span of the vectors (2.3) is not even dense.] The construction above, which uses the Zak transform, will not work for arbitrary a and b; if $b \neq 2\pi/a$, then Eq. (2.5) will no longer be true in general. This is due to the fact that in general the operators $W(ma,nb)$ do not mutually commute. It is, however, possible to use again the same construction in the case where the density is an integral multiple of the density for the von Neumann lattice. For the sake of definiteness, we shall consider the case where $ab = \pi$.

We now have to study the operator

$$P = \sum_m \sum_n \left| ma,\frac{n\pi}{a};\Omega\right\rangle \left\langle ma,\frac{n\pi}{a};\Omega\right| .$$

For $n = 2l$, it is clear from (2.3) that

$$|m,2l;\ a,\pi/a\rangle = W(ma,12\pi/a)\Omega = |ma,2\pi l/a;\Omega\rangle .$$

On the other hand,

$$|m,2l+1;\ a,\pi/a\rangle = e^{imab/4}W(ma,2\pi l/a)W(0,\pi/a)\Omega$$
$$= e^{imab/4}|ma,2\pi l/a;\ W(0,\pi/a)\Omega\rangle .$$

Hence

$$P = \sum_m \sum_n \left| ma,\frac{2\pi n}{a};\Omega\right\rangle\left\langle ma,\frac{2\pi n}{a};\Omega\right|$$
$$+ \sum_m \sum_n \left| ma,\frac{2\pi n}{a};\ W\left(0,\frac{\pi}{a}\right)\Omega\right\rangle$$
$$\times\left\langle ma,\frac{2\pi n}{a};W\left(0,\frac{\pi}{a}\right)\Omega\right| .$$

Using (2.5) again, we then see that

$$\langle f,UPU^{-1}f\rangle = \int dk\int dq|f(k,q)|^2(|U\Omega(k,q)|^2$$
$$+ |[UW(0,\pi/a)\Omega](k,q)|^2),$$

where U is again the Zak transform as defined above, in Sec. II B. A calculation of $UW(0,\pi/a)\Omega$ gives

$$[UW(0,\pi/a)\Omega](k,q)$$
$$= 2^{-1/2}\pi^{-3/4}a^{1/2}e^{i\pi q/a} \exp(-q^2/2)$$
$$\times\theta_3[(ak - aiq - \pi)/2, \exp(-a^2/2)],$$

hence

$$|U\Omega(k,q|^2 + |[UW(0,\pi/a)\Omega](k,q)|^2$$
$$= 2^{-1}\pi^{-3/2}\exp(-q^2)a$$
$$\times\{|\theta_3[a(k - iq)/2, \exp(-a^2/1)]|^2$$
$$+ \theta_3[(ak - iqa - \pi)/2, \exp(-a^2/2)]|^2\} .$$

This function is continuous and has no zeros, since the zeros of $\theta_3[u,\exp(-a^2/2)]$ occur only at $u = \pi(m + \frac{1}{2} + ia^2(n + \frac{1}{2}))$. There exist therefore $A,B > 0$ such that

$$A \leqslant |U\Omega(k,q)|^2 + |[UW(0,\pi/a)\Omega](k,q)|^2 \leqslant B;$$

this implies that the set of canonical coherent states $\{|ma,n\pi/a;\ \Omega\rangle\}$ $(m,n\in\mathbf{Z})$ is a frame, with

$$A \leqslant \sum_m \sum_n \left| ma,\frac{n\pi}{a};\Omega\right\rangle \left\langle ma,\frac{n\pi}{a};\Omega\right| \leqslant B .$$

A numerical estimate of A and B gives, in the case $a = 2$,

$$A \geqslant 1.60 ,$$

$$B \leqslant 2.43 .$$

Remark: The above analysis also works if the density of the chosen lattice is another, higher multiple of the critical von Neumann density, i.e., for $ab = 2\pi/n$, where $n = 3,4,\ldots$. The ratio B/A of the upper and lower bound of the frame is clearly a decreasing function of n.

D. Lattices with analyzing wavelets of compact support

We shall now consider families of the type $|na,mb;h\rangle$, where $h(x)$ is a function of compact support.

As an example, consider first the case where $h(x)$ is the characteristic function of an interval $[-L/2,L/2]$, i.e., $h(x) = 1$ if x belongs to this interval, and is zero otherwise. It is then easy to see that, with the choice $a = L$ and $b = 2\pi L$ (hence again $ab = 2\pi$), the family $\{|ma,nb;h\rangle\}$ $(m,n\in\mathbf{Z})$ consisting of the functions $\exp[2\pi inx/L]h(x)$ is an orthonormal basis of $L^2(\mathbf{R})$ and therefore certainly a frame.

For reasons explained in the Introduction, however, we prefer to work with smoother functions h. We shall see that under fairly general conditions, a lattice based on continuous functions of compact support also gives rise to a frame. The price to be paid is a higher density of the lattice; furthermore, the frame will not be tight in general.

Theorem 1: Let $h(x)$ be a continuous function on $\mathbf{R}$, with support in the interval $[-L/2,L/2]$. Assume that $h(x)$ is bounded away from zero in a subinterval $[-\mu L/2,\mu L/2]$ $(0 < \mu < 1)$:

$$|h(x)| \geqslant k,\quad \text{if}|x| \leqslant \mu L/2 \quad (\mu < 1) .$$

Define now a lattice in phase space by taking $a = \mu L$ and $b = 2\pi/L$ (hence $ab = 2\pi\mu$, but the "oversampling parameter" μ^{-1} need not be an integer, contrary to Sec. II C). Consider the set of states

$$\{|ma,nb;h\rangle\} \quad (m,n\in\mathbf{Z});$$

Downloaded 14 Oct 2004 to 128.112.156.90. Redistribution subject to AIP license or copyright, see http://jmp.aip.org/jmp/copyright.jsp

then this set is a frame, with

$$A \geq L \inf_{|x| \leq \mu L/2} |h(x)|^2 \geq k$$

and

$$B \leq L \sup_{x \in \mathbf{R}} \left[\sum_m |h(x+m\mu)|^2 \right]$$

$$\leq L(1+2[\mu^{-1}])(\|h\|_\infty)^2,$$

where $[\mu^{-1}]$ is the largest integer not exceeding μ^{-1}.

Proof: For typographical convenience, write $\Delta = [-L/2, L/2]$. Let f be any element of $L^2(\mathbf{R})$. Then

$$\langle ma,nb;h|f\rangle = e^{-imn\mu\pi} \int_\Delta dx\, h(x) \times e^{-2i\pi nx/L} f(x+m\mu L).$$

Hence, by considering the above integral as $L^{1/2}$ times the nth Fourier coefficient of the function $h(x)f(x+m\mu L)$ defined on the interval Δ,

$$\sum_n |\langle ma,nb;h|f\rangle|^2 = L \int_\Delta dx |h(x)|^2 |f(x+\mu mL)|^2$$

$$\geq k^2 L \int_{\mu\Delta} dx |f(x+\mu mL)|^2.$$

This implies

$$\sum_m \sum_n |\langle ma,nb;h|f\rangle|^2 \geq k|L \int dx |f(x)|^2.$$

On the other hand, we clearly have

$$\sum_m \sum_n |\langle ma,nb;h|f\rangle|^2$$

$$= L \int_\Delta dx \Big(\sum_m |h(x+m\mu L)|^2\Big) |f(x)|^2$$

$$\leq bL \int dx |f(x)|^2,$$

with

$$b = \sup_{x \in \mathbf{R}} \left[\sum_m |h(x+m\mu L)|^2 \right]$$

$$\leq (2[\mu^{-1}]+1)(\|h\|_\infty)^2,$$

and so our assertions are proved.

E. Tight frames with analyzing wavelets of compact support

We keep the assumptions and notations of Sec. II D. The arguments of that subsection show that

$$\sum_n \left| \left\langle m\mu L, \frac{2\pi n}{L}; h \right| f \right\rangle \Big|^2$$

$$= L \int_\Delta dx |h(x)|^2 |f(x+\mu mL)|^2$$

$$= L \int dx |h(x-\mu mL)|^2 |f(x)|^2.$$

Consequently we have

$$\sum_n \sum_m \left| \left\langle m\mu L, \frac{2\pi n}{L}; h \right| f \right\rangle \Big|^2$$

$$= L \int dx |f(x)|^2 \left[\sum_m |h(x+\mu mL)|^2 \right],$$

and we obtain the following result.

Theorem 2: Let $h(x)$ be continuous on $\mathbf{R}$, with support in $[-L/2, L/2]$, and bounded away from zero on $[-\mu L/2, \mu L/2]$, where $0<\mu<1$. Assume furthermore that the function $\Sigma_m |h(x+\mu mL)|^2$ is a constant, i.e., independent of x. Then the family $\{|m\mu L, 2\pi n/L; h\rangle\}$ $(m,n \in \mathbf{Z})$ is a tight frame.

Remark: By the assumptions on h, the sum $\Sigma_m |h(x+\mu mL)|^2$ has only finitely many nonzero terms, and defines a continuous function of x.

We shall now give a procedure for constructing functions h that are k times continuously differentiable and satisfy the condition in Theorem 2:

$$\sum_m |h(x+m\mu L)|^2 = \text{const.} \tag{2.6}$$

Here k may be any positive integer or even ∞. We start by choosing a function g that is $2k$ times continuously differentiable and such that $g(x)=0$ for $x \leq 0$, and $g(x)=1$ for $x \geq 1$. Assume in addition that g is everywhere increasing.

For the sake of simplicity, we shall now assume that $\mu \geq \frac{1}{2}$. We then define h as follows:

$$h(x) = \begin{cases} 0, & \text{for } x \leq -L/2, \\ \{g[(x/L+1/2)/(1-\mu)]\}^{1/2}, & \text{for } -L/2 \leq x \leq -L(2\mu-1)/2, \\ 1, & \text{for } -L(2\mu-1)/2 \leq x \leq L(2\mu-1)/2, \\ \{1-g([x/L-(2\mu-1)/2]/(1-\mu))\}^{1/2}, & \text{for } L(2\mu-1)/2 \leq x \leq L/2, \\ 0, & \text{for } x \geq L/2. \end{cases}$$

The function $h(x)$ defined in this way is non-negative, with support $[-L/2, L/2]$, and equal to 1 on $[-(2\mu-1)L/2, (2\mu-1)L/2]$. Since g is a C^{2k} function, one sees that h is indeed a C^k function. The points $x = \pm(2\mu-1)L/2$, where h becomes constant, have been chosen so that their distance to the furthest edge of supp(f) is exactly μL. It is now easy to check that h fulfills the condition (2.6): for $|x| \leq (2\mu-1)L/2$, one has

$$\sum_m |h(x+m\mu L)|^2 = |h(x)|^2 = 1;$$

and for x in $[(2\mu-1)L/2, L/2]$, one has

$$\sum_m |h(x+m\mu L)|^2$$

$$= |h(x)|^2 + |h(x-\mu L)|^2$$

$$= 1 - g((x/L-(2\mu-1)/2)/(1-\mu))$$

$$+ g(((x-\mu L)/L+1/2)/(1-\mu)) = 1.$$

For x outside $[-(2\mu-1)L/2, L/2]$ the result follows by simple translation. Hence

$$\sum_m |h(x+m\mu L)|^2 = 1,$$

Downloaded 14 Oct 2004 to 128.112.156.90. Redistribution subject to AIP license or copyright, see http://jmp.aip.org/jmp/copyright.jsp

which implies

$$\sum_m \sum_n \left| m\mu L, \frac{n2\pi}{L}; h \right\rangle \left\langle m\mu L, \frac{n2\pi}{L}; h \right| = L1,$$

and we have constructed a tight frame!

The above construction may be clarified by the following easy example.

Example: We define a function h satisfying the condition (2.6) as follows:

$$h(x) = \begin{cases} 0, & \text{if} |x| \geq \pi/2, \\ \cos x, & \text{if} |x| \leq \pi/2. \end{cases}$$

Hence $L = \pi$. We take $\mu = \frac{1}{2}$. Then (see also Fig. 1), with χ the characteristic function of the interval $[-\pi/2, \pi/2]$, one has

$$\sum_m |h(x + \mu mL)|^2 = \sum_m \cos^2\left(x + \frac{m\pi}{2}\right)\chi\left(x + \frac{m\pi}{2}\right)$$
$$= \cos^2 x + \sin^2 x = 1.$$

In this example, the corresponding function g is the function

$$g(x) = \begin{cases} 0, & \text{if } x \geq 0, \\ \sin^2(\pi x/2), & \text{if } 0 \leq x \leq 1, \\ 1, & \text{if } x \geq 1. \end{cases}$$

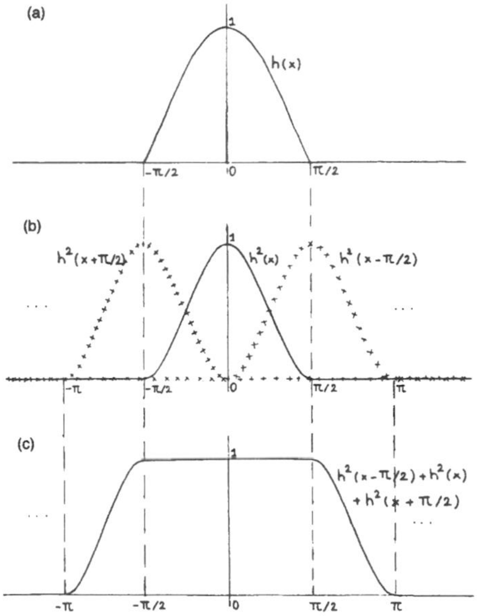

FIG. 1. (a) The function $h(x) = \cos x \, \chi_{[-\pi/2,\pi/2]}(x)$. (b) $h^2(x)$ and two translated copies, $h^2(x+\pi/2)$ and $h^2(x-\pi/2)$. (c) The sum $h^2(x-\pi/2) + h^2(x) + h^2(x+\pi/2)$ is equal to 1, for $-\pi/2 \leq x \leq \pi/2$. Analogously $\Sigma_{l=-N}^{N} h^2(x + l\pi/2) = 1$, for $-N\pi/2 \leq x \leq N\pi/2$, and $\Sigma_{l\in\mathbb{Z}} h^2(x + l\pi/2) = 1$, for all x.

Remark: In our construction, we have assumed that $\mu \geq \frac{1}{2}$. For smaller values of μ, a similar but more complicated construction can be made (see Appendix).

F. A closer look at condition (2.5)

Let h be a function continuous on $\mathbb{R}$, vanishing on the set $\mathbb{R} \setminus [-L/2, L/2]$ and nowhere else. The discussion of the preceding sections shows that the family of functions

$$h_{mn}(x) = e^{2i\pi nx/L} h(x + \mu mL) \quad (0 < \mu < 1; n, m \in \mathbb{Z})$$

is a tight frame if and only if (2.6) holds, i.e., one has

$$\sum_n f(x + na) = \text{const}, \tag{2.7}$$

with $a = \mu L$ and $f(x) = |h(x)|^2$.

In this subsection we shall study the class of functions that satisfy (2.7); at first, we shall not require f to be positive or to have compact support (as opposed to the assumptions on f in the preceding subsections). However, we need to impose some assumptions on f in order to ensure that the left-hand side of (2.7) is well defined. It will be convenient to work with the space $\mathscr{C}$ defined as follows.

Definition:

$\mathscr{C} = \{f: \mathbb{R} \to \mathbb{C};\ f$ is measurable,

and there exists a $C > 0$ and a $K > 1$,

such that $|f(x)| \leq C(1 + |x|)^{-K}\}$.

It is clear that, for f in $\mathscr{C}$, the series $\Sigma_n f(x + na)$ is absolutely convergent, uniformly on the interval $[-a/2, a/2]$. We shall now derive a necessary and sufficient condition for elements of $\mathscr{C}$ to satisfy (2.7). Take f in $\mathscr{C}$. Denote by Δ the interval $[-a/2, a/2]$. Define, for q in Δ,

$$F(q) = \sum_n f(q + na).$$

F is bounded, and hence belongs to $L^2(\Delta)$. We can therefore write its Fourier series as

$$F(q) = \sum_n c_n e^{2\pi i nq/a};$$

this series converges in the L^2 sense and also pointwise almost everywhere. The coefficients c_n are given by

$$c_n = \frac{1}{a} \int_\Delta dq\, e^{-2\pi i nq/a} \sum_n f(q + na)$$
$$= \frac{1}{a} \int dq\, e^{-2\pi i nq/a} f(q) = (2\pi)^{1/2} \frac{1}{a} \hat{f}\left(\frac{2\pi n}{a}\right),$$

where the interchange of integration and summation is justified, since the series converges absolutely. We thus have

$$\sum_n f(q + na) = (2\pi)^{1/2} \frac{1}{a} \sum_n \hat{f}\left(\frac{2\pi n}{a}\right) e^{2\pi i nq/a}. \tag{2.8}$$

This is Poisson's summation formula; see, e.g., Ref. 19, for a derivation of this formula for other classes of functions. It is now clear that the condition $\hat{f}(2\pi n/a) = 0$ for $n \neq 0$ is necessary and sufficient for $F = \text{const}$. Recapitulating, take f in $\mathscr{C}$. Then $\Sigma_n f(q + na)$ is independent of q if and only if, for every nonzero integer n, one has $\hat{f}(2\pi n/a) = 0$.

This motivates the following definition.

Downloaded 14 Oct 2004 to 128.112.156.90. Redistribution subject to AIP license or copyright, see http://jmp.aip.org/jmp/copyright.jsp

Definition:

$$\mathscr{I}_a = \{f\colon \mathbf{R}\to\mathbf{C};\ f\in\mathscr{C} \text{ and } \hat f(2\pi n/a) = 0, \text{ for } n\in\mathbf{Z},\ n\neq 0\}.$$

The set $\mathscr{I}_a$ has many interesting properties. We enumerate a few of them.

(1) $\mathscr{I}_a$ is an ideal under convolution in C, i.e., if $f\in\mathscr{I}_a$, $g\in\mathscr{C}$, then $f*g\in\mathscr{I}_a$.

(2) $\mathscr{I}_a$ is invariant under translations: if $f\in\mathscr{I}_a$ then, for every $y\in\mathbf{R}$, the function $x\to f(y-x)$ also belongs to $\mathscr{I}_a$.

(3) If $f\in\mathscr{I}_a$, then, for every $y>0$, the function $x\to f(yx)$ belongs to $\mathscr{I}_{ya}$.

(4) If $f\in\mathscr{I}_a$ then the integral of f can be replaced by a discrete sum:

$$\int dx\, f(x) = a\sum_n f(na) = a\sum_n f(q+na) \quad (\text{for all } q).$$

Proof:

(1) It is easy to check that for $f,g\in\mathscr{C}$, one has $f*g\in\mathscr{C}$. Since $\mathscr{C}\subset L^1(\mathbf{R})$, we have $(f*\hat g)(k)\hat g(k)$ for all real k; hence $(f*\hat g)(2\pi n/a) = 0$ if $\hat f(2n/a) = 0$.

The assertions (2) and (3) are trivial.

$$(4)\ \int dx\, f(x) = (2\pi)^{1/2}\hat f(0) = a(2\pi)^{1/2}\frac{1}{a}\sum_n \hat f\left(\frac{2\pi n}{a}\right)e^{2\pi i n q/a},$$

since $\hat f(2\pi n/a) = 0$ for $n\neq 0$. Then (2.8) gives

$$\int dx\, f(x) = a\sum_n f(q+na).$$

Remark: Notice that (2.7) can be given an interpretation in terms of Zak transform, defined in Sec. II B. To impose condition (2.7) on a function f amounts to requiring that its Zak transform $(Uf)(k,q)$, defined on $[-\pi/a, \pi/a)\times[-a/2,a/2)$, should be constant along the line $k=0$.

III. THE AFFINE CASE

A. Review of definitions and basic properties

The group of shifts and dilations, or the "$ax+b$ group," is the set $\mathbf{R}^*\times\mathbf{R}$ (where $\mathbf{R}^*$ is the set of nonzero real numbers) with the group law

$$(a,b)(a',b') = (aa',ab'+b).$$

We shall here be concerned with the following representation of this group on $L^2(\mathbf{R})$:

$$[U(a,b)f](x) = |a|^{-1/2}f((x-b)/a). \tag{3.1}$$

This representation is irreducible and square integrable, so there exists a dense set of admissible vectors. The admissibility condition (1.5) can in this case be rewritten as

$$c_g = 2\pi\int dp|p|^{-1}|\hat g(p)|^2 < \infty, \tag{3.2}$$

where $\hat g$ is the Fourier transform of g:

$$\hat g(p) = (2\pi)^{-1/2}\int dx\, e^{-ipx}g(x)dx.$$

The fact that not every element of $L^2(\mathbf{R})$ is admissible with respect to the representation (3.1) stems from the nonunimodularity of the $ax+b$ group. The left-invariant measure on the $ax+b$ group is $a^{-2}\,da\,db$; the right-invariant measure is $|a|^{-1}\,da\,db$.

If g is an admissible vector, we define

$$|a,b;g\rangle = U(a,b)g;$$

such families of vectors can be called "affine coherent states."[11,12] The notation just used does not differ from the notation (2.2), used for the Weyl–Heisenberg group. However, it should be clear from the context which family is used at any one time. The general expression (1.4) in the Introduction can then be written for the $ax+b$ group in the following form:

$$\int a^{-2}\,da\,db\,|a,b;g\rangle\langle a,b;g| = c_g\mathbf{1},$$

where c_g is defined by (3.2).

B. Frames of affine coherent states, based on band-limited analyzing wavelets

The families that we shall consider are defined as

$$\{|a_n^+,b_{mn};g\rangle, |a_n^-,b_{mn};g\rangle\} \quad (m,n\in\mathbf{Z}),$$

where

$$a_n = \exp(\alpha n), \quad b_{mn} = \beta m a_n,$$

for some positive numbers α,β. We shall now derive restrictions on these numbers under which this discrete family is a frame.

The function g is supposed to be band limited, i.e., it is square integrable and its Fourier transform has compact support. We shall also assume that the support of $\hat g$ contains only strictly positive frequencies, i.e., is contained in an interval $[l,L]$, with $0<l<L<\infty$. This will enable us to decouple positive and negative frequencies in our calculations, which will turn out to be very convenient. Note that the requirement $l>0$ automatically guarantees that g is admissible [since the condition (3.2) is trivially satisfied].

Let f be any element of $L^2(\mathbf{R})$. We want to show that, under certain conditions on α,β,g to be derived here, we have

$$A\|f\|^2 \leqslant \sum_m\sum_n\{|\langle a_n^+,b_{nm};g|f\rangle|^2 + |\langle a_n^-,b_{mn};g|f\rangle|^2\} \leqslant B\|f\|^2,$$

with $A>0$, $B<\infty$.

An easy calculation leads to

$$\sum_m |\langle a_n^+,b_{mn};g|f\rangle|^2 = \frac{1}{a_n^+}\sum_m\left|\left(\int_l^L dw\, e^{-iw\beta m}\hat g(w)\hat f\left(\frac{w}{a_n^+}\right)\right)\right|^2.$$

If we impose on β the condition

$$\beta = 2\pi/(L-l),$$

this simplifies to

Downloaded 14 Oct 2004 to 128.112.156.90. Redistribution subject to AIP license or copyright, see http://jmp.aip.org/jmp/copyright.jsp

$$\sum_m |\langle a_n^+, b_{nm}; g | f\rangle|^2 = \frac{1}{a_n^+}\int dw \left|\hat{g}(w)|^2 |\hat{f}\left(\frac{w}{a_n}\right)\right|^2$$
$$= \int dw |\hat{g}(a_n^+ w)|^2 . \quad (3.3)$$

Define now

$F_+(s) = f(e^s)$,

$G(s) = \hat{g}(e^s)$.

Since $a_n^+ = \exp(\alpha n) > 0$ for all $n \in \mathbf{Z}$, and since supp $\hat{g} \subset \mathbf{R}_+$ we can make the substitution $t = e^s$ in the integral (3.3) and write

$$\sum_n \sum_m |\langle a_n^+, b_{mn}; g | f\rangle|^2$$
$$= \int ds \left[\sum_n |G(s+\alpha n)|^2\right] e^s |F+(s)|^2 . \quad (3.4)$$

Since supp $G = [\log l, \log L]$ is compact, only a finite number of terms contribute in the sum $\Sigma_n |G(s+\alpha n)|^2$ for any s. If we define now

$$A = \inf_{s\in\mathbf{R}} \left[\sum_n |G(s+\alpha n)|^2\right],$$

$$B = \sup_{s\in\mathbf{R}} \left[\sum_n |G(s+\alpha n)|^2\right],$$

then clearly

$$A \int_0^\infty dw |\hat{f}(w)|^2$$
$$<\sum_n \sum_m |\langle a_n^+, b_{mn}; g | f\rangle|^2 < B \int_0^\infty dw |\hat{f}(w)|^2 , \quad (3.5)$$

where we have used

$$\int ds\, e^s |F_+(s)|^2 = \int_0^\infty dw |\hat{f}(w)|^2 .$$

A similar calculation can be made for vectors involving a_n^-. Introducing $F_-(s) = \hat{f}(-e^s)$, one finds

$$\sum_n \sum_m |\langle a_n^-, b_{mn}; g | f\rangle|^2$$
$$= \int ds \left[\sum_n |G(s+\alpha n)|^2\right] e^s |F_-(s)|^2 ,$$

hence

$$A \int_{-\infty}^0 dw |\hat{f}(w)|^2$$
$$<\sum_n \sum_m |\langle a_n^-, b_{mn}; g | f\rangle|^2 < B \int_{-\infty}^0 dw |\hat{f}(w)|^2 .$$

Combining this with (3.5) we find thus

$$A \|f\|^2 < \sum_n \sum_m \{|\langle a_n^+, b_{mn}; g | f\rangle|^2$$
$$+ |\langle a_n^-, b_{mn}; g | f\rangle|^2\} < B \|f\|^2 . \quad (3.6)$$

If we can derive conditions on α, β, g, ensuring that $A > 0$, $B < \infty$, then (3.6) implies that under these conditions the set $\{|a^\pm, b_{mn}; g\rangle;\ m,n \in \mathbf{Z}\rangle\}$ is a frame. Since supp $\hat{g} = [\log l, \log L]$, it is clear that A is zero unless $\alpha < \log(L/l)$. If we assume that $\hat{g}(w)$ is a continuous function without zeros in the interior of its support, then this condition is also sufficient to ensure that $A > 0$. Indeed, we then have

$$A > \inf\{|G(s)|^2;\ \log l + (\log (L/l) - \alpha)/2$$
$$< s < \log L\, (\log - (L/l) - \alpha)/2\} .$$

As for B, it is not hard to show that

$$B < \{2[\alpha^{-1} \log(L/l)] + 1\} \|\hat{g}\|_\infty^2 < \infty ,$$

where again we have used the notation $[\mu]$ for the largest integer not exceeding μ.

We have thus derived a set of sufficient conditions ensuring that our construction leads to a frame. The theorem below brings all these conditions together, rewritten in a slightly different form, and states our main conclusion.

Theorem: Let g: $\mathbf{R} \to \mathbf{C}$ satisfy the following conditions: (i) $\hat{g}$ has compact support $[l, kl]$, with $l > 0, k > 1$; and (ii) $|\hat{g}|$ is a continuous function, without zeros in the open interval (l, kl). Take $a \in (0, k)$. Define, for $m, n \in \mathbf{Z}$,

$$a_n^\pm = \pm a^n ,$$

$$b_{mn} = 2\pi / [(k-1) l] m\, a^n .$$

Then the set $\{|a_n^\pm, b_{mn}; g\rangle;\ m,n \in \mathbf{Z}\}$ is a frame, i.e.,

$$A \mathbf{1} < \sum_n \sum_m \{|a_n^+, b_{mn}; g\rangle \langle a_n^+, b_{mn}; g|$$
$$+ |a_n^-, b_{mn}; g\rangle \langle a_n^-, b_{mn}; g|\} < B \mathbf{1} .$$

The lower and upper bounds A and B are given by

$$A = \inf_{w\in\mathbf{R}} \sum_n |\hat{g}(a^n w)|^2$$
$$> \inf\{|\hat{g}(w)|^2;\ w \in [l(k/a)^{1/2}, l(ka)^{1/2}]\} ,$$

$$B = \sup_{w\in R} \sum_n |\hat{g}(a^n w)|^2$$
$$< \{2[\log(k/a)] + 1\} \|\hat{g}\|_\infty^2 .$$

Remarks: (1) The same conclusions can be drawn under slightly less restrictive conditions on g. Strictly speaking we only need $\|\hat{g}\|_\infty < \infty$ and $\inf_{w\in\Delta} |\hat{g}(w)| > 0$ for any closed interval Δ contained in (l, kl); both these conditions are of course satisfied if $\hat{g}$ is continuous and has no zeros in (l, kl).

(2) As the calculations preceding the above theorem show, the positive and negative frequencies decouple neatly. It is therefore possible to choose a different function g_- (and accordingly, also a different lattice a_n, b_{mn}) for the negative frequency domain than for the positive frequency domain.

We have thus constructed a frame, based on a band-limited function g, under fairly general conditions on g. In general, the ratio B/A, comparing the upper with the lower bound, will be larger than 1. Again, however, as in the Weyl–Heisenberg case, it is possible to choose g in such a way that the frame becomes tight, i.e., $B/A = 1$; such tight frames have been used previously by one of us (Y. M.) in Ref. 13(a); they were also used in Ref. 13(b). The construction of such a frame follows more or less the same lines as in the Weyl–Heisenberg case (see Sec. II E); we shall show in the next subsection how the construction works in the present case.

Downloaded 14 Oct 2004 to 128.112.156.90. Redistribution subject to AIP license or copyright, see http://jmp.aip.org/jmp/copyright.jsp

C. Tight frames based on band-limited functions

We shall stick to the same construction as in the preceding subsection, and try to find a function g such that the frame based on g is quasiorthogonal.

Going back to (3.4), it is clear that the frame will be quasiorthogonal if and only if

$$\sum_n |G(s+\alpha n)|^2 = \text{const}, \tag{3.7}$$

where $G(s) = \hat{g}(e^s)$, and $\alpha = \log(a)$ with $a<k$, $\operatorname{supp}\hat{g} = [l,kl]$ $(l>0,\ k>1)$.

This condition (3.7) is exactly the same as the condition (2.6) in the Weyl–Heisenberg case; the analog of L is here $\log(kl) - \log(l) = \log(k)$, while the role of μ is played by $\alpha/\log k = \log(a)/\log(k) < 1$. The only difference is that the function G need not be centered around zero, as was supposed in Sec. II E.

We can therefore copy the construction made in Sec. II E to define a suitable G, hence a suitable g. Explicitly, and directly in terms of $\hat{g}$ rather than in terms of G, this gives

$$\hat{g}(w) = \begin{cases} 0, & \text{for } w<l, \\ [q(\log(w/l)/\log(k/a))]^{1/2}, & \text{for } l<w<lk/a, \\ 1, & \text{for } lk/a<w<al \\ [1-q(\log(w/l)/\log(k/a))]^{1/2} & \text{for } al<w<kl, \\ 0, & \text{for } w>kl, \end{cases} \tag{3.8}$$

where q is a C^{2k} function such that

$$q(x) = \begin{cases} 0, & \text{for } x<0, \\ 1, & \text{for } x>1, \end{cases} \tag{3.9}$$

a strictly increasing between 0 and 1.

Notice that we have assumed that $a^2>k$; this is equivalent to the assumption $\mu>\frac{1}{2}$ in Sec. II E. If $a^2<k$, a similar but more complicated construction can be made.

For $\hat{g}$ constructed as above, the condition (3.7) is satisfied;

$$\sum_n |G(s+\alpha n)|^2 = 1,$$

which implies that the corresponding frame is tight. We thus have proved the following theorem.

Theorem: Let the function $\hat{g}$, with compact support $[l,kl]$ $(l>0, k>1)$, be constructed according to (3.8), with $a>k^{1/2}$ where q is a function satisfying (3.9). Then the set of vectors

$$\{|a^n, 2\pi m a^n/(k-1)l; g\rangle,$$
$$|-a^n, 2\pi m a^n/(k-1)l; g\rangle;\ m,n\in\mathbb{Z}\}$$

(i.e., the set of functions

$$|a|^{-n/2} g[a^{-n}x + 2\pi m/(k-1)l],$$
$$|a|^{-n/2} g[-a^{-n}x + 2\pi m/(k-1)l])$$

is a quasiorthogonal frame in $L^2(\mathbb{R})$, with $A=B=1$. This means the following: If f is any function in $L^2(\mathbb{R})$ and if we define coefficients $f_{mn}^{(\pm)}$ $(m,n\in\mathbb{Z})$ by

$$f_{mn}^{(+)} = |a|^{-n/2}\int dx\, g\Big[a^{-n}x + \frac{2\pi m}{(k-1)l}\Big]^* f(x),$$

$$f_{mn}^{(-)} = |a|^{-n/2}\int dx\, g\Big[-a^{-n}x + \frac{2\pi m}{(k-1)l}\Big]^* f(x),$$

then

$$f(x) = \sum_m\sum_n f_{mm}^{(+)}|a|^{-n/2} g\Big[a^{-n}x + \frac{2\pi m}{(k-1)l}\Big] + \sum_m\sum_n f_{mm}^{(-)}|a|^{-n/2} g\Big[-a^{-n}x + \frac{2\pi m}{(k-1)l}\Big],$$

where the sum converges in $L^2(\mathbb{R})$.

Let us give some specific examples.

Example 1: We take $l=1$, $k=3$, $a=\sqrt{3}$. Define

$$\hat{g}(w) = \begin{cases} 0, & \text{for } w<1, \\ \sin[\pi\log w/\log 3], & \text{for } 1<w<3, \\ 0, & \text{for } w>3. \end{cases}$$

The corresponding g cannot be calculated in closed analytic form. A graph of $\operatorname{Re} g$, $\operatorname{Im} g$ is given in Fig. 2. The corresponding function q is the same as in the example in Sec. II E:

$$q(x) = \begin{cases} 0, & \text{for } x<0, \\ \sin^2(\pi x/2), & \text{for } 0<x<1, \\ 1, & \text{for } x>1. \end{cases}$$

Example 2: We take $l=1$, $k=4$, $a=2$. Define

$$\hat{g}(w) = \begin{cases} 0, & \text{for } w<1, \\ 2\sqrt{2}[\log w/\log 2]^2, & \text{for } 1<w<\sqrt{2}, \\ [1-8(1-\log w/\log 2)^4]^{1/2}, & \text{for } \sqrt{2}<w<2\sqrt{2}, \\ 2\sqrt{2}[2-(\log w/\log 2)]^2, & \text{for } 2\sqrt{2}<w<4, \\ 0, & \text{for } w>4. \end{cases}$$

The corresponding function q is

$$q(x) = \begin{cases} 0, & \text{for } x<0, \\ 8x^4, & \text{for } 0<x<\frac{1}{2}, \\ 1-8(1-x)^4, & \text{for } \frac{1}{2}<x<1, \\ 1, & \text{for } x>1. \end{cases}$$

Graphs of $\operatorname{Re} g$, $\operatorname{Im} g$ are given in Fig. 3.

Because of the correspondence, noted above, between tight frames for the $ax+b$ group and the Weyl–Heisenberg group, all the examples given in the Appendix for the Weyl–Heisenberg group can easily be transposed to the present case.

D. A closer look at the necessary and sufficient condition

The necessary and sufficient condition that a band-limited function g, concentrated on positive frequencies, has to satisfy in order to generate a tight frame, is given by (3.7). This can be rewritten as

$$\sum_n f(a^n w) = \text{const}, \tag{3.10}$$

where $f(w) = |\hat{g}(w)|^2$.

Downloaded 14 Oct 2004 to 128.112.156.90. Redistribution subject to AIP license or copyright, see http://jmp.aip.org/jmp/copyright.jsp

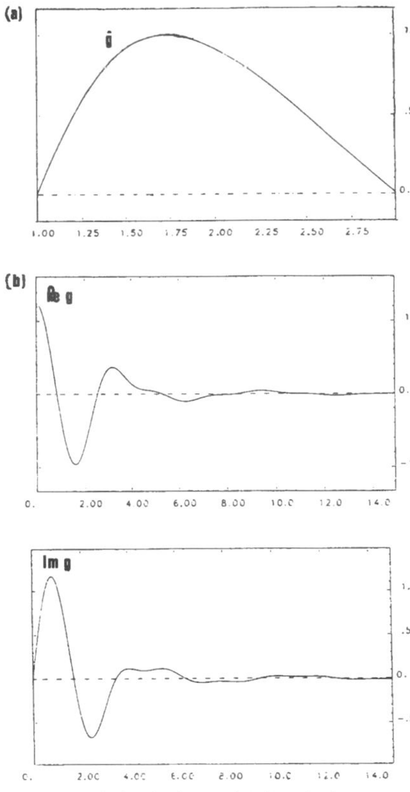

FIG. 2. (a) The function $\hat{g}(w) = \sin(\pi \log w/\log 3)\chi_{[1,3]}(w)$. (b) Real and imaginary parts of the inverse Fourier transform g of $\hat{g}$.

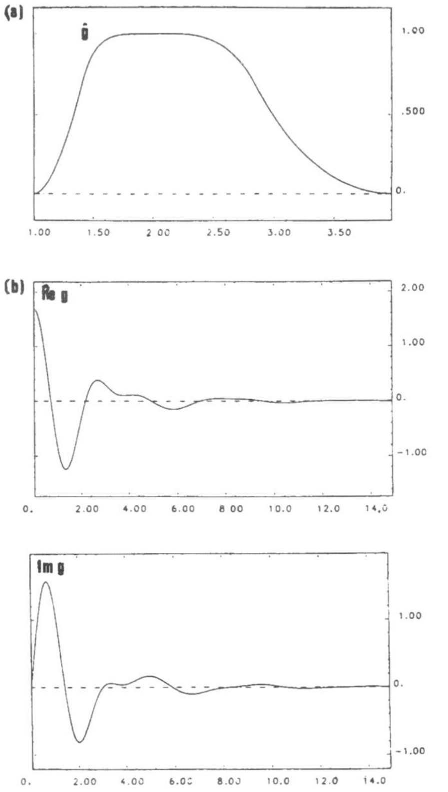

FIG. 3. (a) The function $\hat{g}(w)$ defined in example 2 of Sec. III D. (b) Real and imaginary parts of the inverse Fourier transform g or $\hat{g}$.

In this subsection we shall study the class of functions f: $\mathbf{R}_+ \to \mathbf{C}$, satisfying (3.10); for the purpose of this subsection only, we shall not require f to be positive or to have compact support. This study will be completely analogous to our study in Sec. II F of the functions satisfying condition (2.7); since the arguments run along exactly the same lines, we shall not go into as much detail here. The main difference is that we shall work with the Mellin transform of f rather than with its Fourier transform; this, of course, is due to the difference between (3.10), where the constant enters multiplicatively, and (2.7) where it enters additively.

The Mellin transform F of f: $\mathbf{R}_+ \to \mathbf{C}$ is defined by

$$F(s) = \int_0^\infty w^{s-1} f(w)\,dw\,;$$

the inversion formula is

$$f(w) = \frac{1}{2\pi i}\int ds\; t^{-s} F(s)\,,$$

where the integral is taken from $c - i\infty$ to $c + i\infty$ and where $c > 0$ has to be chosen so that the integral converges.

For the sake of convenience we shall restrict ourselves to functions $f \in \mathscr{C}^\sim$, where $\mathscr{C}^\sim$ is defined below.

Definition:

$$\mathscr{C}^\sim = \{f\colon \mathbf{R}_+ \to \mathbf{C};\ f \text{ is measurable and there exists } C>0 \text{ and } k>1 \text{ such that } |f(w)| < C(1 + |\log w|)^{-k}\}\,.$$

For functions $f \in \mathscr{C}^\sim$ the series $\Sigma_n f(a^n t)$ is absolutely convergent, uniformly on $(1,a]$; the Mellin transform F of f is well defined for purely imaginary arguments, and the inversion formula applies, with $c = 0$.

Notice that if we define g: $\mathbf{R} \to \mathbf{C}$ by $g(x) = f(e^x)$, we immediately find

$$f \in \mathscr{C}^\sim \Leftrightarrow g \in \mathscr{C}$$

(where $\mathscr{C}$ is defined in Sec. II F) and

$$F(ik) = (2\pi)^{1/2}\hat{g}(k) \quad (k \in \mathbf{R})\,.$$

This enables us to translate the results of Sec. II F to the present situation.

Downloaded 14 Oct 2004 to 128.112.156.90. Redistribution subject to AIP license or copyright, see http://jmp.aip.org/jmp/copyright.jsp

We have, for $f \in \mathscr{C}^-$,

$$\sum_n f(a^n w) = \frac{1}{\log a} \sum_n F\left(\frac{2\pi i n}{\log a}\right) w^{2\pi i n/\log a},$$

which leads to the following theorem.

Theorem: Take $f \in \mathscr{C}^-$. Then $\Sigma_n f(a^n w)$ is independent of w if and only if $F(2\pi i n/\log a) = 0$ for all nonzero integers n.

This then motivates the following definition.

Definition:

$$\mathscr{I}_a^- = \{f: \mathbf{R}_+ \to \mathbf{C}, \text{ and } F(2\pi i n/\log a) = 0, \text{ for } n \in \mathbf{Z},\ n \neq 0\}.$$

The following theorem lists a few properties of $\mathscr{I}_a^-$.

Theorem: (1) $\mathscr{I}_a^-$ is an ideal in $\mathscr{C}^-$ under "Mellin convolution," i.e., if $f \in \mathscr{I}_a^-, g \in \mathscr{C}^-$, then the function $f \underset{M}{*} g$ defined by

$$(f \underset{M}{*} g)(w) = \int_0^\infty \frac{du}{u} f(u) g\left(\frac{t}{u}\right),$$

belongs to $\mathscr{I}_a^-$.

(2) $\mathscr{I}_a^-$ is invariant under dilations: if $f \in \mathscr{I}_a^-$, then, for every $u \neq 0$, the function $w \to f(uw)$ also belongs to $\mathscr{I}_a^-$.

(3) If $f \in \mathscr{I}_a$, then the integral of $w^{-1} f(w)$ can be replaced by a discrete sum:

$$\int_0^\infty dw\, w^{-1} f(w) = \log a \sum_n f(a^n)$$

$$= \log a \sum_n f(a^n t), \quad \text{for all } t.$$

Notice that the "Mellin convolution" defined above is exactly the convolution in the sense of Ref. 19, with respect to the multiplicative group of nonzero real numbers. It is obvious that the Mellin transform of a "Mellin convolution" of two functions is given, up to a constant factor, by the product of the two Mellin transforms.

ACKNOWLEDGMENT

One of us (I. D.) would like to thank the Centre de Physique Theorique 2, CNRS-Marseille, where part of this work was done, for its hospitality and support.

APPENDIX: MORE EXAMPLES

We give a few more examples of functions h supported in $[-L/2, L/2]$ and satisfying the condition

$$\sum_{n \in \mathbf{Z}} |h(x + n\mu L)|^2 = \text{const}. \tag{A1}$$

We start by showing how one can extend the construction given in Sec. II E, for the case $\mu \geq \frac{1}{2}$, to more general μ.

For $2^{-(k+1)} \leq \mu \leq 2^{-k}$, with $k \geq 0$, one can always define $\mu^\sim = 2^k \mu$. Obviously $\frac{1}{2} \leq \mu^\sim \leq 1$. If we replace μ by $\mu^\sim$ in the construction of Sec. II E, we obtain a function h satisfying

$$\sum_{n \in \mathbf{Z}} |h(x + n 2^k \mu L)|^2 = \text{const} = C.$$

Hence

$$\sum_{n \in \mathbf{Z}} |h(x + n\mu L)|^2 = 2^k C.$$

Functions h constructed in this way thus obviously satisfy condition (A1). It is clear from the construction that the tight frame $\{|n\mu L, m2\pi/L; h\rangle\}$ generated by h is in this case a superposition of the tight frame $\{|n\mu^\sim L, m2\pi/L; h\rangle\}$ and translated copies of this frame.

There also exist, of course, functions h satisfying (A1), for $\mu < \frac{1}{2}$, which cannot be reduced to the case $\frac{1}{2} \leq \mu \leq 1$. We give here an example of such a function, for the case $\mu = \frac{1}{3}$.

Let g be again a C^{2k} function, strictly increasing, such that

$$g(x) = \begin{cases} 0, & \text{for } x \leq 0, \\ 1, & \text{for } x \geq 1. \end{cases}$$

Define then

$$h(x) = \begin{cases} 0, & \text{for } x \leq -L/2, \\ \{g[(6x + 3L)/4L]\}^{1/2}, & \text{for } -L/2 \leq x \leq L/6, \\ \{1 + g(1/2) - g[(6x + L)/4L] \\ \quad - g[(6x - L)/4L]\}^{1/2}, & \\ \quad \text{for } L/6 \leq x \leq L/2, & \\ 0, & \text{for } x \geq L/2. \end{cases}$$

Obviously h has support $[-L/2, L/2]$. One can check that h is a C^k function satisfying

$$\sum_{n \in \mathbf{Z}} \left| h\left(x + \frac{nL}{3}\right) \right|^2 = 1 + g\left(\frac{1}{2}\right).$$

This example can also be adapted to cover the case $\frac{1}{2} > \mu > \frac{1}{4}$ (instead of only $\mu = \frac{1}{3}$).

Finally, note that another class of examples, for the special cases $\mu = 1/2(k+1)$, with k a positive integer, can be constructed with the help of spline functions. Choose a knot sequence $t = (t_j)_{j \in \mathbf{Z}}$ with equidistant knots, $t_{j+1} - t_j = d > 0$ for all j. Let $B_{j,2k+2,t}$ be the jth B-spline of order $2k + 2$ for the knot sequence t. (For the definition of B-splines, see, e.g., Ref. 20.) Then the $B_{j,2k+2,t}$ are all translated copies of $B_{0,2k+2,t}$:

$$B_j(x) = B_0(x - jd).$$

The B_j are (positive) C^{2k} functions, with support $[t_j, t_{j+2(k+1)}] = [t_j, t_j + 2(k+1)d]$. They have, moreover, the property that

$$\sum_{j \in \mathbf{Z}} B_j(x) = 1, \quad \text{for all } x$$

(only a finite number of terms contribute for any x). It is now easy to check that the function h, defined by

$$h(x) = \{B_{0,k,t}[t_0 + (k+1)d(2x + L)/L]\}^{1/2},$$

is a C^k function with support $[-L/2, L/2]$, satisfying condition (A1) with $\mu = 1/2(k+1)$.

[1] R. M. Young, *An Introduction to Non-harmonic Fourier Series* (Academic, New York, 1980).
[2] R. J. Duffin and A. C. Schaeffer, Trans. Am. Math. Soc. **72**, 341 (1952).

Downloaded 14 Oct 2004 to 128.112.156.90. Redistribution subject to AIP license or copyright, see http://jmp.aip.org/jmp/copyright.jsp

[3]I. Daubechies and A. Grossmann "Frames in the Bargmann space of entire functions" (to be published).
[4]A. Grossmann and J. Morlet, SIAM J. Math. Ann. **15**, 723 (1984); A. Grossmann and J. Morlet, "Decomposition of functions into wavelets of constant shape, and related transforms," to appear in *Mathematics and Physics, Lectures on Recent Results* (World Scientific, Singapore, 1985); P. Goupillaud, A. Grossmann, and J. Morlet, Geoexploration **23**, 85 (1984).
[5]M. Bertero, C. De Mol, and G. Viano, "The stability of inverse problems," in *Inverse Scattering Problems in Optics*, edited by H. P. Baltes (Springer, Berlin, 1980).
[6](a) M. Duflo and C. C. Moore, J. Funct. Anal. **21**, 208 (1976); A. L. Carey, Bull. Austr. Math. Soc. **15**, 1 (1976); (b) S. A. Gaal, *Linear Analysis and Representation Theory* (Springer, Berlin, 1973); (c) A. Grossmann, J. Morlet, and T. Paul, J. Math. Phys. **26**, 2473 (1985).
[7]J. von Neumann, *Mathematical Foundation of Quantum Mechanics* (Princeton U. P., Princeton, NJ, 1985) (English translation).
[8]V. Bargmann, P. Butera, L. Girardello, and J. R. Klauder, Rep. Math. Phys. **2**, 221 (1971).
[9]A. M. Perelomov, Teor. Mat. Fiz. **6**, 213 (1971).
[10](a) H. Bacry, A. Grossmann, and J. Zak, Phys. Rev. B **12**, 1118 (1975); (b) R. Balian, C. R. Acad. Sci. Paris, Ser. 2 **292**, 1357 (1981).
[11]E. W. Aslaksen and J. R. Klauder, J. Math. Phys. **10**, 2267 (1969).
[12]T. Paul, J. Math. Phys. **25**, 3252 (1984); A. Grossmann and T. Paul, "Wave functions on subgroups of the group of affine canonical transformations," in *Lecture Notes in Physics*, Vol. 211 (Springer, Berlin, 1984).
[13](a) Y. Meyer, "La transformation en ondellettes et les nouveaux paraproduits," to be published; (b) M. Frasier and B. Jawerth, "Decomposition of Besov spaces," to be published.
[14]Y. Meyer, "La transformation en ondelettes: de la recherche pétrolière à la géométrie des espaces de Banach," to be published; P. G. Lemarié and Y. Meyer, "Ondelettes et bases hilbertiennes," to be published.
[15]A. Grossmann and T. Paul (in preparation).
[16]J. Klauder and B. S. Skagerstam, *Coherent States. Applications in Physics and Mathematical Physics* (World Scientific, Singapore, 1985).
[17]J. Zak, in *Solid State Physics*, edited by H. Ehrenreich, F. Seitz, and D. Turnbull (Academic, New York, 1972), Vol. 27.
[18]M. Abramowitz and I. Stegun, *Handbook of Mathematical Functions* (Dover, New York, 1964).
[19]E. Hewitt and K. A. Ross, *Abstract Harmonic Analysis, II* (Springer, Berlin, 1970).
[20]C. de Boor, *A Practical Guide to Splines* (Springer, Berlin, 1978).

Downloaded 14 Oct 2004 to 128.112.156.90. Redistribution subject to AIP license or copyright, see http://jmp.aip.org/jmp/copyright.jsp

Decomposition of Besov Spaces

MICHAEL FRAZIER & BJÖRN JAWERTH

1. Introduction. In the last decade, many function or distribution spaces have been found to admit a decomposition, in the sense that every member of the space is a linear combination of basic functions of a particularly elementary form. Such decompositions simplify the analysis of the spaces and the operators acting on them. Here we obtain two types of decompositions for distributions in the homogeneous Besov spaces $\dot{B}_p^{\alpha q}$, $-\infty < \alpha < +\infty$, $0 < p$, $q \leqq +\infty$, and present some applications of these results.

Defining the Fourier transform by $\hat{f}(\xi) = \int f(x)e^{-ix\cdot\xi}dx$, let $\{\varphi_\nu\}_{\nu\in\mathbf{Z}}$ be a family of functions on $\mathbf{R}^n$ satisfying

(1.1) $$\varphi_\nu \in \mathcal{S},$$

(1.2) $$\operatorname{supp} \hat{\varphi}_\nu \subseteq \left\{\xi \in \mathbf{R}^n : \frac{1}{2} \leqq 2^{-\nu}|\xi| \leqq 2\right\},$$

(1.3) $$|\hat{\varphi}_\nu(\xi)| \geqq c > 0 \text{ if } \frac{3}{5} \leqq 2^{-\nu}|\xi| \leqq \frac{5}{3},$$

and

(1.4) $$|\partial^\gamma \hat{\varphi}_\nu(\xi)| \leqq c_\gamma 2^{-\nu|\gamma|} \qquad \text{for every multi-index } \gamma.$$

The Besov space $\dot{B}_p^{\alpha q}$, $-\infty < \alpha < +\infty$, $0 < p$, $q \leqq +\infty$, is the collection of all $f \in \mathcal{S}'/\mathcal{P}$ (tempered distributions modulo polynomials) such that

$$\|f\|_{\dot{B}_p^{\alpha q}} = \left(\sum_{\nu\in\mathbf{Z}} (2^{\nu\alpha}\|\varphi_\nu * f\|_{L^p})^q\right)^{1/q} < +\infty,$$

with the usual interpretation if $q = +\infty$. This definition is independent of the family $\{\varphi_\nu\}$ satisfying (1.1–4); see [20], p. 64.

Our purpose is to show that each $f \in \dot{B}_p^{\alpha q}$ can be decomposed into a sum of simple building blocks. The building blocks in our first decomposition are similar to the atoms in the atomic decomposition of Hardy spaces $H^p(\mathbf{R}^n)$, $0 < p \leqq 1$ ([10], [18], [6], [30]). We define an (α,p)-atom $a(x)$ $(-\infty < \alpha < +\infty$, $0 < p \leqq +\infty)$ to be a function satisfying, for some cube $Q \subseteq \mathbf{R}^n$,

(1.5) $$\operatorname{supp} a \subseteq 3Q,$$

Indiana University Mathematics Journal ©, Vol. 34, No. 4 (1985)

$$(1.6)\qquad |\partial^\gamma a(x)| \leqq |Q|^{\alpha/n-1/p-|\gamma|/n} \qquad \text{if } |\gamma| \leqq K,$$

and

$$(1.7)\qquad \int x^\gamma a(x)\,dx = 0 \qquad \text{if } |\gamma| \leqq N,$$

where $K \geqq ([\alpha]+1)_+$ and $N \geqq \max([n(1/p-1)_+ - \alpha], -1)$ are fixed integers. Here $x_+ = \max(x,0)$, $[x]$ is the greatest integer in x, and $3Q$ is the cube in $\mathbf{R}^n$ concentric with Q but with side length three times the side length $\ell(Q)$ of Q. In (1.7), $N = -1$ means that $a(x)$ is not required to have any vanishing moments.

We write a_Q for an atom satisfying (1.5–7) for a given cube Q, and adopt the convention hereafter that whenever Q appears as a summation index, the sum runs only over dyadic cubes. Our result is the (quasi-) norm equivalence

$$(1.8)\qquad \|f\|_{\dot B_p^{\alpha q}} \approx \inf\left\{\left(\sum_{\nu\in\mathbf{Z}}\left(\sum_{\ell(Q)=2^{-\nu}}|s_Q|^p\right)^{q/p}\right)^{1/q} : f = \lim_{j\to\infty}\sum_{\nu=-j}^{j}\sum_{\ell(Q)=2^{-\nu}} s_Q a_Q \right.$$
$$\left.\text{(in } \mathcal{S}'/\mathcal{P}\text{) and each } a_Q \text{ is an } (\alpha,p)\text{-atom}\right\}.$$

In our other decomposition of $\dot B_p^{\alpha q}$, the building blocks, although not of compact support, are taken from a fixed, explicitly given family of functions which have simple properties. Fix $\psi \in \mathcal{S}$ satisfying supp $\hat\psi(\xi) \subseteq \{\xi \in \mathbf{R}^n : |\xi| \leqq \pi\}$, $\int x^\gamma \psi(x)\,dx = 0$ if $|\gamma| \leqq N$, and $\hat\psi(\xi) \geqq c > 0$ if $1/2 \leqq |\xi| \leqq 2$ (N is the fixed integer above). For each $\nu \in \mathbf{Z}$ and $k = (k_1,\ldots,k_n) \in \mathbf{Z}^n$, set

$$(1.9)\qquad Q_{\nu k} = \{x = (x_1,\ldots,x_n) \in \mathbf{R}^n : k_i 2^{-\nu} \leqq x_i < (k_i+1)2^{-\nu}, i = 1,\ldots,n\},$$

and define

$$(1.10)\qquad \psi_Q(x) = |Q|^{\alpha/n-1/p}\psi(2^\nu x - k) \qquad \text{if } Q = Q_{\nu k}.$$

We will prove that

$$(1.11)\qquad \|f\|_{\dot B_p^{\alpha q}} \approx \inf\left\{\left(\sum_{\nu\in\mathbf{Z}}\left(\sum_{\ell(Q)=2^{-\nu}}|s_Q|^p\right)^{q/p}\right)^{1/q} : f \right.$$
$$\left. = \lim_{j\to\infty}\sum_{\nu=-j}^{j}\sum_{\ell(Q)=2^{-\nu}} s_Q\psi_Q \text{ (in } \mathcal{S}'/\mathcal{P}\text{)}\right\}.$$

In fact, in our representation $f = \sum_Q s_Q\psi_Q$, each s_Q for $Q = Q_{\nu k}$, is a multiple of the "sample value" $\varphi_\nu * f(x_Q)$ for appropriate $\{\varphi_\nu\}$ satisfying (1.1–4), where $x_Q = 2^{-\nu}k$.

After completing the proofs of (1.8) and (1.11) in §2 and §3, we consider the space of functions of bounded mean oscillation (*BMO*). Our decomposition (1.8) of $\dot B_p^{\alpha q}$ corresponds to the decomposition of *BMO* given by Uchiyama ([29], pp. 224–228). In §4, we prove the analogue of (1.11) for *BMO*.

Both of our decomposition methods utilize a discrete version of Calderón's reproducing formula ([7], cf. [20], pp. 52–54), and a classical result of Plancherel-Pólya [23]. The primary difference between these methods is the manner in which a certain convolution is written as a discrete sum. In (1.11) this is done on the Fourier transform, or frequency, side using a Fourier series, while in (1.8) it is done directly on the time side.

Each decomposition has advantages. For example, in §5 we see that the fact that the (α,p)-atoms in (1.8) have compact support is convenient for the consideration of "trace" problems. The well-known result that

$$\operatorname{Tr} \dot{B}_p^{\alpha q}(\mathbf{R}^n) = \dot{B}_p^{\alpha-1/p,q}(\mathbf{R}^{n-1}) \qquad \text{if } \alpha - \frac{1}{p} > (n-1)\left(\frac{1}{p} - 1\right)_+$$

is immediate, and the result that

$$\operatorname{Tr} \dot{B}_p^{1/p,1}(\mathbf{R}^n) = L^p(\mathbf{R}^{n-1}) \qquad \text{if } 1 \leqq p < +\infty$$

([21], [15], [5]) is extended to show that Tr $\dot{B}_p^{1/p,q}(\mathbf{R}^n) = L^p(\mathbf{R}^{n-1})$ whenever $0 < p < +\infty$, $q \leqq \max(1,p)$.

On the other hand, one advantage of (1.11) is the simplicity of the building blocks ψ_Q on the frequency side. In §6, this is exploited to give a very simple proof that $\dot{B}_p^{\alpha q}$ has the lower majorant property if $0 < p \leqq 1$.

There are results for the inhomogeneous Besov spaces $B_p^{\alpha q}$, corresponding to those described above for $\dot{B}_p^{\alpha q}$. These are stated in §7. In conclusion, in §8, some possible extensions of the ideas in this paper are mentioned.

There are many antecedents in the literature of our methods and results. Calderón's formula plays a key role in many decomposition results; in particular it is crucial to the simplest known proofs of the atomic decomposition of $H^p(\mathbf{R}^n)$, $0 < p \leqq 1$ ([6], [30]). Uchiyama [29], following Chang-Fefferman [8], used a similar formula to prove the *BMO* decomposition mentioned above. Our application of Calderón's formula to $\dot{B}_p^{\alpha q}$ in general was prompted by recent work of Wilson [31], who used it to obtain (1.8) for the case of the "special atoms" space $\dot{B}_1^{01}(\mathbf{R}^1)$. Another application was made by Cohen [9], who decomposed the spaces $H^{(p,q)}$, which are similar to the Besov spaces. Methods not relying on Calderón's formula have been used by de Souza et al. [12] to obtain (1.8) for $\dot{B}_1^{\alpha 1}$, $0 < \alpha < 1$, on the circle.

A motivation for our second decomposition is the work of Coifman-Rochberg [11], where Bergman spaces are decomposed into building blocks obtained from the Bergman kernel. Rochberg-Semmes [26] obtain similar results for *BMO*. Ricci-Taibleson [24] employ related ideas to prove (1.8) in $\mathbf{R}^1$ for $\alpha < 0$; their work is extended to $\mathbf{R}^n$ by Bui [4]. Further references to related work can be found in Rochberg's survey [25].

Throughout the paper, "c" denotes constants which may differ at each appearance, possibly depending on the dimension or other parameters.

Acknowledgements. The first-named author would like to thank M. Wilson for his notes on his recent work, R. Johnson for providing a difficult-to-obtain reference, and M. Taibleson for background during the early stages of this investigation. Both authors would like to thank Y.-S. Han and our other colleagues at Washington University in St. Louis for their interest.

2. Decomposition of $\dot{B}_p^{\alpha q}$. The proofs of both of our decomposition results depend on two lemmas, which are of independent interest. The first is a general result about distributions.

Lemma 2.1. *Suppose $f \in \mathcal{S}'/\mathcal{P}$ and that φ and ψ are functions satisfying*

$$\varphi, \psi \in \mathcal{S}, \tag{2.1}$$

$$\operatorname{supp} \hat{\varphi} \subseteq \{\xi : |\xi| < \pi\}, \qquad \operatorname{supp} \hat{\psi} \subseteq \{\xi : |\xi| \leqq \pi\}, \tag{2.2}$$

and

$$\sum_{\nu \in \mathbf{Z}} \hat{\varphi}(2^\nu \xi)\hat{\psi}(2^\nu \xi) = 1 \qquad \textit{if } \xi \in \mathbf{R}^n \setminus \{0\}. \tag{2.3}$$

If $\varphi_\nu(x) = 2^{\nu n}\varphi(2^\nu x)$ and $\psi_\nu(x) = 2^{\nu n}\psi(2^\nu x)$, then

$$f(\cdot) = \sum_{\nu \in \mathbf{Z}} 2^{-\nu n} \sum_{k \in \mathbf{Z}^n} \varphi_\nu * f(2^{-\nu}k)\psi_\nu(\cdot - 2^{-\nu}k). \tag{2.4}$$

In (2.4), the convergence of the right-hand side, as well as the equality, is in $\mathcal{S}'/\mathcal{P}$.

Proof. By (2.3), we have

$$f = \sum_{\nu \in \mathbf{Z}} \psi_\nu * \varphi_\nu * f. \tag{2.5}$$

(See Remark 2.2 below for details about convergence in (2.5).) Hence, (2.4) will follow from

$$\psi_\nu * \varphi_\nu * f(x) = 2^{-\nu n} \sum_{k \in \mathbf{Z}^n} \varphi_\nu * f(2^{-\nu}k)\psi_\nu(x - 2^{-\nu}k). \tag{2.6}$$

To prove (2.6), we note that $\varphi_\nu * f$ is slowly increasing, and hence,

$$f_{\nu,\delta}(x) = \varphi_\nu * f(x) \prod_{i=1}^n \left(\frac{\sin(\delta x_i)}{\delta x_i}\right)^j \in L^2$$

if j is large enough. Also, by (2.2) and the fact that

$$\operatorname{supp}[(\sin \delta x)/\delta x]\hat{} \subseteq [-\delta, \delta], \operatorname{supp} \hat{f}_{\nu,\delta}(\xi) \subseteq \{\xi : |\xi| \leqq 2^\nu \pi\}$$

if δ is sufficiently small. We have

$$f_{\nu,\delta} * \psi_\nu(x) = (2\pi)^{-n} \int \hat{f}_{\nu,\delta}(\xi)\hat{\psi}_\nu(\xi)e^{ix\cdot\xi}d\xi. \tag{2.7}$$

Extending $\hat{\psi}_\nu(\xi)e^{ix\cdot\xi}$ periodically with period $2^{\nu+1}\pi$ in each variable and representing $\hat{\psi}_\nu(\xi)e^{ix\cdot\xi}$ by its Fourier series, we obtain, by Fourier inversion,

$$\hat{\psi}_\nu(\xi)e^{ix\cdot\xi} = 2^{-\nu n}\sum_{k\in\mathbf{Z}^n}\psi_\nu(x-2^{-\nu}k)e^{i2^{-\nu}k\cdot\xi} \tag{2.8}$$

if $|\xi| \leqq 2^\nu\pi$. Inserting (2.8) in (2.7) and using Fourier inversion again yields

$$f_{\nu,\delta} * \psi_\nu(x) = 2^{-\nu n}\sum_{k\in\mathbf{Z}^n}\psi_\nu(x-2^{-\nu}k)f_{\nu,\delta}(2^{-\nu}k). \tag{2.9}$$

Letting $\delta \to 0$ in (2.9) and applying the dominated convergence theorem, (2.6) follows. □

Remark 2.2. The convergence of (2.5) under the assumptions (2.1–3) is considered by Peetre ([20], pp. 52–54). Suppose $f \in \mathcal{S}'$ is of order M, and let

$$L = \inf\left\{j : \int x^\gamma\psi * \varphi(x)\,dx \neq 0 \text{ for some } \gamma \text{ with } |\gamma| = j\right\}$$

(note $L \neq 0$ by (2.3)). Peetre's discussion shows in fact that $\sum_{\nu=0}^{\infty}\psi_\nu * \varphi_\nu * f$ converges in $\mathcal{S}'$, and that there exist polynomials $\{P_k\}_{k=1}^{\infty}$ with deg $P_k \leqq M$ and P with deg $P < \max(M+1, L)$ such that

$$f = \lim_{k\to\infty}\left(\sum_{\nu=-k}^{\infty}\psi_\nu * \varphi_\nu * f + P_k\right) + P \qquad (\text{in } \mathcal{S}').$$

Remark 2.3. Let $\tilde{\varphi}_{\nu k}(x) = \overline{\varphi_\nu(2^{-\nu}k - x)}$ and $\psi_{\nu k}(x) = \psi_\nu(x - 2^{-\nu}k)$, $\nu \in \mathbf{Z}$, $k \in \mathbf{Z}^n$. Then (2.4) means that $f = \sum_{\nu\in\mathbf{Z}} 2^{-\nu n}\sum_{k\in\mathbf{Z}^n}\langle f,\tilde{\varphi}_{\nu k}\rangle\psi_{\nu k}$. Suppose in particular that $\psi \in \mathcal{S}$ satisfies supp $\hat{\psi} \subseteq \{\xi : 1/2 \leqq |\xi| \leqq 2\}$ and $\sum_{\nu\in\mathbf{Z}}|\hat{\psi}(2^\nu\xi)|^2 = 1$ if $\xi \neq 0$. Then by putting $\hat{\varphi}(\xi) = \bar{\hat{\psi}}(\xi)$, or equivalently $\varphi(x) = \overline{\psi(-x)}$, we obtain the "Fourier series expansion"

$$f = \sum_{\nu\in\mathbf{Z}} 2^{-\nu n}\sum_{k\in\mathbf{Z}^n}\langle f,\psi_{\nu k}\rangle\psi_{\nu k}.$$

The functions $\psi_{\nu k}$, $\nu \in \mathbf{Z}$, $k \in \mathbf{Z}^n$, are almost orthogonal since $\langle\psi_{\nu k},\psi_{\mu\ell}\rangle = 0$ if $|\nu - \mu| > 1$, and $\langle\psi_{\nu k},\psi_{\mu\ell}\rangle$ is small if $|\nu-\mu| \leqq 1$ and $|k - \ell|$ is large.

The second lemma leading to our decomposition is a classical result of Plancherel-Pólya [23]. Although there are many proofs in the literature ([3], p. 101, [16], and [28], p. 19), we supply a proof based on Lemma 2.1.

Lemma 2.4. *Let $0 < p \leqq +\infty$ and $\nu \in \mathbf{Z}$. Suppose $g \in \mathcal{S}'$ and* supp $\hat{g} \subseteq \{\xi : |\xi| \leqq 2^{\nu+1}\}$. *If $Q_{\nu k}$ is defined by (1.9), then*

$$\left(\sum_{k\in\mathbf{Z}^n}\sup_{z\in Q_{\nu k}}|g(z)|^p\right)^{1/p} \leqq c_{n,p}2^{\nu(n/p)}\|g\|_{L^p}. \tag{2.10}$$

Proof. By the Paley-Wiener theorem, g is a function of exponential type $2^{\nu+1}$. Also, g is slowly increasing. Let $\psi \in \mathcal{S}$ satisfy supp $\hat{\psi} \subseteq \{\xi : |\xi| \leqq \pi\}$, and $\hat{\psi}(\xi) = 1$ if $|\xi| \leqq 2$. If $\psi_\nu(x) = 2^{\nu n}\psi(2^\nu x)$, and $g(x + y) = g^y(x)$, then, exactly as in the proof of (2.6),

$$g(x + y) = \psi_\nu * g^y(x) = 2^{-\nu n} \sum_{\ell \in \mathbf{Z}^n} g(2^{-\nu}\ell + y)\psi_\nu(x - 2^{-\nu}\ell).$$

Therefore, for any $y \in Q_{\nu k}$,

$$\sup_{z \in Q_{\nu k}} |g(z)| \leqq \sup_{|x| \leqq 2^{-\nu}\sqrt{n}} |g(x + y)| \leqq 2^{-\nu n} \sum_{\ell \in \mathbf{Z}^n} |g(2^{-\nu}\ell + y)| \sup_{|x| \leqq 2^{-\nu}\sqrt{n}} |\psi_\nu(x - 2^{-\nu}\ell)|.$$

Since $\psi \in \mathcal{S}$,

$$\sup_{|x| \leqq 2^{-\nu}\sqrt{n}} |\psi_\nu(x - 2^{-\nu}\ell)| \leqq c_M 2^{\nu n}(1 + |\ell|)^{-M},$$

for any M. Taking M sufficiently large and applying the p-triangle inequality $|a + b|^p \leqq |a|^p + |b|^p$ if $0 < p \leqq 1$ or Hölder's inequality if $p > 1$, we obtain

$$\sup_{z \in Q_{\nu k}} |g(z)|^p \leqq c_p \sum_{\ell \in \mathbf{Z}^n} |g(2^{-\nu}\ell + y)|^p (1 + |\ell|)^{-n-1},$$

for any $y \in Q_{\nu k}$. Integrating with respect to y over $Q_{\nu k}$ yields

$$2^{-\nu n} \sup_{z \in Q_{\nu k}} |g(z)|^p \leqq c_p \sum_{\ell \in \mathbf{Z}^n} (1 + |\ell|)^{-n-1} \int_{Q_{\nu, k+\ell}} |g(x)|^p \, dx. \tag{2.11}$$

Summing over $k \in \mathbf{Z}^n$, we obtain

$$2^{-\nu n} \sum_{k \in \mathbf{Z}^n} \sup_{z \in Q_{\nu k}} |g(z)|^p \leqq c_p \|g\|_{L^p}^p \sum_{\ell \in \mathbf{Z}^n} (1 + |\ell|)^{-n-1} \leqq c_{p,n} \|g\|_{L^p}^p,$$

which proves (2.10). □

Remark 2.5. An alternate way of proving Lemma 2.4, which has the advantage of making the connection with Hardy spaces more obvious, is to observe that

$$\sup_{y \in Q_{\nu k}} |g(y)| \leqq \inf_{z \in Q_{\nu k}} N_\nu(g)(z),$$

where $N_\nu(g)(z) = \sup_{|z-y| < 2^{-\nu}\sqrt{n}} |g(y)|$. Hence

$$\left(\sum_{k \in \mathbf{Z}^n} \sup_{y \in Q_{\nu k}} |g(y)|^p \right)^{1/p} \leqq 2^{\nu(n/p)} \|N_\nu(g)\|_{L^p}.$$

Since supp $\hat{g} \subseteq \{\xi : |\xi| \leqq 2^{\nu+1}\}$, (2.10) follows from the well-known inequality $\|N_\nu(g)\|_{L^p} \leqq c\|g\|_{L^p}$ ([13], [22], [16], [28]); if in addition $\hat{g}(\xi) = 0$ for $|\xi| < 2^{\nu-1}$, this is essentially a restatement of the fact that $\|g\|_{L^p} \approx \|g\|_{H^p}$.

In addition to (2.10), Plancherel-Pólya [23] prove that the inequality

$$\|g\|_{L^p} \leqq c_{n,p} 2^{-\nu n/p} \left(\sum_{k \in \mathbf{Z}^n} \inf_{z \in Q_{\nu k}} |g(z)|^p \right)^{1/p} \tag{2.12}$$

holds as well if supp $\hat{g} \subseteq \{\xi : |\xi| < \varepsilon 2^{\nu+1}\}$ for ε sufficiently small. A proof of this can be given, using Lemma 2.1 in a manner similar to our proof of Lemma 2.4. A more precise statement can be obtained from an interpolation formula in Boas, [3], p. 192.

The relevance of the Plancherel-Pólya result to Besov spaces was, as far as we know, first pointed out by Peetre [20], p. 227.

Our main decomposition theorem for $\dot{B}_p^{\alpha q}$ now follows readily from Lemmas 2.1 and 2.4.

Theorem 2.6. *Let $-\infty < \alpha < +\infty$, $0 < p, q \leqq +\infty$. Then each $f \in \dot{B}_p^{\alpha q}$ can be decomposed as follows:*

i).
$$f = \sum_{\nu \in \mathbf{Z}} \sum_{\ell(Q) = 2^{-\nu}} s_Q \psi_Q,$$

where the ψ_Q's are defined by (1.10), and

ii).
$$f = \sum_{\nu \in \mathbf{Z}} \sum_{\ell(Q) = 2^{-\nu}} s_Q a_Q,$$

where the a_Q's are (α,p)-atoms.

In both cases the convergence is in $\mathcal{S}'/\mathcal{P}$ and the numbers s_Q satisfy

$$\left(\sum_{\nu \in \mathbf{Z}} \left(\sum_{\ell(Q) = 2^{-\nu}} |s_Q|^p \right)^{q/p} \right)^{1/q} \leqq c \|f\|_{\dot{B}_p^{\alpha q}}, \tag{2.13}$$

for some constant c independent of f.

Proof. i). Our assumptions on ψ imply that there is a function φ satisfying (1.1–4) for $\nu = 0$, and such that (2.3) holds. By Lemma 2.1,

$$f(\cdot) = \sum_{\nu \in \mathbf{Z}} \sum_{k \in \mathbf{Z}^n} \varphi_\nu * f(2^{-\nu}k) 2^{\nu n(\alpha/n - 1/p)} \psi_{Q_{\nu k}}(\cdot).$$

For $Q = Q_{\nu k}$, define

$$s_Q = 2^{\nu n(\alpha/n - 1/p)} \varphi_\nu * f(2^{-\nu}k). \tag{2.14}$$

Clearly $f = \sum_Q s_Q \psi_Q$, and (2.13) follows easily by applying (2.10) to $\varphi_\nu * f$.

ii). Select a function $\theta \in \mathcal{S}$ satisfying supp $\theta \subseteq \{x : |x| \leqq 1\}$, $\int x^\gamma \theta(x)\,dx = 0$ if $|\gamma| \leqq N$, and $\hat{\theta}(\xi) \geqq c > 0$ if $1/2 \leqq |\xi| \leqq 2$. (Such a θ is easy to construct: let $\Theta \in \mathcal{S}$ be a real-valued radial function satisfying supp $\Theta \subseteq \{x : |x| \leqq 1\}$ and $\hat{\Theta}(0) = 1$. Then for some $\varepsilon > 0$, $\hat{\Theta}(\xi) \geqq 1/2$ for all ξ satisfying $|\xi| < 2\varepsilon < 1$. Then $\theta(x) = (-\Delta)^N (\varepsilon^{-n} \Theta(x/\varepsilon))$ satisfies all of the requirements.) Our conditions on θ guarantee that a function φ exists which satisfies (1.1–4) for $\nu = 0$ and so that (2.3) holds. Therefore, if $\theta_\nu(x) = 2^{\nu n}\theta(2^\nu x)$,

$$f = \sum_{\nu\in\mathbf{Z}} \theta_\nu * \varphi_\nu * f = \sum_{\nu\in\mathbf{Z}} \sum_{\ell(Q)=2^{-\nu}} \int_Q \theta_\nu(x-y)\varphi_\nu * f(y)\,dy.$$

Define, for $Q = Q_{\nu k}$,

$$s_Q = C2^{\nu n(\alpha/n - 1/p)} \sup_{y\in Q} |\varphi_\nu * f(y)|, \tag{2.15}$$

and

$$a_Q(x) = \frac{1}{s_Q}\int_Q \theta_\nu(x-y)\varphi_\nu * f(y)\,dy,$$

where C is a constant, picked large enough so that every a_Q satisfies (1.6). The a_Q's are (α,p)-atoms, since (1.5) and (1.7) are consequences of our requirements on θ. Finally, (2.13) follows from Lemma 2.4 exactly as in i). □

Remark 2.7. There is a simpler proof for $p \geqq 1$ of Theorem 2.6 ii), that does not depend on Lemma 2.4. Replace (2.15) by

$$s_Q = C|Q|^{-\alpha/n}\left(\int_Q |\varphi_\nu * f(y)|^p\,dy\right)^{1/p} \tag{2.16}$$

(cf. [29], p. 225), and continue with $a_Q(x) = (1/s_Q)\int_Q \theta_\nu(x-y)\varphi_\nu * f(y)\,dy$ as above. Then (1.6) follows by Hölder's inequality if C is a large enough constant, and (1.5) and (1.7) follow as above. With this definition of s_Q, (2.13) is trivial.

In each of our decompositions, s_Q, for $Q = Q_{\nu k}$, is determined by the values of $\varphi_\nu * f$ on $Q_{\nu k}$. Up to multiple factors, s_Q in (2.15) is the sup of $|\varphi_\nu * f|$ on $Q_{\nu k}$, s_Q in (2.16) is the L^p-average of $\varphi_\nu * f$ on $Q_{\nu k}$, and in (2.14) s_Q is the sample value $\varphi_\nu * f(2^{-\nu}k)$. By Plancherel-Pólya, (2.10) and (2.12), these values are roughly interchangeable.

3. Converse of the decomposition theorem. In this section we prove a theorem which contains the converses to the results in Theorem 2.6.

We call a function m an (α,p)-molecule if there exist $\mu \in \mathbf{Z}$ and a point $x_0 \in \mathbf{R}^n$ such that

$$|\partial^\gamma m(x)| \leqq 2^{\mu(n/p-\alpha+|\gamma|)}(1 + 2^\mu|x-x_0|)^{-M-|\gamma|} \qquad \text{if } |\gamma| \leqq K \tag{3.1}$$

and

$$\int x^\gamma m(x)\,dx = 0 \qquad \text{if } |\gamma| \leqq N, \tag{3.2}$$

where M is a large, fixed number; $M \geqq N + 10n\max(1/p,1)$ is certainly enough. We recall that $K \geqq ([\alpha]+1)_+$ and $N \geqq \max([n(1/p-1)_+ - \alpha],-1)$ are fixed integers.

For each $\mu \in \mathbf{Z}$, let $\{x_{\mu,j}\}_j$ be an arbitrary sequence of points in $\mathbf{R}^n$. We will write $m = m_{\mu,j}$ if m satisfies (3.1–2) for $x_0 = x_{\mu,j}$.

We will use the notation m_Q for an (α,p)-molecule which is in fact concentrated on the dyadic cube $Q = Q_{\mu,\ell}$ (defined by (1.9)); i.e. m_Q satisfies (3.1–2) with $\ell(Q) = 2^{-\mu}$ and $x_0 = x_Q = 2^{-\mu}\ell$.

This distinction in notation is adopted to emphasize the fact that our estimates require the (α,p)-molecules to be in correspondence with the dyadic cubes only if $1 < p \leqq +\infty$.

Theorem 3.1. *Let $-\infty < \alpha < +\infty$, $0 < q \leqq +\infty$.*

a). Let $0 < p \leqq 1$. Suppose $f = \sum_{\mu\in\mathbf{Z}} \sum_j s_{\mu,j} m_{\mu,j}$, where the $m_{\mu,j}$'s are (α,p)-molecules, indexed as above. Then

$$\|f\|_{\dot{B}_p^{\alpha q}} \leqq c\left(\sum_{\mu\in\mathbf{Z}}\left(\sum_j |s_{\mu,j}|^p\right)^{q/p}\right)^{1/q}.$$

b). Let $1 < p \leqq +\infty$. Suppose $f = \sum_{\mu\in\mathbf{Z}} \sum_{\ell(Q)=2^{-\mu}} s_Q m_Q$, where each m_Q is an (α,p)-molecule concentrated on Q. Then

$$\|f\|_{\dot{B}_p^{\alpha q}} \leqq c\left(\sum_{\mu\in\mathbf{Z}}\left(\sum_{\ell(Q)=2^{-\mu}} |s_Q|^p\right)^{q/p}\right)^{1/q}.$$

In both cases the constant c is independent of f.

Remark 3.2. In general, the sums in Theorem 3.1 converge in $\mathcal{S}'/\mathcal{P}$. In addition, though, (2.2) and (2.1) with $|\gamma| = 0$ imply that the sums for $\mu \geqq 0$ in a) and b) converge in $\mathcal{S}'$. For $\mu \leqq 0$, the sums in a) and b) converge in L^∞ for $\alpha < n/p$ (or $\alpha = n/p$ if $q \leqq 1$). For $\alpha > n/p$ (or $\alpha = n/p$ if $q > 1$), the sums for $\mu \leqq 0$ may diverge in $\mathcal{S}'$, and in fact may diverge pointwise to $+\infty$ uniformly on an open set. In this case, however, the γ derivatives of the sums converge in L^∞ whenever $\alpha - |\gamma| < n/p$ (or $\alpha - |\gamma| \leqq n/p$ if $q \leqq 1$). Hence, there exists a sequence $\{P_m\}_{m=1}^\infty$ of polynomials of degree $\leqq \alpha - n/p$ (or $< \alpha - n/p$ if $q \leqq 1$) such that in a) $\sum_{\mu=-m}^\infty \sum_j s_{\mu,j} m_{\mu,j} + P_m$, or in b) $\sum_{\ell(Q)\leqq 2^m} s_Q m_Q + P_m$, converges in $\mathcal{S}'$ as $m \to +\infty$; cf. [20] pp. 52–56.

Theorem 3.1 is a consequence of two elementary lemmas which we prove first. In this section, $\{\varphi_\nu\}_{\nu\in\mathbf{Z}}$ will always denote the family of test functions (satisfying (1.1–4)) used to define $\dot{B}_p^{\alpha q}$.

Lemma 3.3. *Let $m_{\mu,j}$ be an (α,p)-molecule. Then*

(3.3) $$|\varphi_\nu * m_{\mu,j}(x)| \leqq c 2^{\mu(n/p-\alpha)} 2^{-(\mu-\nu)(N+1+n)} (1 + 2^\nu |x - x_{\mu,j}|)^{N+1+n-M}$$

if $\nu \leqq \mu$, and

(3.4) $$|\varphi_\nu * m_{\mu,j}(x)| \leqq c 2^{\mu(n/p-\alpha)} 2^{-(\nu-\mu)K} (1 + 2^\mu |x - x_{\mu,j}|)^{N+1+n-M}$$

if $\mu \leqq \nu$.

Proof. Consider (3.3) first. By translation and dilation invariance, we may assume $\nu = 0$ and $x_{\mu,j} = 0$. Put $m = m_{\mu,j}$ and $\varphi = \varphi_0$. By (3.2),

$$\varphi * m(x) = \int m(x-y)\Bigl(\varphi(y) - \sum_{|\beta|\leqq N} \partial^\beta \varphi(x)(y-x)^\beta/\beta!\Bigr)\,dy.$$

Hence,

$$|\varphi * m(x)| \leqq \Bigl(\int_{|x-y|\leqq|x|/2} + \int_{|x-y|>|x|/2}\Bigr)|m(x-y|\,|x-y|^{N+1}\Phi(x,y)\,dy = I + II,$$

where $\Phi(x,y) = \sup_{|\beta|=N+1}\sup_{0<\varepsilon<1} |\partial^\beta\varphi(x+\varepsilon(y-x))|/\beta!$. Since $\varphi \in \mathcal{S}$, $\Phi(x,y) \leqq c(1+|x|)^{N+1+n-M}$ if $|x-y| \leqq |x|/2$. Using this and (3.1),

$$\begin{aligned} I &\leqq c\int |m(x-y)|\,|x-y|^{N+1}dy(1+|x|)^{N+1+n-M} \\ &\leqq c2^{\mu(n/p-\alpha)}\int (1+2^\mu|y|)^{-M}|y|^{N+1}dy(1+|x|)^{N+1+n-M} \\ &\leqq c2^{\mu(n/p-\alpha)}2^{-\mu(N+1+n)}(1+|x|)^{N+1+n-M}. \end{aligned}$$

Since Φ is bounded,

$$\begin{aligned} II &\leqq \int_{|x-y|>|x|/2} |m(x-y)|\,|x-y|^{N+1}dy \\ &\leqq c2^{\mu(n/p-\alpha)}\int_{|x-y|>|x|/2} (1+2^\mu|x-y|)^{-M}|x-y|^{N+1}dy \\ &\leqq c2^{\mu(n/p-\alpha)}2^{-\mu(N+1+n)}(1+|x|)^{N+1+n-M}, \end{aligned}$$

by (3.1). This proves (3.3).

Now (3.4) follows similarly by reversing the roles of φ and m in the above proof. By (1.2), φ has moments of arbitrary order; we subtract a Taylor polynomial of degree $K-1$ from m in the convolution $\varphi * m(x)$ and use (3.1) with $|\gamma| = K$.

Lemma 3.4. *Let* $1 \leqq p \leqq +\infty$ *and* $\mu, \eta \in \mathbf{Z}$, $\eta \leqq \mu$. *Suppose* $F(x) = \sum_{\ell(Q)=2^{-\mu}} s_Q f_Q(x)$, *where*

$$|f_Q(x)| \leqq 2^{\mu(n/p-\alpha)}(1+2^\eta|x-x_Q|)^{-n-1}.$$

Then

$$\|F\|_{L^p} \leqq c2^{-\mu\alpha}\Bigl(\sum_{\ell(Q)=2^{-\mu}} |s_Q|^p\Bigr)^{1/p} 2^{(\mu-\eta)n}.$$

Proof. By our assumption on the f_Q's,

$$\|F\|_{L^p}^p \leqq c \sum_{\ell(S)=2^{-\mu}} 2^{-\mu n}\left(2^{\mu(n/p-\alpha)} \sum_{\ell(Q)=2^{-\mu}} |s_Q|(1+2^{\eta}|x_S-x_Q|)^{-n-1}\right)^p,$$

where $\{S\}$ are the dyadic cubes of side length $2^{-\mu}$. Determine $k, \ell \in \mathbf{Z}^n$ such that $x_Q = 2^{-\mu}k$ and $x_S = 2^{-\mu}\ell$, and set $s_Q = s_k$ in this case. Then by Young's inequality,

$$\begin{aligned}\|F\|_{L^p}^p &\leqq c2^{-\mu\alpha p} \sum_{\ell\in\mathbf{Z}^n}\left(\sum_{k\in\mathbf{Z}^n} |s_k|(1+2^{\eta-\mu}|k-\ell|)^{-n-1}\right)^p \\ &\leqq c2^{-\mu\alpha p}\left(\sum_{k\in\mathbf{Z}^n}|s_k|^p\right)\left(\sum_{\ell\in\mathbf{Z}^n}(1+2^{\eta-\mu}|\ell|)^{-n-1}\right)^p \\ &\leqq c2^{-\mu\alpha p}\left(\sum_{\ell(Q)=2^{-\mu}}|s_Q|^p\right)2^{(\mu-\eta)np},\end{aligned}$$

which proves the lemma. □

Finally, we come to the

Proof of Theorem 3.1. To prove a), we write

$$\varphi_\nu * \sum_{\mu\in\mathbf{Z}}\sum_j s_{\mu,j}m_{\mu,j} = \varphi_\nu * \left(\sum_{\mu=-\infty}^{\nu} + \sum_{\mu=\nu+1}^{+\infty}\right)\sum_j s_{\mu,j}m_{\mu,j}$$

and use Lemma 3.3 and the p-triangle inequality $|a+b|^p \leqq |a|^p + |b|^p$ to see that

$$\begin{aligned}\|f\|_{\dot B_p^{\alpha q}}^q \leqq c\sum_{\nu\in\mathbf{Z}}&\left(\sum_{\mu=-\infty}^{\nu} 2^{-(\nu-\mu)(K-\alpha)p}\sum_j|s_{\mu,j}|^p\right)^{q/p} \\ &+ c\sum_{\nu\in\mathbf{Z}}\left(\sum_{\mu=\nu+1}^{+\infty} 2^{-(\mu-\nu)(N+1+n-n/p+\alpha)p}\sum_j|s_{\mu,j}|^p\right)^{q/p}.\end{aligned}$$

Note that $K-\alpha > 0$ and $N+1+n-n/p+\alpha > 0$ by definition. Hence a) follows by applying Young's inequality if $q \geqq p$, and by using the (q/p)-triangle inequality and $\|s * r\|_{\ell^1} \leqq \|s\|_{\ell^1}\|r\|_{\ell^1}$ if $q \leqq p$.

The proof of b) is similar. If $\mu > \nu$ we apply (3.3) and Lemma 3.4 with $\eta = \nu$, and if $\mu \leqq \nu$ we apply (3.4) and Lemma 3.4 with $\eta = \mu$, to obtain

$$\begin{aligned}\|f\|_{\dot B_p^{\alpha q}}^q \leqq c\sum_{\nu\in\mathbf{Z}}&\left(\sum_{\mu=-\infty}^{\nu} 2^{-(\nu-\mu)(K-\alpha)}\left(\sum_{\ell(Q)=2^{-\mu}}|s_Q|^p\right)^{1/p}\right)^q \\ &+ c\sum_{\nu\in\mathbf{Z}}\left(\sum_{\mu=\nu+1}^{\infty} 2^{-(\mu-\nu)(N+1+\alpha)}\left(\sum_{\ell(Q)=2^{-\mu}}|s_Q|^p\right)^{1/p}\right)^q.\end{aligned}$$

Now $K-\alpha > 0$ and $N+1+\alpha > 0$ by definition, so b) follows by considering $q \geqq 1$ and $q < 1$ separately, similarly to the proof of a). □

We should perhaps note that an immediate consequence of Theorem 3.1 is that if $q < \infty$, our representations of $f \in \dot{B}_p^{\alpha q}$ in Theorem 2.6 converge to f in $\dot{B}_p^{\alpha q}$ (quasi-) norm. This may also be seen directly from the proof of the decomposition.

The functions ψ_Q and a_Q in Theorem 2.6 may be taken to have any fixed number of vanishing moments (or infinitely many, in the case of ψ_Q). Conversely, though, in Theorem 3.1 the assumption that $m_{\mu,j}$, or m_Q, has at least $N = \max([n(1/p - 1)_+ - \alpha], -1)$ vanishing moments cannot be improved. For $0 < p \leqq 1$, this follows from the general fact that any $f \in \dot{B}_p^{\alpha q}$ satisfies $|\hat{f}(\xi)| \leqq c|\xi|^{n(1/p-1)-\alpha}$; the same holds with $\alpha = 0$ for $f \in H^p$ ([27], p. 105). For $p > 1$, the sharpness of the moment assumption (3.2) in Theorem 3.1 can be demonstrated by examples similar to those considered below in §5.

A consideration of (3.1) leads to a remark about the atomic decomposition of $H^p(\mathbf{R}^n)$, $0 < p \leqq 1$. A distribution $f \in H^p(\mathbf{R}^n)$ satisfies $f = \sum_i \lambda_i a_i$ where $\sum_i |\lambda_i|^p \leqq c\|f\|_{H^p}^p$ and each "p-atom" a_i satisfies $\int x^\gamma a_i(x)\,dx = 0$ if $|\gamma| \leqq [n(1/p - 1)]$, supp $a_i \subseteq Q_i$, and $\|a_i\|_{L^\infty} \leqq |Q_i|^{-1/p}$, for some associated cube $Q_i \subseteq \mathbf{R}^n$. Clearly, if a_i also satisfies

$$(3.9) \qquad |\partial^\gamma a_i(x)| \leqq |Q_i|^{-1/p-|\gamma|/n}$$

for $|\gamma| = 1$, then a_i is a $(0,p)$-atom. The condition $\sum_i |\lambda_i|^p < \infty$ is precisely the summation condition on the coefficients $\{s_{\mu,j}\}$ in our decomposition of $\dot{B}_p^{0p}$. Since $\dot{B}_p^{0p} \subsetneqq H^p$ for $0 < p \leqq 1$, we see that it is not possible in general to obtain the smoothness estimate (3.9) for $|\gamma| = 1$ in the atomic decomposition of $H^p(\mathbf{R}^n)$. In our decompositions of $\dot{B}_p^{\alpha q}$, however, our building blocks are C^∞ and may be taken to satisfy (3.1) for an arbitrarily large K. In particular, then, for $0 < p \leqq 1$, the space

$$\left\{\sum_i \lambda_i a_i : \sum_i |\lambda_i|^p < \infty, \text{ each } a_i \text{ is a } p\text{-atom satisfying (3.9) for } |\gamma| \leqq J\right\}$$

is either H^p, if $J = 0$, or $\dot{B}_p^{0p}$, if $J \geqq 1$.

4. Decomposition of $BMO(\mathbf{R}^n)$. The space of functions of bounded mean oscillation is defined by

$$BMO = \left\{f \in L^1_{\text{loc}}/\mathbf{C} : \|f\|_{BMO} = \sup_I \frac{1}{|I|}\int_I |f - f_I| < +\infty\right\},$$

where the sup is taken over all cubes $I \subseteq \mathbf{R}^n$ (not necessarily dyadic), and $f_I = (1/|I|)\int_I f$.

For a sequence $\{s_Q\}_Q$, indexed by the dyadic cubes, define

$$\|\{s_Q\}\|_{\mathscr{C}} = \sup_{J \text{ dyadic}} \left(\frac{1}{|J|}\sum_{Q \subseteq J} |s_Q|^2 |Q|\right)^{1/2};$$

$\|\{s_Q\}\|_{\mathscr{C}}^2$ is equivalent to the Carleson norm of the measure $\sum_Q |s_Q|^2 \delta_{(x_Q,\ell(Q))}$, where $\delta_{(x,t)}$ is the point mass at $(x,t) \in \mathbf{R}_+^{n+1}$.

Uchiyama's decomposition of BMO functions into $(0,\infty)$-atoms ([29], pp. 224–

228) shows the close relation between the $\mathscr{C}$- and BMO-norms. Recall that a $(0,\infty)$-atom a_Q, for some cube Q, is a function satisfying supp $a_Q \subseteq 3Q$, $|\partial^\gamma a_Q(x)| \leqq \ell(Q)^{-|\gamma|}$ for $|\gamma| \leqq K$, and $\int x^\gamma a_Q(x)dx = 0$ for $|\gamma| \leqq N$, where $K \geqq 1$ and $N \geqq 0$ are fixed integers. Uchiyama's result is that

$$(4.1)\quad \|f\|_{BMO} \approx \inf\left\{ \|\{s_Q\}\|_{\mathscr{C}} : f = \sum_Q s_Q a_Q (\text{in } \mathscr{S}'/\mathbf{C}), \right.$$
$$\left. \text{where each } a_Q \text{ is a } (0,\infty)\text{-atom} \right\}.$$

(In [29], it is in fact the L^2-, rather than the $(\mathscr{S}'/\mathbf{C})$-, version of (4.1) that is stated.) Clearly, (4.1) is the BMO-analogue of Theorem 2.6 ii).

The following theorem gives a decomposition of BMO corresponding to Theorem 2.6 i).

Theorem 4.1. *a). If $f \in BMO(\mathbf{R}^n)$, then there exists a sequence $\{s_Q\}_Q$ satisfying $\|\{s_Q\}\|_{\mathscr{C}} \leqq c\|f\|_{BMO}$ such that $f = \sum_Q s_Q \psi_Q$ (in $\mathscr{S}'/\mathbf{C}$), where ψ_Q is defined by (1.10) with $\alpha = 0$ and $p = \infty$.*

b). Conversely, suppose m_Q satisfies $\int m_Q(x)dx = 0$ and

$$|\partial^\gamma m_Q(x)| \leqq \ell(Q)^{-|\gamma|}(1 + \ell(Q)^{-1}|x - x_Q|)^{-n-1-|\gamma|}$$

if $|\gamma| \leqq 1$, for each cube $Q \subseteq \mathbf{R}^n$. If $\|\{s_Q\}\|_{\mathscr{C}} < \infty$, then $\sum_Q s_Q m_Q$ converges in $\mathscr{S}'/\mathbf{C}$ and weak-$*$ in BMO (regarded as $(H^1)^*$), with $\left\|\sum_Q s_Q m_Q\right\|_{BMO} \leqq c\|\{s_Q\}\|_{\mathscr{C}}$.

Proof. The proof of a) is a direct application of the methods of §2. By Lemma 2.1, $f = \sum_{\nu\in\mathbf{Z}} \sum_{k\in\mathbf{Z}^n} \varphi_\nu * f(2^{-\nu}k)\psi_{Q_{\nu k}}(x)$ (in $\mathscr{S}'/\mathbf{C}$). Set $s_Q = \varphi_\nu * f(2^{-\nu}k)$ if $Q = Q_{\nu k}$. It is enough to prove that

$$(4.2)\qquad \sum_{\nu,k:Q_{\nu k}\subseteq J} 2^{-\nu n}|\varphi_\nu * f(2^{-\nu}k)|^2 \leqq c|J|\|f\|^2_{BMO},$$

for every dyadic cube J.

In (4.2), we may clearly assume $J = [0,1]^n$. We apply the estimate (2.11) of Lemma 2.4 to $\varphi_\nu * f$. (This is appropriate here because (2.11) is a nearly localized version of Lemma 2.4.) This gives

$$\sum_{\nu=0}^\infty 2^{-\nu n} \sum_{k\in[0,2^\nu)^n} |\varphi_\nu * f(2^{-\nu}k)|^2$$
$$\leqq c\sum_{\nu=0}^\infty \sum_{k\in[0,2^\nu)^n} \sum_{\ell\in\mathbf{Z}^n} (1 + |\ell|)^{-n-1} \int_{Q_{\nu,k+\ell}} |\varphi_\nu * f|^2$$
$$= c\sum_{\nu=0}^\infty \sum_{k\in[0,2^\nu)^n} \sum_{r\in\mathbf{Z}^n} (1 + |k-r|)^{-n-1} \int_{Q_{\nu r}} |\varphi_\nu * f|^2$$
$$= c\sum_{\nu=0}^\infty \sum_{k\in[0,2^\nu)^n} \sum_{m\in\mathbf{Z}^n} \sum_{r:Q_{\nu r}\subseteq Q_{0m}} (1 + |k-r|)^{-n-1} \int_{Q_{\nu r}} |\varphi_\nu * f|^2.$$

We claim that if $Q_{\nu r} \subseteq Q_{0m}$, then $\sum_{k\in[0,2^\nu)^n}(1+|k-r|)^{-n-1} \leqq c(1+|m|)^{-n-1}$. For $|m|$ small, say $|m| \leqq 10\sqrt{n}$, this is trivial; if $|m| > 10\sqrt{n}$, then

$$|r| > 2^{\nu-1}|m| \geqq 5\cdot 2^\nu\sqrt{n} > 5|k|,$$

so that

$$\sum_{k\in[0,2^\nu)^n}(1+|k-r|)^{-n-1} \leqq c2^{\nu n}(2^\nu|m|)^{-n-1} \leqq c(1+|m|)^{-n-1}.$$

Hence, using this in the inequality above,

$$\sum_{\nu=0}^{\infty} 2^{-\nu n}\sum_{k\in[0,2^\nu)^n}|\varphi_\nu * f(2^{-\nu}k)|^2 \leqq c\sum_{\nu=0}^{\infty}\sum_{m\in\mathbf{Z}^n}\sum_{r:Q_{\nu r}\subseteq Q_{0m}}(1+|m|)^{-n-1}\int_{Q_{\nu r}}|\varphi_\nu * f|^2$$

$$\leqq c\sum_{\nu=0}^{\infty}\sum_{m\in\mathbf{Z}^n}(1+|m|)^{-n-1}\int_{Q_{0m}}|\varphi_\nu * f|^2.$$

Therefore (4.2) is reduced to

$$\sum_{\nu=0}^{\infty}\int_Q|\varphi_\nu * f|^2 \leqq c\|f\|_{BMO}^2, \qquad \text{for } Q = Q_{0m},\ m\in\mathbf{Z}^n. \tag{4.3}$$

The proof of (4.3) is standard (see Fefferman-Stein [13], p. 146). Writing

$$f = f_{3Q} + (f-f_{3Q})\chi_{3Q} + (f-f_{3Q})\chi_{\mathbf{R}^n\setminus 3Q} = f_1 + f_2 + f_3,$$

f_1 contributes nothing to (4.3), while

$$\sum_{\nu=0}^{\infty}\int_Q|\varphi_\nu * f_2|^2 \leqq \int_{\mathbf{R}^n}\sum_{\nu=0}^{\infty}|\hat{\varphi}_\nu|^2|\hat{f}_2|^2 \leqq c\|f_2\|_{L^2}^2 \leqq c\|f\|_{BMO}^2.$$

For $x \in Q$, we have the pointwise estimate

$$|\varphi_\nu * f_3(x)| \leqq \int_{\mathbf{R}^n\setminus 3Q} 2^{\nu n}|f-f_{3Q}|(1+2^\nu|x-y|)^{-n-1}\,dy \leqq c2^{-\nu}\|f\|_{BMO}$$

([13], p. 142). Altogether, this implies (4.3) and completes the proof of a).

In b), the convergence of $f_m = \sum_{\ell(Q)\leqq 2^m} s_Q m_Q$ in $\mathcal{S}'/\mathbf{C}$, to some f, is established as in Remark 3.2. If $\sup_m\|f_m\|_{BMO} = A < +\infty$, then $|\int f_m h| \leqq cA\|h\|_{H^1}$ for any $h \in H^1 \cap \mathcal{S}$ such that $\hat{h} = 0$ in a neighborhood of the origin. It follows that $|\int fh| \leqq c\|h\|_{H^1}$ for these h, so that $\|f\|_{BMO} \leqq cA$ and f_m converges to f weak-* in BMO, by the H^1-BMO duality theorem ([13], p. 145). Therefore we need only to prove $\|f_m\|_{BMO} \leqq c\|\{s_Q\}\|_{\mathscr{C}}$.

The proof of this is contained in Uchiyama's work in [29]. By Lemma 3.5 of [29], each m_Q may be written $m_Q = \sum_{j=0}^{\infty} 2^{-j(n+1)} m_{Q,j}$, where

$$\operatorname{supp} m_{Q,j} \subseteq 2^jQ, \qquad \|m_{Q,j}\|_{\mathrm{Lip}\,1} \leqq c2^{-j}\ell(Q)^{-1}, \qquad \text{and} \int m_{Q,j}(x)\,dx = 0.$$

Then Lemma 3.4 of [29] implies that $\left\|\sum_{\ell(Q)\leq 2^m} s_Q m_{Q,j}\right\|_{BMO} \leq c2^{jn}\|\{s_Q\}\|_{\mathscr{C}}$. Therefore

$$\left\|\sum_{\ell(Q)\leq 2^m} s_Q m_Q\right\|_{BMO} \leq \sum_{j=0}^{\infty} 2^{-j(n+1)}\left\|\sum_Q s_Q m_{Q,j}\right\|_{BMO} \leq c\sum_{j=0}^{\infty} 2^{-j}\|\{s_Q\}\|_{\mathscr{C}} = c\|\{s_Q\}\|_{\mathscr{C}}. \quad \square$$

We may note that the proof of Theorem 4.1 shows that

$$\|f\|_{BMO} \approx \sup_{S \text{ dyadic}}\left(\frac{1}{|S|}\sum_{Q_{\nu k}\subseteq S} 2^{-\nu n}\sup_{Q_{\nu k}}|\varphi_\nu * f|^2\right)^{1/2} \tag{4.4}$$

$$\approx \sup_{S \text{ dyadic}}\left(\frac{1}{|S|}\sum_{\nu=-\log_2\ell(S)}^{\infty}\int_S|\varphi_\nu * f|^2\right)^{1/2}$$

This should be compared with (2.10), with Theorem 3 iii of Fefferman-Stein [13], and with the definition of the $\dot{B}_p^{\alpha q}$ norm.

5. Application to trace theorems. Let $f \in \dot{B}_p^{\alpha q}$. In Theorem 2.6 ii) we obtained s_Q and a_Q such that $f = \sum_Q s_Q a_Q$ where supp $a_Q \subseteq 3Q$, $|\partial^\gamma a_Q(x)| \leq |Q|^{\alpha/n-1/p-|\gamma|/n}$ if $|\gamma| \leq K$, $\int x^\gamma a_Q(x)dx = 0$ if $|\gamma| \leq N$, and

$$\left(\sum_{\nu=\mathbf{Z}}\left(\sum_{\ell(Q)=2^{-\nu}}|s_Q|^p\right)^{q/p}\right)^{1/p} \leq c\|f\|_{\dot{B}_p^{\alpha q}}.$$

Write $x \in \mathbf{R}^n$ as $x = (x',x_n)$, $x' \in \mathbf{R}^{n-1}$, $x_n \in \mathbf{R}$, and let $\pi:\mathbf{R}^n \to \mathbf{R}^{n-1}$ be the natural projection $\pi(x) = x'$. For each dyadic cube J in $\mathbf{R}^{n-1}$ we set

$$t_J = \sum_{\substack{Q:\pi(Q)=J\\ 3Q\cap J\neq\varnothing}}|s_Q| \quad\text{and}\quad h_J(x') = \sum_{\substack{Q:\pi(Q)=J\\ 3Q\cap J\neq\varnothing}} s_Q a_Q(x',0)/t_J.$$

The restriction, or trace, of f to $\mathbf{R}^{n-1}$ is now

$$\operatorname{Tr} f(x') = \sum_{\mu\in\mathbf{Z}}\sum_{\ell(J)=2^{-\mu}} t_J h_J(x') \tag{5.1}$$

whenever the sum converges in $\mathscr{S}'/\mathscr{P}$. Clearly,

$$\left(\sum_{\mu\in\mathbf{Z}}\left(\sum_{\ell(J)=2^{-\mu}}|t_J|^p\right)^{q/p}\right)^{1/q} \leq c\|f\|_{\dot{B}_p^{\alpha q}}, \tag{5.2}$$

$$\operatorname{supp} h_J \subseteq 3J, \tag{5.3}$$

and

$$|\partial^\gamma h_J(x')| \leq |J|^{(\alpha-1/p/(n-1))-1/p-|\gamma|/(n-1)} \qquad \text{if } |\gamma| \leq K, \tag{5.4}$$

since $|Q| = |J|^{n/(n-1)}$ if $\pi(Q) = J$. In other words, each h_J satisfies all the requirements for an $(\alpha - 1/p,p)$-atom except possibly the moment condition (1.7). However, for $\alpha - 1/p > (n-1)(1/p-1)_+$, an $(\alpha - 1/p,p)$-atom in $\mathbf{R}^{n-1}$ is not required to have any vanishing moments. In these cases, then, by Theorem 3.1 and (5.2),

$$\|\operatorname{Tr} f\|_{\dot{B}_p^{\alpha-1/p,q}(\mathbf{R}^{n-1})} \leqq c\|f\|_{\dot{B}_p^{\alpha q}(\mathbf{R}^n)}. \tag{5.5}$$

Although $\operatorname{Tr} f$ in (5.1) is expressed in terms of the decomposition of $f \in \dot{B}_p^{\alpha q}(\mathbf{R}^n)$ given in Theorem 2.6 ii), it is clear that $\operatorname{Tr} f(x') = f(x',0)$ if that decomposition converges absolutely and uniformly. This is the case, for instance, if $f \in \dot{B}_\infty^{01}$. In particular, if f is any function of the form $f = \sum_{\nu=-M}^{N} \sum_{\ell(Q)=2^{-\nu}} \lambda_Q b_Q$, where $\sup_Q |\lambda_Q| < +\infty$ and the b_Q's are (α,p)-atoms satisfying $\int b_Q(x)dx = 0$ and $|\partial^\gamma b(x)| \leqq \ell(Q)^{(\alpha-n/p-|\gamma|)}$, $|\gamma| \leqq 1$, for some α, p, M, N, then $f \in \dot{B}_\infty^{01}$ and hence $\operatorname{Tr} f(x') = f(x',0)$. Since such sums are dense in $\dot{B}_p^{\alpha q}$ for $q < +\infty$, by (5.5) the map $\operatorname{Tr}: \dot{B}_p^{\alpha q}(\mathbf{R}^n) \to \dot{B}_p^{\alpha-1/p,q}$ is the unique continuous linear extension to $\dot{B}_p^{\alpha q}$ of the pointwise restriction operator if $\alpha - 1/p > (n-1)(1/p-1)_+$. Moreover, Tr extends the restriction operator for $\dot{B}_p^{\alpha\infty}$ as well, since the inclusion $\dot{B}_p^{\alpha\infty} \subseteq \dot{B}_p^{\alpha_0 1} + \dot{B}_p^{\alpha_1 1}$ holds whenever $\alpha_0 < \alpha < \alpha_1$.

It is easy to see that the trace map Tr above is onto $\dot{B}_p^{\alpha-1/p,q}(\mathbf{R}^{n-1})$, since any $h_J(x')$ satisfying (5.3–4) can be obtained as the restriction of an (α,p)-atom $a_Q(x)$. Hence, we obtain the known fact (which is classical if $p \geqq 1$) that $\operatorname{Tr} \dot{B}_p^{\alpha q}(\mathbf{R}^n) = \dot{B}_p^{\alpha-1/p,q}(\mathbf{R}^{n-1})$ when $\alpha - 1/p > (n-1)(1/p-1)_+$ (cf. [17], [28]). The failure of this result for $\alpha - 1/p \leqq (n-1)(1/p-1)_+$ is clearly due to the failure of the h_J's to have vanishing moments, which is first necessary at this critical index.

More generally, the existence, or non-existence, of the trace of $\dot{B}_p^{\alpha q}$ is equivalent to the question whether we can make sense of the sums in (5.1) whenever (5.2–4) hold, since any such expression can arise from a suitable $f \in \dot{B}_p^{\alpha q}(\mathbf{R}^n)$. In particular, it is not difficult to see that the sums in (5.1) always converge, and thus the trace exists, in $\mathcal{S}'/\mathcal{P}$, if and only if $\alpha - 1/p > (n-1)(1/p-1)_+$ or $\alpha - 1/p = (n-1)(1/p-1)_+$ and $0 < q \leqq 1$. This is also previously known ([21], [28]).

Suppose now $0 < p < 1$. When $0 < \alpha - 1/p < (n-1)(1/p-1)$, $\alpha - 1/p = (n-1)(1/p-1)$ and $q > 1$, or $\alpha = 1/p$, $q \leqq p$, (5.1) does not necessarily converge in $\mathcal{S}'/\mathcal{P}$, but does converge in $L^p + L^\infty(\mathbf{R}^{n-1})$. This was observed in [17], and may be seen readily from (5.1–4). This is best possible in the sense that the sums (5.1) do not necessarily converge in $L^p + L^\infty$ when $\alpha - 1/p < 0$ or $\alpha = 1/p$, $q > p$. Let us show this, for example, in the case $\alpha = 1/p$, $q > p$.

Pick a sequence $\{t_\mu\}_{\mu=2}^\infty \in \ell^q \setminus \ell^p$ (and to be precise, since this is not our usual convention, in $c_0 \setminus \ell^p$ if $q = +\infty$) and a collection $\{J_\mu\}_{\mu=2}^\infty$ of dyadic cubes satisfying $J_\mu \subseteq [0,1]^{n-1}$, $\ell(J_\mu) = 2^{-\mu}$, and $3J_\mu \cap 3J_\nu = \varnothing$ if $\mu \neq \nu$. Set $t_J = t_\mu$ if $J = J_\mu$ and $t_J = 0$ for any $J \notin \{J_\mu\}_{\mu=2}^\infty$. Let $\{h_J\}_J$ be functions satisfying (5.3–4) and, in addition, $h_J(x') \geqq c|J|^{-1/p}$ if $x' \in J$, for some small constant c. Then

$$\left(\sum_{\mu\in\mathbf{Z}}\left(\sum_{\ell(J)=2^{-\mu}} |t_J|^p\right)^{q/p}\right)^{1/q} = \left(\sum_{\mu=2}^{\infty} |t_\mu|^q\right)^{1/q} < +\infty,$$

and it is clear that $\sum_{\mu\in\mathbf{Z}}\sum_{\ell(J)=2^{-\mu}} t_J h_J$ would arise as the trace of a suitable $f \in \dot{B}_p^{(1/p)q}(\mathbf{R}^n)$ if the trace operator were continuous. But if we let $h_N = \sum_{\mu=2}^N \sum_{\ell(J)=2^{-\mu}} t_J h_J$ for large N, then supp $h_N \subset [0,1]^{n-1}$. Hence, $\|h_n\|_{L^p+L^\infty} \geqq c\|h_N\|_{L^p} \geqq c\left(\sum_{\mu=2}^N |t_\mu|^p\right)^{1/p}$, which can be made arbitrarily large. Therefore the sum $\sum_J t_J h_J$ cannot converge in $L^p + L^\infty$.

We know from above that the trace of $\dot{B}_p^{\alpha q}(\mathbf{R}^n)$ in the limiting case $\alpha = 1/p$ exists in $L^p + L^\infty(\mathbf{R}^{n-1})$ only if $0 < q \leqq \min(1,p)$. In fact, we then have the following result.

Theorem 5.1. *Let* $0 < p < +\infty$, $0 < q \leqq \min(1,p)$. *Then*

$$\operatorname{Tr} \dot{B}_p^{(1/p)q}(\mathbf{R}^n) = L^p(\mathbf{R}^{n-1}).$$

Proof. It is easy to verify that the sums in (5.1) converge in $L^p(\mathbf{R}^{n-1})$ if $f \in \dot{B}_p^{(1/p)q}(\mathbf{R}^n)$, $0 < q \leqq \min(1,p)$, and that Tr is bounded from $\dot{B}_p^{(1/p)q}(\mathbf{R}^n)$ to $L^p(\mathbf{R}^{n-1})$.

To show that Tr is onto L^p, it is sufficient to show that each $h \in L^p(\mathbf{R}^{n-1})$ has a decomposition

$$h(x') = \sum_{\mu\in\mathbf{Z}}\sum_{\ell(J)=2^{-\mu}} t_J h_J(x'),$$

where the h_J's satisfy (5.3–4) with $\alpha = 1/p$, and

$$\left(\sum_{\mu\in\mathbf{Z}}\left(\sum_{\ell(J)=2^{-\mu}} |t_J|^p\right)^{q/p}\right)^{1/q} \leqq c\|h\|_{L^p(\mathbf{R}^{n-1})}.$$

To prove such a decomposition, start by picking a $\Phi \in C_0^\infty$ satisfying supp $\Phi \subseteq [0,1]^{n-1}$, $0 \leqq \Phi \leqq 1$, and $\|1 - \Phi\|_{L^p([0,1]^{n-1})} \leqq \min(1/5,(1/5)^{1/p})$. If

$$J = \{x : k_i 2^{-\mu} \leqq x_i < (k_i + 1)2^{-\mu}, i = 1,\ldots,n-1\},$$

put

$$h_J(x') = C\Phi(2^\mu x' - k)2^{\mu(n-1)/p}, \tag{5.6}$$

where $k = (k_1,\ldots,k_{n-1})$ and C is chosen small enough for h_J to satisfy (5.4).

Fix a non-negative $h \in L^p(\mathbf{R}^{n-1})$; it is enough to prove the decomposition for such functions. By choosing the side length $2^{-\mu_1}$ small enough, it is possible to find a simple function

$$e_1(x') = \sum_{\ell(J)=2^{-\mu_1}} r_J \chi_J(x')$$

such that $e_1 \geqq 0$ and $\|h - e_1\|_{L^p} \leqq \min(1/4,(1/4)^{1/p})\|h\|_{L^p}$. We define the smooth version

$$\tilde{e}_1(x') = \sum_{\ell(J)=2^{-\mu_1}} t_J h_J(x'),$$

where the h_J's are given by (5.6) and $t_J = r_J 2^{-\mu_1(n-1)/p}/C$, with the same constant C. If we set $D = C/\max(5/4,(5/4)^{1/p})$, then

$$\left(\sum_{\ell(J)=2^{-\mu_1}} |t_J|^p\right)^{1/p} = \|e_1\|_{L^p}/C \leqq \|h\|_{L^p}/D, \tag{5.7}$$

and we picked Φ so that $\|e_1 - \tilde{e}_1\|_{L^p} \leqq \min(1/5,(1/5)^{1/p})\|e_1\|_{L^p}$. Hence,

$$\|h - \tilde{e}_1\|_{L^p} \leqq \|h\|_{L^p}/2. \tag{5.8}$$

If this process is repeated with h replaced by $h - \tilde{e}_1$, we obtain $\tilde{e}_2 = \sum_{\ell(J)=2^{-\mu_2}} t_J h_J$ such that

$$\left(\sum_{\ell(J)=2^{-\mu_2}} |t_J|^p\right)^{1/p} \leqq \|h - \tilde{e}_1\|_{L^p}/D \leqq \|h\|_{L^p}/2D$$

and

$$\|h - \tilde{e}_1 - \tilde{e}_2\|_{L^p} \leqq \|h - \tilde{e}_1\|_{L^p}/2 \leqq \|h\|_{L^p}/4,$$

by (5.8). We can also arrange so that $\mu_2 > \mu_1$. Continuing this process inductively, we obtain the functions $\tilde{e}_i = \sum_{\ell(J)=2^{-\mu_i}} t_J h_J$, $i = 1, 2, \ldots$, satisfying

$$\left(\sum_{\ell(J)=2^{-\mu_i}} |t_J|^p\right)^{1/p} \leqq \|h\|_{L^p}/2^{i-1}D, \tag{5.9}$$

$$\left\| h - \sum_{i=1}^{m} \tilde{e}_i \right\|_{L^p} \leqq 2^{-m}\|h\|_{L^p}, \qquad m = 1, 2, \ldots, \tag{5.10}$$

and $\mu_{i+1} > \mu_i$ for every i. The required decomposition of h is $h(x') = \sum_{i=1}^{\infty} \tilde{e}_i(x')$. By (5.10) this sum converges in L^p and by (5.9),

$$\left(\sum_{i=1}^{\infty}\left(\sum_{\ell(J)=2^{-\mu_i}} |t_J|^p\right)^{q/p}\right)^{1/q} \leqq c\|h\|_{L^p}. \qquad \square$$

Theorem (5.1) was previously known when $1 \leqq p < +\infty$ and $q = 1$; see [1], [21], [15], and [5].

6. The lower majorant property for $\dot{B}_p^{\alpha q}$, $0 < p \leqq 1$. A space X of tempered distributions on $\mathbf{R}^n$ is said to have the lower majorant property if, for each $f \in X$, there is a $g \in X$ such that

$$|\hat{f}(\xi)| \leqq \hat{g}(\xi), \qquad \text{if } \xi \in \mathbf{R}^n, \tag{6.1}$$

and

$$\|g\|_X \leqq c\|f\|_X,$$

with c independent of f. For instance, the Hardy spaces $H^p(\mathbf{R}^n)$, $0 < p \leqq 1$, are known to have this property; $L^p(\mathbf{R}^n)$ ($p \geqq 1$) has the lower majorant property if and only if $p = 1$ or $p = 2k/(2k-1)$, $k = 1, 2, 3, \ldots$. References can be found in [19] and [2].

For a space $X \subseteq \mathcal{S}'/\mathcal{P}$, the definition of the lower majorant property is as above, except that the origin is excluded in (6.1). For $\dot{B}_p^{\alpha q}(\mathbf{R}^n)$, the following theorem is an immediate consequence of Theorem 2.6 i) and Theorem 3.1 a).

Theorem 6.1. *Let* $-\infty < \alpha < +\infty$, $0 < q \leqq +\infty$. *If* $0 < p \leqq 1$, *then* $\dot{B}_p^{\alpha q}$ *has the lower majorant property.*

Proof. By Theorem 2.6 i), each $f \in \dot{B}_p^{\alpha q}$ has a representation $f = \sum_Q s_Q \psi_Q$ where

$$\left(\sum_{\nu\in\mathbf{Z}}\left(\sum_{\ell(Q)=2^{-\nu}} |s_Q|^p\right)^{q/p}\right)^{1/q} \leqq c\|f\|_{\dot{B}_p^{\alpha q}}.$$

The ψ_Q's are defined by (1.10); up to a multiple, they are obtained by translating the dilates of a fixed ψ by $x_Q = 2^{-\nu}k$. We choose, as we may, ψ so that $0 \leqq \hat{\psi} \leqq 1$. If we put $g(x) = \sum_{\nu\in\mathbf{Z}} \sum_{\ell(Q)=2^{-\nu}} |s_Q|\psi_Q(x + x_Q)$ we then have $|\hat{f}(\xi)| \leqq \hat{g}(\xi)$.

Now since $0 < p \leqq 1$, it does not matter that the building blocks $\psi_Q(x + x_Q)$ are not evenly scattered. Indeed, by Theorem 3.1 a), we obtain

$$\|g\|_{\dot{B}_p^{\alpha q}} \leqq c\left(\sum_{\nu\in\mathbf{Z}}\left(\sum_{\ell(Q)=2^{-\nu}} |s_Q|^p\right)^{q/p}\right)^{1/q} \leqq c\|f\|_{\dot{B}_p^{\alpha q}}. \qquad \square$$

7. Inhomogeneous Besov spaces. It is not difficult to prove analogues of Theorems 2.6 and 3.1 for the inhomogeneous Besov spaces $B_p^{\alpha q}(\mathbf{R}^n)$, $-\infty < \alpha < +\infty$, $0 < p, q \leqq +\infty$. To define these spaces, let $\{\varphi_\nu\}_{\nu=0}^\infty$ satisfy (1.1–4) and let $\Phi \in \mathcal{S}$ satisfy

$$\text{(7.1)} \qquad \operatorname{supp} \hat{\Phi} \subseteq \{\xi : |\xi| \leqq 1\} \quad \text{and} \quad \hat{\Phi}(\xi) \geqq c > 0 \qquad \text{if } |\xi| \leqq \frac{5}{6}.$$

Then $B_p^{\alpha q}$ is the set of $f \in \mathcal{S}'(\mathbf{R}^n)$ such that

$$\|f\|_{B_p^{\alpha q}} = \|\Phi * f\|_{L^p} + \left(\sum_{\nu=0}^{\infty} (2^{\nu\alpha}\|\varphi_\nu * f\|_{L^p})^q\right)^{1/q} < +\infty.$$

This definition is independent of the choice of Φ and $\{\varphi_\nu\}_{\nu=0}^\infty$ ([20], p. 49).

Suppose $\Psi \in \mathcal{S}$ satisfies $\operatorname{supp} \hat{\Psi} \subseteq \{\xi : |\xi| \leqq \pi\}$ and $\hat{\Psi}(\xi) \geqq c > 0$ if $|\xi| \leqq 1$. We have then the following decomposition results.

Theorem 7.1. *Let* $-\infty < \alpha < +\infty$ *and* $0 < p, q \leqq +\infty$.

a). Each $f \in B_p^{\alpha q}$ *can be decomposed as follows:*

i). $f(\cdot) = \sum_{k\in\mathbf{Z}^n} s_k \Psi(\cdot - k) + \sum_{\nu=0}^{\infty} \sum_{\ell(Q)=2^{-\nu}} s_Q \psi_Q(\cdot)$, *where the* ψ_Q*'s are defined by (1.10), or*

$$ii).\ f = \sum_{k\in\mathbf{Z}^n} s_k b_k + \sum_{\nu=0}^{\infty} \sum_{\ell(Q)=2^{-\nu}} s_Q a_Q,$$

where the a_Q*'s are* (α,p)*-atoms, and the* b_k*'s satisfy* supp $b_k \subseteq 3Q_{0k}$ *and* $|\partial^\gamma b_k(x)| \leqq 1$ *if* $|\gamma| \leqq K$*. In both cases the convergence is in* $\mathcal{S}'$*, and*

$$\left(\sum_{k\in\mathbf{Z}^n} |s_k|^p\right)^{1/p} + \left(\sum_{\nu=0}\left(\sum_{\ell(Q)=2^{-\nu}} |s_Q|^p\right)^{q/p}\right)^{1/q} \leqq c\|f\|_{B_p^{\alpha q}},$$

with c independent of f.

b). Conversely, suppose $f = \sum_{k\in\mathbf{Z}^n} s_k m_k + \sum_{\nu=0}^{\infty} \sum_{\ell(Q)=2^{-\nu}} s_Q m_Q$*, where each* m_Q *is an* (α,p)*-molecule concentrated on* Q*, and each* m_k *satisfies*

$$|\partial^\gamma m_k(x)| \leqq (1 + |x-k|)^{-M-|\gamma|} \qquad \text{if } |\gamma| \leqq K,$$

for some sufficiently large M*. Then*

$$\|f\|_{B_p^{\alpha q}} \leqq c\left(\sum_{k\in\mathbf{Z}^n} |s_k|^p\right)^{1/p} + c\left(\sum_{\nu=0}^{\infty}\left(\sum_{\ell(Q)=2^{-\nu}} |s_Q|^p\right)^{q/p}\right)^{1/q}$$

For $0 < p \leqq 1$, the conclusion of b) holds even if the m_k's and m_Q's are not centered near k and Q, respectively.

To prove a), one obtains $\Phi \in \mathcal{S}$ satisfying (7.1) and φ satisfying (1.1–4) for $\nu = 0$ such that

$$\hat{\Phi}(\xi)\hat{\Psi}(\xi) + \sum_{\nu=0}^{\infty} \hat{\bar{\varphi}}(2^\nu\xi)\hat{\psi}(2^\nu\xi) = 1 \qquad \text{for all } \xi \in \mathbf{R}^n,$$

and proceeds as in Lemma 2.1 and Theorem 2.6. The proof of b) uses the inequalities of §3 and similar estimates, not requiring the assumption of vanishing moments, however, for Φ and m_k.

The results of §5–6 also have analogues for $B_p^{\alpha q}$. The standard result that $\operatorname{Tr} B_p^{\alpha q}(\mathbf{R}^n) = B_p^{\alpha-1/p,q}(\mathbf{R}^{n-1})$ if $\alpha - 1/p > (n-1)(1/p-1)_+$, and the analogue of Theorem 5.1 that $\operatorname{Tr} B_p^{(1/p)q}(\mathbf{R}^n) = L^p(\mathbf{R}^{n-1})$ if $0 < p < +\infty$ and $0 < q \leqq \min(1,p)$, follow from Theorem 7.1 by the techniques of §5. Also, remarks about the non-existence of the trace in $\mathcal{S}'$ or in L^p are analogous to those in §5. Further, $B_p^{\alpha q}$ is proved to have the lower majorant property if $0 < p \leqq 1$ exactly as in §6.

8. Conclusion. The decompositions in Theorem 2.6 provide a natural approach to many of the well-known properties of the Besov spaces, including the standard embedding and interpolation results. Also, they yield a way of comparing the Besov spaces to other spaces known to have a decomposition, such as H^p, $0 < p \leqq 1$, or L^p. We have seen, for example, that the atomic decompositions of $\dot{B}_p^{0p}$ and H^p differ only in the smoothness assumption on the atoms. On the other hand, the main distinction between the building blocks obtained in the decomposition of L^p (Theorem 5.1) and the p-atoms for H^p, or the $(0,p)$-atoms for

$\dot{B}_p^{0p}$, is that no vanishing moments are assumed in the case of L^p. It would be interesting to clarify the relation between $\dot{B}_p^{0p}$ and L^p, H^p and L^p, and $\dot{B}_p^{0p}$ and H^p, by determining the interpolation spaces between each of these couples.

There are a number of directions in which the methods of this paper could possibly be extended. It is straightforward to obtain decompositions similar to (1.8) and (1.11) for Besov spaces defined with respect to a measure satisfying the doubling property. In the case of the polydisk, as well, our results generalize in an obvious way. Since the machinery necessary for Calderón's representation formula has been developed by Folland-Stein in [14] for appropriate homogeneous groups, it should be possible to extend our approach to this setting. In the case of more general domains in $\mathbf{R}^n$, for example Lipschitz domains, it may be natural to define Besov spaces for $\alpha > 0$ via the atomic decomposition ($\alpha > 0$ is taken here because (α,p)-atoms are not required to have vanishing moments in this range). This point of view might be useful in the study of differential equations on these domains (cf. [28], Chapter 3), especially since trace theorems are easy in the atomic context. In the proof of Theorem 2.6 i), it would be interesting to replace the Fourier series expansion with a representation in terms of other bases in L^2, for example certain sets $\{e^{i\lambda_k x}\}_k$, or the eigenfunctions of some differential operator other than the Laplacian. Similarly, it may be possible to replace Fourier series by appropriate group representations in more abstract settings.

Finally, we would like to draw attention to some features relating to the form of the expansion described in Remark 2.3. After a normalization and reindexing, we obtain an expansion of the form $f = \sum_i \langle f,\psi_i\rangle\psi_i$, with $\|\psi_i\|_{L^2} \leqq c$. The key aspect of this decomposition in our treatment of Besov spaces is that the norm of f is equivalent to the appropriate sequence space norm of the coefficients $\langle f,\psi_i\rangle$. Although the expansion is not orthonormal, it nevertheless has many of the advantages of an orthonormal expansion. For instance, it follows directly from the identity $f = \sum_i \langle f,\psi_i\rangle\psi_i$ that $\|f\|_{L^2} = \left(\sum_i |\langle f,\psi_i\rangle|^2\right)^{1/2}$. Applying this to ψ_j gives $\sup_j \sum_i |\langle\psi_j,\psi_i\rangle|^2 \leqq c$, which is an almost orthogonality property. Moreover, writing an operator T in the form

$$Tf = \sum_i \langle Tf,\psi_i\rangle\psi_i = \sum_{i,j} \langle f,\psi_j\rangle\langle T\psi_j,\psi_i\rangle\psi_i$$

effectively reduces the study of T to the study of the matrix $\{\langle T\psi_j,\psi_i\rangle\}$.

More generally, suppose $\{\psi_i\}_i$ is a quasi-orthogonal family, or that the matrix $\{\langle\psi_i,\psi_j\rangle\}$ is bounded on ℓ^2, and in addition that $\psi_j = \sum_i \langle\psi_j,\psi_i\rangle\psi_i$, for each j. Then clearly the operator P defined by $Pf = \sum_i \langle f,\psi_i\rangle\psi_i$ is a bounded projection onto $\mathcal{H}$, the closure of the span of $\{\psi_i\}_i$. Also we can write

$$Pf(x) = \sum_i \int f(y)\overline{\psi_i(y)}\,dy\,\psi_i(x) = \int K(x,y)f(y)\,dy$$

for $K(x,y) = \sum_i \overline{\psi_i(y)}\psi_i(x)$. This is reminiscent of the Bergman and Szegö kernels except that the ψ_i's are not necessarily orthonormal. If the ψ_i's are sufficiently localized, as in Lemma 2.4, then for $f \in \mathcal{H}$, the identity $f = \sum_i \langle f,\psi_i\rangle\psi_i$ can be used to prove an analogue of Plancherel-Pólya.

References

1. S. Agmon & L. Hörmander, *Asymptotic properties of solutions of differential equations with simple characteristics,* J. Analyse Math. **30** (1976), 1–38.
2. A. Baernstein & E. Sawyer, *Embedding and multiplier theorems for* $H^p(\mathbf{R}^n)$, Mem. Amer. Math. Soc. (to appear).
3. R. Boas, *Entire Functions,* Academic Press, New York, 1954.
4. H.-Q. Bui, *Representation theorems and atomic decomposition of Besov spaces,* (preprint).
5. V. I. Burenkov & M. L. Gol'dman, *On the extension of functions of* L_p, (Russian), Trudy Mat. Inst. Steklov. **150** (1979), 31–51.
6. A. P. Calderón, *An atomic decomposition of distributions in parabolic* H^p *spaces,* Adv. in Math. **25** (1977), 216–225.
7. A. P. Calderón & A. Torchinsky, *Parabolic maximal functions associated with a distribution I,* Adv. in Math. **16** (1975), 1–64.
8. S. Y. Chang & R. Fefferman, *A continuous version of duality of* H^1 *and BMO on the bidisc,* Ann. of Math. **112** (1980), 179–201.
9. G. Cohen, $H^{(p,q)}$*: some new spaces of distributions with atomic decomposition,* (preprint).
10. R. R. Coifman, *A real variable characterization of* H^p, Studia Math. **51** (1974), 269–274.
11. R. R. Coifman & R. Rochberg, *Representation theorems for holomorphic and harmonic functions in* L^p, Astérisque **77** (1980), 11–66.
12. G. S. de Souza, R. O'Neill & S. Sampson, *An analytic characterization of the special atom spaces,* (preprint).
13. C. Fefferman & E. M. Stein, H^p *spaces of several variables,* Acta Math. **129** (1972), 137–193.
14. G. B. Folland & E. M. Stein, *Hardy Spaces on Homogeneous Groups,* Princeton University Press, Princeton, N.J., 1982.
15. M. L. Gol'dman, *On the extension of functions of* $L_p(\mathbf{R}_m)$ *in spaces with a greater number of dimensions,* (Russian), Mat. Zametki **25** (1979), 513–520.
16. B. Jawerth, *Weighted norm inequalities for functions of exponential type,* Ark. Mat. **15** (1977), 223–228.
17. B. Jawerth, *The trace of Sobolev and Besov spaces if* $0 < p < 1$, Studia Math. **62** (1978), 65–71.
18. R. Latter, *A characterization of* $H^p(\mathbf{R}^n)$ *in terms of atoms,* Studia Math. **62** (1978), 93–101.
19. E. T. Y. Lee & G.-I. Sunouchi, *On the majorant properties of* $L^p(G)$, Tôhoku Math. J. **31** (1979), 41–48.
20. J. Peetre, *New Thoughts on Besov Spaces,* Duke University Math. Series **1**, Dept. Math., Duke Univ., Durham, N.C., 1976.
21. J. Peetre, *The trace of Besov space—a limiting case,* Technical Report, Lund, 1975.
22. J. Peetre, *On spaces of Triebel-Lizorkin type,* Ark. Mat. **13** (1975), 123–130.
23. M. Plancherel & G. Pólya, *Fonctions entières et intégrales de Fourier multiples,* Comment. Math. Helv. **9** (1937), 224–248.
24. F. Ricci & M. Taibleson, *Boundary values of harmonic functions in mixed norm spaces and their atomic structure,* Ann. Scuola Norm. Sup. Pisa Cl. Sci. (4) **10** (1983), 1–54.
25. R. Rochberg, *Decomposition theorems for Bergman spaces and their applications,* (preprint).
26. R. Rochberg & S. Semmes, *A decomposition theorem for BMO and applications,* (preprint).
27. M. Taibleson & G. Weiss, *The molecular characterization of certain Hardy spaces,* Astérisque **77** (1980), 67–151.
28. H. Triebel, *Theory of Function Spaces,* Akademische Verlagsgesellschaft Geest and Portig K. G., Leipzig, 1983.

29. A. UCHIYAMA, *A constructive proof of the Fefferman-Stein decomposition of* $BMO(\mathbf{R}^n)$, Acta Math. **148** (1982), 215–241.
30. M. WILSON, *On the atomic decomposition for Hardy spaces*, Pacific J. Math. **116** (1985), 201–207.
31. M. WILSON, personal communication.

The second author was partially supported by National Science Foundation Grant DMS-840324.

Washington University—ST. LOUIS, MO 63130

Received September 17, 1984

JOURNAL OF FUNCTIONAL ANALYSIS 86, 307–340 (1989)

Banach Spaces Related to Integrable Group Representations and Their Atomic Decompositions, I

HANS G. FEICHTINGER AND K. H. GRÖCHENIG*

Institut für Mathematik, Universität Wien, Strudlhofgasse 4, A-1090 Vienna, Austria, and Department of Mathematics, U-9, University of Connecticut, Storrs, CT 06269, USA

Communicated by the Editors

Received December 1987

We present a general theory of Banach spaces which are invariant under the action of an integrable group representation and give their atomic decompositions with respect to coherent states, i.e., the atoms arise from a single element under the group action. Several well-known decomposition theories are contained as special examples and are unified under the aspect of group theory. © 1989 Academic Press, Inc.

1. INTRODUCTION

The aim of an atomic decomposition for a space of functions or distributions is to represent every element as a sum of "simple functions," usually called atoms. If this is possible, properties of these function spaces, such as duality, interpolation, or operator theory for them, can be understood better by means of the atomic decomposition. Of course, the meaning of "simple function" depends on the point of view. Thus, for example, the atoms in the decomposition of Hardy spaces are subject to support and moment conditions (cf. [CW]). The atoms for the spaces of Besov–Triebel–Lizorkin type are transforms of a single function, where the transformations are given by a certain unitary group representation (cf. [FJ1, FJ2]). This type of atom is called "generalized coherent state" in theoretical physics [KS], where it is used in quite different contexts. From our point of view, the Gabor-type expansions of the modulation spaces as given in [F4] and the atomic decompositions for Bergman spaces (see [R], [RT]) are further examples of such decompositions with respect to generalized coherent states.

* The second author was supported by the Österreichische Forschungsgemeinschaft under Project No. 09/0010.

0022-1236/89 $3.00

Copyright © 1989 by Academic Press, Inc.
All rights of reproduction in any form reserved.

580/86/2-7*

For each of these families of spaces, specific methods have been developed—either a Fourier analytic approach or complex variable theory—and the proofs in the papers just mentioned depend heavily on particular features of these function spaces.

In this paper we present a general theory of atomic decompositions with respect to generalized coherent states. It contains the above-mentioned examples as special cases and many new theories of atomic decompositions. Furthermore it reveals that atomic decompositions for these spaces are consequences of a single phenomenon, namely the action of a suitable group on these spaces. Thus it is possible to treat all known theories of atomic decompositions with respect to coherent states in a unified way.

In the abstract theory we start with an integrable, irreducible, unitary representation π of a locally compact group $\mathscr{G}$ on a Hilbert space $\mathscr{H}$. Within this context we construct a scale of Banach spaces which are related to $\mathscr{H}$ and which we call coorbit spaces. They are defined by the behaviour of the extended representation coefficients $V_g(f)(x) := \langle \pi(x)\, g, f \rangle$ (also called wavelet transform or voice transform in special cases; cf. [GMP1]). Thus the distribution f belongs to the coorbit space $\mathscr{C}o\ Y$ if and only if the transform $V_g(f)$ of f with respect to a suitable analyzing vector g belongs to a function space Y on $\mathscr{G}$. According to the theory of square integrable representations these extended representation coefficients satisfy a reproducing formula. Thus all questions about coorbit spaces which may contain rather wild functions or distributions can be transferred to related questions concerning well-behaved functions on the group $\mathscr{G}$.

In our opinion the stronger condition of integrability of the representation is essential to the theory because in that case there exist a minimal and a maximal space (cf. [F2] for some general ideas) and one may go beyond the Hilbert space. The integrability is implicitly used in the theory of wavelets although it claims to use only square integrable representations (cf. [GMP2]).

Once the relation between coorbit spaces and corresponding function spaces on the group $\mathscr{G}$ is understood we are left to study convolution operators between various spaces on $\mathscr{G}$ and to do non-commutative harmonic analysis. The construction of an atomic decomposition turns out to be equivalent to the discretization of a certain convolution operator. The atoms are of the form $\pi(x_i)\, g$ for a suitable analyzing vector g and a suitable well-spread point set $(x_i)_{i \in I}$ in $\mathscr{G}$. The choice of the atoms is very flexible and thus the atomic decompositions which are obtained by this method may be adjusted to a wide range of applications. The technical tools for this task will be furnished by the theory of Wiener-type spaces and their convolution relations (cf. [F1]).

The paper is organized as follows. Section 2 collects some facts concerning square integrable representations, and Section 3 provides the terminology of Banach spaces and Wiener-type spaces and contains several auxiliary results, some of which will only be used in Part II. In Section 4, the central part of this paper, we define coorbit spaces and investigate several basic properties. The technique for the discretization of convolution operators relevant for the atomic decompositions is given in Section 5. Section 6 contains the main result concerning the atomic decompositions of coorbit spaces and some direct consequences. Finally, we show the stability of this method.

In Part II of this series we shall study the properties of the coorbit spaces as Banach spaces, such as duality results, embeddings, interpolation, unconditional bases, and related questions. This investigation will be based entirely on the existence of an atomic decomposition for these spaces.

In Part III we shall apply the general theory to more examples. Even in those situations where atomic decompositions are known to exist the general approach can provide new insights. Due to the flexibility of our theory the class of possible atoms is much larger than it was supposed to be in concrete cases. Among the new examples we mention the atomic decompositions for function spaces on the Heisenberg groups which are virtually contained in this theory. It remains to check some conditions and to write down explicitly the decompositions, thus no new theory or method is required.

2. Square Integrable Group Representations

We have already demonstrated the importance of (square) integrable representations for the theory of coorbit spaces in [FG, F2]. For the sake of completeness and for the convenience of the reader we list briefly some of the relevant facts and formulas to be used in the sequel. For a general orientation concerning related matters the reader may consult, e.g., [GMP1], [C], [DM].

2.1. An irreducible, unitary continuous representation π of a locally compact (=: l.c.) group $\mathscr{G}$ on a Hilbert space $\mathscr{H}$ is called *square integrable*, if at least one of the representation coefficients $\langle \pi(x) g, g \rangle$, $g \neq 0$, $g \in \mathscr{H}$, is square integrable with respect to the Haar measure on $\mathscr{G}$. We shall assume that the inner product in $\mathscr{H}$ is conjugate linear in the first and linear in the second factor. It follows that the wavelet transform V_g defined by $V_g(f)(x) := \langle \pi(x) g, f \rangle$ is a linear mapping from $\mathscr{H}$ to $C^b(\mathscr{G})$.

2.2. There is a unique positive, selfadjoint, densely defined operator A on $\mathscr{H}$ such that $V_g(g) \in L^2(\mathscr{G})$ iff $g \in \operatorname{dom} A$ and the *orthogonality relations* hold:

$$\int \overline{V_{g_1}(f_1)(x)}\, V_{g_2}(f_2)(x)\, dx = \langle Ag_2, Ag_1 \rangle \langle f_1, f_2 \rangle \tag{2.1}$$

$$\forall f_i \in \mathscr{H},\ g_i \in \operatorname{dom} A,\ i = 1, 2.$$

As an important consequence one has

$$V_{g_1}(f_1) * V_{g_2}(f_2) = \langle Ag_1, Af_2 \rangle\, V_{g_2}(f_1) \tag{2.2}$$

$$\forall g_1, f_2 \in \operatorname{dom} A,\ g_2, f_1 \in \mathscr{H}.$$

For unimodular groups A is just a scalar multiple of the identity operator, in particular $\operatorname{dom} A = \mathscr{H}$.

2.3. Choosing now $g_1 = g_2 = f_2 =: g \in \operatorname{dom} A$, normalized such that $\|Ag\| = 1$, one obtains the *reproducing formula*

$$V_g(f) = V_g(f) * V_g(g) \qquad \forall f \in \mathscr{H}. \tag{2.3}$$

The mapping $V_g : f \to V_g(f)$ from $\mathscr{H}$ into $L^2(\mathscr{G})$ is isometric and satisfies

$$V_g(\pi(x) f) = L_x V_g(f). \tag{2.4}$$

Thus V_g is the intertwining operator between π and the left regular representation L, and π is equivalent to a subrepresentation of L.

2.4. The orthogonal projection from $L^2(\mathscr{G})$ onto the range of V_g is given by the convolution operator $F \mapsto F * V_g(g)$. Consequently, a function $F \in L^2(\mathscr{G})$ belongs to the range of V_g, i.e., $F = V_g(f)$ for some $f \in \mathscr{H}$ if and only if $F * V_g(g) = F$.

2.5. An irreducible unitary representation π is called integrable if at least one representation coefficient $V_g(g)$ is integrable. Integrable representations are obviously square integrable and thus the relations (2.1)–(2.3) hold.

3. Banach Function Spaces and Wiener-Type Spaces on Groups

There is a natural parameterization of the general coorbit spaces to be defined in Section 4 through *Banach function spaces* (also called *solid BF-spaces*) on the l.c. group $\mathscr{G}$ (not only rearrangement invariant Banach function spaces, or weighted L^p – spaces, for example). For this reason we collect a number of definitions and basic facts concerning these spaces. In order to have a better separation of *local* and *global* properties of the norms involved it will be important to include also *Wiener-type spaces* in our discussion. These spaces have been introduced in full generality in [F1], using "continuous" control functions. The convolution relations among these spaces will be relevant for our atomic decompositions.

Our *general setting* will be that of a *solid BF-space* (or Banach function spaces) on a l.c. group $\mathscr{G}$. By this we mean a Banach space $(Y, \| \|_Y)$ of measurable functions on $\mathscr{G}$, which is continuously embedded into $L^1_{\text{loc}}(\mathscr{G})$, and satisfies the *solidity* condition (cf. [Z], Chap. 15])

$$f \in Y,\ g \in L^1_{\text{loc}} \text{ with } |g(x)| \leqslant |f(x)| \text{ l.a.e.} \Rightarrow g \in Y \text{ and } \|g\|_Y \leqslant \|f\|_Y.$$

Of course, L^p-spaces and mixed norm spaces are the prototypical examples for such spaces, as well as arbitrary rearrangement invariant Banach function spaces, including Lorentz or Orlicz spaces and the like (cf. [LT]). Weighted versions of L^p-spaces have been used as demonstration objects in [FG].

We shall need the following operators on $L^1_{\text{loc}}(\mathscr{G})$: The left and right translation operators L_x and R_x, given by $L_x f(y) := f(x^{-1}y)$, $R_x f(y) := f(yx)$, and the involutions $^\vee$ and $^\nabla$, given by $f^\vee(x) := f(x^{-1})$ and $f^\nabla(x) := \overline{f(x^{-1})}$. A BF-space is called *left* (*right*) *invariant* if $L_x Y \subseteq Y$ (resp. $R_x Y \subseteq Y$) for all $x \in \mathscr{G}$. It follows from the closed graph theorem that the operators L_x and R_x resp. are bounded on $(Y, \| \|_Y)$ for each $x \in \mathscr{G}$, and that the mapping $w: x \mapsto |||L_x|||_Y$ (resp. $|||R_x|||_Y$), using these symbols for the operator norms of these operators on $(Y, \| \|_Y)$, is well defined and submultiplicative, i.e., it satisfies $w(x \circ y) \leqslant w(x)\, w(y)$ for $x, y \in \mathscr{G}$. It is useful to observe that the spaces $L^1_w(\mathscr{G}) := \{F \mid Fw \in L^1\}$, endowed with their natural norm, are Banach convolution algebras (called *Beurling algebras*), continuously embedded into $(L^1, \| \|_1)$, if $w(x) \geqslant \delta_0 > 0$ for all $x \in \mathscr{G}$.

We shall assume throughout this paper that the solid BF-spaces $(Y, \| \|_Y)$ are *two-sided*, i.e., left and right *invariant*. Furthermore we take up the *convention* to use w only for a weight function (considered as fixed for a given Y or a family of such spaces) for which Y is a Banach module over L^1_w, i.e., satisfying

$$Y * L^{1\vee}_w \subseteq Y$$

and

$$\|F * G\|_Y \leqslant \|F\|_Y \|G^\vee\|_{1,w} \quad \text{for} \quad F \in Y, G \in L^1_w. \tag{3.1}$$

If the space $\mathscr{K}(\mathscr{G})$ of continuous complex-valued functions with compact support is dense in $(Y, \| \|_Y)$ or more generally, if right translation is continuous in $(Y, \| \|_Y)$, i.e., the mapping $x \to R_x f$ is continuous from $\mathscr{G}$

into $(Y, \| \|_Y)$ for all $f \in Y$, (3.1) is equivalent to the estimate $|||R_x||| \leqslant Cw(x)$ for all $x \in \mathscr{G}$.

Since the norm of an element of a BF-space Y does not reflect the local behaviour of these functions, the *Wiener-type spaces* (as treated in [F1]) are a more appropriate tool. Given a solid BF-space B, the global behaviour of a given function in $B_{\text{loc}}(\mathscr{G})$ (i.e., belonging locally to B) can be conveniently described using a *control function* given as follows: Let $k \in \mathscr{K}(\mathscr{G})$ be any non-zero *window-function* (one should think of a plateau-like function, satisfying $0 \leqslant k(x) \leqslant 1$ for all $x \in \mathscr{G}$ and $k(z) \equiv 1$ on a compact neighbourhood of the identity) and define the *control function* as

$$K(f, k; B)(x) := \|(L_x k) f\|_B \qquad \text{for} \quad x \in \mathscr{G}.$$

It is clear that for the control function only the local behaviour of f near x, measured in terms of the *local norm* $\| \|_B$, is relevant (note that K depends on a *continuous* parameter). The best way to impose growth or integrability conditions on K (at infinity, i.e., globally) is to choose as *global component* any reasonable two-sided translation invariant solid BF-space $(Y, \| \|_Y)$.

DEFINITION 3.1. (of Wiener-type spaces). Given any solid BF-space $(B, \| \|_B)$ we define the *Wiener-type space*

$$W(B, Y) := \{f \in B_{\text{loc}}, K(f, k; B) \in Y\}. \tag{3.2}$$

It is endowed with the natural norm $\|f \mid W(B, Y)\| := \|K(f, k; B)\|_Y$.

The same definition also makes sense for the space $B = C^0(\mathscr{G})$ (with the sup-norm $\| \|_\infty$) and $B = M(\mathscr{G})$, the space of bounded regular Borel measures, which is considered as the dual of $C^0(\mathscr{G})$ by the Riesz representation theorem.

As has been shown in [F1] these spaces are two-sided invariant Banach spaces, which do not depend on the particular choice of the window-function k. Moreover, different functions k_1 and k_2 define equivalent norms.

In order to prepare the appropriate terminology for the discrete description of these spaces let us give the following definitions:

DEFINITION 3.2. Given some neighborhood U of the identity in $\mathscr{G}$, a family $X = (x_i)_{i \in I}$ in $\mathscr{G}$ is called *U-dense* if the family $(x_i U)_{i \in I}$ covers $\mathscr{G}$.

The family is called *V-separated* if for some relatively compact neighbourhood V of the identity the sets $(x_i V)_{i \in I}$ are pairwise disjoint.

It is called *relatively separated* if it is the finite union of V-separated families. Finally, we shall call a family $(x_i)_{i \in I}$ *well-spread* in $\mathscr{G}$ if it is both U-dense and relatively separated.

Without loss of generality we shall always assume that the group $\mathscr{G}$ is σ-compact, therefore all index sets I, partitions of unity, and coverings of the group used in this paper will be countable.

LEMMA 3.3. *The following properties for* $X=(x_i)_{i\in I}$ *in* $\mathscr{G}$ *are equivalent*:

(i) *The family* $(x_i)_{i\in I}$ *is relatively separated*;

(ii) *For any compact set* $K\subseteq\mathscr{G}$, *there exists a finite partition of the index set* I, $I=\bigcup_{r=1}^{s} I_r$, *such that each of the families* $(x_iK)_{i\in I_r}$ *consists of pairwise disjoint sets (and conversely, any relatively separated family is obtained in this way).*

(iii) *Given any relatively compact set* W *with non-void interior*

$$\sup_{i\in I} \#\{k \mid x_k W\cap x_i W\neq\emptyset\}<\infty.$$

Proof. The reader who does not want to check the details is referred to [FGr], Lemma 2.9]. ▮

Using this terminology, any given solid BF-space Y may be quite naturally associated with a corresponding sequence space $Y_d(X)$ (sometimes called solid BK-space).

DEFINITION 3.4. Given a discrete family $X=(x_i)_{i\in I}$ in $\mathscr{G}$ and a solid, translation invariant BF-space $(Y, \|\ \|_Y)$ we define the *associate discrete BK-space* $Y_d(X)$ as $\{\Lambda \mid \Lambda=(\lambda_i)_{i\in I}$ with $\sum_{i\in I}\lambda_i c_{x_iW}\in Y\}$, with natural norm $\|\Lambda \mid Y_d\|:=\|\sum_{i\in I}|\lambda_i|\, c_{x_iW}\mid Y\|$. Whenever convenient we omit the indication of the dependence on X.

For a well-spread family X this definition does not depend on the choice of W, i.e., different sets W define the same space Y_d and equivalent norms (over a fixed system $X=(x_i)_{i\in I}$). The following lemma collects several basic properties of Y_d (which follow by [F1, FGr] or easy direct computations):

LEMMA 3.5. (a) *If the functions with compact support are dense in* Y, *the finite sequences form a dense subspace of* Y_d.

(b) *With* $w_0(x):=|\!|\!|L_x|\!|\!|_Y$ *and* $w(i):=w_0(x_i)$ *the inclusions* $l^1_w\subseteq Y_d\subseteq l^\infty_{1/w}$ *hold.*

(c) *Given two well-spread families* X *and* X', $Y_d(X)\subseteq Z_d(X)$ *if and only if* $Y_d(X')\subseteq Z_d(X')$. *The last statement allows us to write unambiguously* $Y_d\subseteq Z_d$ *in the sequel.*

(d) *Dependence of* $Y_d(X)$ *on* X: *Assume that for two well-spread families* X *and* X' *over the same index set there is a compact set* Q *such that*

$x_i^{-1}x_i' \in Q$ for all $i \in I$. Then $Y_d(X) = Y_d(X')$, and the corresponding norms are equivalent.

(e) *Given a weighted L^p-space L^p_m, the associated sequence space over X is the appropriate weighted l^p-space l^p_m, the discrete weight m being given by $m(i) := m(x_i)$ for $i \in I$. The same is true for general rearrangement invariant solid BF-space on $\mathscr{G}$ instead of $L^p(\mathscr{G})$.*

(f) *Let $X = (x_i)_{i \in I}$ be a U_0-dense and relatively separated family in $\mathscr{G}$. Then for any partition $(I_r)_{r=1}^s$ of the index set I the projections*

$$P_r \colon \Lambda \mapsto \Lambda_r := \begin{cases} \lambda_i & \text{for} \quad i \in I_r \\ 0 & \text{else} \end{cases}$$

define a partition of unity on Y_d.

Thus $\sum_{r=1}^s \|\sum_{i \in I_r} \lambda_i c_{x_i W} \mid Y\|$ defines an equivalent norm on Y_d.

Establishing discrete descriptions of these spaces involves the use of suitable partitions of unity.

DEFINITION 3.6. Given any compact neighbourhood of e in $\mathscr{G}$, a family $\Psi = (\psi_i)_{i \in I}$ in $C^0(\mathscr{G})$ is called a *bounded uniform partition of unity of size* U (we shall use the acronym U-BUPU in the sequel, or BUPU if the size of U is not relevant) if the following properties hold:

(B1) $0 \leqslant \psi_i(x) \leqslant 1$ for all $i \in I$ (hence Ψ is bounded in $(C^0, \| \ \|_\infty)$).

(B2) There is a well-spread family $(x_i)_{i \in I}$ in $\mathscr{G}$ such that $\operatorname{supp} \psi_i \subseteq x_i U \ \forall i \in I$.

(B3) $\sum_{i \in I} \psi_i(x) \equiv 1$.

It is possible to construct arbitrary fine BUPUs, i.e., U-BUPUs for any given U for arbitrary l.c. groups (cf. [F3], for example).

Using BUPU's one can give the following discrete characterization of Wiener-type spaces:

PROPOSITION 3.7. (Theorem 2 of [F1]). *$f \in W(B, Y)$ if and only if $(\|f\psi_i\|_B)_{i \in I} \in Y_d(X)$ for some BUPU Ψ (and the norm of this sequence in Y_d defines an equivalent norm). In particular, $W(B, Y^1) \subseteq W(B, Y^2)$ if and only if $Y_d^1 \subseteq Y_d^2$.*

The above discrete characterization also tells us that $f = \sum_{i \in I} f\psi_i$ and that the series is norm convergent if the finite sequences are dense in Y_d. Here the building blocks $f\psi_i$ satisfy certain support and summability conditions, but on the other hand any function of the form $f = \sum_{i \in I} f_i$ with $(f_i)_{i \in I}$ satisfying the same conditions belongs to $W(B, Y)$.

LEMMA 3.8. (a) *A continuous function f belongs to* $W(C^0, Y)$ *if and only if* $(f(x_i))_{i\in I}$ *belongs to* Y_d *for every relatively separated family* $X=(x_i)_{i\in I}$.

(b) *The discrete measure* $\sum_{i\in I}\lambda_i\delta_{x_i}$ *belongs to* $W(M, Y)$ *if and only if* $(\lambda_i)_{i\in I}$ *belongs to* Y_d *for any (one) such X and the corresponding norms are equivalent.*

In the general situation we need "right" Wiener-type spaces $W^R(B, Y)$. They can be obtained by replacing (left translation operators) L_z by right translation operators R_z (or by applying the involution $^\vee$). They enjoy analogous properties.

The following lemma relates a solid BF-space to other (solid) Wiener-type spaces having the same global behaviour:

LEMMA 3.9. *For any translation invariant BF-space* $(Y, \| \|_Y)$ *let us denote by* $w_0(x)$ *the norm of the left translation operator*: $w_0(x) := |||L_x|||_Y$. *Then*

(a) $W(C^0, Y) \subseteq Y \subseteq W(L^1, Y)$,

(b) $Y \subseteq W(L_1, L^\infty_{1/w_0^\vee})$.

Proof. (a) Let $k\in\mathscr{K}(\mathscr{G})$ be a positive window-function, satisfying $k(x)\equiv 1$ on a neighbourhood of e. The first inclusion then follows from the obvious inequality $|F(y)| \leqslant \|(L_y k)F\|_\infty$. Taking the norm in Y one has $\|F|\, Y\| \leqslant \|F|\, W(C^0, Y)\|$, as a consequence of the solidity of Y.

For the second embedding note first that (due to the solidity of Y) $F\in Y$ may be assumed to be positive. It then follows that

$$K(F, k; L^1)(x) = \|(L_x k)F\|_1 = \int_{\mathscr{G}} |(L_x k)(y)\, F(y)|\, dy = F * k^\vee(x).$$

By (3.1) we have

$$\|F|\, W(L^1, Y)\| \leqslant \|F\|_Y\, \|k\|_{1,w}.$$

(b) Because Y is embedded into $L^1_{\mathrm{loc}}(\mathscr{G})$, it follows that for any $k\in\mathscr{K}(\mathscr{G})$ there exists $C_k>0$ such that $\|kF\| \leqslant C_k \|F\|_Y$ for all $F\in Y$. Then the estimate for the control function

$$\begin{aligned} K(F, k, L^1)(x) &= \|(L_x k)F\|_1 = \|k(L_{x^{-1}}F)\|_1 \\ &\leqslant C_k\, \|L_{x^{-1}}F\|_Y \leqslant C_k w_0(x^{-1})\, \|F\|_Y \end{aligned}$$

implies

$$\|F|\, W(L^1, L^\infty_{1/w_0^\vee})\| = \|K(F, k, L^1)|\, L^\infty_{1/w_0^\vee}\| \leqslant C_k\, \|F\|_Y. \quad \blacksquare$$

The convolution relation below will serve as a substitute for Cotlar's lemma. The properties of $W(L^\infty, L^1_w)$ allow one to decompose a convolution operator into pieces which are easily estimated and then to reconstruct the operator as an (absolutely) convergent sum. The next proposition enables us to analyze convolution operators on a wide range of function spaces on $\mathscr{G}$.

PROPOSITION 3.10. *Under the assumption* (3.1) (*relating* Y *and* w) *one has*

$$W(M, Y) * W^R(C^0, L^1_w) \subseteq Y. \tag{3.3}$$

Proof. Recall that $G \in W^R(C^0, L^1_w)$ has a decomposition $G = \sum_{n \geq 1} R_{z_n} G_n$ with $G_n \in \mathscr{K}(\mathscr{G})$, supp $G_n \subseteq Q = Q^{-1}$ (compact), and

$$\sum_{n \geq 1} \|G_n\|_\infty \, w(z_n) \leq C \|G \mid W^R(C^0, L^1_w)\| < \infty.$$

Thus let us consider first the effect of the convolution of $\mu \in W(M, Y)$ with the building blocks G_n. Choosing a function $k \in \mathscr{K}(\mathscr{G})$, with $k \geq 0$ and $k(x) \equiv 1$ on Q, we obtain

$$\begin{aligned} |\mu * G_n(x)| &= \left| \int_{\mathscr{G}} G_n(y^{-1}x) \, d\mu(y) \right| \\ &= \left| \int_{\mathscr{G}} L_x k(y) \, G_n(y^{-1}x) \, d\mu(y) \right| \\ &\leq \|G_n\|_\infty \, \langle L_x k, \mu \rangle = \|G_n\|_\infty \, \|L_x k \cdot \mu\|_M \\ &= \|G_n\|_\infty \, K(\mu, k; M)(x) \qquad \text{for} \quad x \in \mathscr{G} \text{ a.e.} \end{aligned}$$

From the assumption $\mu \in W(M, Y)$ we see that the convolution $\mu * G_n$ is well-defined. An application of the Y-norm on both sides yields

$$\|\mu * G_n\|_Y \leq \|G_n\|_\infty \, \|\mu \mid W(M, Y)\|.$$

To finish we put together all the pieces of G and arrive at

$$\begin{aligned} \|\mu * G\|_Y &\leq \sum_{n \geq 1} \|\mu * R_{z_n} G_n\|_Y \leq \sum_{n \geq 1} \|R_{z_n}(\mu * G_n)\| \\ &\leq \sum_{n \geq 1} w(z_n) \, \|\mu * G_n\|_Y \\ &\leq \left(\sum_{n \geq 1} w(z_n) \, \|G_n\|_\infty \right) \|\mu \mid W(M, Y)\| \\ &\leq C \, \|G \mid W^R(C^0, L^1_w)\| \, \|\mu \mid W(M, Y)\|. \quad \blacksquare \end{aligned}$$

4. The General Theory of Coorbit Spaces

In this central section we introduce coorbit spaces with respect to a given group representation. We explore the basic properties as far as they are needed for their characterization through atomic decompositions. The detailed study of their properties as Banach spaces will be pursued in Part II. Some of these facts were already presented in [FG] without rigorous proof, hence we shall provide detailed arguments here wherever necessary and further explore the general theory. The main components for our theory are described below.

We start with an irreducible, unitary, continuous representation π of a locally compact group $\mathscr{G}$ on a Hilbert space $\mathscr{H}$ which is at least integrable. Given a weight function w on $\mathscr{G}$ the set of analyzing vectors $\mathscr{A}_w$ is given by

$$\mathscr{A}_w := \{g \in \mathscr{H},\ V_g(g) \in L^1_w(\mathscr{G})\}.$$

We shall always assume $\mathscr{A}_w$ is non-trivial. Since π is irreducible $\mathscr{A}_w$ is a dense linear subspace of $\mathscr{H}$. Fixing an arbitrary non-zero element $g \in \mathscr{A}_w$, the space $\mathscr{H}^1_w$ is defined as

$$\mathscr{H}^1_w := \{f \in \mathscr{H},\ V_g(f) \in L^1_w\}.$$

$\mathscr{H}^1_w$ is a π-invariant Banach space which is dense in $\mathscr{H}$ and minimal in a certain sense (cf. [FG, Corollary 4.7]). The set $\{\pi(x)g, x \in \mathscr{G}\}$ is a total subset of $\mathscr{H}^1_w$ (cf. [FG, Corollary 4.8]).

As an appropriate reservoir within which coorbit spaces are obtained by way of suitable selection we shall take the space $(\mathscr{H}^1_w)^\urcorner$ of all continuous conjugate-linear functionals (="antifunctionals") on $\mathscr{H}^1_w$. This allows us to preserve the notation of the inner product on $\mathscr{H}$ and to write $\langle f, h\rangle$ for the result of the action of the antifunctional $h \in (\mathscr{H}^1_w)^\urcorner$ on $f \in \mathscr{H}^1_w$. One has the inclusions

$$\mathscr{H}^1_w \subsetneq \mathscr{H} \subsetneq (\mathscr{H}^1_w)^\urcorner \tag{4.1}$$

and for $f, h \in \mathscr{H}^1_w \subseteq (\mathscr{H}^1_w)^\urcorner$: $\langle f, h\rangle = \overline{\langle h, f\rangle}$. Correspondingly the bracket $\langle F, H\rangle$ for functions F, H on the group will always mean the antidual pairing $\langle F, H\rangle := \int_G \overline{F(y)}\, H(y)\, dy$ whenever the integral exists (e.g., for the $L^1_w - L^\infty_{1/w}$-antiduality). The action of π on $\mathscr{H}^1_w$ can be extended to $(\mathscr{H}^1_w)^\urcorner$ by the usual rule:

$$\langle f, \pi(x)h\rangle := \langle \pi(x^{-1})f, h\rangle \qquad \text{for } f \in \mathscr{H}^1_w,\ h \in (\mathscr{H}^1_w)^\urcorner.$$

Therefore it is reasonable to consider the extended representation coefficients (wavelet transform) $V_g(f)(x) := \langle \pi(x)g, f\rangle$ for $f \in (\mathscr{H}^1_w)^\urcorner$ (for fixed $g \in \mathscr{A}_w$).

THEOREM 4.1 (Properties of $(\mathscr{H}^1_w)^\urcorner$ and the Wavelet Transform). (i) *The inner product on* $\mathscr{H}$ *extends to a sesquilinear* π*-invariant pairing* $\langle \cdot, \cdot \rangle$ *on* $\mathscr{H}^1_w \times (\mathscr{H}^1_w)^\urcorner$. *For any element* $f \in (\mathscr{H}^1_w)^\urcorner$ *the wavelet-transform* $V_g(f)(x) := \langle \pi(x)\, g, f \rangle$ *is a continuous function in* $L^\infty_{1/w}(\mathscr{G})$.

(ii) $V_g : (\mathscr{H}^1_w)^\urcorner \to L^\infty_{1/w}(\mathscr{G})$ *is one-to-one from* $(\mathscr{H}^1_w)^\urcorner$ *into* $L^\infty_{1/w}(\mathscr{G})$ *and intertwines* π *and* L, *i.e., we have*

$$V_g(\pi(x) f) = L_x V_g(f) \qquad \forall f \in (\mathscr{H}^1_w)^\urcorner. \tag{4.2}$$

(iii) *If* g *is normalized by* $\|Ag\| = 1$ *the reproducing formula holds true, i.e.,*

$$V_g(f) = V_g(f) * V_g(g) \qquad \text{for all } f \in (\mathscr{H}^1_w)^\urcorner. \tag{4.3}$$

(iv) *Conversely, for every* $F \in L^\infty_{1/w}(\mathscr{G})$ *satisfying the relation* $F * V_g(g) = F$ *there exists a unique element* $f \in (\mathscr{H}^1_w)^\urcorner$ *with* $V_g(f) = F$.

(v) *A bounded net* $(f_\alpha)_{\alpha \in I}$ *in* $(\mathscr{H}^1_w)^\urcorner$ *is* w^**-convergent to an element* $f \in (\mathscr{H}^1_w)^\urcorner$ *if and only if* $V_g(f_\alpha)$ *converges pointwise to* $V_g(f)$ *if and only if* $V_g(f_\alpha)$ *converges uniformly on compact sets to* $V_g(f)$.

Proof. (i) The properties of $\langle \cdot, \cdot \rangle$ on $\mathscr{H}^1_w \times (\mathscr{H}^1_w)^\urcorner$ follow from the definition. $V_g(f) \in L^\infty_{1/w}(\mathscr{G})$ follows from the estimate

$$\begin{aligned} |V_g(f)(x)| &= |\langle \pi(x) g, f \rangle| \\ &\leqslant \|\pi(x) g \mid \mathscr{H}^1_w\| \; \|f \mid (\mathscr{H}^1_w)^\urcorner\| \\ &\leqslant w(x)\, \|g \mid \mathscr{H}^1_w\| \; \|f \mid (\mathscr{H}^1_w)^\urcorner\|. \end{aligned} \tag{4.4}$$

(ii) $V_g(f)(x) = \langle \pi(x) g, f \rangle = 0$ for all $x \in \mathscr{G}$ implies $f = 0$ because the set of atoms $\pi(x) g$, $x \in \mathscr{G}$ is total in $\mathscr{H}^1_w$. Formula (4.2) is obvious.

(iii) The reproducing formula (2.3) for $f \in \mathscr{H}$ written out in full yields (taking into account the conjugate linearity in the first factor)

$$\begin{aligned} \langle \pi(x) g, f \rangle &= \int V_g(g)(y^{-1}x) \langle \pi(y) g, f \rangle \, dy \\ &= \left\langle \int \overline{V_g(g)(y^{-1}x)}\, \pi(y)\, g \, dy, f \right\rangle \end{aligned}$$

and consequently

$$\pi(x) g = \int \langle \pi(y) g, \pi(x) g \rangle\, \pi(y) g \, dy. \tag{4.5}$$

For $g \in \mathscr{A}_w$ this identity is valid even in $\mathscr{H}^1_w$. Applying an element $f \in (\mathscr{H}^1_w)^\urcorner$ to (4.5) implies the same reproducing formula for $(\mathscr{H}^1_w)^\urcorner$, because we have the same rules of calculation as for the inner product (whereas a bilinear extension of the inner product would result in a different reproducing formula!).

(iv) Let us first calculate the adjoint $V_g^*: L_{1/w}^\infty \mapsto (\mathscr{H}_w^1)^\urcorner$. For $F \in L_{1/w}^\infty(\mathscr{G})$, $h \in \mathscr{H}_w^1$ one has by definition

$$\langle h, V_g^*(F) \rangle = \langle V_g(h), F \rangle = \int \overline{\langle \pi(y)g, h \rangle}\, F(y)\, dy$$

$$= \left\langle h, \int F(y)\, \pi(y)g\, dy \right\rangle \tag{4.6}$$

and thus

$$V_g^*(F) = \int F(y)\, \pi(y)g\, dy \in (\mathscr{H}_w^1)^\urcorner. \tag{4.7}$$

This integral is well defined in the weak sense because the representation is integrable and $\mathscr{A}_w \neq \{0\}$.

Next we remark that for $f \in L_{1/w}^\infty$ one has

$$V_g(V_g^*(F))(x) = \langle L_x V_g(g), F \rangle = F * V_g(g)(x), \tag{4.8}$$

as a consequence of (4.6) and the symmetry condition $V_g(g) = V_g(g)^\nabla$.

Assertion (iv) follows immediately. Applying these formulas again, the relation

$$V_g(V_g^*(V_g(f))) = V_g(f) * V_g(g) = V_g(f) \tag{4.9}$$

shows that $V_g^* \circ V_g: (\mathscr{H}_w^1)^\urcorner \mapsto (\mathscr{H}_w^1)^\urcorner$ is the identity operator.

(v) Suppose that a net (f_α) converges to f in $(\mathscr{H}_w^1)^\urcorner$ in the w^*-sense. Then a fortiori $V_g(f_\alpha)(x) = \langle \pi(x)\, g, f_\alpha \rangle \to \langle \pi(x)\, g, f \rangle$ for all $x \in \mathscr{G}$. Since for a compact set $K \subseteq \mathscr{G}$, the set $\{\pi(x)\, g: x \in K\}$ is compact in $\mathscr{H}_w^1$ (being the image of a compact set under a continuous mapping), pointwise convergence implies uniform convergence of $V_g(f_\alpha)$ on K.

On the other hand, pointwise convergence of $\langle \pi(x)\, g, f_\alpha \rangle$ implies w^*-convergence whenever the net (f_α) is bounded in $(\mathscr{H}_w^1)^\urcorner$ because the set $\{\pi(x)\, g, x \in \mathscr{G}\}$ is total in $\mathscr{H}_w^1$. ∎

Remark. The use of the antidual $(\mathscr{H}_w^1)^\urcorner$ instead of the dual $(\mathscr{H}_w^1)'$ is perhaps surprising, but very convenient. It allows us to carry over the notations and formulas from Hilbert space without modifications, e.g., the reproducing formula for $f \in \mathscr{H}$ or $f \in (\mathscr{H}_w^1)^\urcorner$, whereas a *bi*linear extension of the scalar product would have led us to use both π and its conjugate representation $\bar{\pi}$. The reproducing formula would then have the form

$$V_g(f) * V_g(g) = V_g(f) \qquad \text{for} \quad f \in \mathscr{H},$$

but

$$V_g(f) * \overline{V_g(g)} = V_g(f) \qquad \text{for } f \in (\mathscr{H}^1_w)^\urcorner,$$

and thus reveal an undesirable difference between the Hilbert space concerned and the extended situation.

Since the antidual $(\mathscr{H}^1_w)^\urcorner$ can always be „identified" with $(\mathscr{H}^1_w)'$ using the correspondence between $f \in (\mathscr{H}^1_w)^\urcorner$ and $\bar{f} \in (\mathscr{H}^1_w)'$, $\langle h, \bar{f} \rangle := \overline{\langle h, f \rangle}$ this technical detail is of no consequence.

In order to introduce coorbit spaces in full generality we consider as our next main ingredient the family of translation invariant solid BF-spaces on the group $\mathscr{G}$ (cf. Section 3).

To every such space Y on $\mathscr{G}$ are associated the two submultiplicative functions, describing the asymptotic behaviour of left and right translation operators: $x \to |||L_x|||_Y$ and $x \to |||R_x|||_Y$. We shall work in the sequel with the weight function w given by

$$w(x) := \max(|||L_x|||_Y, |||L_{x^{-1}}|||_Y, |||R_x|||_Y, |||R_{x^{-1}}|||_Y \Delta^{-1}(x)). \tag{4.10}$$

By this choice it is clear that the convolution relations (3.1) and, by Proposition 3.10, also (3.3) are valid.

We shall always work under the hypothesis that the space $\mathscr{A}_w$ is nontrivial and only function spaces Y with such weight functions will be considered. Thus the spaces $\mathscr{H}^1_w$ and $(\mathscr{H}^1_w)^\urcorner$ are well defined, and the following definition is reasonable:

DEFINITION 4.1.

$$\mathscr{C}o\, Y := \{ f \in (\mathscr{H}^1_w)^\urcorner \text{ with } V_g(f) \in Y \}.$$

As a natural norm we take $\|f\|_{\mathscr{C}o\, Y} := \|V_g(f)\|_Y$. $\mathscr{C}o\, Y$ will be called the *coorbit of Y under the representation* π.

This terminology is strongly influenced by the work of J. Peetre (cf. [P1, p. 200]). In the definition $\mathscr{C}o\, Y$ seems to depend not only on the representation π, but also on the choice of the weight w and the analyzing vector g. The independence of these ingredients will be stated in the next theorem. Thus we need not indicate these parameters in the notation. We shall also suppress the dependence of $\mathscr{C}o\, Y$ on π in the notation whenever convenient (cf. Theorem 4.6 below).

We shall use the following convention: lowercase letters denote elements in the coorbit spaces and the corresponding capital letters are their V_g-transforms, e.g., $G := V_g(g)$, $F := V_g(f)$, etc., where an admissible vector g is fixed throughout the discussion.

THEOREM 4.2 (Basic Properties of Coorbit Spaces). (i) *$\mathcal{C}o\, Y$ is a π-invariant Banach space which is continuously embedded into $(\mathcal{H}^1_w)^\urcorner$.*

(ii) *The definition of $\mathcal{C}o\, Y$ is independent of the choice of the analyzing vector $g \in \mathcal{A}_w$, i.e., different vectors $g \in \mathcal{A}_w$ define the same space and equivalent norms.*

(iii) *$\mathcal{C}o\, Y$ is independent of the reservoir $(\mathcal{H}^1_w)^\urcorner$, i.e., if w_2 is another weight with $w(x) \leqslant Cw_2(x)$ for all $x \in \mathcal{G}$ and $\mathcal{A}_{w_2} \neq \{0\}$, then*

$$\mathcal{C}o\, Y = \{f \in (\mathcal{H}^1_w)^\urcorner \text{ with } V_g(f) \in Y\}$$
$$= \{f \in (\mathcal{H}^1_{w_2})^\urcorner \text{ with } V_g(f) \in Y\}.$$

The proof of the theorem and all further developments will rely on the following

PROPOSITION 4.3. (i) *Given $g \in \mathcal{A}_w$, a function $F \in Y$ is of the form $V_g(f)$ for some $f \in \mathcal{C}o\, Y$ if and only if F satisfies the reproducing formula, i.e., $F = F * V_g(g)$. It follows that*

(ii) *$V_g: \mathcal{C}o\, Y \to Y$ establishes an isometric isomorphism between $\mathcal{C}o\, Y$ and the closed subspace $Y * V_g(g)$ of Y, whereas $F \mapsto F * V_g(g)$ defines a bounded projection from Y onto this subspace.*

(iii) *Every function $F = F * V_g(g)$ is continuous, belongs to $L^\infty_{1/w}(\mathcal{G})$, and the evaluation mapping $F \mapsto F(x)$ may also be written as $F(x) = \langle L_x G, F \rangle$.*

This proposition will be our principal instrument because it allows us to translate all problems concerning coorbit spaces into problems about continuous functions on the group $\mathcal{G}$ and to make use of the convolution relations for Wiener-type spaces there. From another point of view the proposition tells us that $\mathcal{C}o\, Y$ is (isomorphic to) a Banach space with a reproducing kernel.

In the proof of the proposition we make use of a restricted class of analyzing vectors $\mathcal{B}_w$ (better vectors), which we encountered already in [FG] as atoms for $\mathcal{C}o\, L^p_m$. We define

$$\mathcal{B}_w := \{g \in \mathcal{H}: V_g(g) \in W^R(C^0, L^1_w)\}.$$

Then $\mathcal{B}_w \subseteq \mathcal{A}_w$ and $\mathcal{B}_w$ is still dense in $\mathcal{H}^1_w$. Moreover, every $V_g(g)$, with $g \in \mathcal{B}_w$, has representations of the form

$$V_g(g) = \sum_{n=1}^{\infty} L_{z_n} H_n = \sum_{n=1}^{\infty} R_{z_n} G_n \tag{4.11}$$

for suitable sequences of functions G_n, H_n in $\mathscr{K}(\mathscr{G})$, with $\operatorname{supp} G_n \cup \operatorname{supp} H_n \subseteq Q$ (a fixed compact set in $\mathscr{G}$) for all n and

$$\sum_{n=1}^{\infty} w(z_n)\, \|G_n\|_\infty \cong \|V_g(g) \mid W^R(C^0, L^1_w)\| \tag{4.12}$$

(and $\cong \sum_{n=1}^{\infty} w(z_n)\, \|H_n\|_\infty$, respectively).

The first representation in (4.11) follows from Proposition 3.7 and the second one from the relations

$$V_g(g)^\nabla = V_g(g) \qquad \text{and} \qquad (L_x G)^\vee := R_x(G^\vee). \tag{4.13}$$

Proof of Proposition 4.3. The reproducing formula for $f \in \mathscr{C}o\, Y \subseteq (\mathscr{H}^1_w)^\urcorner$ has already been stated and proved in Theorem 4.1(iii). Moreover, by relation (3.1) it is now true as a convolution in Y.

For the converse take $F \in Y$ satisfying $F * V_g(g) = F$. As soon as we have shown that $F \in L^\infty_{1/w}(\mathscr{G})$ we can apply Theorem 4.1(iv) and find that $f := V_g^*(F) \in (\mathscr{H}^1_w)^\urcorner$ satisfies $V_g(f) = F \in L^\infty_{1/w} \cap Y$ and consequently $f \in \mathscr{C}o\, Y$.

In order to verify that $F * V_g(g) = F$ implies $F \in L^\infty_{1/w}$ we assume first $g \in \mathscr{B}_w$, whence $V_g(g) \in W^R(C^0, L^1_w)$. By Lemma 3.9(b) any $F \in Y$ belongs to $W(L_1, L^\infty_{1/w_0^\vee})$. Since $w_0^\vee(x) \leqslant w(x)$, by (4.10) Proposition 3.10 applies (with Y replaced by $L^\infty_{1/w}$) and yields $F = F * V_g(g) \subseteq L^\infty_{1/w}$.

Now take an arbitrary (normalized $g \in \mathscr{A}_w$. We know that for any fixed non-zero $g_0 \in \mathscr{B}_w$, g is of the form $g = V^*_{g_0}(\phi)$ for some $\phi \in L^1_{w^*}$, where $w^* := w + w^\vee \Delta^{-1}$ [FG, Lemma 4.2], and that $V_g(g) = \phi * V_{g_0}(g_0) * \phi^\nabla$ (by an elementary calculation). Then $F = F * V_g(g) = F * \phi * V_{g_0}(g_0) * \phi^\nabla$, where $F * \phi \in Y \subseteq W(L^1, L^\infty_{1/w})$ by (3.1) and 3.9(b), hence $F * \phi * V_{g_0}(g_0) \in L^\infty_{1/w}(\mathscr{G})$ by Proposition 3.10 and finally $F \in L^\infty_{1/w} * L^{1\vee}_w \subseteq L^\infty_{1/w}$ as desired.

Altogether we have found for every $F = F * V_g(g) \in Y$ some $f \in \mathscr{C}o\, Y$ with $V_g(f) = F$ which is even uniquely determined by the injectivity of V_g. Furthermore

$$F(x) = F * G(x) = \int F(y)\, G(y^{-1}x)\, dy$$

$$= \int \overline{G(x^{-1}y)}\, F(y)\, dy = \langle L_x G, F \rangle,$$

where we have used that $G = G^\nabla$ and the fact that for $F \in L^\infty_{1/w}$ and $G \in L^1_w$ the convolution can be written pointwise in the above way. The remaining assertions are now obvious. ∎

Proof of Theorem 4.2. (i) and (ii): After having clarified the technical details concerning the reservoir and the isomorphism between $\mathscr{C}o\, Y$ and

$Y * V_g(g)$, the proof, based on the orthogonality and convolution relations, is literally the same as in [FG, Theorem 5.2].

For the embedding $\mathcal{C}o\, Y \hookrightarrow (\mathcal{H}^1_w)^\urcorner$ we observe first that

$$\|f \mid (\mathcal{H}^1_w)^\urcorner\| \cong \|V_g(f) \mid L^\infty_{1/w}\|. \tag{4.14}$$

The estimate $\|V_g(f) \mid L^\infty_{1/w}\| \leqslant \|f \mid (\mathcal{H}^1_w)^\urcorner\| \; \|g \mid \mathcal{H}^1_w\|$ was established in (4.4). For the converse we use the fact that V_g^* is an isometry from $L^1_w * G$ onto $\mathcal{H}^1_w$ (Proposition 4.3). Therefore

$$\begin{aligned}\|f \mid (\mathcal{H}^1_w)^\urcorner\| &= \sup_{\|h \mid \mathcal{H}^1_w\| = 1} \langle h, f\rangle = \sup_{\|H * G \mid L^1_w\| = 1} \langle V_g^*(H * G), f\rangle \\ &= \sup_{\|H * G \mid L^1_w\| = 1} \langle H * G, V_g(f)\rangle \leqslant \sup_{\|H \mid L^1_w\| = 1} \langle H, V_g(f)\rangle \\ &= \|V_g(f) \mid L^\infty_{1/w}(\mathcal{G})\|.\end{aligned}$$

Then one obtains with the help of the reproducing formula, Proposition 3.10, and Lemma 3.9(b),

$$\begin{aligned}\|f \mid (\mathcal{H}^1_w)^\urcorner\| &\leqslant \|V_g(f) \mid L^\infty_{1/w}\| \\ &\leqslant \|V_g(f) \mid W(L^1, L^\infty_{1/w})\| \; \|G \mid W^R(C^0, L^1_w)\| \leqslant C\, \|V_g(f) \mid Y\|\end{aligned}$$

and the continuity of the embedding of $\mathcal{C}o\, Y$ into $(\mathcal{H}^1_w)^\urcorner$ is proved.

(iii) If $w(x) \leqslant C w_2(x)$ for all $x \in \mathcal{G}$ it is obvious that $\mathcal{H}^1_{w_2} \subseteq \mathcal{H}^1_w$ as a dense subspace, and therefore $(\mathcal{H}^1_w)^\urcorner \subseteq (\mathcal{H}^1_{w_2})^\urcorner$ (in the sense of continuous embeddings). Fixing $g \in \mathcal{A}_{w_2} \subseteq \mathcal{A}_w$ we suppose that for some $f \in (\mathcal{H}^1_{w_2})^\urcorner$, $V_g(f) \in Y$. We have to verify that f belongs already to $(\mathcal{H}^1_w)^\urcorner$ respectively that $V_g(f) \in L^\infty_{1/w}(\mathcal{G})$. But this has been proved in Proposition 4.3. The embedding $Y \hookrightarrow W(L^1, L^\infty_{1/w_0^\vee})$ has nothing to do with the weight function w_2, so the reproducing formula for $V_g(f)$ and the same convolution arguments as in Proposition 4.3 lead to the desired conclusion. ▌

COROLLARY 4.4. (a) $\mathcal{C}o\, L^\infty_{1/w} = (\mathcal{H}^1_w)^\urcorner$.

(b) $\mathcal{C}o\, L^2 = \mathcal{H}$.

Proof. For (a) there is nothing more to prove. In order to verify (b) note that for $Y = L^2(\mathcal{G})$ the minimal reservoir is $(\mathcal{H}^1_w)^\urcorner$, with $w(x) = 1 + \Delta^{-1/2}(x)$. Since (cf. Section 2.3) $V_g(f) \in L^2(\mathcal{G})$ for $f \in \mathcal{H}$ the inclusion

$\mathscr{H} \subseteq \mathscr{C}o\, L^2$ is clear. On the other hand, if for $f \in (\mathscr{H}^1_w)^\urcorner$, $V_g(f) \in L^2$, then there exists as a consequence of Section 2.4 some $f' \in \mathscr{H}$ such that $V_g(f') = V_g(f)$. By the injectivity of V_g on $(\mathscr{H}^1_w)^\urcorner$ it follows that $f' = f \in \mathscr{H}$ and thus $\mathscr{C}o\, L^2 \subseteq \mathscr{H}$. ∎

Our next aim is to introduce the notation of an orbit space (as opposed to the coorbit spaces considered so far) and to show that orbit and coorbit spaces coincide. Recall to this end that the mapping

$$(F, g) \to V_g^*(F) = \int F(y)\, \pi(y) g\, dy$$

is bilinear from $L^\infty_{1/w} \times \mathscr{A}_w$ onto $(\mathscr{H}^1_w)^\urcorner$. For a subset Z in $L^\infty_{1/w}(\mathscr{G})$ and a fixed vector $g \in \mathscr{A}_w$ one may speak of the *Z-orbit* of g, given by

$$\mathcal{O}_g(Z) := \{V_g^*(F), F \in Z\} \subseteq (\mathscr{H}^1_w)^\urcorner. \tag{4.15}$$

For a fixed function F, $g \mapsto V_g^*(F)$ is a (densely defined linear) operator from $\mathscr{A}_w \to (\mathscr{H}^1_w)^\urcorner$, which is usually denoted by $\pi(F)$ and which may be reasonable for other functions F than those in $L^\infty_{1/w}$, e.g., for $F \in L^p_{w^s}$, with $s := 1/p - 1/p'$.

If $g \in \mathscr{B}_w$, then one can show that, for $F \in Y$ (with canonically associated weight w (4.10)), $V_g^*(F) = \int F(y)\, \pi(y) g\, dy$ makes sense as a weak integral. For general $g \in \mathscr{A}_w$ we may define

$$V_g^*(F) := V_g^*(F * V_g(g)). \tag{4.16}$$

This is possible due to Proposition 4.3, and is consistent with the weak integral definition because one can check that $V_g^*(F * H) = V^*_{V_g^*(H)}(F)$ for $H \in L^1_{w^*}$ (whenever reasonable). The above formula is perhaps easier to remember if it is rewritten in the form $\pi(F * H) = \pi(F)\, \pi(H)$. Finally, $V_g^*(V_g(g)) = g$ (by (4.9)).

In the light of this definition Proposition 4.3 gives the following

COROLLARY 4.5. *For every* $g \in \mathscr{A}_w$

$$\mathscr{C}o\, Y = \mathcal{O}_g(Y)$$

and the orbit-norm

$$\|f \mid \mathcal{O}_g(Y)\| := \inf\{\|F \mid Y\|, f = V_g^*(F)\} \tag{4.17}$$

is an equivalent norm on $\mathscr{C}o\, Y$.

Proof. By Proposition 4.3 every $f \in \mathcal{C}o\, Y$ is of the form $V_g^*(F)$, $F \in Y$ (hence $\mathcal{C}o\, Y \subseteq \mathcal{O}(Y)$). On the other hand, $V_g^*(F)$, $F \in Y$ is an element of $\mathcal{C}o\, Y$ by formula (4.8). The equivalence of the norms follows from

$$\|f \mid \mathcal{O}(Y)\| = \|V_g^*(V_g(f)) \mid \mathcal{O}(Y)\| \leqslant \|V_g(f) \mid Y\| = \|f \mid \mathcal{C}o\, Y\|.$$

Conversely one can find for $f \in \mathcal{O}(Y)$ and $C > 1$ some $F \in Y$ such that

$$f = V_g^*(F) = V_g^*(F * V_g(g)) \quad \text{and} \quad \|F \mid Y\| \leqslant C\, \|f \mid \mathcal{O}(Y)\|.$$

Consequently,

$$\begin{aligned}\|f \mid \mathcal{C}o\, Y\| &= \|V_g(f) \mid Y\| \\ &= \|V_g(V_g^*(F * V_g(g))) \mid Y\| = \|F * V_g(g) \mid Y\| \\ &\leqslant \|F \mid Y\|\, \|V_g(g) \mid L_w^1\| \leqslant C'\, \|f \mid \mathcal{O}(Y)\|. \quad \blacksquare\end{aligned}$$

COROLLARY 4.6. *$\mathcal{C}o\, Y$ is a retract of $(Y, \|\ \|_Y)$.*

Proof. The linear mappings $V_g: \mathcal{C}o\, Y \to Y$ and $V_g^*: Y \to \mathcal{C}o\, Y$ are bounded by definition and Corollary 4.5, respectively, and $V_g^* \circ V_g = \mathrm{Id}_{\mathcal{C}o\, Y}$ (cf. (4.9)). ∎

THEOREM 4.7. (i) *Given a weight function w the family of coorbit spaces $\mathcal{C}o\, Y$ with Y satisfying* (3.1) *is closed with respect to arbitrary interpolation methods.*

(ii) *The subfamily of coorbit spaces with respect to weighted L^p-spaces is closed with respect to complex interpolation.*

Proof. It is clear that the family of spaces Y under consideration is closed with respect to interpolation. Since interpolation functors commute with taking retracts, (i) is proved and (ii) is an obvious consequence. ∎

Next we describe to what extent coorbit spaces depend on the realization of a representation π and how intertwining operators between equivalent representations can be extended in a canonical way to isomorphisms between corresponding coorbit spaces.

THEOREM 4.8 (Automatic Extension of Intertwining Operators, Dependence on π). (i) *Assume that $(\pi_1, \mathcal{H}_1)$ and $(\pi_2, \mathcal{H}_2)$ are two equivalent integrable (irreducible, unitary, continuous) representations of $\mathcal{G}$, i.e., that there is an isometry $T: \mathcal{H}_1 \to \mathcal{H}_2$ such that $\pi_2(x) T = T \pi_1(x)$ for all $x \in \mathcal{G}$. Then T can be uniquely extended to a bounded invertible intertwining operator between $\mathcal{C}o_{\pi_1} Y$ and $\mathcal{C}o_{\pi_2} Y$.*

(ii) *Let $\alpha: \mathcal{G}_1 \mapsto \mathcal{G}_2$ be a surjective homomorphism, and π_1 and π_2 integrable (irreducible, unitary) representations of $\mathcal{G}_1$ and $\mathcal{G}_2$ such that*

$\pi_2 \circ \alpha(x) T = T\pi_1(x)$ for all $x \in \mathscr{G}_1$. For any solid BF-space Y on $\mathscr{G}_1$ we denote by $\alpha^* Y$ the solid BF-space on $\mathscr{G}_2$ given by $\alpha^* Y = \{F$, such that $F \circ \alpha \in Y\}$ with the norm $\|F \mid \alpha^* Y\| = \|F \circ \alpha \mid Y\|$. Then the extension of the intertwining operator T maps $\mathscr{C}o_{\pi_1} Y$ onto $\mathscr{C}o_{\pi_2} \alpha^* Y$.

(iii) *If, in particular, $\alpha: \mathscr{G} \mapsto \mathscr{G}$ is an automorphism of $\mathscr{G}$ and π an integrable representation of $\mathscr{G}$ such that $\pi \circ \alpha(x) T = T\pi(x)$ then $\mathscr{C}o\, Y$ is invariant under T whenever $\alpha^* Y = Y$.*

Proof. (i) To indicate the dependence of the spaces involved on the representation we add π_1, π_2 to our notation. We consider first the effect of $T: \mathscr{H}_1 \mapsto \mathscr{H}_2$ on the analyzing vectors $g \in \mathscr{A}_w(\pi_1)$,

$$\int_{\mathscr{G}} \langle \pi_1(x) g, g \rangle\, w(x)\, dx = \int_{\mathscr{G}} \langle T\pi_1(x) g, Tg \rangle\, w(x)\, dx$$

$$= \int_{\mathscr{G}} \langle \pi_2(x)\, Tg, Tg \rangle\, w(x)\, dx,$$

which implies that T maps $\mathscr{A}_w(\pi_1)$ onto $\mathscr{A}_w(\pi_2)$. Now we take arbitrary normalized vectors $g \in \mathscr{A}_w(\pi_1), g_2 \in \mathscr{A}_w(\pi_2)$, and $f \in \mathscr{H}_1$ and compute the following convolution with the help of (2.2) (with $V^i_{g_i}(f_i)(x) := \langle \pi_i(x) g_i, f_i \rangle$):

$$V^2_{g_2}(Tf) = V^2_{Tg}(Tf) * V^2_{g_2}(Tg) = V^1_g(f) * V^2_{g_2}(Tg) \tag{4.18}$$

or equivalently $Tf = V^{2*}_{g_2} \circ C \circ V^1_g(f)$, where C is the right convolution with $\langle \pi_2(\cdot)\, g_2, Tg \rangle$. This relation means that the intertwining operator $T: \mathscr{H}_1 \to \mathscr{H}_2$ can be obtained by (1) taking the V_g-transform with respect to π_1, (2) taking a right convolution with $\langle \pi_2(\cdot)\, g_2, Tg \rangle$ which is in $L^1_w(\mathscr{G}_1)$ whenever $g \in \mathscr{A}_w(\pi_1)$, $g_2 \in \mathscr{A}_w(\pi_2)$, and (3) reversing the V_g-transform with respect to π_2. Since $\langle \pi_2(\cdot)\, g_2, Tg \rangle \in L^1_w(\mathscr{G}_1)$ this version of T can be extended to $f \in \mathscr{C}o_{\pi_1} Y$. Consequently T is a bounded operator from $\mathscr{C}o_{\pi_1} Y$ onto $\mathscr{C}o_{\pi_2} Y$ and the intertwining property follows from (4.2).

T is invertible because we may proceed the same way with $T^{-1}: \mathscr{H}_2 \to \mathscr{H}_1$. The uniqueness follows from the fact that T is completely known from the images of $\pi_1(x) g$ (which, however, takes place within the Hilbert spaces).

Another equivalent extension procedure would have been to restrict $T^{-1}: \mathscr{H}^1_w(\pi_2) \to \mathscr{H}^1_w(\pi_1)$, then to consider the adjoint mapping $T^{-1*}: \mathscr{H}^1_w(\pi_1) \to \mathscr{H}^1_w(\pi_2)^\urcorner$ and to show that T^{-1*} maps $\mathscr{C}o_{\pi_1} Y$ onto $\mathscr{C}o_{\pi_2} Y$ via the identification of orbits and coorbits (Corollary 4.5).

(ii) This is an immediate consequence of (i). Since $\pi_2 \circ \alpha(x) T = T\pi_1(x)$ one obtains by the same calculation as above $\mathscr{A}_w(\pi_1) = \mathscr{A}_w(\pi_2 \circ \alpha) =$

$\mathscr{A}_{w \circ \alpha^{-1}}(\pi_2)$. (Note that the integrability of π_1 and the intertwining of π_1 and π_2 imply the kernel of α to be compact and the ambiguity of $w \circ \alpha^{-1}$ disappears by defining it to be constant on the cosets of $\ker \alpha$). Furthermore, T extends from $\mathscr{C}o_{\pi_1} Y$ onto $\mathscr{C}o_{\pi_2 \circ \alpha} Y$, i.e., if $f \in \mathscr{C}o\, Y$, then $Tf \in \mathscr{C}o_{\pi_2 \circ \alpha} Y \Leftrightarrow \langle \pi_2(\alpha(x))\, g_2, Tf \rangle \in Y$, which means by definition that $\langle \pi_2(y)\, g_2, Tf \rangle \in \alpha^* Y \Leftrightarrow Tf \in \mathscr{C}o_{\pi_2} \alpha^* Y$.

(iii) This is now trivial. ▌

Remark. Parts (ii) and (iii) are, on this abstract level, mere tautologies, but they will allow us in a concrete situation to conclude without any further effort that certain integral operators (the so-called metaplectic representation) leave a large scale of function spaces on $\mathbb{R}^n$ (certain modulation spaces) invariant, whereas a direct proof involves long calculations (cf. [P2]).

THEOREM 4.9. *Assume that $(Y, \| \ \|_Y)$ has an absolutely continuous norm (which is equivalent to the assumption that the Banach dual Y' coincides with the Köthe-dual $Y^\alpha := \{H \in L^1_{\text{loc}} : HF \in L^1(\mathscr{G})\ \forall F \in Y\}$.). Then*

$$(\mathscr{C}o\, Y)' = \mathscr{C}o\, Y^\alpha = \mathscr{C}o\, Y'.$$

Proof. Since $|||L_x | Y^\alpha||| = |||L_{x^{-1}} | Y|||$ and $|||R_x | Y^\alpha||| = |||R_{x^{-1}} | Y|||\, \Delta^{-1}(x)$ the canonical weights w defined in (4.10) for Y^α and Y coincide. Therefore $\mathscr{C}o\, Y$ and $\mathscr{C}o\, Y^\alpha$ are selected from the same reservoir $(\mathscr{H}^1_w)^\urcorner$ and both spaces have the same set of analyzing vectors. Thus the following arguments are consistent. Let $i: \mathscr{C}o\, Y^\alpha \mapsto (\mathscr{C}o\, Y)'$ be the mapping

$$i(h)(f) = \langle V_g(h), V_g(f) \rangle, \tag{4.19}$$

Then $|i(h)(f)| \leqslant \|V_g(h) | Y^\alpha\|\ \|V_g(f) | Y\| = \|h | \mathscr{C}o\, Y^\alpha\|\ \|f | \mathscr{C}o\, Y\|$ and the norm of the functional $i(h)$ is equivalent to $\|h | \mathscr{C}o\, Y^\alpha\|$ by an argument similar to that in (4.14). In particular, i is one-to-one.

For the converse we have to show that i is onto. We identify $\mathscr{C}o\, Y$ with the closed subspace $Y * G$ (Proposition 4.3(ii)). By the Hahn–Banach theorem we can extend any given $k \in (\mathscr{C}o\, Y)' \simeq (Y * G)'$ to a functional $K \in Y'$ such that

$$K(V_g(f)) = \langle k, f \rangle \qquad \text{for all} \quad f \in \mathscr{C}o\, Y. \tag{4.20}$$

Since $Y' = Y^\alpha$ by assumption there exists $H \in Y^\alpha$ such that

$$K(F) = \int \overline{H(x)}\, F(x)\, dx = \langle H, F \rangle \qquad \text{for all} \quad F \in Y.$$

By the relation $\langle H, V_g(f)\rangle = \langle H, V_g(f) * G\rangle = \langle H * G, V_g(f)\rangle$ the functional $F \to \langle H * G, F\rangle$ is another admissible extension of K. Let h be the associated element in $\mathscr{C}o\, Y^\alpha$ with $V_g(h) = H$ (by Proposition 4.3(i)), then

$$i(h)(f) = \langle H * G, V_g(f)\rangle = \langle H, V_g(f)\rangle = \langle k, f\rangle. \quad \blacksquare$$

COROLLARY 4.10. *$\mathscr{C}o\, Y$ is a reflexive Banach space, if Y is reflexive.*

Proof. For a reflexive solid BF-space the Banach dual coincides with its Köthe-dual Y^α and $Y = Y'' = Y^{\alpha\alpha}$ (cf. [Z]), therefore $(\mathscr{C}o\, Y)'' = \mathscr{C}o\, Y$. $\blacksquare$

5. DISCRETIZATION OF CONVOLUTIONS

In this section the relevant techniques leading to the atomic decompositions are presented. The basic idea is—not unfamiliar—to replace certain convolution products by sums of translates of one convolution factor. Since such results seem to be of independent interest we state them separately here. Related methods can also be used to derive various results on the complete reconstruction of band-limited functions on $\mathbb{R}^m$ from an irregular sampling (such questions will be discussed in detail elsewhere), in a way similar to Shannon's sampling theorem.

As in the preceding sections, Y is a translation invariant, solid BF-space and w a weight function such that (3.1) holds true.

PROPOSITION 5.1. *There is a constant $C_d > 0$ such that for any U_0-dense and relatively separated family $X = (x_i)_{i \in I}$ and for any U_0-BUPU Ψ the linear coefficient mapping $F \mapsto \Lambda = (\lambda_i)_{i \in I} := (\langle \psi_i, F\rangle)_{i \in I}$ satisfies the estimate*

$$\|\Lambda \mid Y_d(X)\| \leqslant C_d \|F \mid Y\|.$$

Proof. Let $F \in Y$ and $k \in \mathscr{K}(\mathscr{G})$ be a plateau function with $k(x) \equiv 1$ on QU_0. Then for every $y \in \mathscr{G}$ the control function

$$K(F, y) = \sum_{i \in I} \langle \psi_i, |F|\rangle\, c_{x_i Q}(y)$$

is a finite sum over the index set $I_y := \{i \mid x_i \in yQ\}$ hence

$$K(F, y) = \sum_{i \in I_y} \langle \psi_i, |F|\rangle \leqslant \langle L_y k, |F|\rangle.$$

As in Lemma 3.9(a) we derive

$$\|\Lambda \mid Y_d\| = \left\| \sum_{i \in I} |\langle \psi_i, F\rangle| \, c_{x_i} Q \mid Y \right\|$$
$$\leqslant \|K(F, y) \mid Y\| \leqslant \|F \mid Y\| \; \|k \mid L^1_w\|. \quad \blacksquare$$

PROPOSITION 5.2. *Let $X = (x_i)_{i \in I}$ be a relatively separated family in $\mathscr{G}$, and let $G \in W^R(C^0, L^1_w)$ be given. Then the mapping*

$$\Lambda = (\lambda_1)_{i \in I} \mapsto \sum_{i \in I} \lambda_i L_{x_i} G$$

is a bounded, linear operator from $Y_d(X)$ into Y, satisfying (for some constant $C_s > 0$, independent of X and G)

$$\left\| \sum_{i \in I} \lambda_i L_{x_i} G \mid Y \right\| \leqslant C_s \, \|G \mid W^R(C^0, L^1_w)\| \; \|\Lambda \mid Y_d\|. \tag{5.1}$$

The convergence of the sum has to be understood with respect to the Y-norm if the finite sequences are dense in Y_d and in the pointwise sense otherwise.

Proof. By Lemma 3.4(b) the measure $\mu_\Lambda := \sum_{i \in I} \lambda_i \delta_{x_i}$ belongs to $W(M, Y)$ and $\|\mu_\Lambda \mid W(M, Y)\| \leqslant C' \, \|\Lambda \mid Y_d\|$ with a C' independent of X. Thus the proposition is a consequence of Proposition 3.10.

Furthermore, $F(x) = \sum_{i \in I} \lambda_i L_{x_i} G(x)$ is defined pointwise: since $Y_d(X) \subseteq l^\infty_{1/w}$ (Lemma 3.5(b)) and $(G(x_i^{-1} x))_{i \in I} \in l^1_w$ (by Lemma 3.8(a)) for all $x \in \mathscr{G}$, the partial sums of F converge pointwise by the $l^1_w - l^\infty_{1/w}$-duality. If the finite sequences are dense in Y_d the norm convergence of the sum follows directly from (5.1). $\blacksquare$

In order to obtain an atomic decomposition of the elements of coorbit spaces we want to discretize the reproducing formula. We do so by approximating the (right) convolution operator

$$T\colon Y \to Y, \qquad F \mapsto F * G \tag{5.2}$$

by the operator

$$T_\Psi F := \sum_{i \in I} \langle \psi_i, F\rangle \, L_{x_i} G, \tag{5.3}$$

where $\Psi = (\psi_i)_{i \in I}$ is an arbitrary U_0-BUPU. Writing $TF = F * G$ as a vector-valued integral $TF = \int F(y) \, L_y G \, dy$ we see that T_Ψ may be interpreted as a Riemannian sum for this integral.

PROPOSITION 5.3. *For G, $X=(x_i)_{i\in I}$, and Y as above, every T_Ψ maps Y into $Y*G$ and the family $\{T_\Psi\}$ (where Ψ runs through the system of U_0-BUPUs) is uniformly bounded by $C_d C_s \|G \mid W^R(C^0, L^1_w)\|$.*

Proof. By Propositions 5.1 and 5.2

$$\|T_\Psi F \mid Y\| = \left\| \sum_{i\in I} \langle \psi_i, F\rangle \, L_{x_i} G \mid Y \right\| \leqslant \|(\langle \psi_i, F\rangle)_{i\in I} \mid Y_d\|$$

$$\leqslant C_s \|G \mid W^R(C^0, L^1_w)\| \, \|F \mid Y\|.$$

The constant $C_s C_d$ is independent of G, X, and Ψ. ∎

The following result is fundamental for the theory of atomic decompositions developed in Section 6, but it should also be of independent interest. It confirms the intuition that a refinement of the partition of unity Ψ will increase the degree of approximation of T_Ψ to T. This is not difficult to verify in the strong operator topology, but the point is that one can verify convergence in the norm topology, because the rate of approximation depends only on the smoothness of the right convolution factor $G \in W^R(C^0, L^1_w)$, but not on the domain Y of the convolution operator (as long as (3.1) holds true).

For the following we consider the set of BUPUs as a directed set which is ordered by the inclusion of the corresponding neighbourhoods. We write $\Psi \to \infty$ if these neighbourhoods run through a neighbourhood basis of e. Thus $\Psi \to T_\Psi$ is a net and convergence of this net will be understood in that sense.

PROPOSITION 5.4. *Let Y be a translation invariant solid BF-space, w its canonical weight, and $G \in W^R(C^0, L^1_w)$. Then the net $\{T_\Psi\}$ of approximating operators T_Ψ converges in the norm to the convolution operator T, i.e.,*

$$\lim_{\Psi\to\infty} |||T - T_\Psi \mid Y||| = 0.$$

Proof. Again, we first give the corresponding estimate for the pieces G_n of $G \in W^R(C^0, L^1_w)$. Let Ψ be a U-BUPU (for some $U \subseteq U_0$). Then

$$\mathcal{R}_n := \left\| F * G_n - \sum_{i\in I} \langle \psi_i, F\rangle \, L_{x_i} G_n \mid Y \right\|$$

$$= \left\| \sum_{i\in I} \int_{x_i U} \psi_i(z) \, F(z)(L_z G_n - L_{x_i} G_n) \, dz \mid Y \right\|$$

(as a vector-valued integral).

Since for each $y \in \mathcal{G}$ the integrand vanishes for $i \notin I_y := \{i \in I, x_i \in yU_0Q\}$ we obtain, taking the norm with respect to the variable y,

$$\begin{aligned}\|\mathcal{R}_n \mid Y\| &= \left\| \sum_{i \in I_y} \int_{x_iU} \psi_i(z)\, F(z)(L_zG_n(y) - L_{x_i}G_n(y))\, dz \mid Y \right\| \\ &\leqslant \left\| \sum_{i \in I_y} \sup_{z \in x_iU} \|L_zG_n - L_{x_i}G_n\|_\infty \langle \psi_i, |F| \rangle \mid Y \right\| \\ &\leqslant (\sup_{u \in U} \|L_uG_n - G_n\|_\infty) \left\| \sum_{i \in I_y} \langle \psi_i, |F| \rangle \mid Y \right\| \\ &\leqslant \omega_U(G_n)\, C_d\, \|F \mid Y\|\end{aligned}$$

according to the proof of Proposition 5.1 and the notation

$$\omega_U(H) := \sup_{u \in U} \|L_uH - H\|_\infty. \tag{5.4}$$

Summing up over n and interchanging the order of summation one obtains

$$\begin{aligned}\|TF - T_\Psi F \mid Y\| &= \left\| \sum_n R_{z_n}(F * G_n - \sum_{i \in I} \langle \psi_i, F \rangle L_{x_i}G_n) \mid Y \right\| \\ &\leqslant \sum_n w(z_n)\, \mathcal{R}_n \leqslant C_d \|F \mid Y\| \sum_n w(z_n)\, \omega_U(G_n) \\ &:= C_d \|F \mid Y\|\, \Omega_U(G).\end{aligned}$$

Observing that $\omega_U(G_n) \leqslant 2 \|G_n\|_\infty$ implies

$$\Omega_U(G) \leqslant 2 \sum_n w(z_n)\, \|G_n\|_\infty = 2C_s \|G \mid W^R(C^0, L^1_w)\| < \infty \tag{5.5}$$

and one finds for $\varepsilon > 0$ some finite set $E \subseteq I$ such that $\sum_{n \notin E} w(z_n) \|G_n\|_\infty \leqslant \varepsilon/4C_d$, hence $\sum_{n \notin E} w(z_n)\, \omega_U(G_n) \leqslant \varepsilon/2C_d$. G_n being uniformly continuous we can choose $U_1 \subseteq U_0$ such that $\omega_U(G_n) \leqslant \varepsilon/(2\,|E|\, C_d)$ for all $n \in E$. Consequently one has $\Omega_U(G) \leqslant \varepsilon/C_d$ and therefore

$$|||TF - T_\Psi F \mid Y||| \leqslant \varepsilon \|F \mid Y\|$$

for every U_1-BUPU Ψ and $\varepsilon > 0$. ∎

Remark. $\Omega_U(G)$ can be viewed as a modulus of continuity of G.

6. The Atomic Decomposition Theorem and Stability Results

This section contains our main result, the atomic decomposition of the coorbit spaces. For the proof we shall combine the methods of the preceding sections. The identification of $\mathscr{C}o\, Y$ with $Y * G$ allows to work exclusively with functions on the group and subject to the reproducing formula. Then we apply the discretization technique of Section 5 and argue along the same lines as in [FG] in order to arrive at the decomposition of $\mathscr{C}o\, Y$. In contrast to the direct methods in [FJ1, FJ2, F4, R] the admissible atoms satisfy rather mild conditions and form a dense subspace $\mathscr{B}_w$ of $\mathscr{H}$.

THEOREM 6.1. (The Atomic Decomposition in $\mathscr{C}o\, Y$). *For any $g \in \mathscr{B}_w$ there exist positive constants C_0 and C_1 (depending only on g) and a neighbourhood U of e such that for an arbitrary U-dense and relatively separated family $X = (x_i)_{i \in I} \subseteq \mathscr{G}$ the following is true:*

(i) *Analysis: There exists a bounded linear operator $A: \mathscr{C}o\, Y \to Y_d(X)$, i.e., writing $\Lambda := (\lambda_i)_{i \in I} := A(f)$ one has*

$$\|\Lambda \mid Y_d(X)\| \leqslant C_0 \|f \mid \mathscr{C}o\, Y\|, \tag{6.1}$$

such that every $f \in \mathscr{C}o\, Y$ can be represented as

$$f = \sum_{i \in I} \lambda_i \pi(x_i) g, \tag{6.2}$$

(ii) *Synthesis: Conversely, assuming that $X = (x_i)_{i \in I}$ is relatively separated, every $\Lambda \in Y_d$ defines an element $f = \sum_{i \in I} \lambda_i \pi(x_i) g$ in $\mathscr{C}o\, Y$ with*

$$\|f \mid \mathscr{C}o\, Y\| \leqslant C_1 \|\Lambda \mid Y_d(X)\|. \tag{6.3}$$

In both cases convergence takes place in the norm of $\mathscr{C}o\, Y$, if the finite sequences are norm dense in Y_d, and in the w^*-sense of $(\mathscr{H}^1_w)^\neg$ otherwise.

Proof. We may work with a normalized $g \in \mathscr{B}_w$, $\|Ag\| = 1$, for then the operator $T: F \mapsto F * G$ $(G := V_g(g))$ is a bounded projection from Y onto $Y * G$ (because $G * G = G$ from the orthogonality relations and Proposition 4.3(ii)). Remember that V_g is an isometrical isomorphism from $\mathscr{C}o\, Y$ onto $Y * G$ that intertwines π and L. Thus for any $F \in Y * G$ there is a unique $f \in \mathscr{C}o\, Y$ with $V_g(f) = F$ and to $L_{x_i} G$ correspond exactly the elements $\pi(x_i) g$. Consequently, in order to obtain the atomic decomposition it suffices to discretize the convolution $F * G$.

This has been done in Proposition 5.4: Since T acts on $Y * G$ as the identity operator and since the range of T_Ψ is always contained in $Y * G$ we can choose a neighbourhood U such that for every U-dense family $X = (x_i)_{i \in I}$ and corresponding U-BUPU Ψ

$$|\!|\!|\mathrm{Id} - T_\Psi \mid Y * G|\!|\!| < a < 1$$

by Proposition 5.4. This means that T_Ψ is invertible on $Y * G$, more precisely, T_Ψ^{-1} can be represented by the Neumann series $T_\Psi^{-1} = \sum_{n=0}^{\infty} (\mathrm{Id} - T_\Psi)^n$ and $|||(T_\Psi|_{Y * G})^{-1} | Y * G||| \leqslant (1-a)^{-1}$.

It follows that any $F \in Y * G$ has the expansion

$$F = T_\Psi(T_\Psi^{-1} F) = \sum_{i \in I} \langle \psi_i, T_\Psi^{-1} F \rangle L_{x_i} G \quad \text{in} \quad Y * G. \tag{6.4}$$

Pulling back to $\mathscr{C}o\, Y$ (cf. Proposition 4.3(ii)) we obtain for $f \in \mathscr{C}o\, Y$

$$f = \sum_{i \in I} \langle \psi_i, T_\Psi^{-1} V_g(f) \rangle \, \pi(x_i)\, g. \tag{6.5}$$

Since $T_\Psi^{-1} V_g(f) \in Y * G \subseteq Y$ the coefficients $\lambda_i := \langle \psi_i, T_\Psi^{-1} V_g(f) \rangle$ fulfill

$$\begin{aligned} \|\Lambda \mid Y_d\| &\leqslant C_d \, \|T_\Psi^{-1} V_g(f) \mid Y\| \\ &\leqslant C_d \, |||T_\Psi^{-1}||| \, \|V_g(f) \mid Y\| \leqslant C_d (1-a)^{-1} \, \|f \mid \mathscr{C}o\, Y\| \end{aligned}$$

after an application of Proposition 5.1. The constant $C_0 := C_d(1-a)^{-1}$ depends only on the size of U (consequently it depends only on g and the arbitrary choice of a window-function k). The linearity of $f \mapsto (\lambda_i)_{i \in I}$ is obvious from the construction.

In order to prove (ii) we apply V_g and then Proposition 5.2: Since $Y_d(X) \subseteq L^\infty_{1/w}$ (cf. 3.5(b)) and $G \in W^{\mathrm{R}}(C^0, L^1_w)$ by the assumption $g \in \mathscr{B}_w$, the function

$$F(x) = V_g\Big(\sum_{i \in I} \lambda_i \pi(x_i) g\Big)(x) = \sum_{i \in I} \lambda_i L_{x_i} G(x)$$

belongs to $L^\infty_{1/w}(\mathscr{G})$ and thus defines a unique element $f \in (\mathscr{H}^1_w)^\urcorner$ (cf. Corollary 4.4(a)). The pointwise convergence of the partial sums of F implies the w^*-convergence of $f := \sum_{i \in I} \lambda_i \pi(x_i) g$. Once f is identified as an element of $(\mathscr{H}^1_w)^\urcorner$ it belongs to $\mathscr{C}o\, Y$ by Proposition 5.2 (where also the type of convergence is stated) and C_1 in (6.3) equals $C_1 := C_s \, \|G \mid W^{\mathrm{R}}(C^0, L^1_w)\|$. ∎

Remarks. (a) The constant C_0 depends *only* on $g \in \mathscr{B}_w$ and is the same for the family of spaces Y which have the same estimate for the right translation norms. Given $g \in \mathscr{B}_w$ all these spaces have the same set of atoms $\{\pi(x_i) g, i \in I\}$. Furthermore, the size of U of the U-dense family $(x_i)_{i \in I}$ depends only on g via the modulus of continuity $\Omega_U(V_g(g))$. It can be estimated explicitly in concrete examples.

(b) As a special case of the theorem, elements in $(\mathscr{H}^1_w)^\urcorner$ are characterized by the existence of a representation of the form $\sum_{i \in I} \lambda_i \pi(x_i) g$ with

$\sup_i |\lambda_i| \, w(x_i)^{-1} < \infty$. The method which was given in [FG] for the spaces $\mathscr{C}o\, L^p_m$ was not applicable to this situation.

(c) The analysis of atomic decompositions is by no means restricted to coorbit spaces under *irreducible* integrable representations. As soon as one disposes of a reproducing formula $V_g(f) * V_g(g) = V_g(f)$ and $V_g(g) \in L^1_w(\mathscr{G})$ our theory of coorbits applies. Concrete examples indicate that this is true for a larger class of integrable representations than the irreducible ones. It is planned to investigate this point in another paper.

(d) In view of the non-uniqueness of the atomic representation of elements in coorbit spaces it is worth mentioning that our method is optimal in the following sense: Assume $f \in (\mathscr{H}^1_w)^\urcorner$ has a representation $\sum_{i \in I} \lambda_i \pi(x_i') \, g$ with coefficients satisfying certain decay/summability conditions, more precisely, belonging to some space $Y_d(X')$ for some relatively separated family $X' = (x_i')_{i \in I}$. Then the coefficients arising in our construction satisfy the same conditions, i.e., they belong to the corresponding sequence space $Y_d(X)$.

The proof of the above theorem even shows the following:

COROLLARY 6.2. *For any U-dense and relatively separated family $X = (x_i)_{i \in I}$ the coorbit space $\mathscr{C}o\, Y$ is a retract of the solid BK-space $Y_d(X)$.*

Proof. We observe that the mappings $A: \mathscr{C}o\, Y \to Y_d(X)$ of Theorem 6.1(i) and $B: Y_d(X) \to \mathscr{C}o\, Y$, with $B(\Lambda) := \sum_i \lambda_i \pi(x_i) g$, are both bounded linear operators and satisfy $B \circ A = \mathrm{Id}_{\mathscr{C}o\, Y}$. Therefore $\mathscr{C}o\, Y$ is a retract of Y_d.

The atomic decomposition of coorbit spaces allows one to reduce many problems to corresponding problems for sequence spaces. In Part II we shall apply this principle to the investigation of the Banach space theoretical properties of coorbit spaces. Here we treat only the behaviour of coorbit spaces under interpolation, because it is a direct consequence of Corollary 6.2.

COROLLARY 6.3. *A given family of coorbit spaces is closed with respect to a certain family of interpolation methods whenever the corresponding family of sequence spaces $Y_d(X)$ is stable under this family.*

The atomic decomposition theorem 6.1 gives an effective procedure to calculate suitable coefficients for a given function in order to expand it in terms of a given family of atoms $\pi(x_i) g$. We finish this section with an investigation of the properties of these coefficients, their dependence on f, and the particular ingredients of the method. Since one may think of the assertions given below as statements on the stability of the atomic decomposition method described in this paper with respect to small perturbations they should be of relevance in connection with numerical analysis.

For later reference let us denote by $A = A(\Psi, X, g)$ the mapping from $\mathscr{C}o\, Y$ into $Y_d(X)$ which associates to every $f \in \mathscr{C}o\, Y$ its coefficients $((Af)_i)_{i \in I}$ in the atomic decomposition, using a fixed partition Ψ associated with the family $X = (x_i)_{i \in I}$ and some $g \in \mathscr{B}_w$ as basic atom.

According to (6.5)

$$(Af)_i = \langle \psi_i, T_\Psi^{-1} V_g(f) \rangle \tag{6.6}$$

and thus $A = A(\Psi, X, g)$ can be written as a product

$$A(\Psi, X, g) = C(\Psi) \circ T(\Psi, X, g)^{-1} \circ V_g, \tag{6.7}$$

where $V_g : \mathscr{C}o\, Y \to Y$, $T(\Psi, X, g)^{-1}$: $Y * GY * G$ ($T(\Psi, X, g)$ is defined in (5.3)), and $C(\Psi)$: $Y \to Y_d$ is given as $C(\Psi) := (\langle \psi_i, F \rangle)_{i \in I}$.

By Theorem 6.1 the mapping $A: \mathscr{C}o\, Y \to Y_d$ is always norm continuous. Additionally, we have the following weak continuity on $(\mathscr{H}^1_w)^\urcorner$:

PROPOSITION 6.4. *$A(\Psi, X, g)$ is w*-continuous from $(\mathscr{H}^1_w)^\urcorner$ into $l^\infty_{1/w}$, in particular, for any bounded w*-convergent net $f_\alpha \to f$ in $(\mathscr{H}^1_w)^\urcorner$ the coefficients converge pointwise, i.e., $(Af_\alpha)_i \to (Af)_i$ for all $i \in I$.*

Proof. It is our aim to verify that $A = \mathscr{A}^*$ for a bounded operator $\mathscr{A}$ from l^1_w into $\mathscr{H}^1_w$ and that consequently A is w^*-continuous as an operator from $(\mathscr{H}^1_w)^\urcorner$ into $l^\infty_{1/w}$.

If one takes into consideration that in (6.6) we need only the restriction of $C(\Psi)$ and $T(\Psi, X, g)$ to $L^\infty_{1/w} * G$ one is led to the operator

$$\mathscr{A} = \mathscr{V}_g \circ \mathscr{T}(\Psi, X, g)^{-1} \circ \mathscr{C}(\Psi), \tag{6.8}$$

which is composed of the following bounded linear operators:

$$\mathscr{C}(\Psi): l^1_w \to L^1_w * G \quad \text{with} \quad \mathscr{C}(\Psi)(\Lambda) := \sum_{i \in I} \lambda_i \psi_i * G$$

$$\mathscr{T}(\Psi, X, g): L^1_w * G \to L^1_w * G \quad \text{with} \quad \mathscr{T}(\Psi, X, g)F := \sum_{i \in I} \langle L_{x_i} G, F \rangle \psi_i * G \tag{6.9}$$

$$\mathscr{V}_g : L^1_w * G \to \mathscr{H}^1_w \quad \text{with} \quad \mathscr{V}_g(F) := \int F(y) \pi(y) g\, dy.$$

One verifies by routine calculations that $\mathscr{V}^*_g = V_g$,

$$\mathscr{T}(\Psi, X, g)^* = T(\Psi, X, g)|_{L^\infty_{1/w} * G} \quad \text{and} \quad \mathscr{C}(\Psi)^* = C(\Psi)|_{L^\infty_{1/w} * G}.$$

For example, for $F \in L^1_w * G$, $H = H * G \in L^\infty_{1/w} * G$ one obtains

$$\begin{aligned}\langle \mathscr{T}(\Psi, X, g)\, F, H\rangle &= \Big\langle \sum_{i \in I} \langle L_{x_i} G, F\rangle\, \psi_i * G, H \Big\rangle \\ &= \sum_{i \in I} \langle F, L_{x_i} G\rangle \langle \psi_i, H * G\rangle \\ &= \Big\langle F, \sum_{i \in I} \langle \psi_i, H\rangle\, L_{x_i} G \Big\rangle = \langle F, T(\Psi, X, g) H\rangle.\end{aligned}$$

Since $\mathscr{T}$ considered as an operator on $L^1_w * G$ and T on $L^\infty_{1/w} * G$ satisfy

$$|||\mathrm{Id} - \mathscr{T}||| = |||(\mathrm{Id} - \mathscr{T})^*||| = |||\mathrm{Id} - T||| < 1$$

by Proposition 5.4, one can build the Neumann series $\sum_{n=0}^{\infty} (\mathrm{Id} - \mathscr{T})^n$, which is norm convergent to $\mathscr{T}(\Psi, X, g)^{-1}$ and $(\mathscr{T}(\Psi, X, g)^{-1})^* = T(\Psi, X, g)^{-1}$. Consequently,

$$\begin{aligned}\mathscr{A}(\Psi, X, g)^* &= (\mathscr{V}_g \circ \mathscr{T}(\Psi, X, g)^{-1} \circ \mathscr{C}(\Psi))^* = \mathscr{C}(\Psi)^* \circ (\mathscr{T}(\Psi, X, g)^{-1})^* \circ \mathscr{V}_g^* \\ &= C(\Psi) \circ T(\Psi, X, g)^{-1} \circ V_g = A(\Psi, X, g). \quad \blacksquare\end{aligned}$$

Now we introduce for our "parameters" Ψ, X, g the following "metrics" (distance functions) which will allow us to express the continuous dependence of the coefficients from these parameters:

(a) Fixing any $h \in \mathscr{B}_w$ we set for $g, g' \in \mathscr{B}_w$

$$d_0(g, g') := \| V_h(g - g') \,|\, L^1_w\| + \| V_g(g) - V_{g'}(g') \,|\, W^R(C^0, L^1_w)\|. \tag{6.10}$$

(b) Two well-spread sets $X = (x_i)_{i \in I}$ and $X' = (x'_i)_{i \in I}$ with the same index set are called V-close (for some neighbourhood V of e in $\mathscr{G}$) if

$$x_i^{-1} x'_i \in V \text{ for all } i \in I. \tag{6.11}$$

Of course, one could work with a metric d_1 in case of a metric group $\mathscr{G}$.

(c) For two families $\Psi = (\psi_i)_{i \in I}$ and $\Psi' = (\psi'_i)_{i \in I}$ of continuous functions satisfying $\operatorname{supp} \psi_i \cup \operatorname{supp} \psi'_i \subseteq x_i Q$ for some compact set Q we set

$$d_2(\Psi, \Psi') = \sup_{i \in I} \|\psi_i - \psi'_i\|_\infty. \tag{6.12}$$

Using this terminology we may formulate

THEOREM 6.5. *Assume that for $g_0 \in \mathscr{B}_w$, Ψ_0 and X_0 fulfill the conditions allowing one to obtain the atomic decomposition as described above. Then the* *mapping* $(\Psi, X, g) \mapsto A(\Psi, X, g)$ *is continuous at* (Ψ_0, X_0, g_0), *i.e., for* $\varepsilon > 0$ *there exists* $\delta > 0$ *and some* V *such that* $d_0(g_0, g) < \delta$, $d_2(\Psi, \Psi_0) < \delta$ *and* V*-closeness of* X *to* X_0 *implies*

$$|||A(\Psi, X, g) - A(\Psi_0, X_0, g_0)|||_{\mathscr{C}_0 Y \to Y_d} < \varepsilon. \tag{6.13}$$

Proof. Since composition of operators is norm continuous, by (6.7) it suffices in view of (6.7) to verify the norm continuous dependence of the operators C, T^{-1}, and V on their parameters separately.

Step 1. That the mapping $\Psi \to C(\Psi)$ is continuous with respect to the operator norm follows from the following quantitative version of Proposition 4.1:

LEMMA 6.6. *Let a relatively separated family* $X = (x_i)_{i \in I}$ *and a compact set* $Q = Q^{-1}$ *be given. Set* $I_y := \{i: x_i \in yQ\}$ *and* $h := \sup_{y \in \mathscr{G}} \# I_y$ *as usual. Assume that* $H_i \in \mathscr{K}(\mathscr{G})$ *satisfy* $\operatorname{supp} H_i \subseteq x_i Q$ *for all* $i \in I$ *and* $\sup_{i \in I} \|H_i\|_\infty < \infty$. *Then*

$$\|(\langle H_i, F\rangle)_{i \in I} \mid Y_d(X)\| \leqslant c_0 h \sup_{i \in I} \|H_i\|_\infty \|F \mid Y\|. \tag{6.14}$$

Proof. Without loss of generality we may assume F and H_i to be non-negative. Set $K(y) = \sum_{i \in I_y} \langle H_i, F\rangle$ and choose $k \in \mathscr{K}(\mathscr{G})$ with $k(z) \equiv 1$ on Q^2. Then $\|(\langle H_i, F\rangle)_{i \in I} \mid Y_d\| = \|K \mid Y\|$, and

$$K(y) = \left\langle \sum_{i \in I_y} H_i, F \right\rangle = \left\langle \left(\sum_{i \in I_y} H_i \right) L_y k, F \right\rangle$$

$$\leqslant \left\| \sum_{i \in I_y} H_i \right\|_\infty \langle L_y k, F\rangle \leqslant h \sup_i \|H_i\|_\infty \, F * k^\vee(y).$$

Taking the Y-norm on both sides one obtains

$$\|(\langle H_i, F\rangle)_{i \in I} \mid Y_d\| \leqslant \|k \mid L^1_w\| \, h \sup_{i \in I} \|H_i\|_\infty \, \|F \mid Y\|.$$

Step 2. Since $T_0 = T(\Psi_0, X_0, g_0)$ is invertible and operator inversion is a norm continuous mapping in a neighbourhood of T_0, the continuity of $(\Psi, X, g) \to T(\Psi, X, g)^{-1}$ follows from the continuity of $(\Psi, X, g) \to T(\Psi, X, g)$. In order to prove this let us separate variables once more by writing $T(\Psi, X, g) = S(X, G)\, C(\Psi)$, with $S(X, G)\colon Y_d \to Y$, $S(X, G)(\Lambda) = \sum_{i \in I} \lambda_i L_{x_i} G$. Then

$$S(X_0, G_0) - S(X, G) = S(X_0, G_0 - G) + (S(X_0, G) - S(X, G)).$$

We estimate the two terms separately:

(a) The continuity with respect to G follows from Proposition 5.2, showing that

$$\|S(X_0, G_0 - G)(\Lambda) \mid Y\| \leqslant C_s \|G_0 - G \mid W^R(C^0, L^1_w)\| \, \|\Lambda \mid Y\|.$$

(b) Continuity with respect to X, i.e., an estimate for the second term, is shown as follows. Assuming that X and X_0 are V-close we can write

$$S(X_0, G)(\Lambda) - S(X, G)(\Lambda) = \sum_{i \in I} \lambda_i (L_{x_i'} G - L_{x_i} G)$$
$$= \sum_{i \in I} \lambda_i L_{x_i}(L_{u_i} G - G) \qquad \text{with } u_i \in V \text{ for all } i \in I. \tag{6.15}$$

Because of the decomposition (4.11) of G the auxiliary function

$$G^*(x) := \sup_{v \in V} |G(v^{-1}x) - G(x)|$$

satisfies

$$G^*(x) \leqslant \sum_{n=1}^{\infty} R_{z_n} (\sup_{v \in V} |G_n(v^{-1}x) - G_n(x)|) := \sum_{n=1}^{\infty} R_{z_n} H_n(x).$$

Since supp $H_n \subseteq VQ$ and $\|H_n\|_\infty = \omega_V(G_n)$ imply (cf. (5.5))

$$\|G^* \mid W^R(C^0, L^1_w)\| \cong \sum_{n=1}^{\infty} \|H_n\|_\infty \, w(z_n)$$
$$= \sum_{n=1}^{\infty} \omega_V(G_n) \, w(z_n) = \Omega_V(G),$$

and thus $G^* \in W^R(C^0, L^1_w)$ with arbitrarily small norm if X and X_0 are sufficiently close to each other. Now (6.15) can be estimated (writing $|\Lambda|$ for $(|\lambda_i|)_{i \in I}$)

$$|S(X_0, G)(\Lambda) - S(X, G)(\Lambda)| \leqslant S(X, G^*)(|\Lambda|). \tag{6.16}$$

Invoking Proposition 5.2 once more one obtains

$$\|S(X_0, G)(\Lambda) - S(X, G)(\Lambda) \mid Y_d\| \leqslant \|S(X, G^*)(|\Lambda|) \mid Y\|$$
$$\leqslant C \, \|G^* \mid W^R(C^0, L^1_w)\| \, \|\Lambda \mid Y_d\|$$

and thus the estimate for the operator norm

$$|||S(X_0, G) - S(X, G)|||_{Y_d(X) \to Y} \leqslant C_s \Omega_V(G).$$

Step 3. $g \to V_g$ is continuous.

In order to estimate the operator norm $|||V_g - V_{g0}|||_{\mathscr{C}o\, Y \to Y}$ we measure the norms in $\mathscr{C}o\, Y$ and $\mathscr{H}^1_w$ with respect to a fixed vector $h \in \mathscr{B}_w$. Since on $L^1_w \cap L^\infty_{1/w}$, $V_g^* = \mathscr{V}_g$ by the definitions (4.7) and (6.9) and thus $\mathscr{V}_g \circ V_g = \mathrm{Id}$ by (4.9), the formula

$$\begin{aligned} V_{\mathscr{V}_h(G)}(f)(x) &= \langle \pi(x)\, \mathscr{V}_h(G), f \rangle = \langle \mathscr{V}_h(L_x G), f \rangle \\ &= \langle L_x G, V_h(f) \rangle = V_h(f) * G^\nabla(x) \end{aligned} \tag{6.17}$$

is valid for $G \in L^1_w \cap L^\infty_{1/w}$, in particular

$$V_g(f) = V_{\mathscr{V}_h(V_h(g))}(f) = V_h(f) * V_h(g)^\nabla \tag{6.18}$$

is an extension of the orthogonality relation (2.2) to $f \in \mathscr{C}o\, Y$. Therefore

$$\begin{aligned} \|V_g(f) - V_{g0}(f) \mid Y\| &= \|V_h(f) * V_h(g - g_0)^\nabla \mid Y\| \qquad \text{(by (3.1))} \\ &\leqslant \|V_h(f) \mid Y\|\ \|V_h(g - g_0) \mid L^1_w\| = \|f \mid \mathscr{C}o\, Y\|\ \|g - g_0 \mid \mathscr{H}^1_w\|. \quad \blacksquare \end{aligned}$$

Remark. Under mild additional conditions on g it can even be shown that it is sufficient to use the L^1-norm in (6.12) instead of the L^∞-norm in order to verify the continuous dependence of the coefficients from the system $\Psi = (\psi_i)_{i \in I}$. Thus for practical purposes it is no problem to replace the BUPUs Ψ by a family of characteristic functions corresponding to a sufficiently fine partition of the group $\mathscr{G}$.

References

[C] A. L. Carey, Square-integrable representations of non-unimodular groups, *Bull. Austral. Math. Soc.* **15** (1976), 1–12.

[CW] R. R. Coifman and G. L. Weiss, Extensions of Hardy spaces and their use in analysis, *Bull. Amer. Math. Soc.* **83** (1977), 569–645.

[DM] M. Duflo and C. C. Moore, On the regular representation of a non-unimodular locally compact group, *J. Funct. Anal.* **21** (1976), 209–243.

[F1] H. G. Feichtinger, Banach convolution algebras of Wiener's type, *in* "Proceedings, Conference on Functions, Series, Operators, Budapest, 1980," pp. 509–524, Colloq. Math. Soc. János Bolyai, North-Holland, Amsterdam, 1983.

[F2] H. G. Feichtinger, "Modulation Spaces on Locally Compact Abelian Groups," Technical Report, Vienna, 1983.

[F3] H. G. Feichtinger, Minimal Banach spaces and atomic decompositions, *Publ. Math. Debrecen* **33** (1986), 167–168 and **34** (1987), 231–240.

[F4] H. G. Feichtinger, Atomic characterizations of modulation spaces, *in* "Proceedings, Conference on Constructive Function Theory, Edmonton, July 1986." *Rocky Mountain J.*, to appear.

[FGr] H. G. Feichtinger and P. Gröbner, Banach spaces of distributions defined by decomposition methods, I, *Math. Nachr.* **123** (1985), 97–120.

[FG] H. G. Feichtinger and K. Gröchenig, A unified approach to atomic decompositions via integrable group representations, *in* "Proceedings, Conference on Functions, Spaces and Applications, Lund, 1986," Lecture Notes in Mathematics, Springer-Verlag, **1302** (1988), 52–73, New York/Berlin, 1987.

[FJ1] M. Frazier and B. Jawerth, Decomposition of Besov spaces, *Indiana Univ. Math. J.* **34** (1985), 777–799.

[FJ2] M. Frazier and B. Jawerth, The φ-transform and decompositions of distribution spaces, *in* Proceedings, Conference on Functions, Spaces and Applications, Lund, 1986," Lecture Notes in Mathematics, Springer-Verlag, **1302** (1988), Heidelberg, 1987.

[GMP1] A. Grossmann, J. Morlet, and T. Paul, Transforms associated to square integrable group representations, I, *J. Math. Phys.* **26**, No. 10 (1985), 2473–2479.

[GMP2] A. Grossmann, J. Morlet, and T. Paul, Transforms associated to square integrable group representations, II, Examples, *Ann. Inst. H. Poincaré* **45** (1986), 293–309.

[KS] J. Klauder and B. Skagerstam (Eds.), Coherent States—Applications in Physics and Mathematical Physics," World Sci. Publ., Singapore, 1985.

[LT] J. Lindenstrauss and L. Tzafriri, "Classical Banach Spaces," Springer-Verlag, Berlin/Heidelberg/New York, 1977.

[P1] J. Peetre, Paracommutators and minimal spaces, *in* "Operators and Function Theory" (S. C. Powers, Ed.), NATO ASI Series, Reidel, Dordrecht, 1985.

[P2] J. Peetre, Some calculations related to Fock space and the Shale–Weil representation, preprint, 1987.

[RT] F. Ricci and M. Taibleson, Boundary values of harmonic functions in mixed norm spaces and their atomic structure, *Ann. Scuola Norm. Sup. Pisa* (4) **10** (1983), 1–54.

[R] R. Rochberg, Decomposition theorems for Bergman spaces and their applications, *in* "Operators and Function Theory" (S. C. Powers, Eds.), NATO ASI Series, Reidel, Dordrecht, 1985.

[Z] A. Zaanen, "Integration," 2nd ed., North-Holland, Amsterdam, 1967.

IEEE TRANSACTIONS ON INFORMATION THEORY, VOL. 36, NO. 5, SEPTEMBER 1990 961

The Wavelet Transform, Time-Frequency Localization and Signal Analysis

INGRID DAUBECHIES, MEMBER, IEEE

Abstract—Two different procedures are studied by which a frequency analysis of a time-dependent signal can be effected, locally in time. The first procedure is the short-time or windowed Fourier transform, the second is the "wavelet transform," in which high frequency components are studied with sharper time resolution than low frequency components. The similarities and the differences between these two methods are discussed. For both schemes a detailed study is made of the reconstruction method and its stability, as a function of the chosen time-frequency density. Finally the notion of "time-frequency localization" is made precise, within this framework, by two localization theorems.

I. Introduction

A. The Windowed Fourier Transform and Coherent States

IN SIGNAL ANALYSIS one often encounters the so-called short-time Fourier transform, or windowed Fourier transform. This consists of multiplying the signal $f(t)$ with a usually compactly supported window function g, centered around 0, and of computing the Fourier coefficients of the product gf. These coefficients give an indication of the frequency content of the signal f in a neighborhood of $t=0$. This procedure is then repeated with translated versions of the window function (i.e., $g(t)$ is replaced by $g(t \pm t_0)$, $g(t \pm 2t_0), \cdots$, where t_0 is a suitably chosen time translation step). This results in a collection of Fourier coefficients

$$c_{mn}(f) = \int_{-\infty}^{\infty} dt\, e^{im\omega_0 t} g(t - nt_0) f(t) \qquad (m, n \in Z). \quad (1.1)$$

Similar coefficients also occur in a transform first proposed by Gabor [1] for data transmission. The original proposal used a Gaussian function g, and parameters ω_0, t_0 such that $\omega_0 \cdot t_0 = 2\pi$. A Gaussian window function is, of course, not compactly supported, but it has many other qualities. One of these is that it is the function which is optimally concentrated in both time and frequency, and therefore well-suited for an analysis in which both time and frequency localization are important. Gabor's original proposal, with $\omega_0 \cdot t_0 = 2\pi$, leads to unstable reconstruction (see [2], [3]; we shall come back to this in Section II-C-1). The Gabor functions have been used in many different settings in signal analysis, either in discrete lattices (with $\omega_0 \cdot t_0 < 2\pi$ for stable reconstruction) or in the continuous form described next. In many of these applications their usefulness stems from their time-frequency localization properties (see e.g., [4]).

Whatever the choice for g (Gaussian, supported on an interval, etc.), it is interesting to know to which extent the coefficients $c_{mn}(f)$ of (1.1) define the function f. This is one of the main issues of this paper.

The coefficients $c_{mn}(f)$ in (1.1) can also be viewed as inner products of the signal f to be analyzed with a discrete lattice of coherent states. Let us clarify this statement. By "coherent states" we understand here the family of square integrable functions $g^{(p,q)}$, generated from a single $L^2(\mathbb{R})$-function g, by phase space translations (p, q). A "discrete lattice" of coherent states is a discrete subset of the whole family obtained by restricting the labels (p, q) to a regular rectangular lattice in phase space. "Phase space," a term we borrow here from physics, stands for the two-dimensional time-frequency space, considered as one geometric whole. More precisely, $g^{(p,q)}(x)$ is obtained from $g(x)$ by a translation of x by q, and by a similar translation of the Fourier transform $\hat{g}$ by p, i.e.,

$$g^{(p,q)}(x) = e^{ipx} g(x - q). \quad (1.2)$$

Families of coherent states, as defined by (1.2), are used in many different areas of theoretical physics. Their name stems from their use in quantum optics (where the Gaussian choice for g is favored, $g(x) = \pi^{-1/4} \exp(-x^2/2)$) [5], [6], but they have since spilled over to many other fields. See e.g., [7] for a review, including many generalizations of the original concept. In quantum mechanics, they are particularly useful in semiclassical arguments because they make it possible to study quantum phenomena in a phase-space setting. If the original function g is centered around (0,0) in phase space, i.e., if the mean value of position, $\int dx\, x |g(x)|^2$, and of momentum, $\int dk\, k |\hat{g}(k)|^2$, are both zero, then the state $g^{(p,q)}$ will be centered around (p, q), i.e.,

$$\int dx\, x |g^{(p,q)}(x)|^2 = q$$

Manuscript received November 9, 1987; revised December 7, 1988. This work was presented in part at the International Conference on Mathematical Physics, Marseille, France, July 1986.

The author was with the Mathematics Department, University of Michigan, Ann Arbor, MI 48109-1003. She is now with AT&T Bell Laboratories, 600 Mountain Ave., Murray Hill, NJ 07974.

IEEE Log Number 9036002.

0018-9448/90/0900-0961$01.00 ©1990 IEEE

$$\int dk\, k|(g^{(p,q)})^{\wedge}(k)|^2 = p.$$

Here $\hat{g}$ denotes the Fourier transform of g, $\hat{g}(k) = (2\pi)^{-1/2}\int dx e^{ikx} g(x)$. The inner products $(g^{(p,q)}, \phi)$ will therefore "measure" the phase space content of the function ϕ around the phase space point (p,q). See e.g., [8, Section V-A.] and [9] for two different methods of using coherent states in a study of semiclassical approximations to quantum mechanics.

A very important property of the coherent states is the so-called "resolution of the identity." It has been discovered and rediscovered many times (see the earlier papers in [7]). A discussion of its relevance for signal analysis can be found in [10]. For the sake of completeness, we give its (easy) proof in Appendix A. The "resolution of the identity" says that the map Φ from $L^2(\mathbb{R})$ into $L^2(\mathbb{R}^2)$, defined by $\Phi f(p,q) = \langle g^{(p,q)}, f\rangle$, is an isometry (up to a constant factor), i.e.,

$$\int dp \int dq |\langle g^{(p,q)}, f\rangle|^2 = 2\pi \|g\|^2 \|f\|^2. \tag{1.3a}$$

Here, as in the remainder of this paper, $\langle f, g\rangle$ stands for the L^2-inner product of the functions f and g,

$$\langle f, g\rangle = \int dx \overline{f(x)} g(x)$$

(note that this inner product is antilinear in the first argument and linear in the second argument, following the physicists' convention), while $\|f\|$ stands for the L^2-norm of f,

$$\|f\|^2 = \langle f, f\rangle = \int dx |f(x)|^2.$$

Formula (1.3a) implies that

$$\frac{1}{2\pi}\int dp \int dq \langle g^{(p,q)}, f\rangle g^{(p,q)} = f. \tag{1.3b}$$

This means that a function f can be recovered completely, and easily, from the phase space "projections" $\langle g^{(p,q)}, f\rangle$. Note that, since (1.3) holds for any g, one can use this freedom to choose g optimally for the application at hand. This freedom of choice was exploited in e.g., [8, Section V-A]; it will also be important to us here.

If, instead of letting (p,q) roam over all of phase space, in a continuous fashion, we rather restrict ourselves to a discrete sublattice of phase space, then we revert to (1.1). That is, if we choose $p_0, q_0 > 0$, and we define

$$g_{mn}(x) = g^{(mp_0, nq_0)}(x)$$
$$= e^{imp_0 x} g(x - nq_0)$$

then clearly (with x interpreted as "time," $p_0 = \nu_0$, $q_0 = t_0$)

$$c_{mn}(f) = \langle g_{mn}, f\rangle = \langle g^{(mp_0, nq_0)}, f\rangle.$$

This shows that the short-time Fourier transform can indeed be viewed as the computation of inner products with a discrete lattice of coherent states.

B. Phase Space in Signal Analysis

The appearance of a phase space concept, such as the coherent states, in signal analysis, is not altogether surprising. As pointed out by N. G. de Bruijn [11], it is an entirely natural concept in music. Let us consider a time-dependent signal, which is a piece of music played (for the sake of simplicity), by a single instrument. If we disregard problems with high harmonics, with the "attack" of the notes, etc. (this is a simple example!), the musical score corresponding to the piece can be considered as a satisfactory representation of the time-dependent acoustic signal. A musical score indicates which notes have to be played at consecutive time steps. Thus, it gives a frequency analysis, locally in time, and is much closer to a short-time Fourier transform or a coherent states analysis, than to e.g., the Fourier transform, in which all track of time-dependence is lost, or at least not explicitly recognizable. The notation of a musical score, indicating time horizontally, and frequency vertically, is really a phase space notation. Phase space is thus seen to be a natural concept in signal analysis. This is also illustrated by the successful use, in signal analysis, of that other phase space concept, the Wigner distribution, as in [12] or [13].

C. Frames

As long as the continuously labelled coherent states $g^{(p,q)}$ are used, we know, from (1.3), that knowing a signal f is equivalent to knowing the inner products $\langle g^{(p,q)}, f\rangle$. This need not be automatically true if one restricts the labels (p,q) to the discrete sublattice (mp_0, nq_0), $m, n \in \mathbb{Z}$; some conditions on g, p_0, q_0 will be required. Let us define the map T

$$T: L^2(\mathbb{R}) \to l^2(\mathbb{Z}^2)$$
$$(Tf)_{m,n} = \langle g_{mn}, f\rangle = c_{mn}(f). \tag{1.4}$$

This map is the discrete analog of the "continuous" map Φ in Section I-A. For all the cases of interest to us, the map T will be bounded. To have complete characterization of a function f by its coefficients $c_{mn}(f)$ we shall require that T be one-to-one. If the characterization of f by means of the $c_{mn}(f)$ is to be of any use for practical purposes, one needs more than this, however. It is important that the reconstruction of f from the coefficients $c_{mn}(f)$ (which is possible, in principle, if T is one-to-one) be numerically stable: if the sequences $c_{mn}(f), c_{mn}(g)$ are "close" for two given functions f and g, then we want this to mean that f and g are "close" as well. More concretely, we require that

$$A\|f\|^2 \le \sum_{m,n} |\langle g_{mn}, f\rangle|^2 \le B\|f\|^2 \tag{1.5}$$

with $A > 0$, $B < \infty$, and A, B independent of f. This can be rewritten, in operator language, as

$$A\mathbb{1} \le T^*T \le B\mathbb{1}. \tag{1.6}$$

Here the inequality $T_1 \le T_2$, where T_1, T_2 are symmetric operators on $L^2(\mathbb{R})$, stands for $\langle f, T_1 f\rangle \le \langle f, T_2 f\rangle$ for all

$f \in L^2(\mathbb{R})$. A set of vectors $\{\phi_j; j \in J\}$ in a Hilbert space $\mathscr{H}$ for which the sums $\sum_{j \in J} |\langle \phi_j, f\rangle|^2$ yield upper and lower bounds for the norms $\|f\|^2$, as in (1.5), is also called a *frame*. The concept "frame" was introduced by Duffin and Schaeffer [14] in the context of nonharmonic Fourier analysis; see also [15]. We shall thus require that the $\{g_{mn}; m,n \in \mathbb{Z}\}$ constitute a frame; we shall give the name *frame bounds* to the constants A, B.

In a previous paper [16] particular functions g were constructed, for given $p_0, q_0 > 0$ (with $p_0 \cdot q_0 < 2\pi$) such that the corresponding g_{mn} constitute a frame. For this special construction, the frame bounds A, B are known explicitly in terms of g, and explicit inversion formulas for T can be given. In a particular case of this construction one even finds that the inequalities in (1.5), (1.6) becomes equalities, i.e., $A = B$. Whenever $A = B$ the frame is called a *tight frame*. The inversion $c_{mn}(f) \to f$ then becomes trivial; since $T^*T = A\mathbb{1}$, one has

$$f = A^{-1}T^*Tf = A^{-1}\sum_{m,n} g_{mn}\langle g_{mn}, f\rangle$$

where the sum converges strongly, i.e.,

$$\left\| f - A^{-1} \sum_{\substack{m,n \\ |m|,|n| \le K}} g_{mn}\langle g_{mn}, f\rangle \right\| \xrightarrow[K \to \infty]{} 0.$$

Remark: Note that frames, even tight frames, are not bases in general, in spite of the resemblance between the reconstruction formula for a tight frame and the standard expansion with respect to an orthonormal basis. In general, a frame contains "too many" vectors. An example in the finite-dimensional space $\mathbb{C}^2$ is given by $u_1 = e_1$, $u_2 = -1/2e_1 + \sqrt{3}/2e_2$, $u_3 = -1/2e_1 - \sqrt{3}/2e_2$, where $e_1 = (1,0)$ and $e_2 = (0,1)$ constitute the standard basis for $\mathbb{C}^2$. One easily checks that, for all $v \in \mathbb{C}^2$,

$$\sum_{l=1}^{3} |\langle u_l, v\rangle|^2 = \frac{3}{2}\|v\|^2$$

so that the $\{u_l; l = 1,2,3\}$ constitute a tight frame, with inversion formula

$$v = \frac{2}{3}\sum_{l=1}^{3} u_l\langle u_l, v\rangle.$$

The $\{u_l;\ l = 1,2,3\}$ do not constitute a basis because they are not linearly independent. In the infinite-dimensional frames we shall consider in this paper any finite number of vectors will be linearly independent in general, but there will still be "too many" vectors in the sense that any of them lies in the closed linear span of all the others. If the vectors constituting a tight frame are normalized, then the frame constant $A = B$ indicates the "rate of redundancy" of the frame; if $A = B = 1$ then the frame is automatically an orthonormal basis.

While the constructions in [16] lead to satisfactory and easy inversion formulas for T, they have the drawback that the function g cannot be chosen freely, but has to be of the very particular type constructed in [16]. For some applications however, the window function g might be imposed *a priori*, and not be of this particular type. In that case it is of interest to find ranges for the parameters p_0, q_0 such that the g_{mn}, associated with the triplet g, p_0, q_0, constitute a frame. As we shall see next, *snug frames*, which are close to tight frames, i.e., for which the ratio B/A is close to 1, are particularly interesting, because they lead to easy inversion formulas with rapid convergence properties. It is therefore important to have good estimates for the frame bounds A and B. One of the questions we shall address in this paper is therefore the following: given g,

1) find a range R such that for $(p_0, q_0) \in R$ the associated g_{mn} are a frame, and
2) for $(p_0, q_0) \in R$, compute estimates for the frame bounds A, B.

Once good estimates for the frame bounds A, B are obtained, one can construct a dual function $\tilde{g}$ (depending on p_0, q_0 as well as on g; see Section II-A next) which leads to the easy reconstruction formula

$$f(x) = \sum_{m,n} \langle g_{mn}, f\rangle e^{imp_0x}\tilde{g}(x - nq_0).$$

The function $\tilde{g}$ is essentially a multiple of g, with correction terms of the order of $B/A - 1$; the closer B/A is to 1, the faster the series for $\tilde{g}$ converges. Obtaining good frame bounds is important even if different reconstruction procedures are considered, or if only characterization of f by means of the $c_{mn}(f) = \langle g_{mn}, f\rangle$ and not full reconstruction (after e.g., transmission) is the main goal, since *any* stable reconstruction or characterization procedure can exist only if the g_{mn} constitute a frame.

All of these concern expansions with respect to coherent states constructed according to (1.2). We shall also discuss a different type of expansion, the so-called "wavelet expansion" [17], [18], which corresponds to coherent states of a different type.

D. Wavelets—A Different Kind of Coherent States

The coherent states $g^{(p,q)}$ all have the same envelope function g, which is translated by the amount q, and "filled in" with oscillations with frequency p (see Fig. 1a). They typically have all the same width in time and in frequency. "Wavelets" are similar to the $g^{(p,q)}$ in that they also constitute a family of functions, derived from one single function, and indexed by two labels, one for position and one for frequency. More explicitly,

$$h^{(a,b)}(x) = |a|^{-1/2}h\left(\frac{x-b}{a}\right)$$

where h is a square integrable function such that

$$C_h = \int dy\, |y|^{-1}|\hat{h}(y)|^2 < \infty \tag{1.7}$$

and where $a, b \in \mathbb{R}$, $a \neq 0$. Note that if h has some decay at infinity, then (1.7) is equivalent to the requirement $\int dx\, h(x) = 0$. The parameter b in the $h^{(a,b)}$ gives the position of the wavelet, while the dilation parameter a

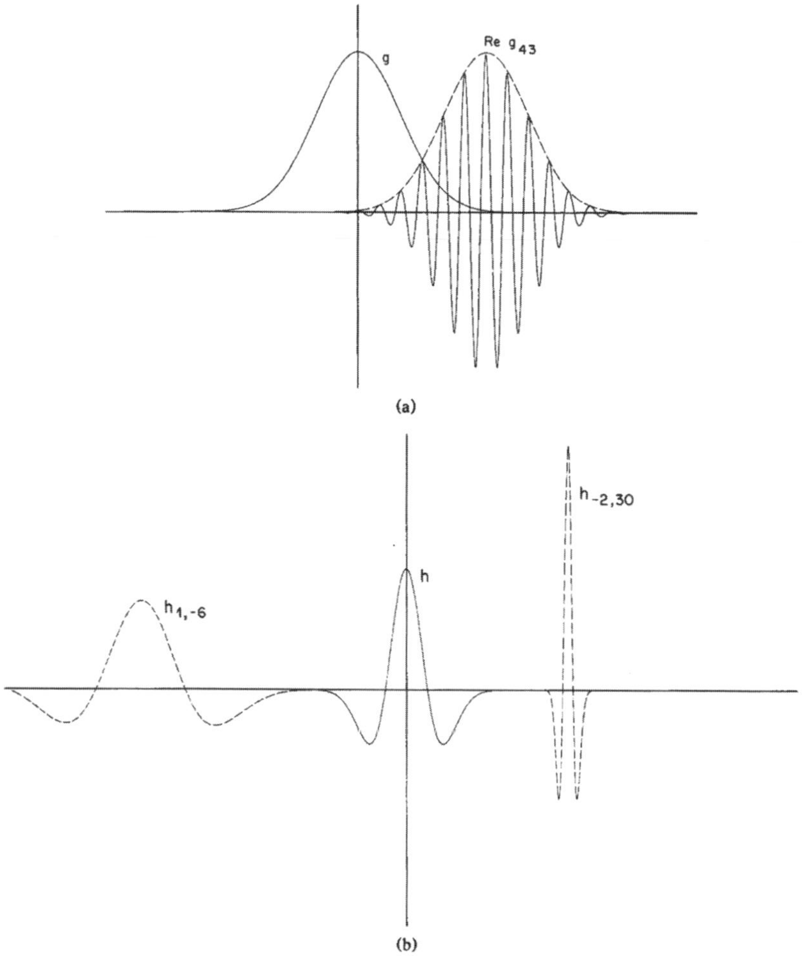

Fig. 1. (a) Typical choice for the window function $g = g_{00}$, and typical g_{mn}. In this case $g(x) = \pi^{-1/4}\exp(-x^2/2)$, $p_0 = \pi$, $q_0 = 1$; figure shows $\operatorname{Re} g_{43}(x) = \pi^{-1/4}\cos(4\pi x)\exp[-(x-3)^2/2]$. (b) Typical choice for basic wavelet $h = h_{00}$, and a few typical h_{mn}. In this case $h(x) = 2/\sqrt{3}\,\pi^{-1/4}(1-x^2)\exp(-x^2/2)$, $a_0 = 2$, $b_0 = 1$.

governs its frequency. For $|a| \ll 1$, the wavelet $h^{(a,b)}$ is a very highly concentrated "shrunken" version of h, with frequency content mostly in the high frequency range. Conversely, for $|a| \gg 1$, the wavelet $h^{(a,b)}$ is very much spread out and has mostly low frequencies (see Fig. 1(b)). As a result of this construction, wavelets will be a better tool than the "canonical" coherent states $g^{(p,q)}$ in situations where better time-resolution at high frequencies than at low frequencies is desirable.

There exists a resolution of the identity for wavelets as well as for the canonical coherent sates. One finds, for all $f_1, f_2 \in L^2(\mathbb{R})$,

$$\int \frac{da}{a^2} \int ab \langle f_1, h^{(a,b)} \rangle \langle h^{(a,b)}, f_2 \rangle = C_h \langle f_1, f_2 \rangle.$$

See Appendix A. Again, this implies that a function f can be recovered easily from the inner products $\langle h^{(a,b)}, f \rangle$, since

$$f = C_h^{-1} \int \frac{da}{a^2} \int db\, \langle h^{(a,b)}, f \rangle h^{(a,b)}. \tag{1.8}$$

The similarities between the $g^{(p,q)}$ associated with the short-time Fourier transform and the wavelets $h^{(a,b)}$ are no accident: both families are special cases of "coherent states associated with a Lie-group," in the first case the Weyl–Heisenberg group, in the second case the "$ax+b$"-group or affine group. Wavelets are therefore also called "affine coherent states." They were first introduced, under this name, in [19]. They are of great interest for their own sake, and have led to interesting applications in quantum mechanics (see, e.g., [20], [21]). They are also related to the use of dilation and translation techniques in harmonic analysis, which have turned out to be very powerful tools. It is likely that these techniques, and affine coherent states, will find interesting applications in quantum mechanics problems (see e.g., [22]). We shall not go into these aspects here. Note that the "resolutions of the identity" (1.3) and (1.8) can be used as starting points for the construction of time-frequency localization or filter operators [23], [24].

E. Discrete Lattices of Wavelets—The Wavelet Transform

In analogy to the lattice of coherent states associated with the short-time Fourier transform, which can be viewed as a discrete subset of the continuously labeled $g^{(p,q)}$, we shall also consider discrete lattices of wavelets. We choose to discretize the dilation parameter a by taking powers of a fixed dilation step $a_0 > 1$, $a = a_0^m$ with $m \in \mathbb{Z}$. For different values of m the wavelets will be more or less concentrated, and we adapt the discretized translation steps to the width of the wavelet by choosing $b = nb_0 a_0^m$, with $n \in \mathbb{Z}$. This leads to

$$h_{mn}(x) = h^{(a_0^m, nb_0a_0^m)}(x) = a_0^{-m/2} h(a_0^{-m}x - nb_0). \quad (1.9)$$

As we shall see, it is also important for the discrete case to have $\int dx\, h(x) = 0$.

The "discrete wavelet transform" is the analog of the map defined by (1.4) for the Weyl–Heisenberg case.[1]

Again one can investigate, as in the Weyl–Heisenberg case, whether the h_{mn} and the associated wavelet transform lead to a "discrete approximation" of the resolution of the identity (1.8), i.e., whether a family of wavelets h_{mn} constitutes a frame. As was shown in [16], it is possible, given a_0, b_0, to construct explicit functions h such that the associated h_{mn} constitute a frame, or even, in particular cases, a tight frame. These functions h are, as in the Weyl–Heisenberg case, of a very particular type. Typically their Fourier transform $\hat{h}$ has a compact support; since $\hat{h}$ may be chosen in C^∞, the function h may have arbitrarily fast decay, or

$$|h(x)| \le C_k(1+|x|)^{-k} \quad (1.10)$$

with $C_k < \infty$ for all $k \in \mathbb{N}$. In practice, however, the constants C_k turn out to be fairly large for functions h of the type constructed in [16]. This means that while h satisfies (1.10), its graph is very spread out (see [16]); typically the distance between the maximum x_0 of h and the point x defined by $\sup\{|h(y)|; y \ge x\} = 10^{-2}|h(x_0)|$ will be an order of magnitude larger than b_0, the translation step parameter. For practical purposes, it is desirable to use functions h that are more concentrated than this, such as e.g., $h(x) = (1-x^2)e^{-x^2/2}$. We shall therefore address the same questions as the Weyl–Heisenberg case, i.e., for given h:

1) find a range R for the parameters such that for $(a_0, b_0) \in R$ the associated h_{mn} constitute a frame; and
2) for $(a_0, b_0) \in R$, compute estimates on the frame bounds A, B.

[1]To distinguish the g_{mn} from the wavelets h_{mn}, we shall call the g_{mn} "Weyl–Heisenberg" coherent states, after the group with which they are associated (see Section I-D).

The wavelet transform can be used, like the short-time Fourier transform, for signal analysis purposes. As in the short-time Fourier transform the two integer indices, m and n, control respectively, the frequency range and the time translation steps. There are however some significant differences between the two transforms. Some of these differences may well make the less widely used wavelet transform a better tool for the analysis of some types of signals (e.g., acoustic signals, such as music or speech) than the short-time Fourier transform. Let us digress a little on a qualitative discussion of these differences.

F. Qualitative Comparison of the Short-Time Fourier Transform and the Wavelet Transform

To illustrate this comparison we give graphs of typical g_{mn} and h_{mn} in Fig. 1, and of the associated phase space lattices in Fig. 2. More specifically, we represent each g_{mn} or h_{mn} by the point in phase space around which that function is mostly concentrated. In the Weyl–Heisenberg case, assuming that $\int dx\, |g(x)|^2 = 1$ and $\int dx\, x|g(x)|^2 = 0 = \int dk\, k|\hat{g}(k)|^2$, the lattice points are given by

$$(nq_0, mp_0) = \left(\int dx\, x|g_{mn}(x)|^2, \int dk\, k\left|(g_{mn})^{\wedge}(k)\right|^2 \right).$$

In the affine case we again associate to every h_{mn} the space localization point $\int dx\, x|h_{mn}(x)|^2 = nb_0 a_0^m$ (assuming that $\int dx\, |h(x)|^2 = 1$ and $\int dx\, x|h(x)|^2 = 0$). Since the function $|\hat{h}|$ and consequently all the $|(h_{mn})^{\wedge}|$ is even in many applications, the choice $\int dk\, k|(h_{mn})^{\wedge}(k)|^2$ is not appropriate for the frequency localization, since this integral is zero. This is due to the fact that the $(h_{mn})^{\wedge}$ have *two* peaks, one for positive and one for negative frequencies. We therefore represent the frequency content of h_{mn} by *two* points, namely

$$\int_0^\infty dk\, k\left|(h_{mn})^{\wedge}(k)\right|^2 \quad \text{and} \quad \int_{-\infty}^0 dk\, k\left|(h_{mn})^{\wedge}(k)\right|^2.$$

The two lattice points corresponding to the positive and negative frequency localizations of h_{mn} are thus

$$(nb_0 a_0^m, a_0^{-m}\omega_\pm) = \left(\int dx\, x|h_{mn}(x)|^2, \int_{0 \le \pm k < \infty} dk\, k\left|(h_{mn})^{\wedge}(k)\right|^2 \right)$$

where $\omega_\pm = \int_{0 \le \pm k < \infty} dk\, k|\hat{h}(k)|^2$. In Fig. 1(a) a few typical Weyl–Heisenberg coherent states were given, in Fig. 1(b) some typical wavelets. Fig. 1 shows one very basic difference between the two approaches; while the size of the support of the g_{mn} is fixed, the support of the h_{mn} is essentially proportional to a_0^m. As a result, high frequency h_{mn}, which correspond to $m \ll 0$, are very much concentrated. This means, of course, that the time-translation step (if x is interpreted as "time") has to be smaller for high-frequency h_{mn}, as shown by the phase space lattice in Fig. 2(b). It also means, however, that the wavelet transform will be able to "zoom in" on singularities, using more and more concentrated h_{mn} corresponding to higher and higher frequencies.

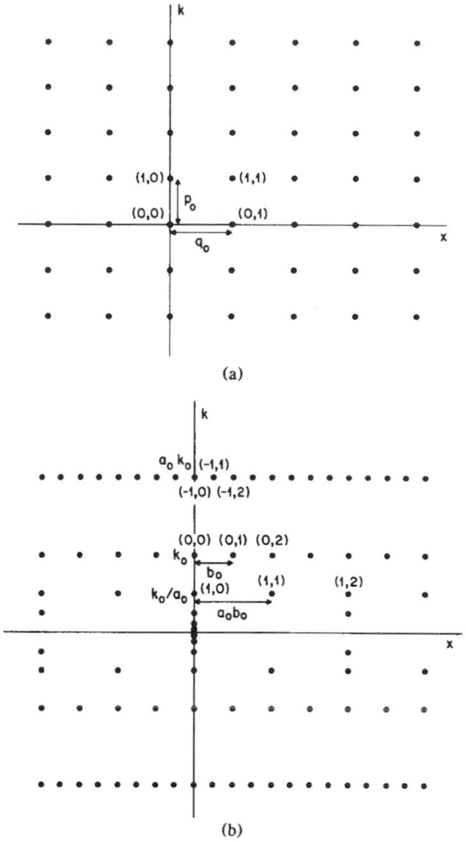

Fig. 2. (a) Phase space lattice corresponding to short-time Fourier transform (see text). (b) Phase space lattice corresponding to wavelet transform (see text). Constant k_0 is given by $k_0 = \int_0^\infty dk\,k^{-1}|\hat{h}(k)|^2$; we have assumed $\hat{h}$ to be even, and we have chosen $a_0 = 2$.

Let us illustrate this by the following simple example, taken from a grossly simplified problem in the synthesis of music. Typically one needs to be able to handle relatively low frequencies corresponding to the lowest notes and very high frequencies corresponding to high harmonics. Suppose one wants to be able to represent tones with frequency of the order of $2\pi/T$. Suppose also that one wants to be able to render faithfully the "attack" of notes. This "attack" consists of very high harmonics at the start of the note which die out very quickly, typically in a time $t_0 \ll T$. We have represented one component of such an "attack" very schematically in Fig. 3. Intuitively, the function $f(t)$ in Fig. 3 seems to correspond to a signal with "frequency" $2\pi 3/t_0$ during the time interval $[0, t_0]$, while its amplitude on $[t_0, T]$ is zero. Let us compare the performances of discrete Weyl–Heisenberg coherent states and of wavelets for this problem. In the first case, the support of g, and hence that of all the g_{mn}, needs to have a width of at least T. The high frequency $6\pi/t_0$ corresponds to a value for m of approximately $6\pi t_0^{-1} p_0^{-1}$. In practice however, since $T \gg t_0$, much higher values of m will be needed to reproduce, by means of the g_{mn} sketched in Fig. 1(a) (and which all have width T), a function f which is nonzero only in the interval $[0, t_0]$. This is not the case if wavelets are used. The high frequency wavelets have very small support, so that the above problem (having to bring in much higher frequencies than intuitively needed) does not occur. Moreover, even for the high frequencies corresponding to f, which correspond in the wavelet transform to very negative values of m and a very small time translation step (see Fig. 2(b)), only a few of the many time-steps necessary to cover $[0, T]$ would be needed, namely only those corresponding to $[0, t_0]$. This is what is meant by the "zooming in" property of the wavelets. For this kind of problem, wavelets thus provide a more efficient way (needing fewer coefficients) of representing the signal.

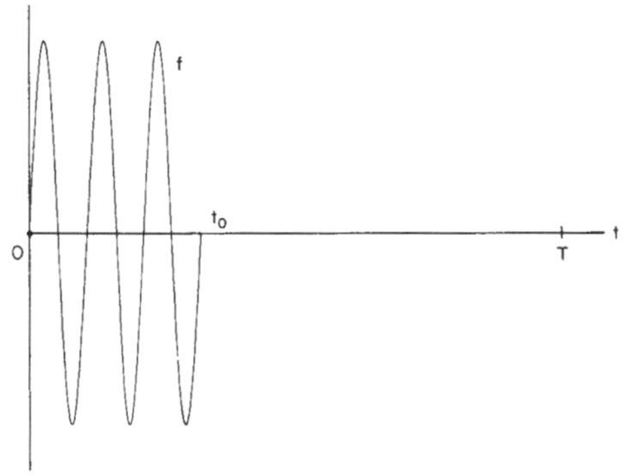

Fig. 3. One component of attack of note (see text). We take, as model,

$$f(t) = \begin{cases} \sin(6\pi t/t_0), & 0 \le t \le t_0 \\ 0 \quad t \le 0, & \text{or } t \ge t_0. \end{cases}$$

Lowest frequency of interest is $2\pi T^{-1}$; typically $t_0 \ll T$.

This example is so much simplified that it is rather unrealistic. The "zooming in" faculty of the wavelets, illustrated by this example, does however play an important role in more realistic applications; it makes the wavelets a useful tool in the areas of signal analysis such as seismic analysis [25] and music synthesis [26]. This same property also makes the wavelets a choice tool for the detection of singularities [27], [28], which is of great interest to the analysis of vision [29], [38], [40], and for the study of fractals [67].

As a final remark we note that the wavelet transform, unlike the short-time Fourier transform, treats frequency in a logarithmic way (as clearly shown by Fig. 2), which is similar to our acoustic perception. This corresponds to constant Q as opposed to constant bandwidth filters. This is another argument for the use of wavelets for the analysis and/or synthesis of acoustic signals.

G. Short Historical Review of the Wavelet Transform—Orthonormal Bases of Wavelets

As was already pointed out at the end of Section I-E, the use of functions $h^{(a,b)}$ generated from a single initial function h by means of dilations and translations is not new in either mathematics or physics, although their use is not so widespread as the Weyl–Heisenberg coherent states $g^{(p,q)}$ described in Section I-A. The use of the $h^{(a,b)}$ for signal analysis purposes, analogous to the use of the short-time Fourier transform, is much more recent, however. As a tool for signal analysis, the wavelet transform was first proposed by the geophysicist J. Morlet in view of applications for the analysis of seismic data [17], [25]. J. Morlet's original name for the wavelets was "wavelets of constant shape", to contrast them with the analyzing functions in the short-time Fourier transform, which do not have a constant shape (see Fig. 1). The numerical success of J. Morlet's method prompted A. Grossmann to make a more detailed study of the wavelet transform; this resulted in the recognition that the wavelets $h^{(a,b)}$ corresponded to a square integrable representation of the $ax+b$-group. The resolution of the identity (1.8) and associated interpolation techniques were then proposed (and implemented) for reconstruction schemes. This work was presented in a series of papers [17], [18], [30], in which the original longer name was shortened to "wavelet." These papers were concerned with the map associating to a function $f \in L^2(\mathbb{R})$ the function $\Phi f(a,b) = \langle h^{(a,b)}, f\rangle$, where a, b ran continuously over $\mathbb{R}^* \times \mathbb{R}$. (Here $\mathbb{R}^* = \mathbb{R}\backslash\{0\}$.) For closer comparison with the numerical situation, it is necessary to limit oneself to the discrete sublattices described in Section I-F. Again it was A. Grossmann who first realized the importance of the "frame" concept in this connection. It was around this time that I first became involved in the subject. Around the same time Y. Meyer, having learned about A. Grossmann's and J. Morlet's results, pointed out to them that there was a connection between their signal analysis methods and existing, powerful techniques in the mathematical study of singular integral operators. All this resulted in our first construction of a special type of frames [16]. It also was the start of a cross-fertilization between the signal analysis applications and the purely mathematical aspects of techniques based on dilations and translations. The wavelet frames constructed in [16] are based on functions h with compactly supported and C^∞ Fourier transform. A similar construction can be made for Weyl–Heisenberg coherent states [16]. In that case, due to the Balian–Low theorem (see Section II-C-1), there is a tradeoff between redundancy of the frame and smoothness of the frame functions: if one requires the g_{mn} to constitute a basis (i.e., no redundancy), then either $xg(x)$ or $k\hat{g}(k)$ is not square integrable. For the Weyl–Heisenberg case one is thus forced to consider frames rather than bases if the phase space localization properties of g are important. It seemed natural to assume that the same would be true for the affine case, and [16] was written under that implicit assumption (although it is never explicitly stated). Shortly after [16] was written, however, Y. Meyer constructed an orthonormal basis, for $L^2(\mathbb{R})$, of wavelets h_{mn}, based on a function h with compactly supported and C^∞ Fourier transform $\hat{h}$ [31]. This quite amazing basis turned out to be an unconditional basis for all L^p-spaces, $1 < p < \infty$, all the Sobolev spaces, etc, [31]. A basis $\{e_l\}_{l\in J}$ in a Banach space E is called "unconditional" if the convergence of $\Sigma_l \lambda_l e_l$ depends only on the $|\lambda_l|$, and is therefore independent of, e.g., the order in which the terms are summed. For a normalized basis $\{e_l\}_{l\in J}$ ($\|e_l\| = 1$ for all l) in a Hilbert space $\mathscr{H}$, this is equivalent to requiring that there exist $A > 0$, $B < \infty$ so that, for all $\nu \in \mathscr{H}$,

$$A\|\nu\|^2 \leq \sum_{l\in J} |\langle e_l, \nu\rangle|^2 \leq B\|\nu\|^2.$$

In a Hilbert space a normalized basis is thus an unconditional basis if and only if it is also a frame. Such bases are also called Riesz bases. Every orthonormal basis is automatically a Riesz basis; if $(\psi_n)_{n\in\mathbb{N}}$ is an orthonormal basis, then $e_n = (1+n^2)^{-1/2}(n\psi_1 + \psi_n)$ is an example of a basis of normalized vectors that is *not* a Riesz basis. The Meyer basis was generalized to more than one dimension by P. G. Lemarié and Y. Meyer [32]. Even though the function h in Y. Meyer's basis has fast decay in the mathematical sense (it decays faster than any inverse polynomial), the constants involved are very large, so that it is not very well localized in practice. This limits its usefulness for signal analysis. If one is willing to accept functions h with less regularity (C^k instead of C^∞) and to forego the compact support of $\hat{h}$, then bases with better localization can be used. G. Battle [33] and P. G. Lemarié [34] independently, using very different approaches, constructed orthonormal bases with exponential decay. Ph. Tchamitchian [35] constructed bases of compactly supported wavelets, which are however nonorthogonal.

A major breakthrough in the understanding of orthonormal wavelet bases came with the concept of multiresolution analysis, developed by S. Mallat and Y. Meyer [36], [37]. A multiresolution analysis consists in breaking up $L^2(\mathbb{R})$ into a ladder of spaces V_j,

$$\cdots \subset V_2 \subset V_1 \subset V_0 \subset V_{-1} \subset V_{-2} \subset \cdots$$

where $V_{-m} \to L^2(\mathbb{R})$ for $m \to \infty$. The orthonormal projections onto the V_m correspond to approximations with "resolution" 2^m; a typical (but too simple to be practical) example is

$$V_j = \{f \in L^2(\mathbb{R});$$
$$f \text{ is constant on every interval } [l2^j, (l+1)2^j), l \in \mathbb{Z}\}.$$

The V_j are all obtained from V_0 by the appropriate dilation,

$$f(x) \in V_j \Leftrightarrow f(2^j x) \in V_0$$

and V_0 is generated by the integer translates $\phi_{0n}(x) = \phi(x-n)$ of one single function ϕ. (In this example, $\phi(x) = 1$ for $0 \leq x < 1$, $\phi(x) = 0$ otherwise). Then there exists a function ψ such that the $\psi_{jn}(x)=2^{-j/2}\psi(2^{-j}x-n)$, j fixed, are an orthonormal basis of the orthogonal complement W_j of V_j in V_{j-1}. In other words, the inner products $\{\langle\psi_{jn},f\rangle;\ n\in\mathbb{Z}\}$ contain exactly all the information that is present in the approximation to f at level $j-1$ (with resolution 2^{j-1}) but which is lacking in the next coarser level j (resolution 2^j). Because of the "ladder property" of the V_j it follows that the whole collection $\{\psi_{jn};\ j,n\in\mathbb{Z}\}$ is an orthonormal basis of wavelets for $L^2(\mathbb{R})$. Readers who would like to learn more about multiresolution analysis should consult [36], [37], or [38], where explicit recipes are given for the construction of wavelet bases, together with examples. Wavelet decompositions using multiresolution analysis have been implemented in vision analysis [38], [40]. A short review is also included in [39]. Note that there is a connection between orthonormal wavelet bases and quadrature mirror filters, used in subband coding [39], [40].

Incidentally, the discovery of multiresolution analysis is another instance in which signal analysis applications provided the first intuition leading to the mathematical construction. Existing, more rudimentary techniques in vision analysis inspired S. Mallat to view orthonormal wavelet bases as a more refined tool for multiresolution approximations, which led to [36], [37]. S. Mallat has implemented these ideas, using the Battle–Lemarié bases, into an algorithm for decomposition and reconstruction of images [29], [38]. Finally, in February 1987, using Mallat's algorithm as inspiration, I succeeded in constructing orthonormal bases of wavelets with compact support, which are discussed elsewhere [39].

These different families of orthonormal wavelet bases have created quite a stir among mathematicians. Apart from applications to signal analysis, they should be useful in physics also. A first application, to quantum field theory, can be found in [41]. All these orthonormal bases are, however, rather spread out numerically, if one wants h to be reasonably regular ($h\in C^k$, with k large enough). If one is willing to give up the requirement of a basis and to settle for a frame (giving up the linear independence—see the remark in Section I-C), then much better localized C^∞ h can be chosen. This makes frames more interesting than orthonormal bases for certain wavelet applications in signal analysis. Another reason is that, as will be shown later, for a given desired reconstruction precision, frames allow one to calculate the wavelet coefficients with less precision than would be needed in the orthonormal case; the number of coefficients calculated is, of course, higher. This may be useful in some numerical applications. The present paper addresses "frame" questions. We shall formulate criteria under which the h_{mn} constitute a frame, and then derive some properties of these frames. Since similar techniques can be used for Weyl–Heisenberg coherent states, we address the same questions for that case as well. The basic results of this paper were reported, in an abridged version, at two mathematics conferences, in March 1986 [42] and July 1986 [43], respectively.

H. Organization of the Paper

In Section II we study "frame questions." We start with some generalities concerning frames in Hilbert spaces. We review the construction of [16], leading to tight frames. For more general choices of the initial function g or h, we answer the questions 1) and 2) asked previously (sufficient condition for frame, estimates for the frame constants A and B), both for the Weyl–Heisenberg and for the wavelet case. We discuss whether certain ranges for p_0,q_0 or a_0,b_0 are *a priori* excluded, and we give inequalities linking A, B, g, or h and p_0,q_0 or a_0,b_0. We give many examples, and numerical tables of frame constants for these examples. For particular choices of g or h, our results can be translated into estimates on entire functions [44]. A special case of this kind of estimate was already given in [45]. Since numerically one is also interested in convergence in other topologies than only L^2, we address the question of convergence in other spaces as well. To improve the readability of the paper, we have relegated many of the technical proofs to appendices.

In Section III we show how to use these frames for phase space localization. Concretely this means the following. Suppose that a signal f is mostly concentrated in $[-T,T]$ in time, and that its Fourier transform $\hat{f}$ is concentrated mostly between the frequencies Ω_1 and Ω_2. This means that in phase space, f is mostly concentrated on $[-T,T]\times([-\Omega_2,-\Omega_1]\cup[\Omega_1,\Omega_2])$. One would then expect that only those phase space lattice points, in Fig. 2, lying within this box (plus those lying very close to it) would suffice to approximately reconstruct f. It turns out that this intuition is essentially right. Note that such a "box" contains only a finite number of points. We also show how the "over-sampling" inherent to working with frames permits, for fixed desired precision on the reconstruction of f, to compute the coefficients $c_{mn}(f)$ with less precision than would be needed in the orthonormal case.

I. Some Remarks

A first remark we want to make is that while we shall stick, throughout the paper, to one dimension (i.e., a two-dimensional phase space), it is possible to generalize the results to more than one dimension. For the Weyl–Heisenberg case this is trivially true. For the wavelet case two possibilities exist. In the first case dilations and translations are used, independently in the n dimensions, and then again the extension is trivial. In the second case one uses n-dimensional translations, but only one dilation parameter, which acts simultaneously on all dimensions. In this case several (a finite number) h_j have to be introduced. This construction is then similar to the generalization in [32] of Y. Meyer's basis to n dimensions: in that case 2^n-1 different functions h_j are used.

A second remark is that we have restricted ourselves here to regularly spaced lattices, in the Weyl–Heisenberg (WH) case as well as the wavelet case. While it is possible to relax this regularity somewhat (we can also handle the

WH case if one of the two variables, p or q, is not regularly spaced, as long as the other is; likewise, in the wavelet case, we can handle dilations that are more irregular than the geometric sequence a_0^m), the methods presented here are essentially unable to cope with distributions of phase space points that would have the same density (and this probably suffices to define frames, as in nonharmonic Fourier analysis [15]), but which would not be given by lattices. Using different methods, one can indeed show that less regular phase space point distributions with a sufficiently high density do indeed lead to frames. For very special choices of g or h for which the "frame questions" can be formulated as properties of entire function spaces, that was proved in [45], [47]. Using the full power of the underlying group structure, Feichtinger and Gröchenig [48] proved the same result for much more general functions g or h (essentially, they only require that g or h is reasonably "nice"), without recourse to entire function spaces. The results in [46]–[48] are, however, more qualitative than quantitative in that they establish that certain discrete families of coherent states constitute frames without caring too much about the values of the frame constants. Moreover, the phase space densities of the discrete families considered in [46]–[48] seems to be considerably higher than in our examples.

II. Frames and Frame Bounds

A. Generalities Concerning Frames

We start by reviewing some general properties of frames [14], [15].

Suppose that the $(\phi_l;\ l \in J)$ constitute a frame in the Hilbert space $\mathscr{H}$, i.e., there exist $A > 0$, $B < \infty$ such that, for all $f \in \mathscr{H}$,

$$A\|f\|^2 \le \sum_{l \in J} |\langle \phi_l, f\rangle|^2 \le B\|f\|^2. \tag{2.1.1}$$

Define T: $\mathscr{H} \to l^2(J)$ by $(Tf)_l = \langle \phi_l, f\rangle$. Here $l^2(J)$ stands for the space of square summable complex sequences indexed by J. The operator T is clearly bounded, $\|Tf\| \le B^{1/2}\|f\|$. We shall call T the "frame operator" associated with the frame $(\phi_l)_{l \in J}$. Its adjoint operator T^* maps $l^2(J)$ onto $\mathscr{H}$; it is defined by $T^*c = \sum_{l \in J} c_l \phi_l$, where $c = (c_l)_{l \in J} \in l^2(J)$. The frame inequalities (2.1.1) can be rewritten as

$$A\mathbb{1} \le T^*T \le B\mathbb{1}.$$

Since the symmetric operator T^*T is bounded below by a strictly positive constant, it is invertible, with a bounded inverse. This inverse satisfies

$$B^{-1}\mathbb{1} \le (T^*T)^{-1} \le A^{-1}\mathbb{1}. \tag{2.1.2}$$

Define $\tilde{\phi}_l = (T^*T)^{-1}\phi_l$. Then the family $(\tilde{\phi}_l)_{l \in J}$ constitutes another frame. More precisely, we have

Proposition 2.1:

1) The family $(\tilde{\phi}_l)_{l \in J}$, with $\tilde{\phi}_l = (T^*T)^{-1}\phi_l$, constitutes a frame with bounds B^{-1} and A^{-1}.

2) The associated frame operator $\tilde{T}$ is given by $\tilde{T} = T(T^*T)^{-1}$ and satisfies

$$\tilde{T}^*\tilde{T} = (T^*T)^{-1}$$
$$\tilde{T}^*T = \mathbb{1} = T^*\tilde{T} \tag{2.1.3}$$

where $\tilde{T}T^* = T\tilde{T}^*$ is the orthogonal projection operator, in $l^2(J)$, onto the range of T.

Proof:

1) For any $f \in \mathscr{H}$, we have

$$\langle \tilde{\phi}_l, f\rangle = \langle (T^*T)^{-1}\phi_l, f\rangle = \langle \phi_l, (T^*T)^{-1}f\rangle.$$

Hence

$$\sum_l |\langle \tilde{\phi}_l, f\rangle|^2 = \|T(T^*T)^{-1}f\|^2$$
$$= \langle (T^*T)^{-1}f, T^*T(T^*T)^{-1}f\rangle$$
$$= \langle f, (T^*T)^{-1}f\rangle.$$

It then follows from (2.1.2) that the $(\tilde{\phi}_l)_{l \in J}$ constitute a frame, with frame bounds B^{-1} and A^{-1}.

2) By the definition of the $\tilde{\phi}_l$ we have

$$(\tilde{T}f)_l = \langle \tilde{\phi}_l, f\rangle = \langle (T^*T)^{-1}\phi_l, f\rangle$$
$$= \langle \phi_l, (T^*T)^{-1}f\rangle = \left(T(T^*T)^{-1}f\right)_l,$$

hence $\tilde{T} = T(T^*T)^{-1}$. It follows that

$$\tilde{T}^*\tilde{T} = (T^*T)^{-1}T^*T(T^*T)^{-1} = (T^*T)^{-1}$$

and

$$\tilde{T}^*T = (T^*T)^{-1}T^*T = \mathbb{1},$$
$$T^*\tilde{T} = T^*T(T^*T)^{-1} = \mathbb{1}.$$

3) Finally $\tilde{T}T^* = T(T^*T)^{-1}T^* = T\tilde{T}^*$; it remains to prove that this is the orthogonal projection operator P onto the range of T. Since $T^*c = 0$ for any c orthogonal to the range of T, it suffices to prove that $T(T^*T)^{-1}T^*c = c$ for c in the range of T. If c is in the range of T, $c = Tf$, then

$$T(T^*T)^{-1}T^*c = T(T^*T)^{-1}T^*Tf = Tf = c.$$

This proves $\tilde{T}T^* = P = T\tilde{T}^*$. □

The operation $\{\phi_l;\ l \in J\} \to \{\tilde{\phi}_l;\ l \in J\}$ defines, in a sense, a duality operation. The same procedure, applied to the frame $\{\tilde{\phi}_l;\ l \in J\}$, gives the original frame $\{\phi_l;\ l \in J\}$ back again. We shall therefore call $\{\tilde{\phi}_l;\ l \in J\}$ the *dual frame* of $\{\phi_l;\ l \in J\}$. The duality $\phi_l \leftrightarrow \tilde{\phi}_l$ is also expressed by (2.1.3); for any $f, g \in \mathscr{H}$ we have

$$\langle f, g\rangle = \sum_{l \in J} \langle f, \tilde{\phi}_l\rangle\langle \phi_l, g\rangle$$
$$= \sum_{l \in J} \langle f, \phi_l\rangle\langle \tilde{\phi}_l, g\rangle.$$

In what follows we shall so often encounter T^*T (rather than T alone) that it makes sense to introduce a new notation for this operator. We define

$$\mathbb{T} = T^*T, \qquad \tilde{\mathbb{T}} = \tilde{T}^*\tilde{T} = \mathbb{T}^{-1}.$$

In particular

$$\mathbb{T} = \sum_{l \in J} \phi_l \langle \phi_l, \cdot \rangle$$
$$A\mathbb{1} \le \mathbb{T} \le B\mathbb{1}$$
$$\tilde{\phi}_l = \mathbb{T}^{-1}\phi_l.$$

Proposition 2.1 gives us an inversion formula for T. If the elements $f \in \mathscr{H}$ are characterized by means of the inner products $(\langle f, \phi_l \rangle;\ l \in J)$, then f can be reconstructed from these by means of

$$f = \sum_l \tilde{\phi}_l \langle \phi_l, f \rangle. \tag{2.1.4}$$

In (2.1.4) the vectors $\tilde{\phi}_l$ are defined by $\tilde{\phi}_l = \mathbb{T}^{-1}\phi_l$. Note that, in general, if the frame is redundant (i.e., contains "more" vectors than a basis would), there exist other vectors in $\mathscr{H}$ that could equally well play the role of the $\tilde{\phi}_l$ and lead to a reconstruction formula. This is due to the fact that the ϕ_l are not linearly independent in the general case. This phenomenon can be illustrated with the two-dimensional example of Section I-C. Take $\mathscr{H} = \mathbb{C}^2$, and define

$$\phi_1 = e_1, \qquad \phi_2 = -\frac{1}{2}e_1 + \frac{\sqrt{3}}{2}e_2, \qquad \phi_3 = -\frac{1}{2}e_1 - \frac{\sqrt{3}}{2}e_2$$

where $e_1 = (1,0)$, $e_2 = (0,1)$ constitute the standard orthonormal basis in $\mathbb{C}^2$. One can check very easily that for all $u \in \mathscr{H}$,

$$\sum_{l=1}^{3} |\langle \phi_l, u \rangle|^2 = \frac{3}{2}\|u\|^2$$

so that $\mathbb{T} = 3/2\mathbb{1}$, hence $\tilde{\phi}_l = \mathbb{T}^{-1}\phi_l = 2/3\phi_l$, and (2.1.4) becomes

$$u = \frac{2}{3}\sum_{l=1}^{3} \phi_l \langle \phi_l, u \rangle.$$

Since $\Sigma_{l=1}^{3}\phi_l = 0$ in this example, it is clear, however, that for any choice of $a \in \mathscr{H}$, an equally valid reconstruction formula is given by

$$u = \frac{2}{3}\sum_{l=1}^{3} (\phi_l + a)\langle \phi_l, u \rangle.$$

The choice $a = 0$, corresponding to (2.1.4), is the "minimal solution" in the following sense. The image, in $\mathbb{C}^3$, of $\mathbb{C}^2$ under the frame operator T is the two-dimensional subspace with equation $x_1 + x_2 + x_3 = 0$; we denote this subspace by *RanT*. Vectors in $\mathbb{C}^3$ orthogonal to *RanT* are all of the type $c = \lambda(1,1,1)$. When the components of such a vector are substituted for the $\langle \phi_l, u \rangle$ in (2.1.4), then the "reconstruction" leads to 0, since

$$\sum_{l=1}^{3} \tilde{\phi}_l c_l = \frac{2}{3}\lambda \sum_{l=1}^{3} \phi_l = 0.$$

This is no longer the case if one of the other "equally valid" choices is used, where $\tilde{\phi}_l$ is replaced by $2/3(\phi_l + a)$. Of all the possible "quasi-inverses" for T, $\tilde{T}^*$ is therefore the only one that automatically projects sequences in $(RanT)^{\perp}$ to zero while effecting the reconstruction.

A similar phenomenon occurs with the infinite-dimensional frames we shall consider. While no finite number of them will be linearly dependent, there generally do exist convergent linear combinations, involving infinitely many ϕ_l, which sum to zero. This again causes "reconstruction formulas" to be nonunique, but again the $\tilde{\phi}_l = \mathbb{T}^{-1}\phi_l$ will offer the optimal solution, in the sense that they are the *only* choice leading to

$$\sum_l c_l \tilde{\phi}_l = 0, \qquad \text{for all } c = (c_l)_{l \in J} \perp RanR.$$

This is easy to check by the following argument. Write $\tilde{\phi}_l = \mathbb{T}^{-1}\phi_l + u_l$, and suppose that (2.1.4) holds. It follows that $\Sigma u_l \langle \phi_l, f \rangle = 0$, for all $f \in \mathscr{H}$, or $\Sigma u_l c_l = 0$, for all $c \in RanT$. The optimality condition implies $\Sigma u_l c_l = 0$ for all $c \perp RanT$. It follows that $\Sigma u_l c_l = 0$ for all $c \in l^2(J)$, hence $u_l = 0$ for all l.

To apply (2.1.4) it is, of course, necessary to construct the $\tilde{\phi}_l$ first. Writing $\mathbb{T}$ as $2(A+B)^{-1}[\mathbb{1} - (2(A+B)^{-1}\mathbb{T})]$, one has the following series for the $\tilde{\phi}_l$,

$$\tilde{\phi}_l = \frac{2}{A+B}\sum_{k=0}^{\infty}\left(\mathbb{1} - \frac{2\mathbb{T}}{A+B}\right)^k \phi_l. \tag{2.1.5}$$

Since

$$\frac{A-B}{A+B}\mathbb{1} \le \mathbb{1} - \frac{2\mathbb{T}}{A+B} \le \frac{B-A}{B+A}\mathbb{1}$$

or

$$\left\| \mathbb{1} - \frac{2\mathbb{T}}{A+B} \right\| \le \frac{B/A - 1}{B/A + 1} < 1 \tag{2.1.6}$$

the series in (2.1.5) always converges. The closer B/A is to 1, the faster the convergence, of course.

Remark: For the reconstruction formula (2.1.4) for f from its coefficients $\langle \phi_l, f \rangle$, we have used (2.1.3), namely that $\tilde{T}^*T = \mathbb{1}$. One can also interpret (2.1.3) differently. From $T^*\tilde{T} = \mathbb{1}$ it follows that, for all $f \in \mathscr{H}$,

$$f = \sum_l \langle \tilde{\phi}_l, f \rangle \phi_l.$$

This tells us that any function f can be expanded in the ϕ_l, and gives us a recipe for calculating appropriate coefficients. This is the point of view taken in so-called "atomic decompositions." It is thus clear that frames and atomic decompositions are dual notions [48].

All the previous formulas become much simpler if the frame is *tight*, i.e., if $B = A$. In that case

$$\mathbb{T} = A\mathbb{1}, \qquad \tilde{\mathbb{T}} = A^{-1}\mathbb{1}$$
$$\tilde{\phi}_l = A^{-1}\phi_l$$
$$f = A^{-1}\sum_{l \in J} \phi_l \langle \phi_l, f \rangle.$$

If the frame constant for a tight frame is equal to one, $A = 1$, then the reconstruction formula looks exactly like the decomposition of a function with respect to an orthonormal basis,

$$f = \sum_{l \in J} \phi_l \langle \phi_l, f \rangle .$$

In fact, if the vectors in a tight frame with frame constant 1 are all normalized, then the frame constitutes an orthonormal basis, as can easily be seen by the following argument:

$$\|\phi_k\|^2 = \sum_{l \in J} |\langle \phi_l, \phi_k \rangle|^2$$
$$= \|\phi_k\|^4 + \sum_{l \notin J} |\langle \phi_l, \phi_k \rangle|^2$$

which implies $\langle \phi_l, \phi_k \rangle = 0$ for $l \neq k$ if $\|\phi_k\| = 1$. As we explained in the introduction, there are circumstances in which one prefers to work with nontight frames. In that case the $\tilde{\phi}_l$ have to be constructed explicitly from the ϕ_l. Formula (2.1.5) then gives an algorithm for the computation of the $\tilde{\phi}_l$; as (2.1.6) shows, it pays to have a ratio B/A as close to 1 as possible. Frames that are almost tight (i.e., B/A close to 1) will be called *snug frames*. The "snugness" can be measured by $S = [B/A - 1]^{-1}$. In the examples below we shall encounter values of $S \geq 100$, corresponding to B/A of the order of 1.01 or even smaller, for quite realistic frames (i.e., reasonably large values of p_0, q_0 or a_0, b_0—see Introduction).

Note also that, while in principle all the different $\tilde{\phi}_l$, $l \in J$, have to be calculated separately, in practice simplifications may occur. Let us show what happens for the discrete Weyl–Heisenberg coherent state frames and for the wavelet frames defined in the introduction. We take therefore $\mathscr{H} = L^2(\mathbb{R})$, $J = \mathbb{Z}^2$.

In the Weyl–Heisenberg case, we can rewrite $\mathbb{T}$ as

$$\mathbb{T} = \sum_{m,n \in \mathbb{Z}} W(mp_0, nq_0) g \langle W(mp_0, nq_0) g, \cdot \rangle$$

where the operator $W(p,q)$ is defined by $[W(p,q)g](x) = e^{ipx} g(x-q)$. One easily checks that $W(p,q)W(p',q') = e^{-ip'q} W(p+p', q+q')$. Using this multiplication formula for the operators W, one easily finds that

$$W(mp_0, nq_0)\mathbb{T} = \mathbb{T} W(mp_0, nq_0).$$

Hence

$$W(mp_0, nq_0)\mathbb{T}^{-1} = \mathbb{T}^{-1} W(mp_0, nq_0).$$

This implies

$$(g_{mn})^{\sim} = \mathbb{T}^{-1} g_{mn}$$
$$= \mathbb{T}^{-1} W(mp_0, nq_0) g$$
$$= W(mp_0, nq_0) \mathbb{T}^{-1} g,$$

or $(g_{mn})^{\sim}(x) = e^{imp_0x}(g_{00})^{\sim}(x - nq_0) = \tilde{g}_{mn}(x)$, where $\tilde{g} = (g_{00})^{\sim} = \mathbb{T}^{-1}g$. We have thus only one function to compute, i.e., $\tilde{g}_{00} = \mathbb{T}^{-1}g$, instead of a double infinity of different $(g_{mn})^{\sim}(m, n \in \mathbb{Z})$. If moreover B/A is close to 1, then the rapidly converging series (2.1.5) can be used for this computation.

In the affine case, we find

$$\mathbb{T} = \sum_{m,n \in \mathbb{Z}} U(a_0^m, a_0^m nb_0) h \langle U(a_0^m, a_0^m nb_0) h, \cdot \rangle$$

where $[U(a,b)f](x) = |a|^{-1/2} f(x - b/a)$. Again we can use the composition law

$$U(a',b')U(a'',b'') = U(a'a'', b' + a'b'')$$

to show that

$$U(a_0^m, 0)\mathbb{T} = \mathbb{T} U(a_0^m, 0).$$

Note that we have no translation in these U-operators; it turns out that $\mathbb{T}$ does *not* commute with $U(a_0^m, nb_0 a_0^m)$ if $n \neq 0$. Since $U(a_0^m, a_0^m nb_0) = U(a_0^m, 0) U(1, nb_0)$, we find

$$(h_{mn})^{\sim} = U(a_0^m, 0)\mathbb{T}^{-1} U(1, nb_0) h$$

or

$$(h_{mn})^{\sim}(x) = a_0^{-m/2}(h_{0n})^{\sim}(a_0^{-m}x).$$

The simplification is less drastic than in the WH case, since only one of the two indices is eliminated, and one still has to compute the infinite number of $(h_{0n})^{\sim} = \mathbb{T}^{-1}h_{0n}$. In practice, however, only a finite number are needed, and for the computation of these it is again a big help if B/A is close to 1.

Note that if one stops at the first term ($k = 0$) in (2.1.5), one obtains the approximate inversion formula

$$f_{\text{approx}} = \frac{2}{A+B} \sum_{m,n} \phi_{mn} \langle \phi_{mn}, f \rangle . \qquad (2.1.7)$$

If B/A is close enough to one, this is a fairly good approximation. In the first calculations with wavelets, the formula used for analysis and reconstruction was very similar to (2.1.7) [25], without theoretical basis. It turns out that in those frames B/A was indeed very close to 1, so that f_{approx} is a good approximation to f.

B. Tight Frames

1) Explicitly Constructed Tight Frames: In this subsection we review shortly, for the sake of completeness, the explicit construction of tight frames given in [16]. We have drawn inspiration from [31] to make the construction more elegant. For examples, graphs and a more detailed discussion we refer to [16].

In both cases an auxiliary function ν of the following type is used

$$\nu: \mathbb{R} \to \mathbb{R}$$
$$\nu(x) = \begin{cases} 0, & \text{if } x \leq 0 \\ 1, & \text{if } x \geq 1. \end{cases}$$

If ν is chosen to be a C^k (or C^∞) function, then the resulting frames will consist of C^k (C^∞) functions.

a) The Weyl–Heisenberg case: Choose $p_0, q_0 > 0$, with $p_0 \cdot q_0 < 2\pi$. (In Section II-C we shall have more to say about this restriction.) We shall assume here that $p_0 \cdot q_0 \geq \pi$. (See [16, Appendix] for indications how the construction can be adapted if $p_0 \cdot q_0 < \pi$). Define $\lambda = 2\pi/p_0 - q_0$; one has $0 < \lambda \leq \pi/p_0$. The function g is then constructed as follows:

$$g(x) = q_0^{-1/2}\begin{cases} 0, & x \le -\pi/p_0 \\ \sin\left[\frac{\pi}{2}\nu\left(\lambda^{-1}\left(\frac{\pi}{p_0}+x\right)\right)\right], & -\pi/p_0 \le x \le -\frac{\pi}{p_0}+\lambda \\ 1, & -\frac{\pi}{p_0}+\lambda \le x \le \frac{\pi}{p_0}-\lambda \\ \cos\left[\frac{\pi}{2}\nu\left(\lambda^{-1}\left(x-\frac{\pi}{p_0}+\lambda\right)\right)\right], & \frac{\pi}{p_0}-\lambda \le x \le \pi/p_0 \\ 0, & x \ge \pi/p_0. \end{cases}$$

The overall factor $q_0^{-1/2}$ normalizes g; one easily checks that $\int dx |g(x)|^2 = 1$. The following two properties of g are crucial.

$$\begin{aligned} &1) \quad |\text{support } g| = 2\pi/p_0, \\ &2) \quad \sum_{k\in\mathbb{Z}} |g(x-kq_0)|^2 = 1/q_0. \end{aligned} \qquad (2.2.1)$$

These two properties together ensure that the g_{mn}, $g_{mn}(x) = e^{imp_0x}g(x-nq_0)$, constitute a tight frame. By Plancherel's theorem we have indeed for all $f \in L^2(\mathbb{R})$,

$$\sum_{m,n\in\mathbb{Z}} |\langle g_{mn}, f\rangle|^2$$

$$= \sum_{m,n\in\mathbb{Z}} \left| \int_{|x-nq_0|\le\pi/p_0} dx\, e^{imp_0x} g(x-nq_0)^* f(x)\right|^2$$

$$= \sum_{n\in\mathbb{Z}} \frac{2\pi}{p_0} \int dx |g(x-nq_0)|^2 |f(x)|^2$$

$$= \frac{2\pi}{p_0 q_0} \int dx |f(x)|^2$$

where we have used the notation $g(y)^*$ to denote the complex conjugate of $g(y)$. Note that the same calculation can be made whenever $|\text{support } g| \le 2\pi/p_0$, even if (2.2.1) is not satisfied. In that case the operator $\mathbb{T}$ of Section II-A reduces to multiplication by the periodic function $G(x) = 2\pi p_0^{-1}\sum_{n\in\mathbb{Z}}|g(x-nq_0)|^2$ (see also [50]).

One finds thus $\tilde{g}_{mn}(x) = G(x)^{-1}g_{mn}(x)$, and the inversion formula becomes

$$f = G^{-1}\sum_{m,n} g_{mn}\langle g_{mn}, f\rangle.$$

Remark: If the function ν is chosen to be C^∞, then g is a compactly supported C^∞-function. While the uncertainty principle can of course not be violated, this nevertheless allows fairly good localization of g in both time and frequency. The coefficients $\langle g_{mn}, f\rangle$ do therefore correspond to a reasonably accurate time-frequency localized picture of the signal f. Moreover, there exists a constant $A = 2\pi/p_0 \cdot q_0$ such that

$$\int dx |f(x)|^2 = A^{-1}\sum_{m,n} |\langle g_{mn}, f\rangle|^2.$$

The role of this constant A is crucial. If one restricts oneself to $A = 1$, then this forces the g_{mn} to be an orthonormal basis, which leads to functions g that are badly localized in either time or frequency. This is a consequence of the Balian–Low theorem [51], [52] (Theorem 2.3 next), rediscovered in [53]. As soon, however, as one allows $A \ne 1$, the picture changes drastically, and much better time-frequency localization is attainable.

b) The wavelet case: In this case there are no *a priori* restrictions on the choice of the parameters a_0, b_0, other than $a_0 \ne 0$ or 1 and $b_0 \ne 0$. We may choose, without loss of generality, $a_0 > 1$, and $b_0 > 0$. Define $l = 2\pi/[b_0(a_0^2 - 1)]$. The tight frame of wavelets will be based on the function h with Fourier transform $\hat{h}$ constructed as follows [16]:

$$\hat{h}(y) = (\log a_0)^{-1/2}\begin{cases} 0, & y \le l \\ \sin\left[\frac{\pi}{2}\nu\left(\frac{y-l}{l(a_0-1)}\right)\right], & l \le y \le a_0 l \\ \cos\left[\frac{\pi}{2}\nu\left(\frac{y-a_0 l}{l a_0(a_0-1)}\right)\right], & a_0 l \le y \le a_0^2 l \\ 0, & y \ge a_0^2 l. \end{cases} \qquad (2.2.2)$$

The function h itself is given by the inverse Fourier transform,

$$h(x) = \frac{1}{\sqrt{2\pi}} \int dy\, e^{-ixy} \hat{h}(y).$$

The normalization of h has been chosen so that

$$\int dy |y|^{-1} |\hat{h}(y)|^2 = 1.$$

The following two properties of h are again crucial:

$$1)\quad |\text{support}\, \hat{h}| = (a_0^2 - 1)l = 2\pi / b_0,$$
$$2)\quad \sum_{k \in \mathbb{Z}} |\hat{h}(a_0^k y)|^2 = (\log a_0)^{-1} \chi_{[0,\infty)}(y), \tag{2.2.3}$$

where $\chi_{[0,\infty)}$ is the indicator function of the right half line, $\chi_{[0,\infty)}(y) = 1$ if $y \geq 0$, $\chi_{[0,\infty)}(y) = 0$, otherwise. These properties together enable us to construct a tight frame based on h. By Parseval's theorem, we have, for all $f \in L^2(\mathbb{R})$,

$$\begin{aligned}\sum_{m,n \in \mathbb{Z}} |\langle h_{mn}, f \rangle|^2 &= \sum_{m,n \in \mathbb{Z}} a_0^m \left| \int dy\, e^{inb_0 a_0^m y} \hat{h}(a_0^m y)^* \hat{f}(y) \right|^2 \\ &= \sum_{m \in \mathbb{Z}} \frac{2\pi}{b_0} \int dy\, |\hat{h}(a_0^m y)|^2 |\hat{f}(y)|^2 \\ &= \frac{2\pi}{b_0 \log a_0} \int_0^\infty dy\, |\hat{f}(y)|^2. \end{aligned} \tag{2.2.4}$$

If we define $h^+ = h$, $h^- = h^*$ (i.e., $(h^-)^\wedge(k) = \hat{h}(-k)$), then this calculation shows that the $(h^{\pm}_{mn};\ m,n \in \mathbb{Z})$ constitute a tight frame. Specifically,

$$\sum_{\epsilon = + \text{ or } -} \sum_{m,n \in \mathbb{Z}} |\langle h^{\epsilon}_{mn}, f \rangle|^2 = \frac{2\pi}{b_0 \log a_0} \|f\|^2. \tag{2.2.5}$$

Similarly the $(h^{(\lambda)}_{mn};\ m,n \in \mathbb{Z},\ \lambda = 1 \text{ or } 2)$, with $h^{(1)} = \operatorname{Re} h$, $h^{(2)} = \operatorname{Im} h$, constitute a tight frame. Strictly speaking, the frames $\{h^{\epsilon}_{mn};\ \epsilon = + \text{ or } -,\ m,n \in \mathbb{Z}\}$ or $\{h^{(\lambda)}_{mn};\ \lambda = 1 \text{ or } 2,\ m,n \in \mathbb{Z}\}$ are not quite of the type described in the introduction, since they are not obtained by dilating and translating one single function. As shown by the computation leading to (2.2.4), the operator $\mathbb{T} = T^*T$ handles the positive and negative frequency domains independently. Since the Fourier transform $\hat{h}$ of h, defined by (2.2.2), is entirely supported on the positive half line, the use of a second function, which will handle the negative frequencies, is therefore unavoidable. In other examples (see e.g., Section II-C-2) the function h will be chosen such that $\hat{h}$ is supported on both the negative and the positive half lines, and one function suffices.

Let us return to the construction (2.2.2). For a special set of signals f to be analyzed, one may restrict oneself, even if support $\hat{h} \subset [0,\infty)$, to only one basic function h, and the associated wavelets h_{mn}. This happens if one knows *a priori*, as is often the case, that the signal f is real. Since then $\hat{f}(-y) = \hat{f}(y)^*$, one finds

$$\sum_{m,n \in \mathbb{Z}} |\langle h_{mn}, f \rangle|^2 = \frac{\pi}{b_0 \log a_0} \|f\|^2.$$

As for the Weyl–Heisenberg case, a calculation similar to (2.2.4) can be carried out whenever the support width of $\hat{h}$ is smaller than $2\pi / b_0$, even if (2.2.3) is not satisfied. The operator $\mathbb{T}$ becomes then

$$(\mathbb{T}f)^\wedge(y) = \hat{H}(y)\hat{f}(y)$$

with

$$\hat{H}(y) = \frac{2\pi}{b_0} \sum_m \left[|h(a_0^m y)|^2 + |h(-a_0^m y)|^2 \right]$$

or, equivalently,

$$\begin{aligned}(\mathbb{T}f)(x) &= \frac{1}{\sqrt{2\pi}} (H * f)(x) \\ &= \frac{1}{\sqrt{2\pi}} \int dx'\, H(x - x') f(x').\end{aligned}$$

Hence

$$\left[(h^{\pm}_{mn})^{\sim} \right]^\wedge(y) = \left[\hat{H}(y) \right]^{-1} (h^{\pm}_{mn})^\wedge(y).$$

In the construction of the orthonormal basis, in [31], the procedure starts in the same way. The function ν is chosen to be C^∞, and has one additional property,

$$\nu(x) + \nu(1 - x) = 1, \qquad \text{for all } x \in \mathbb{R}. \tag{2.2.6}$$

The "doubling" (using superscripts $\pm$ to be able to cover, in Fourier space, the whole real line instead of only the half line) is avoided in [31] by incorporating also the mirror image of h into the basis function ψ. Explicitly,

$$\hat{\psi}(y) = \left(\frac{b_0 \log a_0}{2\pi} \right)^{1/2} e^{ib_0 y/2} \left[\hat{h}(y) + \hat{h}(-y) \right].$$

Due to the extra property (2.2.6) of ν, one easily checks that $\|\psi\|^2 = \int dx\, |\psi(x)|^2 = 1$. Note that the size of the support of $\hat{\psi}$ is no longer equal to $2\pi / b_0$, so that the straightforward calculation made previously no longer works. In fact, the set of functions $\psi_{mn}(x) = a_0^{-m/2} \psi(a_0^{-m} x - nb_0)$, $m,n \in \mathbb{Z}$, turns out to be an orthonormal basis of $L^2(\mathbb{R})$ as the result of some miraculous cancellations, for which the property (2.2.6) of the function ν turns out to be quite crucial [31]. These cancellations occur only when a_0 takes on a very special value, namely when

$$a_0 = \frac{k+1}{k}, \qquad \text{with } k \in \mathbb{N}.$$

The case $k = 1$, or $a_0 = 2$, corresponds to Y. Meyer's construction in [31]. With $b_0 = 1$ (as in [31]), Y. Meyer's basic wavelet is thus given by

$$\hat{\psi}(y) = \frac{1}{\sqrt{2\pi}} e^{iy/2} [\omega(y) + \omega(-y)] \tag{2.2.7}$$

with

$$\omega(y)=\begin{cases}0, & y\le \frac{2\pi}{3}\\ \sin\left[\frac{\pi}{2}\nu\left(\frac{3y}{2\pi}-1\right)\right], & \frac{2\pi}{3}\le y\le\frac{4\pi}{3}\\ \cos\left[\frac{\pi}{2}\nu\left(\frac{3y}{4\pi}-1\right)\right], & \frac{4\pi}{3}\le y\le\frac{8\pi}{3}\\ 0, & y\ge\frac{8\pi}{3}\end{cases}$$

where ν is a C^∞ function from $\mathbb{R}$ to $[0,1]$ such that $\nu(y)=0$ for $y\le 0$, $\nu(y)=1$ for $y\ge 1$ and $\nu(y)+\nu(1-y)=1$ for all y. As pointed out in the introduction, the concept of multiscale analysis allows one to understand more deeply why this construction works, so that the "miraculous cancellations" just mentioned become less so.

2) Relations Between the Frame Parameters and the Frame Bounds: In the explicitly constructed tight frames, the frame constant A is given by $A=2\pi/p_0q_0$, for the Weyl–Heisenberg case, and by $A=\pi/b_0\log a_0$, for the affine or wavelet case. This is no coincidence. We show in this subsection that these values for A are imposed by the normalizations chosen for g,h, and are independent of the details of the construction, i.e., they are generally true for *all* tight frames. More generally, we prove inequalities for the frame bounds A,B for *all* Weyl–Heisenberg frames or wavelet frames, tight or not.

a) The Weyl–Heisenberg case: Let us assume that the $(\phi_{mn};\ m,n\in\mathbb{Z})$ are an arbitrary frame of discrete Weyl–Heisenberg coherent states, with frame constants A,B, and lattice spacings p_0,q_0, i.e.,

$$\phi_{mn}(x)=e^{imp_0x}\phi(x-nq_0),$$
$$A\|f\|^2\le\sum_{m,n}|\langle\phi_{mn},f\rangle|^2\le B\|f\|^2.$$

We shall see next that a frame is possible only if $p_0\cdot q_0\le 2\pi$. Let us therefore restrict ourselves to this case. Then there exists, for the same values p_0,q_0, a tight frame $\{g_{mn};\ m,n\in\mathbb{Z}\}$ with $\|g\|=1$. (For $p_0\cdot q_0<2\pi$, we take g as constructed in Section II-B-1a); for $p_0\cdot q_0=2\pi$, take $g(x)=q_0^{-1/2}$ if $|x|\le q_0/2$, $g(x)=0$, otherwise). One easily checks that

$$\langle g_{mn},\phi\rangle=e^{-imnp_0q_0}\langle g,\phi_{-m-n}\rangle. \tag{2.2.8}$$

It follows that

$$A\|g\|^2\le\sum_{m,n}|\langle\phi_{mn},g\rangle|^2$$
$$=\sum_{m,n}|\langle g_{mn},\phi\rangle|^2=\frac{2\pi}{p_0q_0}\|\phi\|^2.$$

Similarly $B\|g\|^2\ge 2\pi/p_0q_0\|\phi\|^2$. Since $\|g\|=1$, we find

$$A\le\frac{2\pi}{p_0q_0}\|\phi\|^2\le B. \tag{2.2.9}$$

This is true for any Weyl–Heisenberg-frame with lattice spacings p_0,q_0. In particular, if the ϕ_{mn} constitute a tight frame (i.e., $A=B$), then necessarily

$$A=\frac{2\pi}{p_0q_0}\|\phi\|^2.$$

a) The wavelet case: In this case also we shall derive inequalities similar to (2.2.9). Since there is no equivalent to (2.2.8) for wavelet frames, the derivation will be slightly more complicated. Let us assume, again, that $(\phi_{mn};\ m,n\in\mathbb{Z})$ is an arbitrary frame of wavelets, with lattice spacings determined by a_0,b_0, and with frame constants A,B, i.e.,

$$\phi_{mn}(x)=a_0^{-m/2}\phi(a_0^{-m}x-nb_0)$$
$$A\|f\|^2\le\sum_{m,n}|\langle\phi_{mn},f\rangle|^2\le B\|f\|^2. \tag{2.2.10}$$

Take now any positive operator C which is trace-class, i.e., which is of the form

$$C=\sum_l c_lu_l\langle u_l,\cdot\rangle$$

where the u_l are orthonormal, $c_l>0$ and $\operatorname{tr}C=\Sigma_lc_l<\infty$. Then (2.2.10) implies

$$\sum_l c_lA\|u_l\|^2\le\sum_l\sum_{m,n}c_l|\langle u_l,\phi_{mn}\rangle|^2\le\sum_l c_lB\|u_l\|^2$$

or

$$ATrC\le\sum_{m,n}\langle\phi_{mn},C\phi_{mn}\rangle\le BTrC. \tag{2.2.11}$$

We shall apply this to the following operator,

$$C=\int\int\frac{da\,db}{a^2}h^{(a,b)}\langle h^{(a,b)},\cdot\rangle t(a,b) \tag{2.2.12}$$

constructed with the help of the affine coherent states introduced in Section I-D. The function $h\in L^2(\mathbb{R})$ should satisfy $\int dy\,|y|^{-1}|\hat{h}(y)|^2<\infty$ (see Section I-D). Here $t(a,b)$ is a positive function in $L^1(\mathbb{R}^*\times\mathbb{R};\ a^{-2}da\,db)$. The operator (2.2.12) is positive and trace-class, with

$$\operatorname{tr}C=\int\frac{da\,db}{a^2}t(a,b)$$

where we have assumed $\|h\|=1$. On the other hand,

$$\langle\phi_{mn},C\phi_{mn}\rangle$$
$$=\langle\phi,U(a_0^{-m},a_0^{-m}nb_0)^{-1}CU(a_0^{-m},a_0^{-m}nb_0)\phi\rangle$$
$$=\int\frac{da\,db}{a^2}\left|\langle\phi,U(a_0^{-m},a_0^{-m}nb_0)^{-1}U(a,b)h\rangle\right|^2t(a,b)$$
$$=\int\frac{da\,db}{a^2}|\langle\phi,U(a,b)h\rangle|^2t(a_0^ma,a_0^m(b+nb_0)) \tag{2.2.13}$$

where we have used the composition law of the U-operators. We now restrict ourselves to functions t of the form

$$t(a,b)=\chi_{[1,a_0)}(|a|)\cdot t_1(b/|a|)$$

where $\chi_{[1,a_0)}$ is the indicator function of the half open interval $[1, a_0)$, i.e., $\chi_{[1,a_0)}u = 1$ if $1 \le u < a_0$, 0 otherwise. Since $\sum_m \chi_{[1,a_0)}(a_0^m|a|) = 1$, (2.2.13) then leads to

$$\sum_{m,n} \langle \phi_{mn}, C\phi_{mn} \rangle = \int \frac{da\,db}{a^2} |\langle \phi, U(a,b)h \rangle|^2 \sum_{n \in \mathbb{Z}} t_1\left(\frac{b + nb_0}{|a|}\right). \quad (2.2.14)$$

This sum over n can be estimated by the following lemma.

Lemma 2.2: Let f be a positive, continuous, bounded function on $\mathbb{R}$, with $f(x) \to 0$ as $|x| \to \infty$. Assume that f has a finite number of local maxima, at x_j, $j = 1, \cdots, N$. Define

$$\Delta_j = \sup_{\delta \in [0,1]} \int_{x_j - \delta}^{x_j - \delta + 1} dx f(x).$$

Then

$$\int_{-\infty}^{\infty} dx f(x) - \sum_{j=1}^{N} \Delta_j \le \sum_{n \in \mathbb{Z}} f(n) \le \int_{-\infty}^{\infty} dx f(x) + \sum_{j=1}^{N} f(x_j).$$

For the sake of completeness we provide a proof for the case $N = 1$; the case for general N can be proved analogously, though the inequalities can be sharpened if some of the x_j are within a distance 1 of each other.

Proof: Let n_0 be the largest integer not exceeding x_1, the point where f reaches its maximum. Since f is increasing on $(-\infty, x_1]$ and decreasing on $[x_1, \infty)$, and since $n_0 \le x_1 < n_0 + 1$:

$$\sum_{n=-\infty}^{n_0 - 1} f(n) \le \sum_{n=-\infty}^{n_0 - 1} \int_n^{n+1} dx f(x) = \int_{-\infty}^{n_0} dx f(x),$$

$$\sum_{n=n_0+1}^{\infty} f(n) \le \sum_{n=n_0+1}^{\infty} \int_{n-1}^{n} dx f(x) = \int_{n_0}^{\infty} dx f(x).$$

Hence

$$\sum_{n=-\infty}^{\infty} f(n) \le \int_{-\infty}^{\infty} dx f(x) + f(n_0) \le \int_{-\infty}^{\infty} dx f(x) + f(x_1).$$

On the other hand,

$$\sum_{n=-\infty}^{n_0} f(n) \ge \sum_{n=-\infty}^{n_0} \int_{n-1}^{n} dx f(x) = \int_{-\infty}^{n_0} dx f(x),$$

$$\sum_{n=n_0+1}^{\infty} f(n) \ge \sum_{n=n_0+1}^{\infty} \int_{n}^{n+1} dx f(x) = \int_{n_0+1}^{\infty} dx f(x).$$

Hence

$$\sum_{n=-\infty}^{\infty} f(n) \ge \int_{-\infty}^{\infty} dx f(x) - \int_{n_0}^{n_0+1} dx f(x) \ge \int_{-\infty}^{\infty} dx f(x) - \Delta_1. \quad \blacksquare$$

In particular, if f in Lemma 2.2 has only one local maximum, at $x = x_1$, then

$$\int_{-\infty}^{\infty} dx f(x) - f(x_1) \le \sum_{n=-\infty}^{\infty} f(n) \le \int_{-\infty}^{\infty} dx f(x) + f(x_1).$$

Let us apply this to (2.2.14). Choose F to be any positive, continuous L^1 function on $\mathbb{R}$, tending to zero at infinity, with one local maximum, at $x = 0$. Choose $\lambda > 0$, and define

$$t_1(x) = F(\lambda x).$$

Applying Lemma 2.2 to (2.2.14) then leads to

$$\left| \sum_{m,n} \langle \phi_{mn}, C\phi_{mn} \rangle - \frac{1}{\lambda b_0} \left[\int dx F(x) \right] \cdot \int \frac{da\,db}{|a|} |\langle \phi, U(a,b)h \rangle|^2 \right| \le F(0) \int \frac{da\,db}{a^2} |\langle \phi, U(a,b)h \rangle|^2$$

or

$$\left| \sum_{m,n} \langle \phi_{mn}, C\phi_{mn} \rangle - \frac{1}{\lambda b_0} \left[\int dx F(x) \right] 2\pi \|h\|^2 \cdot \left[\int dy\, |y|^{-1} |\hat{\phi}(y)|^2 \right] \right| \le F(0) \cdot 2\pi \|\phi\|^2 \cdot \left[\int dy\, |y|^{-1} |\hat{h}(y)|^2 \right]. \quad (2.2.15)$$

On the other hand,

$$TrC = 2 \int_1^{a_0} \frac{da}{a^2} \int db\, F\left(\lambda \frac{b}{a}\right) = \frac{2}{\lambda} \log a_0 \cdot \int dx F(x). \quad (2.2.16)$$

Inserting (2.2.15) and (2.2.16) into (2.2.11), and taking the limit for $\lambda \to 0$ leads to

$$A \le \frac{\pi}{b_0 \log a_0} \int dy\, |y|^{-1} |\hat{\phi}(y)|^2 \le B. \quad (2.2.17)$$

These inequalities hold for *any* frame of wavelets ϕ_{mn}. Incidentally, (2.2.17) shows that the basic function ϕ for a frame of wavelets must satisfy the same "admissibility condition," i.e., $\int dy |y|^{-1} |\hat{\phi}(y)|^2 < \infty$, as the functions from which continuously labeled affine coherent states are constructed (see Section I-D). In particular, if the ϕ_{mn} constitute a tight frame of wavelets, then

$$A = \frac{\pi}{b_0 \log a_0} \int dy\, |y|^{-1} |\hat{\phi}(y)|^2.$$

In the explicitly constructed tight frames in Section II-B-1b, two functions (either $h^{\pm}$, with $h^+ = h$, $h^- = h^*$, or $h^{(\lambda)}$, $\lambda = 1, 2$, with $h^{(1)} = \mathrm{Re}\, h$, $h^{(2)} = \mathrm{Im}\, h$) were needed to construct tight frames of wavelets (i.e., the $h_{mn}^{\pm}$, or the $h_{mn}^{(\lambda)}$, $\lambda = 1, 2$). For such frames the arguments lead to the frame constant

$$A = \frac{2\pi}{b_0 \log a_0} \int dy\, |y|^{-1} |\hat{h}(y)|^2.$$

This agrees with the value $A = 2\pi / b_0 \log a_0$ obtained in Section II-B-Ib (see (2.2.5)), since we chose the normalization of h such that $\int dy\, |y|^{-1} |\hat{h}(y)|^2 = 1$.

C. General (Not Necessarily Tight) Frames in $L^2(\mathbb{R})$—Ranges for the Lattice Spacings—Frame Bounds

As we already explained in the introduction, it may be necessary in some applications to resort to nontight frames. This can be the case if the basic function g or h is imposed *a priori* (because of its adaptation to the problem at hand), or in the case of wavelets, when the explicit examples of functions h leading to tight frames are too spread out.

In this section we treat the following questions:

1) Is there a range of parameters that is excluded *a priori* (i.e., independently of the choice of g or h)?
2) Given g (or h), determine a range R for the parameters p_0, q_0 (resp. a_0, b_0) such that if $(p_0, q_0) \in R$ (resp. $(a_0, b_0) \in R$), then the associated g_{mn} (resp. h_{mn}) constitute a frame.
3) For g, p_0, q_0 (or h, a_0, b_0) chosen as in 2), compute estimates for the frame bounds A and B.

In order to interrupt the flow of the exposition as little as possible, we relegate all the technical proofs for this section to Appendix C.

1) The Weyl–Heisenberg Case:

a) Critical value for the product $p_0 \cdot q_0$: In the Weyl–Heisenberg case there exists a critical value, 2π, for the product $p_0 \cdot q_0$. This is already illustrated by the construction in Section II-A-1a), which only works if $p_0 \cdot q_0 < 2\pi$. The following theorem states that at the critical value $p_0 \cdot q_0 = 2\pi$, only functions g that are either not very smooth or do not decay very fast can give rise to a frame.

Theorem 2.3 (Balian–Low–Coifman–Semmes): Choose $g \in L^2(\mathbb{R})$, $p_0 > 0$. If the g_{mn} associated with $g, p_0, q_0 = 2\pi / p_0$, constitute a frame, then either $xg \notin L^2$ or $g' \notin L^2$.

Remark: This theorem was first published by R. Balian [51] for the case where the g_{mn} are an orthonormal basis. In the 1985 Festschrift for the 60th birthday of the physicist G. Chew, F. Low also discusses this problem [52]. He gives (independently) essentially the same proof as in [51]. The proof presented in [51], [52] contains a technical gap that was filled by R. Coifman and S. Semmes. The proof extends easily from the basis case to the frame case. We give here the proof as completed by R. Coifman and S. Semmes.

The proof of Theorem 2.3 uses the Zak transform U_Z. This transform maps $L^2(\mathbb{R})$ unitarily onto $L^2([0,1]^2)$; it was first systematically studied by J. Zak [54], in connection with solid state physics. Some of its properties were known long before Zak's work, however. In [55] the same transform is called the Weil–Brezin map, and it is claimed that the same transform was already known to Gauss. It was also used by Gel'fand (see, e.g., Ch. XIII in [56]). J. Zak seems, however, to have been the first to recognize it as the versatile tool it is and to have studied it systematically. It has many very interesting properties; its applications range from solid state physics to the derivation, in [57], of new relationships between Jacobi's theta functions. Before embarking on the proof of Theorem 2.3, we briefly review the definition and some of the properties of U_Z. The Zak transform U_Z is defined by

$$(U_Z f)(t,s) = \lambda^{1/2} \sum_{l \in \mathbb{Z}} e^{2\pi i t l} f(\lambda(s-l)) \qquad (2.3.1)$$

where the parameter $\lambda > 0$ can be adjusted to the problem at hand. For the proof of Theorem 2.3 we shall take $\lambda = q_0$.

Strictly speaking, the definition (2.3.1) does only make sense for the subspace of the L^2-functions for which the series converges. It is, however, easy to extend (2.3.1) from those functions for which it is well-defined, to all of $L^2(\mathbb{R})$. One way of doing this is to observe that the images under (2.3.1) of the orthonormal basis e_{mn} of $L^2(\mathbb{R})$,

$$e_{mn}(x) = \lambda^{-1/2} e^{2\pi i m x/\lambda} \chi_{[0,\lambda)}(x - n\lambda)\,(m, n \in \mathbb{Z})$$

are well-defined, and constitute again an orthonormal basis of $L^2([0,1]^2)$,

$$(U_Z e_{mn})(t,s) = e^{-2\pi i t n} e^{2\pi i m s}.$$

It follows that (2.3.1) defines a unitary map from $L^2(\mathbb{R})$ to $L^2([0,1]^2)$. On the other hand, (2.3.1) can also be extended to values of (t,s) outside $[0,1]^2$. For $f \in L^2(\mathbb{R})$, the resulting function is in $L^2_{loc}(\mathbb{R}^2)$, and satisfies

$$(U_Z f)(t+1,s) = (U_Z f)(t,s)$$
$$(U_Z f)(t,s+1) = e^{2\pi i t}(U_Z f)(t,s) \qquad (2.3.2)$$

almost everywhere (a.e.) with respect to Lebesgue measure on $\mathbb{R}^2$. A rather remarkable consequence of (2.3.2) is the following. Suppose that f is such that $U_Z f$ is continuous (on $\mathbb{R}^2$). Then $U_Z f$ must necessarily have a zero in $[0,1]^2$. The proof of this fact, first pointed out in [58], is quite simple. The presentation given here is borrowed from [59]. If $U_Z f$ had no zero in $[0,1]^2$, then $\log U_Z f$ would be a univalued, continuous function on $[0,1]^2$ extending to a continuous function in $\mathbb{R}^2$ by (2.3.2). On the other hand, (2.3.2) implies that

$$(\log U_Z f)(t+1,s) = (\log U_Z f)(t,s) + 2\pi i k$$
$$(\log U_Z f)(t,s+1) = (\log U_Z f)(t,s) + 2\pi i t + 2\pi i l \qquad (2.3.3)$$

where k, l are integers, independent of t, s because of the continuity of $\log U_Z f$. But (2.3.3) leads immediately to the following contradiction:

$$\begin{aligned} 0 &= (\log U_Z f)(0,0) - (\log U_Z f)(1,0) \\ &\quad + (\log U_Z f)(1,0) - (\log U_Z f)(1,1) \\ &\quad + (\log U_Z f)(1,1) - (\log U_Z f)(0,1) \\ &\quad + (\log U_Z f)(0,1) - (\log U_Z f)(0,0) \\ &= -2\pi i k - 2\pi i - 2\pi i l + 2\pi i k + 2\pi i l = -2\pi i \neq 0. \end{aligned}$$

This proves that we started from a false premise, i.e., that $U_Z f$ has a zero in $[0,1]^2$. A similar argument proves Theorem 2.3. The above is only one of the many properties of the Zak transform. For more of these properties, and interesting applications of the Zak transform to signal analysis, we urge the reader to consult [60].

The images, under the Zak transform, of functions g_{mn} (with $p_0 \cdot q_0 = 2\pi$), are remarkably simple. One easily checks that, for the choice $\lambda = q_0$,

$$(U_Z g_{mn})(t,s) = e^{2\pi ims} e^{-2\pi itn} (U_Z g)(t,s). \quad (2.3.4)$$

An immediate consequence of (2.3.4) is

$$\sum_{m,n} |\langle g_{mn}, f\rangle|^2 = \sum_{m,n} |\langle U_Z g_{mn}, U_Z f\rangle|^2$$
$$= \sum_{m,n} \left| \int_0^1 dt \int_0^1 ds\, e^{-2\pi ims} e^{2\pi itn} \cdot (U_Z g)^*(t,s)(U_Z f)(t,s) \right|^2$$
$$= \int_0^1 dt \int_0^1 ds |(U_Z g)(t,s)|^2 |(U_Z f)(t,s)|^2.$$

It follows that the g_{mn} constitute a frame, with frame bounds A, B, if and only if, for $(t,s) \in [0,1]^2$ a.e. (and hence, by (2.3.2), for $(t,s) \in \mathbb{R}^2$ a.e.)

$$A^{1/2} \le |(U_Z g)(t,s)| \le B^{1/2}.$$

This is another crucial ingredient of the proof of Theorem 2.3, to which we now turn.

Proof of Theorem 2.3: Suppose that the g_{mn} constitute a frame, and assume also that xg, $g' \in L^2(\mathbb{R})$. We want to show that this leads to a contradiction. Define $G(t,s) = (U_Z g)(t,s)$. Since the g_{mn} constitute a frame, we have

$$a \le |G(t,s)| \le b \quad (2.3.5)$$

for some $a > 0$, $b < \infty$, and for $(t,s) \in \mathbb{R}^2$, a.e. For compactly supported f one finds

$$[U_Z(xf)](t,s)$$
$$= q_0^{3/2} s (U_Z f)(t,s) - q_0^{3/2} \sum_l e^{2\pi itl} lf(q_0(s-l))$$
$$= q_0^{3/2} s (U_Z f)(t,s) - q_0^{3/2} (2\pi i)^{-1} \partial_t (U_Z f)(t,s).$$

This shows that $xg \in L^2$ implies $\partial_t G = \partial_t (U_Z g) \in L^2_{loc}(\mathbb{R}^2)$. Similarly $g' \in L^2$ implies $\partial_s G \in L^2_{loc}(\mathbb{R}^2)$.

If the square integrability of the partial derivatives of G implied that G was continuous (which it does not, since we are in more than one dimension), then the proof would be finished. By the argument given before the proof, $\inf |G|$ would then be zero, which is in contradiction with (2.3.5). This is essentially the argument of Balian in [51], where the implicit assumption that G is continuous seems to be made. Lemma 2.4, due to R. Coifman and S. Semmes, which we state below, shows how the boundary conditions (2.3.2), together with the bounds (2.3.5), lead to a contradiction, without assuming continuity for G. The main idea is to use an averaged version of G. This averaged function is automatically continuous, and, if the averaging is done on a small enough scale, close enough to G so that the properties inherited from (2.3.5) and (2.3.2) still lead to a contradiction. This then proves Theorem 2.3. □

Lemma 2.4: Assume that G is a bounded function on $\mathbb{R}^2$ that is locally square integrable and which satisfies

$$G(t+1,s) = G(t,s)$$
$$G(t,s+1) = e^{2\pi it} G(t,s).$$

If both $\partial_t G$ and $\partial_s G$ are locally square integrable, then $\text{ess inf}_{0 \le t,s \le 1} |G(t,s)| = 0$.

Here the "essential infimum" of a measurable function f is defined by

$$\text{ess inf} f = \inf\{\lambda; |\{x; f(x) \le \lambda\}| > 0\}$$

where $|V|$ denotes the Lebesgue measure of the set V. This definition avoids values taken by f on sets of measure zero. For instance, for $f(x) = 1$ for $x \ne 0$, $f(0) = 0$, one has $\inf f = 0$, but $\text{ess inf} f = 1$.

Proof of Lemma 2.4: This proof is rather technical; it is given in Appendix B. □

Note added in proof: After this paper was written, G. Battle produced a very elegant new proof of Theorem 2.3 which avoids the use of the Zak transform [69]. Battle's paper only treats the case where the g_{mn} are an orthonormal basis (as did the original papers by Balian and Low). His argument was extended to frames by A. J. E. M. Janssen and I. Daubechies [70].

For the critical value $p_0 \cdot q_0 = 2\pi$ we see thus that not much regularity and/or decay can be expected from a function g leading to a frame. This is in marked contrast with the case $p_0 \cdot q_0 < 2\pi$. In that case, as shown in Section II-B-1a, there even exist C^∞-functions g with compact support such that the associated g_{mn} constitute a tight frame.

This critical value $p_0 \cdot q_0 = 2\pi$, has a physical meaning. As shown by Fig. 1(a), $(p_0 \cdot q_0)^{-1}$ is the density, in phase space, of the discrete lattice of functions g_{mn}. The density $(2\pi)^{-1}$ is nothing but the Nyquist density; it is well-known in information theory that time-frequency densities at least as high as the Nyquist density are needed for a full transmission of information. It is therefore not surprising to encounter this same critical value here. One encounters the same argument in quantum physics, often used in semiclassical approximations, where a complete set of independent states (i.e., a set of linearly independent functions in $L^2(\mathbb{R})$ whose linear combinations span a dense subspace in $L^2(\mathbb{R})$) heuristically corresponds to a density $(2\pi)^{-1}$ in phase space. In other words, every state "occupies" a "cell" of area 2π in phase space.

For $p_0 \cdot q_0 > 2\pi$, the intuition from physics or information theory suggests that the associated phase space lattice is "too loose," i.e., that the g_{mn} cannot span the whole Hilbert space. More precisely, we expect that for *any* $g \in L^2(\mathbb{R})$, there exists at least one $f \in L^2(\mathbb{R})$, $f \neq 0$, such that $\langle g_{mn}, f \rangle = 0$ for all $m, n \in \mathbb{Z}$. This is indeed the case.

If $p_0 \cdot q_0 / 2\pi > 1$ is rational, then the following argument, again using the Zak transform, shows how to construct a function f.

By a dilation argument we can restrict ourselves to the case $p_0 = 2\pi$, $q_0 = K/L$, with $K, L \in \mathbb{N}$, $K > L > 0$. Let F, G, G_{mn} be the Zak transforms of respectively f, g, g_{mn}, as defined by (2.3.1), where we take $\lambda = 1$. Then

$$G_{mn}(t,s) = e^{2\pi i m s} G\left(t, s - n\frac{K}{L}\right).$$

For $n = kL + l$, with $l, k \in \mathbb{Z}$, $0 \leq l < L$, this reduces to

$$G_{mn}(t,s) = e^{2\pi i m s} e^{-2\pi i k K t} G\left(t, s - \frac{l}{L}K\right)$$

where we have used (2.3.2). The function F will be orthogonal in $L^2([0,1]^2)$ to each G_{mn} if, for all $s \in [0,1]$ and for all $k, l \in \mathbb{Z}$ with $0 \leq l < L$,

$$\int_0^1 dt\, \overline{F(t,s)} e^{-2\pi i k K t} G\left(t, s - \frac{l}{L}K\right) = 0. \quad (2.3.6)$$

We can rewrite (2.3.6) as

$$\int_0^{1/K} dt\, e^{-2\pi i k K t} \sum_{m=0}^{K-1} G\left(t + \frac{m}{K}, s - \frac{l}{L}K\right) \overline{F\left(t + \frac{m}{K}, s\right)} = 0. \quad (2.3.7)$$

We are thus led to the linear system of equations

$$\sum_{m=0}^{K-1} A_{lm}(t,s) \phi_m(t,s) = 0 \qquad 0 \leq l < L \quad (2.3.8)$$

where

$$A_{lm}(t,s) = G\left(t + \frac{m}{K}, s - \frac{l}{L}K\right) \quad (2.3.9)$$

$$\phi_m(t,s) = \overline{F\left(t + \frac{m}{K}, s\right)}. \quad (2.3.10)$$

Since the system (2.3.8) has L equations for $K > L$ unknowns, it always has a nonzero solution, for every pair $(t,s) \in [0, 1/K] \times [0,1]$. The $\phi_m(t,s)$ solving (2.3.8) can, moreover, be chosen in $L^2([0,1/K] \times [0,1])$. One way of doing this is to choose a fixed $u \in \mathbb{C}^K$, and define $\phi_m(t,s) = \lim_{\tau \to \infty} u_m(t,s;\tau)$, where

$$u(t,s;\tau) = \exp[-\tau A^*(t,s) A(t,s)] u.$$

Clearly $\|u(t,s;\tau)\|^2 \leq \|u\|^2$; hence $\sum_{m=1}^K |\phi_m(t,s)|^2 \leq \|u\|^2$. On the other hand, since all the A_{lm} are in $L^2([0,1/K] \times [0,1])$, the $u(t,s;\tau)$ are clearly measurable in t, s; as pointwise limits of measurable functions, the ϕ_m are measurable too. Putting the ϕ_m together according to (2.3.10) then defines a function F in $L^2([0,1]^2)$ which is orthogonal to all the G_{mn}. Note that while F may be zero a.e. for some choices of u, it is impossible that the functions $F(u_k)$ associated with K linearly independent vectors $u_1, \cdots, u_K$ in $\mathbb{C}^K$ all be zero a.e., since this would mean that $A^*(t,s)A(t,s) > 0$ a.e. in t, s, which contradicts rank $[A(t,s)] < K$. For appropriately chosen $u \in \mathbb{C}^K$, the previous construction leads thus to a nontrivial function $F \in L^2([0,1]^2)$ orthogonal to all the G_{mn}.

This argument does not work if $p_0 \cdot q_0 / 2\pi$ is irrational. It is nevertheless still true that the g_{mn} do not span $L^2(\mathbb{R})$, whatever g in $L^2(\mathbb{R})$ is chosen, even for irrational values of $p_0 \cdot q_0 / 2\pi$. The only proof that I know of this fact uses von Neumann algebras; it was pointed out to me by R. Howe and T. Steger. The proof consists in the computation of the coupling constant of the von Neumann algebra spanned by the Weyl operators $(W(mp_0, nq_0);\ m, n \in \mathbb{Z})$. The coupling constant for this von Neumann algebra was computed explicitly by M. Rieffel [49]; for $p_0 \cdot q_0 > 2\pi$ it is larger than 1, which implies that the von Neumann algebra has no cyclic element. This means that for any $g \in L^2(\mathbb{R})$, the closed linear span of the g_{mn} is a proper subspace of $L^2(\mathbb{R})$, which was the desired result. Unfortunately, this proof does not seem very illuminating from the signal analyst's point of view.

Note added in proof: Recently H. Landau [71] found a different, intuitively much more appealing argument to prove that the g_{mn} cannot constitute a frame if $p_0 \cdot q_0 > 2\pi$. His proof works for all g which are reasonably "nice" (decaying in both time and frequency).

This concludes what we have to say about the critical value $(p_0 \cdot q_0 = 2\pi)$ in the Weyl–Heisenberg case. We shall always assume $p_0 \cdot q_0 \leq 2\pi$ in what follows.

b) Ranges for the parameters p_0, q_0, and estimates for the frame bounds: In the preceding subsection we have excluded parameters p_0, q_0 for which $p_0 \cdot q_0 > 2\pi$. Even if $p_0 \cdot q_0 \leq 2\pi$, however, we do not automatically have a frame for arbitrary functions g. When, for example, $g(x) = 1$ for $0 \leq x \leq 1$, and $g(x) = 0$, otherwise, the g_{mn} cannot constitute a frame if $q_0 > 1$, even if $p_0 \cdot q_0 < 2\pi$. Indeed, for $q_0 > 1$, one finds that $\langle g_{mn}, f \rangle = 0$, for all $m, n \in \mathbb{Z}$ if the support of $f \subset [1, q_0]$, independently of the choice of p_0. This is thus a case where an inappropriate choice of q_0 excludes the possibility of a frame, for all values of p_0.

The theorem below gives sufficient conditions on g and q_0 under which this cannot happen, i.e., there always exist some $p_0 > 0$ (in fact, a whole interval) leading to a frame.

Theorem 2.5: If

1) $$m(g;q_0) = \operatorname*{ess\,inf}_{x \in [0,q_0]} \sum_n |g(x - nq_0)|^2 > 0 \quad (2.3.11)$$

2) $$M(g;q_0) = \operatorname*{ess\,sup}_{x \in [0,q_0]} \sum_n |g(x - nq_0)|^2 < \infty \quad (2.3.12)$$

and

3) $$\sup_{s \in \mathbb{R}} \left[(1+s^2)^{(1+\epsilon)/2} \beta(s)\right] = C_\epsilon < \infty \qquad \text{for some } \epsilon > 0$$

where

$$\beta(s) = \sup_{x \in [0,q_0]} \sum_{n \in \mathbb{Z}} |g(x - nq_0)|\,|g(x + s - nq_0)| \quad (2.3.13)$$

then there exists a $P_0^c > 0$ such that

$$\forall p_0 \in (0, P_0^c): \text{the } g_{mn} \text{ associated with } g, p_0, q_0 \text{ are a frame}$$
$$\forall \delta > 0: \exists p_0 \text{ in } [P_0^c, P_0^c + \delta] \text{ such that the } g_{mn} \text{ associated to } g, p_0, q_0 \text{ are not a frame.}$$

Proof: see Appendix C. □

The conditions (2.3.11)–(2.3.13) may seem rather technical. They are, in fact, extremely reasonable. Condition (2.3.11) specifies that the collection of g and its translates should not have any "gaps." This already excludes the example given at the start of this subsection. The conditions (2.3.12), (2.3.13) are satisfied if g has sufficient decay at ∞, in particular, if $|g(x)| \le C[1 + x^2]^{-3/2}$.

Note that both (2.3.11) and (2.3.12) are necessary conditions. If (2.3.11) is not satisfied, then for every $\epsilon > 0$ one can find a nonzero f in $L^2(\mathbb{R})$ such that

$$\sum_{m,n} |\langle g_{mn}, f\rangle|^2 \le \epsilon \|f\|^2$$

which means there exists no nonzero lower frame bound A for the g_{mn}. Similarly there exists no finite upper frame bound B for the g_{mn} if (2.3.12) is not satisfied.

Remarks

1) At the end of the preceding subsection we showed that the g_{mn} can constitute a frame only if $p_0 \cdot q_0 \le 2\pi$. Hence necessarily $P_0^c \le 2\pi/q_0$.
2) The set $\{p_0$; the g_{mn} associated to g, p_0, q_0 constitute a frame$\}$ with g and q_0 fixed need not be connected. It is possible that this set contains values of p_0 larger than P_0^c. An example is given by the following construction. Let ϕ be a C^∞ function with support $[0, 1/3]$ such that $|\phi|$ has no zeros in $(0, 1/3)$. Define g by

$$\hat{g}(y) = \begin{cases} 0, & y \le 0 \\ \phi(y), & 0 \le y \le 1/3 \\ \phi(y - 1/3), & 1/3 \le y \le 2/3 \\ \phi(y - 2/3), & 2/3 \le y \le 1 \\ 0, & y \ge 1. \end{cases}$$

Take $q_0 = 2\pi$. Then, since support $\hat{g} = [0, 1]$,

$$\sum_{m,n} |\langle g_{mn}, f\rangle|^2 = \int dy \sum_m |\hat{g}(y - mp_0)|^2 |\hat{f}(y)|^2.$$

This implies that the g_{mn} constitute a frame if and only if $\inf \sum_m |\hat{g}(y - mp_0)|^2 > 0$. Consequently $P_0^c = 1/3$, while the set of all p_0 leading to a frame is $(0, 1/3) \cup (1/3, 2/3)$.

For reasonably smooth g with sufficient decay at ∞, the constants $m(g; q_0), M(g; q_0)$ and the function $\beta(s)$ can easily be computed numerically. These constants are useful in estimations of the frame bounds A, B, as the following theorem shows.

Theorem 2.6: Assume that (2.3.11), (2.3.12), (2.3.13) are satisfied. Define

$$p_0^c = \inf\left\{p_0 \,\middle|\, \sum_{k=1}^{\infty}\left[\beta\left(\frac{2\pi}{p_0}k\right)\beta\left(-\frac{2\pi}{p_0}k\right)\right]^{1/2} \ge \frac{1}{2}m(g; q_0)\right\}.$$

Then $p_0^c \le P_0^c$, and for $0 < p_0 < p_0^c$, the following estimates for the frame bounds A, B holds

$$A \ge \frac{2\pi}{p_0}\left(m(g; q_0) - 2\sum_{k=1}^{\infty}\left[\beta\left(\frac{2\pi}{p_0}k\right)\beta\left(-\frac{2\pi}{p_0}k\right)\right]^{1/2}\right)$$
$$B \le \frac{2\pi}{p_0}\left(M(g; q_0) + 2\sum_{k=1}^{\infty}\left[\beta\left(\frac{2\pi}{p_0}k\right)\beta\left(-\frac{2\pi}{p_0}k\right)\right]^{1/2}\right).$$

Proof: see Appendix C. □

These bounds for A and B can be improved by the following observation. If the $g_{mn}(x) = e^{imp_0x}g(x - nq_0)$ constitute a frame, then so do the

$$(g_{mn})^\wedge(\zeta) = e^{-inq_0\zeta}\hat{g}(\zeta - mp_0),$$

where $\hat{g}$ denotes the Fourier transform of g. It follows that

$$A \ge \max\left(\frac{2\pi}{p_0}\left\{m(g; q_0) - 2\sum_{k=1}^{\infty}\left[\beta\left(\frac{2\pi}{p_0}k\right)\beta\left(-\frac{2\pi}{p_0}k\right)\right]^{1/2}\right\},\right.$$
$$\left.\frac{2\pi}{q_0}\left\{m(\hat{g}; p_0) - 2\sum_{k=1}^{\infty}\left[\hat{\beta}\left(\frac{2\pi}{q_0}k\right)\hat{\beta}\left(-\frac{2\pi}{q_0}k\right)\right]^{1/2}\right\}\right) \tag{2.3.14}$$

$$B \le \min\left(\frac{2\pi}{p_0}\left\{M(g; q_0) + 2\sum_{k=1}^{\infty}\left[\beta\left(\frac{2\pi}{p_0}k\right)\beta\left(-\frac{2\pi}{p_0}k\right)\right]^{1/2}\right\}\right.$$
$$\left.\frac{2\pi}{q_0}\left\{M(\hat{g}; p_0) + 2\sum_{k=1}^{\infty}\left[\hat{\beta}\left(\frac{2\pi}{q_0}k\right)\hat{\beta}\left(-\frac{2\pi}{q_0}k\right)\right]^{1/2}\right\}\right). \tag{2.3.15}$$

Here $M(\hat{g}; p_0)$, $m(\hat{g}; p_0)$ and $\hat{\beta}(s)$ are the obvious extensions of the quantities in (2.3.11), (2.3.12), and (2.3.13). For instance,

$$\hat{\beta}(s) = \sup_{\xi \in [0, p_0]} \sum_{m \in \mathbb{Z}} |\hat{g}(\xi + mp_0)||\hat{g}(\xi + s + mp_0)|.$$

Similarly, one has the following better lower bound for P_0^c,

$$P_0^c \ge p_0^c = \inf\left\{p_0 \,\middle|\, 2\sum_{k=1}^{\infty}\left[\beta\left(\frac{2\pi}{p_0}k\right)\beta\left(-\frac{2\pi}{p_0}k\right)\right]^{1/2} \ge m(g, q_0) \text{ and } 2\sum_{k=1}^{\infty}\left[\hat{\beta}\left(\frac{2\pi}{q_0}k\right)\hat{\beta}\left(-\frac{2\pi}{q_0}k\right)\right]^{1/2} \ge m(\hat{g}; p_0)\right\}. \tag{2.3.16}$$

C) Examples

i) The Gaussian case

In this case

$$g(x)=\pi^{-1/4}e^{-x^2/2}. \tag{2.3.17}$$

This is the basic function for the so-called "canonical coherent states" in physics [5]. It is also the basic function chosen by Gabor [1] in the definition of his expansion. In the notations used in this paper, Gabor's approach amounts to writing an expansion with respect to the g_{mn} associated to g, p_0, q_0, where $p_0 \cdot q_0 = 2\pi$. This choice seems very natural from the point of view of information theory, since it corresponds exactly to the Nyquist density. However, since both xg and g' are square integrable in this case, Theorem 2.3 tells us that the g_{mn} cannot possibly constitute a frame. In fact, the Zak transform $U_Z g$ of g (see Section II-C-1a) can be constructed explicitly in this case; it is one of Jacobi's theta-functions, and it

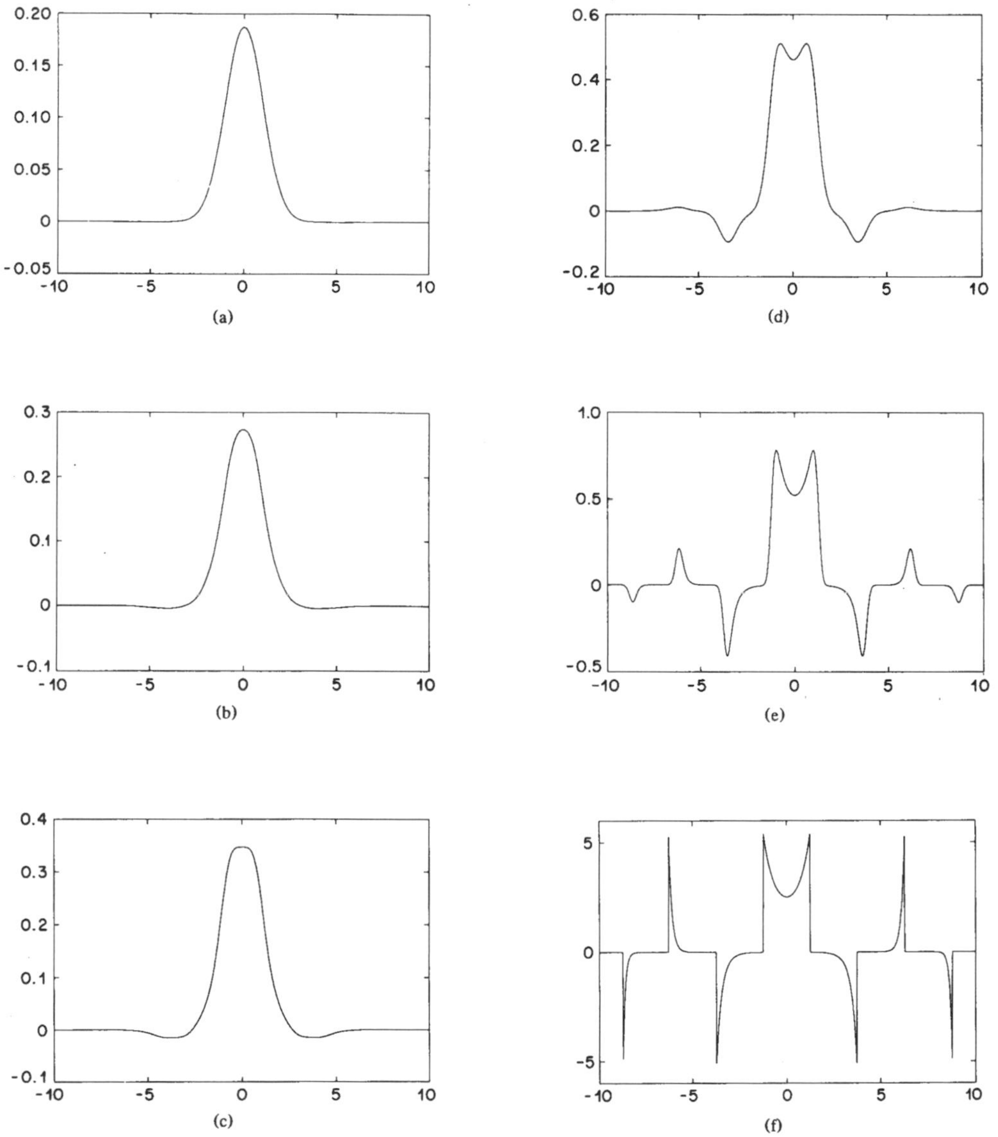

Fig. 4. Function $\tilde{g}$ for different values of $p_0 \cdot q_0$. We have taken $g(x)=\pi^{-1/4}\exp(-x^2/2)$, and $p_0 = q_0$ in every case. (a) $p_0 \cdot q_0 = \pi/2$. (b) $p_0 \cdot q_0 = 3\pi/4$. (c) $p_0 \cdot q_0 = \pi$. (d) $p_0 \cdot q_0 = 3\pi/2$. (e) $p_0 \cdot q_0 = 1.9\pi$. (f) $p_0 \cdot q_0 = 2\pi$. In this last case $\mathbb{T}$ does not have bounded inverse, and $\tilde{g} \notin L^2$. This function $\tilde{g}$ was first computed by Bastiaans [2].

has a zero in $(1/2, 1/2)$ [2], [3], [60], [63]. The operator $\mathbb{T}$ of Section II-A is unitarily equivalent with multiplication by $|U_Z g|^2$ on $L^2([0,1]^2)$, and is therefore one-to-one, but it has an unbounded inverse, i.e.,

$$\inf_{f \in L^2(\mathbb{R})} \|f\|^{-2} \sum_{m,n} |\langle h_{mn}, f \rangle|^2 = 0.$$

One can still write an expansion of type

$$f = \sum_{m,n} \tilde{g}_{mn} \langle g_{mn}, f \rangle$$

but this expansion has, in general, bad convergence properties because $\tilde{g} \notin L^2(\mathbb{R})$ (since $U_Z \tilde{g} = U_Z(\mathbb{T}^{-1} g) = |U_Z g|^{-2} U_Z g$ has a pole at $(1/2, 1/2)$, and is therefore not in $L^2([0,1]^2)$). It is this fact that makes the expansion associated to the Gabor wave functions unstable [2], [3], [60]. The function $\tilde{g}$ was explicitly constructed in [2]; it turns out to be discontinuous as well as nonsquare-integrable (see Fig. 4(f)). Explicit computation of $\tilde{g}$ (via the Zak transform), for $p_0 = q_0 = \sqrt{2\pi}$, leads to [2]

$$\tilde{g}(x) = C e^{x^2/2} \sum_{n + 1/2 \geq x/\sqrt{2\pi}} (-1)^n e^{-\pi(n+1/2)^2}$$

where C is a normalization constant.

For g Gaussian, as in (2.3.17), the fact that the g_{mn} do not span $L^2(\mathbb{R})$ if $p_0 \cdot q_0 > 2\pi$, has been known for quite a while [61], [62]. The proofs in [61], [62] use entire function techniques. These techniques do not work, however, for non-Gaussian g.

Our first table of numerical results, for Gaussian g, lists p_0^c, for different values of q_0 (see Table I). It turns out that p_0^c is always very close to $2\pi/q_0$; for this reason we have tabulated $(2\pi/q_0 \cdot p_0^c) - 1$, rather than p_0^c itself. The difference $(2\pi/q_0 \cdot p_0^c) - 1$ is largest for $q_0 = \sqrt{2\pi}$, where it is about 4×10^{-3}.

TABLE I*

q_0	$(2\pi/q_0 \cdot p_0^c) - 1$
1.0	6×10^{-8}
1.5	3×10^{-7}
2.0	2×10^{-4}
2.5	4×10^{-3}
3.0	4×10^{-4}
3.5	2×10^{-5}
4.0	8×10^{-7}

*The deviation, for $g(x) = \pi^{-1/4} \exp(-x^2/2)$, of the estimated value p_0^c from the optimal value $2\pi/q_0$ (see text).

Note that $P_0^c \leq 2\pi/q_0$. The numerical results show therefore that in this case the estimate p_0^c is remarkably close to the true critical value P_0^c. In fact they suggest the conjecture that $P_0^c = 2\pi/q_0$, for all $q_0 > 0$, in this case. I believe that this conjecture holds for all positive functions g with positive Fourier transform.

Next we list the estimates (2.3.14) and (2.3.15) on the frame bounds for a few values of q_0, $p_0 < p_0^c$. We also tabulate the corresponding B/A, and $r = (B/A - 1)/(B/A + 1)$. This parameter measures how snug the frame is, i.e., how close it is to a tight frame; as explained in Section II-A, this parameter is essential if one wants to apply the inversion formula as indicated in Section II-A. We have grouped together different values of q_0, p_0 corresponding to the same value of $q_0 \cdot p_0$. If $q_0 \cdot p_0 = 2\pi/k$, with k integer, then a different method, based on the Zak transform, enables us to write explicit, exact expressions for the frame bounds A and B. With U_Z defined as in Section II-C-1a (see (2.3.1), with $\lambda = q_0$), one finds indeed that, for $m \in \mathbb{Z}$, $m = m'k + r, m'$, $r \in \mathbb{Z}$, $0 \leq r < k$,

$$(U_Z g_{mn})(t,s) = (U_Z g_{m'k+rn})(t,s) = e^{2\pi irn/k} e^{-2\pi itn} e^{2\pi im's} (U_Z g^r)(t,s)$$

where we have used $p_0 = 2\pi/q_0$, and where

$$g^r(x) = e^{2\pi irxq_0/k} g(x).$$

Hence

$$(U_Z g^r)(t,s) = (U_Z g)\left(t - \frac{r}{k} s\right).$$

This is entirely similar to computations made in Section II-C-1a; see also [45]. Consequently, as in Section II-C-1a (see also [16], [45])

$$\sum_{m,n} |\langle g_{mn}, f \rangle|^2 = \int_0^1 dt \int_0^1 ds \left[\sum_{r=0}^{k-1} \left| (U_Z g)\left(t - \frac{r}{k}, s\right) \right|^2 \right] |(U_Z f)(t,s)|^2$$

which implies

$$A = \inf \left[\sum_{r=0}^{k-1} \left| (U_Z g)\left(t - \frac{r}{k}, s\right) \right|^2 \right]$$

$$B = \sup \left[\sum_{r=0}^{k-1} \left| (U_Z g)\left(t - \frac{r}{k}, s\right) \right|^2 \right].$$

To find A and B, in the case where $2\pi/(p_0 \cdot q_0)$ is an integer, it suffices therefore to compute $U_Z g$ and these two extrema.

In Table II next we list the estimates, given by (2.3.14), (2.3.15) for A, B, B/A and r, for a few values of q_0, in the cases $p_0 \cdot q_0 = \pi/2$, $3\pi/4$, π, $3\pi/2$, and 1.9π. In the cases $p_0 \cdot q_0 = \pi/2$ and π we also list the exact values for A, B, denoted by A_{exact}, B_{exact}. For values of p_0 close to the critical value $2\pi/q_0$ (see the case $q_0 \cdot p_0 = 1.9\pi$ in Table I), the ratio B/A becomes very large, as was to be expected. The convergence of the formula for $\tilde{g}$, measured by r, will be very slow. For $q_0 = 2.0$, $p_0 = \pi/q_0 = \pi/2$, the ratio r is already of the order of 0.2, while $q_0 = 1.5$, $p_0 = \pi/3$, $q_0 p_0 = \pi/2$ leads to $r = 0.025$. In the latter case a few terms will suffice to obtain an accuracy of 10^{-6} is the computation of $\tilde{g}$. This shows that very good frames can be obtained with lattice spacings that are not very small.

In the two cases where the exact values of A, B can be computed by other means ($q_0 \cdot p_0 = \pi/2$ and $q_0 \cdot p_0 = \pi$), the estimates for A and B, as calculated from (2.3.14) and

TABLE II
VALUES FOR THE FRAME BOUNDS A, B, THEIR RATIO B/A, AND THE CONVERGENCE FACTOR $r=(B/A-1)/(B/A+1)$, FOR THE CASE $g(x)=\pi^{-1/4}\exp(-x^2/2)$, FOR DIFFERENT VALUES OF q_0, p_0*

			$q_0 \cdot p_0 = \pi/2$			
q_0	Aa	A_{exact}	B	B_{exact}	B/A	$r=(B-A)/(B+A)$
0.5	1.203	1.221	7.091	7.091	5.896	0.710
1.0	3.853	3.854	4.147	4.147	1.076	0.037
1.5	3.899	3.899	4.101	4.101	1.052	0.025
2.0	3.322	3.322	4.679	4.679	1.408	0.170
2.5	2.365	2.365	5.664	5.664	2.395	0.411
3.0	1.427	1.427	6.772	6.772	4.745	0.652

			$q_0 \cdot p_0 = 3\pi/4$	
q_0	A	B	B/A	$r=(B-A)/(B+A)$
1.0	1.769	3.573	2.019	0.338
1.5	2.500	2.833	1.133	0.062
2.0	2.210	3.124	1.414	0.172
2.5	1.577	3.776	2.395	0.411
3.0	0.951	4.515	4.745	0.652

				$q_0 \cdot p_0 = \pi$		
q_0	A	A_{exact}	B	B_{exact}	B/A	$r=(B-A)/(B+A)$
1.0	0.601	0.601	3.546	3.546	5.901	0.710
1.5	1.519	1.540	2.482	2.482	1.635	0.241
2.0	1.575	1.600	2.425	2.425	1.539	0.212
2.5	1.172	1.178	2.843	2.843	2.426	0.416
3.0	0.713	0.713	3.387	3.387	4.752	0.652

			$q_0 \cdot p_0 = 3\pi/2$	
q_0	A	B	B/A	$r=(B-A)/(B+A)$
1.0	0.027	3.545	130.583	0.985
1.5	0.342	2.422	7.082	0.753
2.0	0.582	2.089	3.592	0.564
2.5	0.554	2.123	3.834	0.586
3.0	0.393	2.340	5.953	0.712
3.5	0.224	2.656	11.880	0.845
4.0	0.105	3.014	28.618	0.932

			$q_0 \cdot p_0 = 1.9\pi$	
q_0	A	B	B/A	$r=(B-A)/(B+A)$
1.5	0.031	2.921	92.935	0.979
2.0	0.082	2.074	25.412	0.924
2.5	0.092	2.021	22.004	0.913
3.0	0.081	2.077	25.668	0.925
3.5	0.055	2.218	40.455	0.952
4.0	0.031	2.432	79.558	0.975

*Where possible ($q_0 \cdot p_0 = \pi/2$, $q_0 \cdot p_0 = \pi$) the estimates for A, B are compared with the exact values (computed via the Zak transform; see text).

(2.3.15), turn out to be remarkably close to the true values, giving deviations of at most a few percent on A, and less than 0.1% on B. In the next example we shall find similar orders of magnitude for the deviation of our estimates for A, B with respect to the exact values. The formulas (2.3.14) and (2.3.15) seem thus to give quite good results, despite the rather brutal estimating methods used.

For every one of the values of the product $q_0 \cdot p_0$ in Table II we have also computed $\tilde{g}$ by means of the inversion formula (2.1.5). These functions $\tilde{g}$ are plotted in Fig. 4, for the choice $q_0 = p_0$.

For $p_0 = q_0 = (\pi/2)^{1/2}$, we know, from Table II, that the frame is snug ($r \simeq 0.02$ in this case), i.e., that $\mathbb{T}$ is very close to a multiple of the unit operator. Consequently $\tilde{g} = \mathbb{T}^{-1}g$ is very close to a (scaled) Gaussian. For increasing $p_0 = q_0$, several things happen to $\tilde{g}$. The decrease of both frames bounds A and B, which reflects the decrease in the "oversampling ratio" $2\pi(p_0 q_0)^{-1}$, causes the amplitude of $\tilde{g}$ to increase. Moreover, the frame becomes less and less snug (for $p_0 = q_0 = (1.9\pi)^{1/2}$, one even has $r > 0.9$), causing $\tilde{g}$ to deviate more and more from a Gaussian. In all these examples (Figs. 4(a)–4(e)), the function $\tilde{g}$ remains square integrable, however. One can even show that it remains C^∞, with fast decay. This is no longer the case if $p_0 = q_0 = (2\pi)^{1/2}$. In this limiting case, where the g_{mn} no longer constitute a frame, one can still construct $\tilde{g} = \mathbb{T}^{-1}g$, but, as previously shown, $\tilde{g}$ is no longer in L^2. This singular $\tilde{g}$ was first plotted by Bastiaans [2]; we have replotted it here in Fig. 4(f). One clearly sees how the regular $\tilde{g}$, for lower values of $p_0 \cdot q_0$, approaches the singular limiting function as $p_0 = q_0$ increases towards $(2\pi)^{1/2}$.

ii) The exponential case: We take $g(x) = e^{-|x|}$. In this case $m(g; q_0)$ and $M(g; q_0)$ can be calculated explicitly. One finds

$$m(g; q_0) = (\sinh q_0)^{-1}$$

$$M(g; q_0) = \coth q_0.$$

The function β cannot be written in closed analytic form. In Table III we list p_0^c for a few values of q_0; we also list again $2\pi/(q_0 \cdot p_0^c) - 1$, which is much less close to zero in this case. In Table IV we list A, B, B/A and $r = (B/A - 1)/(B/A + 1)$ for several values of q_0, p_0. Again we have grouped together those pairs q_0, p_0 with the same value of $p_0 \cdot q_0$; in the cases $p_0 \cdot q_0 = \pi/2$, $p_0 \cdot q_0 = \pi/4$ we compare the estimates with the exact values.

TABLE III
THE ESTIMATED VALUES p_0^c, AND THEIR DEVIATION FROM $2\pi/q_0$, FOR $g(x) = e^{-|r|}$

q_0	p_0^c	$2\pi/(q_0 \cdot p_0^c) - 1$
0.5	5.32	1.36
1.0	2.99	1.10
1.5	2.52	0.66
2.0	2.21	0.42
2.5	1.92	0.30
3.0	1.68	0.25
3.5	1.48	0.22
4.0	1.33	0.17

In the two cases, in Table IV, where the exact values of A, B can be computed via the Zak transform (for $q_0 \cdot p_0 = \pi/2$ and $q_0 \cdot p_0 = \pi/4$), we see again that the estimates for A and B given by (2.3.14) and (2.3.15) are remarkably close to the exact values. The error on B is negligible, and the error on A does not exceed a few percent. Note also that frames based on the exponential used here are much less "snug" than Gaussian frames (compare e.g., the value of r for $q_0 = 1$, $p_0 = \pi/2$, which is 0.399 in the present case, but 0.037 for a Gaussian).

TABLE IV
VALUES FOR THE FRAME BOUNDS A, B, THEIR RATIO B/A, AND THE CONVERGENCE FACTOR $r=(B/A-1)/(B/A+1)$, FOR THE CASE $g(x)=\exp(-|x|)$, FOR DIFFERENT VALUES OF q_0, p_0*

			$q_0 \cdot p_0 = \pi/2$			
q_0	A	A_{exact}	B	B_{exact}	B/A	$r=(B-A)/(B+A)$
1.0	2.600	2.724	6.056	6.056	2.330	0.399
1.5	2.665	2.692	6.781	6.781	2.544	0.436
2.0	2.179	2.190	8.326	8.326	3.821	0.585
2.5	1.648	1.657	10.140	10.140	6.152	0.720
3.0	1.197	1.206	12.060	12.060	10.074	0.819

		$q_0 \cdot p_0 = 3\pi/8$		
q_0	A	B	B/AQ	$r=(B-A)/(B+A)$
0.5	2.014	8.873	4.405	0.630
1.0	4.203	7.339	1.746	0.272
1.5	3.724	8.872	2.382	0.409
2.0	2.938	11.068	3.767	0.580
2.5	2.204	13.514	6.133	0.720
3.0	1.597	16.080	10.068	0.819

				$q_0 \cdot p_0 = \pi/4$		
q_0	A	A_{exact}	B	B_{exact}	B/A	$r=(B-A)/(B+A)$
1.0	6.757	6.766	10.554	10.554	1.562	0.219
1.5	5.634	5.645	13.259	13.259	2.353	0.404
2.0	4.412	4.426	16.597	16.597	3.762	0.580
2.5	3.306	3.322	20.271	20.271	6.132	0.720
3.0	2.396	2.413	24.119	24.119	10.068	0.819

*Where possible ($q_0 \cdot p_0 = \pi/2$, $q_0 \cdot p_0 = \pi/4$) the estimates for A, B are compared with the exact values (computed via the Zak transform; see text).

2) The Wavelet Case:

a) Ranges for the parameters a_0, b_0—Estimates for the frame bounds: In Section II-B-1b) we construct tight frames for arbitrary choices of $a_0 > 1$, $b_0 > 0$. This shows that there exists no absolute, *a priori* limitation on a_0, b_0—values leading to frames, unlike the Weyl–Heisenberg case, where $p_0 \cdot q_0 \leq 2\pi$ is a necessary condition (see Section II-C-1a)). This freedom in the choice of a_0, b_0 is deceptive, however, because of the behavior of wavelet frames under dilations. If the h_{mn}, based on h, with parameters a_0, b_0, constitute a frame, then so do the $h_{\gamma;mn}$, based on $h_\gamma(x) = \gamma^{1/2} h(\gamma x)$, with frame parameters $a_0, \gamma^{-1} b_0$. This explains, at least partially, why a frame can be constructed for *any* pair a_0, b_0. To eliminate this dilational freedom, let us restrict our attention, in the present discussion, to frames such that $\|h\| = 1$ and $\int dk\, |k|^{-1} |\hat{h}(k)|^2 = 1$. Under this restriction, one might hope again that there exists a critical curve $b_0^c(a_0)$ separating the "frameable" pairs from the "nonframeable," with the orthogonal bases corresponding to the curve itself. This was the situation for the Weyl–Heisenberg case. It turns out however that this picture is not true in the wavelet case. At the end of this subsection, in Theorem 2.10, we establish the following counterexample. We take Y. Meyer's basic wavelet ψ, and look at the ψ_{mn,b_0}, a family of wavelets generated from ψ with $a_0 = 2$, b_0 arbitrary. For $b_0 = 1$, these wavelets constitute an orthonormal basis [31]. If there existed a nice critical curve $b_0^c(a_0)$ separating frameable and nonframeable values, then we would expect that the $\psi_{mn;b_0}$ would not be a frame for $b_0 > 1$ ("not enough" vectors), and might be a frame consisting of nonindependent vectors for $b_0 < 1$ ("too many" vectors). It turns out, however (see Theorem 2.10), that there exists $\epsilon > 0$ such that, for all values of b_0 in $(1-\epsilon, 1+\epsilon)$, the associated $\psi_{mn;b_0}$ constitute a basis for $L^2(\mathbb{R})$. This baffling fact shows that the concept of "phase space density," so well-suited for Weyl–Heisenberg frames, is not well adapted to the wavelet situation.

For the wavelet case this is all we have to say in answer to question 1), as formulated at the start of Section II-C. The following theorem addresses question 2), i.e., the determination for a given function h, of a range R such that the h_{mn} are a frame for all choices $(a_0, b_0) \in R$. The formulation of this theorem is very similar to Theorem 2.5, and so is its proof.

Theorem 2.7: If

1) $$m(h; a_0) = \operatorname*{ess\,inf}_{|x| \in [1, a_0]} \sum_{m \in \mathbb{Z}} \left| \hat{h}(a_0^m x) \right|^2 > 0 \qquad (2.3.18)$$

2) $$M(h; a_0) = \operatorname*{ess\,sup}_{|x| \in [1, a_0]} \sum_{m \in \mathbb{Z}} \left| \hat{h}(a_0^m x) \right|^2 < \infty \qquad (2.3.19)$$

and

3) $$\sup_{s \in \mathbb{R}} \left[(1+s^2)^{(1+\epsilon)/2} \beta(s) \right] = C_\epsilon < \infty \quad \text{for some } \epsilon > 0 \qquad (2.3.20)$$

where

$$\beta(s) = \sup_{|x| \in [1, a_0]} \sum_{m \in \mathbb{Z}} \left| \hat{h}(a_0^m x) \right| \left| \hat{h}(a_0^m x + s) \right|. \qquad (2.3.21)$$

Then there exists a $B_0^c > 0$ such that

$\forall b_0 \in (0, B_0^c)$: the h_{mn} associated to h, a_0, b_0 constitute a frame,

$\forall \delta > 0$ $:\exists b_0$ in $[B_0^c, B_0^c + \delta]$ such that the g_{mn} associated to h, a_0, b_0 are not a frame.

Proof: see Appendix C. □

Remarks:

1) The conditions (2.3.18), (2.3.19) are again necessary conditions. If (2.3.18) is not satisfied, then $\inf_{f \in L^2} \|f\|^{-2} \Sigma_{m,n} |\langle h_{mn}, f \rangle|^2 = 0$, which excludes the existence of a nonzero lower frame bound. Similarly (2.3.19) is necessary for the existence of a finite upper frame bound.

2) The range R of "good" parameters, i.e., the set of (a_0, b_0) such that the h_{mn}, associated to h, a_0, b_0, constitute a frame, need not be connected. It is possible to construct functions h such that, for fixed b_0, there exist $a_{0,1} < a_{0,2} < a_{0,3}$, for which the h_{mn} associated with $h, a_{0,j}; b_0$ constitute a frame if $j = 1$ or 3, but do not if $j = 2$. (The construction is similar to the one given for the Weyl–Heisenberg case. See Remark 2, following Theorem 2.5.)

3) Theorem 2.7 is only useful for choices of h for which the support of $\hat{h}$ contains negative as well as positive frequencies. In some cases one prefers to work with functions h with support $\hat{h} \subset \mathbb{R}_+$. Functions with this property are also called "analytic signals," because they extend to functions analytic on a half-plane. See e.g., [68]. The frame to be used then consists of $\{h^{\pm}_{mn}; m,n \in \mathbb{Z},\ h^+ = h,\ h^- = h^*\}$ or of $\{h^{(\lambda)}_{mn};\ m,n \in \mathbb{Z},\ \lambda = 1,2\ h^{(1)} = \sqrt{2}\ \mathrm{re}\, h,\ h^{(2)} = \sqrt{2}\ \mathrm{im}\, h\}$. For these frames the conclusions of Theorem 2.7 hold under very similar conditions. The only changes to be made concern the definitions of $m(h;a_0)$, $M(h;a_0)$ and $\beta(s)$. In each of these definitions, the condition $|x| \in [1,a_0]$ should be replaced by $x \in [1,a_0]$.

As in the Weyl–Heisenberg case (Theorem 2.5), the conditions (2.3.18)–(2.3.20) may seem very technical. They are however very easy to check on a computer. Good estimates for $m(h;a_0)$, $M(h;a_0)$ and $\beta(s)$ lead again to useful inequalities for the frame bounds A and B.

Theorem 2.8: Under the same conditions as in Theorem 2.7, the following lower bound for B_0^c holds

$$B_0^c \ge b_0^c = \inf\left\{ b_0 \left| 2\sum_{k=1}^{\infty}\left[\beta\left(-\frac{2\pi}{b_0}k\right)\beta\left(\frac{2\pi}{b_0}k\right)\right]^{1/2} \right. \ge m(h;a_0)\right\}. \qquad (2.3.22)$$

For $0 < b_0 < b_0^c$, the following estimates for the frame bounds A, B hold

$$A \ge \frac{2\pi}{b_0}\left\{ m(h;a_0) - 2\sum_{k=1}^{\infty}\left[\beta\left(\frac{2\pi}{b_0}k\right)\beta\left(-\frac{2\pi}{b_0}k\right)\right]^{1/2}\right\}, \qquad (2.3.23)$$

$$B \le \frac{2\pi}{b_0}\left\{ M(h;a_0) + 2\sum_{k=1}^{\infty}\left[\beta\left(\frac{2\pi}{b_0}k\right)\beta\left(-\frac{2\pi}{b_0}k\right)\right]^{1/2}\right\}. \qquad (2.3.24)$$

Proof: see Appendix C. □

Remark: The proof in Appendix C applies for the case where both $\mathbb{R}_+ \cap \mathrm{support}\,\hat{h}$ and $\mathbb{R}_- \cap \mathrm{support}\,\hat{h}$ have nonzero measure. If this is not the case, e.g., if support $\hat{h} \subset \mathbb{R}_+$, and the frame considered is either $\{h^{\pm}_{mn} | m,n \in \mathbb{Z},\ h^+ = h,\ h^- = h^*\}$ or $\{h^{(\lambda)}_{mn} | m, (n \in \mathbb{Z},\ \lambda = 1,2,\ h^{(1)} = \sqrt{2}\ \mathrm{re}\, h,\ h^{(2)} = \sqrt{2}\ \mathrm{im}\, h\}$ see Section II-B-1b)), then the definitions of $m(h;a_0), M(h;a_0), \beta(s)$ have to be slightly changed (the restriction $|x| \in [1,a_0]$ is replaced by $x \in [1,a_0]$—see Remark 3 following Theorem 2.7), and the same formulas (2.3.23), (2.3.24) apply.

In most practical examples the dilation parameter a_0 is equal to 2. In this case the estimates (2.3.23) and (2.3.24) can be sharpened. The following corollary is due to Ph. Tchamitchian.

Theorem 2.9: Choose $a_0 = 2$. Under the same conditions as in Theorem 2.7, the following estimates for the frame bounds A, B hold,

$$A \ge \frac{2\pi}{b_0}\left\{ m(h;2) - 2\sum_{l=0}^{\infty}\left[\beta_1\left(\frac{2\pi}{b_0}(2l+1)\right) \cdot \beta_1\left(-\frac{2\pi}{b_0}(2l+1)\right)\right]^{1/2}\right\} \qquad (2.3.25)$$

$$B \le \frac{2\pi}{b_0}\left\{ M(h;2) + 2\sum_{l=0}^{\infty}\left[\beta_1\left(\frac{2\pi}{b_0}(2l+1)\right) \cdot \beta_1\left(-\frac{2\pi}{b_0}(2l+1)\right)\right]^{1/2}\right\} \qquad (2.3.26)$$

where

$$\beta_1(s) = \sup_x \sum_{m\in\mathbb{Z}} \left| \sum_{m'\ge 0} \hat{h}(2^{m+m'}x)\hat{h}^*[2m'(2^m x + s)] \right|.$$

Proof: See Appendix C. □

Note that the estimates in Theorem 2.9 use some of the phase information contained in $\hat{h}$, which is completely lost in the estimates in Theorem 2.8. It is therefore to be expected that the estimates (2.3.25), (2.3.26) are a significant improvement (2.3.23), (2.3.24) for complex $\hat{h}$. This is illustrated most dramatically by applying both pairs of estimates to the basic wavelet in Y. Meyer's basis (defined at the end of Section II-B-1b)). For $h = \psi$, with ψ defined by (2.2.7), one finds $\Sigma_m |\hat{\psi}(2^m y)|^2 = 1$, $\beta(2\pi) = \beta(-2\pi) = 1/2$, $\beta(4\pi) = \beta(-4\pi) = 1/2$, and $\beta(k2\pi) = 0$ if $|k| \ge 3$, hence $\Sigma_{k=1}^{\infty}[\beta(2\pi k)\beta(-2\pi k)]^{1/2} = 1$. Applying Theorem 2.8 we therefore find $A \ge -1$, $B \le 3$.

This means that if we had only the bounds (2.3.23), (2.3.24) to go by, we wouldn't be able to recognize that the ψ_{mn} constitute a frame. Since, for $a_0 = 2$ and $b_0 = 1$, the ψ_{mn} do in fact constitute an orthonormal base, this is a rather poor performance. Calculating β_1 we find that

$$\beta_1(2\pi k) = 0, \qquad \text{for all odd } k.$$

The estimates in Theorem 2.9 thus lead to $A \ge 1$, $B \le 1$, or equivalently to the optimal estimate $A = B = 1$. (Note that this automatically proves that the ψ_{mn} constitute an orthonormal basis, since $\|\psi_{mn}\| = \|\psi\| = 1$. We showed in Section II-A that a tight frame with frame constant 1, consisting of normalized vectors, necessarily is an orthonormal basis.) Since the phase factor in the definition of Y. Meyer's basic wavelet (2.2.7) is essential for the ψ_{mn} to constitute an orthonormal basis (see e.g., [31]), it is natural that the phase-dependent estimates (2.3.25), (2.3.26) perform much better than the phase-independent estimates (2.3.23), (2.3.24) in this case. Using Theorem 2.9 one can prove the result we just announced, i.e., that the wavelets ψ_{mn}, associated with Y. Meyer's ψ, and with $a_0 = 2$, b_0 arbitrary, still constitute a basis for $b_0 \neq 1$, $|b_0 - 1| < \epsilon$, for some $\epsilon > 0$. The following Theorem was proved in collaboration with Ph. Tchamitchian.

Theorem 2.10: Let ψ be the function defined by (2.2.7). Then there exists $\epsilon > 0$ such that the $\psi_{mn;b_0}$,

$$\psi_{mn;b_0}(x) = 2^{-m/2}\psi(2^{-m}x - nb_0), \qquad m,n \in \mathbb{Z}$$

constitute a basis for $L^2(\mathbb{R})$, for any choice $b_0 \in (1-\epsilon, 1+\epsilon)$.

Proof: See Appendix C. □

This result is quite surprising: it shows, as pointed out at the start of Section II-C-2, that it is not always safe to apply "phase space density intuition" to families of wavelets.

Remark: It follows from the proof that ϵ can be estimated explicitly, from computations of the frame bounds A, B, as given by (2.3.25), (2.3.26) on the one hand, and of

$$\sum_{\substack{|m|<1 \\ m \in \mathbb{Z}}} |\langle \psi_{mn;b_0}, \psi \rangle|$$

on the other hand. This estimate for ϵ depends, of course, on the choice for the function ν (see (2.2.7)). For the (non C^∞) choice $\nu(x) = x$ if $0 \le x \le 1$, $\nu(x) = 0$ otherwise, one finds $\epsilon > 0.02$. The ψ_{mn,b_0} remain a frame for $b_0 \le 1.08$ in this case.

In many practical examples $\hat{h}$ decays very fast, and is real. In those cases the differences between estimates using β (i.e., (2.3.23), (2.3.24)) or β_1, (i.e., (2.3.25), (2.3.26)) are very small, and much less dramatic then for Y. Meyer's wavelet ψ. In fact, if $\hat{h}$ is positive (e.g., the Mexican hat function, the 8th derivative of the Gaussian, the modulated Gaussian, $\cdots$: see the next subsection) the estimates using β perform better than those using β_1, as can easily be checked directly by the formulas (2.3.23), (2.3.24), and (2.3.25), (2.3.26).

As already mentioned above, most practical applications use $a_0 = 2$. This choice makes numerical computations much easier since it means that the translation steps $b_0 \cdot 2^m$ (see Fig. 1(b)), for two different frequency levels $m_1 > m_2$, are multiples of each other. It also makes the different frequency levels correspond to "octaves." On the other hand one likes to use functions h with fairly concentrated Fourier transforms $\hat{h}$, corresponding to a good frequency resolution. This means that $\Sigma_m |\hat{h}(2^m x)|^2$ is bound to have rather large oscillations; since then $m(h;2)$ is much smaller than $M(h;2)$, this leads to high, and therefore unpleasant values of B/A. This can be avoided by the use of several functions h^j, chosen so that the minima of $\Sigma_m |\hat{h}^j(2^m x)|^2$ are compensated by the maxima of $\Sigma_m |\hat{h}^{j'}(2^m x)|^2$, for some $j' \neq j$. This is made explicit by the following corollary of Theorem 2.7.

Corollary 2.11: Let $h^0, \cdots, h^{N-1}$ be N functions satisfying the conditions (2.3.18), (2.3.19), and (2.3.20). Define

$$m(h^0, \cdots, h^{N-1}; a_0) = \operatorname*{ess\,inf}_{|x| \in [1, a_0]} \sum_{j=0}^{N-1} \sum_m |\hat{h}^j(a_0^m x)|^2 \tag{2.3.27}$$

$$M(h^0, \cdots, h^{N-1}; a_0) = \operatorname*{ess\,sup}_{|x| \in [1, a_0]} \sum_{j=0}^{N-1} \sum_m |\hat{h}^j(a_0^m x)|^2 \tag{2.3.28}$$

$$\beta^j(s) = \sup_{|x| \in [1, a_0]} \sum_{m \in \mathbb{Z}} |\hat{h}^j(a_0^m x)| \, |\hat{h}^j(a_0^m x + s)|. \tag{2.3.29}$$

Choose b_0 such that

$$2 \sum_{j=0}^{N-1} \sum_{k=1}^{\infty} \left[\beta^j\left(\frac{2\pi}{b_0}k\right) \beta^j\left(-\frac{2\pi}{b_0}k\right) \right]^{1/2} < m(h^0, \cdots, h^{N-1}; a_0).$$

Then the $\{h^j_{mn}; m,n \in \mathbb{Z},\ j = 0, \cdots, N-1\}$ constitute a frame. The following estimates for the frame bounds A and B hold,

$$A \ge \frac{2\pi}{b_0} \left\{ m(h^0, \cdots, h^{N-1}; a_0) - 2 \sum_{j=0}^{N-1} \sum_{k=1}^{\infty} \cdot \left[\beta^j\left(\frac{2\pi}{b_0}k\right) \beta^j\left(-\frac{2\pi}{b_0}k\right) \right]^{1/2} \right\} \tag{2.3.30}$$

$$B \le \frac{2\pi}{p_0} \left\{ M(h^0, \cdots, h^{N-1}; a_0) + 2 \sum_{j=0}^{N-1} \sum_{k=1}^{\infty} \cdot \left[\beta^j\left(\frac{2\pi}{p_0}k\right) \beta^j\left(-\frac{2\pi}{p_0}k\right) \right]^{1/2} \right\}. \tag{2.3.31}$$

Proof: The proof is a simple variant on the proof of Theorem 2.8. □

In the special case where $a_0 = 2$, one can, of course, replace β^j in (2.3.30), (2.3.31) by β_1^j, with β_1^j defined as in Theorem 2.8, for $j = 0, \cdots, N-1$. Then the sums over $k \neq 0$ also have to be replaced by sums over only odd k.

The number N of functions used is called the number of "voices" per octave [28]. In numerical calculations $N = 4$ seems to be a satisfactory choice.

In practice one often chooses the $h^0, \cdots, h^{N-1}$ to be dilated versions of one function h, i.e.,

$$h^j(x) = 2^{-j/N} h(2^{-j/N} x), \qquad j = 0, \cdots, N-1. \tag{2.3.32}$$

The phase space lattice corresponding to the $\{h^j_{mn};\ m,n \in \mathbb{Z},\ j = 0, \cdots, N-1\}$ is the superposition of N lattices of the type depicted in Fig. 1(b), shifted with respect to each other in frequency. Fig. 5 shows such a lattice, for the case $N = 4$, and for $a_0 = 2$.

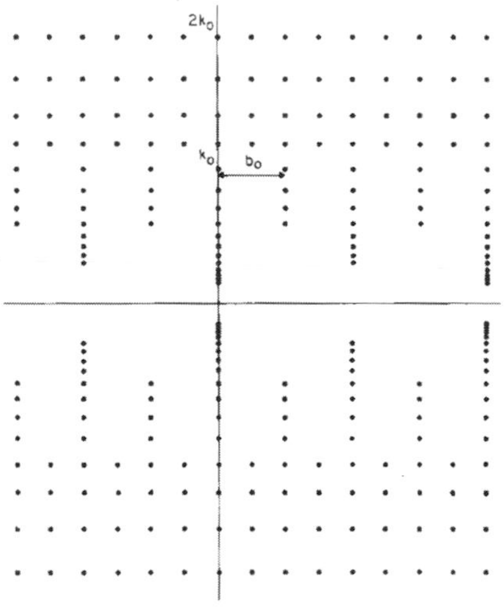

Fig. 5. Phase space lattice corresponding to wavelet frame with $a_0 = 2$, $b_0 = 1$, and 4 voices per octave (see text). The different voice wavelets are obtained by dilating one single function, $h^j(x) = 2^{-j/4}h(2^{-j/4}x)$, $j = 0, 1, 2, 3$. As in Fig. 1(b), $k_0 = \int_0^\infty dk\,|k|\,|\hat{h}(k)|^2$, and $\hat{h}$ is assumed to be even.

TABLE V-A
FRAME BOUNDS FOR WAVELET FRAMES*

$N=1$				$N=2$			
b_0/π	A	B	B/A	b_0/π	A	B	B/A
0.25	4.038	8.409	2.082	0.25	11.950	16.294	1.364
0.50	1.838	4.386	2.387	0.50	5.711	8.411	1.473
0.75	1.412	3.007	2.634	0.75	3.629	5.785	1.594
1.00	0.585	2.527	4.323	1.00	2.410	4.651	1.930
1.25	0.337	2.152	6.380	1.25	1.709	3.939	2.305
				1.50	0.999	3.708	3.713
				1.75	0.556	3.479	6.259
				2.00	0.202	3.328	16.473
$N=3$				$N=4$			
b_0/π	A	B	B/A	b_0/π	A	B	B/A
0.25	20.035	24.331	1.214	0.25	27.986	32.500	1.161
0.50	9.655	12.528	1.298	0.50	13.598	16.645	1.224
0.75	6.185	8.603	1.391	0.75	8.687	11.475	1.321
1.00	4.230	6.861	1.622	1.00	6.042	9.080	1.503
1.25	3.050	5.823	1.909	1.25	4.431	7.666	1.730
1.50	2.033	5.361	2.637	1.50	3.076	7.005	2.278
1.75	1.313	5.025	3.828	1.75	2.031	6.610	3.255
2.00	0.736	4.810	6.539	2.00	1.261	6.300	4.998

*Based on the function

$$h(x) = \begin{cases} \pi^{-1/2} \sin x & |x| \le \pi \\ 0, & \text{otherwise} \end{cases}.$$

The dilation parameter $a_0 = 2$ in all cases; N is the number of "voices" (see text).

Remark: Note that the h^j, constructed by dilations from one function h as in (2.3.32), do not have the same L^2-normalization, $\|h^j\|_{L^2} = 2^{-j/2N}\|h\|_{L^2}$. This change in normalization compensates for the fact that the phase space lattice in Fig. 5 is "denser" than the corresponding lattice (see e.g., Fig. 1(b)) would be for the same basic wavelet h, but with $a_0 = 2^{1/N}$, and with only one function (namely h) per dilation step (instead of N, as in Fig. 5).

b) Examples: We illustrate the different bounds with several examples. In practice, it is by far preferable to use $a_0 = 2$, rather than noninteger values. We have therefore, in all our examples, restricted ourselves to $a_0 = 2$, and introduced several voices (see Corollary 2.11). The different voice-functions h^j are always obtained, in these examples, by dilation of one given function h (see (2.3.32)).

i) One cycle of the sine-function: In this case we take

$$h(x) = \begin{cases} \dfrac{1}{\sqrt{\pi}} \sin x, & \text{if } |x| \le \pi \\ 0, & \text{otherwise} \end{cases}.$$

The Fourier transform $\hat{h}$ of h is given by

$$\hat{h}(y) = \frac{\sqrt{2}}{\pi} i \frac{\sin y\pi}{1 - y^2}.$$

Since this is an oscillating function, one suspects that the formulas analogous to (2.3.30), (2.3.31), but using β_1 rather than β, will perform better than the β-formulas. This turns out to be true. Table V-A lists the estimates for the frame bounds A, B and their ratio, for a few values of b_0, and of N, the number of voices.

ii) The Mexican hat: The Mexican-hat function is the second derivative of the Gaussian (up to a sign),

$$h(x) = \frac{2}{\sqrt{3}} \pi^{-1/4} (1 - x^2) e^{-x^2/2}$$

$$\hat{h}(y) = \frac{2}{\sqrt{3}} \pi^{-1/4} y^2 e^{-y^2/2}.$$

The graph of h looks a bit like a transverse section of a Mexican hat (see Fig. 2(b)), whence the name. Since $\hat{h}$ is positive, the formulas using β rather than β_1 are more effective here. The same will be true in our next examples. Table V-B lists the estimates for the frame bounds A, B and their ratio, for a few values of the translation parameter b_0, and of N, the number of voices.

iii) The eighth derivative of the Gaussian: Functions like the Mexican hat and higher order derivatives of the Gaussian are useful in applications of the wavelet transform to edge detection (see, e.g., [27]). Table V-C lists the estimates for the frame bounds A, B for wavelets based on the eighth derivative of the Gaussian,

$$h(x) = \left(\frac{2^{15} 7!}{15!} \right)^{1/2} \pi^{-1/4} (x^8 - 28x^6 + 210x^4 - 405x^2 + 90) e^{-x^2/2}$$

$$\hat{h}(y) = \left(\frac{2^{15} 7!}{15!} \right)^{1/2} \pi^{-1/4} y^8 e^{-y^2/2}.$$

This is a typical example of a function where, with $a_0 = 2$ fixed, the introduction of several voices is necessary. For $N = 1$ (i.e., only one voice, corresponding to the phase

TABLE V-B
FRAME BOUNDS FOR WAVELET FRAMES*

	$N=1$				$N=2$		
b_0	A	B	B/A	b_0	A	B	B/A
0.25	13.091	14.183	1.083	0.25	27.273	27.278	1.0002
0.50	6.546	7.092	1.083	0.50	13.637	13.639	1.0002
0.75	4.364	4.728	1.083	0.75	9.091	9.093	1.0002
1.00	3.223	3.596	1.116	1.00	6.768	6.870	1.015
1.25	2.001	3.454	1.726	1.25	4.834	6.077	1.257
1.50	0.325	4.221	12.986	1.50	2.609	6.483	2.485
				1.75	0.517	7.276	14.061
	$N=3$				$N=4$		
b_0	A	B	B/A	b_0	A	B	B/A
0.25	40.914	40.914	1.0000	0.25	54.552	54.552	1.0000
0.50	20.457	20.457	1.0000	0.50	27.276	27.276	1.0000
0.75	13.638	13.638	1.0000	0.75	18.184	18.184	1.0000
1.00	10.178	10.279	1.010	1.00	13.586	13.690	1.007
1.25	7.530	8.835	1.173	1.25	10.205	11.616	1.138
1.50	4.629	9.009	1.947	1.50	6.594	11.590	1.758
1.75	1.747	9.942	5.691	1.75	2.928	12.659	4.324

*Based on the Mexican hat function

$$h(x)=2/\sqrt{3}\,\pi^{-1/4}(1-x^2)e^{-x^2/2}.$$

The dilation parameter $a_0=2$ in all cases; N is the number of voices (see text).

TABLE V-C
FRAME BOUNDS FOR WAVELET FRAMES*

b_0	A	B	B/A
	$N=2$		
0.125	25.515	26.569	1.041
0.250	12.758	13.285	1.041
0.375	8.505	8.856	1.041
0.500	6.379	6.642	1.041
0.625	5.101	5.316	1.042
0.750	3.995	4.686	1.173
0.875	1.669	5.772	3.459
	$N=3$		
b_0	A	B	B/A
0.125	39.054	39.073	1.0005
0.250	19.527	19.536	1.0005
0.375	13.018	13.024	1.0005
0.500	9.764	9.768	1.0005
0.625	7.808	7.817	1.001
0.750	6.251	6.770	1.083
0.875	3.563	7.598	2.132
1.000	0.163	9.603	58.929
	$N=4$		
b_0	A	B	B/A
0.125	52.085	52.085	1.0000
0.250	26.042	26.042	1.0000
0.375	17.362	17.362	1.0000
0.500	13.022	13.022	1.0000
0.625	10.414	10.420	1.0005
0.750	8.420	8.941	1.062
0.875	5.313	9.568	1.801
1.000	1.127	11.894	10.550

*Based on the 8th derivative of the Gaussian,

$$h(x)=\left(\frac{2^{15}7!}{15!}\right)^{1/2}$$

$$\cdot\pi^{-1/4}(x^8-28x^6+210x^4-405x^2+90)e^{-x^2/2}.$$

The dilation parameter $a_0=2$ in all cases; N is the number of voices.

space lattice in Fig. 1(b)) one finds that $M(\hat{h};2)/m(\hat{h};2)$ is equal to 3.385. This means that the ratio B/A, which is bounded below by $M(\hat{h};2)/m(\hat{h};2)$, is pretty large, with $N=1$, even for very small values of b_0. As soon as more voices are introduced, the ratio M/m becomes much smaller, and snug frames can be constructed, for appropriate choices of b_0. We have restricted ourselves, in Table Vc, to the choices $N=2,3,4$, excluding $N=1$.

iv) The modulated Gaussian: In this case we take

$$h(x)=\pi^{-1/4}\left(e^{-ikx}-e^{-k^2/2}\right)e^{-x^2/2}$$

$$\hat{h}(y)=\pi^{-1/4}\left(e^{-(y-k)^2/2}-e^{-k^2/2}e^{-y^2/2}\right)$$

where $k=\pi(2/\ln 2)^{1/2}$.

The subtraction term in the definition of $h,\hat{h}$ ensures that $\hat{h}(0)=0$; for the value of k chosen here, this term is negligible in practice. The value of k has been fixed so that the ratio between the highest and the second highest local maxima of $\operatorname{Re} h$ is approximately $1/2$. This wavelet is exactly the wavelet used by J. Morlet in his numerical computations [25], [26]. If only real signals f are decomposed and reconstructed, by means of the $\langle h_{mn}, f\rangle$, then the complex wavelet h consists really of two wavelets, $\operatorname{Re} h$ and $\operatorname{Im} h$. In this case the frame bounds for real signals can be rewritten as

$$A=\frac{2\pi}{b_0}\left\{\frac{1}{2}\min_x\sum_{m\in\mathbb{Z}}\left[\left|\hat{h}(a_0^m x)\right|^2+\left|\hat{h}(-a_0^m x)\right|^2\right]-R\right\}$$

$$B=\frac{2\pi}{p_b}\left\{\frac{1}{2}\max_x\sum_{m\in\mathbb{Z}}\left[\left|\hat{h}(a_0^m x)\right|^2+\left|\hat{h}(-a_0^m x)\right|^2\right]+R\right\}$$

with

$$R=2\sum_{\epsilon=+,-}\sum_{l=1}^{\infty}\left[\beta_\epsilon\left(\frac{2\pi}{b_0}l\right)\beta_\epsilon\left(-\frac{2\pi}{b_0}l\right)\right]^{1/2}$$

and

$$\beta_\epsilon(s)=\frac{1}{4}\sup_x\sum_m\left|\hat{h}(a_0^m x)+\epsilon\hat{h}(-a_0^m x)\right|$$

$$\cdot\left|\hat{h}(a_0^m x+s)+\epsilon\hat{h}(-a_0^m x-s)\right|.$$

Note that $\hat{h}$ is almost completely concentrated on the positive frequency half line; neglecting terms with negative arguments for $\hat{h}$ in the previous formulas leads back to the frame bounds for support $\hat{h}\subset\mathbb{R}_+$. In the first reconstructions with this wavelet, before even the connection with continuously labelled affine coherent states was made (see Sections I-E, I-F), a formula similar to (2.1.7) was used. This reconstruction formula turned out to be extremely precise. Our calculations of frame bounds show why. For $N=4$, $b_0=1$, for instance, which are choices that do correspond to values used in practice (in fact, [25] uses even higher values of N, and smaller values of b_0), we find $B/A=1.0008$. Hence $r=(B-A)/(B+A)\simeq 0.04\%$, which explains why (2.1.7) is such a good approximation to the true reconstruction formula.

As in the previous example, the ratio M/m is rather large for $N=1$, and we have computed the frame bounds only for $N=2$, 3, and 4. They are tabulated in Table V-D.

TABLE V-D
FRAME BOUNDS FOR WAVELET FRAMES*

b_0	A	B	B/A
		$N=2$	
0.5	6.019	7.820	1.299
1.0	3.009	3.910	1.230
1.5	1.944	2.669	1.373
2.0	1.173	2.287	1.950
2.5	0.486	2.282	4.693
		$N=3$	
0.5	10.295	10.467	1.017
1.0	5.147	5.234	1.017
1.5	3.366	3.555	1.056
2.0	2.188	3.002	1.372
2.5	1.175	2.977	2.534
3.0	0.320	3.141	9.824
		$N=4$	
0.5	13.837	13.846	1.0006
1.0	6.918	6.923	1.0008
1.5	4.540	4.688	1.032
2.0	3.013	3.910	1.297
2.5	1.708	3.829	2.242
3.0	0.597	4.017	6.732

*Based on the modulated Gaussian,

$$h(x)=\pi^{-1/4}\left(e^{-ikx}-e^{-k^2/2}\right)e^{-x^2/2},$$

with $k=\pi(2/\ln 2)^{1/2}$.

The dilation constant $a_0=2$ in all cases; N is the number of voices.

Remark: The tables show that extremely snug frames can be obtained for quite reasonable phase space lattices (i.e., N not too large, b_0 not too small). Note, however, that even when B/A s not very close to 1, the frame may still be useful. For $B/A \simeq 1.5$, e.g., the convergence factor $r=(B-A)/(B+A)$ is still of the order of 0.2; while this is insufficient to permit the use of the approximation formula (2.1.7), it does mean that only a few iterations will suffice for the computation of the $(h_{mn})^{\sim}$ up to e.g., 10^{-3}, leading, again, to a very accurate reconstruction formula.

D. Frames in Other Spaces than $L^2(\mathbb{R})$

The results in this section are more technical and specialized than those in the preceding sections. The reader who is mostly interested in L^2-results can safely omit reading this section and go to Section III, where we again discuss L^2-estimates.

1) Motivation: Why Other Spaces than $L^2(\mathbb{R})$?: So far, we have restricted ourselves to studying frames in $L^2(\mathbb{R})$. The preceding section was mainly concerned with the formulation of conditions under which the short-time Fourier expansions (Weyl–Heisenberg case) or the wavelet expansions (affine case) would converge with respect to the L^2-norm. As explained in the introduction, both types of expansions are used in the analysis, e.g., of time-dependent signals. For such signals $f(t)$, the square of the L^2-norm, $\int_{-\infty}^{\infty}dt|f(t)|^2$, is a natural quantity, often called the "energy" of the signal in electrical engineering literature. It is therefore customary to require L^2-convergence for series representations of these signals.

This does not mean, however, that convergence in other topologies is not important. For example, convergence in $L^p(\mathbb{R})$-spaces,

$$L^p(\mathbb{R})=\left\{f;\|f\|_p=\left[\int dx|f(x)|^p\right]^{1/p}<\infty\right\}$$

with $p\neq 2$, where the coefficients are weighted in a way different from $L^2(\mathbb{R})$, would certainly indicate that the series is more "robust" than if it converged in $L^2(\mathbb{R})$ alone. On the other hand, if the signals themselves have some regularity (consider, e.g., the case where not only f but also its first k derivatives are in $L^2(\mathbb{R})$), then one would wish that the partial sums of the series representing f share that regularity, and that the convergence respects this (in the previous example, this would mean L^2-convergence also for the first k derivatives). It would therefore be interesting to have convergence in the Sobolev spaces

$$\mathscr{H}_s(\mathbb{R})=\left\{f;\|f\|^2_{\mathscr{H}_s}=\int dy(1+y^2)^s|\hat{f}(y)|^2<\infty\right\}$$

for at least some $s>0$.

In the two subsections following this one, we shall show that for suitably chosen g or h, and appropriate parameters p_0, p_0 or a_0, b_0, the resulting frames are frames in $\mathscr{H}_s(\mathbb{R})$, at least for some strip $|s|<s_0$. In the remainder of this subsection we give some examples illustrating "what can go wrong." The examples given here all pertain to the wavelet case. I want to thank Y. Meyer and Ph. Tchamitchian for having pointed them out to me.

If the h_{mn} $(h_{mn}(x)=a_0^{-m/2}h(a_0^{-m}x-nb_0))$ constitute a frame in $L^2(\mathbb{R})$, then, as shown in Section II-A, there exists a dual frame $\{(h_{mn})^{\sim};\ m,n\in\mathbb{Z}\}$ such that, for all f in $L^2(\mathbb{R})$,

$$f=\sum_{m,n}(h_{mn})^{\sim}\langle h_{mn},f\rangle \qquad (2.4.1)$$

where the series converges in $L^2(\mathbb{R})$. If the frame is not tight, then the $(h_{mn})^{\sim}$ are not multiples of the h_{mn}, and in general they will not be generated by dilations and translations of a single function (see Section II-A).

There are several ways in which (2.4.1) may fail to extend to $L^p(\mathbb{R})$, with $p\neq 2$, or to $\mathscr{H}_s(\mathbb{R})$, with $s\neq 0$. One possibility is that the coefficients $\langle h_{mn},f\rangle$ are not well defined for all f in the space under consideration, i.e., that h does not belong to the dual of this space. This is not really a problem: it is enough to impose some regularity and/or decay conditions on h to avoid this. Something much more pernicious may happen, however. It is possible that even though h is a "nice" function, the elements $(h_{mn})^{\sim}$ of the dual frame are not, so that (2.4.1) fails. The examples we shall give here illustrate this.

The first example is due to P.-G. Lemarié. It was communicated to me by Y. Meyer. Let ψ be Y. Meyer's wavelet, as As already mentioned, the $\psi_{mn}(x)=$

$2^{-m/2}\psi(2^{-m}x-n)$ constitute an unconditional basis for many function spaces, including all the $L^p(\mathbb{R})$, $1<p<\infty$. The wavelet coefficients $\langle\psi_{mn},f\rangle$ can be used to characterize these spaces and their duals [31]. In particular, L^p-spaces are characterized as follows: for any function f one has the equivalence

$$f\in L^p(\mathbb{R})\Leftrightarrow\left[\sum_{m,n}2^{-m}|\langle\psi_{mn},f\rangle|^2\chi_{mn}\right]^{1/2}\in L^p(\mathbb{R}) \tag{2.4.2}$$

where χ_{mn} is the indicator function of the interval $I_{mn}=[2^m n,2^m(n+1))$. The basic wavelet in Lemarié's construction is the following perturbation of ψ:

$$h(x)=\psi(x)+\sqrt{2}\,r\psi(2x).$$

This function h is clearly in C^∞, and has fast decay at ∞. We use the same dilation and translation parameters as for Y. Meyer's basis, i.e., $h_{mn}(x)=2^{-m/2}h(2^{-m}x-n)$. Then $h_{mn}=\psi_{mn}-r\psi_{m-1\,2n}=(1-rU)\psi_{mn}$, where U is the partial isometry defined by $U\psi_{mn}=\psi_{m-1\,2n}$. For $|r|<1$, the operator $(1-rU)$ is one-to-one and onto, which means that the h_{mn} still constitute a basis for $L^2(\mathbb{R})$ if $|r|<1$. The dual frame $(h_{mn})^\sim$ is given by

$$\begin{aligned}(h_{mn})^\sim&=[(1-rU)(1-r^*U^*)]^{-1}h_{mn}\\&=(1-r^*U^*)^{-1}\psi_{mn}\\&=\sum_{k=0}^{\infty}r^{*k}U^{*k}\psi_{mn}\end{aligned} \tag{2.4.3}$$

where $U^*\psi_{m\,2n+1}=0$, and $U^*\psi_{m\,2n}=\psi_{m+1\,n}$. In particular

$$(h_{00})^\wedge=\sum_{j=0}^{\infty}r^{*k}\psi_{-k0}.$$

It turns out that if $|r|>1/\sqrt{2}$, this function does not belong to $L^p(\mathbb{R})$ for large p, as shown by the following argument. If we apply (2.4.2) to $(h_{00})^\sim$, we find

$$\begin{aligned}(h_{00})^\sim\in L^p(\mathbb{R})&\Leftrightarrow\left[\sum_{m=0}^{\infty}(2|r|^2)^m\chi_{-m0}\right]^{1/2}\in L^p(\mathbb{R})\\&\Leftrightarrow\sum_{m=0}^{\infty}\int_{2^{-(m+1)}}^{2^{-m}}dx\left[\sum_{j=0}^{m}(2|r|^2)^j\right]^{p/2}<\infty\\&\Leftrightarrow\sum_{m=0}^{\infty}2^{-(m+1)}\left|\frac{(2|r|^2)^{m+1}-1}{2|r|^2-1}\right|^{p/2}<\infty.\end{aligned}$$

If $|r|>1/\sqrt{2}$, this shows that $(h_{00})^\sim\notin L^p(\mathbb{R})$ for all $p>2\ln 2/[\ln 2+2\ln|r|]$. This implies that, even though the h_{mn} themselves are C^∞ functions with fast decay, and constitute a frame for $L^2(\mathbb{R})$ (with $A=(1-|r|^2)^{1/2}$, $B=(1+|r|^2)^{1/2}$), the frame expansion (2.4.1) does not extend to all $L^p(\mathbb{R})$-spaces, $1<p<\infty$, if $|r|>1/\sqrt{2}$.

Note that, in the example, the $(h_{m0})^\sim$ are the only functions in the dual frame causing problems. For $n\neq 0$, only a finite number of terms in the series (2.4.3) contribute, and $(h_{mn})^\sim$ is still infinitely differentiable and decaying rapidly. This might lead one to believe that the problems are caused by the fact that the $(h_{mn})^\sim$ are not given, in general, by dilations and translations of a single function $\tilde h$. It is true, also, that this phenomenon ($h\in\mathscr{S}$, and some $(h_{mn})^\sim\notin L^p$) does not occur for the tight frames constructed in Section II-B-2b). For these frames, results analogous to those for Y. Meyer's basis hold, and the expansions

$$f=A^{-1}\sum_{m,n}h_{mn}\langle h_{mn},f\rangle$$

are valid, and converge, in particular, for all L^p-spaces $(1<p<\infty)$ and all $\mathscr{H}_s(\mathbb{R})$. Nevertheless, there exist (non-tight) frames for which the $(h_{mn})^\sim$ happen to be dilations and translations of a single function $\tilde h$, and where, as in our first example, h is "nice" and $\tilde h$ is not, causing (2.4.1) to fail in at least some function spaces. Our second example illustrates this; it is a special case of a construction made by Ph. Tchamitchian [35].

In [35], Ph. Tchamitchian constructs functions σ,τ such that the dyadic (i.e., $a_0=2$, $b_0=1$) lattices of wavelets generated by σ and τ constitute biorthogonal bases for $L^2(\mathbb{R})$, i.e.,

$$\sum_{m,n}\sigma_{mn}\langle\tau_{mn},\cdot\rangle=\mathbb{1} \tag{2.4.4}$$

where the sum converges strongly in $L^2(\mathbb{R})$, and with $\sigma_{mn}(x)=2^{-m/2}\sigma(2^{-m}x-n)$, $\tau_{mn}(x)=2^{-m/2}\tau(2^{-m}x-n)$. The details of this construction, together with proofs and applications, are given in [35b]. It is possible to choose both σ and τ with compact support. One such example is the following (see [35b]):

$$\hat\sigma(y)=y^{-1}e^{iy/2}\left(\frac{1}{2}-\frac{9}{16}\cos\frac{y}{2}+\frac{1}{16}\cos\frac{3y}{2}\right)$$

or

$$\sigma(x)=\frac{1}{4}\sqrt{2\pi}\begin{cases}0, & x\le -1\\ 1/8, & -1\le x\le 0\\ -1, & 0\le x\le 1/2\\ 1, & 1/2\le x\le 1\\ -1/8, & 1\le x\le 2\\ 0, & x\ge 2\end{cases}$$

and

$$\hat\tau(y)=ye^{iy/2}\prod_{j=1}^{\infty}P\left[\cos^2\left(2^{-j}\frac{y}{4}\right)\right]$$

where $P(u)=3u^2-2u^3$. Since $P(\cos y)=1+O(y^4)$, the infinite product in the definition of $\hat\tau$ is well defined. One finds that $\hat\tau$ is entire and of exponential type, implying that τ has compact support. Moreover, since $P(u)\le u^2$ on $[0,1]$,

$$\begin{aligned}|\hat\tau(y)|&\le|y|\prod_{j=1}^{\infty}\cos^4[2^{-j}y/4]\\&=|y|\left[\frac{\sin y/4}{y/4}\right]^4\le C\left(1+|y|^2\right)^{-3/2}.\end{aligned} \tag{2.4.5}$$

Using this decay of $\hat{\tau}$, the compactness of support τ, and $\int dx\, \tau(x) = \hat{\tau}(0) = 0$, one can easily prove that $\Sigma_{m,n}|\langle \tau, \tau_{mn}\rangle| < \infty$. The decay properties of $\hat{\sigma}$ are rather weak, but one can use $\int dx\, \sigma(x) = 0$, the compactness of support σ and the fact that σ is piecewise constant in an easy proof of $\Sigma_{m,n}|\langle \sigma, \sigma_{mn}\rangle| < \infty$. This implies that, for all sequences $(c_{mn})_{m,n \in \mathbb{Z}}$ in $l^2(\mathbb{Z})$,

$$\Big\| \sum_{m,n} c_{mn}\sigma_{mn} \Big\|^2 \le \Big(\sum_{m,n} |\langle \sigma, \sigma_{mn}\rangle| \Big) \cdot \Big(\sum_{m,n} |c_{mn}|^2 \Big) \le C \sum_{m,n} |c_{mn}|^2$$

$$\Big\| \sum_{m,n} c_{mn}\tau_{mn} \Big\|^2 \le \Big(\sum_{m,n} |\langle \tau, \tau_{mn}\rangle| \Big) \cdot \Big(\sum_{m,n} |c_{mn}|^2 \Big) \le C \sum_{m,n} |c_{mn}|^2 .$$

Together with (2.4.4) (for the proof of which we refer to [35b]), this implies that the σ_{mn} and the τ_{mn} both constitute frames. Moreover (see [35]), both the σ_{mn} and the τ_{mn} constitute bases in $L^2(\mathbb{R})$.

It is now easy to construct the example previously announced. Take $h(x) = \tau(x)$. Then h is of compact support, and h belong to $\mathcal{H}_{5/2-\epsilon}(\mathbb{R})$, for all $\epsilon > 0$, because of (2.4.5). Since the $h_{mn}(x) = 2^{-m/2}h(2^{-m}x - n)$ constitute a frame, we can construct the dual frame, $(h_{mn})^{\sim}$. Since the h_{mn} are linearly independent, however, and because of (2.4.4), we find $(h_{mn})^{\sim} = \sigma_{mn}$. Since $\sigma \in \mathcal{H}_s(\mathbb{R})$ only if $s < 1/2 - \epsilon$, we see that *all* the functions in the dual frame are much less "nice" than those in the original frame; in particular, they are discontinuous, while h itself, together with its first derivative, is continuous.

2) Weyl–Heisenberg Frames in Sobolev Spaces: The examples in the preceding subsection show that one must be wary when trying to generalize frames, and the associated expansions, to other function spaces than $L^2(\mathbb{R})$. One can however apply the techniques used in Section II-C to extend the notion of frame to a "strip" of Sobolev spaces $\mathcal{H}_s$, with $|s| < s_0$.

Proposition 2.11: Define (as in Section II-C-1)

$$m(\hat{g}; p_0) = \inf_{x \in \mathbb{R}} \sum_{m \in \mathbb{Z}} |\hat{g}(x + mp_0)|^2,$$

$$M(\hat{g}; p_0) = \sup_{m \in \mathbb{Z}} \sum_{m \in \mathbb{Z}} |\hat{g}(x + mp_0)|^2.$$

Assume that $m(\hat{g}; p_0) > 0$, $M(\hat{g}; p_0) < \infty$. Define, for $s \ge 0$,

$$\beta_s^{\pm}(y) = \sup_x (1 + x^2)^{\mp s/2}\big[1 + (x + y)^2\big]^{\pm s/2} \cdot \sum_{m \in \mathbb{Z}} |\hat{g}(x + mp_0)|\,|\hat{g}(x + mp_0 + y)|.$$

If

$$\sum_{k \ne 0} \left[\beta_s^{+}\left(\frac{2\pi}{q_0}k\right)\beta_s^{-}\left(-\frac{2\pi}{q_0}k\right)\right]^{1/2} < m(\hat{g}; p_0) \quad (2.4.6)$$

then the operator $\mathbb{T}$,

$$\mathbb{T} = \sum_{m,n} g_{mn}\langle g_{mn}, \cdot\rangle$$

is bounded, with a bounded inverse, on both $\mathcal{H}_s$ and $\mathcal{H}_{-s}$. In particular, for all $f_1 \in \mathcal{H}_s$, $f_2 \in \mathcal{H}_{-s}$,

$$A_s\|f_1\|_s \le \|\mathbb{T}f_1\|_s \le B_s\|f_1\|_s \quad (2.4.7)$$

$$A_s\|f_2\|_{-s} \le \|\mathbb{T}f_2\|_{-s} \le B_s\|f_2\|_{-s} \quad (2.4.8)$$

where

$$A_s = \frac{2\pi}{q_0}\left\{m(\hat{g}; p_0) - \sum_{k \ne 0}\left[\beta_s^{+}\left(\frac{2\pi}{q_0}k\right)\beta_s^{-}\left(-\frac{2\pi}{q_0}k\right)\right]^{1/2}\right\} \quad (2.4.9)$$

and

$$B_s = \frac{2\pi}{q_0}\left\{M(\hat{g}; p_0) + \sum_{k \ne 0}\left[\beta_s^{+}\left(\frac{2\pi}{q_0}k\right)\beta_s^{-}\left(-\frac{2\pi}{q_0}k\right)\right]^{1/2}\right\}. \quad (2.4.10)$$

Proof: We shall show that, for all $f \in \mathcal{H}_s$,

$$A_s\|f\|_s\|f\|_{-s} \le \langle f, \mathbb{T}f\rangle \le B_s\|f\|_s\|f\|_{-s}. \quad (2.4.11)$$

Here, as before, we use the notation $\langle\ ,\ \rangle$ for the duality extending the L^2-inner product,

$$\langle f_2, f_1\rangle = \int dx\, \overline{f_2(x)} f_1(x).$$

With respect to $\langle\ ,\ \rangle$, the spaces $\mathcal{H}_s$ and $\mathcal{H}_{-s}$ are each other's dual. On the other hand $\mathbb{T}$ is symmetric with respect to $\langle\ ,\ \rangle$. Since $\mathcal{H}_s$ is dense in $\mathcal{H}_{-s}$, it follows therefore from the upper bound in (2.4.11) that $\mathbb{T}$ is bounded on both $\mathcal{H}_s$ and $\mathcal{H}_{-s}$, with norm smaller than B_s. Similarly the lower bound in (2.4.11) implies that, for all $f_1 \in \mathcal{H}_s$, $f_2 \in \mathcal{H}_{-s}$, $\|\mathbb{T}f_1\|_s \ge A_s\|f_1\|_s$, and $\|\mathbb{T}f_2\|_{-s} \ge A_s\|f_2\|_{-s}$. Again using the symmetry of $\mathbb{T}$ with respect to $\langle\ ,\ \rangle$ one easily checks that this implies that $\mathbb{T}$ is invertible, with a bounded inverse with norm smaller than A_s^{-1}, both on $\mathcal{H}_s$ and $\mathcal{H}_{-s}$.

It remains to prove (2.4.11). This is done along the same lines as in Section II-C.

$$\langle f, \mathbb{T}f\rangle = \frac{2\pi}{q_0}\sum_{m,k}\int dx\, \hat{g}(x + mp_0)\hat{g}\left(x + mp_0 + \frac{2\pi}{q_0}k\right)^* \cdot \hat{f}(x)^*\hat{f}\left(x + \frac{2\pi}{q_0}k\right)$$

$$= \frac{2\pi}{q_0}\int dx\left[(1 + x^2)^{s/2}|\hat{f}(x)|\right] \cdot \left[(1 + x^2)^{-s/2}|\hat{f}(x)|\right]\sum_m |\hat{g}(x + mp_0)|^2 + r,$$

with

$$|r| \le \frac{2\pi}{q_0} \sum_{k \neq 0} \left\{ \int dx\,(1+x^2)^s |\hat{f}(x)|^2 \sum_m (1+x^2)^{-s/2} \right.$$
$$\left. \cdot \left[1+\left(x+\frac{2\pi}{q_0}k\right)^2\right]^{s/2} |\hat{g}(x+mp_0)| \left|\hat{g}\left(x+mp_0+\frac{2\pi}{q_0}k\right)\right| \right\}^{1/2}$$
$$\cdot \left\{ \int dx\,(1+x^2)^{-s} |\hat{f}(x)|^2 \sum_m (1+x^2)^{s/2} \right.$$
$$\left. \cdot \left[1+\left(x-\frac{2\pi}{q_0}k\right)^2\right]^{-s/2} |\hat{g}(x+mp_0)| \left|\hat{g}\left(x+mp_0-\frac{2\pi}{q_0}k\right)\right| \right\}^{1/2}$$
$$\le \frac{2\pi}{q_0} \|f\|_s \|f\|_{-s} \sum_{k\neq 0} \left[\beta_s^+\left(\frac{2\pi}{q_0}k\right)\beta_s^-\left(\frac{2\pi}{q_0}k\right)\right]^{1/2}.$$

The bounds (2.4.11) then follow immediately. □

Remarks:

1) If the conditions of Proposition 2.11 are satisfied, then, since $\mathbb{T}$ is invertible, and has bounded inverse on $\mathscr{H}_s$, we have
$$\tilde{g}_{mn} = \mathbb{T}^{-1} g_{mn} \in \mathscr{H}_s$$
and, for all $f \in \mathscr{H}_s$,
$$f = \sum_{m,n} \tilde{g}_{mn} \langle g_{mn}, f\rangle$$
where the series converges in $\mathscr{H}_s$: the frame expansion holds on $\mathscr{H}_s$.

2) If $\mathbb{T}$ is a bounded and invertible operator on both $\mathscr{H}_s$ and $\mathscr{H}_{-s}$, then, by standard (highly nontrivial) interpolation theorems (see e.g., Section IX-4 in [64]), it is automatically bounded and invertible on any $\mathscr{H}_{s'}$ with $|s'| \le s$. Proposition 2.11 gives thus a sufficient condition for the frame expansion formula to hold on a "strip" of Sobolev spaces. To obtain a wide "strip," for fixed p_0, it is clear from (2.4.6) that we have to choose q_0 sufficiently small. Typically, we would expect the critical value $q_0^c(s)$ (the value of q_0 for which equality occurs in (2.4.6)) to be a decreasing function of s. Note that one needs to impose a decay condition on $\beta^{\pm}(s)$, similar to (2.3.13), to ensure the existence of $q_0^c(s)$.

3) Since $\sum_{m\in\mathbb{Z}} |\hat{g}(x+mp_0)||\hat{g}(x+mp_0+y)|$ is periodic in x, with period p_0, and since $[1+(x+y)^2]/(1+x^2)$ has its maximum, resp. minimum, at $x = -y/2 \pm \operatorname{sgn}(y)\sqrt{1+y^2/4}$, it suffices, for numerical computation of $\beta_s^{\pm}(y)$, to take the supremum over values of x in an interval of length p_0 around $-y/2 \pm \operatorname{sgn}(y)\sqrt{1+y^2/4}$.

Example: For $g(x) = \pi^{-1/4}\exp(-x^2/2)$, Table VI-A gives the values of A_s and B_s for a few values of s, for $p_0 = \pi/2$, $q_0 = 1$. Table VI-B shows how $q_0^c(s)$ changes with s, for fixed $p_0 = \pi/2$.

TABLE VI-A
FRAME BOUNDS*

s	A_s	B_s	B_s/A_s
0.0	3.853	4.147	1.076
1.0	3.852	4.148	1.077
2.0	3.849	4.151	1.079
3.0	3.836	4.164	1.086
4.0	3.787	4.213	1.112
5.0	3.600	4.400	1.222
6.0	2.865	5.135	1.793

*In the Sobolev spaces $\mathscr{H}_s$ for Weyl–Heisenberg frames based on the Gaussian $g(x) = \pi^{1/4}\exp(-x^2/2)$, with $p_0 = \pi/2$, $q_0 = 1.0$, for changing values of s. For $s = 7$, our estimate for A_s becomes negative: the frame breaks down.

TABLE VI-B*

s	$q_0^c(s)$
0.0	3.99
1.0	2.28
2.0	1.72
3.0	1.42
4.0	1.24
5.0	1.13
6.0	1.06
7.0	0.99

*The critical value $q_0^c(s)$ for the translation parameter q_0, as a function of the index s of the Sobolev space in which the frame is consdered. Values of q_0 smaller than $q_0^c(s)$ lead to a frame in $\mathscr{H}_s$ and $\mathscr{H}_{-s}$. The basic function here is $g(x) = \pi^{1/4}\exp(-x^2/2)$; $p_0 = \pi/2$ is fixed.

3) Wavelet Frames in Sobolev Spaces: Since the techniques used for proving frame bounds in $L^2(\mathbb{R})$ were essentially the same for the wavelet case as for the Weyl–Heisenberg case, it is not surprising that we can prove the following proposition.

Proposition 2.12: Define (as in Theorem 2.7)

$$m(h;a_0) = \inf_x \sum_{m\in\mathbb{Z}} |\hat{h}(a_0^m x)|^2,$$
$$M(h;a_0) = \sup_x \sum_{m\in\mathbb{Z}} |\hat{h}(a_0^m x)|^2.$$

Assume that $m(h;a_0) > 0$, $M(h;a_0) < \infty$. Define, for $s \ge 0$,

$$\beta_s^{\pm}(y) = \sup_x (1+x^2)^{\mp s/2} \sum_{m\in\mathbb{Z}} |\hat{h}(a_0^m x)||\hat{h}(a_0^m x + y)| \cdot \left[1+(x+a_0^{-m}y)^2\right]^{\pm s/2}.$$

If

$$\sum_{k\neq 0}\left[\beta_s^+\left(\frac{2\pi}{b_0}k\right)\beta_s^-\left(-\frac{2\pi}{b_0}k\right)\right]^{1/2} < m(h;a_0) \quad (2.4.12)$$

then the operator $\mathbb{T}$,

$$\mathbb{T} = \sum_{m,n} h_{mn}\langle h_{mn},\cdot\rangle$$

is bounded, with a bounded inverse, on both $\mathscr{H}_s$ and

$\mathscr{H}_{-s}$. In particular, for all $f_1 \in \mathscr{H}_s$, $f_2 \in \mathscr{H}_{-s}$,

$$A_s \|f_1\|_s \le \|\mathbb{T} f_1\|_s \le B_s \|f_1\|_s \tag{2.4.13}$$

$$A_s \|f_2\|_{-s} \le \|\mathbb{T} f_2\|_{-s} \le B_s \|f_2\|_{-s} \tag{2.4.14}$$

where

$$A_s = \frac{2\pi}{b_0}\left\{ m(h;a_0) - \sum_{k \ne 0}\left[\beta_s^+\left(\frac{2\pi}{b_0}k\right)\beta_s^-\left(-\frac{2\pi}{b_0}k\right)\right]^{1/2}\right\}$$

$$B_s = \frac{2\pi}{p_0}\left\{ M(h;a_0) - \sum_{k \ne 0}\left[\beta_s^+\left(\frac{2\pi}{b_0}k\right)\beta_s^-\left(-\frac{2\pi}{b_0}k\right)\right]^{1/2}\right\}.$$

Proof: The proof is entirely analogous to the proof of Proposition 2.11. □

Remarks:

1) If $\mathbb{T}$ is bounded, with a bounded inverse, on $\mathscr{H}_s$, then

$$(h_{mn})^{\sim} = \mathbb{T}^{-1} h_{mn} \in \mathscr{H}_s$$

and, for all $f \in \mathscr{H}_s$,

$$f = \sum_{m,n} (h_{mn})^{\sim} \langle h_{mn}, f\rangle \tag{2.4.15}$$

where the series converges in $\mathscr{H}_s$. This means that the phenomenon illustrated by the examples in Section II-D-1 cannot happen, at least in $\mathscr{H}_s$, if h satisfies the conditions in Proposition 2.12.

2) By interpolation (see e.g., Section IX-4 in [64]), one finds that if (2.4.13), (2.4.14) hold in $\mathscr{H}_s$, $\mathscr{H}_{-s}$, respectively, then $\mathbb{T}$ is bounded, with a bounded inverse, on all $\mathscr{H}_{s'}$ with $|s'| \le s$. Under the conditions in Proposition 2.12, the frame expansion (2.4.15) is therefore valid in all $\mathscr{H}_{s'}$ with $|s'| \le s$.

3) Typically (as in the Weyl–Heisenberg case), we expect that, for fixed a_0, the critical value $b_0^c(s)$ (i.e., the value of b_0 for which equality holds in (2.4.12)) decreases with s. It is a remarkable fact that for Y. Meyer's basis, the wavelet expansion is valid in $\mathscr{H}_s$, for all $s \in \mathbb{R}$, for $a_0 = 2$, and for *fixed* $b_0 = 1$. Something similar is true for the tight frames constructed in Section II-B-2b). However, for general functions h, leading to nontight frames, we rather expect $b_0^c(s)$ to decrease with s.

4) For a given function h, the strip of Sobolev spaces for which the h_{mn} constitute a frame, i.e., the possible values of s for which (2.4.13), (2.4.14) holds, is constrained by the behavior of $\hat{h}$ around zero. If $\hat{h}(x) = O(x^a)$ for $x \to 0$, then clearly (see the definition of β_s^+)

$$\beta_s^+(y) \ge C\left[\inf_{|u| \le a_0^{-M}} |\hat{h}(y+u)|\,|y+u|^s\right] \sum_{m=-\infty}^{-M} a_0^{m(\alpha - s)}.$$

This diverges for $s \ge \alpha$.

TABLE VII
FRAME BOUNDS*

	$b_0 = 1.0$				$b_0 = 1.25$		
s	A_s	B_s	B_s/A_s	s	A_s	B_s	B_s/A_s
0.00	3.223	3.596	1.116	0.00	2.001	3.454	1.726
0.25	3.223	3.596	1.116	0.25	1.995	3.460	1.734
0.50	3.222	3.597	1.116	0.50	1.985	3.470	1.748
0.75	3.221	3.597	1.117	0.75	1.971	3.484	1.768
1.00	3.220	3.599	1.118	1.00	1.953	3.502	1.793
1.25	3.218	3.600	1.119	1.25	1.926	3.529	1.832
1.50	3.216	3.602	1.120	1.50	1.877	3.578	1.906
1.75	3.214	3.605	1.122	1.75	1.794	3.661	2.041

In the Sobolev spaces $\mathscr{H}_s$ for wavelet frames based on the Mexican hat function $h(x) = 23^{-1/2}\pi^{-1/4}(1-x^2)\exp(-x^2/2)$, with $a_0 = 2.0$, $b_0 = 1.0$ and 1.25 for changing values of s.

Example: Table VII gives A_s and B_s, for a few values of s, for the Mexican hat function $h(x) = 2/\sqrt{3}\,\pi^{-1/4}(1-x^2)e^{-x^2/2}$, for $a_0 = 2.0$, and for $b_0 = 1.0$ and 1.25.

III. Phase Space Localization

Let us recall the intuition, already mentioned in the introduction, which leads us to expect the phase space localization results we shall give here.

Both the Weyl–Heisenberg coherent states and the wavelets can be used to give a representation in the time-frequency plane of time-dependent signals, provided the basic functions g or h and the parameters p_0, q_0 or a_0, b_0 are suitably chosen (see Section II). A graphical picture of these representations is given in Fig. 1. Ideally, one would like these representations to be reasonably "sharp." If e.g., the pair (m_0, n_0) corresponds to the time t_0 and the frequency ω_0 (see Fig. 1), then we would wish that the frequency content, in a band around ω_0, of the signal f, during a time interval around t_0, is essentially mirrored by the coefficients $\langle g_{mn}, f\rangle$ or $\langle h_{mn}, f\rangle$ with m close to m_0, n close to n_0. It is intuitively clear that the basic analyzing functions g or h have to be themselves well localized in time and frequency for such "sharpness" of the associated representations to be attainable.

In this section we shall try to make these qualitative statements more precise. We shall do this by giving a sense to the "sharpness" of the time-frequency representation, and by showing how the localization, in time and frequency, of the analyzing functions g or h matters. The Weyl–Heisenberg case shall be discussed in Section III-A, the wavelet case in Section III-B. In both cases we shall see that signals that are essentially limited to a given finite time interval and to a given finite range in frequency can essentially be represented by a finite number of expansion coefficients. (All these qualifications will be made more quantitative next). We shall use this fact to explain, in Section III-C, a phenomenon that was first noticed by J. Morlet in numerical calculations. To reconstruct a signal f with a precision ϵ, it is sufficient, for some frames, to calculate the expansion coefficients with a precision $C\epsilon$, where C turns out to be significantly larger than would be expected for orthonormal bases. We have called this phenomenon the "reduction of calculational noise" (for example, it often reduces round-off errors), and we explain it in Section III-C.

A. The Weyl–Heisenberg Case

Assume that g is normalized, $\int dx\,|g(x)|^2 = 1$, and that

$$\int dx x|g(x)|^2 = 0 = \int dy y|\hat{g}(y)|^2.$$

Then g_{mn} is localized around the phase space point (mp_0, nq_0),

$$\int dx x|g_{mn}(x)|^2 = nq_0, \qquad \int dy y|(g_{mn})^\wedge(y)|^2 = mp_0.$$

Suppose, on the other hand, that we restrict ourselves to the analysis and reconstruction of signals that are essentially time-limited to the interval $[-T,T]$ and essentially band-limited (limited in frequency) to the interval $[-\Omega,\Omega]$. We have introduced the word "essentially" in this statement because, as is well-known, no function f can have both a compact support and a Fourier transform with compact support. A more precise statement of the "essential" time- and band-limitedness is

$$\|(1-Q_T)f\| \ll \|f\|$$
$$\|(1-P_\Omega)f\| \ll \|f\|$$

where

$$(Q_T f)(t) = \chi_{[-T,T]}(t)f(t)$$
$$(P_\Omega f)^\wedge(w) = \chi_{[-\Omega,\Omega]}(w)\hat{f}(w)$$

and where χ_I denotes the indicator function of the interval I. Effectively, these limitations single out a rectangle of phase space as more important than other regions.

We shall assume, in what follows, that the g_{mn} constitute a frame, with frame bounds A and B and with dual frame $\tilde{g}_{mn}$. The reconstruction formula, valid for all functions in $L^2(\mathbb{R})$, and therefore in particular for the functions f of interest here, is (see Section II-A)

$$f = \sum_{m,n\in\mathbb{Z}} \tilde{g}_{mn}\langle g_{mn}, f\rangle. \tag{3.1}$$

If the g_{mn} are "well localized" in phase space around (mp_0, nq_0), then it is to be expected that $\langle g_{mn}, f\rangle$ will be small if the distance, in phase space, from (mp_0, nq_0) to the rectangle $[-\Omega,\Omega]\times[-T,T]$, is large. In other words, one expects that only those m,n for which (mp_0, nq_0) lies in or close to $[-\Omega,\Omega]\times[-T,T]$ will play a significant role in the reconstruction (3.1) of f. Fig. 6 represents the situation. As in Fig. 1(a), the g_{mn} are represented by their "phase space centers" (mp_0, nq_0).

The signal f is essentially concentrated, in phase space, on the rectangle $[-\Omega,\Omega]\times[-T,T]$. Under suitable conditions on g, it turns out that, for all $\epsilon > 0$, there exist $t_\epsilon, \omega_\epsilon$ such that the partial reconstruction of f using only the (finitely many) m,n associated to the "enlarged rectangle" $[-(\Omega+\omega_\epsilon),(\Omega+\omega_\epsilon)]\times[-(T+t_\epsilon),(T+t_\epsilon)]$ (in

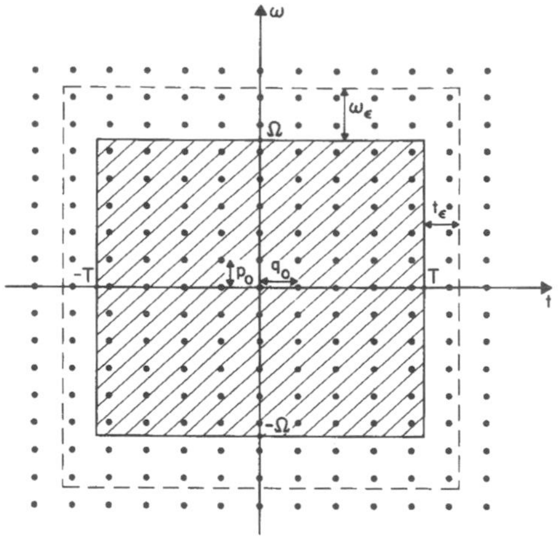

Fig. 6. Rectangular lattice (nq_0, mp_0) indicating localization of g_{mn}, and rectangle in phase space $[-T,T]\times[-\Omega,\Omega]$ on which signal f is mainly concentrated. Coefficients $\langle g_{mn}, f\rangle$ corresponding to (nq_0, mp_0) in enlarged rectangle (in dashed lines) suffice to reconstruct f up to error of order ϵ.

dashed lines in the figure), is equal to f, up to an error of order ϵ. This is essentially the content of the following theorem.

Theorem 3.1: Suppose that the $g_{mn}(x) = e^{imp_0x}g(x - nq_0)$ constitute a frame, with frame bounds A, B. Assume that

$$|g(x)| \le C(1+x^2)^{-\alpha}, \qquad |\hat{g}(y)| \le C(1+y^2)^{-\alpha}$$

for some $C < \infty$, $\alpha > 1/2$. Then, for any $\epsilon > 0$, there exist $t_\epsilon, \omega_\epsilon > 0$ such that, for all $f \in L^2(\mathbb{R})$, and for all $T, \Omega > 0$,

$$\left\| f - \sum_{\substack{|mp_0| \le \Omega + \omega_\epsilon \\ |nq_0| \le T + t_\epsilon}} \tilde{g}_{mn}\langle g_{mn}, f\rangle \right\| \le \sqrt{\frac{B}{A}}\,\big[\|(\mathbb{1}-P_\Omega)f\| + \|(\mathbb{1}-Q_T)f\| + \epsilon\|f\|\big] \tag{3.2}$$

where the $\tilde{g}_{mn}$ constitute the dual frame to the g_{mn}.

Proof: See Appendix D. □

Remarks:

1) In the limit as $\epsilon \to 0$, one finds $t_\epsilon \to \infty$ or $\omega_\epsilon \to \infty$. (With our proof, in Appendix D, both t_ϵ and ω_ϵ tend to ∞ as $\epsilon \to 0$. If g or $\hat{g}$ has compact support, then t_ϵ or ω_ϵ can be kept finite. In all cases at least one of $t_\epsilon, \omega_\epsilon$ must diverge as $\epsilon \to 0$.) This is natural; infinite precision cannot be obtained when using only a finite set of g_{mn}. In particular, for $g(x) = \pi^{-1/4}\exp(-x^2/2)$, one finds $\omega_\epsilon, t_\epsilon \sim_{\epsilon\to 0} C|\log\epsilon|^{1/2}$.
2) Note that the decay conditions on g and $\hat{g}$ exclude all orthonormal bases g_{mn}, by Theorem 2.3.
3) The important fact about Theorem 3.1 is that $t_\epsilon, \omega_\epsilon$ are *independent* of T, Ω: the "enlargement proce-

dure" depends only on ϵ, the desired precision of the finite construction. For fixed ϵ, the number of points $N_\epsilon(T,\Omega)$ used in this finite reconstruction is $4(T+t_\epsilon)(\Omega+\omega_\epsilon)/p_0\cdot q_0$. Hence

$$\lim_{T,\Omega\to\infty}(4T\Omega)^{-1}N_\epsilon(T,\Omega)=(p_0\cdot q_0)^{-1}, \quad (3.3)$$

which is independent of ϵ. This can be compared with an analogous result involving prolate spheroidal wave functions (see [65], [66]; the formula discussed here is explained very clearly in [66]). The prolate spheroidal wave functions are the eigenfunctions of the compact operator $P_\Omega Q_T P_\Omega$ (which also describes localization in phase space, singling out the rectangle $[-\Omega,\Omega]\times[-T,T]$). Signals f that are essentially time-limited to $[-T,T]$ and bandlimited to $[-\Omega,\Omega]$ can be expanded in prolate spheroidal wave functions. Again, if a reconstruction up to a finite error ϵ is desired, one can truncate the expansion to a finite sum. The number, $n_\epsilon(T,\Omega)$, of terms needed is the number of eigenfunctions of $P_\Omega Q_T P_\Omega$ with eigenvalue larger than ϵ. One has [65], [66],

$$n_\epsilon(T,\Omega)=\frac{2T\Omega}{\pi}+C\epsilon\log(T\Omega)+o(\epsilon).$$

In this case

$$\lim_{T,\Omega\to\infty}(4T\Omega)^{-1}n_\epsilon(T,\Omega)=(2\pi)^{-1}. \quad (3.4)$$

This limit is exactly the Nyquist density. In fact, this result was the first rigorous formulation of the intuitive Nyquist density idea [65]. If we compare (3.3) with (3.4), then we see that in our present approach, the density $(p_0\cdot q_0)^{-1}$ is higher than the Nyquist density $(2\pi)^{-1}$. We know indeed (see Section II-C-1a)) that $p_0\cdot q_0\le 2\pi$ for all frames $\{g_{mn};\ m,n\in\mathbb{Z}\}$. For the examples with good phase space localization (measured by e.g., higher moments of $|g|$ and $|\hat{g}|$), we even have, by Theorem 2.3, that $p_0\cdot q_0<2\pi$. The "oversampling" of our phase space density with respect to the optimal Nyquist density, measured by the ratio $2\pi(p_0\cdot q_0)^{-1}>1$, is the price we have to pay for the fact that the frames of interest to us are, in general, not orthonormal. We also gain something with respect to the prolate spheroidal wave functions, however. The construction of the different g_{mn}, obtained from one function g by translations in phase space, is simpler than the construction of the (orthonormal) prolate spheroidal wave functions.

Note that (3.3) can in fact be considered as a *definition* of the phase space density for the frame under consideration. While we have used the term "phase space density" before in heuristic discussions, (3.3) is the first mathematically precise statement justifying this terminology.

Example: In the Table VIII we give the values of $t_\epsilon=\omega_\epsilon$ corresponding to given ϵ for $g(x)=\pi^{-1/4}\exp(-x^2/2)$, in the cases $p_0=q_0=\pi^{1/2}$ and $p_0=q_0=(\pi/2)^{1/2}$. In each case we have used (D.6) in the estimates.

TABLE VIII
THE "ENLARGEMENT PARAMETER" t_ϵ*

ϵ	$t_\epsilon=\omega_\epsilon$
$p_0=q_0=\sqrt{\pi/2}=1.25$	
0.1	2.52
0.05	2.78
0.01	3.31
0.005	3.51
0.001	3.94
0.0005	4.11
0.0001	4.48
$p_0=q_0=\sqrt{\pi}=1.77$	
0.1	2.55
0.05	2.81
0.01	3.33
0.005	3.54
0.001	3.97
0.0005	4.14
0.0001	4.50

*As function of the error ϵ, for the Weyl-Heisenberg frame of Gaussians, respectively for $p_0=q_0=\sqrt{\pi/2}$, $p_0=q_0=\sqrt{\pi}$ (see text)

B. The Wavelet Case

In this case we assume that h is normalized so that $\int dx\,|h(x)|^2=1$, and that $\int dx\,x|h(x)|^2=0$. Let us assume, for the sake of simplicity, that $|\hat{h}|$ is even (in practice, $\hat{h}$ is either even or odd; even if the frame is constructed by means of a function h with support $\hat{h}\subset\mathbb{R}_+$, then the effective basic wavelets $h^1=\mathrm{Re}\,h$, $h^2=\mathrm{Im}\,h$ satisfy the condition that $|\hat{h}^j|$ is even—see the remarks following Theorem 2.7 and 2.8). Let us also assume that

$$\int_0^\infty dk\,k|\hat{h}(k)|^2=1$$

(this does not imply any loss of generality: it can easily be achieved by dilating h). Then h_{mn} is concentrated around the phase space points $(\pm a_0^{-m}, a_0^m nb_0)$

$$\int dx\,x|h_{mn}(x)|^2=a_0^m nb_0$$

$$\int_0^\infty dk\,k|(h_{mn})^\wedge(k)|^2=a_0^{-m}=-\int_{-\infty}^0 dk\,k|(h_{mn})^\wedge(k)|^2.$$

Again, we suppose we are mainly interested in functions f localized in phase space. In this case, we assume that they are essentially time-limited to $[-T,T]$, and with frequencies $|\omega|$ mainly concentrated in $[\Omega_0,\Omega_1]$, where $0<\Omega_0<\Omega_1<\infty$. The need for a lower bound Ω_0 on the frequencies $|\omega|$, as opposed to the Weyl–Heisenberg case, where we only introduced an upper bound, stems from the logarithmic rather than linear treatment of frequencies by the wavelet transform.

Fig. 7 represents the situation graphically. The dots represent the "phase space localization centers" of the h_{mn}; for the sake of simplicity we have taken $a_0=2$. The

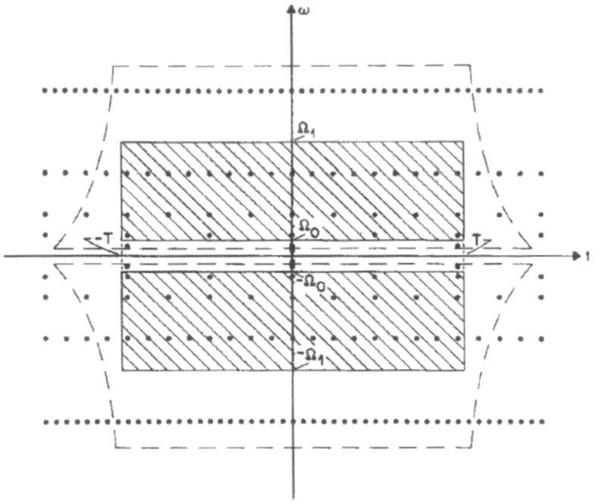

Fig. 7. Lattice $(nb_0a_0^m, \pm a_0^{-m}k_0)$, indicating localization of h_{mn} (see Fig. 1(b)), and two rectangles $[-T,T]\times[\Omega_0,\Omega_1]$, $[-T,T]\times[-\Omega_1,-\Omega_0]$ on which signal f is mainly concentrated. Coefficients $\langle h_{mn}, f\rangle$ corresponding to lattice points within set B_ϵ (in dashed lines) suffice to reconstruct f up to error ϵ.

signals f of interest are essentially concentrated in the set $[-T,T]\times([-\Omega_1,-\Omega_0]\cup[\Omega_0,\Omega_1])$, marked with full lines in the figure.

We shall assume, of course, that the h_{mn} constitute a frame, with frame bounds A, B and with dual frame $(h_{mn})^\sim$. The associated reconstruction formula is then

$$f=\sum_{m,n\in\mathbb{Z}}(h_{mn})^\sim\langle h_{mn},f\rangle. \tag{3.5}$$

This formula is valid for all functions in $L^2(\mathbb{R})$. For signals essentially concentrated in $[-T,T]\times([-\Omega_1,-\Omega_0]\cup[\Omega_0,\Omega_1])$ we can again restrict ourselves to a finite subset $B_\epsilon(T,\Omega_0,\Omega_1)$ of indexes, provided the basic wavelet h is itself sufficiently concentrated. Partial reconstruction of such a signal f, using only the pairs $(m,n)\in B_\epsilon$, is then an approximation of f with an error at most equal to ϵ. The set B_ϵ includes all the (m,n) for which $\Omega_0\le a_0^{-m}\le\Omega_1$, and $|a_0^m nb_0|\le T$; it is indicated in dashed lines in Fig. 7. The following theorem gives an exact statement of this result.

Theorem 3.2: Suppose that the

$$h_{mn}(x)=a_0^{-m/2}h(a_0^{-m}x-nb_0)$$

constitute a frame, with frame bounds A, B, and dual frame $(h_{mn})^\sim$. Assume that

$$|\hat{h}(y)|\le C|y|^\beta(1+y^2)^{-(\alpha+\beta)/2}$$

where $\beta>0$, $\alpha>1$, and that, for some $\gamma>1/2$

$$\int dx(1+x^2)^\gamma|h(x)|^2<\infty.$$

Fix $T>0$, $0<\Omega_0<\Omega_1$. Then, for any $\epsilon>0$, there exists a finite subset $B_\epsilon(T,\Omega_1,\Omega_2)$ of $\mathbb{Z}^2$ such that, for all $f\in L^2(\mathbb{R})$,

$$\Big\|f-\sum_{(m,n)\in B_\epsilon(T,\Omega_1,\Omega_2)}(h_{mn})^\sim\langle h_{mn},f\rangle\Big\|$$
$$\le(B/A)^{1/2}\Big[\|(\mathbb{1}-Q_T)f\|+\|(\mathbb{1}-P_{\Omega_1}+P_{\Omega_0})f\|+\epsilon\|f\|\Big]. \tag{3.6}$$

Proof: See Appendix D. □

Remarks:

1) The "enlargement procedure," i.e., the transition from the original "box" $[-T,T]\times([-\Omega_1,-\Omega_0]\cup[\Omega_0,\Omega_1])$ to $B_\epsilon(T,\Omega_0,\Omega_1)$, is more complicated here than in the Weyl–Heisenberg case; the parameters m_0, m_1, and t also depend on Ω_0,Ω_1 and not only on ϵ (see the proof in Appendix D).
2) The construction of the set B_ϵ (see also Fig. 7) exhibits the fact that wavelets give a higher resolution at high frequencies than at low frequencies. For fixed m, the "extension" of the box in the time-variable, as defined by (D.7), corresponds to consideration of all the pairs (m,n) with time-localization $|a_0^m nb_0|\le T+a_0^m t$. The "extension" $a_0^m t$ thus depends on m and is small for large negative values of m, which correspond to high frequencies.
3) A crude estimate leads to

$$\#B_\epsilon(T,\Omega_0,\Omega_1)$$
$$\le C\frac{2T}{b_0\ln a_0}\left(\epsilon^{-1/\beta}\Omega_1-\epsilon^{1/\alpha-\kappa}\Omega_0\right)$$
$$+C\frac{\epsilon^{-2}}{b_0\ln a_0}\left(C+(\ln\Omega_1-\ln\Omega_0)^2\right). \tag{3.7}$$

If $\Omega_0\to0$, T, and $\Omega_1\to\infty$, then

$$(4T\Omega_1)^{-1}[\#B_\epsilon(T,\Omega_0,\Omega_1)]\underset{\to}{\le}\frac{C\epsilon^{-1/\beta}}{2b_0\ln a_0} \tag{3.8}$$

which is not independent of ϵ. While the estimate (3.7) is admittedly crude, finer estimates lead to a similar result. This is in contrast with the analogous result for the Weyl–Heisenberg case (see (3.3)) where the corresponding limit was independent of ϵ, and gave the phase space density of the frame. In fact, it gave a procedure to *define* the phase space density of the frame. Because of the ϵ-dependence of the right-hand side of (3.7), we cannot use the same procedure to define a phase space density corresponding to wavelet frames. This illustrates again (see also the discussion in Section II-C-2a)) that the concept "phase space density" is not well-suited to the wavelet representation.
4) The estimates made in the proof of Theorem 3.2 lead to rather crude bounds. As in the Weyl–Heisenberg case, it is possible to write more complicated, but less coarse estimates. For the choice

$h(x) = 2.3^{-1/2}\pi^{-1/4}(1-x^2)e^{-x^2/2}$, $a_0 = 2.0$, $b_0 = 0.5$, with $\Omega_0 = 10$, $\Omega_1 = 1000$, and $T = 100$, one finds

ϵ	m_0	m_1	t
0.1	−11	−2	4.95
0.05	−12	−2	4.97
0.01	−13	−2	5.75.

C. The Reduction of Calculational Noise

J. Morlet noticed, some time ago, that in numerical wavelet calculations, it often sufficed to calculate the wavelet coefficients to a precision of, say, 10^{-2}, to be able to reconstruct the original signal with a precision of, say, 10^{-3}. This rather surprising fact can be explained as a consequence of both phase space localization, and "oversampling."

Phase space localization is necessary to restrict oneself to a finite number of coefficients. We cannot hope to control an infinite number of coefficients if they can all induce an error of the same order. The role of "oversampling" is the following. Let us go back to the frame operator T (not to be confused with the time-limit T in Sections III-A and III-B) defined in Sections I-C and II-A. We have

$$T: L^2(\mathbb{R}) \to l^2(\mathbb{Z}^2)$$
$$(Tf)_{m,n} = \langle \phi_{mn}, f \rangle$$

where ϕ_{mn} stands for either h_{mn} or g_{mn}. Since the ϕ_{mn} constitute a frame, this operator is bounded and has a bounded inverse on its closed range. The operator T is onto (its range is all of $l^2(\mathbb{Z}^2)$) if and only if the ϕ_{mn} constitute a basis. In general, however, the ϕ_{mn} are not independent, and the range of T ($Ran T$) is a proper subspace of $l^2(\mathbb{Z}^2)$. The inversion procedure,

$$f = \sum_{m,n} (\phi_{mn})^{\sim} \langle \phi_{mn}, f \rangle$$

when applied to elements c of $l^2(\mathbb{Z}^2)$ not necessarily in $Ran\ T$:

$$\sum_{m,n} (\phi_{mn})^{\sim} c_{mn}$$

in fact consists of 1) a projection of $l^2(\mathbb{Z}^2)$ onto $RanT$, 2) the inversion of T on its range (see Section II-A). We shall model the finite precision of numerical calculations by adding random "noise" to the coefficients $\langle \phi_{mn}, f \rangle$, thus leading to modified coefficients $c_{mn}(f)$. The "noise" component of these coefficients "lives" on all of $l^2(\mathbb{Z}^2)$. If we apply the inversion procedure, this component will therefore be reduced in norm by the projection onto $RanT$. This reduction will be the more pronounced the "smaller" $RanT$ is, as a subspace of $l^2(\mathbb{Z}^2)$, i.e., the more pronounced the oversampling of redundancy in the frame. The calculations below show how this work in practice.

Let us assume that we are interested in signals f that are essentially localised in the time interval $[-T, T]$, and in the frequency range $[-\Omega, \Omega]$ (in the Weyl–Heisenberg case) or $[-\Omega_1, -\Omega_0] \cup [\Omega_0, \Omega_1]$ (in the wavelet case), i.e.,

$$\|(\mathbb{1} - Q_T)f\| \le \epsilon \|f\|$$
$$\|(\mathbb{1} - P_\Omega)f\| \le \epsilon\|f\| \quad \text{or} \quad \|(\mathbb{1} - P_{\Omega_1} + P_{\Omega_0})f\| \le \epsilon\|f\|.$$

Then, by Theorems 3.1 and 3.2, there exists an "enlarged box" B_ϵ such that

$$\left\| f - \sum_{(m,n)\in B_\epsilon} (\phi_{mn})^{\sim} \langle \phi_{mn}, f \rangle \right\| \le 3(B/A)^{1/2} \epsilon \|f\|$$

where ϕ denotes either g or h. Since B_ϵ is a finite subset of $\mathbb{Z}^2$, we restrict ourselves therefore to the finitely many coefficients $\langle \phi_{mn}, f \rangle$.

In practical calculations, the coefficients $\langle \phi_{mn}, f \rangle$ will be computed with finite precision. Let us take the following model for the errors. Assume that the coefficients to be used in the calculations are given by

$$c_{mn}(f) = \langle \phi_{mn}, f \rangle + \gamma_{mn}$$

where the γ_{mn} are independent identically distributed random variables, with mean zero, and with variance α^2,

$$\mathbb{E}(\gamma_{mn}^2) = \alpha^2.$$

This means that the $\langle \phi_{mn}, f \rangle$ are known with "precision" α. Note that our model is only a first approximation. In general the ϕ_{mn}, and hence the $\langle \phi_{mn}, f \rangle$, are not linearly independent, which means that the round-off errors should not be regarded as independent random variables. With this approximation, we find that the estimated error between f and a partial reconstruction, using only the finitely many coefficients associated to $(m,n) \in B_\epsilon$, and even those only with finite precision (i.e., replace $\langle \phi_{mn}, f \rangle$ by $c_{mn}(f)$), is given by

$$\mathbb{E}\left(\left\| f - \sum_{(m,n)\in B_\epsilon} c_{mn}(f)(\phi_{mn})^{\sim} \right\|^2 \right)$$
$$= \mathbb{E}\left(\left\| \left(f - \sum_{(m,n)\in B_\epsilon} (\phi_{mn})^{\sim} \langle \phi_{mn}, f \rangle \right) - \sum_{(m,n)\in B_\epsilon} \gamma_{mn} (\phi_{mn})^{\sim} \right\|^2 \right)$$
$$\le 9(B/A)\epsilon^2\|f\|^2 + A^{-2}\alpha^2 N_\epsilon \tag{3.9}$$

where $N_\epsilon = \# B_\epsilon$, and where we have used $\mathbb{E}(\gamma_{mn}) = 0$, $\mathbb{E}(\gamma_{mn}\gamma_{m'n'}) = \delta_{mm'}\delta_{nn'}\alpha^2$, and $\|(\phi_{mn})^{\sim}\|^2 = \|\mathbb{T}^{-1}\phi_{mn}\|^2 \le A^{-2}\|\phi_{mn}\|^2 = A^{-2}$ (with $\mathbb{T}$ as defined in Section II-A).

The "reduction of calculational noise," observed by J. Morlet, is contained in the second term in (3.9), more particular in the factor $N_\epsilon A^{-2}$. Let us show how.

Assume that we are considering a snug Weyl–Heisenberg frame, g_{mn}. If we assume that B_ϵ is large with respect to the lattice mesh, then (see Section III-A)

$$N_\epsilon = \# B_\epsilon \simeq \frac{4T\Omega}{p_0 \cdot q_0}.$$

On the other hand, if the frame is snug (i.e., $B \simeq A$), we find, by (2.2.9),

$$A \simeq 2\pi(p_0 \cdot q_0)^{-1}$$

(we assume $\|g\| = 1$). Hence

$$N_\epsilon A^{-2} \simeq \pi^{-2} T\Omega(p_0 \cdot q_0). \tag{3.10}$$

If the g_{mn} had constituted an orthonormal basis, then (provided we neglect the loss in phase space localisation due to the use of an orthonormal basis) this factor would have been

$$(N_\epsilon A^{-2})_{\text{orthon. basis}} \simeq \left(\frac{2T\Omega}{\pi}\right). \tag{3.11}$$

The frame gives thus a net gain of $2\pi/p_0 \cdot q_0$ with respect to the orthonormal basis situation.

Something similar happens for wavelets. In this case we don't have such a simple expression for N_ϵ, but we can easily see that the same phenomenon takes place by the following argument. Suppose h, a_0, b_0 are chosen so that the frame is snug (i.e., it is tight for all practical purposes). In particular this means that $m(h; a_0) \simeq M(h; a_0)$ while the β-terms in (2.3.23), (2.3.24) are negligible. Consequently $A \simeq B \simeq 2\pi b_0^{-1} m(h; a_0)$. Consider now the frame with the same h, a_0, but with $b_0' = b_0/2$. This frame will obviously also be snug, with $A' \simeq B' \simeq 2\pi b_0'^{-1} m(h; a_0) = 2A$. On the other hand, there are twice as many points in the graphical representation of this new frame, for every frequency level. Hence $N_\epsilon' = 2N_\epsilon$. Combining these two, we find $N_\epsilon' A'^{-2} = \frac{1}{2} N_\epsilon A^{-2}$, i.e., halving b_0 leads to a gain of 2 in the total error on f, for the same precision on the coefficients.

For the frames used by J. Morlet when he noticed this phenomenon, which were heavily oversampled (e.g., he used up to 15 "voices"—see the end of Section II-C-2a for a definition), a gain factor of 10 or more can be obtained easily. Note, however, that oversampling does not explain completely the observed calculational noise (or quantization noise) reduction. As in vision analysis [38] part of the reduction is a consequence of the fact that, unlike the original signal, the coefficients $c_{mn}(f)$ at every fixed m-level are distributed around zero, with a sharp peak at zero. This makes it possible to drastically reduce the number of quantization steps in the c_{mn}, without significantly altering the quality of the reconstructed signal [38].

IV. Conclusion

We have shown how to characterize functions f by means of a local time-frequency analysis, by computing their inner products with either Weyl–Heisenberg coherent states,

$$c_{mn}(f) = \int dx f(x) g(x - nx_0) e^{im\omega_0 x}$$

or with wavelets,

$$c_{mn}(f) = a_0^{-m/2} \int dx f(x) h(a_0^{-m} x - nb_0).$$

The first approach corresponds to the windowed Fourier transform, the second is the wavelet transform. The wavelet transform handles frequencies in a logarithmic rather than linear way, and seems better adapted to the analysis of acoustic and visual signals than the windowed Fourier transform.

In both cases we have formulated necessary and sufficient conditions for the stable reconstruction of f from the $c_{mn}(f)$. For such a stable reconstruction algorithm to exist, we require that, for some $A > 0$, $B < \infty$, and all $f \in L^2(\mathbb{R})$,

$$A\|f\|^2 \leq \sum_{m,n \in \mathbb{Z}} |c_{mn}(f)|^2 \leq B\|f\|^2. \tag{4.1}$$

We have provided a reconstruction algorithm, which converges at least as fast as a geometric series in $r = B/A - 1$, and we have presented efficient methods for estimating A and B. These methods are illustrated by many examples.

Finally, if g respectively, h is well localized in both time and frequency, and if (4.1) is satisfied, we have shown that the characterization of functions f by their coefficients $c_{mn}(f)$ is truly local in time-frequency. If f is essentially concentrated on a limited region in time-frequency, then only the $c_{mn}(f)$ corresponding to time-frequency points within or close to this region are needed for an approximate reconstruction of f.

Acknowledgment

It is a pleasure for me to thank the many people with whom I have had enjoyable discussions on the subject of this paper, who pointed out relevant references, and who raised interesting questions. First of all, I want to express my gratitude to A. Grossmann, who introduced me to the many properties of coherent states years ago, and more recently stimulated my interest in "frame problems." This paper would not have existed without his continued interest and encouragement. It has been a pleasure to discuss the topics in this paper with him.

The completed proof of the Balian theorem, in Section II-C-1a), is due to R. Coifman and S. Semmes. I want to thank them for letting me include it here. Thanks are also due to R. Howe and T. Steger for solving one of my conjectures and for drawing my attention to reference [49]. The refinements of the frame bounds in Section II-C-2a) are due to Ph. Tchamitchian. Section II-D, on frame bounds and convergence in other spaces than $L^2(\mathbb{R})$, would not have existed without Y. Meyer and Ph. Tchamitchian, who pointed out the problem to me, and who suggested that a modification of the L^2-estimates would lead to useful Sobolev-space estimates.

I would also like to thank P. Deift, H. Feichtinger, P. Jones, H. Landau, P. Lax, R. Littlejohn, M. Sirugue, and D. Thomson for many interesting discussions on one or several subjects of this paper. I am also grateful to C. Tomei and to J. R. Klauder, who suggested the terminology "tight frame" and "snug frame," respectively.

Finally, I would like to thank the referees for making many suggestions that improved the readability of this paper.

Appendix A
Resolutions of the Identity

By explicit computation we have

$$\int dp \int dq \langle f_1, g^{(p,q)}\rangle\langle g^{(p,q)}, f_2\rangle$$

$$= \int dp \int dq \int dx \int dy f_1(x)^* e^{ip(x-y)} g(x-q) g(y-q)^* f_2(y)$$

$$= \int dq \int dx \int dy f_1(x)^* g(x-q) g(y-q)^* f_2(y) 2\pi\delta(x-y)$$

$$= 2\pi \int dq \int dx f_1(x)^* |g(x-q)|^2 f_2(x)$$

$$= 2\pi \int dy \int dx f_1(x)^* f_2(x) |g(y)|^2 = 2\pi \|g\|^2 \langle f_1, f_2\rangle.$$

This proves (1.3a).

In the wavelet case, we work via the Fourier transform,

$$\int \frac{da}{a^2} \int db \langle f_1, h^{(a,b)}\rangle\langle h^{(a,b)}, f_2\rangle$$

$$= \frac{1}{2\pi} \int \frac{da}{a^2} \int db \int dx \int dy \hat{f}_1(x)^* |a| \hat{h}(ax) \hat{h}(ay)^* e^{ib(x-y)} \hat{f}_2(y)$$

$$= \int \frac{da}{a^2} \int dx \int dy \hat{f}_1(x)^* |a| \hat{h}(ax) \hat{h}(ay)^* \hat{f}_2(y) \delta(x-y)$$

$$= \int da \int dx \hat{f}_1(x)^* \hat{f}_2(x) |a|^{-1} |\hat{h}(ax)|^2$$

$$= \int dy \int dx \hat{f}_1(x)^* \hat{f}_2(x) |y|^{-1} |\hat{h}(y)|^2 = C_h \langle f_1, f_2\rangle.$$

Appendix B
Proof of Lemma 2.4

We assume that

$$|G(t,s)| \le b < \infty \tag{B.1}$$

and that G, $\partial_t G$ and $\partial_s G$ are locally square integrable. We show that, together with the "periodicity conditions"

$$G(t+1,s) = G(t,s)$$
$$G(t,s+1) = e^{2\pi i t} G(t,s) \tag{B.2}$$

the extra assumption

$$|G(t,s)| \ge a > 0 \tag{B.3}$$

necessarily leads to a contradiction. To do this we introduce an averaged version G_r of G. More precisely, define, for $r > 0$,

$$G_r(t,s) = \frac{1}{4r^2} \int_{|t'-t|\le r} dt' \int_{|s'-s|\le r} ds' G(t',s').$$

Then G_r is obviously continuous. For $|t-t'| < r$, $|s-s'| < r$ one has

$$|G_r(t,s) - G_r(t',s')| \le br^{-1}(|t-t'| + |s-s'|). \tag{B.4}$$

The function G_r does not satisfy the "periodicity" conditions (B.2), but instead a modified version of them:

$$G_r(t+1,s) = G_r(t,s)$$

$$G_r(t,s+1) = \frac{1}{4r^2} \int_{|t'-t|<r} dt' \int_{|s'-s|<r} e^{2\pi i t'} G(t',s')$$

$$= e^{2\pi i t} G_r(t,s) + \psi(t,s) \tag{B.5}$$

where

$$|\psi(t,s)| \le 2\pi br. \tag{B.6}$$

By choosing r small enough, we can make the deviation of (B.5) from (B.2), as small as desired. The only other ingredient needed is a set of bounds (upper and lower) on $|G_r|$. The upper bound is immediate from (B.1),

$$|G_r(t,s)| \le b.$$

For the lower bound we restrict ourselves to a neighborhood U of $[0,2]^2$. Choose r such that, for all $(t,s) \in U$,

$$\left[\int_{|t'-t|\le 2r} dt' \int_{|s'-s|\le 2r} ds' |\partial_t G(t',s')|^2\right]^{1/2} + \left[\int_{|t'-t|\le 2r} dt' \int_{|s'-s|\le 2r} ds' |\partial_s G(t',s')|^2\right]^{1/2} \le \epsilon$$

the value of ϵ will be fixed later. Then, for any $\alpha > 0$, $\alpha < 1$, and for all $(t,s) \in U$

$$\int_{|t'-t|\le \alpha r} dt' \int_{|s'-s|\le \alpha r} ds' |G(t',s') - G_r(t',s')| \tag{B.7}$$

$$\le \frac{1}{4r^2} \int_{|t'-t|\le \alpha r} dt' \int_{|s'-s|\le \alpha r} ds' \int_{|t''-t'|\le r} dt'' \cdot \int_{|s''-s'|\le r} ds'' \left[\left|\int_{t'}^{t''} dt''' \partial_t G(t''',s'')\right| + \left|\int_{s'}^{s''} ds''' \partial_s G(t',s''')\right|\right]$$

$$\le \frac{1}{4r^2} \int_{|t'''-t|\le(1+\alpha)r} dt''' \int_{|s''-s|\le(1+\alpha)r} ds'' |\partial_t G(t''',s'')| \cdot 8\alpha^2 r^3$$

$$+ \frac{1}{4r^2} \int_{|t'-t|\le \alpha r} dt' \int_{|s'''-s|\le(1+\alpha)r} ds''' |\partial_s G(t',s''')| 8\alpha r^3$$

$$\le 8\alpha r^2 \left\{\left[\int_{|t'-t|\le 2r} dt' \int_{|s'-s|\le 2r} ds' |\partial_t G(t,s)|^2\right]^{1/2} + \left[\int_{|t'-t|\le 2r} dt' \int_{|s'-s|\le 2r} ds' |\partial_s G(t,s)|^2\right]^{1/2}\right\}$$

$$\le 8\alpha r^2 \epsilon. \tag{B.8}$$

We can however also compute a lower bound on (B.7). For $|t'-t| \le \alpha r$, $|s'-s| \le \alpha r$, we have, from (B.4),

$$|G(t',s') - G_r(t',s')|$$
$$\ge |G(t',s')| - |G_r(t',s') - G_r(t,s)| - |G_r(t,s)|$$
$$\ge a - 2\alpha b - |G_r(t,s)|. \tag{B.9}$$

Together with (B.8) this leads to

$$8\alpha r^2 \epsilon \ge \int_{|t'-t|\le \alpha r} dt' \int_{|s'-s|\le \alpha r} ds' |G(t',s') - G_r(t',s')|$$
$$\ge 4\alpha^2 r^2 (a - 2\alpha b - |G_r(t,s)|)$$

or

$$|G_r(t,s)| \ge a - 2\alpha b - 2\alpha^{-1}\epsilon. \tag{B.10}$$

If we choose $\alpha = a/8b$, $\epsilon = \alpha a/8 = a^2/64b$, then the right hand side of (B.10) reduces to $a/2$, and we conclude

$$a/2 \le |G_r(t,s)| \le b. \tag{B.11}$$

We have now all the necessary ingredients. Since G_r is continuous, and satisfies (B.11) on U, with $a > 0$, $b < \infty$, $\gamma_r = \log G_r$ can be defined as a continuous, univalued function on U. As a consequence of (B.5) we see that, for all $(t,s) \in V = U \cap (U-(1,0)) \cap (U-(0,1))$,

$$\gamma_r(t+1,s) = \gamma_r(t,s) + 2\pi i k$$
$$\gamma_r(t,s+1) = \gamma_r(t,s) + 2\pi i l + 2\pi i t + \phi(t,s) \tag{B.12}$$

where k, l are integers, constant over all of V, because of the continuity of γ_r, and where

$$|\phi(t,s)| \le -\log\left(1 - \frac{|\psi(t,s)|}{|G_r(t,s)|}\right) \le 2\frac{|\psi(t,s)|}{|G_r(t,s)|}$$

provided $|\psi(t,s) \le |G_r(t,s)|/2$. For this it is sufficient (see (B.6)) to impose $r \le a/(4\pi b)$, which we can do without loss of the previous estimates. One finds then

$$|\phi(t,s)| \le 1, \qquad \text{for all } (t,s) \in V.$$

By construction V is a neighborhood of $[0,1]^2$. In particular (B.12) is valid on $[0,1]^2$, which is enough to again lead to a contradiction. We have

$$\begin{aligned}0 &= \gamma_r(1,1) - \gamma_r(1,0) + \gamma_r(1,0) - \gamma_r(0,0) \\ &\quad + \gamma_r(0,0) - \gamma_r(0,1) + \gamma_r(0,1) - \gamma_r(1,1) \\ &= 2\pi i l + 2\pi i + \phi(1,0) + 2\pi i k - 2\pi i l - \phi(0,0) - 2\pi i k \\ &= 2\pi i + \phi(1,0) - \phi(0,0) \ne 0\end{aligned}$$

since

$$|\phi(1,0)|, |\phi(0,0)| \le 1.$$

This contradiction concludes the proof. □

Appendix C
Proof of the Theorems in Section III-B-3

Proof of Theorems 2.5, 2.6: Using the Poisson formula

$$\sum_{l \in \mathbb{Z}} e^{ilax} = \frac{2\pi}{a} \sum_{k \in \mathbb{Z}} \delta\left(x - \frac{2\pi}{a}k\right)$$

we find

$$\begin{aligned}\sum_{m,n} |\langle g_{mn}, f\rangle|^2 &= \sum_{m,n} \int dx \int dx'\, e^{imp_0(x-x')} \\ &\quad \cdot g(x-nq_0) g(x'-nq_0)^* f(x)^* f(x') \\ &= \frac{2\pi}{p_0} \sum_{n,k} \int dx\, g(x-nq_0) \\ &\quad \cdot g\left(x - nq_0 - \frac{2\pi}{p_0}k\right)^* f(x)^* f\left(x - \frac{2\pi}{p_0}k\right).\end{aligned}$$

Hence, separating the sum over k into the term for $k=0$ and a sum over $k \ne 0$, and applying Cauchy–Schwarz (once on the integral over x, and once on the sum over k),

$$\begin{aligned}\sum_{m,n} |\langle g_{mn}, f\rangle|^2 &\ge \frac{2\pi}{p_0} \int dx \sum_n |g(x-nq_0)|^2 |f(x)|^2 \\ &\quad - \frac{2\pi}{p_0} \sum_{k \ne 0} \left[\sum_n \int dx |\cdot g(x-nq_0)| \right. \\ &\quad \left. \cdot \left|g\left(x - nq_0 - \frac{2\pi}{p_0}k\right)\right| |f(x)|^2\right]^{1/2} \\ &\quad \cdot \left[\sum_n \int dx |g(x-nq_0)| \right. \\ &\quad \left. \cdot \left|g\left(x - nq_0 - \frac{2\pi}{p_0}k\right)\right| \left|f\left(x - \frac{2\pi}{p_0}k\right)\right|^2\right]^{1/2} \\ &\ge \frac{2\pi}{p_0}\left\{m(g;q_0) - \sum_{k \ne 0}\left[\beta\left(\frac{2\pi}{p_0}k\right)\beta\left(-\frac{2\pi}{p_0}k\right)\right]^{1/2}\right\}\|f\|^2.\end{aligned} \tag{C.1}$$

The decay condition (2.3.13) on β implies that the sum over k always converges. It also implies that this sum tends to zero for $p_0 \to 0$, so that the coefficient of $\|f\|^2$ in (C.1) is strictly positive for small enough p_0. On the other hand, we also have

$$\sum_{m,n} |\langle g_{mn}, f\rangle|^2 \le \frac{2\pi}{p_0}\left\{M(g;q_0) + \sum_{k \ne 0}\left[\beta\left(\frac{2\pi}{p_0}k\right)\beta\left(-\frac{2\pi}{p_0}k\right)\right]^{1/2}\right\}\|f\|^2 \tag{C.2}$$

for all $p_0 > 0$. Together, the lower and upper bounds (C.1) and (C.2) imply that $P_0^c > 0$, with

$$P_0^c = \inf\{p_0 | \text{the } g_{mn} \text{ associated to } g, p_0, q_0 \text{ do not constitute a frame}\}.$$

This proves Theorem 2.5. The inequalities (C.1) and (C.2) also immediately prove Theorem 2.6. □

The proofs of Theorems 2.7 and 2.8 are very similar.

Proofs of Theorems 2.7 and 2.8: One uses the unitarity of the Fourier transform to write

$$\sum_{m,n} |\langle h_{mn}, f\rangle|^2 = \sum_{m,n} a_0^m \int dx \int dx'\, e^{inb_0 a_0^m(x-x')} \hat{h}(a_0^m x) \cdot \hat{h}(a_0^m x')^* \hat{f}(x)^* \hat{f}(x'). \tag{C.3}$$

As in the proof of Theorem 2.5, this can be rewritten with the help of the Poisson formula,

$$\sum_{m,n}|\langle h_{mn},f\rangle|^2=\frac{2\pi}{b_0}\sum_{m,k}\int dx\,\hat{h}(a_0^m x)\hat{h}\left(a_0^m x+\frac{2\pi}{b_0}k\right)^*\cdot\hat{f}(x)^*\hat{f}\left(x+\frac{2\pi}{b_0}ka_0^{-m}\right). \quad \text{(C.4)}$$

The same arguments as in the proofs of Theorems 2.5 and 2.6 then lead to the desired conclusions. □

In the case where support $\hat{h}\subset\mathbb{R}_+$, and the frame $\{h_{mn}^{\pm}\,|\,m,n\in\mathbb{Z}\}$ is used (see Section II-C-2a)), one has to be a little more careful. The equivalent of (C.1) is then

$$\sum_{\epsilon=\pm}\sum_{m,n}|\langle h_{mn}^{\epsilon},f\rangle|^2=\frac{2\pi}{b_0}\int_0^\infty dx\sum_m|\hat{h}(a_0^m x)|^2\left[|\hat{f}(x)|^2+|\hat{f}(-x)|^2\right]+r$$

with

$$|r|\le\frac{2\pi}{b_0}\sum_{k\ne 0}\left\{\sum_m\left[\int_0^\infty dx|\hat{h}(a_0^m x)|\left|\hat{h}\left(a_0^m x+\frac{2\pi}{b_0}k\right)\right||\hat{f}(x)|^2\right]^{1/2}\cdot\left[\int_0^\infty dx\left|\hat{h}\left(a_0^m x-\frac{2\pi}{b_0}k\right)\right||\hat{h}(a_0^m x)||\hat{f}(x)|^2\right]^{1/2}\right.$$
$$\left.+\sum_m\left[\int_0^\infty dx|\hat{h}(a_0^m x)|\left|\hat{h}\left(a_0^m x+\frac{2\pi}{b_0}k\right)\right||\hat{f}(-x)|^2\right]^{1/2}\cdot\left[\int_0^\infty dx\left|\hat{h}\left(a_0^m x-\frac{2\pi}{b_0}k\right)\right||\hat{h}(a_0^m x)||\hat{f}(-x)|^2\right]^{1/2}\right\}.$$

This leads again to the same bounds (2.3.23), (2.3.24).

Proof of Theorem 2.9: Applying the Poisson formula gives

$$\sum_{m,n}|\langle h_{mn},f\rangle|^2=\frac{2\pi}{b_0}\int dx\left[\sum_m|\hat{h}(2^m x)|^2\right]|\hat{f}(x)|^2+\frac{2\pi}{b_0}\sum_{k\ne 0}\sum_m\int dx\,\hat{h}(2^m x)\hat{h}\left(2^m x-\frac{2\pi}{b_0}k\right)^*\hat{f}(x)^*\cdot\hat{f}\left(x-\frac{2\pi}{b_0}2^{-m}k\right). \quad \text{(C.5)}$$

Any $k\in\mathbb{Z}$, $k\ne 0$ can be decomposed, in a unique way, as $k=2^{m'}k'$, where $m',k'\in\mathbb{Z}$, $m'\ge 0$ and k' odd. Substituting this into (C.5), and defining $m''=m-m'$, one finds

$$\sum_{m,n}|\langle h_{mn},f\rangle|^2=\frac{2\pi}{b_0}\int dx\left[\sum_m|\hat{h}(2^m x)|^2\right]|\hat{f}(x)|^2+\frac{2\pi}{b_0}\sum_{k\,odd}\sum_{m'\ge 0}\sum_{m''\in\mathbb{Z}}\int dx\,\hat{h}(2^{m'+m''}x)\cdot\hat{h}\left[2^{m'}\left(2^{m''}x-\frac{2\pi}{b_0}k\right)\right]^*\hat{f}(x)^*\hat{f}\left(x-\frac{2\pi}{b_0}2^{-m''}k\right).$$

Applying Cauchy-Schwarz twice, in the second term (once on the sum over m'', and once on the integral over x) leads to the estimates (2.3.25) and (2.3.26). □

Proof of Theorem 2.10: For $b_0=1$, the $\psi_{mn;1}$ are Y. Meyer's basis. We start by showing that the $\psi_{m,n;b_0}$ still constitute a frame for b_0 in a neighborhood of 1. For $b_0=2$, the $\psi_{mn;2}=\psi_{m\,2n;1}$ do not span $L^2(\mathbb{R})$. We therefore restrict our attention to $b_0<2$.

In the computation of β_1, it is sufficient to take the supremum over $x\in[2\pi/3,\,4\pi/3]$. For x in this interval, and for $|s|\ge 2\pi b_0^{-1}\ge\pi$, one finds that only the couples $(m,m')\in\{(-2,2),(-1,1),(-1,2),(0,0),(0,1),(1,0)\}$ lead to nonzero contributions to β_1. As the supremum of the sum of a finite number of continuous functions, β_1 is therefore continuous. On the other hand, since support $\hat{\psi}\subset[-8\pi/3,8\pi/3]$, one finds that $\beta_1(s)=0$ if $|s|\ge 16\pi/3$. For $b_0<2$, this implies that only the choices $l=0$, 1 or 2 lead to nonzero contributions in the sums (2.3.25), (2.3.26). This implies that the right hand sides of (2.3.25), (2.3.26) are continuous in b_0. For $b_0=1$ these two expressions are equal to 1. By continuity we find therefore that (2.3.25) > 0 and (2.3.26) < ∞ on a neighborhood of $b_0=1$.

To show that the $\psi_{mn;b_0}$ constitute a basis, we introduce the operator

$$S(b_0)=\sum_{m,n}\psi_{mn;b_0}\langle\psi_{mn;1},\cdot\rangle.$$

In the terminology of Section II-A, we can write

$$S(b_0)=T(b_0)^*T(1)$$

where $T(b_0)$, $T(1)$ are the frame operators for the frames $\psi_{mn;b_0}$, $\psi_{mn;1}$ respectively. To prove that the $\psi_{mn;b_0}$ constitute a basis, it is sufficient to prove that $S(b_0)$ is one-to-one and onto, since $S(b_0)\psi_{mn;1}=\psi_{mn;b_0}$. Since $T(1)$ is unitary, and $T(b_0)^*$ is onto (see Proposition 2.1), we only need to prove that $T(b_0)^*$ or $S(b_0)$ is one-to-one. We have

$$\|S(b_0)f\|^2=\sum_{m,n}\sum_{m',n'}\langle f,\psi_{mn;1}\rangle\langle\psi_{mn;b_0},\psi_{m'n';b_0}\rangle\langle\psi_{m'n';1},f\rangle$$
$$\ge\|f\|^2-\sum_{\substack{m,n,m',n'\\(m,n)\ne(m',n')}}|\langle f,\psi_{mn;1}\rangle|\,|\langle\psi_{m'n';1},f\rangle|\cdot|\langle\psi_{mn;b_0},\psi_{m'n';b_0}\rangle|$$
$$\ge\|f\|^2\left[1-\sup_{m,n}\sum_{\substack{m',n'\\(m',n')\ne(m,n)}}|\langle\psi_{mn;b_0},\psi_{m'n';b_0}\rangle|\right]$$
$$=\|f\|^2\left[1-\sum_{\substack{m,n\\(m,n)\ne(0,0)}}|\langle\psi_{mn;b_0},\psi\rangle|\right]. \quad \text{(C.6)}$$

Since support $\psi = [-8\pi/3, -2\pi/3] \cup [2\pi/3, 8\pi/3]$, and $a_0 = 2$, $\langle \psi_{mn;b_0}, \psi \rangle = 0$ for $|m| > 1$. One checks easily, using

$$|\psi(x)| \le C_N(1+|x|^2)^{-N}$$

that

$$\sum_{n \in \mathbb{Z}} |\langle \psi_{mn;b_0}, \psi \rangle|$$

converges, and is continuous in b_0, for $m = 0, \pm 1$. Hence the coefficient of $\|f\|^2$ in the right-hand side of (C.6) is continuous in b_0. Since this coefficient is equal to 1 for $b_0 = 1$, it is therefore strictly positive on a neighborhood of $b_0 = 1$. This implies that, on that neighborhood, $S(b_0)$ is one-to-one. This concludes the proof. □

Appendix D
Proofs of the Theorems in Section III

Proof of Theorem 3.1: Since the g_{mn} constitute a frame, with dual frame $\tilde{g}_{mn}$, we have, for all $f_1, f_2 \in L^2(\mathbb{R})$,

$$\langle f_2, f_1 \rangle = \sum_{m,n \in \mathbb{Z}} \langle f_2, \tilde{g}_{mn}, f_1 \rangle.$$

Fix T, $\Omega > 0$. For t, $\omega > 0$ we define

$$B(t,\omega) = \{(m,n) \in \mathbb{Z}^2; |mp_0| \le \Omega + \omega, |nq_0| \le T + t\}.$$

Then, for $f \in L^2(\mathbb{R})$,

$$\Big\| f - \sum_{(m,n) \in B(t,\omega)} \tilde{g}_{mn} \langle g_{mn}, f \rangle \Big\|$$

$$= \sup_{\|f_1\|=1} \Big| \sum_{(m,n) \notin B(t,\omega)} \langle f_1, \tilde{g}_{mn} \rangle \langle g_{mn}, f \rangle \Big|$$

$$\le \sup_{\|f_1\|=1} \sum_{\substack{|mp_0| > \Omega + \omega \\ n \in \mathbb{Z}}} |\langle f_1, \tilde{g}_{mn} \rangle| \big[|\langle g_{mn}, P_\Omega f \rangle|$$

$$+ |\langle g_{mn}, (1 - P_\Omega) f \rangle| \big]$$

$$+ \sup_{\|f_1\|=1} \sum_{\substack{|nq_0| > T + t \\ m \in \mathbb{Z}}} |\langle f_1, \tilde{g}_{mn} \rangle| \big[|\langle g_{mn}, Q_T f \rangle|$$

$$+ |\langle g_{mn}, (1 - Q_T) f \rangle| \big]$$

$$\le (B/A)^{1/2} \Bigg\{ \|(\mathbb{1} - P_\Omega) f\| + \|(\mathbb{1} - Q_T) f\|$$

$$+ B^{-1/2} \Bigg(\sum_{\substack{|mp_0| > \Omega + \omega \\ n \in \mathbb{Z}}} |\langle g_{mn}, P_\Omega f \rangle|^2 \Bigg)^{1/2}$$

$$+ B^{-1/2} \Bigg(\sum_{\substack{|nq_0| > T + t \\ m \in \mathbb{Z}}} |\langle g_{mn}, Q_T f \rangle|^2 \Bigg)^{1/2} \Bigg\} \tag{D.1}$$

where we have used the fact that the g_{mn} and the $\tilde{g}_{mn}$ both constitute frames, with frame constants A, B and B^{-1}, A^{-1}, respectively. Similar to the proof of Theorem 2.5 we use the Poisson formula to estimate the last two terms in (D.1). This leads to

$$\sum_{\substack{|nq_0| > T + t \\ m \in \mathbb{Z}}} |\langle g_{mn}, Q_T f \rangle|^2$$

$$= \frac{2\pi}{p_0} \sum_{\substack{|nq_0| T + t \\ l \in \mathbb{Z}}} \int dx\, g(x - bq_0)^* g\left(x - nq_0 - \frac{2\pi}{p_0} l\right)$$

$$\cdot (Q_T f)(x) (Q_T f)\left(x - \frac{2\pi}{p_0} l\right)^*$$

$$\le \frac{2\pi}{p_0} \|Q_T f\|^2 \cdot \sum_{l \in \mathbb{Z}} \sup_{\substack{|x| \le T \\ |x - \frac{2\pi}{p_0} l| \le T}} \sum_{|nq_0| > T + t} |g(x - nq_0)|$$

$$\cdot \left| g\left(x - nq_0 - \frac{2\pi l}{p_0}\right) \right|. \tag{D.2}$$

From the conditions on g, we have

$$(\text{D.2}) \le C^2 \sum_{|nq_0| > T + t} \sum_{l \in \mathbb{Z}} \sup_{\substack{|x| \le T \\ |x - \frac{2\pi}{p_0} l| \le T}} \left[1 + (x - nq_0)^2\right]^{-\alpha}$$

$$\cdot \left[1 + \left(x - nq_0 - \frac{2\pi}{p_0} l\right)^2\right]^{-\alpha}. \tag{D.3}$$

The contribution for $n > (T+t)/q_0$ is exactly equal to that for $n < -(T+t)/q_0$, so that we may restrict ourselves to negative n, at the price of a factor 2. By redefining $y = x - (2\pi/p_0) l$ if l is positive, we see that we may restrict ourselves to negative l as well. Hence

$$(\text{D.3}) \le 4C^2 \sum_{nq_0 > T + t} \sum_{l \ge 0} \sup_{\substack{|x| \le T \\ |x - \frac{2\pi}{p_0} l| \le T}} \left[1 + (x + nq_0)^2\right]^{-\alpha}$$

$$\cdot \left[1 + \left(x + nq_0 + \frac{2\pi}{p_0} l\right)^2\right]^{-\alpha}$$

$$\le 4C^2 \sum_{nq_0 > T + t} \sum_{l \ge 0} \left[1 + (nq_0 - T)^2\right]^{-\alpha}$$

$$\cdot \left[1 + \left(nq_0 - T + \frac{2\pi}{p_0} l\right)^2\right]^{-\alpha}. \tag{D.4}$$

However, for any $a, b > 0$, we have

$$\sum_{l \ge 0} \left[1 + (a + bl)^2\right]^{-\alpha}$$

$$\le (1 + a^2)^{-\alpha} + \frac{1}{b} \int_a^\infty dx\, (1 + x^2)^{-\alpha}$$

$$\le \left[1 + 2^{2\alpha - 1} b^{-1} \left(1 + \frac{1}{2\alpha - 1}\right)\right] (1 + a^2)^{-\alpha + 1/2},$$

if $\alpha > 1/2$.

Consequently

$$(D.4) \le 4C^2\left[1+2^{2\alpha-1}\frac{\alpha p_0}{\pi(2\alpha-1)}\right] \cdot \sum_{nq_0>T+t}\left[1+(nq_0-T)^2\right]^{-2\alpha+1/2}.$$

A similar argument shows that, for $t \ge 2q_0$,

$$\sum_{nq_0>T+t}\left[1+(nq_0-T)^2\right]^{-2\alpha+1/2} \le \frac{1}{q_0}\left[1+\frac{1}{2(2\alpha-1)}\right]2^{2(2\alpha-1)}\left[1+(t-2q_0)^2\right]^{-2\alpha+1}.$$

We find therefore that there exists a constant $\kappa(p_0,q_0)$, independent of T or t, such that

$$(D.2) \le \kappa(p_0,q_0)\left[1+t^2\right]^{-2\alpha+1}\|f\|.$$

The term in $P_\Omega f$ is estimated similarly; putting everything together leads to

$$\left\|f-\sum_{(m,n)\in B(t,\omega)}\tilde{g}_{mn}\langle g_{mn},f\rangle\right\| \le (B/A)^{1/2}\Big\{\|(\mathbb{1}-P_\Omega)f\|+\|(\mathbb{1}-Q_T)f\| + B^{-1/2}\Big[\kappa(p_0,q_0)^{1/2}(1+t^2)^{-2\alpha+1/2} + \kappa(q_0,p_0)^{1/2}(1+\omega^2)^{-2\alpha+1/2}\Big]\|f\|\Big\}, \qquad (D.5)$$

since $(1+t^2)^{-2\alpha+1}$, $(1+\omega^2)^{-2\alpha+1}\to 0$ for $t,\omega\to 0$, this proves the theorem. □

Remark: The estimates in this proof cause $t_\epsilon,\omega_\epsilon$, when calculated using (D.5), to be much larger for a given ϵ, and e.g., for Gaussian g, than observed in numerical calculations. The intermediate estimate (D.2) leads to much better values of t_ϵ (a similar formula can of course be written for $P_\Omega f$, leading to estimates for ω_ϵ). If we define

$$\zeta^{\pm}(t;y)=\sup_{\substack{\pm x\ge t\\ \pm(x-y)\ge t}}\sum_{n=0}^{\infty}|g(x\pm nq_0)|\,|g(x-y\pm nq_0)|$$

then $\kappa(p_0,q_0)^{1/2}$, in the estimate (D.5), can be replaced by

$$\left(\frac{2\pi}{p_0}\right)^{1/2}\left\{\sum_{l\in\mathbb{Z}}\left[\zeta^{+}\left(t;\frac{2\pi}{p_0}l\right)+\zeta^{-}\left(t;\frac{2\pi}{p_0}l\right)\right]\right\}^{1/2}. \qquad (D.6)$$

The same thing can be done for $\kappa(q_0,p_0)^{1/2}$.

Proof of Theorem 3.2: We define the set B_ϵ as

$$B_\epsilon(T,\Omega_0,\Omega_1)=\{(m,n);\ m_0\le m\le m_1,\ |nb_0|\le a_0^{-m}T+t\} \qquad (D.7)$$

where m_0, m_1, and t, to be defined next, depend on Ω_0, Ω_1, and ϵ.

One then has (see the proof of Theorem 3.1)

$$\left\|f-\sum_{(m,n)\in B_\epsilon(T,\Omega_0,\Omega_1)}(h_{mn})^{\sim}\langle h_{mn},f\rangle\right\|$$

$$\le \sup_{\|f\|=1}\sum_{\substack{n\in\mathbb{Z}\\ m<m_0\text{ or }m>m_1}}|\langle f_1,(h_{mn})^{\sim}\rangle| \cdot\Big[|\langle h_{mn},(P_{\Omega_1}-P_{\Omega_0})f\rangle| + |\langle h_{mn},(1-P_{\Omega_1}+P_{\Omega_0})f\rangle|\Big]$$

$$+\sup_{\|f_1\|=1}\sum_{\substack{m_0\le m\le m_1\\ |nb_0|\ge a_0^{-m}T+t}}|\langle f_1,(h_{mn})^{\sim}\rangle| \cdot\Big[|\langle h_{mn},Q_Tf\rangle|+|\langle h_{mn},(1-Q_T)f\rangle|\Big]$$

$$\le (B/A)^{1/2}\left\{\|(\mathbb{1}-P_{\Omega_1}+P_{\Omega_0})f\|+\|(\mathbb{1}-Q_T)f\| + B^{-1/2}\left(\sum_{\substack{n\in\mathbb{Z}\\ m<m_0\\ \text{or }m>m_1}}|\langle h_{mn},(P_{\Omega_1}-P_{\Omega_0})f\rangle|^2\right)^{1/2} + B^{-1/2}\left(\sum_{\substack{m_0\le m\le m_1\\ |nb_0|\ge a_0^{-m}T+t}}|\langle h_{mn},Q_Tf\rangle|^2\right)^{1/2}\right\}. \qquad (D.8)$$

One has

$$\sum_{\substack{m>m_1\text{ or }m<m_0\\ n\in\mathbb{Z}}}|\langle h_{mn},(P_{\Omega_1}-P_{\Omega_0})f\rangle|^2$$

$$\le \frac{2\pi}{b_0}\sum_{\substack{m>m_1\text{ or }m<m_0\\ l\in\mathbb{Z}}}\int_{\substack{\Omega_0\le|y|\le\Omega_1\\ \Omega_0\le|y-a_0^{-m}\frac{2\pi}{b_0}l|\le\Omega_1}}dy \cdot|\hat{h}(a_0^m y)|\left|\hat{h}\left(a_0^m y-\frac{2\pi}{b_0}l\right)\right||\hat{f}(y)|\left|\hat{f}\left(y-a_0^{-m}\frac{2\pi}{b_0}l\right)\right|$$

$$\le \frac{2\pi}{b_0}\|(P_{\Omega_1}-P_{\Omega_0})f\|^2\cdot\sum_{l\in\mathbb{Z}}\phi\left(\frac{2\pi}{b_0}l\right)^{\kappa/2} \cdot\left[\sup_{\Omega_0\le|y|\le\Omega_1}\sum_{\substack{m>m_1\\ \text{or}\\ m<m_0}}\left[1+(a_0^m y)^2\right]^{\kappa}|\hat{h}(a_0^m y)|\right]^2$$

where we have used that, for all $x \in \mathbb{R}$,

$$(1+x^2)\left[1+(x-s)^2\right]\phi(s) \geq 1$$

with ϕ defined by

$$\phi(s)=\begin{cases}\left[1+(s/2)^2\right]^{-2}, & \text{if } |s|\leq 2\\ s^{-2}, & \text{if } |s|\geq 2.\end{cases}$$

If we choose $1<\kappa<\alpha$, then

$$K_\kappa(b_0)=\sum_{l\in\mathbb{Z}}\phi\left(\frac{2\pi}{b_0}l\right)^{\kappa/2} \quad \text{converges}$$

and

$$\sup_{\Omega_0\leq|y|\leq\Omega_1}\sum_{\substack{m>m_1\\ \text{or}\\ m<m_0}}\left[1+(a_0^m y)^2\right]^\kappa|\hat{h}(a_0^m y)|^2$$

$$\leq C^2\sup_{\Omega_0\leq|y|\leq\Omega_1}\sum_{\substack{m>m_1\\ \text{or}\\ m<m_0}}(a_0^m|y|)^{2\beta}\left[1+(a_0^m y)^2\right]^{-\alpha-\beta+s\kappa}.$$

Using the same techniques as in the proof of Lemma 2.2, one finds, for $\Omega_0\leq|y|$,

$$\sum_{m>m_1}(a_0^m|y|)^{2\beta}\left[1+(a_0^m y)^2\right]^{-\alpha-\beta+\kappa}$$

$$\leq\frac{1}{\ln a_0}\int_{a_0^{m_1}} dt\, t^{2\beta-1}(1+t^2)^{-\alpha-\beta+\kappa}$$

$$\leq\frac{1}{\ln a_0}\frac{1}{2(\alpha-\kappa)}\left[a_0^{m_1}\Omega_0\right]^{-2(\alpha-\kappa)}$$

where we have assumed that $m_1\geq(\ln\beta-\ln(\alpha-\kappa)-\ln\Omega_0)/\ln a_0$. Similarly, for $|y|\leq\Omega_1$,

$$\sum_{m<m_0}(a_0^m|y|)^{2\beta}\left[1+(a_0^m y)^2\right]^{-\alpha-\beta+\kappa}$$

$$\leq\frac{1}{\ln a_0}\int_0^{a_0^{m_0}\Omega_1}dt\, t^{2\beta-1}(1+t^2)^{-\alpha-\beta+\kappa}$$

$$\leq\frac{1}{\ln a_0}\frac{1}{2\beta}\left[a_0^{m_0}\Omega_1\right]^{2\beta}$$

provided $m_0\leq(\ln\beta-\ln(\alpha-\kappa)-\ln\Omega_0)/\ln a_0$. It is clear from this that, for Ω_0, Ω_1, ϵ given, we can choose m_0, m_1 so that

$$\sum_{\substack{n\in\mathbb{Z}\\ m>m_1\text{ or } m<m_0}}\left|\langle h_{mn},(P_{\Omega_1}-P_{\Omega_0})f\rangle\right|^2\leq B\epsilon^2/4\|f\|^2. \quad \text{(D.9)}$$

On the other hand,

$$\sum_{b_0|n|>a_0^{-m}T+t}|\langle h_{mn},Q_T f\rangle|^2$$

$$\leq\sum_{b_0|n|>a_0^{-m}T+t}\|Q_T h_{mn}\|^2\cdot\|f\|^2$$

$$\leq\|f\|^2\int_{|x|\leq T}dx\,a_0^{-m}\sum_{b_0|n|>a_0^{-m}T+t}|h(a_0^{-m}x-nb_0)|^2$$

$$\leq\|f\|^2\left\{\int_{-\infty}^{-t}du\sum_{l=0}^{\infty}|h(u-lb_0)|^2\right.$$

$$\left.+\int_t^\infty du\sum_{l=0}^\infty|h(u+lb_0)|^2\right\}$$

$$\leq\|f\|^2\sum_{l=0}^\infty\left[1+(t+lb_0)^2\right]^{-\gamma}\int_{|u|\geq t}du(1+u^2)^\gamma|h(u)|^2. \quad \text{(D.10)}$$

Since $\gamma>1/2$, this converges. It is moreover clear that t can be chosen so that the coefficient of $\|f\|^2$ in (D.10) is smaller than $B\epsilon^2/4(m_1-m_0+1)$, i.e., such that

$$\sum_{\substack{m_0\leq m\leq m_1\\ b_0|n|>a_0^{-m}T+t}}|\langle h_{mn},Q_T f\rangle|^2\leq B\epsilon^2/4\|f\|^2. \quad \text{(D.11)}$$

The statement (3.6) now follows directly from (D.8), (D.9) and (D.11). □

References

[1] D. Gabor, "Theory of communication," *J. Inst. Elect. Eng.* (London), vol. 93, no. III, pp. 429–457, 1946.

[2] M. J. Bastiaans, "Gabor's signal expansion and degrees of freedom of a signal," *Proc. IEEE*, vol. 68, pp. 538–539, 1980.
____, "A sampling theorem for the complex spectrogram and Gabor's expansion of a signal in Gaussian elementary signals," *Optical Eng.*, vol. 20, pp. 594–598, 1981.

[3] A. J. E. M. Janssen, "Gabor representation of generalized functions," *J. Math. Appl.*, vol. 80, pp. 377–394, 1981.
____, "Gabor representation and Wigner distribution of signals," in *Proc. IEEE*, 1984, pp. 41.B.2.1–41.B.2.4.

[4] M. J. Davis and E. J. Heller, *J. Chem. Phys.*, vol. 71, pp. 3383–3395, 1979.

[5] J. R. Klauder and E. Sudarshan, *Fundamentals of Quantum Optics*, New York: Benjamin, 1968.

[6] R. J. Glauber, "The quantum theory of coherence," *Phys. Rev.*, vol. 130, pp. 2529–2539, 1963; "Coherent and incoherent states of the radiation field," *Phys. Rev.*, vol. 131, pp. 2766–2788, 1963.

[7] J. R. Klauder and B.-S. Skagerstam, "Coherent states," Singapore: World Scientific, 1985.

[8] E. Lieb, "Thomas-Fermi theory and related theories of atoms and molecules," *Rev. Mod. Phys.*, vol. 53, pp. 603–641, 1981.

[9] G. A. Hagedorn, "Semiclassical quantum mechanics I: The $\hbar\to0$ limit for coherent states," *Comm. Math. Phys.*, vol. 71, pp. 77–93, 1980.

[10] C. W. Helstrom, "An expansion of a signal in Gaussian elementary signals," *IEEE Trans. Inform. Theory*, vol. 12, pp. 81–82, 1966.

[11] N. G. de Bruijn, "Uncertainty principles in Fourier analysis," in *Inequalities*, O. Shisha, Ed. New York: Academic Press, pp. 57–71, 1967.

[12] T. A. C. M. Claassen and W. F. G. Mecklenbräuker, "The Wigner distribution—A tool for time-frequency signal analysis," *Philips J. Res.*, vol. 35, pp. 217–250, 276–300, 372–389, 1980.

[13] C. P. Janse and A. J. M. Kaiser, "Time-frequency distributions of loudspeakers: The application of the Wigner distribution," *J. Audio Eng. Soc.*, vol. 31, pp. 198–223, 1983.

[14] R. J. Duffin and A. C. Schaeffer, "A class of nonharmonic Fourier series," *Trans. Am. Math. Soc.*, vol. 72, pp. 341–366, 1952.

[15] R. M. Young, *An Introduction to Nonharmonic Fourier Series*. New York: Academic Press, 1980.

[16] I. Daubechies, A. Grossmann, and Y. Meyer, "Painless nonorthogonal expansions," *J. Math. Phys.*, vol. 27, pp. 1271–1283, 1986.

[17] A. Grossmann and J. Morlet, "Decomposition of Hardy functions ito square integrable wavelets of constant shape," *SIAM J. Math. Anal.*, vol. 15, pp. 723–736, 1984.
P. Goupillaud, A. Grossmann, and J. Morlet, "Cycle-octave and related transforms in seismic signal analysis," *Geoexploration*, vol. 23, p. 85, 1984.

[18] A. Grossmann, J. Morlet, and T. Paul, "Transforms associated to square integrable group representations," *J. Math. Phys.*, vol. 26, pp. 2473–2479, 1985; *Ann. Inst. Henri Poincaré*, vol. 45, 293–309, 1986.

[19] E. W. Aslaksen and J. R. Klauder, "Unitary representations of the affine group," *J. Math. Phys.*, vol. 9, pp. 206–211, 1968; "Continuous representation theory using the affine group," *J. Math. Phys.*, vol. 10, pp. 2267–2275, 1969.

[20] T. Paul, "Functions analytic on the half-plane as quantum mechanical states," *J. Math. Phys.*, vol. 25, pp. 3252–3263, 1984.

[21] ——, "Affine coherent states and the radial Schrödinger equation. I. Radial harmonic oscillator and hydrogen atom," to be published.

[22] C. Fefferman and R. de la Llave, "Relativistic stability of matter," *Rev. Math. Iberoamericana*, vol. 2, pp. 119–213, 1986.

[23] I. Daubechies, "Time-frequency localization operators—A geometric phase space approach," *IEEE Trans. Inform. Theory*, vol. 34, pp. 605–612, 1988.

[24] I. Daubechies and T. Paul, "Time-frequency localization operators —A geometric phase space approach II. The use of dilations," *Inverse Problems*, vol. 4, pp. 661–680, 1988.

[25] J. Morlet, G. Arens, I. Fourgeau, and D. Giard, "Wave propagation and sampling theory," *Geophys.*, vol. 47, pp. 203–236, 1982; "Sampling theory and wave propagation," NATO ASI Series, vol. 1, *Issues in Acoustic Signal / Image Processing and Recognition*, C. H. Chen, Ed. Berlin: Springer, pp. 233–261.

[26] R. Kronland-Martinet, J. Morlet, and A. Grossmann, "Analysis of sound patterns through wavelet transforms," preprint CTP-87/p. 1981, Centre de Physique Théorique, CNRS, Marseille, France, 1987.

[27] A. Grossmann, "Wavelet transforms and edge detection," to be published in *Stochastic Processes in Physics and Engineering*, Ph. Blanchard, L. Streit, and M. Hasewinkel, Eds.

[28] A. Grossmann, M. Holschneider, R. Kronland-Martinet, and J. Morlet, "Detection of abrupt changes in sound signals with the help of wavelet transforms," in *Inverse Problems: An Interdisciplinary Study, Advances in Electronics and Electron Physics*, Supp. 19. New York: Academic, 1987.

[29] S. Mallat, "A theory for multiresolution signal decomposition," Ph.D. thesis, Univ. of Pennsylvania, Depts. of Elect. Eng. and Comput. Sci., 1988.

[30] A. Grossmann and J. Morlet, "Decomposition of functions into wavelets of constant shape and related transforms," in *Mathematics and Physics, Lectures on Recent Results*. Singapore: World Scientific, 1985.

[31] Y. Meyer, "Principe d'incertitude, bases hilbertiennes et algèbres d'opérateurs," *Séminaire Bourbaki*, nr. 662, 1985–1986.

[32] P. G. Lemarié and Y. Meyer, "Ondelettes et bases hilbertinennes," *Rev. Math. Iberoamericana*, vol. 2, pp. 1–18, 1986.

[33] G. Battle, "A block spin construction of ondelettes. Part I: Lemarié functions," *Comm. Math. Phys.*, vol. 110, pp. 601–615, 1987.

[34] P. G. Lemarié, "Une nouvelle base d'ondelettes de $L^2(\mathbb{R}^n)$," to be published in *J. de Math. Pures et Appl.*

[35] a) Ph. Tchamitchian, "Calcul symbolique sur les opérateurs de Calderón-Zygmund et bases inconditionnelles de $L^2(\mathbb{R}^n)$," C. R. Acad. Sc. Paris, vol. 303, série 1, pp. 215–218, 1986; b) Ph. Tchamitchian, "Biorthogonalité et théorie des opérateurs," to be published in *Rev. Math. Iberoamericana*.

[36] S. Mallat, "Multiresolution approximation and wavelets," *Trans. Am. Math. Soc.*, vol. 135, pp. 69–88, 1989.

[37] Y. Meyer, "Ondelettes, function splines, et analyses graduées," Univ. of Torino, 1986.

[38] S. Mallat, "A Theory for multiresolution signal decomposition: The wavelet representation," *IEEE Trans. Pattern Anal. Machine Intell.*, vol. 31, pp. 674–693, 1989.

[39] I. Daubechies, "Orthonormal basis of compactly supported wavelets," *Comm. Pure Applied Math.*, vol. 41, pp. 909–996, 1988.

[40] S. G. Mallat, "Multifrequency channel decompositions of images and wavelet models," *IEEE Trans. Acoust. Speech Signal Processing*, vol. 37, no. 12, pp. 2091–2110, Dec. 1987.

[41] G. Battle and P. Federbush, "Ondelettes and phase cell cluster expansions: A vindication," *Comm. Math. Phys.*, 1987.

[42] I. Daubechies, "Discrete sets of coherent states and their use in signal analysis," in *Proc. 1986 Int. Conf. Diff. Equations Math. Phys.*, Birmingham, AL, Springer Lecture Notes in Mathematics nr. 1285, pp. 73–82.

[43] I. Daubechies and T. Paul, "Wavelets—Some applications," *Proc. Int. Conf. Math. Phys.*, M. Mebkhout and R. Sénéor, Eds. Singapore: World Scientific, 1987.

[44] I. Daubechies, "How to frame Bargmann and Hardy—Frame bounds for Hilbert spaces of entire functions," unpublished memorandom, AT&T Bell Laboratories, 1987.

[45] I. Daubechies and A. Grossmann, "Frames of entire functions in the Bargmann space," *Comm. Pure Appl. Math.*, vol. 41, pp. 151–164, 1988.

[46] R. R. Coifman and R. Rochberg, "Representation theorems for holomorphic and harmonic functions in L^p," *Astérisque*, vol. 77, pp. 11–66, 1980.

[47] S. Janson, J. Peetre, and R. Rochberg, "Hankel forms and the Fock space," Uppsala Univ., Math. report, 1986.

[48] H. G. Feichtinger and K. Gröchenig, "A unified approach to atomic decompositions via integrable group representations," in *Proc. Function Spaces Applications*, Lund, 1986, to appear in *Springer Lect. Notes in Math.* no. 1302, pp. 52–73; "Banach spaces related to integrable group representations and their atomic decomposition I, II, and III," to be published, *J. Funct. Anal.*, vol. 83, 1989.

[49] M. Rieffel, "Von Neumann algebras associated with pairs of lattices in Lie groups," *Math. Ann.*, vol. 257, pp. 403–418, 1981.

[50] G. Kaiser, "A sampling theorem for signals in the joint time-frequency domain," to be published.

[51] R. Balian, "Un principe d'incertitude fort en théorie du signal on en mécanique quantique," C. R. Acad. Sc. Paris, vol. 292, série 2, 1981.

[52] F. Low, "Complete sets of wave-packets," in *A Passion for Physics — Essays in Honor of Geoffrey Chew*," pp. 17–22. Singapore: World Scientific, 1985.

[53] H. Elbrond Jensen, T. Hoholdt, and J. Justesen, "Double series representation of bounded signals," *IEEE Trans. Inform. Theory*, vol. 34, pp. 613–624, 1988.

[54] J. Zak, "Finite translations in solid state physics." *Phys. Rev. Lett.*, vol. 19, pp. 1385–1397, 1967; "Dynamics of electrons in solids in external fields," *Phys. Rev.*, vol. 168, pp. 686–695; "The *kq*-representation in the dynamics of electrons in solids," *Solid State Phys.*, vol. 27, pp. 1–62, 1972.

[55] W. Schempp, "Radar ambiguity functions, the Heinsenberg group, and holomorphic theta series," *Proc. of the Am. Math. Soc.*, vol. 92, 1984.

[56] M. Reed and B. Simon, *Methods and Modern Mathematical Physics. IV Analysis of Operators*. New York: Academic Press, 1978.

[57] M. Boon, J. Zak, and J. Zucker, "Rational von Neumann lattices," *J. Math. Phys.*, vol. 24, pp. 316–323, 1983.

[58] J. Zak, "Lattice operators in crystals for Bravais and reciprocal vectors," *Phys. Rev.*, vol. B 12, pp. 3023–3026, 1975.

[59] A. J. E. M. Janssen, "Bargmann transform, Zak transform, and coherent states," *J. Math. Phys.*, vol. 23, pp. 720–731, 1982.

[60] A. J. E. M. Janssen, "The Zak transform: A signal transform for sampled time-continuous signals," *Philips J. Res.*, vol. 42, 1987.

[61] V. Bargmann, P. Butera, L. Girardello, and J. R. Klauder, "On the completeness of coherent states," *Rep. Math. Phys.*, vol. 2, pp. 221–228, 1971.

[62] A. M. Perelomov, "Note on the completeness of systems of coherent states," *Teor. i. Matem. Fis.*, vol. 6, pp. 213–224, 1971.

[63] H. Bacry, A. Grossmann, and J. Zak, "Proof of the completeness of lattice states in the *kq*-representation," *Phys. Rev.*, vol. B 12, pp. 1118–1120, 1975.

[64] M. Reed and B. Simon, *Methods of Modern Mathematical Physics, II. Fourier Analysis and Self-Adjointness*. New York: Academic Press, 1975.

[65] D. Slepian and H. O. Pollak, "Prolate spheroidal wave functions, Fourier analysis and uncertainty," I, *Bell. Syst. Tech. J.*, vol. 40,

pp. 43–64, 1961; H. J. Landau and H. O. Pollak, II, *Bell. Syst. Techn. J.*, vol. 40, pp. 65–84, 1961; H. J. Landau and H. O. Pollak, III, *Bell. Syst. Techn. J.*, vol. 41, pp. 1295–1336, 1962.

[66] D. Slepian, "On bandwidth," *Proc. IEEE*, vol. 64, pp. 292–300, 1976.

[67] A. Annéodo, G. Grasseau, and M. Holschneider, "Wavelet transform analysis of invariant measures for some dynamical systems," *Phys. Rev. Lett.*, vol. 61, pp. 2281–2287, 1988.

[68] J. Ville, "Théorie et applications de la notion de signal analytique," *Revue Cables et Transmissions*, vol. 1, pp. 61–74, 1948.

[69] G. Battle, "Heisenberg proof of the Balian-low theorem," *Lett. Math. Phys.*, vol. 15, pp. 175–177, 1988.

[70] I. Daubechies and A. J. E. M. Janssen, "Two theorems on lattice expansions," submitted for publication.

[71] H. Landau, "On the density of phase space functions," submitted for publication.

SECTION V
Multiresolution Analysis

Introduction

Guido Weiss

The "classical" one-dimensional *orthonormal wavelets* are those functions $\psi \in L^2(\mathbb{R})$ for which the system $\{\psi_{j,k}(x) = 2^{j/2}\psi(2^j x - k)\}$, $j, k \in \mathbb{Z}$, is an orthonormal basis for $L^2(\mathbb{R})$. There is a relatively simple characterization of these functions:

Theorem 1 *$\psi \in L^2(\mathbb{R})$ is an orthonormal wavelet if and only if $\|\psi\|_2 = 1$,*

$$\sum_{j\in\mathbb{Z}} |\hat{\psi}(2^j\xi)|^2 = 1 \text{ for a.e. } \xi \in \mathbb{R} \tag{I}$$

and

$$t_q(\xi) \equiv \sum_{j=0}^{\infty} \hat{\psi}(2^j\xi)\overline{\hat{\psi}(2^j(\xi + 2m\pi))} = 0 \tag{II}$$

for a.e. $\xi \in \mathbb{R}$, $m \in 2\mathbb{Z} + 1$.

We are using the *Fourier transform* $\hat{f}$ of $f \in L^2(\mathbb{R})$ that has the form

$$\hat{f}(\xi) = \int_{\mathbb{R}} f(t)e^{-i\xi t}dt.$$

Theorem 1 was obtained independently by G. Gripenberg and X. Wang ([Gri95], [Wan95]). Equalities (I) and (II) were known since the beginning of the development of wavelet theory. In particular, (I) is a variant of the Calderón reproducing formula that has been discussed in the introduction and in the articles of Section IV. Independently of their theoretical importance, these formulae are useful for the construction of wavelets (see [HW96, chap. 7] for examples of such uses).

Perhaps the best known method for the construction of wavelets is known as the *multiresolution analysis* (MRA) method. It was developed and formulated in the "modern form" by S. Mallat. His method is based on ideas that have been introduced in Signal Analysis. In order to be able to discuss this method and its properties, let us present the defining properties of an MRA:

An MRA is a sequence of closed subspaces $V_j, j \in \mathbb{Z}$, of $L^2(\mathbb{R})$ satisfying

(a) $V_j \subset V_{j+1}$ for all $j \in \mathbb{Z}$;

(b) $f \in V_j$ if and only if $f(2\cdot) \in V_{j+1}$ for all $j \in \mathbb{Z}$;

(c) $\overline{\bigcup_{j\in\mathbb{Z}} V_j} = L^2(\mathbb{R})$;

(d) There exists a function $\varphi \in V_0$ such that $\{\varphi(\cdot - k) : k \in \mathbb{Z}\}$ is an orthonormal basis for V_0.

The function φ whose existence is asserted in (d) is called a *scaling function* of the given MRA.

The construction of an orthonormal wavelet ψ when we are given a scaling function φ of an MRA is really rather simple to explain and to establish. This is done in the first paper of this section. One first shows that there exists a unique 2π-periodic function $m_0 \in L^2(\mathbb{T}) \equiv L^2([-\pi, \pi))$ such that $\hat{\varphi}(2\xi) = m_0(\xi)\hat{\varphi}(\xi)$. The function m_0 is called the *low pass filter* associated with the scaling function φ. It is not hard to show that m_0 satisfies the equality

$$|m_0(\xi)|^2 + |m_0(\xi + \pi)|^2 = 1 \tag{1}$$

for a.e. $\xi \in \mathbb{R}$. This property is often referred to as the *Smith-Barnwell equation* (see the fourth paper in Section I in order to obtain an idea of the signal analysis origin of many of the ideas in the theory of wavelets).

The most general wavelet ψ associated with this MRA has the form

$$\hat{\psi}(2\xi) = e^{i\xi}\nu(2\xi)\overline{m_0(\xi + \pi)}\hat{\varphi}(\xi) \tag{2}$$

for any 2π-periodic measurable function ν satisfying

$$|\nu(\xi)| = 1 \text{ a.e. on } \mathbb{T}.$$

It is not difficult to show (from the properties we have described) that such a function ψ belongs to the space $W_0 = V_1 \cap V_0^{\perp}$ and its integer translates form an orthonormal basis of W_0. Dilating the elements of W_0 by 2^j we obtain a closed subspace W_j of V_{j+1} such that $V_{j+1} = V_j \oplus W_j$ for each $j \in \mathbb{Z}$. A consequence of the properties of an MRA is the fact that $V_j \to \{0\}$ as $j \to -\infty$; consequently,

$$V_{j+1} = V_j \oplus W_j = \bigoplus_{\ell=-\infty}^{j} W_\ell$$

for all $j \in \mathbb{Z}$. From property (c), we then have

$$L^2(\mathbb{R}) = \bigoplus_{j=-\infty}^{\infty} W_j \tag{3}$$

Thus, $\{\psi_{jk}\}, j, k \in \mathbb{Z}$, is an orthonormal basis for $L^2(\mathbb{R})$.

All this (and much more) is derived in the two articles by Mallat reproduced in this chapter.

Perhaps the simplest example of an MRA wavelet is the *Haar wavelet:*

$$\psi(x) = \begin{cases} 1 & \text{if } -1 \leqq x < -\frac{1}{2}, \\ -1 & \text{if } -\frac{1}{2} \leqq x < 0, \\ 0 & \text{otherwise.} \end{cases}$$

In this case, simple calculations give us the scaling function $\varphi(x) = \chi_{[-1,0]}(x)$. Thus, $\frac{1}{2}\varphi(\frac{1}{2}x) = \frac{1}{2}\varphi(x) + \frac{1}{2}\varphi(x+1)$ and $m_0(\xi) = \frac{1+e^{i\xi}}{2}$. Another simple example of an MRA wavelet is the *Shannon wavelet* ψ whose Fourier transform satisfies

$$\hat{\psi}(\xi) = e^{i\frac{\xi}{2}}\chi_I(\xi),$$

where

$$I = [-2\pi, -\pi) \cup (\pi, 2\pi].$$

The scaling function φ that produces this wavelet is easily seen to satisfy $\hat{\varphi}(\xi) = \chi_{[-\pi,\pi)}(\xi)$. From this we see that the low pass filter m_0 is the 2π-periodic function whose values on $[-\pi, \pi)$ are given by

$$m_0(\xi) = \begin{cases} 1 & \text{if } \frac{-\pi}{2} \leqq \xi < \frac{\pi}{2}, \\ 0 & \text{if } -\pi \leqq \xi < \frac{-\pi}{2} \text{ or } \frac{\pi}{2} \leqq \xi < \pi. \end{cases}$$

The Haar wavelet ψ produces the orthonormal basis $\{\psi_{j,k}\}$ for $L^2(\mathbb{R})$, usually called the Haar system. It was introduced in 1910 by A. Haar (see the first paper in Section III). The Shannon wavelet is closely connected with the *Shannon sampling theorem*, which dates back to 1949 [Sha49]. In a sense, these two wavelets are "opposites:" the Shannon wavelet is "band limited" (its Fourier transform is compactly supported), while the Haar wavelet is compactly supported. The Haar wavelet is not smooth while the Shannon wavelet belongs to $C^\infty(\mathbb{R})$; however, the latter decreases to 0 at infinity rather slowly (it is $O(\frac{1}{x})$ at ∞). These properties make the Haar wavelet "desirable" in applications because of its compactness. It is also desirable to combine this last property with a certain degree of smoothness. Since, clearly, there do not exist compact functions that are bandlimited, one is led to the problem of constructing compact wavelets having higher degrees of smoothness. This was achieved by I. Daubechies in the last paper of this section. One can see there and in Meyer's paper (the second one in this section) the important role that the MRA method plays in the solution to this problem. A general statement of the principal result of these two papers is

Theorem 2 *For any integer $r = 0, 1, 2, \ldots$, there exists an orthonormal wavelet ψ with compact support such that ψ has bounded derivatives up to order r. Moreover, ψ can be obtained from an MRA whose scaling function φ also has compact support and the same smoothness as ψ.*

Perhaps the most important feature of the MRA method is that the scaling function, and, consequently, the wavelets, can be constructed from the low pass filter m_0. Given appropriate assumptions, one shows that the equality

$$\hat{\varphi}(\xi) = \prod_{j=1}^{\infty} m_0(2^{-j}\xi) \tag{4}$$

presents us the scaling function φ in terms of m_0.

In the second and sixth papers of this section, it is shown that the compactness of φ (and ψ) is assured when m_0 is a trigonometric polynomial (a finite linear combination of the exponentials $e^{im\xi}$, $m \in \mathbb{Z}$). Thus, the construction of compactly supported wavelets leads to the problem of finding those trigonometric polynomials m_0 that satisfy the Smith-Barnwell equality (1) as well as properties that guarantee that φ, given by (4), is a scaling function. This is achieved in these papers by Daubechies and Meyer.

Not all wavelets are MRA wavelets (in the third paper of this section, Mallat exhibits an example, discovered by J.-L. Journé, of an orthonormal wavelet that is not in the MRA class). Both Gripenberg and Wang, again independently, characterized the MRA wavelets in terms of a simple equality: *If* $\psi \in L^2(\mathbb{R})$ *is an orthonormal wavelet, then it is an* MRA *wavelet if and only if*

$$D(\xi) \equiv D_\psi(\xi) \equiv \sum_{j=1}^{\infty} \sum_{k \in \mathbb{Z}} |\hat{\psi}(2^j(\xi + 2k\pi))|^2 = 1$$

for a.e. $\xi \in \mathbb{R}$. (See [HW, Chap. 7, Section 3] for a proof and explanation of this result).

The MRA wavelets have many other important properties. For example, one can show the following rather surprising result:

Theorem 3 *The class of MRA wavelets, as a subset of the unit sphere of* $L^2(\mathbb{R})$, *is pathwise connected.*

We refer the reader to [Wut98] for a proof of this result, as well as other MRA properties. One can see in this paper what is the role played by the lowpass filters in establishing these results. It is, then, a natural problem of finding those properties, in addition to equality (1), that determines which 2π-periodic $m_0 \in L^2([-\pi, \pi))$ are lowpass filters. The paper by A. Cohen in this chapter makes substantial progress in this direction. These lowpass filters are now comopletely characterized in [PSW99]; there, one can see that the result of Cohen is most important for this characterization.

Finally, let us say a few words about the paper of W. Lawton, the fifth one in this section. In the previous chapters, besides orthonormal bases, other reproducing systems are discussed that are produced by the action of translations and dilations on a generating function ψ. There are good mathematical reasons, as well as useful applications, for considering frames, rather than orthonormal bases. Such frames can be obtained by the methods we used for wavelet expansions. Let us point out, for example, that, if the condition $\|\psi\|_2 = 1$ is replaced by the condition $\|\psi\|_2 \leqq 1$, one obtains the fact that (I) and (II) characterize those $\psi \in L^2(\mathbb{R})$ for which $\{\psi_{jk}\}, j, k \in \mathbb{Z}$ is a *normalized tight frame* (NTF) for $L^2(\mathbb{R})$. That is,

$$f = \sum_{j,k \in \mathbb{Z}} \langle f, \psi_{j,k} \rangle \psi_{j,k}$$

for all $f \in L^2(\mathbb{R})$ (with convergence in $L^2(\mathbb{R})$) or, equivalently,

$$\|f\|_2^2 = \sum_{j,k \in \mathbb{Z}} |\langle f, \psi_{j,k} \rangle|^2$$

for all $f \in L^2(\mathbb{R})$. Recently, the term *Parseval frame* has been introduced for such expansions. It is only natural to extend the MRA method to this more general situation. This is carried out in the paper by Lawton (the fifth in this section) in which a larger class of functions $\psi \in L^2(\mathbb{R})$, with compact support, is constructed so that the "affine system" $\{\psi_{j,k}\}, j, k \in \mathbb{Z}$, is a Parseval frame for $L^2(\mathbb{R})$. Several authors have continued similar investigations and the study of the MRA method(s) is still an active area of research in the theory of wavelets. A sample of this activity is given by the following articles: [Aus92], [Aus95], [BL98], [BT01], [PSW99], [PSWX01].

References

[Aus95] P. Auscher. *Solution of two problems on wavelets.* J. Geom. Anal. **5**, (1995), 181–236.

[Aus92] P. Auscher. *Toute base d'ondlettes régulières de $L^2(\mathbb{R})$ est issue d'une analyse multirésolution régulière*, C.R. Acad. Sci. Paris(1) **315** (1992), 769–772.

[BL98] J. Benedetto and S. Li, *The theory of multiresolution analysis frames and application to filter banks*, Appl. & Comp. Harm. Anal. **5** (1998), 389–427.

[BT01] J. Benedetto and O. Treiber, *Wavelet frames: multiresolution analysis and extension principles*, Wavelet transforms and time-frequency signal analysis, Applied Numer. Harm. Analysis, Birkhäuser Boston, 2001, 3–36.

[Gri95] G. Gripenberg, *A necessary and sufficient condition for the existence of a father wavelet*, Studia Math. **114** (1995), 207–226.

[HW96] E. Hernández and G. Weiss, *A First Course on Wavelets*, CRC Press, Boca Raton, Fl., 1996, 1–489.

[PSW99] M. Papadakis, H. Sikic and G. Weiss, *The characterization of low pass filters and some basic properties of wavelets*, J. Four. Anal. Appl. **5** (1999), 495–521.

[PSWX01] M. Paluszynski, H. Sikic, G. Weiss and S. Xiao, *Generalized low pass filters and MRA frame wavelets*, J. Geom. Anal. **11** (2001) 311–371.

[Sha49] C.E. Shannon, *Communications in the presence of noise*, Proc. Inst. Radio Eng. **37** (1949), 10–21.

[Wan95] X. Wang, *The study of wavelets from the properties of their Fourier transform*, Ph.D. thesis, Washington University in St. Louis, 1995.

[Wut98] Wutam Consortium, *Basic properties of wavelets*, J. Fourier Anal. Appl. **4** (1998), 575–594.

674 IEEE TRANSACTIONS ON PATTERN ANALYSIS AND MACHINE INTELLIGENCE, VOL. 11, NO. 7, JULY 1989

A Theory for Multiresolution Signal Decomposition: The Wavelet Representation

STEPHANE G. MALLAT

***Abstract*—Multiresolution representations are very effective for analyzing the information content of images. We study the properties of the operator which approximates a signal at a given resolution. We show that the difference of information between the approximation of a signal at the resolutions 2^{j+1} and 2^j can be extracted by decomposing this signal on a wavelet orthonormal basis of $L^2(R^n)$. In $L^2(R)$, a wavelet orthonormal basis is a family of functions $(\sqrt{2^j}\,\psi(2^j x - n))_{(j,n)\in Z^2}$, which is built by dilating and translating a unique function $\psi(x)$. This decomposition defines an orthogonal multiresolution representation called a wavelet representation. It is computed with a pyramidal algorithm based on convolutions with quadrature mirror filters. For images, the wavelet representation differentiates several spatial orientations. We study the application of this representation to data compression in image coding, texture discrimination and fractal analysis.**

***Index Terms*—Coding, fractals, multiresolution pyramids, quadrature mirror filters, texture discrimination, wavelet transform.**

I. Introduction

In computer vision, it is difficult to analyze the information content of an image directly from the gray-level intensity of the image pixels. Indeed, this value depends upon the lighting conditions. More important are the local variations of the image intensity. The size of the neighborhood where the contrast is computed must be adapted to the size of the objects that we want to analyze [41]. This size defines a resolution of reference for measuring the local variations of the image. Generally, the structures we want to recognize have very different sizes. Hence, it is not possible to define a priori an optimal resolution for analyzing images. Several researchers [18], [31], [42] have developed pattern matching algorithms which process the image at different resolutions. For this purpose, one can reorganize the image information into a set of details appearing at different resolutions. Given a sequence of increasing resolutions $(r_j)_{j\in Z}$, the details of an image at the resolution r_j are defined as the difference of information between its approximation at the resolution r_j and its approximation at the lower resolution r_{j-1}.

Manuscript received July 30, 1987; revised December 23, 1988. This work was supported under the following Contracts and Grants: NSF grant 10DCR-8410771, Air Force Grant AFOSR F49620-85-K-0018, Army DAAG-29-84-K-0061, NSF-CER/DC82-19196 Ao2, and DARPA/ONR ARPA N0014-85-K-0807.

The author is with the Department of Computer Science Courant Institute of Mathematical Sciences, New York University, New York, NY 10012.

IEEE Log Number 8928052.

A multiresolution decomposition enables us to have a scale-invariant interpretation of the image. The scale of an image varies with the distance between the scene and the optical center of the camera. When the image scale is modified, our interpretation of the scene should not change. A multiresolution representation can be partially scale-invariant if the sequence of resolution parameters $(r_j)_{j\in Z}$ varies exponentially. Let us suppose that there exists a resolution step $\alpha \in R$ such that for all integers j, $r_j = \alpha^j$. If the camera gets α times closer to the scene, each object of the scene is projected on an area α^2 times bigger in the focal plane of the camera. That is, each object is measured at a resolution α times bigger. Hence, the details of this new image at the resolution α^j correspond to the details of the previous image at the resolution α^{j+1}. Rescaling the image by α translates the image details along the resolution axis. If the image details are processed identically at all resolutions, our interpretation of the image information is not modified.

A multiresolution representation provides a simple hierarchical framework for interpretating the image information [22]. At different resolutions, the details of an image generally characterize different physical structures of the scene. At a coarse resolution, these details correspond to the larger structures which provide the image "context". It is therefore natural to analyze first the image details at a coarse resolution and then gradually increase the resolution. Such a coarse-to-fine strategy is useful for pattern recognition algorithms. It has already been widely studied for low-level image processing such as stereo matching and template matching [16], [18].

Burt [5] and Crowley [8] have each introduced pyramidal implementation for computing the signal details at different resolutions. In order to simplify the computations, Burt has chosen a resolution step α equal to 2. The details at each resolution 2^j are calculated by filtering the original image with the difference of two low-pass filters and by subsampling the resulting image by a factor 2^j. This operation is performed over a finite range of resolutions. In this implementation, the difference of low-pass filters gives an approximation of the Laplacian of the Gaussian. The details at different resolutions are regrouped into a pyramid structure called the Laplacian pyramid. The Laplacian pyramid data structures, as studied by Burt and Crowley, suffer from the difficulty that data at separate levels are correlated. There is no clear model

0162-8828/89/0700-0674$01.00 © 1989 IEEE

which handles this correlation. It is thus difficult to know whether a similarity between the image details at different resolutions is due to a property of the image itself or to the intrinsic redundancy of the representation. Furthermore, the Laplacian multiresolution representation does not introduce any spatial orientation selectivity into the decomposition process. This spatial homogeneity can be inconvenient for pattern recognition problems such as texture discrimination [21].

In this article, we first study the mathematical properties of the operator which transforms a function into an approximation at a resolution 2^j. We then show that the difference of information between two approximations at the resolutions 2^{j+1} and 2^j is extracted by decomposing the function in a wavelet orthonormal basis. This decomposition defines a complete and orthogonal multiresolution representation called the wavelet representation. Wavelets have been introduced by Grossmann and Morlet [17] as functions $\psi(x)$ whose translations and dilations $(\sqrt{s}\,\psi(sx - t))_{(s,t)\in R^+ \times R}$ can be used for expansions of $L^2(R)$ functions. Meyer [35] showed that there exists wavelets $\psi(x)$ such that $(\sqrt{2^j}\,\psi(2^j x - k))_{(j,k)\in Z^2}$ is an orthonormal basis of $L^2(R)$. These bases generalize the Haar basis. The wavelet orthonormal bases provide an important new tool in functional analysis. Indeed, before then, it had been believed that no construction could yield simple orthonormal bases of $L^2(R)$ whose elements had good localization properties in both the spatial and Fourier domains.

The multiresolution approach to wavelets enables us to characterize the class of functions $\psi(x) \in L^2(R)$ that generate an orthonormal basis. The model is first described for one-dimensional signals and then extended to two dimensions for image processing. The wavelet representation of images discriminates several spatial orientations. We show that the computation of the wavelet representation may be accomplished with a pyramidal algorithm based on convolutions with quadrature mirror filters. The signal can also be reconstructed from a wavelet representation with a similar pyramidal algorithm. We discuss the application of this representation to compact image coding, texture discrimination and fractal analysis. In this article, we omit the proofs of the theorems and avoid mathematical technical detals. Rather, we try to illustrate the practical implications of the model. The mathematical foundations are more thoroughly described in [28].

A. Notation

Z and R denote the set of integers and real numbers respectively. $L^2(R)$ denotes the vector space of measurable, square-integrable one-dimensional functions $f(x)$. For $f(x) \in L^2(R)$ and $g(x) \in L^2(R)$, the inner product of $f(x)$ with $g(x)$ is written

$$\langle g(u), f(u)\rangle = \int_{-\infty}^{+\infty} g(u) f(u)\, du.$$

The norm of $f(x)$ in $L^2(R)$ is given by

$$\|f\|^2 = \int_{-\infty}^{+\infty} |f(u)|^2\, du.$$

We denote the convolution of two functions $f(x) \in L^2(R)$ and $g(x) \in L^2(R)$ by

$$f * g(x) = (f(u) * g(u))(x) = \int_{-\infty}^{+\infty} f(u)\, g(x-u)\, du.$$

The Fourier transform of $f(x) \in L^2(R)$ is written $\hat{f}(\omega)$ and is defined by

$$\hat{f}(\omega) = \int_{-\infty}^{+\infty} f(x)\, e^{-i\omega x}\, dx.$$

$l^2(Z)$ is the vector space of square-summable sequences

$$l^2(Z) = \left\{ (\alpha_i)_{i\in Z} : \sum_{i=-\infty}^{+\infty} |\alpha_i|^2 < \infty \right\}.$$

$L^2(R^2)$ is the vector space of measurable, square-integrable two dimensional functions $f(x, y)$. For $f(x, y) \in L^2(R^2)$ and $g(x, y) \in L^2(R^2)$, the inner product of $f(x, y)$ with $g(x, y)$ is written

$$\langle f(x, y), g(x, y)\rangle = \int_{-\infty}^{+\infty}\int_{-\infty}^{+\infty} f(x, y)\, g(x, y)\, dx\, dy.$$

The Fourier transform of $f(x, y) \in L^2(R^2)$ is written $\hat{f}(\omega_x, \omega_y)$ and is defined by

$$\hat{f}(\omega_x, \omega_y) = \int_{-\infty}^{+\infty}\int_{-\infty}^{+\infty} f(x, y)\, e^{-i(\omega_x x + \omega_y y)}\, dx\, dy.$$

II. Multiresolution Transform

In this section, we study the concept of multiresolution decomposition for one-dimensional signals. The model is extended to two dimensions in Section IV.

A. Multiresolution Approximation of $L^2(R)$

Let A_{2^j} be the operator which approximates a signal at a resolution 2^j. We suppose that our original signal $f(x)$ is measurable and has a finite energy: $f(x) \in L^2(R)$. Here, we characterize A_{2^j} from the intuitive properties that one would expect from such an approximation operator. We state each property in words, and then give the equivalent mathematical formulation.

1) A_{2^j} is a linear operator. If $A_{2^j} f(x)$ is the approximation of some function $f(x)$ at the resolution 2^j, then $A_{2^j} f(x)$ is not modified if we approximate it again at the resolution 2^j. This principle shows that $A_{2^j} \circ A_{2^j} = A_{2^j}$. The operator A_{2^j} is thus a projection operator on a particular vector space $V_{2^j} \subset L^2(R)$. The vector space V_{2^j} can be interpreted as the set of all possible approximations at the resolution 2^j of functions in $L^2(R)$.

2) Among all the approximated functions at the resolution 2^j, $A_{2^j} f(x)$ is the function which is the most similiar to $f(x)$.

$$\forall g(x) \in V_{2^j}, \quad \|g(x) - f(x)\| \geq \|A_{2^j}f(x) - f(x)\|. \tag{1}$$

Hence, the operator A_{2^j} is an orthogonal projection on the vector space V_{2^j}.

3) The approximation of a signal at a resolution 2^{j+1} contains all the necessary information to compute the same signal at a smaller resolution 2^j. This is a causality property. Since A_{2^j} is a projection operator on V_{2^j} this principle is equivalent to

$$\forall j \in Z, \quad V_{2^j} \subset V_{2^{j+1}}. \tag{2}$$

4) An approximation operation is similar at all resolutions. The spaces of approximated functions should thus be derived from one another by scaling each approximated function by the ratio of their resolution values

$$\forall j \in Z, \quad f(x) \in V_{2^j} \Leftrightarrow f(2x) \in V_{2^{j+1}}. \tag{3}$$

5) The approximation $A_{2^j}f(x)$ of a signal $f(x)$ can be characterized by 2^j samples per length unit. When $f(x)$ is translated by a length proportional to 2^{-j}, $A_{2^j}f(x)$ is translated by the same amount and is characterized by the same samples which have been translated. As a consequence of (3), it is sufficient to express the principle 5) for the resolution $j = 0$. The mathematical translations consist of the following.

• Discrete characterization:

$$\text{There exists an isomorphism } I \text{ from } V_1 \text{ onto } I^2(Z). \tag{4}$$

• Translation of the approximation:

$$\forall k \in Z, \; A_1 f_k(x) = A_1 f(x-k), \text{ where } f_k(x) = f(x-k). \tag{5}$$

• Translation of the samples:

$$I(A_1 f(x)) = (\alpha_i)_{i \in Z} \Leftrightarrow I(A_1 f_k(x)) = (\alpha_{i-k})_{i \in Z}. \tag{6}$$

6) When computing an approximation of $f(x)$ at resolution 2^j, some information about $f(x)$ is lost. However, as the resolution increases to $+\infty$ the approximated signal should converge to the original signal. Conversely as the resolution decreases to zero, the approximated signal contains less and less information and converges to zero.

Since the approximated signal at a resolution 2^j is equal to the orthogonal projection on a space V_{2^j}, this principle can be written

$$\lim_{j \to +\infty} V_{2^j} = \bigcup_{j=-\infty}^{+\infty} V_{2^j} \text{ is dense in } L^2(R) \tag{7}$$

and

$$\lim_{j \to -\infty} V_{2^j} = \bigcap_{j=-\infty}^{+\infty} V_{2^j} = \{0\}. \tag{8}$$

We call any set of vector spaces $(V_{2^j})_{j \in Z}$ which satisfies the properties (2)–(8) a *multiresolution approximation* of $L^2(R)$. The associated set of operators A_{2^j} satisfying 1)–6) give the approximation of any $L^2(R)$ function at a resolution 2^j. We now give a simple example of a multiresolution approximation of $L^2(R)$.

Example: Let V_1 be the vector space of all functions of $L^2(R)$ which are constant on each interval $]k, k+1[$, for any $k \in Z$. Equation (3) implies that V_{2^j} is the vector space of all the functions of $L^2(R)$ which are constant on each interval $]k2^{-j}, (k+1)2^{-j}[$, for any $k \in Z$. The condition (2) is easily verified. We can define an isomorphism I which satisfies properties (4), (5), and (6) by associating with any function $f(x) \in V_1$ the sequence $(\alpha_k)_{k \in Z}$ such that α_k equals the value of $f(x)$ on the interval $]k, k+1[$. We know that the vector space of piecewise constant functions is dense in $L^2(R)$. Hence, we can derive that $\bigcup_{j=-\infty}^{+\infty} V_{2^j}$ is dense in $L^2(R)$. It is clear that $\bigcap_{j=-\infty}^{+\infty} V_{2^j} = \{0\}$, so the sequence of vector spaces $(V_{2^j})_{j \in Z}$ is a multiresolution approximation of $L^2(R)$. Unfortunately, the functions of these vector spaces are neither smooth nor continuous, making this multiresolution approximation rather inconvenient. For many applications we want to compute a smooth approximation. In Appendix A, we describe a class of multiresolution approximations where the functions of each space V_{2^j} are n times continuously differentiable.

We saw that the approximation operator A_{2^j} is an orthogonal projection on the vector space V_{2^j}. In order to numerically characterize this operator, we must find an orthonormal basis of V_{2^j}. The following theorem shows that such an orthonormal basis can be defined by dilating and translating a unique function $\phi(x)$.

Theorem 1: Let $(V_{2^j})_{j \in Z}$ be a multiresolution approximation of $L^2(R)$. There exists a unique function $\phi(x) \in L^2(R)$, called a *scaling function*, such that if we set $\phi_{2^j}(x) = 2^j\phi(2^jx)$ for $j \in Z$, (the dilation of $\phi(x)$ by 2^j), then

$$\left(\sqrt{2^{-j}}\, \phi_{2^j}(x - 2^{-j}n)\right)_{n \in Z} \text{ is an orthonormal basis of } V_{2^j}. \quad \blacksquare \tag{9}$$

Indications for the proof of this theorem can be found in Appendix B. The theorem shows that we can build an orthonormal basis of any V_{2^j} by dilating a function $\phi(x)$ with a coefficient 2^j and translating the resulting function on a grid whose interval is proportional to 2^{-j}. The functions $\phi_{2^j}(x)$ are normalized with respect to the $L^1(R)$ norm. The coefficient $\sqrt{2^{-j}}$ appears in the basis set in order to normalize the functions in the $L^2(R)$ norm. For a given multiresolution approximation $(V_{2^j})_{j \in Z}$, there exists a unique scaling function $\phi(x)$ which satisfies (9). However, for different multiresolution approximations, the scaling functions are different. One can easily show that the scaling function corresponding to the multiresolution described in the previous example is the indicator function of the interval $[0, 1]$. In general, we want to have a smoother scaling function. Fig. 1 shows an example of a continuously differentiable and exponentially decreasing scaling function. Its Fourier transform has the shape of a

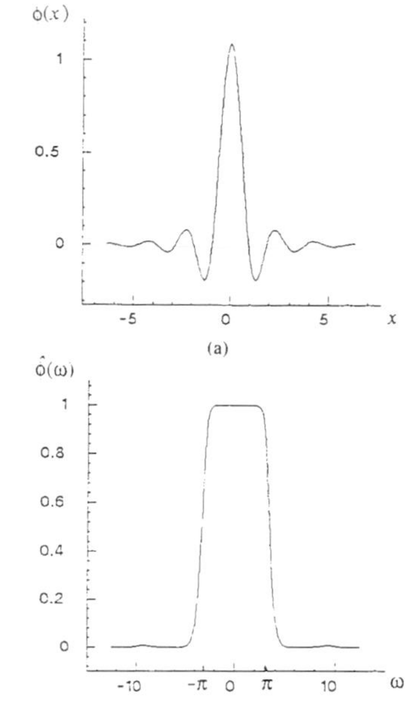

Fig. 1. (a) Example of scaling function $\phi(x)$. This function is computed in Appendix A. (b) Fourier transform $\hat{\phi}(\omega)$. A scaling function is a low-pass filter.

low-pass filter. The corresponding multiresolution approximation is built from cubic splines. This scaling function is described further in Appendix A.

The orthogonal projection on V_{2^j} can now be computed by decomposing the signal $f(x)$ on the orthonormal basis given by Theorem 1. Specifically,

$$\forall f(x) \in L^2(\mathbf{R}),\ A_{2^j} f(x) = 2^{-j} \sum_{n=-\infty}^{+\infty} \langle f(u), \phi_{2^j}(u - 2^{-j}n) \rangle \, \phi_{2^j}(x - 2^{-j}n). \tag{10}$$

The approximation of the signal $f(x)$ at the resolution 2^j, $A_{2^j} f(x)$, is thus characterized by the set of inner products which we denote by

$$A_{2^j}^d f = \left(\langle f(u), \phi_{2^j}(u - 2^{-j}n) \rangle \right)_{n \in \mathbf{Z}}. \tag{11}$$

$A_{2^j}^d f$ is called a *discrete approximation* of $f(x)$ at the resolution 2^j. Since computers can only process discrete signals, we must work with discrete approximations. Each inner product can also be interpreted as a convolution product evaluated at a point $2^{-j}n$

$$\langle f(u), \phi_{2^j}(u - 2^{-j}n) \rangle = \int_{-\infty}^{+\infty} f(u)\, \phi_{2^j}(u - 2^{-j}n)\, du = \left(f(u) * \phi_{2^j}(-u) \right)(2^{-j}n).$$

Hence, we can rewrite $A_{2^j}^d f$:

$$A_{2^j}^d f = \left(\left(f(u) * \phi_{2^j}(-u) \right)(2^{-j}n) \right)_{n \in \mathbf{Z}}. \tag{12}$$

Since $\phi(x)$ is a low-pass filter, this discrete signal can be interpreted as a low-pass filtering of $f(x)$ followed by a uniform sampling at the rate 2^j. In an approximation operation, when removing the details of $f(x)$ smaller than 2^{-j}, we suppress the highest frequencies of this function. The scaling function $\phi(x)$ forms a very particular low-pass filter since the family of functions $(\sqrt{2^{-j}}\ \phi_{2^j}(x - 2^{-j}n))_{n \in \mathbf{Z}}$ is an orthonormal family.

In the next section we show that the discrete approximation of $f(x)$ at the resolution 2^j can be computed with a pyramidal algorithm.

B. Implementation of a Multiresolution Transform

In practice, a physical measuring device can only measure a signal at a finite resolution. For normalization purposes, we suppose that this resolution is equal to 1. Let $A_1^d f$ be the discrete approximation at the resolution 1 that is measured. The causality principle says that from $A_1^d f$ we can compute all the discrete approximations $A_{2^j}^d f$ for $j < 0$. In this section, we describe a simple iterative algorithm for calculating these discrete approximations.

Let $(V_{2^j})_{j \in \mathbf{Z}}$ be a multiresolution approximation and $\phi(x)$ be the corresponding scaling function. The family of functions $(\sqrt{2^{-j-1}}\ \phi_{2^{j+1}}(x - 2^{-j-1}k))_{k \in \mathbf{Z}}$ is an orthonormal basis of $V_{2^{j+1}}$. We know that for any $n \in \mathbf{Z}$, the function $\phi_{2^j}(x - 2^{-j}n)$ is a member of V_{2^j} which is included in $V_{2^{j+1}}$. It can thus be expanded in this orthonormal basis of $V_{2^{j+1}}$:

$$\phi_{2^j}(x - 2^{-j}n) = 2^{-j-1} \sum_{k=-\infty}^{+\infty} \langle \phi_{2^j}(u - 2^{-j}n), \phi_{2^{j+1}}(u - 2^{-j-1}k) \rangle \cdot \phi_{2^{j+1}}(x - 2^{-j-1}k). \tag{13}$$

By changing variables in the inner product integral, one can show that

$$2^{-j-1} \langle \phi_{2^j}(u - 2^{-j}n), \phi_{2^{j+1}}(u - 2^{-j-1}k) \rangle = \langle \phi_{2^{-1}}(u), \phi(u - (k - 2n)) \rangle. \tag{14}$$

When computing the inner products of $f(x)$ with both sides of (13) we obtain

$$\langle f(u), \phi_{2^j}(u - 2^{-j}n) \rangle = \sum_{k=-\infty}^{+\infty} \langle \phi_{2^{-1}}(u), \phi(u - (k - 2n)) \rangle \cdot \langle f(u), \phi_{2^{j+1}}(u - 2^{-j-1}k) \rangle.$$

Let H be the discrete filter whose impulse response is given by

$$\forall n \in \mathbf{Z},\ h(n) = \langle \phi_{2^{-1}}(u), \phi(u - n) \rangle. \tag{15}$$

Let $\tilde{H}$ be the mirror filter with impulse response $\tilde{h}(n) = h(-n)$. By inserting (15) in the previous equation, we obtain

$$\langle f(u), \phi_{2^j}(u - 2^{-j}n) \rangle = \sum_{k=-\infty}^{+\infty} \tilde{h}(2n - k) \langle f(u), \phi_{2^{j+1}}(u - 2^{-j-1}k) \rangle. \tag{16}$$

Equation (16) shows that $A_{2^j}^d f$ can be computed by convolving $A_{2^{j+1}}^d f$ with $\tilde{H}$ and keeping every other sample of the output. All the discrete approximations $A_{2^j}^d f$, for $j < 0$, can thus be computed from $A_1^d f$ by repeating this process. This operation is called a pyramid transform. The algorithm is illustrated by a block diagram in Fig. 5.

In practice, the measuring device gives only a finite number of samples: $A_1^d f = (\alpha_n)_{1 \le n \le N}$. Each discrete signal $A_{2^j}^d f$ $(j < 0)$ has $2^j N$ samples. In order to avoid border problems when computing the discrete approximations $A_{2^j}^d f$, we suppose that the original signal $A_1^d f$ is symmetric with respect to $n = 0$ and $n = N$

$$\alpha_n = \begin{cases} \alpha_{-n} & \text{if } -N < n < 0 \\ \alpha_{2N-n} & \text{if } 0 < n < 0. \end{cases}$$

If the impulse response of the filter $\tilde{H}$ is even ($\tilde{H} = H$), each discrete approximation $A_{2^j}^d f$ will also be symmetric with respect to $n = 0$ and $n = 2^{-j}N$. Fig. 2(a) shows the discrete approximated signal $A_{2^j}^d f$ of a continuous signal $f(x)$ at the resolutions 1, 1/2, 1/4, 1/8, 1/16, and 1/32. These discrete approximated signals have been computed with the algorithm previously described. Appendix A gives the coefficients of the filter H that we used. The continuous approximated signals $A_{2^j} f(x)$ shown in Fig. 2(b) have been calculated by interpolating the discrete approximations with (10). As the resolution decreases, the smaller details of $f(x)$ gradually disappear.

Theorem 1 shows that a multiresolution approximation $(V_{2^j})_{j \in Z}$ is completely characterized by the scaling function $\phi(x)$. A scaling function can be defined as a function $\phi(x) \in L^2(R)$ such that, for all $j \in Z$, $(\sqrt{2^{-j}}\, \phi_{2^j}(x - 2^{-j}n))_{n \in Z}$ is an orthonormal family, and if V_{2^j} is the vector space generated by this family of functions, then $(V_{2^j})_{j \in Z}$ is a multiresolution approximation of $L^2(R)$. We also impose a regularity condition on scaling functions. A scaling function $\phi(x)$ must be continuously differentiable and the asymptotic decay of $\phi(x)$ and $\phi'(x)$ at infinity must satisfy

$$|\phi(x)| = O(x^{-2}) \quad \text{and} \quad |\phi'(x)| = O(x^{-2}).$$

The following theorem gives a practical characterization of the Fourier transform of a scaling function.

Theorem 2: Let $\phi(x)$ be a scaling function, and let H be a discrete filter with impulse response $h(n) = \langle \phi_{2^{-1}}(u), \phi(u - n) \rangle$. Let $H(\omega)$ be the Fourier series defined by

$$H(\omega) = \sum_{n=-\infty}^{+\infty} h(n) e^{-in\omega}. \tag{17}$$

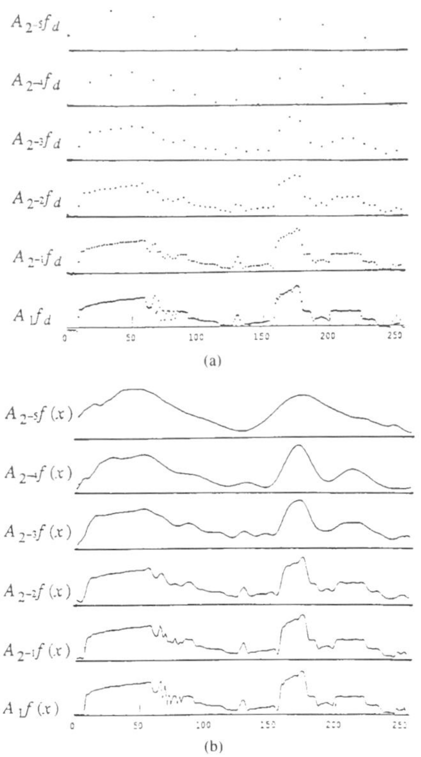

Fig. 2. (a) Discrete approximations $A_{2^j}^d f$ at the resolutions 1, 1/2, 1/4, 1/8, 1/16, and 1/32. Each dot gives the amplitude of the inner product $\langle f(u), \phi_{2^j}(u - 2^{-j}n) \rangle$ depending upon $2^{-j}n$. (b) Continuous approximations $A_{2^j} f(x)$ at the resolutions 1, 1/2, 1/4, 1/8, 1/16, and 1/32. These approximations are computed by interpolating the discrete approximations with (10).

$H(\omega)$ satisfies the following two properties:

$$|H(0)| = 1 \quad \text{and} \quad h(n) = O(n^{-2}) \text{ at infinity}. \tag{17a}$$

$$|H(\omega)|^2 + |H(\omega + \pi)|^2 = 1. \tag{17b}$$

Conversely let $H(\omega)$ be a Fourier series satisfying (17a) and (17b) and such that

$$|H(\omega)| \neq 0 \quad \text{for} \quad \omega \in [0, \pi/2]. \tag{17c}$$

The function defined by

$$\hat{\phi}(\omega) = \prod_{p=1}^{+\infty} H(2^{-p}\omega) \tag{18}$$

is the Fourier transform of a scaling function. ■

Indication for the proof of this theorem are given in Appendix C. The filters that satisfy property (17b) are called *conjugate* filters. One can find extensive descriptions of such filters and numerical methods to synthesize them in the signal processing literature [10], [36], [40]. Given a conjugate filter H which satisfies (17a)–(17c), we can then compute the Fourier transform of the corresponding scaling function with (16). It is possible to choose $H(\omega)$ in order to obtain a scaling function $\phi(x)$ which has good localization properties in both the frequency and spatial domains. The smoothness class of $\phi(x)$ and its asymtotic decay at infinity can be estimated from the properties of $H(\omega)$ [9]. In the multiresolution approximation given in the example of Section II-A, we saw that the scaling function is the indicator function of the interval [0, 1]. One can easily show that the corresponding function $H(\omega)$ satisfies

$$H(\omega) = e^{-i\omega} \cos\left(\frac{\omega}{2}\right).$$

Appendix A describes a class of symmetric scaling functions which decay exponentially and whose Fourier transforms decrease as $1/\omega^n$, for some $n \in N$. Fig. 3 shows the filter H associated with the scaling function given in Fig. 1. This filter is further described in Appendix A.

III. The Wavelet Representation

As explained in the introduction, we wish to build a multiresolution representation based on the differences of information available at two successive resolutions 2^j and 2^{j+1}. This section shows that such a representation can be computed by decomposing the signal using a wavelet orthonormal basis.

A. *The Detail Signal*

Here, we explain how to extract the difference of information between the approximation of a function $f(x)$ at the resolutions 2^{j+1} and 2^j. This difference of information is called the *detail signal* at the resolution 2^j. The approximation at the resolution 2^{j+1} and 2^j of a signal are respectively equal to its orthogonal projection on $V_{2^{j+1}}$ and V_{2^j}. By applying the projection theorem, we can easily show that the detail signal at the resolution 2^j is given by the orthogonal projection of the original signal on the orthogonal complement of V_{2^j} in $V_{2^{j+1}}$. Let O_{2^j} be this orthogonal complement, i.e.,

$$O_{2^j} \text{ is orthogonal to } V_{2^j},$$

$$O_{2^j} \oplus V_{2^j} = V_{2^{j+1}}.$$

To compute the orthogonal projection of a function $f(x)$ on O_{2^j}, we need to find an orthonormal basis of O_{2^j}. Much like Theorem 1, Theorem 3 shows that such a basis can be built by scaling and translating a function $\psi(x)$.

Theorem 3: Let $(V_{2^j})_{j\in Z}$ be a multiresolution vector space sequence, $\phi(x)$ the scaling function, and H the corresponding conjugate filter. Let $\psi(x)$ be a function whose Fourier transform is given by

$$\hat{\psi}(\omega) = G\left(\frac{\omega}{2}\right)\hat{\phi}\left(\frac{\omega}{2}\right)$$

$$\text{with } G(\omega) = e^{-i\omega}\,\overline{H(\omega + \pi)}. \quad (19)$$

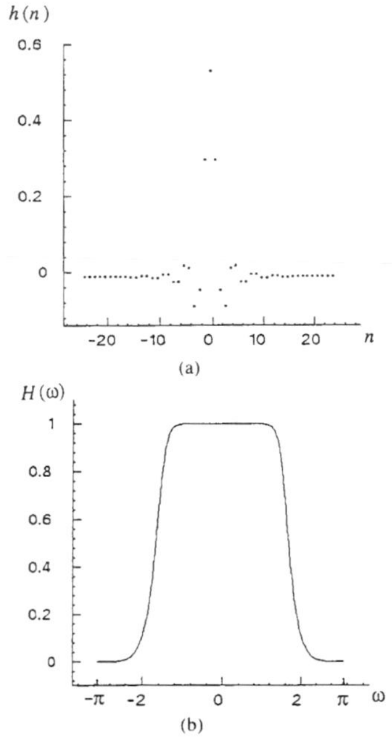

Fig. 3. (a) Impulse response of the filter H associated to the scaling function shown in Fig. 1. The coefficients of this filter are given in Appendix A. (b) Transfer function $H(\omega)$ of the filter H.

Let $\psi_{2^j}(x) = 2^j\psi(2^jx)$ denote the dilation of $\psi(x)$ by 2^j. Then

$$\left(\sqrt{2^{-j}}\,\psi_{2^j}(x - 2^{-j}n)\right)_{n\in Z} \text{ is an orthonormal basis of } O_{2^j}$$

and

$$\left(\sqrt{2^{-j}}\,\psi_{2^j}(x - 2^{-j}n)\right)_{(n,j)\in Z^2}$$

is an orthonormal basis of $L^2(R)$.

$\psi(x)$ is called an *orthogonal wavelet.* ∎

Indications for the proof of this theorem can be found in Appendix D. An orthonormal basis of O_{2^j} can thus be computed by scaling the wavelet $\psi(x)$ with a coefficient 2^j and translating it on a grid whose interval is proportional to 2^{-j}. The wavelet function corresponding to the example of multiresolution given in Section II-A is the Haar wavelet

$$\psi(x) = \begin{cases} 1 & \text{if } 0 \le x < \frac{1}{2} \\ -1 & \text{if } \frac{1}{2} \le x < 1 \\ 0 & \text{otherwise} \end{cases}.$$

This wavelet is not even continuous. In many applications, we want to use a smooth wavelet. For computing a wavelet, we can define a function $H(\omega)$ which satisfies the conditions (17a)–(17c) of Theorem 2, compute the corresponding scaling function $\phi(x)$ with equation (18) and the wavelet $\psi(x)$ with (19). Depending upon choice of $H(\omega)$, the scaling function $\phi(x)$ and the wavelet $\psi(x)$ can have good localization both in the spatial and Fourier domains. Daubechies [9] studied the properties of $\phi(x)$ and $\psi(x)$ depending upon $H(\omega)$. The first wavelets found by Meyer [35] are both C^{∞} and have an asymptotic decay which falls faster than the multiplicative inverse of any polynomial. Daubechies shows that for any $n > 0$, we can find a function $H(\omega)$ such that the corresponding wavelet $\psi(x)$ has a compact support and is n times continuously differentiable [9]. The wavelets described in Appendix A are exponentially decreasing and are in C^n for different values of n. These particular wavelets have been studied by Lemarie [24] and Battle [3].

The decomposition of a signal in an orthonormal wavelet basis gives an intermediate representation between Fourier and spatial representations. The properties of the wavelet orthonormal bases are discussed by Meyer in an advanced functional analysis book [34]. Due to this double localization in the Fourier and the spatial domains, it is possible to characterize the local regularity of a function $f(x)$ based on the coefficients in a wavelet orthonormal basis expansion [25]. For example, from the asymptotic rate of decrease of the wavelet coefficients, we can determine whether a function $f(x)$ is n times differentiable at a point x_0. Fig. 4 shows the wavelet associated with the scaling function of Fig. 1. This wavelet is symmetric with respect to the point $x = 1/2$. The energy of a wavelet in the Fourier domain is essentially concentrated in the intervals $[-2\pi, -\pi] \cup [\pi, 2\pi]$.

Let $P_{O_{2^j}}$ be the orthogonal projection on the vector space O_{2^j}. As a consequence of Theorem 3, this operator can now be written

$$P_{O_{2^j}} f(x) = 2^{-j} \sum_{n=-\infty}^{+\infty} \langle f(u), \psi_{2^j}(u - 2^{-j}n) \rangle \cdot \psi_{2^j}(x - 2^{-j}n). \tag{20}$$

$P_{O_{2^j}} f(x)$ yields to the detail signal of $f(x)$ at the resolution 2^j. It is characterized by the set of inner products

$$D_{2^j} f = \left(\langle f(u), \psi_{2^j}(u - 2^{-j}n) \rangle \right)_{n \in Z}. \tag{21}$$

$D_{2^j} f$ is called the *discrete detail signal* at the resolution 2^j. It contains the difference of information between $A^d_{2^{j+1}} f$ and $A^d_{2^j} f$. As we did in (12), we can prove that each of these inner products is equal to the convolution of $f(x)$ with $\psi_{2^j}(-x)$ evaluated at $2^{-j}n$

$$\langle f(u), \psi_{2^j}(u - 2^{-j}n) \rangle = \left(f(u) * \psi_{2^j}(-u) \right) (2^{-j}n). \tag{22}$$

Equations (21) and (22) show that the discrete detail signal at the resolution 2^j is equal to a uniform sampling of $(f(u) * \psi_{2^j}(-u))(x)$ at the rate 2^j

$$D_{2^j} f = \left(\left(f(u) * \psi_{2^j}(-u) \right) (2^{-j}n) \right)_{n \in Z}.$$

The wavelet $\psi(x)$ can be viewed as a bandpass filter whose frequency bands are approximatively equal to

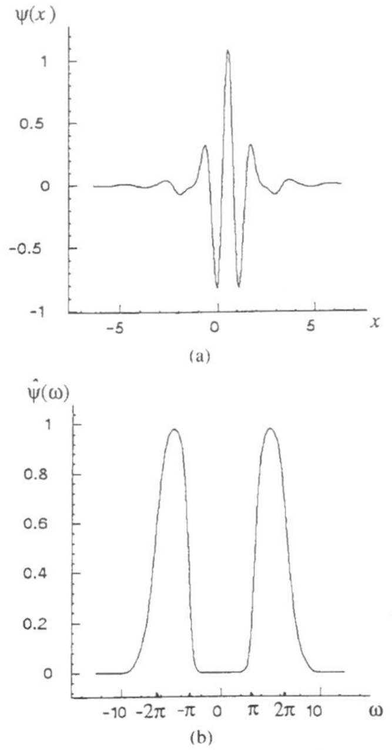

Fig. 4. (a) Wavelet $\psi(x)$ associated to the scaling function of Fig. 1. (b) Modulus of the Fourier transform of $\psi(x)$. A wavelet is a band-pass filter.

$[-2\pi, -\pi] \cup [\pi, 2\pi]$. Hence, the detail signal $D_{2^j} f$ describes $f(x)$ in the frequency bands $[-2^{-j+1}\pi, -2^{-j}\pi] \cup [2^{-j}\pi, 2^{-j+1}\pi]$.

We can prove by induction that for any $J > 0$, the original discrete signal $A^d_1 f$ measured at the resolution 1 is represented by

$$\left(A^d_{2^{-J}} f, (D_{2^j} f)_{-J \le j \le -1} \right). \tag{23}$$

This set of discrete signals is called an *orthogonal wavelet representation*, and consists of the reference signal at a coarse resolution $A^d_{2^{-J}} f$ and the detail signals at the resolutions 2^j for $-J \le j \le -1$. It can be interpreted as a decomposition of the original signal in an orthonormal wavelet basis or as a decomposition of the signal in a set of *independent* frequency channels as in Marr's human vision model [30]. The independence is due to the orthogonality of the wavelet functions.

It is difficult to give a precise interpretation of the model in terms of a frequency decomposition because the overlap of the frequency channels. However, we can control this overlap thanks to the orthogonality of our decomposition functions. That is why the tools of functional analysis give a better understanding of this decomposition. If we ignore the overlapping spectral supports, the interpretation in the frequency domain provides an intuitive approach to the model. In analogy with the Laplacian pyramid data structure, $A^d_{2^{-J}} f$ provides the top-level Gaussian pyramid data, and the $D_{2^j} f$ data provide the successive

Laplacian pyramid levels. Unlike the Laplacian pyramid, however, there is no oversampling, and the individual coefficients in the set of data are independent.

B. Implementation of an Orthogonal Wavelet Representation

In this section, we describe a pyramidal algorithm to compute the wavelet representation. With the same derivation steps as in Section II-B, we show that $D_{2^j}f$ can be calculated by convolving $A^d_{2^{j+1}}f$ with a discrete filter G whose form we will characterize.

For any $n \in \mathbf{Z}$, the function $\psi_{2^j}(x - 2^{-j}n)$ is a member of $\mathbf{O}_{2^j} \subset \mathbf{V}_{2^{j+1}}$. In the same manner as (13), this function can be expanded in an orthonormal basis of $\mathbf{V}_{2^{j+1}}$

$$\psi_{2^j}(x - 2^{-j}n) =$$

$$2^{-j-1} \sum_{k=-\infty}^{+\infty} \langle \psi_{2^j}(u - 2^{-j}n), \phi_{2^{j+1}}(u - 2^{-j-1}k) \rangle \cdot \phi_{2^{j+1}}(x - 2^{-j-1}k). \tag{24}$$

As we did in (14), by changing variables in the inner product integral we can prove that

$$2^{-j-1}\langle \psi_{2^j}(u - 2^{-j}n), \phi_{2^{j+1}}(u - 2^{-j-1}k) \rangle = \langle \psi_{2^{-1}}(u), \phi(u - (k - 2n)) \rangle. \tag{25}$$

Hence, by computing the inner product of $f(x)$ with the functions of both sides of (24), we obtain

$$\langle f(u), \psi_{2^j}(u - 2^{-j}n) \rangle = \sum_{k=-\infty}^{+\infty} \langle \psi_{2^{-1}}(u), \phi(u - (k - 2n)) \rangle \cdot \langle f(u), \phi_{2^{j+1}}(u - 2^{-j-1}k) \rangle. \tag{26}$$

Let G be the discrete filter with impulse response

$$g(n) = \langle \psi_{2^{-1}}(u), \phi(u - n) \rangle, \tag{27}$$

and $\tilde{G}$ be the symmetric filter with impulse response $\tilde{g}(n) = g(-n)$. We show in Appendix D that the transfer function of this filter is the function $G(\omega)$ defined in Theorem 3, equation (19). Inserting (27) to (26) yields

$$\langle f(u), \psi_{2^j}(u - 2^{-j}n) \rangle = \sum_{k=-\infty}^{+\infty} \tilde{g}(2n - k) \langle f(u), \phi_{2^{j+1}}(u - 2^{-j-1}k) \rangle. \tag{28}$$

Equation (28) shows that we can compute the detail signal $D_{2^j}f$ by convolving $A^d_{2^{j+1}}f$ with the filter $\tilde{G}$ and retaining every other sample of the output. The orthogonal wavelet representation of a discrete signal $A^d_1 f$ can therefore be computed by successively decomposing $A^d_{2^{j+1}}f$ into $A^d_{2^j}f$ and $D_{2^j}f$ for $-J \le j \le -1$. This algorithm is illustrated by the block diagram shown in Fig. 5.

In practice, the signal $A^d_1 f$ has only a finite number of samples. One method of handling the border problems

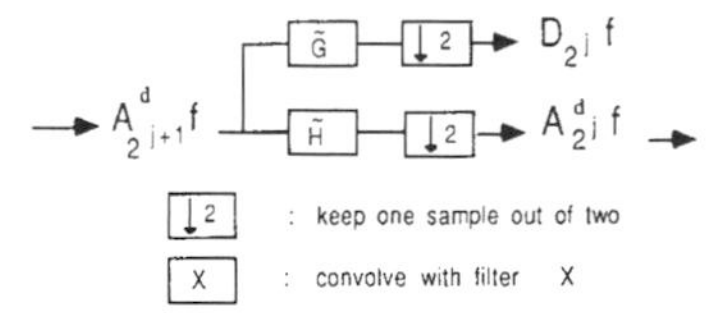

Fig. 5. Decomposition of a discrete approximation $A^d_{2^{j+1}}f$ into an approximation at a coarser resolution $A^d_{2^j}f$ and the signal detail $D_{2^j}f$. By repeating in cascade this algorithm for $-1 \ge j \ge -J$, we compute the wavelet representation of a signal $A^d_1 f$ on J resolution levels.

uses a symmetry with respect to the first and the last sample as in Section II-B.

Equation (19) of Theorem 3 implies that the impulse response of the filter G is related to the impulse response of the filter H by

$$g(n) = (-1)^{1-n} h(1 - n). \tag{29}$$

This equation is provided in Appendix D. G is the *mirror* filter of H, and is a high-pass filter. In signal processing, G and H are called *quadrature mirror* filters [10]. Equation (28) can be interpreted as a high-pass filtering of the discrete signal $A^d_{2^{j+1}}f$.

If the original signal has N samples, then the discrete signals $D_{2^j}f$ and $A^d_{2^j}f$ have 2^jN samples each. Thus, the wavelet representation

$$\left(A^d_{2^{-J}}f, (D_{2^j}f)_{-J \le j \le -1}\right)$$

has the same total number of samples as the original approximated signal $A^d_1 f$. This occurs because the representation is orthogonal. Fig. 6(b) gives the wavelet representation of the signal $A^d_1 f$ decomposed in Fig. 2. The energy of the samples of $D_{2^j}f$ gives a measure of the irregularity of the signal at the resolution 2^{j+1}. Whenever $A_{2^j}f(x)$ and $A_{2^{j+1}}f(x)$ are significantly different, the signal detail has a high amplitude. In Fig. 6, this behavior is observed in the textured area between the abscissa coordinates of 60 and 80.

C. Signal Reconstruction from an Orthogonal Wavelet Representation

We have seen that the wavelet representation is complete. We now show that the original discrete signal can also be reconstructed with a pyramid transform. Since $\mathbf{O}_{2^j}$ is the orthogonal complement of $\mathbf{V}_{2^j}$ in $\mathbf{V}_{2^{j+1}}$, $(\sqrt{2^{-j}}\,\phi_{2^j}(x - 2^{-j}n), \sqrt{2^{-j}}\,\psi_{2^j}(x - 2^{-j}n))_{n\in Z}$ is an orthonormal basis of $\mathbf{V}_{2^{j+1}}$. For any $n > 0$, the function $\phi_{2^{j+1}}(x - 2^{-j-1}n)$ can thus be decomposed in this basis

$$\phi_{2^{j+1}}(x - 2^{-j-1}n) = 2^{-j} \sum_{k=-\infty}^{+\infty} \langle \phi_{2^j}(u - 2^{-j}k), \phi_{2^{j+1}}(u - 2^{-j-1}n) \rangle \cdot \phi_{2^j}(x - 2^{-j}k)$$
$$+ 2^{-j} \sum_{k=-\infty}^{+\infty} \langle \psi_{2^j}(u - 2^{-j}k), \phi_{2^{j+1}}(u - 2^{-j-1}n) \rangle \cdot \psi_{2^j}(x - 2^{-j}k). \tag{30}$$

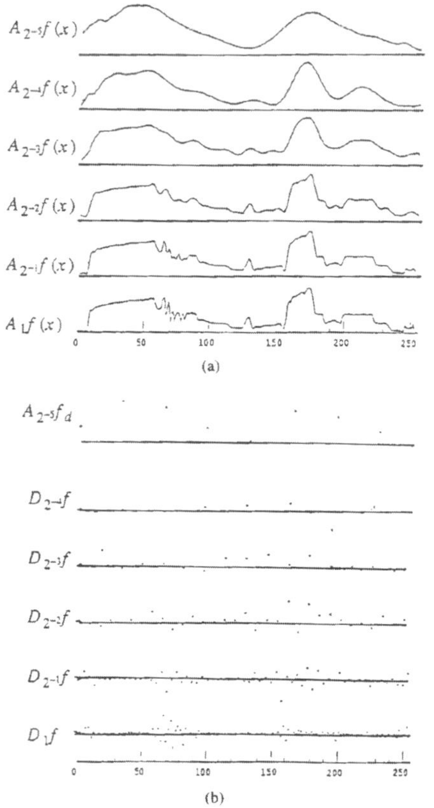

Fig. 6. (a) Multiresolution continuous approximations $A_{2^j} f(x)$. (b) Wavelet representation of the signal $A_1 f(x)$. The dots give the amplitude of the inner products $\langle f(u), \psi_{2^j}(u - 2^{-j}n)\rangle$ of each detail signal $D_{2^j} f$ depending upon $2^{-j}n$. The detail signals samples have a high amplitude when the approximations $A_{2^j} f(x)$ and $A_{2^{j+1}} f(x)$ shown in (a) are locally different. The top graph gives the inner products $\langle f(u), \phi_{2^J}(u - 2^J n)\rangle$ of the coarse discrete approximation $A^d_{2^{-J}} f$.

By computing the inner product of each side of equation (30) with the function $f(x)$, we have

$$\langle f(u), \phi_{2^{j+1}}(u - 2^{-j-1}n)\rangle$$
$$= 2^{-j} \sum_{k=-\infty}^{+\infty} \langle \phi_{2^j}(u - 2^{-j}k), \phi_{2^{j+1}}(u - 2^{-j-1}n)\rangle \cdot \langle f(u), \phi_{2^j}(u - 2^{-j}k)\rangle$$
$$+ 2^{-j} \sum_{k=-\infty}^{+\infty} \langle \psi_{2^j}(u - 2^{-j}k), \phi_{2^{j+1}}(u - 2^{-j-1}n)\rangle \cdot \langle f(u), \psi_{2^j}(u - 2^{-j}k)\rangle. \quad (31)$$

Inserting (14) and (25) in this expression and using the filters H and G, respectively, defined by (15) and (27) yields

$$\langle f(u), \phi_{2^{j+1}}(u - 2^{-j-1}n)\rangle$$
$$= 2 \sum_{k=-\infty}^{+\infty} h(n - 2k) \langle f(u), \phi_{2^j}(u - 2^{-j}k)\rangle$$
$$+ 2 \sum_{k=-\infty}^{+\infty} g(n - 2k) \langle f(u), \psi_{2^j}(u - 2^{-j}k)\rangle. \quad (32)$$

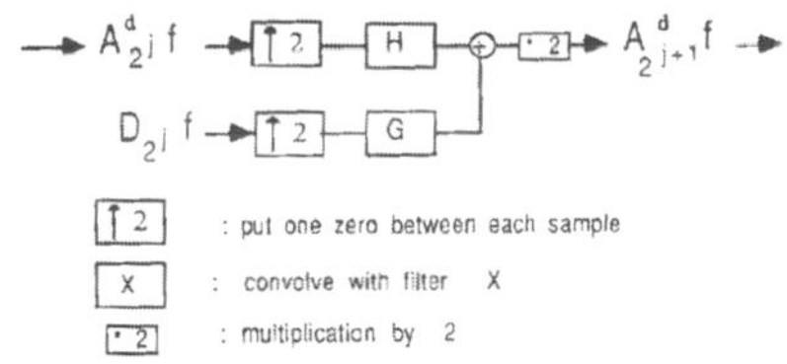

Fig. 7. Reconstruction of a discrete approximation $A^d_{2^{j+1}} f$ from an approximation at a coarser resolution $A^d_{2^j} f$ and the signal detail $D_{2^j} f$. By repeating in cascade this algorithm for $-J \leq j \leq -1$, we reconstruct $A^d_1 f$ from its wavelet representation.

This equation shows that $A^d_{2^{j+1}} f$ can be reconstructed by putting zeros between each sample of $A^d_{2^j} f$ and $D_{2^j} f$ and convolving the resulting signals with the filters H and G, respectively. A quite similar process can be found in the reconstruction algorithm of Burt and Adelson from their Laplacian pyramid [5].

The block diagram shown in Fig. 7 illustrates this algorithm. The original discrete signal $A^d_1 f$ at the resolution 1 is reconstructed by repeating this procedure for $-J \leq j < 0$. From the discrete approximation $A^d_1 f$, we can recover the continuous approximation $A_1 f(x)$ with equation (10). Fig. 8(a) is a reconstruction of the signal $A_1 f(x)$ from the wavelet representation given in Fig. 6(b). By comparing this reconstruction with the original signal shown in Fig. 8(b), we can appreciate the quality of the reconstruction. The low and high frequencies of the signal are reconstructed well, illustrating the numerical stability of the decomposition and reconstruction processes.

IV. Extension of the Orthogonal Wavelet Representation to Images

The wavelet model can be easily generalized to any dimension $n > 0$ [33]. In this section, we study the two-dimensional case for image processing applications. The signal is now a finite energy function $f(x, y) \in L^2(\mathbf{R}^2)$. A multiresolution approximation of $L^2(\mathbf{R}^2)$ is a sequence of subspaces of $L^2(\mathbf{R}^2)$ which satisfies a straightforward two-dimensional extension of the properties (2) to (8). Let $(V_{2^j})_{j \in \mathbf{Z}}$ be such a multiresolution approximation of $L^2(\mathbf{R}^2)$. The approximation of a signal $f(x, y)$ at a resolution 2^j is equal to its orthogonal projection on the vector space V_{2^j}. Theorem 1 is still valid in two dimensions, and one can show that there exists a unique scaling function $\Phi(x, y)$ whose dilation and translation given an or-

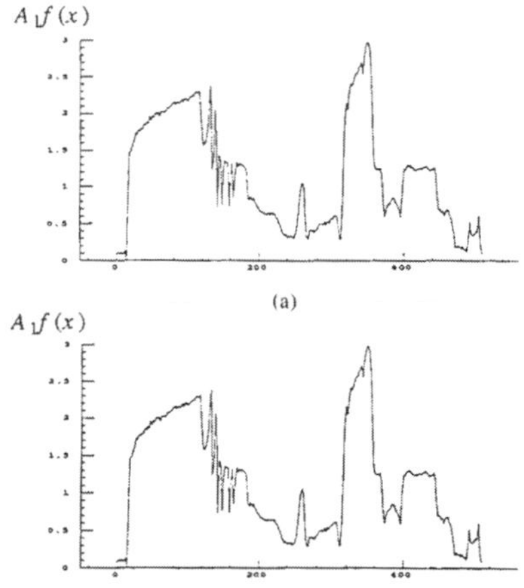

Fig. 8. (a) Original signal $A_1 f(x)$ approximated at the resolution 1. (b) Reconstruction of $A_1 f(x)$ from the wavelet representation shown in Fig. 6(b). By comparing both figures, we can appreciate the quality of the reconstruction.

thonormal basis of each space V_{2^j}. Let $\Phi_{2^j}(x, y) = 2^{2j}\Phi(2^j x, 2^j y)$. The family of functions

$$\left(2^{-j}\Phi_{2^j}(x - 2^{-j}n, y - 2^{-j}m)\right)_{(n,m)\in Z^2}$$

forms an orthonormal basis of V_{2^j}. The factor 2^{-j} normalizes each function in the $L^2(R^2)$ norm. The function $\Phi(x, y)$ is unique with respect to a particular multiresolution approximation of $L^2(R^2)$.

We next describe the particular case of separable multiresolution approximations of $L^2(R^2)$ studied by Meyer [35]. For such multiresolution approximations, each vector space V_{2^j} can be decomposed as a tensor product of two identical subspaces of $L^2(R)$

$$V_{2^j} = V^1_{2^j} \otimes V^1_{2^j}.$$

The sequence of vector spaces $(V_{2^j})_{j\in Z}$ forms a multiresolution approximation of $L^2(R^2)$ if and only if $(V^1_{2^j})_{j\in Z}$ is a multiresolution approximation of $L^2(R)$. One can then easily show that the scaling function $\Phi(x, y)$ can be written as

$$\Phi(x, y) = \phi(x)\,\phi(y)$$

where $\phi(x)$ is the one-dimensional scaling function of the multiresolution approximation $(V^1_{2^j})_{j\in Z}$. With a separable multiresolution approximation, extra importance is given to the horizontal and vertical directions in the image. For many types of images, such as those from man-made environments, this emphasis is appropriate. The orthogonal basis of V_{2^j} is then given by

$$\left(2^{-j}\Phi_{2^j}(x - 2^{-j}n, y - 2^{-j}m)\right)_{(n,m)\in Z^2} = \left(2^{-j}\phi_{2^j}(x - 2^{-j}n)\,\phi_{2^j}(y - 2^{-j}m)\right)_{(n,m)\in Z^2}. \quad (33)$$

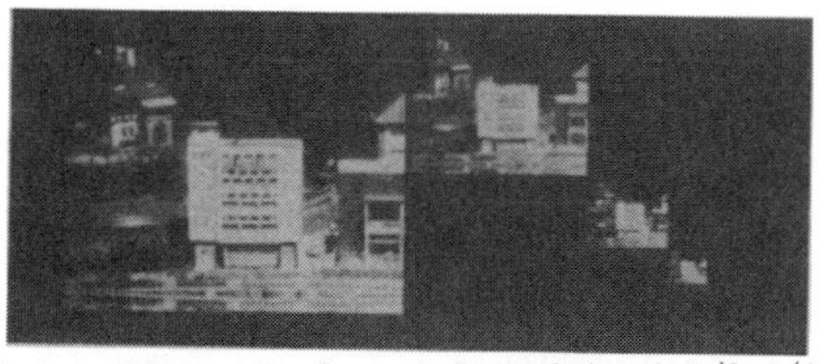

Fig. 9. Approximations of an image at the resolutions 1, 1/2, 1/4, and 1/8 ($j = 0, -1, -2, -3$).

The approximation of a signal $f(x, y)$ at a resolution 2^j is therefore characterized by the set of inner products

$$A^d_{2^j} f = \left(\langle f(x, y), \phi_{2^j}(x - 2^{-j}n)\,\phi_{2^j}(y - 2^{-j}m)\rangle\right)_{(n,m)\in Z^2}.$$

Let us suppose that the camera measures an approximation of the irradiance of a scene at the resolution 1. Let $A^d_1 f$ be the resulting image and N be the number of pixels. One can easily show that for $j < 0$, a discrete image approximation $A^d_{2^j} f$ has $2^j N$ pixels. Border problems are handled by supposing that the original image is symmetric with respect to the horizontal and vertical borders. Fig. 9 gives the discrete approximations of an image at the resolutions 1, 1/2, 1/4, and 1/8.

As in the one-dimensional case, the detail signal at the resolution 2^j is equal to the orthogonal projection of the signal on the orthogonal complement of V_{2^j} in $V_{2^{j+1}}$. Let O_{2^j} be this orthogonal complement. The following theorem gives a simple extension of Theorem 3, and states that we can build an orthonormal basis of O_{2^j} by scaling and translating three wavelets functions, $\Psi^1(x, y)$, $\Psi^2(x, y)$, and $\Psi^3(x, y)$.

Theorem 4: Let $(V_{2^j})_{j\in Z}$ be a separable multiresolution approximation of $L^2(R^2)$. Let $\Phi(x, y) = \phi(x)\,\phi(y)$ be the associated two-dimensional scaling function. Let $\Psi(x)$ be the one-dimensional wavelet associated with the scaling function $\phi(x)$. Then, the three "wavelets"

$$\Psi^1(x, y) = \phi(x)\,\psi(y), \quad \Psi^2(x, y) = \psi(x)\,\phi(y), \quad \Psi^3(x, y) = \psi(x)\,\psi(y)$$

are such that

$$\left(2^{-j}\Psi^1_{2^j}(x - 2^{-j}n, y - 2^{-j}m),\; 2^{-j}\Psi^2_{2^j}(x - 2^{-j}n, y - 2^{-j}m),\; 2^{-j}\Psi^3_{2^j}(x - 2^{-j}n, y - 2^{-j}m)\right)_{(n,m)\in Z^2} \quad (34)$$

is an orthonormal basis of O_{2^j} and

$$\left(2^{-j}\Psi^1_{2^j}(x - 2^{-j}n, y - 2^{-j}m),\; 2^{-j}\Psi^2_{2^j}(x - 2^{-j}n, y - 2^{-j}m),\; 2^{-j}\Psi^3_{2^j}(x - 2^{-j}n, y - 2^{-j}m)\right)_{(n,m)\in Z^3} \quad (35)$$

is an orthonormal basis of $L^2(R^2)$.

Appendix E gives a proof of this theorem. The difference of information between $A^d_{2^{j+1}}f$ and $A^d_{2^j}f$ is equal to the orthonormal projection of $f(x)$ on O_{2^j}, and is characterized by the inner products of $f(x)$ with each vector of an orthonormal basis of O_{2^j}. Theorem 4 says that this difference of information is given by the three detail images

$$D^1_{2^j}f = \left(\langle f(x, y), \Psi^1_{2^j}(x - 2^{-j}n, y - 2^{-j}m)\rangle\right)_{(n,m)\in Z^2}, \tag{36}$$

$$D^2_{2^j}f = \left(\langle f(x, y), \Psi^2_{2^j}(x - 2^{-j}n, y - 2^{-j}m)\rangle\right)_{(n,m)\in Z^2}, \tag{37}$$

$$D^3_{2^j}f = \left(\langle f(x, y), \Psi^3_{2^j}(x - 2^{-j}n, y - 2^{-j}m)\rangle\right)_{(n,m)\in Z^2}. \tag{38}$$

Just as for one-dimensional signals, one can show that in two dimensions the inner products which define $A^d_{2^j}f$, $D^1_{2^j}f$, $D^2_{2^j}f$, and $D^3_{2^j}f$ are equal to a uniform sampling of two-dimensional convolution products. Since the three wavelets $\Psi_1(x, y)$, $\Psi_2(x, y)$, and $\Psi_3(x, y)$ are given by separable products of the functions ϕ and ψ, these convolutions can be written

$$A^d_{2^j}f = \left((f(x, y) * \phi_{2^j}(-x)\,\phi_{2^j}(-y))\,(2^{-j}n, 2^{-j}m)\right)_{(n,m)\in Z^2} \tag{39}$$

$$D^1_{2^j}f = \left((f(x, y) * \phi^j(-x)\,\psi^j(-y))\,(2^{-j}n, 2^{-j}m)\right)_{(n,m)\in Z^2} \tag{40}$$

$$D^2_{2^j}f = \left((f(x, y) * \psi_{2^j}(-x)\,\phi_{2^j}(-y))\,(2^{-j}n, 2^{-j}m)\right)_{(n,m)\in Z^2} \tag{41}$$

$$D^3_{2^j}f = \left((f(x, y) * \psi_{2^j}(-x)\,\psi_{2^j}(-y))\,(2^{-j}n, 2^{-j}m)\right)_{(n,m)\in Z^2}. \tag{42}$$

The expressions (39) through (42) show that in two dimensions, $A^d_{2^j}f$ and the $D^k_{2^j}f$ are computed with separable filtering of the signal along the abscissa and ordinate.

The wavelet decomposition can thus be interpreted as a signal decomposition in a set of independent, *spatially oriented* frequency channels. Let us suppose that $\phi(x)$ and $\psi(x)$ are, respectively, a perfect low-pass and a perfect bandpass filter. Fig. 10(a) shows in the frequency domain how the image $A^d_{2^{j+1}}f$ is decomposed into $A^d_{2^j}f$, $D^1_{2^j}f$, $D^2_{2^j}f$, and $D^3_{2^j}f$. The image $A^d_{2^j}f$ corresponds to the lowest frequencies, $D^1_{2^j}f$ gives the vertical high frequencies (horizontal edges), $D^2_{2^j}f$ the horizontal high frequencies (vertical edges) and $D^3_{2^j}f$ the high frequencies in both directions (the corners). This is illustrated by the decomposition of a white square on a black background explained in Fig. 11(b). The arrangement of the $D^k_{2^j}f$ im-

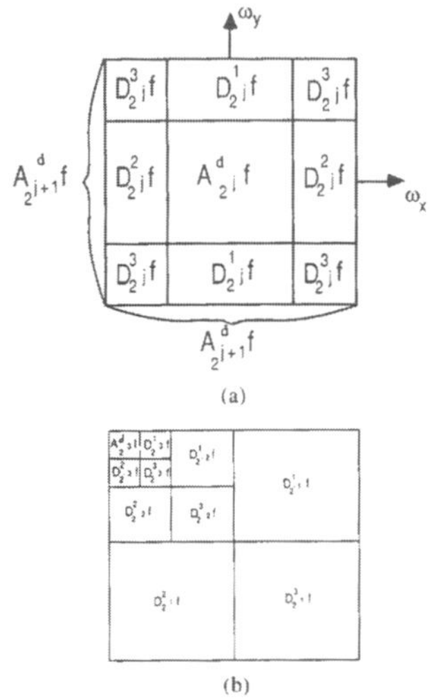

Fig. 10. (a) Decomposition of the frequency support of the image $A^d_{2^{j+1}}f$ into $A^d_{2^j}f$ and the detail images $D^k_{2^j}f$. The image $A^d_{2^j}f$ corresponds to the lower horizontal and vertical frequencies of $A_{2^{j+1}}$. $D^1_{2^j}f$ gives the vertical high frequencies and horizontal low frequencies, $D^2_{2^j}f$ the horizontal high frequencies and vertical low frequencies and $D^3_{2^j}f$ the high frequencies in both horizontal and vertical directions. (b) Disposition of the $D^k_{2^j}f$ and $A^d_{2^j}f$ images of the image wavelet representations shown in this article.

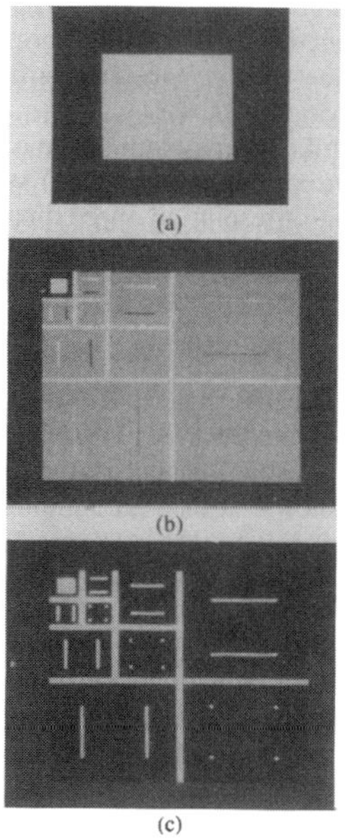

Fig. 11. (a) Original image. (b) Wavelet representation on three resolution levels. The black, grey, and white pixels correspond respectively to negative, zero, and positive wavelet coefficients. The disposition of the detail images is explained in Fig. 10(b). (c) These images show the absolute value of the wavelet coefficients for each detail images $D^k_{2^j}$ shown in (b). Black and white pixels correspond respectively to zero and high amplitude coefficients. The amplitude is high along the edges of the square for each orientation.

ages is shown in Fig. 10(b). The black, grey, and white pixels respectively correspond to negative, zero, and positive coefficients. Fig. 11(c) shows the absolute value of the detail signal samples. The black pixels correspond to zero whereas the white ones have a high positive value. As expected, the detail signal samples have a high amplitude on the horizontal edges, the vertical edges and the corners of the square.

For any $J > 0$, an image $A_1^d f$ is completely represented by the $3J + 1$ discrete images

$$\left(A_{2^{-J}}^d f, \ (D_{2^j}^1 f)_{-J \le j \le -1}, \ (D_{2^j}^2 f)_{-J \le j \le -1}, \ (D_{2^j}^3 f)_{-J \le j \le -1}\right).$$

This set of images is called an *orthogonal wavelet representation* in two dimensions. The image $A_{2^{-J}}^d f$ is the coarse approximation at the resolution 2^{-J} and the $D_{2^j}^k f$ images give the detail signals for different orientations and resolutions. If the original image has N pixels, each image $A_{2^j}^d f$, $D_{2^j}^1 f$, $D_{2^j}^2 f$, $D_{2^j}^3 f$ has $2^j N$ pixels ($j < 0$). The total number of pixels in this new representation is equal to the number of pixels of the original image, so we do not increase the volume of data. Once again, this occurs due to the orthogonality of the representation. In a correlated multiresolution representation such as the Laplacian pyramid, the total number of pixels representing the signal is increased by a factor of 2 in one dimension and of $4/3$ in two dimensions.

A. Decomposition and Reconstruction Algorithms in Two Dimensions

In two dimensions, the wavelet representation can be computed with a pyramidal algorithm similar to the one-dimensional algorithm described in Section III-B. The two-dimensional wavelet transform that we describe can be seen as a one-dimensional wavelet transform along the x and y axes. By repeating the analysis described in Section III-B, we can show that a two-dimensional wavelet transform can be computed with a separable extension of the one-dimensional decomposition algorithm. At each step we decompose $A_{2^{j+1}}^d f$ into $A_{2^j}^d f$, $D_{2^j}^1 f$, $D_{2^j}^2 f$, and $D_{2^j}^3 f$. This algorithm is illustrated by a block diagram in Fig. 12. We first convolve the rows of $A_{2^{j+1}}^d f$ with a one-dimensional filter, retain every other row, convolve the columns of the resulting signals with another one-dimensional filter and retain every other column. The filters used in this decomposition are the quadrature mirror filters $\tilde{H}$ and $\tilde{G}$ described in Sections II-B and III-B.

The structure of application of the filters for computing $A_{2^j}^d$, $D_{2^j}^1$, $D_{2^j}^2$, and $D_{2^j}^3$ is given in Fig. 12. We compute the wavelet transform of an image $A_1^d f$ by repeating this process for $-1 \ge j \ge -J$. This corresponds to a separable conjugate mirror filter decomposition [44].

Fig. 14(b) shows the wavelet representation of a natural scene image decomposed on 3 resolution levels. The pattern of arrangement of the detail images is as explained in Fig. 10(b). Fig. 14(c) gives the absolute value of the wavelet coefficients of each detail image. The wavelet coefficients have a high amplitude around the images edges and in the textured areas within a given spatial orientation.

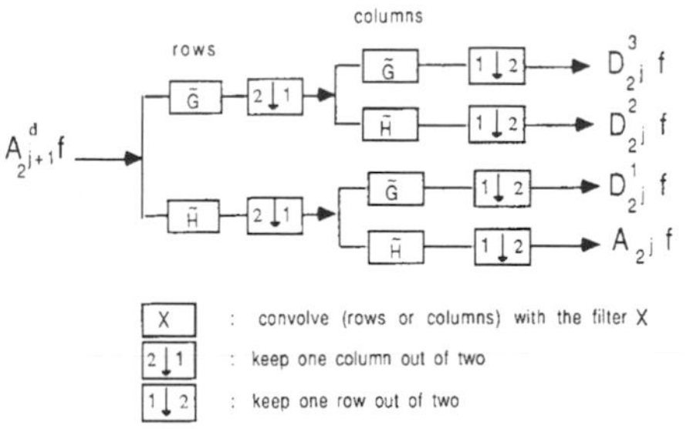

Fig. 12. Decomposition of an image $A_{2^{j+1}}^d f$ into $A_{2^j}^d f$, $D_{2^j}^1 f$, $D_{2^j}^2 f$, and $D_{2^j}^3 f$. This algorithm is based on one-dimensional convolutions of the rows and columns of $A_{2^{j+1}}^d f$ with the one dimensional quadrature mirror filters $\tilde{H}$ and $\tilde{G}$.

The one-dimensional reconstruction algorithm described in Section III-C can also be extended to two dimensions. At each step, the image $A_{2^{j+1}}^d f$ is reconstructed from $A_{2^j}^d f$, $D_{2^j}^1 f$, $D_{2^j}^2 f$, and $D_{2^j}^3 f$. This algorithm is illustrated by a block diagram in Fig. 13. Between each column of the images $A_{2^j}^d f$, $D_{2^j}^1 f$, $D_{2^j}^2 f$, and $D_{2^j}^3 f$, we add a column of zeros, convolve the rows with a one dimensional filter, add a row of zeros between each row of the resulting image, and convolve the columns with another one-dimensional filter. The filters used in the reconstruction are the quadrature mirror filters H and G described in Sections II-B and III-B. The image $A_1^d f$ is reconstructed from its wavelet transform by repeating this process for $-J \le j \le -1$. Fig. 14(d) shows the reconstruction of the original image from its wavelet representation. If we use floating-point precision for the discrete signals in the wavelet representation, the reconstruction is of excellent quality. Reconstruction errors are more thoroughly discussed in the next section.

V. Applications of the Orthogonal Wavelet Representation

A. Compact Coding of Wavelet Image Representations

To compute an exact reconstruction of the original image, we must store the pixel values of each detail image with infinite precision. However, for practical applications, we can allow errors to occur as long as the relevant information is not destroyed for a human observer. In this section, we show how to use the sensitivity of the human visual system as well as the statistical properties of the image to optimize the coding by the wavelet representation. The conjugate mirror filters which implement the wavelet decomposition have also been studied by Woods [44] and Adelson *et al.* [1] for image coding.

Let $(A_{2^{-J}}^d f$, $(D_{2^j}^1 f)_{-J \le j \le -1}$, $(D_{2^j}^2 f)_{-J \le j \le -1}$, $(D_{2^j}^3 f)_{-J \le j \le -1})$ be the wavelet representation of an image $A_1^d f$. Let $(\epsilon_{-J}$, $(\epsilon_j^1)_{-J \le j \le -1}$, $(\epsilon_j^2)_{-J \le j \le -1}$, $(\epsilon_j^3)_{-J \le j \le -1})$ be the mean square errors introduced when coding each image component of the wavelet representa-

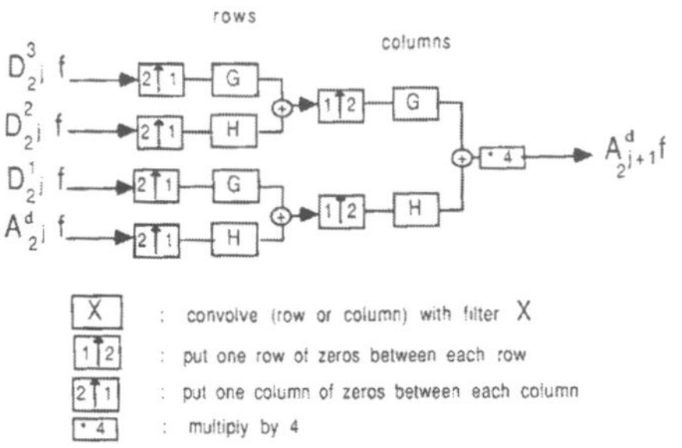

Fig. 13. Reconstruction of an image $A^d_{2^{j+1}}f$ from $A^d_{2^j}f$, $D^1_{2^j}f$, $D^2_{2^j}f$, and $D^3_{2^j}f$. The row and columns of these images are convolved with the one dimensional quadrature mirror filters H and G.

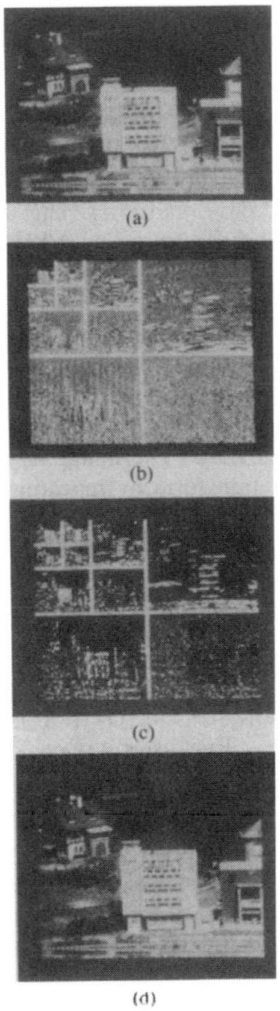

Fig. 14. (a) Original image. (b) Wavelet representation on three resolution levels. The arrangement of the detail images is explained in Fig. 10(b). (c) These images show the absolute value of the wavelet coefficients for each detail images $D^k_{2^j}$ shown in (b). The amplitude is high along the edges and the textured area for each orientation. (d) Reconstruction of the original image from the wavelet representation given in (b).

tion. Let ϵ_0 be the mean square error of the image reconstructed from the coded wavelet representation. Since the wavelet representation is orthogonal in $L^2(R^2)$, one can prove that

$$\epsilon_0 = 2^{2J}\epsilon_{-J} + \sum_{k=1}^{3}\sum_{j=J}^{-1} 2^{-2j}\epsilon_j^k. \tag{43}$$

The factors 2^{-2j} are due to the normalization factor 2^{-j} which appears in (33) and (34). Psychological experiments on human visual sensitivity show that the visible distortion on the reconstructed image will not only depend on the total mean square error ϵ_0, but also on the distribution of this error between the different detail images $D^k_{2^j}f$. The contrast sensitivity function of the visual system [6] shows that the perception of a contrast distortion in the image depends upon the frequency components of the modified contrast. Visual sensitivity also depends upon the orientation of the stimulus. The results of Campbell and Kulikowski [7] show that the human visual system has a maximum sensitivity when the contrast is horizontal or vertical. When the contrast is tilted at 45°, the sensitivity is minimum. A two-dimensional wavelet representation corresponds to a decomposition of the image into independent frequency bands and three spatial orientations. Each detail image $D^k_{2^j}$ gives the image contrast in a given frequency range and along a particular orientation. It is therefore possible to adapt the coding error of each detail image to the sensitivity of human perception for the corresponding frequency band and spatial orientation selectivity. The more sensitive the human visual system, the less coding error we want to introduce in the detail image $D^k_{2^j}f$. Watson [43] has made a particularly detailed study of subband image coding adapted to human visual perception.

Given an allocation of the coding error between the different resolutions and orientations of the wavelet representation, we must then code each detail image with a minimum number of bytes. In order to optimize the coding, one can use the statistical properties of the wavelet coefficients for each resolution and orientation. Natural images are special kinds of two-dimensional signals. This shows up clearly when one looks at the histogram of the detail images $D^k_{2^j}f$. Since the pixels of these detail images are the decomposition coefficients of the original image in an orthonormal family, they are not correlated. The histogram of the detail images could therefore have any distribution. Yet in practice, for all resolutions and orientations, these histograms are symmetrical peaks centered in zero. Natural images must therefore belong to a particular subset of $L^2(R^2)$. The modeling of this subset is a well known problem in image processing [26]. The wavelet orthonormal bases are potentially helpful for this purpose. Indeed, the statistical properties of an image decomposition in a wavelet orthonormal basis look simpler than the statistical properties of the original image. Moreover, the orthogonality of the wavelet functions can simplify the mathematical analysis of the problem.

We have found experimentally that the detail image histograms can be modeled with the following family of histograms:

$$h(u) = Ke^{-(|u|/\alpha)^\beta}. \tag{44}$$

The parameter β modifies the decreasing rate of the peak and α models the variance. This model was built by studing the histograms of seven different images decomposed on four resolution levels each. Our goal was only to define a qualitative histogram model. The constant K is adjusted in order to have $\int_{-\infty}^{\infty} h(u)\, du = N$ where N is the total number of pixels of the given detail image. By changing variables in the integral, one can derive that

$$K = \frac{N\beta}{2\alpha\Gamma\left(\frac{1}{\beta}\right)}, \quad \text{where} \quad \Gamma(t) = \int_0^{\infty} e^{-u} u^{t-1}\, du. \tag{45}$$

The coefficients α and β of the histogram model can be computed by measuring the first and second moment of the detail image histogram:

$$m_1 = \int_{-\infty}^{\infty} |u| h(u)\, du \quad \text{and} \quad m_2 = \int_{-\infty}^{\infty} u^2 h(u)\, du. \tag{46}$$

By inserting (44) of the histogram model and changing variables in these two integrals, we obtain

$$m_1 = 2K \frac{\alpha^2}{\beta} \Gamma\left(\frac{2}{\beta}\right) \quad \text{and} \quad m_2 = 2K \frac{\alpha^3}{\beta} \Gamma\left(\frac{3}{\beta}\right). \tag{47}$$

Thus,

$$\beta = F^{-1}\left(\frac{m_1^2}{m_2 N}\right) \quad \text{where} \quad F(x) = \frac{\Gamma\left(\frac{2}{x}\right)^2}{\Gamma\left(\frac{3}{x}\right)\Gamma\left(\frac{1}{x}\right)} \tag{48}$$

and

$$\alpha = \frac{m_2 \Gamma\left(\frac{1}{\beta}\right)}{N\Gamma\left(\frac{3}{\beta}\right)}. \tag{49}$$

The function $F^{-1}(x)$ is shown in Fig. 15. Fig. 16(a) gives a typical example of a detail image histogram obtained from the wavelet representation of a real image. Fig. 16(b) is the graph of the model derived from (44).

The *a priori* knowledge of the detail signal's statistical distribution given by the histogram model (44) can be used to optimize the coding of these signals. We have developed such a procedure [27] using Max's algorithm [32] to minimize the quantization noise on the wavelet representation. Predictive coding procedures are also effective for this purpose. Results show that one can code an image using such a representation with less than 1.5 bits per pixel with few visible distortions [1], [27], [44].

B. Texture Discrimination and Fractal Analysis

We now describe the application of the wavelet orthogonal representation to texture discrimination and fractal analysis. Using psychophysics, Julesz [21] has developed a texture discrimination theory based on the decomposition of textures into basic primitives called *textons*. These textons are spatially local; they have a particular spatial orientation and narrow frequency tuning. The wavelet representation can also be interpreted as a texton decomposition where each texton is equal to a particular function of the wavelet orthonormal basis. Indeed, these functions have all the discriminative abilities required by the Julesz theory. In the decomposition studied in this article, we have used only three orientation tunings. However, one can build a wavelet representation having as many orientation tunings as desired by using non-separable wavelet orthonormal bases [33].

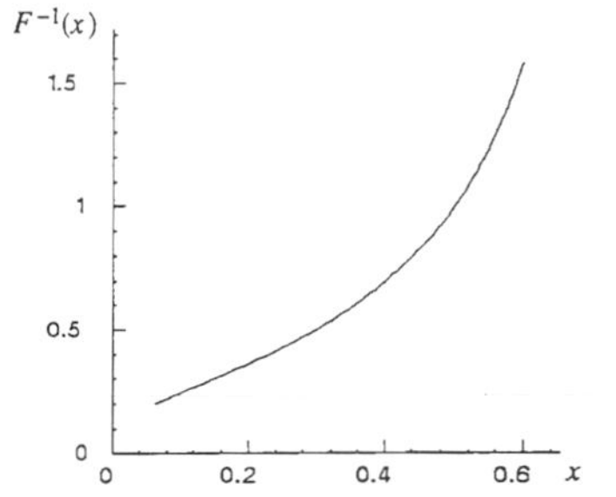

Fig. 15. Graph of the function $F^{-1}(x)$ characterized by (49).

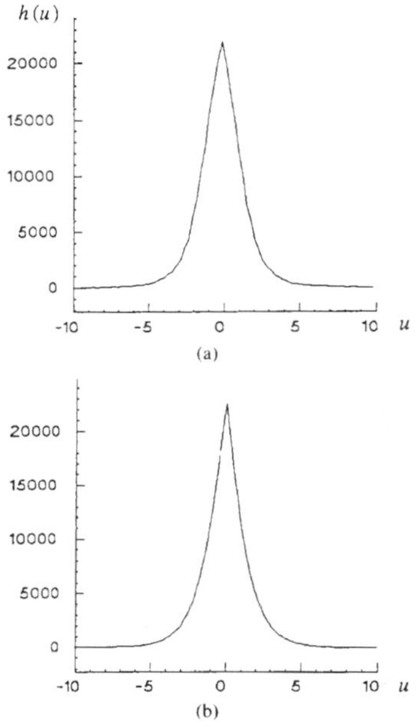

Fig. 16. (a) Typical example of a detail image histogram $h(u)$. (b) Modeling of $h(u)$ obtained from equation (44). The parameters α and β have been computed from the first two moments of the original histogram (α = 1.39 and β = 1.14).

Fig. 17(a) shows three textures synthesized by Beck. Humans cannot preattentively discriminate the middle from the right texture but can separate the left texture from the others. In this example, human discrimination is based mainly on the orientation of these textures as their frequency content is very similar. With a first-order statistical analysis of the wavelet representation shown in Fig. 17(b), we can also discriminate the left texture but not the two others. This example illustrates the ability of our representation to differentiate textures on orientation criteria. This is of course only one aspect of the problem, and a more sophisticated statistical analysis is needed for modeling textures [13]. Although several psychophysical studies have shown the importance of a signal decomposition in several frequency channels [4], [15], there still is no statistical model to combine the information provided by the different channels. From this point of view, the wavelet mathematical model might be helpful to transpose some of the tools currently used in functional analysis to characterize the local regularity of functions [25].

Mandelbrot [29] has shown that certain natural textures can be modeled with Brownian fractal noise. Brownian fractal noise $F(x)$ is a random process whose local differences

$$\frac{|F(x) - F(x + \Delta x)|}{\|\Delta x\|^{H}}$$

has a probability distribution function $g(x)$ which is Gaussian. Such a random process is self-similar, i.e.,

$$\forall r > 0,\ F(x) \text{ and } r^{H}F(rx) \text{ are statistically identical.}$$

Hence, a realization of $F(x)$ looks similar at any scale and for any resolution. Fractals do not provide a general model which can be used for the analysis of any kind of texture, but Pentland [39] has shown that for a fractal texture, the psychophysical perception of roughness can be quantified with the fractal dimension.

Fig. 18(a) shows a realization of a fractal noise which looks like a cloud. Its fractal dimension is 2.5. Fig. 18(b) gives the wavelet representation of this fractal. As expected, the detail signals are similar at all resolutions. The image $A^{d}_{2^{-3}}f$ gives the local dc component of the original fractal image. For a cloud, this would correspond to the local differences of illuminations.

Let us show that the fractal dimension can be computed from the wavelet representation. We give the proof for one-dimensional fractal noise, but the result can be easily extended to two dimensions. The power spectrum of fractal noise is given by [29]

$$P(\omega) = k\omega^{-2H-1}. \tag{50}$$

The fractal dimension is related to the exponent H by

$$D = T + 1 - H \tag{51}$$

where T is the topological dimension of the space in which x varies (for images $T = 2$). Since Brownian fractal noise is not a stationary process, this power spectrum cannot be interpreted in the classical sense. Flandrin [12] has shown how to define precisely this power spectrum formula with a time-frequency analysis. We saw in equation (22) that the detail signals $D_{2^j}f$ are obtained by filtering the signal with $\psi_{2^j}(-x)$ and sampling the output. The power spectrum of the fractal filtered by $\psi_{2^j}(-x)$ is given by

$$P_{2^j}(\omega) = P(\omega)\left|\hat{\psi}(2^{-j}\omega)\right|^2. \tag{52}$$

After sampling at a rate 2^j, the power spectrum of the discrete detail signal becomes [37]

$$P^{d}_{2^j}(\omega) = 2^j \sum_{k=-\infty}^{+\infty} P_{2^j}(\omega + 2^j 2k\pi). \tag{53}$$

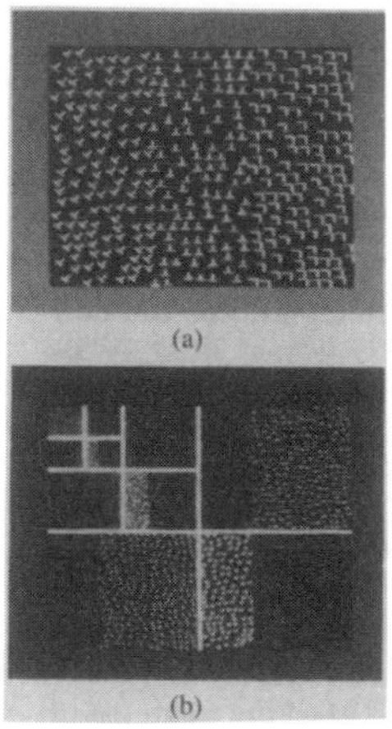

Fig. 17. (a) J. Beck textures: only the left texture is preattentively discriminable by a human observer. (b) These images show the absolute value of the wavelet coefficients of image (a), computed on three resolution levels. The left texture can be discriminated with a first-order statistical analysis of the detail signals amplitude. The two other textures can not be discriminated with such a technic.

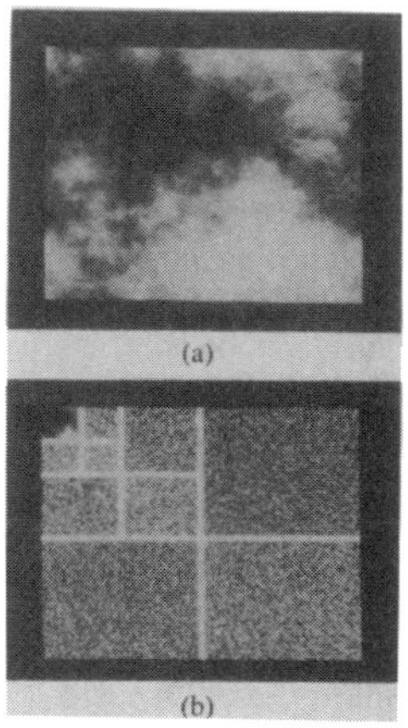

Fig. 18. (a) Brownian fractal image. (b) Wavelet representation on three resolution levels of image (a). As expected, the detail signals are similar at all resolutions.

Let $\sigma_{2^j}^2$ be the energy of the detail signal $D_{2^j}f$

$$\sigma_{2^j}^2 = \frac{2^{-j}}{2\pi} \int_{-2^j\pi}^{+2^j\pi} P_{2^j}^d(\omega)\, d\omega. \tag{54}$$

By inserting (52) and (53) into (54) and changing variables in this integral, we obtain that

$$\sigma_{2^j}^2 = 2^{2H} \sigma_{2^{j+1}}^2. \tag{55}$$

For a fractal, the ratio $\sigma_{2^j}^2/\sigma_{2^{j+1}}^2$ is therefore constant. From the wavelet representation of a Brownian fractal, we compute H from equation (55) and derive the fractal dimension D with (51). This result can easily be extended to two dimensions in order to compute the fractal dimension of fractal images. Analogously, we compute the ratios of the energy of the detail images within each direction, and derive the value of the coefficient H. A similar algorithm has been proposed by Heeger and Pentland for analyzing fractals with Gabor functions [19]. For the fractal shown in Fig. 18(a), we calculated the ratios of the energy of the detail images within each orientation for the resolutions 1/2, 1/4, and 1/8. We recovered the fractal dimension of this image from each of these ratios with a 3 percent maximum error.

Much research work has recently concentrated on the analysis of fractals with the wavelet transform [2]. This topic is promising because multiscale decompositions, such as the wavelet transform, are well adapted to evaluate the self-similarity of a signal and its fractal properties.

VI. Conclusion

This article has described a mathematical model for the computation and interpretation of the concept of a multiresolution representation. We explained how to extract the difference of information between successive resolutions and thus define a new (complete) representation called the wavelet representation. This representation is computed by decomposing the original signal using a wavelet orthonormal basis, and can be interpreted as a decomposition using a set of independent frequency channels having a spatial orientation tuning. A wavelet representation lies between the spatial and Fourier domains. There is no redundant information because the wavelet functions are orthogonal. The computation is efficient due to the existence of a pyramidal algorithm based on convolutions with quadrature mirror filters. The original signal can be reconstructed from the wavelet decomposition with a similar algorithm.

We discussed the application of the wavelet representation to data compression in image coding. We showed that an orthogonal wavelet transform provides interesting insight on the statistical properties of images. The orientation selectivity of this representation is useful for many applications. We reviewed in particular the texture discrimination problem. A wavelet transform is particularly well-suited to analyze the fractal properties of images. Specifically, we showed how to compute the fractal dimension of a Brownian fractal from its wavelet representation. In this article, we emphasized the computer vision applications, but this representation can also be helpful for pattern recognition in other domains. Grossmann and Kronland-Martinet [23] are currently working on speech recognition applications, and Morlet [14] studies seismic signal analysis. The wavelet orthonormal bases are also studied in both pure and applied mathematics [20], [25], and have found applications in *Quantum Mechanics* with the work of Paul [38] and Federbush [11].

Appendix A
An Example Multiresolution Approximation

In this appendix, we describe a class of multiresolution approximations of $L^2(\boldsymbol{R})$ studied by Lemarie [24] and Battle [3]. We explain how to compute the corresponding scaling functions $\phi(x)$, wavelets $\psi(x)$, and quadrature filters H. These multiresolution approximations are built from polynomial splines of order $2p+1$. The vector space V_1 is the vector space of all functions of $L^2(\boldsymbol{R})$ which are p times continuously differentiable and equal to a polynomial of order $2p+1$ on each interval $[k, k+1]$, for any $k \in \boldsymbol{Z}$. The other vector spaces V_{2^j} are derived from V_1 with property (3). Lemarie has shown that the scaling function associated with such a multiresolution approximation can be written

$$\hat{\phi}(\omega) = \frac{1}{\omega^n \sqrt{\Sigma_{2n}(\omega)}} \quad \text{where } n = 2 + 2p, \tag{56}$$

and where the function $\Sigma_n(\omega)$ is given by

$$\Sigma_n(\omega) = \sum_{k=-\infty}^{+\infty} \frac{1}{(\omega + 2k\pi)^n}. \tag{57}$$

We can compute a closed form of $\Sigma_n(\omega)$ by calculating the derivative of order $n-2$ of the equation

$$\Sigma_2(\omega) = \frac{1}{4\sin^2(\omega/2)}.$$

Theorem 2 says that $\hat{\phi}(\omega)$ is related to the transfer function $H(\omega)$ of a quadrature mirror filter by

$$\hat{\phi}(2\omega) = H(\omega)\, \hat{\phi}(\omega).$$

From (56) we obtain

$$H(\omega) = \sqrt{\frac{\Sigma_{2n}(\omega)}{2^{2n}\Sigma_{2n}(2\omega)}}. \tag{58}$$

The Fourier transform of the corresponding orthonormal wavelet can be derived from the property (19) of Theorem 3

$$\hat{\psi}(\omega) = e^{-i(\omega/2)}\overline{H}\left(\frac{\omega}{2} + \pi\right) \hat{\phi}\left(\frac{\omega}{2}\right)$$

$$= \frac{e^{-i(\omega/2)}}{\omega^n} \frac{\sqrt{\Sigma_{2n}\left(\frac{\omega}{2} + \pi\right)}}{\sqrt{\Sigma_{2n}(\omega)\, \Sigma_{2n}\left(\frac{\omega}{2}\right)}}. \tag{59}$$

The wavelet $\psi(x)$ defined by (59) decreases exponentially.

The scaling function shown in Fig. 1 was obtained with $p = 1$, and thus $n = 4$. It corresponds to a multiresolution approximation built from cubic splines. Let

$$N_1(\omega) = 5 + 30\left(\cos\frac{\omega}{2}\right)^2 + 30\left(\sin\frac{\omega}{2}\right)^2\left(\cos\frac{\omega}{2}\right)^2$$

and

$$N_2(\omega) = 2\left(\sin\frac{\omega}{2}\right)^4\left(\cos\frac{\omega}{2}\right)^2 + 70\left(\cos\frac{\omega}{2}\right)^4 + \frac{2}{3}\left(\sin\frac{\omega}{2}\right)^6.$$

The function $\Sigma_8(\omega)$ is given by

$$\Sigma_8(\omega) = \frac{N_1(\omega) + N_2(\omega)}{105\left(\sin\frac{\omega}{2}\right)^8}.$$

For this multiresolution approximation based on cubic splines, the functions $\hat{\phi}(\omega)$ and $\hat{\psi}(\omega)$ are computed from (56) and (59) with $n = 4$. The transfer function $H(\omega)$ of the quadrature mirror filter is given by equation (58). Table I gives the first 12 coefficients of the impulse response $(h(n))_{n\in Z}$. This filter is symmetrical. The impulse response of the mirror filter G is obtained with (29).

TABLE I

n	$h(n)$	n	$h(n)$
0	0.542	6	0.012
1	0.307	7	-0.013
2	-0.035	8	0.006
3	-0.078	9	0.006
4	0.023	10	-0.003
5	-0.030	11	-0.002

Appendix B
Proof of Theorem 1

This appendix gives the main steps of the proof of Theorem 1. More details can be found in [28]. We prove Theorem 1 for $j = 0$. The result can be extended for any $j \in Z$ using the property (3). From the properties (5) and (6) of the isomorphism I from V_1 onto $I^2(Z)$, one can prove that there exists a function $g(x)$ such that $(g(x - k))_{k\in Z}$ is a basis of V_1. We are looking for a function $\phi(x) \in V_1$ such that $(\phi(x - k))_{k\in Z}$ is an orthonormal basis of V_1. Let $\hat{\phi}(\omega)$ be the Fourier transform of $\phi(x)$. With the Poisson formula, we can show that the family of functions $(\phi(x - k))_{k\in Z}$ is orthonormal if and only if

$$\sum_{k=-\infty}^{+\infty} \left|\hat{\phi}(\omega + 2k\pi)\right|^2 = 1. \tag{60}$$

Since $\phi(x) \in V_1$, it can be decomposed in the basis $(g(x - k))_{k\in Z}$:

$$\exists(\alpha_k)_{k\in Z} \in I^2(Z) \quad \text{such that} \quad \phi(x) = \sum_{-\infty}^{+\infty} \alpha_k g(x - k). \tag{61}$$

The Fourier transform of (61) can be written

$$\hat{\phi}(\omega) = M(\omega)\,\hat{g}(\omega) \quad \text{with} \quad M(\omega) = \sum_{k=-\infty}^{+\infty} \alpha_k e^{ik\omega}. \tag{62}$$

By inserting (62) in (60), we obtain

$$M(\omega) = \left(\sum_{k=-\infty}^{+\infty} \left|\hat{g}(\omega + 2k\pi)\right|^2\right)^{-1/2}. \tag{63}$$

One can show that the continuity of the isomorphism I implies that there exists two constants C_1 and C_2 such that

$$C_1 \le \left(\sum_{-\infty}^{+\infty} \left|\hat{g}(\omega + 2k\pi)\right|^2\right)^{1/2} \le C_2.$$

Hence, (63) is defined for any $\omega \in R$. Conversely, it is simple to prove that equations (62) and (63) define the Fourier transform of a function $\phi(x)$ such that $(\phi(x - k))_{k\in Z}$ is an orthonormal family that generates V_1.

Appendix C
Proof of Theorem 2

This appendix gives the main steps of the proof of Theorem 2. More details can be found in [28]. Let us first prove property (17a). Since $\phi_{2^{-1}}(x) \in V_{-1} \subset V_1$, it can be decomposed in the orthogonal basis $(\phi(x - k))_{k\in Z}$

$$\phi_{2^{-1}}(x) = \sum_{k=-\infty}^{+\infty} \langle \phi_{2^{-1}}(u), \phi(u - k)\rangle\, \phi(u - k).$$

The Fourier transform of this equation yields

$$\hat{\phi}(2\omega) = H(\omega)\,\hat{\phi}(\omega) \tag{64}$$

where $H(\omega)$ is the Fourier series defined by (17). One can show that the property (7) of a multiresolution approximation implies that any scaling function satisfies

$$\left|\int_{-\infty}^{+\infty} \phi(x)\,dx\right| = \left|\hat{\phi}(0)\right| = 1. \tag{65}$$

From (64) we obtain $|H(0)| = 1$. Since the asymptotic decay of $\phi(x)$ at infinity satisfies

$$\left|\phi(x)\right| = O(x^{-2})$$

we can also derive that

$$h(n) = \langle \phi^{-1}(u), \phi(u - n)\rangle = O(n^{-2})$$

at infinity.

Let us now prove property (17b). We saw in Appendix A that a scaling function must satisfy

$$\sum_{k=-\infty}^{+\infty} \left|\hat{\phi}(\omega + 2k\pi)\right|^2 = 1. \tag{66}$$

Since $H(\omega)$ is 2π periodic, (64) and (66) yield

$$\left|H(\omega)\right|^2 + \left|H((\omega + \pi))\right|^2 = 1.$$

Let us write

$$\hat{\phi}(\omega) = \prod_{p=1}^{+\infty} H(2^{-p}\omega). \tag{67}$$

We will show that this equation defines the Fourier transform of a scaling function. We need to prove that

(α) $\hat{\phi}(\omega) \in L^2(\mathbf{R})$ and $(\phi(x - n))_{n \in Z}$ is an orthonormal family.

(β) If V_{2^j} is the vector space generated by the family of functions $(\phi_{2^j}(x - 2^{-j}n))_{n \in Z}$, then the sequence of vector spaces $(V_{2^j})_{j \in Z}$ is a multiresolution approximation of $L^2(\mathbf{R})$.

Let us first prove property (α). With the Parseval theorem, we can show that this statement is equivalent to

$$\forall n \in \mathbf{Z}, \int_{-\infty}^{+\infty} \left|\hat{\phi}(\omega)\right|^2 e^{ik\omega}\, d\omega = \begin{cases} 2\pi & \text{if } k = 0 \\ 0 & \text{if } k \neq 0 \end{cases} \tag{68}$$

Let us define the sequence of functions $(g_q(\omega))_{q>1}$ such that

$$g_q(\omega) = \begin{cases} \prod_{p=1}^{q} H(2^{-p}\omega) & \text{for } |\omega| < 2^q\pi \\ 0 & \text{for } |\omega| \geq 2^q\pi. \end{cases} \tag{69}$$

As q tends to $+\infty$, the sequence $(g_q(\omega))_{q>1}$ converges towards $|\phi(\omega)|^2$ almost everywhere. We can also prove that

$$\forall n \in \mathbf{Z}, \int_{-\infty}^{+\infty} g_q(\omega)\, e^{ik\omega}\, d\omega = \begin{cases} 2\pi & \text{if } k = 0 \\ 0 & \text{if } k \neq 0. \end{cases} \tag{70}$$

With hypothesis (17c) of Theorem 2, it is then possible to apply the dominated convergence theorem to the sequence $(g_q(\omega))_{q>1}$ to derive from (70) that $\hat{\phi}(\omega)$ satisfies (68).

Let us now prove property (β). In order to prove that $(V_{2^j})_{j \in Z}$ is a multiresolution approximation of $L^2(\mathbf{R})$, we must show that assertions (2)–(8) apply. The properties (2)–(6) can be derived from the equation

$$\hat{\phi}(2\omega) = H(\omega)\,\hat{\phi}(\omega).$$

Since $H(\omega)$ satisfies (17a), we can show that

$$\left|\hat{\phi}(0)\right| = \left|\int_{-\infty}^{+\infty} \phi(x)\, dx\right| = 1.$$

From this equation, one can prove that the sequence of vector spaces $(V_{2^j})_{j \in Z}$ defined in (β) do satisfy the last two properties (7) and (8) of a multiresolution representation.

Appendix D
Proof of Theorem 3

This appendix gives the main steps of the proof of Theorem 3. More details can be found in [28]. This theorem is proved for $j = -1$. We are looking for a function $\psi(x) \in L^2(\mathbf{R})$ such that $(\sqrt{2^{-1}}\psi_{2^{-1}}(x - 2^{-1}n))_{n \in Z}$ is an orthonormal basis of $\mathbf{O}_{2^{-1}}$. The orthogonality of this family can be expressed with the Poisson formula

$$\sum_{k=-\infty}^{+\infty} \left|\hat{\psi}(2\omega + 2k\pi)\right|^2 = 1. \tag{71}$$

Since $\psi_{2^{-1}}(x) \in \mathbf{O}_{2^{-1}} \subset V_1$, we can decompose it on the orthonormal basis $(\phi(x - n))_{n \in Z}$:

$$\psi_{2^{-1}}(x) = \sum_{n=-\infty}^{+\infty} \langle \psi_{2^{-1}}(u), \phi(u - n)\rangle\, \phi(x - n). \tag{72}$$

Let us define

$$G(\omega) = \sum_{n=-\infty}^{+\infty} \langle \psi_{2^{-1}}(u), \phi(u - n)\rangle\, e^{-in\omega}. \tag{73}$$

The Fourier transform of (73) yields

$$\hat{\psi}(2\omega) = G(\omega)\,\hat{\phi}(\omega). \tag{74}$$

As in Appendix C, (74) and (71) give

$$\left|G(\omega)\right|^2 + \left|G(\omega + \pi)\right|^2 = 1. \tag{75}$$

Since $\mathbf{O}_{2^{-1}}$ is orthogonal to $V_{2^{-1}}$, each function of the family $(\sqrt{2^{-1}}\psi_{2^{-1}}(x - 2^{-1}n))_{n \in Z}$ should be orthogonal to each function of the family $(\sqrt{2^{-1}}\phi_{2^{-1}}(x - 2^{-1}n))_{n \in Z}$. The Poisson formula shows that this property is equivalent to

$$\forall n \in \mathbf{Z}, \sum_{n=-\infty}^{+\infty} \phi(2\omega + 2n\pi)\, \overline{\psi(2\omega + 2n\pi)} = 0. \tag{76}$$

By inserting (64), (66), and (74) in (77), we obtain

$$H(\omega)\,\overline{G(\omega)} + H(\omega + \pi)\,\overline{G(\omega + \pi)} = 0. \tag{77}$$

We can prove that the necessary conditions (75) and (76) on $G(\omega)$ are sufficient to ensure that $(\sqrt{2^{-1}}\psi_{2^{-1}}(x - 2^{-1}n))_{n \in Z}$ is an orthogonal basis of $\mathbf{O}_{2^{-1}}$. An example of such function $G(\omega)$ is given by

$$G(\omega) = e^{-i\omega}\,\overline{H(\omega + \pi)}. \tag{78}$$

The functions $G(\omega)$ and $H(\omega)$ can be viewed as the transfer functions of a pair of quadrature mirror filters. By taking the inverse Fourier transform of (79), we prove that the impulse responses $(g(n))_{n \in Z}$ and $(h(n))_{n \in Z}$ of these filters are related by

$$g(n) = (-1)^{1-n} h(1 - n). \tag{79}$$

By definition, a multiresolution approximation of $L^2(\mathbf{R})$ satisfies

$$\lim_{j \to +\infty} V_{2^j} = L^2(\mathbf{R}) \quad \text{and} \quad \lim_{j \to -\infty} V_{2^j} = \{0\}.$$

Since O_{2^j} is the orthogonal complement of V_{2^j} in $V_{2^{j+1}}$, we can derive that that for any $j \neq k$, O_{2^j} is orthogonal to O_{2^k} and

$$L^2(R) = \bigoplus_{j=-\infty}^{+\infty} O_{2^j}. \tag{80}$$

We proved that for any $j \in Z$, $(\sqrt{2^{-j}}\psi_{2^j}(x - 2^{-j}n))_{n \in Z}$ is an orthonormal basis of O_{2^j}. The family of functions $(\sqrt{2^{-j}}\psi_{2^j}(x - 2^{-j}n))_{(n,j) \in Z^2}$ is therefore an orthonormal basis of $L^2(R)$.

Appendix E
Proof of Theorem 4

This appendix gives the main steps of Theorem 4 proof. More details can be found in [35]. Let $(V_{2^j})_{j \in Z}$ be a multiresolution approximation of $L^2(R)$ such that for any $j \in Z$,

$$V_{2^j} = V^1_{2^j} \otimes V^1_{2^j} \tag{81}$$

where $(V^1_{2^j})_{j \in Z}$ is a multiresolution approximation of $L^2(R)$. We want to prove that the family of functions

$$\begin{aligned}&(2^{-j}\phi_{2^j}(x - 2^{-j}n)\, \psi_{2^j}(y - 2^{-j}m),\\ &2^{-j}\psi_{2^j}(x - 2^{-j}n)\, \phi_{2^j}(y - 2^{-j}m),\\ &2^{-j}\psi_{2^j}(x - 2^{-j}n)\, \psi_{2^j}(y - 2^{-j}m))_{(n,m) \in Z^2}\end{aligned}$$

is an orthonormal basis O_{2^j}. The vector space O_{2^j} is the orthogonal complement of V_{2^j} in $V_{2^{j+1}}$. Let $O^1_{2^j}$ be the orthogonal complement of $V^1_{2^j}$ in $V^1_{2^{j+1}}$. Equation (81) yields

$$\begin{aligned}V_{2^{j+1}} &= V^1_{2^{j+1}} \otimes V^1_{2^{j+1}}\\ &= (O^1_{2^j} \oplus V^1_{2^j}) \otimes (O^1_{2^j} \oplus V^1_{2^j}).\end{aligned}$$

This can be rewritten

$$\begin{aligned}V_{2^{j+1}} = (V^1_{2^j} \otimes V^1_{2^j}) \oplus (V^1_{2^j} \otimes O^1_{2^j}) \oplus (O^1_{2^j} \otimes V^1_{2^j})\\ \oplus (O^1_{2^j} \otimes O^1_{2^j}).\end{aligned}$$

The orthogonal complement of V_{2^j} in $V_{2^{j+1}}$ is therefore given by

$$\begin{aligned}O_{2^j} = (V^1_{2^j} \otimes O^1_{2^j}) \oplus (O^1_{2^j} \otimes V^1_{2^j})\\ \oplus (O^1_{2^j} \otimes O^1_{2^j}).\end{aligned} \tag{82}$$

The family of functions $(\sqrt{2^{-j}}\phi_{2^j}(x - 2^{-j}n))_{j \in Z}$ is an orthonormal basis of $V^1_{2^j}$ and $(\sqrt{2^{-j}}\psi_{2^j}(x - 2^{-j}n))_{j \in Z}$ is an orthonormal basis of $O^1_{2^j}$. Hence, (82) implies that

$$\begin{aligned}&(2^{-j}\phi_{2^j}(x - 2^{-j}n)\, \psi_{2^j}(y - 2^{-j}m),\\ &2^{-j}\psi_{2^j}(x - 2^{-j}n)\, \phi_{2^j}(y - 2^{-j}m),\\ &2^{-j}\psi_{2^j}(x - 2^{-j}n)\, \psi_{2^j}(y - 2^{-j}m))_{(n,m) \in Z^2}\end{aligned}$$

is an orthonormal basis of O_{2^j}.

The vector space $L^2(R^2)$ can be decomposed as a direct sum of the orthogonal spaces O_{2^j}

$$L^2(R^2) = \bigoplus_{j=-\infty}^{+\infty} O_{2^j}.$$

The family of functions

$$\begin{aligned}&(2^{-j}\phi_{2^j}(x - 2^{-j}n)\, \psi_{2^j}(y - 2^{-j}m),\\ &2^{-j}\psi_{2^j}(x - 2^{-j}n)\, \phi_{2^j}(y - 2^{-j}m),\\ &2^{-j}\psi_{2^j}(x - 2^{-j}n)\, \psi_{2^j}(y - 2^{-j}m))_{(n,m) \in Z^3}\end{aligned}$$

is therefore an orthonormal basis of $L^2(R^2)$.

Acknowledgment

I would like to thank particularly R. Bajcsy for her advice throughout this research, and Y. Meyer for his help with some mathematical aspects of this paper. I am also grateful to J.-L. Vila for his comments.

References

[1] E. Adelson and E. Simoncelli, "Orthogonal pyramid transform for image coding," *Proc. SPIE, Visual Commun. Image Proc.*, 1987.

[2] A. Arneodo, G. Grasseau, and H. Holschneider, "On the wavelet transform of multifractals," Tech. Rep., CPT, CNRS Luminy, Marseilles, France, 1987.

[3] G. Battle, "A block spin construction of ondelettes, Part 1: Lemarie functions," *Commun. Math. Phys.*, vol. 110, pp. 601–615, 1987.

[4] J. Beck, A. Sutter, and R. Ivry, "Spatial frequency channels and perceptual grouping in texture segregation," *Comput. Vision Graph. Image Proc.*, vol. 37, 1987.

[5] P. J. Burt and E. H. Adelson, "The Laplacian pyramid as a compact image code," *IEEE Trans. Commun.*, vol. COM-31, pp. 532–540, Apr. 1983.

[6] F. Campbell and D. Green, "Optical and retina factors affecting visual resolution," *J. Physiol.*, vol. 181, pp. 576–593, 1965.

[7] F. Campbell and J. Kulikowski, "Orientation selectivity of the human visual system," *J. Physiol.*, vol. 197, pp. 437–441, 1966.

[8] J. Crowley, "A representation for visual information," Robotic Inst. Carnegie-Mellon Univ., Tech. Rep. CMU-RI-TR-82-7, 1987.

[9] I. Daubechies, "Orthonormal bases of compactly supported wavelets," *Commun. Pure Appl. Math.*, vol. 41, pp. 909–996, Nov. 1988.

[10] D. Esteban and C. Galand, "Applications of quadrature mirror filters to split band voice coding schemes," *Proc. int. Conf. Acoust., Speech, Signal Processing*, May 1977.

[11] P. Federbush, "Quantum field theory in ninety minutes," *Bull. Amer. Math. Soc.*, 1987.

[12] P. Flandrin, "On the spectrum of fractional Brownian motions," UA 346, CNRS, Lyon, Tech. Rep. ICPI TS-8708, France.

[13] A. Gagalowicz, "Vers un modele de textures," dissertation de docteur d'etat, INRIA, May 1983.

[14] P. Goupillaud, A. Grossmann, and J. Morlet, "Cycle octave and related transform in seismic signal analysis," *Geoexploration*, vol. 23, pp. 85–102, 1985/1984.

[15] N. Graham, "Psychophysics of spatial frequency channels," *Perceptual Organization*, Hilldale, NJ, 1981.

[16] W. Grimson, "Computational experiments with a feature based stereo algorithm," *IEEE Trans. Pattern Anal. Machine Intell.*, vol. PAMI-7, pp. 17–34, Jan. 1985.

[17] A. Grossmann and J. Morlet, "Decomposition of Hardy functions into square integrable wavelets of constant shape," *SIAM J. Math.*, vol. 15, pp. 723–736, 1984.

[18] E. Hall, J. Rouge, and R. Wong, "Hierarchical search for image matching," in *Proc. Conf. Decision Contr.*, 1976, pp. 791–796.

[19] D. Heeger and A. Pentland, "Measurement of fractal dimension using Gabor filters," Tech. Rep. TR 391, SRI AI center.

[20] S. Jaffard and Y. Meyer, "Bases d'ondelettes dans des ouverts de Rn," *J. de Math. Pures Appli.*, 1987.

[21] B. Julesz, "Textons, the elements of texture perception and their interactions," *Nature*, vol. 290, Mar. 1981.
[22] J. Koenderink, "The structure of images," in *Biological Cybernetics*. New York: Springer Verlag, 1984.
[23] R. Kronland-Martinet, J. Morlet, and A. Grossmann, "Analysis of sound patterns through wavelet transform," *Int. J. Pattern Recog. Artificial Intell.*, 1988.
[24] P. G. Lemarie, "Ondelettes a localisation exponentielles," *J. Math. Pures Appl.*, to be published.
[25] P. G. Lemarie and Y. Meyer, "Ondelettes et bases Hilbertiennes," *Revista Matematica Ibero Americana*, vol. 2, 1986.
[26] H. Maitre and B. Faust, "The nonstationary modelization of images: Statistical properties to be verified," Conf. Pattern Recogn., May 1978.
[27] S. Mallat, "A compact multiresolution representation: the wavelet model," in *Proc. IEEE Workshop Comput. Vision*, Miami, FL, Dec. 1987.
[28] —, "Multiresolution approximation and wavelet orthonormal bases of L2," *Trans. Amer. Math. Soc.*, June 1989.
[29] B. Mandelbrot, *The Fractal Geometry of Nature*. New York: Freeman, 1983.
[30] D. Marr, *Vision*. New York: Freeman, 1982.
[31] D. Marr and T. Poggio, "A theory of human stereo vision," *Proc. Royal Soc. London*, vol. B 204, pp. 301–328, 1979.
[32] J. Max, "Quantizing for minimum distortion," *Trans. IRE Inform. Theory*, vol. 6, pp. 7–14.
[33] Y. Meyer, "Ondelettes et fonctions splines," *Sem. Equations aux Derivees Partielles*, Ecole Polytechnique, Paris, France, Dec. 1986.
[34] —, *Ondelettes et Operateurs*, Hermann, 1988.
[35] —, "Principe d'incertitude, bases hilbertiennes et algebres d'operateurs," *Bourbaki seminar*, no. 662, 1985–1986.
[36] Millar and C. Paul, "Recursive quadrature mirror filters; Criteria specification and design method," *IEEE Trans. Acoust., Speech, Signal Processing*, vol. 33, pp. 413–420, Apr. 1985.
[37] A. Papoulis, *Probability, Random Variables, and Stochastic Processes*. New York: McGraw-Hill, 1984.
[38] T. Paul, "Affine coherent states and the radial Schrodinger equation. Radial harmonic oscillator and hydrogen atom," Preprint.
[39] A. Pentland, "Fractal based description of natural scenes," *IEEE Trans. Pattern Anal. Machine Intell.*, vol. PAMI-6, pp. 661–674, 1984.
[40] G. Pirani and V. Zingarelli, "Analytical formula for design of quadrature mirror filters," *IEEE Trans. Acoust., Speech, Signal Processing*, vol. ASSP-32, pp. 645–648, June 1984.
[41] A. Rosenfeld and M. Thurston, "Edge and curve detection for visual scene analysis," *IEEE Trans. Comput.*, vol. C-20, 1971.
[42] —, "Coarse-fine template matching," *IEEE Trans. Syst., Man, Cybern.*, vol. SMC-7, pp. 104–107, 1977.
[43] A. Watson, "Efficiency of a model human image code," *J. Opt. Soc. Amer.*, vol. 4, pp. 2401–2417, Dec. 1987.
[44] J. W. Woods and S. D. O'Neil, "Subband coding of images," *IEEE Trans. Acoust., Speech, Signal Processing*, vol. ASSP-34, Oct. 1986.

Stephane G. Mallat was born in Paris, France. He graduated from Ecole Polytechnique, Paris, in 1984 and from Ecole Nationale Superieure des Telecommunications, Paris, in 1985. He received the Ph.D. degree in electrical engineering from the University of Pennsylvania, Philadelphia, PA, in 1988.

Since September 1988, he has been an Assistant Professor in the Department of Computer Science of the Courant Institute of Mathematical Sciences, New York University, New York, NY. His research interests include computer vision, signal processing, and applied mathematics.

Wavelets with Compact Support

Yves Meyer

1. Introduction

The aim of this paper is to give a new proof of a recent and still unpublished result by I. Daubechies. I. Daubechies' theorem is the following.

Theorem 1 *For any $r \geq 1$ there exists a function $\psi = \psi_r$ of the real variable x with the following properties:*

(1.1) *$\psi(x)$ is of class C^r on the real line,*

(1.2) *$\psi(x)$ is compactly supported,*

(1.3) *the collection $2^{j/2}\psi(2^j x - k)$, $j \in \mathbb{Z}$, $k \in \mathbb{Z}$, is an orthonormal basis for $L^2(\mathbb{R})$.*

The only orthonormal basis with the structure given by (1.2) and (1.3) which has been known before is the Haar system for which $\psi(x) = 1$ on $[0, 1/2)$, $\psi(x) = -1$ on $[1/2, 1)$ and $\psi(x) = 0$ elsewhere.

Let us denote by I the dyadic interval $[\frac{k}{2^j}, \frac{k+1}{2^j}]$ and by $\mathcal{J}$ the collection of such dyadic intervals. Then $2^{j/2}\psi(2^j x - k)$ can be rewritten ψ_I and is indeed supported by the "doubled interval" MI where the constant M (M is not 2 as it would be the case for the truly doubled interval) depends on the size of the support of ψ. In the case of the Haar system $M = 1$. The orthonormal basis ψ_I, $I \in \mathcal{J}$, provides a *local Fourier analysis* which has two remarkable features. The formula $f = \sum_{I \in \mathcal{J}} \langle f, \psi_I \rangle \psi_I$ extends to distributions f of order less than r, and for analyzing the behavior of f near x_0 (i.e., on $[x_0 - \varepsilon, x_0 + \varepsilon]$), the only coefficients $\alpha(I) = \langle f, \psi_I \rangle$ that matter are those for which I is small and MI intersects $[x_0 - \varepsilon, x_0 + \varepsilon]$. Indeed if MI does not intersect this small interval $[x_0 - \varepsilon, x_0 + \varepsilon]$ on which we would like to understand the behavior of $f(x)$, then ψ_I vanishes on this interval. On the other hand, if $|I|$ is large, then ψ_I is smooth and the contribution of all I of size $\geq C\varepsilon$ will be a smooth function on $[x_0 - \varepsilon, x_0 + \varepsilon]$. Those simply minded remarks explain the success of the wavelet analysis for discovering the local behavior of functions or distributions. Another remark which explains the striking difference between wavelet analysis and Fourier analysis (or Fourier series) is the fact that the series

$$f = \sum_{I \in \mathcal{J}} \langle f, \psi_I \rangle \psi_I \qquad (*)$$

This manuscript is a typed version of handwritten notes that accompanied the Zygmund Lectures presented by Yves Meyer at the University of Chicago in 1987. The references have been updated.

is unconditionally convergent when f belongs to most of the classical function spaces like L^p $(1 < p < \infty)$, H^1 (the Hardy space à la Stein and Weiss), BMO, the corresponding Riesz potentials, Besov spaces ... (the only limitation is that the index denoting the smoothness s should satisfy $|s| \leq r$).

The fact that $(*)$ is unconditionally convergent allows regroupings of terms. These regroupings are certainly needed (as shown above) for studying the local behavior of $(*)$.

Wavelets play a decisive role in quantum field theory (as G. Battle, Cornell, is showing in his papers) but then we need n-dimensional wavelets and we would like to restrict the wavelet expansion $(*)$ to intervals I (indeed to cubes Q) of size ≤ 1. This can be achieved using the following theorem.

Theorem 2 *For any $r \geq 1$ there exist two functions $\psi = \psi_r$ and $\varphi = \varphi_r$ of the real variable x with the following properties:*

(1.4) *ψ and φ belong to C^r on the real line;*

(1.5) *ψ and φ are compactly supported;*

(1.6) *the collection of all φ_I, $I \in \mathcal{J}$, $|I| = 1$, and of all ψ_I, $|I| \leq 1$, $I \in \mathcal{J}$, is an orthonormal basis for $L^2(\mathbb{R})$.*

As above, $I = [k2^{-j}, (k+1)2^{-j}]$ and $\varphi_I(x) = 2^{j/2}\varphi(2^j x - k)$, $j \geq 0$, $k \in \mathbb{Z}$. It means that if f is a distribution of order less than r, then the first sum

$$S_0(f) = \sum_{|I|-1} \langle f, \varphi_I \rangle \varphi_I \tag{1.7}$$

will somehow give a blurred vision of f, a vision that is too smooth to reveal the local complex behavior of the distribution f. The next steps consist in adding small fluctuations $\alpha_I \psi_I$, $|I| = 1$, of size 1 $(\alpha_I = \langle f, \psi_I \rangle)$ and then smaller and smaller fluctuations in order to recompose the full complexity of our distribution f. The wavelet series expansion is a remarkable way of writing any distribution as a sum of localized fluctuations.

The use of wavelet expansions is not limited to the one-dimensional case. Let us describe, for example, the two-dimensional case.

Theorem 3 *Keeping the same notation as in Theorem 2, the following collection of functions: $2^j\varphi(2^j x - k)\psi(2^j y - \ell)$, $2^j\psi(2^j x - k)\varphi(2^j y - \ell)$, $2^j\psi(2^j x - k)\psi(2^j y - \ell)$, $j \in \mathbb{Z}$, $k \in \mathbb{Z}$, $\ell \in \mathbb{Z}$, is an orthonormal basis of $L^2(\mathbb{R}^2)$.*

Let us denote by $Q = Q(j, k, \ell)$ the "dyadic cube" defined by $k2^{-j} \leq x < (k+1)2^{-j}$, $\ell 2^{-j} \leq y < (\ell + 1)2^{-j}$. Then the wavelets in Theorem 3 can be rewritten $\psi_Q^{(\alpha)}$, where $\alpha \in \{0, 1\}^2$ and $\alpha \neq (0, 0)$. They are supported by MQ and the basis is therefore indexed by $A \times Q$, where $A = \{0, 1\}^2 \backslash (0, 0)$ and Q is the collection of all dyadic cubes Q.

The basis given by Theorem 3 *is not* of tensor product type since the tensor product between the wavelet basis of Theorem 1 with itself would produce bidimensional wavelets supported by rectangles and not by cubes.

Another remarkable feature of wavelet basis is the way they relate to the Calderón-Zygmund operators. A Calderón-Zygmund operator is, by definition, a continuous linear operator $T: L^2(\mathbb{R}^n) \to L^2(\mathbb{R}^n)$ with the following extra property. The distributional kernel $K(x,y)$ of T, once restricted to the open set $\Omega \subset \mathbb{R}^n \times \mathbb{R}^n$ defined by $y \neq x$, $x \in \mathbb{R}^n$, $y \in \mathbb{R}^n$, becomes a function $K(x,y)$ enjoying the following smoothness and size conditions:

(1.8) $|K(x,y)| \leq C|x-y|^{-n}$;

(1.9) there exists an exponent $\theta > 0$ such that, when $0 < \theta < 1$,

$$|K(x',y) - K(x,y)| \leq C|x'-x|^\theta |x-y|^{-n-\theta}$$

when $|x'-x| \leq \frac{1}{2}|x-y|$, and when $m < \theta < m+1$,

$$|\partial^\alpha K(x,y)| \leq C|x-y|^{-n-|\alpha|} \quad \text{for } |\alpha| \leq m$$

while, if $|\alpha| = m$, $\theta - m = \eta$

$$|\partial^\alpha K(x',y) - \partial^\alpha K(x,y)| \leq C|x'-x|^\eta\,|x-y|^{-n-\theta}$$

whenever $|x'-x| \leq \frac{1}{2}|x-y|$;

we make the similar smoothness assumptions for the dependence in y of $K(x,y)$.

A remarkable algebra $\mathcal{A}_\theta$ of Calderón-Zygmund operators has been discovered by P. G. Lemarié. We say that T belongs to $\mathcal{A}_\theta$ if T satisfies the following vanishing moment conditions:

$$\int_{\mathbb{R}^n} K(x,y) y^\alpha \, dy = 0 \quad \text{whenever } |\alpha| \leq m, \tag{1.10}$$

and similarly

$$\int_{\mathbb{R}^n} K(x,y) x^\alpha \, dy = 0 \quad \text{when } |\alpha| \leq m. \tag{1.11}$$

Since these two integrals are highly divergent, we should give a human meaning to (1.10) and (1.11). They mean that, if ψ is a compactly supported function of class $C^\theta(\mathbb{R}^n)$ and if all moments of ψ vanish up to order m, then the moments of $T(\psi)$ and of $T^*(\psi)$ should vanish up to order m. The operator T^* is the adjoint of T and $T(\psi)$ is $\mathcal{O}(|x|^{-n-\theta})$ at infinity by (1.10), while $T^*(\psi)$ is $\mathcal{O}(|x|^{-n-\theta})$ by (1.9). Therefore the moments of $T(\psi)$ and of $T^*(\psi)$ exist up to order m.

Now we have the following and relatively easy characterization of $\mathcal{A}_\theta$ by the entries $\langle T\psi_Q^{(\alpha)}, \psi_R^{(\beta)}\rangle$ of T in the wavelet basis given by Theorem 3. We denote by r the integer $m+1$, by $\mathcal{D}_r$ the space of compactly supported testing functions of class $\mathcal{C}^r$, and by $(\mathcal{D}_r)'$ the corresponding space of distributions.

We start with an operator $T: \mathcal{D}_r \to \mathcal{D}'_r$, which is supposed to be linear and continuous; we would like to decide when this operator T belongs to the operator algebra $\mathcal{B}_m$, which is by definition the union of $\mathcal{A}_\theta$ for $m < \theta < m+1$.

Then $T \in \mathcal{B}_m$ iff there exist an exponent $\theta > m$ and a constant C such that

$$|\langle T\psi_Q^{(\alpha)}, \psi_R^{(\beta)}\rangle| \leq C|Q|^{1/2}|R|^{1/2}(\inf(|Q|,|R|))^{\theta/n}(\operatorname{diam}(Q \cup R))^{-n-\theta}. \tag{1.12}$$

From all our knowledge of the Calderón-Zygmund theory it is relatively easy to show that an operator $T \in \mathcal{A}_\theta$ is bounded on all the classical functional spaces like $L^p(\mathbb{R}^n)$, $1 < p < \infty$,

$L^{p,s}(\mathbb{R}^n)$ when $|s| < \theta$, Besov spaces of smoothness $s < \theta$, the corresponding dual spaces, the Hardy space H^1, its dual BMO, ...

In particular the operators which are bounded on $L^2(\mathbb{R}^n)$ and diagonal in our wavelet basis $\psi_Q^{(\alpha)}$ are also bounded on these function spaces. Then the wavelets $\psi_Q^{(\alpha)}$, $\alpha \in A$, $Q \in \mathcal{Q}$, are an unconditional basis for each one of these spaces.

This approach does not provide the specific growth conditions satisfied by the coefficients $|\langle f, \psi_Q^{(\alpha)} \rangle|$ of the wavelet expansion of f for each one of those spaces. The reader is referred to other papers where those conditions are given explicitly.

After this long introduction, we would like to return to I. Daubechies' theorem. We first treat the one-dimensional case.

2. The Multiscale Analysis Approach to I. Daubechies's Theorem

A multiscale analysis[1] of $L^2(\mathbb{R})$ is, by definition, an increasing sequence V_j, $j \in \mathbb{Z}$, of closed subspaces of $L^2(\mathbb{R})$ with the following extra properties:

(2.1) $\bigcap_{-\infty}^{\infty} V_j = \{0\}$ and $\bigcup_{-\infty}^{\infty} V_j$ is dense in $L^2(\mathbb{R})$;

(2.2) $f(x) \in V_j \iff f(2x) \in V_{j+1}$;

(2.3) $f(x) \in V_0$ and $k \in \mathbb{Z} \implies f(x-k) \in V_0$;

(2.4) there exists an isomorphism $T\colon V_0 \to \ell^2(\mathbb{Z})$ that intertwines the $\mathbb{Z}$-action described by (2.3) and the regular representation of $\mathbb{Z}$ acting on $\ell^2(\mathbb{Z})$.

In other words, (2.4) means the existence of a function $g(x) \in V_0$ such that $g(x-k)$, $k \in \mathbb{Z}$, is an unconditional basis of V_0, which means the following property. Each $f \in V_0$ can be uniquely written $f(x) = \sum_{k=-\infty}^{\infty} \alpha_k g(x-k)$ with $(\sum_{k=-\infty}^{\infty} |\alpha_k|^2)^{1/2} < \infty$, and this ℓ^2-norm is equivalent to $\|f\|_2$.

The multiscale analysis will be called r-regular if the function g is smooth and has a reasonably good decay at infinity: in other words, if we have

$$\begin{aligned} &g(x) \text{ belongs to } C^r(\mathbb{R}) \quad \text{and} \\ &(1+x^2)^r \left(\frac{d}{dx}\right)^m g(x) \to 0 \text{ as } |x| \to \infty \text{ when } 0 \le m \le r. \end{aligned} \tag{2.5}$$

It is an open problem to construct *all* the multiscale analysis of $L^2(\mathbb{R})$. We will later give a partial answer to this problem, which has been raised and partly solved by S. Mallat.

Let us give some examples. In all the examples, V_0 will be first described and V_0 should satisfy (2.3) and (2.4). Then V_j are defined by (2.2). What is to be checked next is the fact that V_j is an increasing sequence. It suffices for that to verify that $V_{(-1)} \subset V_0$. But then we have also to check that $\bigcup V_j$ is dense in $L^2(\mathbb{R})$ and that $\bigcap V_j = \{0\}$. All those properties are implicitly contained in the choice of the space V_0.

Examples are polynomial splines of order r with nodes at $k \in \mathbb{Z}$. In other terms, $f \in V_0$ iff $f(x)$ is a function of class C^{r-1} whose restriction to each interval $[k, k+1]$ coincides with

[1] Author's Note: The term "multiscale analysis" has now been replaced by "multiresolution analysis."

a polynomial of degree $\leq r$. When $r = 0$ we take step functions with joint discontinuities at $k \in \mathbb{Z}$. In all cases f should belong to $L^2(\mathbb{R})$.

Another example is to define $\mathcal{F}V_0$ as being the closed subspace of $L^2(\mathbb{R})$ consisting of all functions supported by $[-\pi, \pi]$. No one of those natural and well-known examples would produce compactly supported wavelets by the procedure which will now be described.

STEP 1. We replace $g(x)$ by $\varphi(x)$ in such a way that the collection $\varphi(x-k)$, $k \in \mathbb{Z}$, is an orthonormal basis of V_0 without losing the smoothness and decay described by (2.5). The optimal choice of φ is given by

$$\hat{\varphi}(\xi) = \hat{g}(\xi) \left(\sum_{-\infty}^{\infty} |\hat{g}(\xi + 2k\pi)|^2 \right)^{-1/2} . \tag{2.6}$$

STEP 2. We define W_j as being the orthogonal complement of V_j in V_{j+1}. Then $f(x) \in W_j$ is equivalent to $f(2x) \in W_{j+1}$. Moreover, $L^2(\mathbb{R}) = \oplus_{-\infty}^{\infty} W_j$. We will now construct (in step 3) a function $\psi \in W_0$ with good smoothness and decay properties such that $\psi(x-k)$, $k \in \mathbb{Z}$, is an orthonormal basis of W_0. Then $2^{j/2}\psi(2^j x - k)$, $k \in \mathbb{Z}$, will be a basis of W_j for every j, and the union of these bases will be the basis we are looking for.

STEP 3. For building ψ we consider the two 2π-periodic functions $m_0(\xi)$ and $m_1(\xi)$, which are, at a formal level, defined by

$$\hat{\varphi}(2\xi) = m_0(\xi)\hat{\varphi}(\xi), \quad \hat{\psi}(2\xi) = m_1(\xi)\hat{\varphi}(\xi), \tag{2.7}$$

and more precisely by

$$\begin{cases} \dfrac{1}{2}\varphi\left(\dfrac{x}{2}\right) = \displaystyle\sum_{-\infty}^{\infty} \alpha_k \, \varphi(x-k) & \text{(since } V_{-1} \subset V_0\text{)}, \\ \dfrac{1}{2}\psi\left(\dfrac{x}{2}\right) = \displaystyle\sum_{-\infty}^{\infty} \beta_k \, \varphi(x-k) & \text{(since } W_{-1} \subset V_0\text{)}, \\ m_0(\xi) = \displaystyle\sum_{-\infty}^{\infty} \alpha_k \, e^{-ik\xi}, & \\ m_1(\xi) = \displaystyle\sum_{-\infty}^{\infty} \beta_k \, e^{-ik\xi}. & \end{cases} \tag{2.8}$$

In the worst case α_k belongs to $\ell^2(\mathbb{Z})$ and so does β_k. Then $m_0(\xi)$ and $m_1(\xi)$ belong to $L^2(\mathbb{T})$. Writing the fact that $\varphi(x-k)$, $k \in \mathbb{Z}$, is orthonormal immediately leads to $\sum_{-\infty}^{\infty} |\hat{\varphi}(\xi + 2k\pi)|^2 = 1$. Since we also have $\sum_{-\infty}^{\infty} |\hat{\varphi}(2\xi + 2k\pi)|^2 = 1$, we deduce $\sum_{-\infty}^{\infty} |m_0(\xi + k\pi)|^2 |\hat{\varphi}(\xi + k\pi)|^2 = 1$. We break this sum into $\sum_{(2k)}$ and $\sum_{(2k+1)}$, which yields $|m_0(\xi)|^2 + |m_0(\xi + \pi)|^2 = 1$. It is now quite easy to see that $\psi(x-k)$, $k \in \mathbb{Z}$, is an orthonormal basis for W_0 iff

$$U(\xi) = \begin{pmatrix} m_0(\xi) & m_1(\xi) \\ m_0(\xi + \pi) & m_1(\xi + \pi) \end{pmatrix} \tag{2.9}$$

is unitary.

In our procedure we know $m_0(\xi)$ through φ. We have to construct ψ through $U(\xi)$. For doing it we decide that $m_1(\xi) = e^{-i\xi}\overline{m_0}(\xi + \pi)$. Other choices would be deduced from this one by multiplication with unimodular π-periodic functions of ξ. For instance, we could as well take $m_1(\xi) = e^{i\xi}\overline{m_0}(\xi + \pi)$, which corresponds to the trivial observation that if ψ is a solution to Theorem 1, so is $\psi(x+1)$.

Then ψ is defined by $\hat{\psi}(\xi) = m_1(\xi/2)\,\hat{\varphi}(\xi/2)$. If we want φ and ψ to be compactly supported it is necessary to assume that $m_0(\xi)$ be a trigonometric polynomial.

A last observation concerns the operator $E_j\colon L^2(\mathbb{R}) \to V_j$ given by the orthogonal projection. Its kernel is $A_j(x,y) = \sum_{-\infty}^{\infty} s^j \varphi(2^j x - k)\bar{\varphi}(2^j y - k) = 2^j B(2^j x, 2^j y)$. We easily observe that $B(x,y)$ is C^r in x and y and that $B(x,y) = 0$ if $|x - y| \geq L$ (assuming from now on that φ is compactly supported). Then it is an easy lemma to observe that $E_j \uparrow \mathbb{1}$ iff $\int_{-\infty}^{\infty} B(x,y)\,dy = 1$, which implies (with $I = \int_{-\infty}^{\infty} \varphi(x)\,dx$) that $\bar{I}\sum_{-\infty}^{\infty} \varphi(x-k) = 1$ and $|I|^2 = 1$. After a renormalization of φ, we can assume $I = 1$ and $\sum_{-\infty}^{\infty} \varphi(x-k) = 1$. It immediately implies that $m_0(0) = 1$.

Collecting all the information we have obtained, we have $m_0(0) = 1$, $m_0(\xi) = \sum_{-N}^{N} \alpha_k e^{ik\xi}$, and $|m_0(\xi)|^2 + |m_0(\xi+\pi)|^2 = 1$. What S. Mallat proposed to do is to start with such an $m_0(\xi)$, called a *quadrature mirror filter* in signal analysis, and to build $\varphi(\xi)$ by

$$\hat{\varphi}(\xi) = \prod_{1}^{\infty} m_0(\xi/2^j). \tag{2.10}$$

Then, if we are lucky enough, $\varphi(x-k)$, $k \in \mathbb{Z}$, would be an orthonormal basis of some closed subspace $V_0 \subset L^2(\mathbb{R})$ and starting from V_0 we should be able to reconstruct the multiresolution analysis. Unfortunately, this program breaks down the way it is stated, since the trivial choice $m_0(\xi) = e^{-i\xi/2}\cos 3\xi/2$ that satisfies the requirements on $m_0(\xi)$ leads to a function $\varphi(x)$ that is $\frac{1}{3}$ on $[-2,1]$ and 0 elsewhere.

Nevertheless the following lemma is true.

Lemma 1 *Let $m_0(\xi)$ be a 2π-periodic function on the real line of class C^1. If we assume*

$$m_0(0) = 1, \quad |m_0(\xi)|^2 + |m_0(\xi+\pi)|^2 = 1, \quad \text{and} \tag{2.11}$$

$$m_0(\xi) \neq 0 \text{ on } (-\pi, \pi), \tag{2.12}$$

then the function φ defined by (2.10) satisfies $\|\varphi\|_2 = 1$. If we further assume $\hat{\varphi}(0) = \mathcal{O}(|\xi|^{-\gamma})$ at infinity for some $\gamma > \frac{1}{2}$, then $\varphi(x-k)$, $k \in \mathbb{Z}$, is an orthonormal sequence and spans a space V_0 leading to a multiscale analysis of $L^2(\mathbb{R})$.

We do not know if (2.12) is necessary. Probably not. Observe that $m_0(\pm\pi) = 0$ by (2.11). We will later give a sufficient condition on $m_0(\xi)$ ensuring that $\hat{\varphi}(\xi) = \mathcal{O}(|\xi|^{-\gamma})$ at infinity for a large γ.

For simplifying the notation we will rewrite $|m_0(\xi)|^2$ as $g(\xi)$. Then we have $0 \leq g(\xi) \leq 1$ and $g(\xi) + g(\xi+\pi) = 1$. This function $g(\xi)$ is 2π-periodic and belongs to $C^1(\mathbb{R})$. We next define $g_k(\xi)$ by $g_k(\xi) = 0$ if $|\xi| \geq \pi 2^k$ and $g_k(\xi) = g(\xi/2)g(\xi/4)\cdots g(\xi/2^k)$ if $|\xi| \leq \pi 2^k$. It is now a trivial computation to check that

$$\int_{-\infty}^{\infty} g_k(\xi) = 2\pi.$$

We next consider the convergent infinite product $G(\xi) = g(\xi/2)g(\xi/4)\cdots$, which makes sense since $g(0) = 1$ and since g is smooth. If $-2^k\pi \leq \xi \leq 2^k\pi$, we have $g(\xi/2^{k+1})\cdots$

$g(\xi/2^j)\cdots \geq c > 0$ by (2.12). Therefore $0 \leq g_k(\xi) \leq \frac{1}{c}G(\xi)$. But $G(\xi) = \lim_{k\to+\infty} g_k(\xi)$ implies $\int_{-\infty}^{\infty} G(\xi) \leq \underline{\lim} \int_{-\infty}^{\infty} g_k(\xi)\, d\xi = 2\pi$ by Fatou's lemma. Therefore, the Lebesgue dominated convergence theorem applies to $g_k(\xi)$ and we have $\int_{-\infty}^{\infty} G(\xi)\, d\xi = 2\pi$. We now assume that $\hat{\varphi}(\xi) = \mathcal{O}(|\xi|^{-\gamma})$ at infinity for some $\gamma > 1/2$. Then the function $\alpha(\xi) = \sum_{-\infty}^{\infty} |\varphi(\xi + 2k\pi)|^2$ is 2π-periodic and continuous. This function satisfies the functional equation

$$\alpha(2\pi) = g(\xi)\,\alpha(\xi) + g(\xi+\pi)\,\alpha(\xi+\pi).$$

We want to prove that $\alpha(\xi) = 1$ identically. Consider $M = \sup_{0\leq\xi\leq 2\pi} \alpha(\xi)$. If $M = \alpha(0)$, we have $M = 1$ since $\alpha(0) = \sum_{-\infty}^{\infty} |\hat{\varphi}(2k\pi)|^2 = |\hat{\varphi}(0)|^2 = 1$. If $M \neq \alpha(0)$, we certainly have $M = \alpha(\xi_0)$ for $0 < \xi_0 < 2\pi$. We next consider $\xi_0/2$ and $\xi_0/2 + \pi$. Then $\alpha(\xi_0) = g(\xi_0/2)\,\alpha(\xi_0/2) + g(\xi_0/2+\pi)\,\alpha(\xi_0/2+\pi)$. Since $g(\xi_0/2) + g(\xi_0/2+\pi) = 1$ and since both are positive, we certainly have $\alpha(\xi_0/2) = \alpha(\xi_0)$. Repeating this argument yields $\alpha(\xi_0/2^m) = M$ and passing to the limit we obtain $M = \alpha(0) = 1$. Similarly we denote by m the minimum of $\alpha(\xi)$ on $[0, 2\pi]$ and we have $m = 1$. Therefore $\alpha(\xi) = 1$. This argument is due to Ph. Tchamitchian.

The fact that $\sum_{-\infty}^{\infty} |\hat{\varphi}(\xi + 2k\pi)|^2 = 1$ is equivalent to saying that the collection $\varphi(x-k)$, $k \in \mathbb{Z}$, is orthonormal. Then V_0 is defined as the span of those vectors. We have $f \in V_0$ iff $\hat{f}(\xi) = m(\xi)\hat{\varphi}(\xi)$, where $m(\xi)$ is 2π-periodic and belongs to L^2, once restricted to $[0, 2\pi]$. Similarly $f \in V_{-1}$ iff $\hat{f}(\xi) = m(\xi)\hat{\varphi}(2\xi)$, where $m(\xi)$ is now a π-periodic function in $L^2([0,\pi])$. Since $\hat{\varphi}(2\xi) = m_0(\xi)\hat{\varphi}(\xi)$ where $m_0(\xi)$ is 2π-periodic we have $\mathcal{F}V_{-1} \subset \mathcal{F}V_0$.

In order to show that $\cup V_j$ is dense in $L^2(\mathbb{R})$ we observe that the kernel $A_j(x,y) = 2^j B(2^j x, 2^j y)$ of the orthogonal projector $E_j\colon L^2(\mathbb{R}) \to V_j$ satisfies $\int_{-\infty}^{\infty} B(x,y)\, dy = 1$, which together with smoothness and decay implies $E_j \uparrow \mathbb{1}$.

Our goal will be to show that for each $r \geq 1$ there is a trigonometric sum $m_0(\xi)$ (depending on r) with the three following properties

(2.13) $m_0(0) = 1$ and $|m_0(\xi)|^2 + |m_0(\xi+\pi)|^2 = 1$;

(2.14) $\hat{\varphi}(\xi) = \prod_1^{\infty} m_0(2^{-j}\xi)$ is $\mathcal{O}(|\xi|^{-r})$ at infinity;

(2.15) $m_0(\xi) \neq 0$ in $(-\pi, \pi)$.

Moreover, we want the degree of $m_0(\xi)$ to be less than Cr.

3. Construction of $m_0(\xi)$

We first construct the trigonometric polynomial $g(\xi) = g_r(\xi)$ which satisfies

(3.1) $g(\xi) \geq 0$, $g(\xi + 2\pi) = g(\xi)$;

(3.2) $g(\xi+\pi) + g(\xi) = 1$ and $g(0) = 1$;

(3.3) $\prod_1^{\infty} g(2^{-j}\xi) = \mathcal{O}(|\xi|^{-r})$ as $|\xi| \to \infty$.

Then we apply a lemma due to F. Riesz which can also be found in Polya-Szegö and tells that *any* nonnegative trigonometric polynomial $g(\xi) = \sum_{-N}^{N} c_k\, e^{ik\xi}$ can be written

$g(\xi) = |P(\xi)|^2$, where $P(\xi) = \sum_0^N a_k\, e^{-ik\xi}$. Moreover $P(\xi)$ is unique if $P(0) = g(0) = 1$. This $P(\xi)$ will be our $m_0(\xi)$. Therefore our φ will be supported by $[0, N]$ and the wavelet ψ will be supported by $\left[-\frac{N}{2}+\frac{1}{2}, \frac{N}{2}+\frac{1}{2}\right]$. If we want a wavelet of causal type (i.e., supported by $[0, +\infty[$) we simply shift ψ by $\frac{N}{2}$ (N even) or $\frac{N-1}{2}$ (N odd) to the right. In the first case, the wavelet will be supported by $\left[\frac{1}{2}, N+\frac{1}{2}\right]$ and in the second case by $[0, N]$. In both cases the length of the support of ψ will be exactly N.

For constructing $g(\xi)$ we make a change of variable $\xi = 2\pi t$, $0 \le t \le 1$ and we then apply the following theorem by J. P. Kahane and Y. Katznelson.

Theorem 4 *For any $\delta \in (0,1)$ and any $C > 1$ there exists an $\alpha > 0$ such that for all continuous and 1-periodic functions $g_m(t)$ satisfying*

(3.4) $0 \le g_m(t) \le 1$ *for all* $t \in \mathbb{R}$;

(3.5) $g_m(t) \le \delta^m$ *for* $\frac{1}{3} \le t \le \frac{2}{3}$;

(3.6) $g_m(t) \le C^m \left|t - \frac{1}{2}\right|^m$ *for* $0 \le t \le 1$,

we have for all $k \ge 1$

$$\sup_{\frac{1}{4}\le t\le\frac{1}{2}} g_m(t)\, g_m(2t) \cdots g_m(2^k t) \le 2^{-\alpha m k}. \tag{3.7}$$

In the application we have in mind C is 2π in such a way that (3.6) is an improvement on (3.5) only when $\left|t - \frac{1}{2}\right| \le \delta/C < \frac{1}{6}$.

Let us take the theorem for granted and construct a trigonometric polynomial $g_m(t)$ satisfying (3.4), (3.5) and (3.6) together with $g_m(t) + g_m(t + \frac{1}{2}) = 1$.

We start with $c_m(\sin 2\pi t)^{2m+1}$ and consider the integral

$$g_m(t) = 1 - c_m \int_0^t (\sin 2\pi s)^{2m+1}\, ds. \tag{3.8}$$

The constant c_m is adjusted so that $g_m(\pi) = 0$ and we obviously have

$$g_m(t) + g_m(t+\pi) = 1, \tag{3.9}$$

$$g_m(t) = \sum_{-2m-1}^{2m+1} \alpha(m,\ell)\, e^{2\pi i \ell t}. \tag{3.10}$$

Moreover $c_m \sim \sqrt{2\pi m}$ as m tends to infinity and if $t = \frac{1}{2} + u$ we have

$$g_m\left(\tfrac{1}{2} + u\right) = -c_m \int_0^u (\sin 2\pi s)^{2m+1}\, ds,$$

which yields

$$\left|g_m\left(\tfrac{1}{2} + u\right)\right| \le (2\pi)^{2m+1} c_m (2m+2)^{-1} |u|^{2m+2}.$$

Then (3.6) is true with $C = 2\pi$ and $2m+2$ instead of m (when m is large enough). On the other hand, $\left|g_m(\frac{1}{2}+u)\right| \le c_m(\sqrt{3}/2)^{2m}$ when $|u| \le \frac{1}{6}$ which yields (3.5).

Returning to the original problem, we compute $|\hat{\varphi}(\xi)|^2 = g(\xi/2)\cdots g(\xi/2^j)\cdots$ for $\pi 2^k \le |\xi| \le 2\pi 2^k$. We can majorize the infinite product by $g(\xi/2)\cdots g(\xi/2^{k+1})$ and then make the change of variable $|\xi| = 2\pi 2^{k+1}t$, $\frac{1}{4} \le t \le \frac{1}{2}$. Then the Kahane-Katznelson theorem gives $|\hat{\varphi}(\xi)|^2 \le 2^{-\alpha mk}$, where $\alpha > 0$ is an absolute constant. Therefore $|\hat{\varphi}(\xi)|^2 = \mathcal{O}(|\xi|^{-\alpha m})$ which gives the required amount of smoothness.

4. Proof of the Kahane-Katznelson Theorem

For simplifying the notation, we will drop the subscript m in g. Next $g(t)$ will be replace by $h(t) = g(t)g(2t)$. If we can prove that $h(t)h(2t)\cdots h(2^k t) \le 2^{-\alpha mk}$ we will certainly have $g(t)g(2t)\cdots g(2^k t) \le 2^{-(\alpha/2)\,mk}$. But now (3.5) is improved into

$$h(t) \le \delta^m \quad \text{for } \frac{1}{6} \le t \le \frac{5}{6} \tag{4.1}$$

and we still have

$$h(t) \le C^m \left| t - \frac{1}{2} \right|^m. \tag{4.2}$$

The next idea is to write $t \in [1/4, 1/2]$ as

$$t = 0.01\alpha_3\alpha_4 \ldots = 0 + 0 \cdot 2^{-1} + 1 \cdot 2^{-2} + \alpha_3 2^{-3} + \cdots,$$

where $\alpha_j = 0$ or 1. Since $h(t)$ is 1-periodic, we have

$$h(2^k t) = h(0.\alpha_{k+1}\alpha_{k+2}\ldots). \tag{4.3}$$

This leads to defining two subsets E and F of $\{2, \ldots, k+2\}$. We write $j \in E$ if $j \ge 3$ and if $\alpha_j = 0$. Similarly $j \in F$ if $\alpha_j = 1$. A *component* of E (or of F) is a maximal arithmetical progression $I = [j, j+1, j+2, \ldots, j+\ell]$ contained in E (or in F). Then E is the disjoint union of its components named $I_1, I_2, \ldots$ and similarly for F. The origin of I_1 will be $a_1, \ldots$ and similarly b_1 will be the origin of $J_1, \ldots$.

If $k = a_q - 2$, a_q being the origin of I_q, then $h(2^k t) = h(0.10\alpha_{k+3}\ldots)$. But $s = 0.10\alpha_{k+3}\cdots \in [1/2, 3/4]$ and $h(s) \le \delta^m$. If the length of I_q (denoted $|I_q|$) is large enough we have a better estimate. Indeed $|s - \frac{1}{2}| \le \frac{1}{4}2^{-|I_q|}$ and therefore $h(s) \le \left(\frac{C}{4}\right)^m 2^{-m|I_q|}$.

We have obtained two competing estimates namely $h(s) \le \delta^m$ (for short I_q) and $h(s) \le \left(\frac{C}{4}\right)^m 2^{-m|I_q|}$ which is better when $|I_q| \ge \log_2 C + \log_2 1/\delta - 2 = A$. Now we decide that the only value of j we are going to use for estimating the product $h(t)\cdots h(2^j t)\cdots h(2^k t) = P_k(t)$ will be either $j = a_q - 2$ or $j = b_q - 2$ (a_q is the origin of I_q and b_q is the origin of J_q). Therefore those values will be distinct and the product $P_k(t)$ will be majorized by the subproduct in which we delete all $h(2^j t)$ for $j \ne a_q - 2$ and $j \ne b_q - 2$. Since $h(t) \le 1$ we are in good shape.

The estimate of $h(2^j t)$ we have obtained above can be rewritten $h(2^j t) \le 2^{-\alpha m|I_q|}$ for some $\alpha > 0$ which can be easily computed in terms of δ and C. This holds if $j = a_q - 2$. If $j = b_q - 2$ a similar computation leads to $h(2^j t) \le 2^{-\alpha m|J_q|}$ and putting all these estimates together we obtain

$$P_k(t) = h(t)\cdots h(2^k t) \le 2^{-\alpha m \text{ Card } E}\, 2^{-\alpha m \text{ Card } F} = 2^{-\alpha mk}.$$

References

[1] I. Daubechies, *Orthonormal bases of compactly supported wavelets*, Comm. Pure Appl. Math. **41** (1988), 909–996.

[2] S. Mallat, *A theory for multiresolution signal decomposition: the wavelet representation*, IEEE Trans. Pattern Anal. Machine Intell. **11** (1989), 674–693.

[3] P. G. Lemarié and Y. Meyer, *Ondelettes et bases hilbertiennes*, Rev. Mat. IberoAmericana **2** (1986), 1–18.

TRANSACTIONS OF THE
AMERICAN MATHEMATICAL SOCIETY
Volume 315, Number 1, September 1989

MULTIRESOLUTION APPROXIMATIONS AND WAVELET ORTHONORMAL BASES OF $\mathbf{L}^2(\mathbf{R})$

STEPHANE G. MALLAT

ABSTRACT. A multiresolution approximation is a sequence of embedded vector spaces $(\mathbf{V}_j)_{j\in\mathbf{Z}}$ for approximating $\mathbf{L}^2(\mathbf{R})$ functions. We study the properties of a multiresolution approximation and prove that it is characterized by a 2π-periodic function which is further described. From any multiresolution approximation, we can derive a function $\psi(x)$ called a wavelet such that $(\sqrt{2^j}\psi(2^j x - k))_{(k,j)\in\mathbf{Z}^2}$ is an orthonormal basis of $\mathbf{L}^2(\mathbf{R})$. This provides a new approach for understanding and computing wavelet orthonormal bases. Finally, we characterize the asymptotic decay rate of multiresolution approximation errors for functions in a Sobolev space $\mathbf{H}^s$.

1. INTRODUCTION

In this article, we study the properties of the multiresolution approximations of $\mathbf{L}^2(\mathbf{R})$. We show how they relate to wavelet orthonormal bases of $\mathbf{L}^2(\mathbf{R})$. Wavelets have been introduced by A. Grossmann and J. Morlet [7] as functions whose translations and dilations could be used for expansions in $\mathbf{L}^2(\mathbf{R})$. J. Stromberg [16] and Y. Meyer [14] have proved independently that there exists some particular wavelets $\psi(x)$ such that $(\sqrt{2^j}\psi(2^j x - k))_{(j,k)\in\mathbf{Z}^2}$ is an orthonormal basis of $\mathbf{L}^2(\mathbf{R})$; these bases generalize the Haar basis. If $\psi(x)$ is regular enough, a remarkable property of these bases is to provide an unconditional basis of most classical functional spaces such as the Sobolev spaces, Hardy spaces, $\mathbf{L}^p(\mathbf{R})$ spaces and others [11]. Wavelet orthonormal bases have already found many applications in mathematics [14, 18], theoretical physics [6] and signal processing [9, 12].

Notation. $\mathbf{Z}$ and $\mathbf{R}$ respectively denote the set of integers and real numbers.

$\mathbf{L}^2(\mathbf{R})$ denotes the space of measurable, square-integrable functions $f(x)$.

The inner product of two functions $f(x) \in \mathbf{L}^2(\mathbf{R})$ and $g(x) \in \mathbf{L}^2(\mathbf{R})$ is written $\langle g(u), f(u)\rangle$.

The norm of $f(x) \in \mathbf{L}^2(\mathbf{R})$ is written $\|f\|$.

Received by the editors February 12, 1988.

1980 *Mathematics Subject Classification* (1985 *Revision*). Primary 42C05, 41A30.

Key words and phrases. Approximation theory, orthonormal bases, wavelets.

This work is supported in part by NSF-CER/DCR82-19196 A02, NSF/DCR-8410771, Air Force/F49620-85-K-0018, ARMY DAAG-29-84-K-0061, and DARPA/ONR N0014-85-K-0807.

©1989 American Mathematical Society
0002-9947/89 $1.00 + $.25 per page

The Fourier transform of any function $f(x) \in \mathbf{L}^2(\mathbf{R})$ is written $\hat{f}(\omega)$.
Id is the identity operator in $\mathbf{L}^2(\mathbf{R})$.
$\mathbf{l}^2(\mathbf{Z})$ is the vector space of square-summable sequences:

$$\mathbf{l}^2(\mathbf{Z}) = \left\{ (\alpha_i)_{i\in\mathbf{Z}} : \sum_{i=-\infty}^{+\infty} |\alpha_i|^2 < \infty \right\}.$$

Definition. A multiresolution approximation of $\mathbf{L}^2(\mathbf{R})$ is a sequence $(\mathbf{V}_j)_{j\in\mathbf{Z}}$ of closed subspaces of $\mathbf{L}^2(\mathbf{R})$ such that the following hold:

(1) $$\mathbf{V}_j \subset \mathbf{V}_{j+1} \quad \forall j \in \mathbf{Z},$$

(2) $$\bigcup_{j=-\infty}^{+\infty} \mathbf{V}_j \text{ is dense in } \mathbf{L}^2(\mathbf{R}) \quad \text{and} \quad \bigcap_{j=-\infty}^{+\infty} \mathbf{V}_j = \{0\},$$

(3) $$f(x) \in \mathbf{V}_j \Leftrightarrow f(2x) \in \mathbf{V}_{j+1} \quad \forall j \in \mathbf{Z},$$

(4) $$f(x) \in \mathbf{V}_j \Rightarrow f(x - 2^{-j}k) \in \mathbf{V}_j \quad \forall k \in \mathbf{Z},$$

(5) There exists an isomorphism **I** from $\mathbf{V}_0$ onto $\mathbf{l}^2(\mathbf{Z})$ which commutes with the action of $\mathbf{Z}$.

In property (5), the action of $\mathbf{Z}$ over $\mathbf{V}_0$ is the translation of functions by integers whereas the action of $\mathbf{Z}$ over $\mathbf{l}^2(\mathbf{Z})$ is the usual translation. The approximation of a function $f(x) \in \mathbf{L}^2(\mathbf{R})$ at a resolution 2^j is defined as the orthogonal projection of $f(x)$ on $\mathbf{V}_j$. To compute this orthogonal projection we show that there exists a unique function $\phi(x) \in \mathbf{L}^2(\mathbf{R})$ such that, for any $j \in \mathbf{Z}$, $(\sqrt{2^j}\phi(2^j x - k))_{k\in\mathbf{Z}}$ is an orthonormal basis of $\mathbf{V}_j$. The main theorem of this article proves that the Fourier transform of $\phi(x)$ is characterized by a 2π-periodic function $H(\omega)$. As an example we describe a multiresolution approximation based on cubic splines.

The additional information available in an approximation at a resolution 2^{j+1} as compared with the resolution 2^j, is given by an orthogonal projection on the orthogonal complement of $\mathbf{V}_j$ in $\mathbf{V}_{j+1}$. Let $\mathbf{O}_j$ be this orthogonal complement. We show that there exists a function $\psi(x)$ such that $(\sqrt{2^j}\psi(2^j x - k))_{k\in\mathbf{Z}}$ is an orthonormal basis of $\mathbf{O}_j$. The family of functions $(\sqrt{2^j}\psi(2^j x - k))_{(k,j)\in\mathbf{Z}^2}$ is a wavelet orthonormal basis of $\mathbf{L}^2(\mathbf{R})$.

An important problem in approximation theory [4] is to measure the decay of the approximation error when the resolution increases, given an a priori knowledge on the function smoothness. We estimate this decay for functions in Sobolev spaces $\mathbf{H}^s$. This result is a characterization of Sobolev spaces.

2. Orthonormal bases of multiresolution approximations

In this section, we prove that there exists a unique function $\phi(x) \in \mathbf{L}^2(\mathbf{R})$ such that for any $j \in \mathbf{Z}$, $(\sqrt{2^j}\phi(2^j x - k))_{k\in\mathbf{Z}}$ is a wavelet orthonormal basis of $\mathbf{V}_j$. This result is proved for $j = 0$. The extension for any $j \in \mathbf{Z}$ is a consequence of property (3).

Let us first detail property (5) of a multiresolution approximation. The operator $\mathbf{I}$ is an isomorphism from $\mathbf{V}_0$ onto $\mathbf{l}^2(\mathbf{Z})$. Hence, there exists a function $g(x)$ satisfying

$$(6) \qquad g(x) \in \mathbf{V}_0 \quad \text{and} \quad \mathbf{I}(g(x)) = \varepsilon(n), \quad \text{where } \varepsilon(n) = \begin{cases} 1 & \text{if } n = 0, \\ 0 & \text{if } n \neq 0. \end{cases}$$

Since $\mathbf{I}$ commutes with translations of integers:

$$\mathbf{I}(g(x-k)) = \varepsilon(n-k).$$

The sequence $(\varepsilon(n-k))_{k\in\mathbf{Z}}$ is a basis of $\mathbf{l}^2(\mathbf{Z})$, hence $(g(x-k))_{k\in\mathbf{Z}}$ is a basis of $\mathbf{V}_0$. Let $f(x) \in \mathbf{V}_0$ and $\mathbf{I}(f(x)) = (\alpha_k)_{k\in\mathbf{Z}}$. Since $\mathbf{I}$ is an isomorphism, $\|f\|$ and $(\sum_{k=-\infty}^{+\infty} |\alpha_k|^2)^{1/2}$ are two equivalent norms on $\mathbf{V}_0$. Let us express the consequence of this equivalence on $g(x)$. The function $f(x)$ can be decomposed as:

$$(7) \qquad f(x) = \sum_{k=-\infty}^{+\infty} \alpha_k g(x-k).$$

The Fourier transform of this equation yields

$$(8) \qquad \hat{f}(\omega) = M(\omega)\hat{g}(\omega), \quad \text{where } M(\omega) = \sum_{k=-\infty}^{+\infty} \alpha_k e^{-ik\omega}.$$

The norm of $f(x)$ is given by

$$\|f\|^2 = \int_{-\infty}^{+\infty} |\hat{f}(\omega)|^2 \, d\omega = \int_0^{2\pi} |M(\omega)|^2 \sum_{k=-\infty}^{+\infty} |\hat{g}(\omega + 2k\pi)|^2 \, d\omega.$$

Since $\|f\|$ and $(\sum_{k=-\infty}^{+\infty} |\alpha_k|^2)^{1/2}$ are two equivalent norms on $\mathbf{V}_0$, it follows that

$$(9) \quad \exists C_1, C_2 > 0 \text{ such that } \forall \omega \in \mathbf{R}, \quad C_1 \leq \left(\sum_{k=-\infty}^{+\infty} |\hat{g}(\omega + 2k\pi)|^2 \right)^{1/2} \leq C_2.$$

We are looking for a function $\phi(x)$ such that $(\phi(x-k))_{k\in\mathbf{Z}}$ is an orthonormal basis of $\mathbf{V}_0$. To compute $\phi(x)$ we orthogonalize the basis $(g(x-k))_{k\in\mathbf{Z}}$. We can use two methods for this purpose, both useful.

The first method is based on the Fourier transform. Let $\hat{\phi}(\omega)$ be the Fourier transform of $\phi(x)$. With the Poisson formula, we can express the orthogonality of the family $(\phi(x-k))_{k\in\mathbf{Z}}$ as

$$(10) \qquad \sum_{k=-\infty}^{+\infty} |\hat{\phi}(\omega + 2k\pi)|^2 = 1.$$

Since $\phi(x) \in \mathbf{V}_0$, equation (8) shows that there exists a 2π-periodic function $M_\phi(\omega)$ such that

$$\hat{\phi}(\omega) = M_\phi(\omega)\hat{g}(\omega). \tag{11}$$

By inserting equation (11) into (10) we obtain

$$M_\phi(\omega) = \left(\sum_{k=-\infty}^{+\infty} |\hat{g}(\omega + 2k\pi)|^2 \right)^{-1/2}. \tag{12}$$

Equation (9) proves that (12) defines a function $M_\phi(\omega) \in \mathbf{L}^2([0, 2\pi])$. If $\phi(x)$ is given by (11), one can also derive from (9) that $g(x)$ can be decomposed on the corresponding orthogonal family $(\phi(x-k))_{k\in\mathbf{Z}}$. This implies that $(\phi(x-k))_{k\in\mathbf{Z}}$ generates $\mathbf{V}_0$.

The second approach for building the function $\phi(x)$ is based on the general algorithm for orthogonalizing an unconditional basis $(e_\lambda)_{\lambda\in\Lambda}$ of a Hilbert space $\mathbf{H}$. This approach was suggested by Y. Meyer. Let us recall that a sequence $(e_\lambda)_{\lambda\in\Lambda}$ is a normalized unconditional basis if there exist two positive constants A and B such that for any sequence of numbers $(\alpha_\lambda)_{\lambda\in\Lambda}$,

$$A\left(\sum_{\lambda\in\Lambda} |\alpha_\lambda|^2\right)^{1/2} \le \left\| \sum_{\lambda\in\Lambda} \alpha_\lambda e_\lambda \right\| \le B\left(\sum_{\lambda\in\Lambda} |\alpha_\lambda|^2\right)^{1/2}. \tag{13}$$

We first compute the Gram matrix $\mathbf{G}$, indexed by $\Lambda \times \Lambda$, whose coefficients are $\langle e_{\lambda 1}, e_{\lambda 2}\rangle$. Equation (13) is equivalent to

$$A^2\mathrm{Id} \le \mathbf{G} \le B^2\mathrm{Id}. \tag{14}$$

This equation shows that we can calculate $\mathbf{G}^{-1/2}$, whose coefficients are written $\gamma(\lambda_1, \lambda_2)$. Let us define the vectors $f_\lambda = \sum_{\lambda\in\Lambda} \gamma(\lambda, \lambda')e_{\lambda'}$. It is well known that the family $(f_\lambda)_{\lambda\in\Lambda}$ is an orthonormal basis of $\mathbf{H}$. This algorithm has the advantage, with respect to the usual Gram-Schmidt procedure, of preserving any supplementary structure (invariance under the action of a group, symmetries) which might exist in the sequence $(e_\lambda)_{\lambda\in\Lambda}$. In our particular case we verify immediately that both methods lead to the same result. The second one is more general and can be used when the multiresolution approximation is defined on a Hilbert space where the Fourier transform does not exist [8].

In the following, we impose a regularity condition on the multiresolution approximations of $\mathbf{L}^2(\mathbf{R})$ that we study. We shall say that a function $f(x) \in \mathbf{L}^2(\mathbf{R})$ is regular if and only if it is continuously differentiable and satisfies:

$$\exists C > 0, \forall x \in \mathbf{R}, \quad |f(x)| \le C(1+x^2)^{-1} \quad \text{and} \quad |f'(x)| \le C(1+x^2)^{-1}. \tag{15}$$

A multiresolution approximation $(\mathbf{V}_j)_{j\in\mathbf{Z}}$ is said to be regular if and only if $\phi(x)$ is regular.

3. Properties of $\phi(x)$

In this section, we study the functions $\phi(x)$ such that for all $j \in \mathbf{Z}$, $(\sqrt{2^j}\phi(2^j x - n))_{n\in\mathbf{Z}}$ is an orthonormal family, and if $\mathbf{V}_j$ is the vector space generated by this family of functions, then $(\mathbf{V}_j)_{j\in\mathbf{Z}}$ is a regular multiresolution approximation of $\mathbf{L}^2(\mathbf{R})$. We show that the Fourier transform of $\phi(x)$ can be computed from a 2π-periodic function $H(\omega)$ whose properties are further described.

Property (2) of a multiresolution approximation implies that

$$\frac{1}{2}\phi\left(\frac{x}{2}\right) \in \mathbf{V}_{-1} \subset \mathbf{V}_0 .$$

The function $\frac{1}{2}\phi(\frac{x}{2})$ can thus be decomposed in the orthonormal basis $(\phi(x-k))_{k\in\mathbf{Z}}$ of $\mathbf{V}_0$:

$$\frac{1}{2}\phi\left(\frac{x}{2}\right) = \sum_{k=-\infty}^{\infty} h_k \phi(x+k), \quad \text{where } h_k = \frac{1}{2}\int_{-\infty}^{\infty} \phi\left(\frac{x}{2}\right)\overline{\phi}(x+k)\,dx . \tag{16}$$

Since the multiresolution approximation is regular, the asymptotic decay of h_k satisfies $|h_k| = O(1+k^2)^{-1}$. The Fourier transform of equation (16) yields

$$\hat{\phi}(2\omega) = H(\omega)\hat{\phi}(\omega), \quad \text{where } H(\omega) = \sum_{k=-\infty}^{\infty} h_k e^{-ik\omega} . \tag{17}$$

The following theorem gives a necessary condition on $H(\omega)$.

Theorem 1. *The function* $H(\omega)$ *as defined above satisfies*:

$$|H(\omega)|^2 + |H(\omega+\pi)|^2 = 1 , \tag{18}$$

$$|H(0)| = 1 . \tag{19}$$

Proof. We saw in equation (10) that the Fourier transform $\hat{\phi}(\omega)$ must satisfy

$$\sum_{k=-\infty}^{+\infty} |\hat{\phi}(\omega + 2k\pi)|^2 = 1 , \tag{20}$$

and therefore

$$\sum_{k=-\infty}^{+\infty} |\hat{\phi}(2\omega + 2k\pi)|^2 = 1 . \tag{21}$$

Since $\hat{\phi}(2\omega) = H(\omega)\hat{\phi}(\omega)$, this summation can be rewritten

$$\sum_{k=-\infty}^{+\infty} |H(\omega + k\pi)|^2 |\hat{\phi}(\omega + k\pi)|^2 = 1 . \tag{22}$$

The function $H(\omega)$ is 2π-periodic. Regrouping the terms for $k \in 2\mathbf{Z}$ and $k \in 2\mathbf{Z}+1$ and inserting equation (20) yields

$$|H(\omega)|^2 + |H(\omega+\pi)|^2 = 1.$$

In order to prove that $|H(0)| = 1$, we show that

$$|\hat{\phi}(0)| = 1. \tag{23}$$

Let us prove that this equation is a consequence of property (2) of a multiresolution approximation. Let $\mathbf{P}_{\mathbf{V}_j}$ be the orthogonal projection on $\mathbf{V}_j$. Since $(\sqrt{2^j}\phi(2^jx-n))_{n\in\mathbf{Z}}$ is an orthonormal basis of $\mathbf{V}_j$, the kernel of $\mathbf{P}_{\mathbf{V}_j}$ can be written:

$$2^jK(2^jx, 2^jy), \quad \text{where } K(x,y) = \sum_{k=-\infty}^{\infty} \phi(x-k)\overline{\phi}(y-k). \tag{24}$$

Property (2) implies that the sequence of operators $(\mathbf{P}_{\mathbf{V}_j})_{j\in\mathbf{Z}}$ tends to Id in the sense of strong convergence for operators. The next lemma shows that the kernel $K(x,y)$ must satisfy $\int_{-\infty}^{\infty} K(x,y)\,dy = 1$.

Lemma 1. *Let* $g(x)$ *be a regular function (satisfying* (15)) *and* $A(x,y) = \sum_{k=-\infty}^{\infty} g(x-k)\overline{g}(y-k)$. *Then the following two properties are equivalent*:

$$\int_{-\infty}^{+\infty} A(x,y)\,dy = 1 \quad \textit{for almost all } x. \tag{25}$$

(26) *The sequence of operators* $(\mathbf{T}_j)_{j\in\mathbf{Z}}$ *whose kernels are* $2^jA(2^jx, 2^jy)$, *tends to* Id *in the sense of strong convergence for operators.*

Proof. Let us first prove that (25) implies (26). Since $g(x)$ is regular, $\exists C > 0$ such that

$$|A(x,y)| \le C(1+|x-y|)^{-2}. \tag{27}$$

Hence, the sequence of operators $(\mathbf{T}_j)_{j\in\mathbf{Z}}$ is bounded over $\mathbf{L}^2(\mathbf{R})$. For proving that

$$\forall j \in \mathbf{Z}, \forall f \in \mathbf{L}^2(\mathbf{R}) \quad \lim_{j\to+\infty} \|f - \mathbf{T}_j(f)\| = 0, \tag{28}$$

we can thus restrict ourselves to indicator functions of intervals. Indeed, finite linear combinations of these indicator functions are dense in $\mathbf{L}^2(\mathbf{R})$. Let $f(x)$ be the indicator function of an interval $[a,b]$,

$$f(x) = \begin{cases} 1 & \text{if } a \le x \le b, \\ 0 & \text{otherwise.} \end{cases}$$

Let us first prove that $\mathbf{T}_jf(x)$ converges almost everywhere to $f(x)$:

$$\mathbf{T}_j(f)(x) = \int_a^b 2^jA(2^jx, 2^jy)\,dy. \tag{29}$$

Equation (27) implies that

$$|\mathbf{T}_j(f)(x)| \le C2^j \int_a^b (1+2^j|x-y|)^{-2}\,dy \le \frac{C'}{1+2^j \operatorname{dist}(x,[a,b])^2}. \tag{30}$$

If x is not a member of $[a,b]$, this inequality implies that

$$\lim_{j\to+\infty} \mathbf{T}_j(f)(x) = 0.$$

Let us now suppose that $x \in]a,b[$,

$$\mathbf{T}_j(f)(x) = \int_{2^j a}^{2^j b} A(2^j x, y)\,dy. \tag{31}$$

By applying property (25), we obtain

$$\mathbf{T}_j(f)(x) = 1 - \int_{-\infty}^{2^j a} A(2^j x, y)\,dy - \int_{2^j b}^{+\infty} A(2^j x, y)\,dy. \tag{32}$$

Since $x \in]a,b[$, inserting (27) in the previous equation yields

$$\lim_{j\to+\infty} \mathbf{T}_j(f)(x) = 1. \tag{33}$$

Equation (30) shows that for $j \ge 0$, there exists $C'' > 0$ such that

$$|\mathbf{T}_j f(x)| \le \frac{C''}{1+x^2}.$$

We can therefore apply the theorem of dominated convergence on the sequence of functions $(\mathbf{T}_j f(x))_{j\in\mathbf{Z}}$ and prove that it converges strongly to $f(x)$.

Conversely let us show that (26) implies (25). Let us define

$$\alpha(x) = \int_{-\infty}^{\infty} A(x,y)\,dy. \tag{34}$$

The function $\alpha(x)$ is periodic of period 1 and equation (27) implies that $\alpha(x) \in \mathbf{L}^\infty(\mathbf{R})$. Let $f(x)$ be the indicator function of $[-1,1]$. Property (26) implies that $\mathbf{T}_j f(x)$ converges to $f(x)$ in $\mathbf{L}^2(\mathbf{R})$ norm. Let $1 > r > 0$ and $x \in [-r,r]$,

$$\mathbf{T}_j(f)(x) = \int_{-1}^{1} 2^j A(2^j x, 2^j y)\,dy.$$

Similarly to equation (32), we show that

$$\mathbf{T}_j(f)(x) = \alpha(2^j x) + O(2^{-j}). \tag{35}$$

Since $\alpha(2^j x)$ is 2^{-j} periodic and converges strongly to 1 in $\mathbf{L}^2([-r,r])$, $\alpha(x)$ must therefore be equal to 1. This proves Lemma 1.

Since $(\mathbf{P}_{\mathbf{V}_j})_{j\in\mathbf{Z}}$ tends to Id in the sense of strong convergence for operators, this lemma shows that the kernel $K(x,y)$ must satisfy $\int_{-\infty}^{\infty} K(x,y)\,dy = 1$. Hence, we have

$$\int_{-\infty}^{\infty} K(x,y)\,dy = \sum_{k=-\infty}^{\infty} \overline{\gamma}\phi(x-k) = 1, \tag{36}$$

with

$$\gamma = \int_{-\infty}^{\infty} \phi(y)\, dy = \hat{\phi}(0). \tag{37}$$

By integrating equation (36) in x on $[0,1]$, we obtain $|\hat{\phi}(0)|^2 = 1$. From equation (17), we can now conclude that $|H(0)| = 1$. This concludes the proof of Theorem 1.

The next theorem gives a sufficiency condition on $H(\omega)$ in order to compute the Fourier transform of a function $\phi(x)$ which generates a multiresolution approximation.

Theorem 2. *Let* $H(\omega) = \sum_{k=-\infty}^{\infty} h_k e^{-ik\omega}$ *be such that*

$$|h_k| = O(1+k^2)^{-1}, \tag{38}$$

$$|H(0)| = 1, \tag{39}$$

$$|H(\omega)|^2 + |H(\omega+\pi)|^2 = 1, \tag{40}$$

$$H(\omega) \neq 0 \quad \text{on } [-\pi/2, \pi/2]. \tag{41}$$

Let us define

$$\hat{\phi}(\omega) = \prod_{k=1}^{\infty} H(2^{-k}\omega). \tag{42}$$

The function $\hat{\phi}(\omega)$ *is the Fourier transform of a function* $\phi(x)$ *such that* $(\phi(x-k))_{k\in\mathbf{Z}}$ *is an orthonormal basis of a closed subspace* $\mathbf{V}_0$ *of* $\mathbf{L}^2(\mathbf{R})$. *If* $\phi(x)$ *is regular, then the sequence of vector spaces* $(\mathbf{V}_j)_{j\in\mathbf{Z}}$ *defined from* $\mathbf{V}_0$ *by* (3) *is a regular multiresolution approximation of* $\mathbf{L}^2(\mathbf{R})$.

Proof. Let us first prove that $\hat{\phi}(\omega) \in \mathbf{L}^2(\mathbf{R})$. To simplify notations we denote $M(\omega) = |H(\omega)|^2$ and denote by $M_k(\omega)$ $(k \geq 1)$ the continuous function defined by

$$M_k(\omega) = \begin{cases} 0 & \text{if } |\omega| > 2^k\pi, \\ M\left(\frac{\omega}{2}\right) M\left(\frac{\omega}{4}\right) \cdots M\left(\frac{\omega}{2^k}\right) & \text{if } |\omega| \leq 2^k\pi. \end{cases}$$

Lemma 2. *For all* $k \in \mathbf{N}$, $k \neq 0$,

$$I_k^n = \int_{-\infty}^{\infty} M_k(\omega) e^{i2n\pi\omega}\, d\omega = \begin{cases} 2\pi & \text{if } n = 0, \\ 0 & \text{if } n \neq 0. \end{cases} \tag{43}$$

Proof. Let us divide the integral I_k^n into two parts:

$$I_k^n = \int_{-2^k\pi}^{0} M_k(\omega) e^{i2n\pi\omega}\, d\omega + \int_{0}^{2^k\pi} M_k(\omega) e^{i2n\pi\omega}\, d\omega.$$

Since $M(2^{-j}\omega + 2^{k-j}\pi) = M(2^{-j}\omega)$ for $0 \leq j < k$ and

$$M(2^{-k}\omega) + M(2^{-k}\omega + \pi) = 1,$$

by changing variables $\omega' = \omega + 2^k\pi$ in the first integral, we obtain

$$I_k^n = \int_0^{2^k\pi} M\left(\frac{\omega}{2}\right)\cdots M\left(\frac{\omega}{2^{k-1}}\right) e^{i2n\pi\omega}\, d\omega.$$

Since $M(\omega)$ is 2π-periodic, this equation implies

$$I_k^n = \int_{-2^{k-1}\pi}^{2^{k-1}\pi} M\left(\frac{\omega}{2}\right)\cdots M\left(\frac{\omega}{2^{k-1}}\right) e^{i2n\pi\omega}\, d\omega = I_{k-1}^n.$$

Hence, we derive that

$$I_k^n = I_{k-1}^n = \cdots = I_1^n = \begin{cases} 2\pi & \text{if } n = 0, \\ 0 & \text{if } n \neq 0. \end{cases}$$

This proves Lemma 2.

Let us now consider the infinite product

$$M_\infty(\omega) = \lim_{k\to\infty} M_k(\omega) = \prod_{j=1}^{\infty} M(2^{-j}\omega) = |\hat{\phi}(\omega)|^2. \tag{44}$$

Since $0 \le M(\omega) \le 1$, this product converges. From Fatou's lemma we derive that

$$\int_{-\infty}^{\infty} M_\infty(\omega)\, d\omega \le \lim_{k\to\infty} \int_{-\infty}^{\infty} M_k(\omega)\, d\omega = 2\pi. \tag{45}$$

Equation (42) thus defines a function $\hat{\phi}(\omega)$ which is in $\mathbf{L}^2(\mathbf{R})$. Let $\phi(x)$ be its inverse Fourier transform. We must show that $(\phi(x-k))_{k\in\mathbf{Z}}$ is an orthonormal family. For this purpose, we want to use Lemma 2 and apply the theorem of dominated convergence on the sequence of functions $(M_k(\omega)e^{i2n\pi\omega})_{k\in\mathbf{Z}}$. The function $M_\infty(\omega)$ can be rewritten

$$M_\infty(\omega) = e^{-\sum_{j=1}^{\infty} \mathrm{Log}(M(2^{-j}\omega))}. \tag{46}$$

Since $H(\omega)$ satisfies both conditions (38) and (39), it follows that $\mathrm{Log}(M(\omega)) = O(\omega)$ in the neighborhood of 0, and therefore

$$\lim_{\omega\to 0} M_\infty(\omega) = M_\infty(0) = 1. \tag{47}$$

As a consequence of (38), $H(\omega)$ is a continuous function. From property (41) together with (47), we derive that

$$\exists C > 0 \text{ such that } \forall\omega \in [-\pi, \pi] \quad M_\infty(\omega) \ge C. \tag{48}$$

For $|\omega| \le 2^k\pi$, we have

$$M_\infty(\omega) = M_k(\omega) M_\infty(\omega/2^k).$$

Hence, equation (48) yields

$$0 \le M_k(\omega) \le \tfrac{1}{C} M_\infty(\omega). \tag{49}$$

Since $M_k(\omega) = 0$ for $|\omega| > 2^k\pi$, inequality (48) is satisfied for all $\omega \in \mathbf{R}$. We proved in (45) that $M_\infty(\omega) \in \mathbf{L}^1(\mathbf{R})$, so we can apply the dominated convergence theorem on the sequence of functions $(M_k(\omega)e^{i2n\pi\omega})_{k\in\mathbf{Z}}$. From Lemma 2, we obtain

$$(50) \qquad \int_{-\infty}^{\infty} M_\infty(\omega)e^{i2n\pi\omega}\,d\omega = \begin{cases} 2\pi & \text{if } n = 0, \\ 0 & \text{if } n \neq 0. \end{cases}$$

With the Parseval theorem applied to the inner products $\langle\phi(x), \phi(x-k)\rangle$, we conclude from (50) that $(\phi(x-k))_{k\in\mathbf{Z}}$ is orthonormal.

Let us call $\mathbf{V}_0$ the vector space generated by this orthonormal family. We suppose now that the function $\phi(x)$ is regular. Let $(\mathbf{V}_j)_{j\in\mathbf{Z}}$ be the sequence of vector spaces derived from $\mathbf{V}_0$ with property (3), $(\sqrt{2^j}\phi(2^jx-k))_{k\in\mathbf{Z}}$ is an orthonormal basis of $\mathbf{V}_j$ for any $j \in \mathbf{Z}$. We must prove that $(\mathbf{V}_j)_{j\in\mathbf{Z}}$ is a multiresolution approximation of $\mathbf{L}^2(\mathbf{R})$. We only detail properties (1) and (2) since the other ones are straightforward.

To prove (1), it is sufficient to show that $\mathbf{V}_{-1} \subset \mathbf{V}_0$. The vector spaces $\mathbf{V}_0$ and $\mathbf{V}_{-1}$ are respectively the set of all the functions whose Fourier transform can be written $M(\omega)\hat{\phi}(\omega)$ and $M(2\omega)\hat{\phi}(2\omega)$, where $M(\omega)$ is any 2π-periodic function such that $M(\omega) \in \mathbf{L}^2([0, 2\pi])$. Since $\hat{\phi}(\omega)$ is defined by (42), it satisfies

$$(51) \qquad \hat{\phi}(2\omega) = H(\omega)\hat{\phi}(\omega),$$

with $|H(\omega)| \leq 1$. The function $M(2\omega)H(\omega)$ is 2π-periodic and is a member of $\mathbf{L}^2([0, 2\pi])$. From equation (51), we can therefore derive that any function of $\mathbf{V}_{-1}$ is in $\mathbf{V}_0$.

Let $\mathbf{P}_{\mathbf{V}_j}$ be the orthogonal projection operator on $\mathbf{V}_j$. To prove (2), we must verify that

$$(52) \qquad \lim_{j\to+\infty} \mathbf{P}_{\mathbf{V}_j} = \mathrm{Id} \quad \text{and} \quad \lim_{j\to-\infty} \mathbf{P}_{\mathbf{V}_j} = 0.$$

Since $(\sqrt{2^j}\phi(2^jx-k))_{k\in\mathbf{Z}}$ is an orthonormal basis of $\mathbf{V}_j$, the kernel of $\mathbf{P}_{\mathbf{V}_j}$ is given by

$$(53) \qquad 2^j \sum_{k=-\infty}^{\infty} \phi(2^jx-k)\bar{\phi}(2^jy-k) = 2^jK(2^jx, 2^jy).$$

Since $(\phi(x-k))_{k\in\mathbf{Z}}$ is an orthogonal family, we have

$$\sum_{-\infty}^{+\infty} |\hat{\phi}(\omega + 2k\pi)|^2 = 1.$$

We showed in (47) that $|\hat{\phi}(0)| = 1$, so for any $k \neq 0$, the previous equation implies that $\hat{\phi}(2k\pi) = 0$. The Poisson formula yields

$$(54) \qquad \sum_{k=-\infty}^{\infty} \phi(x-k) = \int_{-\infty}^{+\infty} \phi(u)\,du = \hat{\phi}(0).$$

We can therefore derive that

$$\int_{-\infty}^{+\infty} K(x,y)\,dy = |\hat{\phi}(0)|^2 = 1 \quad \text{for almost all } x.$$

Lemma 1 enables us to conclude that $\lim_{j\to+\infty} \mathbf{P}_{\mathbf{V}_j} = \mathrm{Id}$. Since $\phi(x)$ is regular, similarly to (27), we have

$$|2^j K(2^j x, 2^j y)| \leq \frac{C2^j}{(1+2^j|x-y|)^2}. \tag{55}$$

From this inequality, we easily derive that $\lim_{j\to-\infty} \mathbf{P}_{\mathbf{V}_j} = 0$. This concludes the proof of Theorem 2.

Remarks. 1. The necessary conditions on $H(\omega)$ stated in Theorem 1 are not sufficient to define a function $\phi(x)$ such that $(\phi(x-k))_{k\in\mathbf{Z}}$ is an orthonormal family. A counterexample is given by $H(\omega) = \cos(3\omega/2)$. The function $\phi(x)$ whose Fourier transform is defined by (42) is equal to $\frac{1}{3}$ in $[-\frac{3}{2}, \frac{3}{2}]$ and 0 elsewhere. It does not generate an orthogonal family. A. Cohen [2] showed that the sufficient condition (41) is too strong to be necessary. He gave a weaker condition which is necessary and sufficient.

2. It is possible to control the smoothness of $\hat{\phi}(\omega)$ from $H(\omega)$. One can show that if $H(\omega) \in \mathbf{C}^q$ then $\hat{\phi}(\omega) \in \mathbf{C}^q$ and

$$\frac{d^n H(0)}{d\omega^n} = 0 \text{ for } 1 \leq n \leq q \Leftrightarrow \frac{d^n \hat{\phi}(0)}{d\omega^n} = 0 \text{ for } 1 \leq n \leq q. \tag{56}$$

I. Daubechies [3] and P. Tchamitchian [17] showed that we can also obtain a lower bound for the decay rate of $\hat{\phi}(\omega)$ at infinity. As a consequence of (40),

$$\frac{d^n H(0)}{d\omega^n} = 0 \quad \text{for } 1 \leq n \leq q$$

implies that

$$\frac{d^n H((2k+1)\pi)}{d\omega^n} = 0 \quad \text{for } 0 \leq n \leq q-1 \text{ and } k \in \mathbf{Z}.$$

Hence, we can decompose $H(\omega)$ into

$$H(\omega) = (\cos(\omega/2))^q\, M_0(\omega), \tag{57}$$

where $M_0(\omega)$ is a 2π-periodic function whose amplitude is bounded by $A > 0$. One can then show that

$$\prod_{j=-1}^{+\infty} |M_0(2^{-j}\omega)| = O(|\omega|^{\mathrm{Log}(A)/\mathrm{Log}(2)}) \tag{58}$$

at infinity. Since

$$\prod_{j=-1}^{+\infty} \cos\left(2^{-j}\frac{\omega}{2}\right) = \frac{\sin(\omega/2)}{\omega/2},$$

it follows that

$$|\hat{\phi}(\omega)| = O(|\omega|^{-q+\mathrm{Log}(A)/\mathrm{Log}(2)}) \quad \text{at infinity}. \tag{59}$$

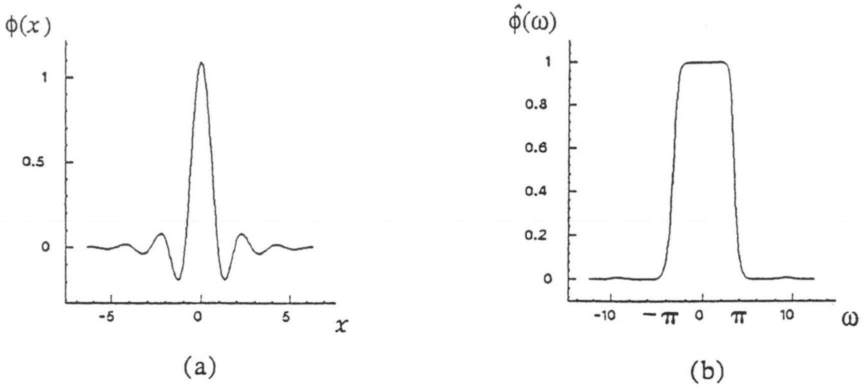

(a) Graph of the function $\phi(x)$ derived from a cubic spline multiresolution approximation. It decreases exponentially.

(b) Graph of $\hat{\phi}(\omega)$. It decreases like $1/\omega^4$ at infinity.

FIGURE 1

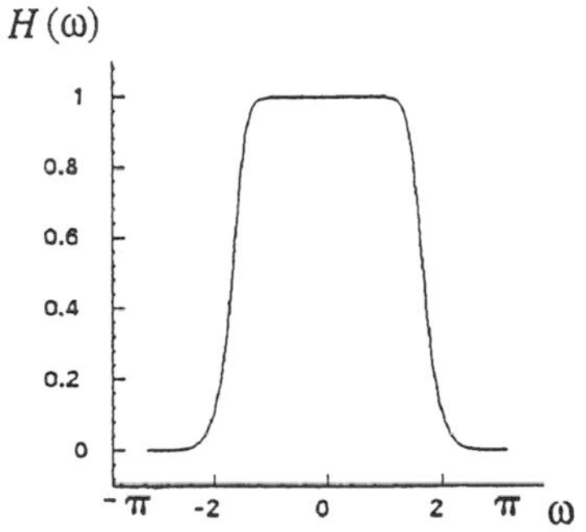

Graph of the function $H(\omega)$ derived from a cubic spline multiresolution approximation.

FIGURE 2

Example. We describe briefly an example of multiresolution approximation from cubic splines found independently by P. Lemarie [10] and G. Battle [1].

The vector space $\mathbf{V}_0$ is the set of functions which are $\mathbf{C}^2$ and equal to a cubic polynomial on each interval $[k, k+1]$, $k \in \mathbf{Z}$. It is well known that there exists a unique cubic spline $g(x) \in \mathbf{V}_0$ such that

$$\forall k \in \mathbf{Z}, \quad g(k) = \begin{cases} 1 & \text{if } k = 0, \\ 0 & \text{if } k \neq 0. \end{cases}$$

The Fourier transform of $g(x)$ is given by

$$\hat{g}(\omega) = \left(\frac{\sin(\omega/2)}{\omega/2}\right)^4 \left(1 - \frac{2}{3}\sin^2\frac{\omega}{2}\right)^{-1}. \tag{60}$$

Any function $f(x) \in \mathbf{V}_0$ can thus be decomposed in a unique way,

$$f(x) = \sum_{k=-\infty}^{\infty} f(k) g(x-k).$$

Hence, for a cubic spline multiresolution approximation, the isomorphism $\mathbf{I}$ of property (5) can be defined as the restriction to $\mathbf{Z}$ of the functions $f(x) \in \mathbf{V}_0$. One can easily show that the sequence of vector spaces $(\mathbf{V}_j)_{j\in\mathbf{Z}}$ built with property (3) is a regular multiresolution approximation of $\mathbf{L}^2(\mathbf{R})$. Let us define

$$\Sigma_8(\omega) = \sum_{-\infty}^{+\infty} \frac{1}{(\omega + 2k\pi)^8}. \tag{61}$$

It follows from equations (60), (12) and (17) that

$$\hat{\phi}(\omega) = \frac{1}{\omega^4\sqrt{\Sigma_8(\omega)}} \quad \text{and} \quad H(\omega) = \sqrt{\frac{\Sigma_8(\omega)}{2^8\Sigma_8(2\omega)}}. \tag{62}$$

We calculate $\Sigma_8(\omega)$ by computing the 6th derivative of the formula

$$\Sigma_2(\omega) = \frac{1}{4\sin^2(\omega/2)}.$$

Figure 1 shows the graph of $\phi(x)$ and its Fourier transform. It is an exponentially decreasing function. Figure 2 shows $H(\omega)$ on $[-\pi, \pi]$.

4. The wavelet orthonormal basis

The approximation of a function at a resolution 2^j is equal to its orthogonal projection on $\mathbf{V}_j$. The additional precision of the approximation when the resolution increases from 2^j to 2^{j+1} is thus given by the orthogonal projection on the orthogonal complement of $\mathbf{V}_j$ in $\mathbf{V}_{j+1}$. Let us call this vector space $\mathbf{O}_j$. In this section, we describe an algorithm, which is now classic [13], in order to find a wavelet $\psi(x)$ such that $(\sqrt{2^j}\psi(2^j x - k))_{k\in\mathbf{Z}}$ is an orthonormal basis of $\mathbf{O}_j$.

We are looking for a function $\psi(x)$ such that $\psi(x/2) \in \mathbf{O}_{-1} \subset \mathbf{V}_0$. Its Fourier transform can thus be written

$$\hat{\psi}(2\omega) = G(\omega)\hat{\phi}(\omega), \tag{63}$$

where $G(\omega)$ is a 2π-periodic function in $\mathbf{L}^2([0, 2\pi])$. Since $\mathbf{V}_0 = \mathbf{V}_{-1} \oplus \mathbf{O}_{-1}$, the Fourier transform of any function $f(x) \in \mathbf{V}_0$ can be decomposed as

$$\hat{f}(\omega) = a(\omega)\hat{\phi}(\omega) = b(\omega)\hat{\phi}(2\omega) + c(\omega)\hat{\psi}(2\omega), \tag{64}$$

where $a(\omega)$ is 2π-periodic and a member of $\mathbf{L}^2([0,\pi])$, and $b(\omega), c(\omega)$ are both π-periodic and members of $\mathbf{L}^2([0,\pi])$. By inserting (17) and (63) in the previous equation, it follows that

$$a(\omega) = b(\omega)H(\omega) + c(\omega)G(\omega). \tag{65}$$

The orthogonality of the decomposition is equivalent to

$$\int_0^{2\pi} |a(\omega)|^2 \, d\omega = \int_0^{\pi} |b(\omega)|^2 \, d\omega + \int_0^{\pi} |c(\omega)|^2 \, d\omega.$$

It is satisfied for any $a(\omega)$ if and only if

$$\begin{cases} |H(\omega)|^2 + |G(\omega)|^2 = 1, \\ H(\omega)\overline{G(\omega)} + H(\omega+\pi)\overline{G(\omega+\pi)} = 0. \end{cases} \tag{66}$$

These equations are necessary and sufficient conditions on $G(\omega)$ to build $\psi(x)$. The functions $b(\omega)$ and $c(\omega)$ are respectively given by

$$\begin{cases} b(\omega) = a(\omega)\overline{H(\omega)} + a(\omega+\pi)\overline{H(\omega+\pi)}, \\ c(\omega) = a(\omega)\overline{G(\omega)} + a(\omega+\pi)\overline{G(\omega+\pi)}. \end{cases} \tag{67}$$

Condition (66) together with (40) can also be expressed by writing that

$$\begin{bmatrix} H(\omega) & G(\omega) \\ H(\omega+\pi) & G(\omega+\pi) \end{bmatrix} \tag{68}$$

is a unitary matrix. A possible choice for $G(\omega)$ is

$$G(\omega) = e^{-i\omega}\overline{H(\omega+\pi)}. \tag{69}$$

Any vector spaces $\mathbf{V}_J$ can be decomposed as

$$\mathbf{V}_J = \bigoplus_{j=-\infty}^{J-1} \mathbf{O}_j. \tag{70}$$

Since $\bigcup_{j=-\infty}^{+\infty} \mathbf{V}_j$ is dense in $\mathbf{L}^2(\mathbf{R})$, the direct sum $\bigoplus_{j=-\infty}^{+\infty} \mathbf{O}_j$ is also dense in $\mathbf{L}^2(\mathbf{R})$. The family of functions $(\sqrt{2^j}\psi(2^j x - k))_{(k,j)\in\mathbf{Z}^2}$ is therefore an orthonormal basis of $\mathbf{L}^2(\mathbf{R})$.

Multiresolution approximations provide a general approach to build wavelet orthonormal bases. We first define a function $H(\omega)$ which satisfies the hypothesis of Theorem 2 and compute the corresponding function $\phi(x)$ with equation (42). From equations (63) and (69), we can also derive the Fourier transform of a wavelet $\psi(x)$ which generates an orthonormal basis. Figure 3 is the graph of the wavelet derived from the cubic spline multiresolution approximation described in the previous section.

The Haar basis is a particular case of wavelet orthonormal basis with

$$\psi(x) = \begin{cases} 1 & \text{if } 0 \le x < 1/2, \\ -1 & \text{if } 1/2 \le x < 1, \\ 0 & \text{otherwise.} \end{cases}$$

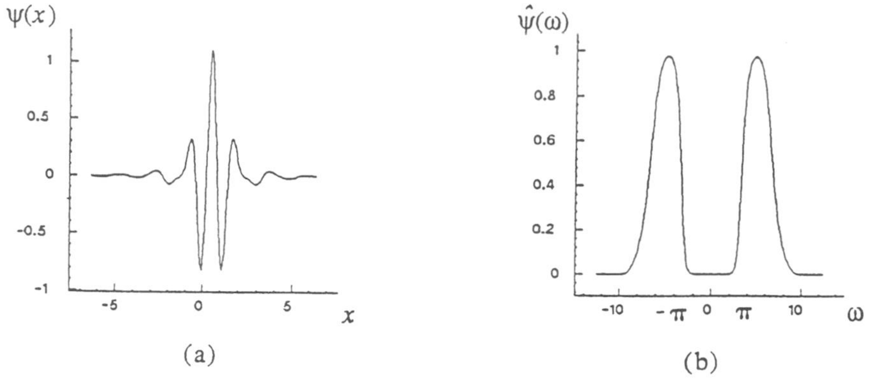

(a) Graph of the function $\psi(x)$ derivedfrom a cubic spline multiresolution approximation. It decreases exponentially.
(b) Graph of $|\hat{\psi}(\omega)|$. It decreases like $1/\omega^4$ at infinity.

FIGURE 3

The corresponding function $\phi(x)$ is the indicator function of $[0,1]$, so the Haar multiresolution approximation is not regular. It is characterized by the function $H(\omega) = e^{-i\omega/2}\cos(\omega/2)$. With some other choice of $H(\omega)$, we can build wavelets which are much more regular than the Haar wavelet.

The smoothness and the asymptotic decay rate of a wavelet $\psi(x)$ defined by (63) and (69) is controlleld by the behavior of $H(\omega)$. The asymptotic decay rate of $\psi(x)$ is estimated by observing that if $H(\omega) \in \mathbf{C}^q$, then $\hat{\psi}(\omega) \in \mathbf{C}^q$ and

(71) $$\frac{d^n H(0)}{d\omega^n} = 0 \text{ for } 1 \leq n \leq q \Leftrightarrow \begin{cases} \dfrac{d^n G(0)}{d\omega^n} = 0 \text{ for } 0 \leq n \leq q-1, \\ \dfrac{d^n \hat{\psi}(0)}{d\omega^n} = 0 \text{ for } 0 \leq n \leq q-1. \end{cases}$$

We can also obtain a lower bound of the asymptotic decay rate of $\hat{\psi}(\omega)$ from the lower bound (59) on the decay rate of $\hat{\phi}(\omega)$:

(72) $$|\hat{\psi}(\omega)| = O(|\omega|^{-q+\mathrm{Log}(A)/\mathrm{Log}(2)}) \text{ at infinity}.$$

Outside the Haar basis, the first classes of wavelet orthonormal bases were found independently by Y. Meyer [14] and J. Stromberg [16]. Y. Meyer's bases are given by the class of functions $H(\omega)$ satisfying the hypothesis of Theorem 2, equal to 1 on $[-\pi/3, \pi/3]$ and continuously differentiable at any order. The Fourier transform of Y. Meyer's wavelets are in $\mathbf{C}^\infty$, so $\psi(x)$ has a decay faster than any power. We can also easily derive that $\hat{\psi}(\omega)$ has a support contained in $[-8\pi/3, 8\pi/3]$ so $\psi(x)$ is in $\mathbf{C}^\infty$.

By using a multiresolution approach, I. Daubechies [3] has recently proved that for any $n \geq 1$, we could find some wavelets $\psi(x) \in \mathbf{C}^n$ having a compact support. Indeed, she showed that we can find trigonometrical polynomials

$$H(\omega) = \sum_{k=-N}^{N} h_k e^{-ik\omega}$$

such that the constant $q - \mathrm{Log}(A)/\mathrm{Log}(2)$ of equation (72) is as large as desired. The corresponding wavelet $\psi(x)$ has a support contained in $[-2N-1, 2N+1]$ and its differentiability is estimated from (72).

If $\psi(x)$ is regular enough, Y. Meyer and P. Lemarie showed that wavelet orthonormal bases provide unconditional bases for most usual functional spaces [11]. We can thus find whether a function $f(x)$ is inside $\mathbf{L}^p(\mathbf{R})$ $(1 < p < \infty)$, a Sobolev space or a Hardy space from its decomposition coefficients in the wavelet basis. As an example, one can prove that if a wavelet $\psi(x)$ satisfies

$$\exists C \geq 0, \quad |\psi(x)| < C(1+|x|)^{-1-q}, \tag{73}$$

$$\int_{-\infty}^{+\infty} x^n \psi(x)\, dx = 0 \quad \text{for } 1 \leq n \leq q, \tag{74}$$

$$\exists C' \geq 0, \quad \left|\frac{d^n \psi(x)}{dx^n}\right| < C'(1+|x|)^{1-q} \quad \text{for } n \leq q, \tag{75}$$

then for any $s \leq q$, the family of functions $(\sqrt{2^j}\psi(2^j x - k))_{(j,k)\in\mathbf{Z}^2}$ is an unconditional basis of the Sobolev space $\mathbf{H}^s$. As a consequence, for any $f(x) \in \mathbf{L}^2(\mathbf{R})$, if $\alpha(k, j) = \langle f(x), \sqrt{2^j}\psi(2^j x - k)\rangle$ then

$$f \in \mathbf{H}^s \Leftrightarrow \left(\sum_{j=-\infty}^{+\infty} \left\{\left(\sum_{k=-\infty}^{+\infty} |\alpha(k, j)|^2\right)^{1/2} 2^{js}\right\}^2\right)^{1/2} < +\infty. \tag{76}$$

Remarks. 1. The couples of functions $H(\omega)$ and $G(\omega)$ which satisfy (69) were first studied in signal processing for multiplexing and demultiplexing a signal on a transmission line [5, 15]. Let $A = (\alpha_n)_{n\in\mathbf{Z}}$ be a discrete time sequence and $a(\omega)$ the corresponding Fourier series. The goal is to decompose A in two sequences B and C each having half as many samples per time unit and such that B and C contain respectively the low and the high frequency components of A. Equations (67) enable us to achieve such a decomposition where $b(\omega)$ and $c(\omega)$ are respectively the Fourier series of B and C. In signal processing, $H(\omega)$ and $G(\omega)$ are interpreted respectively as the transfer functions of a discrete low-pass filter H and a discrete high-pass filter G. They are called *quadrature mirror* filters. The sequences B and C are respectively computed by convolving A with the filters H and G and keeping one element out of two of the resulting sequences.

2. If a function is characterized by N samples uniformly distributed, its decomposition in a wavelet orthonormal basis can be computed with an algorithm of complexity $O(N)$. This algorithm is based on discrete convolutions with the quadrature mirror filters H and G [12].

3. We proved that we can derive a wavelet orthonormal basis from any multiresolution approximation. It is however not true that we can build a multiresolution approximation from any wavelet orthonormal basis. The function $\psi(x)$ whose Fourier transform is given by

$$\hat{\psi}(\omega) = \begin{cases} 1 & \text{if } 4\pi/7 \leq |\omega| \leq \pi \text{ or } 4\pi \leq |\omega| \leq 4\pi + 4\pi/7, \\ 0 & \text{otherwise}, \end{cases} \tag{77}$$

is a counterexample due to Y. Meyer. The translates and dilates

$$(\sqrt{2^j}\,\psi(2^j x - k))_{(k,j)\in\mathbf{Z}^2}$$

of this function constitute an orthonormal basis of $\mathbf{L}^2(\mathbf{R})$. Let $\mathbf{V}_J$ be the vector space generated by the family of functions

$$(\sqrt{2^j}\,\psi(2^j x - k))_{k\in\mathbf{Z},-\infty<j<J}.$$

One can verify that the sequence of vector spaces $(\mathbf{V}_J)_{J\in\mathbf{Z}}$ does not satisfy property (5) of a multiresolution approximation. Hence, this wavelet is not related to a multiresolution approximation. It might however be sufficient to impose a regularity condition on $\psi(x)$ in order to always generate a multiresolution approximation.

4. Multiresolution approximations have been extended by S. Jaffard and Y. Meyer to $\mathbf{L}^2(\Omega)$ where Ω is an open set of $\mathbf{R}^n$ [13]. This enables us to build wavelet orthonormal bases in $\mathbf{L}^2(\Omega)$.

5. Approximation error

When approximating a function at the resolution 2^j, the error is given by $\varepsilon_j = \|f - \mathbf{P}_{\mathbf{V}_j}(f)\|$. Property (2) of a multiresolution approximation implies that $\lim_{j\to+\infty} \varepsilon_j = 0$. A classical problem in approximation theory is to estimate the convergence rate of ε_j given an a priori knowledge on the smoothness of $f(x)$ or derive the smoothness of $f(x)$ from the convergence rate of ε_j [4].

Theorem 3. *Let $(\mathbf{V}_j)_{j\in\mathbf{Z}}$ be a multiresolution approximation such that the associated function $\phi(x)$ satisfies*

$$\exists C \geq 0, \quad |\phi(x)| < C(1+|x|)^{-3-q}, \tag{78}$$

$$\int_{-\infty}^{+\infty} x^n \phi(x)\,dx = 0 \quad \textit{for } 1 \leq n \leq q+1, \tag{79}$$

$$\exists C' \geq 0, \quad \left|\frac{d^n\phi(x)}{dx^n}\right| < C'(1+|x|)^{-1-q} \quad \textit{for } n \leq q. \tag{80}$$

Let $\varepsilon_j = \|f - \mathbf{P}_{\mathbf{V}_j}(f)\|$. For all $f \in \mathbf{L}^2(\mathbf{R})$, if $0 < s \leq q$ then

$$f(x) \in \mathbf{H}^s \Leftrightarrow \sum_{j=-\infty}^{+\infty} \varepsilon_j^2 2^{2sj} < +\infty. \tag{81}$$

Proof. Let $\mathbf{P}_{\mathbf{O}_j}$ denote the orthogonal projection on the vector space $\mathbf{O}_j$ and let $\psi(x)$ be the wavelet defined by (63) and (69). Let $f(x)$ be in $\mathbf{L}^2(\mathbf{R})$ and let $\alpha(k,j) = \langle f(x), \sqrt{2^j}\psi(2^jx-k)\rangle$. The approximation error is given by

$$(82)\quad \varepsilon_J = \|f - \mathbf{P}_{\mathbf{V}_J}(f)\| = \left(\sum_{j=J}^{\infty} \|\mathbf{P}_{\mathbf{O}_j}(f)\|^2\right)^{1/2} = \left(\sum_{j=J}^{+\infty}\sum_{k=-\infty}^{+\infty} |\alpha(k,j)|^2\right)^{1/2}.$$

In order to prove the theorem we show that if the function $\phi(x)$ satisfies conditions (78), (79) and (80) then $\psi(x)$ satisfies conditions (73), (74) and (75). We then apply property (76) to finish the proof.

It follows from (78) that the function $H(\omega) = \sum_{k=-\infty}^{+\infty} h_k e^{-ik\omega}$ satisfies

$$(83)\qquad h_k = O((1+|k|)^{-3-q}).$$

The function $G(\omega)$ defined in (69) can be written

$$G(\omega) = \sum_{k=-\infty}^{\infty} g_k e^{-ik\omega}, \quad \text{with } g_k = h_{-k-1}(-1)^{-k-1}.$$

The wavelet $\psi(x)$ is thus given by

$$\frac{1}{2}\psi\left(\frac{x}{2}\right) = \sum_{k=-\infty}^{+\infty} h_{-k-1}(-1)^{-k-1}\phi(x-k).$$

With the above expression and equations (78), (80) and (83) we can derive that

$$|\psi(x)| = O((1+|x|)^{-1-q}) \quad \text{and} \quad \left|\frac{d^n\psi(x)}{dx^n}\right| = O((1+|x|)^{-1-q}) \quad \text{for } n \le q.$$

Equations (78) and (79) imply that $\hat{\phi}(\omega) \in \mathbf{C}^{q+1}$ and

$$\frac{d^n\hat{\phi}(0)}{d\omega^n} = 0 \quad \text{for } 1 \le n \le q+1.$$

From (56) and (71) it thus follows that $d^n\hat{\psi}(0)/d\omega^n = 0$ for $0 \le n \le q$ and therefore

$$\int_{-\infty}^{+\infty} x^n\psi(x)\,dx = 0 \quad \text{for } 0 \le n \le q.$$

We can now finish the proof of this theorem by applying property (76). Let $\beta_j = \sum_{k=-\infty}^{+\infty} |\alpha(k,j)|^2$; equation (82) yields

$$\sum_{j=J}^{+\infty} \beta_j = \varepsilon_J^2.$$

This implies that $\beta_J = \varepsilon_J^2 - \varepsilon_{J+1}^2$. The right-hand side of property (76) is therefore given by

$$\sum_{j=-\infty}^{+\infty} \beta_j 2^{2js} = (1-2^{-2s}) \sum_{j=-\infty}^{+\infty} \varepsilon_J^2 2^{2sj}.$$

The right-hand side statement of property (76) is thus equivalent to the right-hand side statement of (81). This concludes the proof of Theorem 3.

Acknowledgments. I deeply thank Yves Meyer who helped me all along this research. I am also very grateful to Ruzena Bajcsy for her support and encouragement. I finally would like to thank Ingrid Daubechies for all her comments.

References

1. G. Battle, *A block spin construction of ondelettes, Part* 1: *Lemarie functions*, Comm. Math. Phys. **110** (1987), 601–615.
2. A. Cohen, *Analyse multiresolutions et filtres miroirs en quadrature*, Preprint, CEREMADE, Université Paris Dauphine, France.
3. I. Daubechies, *Orthonormal bases of compactly supported wavelets*, Comm. Pure Appl. Math. (to appear).
4. R. DeVore, *The approximation of continuous functions by positive linear operators*, Lecture Notes in Math., vol. 293, Springer-Verlag, 1972.
5. D. Esteban and C. Galand, *Applications of quadrature mirror filters to split band voice coding schemes*, Proc. Internat. Conf. Acoustic Speech and Signal Proc., May 1977.
6. P. Federbush, *Quantum field theory in ninety minutes*, Bull. Amer. Math. Soc. **17** (1987), 93–103.
7. A. Grossmann and J. Morlet, *Decomposition of Hardy functions into square integrable wavelets of constant shape*, SIAM J. Math. **15** (1984), 723–736.
8. S. Jaffard and Y. Meyer, *Bases d'ondelettes dans des ouverts de Rn*, J. Math. Pures Appl. (1987).
9. R. Kronland-Martinet, J. Morlet and A. Grossmann, *Analysis of sound patterns through wavelet transform*, Internat. J. Pattern Recognition and Artificial Intelligence (1988).
10. P. G. Lemarie, *Ondelettes a localisation exponentielles*, J. Math. Pures Appl. (to appear).
11. P. G. Lemarie and Y. Meyer, *Ondelettes et bases Hilbertiennes*, Rev. Mat. Ibero-Amer. **2** (1986).
12. S. Mallat, *A theory for multiresolution signal decomposition: the wavelet representation* (Tech. Rep. MS-CIS-87-22, Univ. of Pennsylvania, 1987), IEEE Trans. Pattern Analysis and Machine Intelligence, July 1989.
13. Y. Meyer, *Ondelletes et fonctions splines*, Seminaire Equations aux Derivees Partielles, Ecole Polytechnique, Paris, France, 1986.
14. ——, *Principe d'incertitude, bases hilbertiennes et algebres d'operateurs*, Bourbaki Seminar, 1985–86, no. 662.
15. M. J. Smith and T. P. Barnwell, *Exact reconstruction techniques for tree-structured subband coders*, IEEE Trans. Acoust. Speech Signal Process **34** (1986).
16. J. Stromberg, *A modified Franklin system and higher-order systems of R^n as unconditional bases for Hardy spaces*, Conf. in Harmonic Analysis in honor of A. Zygmund, Wadsworth Math. Series, vol. 2, Wadsworth, Belmont, Calif., pp. 475–493.
17. P. Tchamitchian, *Biorthogonalite et theorie des operateurs*, Rev. Mat. Ibero-Amer. **2** (1986).
18. ——, *Calcul symbolique sur les operateurs de Calderon-Zygmund et bases inconditionnelles de L2*, C.R. Acad. Sci. Paris Sér. I Math. **303** (1986), 215–218.

Courant Institute of Mathematical Sciences, New York University, New York, New York 10012

Wavelets, Multiresolution Analysis, and Quadrature Mirror Filters

A. Cohen
Translated by Robert D. Ryan.

ABSTRACT. In this paper, we give a complete characterization of the "quadrature mirror filters" associated with a multiresolution analysis. As an application, new types of wavelets belonging to the Schwartz class are constructed.

1. Introduction

The purpose of this article is to elucidate the mathematical relations between the theory of wavelets and the theory of quadrature mirror filters (QMF).

Since the work of J. O. Stromberg (1981), we have known how to construct orthonormal bases for $L^2(\mathbb{R})$ of the form $\{\psi(2^j x - k)\}_{j\in\mathbb{Z},k\in\mathbb{Z}}$. This discovery was foreshadowed by the Haar system (1909) and later by Franklin's system (1927).

Several years later, Y. Meyer ([1], [2], [3]) discovered a basis of the type $\{\psi(2^j x - k)\}_{j\in\mathbb{Z},k\in\mathbb{Z}}$ where the wavelet ψ belongs to the Schwartz class. He then introduced with S. Mallat ([4], [5]) the algebraic framework for multiresolution analyses, which proved to be particularly fruitful for the construction of wavelet bases. It was in this context that S. Mallat noticed the presence of QMFs, which appear as the natural tools for passing from one resolution scale to the next.

QMFs were introduced by Esteban and Galand [10] in 1977 for coding speech. They were later perfected by Smith, Barnwell, and Adelson ([11], [12]), who gave them the exact form of the filters that appear in multiresolution analyses.

Since then, two questions have arisen: Which QMFs are associated with multiresolution analyses (and thus with wavelets), and can one analyze, using the these QMFs, the characteristics of the wavelets that come from the multiresolution analyses based on these filters? I. Daubechies and S. Mallat made the first contributions to the solution of these problems ([5], [6]).

In this article, after some preliminaries, Theorem 2 provides a complete answer to the first question. Some examples are then presented to illustrate the meaning of this result. The second question, which concerns the classification of QMFs according to the regularity of the corresponding wavelets, is broached in the Appendix, but here there are still many open problems.

Translator's Note: The references to the article have been updated. A few typos were corrected with the author's concurrence.

2. Review: Multiresolution Analyses

We begin by recalling the definition of a multiresolution analysis as introduced by Y. Meyer ([1], [2], [3]) and S. Mallat ([4], [5]).

We work in the space $L^2(\mathbb{R}^n)$ of square-summable functions endowed with its natural Hilbert space structure. If $f \in L^2(\mathbb{R}^n)$, we will denote its Fourier transform by $\hat{f}$ and its norm by $\|f\|$. We denote the canonical basis of $\mathbb{R}^n$ by $\{e_i\}_{i=1}^n$; $\{f_j\}_{j=0}^{2^n-1}$ denotes the family of all expressions $\sum_{i=1}^n a_i e_i$, where $a_i \in \{0,1\}$ and $f_0 = 0$. If x and y are in $\mathbb{R}^n$, then xy denotes their scalar product. Finally, x_i is the ith coordinate of x.

A multiresolution analysis is a sequence $\{V_j\}_{j\in\mathbb{Z}}$ of closed vector subspaces of $L^2(\mathbb{R}^n)$ that satisfy the following properties:

(P1) $V_j \subset V_{j+1}$.

(P2) $f(x) \in V_j \Longleftrightarrow f(2x) \in V_{j+1}$.

(P3) $\bigcap_{j\in\mathbb{Z}} V_j = \{0\}$.

(P4) $\bigcup_{j\in\mathbb{Z}} V_j$ is dense in $L^2(\mathbb{R}^n)$.

(P5) There exists a function $g(x)$ in V_0 such that $\{g(x-k)\}_{k\in\mathbb{Z}^n}$ is a Riesz basis for V_0.

Recall that a family $\{e_j\}$ is a Riesz basis for a Hilbert space H if and only if it satisfies the following properties:

(a) The finite linear combinations $\sum \beta_i e_i$ are dense in H.

(b) There are two strictly positive constants C_1 and C_2 such that for all finite sequences $\{\beta_i\}$ one has

$$C_1 \left(\sum |\beta_i|^2\right)^{1/2} \leq \left\|\sum \beta_i e_i\right\| \leq C_2 \left(\sum |\beta_i|^2\right)^{1/2}.$$

We see then that a consequence of (P5) is that

$$f(x) \in V_0 \Longleftrightarrow f(x-k) \in V_0 \text{ for all } k \in \mathbb{Z}^n.$$

Given these conditions, we know how to construct a function φ such that the family $\{\varphi(x-k)\}_{k\in\mathbb{Z}^n}$ is an orthonormal basis for V_0. The function φ is defined in terms of its Fourier transform:

$$\hat{\varphi}(x) = \hat{g}(x) \Big/ \sqrt{\sum_{k\in\mathbb{Z}^n} |\hat{g}(x+2k\pi)|^2}.$$

We see that the condition $\sum_{k\in\mathbb{Z}^n} |\hat{\varphi}(x+2k\pi)|^2 = 1$ is equivalent to

$$\int_{\mathbb{R}^n} |\hat{\varphi}(x)|^2 e^{ikx} dx = (2\pi)^n \delta_{0,k},$$

which means that the family $\{\varphi(x-k)\}_{k\in\mathbb{Z}^n}$ is an orthonormal system.

We are now going to introduce the function m_0, which plays a crucial role in the construction of multiresolution analyses and wavelet bases. For this, we make the following observation: The function $\varphi(x/2)$ is in V_{-1} and thus in V_0. Hence, it can be written in terms of the $\varphi(x-k)$. We have $\varphi(x/2) = \sum \beta_k \varphi(x-k)$, which by taking Fourier transforms is $\hat{\varphi}(2x) = m_0(x)\hat{\varphi}(x)$. The function m_0 belongs to $L^2([0,2\pi]^n)$ and is $2\pi\mathbb{Z}^n$-periodic, which means that $m_0(x) = m_0(x+2\pi e_i)$ for all i between 1 and n. On the other hand, the condition $\sum_{k\in\mathbb{Z}^n} |\hat{\varphi}(x+2k\pi)|^2 = 1$ implies that $\sum_{j=0}^{2^n-1} |m_0(x+\pi f_j)|^2 = 1$, which is the QMF property generalized to the multidimensional case ([11], [12]).

The wavelets associated with our multiresolution analysis are then defined by relations of the type $\hat{\psi}_p(2x) = m_p(x)\hat{\varphi}(x)$. The index p runs between 1 and 2^n-1, and the functions m_p are such that the matrix $(a_{ij}) = (m_i(x+\pi f_j))$ is unitary for all x. A general solution for this problem was given by K. Gröchenig [9].

The family $\{2^{nj/2}\psi_p(2^j x - k)\}_{j\in\mathbb{Z},k\in\mathbb{Z}^n}$, p running between 1 and 2^n-1, is then an orthonormal basis for $L^2(\mathbb{R}^n)$. This approach for constructing wavelets from a multiresolution analysis was introduced by Y. Meyer ([1], [2], [3]).

From now on, we will be concerned with a specific class of multiresolution analyses: a multiresolution analysis is said to be regular if and only if, for all m in $\mathbb{N}$, $(1+|x|)^m\varphi(x) \in L^2(\mathbb{R}^n)$. This is equivalent to assuming that $\hat{\varphi}(x) \in H^m(\mathbb{R}^n)$ for all m in $\mathbb{N}$, where $H^m(\mathbb{R}^n)$ denotes the usual Sobolev space. No regularity condition is imposed on φ.[1]

We can estimate the tails of the integrals of $|\varphi(x)|^2$ by

$$\int_{|x|>A} |\varphi(x)|^2\,dx \le (1+A)^{-m}\int (1+|x|)^m |\varphi(x)|^2\,dx \quad (A>0),$$

so $\int_{|x|>A} |\varphi(x)|^2\,dx \le C_m(1+A)^{-m}$ for all m in $\mathbb{N}$.

We can now justify the term "regular analysis." Using Schwarz's inequality, we have the following estimate:

$$\left|\int \varphi(x/2)\overline{\varphi}(x-k)\,dx\right| \le C_1\left(\int_{|x|>k/2} |\varphi(x/2)|^2\,dx\right)^{1/2} + C_2\left(\int_{|x|<k/2} |\varphi(x-k)|^2\,dx\right)^{1/2},$$

which implies that

$$\left|\int \varphi(x/2)\overline{\varphi}(x-k)\,dx\right| \le C_m(1+|k|)^{-m}$$

for all m in $\mathbb{N}$.

The Fourier coefficients of m_0 are rapidly decreasing, and thus m_0 is regular. We will see later that this condition is sufficient to ensure that $(1+|x|)^m\varphi(x) \in L^2(\mathbb{R}^n)$ for all m in $\mathbb{N}$.

3. Aspects of the Problem

As an initial problem, we can ask the following question: Given a function φ such that $\hat{\varphi}(x) \in H^m(\mathbb{R}^n)$ for all m in $\mathbb{N}$ and such that the sequence $\{\varphi(x-k)\}_{k\in\mathbb{Z}^n}$ is orthonormal, is

[1]Translator's note: In the version of the present work published later in the book "Wavelets and Multiscale Signal Processing" (Chapman and Hall, 2005), Cohen used the term "localized analysis" rather than "analysis with a regular filter" (analyse à filtre régulier). Here, for brevity, I will use the term "regular analysis." This should not be confused with the term "r-regular analysis" used by Meyer [1].

it possible to create a multiresolution analysis by letting V_j be the Hilbert space generated by the system $\{2^{nj/2}\varphi(2^j x - k)\}_{k\in\mathbb{Z}^n}$?

Properties (P2) and (P5) are trivially satisfied, but what about the other three? The following theorem clarifies this point.

Theorem 1 *Let φ be such that $\hat{\varphi}(x) \in H^m(\mathbb{R}^n)$ for all m in $\mathbb{N}$ and suppose that $\{\varphi(x-k)\}_{k\in\mathbb{Z}^n}$ is an orthonormal sequence. Let V_j be the Hilbert space generated by the system $\{2^{nj/2}\varphi(2^j x - k)\}_{k\in\mathbb{Z}^n}$. Then*

- (*a*) (P1) $\Leftrightarrow$ *There exists a $2\pi\mathbb{Z}^n$-periodic function m_0 in $L^2([0,2\pi]^n)$ such that $\hat{\varphi}(2x) = m_0(x)\hat{\varphi}(x)$;*
- (*b*) (P3) *is always true;*
- (*c*) (P4) $\Leftrightarrow$ $|\hat{\varphi}(0)| = |\int_{\mathbb{R}^n} \varphi(x)\,dx| = 1.$

Proof. (a) It is clear that the existence of m_0 satisfying $\hat{\varphi}(2x) = m_0(x)\hat{\varphi}(x)$ is equivalent to $\varphi(x/2) \in V_0$ and hence equivalent to $V_{-1} \subset V_0$. But by the definition of V_j in the hypothesis, this means that $V_j \subset V_{j+1}$ for all j.

(b) Let T_j be the projection operator that maps $L^2(\mathbb{R}^n)$ onto V_j. We wish to show that, for all f in $L^2(\mathbb{R}^n)$, $\|T_j(f)\|_{L^2(\mathbb{R}^n)}$ converges to 0 as j tends to $-\infty$. We note that an element f of $L^2(\mathbb{R}^n)$ can be written $f = g + h$, where $\|h\|$ is arbitrarily small and g is a finite linear combination of the characteristic functions of bounded hypercubes. (We know that these functions are dense in $L^2(\mathbb{R}^n)$.) Since T_j is uniformly bounded ($\|T_j\| = 1$), it is thus sufficient to show that $\|T_j(\mathbf{1}_{[a,b]^n})\| \to 0$ for all $(a,b) \in \mathbb{R}$. We write $A = [a,b]^n$. We have

$$\begin{aligned}
\|T_j(\mathbf{1}_A)\|^2 &= \sum_{k\in\mathbb{Z}^n} |\langle \mathbf{1}_A, 2^{nj/2}\varphi(2^j x - k)\rangle|^2 \\
&= 2^{nj} \sum_{k\in\mathbb{Z}^n} \left| \int_A \varphi(2^j x - k)\,dx \right|^2 \\
&= 2^{-nj} \sum_{k\in\mathbb{Z}^n} \left| \int_{2^j A} \varphi(x - k)\,dx \right|^2 \\
&\le (b-a)^n \sum_{k\in\mathbb{Z}^n} \int_{2^j A} |\varphi(x-k)|^2\,dx.
\end{aligned}$$

We may assume that $|2^j a|$ and $|2^j b|$ are less than 1. The bounds on the tails of the integral of $|\varphi(x)|^2$ allow us to write

$$\|T_j(\mathbf{1}_{[a,b]^n})\|^2 \le (b-a)^n \sum_{|k|<K} \int_{2^j A} |\varphi(x-k)|^2\,dx + \varepsilon(K),$$

where $\varepsilon(K) \to 0$ as $K \to +\infty$. This clearly implies that $\|T_j(\mathbf{1}_{[a,b]^n})\|$ tends to 0 and proves the result.

(c) First assume (P4), or equivalently, that $\|T_j(f) - f\| \to 0$ for all f. Take $f = \mathbf{1}_{[-1,1]^n} = \mathbf{1}_B$. We then have

$$T_j(f)(x) = 2^{nj} \int_B \sum_{k \in \mathbb{Z}^n} \varphi(2^j x - k)\overline{\varphi}(2^j y - k)\, dy$$
$$= \int_{2^j B} \sum_{k \in \mathbb{Z}^n} \varphi(2^j x - k)\overline{\varphi}(y - k)\, dy.$$

This quantity tends to 1 in $L^2([-1,1]^n)$, thus a fortiori in $L^2([-r,r]^n)$ for $0 < r < 1$. We are going to show that the expression

$$\int_{\mathbb{R}^n} \sum_{k \in \mathbb{Z}^n} \varphi(2^j x - k)\overline{\varphi}(y - k)\, dy$$

also tends to 1 in $L^2([-r,r]^n)$. Let $\|\cdot\|$ denote the norm in $L^2([-r,r]^n)$ and write

$$R_j = \left\| \sum_{k \in \mathbb{Z}^n} \int_{|y_i| > 2^j} \varphi(2^j x - k)\overline{\varphi}(y - k)\, dy \right\|$$
$$\leq \sum_{k \in \mathbb{Z}^n} \left\| \int_{|y_i| > 2^j} \varphi(2^j x - k)\overline{\varphi}(y - k)\, dy \right\|$$
$$\leq \sum_{k \in \mathbb{Z}^n} \left| \int_{|y_i| > 2^j} \varphi(y - k)\, dy \right| \|\varphi(2^j x - k)\|,$$

so

$$R_j \leq 2^{-nj/2} \sum_{k \in \mathbb{Z}^n} \left| \int_{|y_i| > 2^j} \varphi(y - k)\, dy \right| \left(\int_{|x_i| < r2^j} |\varphi(x - k)|^2\, dx \right)^{1/2}.$$

At this point, the following observation is useful. We have $(1 + |x|)^m \varphi \in L^1(\mathbb{R}^n)$ for all m (which is easily shown using Schwarz's inequality), and we can estimate the tails of the integral of $|\varphi|$ just as we did for $|\varphi|^2$.

We then separate the sum as follows: $2^{nj/2} R_j \leq A_j + B_j$, where

$$A_j = \sum_{|k_i| \leq (r+1)2^{j-1}} \left| \int_{|y_i| > 2^j} \varphi(y - k)\, dy \right| \quad \text{and}$$
$$B_j = \sum_{|k_i| > (r+1)2^{j-1}} \int_{\mathbb{R}^n} |\varphi(y)|\, dy \left(\int_{|x_i| < r2^j} |\varphi(x - k)|^2\, dx \right)^{1/2}.$$

In both cases, the points k are at least the distance $(1 - r)2^{j-1}$ from the domains of integration of $|\varphi(x - k)|$ and $|\varphi(x - k)|^2$. This allows us to show that A_j and B_j are bounded by $C_m 2^{-mj}$ for all m, which is sufficient to deduce that R_j tends to 0.

Thus, the sequence defined by $h_j(x) = \int_{\mathbb{R}^n} \sum_{k\in\mathbb{Z}^n} \varphi(2^j x - k)\overline{\varphi}(y-k)\,dy$ tends to 1 in $L^2([-r,r]^n)$. Notice then that $h_j(x) = h(2^j x)$, where

$$h(x) = \int_{\mathbb{R}^n} \sum_{k\in\mathbb{Z}^n} \varphi(x-k)\overline{\varphi}(y-k)\,dy,$$

and it is clear that h is $\mathbb{Z}^n$-periodic. This means that h_j can converge only if $h(x) = 1$. Integrating over $E = [0,1]^n$ shows that

$$1 = \int_E h(x)\,dx = \int_{\mathbb{R}^n} \varphi(x)\,dx \int_{\mathbb{R}^n} \overline{\varphi}(y)\,dy \quad \text{and so} \quad \left|\int_{\mathbb{R}^n} \varphi(y)\,dy\right|^2 = 1.$$

To prove the other implication, suppose that $\left|\int_{\mathbb{R}^n} \varphi(y)\,dy\right| = 1$. As before, it is sufficient to consider only functions of the form $\mathbf{1}_{[a,b]^n} = \mathbf{1}_A$ to prove that $\|T_j(f) - f\|$ tends to 0 as j tends to $+\infty$. Note that $\|T_j(f) - f\|^2 = \|f\|^2 - \|T_j(f)\|^2$ by the Pythagorean theorem. Thus we only need to show that $\|T_j(f)\|^2$ tends to $(b-a)^n$.

We have already established that $\|T_j(\mathbf{1}_A)\|^2 = 2^{-nj} \sum_{k\in\mathbb{Z}^n} \left|\int_{2^j A} \varphi(x-k)\,dx\right|^2$. We divide the sum into three parts: $\|T_j(\mathbf{1}_A)\|^2 = A_j + B_j + C_j$, with

$$\begin{aligned}
A_j &= 2^{-nj} \sum_{k\in Z_{ja}} \left|\int_{2^j A} \varphi(x-k)\,dx\right|^2, \\
Z_{ja} &= [2^j a + 2^{j/2}, 2^j b - 2^{j/2}]^n \cap \mathbb{Z}^n; \\
B_j &= 2^{-nj} \sum_{k\in Z_{jb}} \left|\int_{2^j A} \varphi(x-k)\,dx\right|^2, \\
Z_{jb} &= \mathbb{Z}^n \setminus [2^j a - 2^{j/2}, 2^j b + 2^{j/2}]^n; \\
C_j &= 2^{-nj} \sum_{k\in Z_{jc}} \left|\int_{2^j A} \varphi(x-k)\,dx\right|^2, \\
Z_{jc} &= \mathbb{Z}^n \setminus (Z_{ja} \cup Z_{jb}).
\end{aligned}$$

We have already shown in thc first part of the proof that B_j tends to 0 (the k are at least the distance $2^{j/2}$ from the domain of integration).

We can estimate C_j from the hypothesis $\left|\int_{\mathbb{R}^n} \varphi(y)\,dy\right|^2 = 1$:

$$C_j \le 2^{-nj}(\mathrm{card}(Z_{jc})) \le 2^{-nj}(n2^{(j+1)(n-1)}2^{j/2+1}) = n2^{-j/2+n}.$$

This shows us that C_j also tends to 0.

It remains to examine A_j. For any $\varepsilon > 0$, $\left|\left|\int_{2^j A} \varphi(y-k)\,dy\right|^2 - 1\right| < \varepsilon$ uniformly for k in Z_{ja} for all sufficiently large j, since k is inside the domain of integration by at least the distance $2^{j/2}$ from the boundary. Furthermore, it is clear that $2^{-nj}(\mathrm{card}(Z_{ja}))$ tends to $(b-a)^n$, and thus that $\|T_j(\mathbf{1}_{[b-a]^n})\|^2$ tends to $(b-a)^n$, which proves the theorem. □

We thus have $|\hat{\varphi}(0)| = 1$ in the case of a regular analysis. The function φ being defined up to a factor of modulus 1, we can take $\hat{\varphi}(0) = 1$. It follows that $m_0(0) = 1$.

By iterating the relation $\hat{\varphi}(x) = m_0(x/2)\hat{\varphi}(x/2)$, we obtain the equality

$$\hat{\varphi}(x) = \prod_{k=1}^{+\infty} m_0\left(\frac{x}{2^k}\right).$$

(The product converges since $m_0(0) = 1$ and m_0 is regular, and thus satisfies a Lipschitz condition, at 0.)

A second problem, raised by S. Mallat ([4],[5]), is the following:

Given a regular $2\pi\mathbb{Z}^n$-periodic function m_0 satisfying $\sum_{j=0}^{2^n-1} |m_0(x + f_j\pi)|^2 = 1$ and $m_0(0) = 1$, does the equality $\hat{\varphi}(x) = \prod_{k=1}^{+\infty} m_0(x/2^k)$ allow us to construct a multiresolution analysis? This problem is particularly attractive, since it reduces the rather mysterious notion of a multiresolution analysis to a single function having simple properties. Furthermore, in practice, it is the function m_0 that is used in wavelet decomposition algorithms. The answer, however, is negative, counterexamples having been given. Thus, the quadrature mirror filter must have additional properties. We will identify these properties under the assumption that the analysis is regular.

4. Which Filters to Choose?

We use the notation $P = [-\pi, \pi]^n$ and begin by introducing a useful definition.

Definition 1 *A compact set K is said to be congruent to $P = [-\pi, \pi]^n$ modulo 2π if and only if for almost all x in P, there exists a unique y in K such that $x - y \in 2\pi\mathbb{Z}^n$.*

An immediate property of K is then the fact that if f is in L^1_{loc} and is $2\pi\mathbb{Z}^n$-periodic, then $\int_K f = \int_P f$. We will be interested in functions m_0 that have the following properties:

(Q1) There exists a compact set K congruent to P, with $0 \in \text{int}(K)$, and such that for all j in $\mathbb{N}^*$, we have $m_0(x/2^j) \neq 0$ for all x in K. Notice that, since K is bounded and since $m_0(0) = 1$, it is sufficient to verify this property for a finite number of integers j. We assume in addition that m_0 satisfies the properties already mentioned:

(Q2) m_0 is regular and $2\pi\mathbb{Z}^n$-periodic.

(Q3) $m_0(0) = 1$.

(Q4) $\sum_{j=0}^{2^n-1} |m_0(x + f_j\pi)|^2 = 1$.

We can now state and prove the following result:

Theorem 2 *A function m_0 generates a regular multiresolution analysis if and only if it satisfies properties (Q1), (Q2), (Q3), and (Q4).*

Proof. We first assume that m_0 satisfies the four conditions. Then, for all m in $\mathbb{N}^*$, define

$$h_m(x) = \prod_{k=1}^{m} m_0\left(\frac{x}{2^k}\right)\mathbf{1}_{2^m K},$$

where K is the compact set in (Q1) and $\mathbf{1}_{2^m K}$ is the characteristic function of this set dilated by the factor 2^m. Since $0 \in \operatorname{int}(K)$, properties (Q2) and (Q3) clearly imply the pointwise convergence of h_m as m tends to $+\infty$. Denote this pointwise limit by h.

We now evaluate the quantity $\int_{\mathbb{R}^n} |h_m(x)|^2 e^{ikx}\, dx$ for all k in $\mathbb{Z}^n$.

$$\begin{aligned}
\int_{\mathbb{R}^n} |h_m(x)|^2 e^{ikx}\, dx &= \int_{2^m K} \prod_{j=1}^{m} \left| m_0\left(\frac{x}{2^j}\right)\right|^2 e^{ikx}\, dx \\
&= 2^{mn} \int_K \prod_{j=0}^{m-1} |m_0(2^j x)|^2 e^{i2^m kx}\, dx \\
&= 2^{mn} \int_P \prod_{j=0}^{m-1} |m_0(2^j x)|^2 e^{i2^m kx}\, dx \\
&= 2^{mn} \int_{P/2} \prod_{j=1}^{m-1} |m_0(2^j x)|^2 \left(\sum_{j=0}^{2^n-1} |m_0(x + f_j \pi)|^2 \right) e^{i2^m kx}\, dx \\
&= 2^{n(m-1)} \int_P \prod_{j=0}^{m-1} |m_0(2^j x)|^2 e^{i2^{m-1}kx}\, dx \\
&= \int_{\mathbb{R}^n} |h_{m-1}(x)|^2 e^{ikx}\, dx \\
&= \cdots = \int_P e^{ikx}\, dx = (2\pi)^n \delta_{0,k},
\end{aligned}$$

where $\delta_{0,k} = 1$ if $k = 0$ and 0 otherwise.

This tells us that for all m in $\mathbb{N}^*$, $\|h_m\|^2 = (2\pi)^n$. By Fatou's lemma, the pointwise limit h is also in $L^2(\mathbb{R}^n)$ and we have $\|h\|^2 \leq (2\pi)^n$.

On the other hand, (Q1) implies that h_m vanishes outside $2^m K$ and satisfies the equality $|h_m(x)|^2 = |h(x)|^2/|h(x/2^m)|^2$ on $2^m K$. Thus, the inequality

$$|h_m(x)|^2 \leq |h(x)|^2/C$$

is true everywhere, and this allows us to apply Lebesgue's dominated convergence theorem. Hence, we have

$$\lim_{m\to\infty} \|h - h_m\|_{L^2(\mathbb{R}^n)} = 0.$$

Now by defining $\hat{\varphi} = h$, it follows from the computation above that

$$\int_{\mathbb{R}^n} |\hat{\varphi}(x)|^2 e^{ikx}\, dx = (2\pi)^n \delta_{0,k}.$$

Thus, $\{\varphi(x-k)\}_{k\in\mathbb{Z}^n}$ is an orthonormal system.

We are going to show that $\hat{\varphi}$ is in H^m for all m in $\mathbb{N}$. Let J_m be a sequence of functions of a real variable. We assume that J_m is even and regular, is 1 on $[0, 2^m\pi]$, is zero outside $[0, 2^m\pi + 1]$, and is between 0 and 1 on $[2^m\pi, 2^m\pi + 1]$. Then the successive derivatives of J_m are bounded uniformly in m. We then define the function I_m by $I_m(x) = \prod_{i=1}^n J_m(x_i)$.

Let $g_m(x) = \prod_{k=1}^{m} m_0(x/2^k) \cdot I_m(x)$. It is clear that $(d/dx)^k(g_m)$ tends pointwise to $(d/dx)^k(\hat{\varphi})$ (here, k is a multi-index).

We are first going to estimate the L^2 norm of $(d/dx_i)(g_m)$. We have

$$\begin{aligned}
\left\|\frac{d}{dx_i}(g_m)\right\| &\leq \left\|\prod_{k=1}^{m} m_0\Big(\frac{x}{2^k}\Big)\frac{d}{dx}(I_m)\right\| \\
&\quad + \sum_{l=1}^{m}\left\|\sum_{k=1,k\neq l}^{m} m_0\Big(\frac{x}{2^k}\Big)I_m(x)\frac{d}{dx_i}\Big(m_0\Big(\frac{x}{2^i}\Big)\Big)\right\| \\
&\leq \sup\left(\frac{dI_m}{dx_i}\right)\left(\int_{2^{m+1}P}\prod_{k=1}^{m}\left|m_0\Big(\frac{x}{2^k}\Big)\right|^2 dx\right)^{1/2} \\
&\quad + \sum_{l=1}^{m}\sup\left(\frac{dm_0}{dx_i}\right)\left(\int_{2^{m+1}P}\prod_{k=1,k\neq l}^{m}\left|m_0\Big(\frac{x}{2^k}\Big)\right|^2 dx\right)^{1/2} \\
&\leq 2^n\sup\left(\frac{dI_m}{dx_i}\right)\left(\int_{2^{m}P}\prod_{k=1}^{m}\left|m_0\Big(\frac{x}{2^k}\Big)\right|^2 dx\right)^{1/2} \\
&\quad + 2^n\sum_{l=1}^{m}\sup\left(\frac{dm_0}{dx_i}\right)\left(\int_{2^{m}P}\prod_{k=1,k\neq l}^{m}\left|m_0\Big(\frac{x}{2^k}\Big)\right|^2 dx\right)^{1/2}
\end{aligned}$$

We are now going to evaluate the norm in $L^2(2^m P)$ of the truncated products $\prod_{k=1,k\neq l}^{m} m_0(x/2^k)$. We have

$$\begin{aligned}
\int_{2^m P}\prod_{k=0,k\neq l}^{m}\left|m_0\Big(\frac{x}{2^k}\Big)\right|^2 dx &= 2^{nm}\int_P \prod_{k=1,k\neq m-l}^{m-1}|m_0(2^k x)|^2\,dx \\
&= 2^{n(m-1)}\int_P \prod_{k=0,k\neq m-l-1}^{m-2}|m_0(2^k x)|^2\,dx \\
&= \cdots = 2^{nl}\int_P \prod_{k=1}^{l-1}|m_0(2^k x)|^2\,dx \\
&= 2^{n(l-1)}\int_{2P}\prod_{k=0}^{l-2}|m_0(2^k x)|^2\,dx \\
&= 2^{nl}\int_P \prod_{k=0}^{l-2}|m_0(2^k x)|^2\,dx \\
&= \cdots = (4\pi)^n.
\end{aligned}$$

(We have used the same techniques we used in the computation of $\int_{\mathbb{R}^n}|h_m(x)|^2\,dx$.)

By substituting this result in the inequality obtained before, it is clear that $\|d/dx_i(g_m)\|_{L^2(\mathbb{R}^n)}$ is bounded independently of m. It is clear that by further differentiation, one will obtain a similar result: for a multi-index k, $\|(d/dx)^k(g_m)\|_{L^2(\mathbb{R}^n)}$ is bounded by a constant that depends only on the sup norms of the successive derivatives of m_0 and those of I_m, which are independent of m.

Fatou's lemma guarantees that $(d/dx)^k(\hat{\varphi})$ is in $L^2(\mathbb{R}^n)$ for all multi-indices k in $\mathbb{N}^n$, in other words, $\hat{\varphi}$ is in H^m for all m.

By Theorem 1, this means that the multiresolution analysis generated by m_0 is regular. This proves the theorem in one direction.

To prove it in the other direction, assume that we are given a regular analysis. We have already seen that the function m_0 satisfies (Q2), (Q3), and (Q4), so it remains to show that m_0 satisfies (Q1). This proves to be a consequence of the facts that $\hat{\varphi} \in H^{n/2}$ and $\sum_{k\in\mathbb{Z}^n} |\hat{\varphi}(x + 2k\pi)|^2 = 1$.[2]

We are going to "localize" this equality uniformly in x, which means that we are going to establish the following inequality:

$$\sum_{|x+2k\pi|<A(\varepsilon)} |\hat{\varphi}(x + 2k\pi)|^2 > 1 - \varepsilon \quad \text{for all } \varepsilon > 0.$$

For this, we will use the first part of the following classic lemma.

Lemma 1 *Assume that $h \in H^m(\mathbb{R}^n)$ and that $g \in \mathcal{D}(\mathbb{R}^n)$. Then the sequence defined by $a_k = \|h(x)g(x - k)\|_{H^m(\mathbb{R}^n)}$ belongs to $l^2(\mathbb{Z}^n)$. Conversely, if $a_k \in l^2(\mathbb{Z}^n)$ for a function g in $\mathcal{D}(\mathbb{R}^n)$ such that $\sum_{k\in\mathbb{Z}^n} |g(x - k| \neq 0$ for all x in $\mathbb{R}^n$, then the function h belongs to $H^m(\mathbb{R}^n)$.*

For our purposes, we take g in $\mathcal{D}(\mathbb{R}^n)$ such that $g = 1$ on P. For all m in $\mathbb{N}$ and all $\varepsilon > 0$, there exists an $A(m, \varepsilon)$ such that

$$\sum_{|k|>A(m,\varepsilon)} (\|\hat{\varphi}(x)g(x - 2k\pi)\|_{H^m(\mathbb{R}^n)})^2 < \varepsilon.$$

Recall that the $H^{n/2}(\mathbb{R}^n)$ norm dominates, up to a constant, the $(\sup |f|^2)^{1/2}$ norm. Thus, letting $A(\varepsilon) = A(n/2, \varepsilon)$, we have

$$\sum_{|x+2k\pi|<A(\varepsilon)} |\hat{\varphi}(x + 2k\pi)|^2 > 1 - \varepsilon,$$

and this is uniform in x.

We can select A such that $\sum_{|x+2k\pi|<A(\varepsilon)} |\hat{\varphi}(x + 2k\pi)|^2 > 1/2$. We know that, for all x in $P = [-\pi, \pi]^n$, there exists $x + 2k_x\pi$ in $B(0, A)$ such that $|\hat{\varphi}(x + 2k_x\pi)| > C > 0$. The inequality $|\hat{\varphi}(x + 2k_x\pi)| > C > 0$ holds in a cubic neighborhood U_x of x. Since $\hat{\varphi}(0) = 1$, we can choose $k_0 = 0$. Now cover P with a finite family of cubes U_x that includes U_0.

[2] Author's and translator's note: The following is the original proof. In fact, there is a small gap here: It is necessary to show that $\sum_{k\in\mathbb{Z}^n} |\hat{\varphi}(x + 2k\pi)|^2 = 1$ everywhere, since the equations $\int_{\mathbb{R}^n} |\hat{\varphi}(x)|^2 e^{ikx} dx = (2\pi)^n \delta_{0,k}$ only imply that the condition holds almost everywhere. We refer to "Wavelets and Multiscale Signal Processing " by Cohen and Ryan, Chapman & Hall, 1995, for the correct argument.

We can extract from this covering U_i a partition of $[-\pi,\pi]^n$, up to a set of measure zero: It is sufficient to define by induction $R_0 = U_0$, $R_1 = U_1 \setminus R_0, \ldots, R_i = U_i \setminus \cup_{j=0}^{i-1} R_j$. The R_i are then finite unions of rectangles, and for all x in R_i we have $|\hat{\varphi}(x + 2k_x\pi)| > C$.

Property (Q1) is then established by defining the compact set K to be $K = \cup_i \operatorname{closure}(R_i + 2k_i\pi)$. Then $0 \in \operatorname{interior}(U_0) \subset \operatorname{interior}(K)$, and for all $n > 0$ and all x in K, $m_0(x/2^n) \geq C$. This proves the theorem. □

This result establishes a one-to-one correspondence between regular multiresolution analyses and functions that satisfy (Q1), (Q2), (Q3), and (Q4). We do not yet know, however, how to characterize the regularity of the function φ in terms of m_0. For example, this is not known in the case of r-regular analyses, where the function φ satisfies $|(d/dx)^k(\varphi)| < C_m(1+|x|)^{-m}$ for all m and for $0 \leq |k| \leq r$.

5. Applications

Condition (Q1) may appear difficult to use. We are going to give two examples that should shed some light on its significance.

5.1 A family of filters that does not generate a multiresolution analysis

The example in dimension 1. We already know that the functions $m_0(x) = \cos((2n+1)x/2)\, e^{-ix/2}$ for n in $\mathbb{N}^*$ do not work. In general, we will see that if $m_0((2k+1)\pi/(2n+1)) = 0$ for all k in $\mathbb{Z}$, where n is a fixed nonzero integer, then condition (Q1) is not satisfied.

Assume that it was satisfied. Then the corresponding set K would contain an element of the form

$$\frac{2\pi}{2n+1} + 2j\pi = \frac{2\pi((2n+1)j+1)}{2n+1} = \frac{m\pi}{2n+1} = x_m,$$

with $m = 2((2n+1)j+1)$; m is even and nonzero because $n \in \mathbb{N}^*$.

Thus there exists l in $\mathbb{N}^*$ and k in $\mathbb{Z}$ such that $m = 2^l(2k+1)$. In other words, $x_m/2^l = (2k+1)\pi/(2n+1)$, and hence $m_0(x_m/2^l) = 0$, which contradicts the hypothesis (Q1). We thus have a family of quadrature mirror filters that do not generate multiresolution analyses. A similar result exists for the multidimensional cases.

5.2 A family of filters that satisfies (Q1)

Assume that m_0 does not vanish on $P/3 = [-\pi/3, \pi/3]^n$. We will show that (Q1) is then satisfied.

Let Y be the compact set of zeros of m_0 in $(P/2) \setminus (P/3)$. Let y be an element of Y. By using condition (Q4), we can find f_y in the family $\{f_y\}$ such that $m_0(y + \varepsilon f_y \pi) > 2^{-n/2}$, for ε equal to 0 or 1. On the other hand, it is clear that we can choose ε_y in $\{0,1\}$ such that $y + \varepsilon_y f_y \pi$ is in $2P/3 = [-2\pi/3, 2\pi/3]^n$.

We can find a cubic neighborhood U_y of y in $P/2$ such that these two properties hold in all of U_y. Then, using the same techniques as in the proof of Theorem 2, we construct a finite partition R_j that covers Y in $P/2$. Then define K as follows:

$$\frac{K}{2} = \operatorname{closure}\left(\left(\frac{P}{2} \setminus (\cup_j(R_j))\right) \bigcup (\cup_j(R_j + \varepsilon_i f_i \pi))\right).$$

It is clear that K is congruent to P modulo 2π and that 0 is in the interior of K. By hypothesis, m_0 does not vanish on $K/2$ and thus $m_0(x/2)$ does not vanish on K. By construction, $K/2 \subset 2P/3$. Since m_0 does not vanish on $P/3$, we deduce that $m_0(x/2^n) \neq 0$ for all $n \geq 1$. Condition (Q1) is thus satisfied and hence these filters generate multiresolution analyses.

6. Quasi-progressive Wavelet Bases

We begin with the one-dimensional case. We are going to use Theorem 2 to construct an orthonormal basis of wavelets belonging to the Schwartz class $\mathcal{S}(\mathbb{R})$ and such the support of the function $\hat{\psi}$ is in the interval $[-\varepsilon, +\infty)$, where the real ε is strictly positive and arbitrarily small. This application is particularly interesting for the analysis of acoustic signals, where one seeks to isolate the progressive part of the signal.

J. Morlet, A. Grossman, and R. Kronland–Martinet ([6],[7]) used nonorthogonal wavelets $g((t-b)/a)$, $a > 0$, $b \in \mathbb{R}$, with $g(x) = e^{icx} \exp(x^2/2)$. When c is greater than 5, g acts like a progressive wavelet, except for a very small numerical error. Thus take ε such that $0 < \varepsilon < \pi$.

Here is how we define the function m_0: m_0 is regular and ranges between 0 and 1 (conditions (Q2) and (Q3)). The function m_0 is identically equal to 1 on the interval $[0, \pi - \varepsilon/2]$, strictly between 0 and 1 on $(\pi - \varepsilon/2, \pi)$ and zero at π; m_0 is defined on $[-\pi, 0]$ by condition (Q4) ($|m_0(x)|^2 + |m_0(x+\pi)|^2 = 1$); finally, m_0 is extended to all of $\mathbb{R}$ by 2π periodicity (Q2). It is easy to verify that condition (Q1) is also satisfied: We take K to be the interval $[-\varepsilon/2, 2\pi - \varepsilon/2]$, which works perfectly. We can thus generate a multiresolution analysis. It is clear that $\hat{\varphi}$ is zero on the interval $(-\infty, -\varepsilon]$ (m_0 is zero on $[-\varepsilon, -\varepsilon/2]$ since $\varepsilon < \pi$); thus, the function $\hat{\psi}$ will have the same property on $(-\infty, -2\varepsilon]$. In terms of signal theory, we can say that the nonprogressive energy of each wavelet is less than 2ε.

We are now going to prove the following result:

Proposition 1 *The functions φ and ψ belong to the Schwartz class.*

Proof. We will in fact show that $\hat{\varphi}$ is in the Schwartz class, which is equivalent to showing that φ is in the Schwartz class. We begin by proving the following lemma.

Lemma 2 *The function $\hat{\varphi}$ vanishes on all of the intervals $[2^n\pi, 2^{n+1}\pi - \varepsilon]$, where n is in $\mathbb{N}^*$.*

The proof is by induction on n. Since $m_0(x/2)$ is zero on $[2\pi, 4\pi - \varepsilon]$, $\hat{\varphi}$ is also zero there. Assume that $\hat{\varphi}$ is zero on the interval $[2^{n-1}\pi, 2^n\pi - \varepsilon]$, $n > 2$. Note that $m_0(x/2)$ is zero on $[2^{n+1}\pi - 2\pi, 2^{n+1}\pi - \varepsilon]$ and that $\hat{\varphi}(x/2)$ is zero on the interval $[2^n\pi, 2^{n+1}\pi - 2\varepsilon]$. Since $\varepsilon < \pi$, this implies that $\hat{\varphi}(x) = m_0(x/2)\hat{\varphi}(x/2)$ is zero on $[2^n\pi, 2^{n+1}\pi - \varepsilon]$. This proves the lemma.

On the other hand, it is clear that the function $\hat{\varphi}$ is not zero at the points $x_n = 2^n\pi - x$ if $0 < x < \varepsilon$. We are going to examine the decay of $\hat{\varphi}$ and its successive derivatives at these points. Recall that for all m in $\mathbb{N}$ there exist constants C_m and $C_{m,k}$ such that we have the following estimates: For all x in $\mathbb{R}$ and all n in $\mathbb{N}^*$,

$$\left| \left(\frac{d}{dx}\right)^m (\hat{\varphi})(x) \right| \leq C_m \sum_{k=0}^{m} \left| \left(\frac{d}{dx}\right)^k (m_0)\left(\frac{x}{2^n}\right) \right|; \qquad \text{(M1)}$$

For all x in $\mathbb{R}$ and all k in $\mathbb{N}$,

$$\left|\left(\frac{d}{dx}\right)^m (m_0)(\pi - x)\right| \leq C_{m,k}|x|^k. \tag{M2}$$

The estimate (M1) follows easily from Leibniz's formula, the fact that $|m_0|$ is less than or equal to 1, and that the derivatives of m_0 are bounded on $\mathbb{R}$. (M2) means that the zero at π is of infinite order.

If we apply (M1) and then (M2) at the points $x_n = 2^n\pi - x$, then we have the inequality

$$\left|\left(\frac{d}{dx}\right)^m (\hat{\varphi})(x_m)\right| \leq B_{m,k}\varepsilon^k 2^{-nk} \leq D_{m,k}|x_m|^{-k}.$$

Thus, the function $\hat{\varphi}$ belongs to the Schwartz class, and hence so do φ and ψ. □

We have thus used Theorem 2 to construct new wavelets that belong to the Schwartz class and that are progressive, up to an ε energy term.

7. The Multidimensional Case

If we wish to generalize the last result for analyses of $L^2(\mathbb{R}^n)$, the first, or "naive," method would be to choose the function m_0 as follows: In $P = [-\pi, \pi]^n$, m_0 is zero outside $[-\varepsilon, \pi]^n$, identically 1 in $[0, \pi - \varepsilon]^n$, and adjusted between these two cubes so that the quadrature relation (Q4) is satisfied. The function is then extended to be $2\pi\mathbb{Z}^n$-periodic. If we proceed this way, we construct a function φ whose Fourier transform is supported, up to ε, by the part of the space $\{x_i > 0\}$.

This produces nothing new, in the sense that this property is satisfied when we take the tensor product of the analysis constructed in the last section n times with itself. We then have $m_0(x) = \prod_{i=1}^n m_0(x_i)$.

We can, in fact, obtain a better result. We work in dimension 2 to simplify the construction, which will be similar in higher dimensions.

We take the function m_0 to be identically equal to 1 on a parallelogram

$$P_1 = \{(0,0), (\pi - \varepsilon, 0), (a(\pi - \varepsilon), \pi - \varepsilon), (a(\pi - \varepsilon) + \pi - \varepsilon, \pi - \varepsilon)\},$$

where $a > 0$ (if $a = 0$ we are in the previous case), and identically zero around

$$P_2 = \{(-(a+1)\varepsilon, -\varepsilon), (\pi - a\varepsilon, -\varepsilon), (a\pi - \varepsilon, \pi), ((a+1)\pi, \pi)\}.$$

We adjust $0 < m_0 < 1$ in the border $\operatorname{int}(P_2 \setminus P_1)$ so that condition (Q4) is satisfied, and then we extend m_0 to all of $\mathbb{R}^2$ periodically using the translation group $2\pi\mathbb{Z}^2$.

The function so defined satisfies condition (Q1): This follows from the facts that the parallelogram $Q = (P_1 + P_2)/2$ is congruent to P modulo 2π, that $m_0(x/2) \neq 0$ on Q, and that 0 is in the interior of Q. On the other hand, by having assumed at the outset that $\varepsilon < \pi/2$, we note that, up to an ε energy term, $\hat{\varphi}$ is zero outside the homogeneous cone defined by the x-axis and the half-line at the angle whose tangent is $1/a$ (see figures 1 and 2). The result is indeed the same in higher dimensions: We can select the directions in an arbitrarily small solid angle, the parameter a can be as large as we wish.

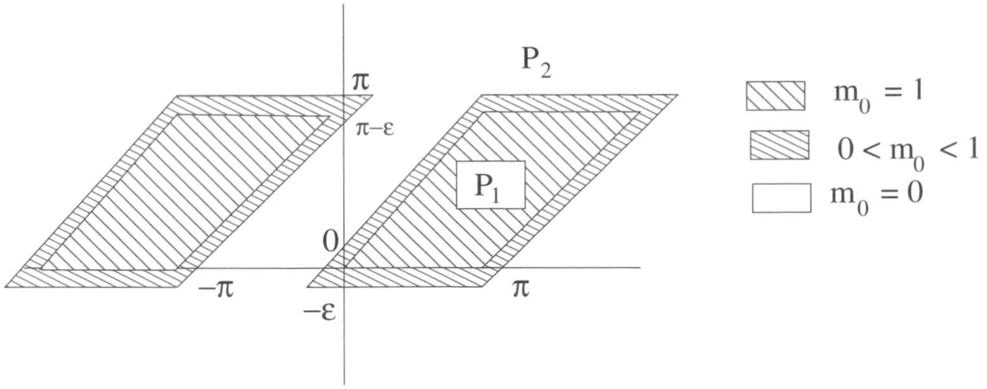

Figure 1. Appearance of the function m_0.

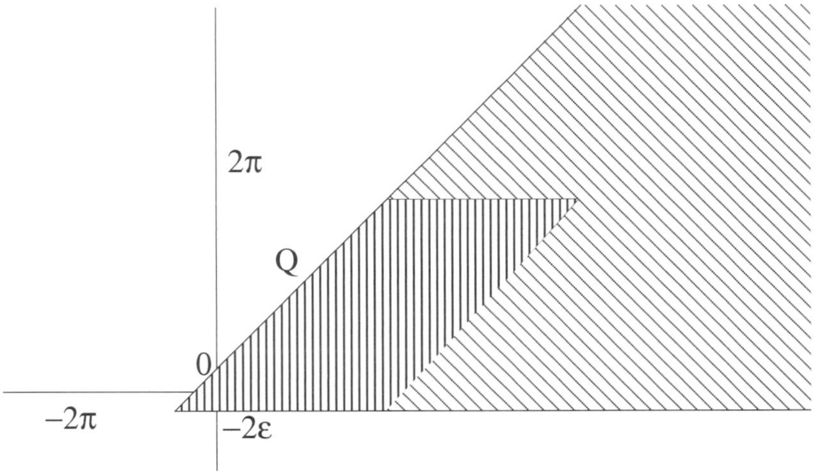

Figure 2. Support of the function $\hat{\varphi}(x) = \prod_{n=1}^{\infty} m_0(x/2^n)$; Q is congruent to P modulo 2π.

We conclude with the following remark: This conic geometry also appears if we consider a multiresolution analysis based on a translation lattice G of $\mathbb{R}^n$ other than $\mathbb{Z}^n$. All of the results presented in this article can be generalized to this framework.

Condition (Q1) is then expressed by replacing P with the fundamental unit of the dual lattice G^*, and the congruence modulo 2π with the congruence modulo G^* ($x \in G^* \Leftrightarrow$ for all y in G, $xy \in 2\pi\mathbb{Z}$). Since the lattice G^* is not necessarily cubic, we get the geometry of our construction using parallelograms, but the translation group is on longer $\mathbb{Z}^n$. Our multidimensional wavelets will be in the Schwartz class by arguments similar to those given in preceding sections.

8. Appendix: Regularity Conditions

We have seen a strong connection between the regularity of the function m_0 and the condition $(1 + |x|)^m \varphi \in L^2(\mathbb{R}^n)$ for all m in $\mathbb{N}$. We can also try to characterize the regularity of the function φ in terms of properties of the filter m_0. Currently, these relations seem less clear, and we will be content to "bracket" these properties with necessary conditions and other conditions that are sufficient for the filter m_0 to generate an analysis that is more or less regular. A more complete analysis is given in [13]. We will work in one dimension.

8.1 Sufficient conditions

In her construction of orthonormal wavelets with compact support, I. Daubechies [8] used the following result due to P. Thamitchian.

Suppose that $m_0(x) = ((1+e^{ix})/2)^n f(x)$ and $f(x) = \sum_{k\in\mathbb{Z}} f_k e^{ikx}$ is such that there exists an $\varepsilon > 0$ such that $\sum_{k\in\mathbb{Z}} |f_k||k|^\varepsilon < +\infty$.

If $B_m = \sup_{x\in\mathbb{R}} |f(x)f(x/2)\cdots f(x/2^{m+1})|$, then

$$|\hat{\varphi}(x)| = \left|\prod_{j=1}^{+\infty} m_0(2^{-j}x)\right| \leq C(1+|x|)^{-n+\log(B_m)/(m\log 2)}.$$

A consequence of this result is that, if for some m we have $B_m < 2^{m(n-1-k)}$, then, clearly, $|x|^k|\hat{\varphi}(x)|$ is in $L^1(\mathbb{R})$, and this gives us a sufficient condition to have $\varphi \in C^k$.

In the case of wavelets belonging to the Schwartz class, we cite the construction due to Y. Meyer [1], which gives us a sufficient condition on the function m_0. Here the filter satisfies the conditions (Q2),(Q3), and (Q4), and we assume further that it vanishes on $[2\pi/3, 4\pi/3]$. (Condition (Q1) is thus satisfied automatically.) Under these conditions, the support of $\hat{\varphi}$ is contained in $[-4\pi/3, 4\pi/3]$, and thus the wavelets belong to the Schwartz class.

The results of section 6 give us another example, plus a method that can be used in other possible cases to show that $\hat{\varphi}$ decreases rapidly.

8.2 Necessary conditions

We are going to see here that the conditions (Q1), (Q2), (Q3), and (Q4) are indeed insufficient to deduce the regularity of the function φ. We can try, for example, to generate r-regular analyses, which means that the function φ is in the class C^{r-1} and satisfies, for all m in $\mathbb{N}$ and for all k between 0 and r, the following property:

$$(1+|x|)^m \left|\left(\frac{d}{dx}\right)^k (\varphi)\right| < C_{m,k}.$$

This is the case, among others, for splines of order r. The next proposition provides a necessary condition for the realization of this program.

Proposition 2 *Let* $C = \frac{1}{2\pi}\int_0^{2\pi} \log(|m_0(x)|)\,dx$. *(*$C$ *can equal* $-\infty$*.) Then*

$$\left|\left(\frac{d}{dx}\right)^k (\varphi)\right| \in L^2(\mathbb{R}) \Longrightarrow C \leq -\left(k+\frac{1}{2}\right)\log 2.$$

Proof. Assume that $C > -(k+1/2)\log 2$, let $I = [2^{-M-1}\pi, 2^{-M}\pi]$, and let $J = I \cup (-I)$. M is assumed to be large enough so that $|\varphi(x)| > B > 0$ for all x in J. We have

$$\left\| \left(\frac{d}{dx}\right)^k (\varphi) \right\|^2 \geq \sum_{n>0} \int_{2^n J} |x|^{2k} |\hat{\varphi}|^2\, dx$$
$$\geq \sum_{n>0} 2^{(2k+1)n} \int_J |x|^{2k} |\hat{\varphi}(2^n x)|^2\, dx$$
$$\geq K \sum_{n>0} 2^{(2k+1)n} \int_J |\hat{\varphi}(2^n x)|^2\, dx.$$

Furthermore, since the transformation $x \mapsto 2x$ on the circle S_1 is ergodic, we know that, for almost every x in J, we have $\lim_{N\to+\infty}(\sum_{n=0}^{N-1} \log(|m_0(2^n x)|)/N) = C$. By Egorov's theorem, this limit is uniform on some subset A of J that has nonzero measure. We can thus find an N_0 in $\mathbb{N}$ such that for all x in A and for all N greater than N_0, we have

$$\sum_{n=0}^{N-1} \frac{\log(|m_0(2^n x)|)}{N} \geq -\left(k + \frac{1}{2}\right) \log 2,$$

which by taking the exponential is $\prod_{n=0}^{N-1} |m_0(2^n x)| \geq 2^{-N(k+1/2)}$. Consequently, it follows that

$$\int_J |\hat{\varphi}(2^N x)|^2\, dx \geq B|A| 2^{-N(2k+1)},$$

where $|A|$ denotes the measure of A.

Thus, the series $\sum_{n>0} \int_{2^n J} |x|^{2k} |\hat{\varphi}|^2\, dx$ is divergent, which tells us that $(d/dx)^k(\varphi)$ cannot belong to $L^2(\mathbb{R})$. □

This result implies that, if φ is in $S(\mathbb{R})$, then we necessarily have

$$\int_0^{2\pi} \log(|m_0(x)|)\, dx = -\infty.$$

Belonging to the Schwartz class implies other properties of the function m_0. In this case, m_0 has zeros at an infinite number of points of the form $2q\pi$, where q is a rational between 0 and 1, and these are zeros of infinite order (one can easily show that, if this were not the case, then φ would not decrease rapidly at values near $2^n q\pi$).

We have thus seen several approaches to the problem of deducing regularity properties of φ from properties of the filter m_0. These results lend hope to finding a complete characterization similar to the one we have for a regular analyses.

References

[1] Y. Meyer, *Wavelets and operators*, Cambridge University Press, Cambridge, 1992.

[2] Y. Meyer, *Ondelettes, fonctions splines et analyses graduées*, Rend. Sem. Mat. Univ. Politec. Torino, **45** (1987), 1–42.

[3] Y. Meyer, *Construction de bases orthonormées d'ondelettes*, Colloq. Math., **60/61** (1990), 141–149.

[4] S. Mallat, *A theory for multiresolution signal decomposition: the wavelet representation*, IEEE Trans. Pattern Anal. Machine Intell., **11** (1989), pp. 674–693.

[5] S. Mallat, *Multiresolution approximation and wavelets orthonormal bases of* $L^2(\mathbb{R})$, Trans. Amer. Math. Soc., **315** (1989), pp. 69–87.

[6] R. Kronland–Martinet, J. Morlet, and A. Grossman, *Analysis of sound patterns through wavelet transforms*, Int. J. Pattern Recognition Artificial Intelligence, special issue on expert systems and pattern analysis, vol. 1, no. 2 (1987), pp. 237–301.

[7] A. Grossman, R. Kronland–Martinet, and J. Morlet, *Reading and understanding continuous wavelet transforms*, Wavelets (Marseille, 1987), Inverse Probl. Theoret. Imaging, Springer, Berlin, 1989, pp. 2–20,

[8] I. Daubechies, *Orthonormal basis of compactly supported wavelets*, Comm. Pure Appl. Math., **41** (1988), 909–996.

[9] K. Gröchenig, *Analyse multi-échelles at bases d'ondelettes*, C. R. Acad. Sci. Paris Série I, **305** (1987), pp. 13–17.

[10] D. Esteban and C. Galand, *Application of quadrature mirror filters to split-band voice coding schemes*, ICASSP '77, IEEE Internat. Conf. on Acoustics, Speech, and Signal Processing, **2**, April 1977, pp. 191–195.

[11] M. J. T. Smith and T. P. Barnwell III, *Exact reconstruction techniques for tree-structured subband coders*, IEEE Trans. Acoustics, Speech, and Signal Processing, **34** (1986), pp. 434–441.

[12] E. H. Adelson, E. Simoncelli, and R. Hingorani, *Orthogonal Pyramid Transforms for Image Coding*, Visual Communications and Image Processing II, Proc. SPIE, Vol. 845 (1987), pp. 50–58.

[13] A. Cohen, *Construction de bases d'ondelettes* α*-hölderiennes*, Rev. Mat. Iberoamericana, **6** (1990), pp. 91–108.

(Manuscript received 25 May 1989.)

Tight frames of compactly supported affine wavelets

Wayne M. Lawton
AWARE, Inc., 124 Mount Auburn Street, Suite 310, Cambridge, Massachusetts 02138

(Received 12 October 1989; accepted for publication 28 March 1990)

This paper extends the class of orthonormal bases of compactly supported wavelets recently constructed by Daubechies [Commun. Pure Appl. Math. **41**, 909 (1988)]. For each integer $N \geq 1$, a family of wavelet functions ψ having support $[0,2N-1]$ is constructed such that $\{\psi_{jk}(x) = 2^{j/2}\psi(2^j x - k) | j,k \in Z\}$ is a tight frame of $L^2(R)$, i.e., for every $f \in L^2(R)$, $f = c\Sigma_{jk} \langle \psi_{jk} | f \rangle \psi_{jk}$ for some $c > 0$. This family is parametrized by an algebraic subset V_N of R^{4N}. Furthermore, for $N \geq 2$, a proper algebraic subset W_N of V_N is specified such that all points in V_N outside of W_N yield orthonormal bases. The relationship between these tight frames and the theory of group representations and coherent states is discussed.

I. INTRODUCTION

This paper discusses families of functions $h_{a,b}(x) = a^{-1/2}h((x-b)/a), a \neq 0$, called *wavelets*,[1] which are generated from a single function $h \in L^2(R)$ by dilation, translation, and possibly reflection. Wavelets provide a means of representing functions as a continuous linear superposition that is analogous to the Fourier transform. This representation is related to group representations. Let G denote the *affine group* $\{x \to ax + b | a \neq 0\}$ and let U denote the unitary representation of G on $L^2(R)$ defined by $U(a,b)h = h_{a,b}$. In this framework wavelets are subsets of $L^2(R)$ having the form $D(h) = \{h_{a,b} | (a,b) \in D\}$ for some subset D of G.

Let G_s denote either G or the connected affine subgroup $G_c = \{(a,b), a > 0\}$ and let $L^2(G_s)$ denote the Hilbert space of measurable functions on G_s that are square integrable with respect to the left-invariant Haar measure $d\mu(a,b) = a^{-2}\, da\, db$ on G_s. A function $h \in L^2(R)$ is *admissible* if $h \neq 0$ and if

$$\int_{G_s} d\mu(a,b) | \langle h_{a,b}, h \rangle |^2 = c_h < \infty.$$

This condition is independent of the choice of G_s and is equivalent to the condition $\int dy |y|^{-1} |\hat{h}(y)|^2 < \infty$, where $\hat{h}(y) = \int dx\, h(x) \exp(-2\pi i x y)$ is the Fourier transform of h. The following result is a consequence of standard results in the theory of unitary group representations:[2] If $G_s = G$ or $G_s = G_c$ and $h \in L^2(R)$ is admissible, then

$$c_h^{-1} \int_{G_s} d\mu(a,b) (U_h f)(a,b) h_{a,b} = P(f), \quad \text{for all } f \in L^2(R), \tag{1}$$

where $P(f)$ denotes the orthogonal projection of f onto the closed subspace H spanned by the wavelets $G_s(h)$.

The wavelets $G_s(h)$ constitute a set of *coherent states* for the action of G_s on H if h is admissible and if H is irreducible under the action of G_s. Note that $L^2(R)$ is irreducible under the action of G but not irreducible under the action of G_c since the subspaces H^+ and H^- of $L^2(R)$ consisting of functions whose Fourier transforms are supported on the set of positive, negative real numbers that are closed and invariant under G_c. However, these subspaces are irreducible under the action of G_c. Therefore, wavelets generated by admissible functions are generalizations of coherent states. Coherent states have extensive applications in physics.[3]

Computational considerations suggest the utility of representing functions by expansions involving discrete wavelets. A *tight frame* of a Hilbert space H is a countable subset $\{h_\alpha\}$ of vectors such that for every $f \in H$, $f = c\Sigma_\alpha \langle h_\alpha | f \rangle h_\alpha$ for some $c > 0$. Therefore, the discrete wavelets $\{h_{a,b} | (a,b) \in D\}$ form a *tight frame* if:

$$f = c \sum_{(a,b) \in D} \langle h_{a,b} | f \rangle h_{a,b}, \quad \text{for every } f \in L^2(R). \tag{2}$$

This relation is a discrete analog of relation (1) in which $P(f) = f$. Daubechies *et al.*[4] constructed tight frames for $L^2(R)$ that consist of discrete subsets $D(h)$ of coherent states for the action of G. In their construction, the set D is a lattice in G having the form $L_{\alpha,\beta} = \{(\pm \exp(\alpha n), \beta m \exp(\alpha n)) | m,n \in Z\}$ for $\beta \neq 0$. The function h is an admissible bandlimited function contained in either H^+ or in H^- satisfying additional restrictions. Since D contains reflections, it is not a subset of G_c and therefore $D(h)$ is not a set of wavelets for G_c. Furthermore, h does not have compact support.

The first bases consisting of wavelets having compact support were orthonormal bases for which $D = \{(2^{-j}, 2^{-j}k) | j,k \in Z\} = L_{\alpha,\beta} \cap G_c$ where $\alpha = \ln(2)$ and $\beta = 1$. In 1910, Haar[5] constructed the following basis for $L^2(R)$: $\{2^{j/2}\psi(2^j x - k) | j,k \in Z\}$ where $\psi = 1$ for $[0,\frac{1}{2}]$, $\psi = -1$ on $[\frac{1}{2},1)$, and $\psi = 0$ on $(-\infty,0) \cup [1,\infty)$. In 1988, Daubechies[6] constructed, for each integer $N \geq 2$, a function ψ_N satisfying: (i) the support of ψ_N has length $2N-1$; (ii) the $0,\ldots,N-1$ moments of ψ_N equal 0; (iii) each ψ_N is continuous and its regularity increases indefinitely as N increases; (iv) $\{2^{j/2}\psi_N(2^j x - k) | j,k \in Z\}$ is an orthonormal wavelet basis for $L^2(R)$. Daubechies constructs each wavelet ψ_N from $2N$ parameters, called *scaling parameters*, which are constrained by a set of N linear and N quadratic equations. These constraints are necessary and sufficient to ensure that all four properties above are satisfied. However, they imply that there exist at most 2^N different wavelets and, furthermore, all these wavelets have identical Fourier moduli. In this paper we prove that if $N-1$ of the

 0022-2488/90/081898-04$03.00 © 1990 American Institute of Physics

Downloaded 14 Oct 2004 to 128.112.156.90. Redistribution subject to AIP license or copyright, see http://jmp.aip.org/jmp/copyright.jsp

linear constraints are removed, then Daubechies' construction results in tight frames of compactly supported wavelets. Furthermore it is proved that for "almost all" choices of scaling parameters, the resulting frame is in fact an orthonormal basis.

II. STATEMENT OF RESULTS

The Haar wavelet function can be constructed as $\psi(x) = \phi(2x) - \phi(2x-1)$, where ϕ is the characteristic function of the set $[0,1)$. The function ϕ satisfies and is uniquely defined in $L^2(R)$, up to normalization, by the scaling relation $\phi(x) = \phi(2x) + \phi(2x-1)$. This approach is generalized below.

Step 1. Choose an integer N and construct a sequence of complex numbers $\{C_0,...,C_{2N-1}\}$, called scaling parameters, that satisfy the following two equations:

$$\sum C_m C^*_{m+2n} = 2\delta_{0n}, \quad \text{for all } n, \tag{3}$$

$$\sum C_m = 2. \tag{4}$$

Here, $C_m = 0$ for $m<0$ and for $m \geqslant 2N$, * denotes complex conjugation, and $\delta_{00} = 1$ and $\delta_{0n} = 0$ for $n \neq 0$. Let V_N denote the set of all such sequences. Then V_N is an algebraic subset of $C^N = R^{4N}$, $V_1 = \{(1\ 1)\}$, and V_N is uncountably infinite for $N \geqslant 2$.

Step 2: Construct a scaling function $\phi \in L^2(R)$ that satisfies the scaling relation

$$\phi(x) = \sum C_m \phi(2x-m). \tag{5}$$

Step 3: Construct the wavelet function $\psi \in L^2(R)$ by

$$\Psi(x) = \sum (-1)^m C^*_{1-m} \phi(2x-m). \tag{6}$$

The main objective of this paper is to prove the following three results.

Theorem 1: If $\{C_0,...,C_{2N-1}\} \in V_N$ and ϕ^0 denotes the characteristic function of $[0,1)$, then the sequence ϕ^j defined by

$$\phi^{j+1}(x) = \sum C_m \phi^j(2x-m), \tag{7}$$

converges weakly in $L^2(R)$ to a function ϕ. Furthermore, ϕ is the unique function that satisfies (5) and for which $\int \phi = 1$. Also, support $(\phi) = [0,2N-1]$.

Theorem 2: If $\{C_0,...,C_{2N-1}\} \in V_N$, $\phi \in L^2(R)$ satisfies (5), and ψ is constructed by (6), then:

(i) $\{\phi(x-m) | m \in Z\}$ forms a partition of unity, i.e., for almost all $x, \Sigma m \phi(x-m) = 1$,

(ii) $\{2^{-j/2}\psi(2^j x - k) | j,k \in Z\}$ is a tight frame of $L^2(R)$; in particular,

$$f(y) = \sum_{j,k} \langle 2^{-j/2}\psi(2^j x - k) | f \rangle 2^{-j/2}\psi 2^j y - k),$$

for all $f \in L^2(R)$.

The next result expresses the fact that "most" of the tight frames constructed above are orthonormal bases. Fix an integer $N>0$, let L_N denote the $4N-3$ dimensional subspace of $L^2(Z)$ consisting of all sequences $\{B_m\}$ such that $B_m = 0$ if $m < -2N+2$ or if $m > 2N-2$. For any $v = \{C_0,...,C_{2N-1}\} \in V_N$, construct the following:

(i) a linear mapping $S_v: L_N \to L_N$ defined by

$$S_v(A)_m = \sum_n \left[\frac{1}{2} C^{\sim} * C\right]_n A_{2m-n}, \tag{8}$$

where $C^{\sim}(k) = C^*(-k)$ for all k and $C^{\sim} * C$ denotes the convolution of $C^{\sim}$ and C;

(ii) a sequence $A_v \in L_N$ by

$$A_{v,m} = \int \phi^*_v(x)\phi_v(x-m)dx, \tag{9}$$

where $\phi_v \in L^2(R)$ is the unique function, constructed as in Theorem 1, that satisfies Eq. (5) and the condition $\int \phi_v = 1$,

(iii) a subset

$$W_N = \{v = \{C_0,...,C_{2N-1}\} \in V_N | DP_v(1) = 0\}, \tag{10}$$

where DP_v denotes the derivative of the characteristic polynomial P_v of S_v. The following result shows that "most," but for $N \geqslant 2$ not all, of the tight frames constructed above are orthonormal bases.

Theorem 3: For all $v \in V_N$, $S_v(\delta_{0m}) = \delta_{0m}$ and $S_v(A_v) = A_v$. The subset of W_N of V_N is a proper algebraic subset of V_N. For $N \geqslant 2$, $v = \{1,0,...,0,1\} \in V_N$, $v \in W_N$, and the set $\{\psi_{jk} | j,k \in Z\}$ is not orthonormal. If $v \in V_N$ and $v \notin W_N$ then $\{\psi_{jk} | j,k \in Z\}$ is an orthonormal basis for $L^2(R)$.

III. DERIVATIONS

Several intermediate results are required for the proofs.

Lemma 1: The sequence $\{\phi^j\}$ in Theorem 1 converges to a distribution ϕ whose support is $[0,2N-1]$.

Proof: The Fourier transforms Φ^j of ϕ^j satisfy $\Phi^{j+1}(y) = \Phi^j(y/2) \quad m_0(y/2) \quad \text{for} \quad j \geqslant 1 \quad \text{and} \quad \Phi^0(y) = \exp(\pi i y)\text{sinc}(\pi y)$, where $m_0(y) = \frac{1}{2}\Sigma_n C_n \exp(2\pi i n y)$. Equation (4) implies $m_0(y) = 1$, therefore the sequence Φ^j converges uniformly to the function $\Phi(y) = \Pi_{j \geqslant 1} m_0(2^{-j}y)$ on compact subsets of R. Furthermore, for $|y| \geqslant 1$, each Φ^j satisfies the inequality $|\Phi^j(y)| \leqslant CB^{1+\log_2(y)}$, where $C = \max\{|\Phi(y)|, |y| \leqslant 1\}$ and $B = \max\{|m_0(y)|\}$. Thus Φ^j and Φ are bounded by polynomial growth and $\{\phi^j\}$ converges to a distribution ϕ such that $\hat{\phi} = \Phi$. Clearly, $\int_\phi = \Phi(0) = 1$. Also, $\text{support}(\phi) = \lim \text{support}(\phi^j) = \lim[0,(N-1)(1-2^{-j})] = [0,N-1]$.

Lemma 2: For all $j \geqslant 0$, $\langle \phi^j | \phi^j \rangle = 1$. The distribution ϕ is in $L^2(R)$.

Proof: Clearly for any integer n, $\langle \phi^0(x) | \phi^0(x-n) \rangle = \delta_{0n}$. By (7), (3) and induction

$$\begin{aligned}
&\langle \phi^{j+1}(x) | \phi^{j+1}(x-n) \rangle \\
&= \langle \sum_m C_m \phi^j(2x-m) | \sum_k C_k \phi^j(2x-2n-k) \rangle \\
&= \sum_m \sum_k C^*_m C_k \langle \phi^j(2x-m) | \phi^{j+1}(2x-2n-k) \rangle \\
&= \frac{1}{2} \sum_m C^*_m C_{m-2k} = \delta_{0n}.
\end{aligned}$$

Therefore, since the unit ball in $L^2(R)$ is weakly compact, any subsequence $\{g_k\}$ of $\{\phi^j\}$ has at least one accumulation

Downloaded 14 Oct 2004 to 128.112.156.90. Redistribution subject to AIP license or copyright, see http://jmp.aip.org/jmp/copyright.jsp

point, say g, in the weak topology. Then there exists a subsequence of $\{g_k\}$ that converges weakly to g. Since the weak topology on $L^2(R)$ is stronger than the distribution topology, $g = \phi$. Therefore, $\{\phi^j\}$ converges weakly to $\phi \in L^2(R)$.

Proof of Theorem 1: It suffices, from Lemmas 1 and 2, to demonstrate the uniqueness of ϕ. Assume ϕ is any distribution satisfying the scaling recursion (7) and the condition $\int \phi = 1$. The Fourier transform satisfies $\hat{\phi}(y) = \Phi(y)\hat{\phi}^{(0)} = \Phi(y)$ where Φ is the function represented by the infinite product expansion in Lemma 2. This concludes the proof.

Lemma 3: The set $\{\phi(x-m)\,|\,m \in Z\}$ forms a partition of unity.

Proof: Let $\{\phi^j\}$ be the sequence in Theorem 1. Since $\phi = \lim \phi^j$ it suffices to prove $\{\phi^j(x-m)\,|\,m\in Z\}$ forms a partition of unity for all $j = 1$. For $j = 1$ this holds since $\phi^0 =$ characteristic function of $[0,1)$. The result for $j \geqslant 1$ follows by induction.

For any vectors g and h in a Hilbert space H let $|g\rangle\langle h|$ denote the operator, called the *outer product* of g and h, defined by $|g\rangle\langle h|(f) = \langle h|f\rangle g$. Therefore, a subset $\{h_\alpha\}$ of H is a tight frame if and only if $\Sigma_\alpha\, |h_\alpha\rangle\langle h_\alpha| = I$, where I denotes the identity operator on H and the limit is in the weak sense.

Lemma 4: Let ϕ be a scaling function with support $[0,2N-1]$ and for any integer J let I_J denote the operator $\Sigma_m |2^{J/2}\phi(2^Jx-m)\rangle\langle 2^{J/2}\phi(2^Jx-m)|$. Then limit $I_J = I$.

Proof: The set of operators $\{I_J\}$ is uniformly bounded since for any $f \in L^2(R)$, the orthonormality of $\{2^{J/2}\phi(2^Jy - n(2N-1))/\|\phi\|\,|n\in Z\}$ and Bessel's inequality[7] imply

$$\|I_J(f)\| = \Big\| \sum_{n=0}^{n=2N-2} \sum_{m=n\,\mathrm{mod}(2N-1)} \langle 2^{J/2}\phi(2^Jy-m)|f(y)\rangle \times 2^{J/2}\phi(2^Jx-m)\Big\| \leqslant (2N-1)\|\phi\|^2\|f\|.$$

Therefore, it suffices to prove that limit $I_J(f) = f$ for all f in a dense subset of $L^2(R)$. Assume f is continuous and has compact support. Then if $m\to\infty$ and $J\to\infty$ such that $2^{-J}m \to x$, $\lim 2^J\langle\phi(2^Jy-m)|f(y)\rangle = f(x)$ uniformly in x. The result now follows since, by Lemma 3, $\{\phi(2^Jy-m)\,|\,m\in Z\}$ forms a partition of unity.

Lemma 5: Let ϕ and ψ denote the scaling function and wavelet corresponding to a $\{C_0,\ldots,C_{2N-1}\}\in V_N$. For any integer J define the operators I_J as in Lemma 4 and $F_J = \Sigma_m |2^{J/2}\psi(2^Jx-m)\rangle\langle 2^{J/2}\psi(2^Jx-m)|$. Then for all J, $I_J + F_J = I_{J+1}$.

Proof: Substitute the expressions for ϕ and ψ in Eqs. (5) and (6) to obtain

$$\begin{aligned} I_J + F_J &= \frac{1}{2}\sum_p C_p^* \sum_q C_q \sum_m |2^{(J+1)/2}\phi(2^{J+1}x-2m-p)\rangle\langle 2^{(J+1)/2}\phi(2^{J+1}x-2m-q)| \\ &\quad + \frac{1}{2}\sum_p(-1)^p C_{1-p}\sum_q(-1)^q C_{1-q}^*\sum_m |2^{(J+1)/2}\phi(2^{J+1}x-2m-p)\rangle\langle 2^{(J+1)/2}\phi(2^{J+1}x-2m-q)| \\ &= \sum_i\sum_k C(i,k)|2^{(J+1)/2}\phi(2^{J+1}x-i)\rangle\langle 2^{(J+1)/2}\phi(2^{J+1}x-k)|, \end{aligned}$$

where

$$C(i,k) = \frac{1}{2}\sum_m [C_{2m+i}^* C_{2m+k} + (-1)^{2m+i} \times C_{1-2m-i}(-1)^{2m+k}C_{1-2m-k}^*].$$

It suffices to prove that $C(i,k) = \delta_{ik}$. Assume i is even and k is odd. Then replace m by $m - i/2$ in the first expression and replace m by $-m-(1-k)/2$ in the second expression (within brackets) above to obtain

$$C(i,k) = \frac{1}{2}\sum_m [C_{2m}^* C_{2m-i+k} - C_{2m-i+k}C_{2m}^*] = 0.$$

If i is odd and k is even it follows similarly that $C(i,k) = 0$. Now assume i is even and k is even. Then replace m by $m-k/2$, respectively $-m-i/2$ in the first and second expression above to obtain

$$C(i,k) = \frac{1}{2}\sum_m [C_{2m-k+i}^* C_{2m} + C_{1+2m}C_{1+2m-k+i}^*].$$

Since $2m$ ranges over all even integers and $1+2m$ ranges over all odd integers, this expression is equivalent to $\frac{1}{2}\Sigma_m C_{m-k+i}^* C_{2m}$. Equation (3) implies this expression equals δ_{ik} because $-k+1$ is even. Finally, if both i and k are odd, replace m by $-m-(i+k)/2$ in the second expression to obtain

$$\frac{1}{2}\sum_m [C_{2m+i}^* C_{2m+k} + C_{1+2m+k}C_{1+2m+i}^*] = \delta_{ik}.$$

Proof of Theorem 2: The first conclusion follows from Lemma 3. The second conclusion follows from Lemma 4 and the repeated application of Lemma 5 to obtain

$$I = \lim_{J\to\infty} I_J = \lim_{J\to\infty} F_{J-1} + F_{J-2} + \cdots + F_{-J} + I_{-J} = \sum_j F_j.$$

Proof of Theorem 3: Let $N \geqslant 1$ be an integer, let $v = \{C_0,\ldots,C_{2N-1}\}\in V_N$, and let S_v denote the operator defined by (8). Then $S_v(\delta_{0m}) = \delta_{0m}$ follows from (3) and

Downloaded 14 Oct 2004 to 128.112.156.90. Redistribution subject to AIP license or copyright, see http://jmp.aip.org/jmp/copyright.jsp

$S_v(A_v) = A_v$ is obtained by substituting the scaling relation (5) into the integrand in Eq. (9). Clearly, $DP_v(1)$ is a polynomial in terms of the scaling parameters $\{C_m\}$ and their complex conjugates. Therefore, W_N is an algebraic subset of V_N. Also, $W_N \neq V_N$ because the scaling parameters constructed by Daubechies are in V_N and not in W_N, therefore W_N is a proper algebraic subset of V_N. Clearly, for $N \geqslant 2$, $v = \{1,0,...,0,1\} \in W_N$ and also the set $\{\psi_{jk} | j,k \in Z\}$ is not orthonormal. Assume $\{\psi_{jk} | j,k \in Z\}$ corresponding to $v \in V_N$ is not an orthonormal basis for $L^2(R)$. It suffices to show $v \in W_N$. Consider the sequence A_v defined by Eq. (9). If A_v is a linear multiple of the sequence δ_{0m} then the functions $\{\phi(x-m) | m \in Z\}$ are orthogonal. However, by construction $\int_\phi = 1$, and by Lemma 3, $\{\phi(x-m) | m \in Z\}$ forms a partition of unity, therefore,

$$A_v(0) = \int \phi^*(x)\phi(x)dx = \int \phi^*(x) \sum_k \phi(x-k)dx$$

$$= \int \phi^*(x)dx = 1.$$

Then Eq. (8), describing the construction of the wavelet ψ in terms of ϕ, implies the functions $\{\psi_{jk} | j,k \in Z\}$ are orthogonal. Since by Theorem 2, this set is a also a quasiorthonormal basis for $L^2(R)$,

$$\psi_{jk} = \sum_{m,n} \langle \psi_{mn} | \psi_{jk} \rangle \psi_{mn} = \langle \psi_{jk} | \psi_{jk} \rangle \psi_{jk},$$

for all j, $k \in Z$, hence $\{\Psi_{jk} | j,k \in Z\}$ is an orthonormal basis. Since this contradicts the original assumption, it follows that A_v is not a linear multiple of δ_{0m}. Therefore, A_v and δ_{0m} are linearly independent eigenvectors in L_N, having eigenvalue 1, for the linear operator S_v defined by Eq. (8). Therefore the characteristic polynomial P_v of S_v has the root 1 occur with multiplicity $\geqslant 2$. Hence its derivative DP_v satisfies $DP_v(1) = 0$ and $v \in W_N$. The proof is complete.

Remark: It follows from Pollen's factorization theorem[8] for unitary matrices over the ring of Laurent polynomials that V_N is an irreducible algebraic set. Therefore, the subset W_N of V_N has zero measure.

Open Problem: Does every $v \in W_N$ give rise to a nonorthonormal basis?

ACKNOWLEDGMENTS

The author is indebted to Ingrid Daubechies for permission to include her proof of Theorem 1, communicated to me in a letter of April 1989, in this paper and for encouragement to publish these results. Special thanks are given to my colleages at AWARE, Inc., in particular to Ramesh Gopinath, who uncovered several errors in earlier versions of the manuscript and provided helpful editorial comments; to David Linden, who discovered the tight frame properties of wavelet bases through numerical experiments prior to theoretical investigations; to David Pollen, who elucidated the detailed structure of the parameter set for wavelets; and to Howard Resnikoff, who first suggested the examination of the eigenvalues of linear operators derived from the scaling relation as used in the proof of Theorem 2.

This research was supported in part by the Advanced Research Projects Agency of the Department of Defense and was monitored by the Air Force Office of Scientific Research under Contract No. F49620-89-C-0125. The United States Government is authorized to reproduce and distribute reprints for governmental purposes notwithstanding any copyright notation hereon.

[1] A. Grossman and J. Morlet, SIAM J. Math. Anal. **15**, 723 (1984).
[2] A. Grossman and J. Morlet, J. Math. Phys. **26**, 2473 (1985).
[3] J. R. Klauder and B. S. Skagerstam, *Coherent States. Applications in Physics and Mathematical Physics* (World Scientific, Singapore, 1984).
[4] I. Daubechies, A. Grossman, and Y. Meyer, J. Math. Phys. **27**, 1271 (1985).
[5] Alfred Haar, Math. Ann. **69**, 336 (1910).
[6] I. Daubechies, Commun. Pure Appl. Math. **41**, 909 (1988).
[7] G. Dahlquist and A. Bjorck, *Numerical Methods* (Prentice-Hall, Englewood Cliffs, NJ, 1979).
[8] D. Pollen, "$SU_I(2,F[z,1/z])$ for F a subfield of C," J. Am. Math. Soc. (to be published).

Downloaded 14 Oct 2004 to 128.112.156.90. Redistribution subject to AIP license or copyright, see http://jmp.aip.org/jmp/copyright.jsp

Orthonormal Bases of Compactly Supported Wavelets

INGRID DAUBECHIES
AT&T Bell Laboratories

Abstract

We construct orthonormal bases of compactly supported wavelets, with arbitrarily high regularity. The order of regularity increases linearly with the support width. We start by reviewing the concept of multiresolution analysis as well as several algorithms in vision decomposition and reconstruction. The construction then follows from a synthesis of these different approaches.

1. Introduction

In recent years, families of functions $h_{a,b}$,

$$h_{a,b}(x) = |a|^{-1/2} h\left(\frac{x-b}{a}\right), \qquad a, b \in \mathbb{R},\ a \neq 0, \tag{1.1}$$

generated from one single function h by the operation of dilations and translations, have turned out to be a useful tool in many different fields of mathematics, pure as well as applied. Following Grossmann and Morlet [1], we shall call such families "wavelets".

Techniques based on the use of translations and dilations are certainly not new. They can be traced back to the work of A. Calderón [2] on singular integral operators, or to renormalization group ideas (see [3]) in quantum field theory and statistical mechanics. Even in these two disciplines, however, the explicit introduction of special families of wavelets seems to have led to new results (see, e.g. [4], [5], [6]). Moreover, wavelets are useful in many other applications as well. They are used for e.g. sound analysis and reconstruction in [7], and have led to a new algorithm, with many attractive features, for the decomposition of visual data in [8]. They seem to hold great promise for the detection of edges and singularities; see [9]. It is therefore fair to surmise that they will have applications in yet other directions.

Depending on the type of application, different families of wavelets may be chosen. One can choose, e.g., to let the parameters a, b in (1.1) vary continuously on their range $\mathbb{R}^* \times \mathbb{R}$ (where $\mathbb{R}^* = \mathbb{R} \setminus \{0\}$). One can then, for instance, represent functions $f \in L^2(\mathbb{R})$ by the functions Uf,

$$(Uf)(a, b) = \langle h_{a,b}, f\rangle = |a|^{-1/2} \int dx\, \overline{h\left(\frac{x-b}{a}\right)} f(x). \tag{1.2}$$

Communications on Pure and Applied Mathematics, Vol. XLI 909–996 (1988)
© 1988 John Wiley & Sons, Inc. CCC 0010-3640/88/070909-88$04.00

If h satisfies the condition

$$\int d\xi\, |\xi|^{-1} |\hat{h}(\xi)|^2 < \infty, \tag{1.3}$$

where ˆ denotes the Fourier transform,

$$\hat{h}(\xi) = \frac{1}{\sqrt{2\pi}} \int dx\, e^{ix\xi} h(x),$$

then U (as defined by (1.2)) is an isometry (up to a constant) from $L^2(\mathbb{R})$ into $L^2(\mathbb{R}^* \times \mathbb{R};\, a^{-2}\, da\, db)$. The map U is called the "continuous wavelet transform"; see [1], [10]. In this form, wavelets are closest to the original work of Calderón. The continuous wavelet transform is also closely related to the "affine coherent state representation" of quantum mechanics (first constructed in [11], see also [12]); in fact, for appropriate choices of h, the $h_{a,b}$ are "affine coherent states", and have been used in the study of some quantum mechanics problems in [11], [12].

Note that the "admissibility condition" (1.3) implies, if h has sufficient decay which we shall always assume in practice, that h has mean zero,

$$\int dx\, h(x) = 0. \tag{1.4}$$

Typically, the function h will therefore have at least some oscillations. A standard example is

$$h(x) = (2/\sqrt{3})\pi^{-1/4}(1 - x^2)e^{-x^2/2}. \tag{1.5}$$

For other applications, including those in signal analysis, one may choose to restrict the values of the parameters a, b in (1.1) to a discrete sublattice. In this case one fixes a dilation step $a_0 > 1$, and a translation step $b_0 \neq 0$. The family of wavelets of interest becomes then, for $m, n \in \mathbb{Z}$,

$$h_{mn}(x) = a_0^{-m/2} h(a_0^{-m}x - nb_0). \tag{1.6}$$

Note that this corresponds to the choices

$$a = a_0^m,$$

$$b = nb_0 a_0^m,$$

indicating that the translation parameter b depends on the chosen dilation rate. For m large and positive, the oscillating function h_{m0} is very much spread out, and the large translation steps $b_0 a_0^m$ are adapted to this wide width. For large but negative m the opposite happens; the function h_{m0} is very much concentrated, and the small translation steps $b_0 a_0^m$ are necessary to still cover the whole range.

A "discrete wavelet transform" T is associated with the discrete wavelets (1.6). It maps functions f to sequences indexed by $\mathbb{Z}^2$,

$$(1.7) \qquad \begin{aligned} (Tf)_{mn} &= \langle h_{mn}, f \rangle \\ &= a_0^{-m/2} \int dx \, \overline{h(a_0^{-m}x - nb_0)}\, f(x). \end{aligned}$$

If h is "admissible", i.e., if h satisfies the condition (1.3), and if h has sufficient decay, then T maps $L^2(\mathbb{R})$ into $l^2(\mathbb{Z}^2)$. In general, T does not have a bounded inverse on its range. If it does, i.e., if, for some $A > 0$, $B < \infty$,

$$A\|f\|^2 < \sum_{m, n \in \mathbb{Z}} |\langle h_{mn}, f \rangle|^2 < B\|f\|^2,$$

for all f in $L^2(\mathbb{R})$, then the set $\{h_{mn};\ m, n \in \mathbb{Z}\}$ is called a "frame". In this case one can construct numerically stable algorithms to reconstruct f from its wavelet coefficients $\langle h_{mn}, f \rangle$. In particular,

$$(1.8) \qquad f = \frac{2}{A + B} \sum_{m, n} h_{mn} \langle h_{mn}, f \rangle + R,$$

with

$$\|R\| \leqq O\left(\frac{B}{A} - 1\right)\|f\|.$$

For B/A close to 1, which is the case in the decompositions and reconstructions of music and other sound signals, as done by A. Grossmann, R. Kronland and J. Morlet [7], the "error term" R can be omitted. In practice, with e.g. the basic wavelet (1.5), and with $a_0 = 2^{1/4}$, $b_0 = .5$, one finds $B/A - 1 < 10^{-5}$, and the reconstruction formula (1.8) restricted to its first term gives excellent results. In fact, even for the larger value $a_0 = 2$, corresponding to $B/A - 1 \cong .08$, the truncated reconstruction formula, when applied to the wavelet decomposition of speech signals, leads to a clearly understandable reconstruction; see [13].

In the use of wavelet frames for sound analysis, and reconstruction, as studied by the Marseilles group [7], the families of wavelets h_{mn} considered are highly redundant, i.e., they are not independent, in the sense that any finite number of them lies in the closed linear span generated by the others. Consequently, the range of the discrete wavelet transform T is a proper subspace of $l^2(\mathbb{Z}^2)$. The higher the redundancy of the frame, the smaller this subspace, which is a desirable feature for some purposes (e.g. the reduction of calculational noise). If a_0, b_0 are chosen very close to 1, 0, respectively, then the resulting frame is very redundant and very close to the continuous family of wavelets (1.1); this type of frame was used in the "edge detection" study mentioned earlier; see [9].

For other applications, as e.g. in S. Mallat's vision decomposition algorithm in [8], it is preferable to work with the other extremum, and to reduce redundancy as much as possible. In this case, one can turn to choices of h and a_0, b_0 (typically $a_0 = 2$) for which the h_{mn} constitute an orthonormal basis. This is the case to which we shall be restricting ourselves in the remainder of this paper. For a more detailed study of general (non-orthonormal) wavelet frames, and a discussion of the similarities and the differences between wavelet transform and windowed Fourier transform, the reader is referred to [14], [15].

One example of an orthonormal basis of wavelets[1] for $L^2(\mathbb{R})$ is the well-known Haar basis. For the Haar basis one chooses

$$h(x) = \begin{cases} 1, & 0 \leqq x < \frac{1}{2}, \\ -1, & \frac{1}{2} \leqq x < 1, \\ 0, & \text{otherwise}, \end{cases} \tag{1.9}$$

and $a_0 = 2$, $b_0 = 1$. The resulting h_{mn},

$$h_{mn}(x) = 2^{-m/2} h(2^{-m}x - n), \qquad m, n \in \mathbb{Z}, \tag{1.10}$$

constitute an orthonormal basis for $L^2(\mathbb{R})$. The h_{mn} also constitute an unconditional basis for all $L^p(\mathbb{R})$, $1 < p < \infty$.

Recently, some much more surprising examples of orthonormal wavelet bases have surfaced. The first one was constructed by Y. Meyer [4] in the summer of 1985. He constructed a C^∞-function h of rapid decay (in fact $\hat{h}$, in his example, is a compactly supported C^∞-function) such that the h_{mn}, as defined by (1.10) (i.e., with $a_0 = 2$, $b_0 = 1$), constitute an orthonormal basis for $L^2(\mathbb{R})$. As in the case of the Haar basis, Y. Meyer's basis is also an unconditional basis for all the L^p spaces, $1 < p < \infty$. Much more is true, however. The Meyer basis turns out to be an unconditional basis for all the Sobolev spaces, for the Hardy-Littlewood space H_1, for the Besov spaces, etc.; see [4]. The Meyer basis is therefore a much more powerful tool than the Haar basis.

Some time later, in 1986, another interesting orthonormal basis of wavelets was constructed, independently, by P. G. Lemarié [17] and G. Battle [18]. In their construction the function h is only C^k, but it has exponential decay (as compared

[1] Following Grossmann and Morlet [1] we call "wavelet" any L^2-function h satisfying the admissibility condition (1.3). This is less restrictive than Y. Meyer [16], who, in keeping with the tradition in harmonic analysis, also imposes some regularity. In the terminology of [16], the Haar basis function (1.9) is not a wavelet.

with decay faster than any power in Y. Meyer's case). It also has k vanishing moments, i.e.,

$$\int dx\, x^j h(x) = 0, \qquad j = 0, 1, \cdots, k-1,$$

which makes these h_{mn} an unconditional basis for all the Sobolev spaces $\mathcal{H}_s$, with $s < k - 1$.

In all these constructions the choices $a_0 = 2$, $b_0 = 1$ were made. The choice for b_0 is of course arbitrary, a simple dilation of the function h allows one to fix any non-zero choice for b_0; it is convenient to choose $b_0 = 1$. The choice of a_0 is far less arbitrary. We shall restrict ourselves here to $a_0 = 2$, although it is possible to consider other, though by no means arbitrary, choices for a_0 (see [4], [21]).

In the fall of 1986, S. Mallat and Y. Meyer [16], [19] realized that these different wavelet basis constructions can all be realized by a "multiresolution analysis". This is a framework in which functions $f \in L^2(\mathbb{R}^d)$ can be considered as a limit of successive approximations, $f = \lim_{m \to -\infty} P_m f$, where the different $P_m f$, $m \in \mathbb{Z}$, correspond to smoothed versions of f, with a "smoothing out action radius" of the order of 2^m. The wavelet coefficients $\langle h_{mn}, f \rangle$, with fixed m, then correspond to the difference between the two successive approximations $P_{m-1} f$ and $P_m f$. A more detailed description of multiresolution analysis will be given in Section 2.

The concept of multiresolution analysis plays a central role in S. Mallat's algorithm for the decomposition and reconstruction of images in [8]. In fact, ideas related to multiresolution analysis (a hierarchy of averages, and the study of their differences) were already present in an older algorithm for image analysis and reconstruction, namely the Laplacian pyramid scheme of P. Burt and E. Adelson [20]. The Laplacian pyramid ideas triggered S. Mallat to view the orthonormal bases of wavelets as a vehicle for multiresolution analysis. Together, S. Mallat and Y. Meyer then carried out a more detailed mathematical analysis, showing how all the "accidental" previous constructions found their natural framework in multiresolution analysis; see [16], [19]. By the use of multiresolution analysis and orthonormal wavelet bases, S. Mallat constructed an algorithm that is both more economical and more powerful in its orientation selectivity. On the other hand, by a curious feedback, the combination of Mallat's ideas and of the restrictions on "filters" imposed in [20] led to my construction of orthonormal wavelet bases of compact support, which is the main topic of this paper.

Because of the important role, in the present construction, of the interplay of all these different concepts, and also to give a wider publicity to them, an extensive review will be given in Section 2 of multiresolution analysis (subsection 2A), of the Laplacian pyramid scheme (subsection 2B) and of Mallat's algorithm (subsection 2C).

Sections 3 and 4 contain the new results of this paper. A closer look at Mallat's work shows that he uses the intermediary of orthonormal wavelet bases for *function* spaces to build an essentially *discrete* algorithm. It seemed therefore natural to wonder whether similar, and as powerful, discrete algorithms could be built *directly*, without using function spaces as an intermediate step. It turns out that it is very easy to write a set of necessary and sufficient conditions, on the "discrete side", ensuring that an algorithm similar to Mallat's works. This is done in subsection 3A. In order to have a useful algorithm, however, an extra regularity condition has to be imposed (this condition is already satisfied in e.g. Burt and Adelsen's Laplacian pyramid scheme). This is done in subsection 3B. The combination of the *discrete* conditions and the regularity condition on the *discrete* algorithm turns out, however, to be strong enough to *impose* an underlying multiresolution analysis of *functions*, with associated orthonormal wavelet basis. Provided the regularity condition is satisfied, there is therefore a one-to-one correspondence between orthonormal wavelet bases and discrete multiresolution decompositions, in the sense of Mallat's algorithm. This equivalence is proved in subsection 3C. Another proof of the same result, using different techniques, can be found in [19]; the proof presented here is more "graphical", and closer to the "filter" point of view of [20].

In Section 4, we exploit the equivalence between discrete and function schemes to build orthonormal bases of wavelets with compact support. Using this equivalence, it turns out that it is sufficient to build a discrete scheme using filters with a finite number of taps. This can be done explicitly, as shown in subsection 4B. As a result one can construct, for any $k \in \mathbb{N}$, a C^k-function ψ with compact support, such that the corresponding ψ_{mn},

$$\psi_{mn}(x) = 2^{-m/2}\psi(2^{-m}x - n),$$

constitute an orthonormal basis. The size of the support increases linearly with the regularity. Moreover, ψ has K consecutive moments equal to zero,

$$\int dx\, x^j \psi(x) = 0, \qquad j = 0, 1, \cdots, K - 1,$$

where K also increases linearly with k. All these properties of the construction are proved in subsection 4C. Finally, the "graphical" approach which, as explained in subsection 3B, was the guideline to the proof of the link between the "regularity" condition of Burt and Adelson (see subsection 2B) and multiresolution analysis, can also be used to plot the functions ψ. We conclude this paper with the plots of a few of the compactly supported wavelets constructed here.

2. Multiresolution Analysis and Image Decomposition and Reconstruction

2.A. A review of multiresolution analysis and orthonormal wavelet bases. In this subsection we review the definition of multiresolution analysis, and show how orthonormal bases of wavelets can be constructed starting from a multiresolution analysis. We illustrate this construction with examples. No proofs will be given; for proofs, more details and generalizations we refer to [16], [19] or [21].

The idea of a multiresolution analysis is to write L^2-functions f as a limit of successive approximations, each of which is a smoothed version of f, with more and more concentrated smoothing functions. The successive approximations thus use a different resolution, whence the name multiresolution analysis. The successive approximation schemes are also required to have some translational invariance. More precisely, a multiresolution analysis consists of

(i) a family of embedded closed subspaces $V_m \subset L^2(\mathbb{R})$, $m \in \mathbb{Z}$,

$$\cdots \subset V_2 \subset V_1 \subset V_0 \subset V_{-1} \subset V_{-2} \subset \cdots \tag{2.1}$$

such that (ii)

$$\bigcap_{m \in \mathbb{Z}} V_m = \{0\}, \qquad \overline{\bigcup_{m \in \mathbb{Z}} V_m} = L^2(\mathbb{R}), \tag{2.2}$$

and (iii)

$$f \in V_m \Leftrightarrow f(2\,\cdot) \in V_{m-1}; \tag{2.3}$$

moreover, there is a $\phi \in V_0$ such that, for all $m \in \mathbb{Z}$, the ϕ_{mn} constitute an unconditional basis for V_m, that is, (iv)

$$V_m = \overline{\text{linear span}\,\{\phi_{mn},\ n \in \mathbb{Z}\}} \tag{2.4a}$$

and there exist $0 < A \leqq B < \infty$ such that, for all $(c_n)_{n \in \mathbb{Z}} \in l^2(\mathbb{Z})$,

$$A \sum_n |c_n|^2 \leqq \left\| \sum_n c_n \phi_{mn} \right\|^2 \leqq B \sum_n |c_n|^2. \tag{2.4b}$$

Here $\phi_{mn}(x) = 2^{-m/2}\phi(2^{-m}x - n)$. Let P_m denote the orthogonal projection onto V_m. It is then clear from (2.1), (2.2) that $\lim_{m \to -\infty} P_m f = f$, for all $f \in L^2(\mathbb{R})$. The condition (2.3) ensures that the V_m correspond to different scales, while the translational invariance

$$f \in V_m \to f(\cdot - 2^m n) \in V_m \quad \text{for all} \quad n \in \mathbb{Z}$$

is a consequence of (2.4).

EXAMPLE 2.1. A typical though crude example is the following. Take the V_m to be spaces of piecewise constant functions,

$$V_m = \left\{ f \in L^2(\mathbb{R});\ f \text{ constant on } \left[2^m n, 2^m(n+1)\right[\text{ for all } n \in \mathbb{Z} \right\}.$$

The conditions (2.1)–(2.3) are clearly satisfied. The projections P_m are defined by

$$P_m f|_{[2^m n, 2^m(n+1)[} = 2^{-m} \int_{2^m n}^{2^m(n+1)} dx\, f(x).$$

The successive $P_m f$ (as m decreases) do therefore correspond to approximations of f on a finer and finer scale. Finally, we can choose for ϕ the characteristic function of the interval $[0,1[$,

$$\phi(x) = \begin{cases} 1, & 0 \leqq x < 1, \\ 0, & \text{otherwise.} \end{cases}$$

Clearly, $\phi \in V_0$ and $V_m = \overline{\text{span}\{\phi_{mn}\}}$.

In what follows, we shall revisit this example to illustrate the construction of an orthonormal wavelet basis from multiresolution analysis.

Note that, in view of (2.3), the condition (2.4a) may be replaced by the weaker condition $V_0 = \overline{\text{span}\{\phi_{0n}\}}$. Moreover, one may, without loss of generality, assume that the ϕ_{0n} are orthonormal (which automatically implies that the ϕ_{mn} are orthonormal for every fixed m). If the ϕ_{0n} are not orthonormal to start with, then one defines $\tilde{\phi}$ by

$$(\tilde{\phi})^{\wedge}(\xi) = C\hat{\phi}(\xi)\Big(\sum_{k\in\mathbb{Z}} |\hat{\phi}(\xi + 2k\pi)|^2\Big)^{-1/2} \tag{2.5}$$

(where we implicitly assume that $\hat{\phi}$ has sufficient decay to make the infinite sum converge). One finds that

$$\overline{\text{span}\{\phi_{0n}\}} = \overline{\text{span}\{\tilde{\phi}_{0n}\}},$$

while, moreover, the $\tilde{\phi}_{0n}$ are orthonormal. See [16] for a detailed proof.

EXAMPLE 2.1 (continued). In this case the ϕ_{0n} are orthonormal from the start. If we define

$$c_{mn}(f) = \langle \phi_{mn}, f \rangle = 2^{-m/2} \int_{2^m n}^{2^m(n+1)} dx\, f(x), \tag{2.6}$$

then

$$P_m f = \sum_n c_{mn}(f)\phi_{mn}.$$

Let us look at the difference between $P_m f$ and the next coarser approximation $P_{m+1} f$. One easily checks that

$$\phi_{m+1\,n} = \frac{1}{\sqrt{2}}\left(\phi_{m\,2n} + \phi_{m\,2n+1}\right);$$

hence

$$c_{m+1\,n}(f) = \frac{1}{\sqrt{2}}\left[c_{m\,2n}(f) + c_{m\,2n+1}(f)\right].$$

This again exhibits $P_{m+1}f$ as an averaged version of $P_m f$, i.e., as a larger scale approximation. The difference between these two successive approximations is given by

$$P_m f - P_{m+1} f = \frac{1}{2}\sum_n \left[c_{m\,2n}(f) - c_{m\,2n+1}(f)\right]\left[\phi_{m\,2n} - \phi_{m\,2n+1}\right].$$

The remarkable fact about this expression is that it can be rewritten under a form very similar to (2.6). Define

$$\psi(x) = \phi(2x) - \phi(2x-1) = \begin{cases} 1, & 0 \leqq x < \frac{1}{2}, \\ -1, & \frac{1}{2} \leqq x < 1, \\ 0, & \text{otherwise}. \end{cases} \tag{2.7}$$

Then

$$\begin{aligned} \psi_{mn}(x) &= 2^{-m/2}\psi(2^{-m}x - n) \\ &= \frac{1}{\sqrt{2}}\left(\phi_{m-1\,2n} - \phi_{m-1\,2n+1}\right), \end{aligned} \tag{2.8}$$

and

$$\begin{aligned} Q_{m+1} f &= P_m f - P_{m+1} f \\ &= \sum_n d_{m+1\,n}(f)\,\psi_{m+1\,n}, \end{aligned} \tag{2.9}$$

where

$$d_{m+1\,n}(f) = \langle \psi_{m+1\,n}, f\rangle = \frac{1}{\sqrt{2}}\left[c_{m\,2n}(f) - c_{m\,2n+1}(f)\right].$$

What is so remarkable about this? Note first, as can easily be checked from (2.7), that for fixed m the ψ_{mn} are orthonormal. The decomposition (2.9) is thus the expansion, with respect to an orthonormal basis, of $Q_{m-1}f$, the orthogonal projection of f onto $W_{m+1} = P_m L^2 - P_{m+1} L^2$, i.e., onto the orthogonal complement of V_{m+1} in V_m. The surprising fact is that, as is clear from (2.9), the W_m are also (as are the V_m) generated by the translates and dilates ψ_{mn} of a single function ψ. Once this is realized, building a wavelet basis becomes trivial. Clearly (2.1)–(2.2), together with $W_m \perp V_m$, $V_{m-1} = V_m \oplus W_m$, imply that the W_m are all mutually orthogonal, and that their direct sum is $L^2(\mathbb{R})$. Since, for each m, the set $\{\psi_{mn};\ n \in \mathbb{Z}\}$ constitutes an orthonormal basis for W_m, it follows that the whole collection $\{\psi_{mn};\ m, n \in \mathbb{Z}\}$ is an orthonormal wavelet basis for $L^2(\mathbb{R})$.

In the example above the function ψ is nothing but the Haar function (see (1.9)), and it is therefore no surprise that the ψ_{mn} constitute an orthonormal basis. The example does however clearly show how this basis can be constructed from a multiresolution analysis. Let us sketch now how the general case works:

For a multiresolution analysis, i.e., a family of spaces V_m and a function ϕ satisfying (2.1)–(2.4), one defines (as in Example 2.1) W_m as the orthogonal complement, in V_{m-1}, of V_m,

$$V_{m-1} = V_m \oplus W_m, \qquad W_m \perp V_m. \tag{2.10}$$

Equivalently,

$$W_m = Q_m L^2(\mathbb{R}) \quad \text{with} \quad Q_m = P_{m-1} - P_m. \tag{2.11}$$

It follows immediately that all the W_m are scaled versions of W_0,

$$f \in W_m \Leftrightarrow f(2^m \cdot) \in W_0, \tag{2.12}$$

and that the W_m are orthogonal spaces which sum to $L^2(\mathbb{R})$,

$$L^2(\mathbb{R}) = \dot{\bigoplus_{m\in\mathbb{Z}}} W_m. \tag{2.13}$$

Because of the properties (2.1)–(2.4) of the V_m, it turns out (see [16], [19]) that in W_0 also (as in V_0) there exists a vector ψ such that its integer translates span W_0, i.e.,

$$\overline{\text{span}\{\psi_{0n}\}} = W_0, \tag{2.14}$$

where, as before, $\psi_{mn}(x) = 2^{-m/2}\psi(2^{-m}x - n)$ for $m, n \in \mathbb{Z}$. It follows immediately from (2.12) that then

$$\overline{\text{span}\{\psi_{mn}\}} = W_m,$$

for all $m \in \mathbb{Z}$.

Intuitively one may understand this similarity between W_0 and V_0 by the fact that V_{-1} is "twice as large" as V_0, since V_0 is generated by the integer translates of a single function ϕ_{00}, while V_{-1} is generated by the integer translates of *two* functions, namely ϕ_{-10} and ϕ_{-11}. It therefore seems natural that the orthogonal complement W_0 of V_0 in V_{-1} is also generated by the integer translates of a single function. This hand-waving argument can easily be made rigorous by using group representation arguments. Mere proof of *existence* of a function ψ satisfying (2.14) would however not be enough for practical purposes. A more detailed analysis leads to the following algorithm for the *construction* of ψ (see [16], [19]). We start from a function ϕ such that the ϕ_{0n} are an orthonormal basis for V_0 (if necessary, we apply (2.5)). Since $\phi \in V_0 \subset V_{-1} = \overline{\text{span}\{\phi(2\cdot - n)\}}$, there exist c_n such that

$$\phi(x) = \sum_n c_n \phi(2x - n). \tag{2.15}$$

Define then

$$\psi(x) = \sum_n (-1)^n c_{n+1} \phi(2x + n). \tag{2.16}$$

The corresponding ψ_{0n} will constitute an orthonormal basis of W_0; see [16], [19]. Consequently the ψ_{mn}, for fixed m, will constitute an orthonormal basis of W_m. It follows then from (2.12) that the $\{\psi_{mn},\ m, n \in \mathbb{Z}\}$ constitute an orthonormal basis of wavelets for $L^2(\mathbb{R})$. This completes the explicit construction, in the general case, of an orthonormal wavelet basis from a multiresolution analysis.

EXAMPLE 2.1 (final visit). As we already noted above, the ϕ_{0n} are orthonormal in this example, and

$$\phi(x) = \phi(2x) + \phi(2x - 1).$$

Applying the recipe (2.15)–(2.16) then leads to

$$\psi(x) = \phi(2x) - \phi(2x - 1),$$

which corresponds to (2.7).

Remarks. 1. One can show (see [16]) that the functions ϕ, ψ having all the above properties necessarily satisfy

$$\int dx\, \psi(x) = 0 \tag{2.17}$$

and

$$\int dx\, \phi(x) \neq 0,$$

where we implicitly assume that ϕ, ψ are sufficiently well-behaved for these integrals to make sense (in all examples of practical interest, $\phi, \psi \in L^1$). In fact, one does not even need to assume that the ϕ_{0n} or ψ_{0n} are orthonormal to derive (2.17)–(2.18). In [15] it is shown that (2.17) has to be satisfied even if the ψ_{mn} constitute only a frame (see the introduction). Note also that the transition (2.5) from ϕ to $\tilde{\phi}$, orthonormalizing the ϕ_{0n}, preserves $\int dx\, \phi(x) \neq 0$.

2. If one restricts oneself to the case where ϕ is a *real* function (as in all the examples above), then ϕ is determined uniquely, up to a sign, by the requirement

that the ϕ_{0n} be orthonormal. One then also has $\int dx\, \phi(x) = \pm 1$; we shall fix the sign of ϕ so that

$$\int dx\, \phi(x) = 1. \tag{2.18}$$

In practice, one can often start the whole construction by choosing an appropriate ϕ, i.e., a function ϕ satisfying (2.15) for some c_n. Provided ϕ is "reasonable" (it suffices, e.g., that $\inf_{|\xi| \leqq \pi} |\hat{\phi}(\xi)| > 0$ and that $\Sigma_{k \in \mathbf{Z}} |\hat{\phi}(\xi + 2k\pi)|^2$ is bounded), the closed linear spans V_m of the ϕ_{mn} (m fixed) then automatically satisfy (2.1)–(2.4) and there exists an associated orthonormal basis of wavelets. Two typical examples are

EXAMPLE 2.2.

$$\phi(x) = \begin{cases} x, & 0 \leqq x \leqq 1, \\ 2 - x, & 1 \leqq x \leqq 2, \\ 0, & \text{otherwise.} \end{cases}$$

This is a linear spline function; the spaces V_m consist of continuous, piecewise linear functions. The c_n are given by

$$\phi(x) = \tfrac{1}{2}\phi(2x) + \phi(2x - 1) + \tfrac{1}{2}\phi(2x - 2).$$

EXAMPLE 2.3.

$$\phi(x) = \begin{cases} x^2, & 0 \leqq x \leqq 1, \\ -2x^2 + 6x - 3, & 1 \leqq x \leqq 2, \\ (3 - x)^2, & 2 \leqq x \leqq 3, \\ 0, & \text{otherwise.} \end{cases}$$

This is a quadratic spline function; the spaces V_m consist of C^1, piecewise quadratic functions. The c_n are given by

$$\phi(x) = \tfrac{1}{4}\phi(2x) + \tfrac{3}{4}\phi(2x - 1) + \tfrac{3}{4}\phi(2x - 2) + \tfrac{1}{4}\phi(2x - 3).$$

In these last two examples the corresponding ψ will be, respectively, continuous and piecewise linear, or C^1 and piecewise quadratic. Starting from spline functions one can, in fact, construct orthonormal bases of wavelets with an arbitrarily high number of continuous derivatives. These bases are the Battle-Lemarié bases (see [17], [18], [16]). In these constructions the initial function ϕ is compactly supported, but the ϕ_{0n} are not orthogonal, as illustrated by the two examples above. One therefore has to apply (2.5) before using (2.15), (2.16); the transition $\phi \to \tilde{\phi}$ in (2.5) leads to a non-compactly supported $\tilde{\phi}$, resulting in a non-compactly supported ψ. Typically the Battle-Lemarié wavelets have exponential decay.

We shall see below that for the construction of orthonormal bases of *compactly supported* wavelets it is more natural to start from the coefficients c_n than from the function ϕ.

Up to now, we have restricted ourselves to one dimension. It is very easy, however, to extend the multiresolution analysis to more dimensions. As pointed out by R. Coifman and Y. Meyer [22], this extension was already inherent in the first construction by P. G. Lemarié and Y. Meyer [23] of an n-dimensional wavelet basis. It becomes much more transparent, however, from the multiresolution analysis point of view. Let us illustrate this for e.g. two dimensions. The case of n dimensions, n arbitrary, is completely similar. Assume that we dispose of a one-dimensional multiresolution analysis, i.e., we have at hand a ladder of spaces V_m, and functions ϕ, ψ satisfying (2.1)–(2.4) and (2.14), where the ϕ_{0n} and the ψ_{0n} are assumed to be orthonormal. Define

$$\mathbf{V}_m = V_m \oplus V_m.$$

Clearly, the $\mathbf{V}_m$ define a ladder of subspaces of $L^2(\mathbb{R}^2)$, satisfying (2.1) and the equivalent, for $\mathbb{R}^2$, of (2.2). Moreover, (2.3) holds, and if we define

$$\Phi(x_1, x_2) = \phi(x_1)\phi(x_2),$$

then this two-dimensional function satisfies the analogue of (2.4),

$$\mathbf{V}_m = \overline{\text{linear span}\left\{\Phi_{mn};\ n \in \mathbb{Z}^2\right\}},$$

where Φ_{mn} is defined by

$$\begin{aligned}\Phi_{mn}(x_1, x_2) &= 2^{-m}\Phi(2^{-m}x_1 - n_1, 2^{-m}x_2 - n_2)\\ &= \phi_{mn_1}(x_1)\phi_{mn_2}(x_2).\end{aligned}$$

Note that we use the *same* dilation for both arguments. Because of the definition (2.10) of W_m, we find immediately that

$$\mathbf{V}_{m-1} = \mathbf{V}_m \oplus \left[(V_m \oplus W_m) \oplus (W_m \oplus V_m) \oplus (W_m \oplus W_m)\right].$$

This implies that an orthonormal basis for the orthogonal complement $\mathbf{W}_m$ of $\mathbf{V}_m$ in $\mathbf{V}_{m-1}$ is given by the functions $\phi_{mn_1}\psi_{mn_2}$, $\psi_{mn_1}\phi_{mn_2}$, $\psi_{mn_1}\psi_{mn_1}$, with $n_1, n_2 \in \mathbb{Z}$, or equivalently, by the two-dimensional wavelets Ψ^l_{mn},

$$\Psi^l_{mn}(x_1, x_2) = 2^{-m}\Psi^l(2^{-m}x_1 - n_1, 2^{-m}x_2 - n_2), \tag{2.19}$$

where $l = 1, 2, 3$, $n \in \mathbb{Z}^2$, and

$$\Psi^1(x_1, x_2) = \phi(x_1)\psi(x_2), \tag{2.20}$$

$$\Psi^2(x_1, x_2) = \psi(x_1)\phi(x_2), \tag{2.21}$$

$$\Psi^3(x_1, x_2) = \psi(x_1)\psi(x_2). \tag{2.22}$$

It follows that the Ψ^l_{mn}, $l = 1, 2, 3$, $m \in \mathbb{Z}$, $n \in \mathbb{Z}^2$, constitute an orthonormal basis of wavelets for $L^2(\mathbb{R}^2)$.

The above construction shows how any multiresolution analysis + associated wavelet basis in one dimension can be extended to d dimensions. The decomposition + reconstruction algorithm constructed by S. Mallat for visual data in [8] uses such a two-dimensional basis.

2.B. The Laplacian pyramid scheme of P. Burt and E. Adelson. In this subsection we review some aspects, relevant for the present paper, of Burt and Adelson's algorithm for the decomposition and reconstruction of images. For a more detailed presentation, and for applications, we refer the reader to [20].

One of the goals of a decomposition scheme for images is to remove the very high correlations existing between neighboring pixels, in order to achieve data compression. Several different schemes have been proposed to achieve this. Typically they use a prediction method, in which the value at a pixel is predicted (by a weighted average) from either previously encoded or neighboring pixels, and only the difference between the actual pixel value and the predicted value is encoded. Using the neighboring pixels for prediction is more natural and should lead to more accurate prediction (and therefore to greater data compression), but is much harder to implement than the easy causal prediction scheme, using only previously encoded pixels. The scheme proposed by Burt and Adelson combines the ease of computation of a causal scheme with the advantages and elegance of a neighborhood-based (noncausal) scheme. The result is—we quote directly from [20a]—

> "$\cdots$ a technique for image encoding in which local operators of many scales but identical shape serve as the basis functions."

The analogy with multiresolution analysis is evident from this quote.

Images are two-dimensional, and the Laplacian pyramid scheme is a two-dimensional algorithm. For simplicity, the review below will be restricted to one dimension. This does not really matter, except in details (which will be pointed out). Moreover, the two-dimensional schemes used in [20] are in fact obtained (for simplicity reasons) as a tensor product of two one-dimensional schemes.

Our presentation will be already adapted to later use in this paper, and slightly different in notation from [20]. Except for this difference, what follows is the construction in [20].

The original (one-dimensional) data can be represented as a sequence of real numbers, $(c_n)_{n \in \mathbf{Z}}$, representing the pixel values. For later convenience, we give this sequence the index 0,

$$c_n^0 = c_n.$$

The main idea is to decompose c^0 into different sequences corresponding to distinct ranges of spatial frequency. The highest level, with only the high frequency content of c^0, is obtained by computing the difference between c^0 and a blurred version $\tilde{c}^0$. The remainder, i.e., the blurred version, contains only lower spatial frequencies, and can therefore be sampled more sparsely than c^0 itself, without loss of information. The Laplacian pyramid algorithm provides an elegant and easily implementable scheme for doing all this. The whole process is repeated several times in order to achieve the desired decomposition.

One starts by transforming the sequence c^0 into a sequence c^1 by means of an operator which both averages and decimates,

$$c_k^1 = \sum_n w(n - 2k) c_n^0. \tag{2.23}$$

The weighing coefficients $w(n)$ are always real; in [20] they are chosen to be symmetric and normalized, i.e.,

$$\begin{aligned} &w(n) = w(-n), \\ &\sum_n w(n) = 1. \end{aligned} \tag{2.24}$$

They are also required to satisfy an "equal contribution constraint", stipulating that the sum of all the contributions from a given node n is independent of n, i.e., all the nodes contribute the same total amount,

$$\sum_k w(n - 2k) \text{ is independent of } n.$$

This can be rewritten as

$$\sum_n w(2n) = \sum_n w(2n + 1). \tag{2.25}$$

We shall come back below to the mathematical significance of this requirement. Examples given in [20] are

$$\begin{aligned} &w(n) = 0 \quad \text{if} \quad |n| > 2, \\ &w(2) = w(-2) = \tfrac{1}{4} - \tfrac{1}{2}a, \\ &w(1) = w(-1) = \tfrac{1}{4}, \\ &w(0) = a; \end{aligned} \tag{2.26}$$

the different values considered for a are $a = .6$, $.5$, $.4$ and $.3$.

The sequence c^1 plays a double role: it will serve as the input sequence (instead of c^0) for the next level of the pyramid, and it is also an intermediate step for the computation of the blurred version $\tilde{c}^0$ to be subtracted from c^0. (Note that we cannot have c^1 itself as this blurred version: c^1 and c^0 "live" on different scales—see Figure 1). More precisely,

$$\tilde{c}_n^0 = \sum_k w(n-2k)c_k^1, \tag{2.27}$$

or

$$\tilde{c}_n^0 = \sum_l \tilde{w}(n,l)c_l^0,$$

where

$$\tilde{w}(n,l) = \sum_k w(n-2k)w(l-2k). \tag{2.28}$$

Notice that this does not quite define a convolution; from (2.28) one sees that $\tilde{w}(n,l) = \tilde{w}(n-2, l-2)$, but $\tilde{w}(n,l) \neq \tilde{w}(n-1, l-1)$ in general. The sequence $\tilde{c}^0$ is clearly a blurred version of c^0; one then defines the difference d^0 by

$$d_n^0 = c_n^0 - \tilde{c}_n^0. \tag{2.29}$$

Knowing this difference sequence (the high spatial frequency content of c^0) and c^1 (a low-pass filtered version of c^0, sampled at a sparser rate) is clearly sufficient to reconstruct the data c^0, since

$$c_n^0 = d_n^0 + \sum_k w(n-2k)c_k^1.$$

The whole process is then iterated. From c^1 one computes c^2 and $\tilde{c}^1$, d^1 is the difference $c^1 - \tilde{c}^1$, etc. A graphical representation of the transitions $c^0 \to c^1 \to c^2 \to \cdots$ and $c^1 \to \tilde{c}^0$ is given in Figures 1a and 1b.

A more condensed notation for all the above is the following. Define the operator F: $l^2(\mathbb{Z}) \to l^2(\mathbb{Z})$ (F for "filter") by

$$(Fa)_k = \sum_n w(n-2k)a_k. \tag{2.30}$$

Then (2.23), (2.27) and (2.29) become

$$c^1 = Fc^0, \tag{2.31}$$

$$\tilde{c}^0 = F^*c^1 = F^*Fc^0, \tag{2.32}$$

$$d^0 = c^0 - \tilde{c}^0 = (\mathbb{1} - F^*F)c^0. \tag{2.33}$$

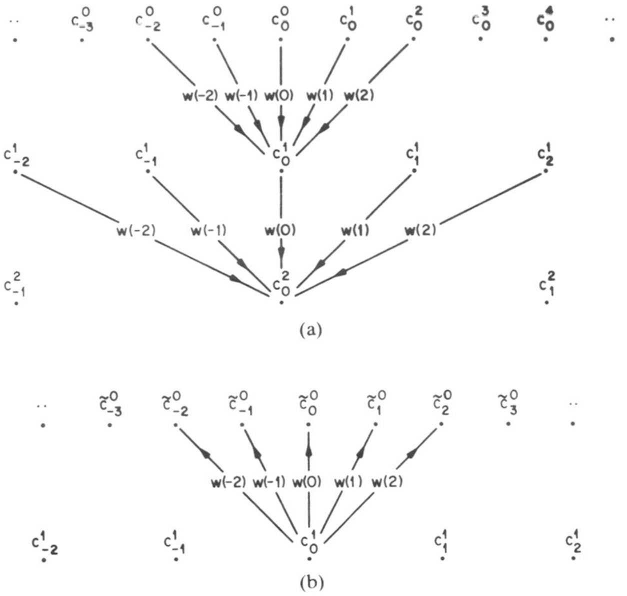

Figure 1. Graphical representation of the Laplacian pyramid scheme (redrawn from [20]).
a. The transition $c^0 \to c^1 \to c^2$. For simplicity's sake we have restricted ourselves to the case $w(n) = 0$ if $|n| > 2$, and only the computation of c_0^1 and c_0^2 are depicted.
b. The transition $c^1 \to \tilde{c}^0$.

Here we use the standard notation F^* for the adjoint of the (bounded) operator F. Note that we implicitly assume that $c^0 \in l^2(\mathbb{Z})$, or, in signal analysis terms, that the data sequence c^0 has finite energy. In practice, the sequence c^0 is finite, and this constraint does not matter.

The total decomposition consists thus in L consecutive steps (in practice $L = 5$ or 6), with

$$c^l = Fc^{l-1}, \qquad l = 1, \cdots, L,$$
$$d^{l-1} = c^{l-1} - F^* c^l = (\mathbb{1} - F^*F)c^{l-1}. \tag{2.34}$$

From the sequences $d^0, \cdots, d^{L-1}, c^L$ one then reconstructs c^0 recursively by

$$c^{l-1} = d^{l-1} + F^*c^l. \tag{2.35}$$

At every step, in the decomposition (2.34) as well as in the reconstruction (2.35) the same filter coefficients are used, and all the operations involved are direct and linear (no solving of complicated systems of equations!). This makes this algorithm very easy to implement. The decimation aspect in the operator F, which reduces the number of entries in the c^l by a factor 2 at every step, makes the whole decomposition algorithm as fast as a fast Fourier transform (see [20]).

Let N be the total number of non-zero entries in c^0. Then the total number of entries in $d^0, \cdots, d^{L-1}, c^L$ (except for edge effects) is

$$N + N/2 + \cdots + N/2^{L-1} + N/2^L = 2N(1 - 2^{-L-1}).$$

After the Laplacian pyramid decomposition there is thus a larger number of entries (almost twice as many) than in the original data sequence. However, it turns out that, because of the removal of correlations, the decomposed data can be greatly compressed (see [20a]). The net effect is still an appreciable data compression. We shall not go into this here, however. Note that the increase of the number of entries is less pronounced in two dimensions (a factor $\frac{4}{3}$ instead of 2).

The similarity between the Laplacian pyramid algorithm and a multiresolution analysis is now clear. In both approaches, the data (a function in multiresolution analysis, a sequence in the Laplacian pyramid) are decomposed into a "pyramid" of approximations, corresponding to less and less detail. Moreover, the differences between each two successive approximations are computed (corresponding to the wavelet decomposition in the multiresolution analysis). However, it is also clear that the schemes are quite different in the details of the computation of the decomposition. The algorithm developed by S. Mallat, described in the next subsection, retains the attractive features of the Laplacian pyramid scheme, but is much closer to the analysis described in subsection 2A.

The filter coefficients $w(n)$, or equivalently the filter operator F, are associated in [20] with "equivalent weighting functions". Only the limit of these functions will be relevant for us; we conclude this subsection by its definition and a few of its properties. One may wonder which kind of input sequence c^0 corresponds to the "simplest" decomposition sequence, i.e., to $d^0 = \cdots = d^{L-1} = 0$, and $(c^L)_n = \delta_{n0}$. The answer is obviously (use the reconstruction algorithm)

$$c^0 = (F^*)^L e, \tag{2.36}$$

where e is the sequence $e_n = \delta_{n0}$. If, e.g., $L = 1$, then the entries of c^0 are exactly the $w(n)$. Since any sequence can be considered as a sum of translated versions of e, the sequence $c^0 = (F^*)^L e$ gives the basic building block for the subspace $(F^*)^L l^2(\mathbb{Z})$, i.e., for the L-th level component sequences. It is therefore important that these sequences c^0 do not look messy, which they well might, for L large enough (for a "messy" example, see Figure 4 in subsection 3B). One can make a graphical representation of the c^0 defined by (2.36), for successive L. We represent the sequence e by a simple histogram, with value 1 for $-\frac{1}{2} \leqq x < \frac{1}{2}$, 0 otherwise (see Figure 2). The sequence F^*e "lives" on a scale twice as small, and will therefore be represented by a histogram with step widths $\frac{1}{2}$ (as opposed to 1 for e); its different amplitudes are given by $2(F^*e)_n = 2w(n)$. Similarly, $(F^*)^l e$ is represented by a histogram with step width 2^{-l}; the successive amplitudes are given by $2^l((F^*)^l e)_n$. We have introduced an extra factor 2 at every step in our representation for normalization purposes: the area under each histogram is always 1. This normalization will be convenient in Section 3. The example plotted in Figure 2 corresponds to the $w(n)$ given by (2.26), with $a = .375$. Plots for $a = .6, .5, .4$ and $.3$, with slightly different conventions, can be found in [20]; our choice $a = .375$ shall become clear below. One finds a very rapid convergence of these histograms to a rather nice function. This surprising feature is in fact due, in large part, to the special form (2.26) of the coefficients $w(n)$, and in particular, to condition (2.25). The following argument shows why. The "representation" of e in Figure 2 is the characteristic function of the interval $[-\frac{1}{2}, \frac{1}{2}[$, which we

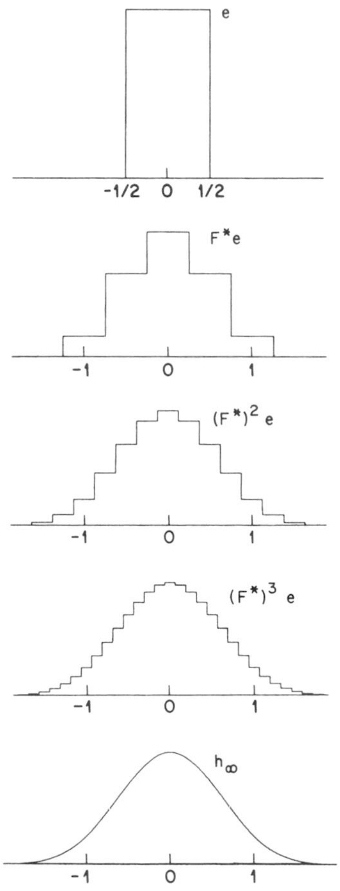

Figure 2. The successive sequences e, F^*e, $(F^*)^2e$, $(F^*)^3e$ represented by histograms, and the limit function h_∞ (see text). We have taken the $w(m)$ as defined by (2.26), with $a = .375$.

denote by h_0. The representations h_1, h_2 of, respectively, $F^*e, (F^*)^2e$ are given by

$$h_1(x) = 2\sum_n w(n)\chi_{[-1/4,1/4[}(x - \tfrac{1}{2}n) \tag{2.37}$$

and

$$h_2(x) = 4\sum_n w(n)\Big[\sum_m w(m)\chi_{[-1/8,1/8[}(x - \tfrac{1}{2}n - \tfrac{1}{4}m)\Big]. \tag{2.38}$$

To make the transition from h_{j-1} to h_j one

(i) divides h_{j-1}, a step function with step width $2^{-(j-1)}$, into its components

$$h_{j-1} = \sum_k a_{j-1,k}\chi_{[2^{-(j-1)}(k-1/2),\,2^{-(j-1)}(k+1/2)[} \tag{2.39}$$

(see Figure 3c),

(ii) replaces every component by a suitably scaled and recentered version of h_1,

$$\chi_{[2^{-j+1}(k-1/2),\,2^{-j+1}(k+1/2)[} \to h_1(2^{j-1}x - k)$$
$$= 2\sum_n w(n)\chi_{[-1/4,1/4[}(2^{j-1}x - k - \tfrac{1}{2}n),$$

(see Figure 3d),

(iii) sums it all up,

$$h_j(x) = 2\sum_k a_{j-1,k}\sum_n w(n)\chi_{[2^{-j}(2k+n-1/2),\,2^{-j}(2k+n+1/2)[}$$

(see Figure 3e).

These different steps are illustrated by Figure 3. The construction amounts to defining

$$h_j = \tilde{T}_j h_{j-1} \quad \text{or} \quad h_j = \tilde{T}_j\tilde{T}_{j-1}\cdots\tilde{T}_1 h_0, \tag{2.40}$$

where

(2.41)

$$(\tilde{T}_l f)(x) = 2\sum_k\sum_n w(n)\big(\chi_{[2^{-l+1}(k-1/2),\,2^{-l+1}(k+1/2)[}f\big)\big(2x - 2^{-l+1}(k+n)\big).$$

The iterative algorithm (2.40), (2.41) is extremely easy to implement numerically.

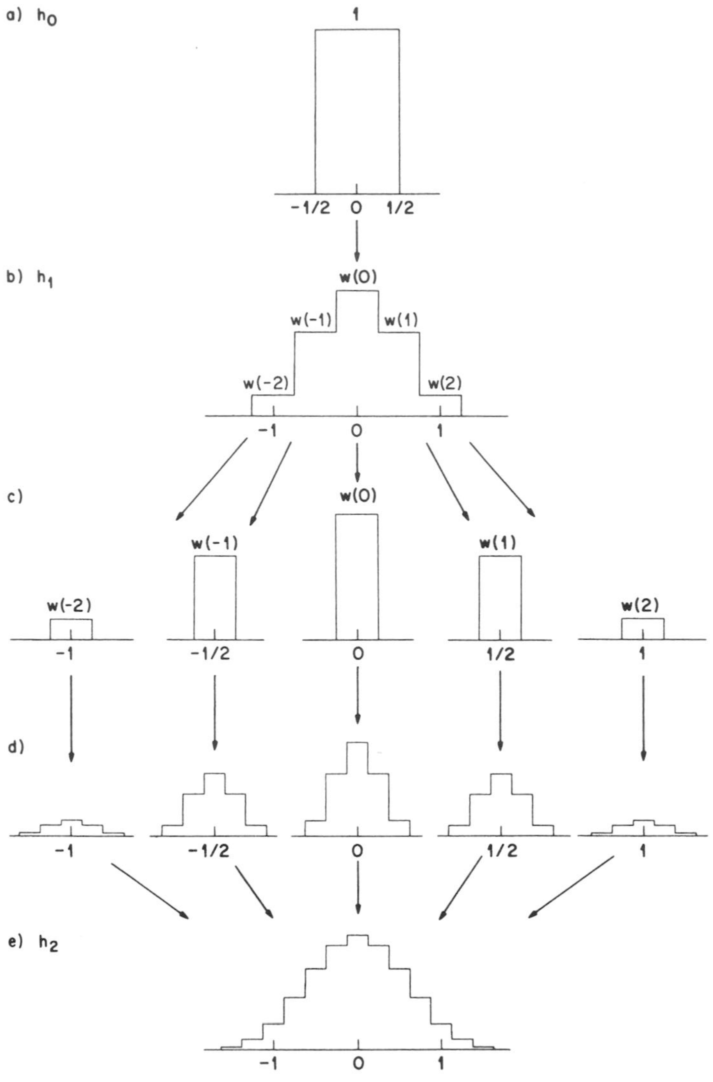

Figure 3. a) $h_0(x) = \chi_{[-1/2,1/2[}(x)$;

b) $h_1(x) = 2\Sigma w(n)\chi_{[n/2-1/4,\, n/2+1/4]}(x)$,

c) h_1 is decomposed into its "components"; each component is a multiple of the characteristic function of an interval of length $\frac{1}{2}$. (The "components" of h_j would have width 2^{-j}).

d) Each "component" is replaced by a proportional version of h_1, centered around the same point as the component, and scaled down by a factor $\frac{1}{2}$. (This scaling factor would be 2^{-j} for h_j).

e) The functions in d) are added to constitute h_2 (or h_{j+1}, if one starts from h_j in c)).

The h_j can, however, also be written differently. Let us go back to expressions (2.36), (2.37) for h_1, h_2. These can be rewritten as

$$h_1(x) = 2\sum_n w(n)h_0(2x-n), \tag{2.42}$$

$$\begin{aligned} h_2(x) &= 4\sum_n w(n)\sum_m w(m)\chi_{[-1/4,1/4[}(2x-n-\tfrac{1}{2}m) \\ &= 2\sum_n w(n)h_1(2x-n). \end{aligned} \tag{2.43}$$

This suggests

$$h_j = Th_{j-1} = \cdots = T^j h_0, \tag{2.44}$$

where

$$(Tf)(x) = 2\sum_n w(n)f(2x-n). \tag{2.45}$$

The following argument shows that (2.44) is indeed true:

$$\begin{aligned} &(T\tilde{T}_l f)(x) \\ &\quad = 4\sum_{m,n} w(m)w(n) \\ &\qquad \cdot \sum_k \left(\chi_{[2^{-l+1}(k-1/2),2^{-l+1}(k+1/2)[}f\right)\left(4x-2m-2^{-l+1}(n+k)\right), \\ &(\tilde{T}_{l+1}Tf)(x) \\ &\quad = 4\sum_{m,n} w(m)w(n)\sum_k \chi_{[2^{-l}(k-1/2),2^{-l}(k+1/2)[}\left(2x-2^{-l}(n+k)\right) \\ &\qquad \cdot f\left(4x-m-2^{-l+1}(n+k)\right). \end{aligned}$$

Substituting $k = k' + 2^{l-1}m$ into this last sum, we find

$$\begin{aligned} &(\tilde{T}_{l+1}Tf)(x) \\ &\quad = 4\sum_{m,n} w(m)w(n)\sum_{k'}\Big[\chi_{[2^{-l}(k'-1/2),2^{-l}(k'+1/2)[}\left(2x-2^{-l}(n+k')-m\right) \\ &\qquad\qquad \cdot f\left(4x-2m-2^{-l+1}(n+k')\right)\Big] \\ &\quad = (T\tilde{T}_l f)(x). \end{aligned}$$

Since (see (2.42), (2.43)) $h_1 = Th_0$, $h_2 = T^2h_0$, it follows that

$$\begin{aligned} h_j &= \tilde{T}_j \cdots \tilde{T}_3 h_2 = \tilde{T}_j \cdots \tilde{T}_3 T h_1 \\ &= \tilde{T}_j \cdots \tilde{T}_3 T \tilde{T}_1 h_0 = \tilde{T}_j \cdots \tilde{T}_4 T \tilde{T}_2 \tilde{T}_1 h_0 \\ &= \cdots = T\tilde{T}_{j-1} \cdots \tilde{T}_1 h_0 = Th_{j-1}, \end{aligned}$$

which proves (2.44).

We have thus two different ways, (2.44) and (2.40), to compute the h_j. Figure 2 shows that, at least for some choices of the $w(n)$, the functions h_j converge, for $j \to \infty$, to a "nice" function h_∞. The explicit proofs which will be given in subsection 3B show that, at least for the examples (2.26) with $.125 < a < .625$, the function h_∞ is continuous (see (2.46) below), has compact support, and that the convergence $h_j \to h_\infty$ is uniform. Let us just accept these facts for the moment.

The two formulas (2.40) and (2.44) are both extremely useful in the study (and the proof) of this convergence. The construction of $h_\infty = \lim_{j\to\infty} h_j$ via (2.40) has the following nice localization feature. To compute the value of $h_{j+1}(x)$ the recursion $h_{j+1} = \tilde{T}_j h_j$ uses only values $h_j(y)$ for $|y - x| \leqq 2^{-j-1}(N+1)$, where we assume $w(n) = 0$ for $n \geqq 2N$. Consequently, the value of $h_\infty(x)$ can be computed using only the values of $h_j(y)$ for $|y - x| \leqq 2^{-j}(N+1)$. For increasing j, this lends a "zoom-in" quality to the graphical construction of which Figure 2 is an example. This is extremely useful when one wants to focus on details of the behavior of h_∞ (see, e.g. Figure 6 in subsection 4B). This localization feature is not present in (2.44). The formula $h_j(x) = (Th_{j-1})(x)$ uses values of h_{j-1} at points which stay at fixed distance of each other (i.e., $2x$, $2(x \pm \frac{1}{2})$, $2(x \pm 1), \cdots$), independently of how large j is. The usefulness of (2.44) is therefore not "graphical". It is, however, this less local formula which will be most useful in proving convergence of the h_j, continuity of h_∞, etc.

Introducing Fourier transforms, (2.45) can be rewritten as

$$(Tf)^\wedge(\xi) = W(\tfrac{1}{2}\xi)\hat{f}(\tfrac{1}{2}\xi),$$

where

$$W(\xi) = \sum_n w(n)e^{in\xi}.$$

Consequently, from (2.45), one obtains

$$\hat{h}_l(\xi) = (2\pi)^{-1/2}\left[\prod_{j=1}^{l} W(2^{-j}\xi)\right]\frac{\sin(2^{-l-1}\xi)}{2^{-l-1}\xi}.$$

For $l \to \infty$, this converges, pointwise, to

$$\hat{h}_\infty(\xi) = (2\pi)^{-1/2} \prod_{j=1}^{\infty} W(2^{-j}\xi),$$

provided this infinite product makes sense. (We shall come back to this, and other convergence problems, in subsection 3B. It turns out that, for $w(n)$ as chosen in (2.26), the convergence $h_l \to h_\infty$ holds in all L^p-spaces, $1 \leqq p \leqq \infty$.) Because of the constraint (2.25), one finds that $W(\xi)$ is divisible by $(1 + e^{i\xi})$,

$$\begin{aligned} W(\xi) &= \tfrac{1}{2}(1 + e^{i\xi})Q(\xi) \\ &= e^{i\xi/2}\cos\tfrac{1}{2}\xi Q(\xi). \end{aligned}$$

Combining this with

$$\prod_{j=1}^{\infty} \cos(2^{-j}\xi) = \frac{\sin\xi}{\xi},$$

we find

$$\hat{h}_\infty(\xi) = (2\pi)^{-1/2} e^{i\xi/2} \frac{2\sin\frac{1}{2}\xi}{\xi} \prod_{j=1}^{\infty} Q(2^{-j}\xi).$$

The constraint (2.25) leads thus to a factor ξ^{-1} in $\hat{h}_\infty$, i.e., to some regularity in h_∞! Without this constraint, as can be easily checked, the graphical procedure in Figure 2 can lead to rather horrible (fractal) functions h_∞ (see e.g. Figure 4). In fact, for the examples (2.26) one even finds two factors $\cos\frac{1}{2}\xi$,

$$W(\xi) = \left(\cos\tfrac{1}{2}\xi\right)^2\left[(8a - 3) + 4(1 - 2a)\left(\cos\tfrac{1}{2}\xi\right)^2\right].$$

Using an estimation technique due to P. Tchamitchian (see Lemma 3.2 below) one finds that this leads to

$$|\hat{h}_\infty(\xi)| \leqq C(1 + |\xi|)^{-2+\log_2[\max(1,\,|8a-3|)]}. \tag{2.46}$$

For $.125 < a < .625$, which includes all the choices in [20], this implies that h_∞ is continuous. For $a = \frac{3}{8} = .375$ (the example chosen in Figure 2), the decay of $\hat{h}_\infty$ is even stronger,

$$\begin{aligned} W(\xi) &= \left(\cos\tfrac{1}{2}\xi\right)^4, \\ \hat{h}_\infty(\xi) &= (2\pi)^{-1/2}\left(\frac{\sin\frac{1}{2}\xi}{\frac{1}{2}\xi}\right)^4. \end{aligned}$$

In this case, h_∞ is thus a fourth-order convolution of $\chi_{[0,1[}$ with itself, which results in a $C^{3-\varepsilon}$ function.

The above remarks show that constraints on the $w(n)$, corresponding to divisibility of $W(\xi)$ by $(1 + e^{i\xi})$, result in regularity of h_∞. Constructions similar to (2.44) will be used in Section 3, where the above "trick" for imposing regularity on h_∞ will turn up again.

This concludes our review of the Laplacian pyramid scheme. The above is by no means a complete review; only those aspects relevant to the present paper have been highlighted. For more details, and especially for applications (data compression, image splining) the reader should consult [20a] and [20b].

Remark. During the last revision of this paper before publication, Y. Meyer drew my attention to related work by G. Deslauriers and S. Dubuc [29]. They are interested in functions defined recursively by the following interpolation scheme. At the l-th step, the values of f at the points $2^{-l}(2k + 1)$, $k \in \mathbb{Z}$, are computed from the $f(k2^{-l+1})$ via the formula

$$f\big((2k+1)2^{-l}\big) = \sum_{m \in \mathbb{Z}} a_m f\big((k-m)2^{-l+1}\big).$$

In many applications considered by Deslauriers and Dubuc, the interpolation procedure is symmetric, i.e., $a_{-m} = a_{m+1}$ for all $m \in \mathbb{Z}$. For suitable choices of the a_m, the functions f constructed via this dyadic interpolation scheme, starting from the $f(k)$, $k \in \mathbb{Z}$, are continuous, and are therefore completely characterized by their values at the dyadic rational points $x = k2^{-l}$, $k \in \mathbb{Z}$, $l \in \mathbb{N}$. A typical function f can be written as

$$f(x) = \sum_{k \in \mathbb{Z}} f(k) g(x-k),$$

where g is the function obtained by interpolation from $g(0) = 1$, $g(k) = 0$ for $k \in \mathbb{Z} \setminus \{0\}$. The definition of g via the dyadic interpolation scheme is exactly the same as our "graphical recursion" (2.40), with the choice $w(0) = \frac{1}{2}$, $w(2n) = 0$ for $n \neq 0$, $w(2n+1) = \frac{1}{2}a_{n+1}$, $n \in \mathbb{Z}$. The analysis of the properties of g in [29] is then carried out by means of the same correspondence between "graphical recursion" and the iterative formula (2.44). Imposing $\Sigma a_m = 1$ (i.e., $\Sigma w(n) = 1$) immediately leads to $w(\xi) = \left[\frac{1}{2}(1 + e^{i\xi})\right]^2 Q(\xi)$, which is then exploited, in [29], to impose continuity on g. There is therefore a clear similarity between the techniques used here and those exposed in [29]. The applications are different, however. Moreover, the proofs given in Section 3 apply to more general cases than those in [29], since we do not impose $w(2n) = 0$ for $n \neq 0$, nor $w(2n+1) = w(-2n-1)$.

2.C. The wavelet based decomposition and reconstruction algorithm of S. Mallat. In [8], Stéphane Mallat exploits the attractive features of multiresolution analysis to construct a decomposition and reconstruction algorithm for

2-d-images that has the same philosophy as the Laplacian pyramid scheme, but is more efficient and orientation selective. It is interesting to remark that the development of the concept of multiresolution analysis was triggered by the multiresolution methods, and in particular by the Laplacian pyramid. The full mathematical study of the concept, by S. Mallat and Y. Meyer, was done more or less simultaneously with the practical development, by S. Mallat, of his algorithm for vision analysis and reconstruction. This is not the only instance in which theoretical developments concerning wavelets find their inspiration in applications: the last few years have seen a constant feedback between theory and applications. In fact, this paper is another such instance.

Let us start by a review of the algorithm in one dimension. As in the previous subsection, we want to decompose a sequence $c^0 = (c_n^0)_n \in l^2(\mathbb{Z})$ into levels corresponding to different spatial frequency bands. To achieve this, we shall use a multiresolution analysis, which can be chosen freely (as long as (2.1)–(2.4) are satisfied), but has to be kept fixed for the whole algorithm. We suppose thus that we have chosen spaces V_m and a function ϕ such that (2.1)–(2.4) are satisfied. We assume (if necessary, we apply (2.5) first) that the ϕ_{0n} are orthonormal. Let $\{\psi_{mn};\ m, n \in \mathbb{Z}\}$ be the associated orthonormal wavelet basis (we shall keep the same notations as in subsection 2A). The multiresolution analysis and orthonormal basis chosen in [8] is one in which the V_m consist of cubic spline functions (cf. Examples 2.2 and 2.3, corresponding to linear and quadratic splines, respectively); the corresponding orthonormal basis is one of the Battle-Lemarié bases. In what follows we shall assume that both ϕ and ψ are real, as they are in [8] and indeed in most practical examples.

Form the data sequence $c^0 \in l^2(\mathbb{Z})$ we construct a function f,

$$f = \sum_n c_n^0 \phi_{0n},$$

or

$$f(x) = \sum_n c_n^0 \phi(x-n).$$

This function is clearly an element of V_0. We can now use the whole multiresolution analysis apparatus on this function. We shall compute the successive $P_j f$, corresponding to more and more "blurred" versions of f (and hence of the data sequence c^0), and also the $Q_j f$, corresponding to the difference in information between the "versions" of f at two successive resolution levels. Eventually, of course, this has to be translated back to a "sequence" (as opposed to a "function") language, but this turns out to be very easy.

As element of $V_0 = V_1 \oplus W_1$, f can be decomposed into its components along V_1 and W_1,

$$f = P_1 f + Q_1 f.$$

Each of these components can be expanded with respect to the orthonormal bases ϕ_{1n}, ψ_{1n}, respectively,

$$P_1 f = \sum_k c_k^1 \phi_{1k},$$

$$Q_1 f = \sum_k d_k^1 \psi_{1k}.$$

The sequence c^1 represents a smoothed version of the original data sequence c^0, while d^1 represents the difference in information between c^0 and c^1 (cf. the discussion of P_j, Q_j in subsection 2A). The sequences c^1, d^1 can be computed as a function of c^0 in the following way. Since the ϕ_{1n} are orthonormal bases of V_1, one has

$$\begin{aligned} c_k^1 &= \langle \phi_{1k}, P_1 f \rangle = \langle \phi_{1k}, f \rangle \\ &= \sum c_n^0 \langle \phi_{1k}, \phi_{0n} \rangle, \end{aligned}$$

where

$$\begin{aligned} \langle \phi_{1k}, \phi_{0n} \rangle &= 2^{-1/2} \int dx\, \phi(\tfrac{1}{2}x - k) \phi(x - n) \\ &= 2^{-1/2} \int dx\, \phi(\tfrac{1}{2}x) \phi(x - (n - 2k)). \end{aligned}$$

This can be rewritten as

$$c_k^1 = \sum_n h(n - 2k) c_n^0 \tag{2.47}$$

with

$$h(n) = 2^{-1/2} \int dx\, \phi(\tfrac{1}{2}x) \phi(x - n).$$

Note that these $h(n)$ are, up to a normalization factor $2^{-1/2}$, exactly the coefficients $c(n)$ appearing in (2.15). Similarly,

$$d_k^1 = \sum_n g(n - 2k) c_n^0 \tag{2.48}$$

with

$$g(n) = 2^{-1/2} \int dx\, \psi(\tfrac{1}{2}x) \phi(x - n).$$

It follows that the expressions for c^1, d^1 as a function of c^0 are of *exactly* the same type as (2.23) in the Laplacian pyramid scheme. The main difference between the two schemes is that *both* the blurred, lower resolution c^1 and the "difference" sequence d^1 are now obtained via a filter of type (2.23). The filter coefficients $h(n)$, $g(n)$ are fixed by the chosen multiresolution analysis framework. It turns out that the $h(n)$ have many properties in common with the $w(n)$ in subsection 2B; for instance, the $h(n)$ satisfy a normalization condition, i.e., $\Sigma_n h(n) = \sqrt{2}$ (see subsection 3A for an explanation of the difference in normalization with the $w(n)$). The requirement $\Sigma_n h(2n) = \Sigma_n h(2n+1)$ is also satisfied by most interesting examples, and in particular in [8] (we shall come back to this later). The filter coefficients $g(n)$ are of a different nature, as one would expect; in particular, $\Sigma_n g(n) = 0$.

Introducing a shorthand notation similar to (2.30), we rewrite (2.47), (2.48) as

$$c^1 = Hc^0,$$

$$d^1 = Gc^0,$$

where H, G are bounded operators from $l^2(\mathbb{Z})$ to itself,

$$(Ha)_k = \sum_n h(n-2k)a_n,$$
$$(Ga)_k = \sum_n g(n-2k)a_n. \tag{2.49}$$

The procedure can now be iterated; since $P_1 f \in V_1 = V_2 \oplus W_2$, we have

$$P_1 f = P_2 f + Q_2 f,$$

$$P_2 f = \sum_k c_k^2 \phi_{2k},$$

$$Q_2 f = \sum_k d_k^2 \psi_{2k}.$$

One finds then

$$c_k^2 = \langle \phi_{2k}, P_{2f} \rangle = \langle \phi_{2k}, P_1 f \rangle$$
$$= \sum_n c_n^1 \langle \phi_{2k}, \phi_{1n} \rangle.$$

It is very easy to check, however, that

$$\langle \phi_{j+1\,k}, \phi_{jn} \rangle = h(n-2k),$$

independently of j. It follows that

$$c_k^2 = \sum_n h(n-2k)c_n^1$$

or

$$c^2 = Hc^1.$$

Similarly,

$$d^2 = Gc^1.$$

Clearly this can now be iterated as many times as wanted. At every step one finds

$$\begin{aligned} P_{j-1}f &= P_j f + Q_j f \\ &= \sum_k c_k^j \phi_{jk} + \sum_k d_k^j \psi_{jk} \end{aligned}$$

with

$$c^j = Hc^{j-1}, \tag{2.50}$$

$$d^j = Gc^{j-1}. \tag{2.51}$$

This is the desired decomposition. The successive c^j are lower and lower resolution versions of the original c^0, each sampled twice as sparsely as their predecessor (due to the factor 2 in the filter coefficients in (2.47)), and the d^j contain the difference in information between c^{j-1} and c^j. Moreover, the c^j, d^j are computed via a tree algorithm (2.50), (2.51). This computation is therefore as easy to implement as the Laplacian pyramid scheme.

Note that Mallat's algorithm is more economical than the Laplacian pyramid scheme. In practice, one will again stop the decomposition after a finite number L of steps, i.e., c^0 will be decomposed into $d^1, \cdots, d^L$ and c^L. If c^0 has initially N non-zero entries, then (neglecting edge effects) the total number of non-zero entries in the decomposition is $N/2 + N/4 + \cdots + N/2^{L-1} + N/2^L + N/2^L = N$. This shows that, unlike the Laplacian pyramid scheme (see subsection 2B), Mallat's algorithm preserves, at every step, the number of non-zero entries (as was to be expected from an algorithm based on an orthonormal basis decomposition).

So far we have only described the decomposition part of the algorithm. The reconstruction part is just as easy. Suppose we know c^j and d^j. Then

$$\begin{aligned} P_{j-1}f &= P_j f + Q_j f \\ &= \sum_k c_k^j \phi_{jk} + \sum_k d_k^j \psi_{jk}, \end{aligned}$$

and hence

$$\begin{aligned} c_n^{j-1} &= \langle \phi_{j-1\,n}, P_{j-1} f \rangle \\ &= \sum_k c_k^j \langle \phi_{j-1\,n}, \phi_{jk} \rangle + \sum d_k^j \langle \phi_{j-1\,n}, \psi_{jk} \rangle \\ &= \sum_k h(n-2k) c_k^j + \sum_k g(m-2k) d_k^j, \end{aligned}$$

or

$$c^{j-1} = H^* c^j + G^* d^j. \tag{2.52}$$

The reconstruction algorithm is therefore also a tree algorithm, using the same filter coefficients as the decomposition.

Remark. In fact, the transition $c^{j-1} \to c^j, d^j$ corresponds to a change of basis in V^{j-1}, $\{\phi_{j-1\,k};\ k \in \mathbb{Z}\} \to \{\phi_{jk}, \psi_{jk};\ k \in \mathbb{Z}\}$. Because of the underlying wavelet structure the orthogonal matrix associated to this basis change has a peculiar structure. The transition $c^j, d^j \to c^{j-1}$ is given by the transposed matrix; this is the reason why the adjoints H^*, G^* of H and G turn up in (2.52).

All the above is one-dimensional. As an image decomposition and reconstruction algorithm, Mallat's scheme is of course two-dimensional, and corresponds to a two-dimensional multiresolution analysis (see subsection 2A). Since the corresponding wavelet basis vectors can all be written as products of one-dimensional ψ_{jk}, ϕ_{jk} (see (2.19)–(2.22)), the two-dimensional algorithm itself can also be generated by a "tensor product" of the one-dimensional algorithm (see [8]). More specifically, the sequences to be decomposed are now elements of $l^2(\mathbb{Z}^2)$,

$$c^0 = \left(c^0_{mn}\right)_{m,\,n \in \mathbb{Z}},$$

and one defines G_r, H_r and G_c, H_c as the filters G, H defined by (2.49), but acting only on the first, respectively, the second, coefficient (r for "rows", c for "columns"). Then c^0 is decomposed into c^1 and *three* difference sequences (corresponding to the Ψ^j, $j = 1, 2, 3,$—see (2.20)–(2.22)) $d^{1,1}$, $d^{1,2}$ and $d^{1,3}$,

$$\begin{aligned} c^1 &= H_c H_r c^0, \\ d^{1,1} &= G_c H_r c^0, \\ d^{1,2} &= H_c G_r c^0, \\ d^{1,3} &= G_c G_r c^0. \end{aligned}$$

The operator $G_c H_r$ "smooths" over the column index, and looks at the "difference" ($\to$ high frequency information) for the row index; typically, $d^{1,1}$ will be large when a horizontal edge is present. Similarly, $d^{1,2}$ detects vertical edges. It follows that, at no extra cost, Mallat's algorithm is orientation sensitive, which the two-dimensional Laplacian pyramid scheme of [20] was not. In [8], S. Mallat gives a very striking graphical representation of the whole two-dimensional scheme, illustrated with several examples, which clearly show, in particular, the orientation specificity of his algorithm.

3. Equivalence Between Mallat's Discrete Algorithm and Multiresolution Analysis

3.A. Weaning Mallat's algorithm from its multiresolution parent. Ultimately, Mallat's decomposition and reconstruction algorithm, i.e., (2.50), (2.51) and (2.52), deals only with sequences; the underlying multiresolution analysis is only used in the computation of the filter operators H and G. In this subsection we extract the properties of H and G that make the scheme work, without reference to multiresolution analysis.

These properties are very easy to deduce from subsection 2C. First of all, we impose

$$\sum_n |h(n)| < \infty, \qquad \sum_n |g(n)| < \infty. \tag{3.1}$$

This implies that the operators H, G, defined by

$$(Ha)_k = \sum_n h(n-2k)a_n,$$

$$(Ga)_k = \sum_n g(n-2k)a_n,$$

are bounded operators on $l^2(\mathbb{Z})$. This condition is satisfied by the $h(n)$, $g(n)$ in subsection 2C; it corresponds to a rather weak decay condition on ϕ. At later stages, we shall impose much stronger decay conditions on the $h(n)$.

A second condition follows from the decomposition formulas (2.50), (2.51) and the reconstruction formula (2.52). The scheme will only work if

$$H^*H + G^*G = \mathbb{1}. \tag{3.2}$$

The third condition expresses orthogonality. Essentially, the decomposition splits the original $l^2(\mathbb{Z})$ into a sum of subspaces. After the first step, we have

$$l^2(\mathbb{Z}) = H^*l^2(\mathbb{Z}) \oplus G^*l^2(\mathbb{Z});$$

after L iterations, one finds

$$l^2(\mathbb{Z}) = \bigoplus_{j=0}^{L-1} (H^*)^j G^* l^2(\mathbb{Z}) + (H^*)^L l^2(\mathbb{Z}).$$

In order to make the decomposition as sharp as possible, i.e., to remove correlations in the original sequence as much as possible, we require that these subspaces be orthogonal. That is, we require

$$HG^* = 0. \tag{3.3}$$

This condition is verified by the filter operators in subsection 2C. One finds

$$\begin{aligned}(HG^*)_{kl} &= \sum_n h(n-2k)g(n-2l) \\ &= \sum_n \langle \phi_{1k}, \phi_{0n} \rangle \langle \phi_{0n}, \psi_{1l} \rangle \\ &= \langle \phi_{1k}, \psi_{1l} \rangle = 0.\end{aligned}$$

So far, H and G play symmetrical roles in our conditions. The final condition will break that symmetry, and identify G as a "difference" operator, and H as an "averaging" operator. Let a be the sequence

$$a_n = \begin{cases} 1 & \text{for} \quad |n| \leqq N, \\ 0 & \text{for} \quad |n| > N, \end{cases}$$

where N is large compared to n_0, with

$$\sum_{|n| \geqq n_0} |h(n)| \leqq \varepsilon,$$

$$\sum_{|n| \geqq n_0} |g(n)| \leqq \varepsilon,$$

for some small ε. If H averages, i.e., corresponds to a low pass filter, and G corresponds to a band pass filter, then we expect (in regions away from the "edges" of a)

$$(Ha)_k \approx \begin{cases} C & \text{for} \quad |k| \leqq \frac{1}{2}N - n_0, \\ 0 & \text{for} \quad |k| \geqq \frac{1}{2}N + n_0, \end{cases}$$

$$(Ga)_k \approx 0 \quad \text{for} \quad |k| \leqq \tfrac{1}{2}N - n_0 \quad \text{and} \quad \text{for} \quad |k| \geqq \tfrac{1}{2}N + n_0.$$

This implies that we require

$$\sum_n g(n) = 0,$$

$$\sum_n h(n) = C.$$

The constant C can be determined as follows. For $N \to \infty$, the edge effects become negligible, and

$$\|Ga\|^2/\|a\|^2 \to 0,$$

$$\|Ha\|^2/\|a\|^2 \sim \sum_{|k| \leq N/2} C^2/2N \to \tfrac{1}{2}C^2.$$

But

$$\|Ha\|^2 + \|Ga\|^2 = \langle a, (H^*H + G^*G)a \rangle = \|a\|^2,$$

hence $C = \sqrt{2}$. Thus our final conditions read

$$\sum_n h(n) = \sqrt{2},$$

(3.4)

$$\sum_n g(n) = 0.$$

These conditions are satisfied in subsection 2C. One has

$$\phi_{10} = \sum_n h(n) \phi_{0n};$$

hence, by integration,

$$2^{-1/2} \int dx\, \phi(\tfrac{1}{2}x) = \left[\sum_n h(n)\right] \int dx\, \phi(x),$$

or

$$\sum_n h(n) = \sqrt{2}, \quad \text{since} \quad \int dx\, \phi(x) \neq 0$$

(see (2.18)). Similarly,

$$\psi_{10} = \sum_n g(n) \phi_{0n};$$

since (see (2.17)) $\int dx\, \psi(x) = 0$, it follows that $\sum_n g(n) = 0$.

We have identified four conditions, (3.1)–(3.4), which guarantee that an algorithm "à la Mallat" works, and corresponds to averaging, respectively difference operations, followed by exact reconstruction. In terms of the $h(n)$, $g(n)$, conditions (3.2) and (3.3) can be rewritten as

(3.5) $$\sum_k [h(m - 2k)h(n - 2k) + g(m - 2k)g(n - 2k)] = \delta_{mn}$$

and

$$\sum_n h(n-2k)g(n-2l) = 0. \tag{3.6}$$

In the remainder of this subsection, we shall rewrite the conditions (3.1)–(3.4) in various ways which make them more tractable to analysis.

In order to get rid of the factors 2 in (3.5), (3.6), we define

$$\begin{aligned} a(n) &= h(2n), \\ b(n) &= h(2n+1), \\ c(n) &= g(2n), \\ d(n) &= g(2n+1). \end{aligned} \tag{3.7}$$

Rewriting (3.5), (3.6) in terms of functions of a, b, c, d leads to

$$\sum_k [a(m-k)a(n-k) + c(m-k)c(n-k)] = \delta_{mn}, \tag{3.8a}$$

$$\sum_k [b(m-k)b(n-k) + d(m-k)d(n-k)] = \delta_{mn}, \tag{3.8b}$$

$$\sum_k [a(m-k)b(n-k) + c(m-k)d(n-k)] = 0, \tag{3.8c}$$

$$\sum_n [a(n-k)c(n-l) + b(n-k)d(n-l)] = 0. \tag{3.8d}$$

In this form the conditions are completely expressed in terms of convolutions of the sequences a, b, c, d. It is therefore natural to introduce the 2π-periodic functions

$$\begin{aligned} \alpha(\xi) &= \sum_n a(n)e^{in\xi}, \\ \beta(\xi) &= \sum_n b(n)e^{in\xi}, \\ \gamma(\xi) &= \sum_n c(n)e^{in\xi}, \\ \delta(\xi) &= \sum_n d(n)e^{in\xi}, \end{aligned} \tag{3.9}$$

and to rewrite the conditions in terms of these functions. We obtain

$$|\alpha(\xi)|^2 + |\gamma(\xi)|^2 = 1, \tag{3.10a}$$

$$|\beta(\xi)|^2 + |\delta(\xi)|^2 = 1, \tag{3.10b}$$

$$\alpha(\xi)\overline{\beta(\xi)} + \gamma(\xi)\overline{\delta(\xi)} = 0, \tag{3.10c}$$

$$\alpha(\xi)\overline{\gamma(\xi)} + \beta(\xi)\overline{\delta(\xi)} = 0. \tag{3.10d}$$

These conditions are obviously not independent. Except for trivial solutions, which would be in contradiction with (3.4), i.e., with

$$\alpha(0) + \beta(0) = \sqrt{2}, \tag{3.11a}$$

$$\gamma(0) + \delta(0) = 0, \tag{3.11b}$$

we find from (3.10c) and (3.10d)

$$\begin{aligned} \gamma(\xi) &= e^{i\lambda(\xi)}\overline{\beta(\xi)}, \\ \delta(\xi) &= -e^{i\lambda(\xi)}\overline{\alpha(\xi)}, \end{aligned} \tag{3.12}$$

where λ is a real function such that $\lambda(\xi + 2\pi) - \lambda(\xi) \in 2\pi\mathbb{Z}$ for all ξ. For the sake of simplicity we shall restrict ourselves to $\lambda(\xi) = 0$ for the moment. We thus choose

$$\begin{aligned} \gamma(\xi) &= \overline{\beta(\xi)}, \\ \delta(\xi) &= -\overline{\alpha(\xi)}. \end{aligned} \tag{3.13}$$

The only equation remaining from the system (3.10) is then

$$|\alpha(\xi)|^2 + |\beta(\xi)|^2 = 1. \tag{3.14}$$

The choice (3.13), together with (3.11b), also implies

$$\alpha(0) - \beta(0) = 0.$$

Hence (from (3.11a)),

$$\alpha(0) = \beta(0) = 2^{-1/2}, \tag{3.15}$$

which agrees with (3.14) for $\xi = 0$. It follows that *any* choice of 2π-periodic functions α and β satisfying (3.14), (3.15) and $\Sigma|a_n| < \infty$, $\Sigma_n|b_n| < \infty$, leads, via (3.13), (3.9) and (3.7), to two filter operators H and G satisfying (3.1)–(3.4). These filter operators can then be used for a decomposition and reconstruction algorithm "à la Mallat", without reference to multiresolution analysis.

Remarks. 1. The system of equations (3.10) can also be rewritten as one matrix equation. If we define the 2×2 matrix-valued 2π-periodic function $M(\xi)$ by

$$M(\xi) = \begin{pmatrix} \alpha(\xi) & \gamma(\xi) \\ \beta(\xi) & \delta(\xi) \end{pmatrix}, \tag{3.16}$$

then (3.10) states that $M(\xi)$ should be unitary, for all ξ.

2. Note that, in view of (3.9) and (3.7), the choice (3.13) is equivalent to

$$g(n) = (-1)^n h(-n+1). \tag{3.17}$$

The equations (3.14) and (3.15) involve only the $h(n)$. They can be rewritten as

$$\sum_n h(n-2k)h(n-2l) = \delta_{kl} \tag{3.18}$$

and

$$\sum_n h(2n) = \sum_n h(2n+1) = 2^{-1/2}.$$

This last condition is implied by (3.18) and

$$\sum h(n) = 2^{1/2}. \tag{3.19}$$

3. If one introduces the 2π-periodic function $H(\xi)$,

$$H(\xi) = \sum_n h(n) e^{in\xi},$$

then the conditions (3.14), (3.15) can also be written in terms of H. Clearly,

$$H(\xi) = \alpha(2\xi) + e^{i\xi}\beta(2\xi),$$

or

$$\alpha(2\xi) = \tfrac{1}{2}[H(\xi) + H(\xi + \pi)],$$

$$\beta(2\xi) = \tfrac{1}{2}e^{-i\xi}[H(\xi) - H(\xi + \pi)].$$

Then (3.14), (3.15) are equivalent with

$$|H(\xi)|^2 + |H(\xi + \pi)|^2 = 2 \tag{3.20}$$

and

$$H(0) = \sqrt{2}\,. \tag{3.21}$$

Under the form (3.20) this condition is not new. It can be found in [16], where $m_0(\xi) = 2^{-1/2}H(\xi)$ is used, rather than H. While this paper was being written, S. Mallat pointed out to me that (3.20) is very similar to a condition derived by M. Smith and T. Barnwell [24] in the construction of "conjugate quadrature filters". In fact, (3.20) is identical to their condition. Smith and Barnwell were looking for, and found, a tree-structured two-band coding scheme with exact reconstruction, which is exactly what this subsection is about! The constructions given later (at least insofar as they describe discrete filters) are therefore, in fact, special cases of their construction. Ultimately, however, our aim here is to construct orthonormal wavelet bases of compact support, which is a very different point of view. Even from the filter point of view, our results go further than Smith and Barnwell's, in that we give complete characterization of the possible filters. We shall however not go into this here.

4. Similarly one can introduce $G(\xi) = \Sigma_n g(n)e^{in\xi}$. The matrix statement (3.16) is then equivalent to the requirement that the matrix

$$\frac{1}{\sqrt{2}}\begin{pmatrix} H(\xi) & G(\xi) \\ H(\xi+\pi) & G(\xi+\pi) \end{pmatrix} \tag{3.22}$$

be unitary. This is the form under which this requirement appears in [16]. Depending on what one wants to do, (3.22) and (3.20) may or may not be more useful than (3.16) and (3.14). The advantage of (3.14), (3.16) is that no correlations are introduced, as in (3.20), (3.22), linking the behavior of H at $\xi + \pi$ with its values at ξ. The conditions (3.16) or (3.22) can be generalized to situations where three or more band filters are considered (corresponding to decimations with factors $3, 4, \cdots$ rather than 2), or even more complicated structures, in more than one dimension (associated with lattices in $\mathbb{Z}^d$; see [21]). It was pointed out to me by P. Auscher [25] that in these cases the generalization of (3.16) is more useful, for practical construction, than the generalization of (3.22), precisely because it avoids introducing correlations.

5. Note that $\Sigma_n h(2n) = \Sigma_n h(2n+1)$, which is a consequence of (3.18)–(3.19) (see Remark 2 above) implies that *all* the possible $H(\xi)$, satisfying all the above conditions, necessarily are divisible by $(1 + e^{i\xi})$ (see subsection 2B).

Finally, let us conclude this subsection with some simple examples.

EXAMPLE 3.1. The simplest possible example is

$$\alpha(\xi) = \beta(\xi) = 2^{-1/2},$$

corresponding to

$$h(0) = 2^{-1/2}, \qquad g(0) = 2^{-1/2},$$

$$h(1) = 2^{-1/2}, \qquad g(1) = -2^{-1/2},$$

all the other $h(n)$, $g(n)$ being zero.

EXAMPLE 3.2. The next simplest example is

$$\alpha(\xi) = 2^{-1/2}\left[\nu(\nu - 1) + (\nu + 1)e^{i\xi}\right]/(\nu^2 + 1),$$

$$\beta(\xi) = 2^{-1/2}\left[(1 - \nu) + \nu(\nu + 1)e^{i\xi}\right]/(\nu^2 + 1),$$

where ν is an arbitrary real number. This corresponds to

$$h(0) = 2^{-1/2}\nu(\nu - 1)/(\nu^2 + 1), \qquad g(0) = 2^{-1/2}\nu(\nu + 1)/(\nu^2 + 1),$$

$$h(1) = 2^{-1/2}(1 - \nu)/(\nu^2 + 1), \qquad g(1) = -2^{-1/2}(\nu + 1)/(\nu^2 + 1),$$

$$h(2) = 2^{-1/2}(\nu + 1)/(\nu^2 + 1), \qquad g(2) = 2^{-1/2}(1 - \nu)/(\nu^2 + 1),$$

$$h(3) = 2^{-1/2}\nu(\nu + 1)/(\nu^2 + 1), \qquad g(3) = -2^{-1/2}\nu(\nu - 1)/(\nu^2 + 1),$$

all the other $h(n)$, $g(n)$ being zero.

Note. We have here taken

$$g(n) = (-1)^n h(3 - n)$$

rather than (3.17); this shift corresponds simply to choosing $\lambda(\xi) = \xi$ instead of 0 in (3.12).

3.B. Introducing a regularity condition. In the preceding subsection we derived and discussed a set of necessary and sufficient conditions, directly on the filter operators, for Mallat's algorithm to work. All these conditions concerned only what happened in one step of decomposition/reconstruction. In the discussion, in subsection 2B, of the Laplacian pyramid scheme, we saw that it is also important that the iterated reconstruction, applied to a sequence consisting of only one non-zero entry, looks still reasonably nice, even after several iterations.

In Mallat's algorithm, a sequence c^0 is decomposed into $d^1, \cdots, d^L, c^L$, with $d^j = GH^{j-1}c^0$, and $c^L = H^L c^0$; the reconstruction formula is then (cf. (2.52))

$$c^0 = \sum_{j=1}^{L} (H^*)^{j-1} G^* d^j + (H^*)^L c^L.$$

The iterated filter operator is thus H^*. It is therefore important (see subsection 2B) to study the behavior of $(H^*)^l e$, for large l, where e is a sequence with only one non-zero entry, e.g. $e_n = \delta_{n0}$. Ideally we want the graphical representation (with histograms—see Figures 2, 3 in subsection 2B) of $(H^*)^l e$ to look "nice", which expresses itself by convergence, for $l \to \infty$, to a reasonably regular function.

To show that this is a genuine concern, we have plotted, in Figure 4, the histogram representation of $(H^*e), \cdots, (H^*)^6 e$, for H^* chosen as in Example 3.2

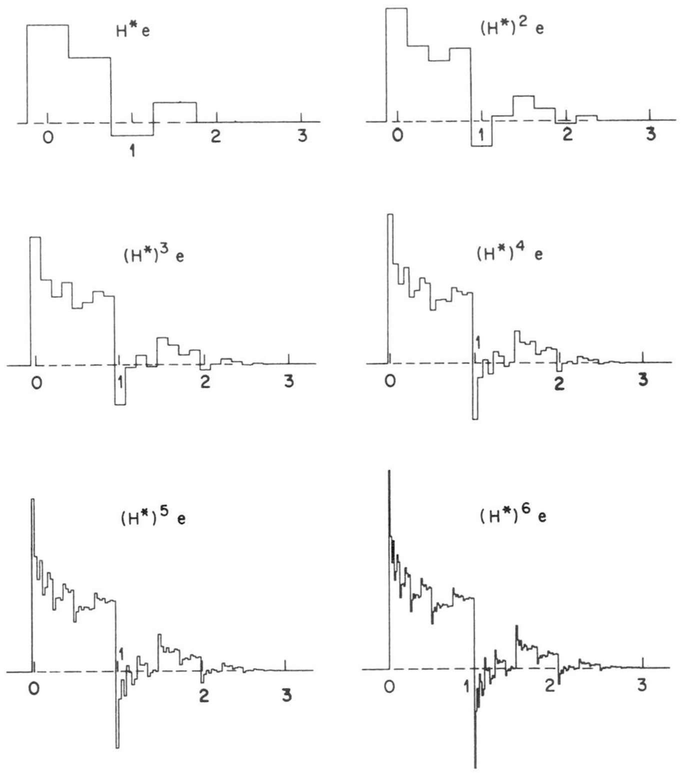

Figure 4. The histogram representations of $(H^*)^j e$, $j = 1, \cdots, 6$, for $h(n)$ which do not satisfy a regularity condition (see text).

(subsection 3A), with $\nu = -1.5$. For increasing l, $(H^*)^l e$ becomes increasingly messy; in fact, $(H^*)^l e$ converges, for $l \to \infty$, to a discontinuous, fractal function.

As in subsection 2B, we represent $(H^*)^l e$ by a histogram η_l with step width 2^{-l}, and with amplitudes given by the successive $2^{l/2}((H^*)^l e)_n$. The normalization, different from that in subsection 2B (because $\Sigma h(n) = \sqrt{2}$ and not 1), is again chosen so that the area under the histogram remains 1 for every l. The stepfunction η_l can be written as (see subsection 2B)

$$\eta_l(x) = \left(T_H^l \chi_{[-1/2,\,-1/2[}\right)(x), \tag{3.23}$$

where

$$(T_H f)(x) = \sqrt{2}\sum_n h(n) f(2x - n). \tag{3.24}$$

By taking Fourier transforms, (3.23) and (3.24) lead to

$$\hat{\eta}_l(\xi) = (2\pi)^{-1/2}\left[\prod_{j=1}^{l} m_0(2^{-j}\xi)\right]\frac{\sin(2^{-l-1}\xi)}{2^{-l-1}\xi}, \tag{3.25}$$

where $m_0(\xi) = 2^{-1/2}\Sigma_n h(n) e^{in\xi}$. Hence, at least in a formal sense, $\eta_l \to \eta_\infty$ for $l \to \infty$, with

$$\hat{\eta}_\infty(\xi) = (2\pi)^{-1/2}\prod_{j=1}^{\infty} m_0(2^{-j}\xi). \tag{3.26}$$

The following lemma ensures that $\hat{\eta}_\infty$ is well defined, i.e., that the infinite product in (3.26) converges, at least pointwise.

LEMMA 3.1. *Suppose that, for some* $\varepsilon > 0$,

$$\sum_n |h(n)|\,|n|^{\varepsilon} < \infty. \tag{3.27}$$

Then (3.26) *converges pointwise, for all* $\xi \in \mathbb{R}$. *The convergence is uniform on compact sets.*

Proof: Since $\Sigma h(n) = \sqrt{2}$, we have $m_0(\xi) = 1 + 2^{-1/2}\Sigma_n h(n)(e^{in\xi} - 1)$, hence $|m_0(\xi) - 1| \leqq \sqrt{2}\,\Sigma_n |h(n)|\,|\sin\frac{1}{2}n\xi|$. For any $0 < \delta \leqq 1$ there exists C_δ such that, for all $\alpha \in \mathbb{R}$, $|\sin\alpha| \leqq C_\delta|\alpha|^\delta$. It follows that

$$|m_0(\xi) - 1| \leqq C\left[\sum_n |h(n)|\,|n|^{\min(1,\varepsilon)}\right]\cdot|\xi|^{\min(1,\varepsilon)};$$

hence

$$|m_0(2^{-j}\xi) - 1| \leqq C\lambda^{-j}|\xi|^{\min(1,\varepsilon)}, \tag{3.28}$$

where $\lambda = 2^{\min(1,\varepsilon)} > 1$. This is sufficient to ensure convergence of (3.26), for any $\xi \in \mathbb{R}$. It immediately follows from (3.28) that the convergence is uniform on compact sets.

Remark. While being more restrictive than (3.1), the condition (3.27) is still very mild. In practice one requires much stronger decay for the $h(n)$. For filter construction purposes, one even restricts oneself to the case where only finitely many $h(n)$ are different from zero.

It is however not sufficient to know that $\hat{\eta}_\infty$ is well defined. In order to avoid situations such as depicted in Figure 4, we require that (i) $\hat{\eta}_\infty$ has sufficient decay, so that η_∞ is sufficiently regular (at least continuous), and (ii) η_l converges to η_∞, pointwise, for $l \to \infty$.

To ensure the decay, for $|\xi| \to \infty$, of $\hat{\eta}_\infty(\xi)$, we shall use the same trick as in subsection 2B, i.e., we shall require that $m_0(\xi)$ is divisible by $(1 + e^{i\xi})^N$, for some $N > 0$. The precise statement is given in the following lemma, using an estimation technique of P. Tchamitchian [5].

LEMMA 3.2. *If* $m_0(\xi) = (1 + e^{i\xi})]^N \mathscr{F}(\xi)$, *where* $\mathscr{F}(\xi) = \Sigma_n f(n) e^{in\xi}$ *satisfies*

$$\sum |f(n)|\,|n|^{\varepsilon} < \infty \quad \textit{for some} \quad \varepsilon > 0 \tag{3.29}$$

and

$$\sup_{\xi \in \mathbb{R}} |\mathscr{F}(\xi)| = B, \tag{3.30}$$

then there exists $C > 0$ *such that, for all* $\xi \in \mathbb{R}$,

$$\left| \prod_{j=1}^{\infty} m_0(2^{-j}\xi) \right| \leqq C(1 + |\xi|)^{-N + (\log B)/(\log 2)}. \tag{3.31}$$

Remarks. 1. It follows from (3.31) that η_∞ is continuous if H satisfies all the above conditions, and if $B < 2^{N-1}$.

2. The condition (3.29) will automatically be satisfied if

$$\sum_n |h(n)|\,|n|^{N+\varepsilon} < \infty. \tag{3.32}$$

Proof: Since $\prod_{j=1}^{\infty}\cos(2^{-j}x) = x^{-1}\sin x$, we have

$$\prod_{j=1}^{\infty} m_0(2^{-j}\xi) = \left[e^{i\xi/2}\prod_{j=1}^{\infty}\cos(2^{-j-1}\xi)\right]^N \prod_{j=1}^{\infty}\mathscr{F}(2^{-j}\xi)$$

(3.33)

$$= e^{iN\xi/2}\left(\frac{\sin\frac{1}{2}\xi}{\frac{1}{2}\xi}\right)^N \prod_{j=1}^{\infty}\mathscr{F}(2^{-j}\xi),$$

where the right-hand side converges uniformly on compact sets because of (3.29). There exists therefore a constant C such that, for all $|\xi| \leqq 1$,

$$\left|\prod_{j=1}^{\infty} m_0(2^{-j}\xi)\right| \leqq C. \tag{3.34}$$

Take now $|\xi| > 1$. Determine $j_0 \in \mathbb{N}$ such that

$$2^{-j_0}|\xi| < 1 \leqq 2^{-j_0+1}|\xi|,$$

i.e.,

$$\log|\xi|/\log 2 < j_0 \leqq 1 + \log|\xi|/\log 2.$$

Then

$$\left|\prod_{j=1}^{\infty}\mathscr{F}(2^{-j}\xi)\right| = \left|\prod_{j=1}^{j_0}\mathscr{F}(2^{-j}\xi)\right| \cdot \left|\prod_{j=1}^{\infty}\mathscr{F}(2^{-j}2^{-j_0}\xi)\right|$$

(3.35)

$$\leqq B^{j_0}\prod_{j=1}^{\infty}\left(1 + 2^{-j\varepsilon}\sum_n |f(n)|\,|n|^{\varepsilon}\right)$$

$$\leqq C\exp\{\log B \cdot \log|\xi|/\log 2\},$$

To estimate $\prod_{j=1}^{\infty}\mathscr{F}(2^{-j}2^{-j_0}\xi)$ we have used the same argument as in the proof of Lemma 3.1. This is allowed since $\Sigma f(n) = \mathscr{F}(0) = m_0(0) = 1$. Together, (3.35), (3.34) and (3.33) imply (3.31).

In our search for "regularity" we have, so far, only used one of the special conditions on the $h(n)$, derived in subsection 3A, namely (3.4), $\Sigma_n h(n) = \sqrt{2}$. And even that has not played a critical role, since it was only used for normalization purposes, and we could have as easily normalized by any other constant which happened to be the sum of the $h(n)$. For our last step, the proof that the histograms η_l converge pointwise to the continuous function η_∞ (assuming B is not too large), we need an extra ingredient, namely $|m_0(\xi)| \leqq 1$. Since, however (see (3.20)), as a consequence of (3.2)–(3.3), $|m_0(\xi)|^2 + |m_0(\xi + \pi)|^2 = 1$, this condition is automatically fulfilled for $h(n)$ satisfying (3.18)–(3.19).

PROPOSITION 3.3. *Define* $m_0(\xi) = 2^{-1/2}\Sigma_n h(n)e^{in\xi}$, *where the* $h(n)$ *satisfy* (3.18), (3.19). *Suppose moreover that*

$$m_0(\xi) = \left[\tfrac{1}{2}(1 + e^{i\xi})\right]^N \mathscr{F}(\xi), \tag{3.36}$$

with $\mathscr{F}(\xi) = \Sigma_n f(n)e^{in\xi}$ *such that*

$$\sum_n |f(n)|\,|n|^\varepsilon < \infty \quad \textit{for some} \quad \varepsilon > 0 \tag{3.37}$$

and

$$\sup_{\xi \in \mathbf{R}} |\mathscr{F}(\xi)| = B < 2^{N-1}. \tag{3.38}$$

Then the piecewise constant functions η_l, *defined recursively by*

$$\eta_l(x) = \sqrt{2}\sum_n h(n)\eta_{l-1}(2x - n), \tag{3.39}$$

with

$$\eta_0(x) = \chi_{[-1/2,1/2[}(x),$$

converge pointwise to the continuous function η_∞ *defined by*

$$\hat{\eta}_\infty(\xi) = (2\pi)^{-1/2}\prod_{j=1}^{\infty} m_0(2^{-j}\xi).$$

Proof: 1. As an intermediate step, we prove $\mu_l \to \eta_\infty$, pointwise, where the μ_l are defined in the same recursive way as the η_l, but starting from a different initial function,

$$\mu_0(x) = \begin{cases} 1 + x, & -1 \leqq x \leqq 0, \\ 1 - x, & 0 \leqq x \leqq 1, \\ 0, & \text{otherwise}. \end{cases}$$

2. Taking Fourier transforms, we find

$$\hat{\mu}_l(\xi) = (2\pi)^{-1/2}\left[\prod_{j=1}^{l} m_0(2^{-j}\xi)\right]\left[\frac{\sin(2^{-l-1}\xi)}{2^{-l-1}\xi}\right]^2.$$

From Lemma 3.1 it follows that $\hat{\mu}_l \to \hat{\eta}_\infty$, uniformly on compact sets. This implies that, for all $\delta > 0$, and for all $R > 0$, we can find l_0 such that, for all $l \geqq l_0$,

$$\int_{|\xi| \leqq R} d\xi \, |\hat{\mu}_l(\xi) - \hat{\eta}_\infty(\xi)| \leqq \delta.$$

On the other hand, $\hat{\eta}_\infty \in L^1$ since $B < 2^{N-1}$. It follows that for all $\delta > 0$ there exists R such that

$$\int_{|\xi| \geqq R} d\xi \, |\hat{\eta}_\infty(\xi)| \leqq \delta.$$

L^1-convergence of $\hat{\mu}_l$ to $\hat{\eta}_\infty$, which implies pointwise convergence of μ_l to η_∞, will then follow if we can prove that, for all $\delta > 0$, there exist R and l_0 large enough, so that, for all $l \geqq l_0$,

$$\int_{|\xi| \geqq R} d\xi \, |\hat{\mu}_l(\xi)| \leqq \delta.$$

3. We need thus to evaluate the integral

$$\int_{|\xi| \geqq R} d\xi \, |P_l(\xi)| \left| \frac{\sin(2^{-l-1}\xi)}{2^{-l-1}\xi} \right|^2,$$

where $P_l(\xi) = \prod_{j=1}^{l} m_0(2^{-j}\xi)$. To do this, we split the integral into two parts, namely $|\xi| \geqq 2^l\pi$ and $R \leqq |\xi| \leqq 2^l\pi$. To evaluate these two parts, we shall use the following three properties of P_l:

$$\text{(i)} \quad |P_l(\xi)| \leqq 1, \qquad \left(\text{since } |m_0(\xi)| \leqq 1\right),$$

$$\text{(ii)} \quad |P_l(\xi)| \leqq \left[\prod_{j=1}^{l} |\cos(2^{-j}\xi)| \right]^N \prod_{j=1}^{l} |\mathscr{F}(2^{-j}\xi)|$$

$$\leqq C \left| \frac{2^{-l}\sin\frac{1}{2}\xi}{\sin(2^{-l-1}\xi)} \right|^N (1 + |\xi|)^\beta,$$

where $\beta = \log B / \log 2$ (use the proof of Lemma 3.2) and

(iii) P_l is periodic, with period $2^{l+1}\pi$.

4. We concentrate first on $|\xi| \geqq 2^l\pi$. Using the periodicity of P_l, we find

$$\int_{|\xi|\geqq 2^l\pi} d\xi\, |P_l(\xi)| \left| \frac{\sin(2^{-l-1}\xi)}{2^{-l-1}\xi} \right|^2$$

$$= \sum_{k\neq 0} \int_{|\xi|\leqq 2^l\pi} d\xi\, |P_l(\xi)| \frac{|\sin(2^{-l-1}\xi)|^2}{|2^{-l-1}\xi + k\pi|^2}$$

$$\leqq C\int_{|\xi|\leqq 2^l\pi} d\xi\, |P_l(\xi)|\, |\sin(2^{-l-1}\xi)|^2.$$

Choose $\lambda = 2^{-\alpha l}$, with $\alpha \in]0,1[$ to be fixed later. Then

$$\int_{|\xi|\leqq 2^l\pi} d\xi\, |P_l(\xi)|\, |\sin(2^{-l-1}\xi)|^2$$

(3.40)

$$\leqq \tfrac{1}{4}\lambda^2 \int_{|\xi|\leqq 2^l\lambda} d\xi\, |P_l(\xi)| + \int_{2^l\lambda\leqq|\xi|\leqq 2^l\pi} d\xi\, |P_l(\xi)|.$$

Now

$$\int_{|\xi|\leqq 2^l\lambda} d\xi\, |P_l(\xi)|$$

$$\leqq \int_{|\xi|\leqq 1} d\xi\, |P_l(\xi)| + C\int_{1\leqq|\xi|\leqq 2^l\lambda} d\xi\, (1+|\xi|)^\beta |2^l \sin(2^{-l-1}\xi)|^{-N}$$

$$\leqq 1 + 2^N C \int_1^\infty dx\, (1+x)^\beta x^{-N} = C_1,$$

where C_1 is finite because $N - \beta > 1$.

On the other hand,

$$\int_{2^l\lambda\leqq|\xi|\leqq 2^l\pi} d\xi\, |P_l(\xi)|$$

$$\leqq C 2^{-lN}(1+2^l\pi)^\beta 2^l \int_\lambda^\pi dx\, |\sin\tfrac{1}{2}x|^{-N}$$

$$\leqq C_2 2^{l(1+\beta-N)}\lambda^{-N}.$$

Putting it all together, and choosing $\alpha = (N-\beta-1)/(N+2) \in]0,1[$, this implies that (3.40) is

(3.41) $$\leqq C_3 2^{-2l(N-\beta-1)/(N+2)}.$$

This clearly tends to zero for $l \to \infty$.

5. We now evaluate the integral of $|\hat{\mu}_l|$ over $R \leqq |\xi| \leqq 2^l\pi$. Since $|\sin x| \geqq 2|x|/\pi$ for $|x| \leqq \frac{1}{2}\pi$, we find

$$\int_{R \leqq |\xi| \leqq 2^l\pi} d\xi \, |P_l(\xi)| \left| \frac{\sin(2^{-l-1}\xi)}{2^{-l-1}\xi} \right|^2$$

$$\leqq 4C \int_{R \leqq |\xi| \leqq 2^l\pi} d\xi \, (1+|\xi|)^{\beta} |\xi|^{-2} |\sin \tfrac{1}{2}\xi|^N \left| \frac{2^{-l}}{\sin(2^{-l-1}\xi)} \right|^{N-2}$$

$$\leqq 4C\pi^{N-2} \int_R^{\infty} dx \, (1+x)^{\beta} x^{-N}.$$

Since $N - \beta - 1 > 0$, this tends to zero for $R \to \infty$, uniformly in l. Together with (3.41) this proves that

$$\int_{|\xi| \geqq R} d\xi \, |\hat{\mu}_l(\xi)|$$

can be made as small as wanted, by choosing l and R large enough. As pointed out in point 2, this proves $\|\hat{\mu}_l - \hat{\eta}_\infty\|_{L^1} \to_{l \to \infty} 0$.

6. We have thus proved that $\mu_l \to \eta_\infty$, pointwise. In fact, we can even show a little bit more. The same arguments (points $2 \to 5$) as above can be stretched a little to prove

$$\int d\xi \, (1+|\xi|)^{\lambda} |\hat{\eta}_\infty(\xi)| < \infty$$

and

$$\int d\xi \, (1+|\xi|)^{\lambda} |\hat{\eta}_\infty(\xi) - \hat{\mu}_l(\xi)| \xrightarrow[l \to \infty]{} 0,$$

where

$$\lambda = \tfrac{1}{2}(N - \beta - 1) > 0.$$

Consequently, η_∞ is λ-Lipschitz,

$$|\eta_\infty(x) - \eta_\infty(y)| \leqq C|x-y|^{\lambda},$$

and the convergence $\mu_l \to \eta_\infty$ is uniform on compact sets.

7. Finally, we only need to show that pointwise convergence of the μ_l implies pointwise convergence of the η_l. The two functions μ_0 and η_0 agree on integers,

$$\mu_0(0) = \eta_0(0) = 1,$$

$$\mu_0(k) = \eta_0(k) = 0 \quad \text{for} \quad k \in \mathbb{Z}, \, k \neq 0.$$

Using the recursion relation (3.39), which both the μ_l and the η_l satisfy, one sees that this implies, for all $l \in \mathbb{N}$,

$$\eta_l(2^{-l}k) = \mu_l(2^{-l}k) \quad \text{for all} \quad k \in \mathbb{Z}.$$

Let $x \in \mathbb{R}$ be arbitrary. For any $\varepsilon > 0$, there exists $\delta > 0$ such that

$$|x - y| \leqq \delta \Rightarrow |\eta_\infty(x) - \eta_\infty(y)| \leqq \tfrac{1}{2}\varepsilon.$$

There also exists l_0 such that, for all $l \geqq l_0$, and all $y \in [x - \delta, x + \delta]$, one has

$$|\eta_\infty(y) - \mu_l(y)| \leqq \tfrac{1}{2}\varepsilon.$$

Choose $l \geqq l_1 = \max(l_0, -\ln\delta/\ln 2)$. Since η_l is piecewise constant, with step width 2^{-l}, it follows that there exists $k \in \mathbb{Z}$ such that

$$|x - 2^{-l}k| \leqq 2^{-l} \leqq \delta$$

and

$$\eta_l(x) = \eta_l(2^{-l}k) = \mu_l(2^{-l}k).$$

Hence

$$|\eta_l(x) - \eta_\infty(x)| \leqq |\mu_l(2^{-l}k) - \eta_\infty(2^{-l}k)| + |\eta_\infty(2^{-l}k) - \eta_\infty(x)| \leqq \varepsilon.$$

Since ε was arbitrary, this shows that η_l converges pointwise to η_∞, for $l \to \infty$.

Remarks. 1. Using only slightly modified arguments, one proves, under the same conditions (in fact, only $B < 2^{N-1/2}$ is needed) that $\eta_l \to \eta_\infty$ in L^2, for $l \to \infty$. One simply replaces the L^1-estimates for $\eta_\infty - \mu_l$ by L^2-estimates for $\eta_\infty - \eta_l$ (no intermediary μ_l are needed).

2. As noted above, it is sufficient that

$$\sum_n |h(n)|\,|n|^{N+\varepsilon} < \infty$$

to ensure (3.37).

3. The $h(n)$ of Example 3.1 do not satisfy the conditions of the proposition, since in this case

$$m_0(\xi) = \tfrac{1}{2}(1 + e^{i\xi}),$$

hence $N = 1$, $B = |\mathscr{F}(\xi)| = 1$, and therefore $B = 2^{N-1}$. However, in this case one checks directly that

$$\eta_l = \chi_{[-2^{-l-1}, 1-2^{-l-1}[}.$$

The limit η_∞ is not continuous in this case, $\eta_\infty = \chi_{[0,1[}$, but the pointwise convergence $\eta_l \to \eta_\infty$ still holds a.e.

4. The coefficients $h(n)$ defined by

$$h(0) = h(3) = 2^{-1/2},$$
$$h(n) = 0 \quad \text{otherwise},$$

satisfy all the "discrete" conditions of subsection 3A, but do not satisfy the conditions in the last proposition (for the same reason as the $h(n)$ of Example 3.1). In this case, however, the pointwise convergence of the η_l fails on a whole interval. It is easy to check that, for any l, the η_l take only two values, 0 and 1. (The easiest way to check this is to use the "graphical" construction (2.40) of the η_l—see subsection 2B and Figure 3.) On the other hand,

$$m_0(\xi) = \tfrac{1}{2}(1 + e^{3i\xi}),$$

hence

$$\hat{\eta}_\infty(\xi) = (2\pi)^{-1/2} \prod_{j=1}^{\infty} m_0(2^{-j}\xi) = (2\pi)^{-1/2} \varepsilon^{3i\xi/2} \frac{\sin\frac{3}{2}\xi}{\frac{3}{2}\xi}$$

or

$$\eta_\infty = \tfrac{1}{3}\chi_{[0,3]}.$$

There is therefore no pointwise convergence for any x between 0 and 3. The L^2-convergence fails too, since $\|\eta_\infty\|^2_{L^2} = \frac{1}{3}$, whereas for all finite l, η_l is the characteristic function of a union of intervals, and hence $\|\eta_l\|^2_{L^2} = \|\eta_l\|_{L^1} = \hat{\eta}_l(0) = 1$.

5. Only two values of ν, in Example 3.2, lead to coefficients $h(n)$ that satisfy the conditions of the proposition. They correspond to $m_0(\xi)$ divisible by $(1 + \varepsilon^{i\xi})^2$. As noted above, all $m_0(\xi)$ satisfying the discrete conditions in subsection 3A are divisible by $(1 + e^{i\xi})$ (see Remark 5 at the end of subsection 3A). In Example 3.2, extra divisibility by another factor $(1 + e^{i\xi})$ leads to the condition

$$h(1) - h(3) = 2h(0),$$

or

$$\nu = \pm 1/\sqrt{3}.$$

The corresponding $h(0), \cdots, h(3)$ are

$$\begin{aligned} h(0) &= (1 \mp \sqrt{3})/(4\sqrt{2}), \\ h(1) &= (3 \mp \sqrt{3})/(4\sqrt{2}), \\ h(2) &= (3 \pm \sqrt{3})/(4\sqrt{2}), \\ h(3) &= (1 \pm \sqrt{3})/(4\sqrt{2}). \end{aligned} \tag{3.42}$$

We shall come back to these $h(n)$ later.

With Proposition 3.3 we have completed our program of writing a set of explicit conditions on the $h(n)$, $g(n)$, without reference to a multiresolution analysis background, which make Mallat's algorithm work, and which moreover lead to filters with sufficient "regularity".

In the case where the $h(n)$, $g(n)$ are calculated starting from a multiresolution analysis (see subsection 2C), one has

$$h(n) = \langle \phi_{10}, \phi_{0n} \rangle,$$

or

$$\phi_{10} = \sum_n h(n) \phi_{0n},$$

i.e.,

$$\phi(\tfrac{1}{2}x) = 2^{1/2} \sum_n h(n) \phi(x-n).$$

This is equivalent to

$$\hat{\phi}(\xi) = 2^{-1/2} \sum_n h(n) e^{in\xi/2} \hat{\phi}(\tfrac{1}{2}\xi) = m_0(\tfrac{1}{2}\xi)\hat{\phi}(\tfrac{1}{2}\xi).$$

It follows that

$$\hat{\phi}(\xi) = \left[\prod_{j=1}^{\infty} m_0(2^{-j}\xi) \right] \hat{\phi}(0), \tag{3.43}$$

or, since $\hat{\phi}(0) = (2\pi)^{-1/2} \int dx\, \phi(x) = (2\pi)^{-1/2}$ (see (2.18)),

$$\phi(x) = \eta_\infty(x). \tag{3.44}$$

As pointed out in subsection 2B, the $\eta_l = T^l \chi_{[-1/2, 1/2[}$ can also be computed via a different recursion, (2.40), which we shall call the "graphical" recursion, and which lies at the basis of the graphical construction technique illustrated by Figure 3. It follows from (3.44) that, in the case where the $h(n)$ are derived from a multiresolution analysis framework, the graphical construction by iteration (see Figure 3, where the $h(n)$ now play the role of the $w(n)$) is therefore nothing but a reconstruction of the function ϕ; in the limit for $l \to \infty$, finer and finer detail is achieved for increasing l.

3.C. Equivalence between the discrete conditions and multiresolution analysis. So far we have formulated conditions, directly on the $h(n)$, which ensure that S. Mallat's algorithm works (with these coefficients), and has regularity (in the sense given to it at the end of subsection 2B, or in subsection 3B). We have seen for every condition how the coefficients $h(n)$ computed from a multiresolution analysis fit into the picture. The main result of this subsection is that these multiresolution-based examples are the *only* ones. It turns out that *any* sequence of $h(n)$ satisfying the conditions in subsections 3A and 3B corresponds to a multiresolution analysis. The function η_∞ defined by (3.26) is then exactly the function ϕ from the multiresolution structure.

To prove this equivalence, we start from a sequence $h(n)$ satisfying (3.18), (3.19) and (3.27). We also assume that the function $m_0(\xi) = 2^{-1/2}\Sigma_n h(n)e^{in\xi}$ satisfies all the conditions in Proposition 3.3. We then define, as in (3.17),

$$g(n) = (-1)^n h(-n+1), \tag{3.45}$$

and, as in (3.44),

$$\phi(x) = \eta_\infty(x),$$

or

$$\hat{\phi}(\xi) = (2\pi)^{-1/2} \prod_{j=1}^{\infty} m_0(2^{-j}\xi). \tag{3.46}$$

From the proof of Proposition 3.3 we know that ϕ is a bounded, uniformly continuous function; since $\hat{\phi} \in L^1 \cap L^\infty$, one also has $\phi \in L^2$. We define, in accordance with (2.16),

$$\psi(x) = \sqrt{2} \sum_n g(n)\phi(2x-n). \tag{3.47}$$

Since $\Sigma_n|g(n)| = \Sigma_n|h(n)| < \infty$, it follows that

$$|\hat{\psi}(\xi)| \leqq 2^{-1/2} \sum_n |h(n)| \cdot |\hat{\phi}(\tfrac{1}{2}\xi)|.$$

All the estimates of subsection 3B on η_∞ carry over, therefore, to ψ, and one finds that ψ is a bounded, uniformly continuous L^2-function. As before, we define $\psi_{jk}(x) = 2^{-j/2}\psi(2^{-j}x-k)$, and $\phi_{jk}(x) = 2^{-j/2}\phi(2^{-j}x-k)$. The definitions (3.46) and (3.47) immediately imply

$$\phi_{jk} = \sum_n h(n-2k)\phi_{j-1\,n}, \tag{3.48}$$

$$\psi_{jk} = \sum_n g(n-2k)\phi_{j-1\,n}. \tag{3.49}$$

We shall prove that the ψ_{jk} constitute an orthonormal basis of $L^2(\mathbb{R})$. In a first step we prove some orthogonality relations.

LEMMA 3.4. *Let* $h(n)$ *satisfy* (3.18), (3.19), (3.28) *and the conditions in Proposition* 3.3. *Let* $g(n)$, ϕ, ψ *be defined by* (3.45), (3.46), (3.47), *respectively.*

Then ϕ, $\psi \in L^2(\mathbb{R})$, *and, for all* j, k, $k' \in \mathbb{Z}$,

(3.50) $$\langle \psi_{jk}, \psi_{jk'} \rangle = \delta_{kk'},$$

(3.51) $$\langle \psi_{jk}, \phi_{jk'} \rangle = 0,$$

(3.52) $$\langle \phi_{jk}, \phi_{jk'} \rangle = \delta_{kk'}.$$

Remark. Note that (3.50)–(3.52) are restricted to one j-level at a time. The orthogonality between j-levels will follow from Lemma 3.5.

Proof: 1. Let η_l be defined as in Proposition 3.3,

$$\eta_l = T^l \chi_{[-1/2,1/2[},$$

with

(3.53) $$(Tf)(x) = \sqrt{2} \sum_n h(n) f(2x - n).$$

For reasons which will become obvious, we add an index 0 to η_l,

$$\eta_{l,0} = \eta_l.$$

For arbitrary $k \in \mathbb{Z}$, we define

$$\eta_{l,k} = (T_k)^l \chi_{[-1/2+k,1/2+k[}$$

with $(T_k f)(x) = \sqrt{2} \sum_n h(n) f(2x - n - k)$. Due to the translations over k, built into $\eta_{0,k}$ as well as into T_k, $\eta_{l,k}$ is just a translated version of $\eta_{l,0}$. This can easily be checked by induction,

$$\eta_{0,k}(x) = \chi_{[-1/2+k,1/2+k[}(x) = \eta_{0,0}(x - k)$$

and

$$\eta_{l,k}(x) = \sqrt{2} \sum_n h(n) \eta_{l-1,k}(2x - n - k)$$

$$= \sqrt{2} \sum_n h(m) \eta_{l-1,0}(2x - 2k - n)$$

$$= \eta_{l,0}(x - k).$$

Since (see Remark 1 following Proposition 3.3) $\|\eta_{l,0} - \phi\|_{L^2} \to 0$ for $l \to \infty$, it follows that $\|\eta_{l,k} - \phi_{0k}\|_{L^2} \to 0$ for $l \to \infty$.

2. Since $\hat{\eta}_{l,0}(\xi) = \left[\prod_{j=1}^{l} m_0(2^{-j}\xi)\right] \hat{\eta}_{0,0}(2^{-l}\xi)$, and since $|m_0(\xi)| \leqq 1$ and $\eta_{0,0} \in L^2$, it follows that all the $\eta_{l,k}$ are in L^2.

3. For fixed l, the different $\eta_{l,k}$ are orthonormal. This can again be proved by induction. By translation invariance, it is sufficient to prove that $\langle \eta_{l,k}, \eta_{l,k'} \rangle = \delta_{kk'}$ for $k' = 0$. We have

$$\langle \eta_{0,k}, \eta_{0,0} \rangle = \int_{-1/2}^{1/2} dx\, \chi_{[-1/2+k,1/2+k[}(x) = \delta_{k0}$$

and

$$\begin{aligned}
\langle \eta_{l,k}, \eta_{l,0} \rangle &= 2 \sum_{n,m} h(n)h(m) \int dx\, \eta_{l-1,k}(2x - n - k)\eta_{l-1,0}(2x - m) \\
&= 2 \sum_{n,m} h(n)h(m) \int dx\, \eta_{l-1,2k+n-m}(2x)\eta_{l-1}(2x) \\
&= \sum_{n,m} h(n)h(m)\, \delta_{0,2k+n-m} = \sum_n h(n)h(m + 2k) \\
&= \delta_{k,0} \qquad \text{(by (3.18))}.
\end{aligned}$$

By induction it follows that $\langle \eta_{l,k}, \eta_{l,k'} \rangle = \delta_{kk'}$ for all l, k, k'.

4. It follows immediately that

$$\begin{aligned}
\langle \phi_{jk}, \phi_{jk'} \rangle &= 2^{-j} \int dx\, \phi(2^{-j}x - k)\phi(2^{-j}x - k') \\
&= \int dx\, \phi(x)\phi(x - k' + k) \\
&= \lim_{l \to \infty} \langle \eta_{l,0}, \eta_{l,k'-k} \rangle = \delta_{kk'}.
\end{aligned}$$

5. With $g(n)$ defined by (3.45), the conditions (3.18), (3.19) on the $h(n)$ imply (see subsection 3A)

$$\sum_n g(n - 2k)h(n - 2l) = 0, \tag{3.54}$$

$$\sum_n g(n - 2k)g(n - 2l) = \delta_{kl}. \tag{3.55}$$

Hence, by (3.48) and (3.49),

$$\begin{aligned}
\langle \psi_{jk}, \phi_{jk'} \rangle &= \sum_{n,n'} g(n - 2k)h(n' - 2k')\langle \phi_{j-1\,n}, \phi_{j-1\,n'} \rangle \\
&= \sum_n g(n - 2k)h(n - 2k') = 0,
\end{aligned}$$

and

$$\langle \psi_{jk}, \psi_{jk'} \rangle = \sum_{m, n'} g(m - 2k) g(n' - 2k') \langle \phi_{j-1\,n}, \phi_{j-1\,n'} \rangle$$

$$= \sum_n g(n - 2k) g(n - 2k') = \delta_{kk'}.$$

The "discrete orthogonality condition" (3.18) plays a crucial role in this proof. In the terminology of subsection 3A, (3.18) is equivalent to $HH^* = \mathbb{1}$, where H^* is the bounded l^2-operator (see subsection 3A)

$$(H^*a)^n = \sum_k h(n - 2k) a_k.$$

This implies that H^*, as an operator from l^2 to l^2, preserves orthogonality of sequences. The operator T_H defined by (3.24) was in fact constructed to exactly reproduce, when acting on $\chi_{[-1/2, 1/2[}$ and its iterates, the action of H^* on the sequence e ($e_n = \delta_{n0}$) and its iterates (see subsection 2B). This implies that repeated application of T_H preserves the orthogonality of the $\eta_{0,k}$. This is what makes the above proof work.

In the following lemma we prove that the ψ_{jk} constitute a tight frame (see Section 1, or (3.57) below). Again, the crucial ingredient will be one of the discrete identities which follow from the conditions on $h(n)$, $g(n)$. From subsection 3A we know that, with $g(n)$ as defined by (3.45), and with $h(n)$ satisfying all the conditions above,

$$\sum_k [h(n - 2k) h(m - 2k) + g(n - 2k) g(m - 2k)] = \delta_{mn}$$

(this can also be derived directly from (3.18) and (3.45)). It follows that (use (3.48), (3.49))

$$\sum_k \left[h(m - 2k) \phi_{jk} + g(m - 2k) \psi_{jk} \right] = \phi_{j-1\,m}. \tag{3.56}$$

This, of course, already points towards multiresolution analysis (see subsection 2A).

LEMMA 3.5. *Let $h(n)$, $g(n)$, ϕ, ψ be as in Lemma 3.4. Then, for all $f \in L^2(\mathbb{R})$,*

$$\sum_{j, k \in \mathbb{Z}} |\langle \psi_{jk}, f \rangle|^2 = \|f\|^2. \tag{3.57}$$

Proof: 1. Take any $f \in C_0^\infty$. Then, since $\phi \in L^2$, $\Sigma_n |\langle \phi_{jn}, f\rangle|^2$ converges, for any $j \in \mathbb{Z}$. Moreover, by (3.56),

$$\sum_n |\langle \phi_{j-1\,n}, f\rangle|^2 = \sum_{n,k,l} \Big[h(n-2k)h(n-2l)\langle \phi_{jk}, f\rangle\langle f, \phi_{jl}\rangle$$

$$+2h(n-2k)g(n-2l)\mathcal{R}e(\langle \phi_{jk}, f\rangle\langle f, \psi_{jl}\rangle)$$

$$+g(n-2k)g(n-2l)\langle \psi_{jk}, f\rangle\langle f, \psi_{jl}\rangle\Big]$$

$$= \sum_k \Big[|\langle \phi_{jk}, f\rangle|^2 + |\langle \psi_{jk}, f\rangle|^2 \Big],$$

where we have used (3.18), (3.54) and (3.55).

2. By iteration, one has, for all $N \in \mathbb{N}$,

$$\sum_n |\langle \phi_{-Nn}, f\rangle|^2 = \sum_k |\langle \phi_{Nk}, f\rangle|^2 + \sum_{j=-N}^{N} \sum_k |\langle \psi_{jk}, f\rangle|^2. \tag{3.58}$$

In this expression we shall let N tend to ∞.

3. We first concentrate on $\Sigma_k |\langle \phi_{Nk}, f\rangle|^2$. Let us suppose, for the sake of definiteness, that $\operatorname{supp} f \subset [-2^{n_0}, 2^{n_0}]$. Take $N \geqq n_0 + 1$, so that the translation steps in the $\phi_{Nk}(x) = \phi_{N0}(x - 2^N k)$ are larger than $|\operatorname{supp} f|$. On the other hand, for any $\varepsilon > 0$ there exists $k_0 \in \mathbb{N}$ such that

$$\int_{|x| \geqq k_0} dx |\phi(x)|^2 \leqq \varepsilon.$$

Then

$$\sum_k |\langle \phi_{Nk}, f\rangle|^2$$

$$= \sum_{|k| \leqq k_0} |\langle \phi_{Nk}, f\rangle|^2 + \sum_{|k| \geqq k_0+1} |\langle \phi_{Nk}, f\rangle|^2$$

$$\leqq (2k_0+1)2^{-N}\|\phi\|_\infty^2 \|f\|_1^2 + \|f\|_2^2 2^{-N} \sum_{|k| \geqq k_0+1} \int_{|x| \leqq 2^{n_0}} dx |\phi(2^{-N}x - k)|^2$$

$$\leqq 2^{-N}(2k_0+1)\|\phi\|_\infty^2 \|f\|_1^2 + \varepsilon \|f\|_2^2.$$

By choosing ε and N appropriately, this can be made arbitrarily small. Hence

$$\sum_k |\langle \phi_{Nk}, f\rangle|^2 \underset{N\to\infty}{\longrightarrow} 0. \tag{3.59}$$

4. We now concentrate on $\Sigma_k|\langle\phi_{-Nk}, f\rangle|^2$. By means of the Poisson formula this can be rewritten as

$$\sum_k |\langle\phi_{-Nk}, f\rangle|^2$$

$$= 2\pi \sum_{l\in\mathbb{Z}} \int d\xi\, \hat{\phi}(2^{-N}\xi)\overline{\hat{\phi}(2^{-N}\xi + 2\pi l)\,\hat{f}(\xi)}\,\hat{f}(\xi + 2\pi l 2^N)$$

$$= 2\pi \int d\xi |\hat{\phi}(2^{-N}\xi)|^2 |\hat{f}(\xi)|^2 + R. \tag{3.60}$$

Here

$$|R| \leqq \sum_{l\neq 0} \int d\xi\, |\hat{f}(\xi)|\,|\hat{f}(\xi + 2\pi l 2^N)|,$$

because $|\hat{\phi}(\xi)| = (2\pi)^{-1/2}\prod_{j=1}^{\infty}|m_0(2^{-j}\xi)| \leqq (2\pi)^{-1/2}$, since $|m_0(\xi)| \leqq 1$. Since $f \in C_0^\infty$, we can find C such that

$$|\hat{f}(\xi)| \leqq C(1 + |\xi|)^{-3}.$$

An easy estimation then leads to

$$|R| \leqq C' 2^{-3N/2}.$$

This tends to zero for $N \to \infty$.

5. We now examine the first term in (3.60). One has

$$|\hat{\phi}(\xi) - \hat{\phi}(0)| = (2\pi)^{-1/2} \left| \prod_{j=1}^{\infty} m_0(2^{-j}\xi) - \prod_{j=1}^{\infty} m_0(0) \right|$$

$$\leqq (2\pi)^{-1/2} \sum_{j=1}^{\infty} |m_0(2^{-j}\xi) - m_0(0)|,$$

since $|m_0(\zeta)| \leqq 1$ for all $\zeta \in \mathbb{R}$. But

$$|m_0(\zeta) - m_0(0)| \leqq 2^{-1/2} \sum_n |h(n)|e^{in\zeta} - 1|$$

$$\leqq C|\zeta|^\varepsilon,$$

where we have used (3.19) and $|e^{i\alpha} - 1| \leqq C_\varepsilon|\alpha|^\varepsilon$ (we assume $0 < \varepsilon \leqq 1$). Hence,

$$|\hat{\phi}(\xi) - \hat{\phi}(0)| \leqq (2\pi)^{-1/2} C \sum_{j=1}^{\infty} |2^{-j}\xi|^\varepsilon \leqq C'|\xi|^\varepsilon.$$

Consequently, using $\hat{\phi}(0) = (2\pi)^{-1/2}$, we find

$$2\pi \int d\xi \, |\hat{\phi}(2^{-N}\xi)|^2 |\hat{f}(\xi)|^2$$

$$\leqq \int d\xi \, |\hat{f}(\xi)|^2 + 2\pi \int d\xi \left[2C'|2^{-N}\xi|^{\varepsilon} + C'^2 |2^{-N}\xi|^{2\varepsilon}\right] |\hat{f}(\xi)|^2$$

$$= \|f\|^2 + C'' 2^{-N\varepsilon} \int d\xi \left(1 + |\xi|^{2\varepsilon}\right) |\hat{f}(\xi)|^2.$$

This converges to $\|f\|^2$ as $N \to \infty$. Hence

$$\sum_k |\langle \phi_{-Nk}, f\rangle|^2 \xrightarrow[N\to\infty]{} \|f\|^2. \tag{3.61}$$

6. Putting together (3.58), (3.60) and (3.61) shows that, for all $f \in C_0^\infty(\mathbb{R})$,

$$\sum_{j,k} |\langle \psi_{jk}, f\rangle|^2 = \|f\|_{L^2}^2. \tag{3.62}$$

Since $C_0^\infty(\mathbb{R})$ is dense in $L^2(\mathbb{R})$, (3.62) extends to all $f \in L^2(\mathbb{R})$.

Since $\|\psi\| = 1$ (this is a special case of (3.50), with $j = k = k' = 0$), (3.57) implies that the ψ_{jk} constitute an orthonormal basis. This completes the proof of the main theorem of this section.

THEOREM 3.6. *Let $h(n)$ be a sequence such that*

(i) $\Sigma_n |h(n)|\,|n|^{\varepsilon} < \infty$ *for some* $\varepsilon > 0$,

(ii) $\Sigma_n h(n-2k)h(n-2l) = \delta_{kl}$,

(iii) $\Sigma h(n) = 2^{1/2}$.

Suppose also that $m_0(\xi) = 2^{-1/2}\Sigma_n h(n)e^{in\xi}$ *can be written as*

$$m_0(\xi) = \left[\tfrac{1}{2}(1 + e^{i\xi})\right]^N \left[\sum_n f(n)e^{in\xi}\right],$$

where

(iv) $\Sigma_n |f(n)|\,|n|^{\varepsilon} < \infty$ *for some* $\varepsilon > 0$,

(v) $\sup_{\xi\in\mathbb{R}} |\Sigma_n f(n)e^{in\xi}| < 2^{N-1}$.

Define

$$g(n) = (-1)^n h(-n+1),$$

$$\hat{\phi}(\xi) = (2\pi)^{-1/2} \prod_{j=1}^{\infty} m_0(2^{-j}\xi),$$

$$\psi(x) = 2^{1/2} \sum_n g(n)\phi(2x - n).$$

Then the $\phi_{jk}(x) = 2^{-j/2}\phi(2^{-j}x - k)$ *define a multiresolution analysis (in the sense of subsection* 2A); *the* ψ_{jk} *are the associated orthonormal wavelet basis.*

Remarks. 1. As we already said in the introduction, this theorem is also proved in [19], under slightly different conditions. The growth restrictions (3.37) and (3.38) on the $h(n)$ are replaced, in [19], by the condition that $\inf_{|\xi| \leq \pi/2}|m_0(\xi)| > 0$. Together with $|m_0(\xi)|^2 + |m_0(\xi + \pi)|^2 = 1$, $m_0(0) = 1$, this condition implies that the ϕ_{jk}, with ϕ defined as above, define a multiresolution analysis. The function ϕ may, however, still be very irregular; the coefficients $h(n)$ used in Figure 4, e.g., satisfy the positivity condition of [19], but are clearly not associated with a regular ϕ. In the present paper, we emphasized regularity of the discrete filters; once regularity is ensured by means of conditions (3.36)–(3.38), equivalence with regular multiresolution analysis follows. Consequently, the techniques of our proofs and the proofs in [19] are quite different. The basic intuition for the present proof was mainly graphical. As explained above, the orthogonality of the ϕ_{0k} follows naturally, given our "graphical" construction, from the discrete conditions. Similarly, (3.60) can be understood graphically.

2. At the end of subsection 3B (Remark 3) we mentioned the link between the present construction and the "conjugated quadrature filters" of Smith and Barnwell [24]. Any of their conjugated quadrature filters will satisfy all the conditions in subsection 3A. Provided they also satisfy the regularity condition in subsection 3B, they can be used to construct orthonormal wavelet bases. Since the goals of [24] are completely different however, most of the examples in [24] do not satisfy our regularity condition.

4. Orthonormal Bases of Wavelets with Compact Support

In subsection 2A we reviewed how orthonormal bases of wavelets can be constructed, starting from a multiresolution analysis framework. The basic ingredient there was a function ϕ such that (2.15) held, for some c_n, without even requiring the ϕ_{0n} to be orthogonal. Theorem 3.6 gives another recipe for constructing an orthonormal basis of wavelets (and the associated multiresolution analysis), this time from a sequence $(h(n))_{n \in \mathbf{Z}}$.

If this sequence has finite length, $h(n) = 0$ for $n < N_-$, or $n > N_+$, then the corresponding basic wavelet has compact support. This can be checked very easily from the graphical construction of ϕ (see Figures 2, 4), or from the recursive definition of the η_l,

$$\phi(x) = \lim_{l \to \infty} \eta_l(x), \tag{4.1}$$

$$\eta_l(x) = \sqrt{2}\sum_n h(n)\eta_{l-1}(2x - n), \tag{4.2}$$

$$\eta_0 = \chi_{[-1/2,1/2[}. \tag{4.3}$$

The recursive definition of the η_l implies that all the η_l have compact support, $\operatorname{supp} \eta_l \subset [N_{l,-}, N_{l,+}]$, with $N_{l,-} = \frac{1}{2}(N_{l-1,-} + N_-)$, and $N_{l,+} = \frac{1}{2}(N_{l-1,+} + N_+)$, while $N_{0,-} = -\frac{1}{2}$, $N_{0,+} = \frac{1}{2}$. Hence $H_{l,-} \to N_-$, $N_{l,+} \to N_+$ for $l \to \infty$, which implies that ϕ has compact support $\subset [N_-, N_+]$. Since only finitely many $g(n)$ are non-zero ($g(n) = 0$ for $n < -N_+ + 1$ or $n > -N_- + 1$), ψ also has compact support,

$$\operatorname{supp} \psi \subset \left[\tfrac{1}{2}(1 - N_+ - N_-), \tfrac{1}{2}(1 + N_+ - N_-)\right].$$

In order to construct orthonormal bases of compactly supported wavelets, it suffices, therefore, to construct finite-length sequences $h(n)$ satisfying all the conditions of Theorem 3.6. An example of such a finite-length sequence is Example 3.2, with $\nu = \pm 1/\sqrt{3}$ (see Remark 5 following Proposition 3.3). In this case one finds (see (3.42)) $N_- = 0$, $N_+ = 3$, and

$$m_0(\xi) = \left[\tfrac{1}{2}(1 + e^{i\xi})\right]^2 \tfrac{1}{2}\left[(1 \mp \sqrt{3}) + (1 \pm \sqrt{3})e^{i\xi}\right]. \tag{4.4}$$

Since

$$\sup_{\xi \in \mathbb{R}} \tfrac{1}{2}\left|(1 \mp \sqrt{3}) + (1 \pm \sqrt{3})e^{i\xi}\right| = \sqrt{3} < 2,$$

the example (3.42) satisfies all the required conditions. The $h(n)$ given by (3.42) correspond, therefore, to an orthonormal basis of continuous wavelets. The basic wavelet has support width equal to $N_+ - N_- = 3$. Figure 5 shows the graphs of ϕ, ψ and their Fourier transforms, for this example. There are several striking features in Figure 5. First of all, it is obvious that even though ϕ and ψ are continuous, they are not very regular. There exist other constructions of compactly supported wavelet bases, in which ϕ and ψ have more regularity, at the cost of larger numbers of non-zero coefficients $h(n)$, which results in larger support widths for ψ, ϕ. For the family of examples we shall examine below, the support width of ψ, ϕ increases linearly with their regularity. Another striking feature of Figure 5 is the lack of any symmetry or antisymmetry axis for ψ, ϕ. This is quite unlike the Meyer wavelets (see [4]) or the Battle-Lemarié wavelets (see [16]). In all these (non-compactly supported) examples, ϕ is an even function, and ψ is symmetric around $x = \frac{1}{2}$. We shall see below that, except for the Haar basis (see (1.9) or Example 3.1), there exist *no* compactly supported wavelet bases in which ϕ is either symmetric or antisymmetric around any axis.

The plots of ψ and ϕ in Figure 5 (and later figures, for other examples) are made by direct implementation of the "graphical recursion algorithm" equivalent with (4.1)–(4.3) (see subsection 2B). This is much more efficient than Fourier transform of the infinite product (3.46) (see [26]). To plot Figure 5, only 8 iterations of type (2.40) were needed (i.e., η_8 is plotted rather than ϕ; the difference is not detectable at the scale of the figure). If more detail is wanted at any point (see Figure 6), it is possible to restrict to a neighborhood, and to locally iterate a few times more to obtain this detail.

In the following subsections we describe families of examples of compactly supported wavelet bases, and their properties. Henceforth, we shall always assume that only finitely many $h(n)$ are non-zero.

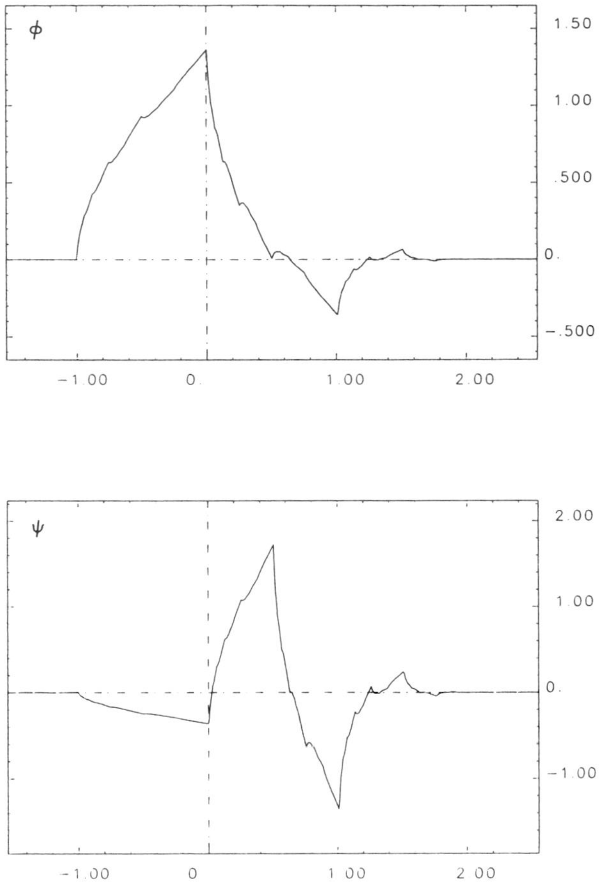

(a)

Figure 5. The functions ϕ, ψ, and the modulus of their Fourier transforms, $|\hat{\phi}|$, $|\hat{\psi}|$, for the orthonormal basis of compactly supported wavelets corresponding to the $h(n)$ in (3.42) (see text). Out of the two possibilities in (3.42) we choose the one corresponding to $\nu = -1/\sqrt{3}$ (i.e., $h(0) = (1 + \sqrt{3})/4\sqrt{2}$, etc.)

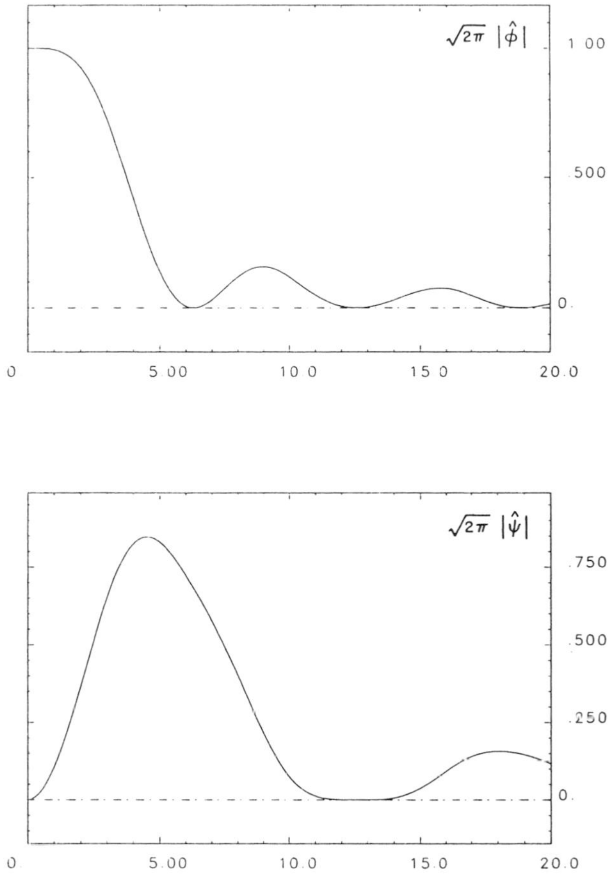

(b)

Figure 5. Continued

4.A. Lack of symmetry. Here we shall use again the notations $a(n), \cdots, d(n)$ (see (3.7)) and $\alpha(\xi), \cdots, \delta(\xi)$ (see (3.9)) introduced in subsection 3A. Let us define, for any trigonometric polynomial $P(\xi) = \Sigma_n p_n e^{in\xi}$, the two numbers

$$N_+(P) = \max\{n;\ p_n \neq 0\},$$
$$N_-(P) = \min\{n;\ p_n \neq 0\}.$$

One easily checks that

$$N_+(|P|^2) = -N_-(|P|^2) = N_+(P) - N_-(P).$$

Since $|\alpha(\xi)|^2 + |\beta(\xi)|^2 = 1$ (see (3.14)), and $\alpha \neq 0$, $\beta \neq 0$ (see (3.15)), this implies

$$N_+(\alpha) - N_-(\alpha) = N_+(\beta) - N_-(\beta). \tag{4.5}$$

On the other hand, the definition (3.7) of the $a(n)$, $b(n)$, gives

$$N_+(m_0) = \max(2N_+(\alpha), 2N_+(\beta) + 1),$$

$$N_-(m_0) = \min(2N_-(\alpha), 2N_-(\beta) + 1).$$

Together with (4.5) this leads to

$$\begin{aligned} &N_+(m_0) - N_-(m_0) \\ &\quad = \max(2N_+(\alpha) - 2N_-(\beta) - 1, 2N_+(\beta) - 2N_-(\alpha) + 1). \end{aligned} \tag{4.6}$$

In any case, $N_+(m_0) - N_-(m_0)$ is an odd number.

If the function ϕ were symmetric around zero, $\phi(x) = \phi(-x)$, then $h(n) = h(-n)$ would follow. This would however imply $N_+(m_0) = -N_-(m_0)$, i.e., $N_+(m_0) - N_-(m_0) = 2N_+(m_0)$ would be even. Since this is in contradiction with (4.6), it follows that the function ϕ, associated with an orthonormal basis of wavelets with compact support, can never be an even function.

What about symmetry with respect to another point $\lambda \neq 0$? Suppose

$$\phi(\lambda + x) = \phi(\lambda - x),$$

where we can, without loss of generality, shift λ to the interval $]0, 1[$. Then it follows that

$$\hat{\phi}(\xi) = e^{2i\lambda\xi}\hat{\phi}(-\xi).$$

Because of the definition of $\hat{\phi}$ as the infinite product (3.46), this implies

$$m_0(\xi) = e^{2i\lambda\xi}m_0(-\xi).$$

Since both $m_0(\xi)$ and $m_0(-\xi)$ are trigonometric polynomials, this leaves only one possible value for λ, namely $\lambda = \frac{1}{2}$. Let us, therefore, assume that ϕ is symmetric with respect to $\frac{1}{2}$,

$$\phi(x + 1) = \phi(-x).$$

Then

$$h(2n + 1) = h(-2n),$$

or

$$b(n) = a(-n).$$

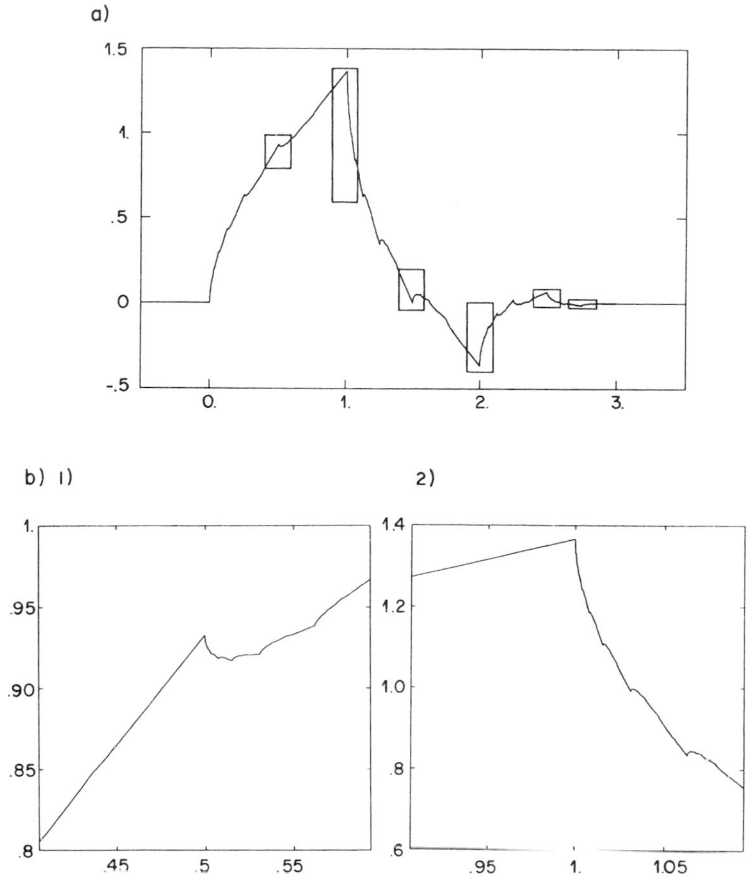

(a)

Figure 6. The function ϕ of Figure 5, and 6 local blow-ups

(a) The different zoom-in zones are shown on the graph of ϕ

(b) The blow-ups around 1) $x = .5$, 2) $x = 1$, 3) $x = 1.5$, 4) $x = 2.$, 5) $x = 2.5$, 6) $x + 2.75$.

The detail in these blow-ups illustrates the fractal, self-similar nature of this function ϕ.

Hence

$$\beta(\xi) = \overline{\alpha(\xi)}\,.$$

Together with (3.14) this implies

$$2|\alpha(\xi)|^2 = 1,$$

or

$$a(n) = \pm 2^{-1/2}\delta_{nk} = b(-n) \quad \text{for some} \quad k \in \mathbb{N}.$$

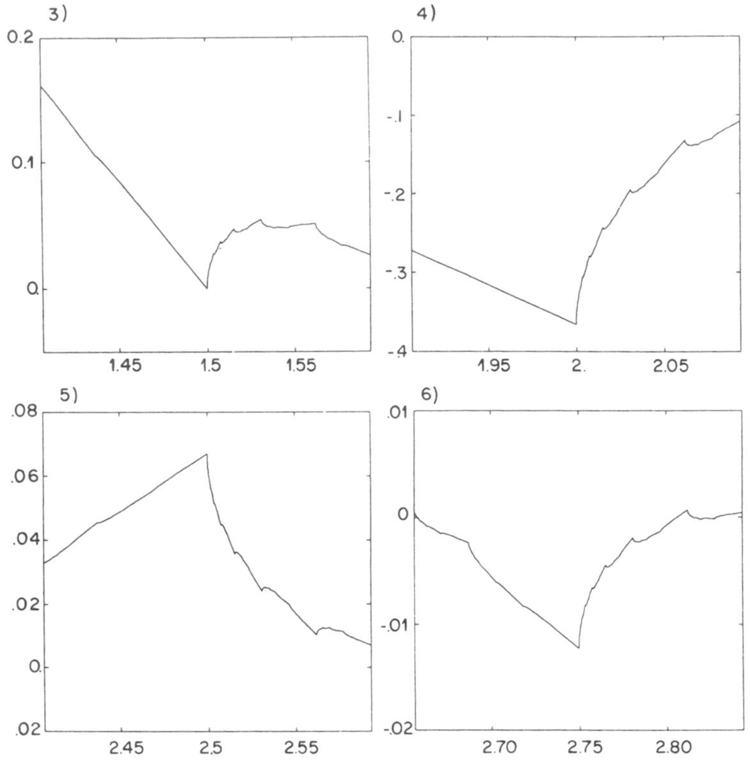

(b)

Figure 6. Continued

We can, again without loss of generality, choose $k = 0$ (this amounts to a translation of ϕ by an integer). The corresponding $h(n)$ are then exactly given by Example 3.1, resulting in $\phi = \chi_{[0,1]}$.

All these arguments prove the following proposition.

PROPOSITION 4.1. *The Haar basis* (1.9) *is the only orthonormal basis of compactly supported wavelets for which the associated averaging function ϕ has a symmetry axis.*

In the following subsection we explicitly characterize all the functions m_0 corresponding to orthonormal wavelet bases with compactly supported basic wavelet.

4.B. Characterization of all orthonormal, compactly supported wavelet bases. The basic condition (3.18) on the $h(n)$ can be rewritten as (see subsection 3A)

$$|m_0(\xi)|^2 + |m_0(\xi + \pi)|^2 = 1. \tag{4.7}$$

On the other hand, we have imposed, in Proposition 3.3, the following structure on m_0:

$$m_0(\xi) = \left[\tfrac{1}{2}(1 + e^{i\xi})\right]^N Q(e^{i\xi}), \tag{4.8}$$

where Q is a polynomial, since only finitely many $h(n)$ are non zero. Moreover, since all the $h(n)$ are real, all the coefficients in Q are real as well. From (4.8) we have

$$|m_0(\xi)|^2 = \left[\cos^2 \tfrac{1}{2}\xi\right]^N |Q(e^{i\xi})|^2.$$

Since $\overline{Q(e^{i\xi})} = Q(e^{-i\xi})$, the polynomial $|Q(e^{i\xi})|^2$ can be rewritten as a polynomial in $\cos \xi$, or, equivalently, as a polynomial in $\sin^2 \frac{1}{2}\xi$. Introducing the shorthand $y = \cos^2 \frac{1}{2}\xi$, (4.7) becomes

$$y^N P(1-y) + (1-y)^N P(y) = 1. \tag{4.9}$$

Any m_0 of type (4.8) which solves (4.7) corresponds therefore to a polynomial P solving (4.9) and satisfying

$$P(y) \geqq 0 \quad \text{for} \quad y \in [0,1]. \tag{4.10}$$

Conversely, every polynomial P satisfying both (4.9) and (4.10) leads to solutions of (4.7), with real coefficients $h(n)$.This is due to the following lemma of Riesz [27].

LEMMA 4.2. *Let A be a positive trigonometric polynomial containing only cosines, $A(\xi) = \sum_{n=0}^N a_n \cos n\xi$ (with $a_n \in \mathbb{R}$). Then there exists a trigonometric polynomial B, of order N, $B(\xi) = \sum_{n=0}^N b_n e^{in\xi}$, with real coefficients b_n, such that*

$$|B(\xi)|^2 = A(\xi). \tag{4.11}$$

The proof of this lemma (see [27]) is simple and elegant. It constructs B explicitly; this construction is now widely used by engineers when designing filters. We include the proof here, because we shall come back to the construction later.

Proof: To

$$A(\xi) = a_0 + \frac{1}{2}\sum_{n=1}^{N} a_n (e^{in\xi} + e^{-in\xi})$$

$$= e^{-iN\xi}\left[\frac{1}{2}\sum_{n=0}^{N-1} a_{N-n}\, e^{in\xi} + a_0\, e^{iN\xi} + \frac{1}{2}\sum_{n=1}^{N} a_n\, e^{i(N+n)\xi}\right]$$

we associate the polynomial

$$P_A(z) = \frac{1}{2}\sum_{n=0}^{N-1} a_{N-n} z^n + a_0 z^N + \frac{1}{2}\sum_{n=1}^{N} a_n z^{N+n}.$$

This polynomial has $2N$ zeros (counting multiplicity). Since $P_A(e^{i\xi}) = e^{iN\xi}A(\xi)$, it follows that the two polynomials $P_A(z)$ and $z^{2N}P_A(z^{-1})$ agree on the unit circle, and therefore on the whole complex plane. They have therefore the same zeros. This means that if z_0 is a zero of $P_A(z)$, $P_A(z_0) = 0$, then so is z_0^{-1}. On the other hand, since the a_n are real, $\overline{P_A(z)} = P_A(\bar{z})$. This implies that if z_0 is a zero of $P_A(z)$, then so is its complex conjugate $\bar{z}_0$. The zeros of $P_A(z)$ therefore come in quadruplets, z_0, $\bar{z}_0$, z_0^{-1} and $\bar{z}_0^{-1}$, or (if $z_0 = r_0$ is real) in duplets, $r_0, r_0^{-1} \in \mathbb{R}$. Let $z_j, \bar{z}_j, z_j^{-1}, \bar{z}_j^{-1}$ be the quadruplets of complex zeros of $P_A(z)$, and r_k, r_k^{-1} the real duplets,

$$P_A(z) = \frac{1}{2}a_N\left[\prod_{k=1}^{K}(z - r_k)(z - r_k^{-1})\right] \cdot \left[\prod_{j=1}^{J}(z - z_j)(z - \bar{z}_j)(z - z_j^{-1})(z - \bar{z}_j^{-1})\right].$$

For $z = e^{i\xi}$ on the unit circle, one finds

$$\begin{aligned}\left|(e^{i\xi} - z_0)(e^{i\xi} - \bar{z}_0^{-1})\right| &= |z_0|^{-1}\left|(e^{i\xi} - z_0)(\bar{z}_0 - e^{-i\xi})\right| \\ &= |z_0|^{-1}|e^{i\xi} - z_0|^2.\end{aligned}$$

Consequently,

$$\begin{aligned}A(\xi) &= |A(\xi)| = |P_A(e^{i\xi})| \\ &= \left[\frac{1}{2}|a_N|\prod_{k=1}^{K}|r_k|^{-1}\prod_{j=1}^{J}|z_j|^{-2}\right]\left|\prod_{k=1}^{K}(e^{i\xi} - r_k)\prod_{j=1}^{J}(e^{i\xi} - z_j)(e^{i\xi} - \bar{z}_j)\right|^2 \\ &= |B(\xi)|^2,\end{aligned}$$

where

$$B(\xi) = \left[\frac{1}{2}|a_N|\prod_{k=1}^{K}|r_k|^{-1}\prod_{j=1}^{J}|z_j|^{-2}\right]^{1/2} \cdot \prod_{k=1}^{K}(e^{i\xi} - r_k)\prod_{j=1}^{J}(e^{2i\xi} - 2e^{i\xi}\,\mathcal{R}e\, z_j + |z_j|^2) \tag{4.12}$$

is clearly a trigonometric polynomial of order N with only real coefficients.

Remarks. 1. Note that B is generally not unique. Out of any quadruplet of zeros z_0, $\bar{z}_0$, z_0^{-1}, $\bar{z}_0^{-1}$ one can choose the pair of zeros to retain, for the construction of B, in four different ways. For every duplet of real zeros of P_A two choices are possible. This results in 2^N different possibilities for B.

2. All these different possibilities, corresponding to different choices of the zeros of P_A to retain for B, constitute, however, the only solutions to (4.11). One can show (see [27]) that, up to an arbitrary phase factor $\pm e^{iK\xi}$, $K \in \mathbb{Z}$, *all* the polynomials B satisfying (4.11) are necessarily of the form (4.12).

If P is a polynomial satisfying (4.9) and (4.10), then Lemma 4.2 tells us that there exists a trigonometric polynomial of the same order such that

$$|Q(e^{i\xi})|^2 = P(\sin^2 \tfrac{1}{2}\xi) = P(\tfrac{1}{2}(1 - \cos \xi)).$$

It follows that $m_0(\xi) = [\frac{1}{2}(1 + e^{i\xi})]^N Q(e^{i\xi})$ satisfies (4.7). If, moreover,

$$\sup_\xi |Q(e^{i\xi})| = \sup_{y \in [0,1]} |P(y)|^{1/2} < 2^{N-1},$$

then all the conditions of Theorem 3.6 are satisfied, and there exists an associated orthonormal wavelet basis.

To construct compactly supported orthonormal wavelet bases, with m_0 of type (4.8), it is therefore necessary and sufficient to find polynomials P solving (4.9) and (4.10), which are moreover strictly bounded above by $2^{2(N-1)}$.

The following two combinatorial lemmas allow one to "guess" a particular solution of (4.9).

LEMMA 4.3.

$$\sum_{j=0}^{k} \binom{n+j}{j} = \binom{n+k+1}{k}.$$

Proof: Define $S_{n,k} = \sum_{j=0}^{k} \binom{n+j}{j}$. Then

$$S_{n+1,k+1} = \frac{(k+n+2)!}{(k+1)!(n+1)!} + \sum_{j=0}^{k} \frac{(n+j)!}{(n+1)!j!}(n+j+1)$$

$$= \binom{k+n+2}{k+1} + S_{n,k} + \sum_{j=1}^{k} \frac{(n+j)!}{(n+1)!(j-1)!}$$

$$= \binom{k+n+2}{n+1} + S_{n,k} + \left[S_{n+1,k+1} - \binom{k+n+2}{k+1} - \binom{k+n+1}{k}\right].$$

Hence

$$S_{n,k} = \binom{n+k+1}{k}.$$

LEMMA 4.4.

$$\sum_{j=0}^{n} \binom{n+j}{j} \left[y^j (1-y)^{n+1} + y^{n+1}(1-y)^j \right] = 1.$$

Proof: Define $A_{n,j} = \binom{n+j}{j}$. Then, by Lemma 4.3, $\sum_{j=0}^{k} A_{n,j} = A_{n+1,k}$. Define

$$S_n(y) = \sum_{j=0}^{n} \binom{n+j}{j} \left[y^j (1-y)^{n+1} + y^{n+1}(1-y)^j \right].$$

Clearly,

$$S_0(a) = (1-a) + a = 1.$$

We shall prove that $S_n(a) = S_{n-1}(a)$, which proves the lemma. By repeatedly inserting factors $[(1-a)+a] = 1$, we find

$$\begin{aligned}
S_{n-1}(a) &= \sum_{j=0}^{n-1} A_{n-1,j} \left[(1-a)^n a^j + a^n (1-a)^j \right] \\
&= A_{n-1,0} \left[(1-a)^{n+1} + a^{n+1} \right] \\
&\quad + (A_{n-1,0} + A_{n-1,1}) \left[(1-a)^n a + a^n (1-a) \right] \\
&\quad + \sum_{j=2}^{n-1} A_{n-1,j} \left[(1-a)^n a^j + a^n (1-a)^j \right] \\
&= \cdots \\
&= \sum_{j-0}^{n-1} \left[\sum_{k=0}^{j} A_{n-1,k} \right] \left[(1-a)^{n+1} a^j + a^{n+1} (1-a)^j \right] \\
&\quad + 2 \left[\sum_{k=0}^{n-1} A_{n-1,k} \right] (1-a)^n a^n \\
&\quad - \sum_{j=0}^{n-1} A_{n,j} \left[(1-a)^{n+1} a^j + a^{n+1} (1-a)^j \right] \\
&\quad + 2A_{n,n-1} \left[(1-a)^{n+1} a^n + a^{n+1} (1-a)^n \right] \\
&= S_n(a) \qquad\qquad (\text{since } 2A_{n,n-1} = A_{n,n}).
\end{aligned}$$

It follows that the polynomial of order $N-1$,

$$P_N(y) = \sum_{j=0}^{N-1} \binom{N-1+j}{j} y^j, \tag{4.13}$$

solves (4.9). Since all the coefficients in this polynomial are positive, (4.10) is clearly also satisfied.

The two explicit examples of compactly supported wavelet bases we have seen so far, i.e., Example 3.1 and (3.42), correspond exactly to a polynomial of type (4.13), with $N = 1, 2$, respectively. For Example 3.1 one has $m_0(\xi) = \frac{1}{2}(1 + e^{i\xi})$, i.e., $N = 1$, and $Q(e^{i\xi}) = 1$, hence $P(y) = 1 = P_1(y)$. For the second example (3.42), we find (see (4.4)) $m_0(\xi) = [\frac{1}{2}(1 + e^{i\xi})]^2 \frac{1}{2}[(1 \mp \sqrt{3})e^{i\xi}]$, corresponding to $N = 2$ and $|Q(e^{i\xi})|^2 = 2 - \cos\xi = 1 + 2\sin^2\frac{1}{2}\xi$; hence $P(y) = 1 + 2y = P_2(y)$.

In fact, for given N, P_N is the *only* polynomial of order less than N which solves (4.9). Even more is true: for *any* polynomial P solving (4.9), the first N terms (orders 0 up till $N-1$) are exactly given by P_N. This is because (4.9) completely determines the first N coefficients $p_0, \cdots, p_{N-1}$ in $P(y) = \sum_{n=0}^{K} p_n y^n$. Since the first term in (4.9) is already of order N, only the second term plays a role in the cancellations for y^k, $k = 0, \cdots, N-1$. This leads to

$$\begin{aligned} &p_0 = 1, \\ &p_k = \sum_{n=0}^{k-1} (-1)^{k-n+1} \binom{N}{k-n} p_n, \qquad k = 1, \cdots, N-1, \end{aligned} \tag{4.14}$$

from which the p_k, $k = 1, \cdots, N-1$, can be determined recursively. Since P_N solves (4.9), it follows from (4.13) that

$$p_k = \binom{N+k-1}{k}.$$

Consequently, *any* polynomial P solving (4.9) is of the form

$$P(y) = P_N(y) + y^N R(y). \tag{4.15}$$

Substitution of (4.15) into (4.9) leads to the following equation for the polynomial R:

$$y^N(1-y)^N R(1-y) + (1-y)^N y^N R(y) = 0,$$

or

$$R(1-y) + R(y) = 0.$$

The polynomial R is therefore antisymmetric with respect to $y = \frac{1}{2}$, or

$$R(y) = \tilde{R}(\tfrac{1}{2} - y),$$

where $\tilde{R}$ is an odd polynomial.

To summarize, we have the following explicit characterization of all solutions m_0 of (4.7), corresponding to only finitely many non-zero $h(n)$.

PROPOSITION 4.5. *Any trigonometric polynomial solution m_0 of (4.7) is of the form*

$$m_0(\xi) = \left[\tfrac{1}{2}(1 + e^{i\xi})\right]^N Q(e^{i\xi}), \tag{4.16}$$

where $N \in \mathbb{N}$, $N \geqq 1$, and where Q is a polynomial such that

$$|Q(e^{i\xi})|^2 = \sum_{k=0}^{N-1} \binom{N-1+k}{k} \sin^{2k} \tfrac{1}{2}\xi + \left[\sin^{2N}\tfrac{1}{2}\xi\right] R\left(\tfrac{1}{2}\cos \xi\right), \tag{4.17}$$

where R is an odd polynomial.

Remarks. 1. Since the proof of Lemma 4.2 shows explicitly how to construct all possible polynomials Q once $|Q(e^{i\xi})|^2$ is known, this proposition is indeed an explicit characterization of all the solutions m_0 of (4.7).

2. In constructing m_0, there are therefore 3 steps at which choices can be made,

(i) choosing $N \in \mathbb{N} \setminus \{0\}$,
(ii) choosing an odd polynomial R (with some restrictions),
(iii) choosing pairs of zeros out of each quadruplet of complex zeros, and one zero out of each duplet of real zeros, of $P_N(z) + z^N R(z - \frac{1}{2})$ (see the proof of Lemma 4.2).

The odd polynomial R cannot be chosen completely freely. One needs, of course, the fact that

$$P_N(y) + y^N R\left(\tfrac{1}{2} - y\right) \geqq 0 \quad \text{for} \quad 0 \leqq y \leqq 1. \tag{4.18}$$

Moreover, condition (v) in Theorem 3.6 requires that

$$\sup_{0 \leqq y \leqq 1} \left[P_N(y) + y^N R\left(\tfrac{1}{2} - y\right)\right] < 2^{2(N-1)}. \tag{4.19}$$

3. For $N = 1$, (4.16), (4.17) and (4.18) reduce to

$$m_0(\xi) = \tfrac{1}{2}(1 + e^{i\xi}) Q(e^{i\xi}) \tag{4.20}$$

with

$$|Q(e^{i\xi}|^2 = 1 + \sin^2 \tfrac{1}{2}\xi R\left(\tfrac{1}{2}\cos \xi\right), \tag{4.21}$$

where R is an odd polynomial such that

$$-\frac{2}{1 - 2|x|} \leqq R(x) \leqq \frac{2}{1 + 2|x|} \quad \text{for} \quad |x| \leqq \tfrac{1}{2}.$$

These conditions can already be found in the construction of conjugate quadrature mirror filters in [24]. The condition (4.19) is impossible to satisfy, however, because $P_1(0) = 1$.

4. Using a different method, Y. Meyer constructs in [28] another polynomial solving (4.7). The solutions to (4.7) proposed in [28] are

$$|m_0(\xi)|^2 = 1 - \frac{(2N-1)!}{[(N-1)!]^2 2^{2N-1}} \int_0^{\xi} \sin^{2N-1} x \, dx. \tag{4.22}$$

This is clearly an even trigonometric polynomial of order $2N - 1$. It turns out to be divisible by $(\frac{1}{2}(1 + \cos\xi))^N = (\cos^2 \frac{1}{2}\xi)^N$. Therefore, by Proposition 4.5, (4.22) is exactly equal to

$$\left(\cos^2 \tfrac{1}{2}\xi\right)^N P_N\left(\sin^2 \tfrac{1}{2}\xi\right).$$

4.C. A family of examples with arbitrarily high regularity. In the remainder of this section, we shall concern ourselves with a special family of functions m_0, and the corresponding wavelet bases. We follow the prescriptions of Remark 2 after Proposition 4.5. For every $N \in \mathbb{N}$, $N \geqq 1$, we choose Q of minimal order, i.e., $R \equiv 0$, $|Q(e^{i\xi})|^2 = P_N(\sin^2 \frac{1}{2}\xi)$. This choice satisfies both the conditions (4.18) and (4.19). From (4.13) the positivity of $P_N(y)$ for $0 \leqq y \leqq 1$ is immediate. Since P_N is strictly increasing for $y \geqq 0$, it follows that

$$\begin{aligned} \sup_{y \in [0,1]} P_N(y) = P_N(1) &= \binom{2N-1}{N-1} = \frac{1}{2}\left[\binom{2N-1}{N-1} + \binom{2N-1}{N}\right] \\ &< \frac{1}{2} \sum_{k=0}^{2N-1} \binom{2N-1}{k} = 2^{2(N-1)}, \end{aligned} \tag{4.23}$$

where we have used Lemma 4.3 in the second equality. This fixes $|Q|^2$. In the construction (via Lemma 4.2) of Q from $|Q|^2$, we systematically retain all the zeros inside the unit circle (this corresponds to a "minimal phase" choice in filter design). For $N \in \mathbb{N}$, $N > 1$ fixed, this determines Q unambiguously, up to a phase factor $e^{iK\xi}$, $K \in \mathbb{Z}$. For the sake of definiteness we fix this phase factor so that Q contains only positive frequencies, starting from zero, i.e.,

$$Q_N(e^{i\xi}) = \sum_{n=0}^{N-1} q_N(n) e^{in\xi} \quad \text{with} \quad q_0 \neq 0. \tag{4.24}$$

These choices uniquely determine Q_N. We shall denote the corresponding m_0 by ${}_N m_0$,

$$\begin{aligned} {}_N m_0(\xi) &= \left[\tfrac{1}{2}(1 + e^{i\xi})\right]^N \sum_{n=0}^{N-1} q_N(n) e^{in\xi} \\ &= 2^{-1/2} \sum_{n=0}^{2N-1} h_N(n) e^{in\xi}. \end{aligned}$$

Table 1 lists the coefficients $h_N(n)$ for the cases $N = 2, 3, \cdots, 10$. For the lowest values of N, $Q_N(\xi)$ can be determined analytically. One has, e.g.,

$$Q_2(\xi) = \tfrac{1}{2}\big[(1+\sqrt{3}) + (1-\sqrt{3})e^{i\xi}\big] \qquad \text{(see (4.4))}$$

and

$$Q_3(\xi) = \tfrac{1}{4}\Big[\Big(1+\sqrt{10}+\sqrt{5+2\sqrt{10}}\Big) + 2(1-\sqrt{10})e^{i\xi} + \Big(1+\sqrt{10}-\sqrt{5+2\sqrt{10}}\Big)e^{2i\xi}\Big].$$

For larger values of N, the coefficients in Table 1 were computed numerically.

Since the ${}_Nm_0$ satisfy all the conditions of Theorem 3.6, there exists an associated orthonormal basis of continuous wavelets with compact support for every ${}_Nm_0$. Let us denote the corresponding ϕ, ψ functions by ${}_N\phi$, ${}_N\psi$. Since $h_N(n) = 0$ for $n < 0$ and $n > 2N-1$, it follows (see the discussion at the start of Section 4) that $\operatorname{supp}({}_N\phi) = [0, 2N-1]$. The support of ${}_N\psi$,

$$({}_N\psi)(x) = \sum_{n=0}^{2N-1} (-1)^n h_N(-n+1)\, {}_N\phi(2x-n),$$

is therefore given by $[-(N-1), N]$. Note that an additional phase factor $e^{iK\xi}$, $K \in \mathbb{Z}$, in (4.24) would amount to shifting the $h_N(n)$ by K, i.e., to shifting the function ${}_N\phi$ by an integer, which does not affect the multiresolution analysis construction. The wavelet ${}_N\psi$ is unaffected by this shift.

From Theorem 3.6, we know that ${}_N\phi$ and ${}_N\psi$ are bounded, continuous functions. For large N, ${}_N\phi$ and ${}_N\psi$ are, in fact, much more regular. To see this, we shall need the following generalization of Lemma 3.2.

LEMMA 4.6. *If $m_0(\xi) = [\frac{1}{2}(1+e^{i\xi})]^N \mathcal{F}(\xi)$, where $\mathcal{F}(\xi) = \sum_n f_n e^{in\xi}$ satisfies*

(4.25) $$\sum_n |f_n|\,|n|^\varepsilon < \infty \quad \textit{for some} \quad \varepsilon > 0,$$

(4.26) $$\sup_\xi \big|\mathcal{F}(\xi)\mathcal{F}(\tfrac{1}{2}\xi)\cdots\mathcal{F}(2^{-k+1}\xi)\big| = B_k,$$

then

(4.27) $$\left|\prod_{j=1}^{\infty} m_0(2^{-j}\xi)\right| \leq C(1+|\xi|)^{-N+\log B_k/(k\log 2)}.$$

Proof: Define

$$\mathcal{F}_k(\xi) = \prod_{j=0}^{k} \mathcal{F}(2^{-j}\xi).$$

Table 1. The coefficients $h_N(n)$ $(n = 0,\cdots,2N-1)$ for $N = 2,3,\cdots,10$.

	n	$h_N(n)$		n	$h_N(n)$
$N = 2$	0	.482962913145	$N = 8$	0	.054415842243
	1	.836516303738		1	.312871590914
	2	.224143868042		2	.675630736297
	3	−.129409522551		3	.585354683654
$N = 3$	0	.332670552950		4	−.015829105256
	1	.806891509311		5	−.284015542962
	2	.459877502118		6	.000472484574
	3	−.135011020010		7	.128747426620
	4	−.085441273882		8	−.017369301002
	5	.035226291882		9	−.044088253931
$N = 4$	0	.230377813309		10	.013981027917
	1	.714846570553		11	.008746094047
	2	.630880767930		12	−.004870352993
	3	−.027983769417		13	−.000391740373
	4	−.187034811719		14	.000675449406
	5	.030841381836		15	−.000117476784
	6	.032883011667	$N = 9$	0	.038077947364
	7	−.010597401785		1	.243834674613
$N = 5$	0	.160102397974		2	.604823123690
	1	.603829269797		3	.657288078051
	2	.724308528438		4	.133197385825
	3	.138428145901		5	−.293273783279
	4	−.242294887066		6	−.096840783223
	5	−.032244869585		7	.148540749338
	6	.077571493840		8	.030725681479
	7	−.006241490213		9	−.067632829061
	8	−.012580751999		10	.000250947115
	9	.003335725285		11	.022361662124
$N = 6$	0	.111540743350		12	−.004723204758
	1	.494623890398		13	−.004281503682
	2	.751133908021		14	.001847646883
	3	.315250351709		15	.000230385764
	4	−.226264693965		16	−.000251963189
	5	−.129766867567		17	.000039347320
	6	.097501605587	$N = 10$	0	.026670057901
	7	.027522865530		1	.188176800078
	8	−.031582039318		2	.527201188932
	9	.000553842201		3	.688459039454
	10	.004777257511		4	.281172343661
	11	−.001077301085		5	−.249846424327
$N = 7$	0	.077852054085		6	−.195946274377
	1	.396539319482		7	.127369340336
	2	.729132090846		8	.093057364604
	3	.469782287405		9	−.071394147166
	4	−.143906003929		10	−.029457536822
	5	−.224036184994		11	.033212674059
	6	.071309219267		12	.003606553567
	7	.080612609151		13	−.010733175483
	8	−.038029936935		14	.001395351747
	9	−.016574541631		15	.001992405295
	10	.012550998556		16	−.000685856695
	11	.000429577973		17	−.000116466855
	12	−.001801640704		18	.000093588670
	13	.000353713800		19	−.000013264203

Then

$$\prod_{j=1}^{\infty} \mathscr{F}(2^{-j}\xi) = \prod_{j=0}^{\infty} \mathscr{F}_k\left[(2^k)^{-j}\tfrac{1}{2}\xi\right].$$

Repeating the proof of Lemma 3.2, with multiplication factor 2^k instead of 2 leads to

$$\left|\prod_{j=0}^{\infty} \mathscr{F}_k\left[(2^k)^{-j}\xi\right]\right| \leqq C\exp\left\{\log B_k \log|\xi|/\log(2^k)\right\}.$$

This implies (4.27).

To interpolate between the standard spaces C^k of k times continuously differentiable functions, we shall use, for $\alpha \notin \mathbb{N}$, $\alpha > 0$, the spaces defined by

$$f \in C^\alpha \Leftrightarrow \int dx\,|\hat{f}(\xi)|(1 + |\xi|)^{1+\alpha} < \infty. \tag{4.28}$$

Note that, for $\alpha = k \in \mathbb{N}$, the condition (4.28) implies $f \in C^k$, but is not necessary.

We then have the following

PROPOSITION 4.7. *There exists* $\lambda > 0$ *such that, for all* $N \in \mathbb{N}$, $N \geqq 2$,

$$_N\phi, {_N\psi} \in C^{\lambda N}. \tag{4.29}$$

Proof: We shall apply Lemma 4.6. Since $Q_N(e^{i\xi})$ has only a finite number of terms, (4.25) is obviously satisfied. We compute

$$B_2 = \sup|Q_N(e^{i\xi})Q_N(e^{i\xi/2})| = \sup_\xi |P_N(\sin^2\tfrac{1}{2}\xi)P_N(\sin^2\tfrac{1}{4}\xi)|$$

$$= \sup_{0\leqq y\leqq 1} |P_N(4y(1-y))P_N(y)|.$$

First, note that (see (4.28))

$$\sup_{0\leqq y\leqq 1} P_N(y) = P_N(1) < 2^{2(N-1)}.$$

Secondly,

$$P_N(y) = \sum_{k=0}^{N-1}\binom{N+k-1}{k}y^k$$

$$\leqq \sum_{k=0}^{N-1} 2^{N+k-1}y^k \leqq 2^{N-1}N\max\left(1,(2y)^N\right).$$

Hence, for $y \leqq \frac{1}{2}$,

$$P_N(y)P_N(4y(1-y)) \leqq N2^{N-1}2^{2(N-1)} = N2^{3(N-1)}.$$

For $y \geqq \frac{1}{4}(2+\sqrt{2})$, or $4y(1-y) \leqq \frac{1}{2}$,

$$P_N(y)P_N(4y(1-y)) \leqq 2^{2(N-1)}N2^{N-1} = N2^{3(N-1)}.$$

Finally, for $\frac{1}{2} \leqq y \leqq \frac{1}{4}(2+\sqrt{2})$,

$$P_N(y)P_N(4y(1-y)) \leqq N^2 2^{4N-2}\Big(\sup_{0\leqq y\leqq 1}\big[4y^2(1-y)\big]\Big)^N$$

$$= N^2 2^{4N-2}\left(\tfrac{16}{27}\right)^N,$$

or

$$B_2 \leqq N2^{2N-1}\left(\tfrac{16}{27}\right)^{N/2}.$$

Consequently,

$$\left|({}_N\phi)^\wedge(\xi)\right| = (2\pi)^{-1/2}\left|\prod_{j=1}^{\infty} m_0(2^{-j}\xi)\right|$$

$$\leqq C(1+|\xi|)^{[\log N - N\log(3\sqrt{3}/4)]/2\log 2}.$$

This exponent is smaller than -1 for $N \geqq 16$. For smaller values of N, one can use the explicit estimate

$$B_1 = \left[\binom{2N-1}{N}\right]^{1/2}$$

to prove that

$$\left|({}_N\phi)^\wedge(\xi)\right| \leqq C(1+|\xi|)^{-1-\kappa N}$$

for some $\kappa > 0$, for all $N \leqq 16$. Hence (4.29) holds for ${}_N\phi$, for some $\lambda > 0$, and for all $N \geqq 2$. Since ${}_N\psi$ is always a finite linear combination of translated and dilated versions of ${}_N\phi$, the same holds for ${}_N\psi$.

Remarks. 1. Since $|\mathrm{supp}({}_N\phi)| = |\mathrm{supp}({}_N\psi)| = 2N-1$, (4.29) shows that the regularity of ${}_N\phi$, ${}_N\psi$ increases linearly with their support width, as announced in the introduction. It turns out that linear increase of the support width with the regularity of ϕ, ψ is the best one can do. More precisely, if a C^k-function ϕ satisfies an equation of the type

$$\phi(x) = \sum_{n=0}^{N} c_n\phi(2x-n)$$

(without necessarily being connected to multiresolution analysis), and if supp $\phi \subset [0, N]$, then $k \leqq N - 2$. For a proof, see [30].

2. The estimate for λ obtained in this proof is, of course, not very good; the argument is too simple. Asymptotically, for large N, one finds

$$_N\phi, {}_N\psi \in C^{(\mu-\varepsilon)N}$$

with

$$\mu \sim \frac{\log\left(\frac{3}{4}\sqrt{3}\right)}{2\log 2} + O\left(\frac{\log N}{N}\right) \geqq .1887 + O\left(\frac{\log N}{N}\right).$$

The same technique, with a little more work, leads to slightly better estimates if larger values of k are used. Using $k = 4$, e.g., leads to

$$\mu \geqq .1936 + O\left(N^{-1}\log N\right).$$

Since the map $y \mapsto 4y(1-y)$ has a fixed point, at $y = \frac{3}{4}$, one finds

$$B_k \geqq \left[P_N\left(\tfrac{3}{4}\right)\right]^{k/2}.$$

One can show that

$$P_N\left(\tfrac{3}{4}\right) \sim C3^N.$$

Even for arbitrarily large k, the values of μ obtained by this method are therefore limited by

$$\mu \leqq 1 - \frac{\log 3}{2\log 2} + O\left(N^{-1}\log N\right) \cong .2075 + O\left(N^{-1}\log N\right).$$

3. Using a more sophisticated method than the brutal estimates above, Y. Meyer [28] showed that, again asymptotically for large N,

$$_N\phi, {}_N\psi \in C^{(\mu-\varepsilon)N}$$

with $\mu = \log(4/\pi)/\log 2 \cong .3485$. His proof uses (4.22) rather than P_N.

4. For small values of N, better estimates can be obtained for the regularity of the $_N\phi$, $_N\psi$ by yet a third method. This method is based on a generalization of a technique used by Riesz in the proof that "Riesz products" can lead to continuous, nowhere differentiable functions. I would like to thank Y. Meyer for introducing me to this technique, and for showing me how to use it to prove $_2\phi, {}_2\psi \in C^{.5-\varepsilon}$. The proof, and a generalization for $N \geqq 3$, are given in the Appendix. It works very well for small values of N, but does not, however, give good asymptotic results. For large N, it leads to logarithmic rather than linear increase of the regularity of the $_N\phi$, $_N\psi$.

Table 2. Regularity estimates.
For $N = 2,\cdots,10$, we give α_N so that ${}_N\phi, {}_N\psi \in C^{\alpha_N}$

N	α_N
2	$.5 - \varepsilon$
3	.915
4	1.275
5	1.596
6	1.888
7	2.158
8	2.415
9	2.661
10	2.902

To conclude this paper, we give in Figure 7 the graphs of ${}_N\phi$, ${}_N\psi$ and their Fourier transforms $({}_N\phi)\hat{}$, $({}_N\psi)\hat{}$, for $N = 3, 5, 7, 9$. (For $N = 2$, these graphs were given in Figure 5.) The graphs were plotted by means of the "graphical algorithm" explained in subsection 2B, using the coefficients $h_N(n)$ of Table 1. One clearly sees that the ${}_N\phi$, ${}_N\psi$ become more regular as N increases. Also noticeable is that $|({}_N\phi)\hat{}|$, $|({}_N\psi)\hat{}|$ become "flatter" as N increases, around 0 and $2\pi \cong 6.28$. This is a direct consequence of (4.7) and (4.8). By (4.3), $({}_Nm_0)(\xi)$ has a zero of order N at $\xi = \pi$. It follows that, by (4.7), $({}_Nm_0)(0) = 1$, and that the first $N - 1$ derivatives $({}_Nm_0)^{(j)}(\xi)$ of ${}_Nm_0$ are zero in $\xi = 0$. Since (this follows from (3.45)) $({}_N\psi)\hat{}(\xi) = \overline{{}_Nm_0(\pi + \frac{1}{2}\xi)}\,({}_N\phi)\hat{}(\frac{1}{2}\xi)$, this means that $[({}_N\psi)\hat{}\,]^{(k)}(0) = 0$ for $k = 0,\cdots, N - 1$, or $\int dx\, x^k({}_N\psi)(x) = 0$ for $k = 0,\cdots, N - 1$. The present construction leads thus also to orthonormal bases of compactly supported wavelets with an arbitrarily high number of zero moments. This property could be useful for quantum field theory (see [18]).

It is also quite striking that the "effective support" (where $|({}_N\psi)(x)| \geq .01\|{}_N\psi\|_\infty$, say) of ${}_N\psi$ is quite a bit smaller than its total support, for N not too small. This is due to the very small value of the $h_N(n)$ for large n (see Table 1). Table 2 lists the estimates for the "regularity index" α_N (where ${}_N\phi, {}_N\psi \in C^{\alpha_N}$) for $N = 2, 3,\cdots, 10$, computed using the method explained in the Appendix.

Remark. Using a different approach (see [30]), these estimates for the regularity index α_N can be sharpened. For $N = 2$ one finds, e.g., $\alpha_2 = 2 - \ln(1 + \sqrt{3})/\ln 2 \simeq .550\cdots$. This is the best possible exponent for $N = 2$ (see [30]).

Appendix
Sharper Regularity Estimates for ${}_N\phi$, ${}_N\psi$

The estimates given here are based on a different way of calculating (4.28). Using the facts that $|({}_N\phi)\hat{}(\xi)| = (2\pi)^{-1/2}\prod_{j=1}^{\infty}|({}_Nm_0)(2^{-j}\xi)|$ is even (because ${}_Nm_0$ is a trigonometric polynomial with real coefficients) and that $|({}_Nm_0)(\xi)| \leq 1$

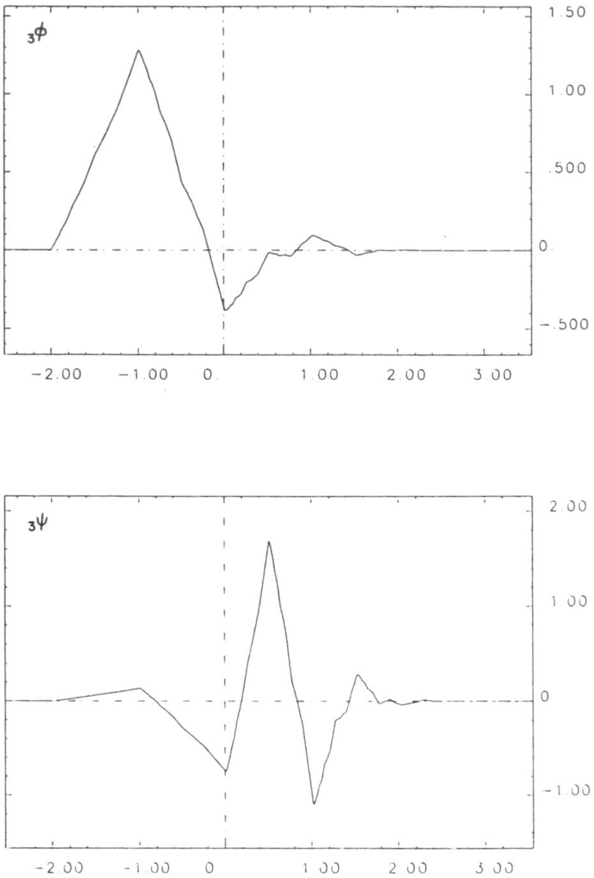

Figure 7. The functions $_N\phi$, $_N\psi$ and the modulus of their Fourier transforms $|(_N\phi)^\wedge|$, $|(_N\psi)^\wedge|$, for increasing values of N (see text). We have each time shifted $_N\phi$ by $N - 1$, so that $\mathrm{supp}(_N\phi) = \mathrm{supp}(_N\psi) = [-(N-1), N]$. One clearly sees that the $_N\phi$, $_N\psi$ become more regular as N increases. The function $_N\phi$ has been plotted using the "graphical construction algorithm" explained in subsection 2B, with the weighting coefficients $_Nh(n)$ given in Table 1. Only 7 iterations were needed. The plot of $_N\psi$ then follows from $(_N\psi)(x) = \sqrt{2}\,\Sigma_n(-1)^n h_N(-n+1)(_N\phi)(2x-n)$.

(see (4.7)), we find

$$\int d\xi |(_N\phi)^\wedge(\xi)|(1+|\xi|)^{1+\alpha} \leqq (2\pi)^{-1/2} 2^{\alpha+1}(1+a)^{\alpha+1} \cdot \left\{ a + \sum_{m=0}^{\infty} 2^{(m+1)(\alpha+1)} \int_{2^m a}^{2^{m+1}a} d\xi \prod_{j=0}^{m} |(_N m_0)(2^{-j}\xi)| \right\}, \tag{A.1}$$

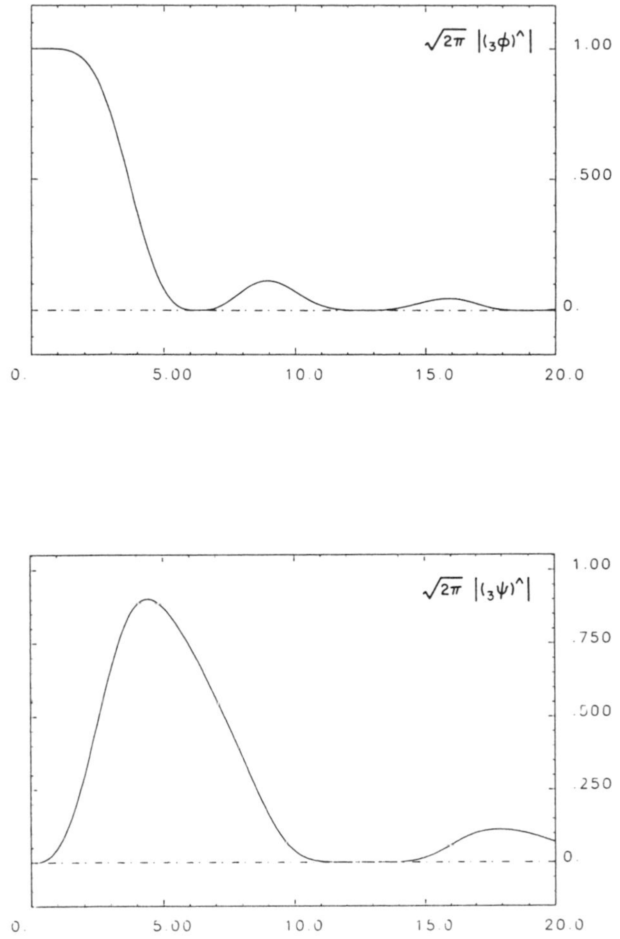

Figure 7. For the plots of $|({}_N\phi)|^{\wedge}$ the infinite product (3.46) was computed (truncated at $j = 10$), $(2\pi)^{1/2}|({}_N\phi)^{\wedge}(\xi)| = \prod_{j=1}^{\infty}|{}_Nm_0(2^{-j}\xi)|$, with ${}_Nm_0(\xi) = 2^{-1/2}\Sigma_n h_N(n)e^{in\xi}$, where the $h_N(n)$ are given in Table 1. The plot of $|({}_N\psi)^{\wedge}(\xi)|$ then follows from

$$|({}_N\psi)^{\wedge}(\xi)| = 2^{-1/2}|\Sigma_n(-1)^n h(-n+1)e^{in\xi/2}|\,|({}_N\phi)^{\wedge}(\tfrac{1}{2}\xi)|.$$

where $a > 0$ is arbitrary for the moment. Using $({}_Nm_0)(\xi) = [\frac{1}{2}(1 + e^{i\xi})]^N Q_N(e^{i\xi})$, we find

$$\prod_{j=0}^{m}\left|({}_Nm_0)(2^{-j}\xi)\right| = \left[\frac{|\sin\xi|}{2^{m+1}\left|\sin(2^{-m-1}\xi)\right|}\right]^N \prod_{j=0}^{m}\left|Q_N\left(e^{-2^{-j}\xi}\right)\right|.$$

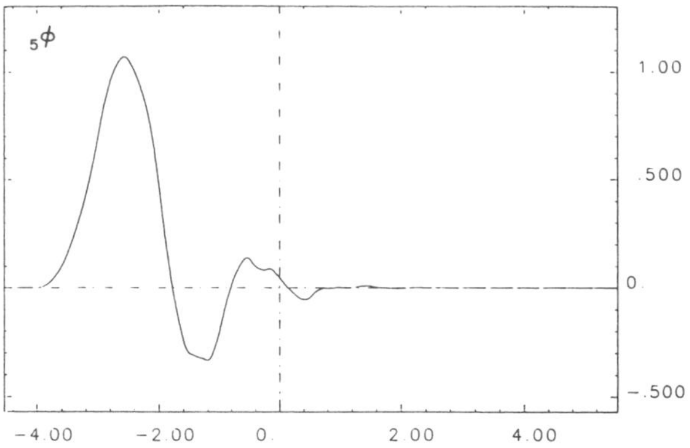

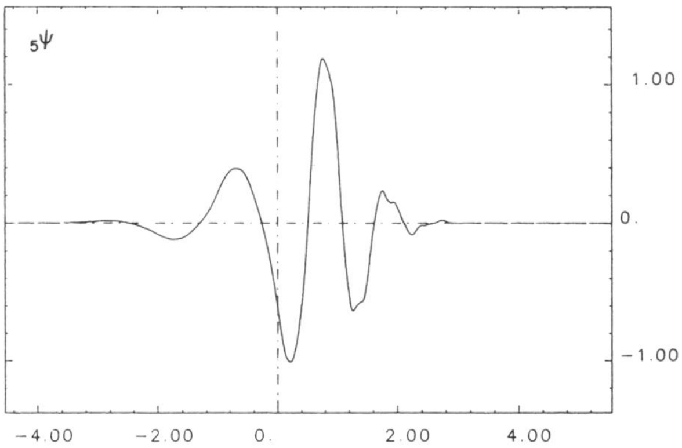

Figure 7. Continued

If we choose $a = \frac{2}{3}\pi$, then $|\sin(2^{-m-1}\xi)| \geqq \sqrt{\frac{3}{2}}$ for $2^m a \leqq |\xi| \leqq 2^{m+1}a$. Hence

$$\int d\xi \left|({}_N\phi)^\wedge(\xi)\right| (1 + |\xi|)^{1+\alpha}$$

$$\leqq C_1 + C_2 \sum_{m=0}^{\infty} 2^{(\alpha+1)m} 2^{-N(m+1)} \int_{2^m 2\pi/3}^{2^{m+1}2\pi/3} \prod_{j=0}^{m} \left| Q_N\left(e^{i2^{-j}\xi}\right)\right|$$

$$\leqq C_1 + C_2 \sum_{m=0}^{\infty} 2^{(\alpha+1)m} 2^{-N(m+1)} (2^{m+1}\pi)^{1/2} \left[\int_0^{2^{m+1}\pi} d\xi \prod_{j=0}^{m} \left| Q_N\left(e^{i2^{-j}\xi}\right)\right|^2 \right]^{1/2}$$

$$\leqq C_1 + C_3 \sum_{m=0}^{\infty} 2^{m(\alpha-N+1)} \left[\int_0^{2\pi} d\xi \prod_{j=0}^{m} P_N\left(\sin^2(2^j \tfrac{1}{2}\xi)\right)\right]^{1/2},$$

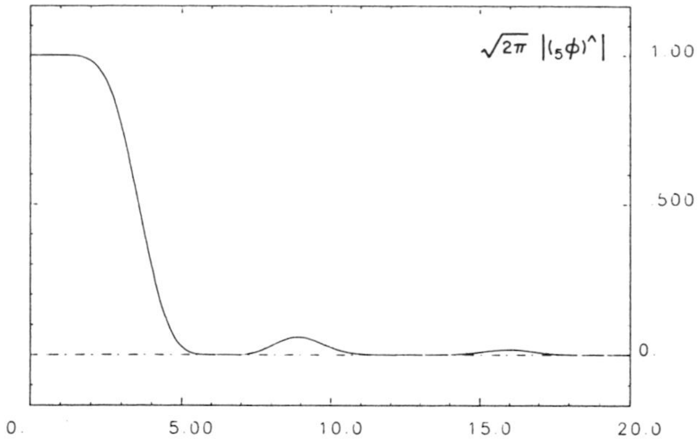

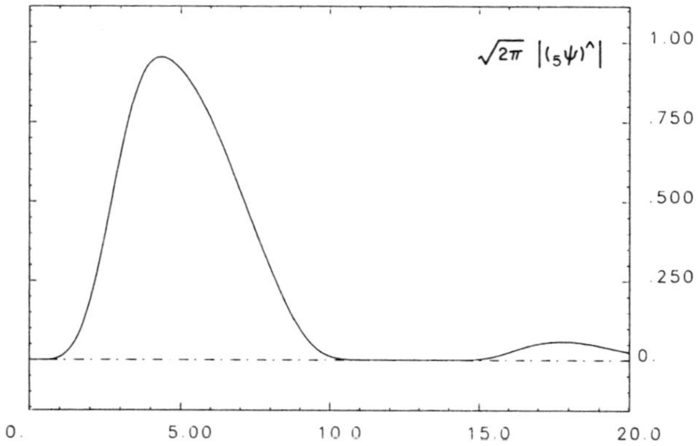

Figure 7. Continued

where we have used $|Q_N(e^{i\xi})|^2 = P_N(\sin^2\frac{1}{2}\xi)$ (see subsection 4C). It follows that (A.1) is convergent, hence ${}_N\phi,\ {}_N\psi \in C^\alpha$, if

$$\text{(A.2)}\quad \limsup_{m\to\infty}\,(2m\log 2)^{-1}\log\left[\int_0^{2\pi} d\xi \prod_0^{2\pi} P_N\left(\sin^2(2^{j-1}\xi)\right)\right] \leqq N-1-\alpha.$$

We know P_N explicitly (see (4.13)),

$$\text{(A.3)}\qquad P_N\left(\sin^2\tfrac{1}{2}\xi\right) = \sum_{k=0}^{N-1}\binom{N+k-1}{k}\left(\sin^2\tfrac{1}{2}\xi\right)^k.$$

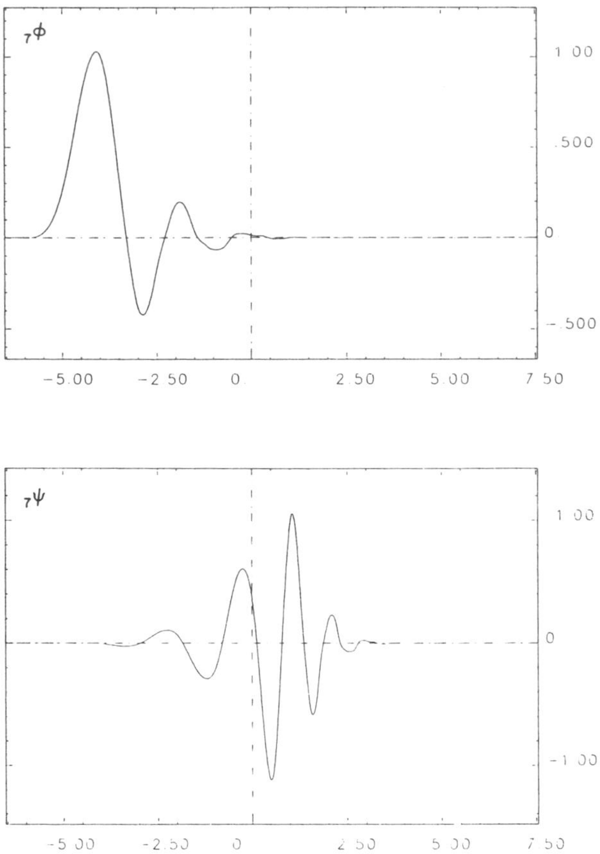

Figure 7. Continued

This can be rewritten as

$$P_N\left(\sin^2\tfrac{1}{2}\xi\right) = \sum_{l=-(N-1)}^{N-1} a_{N,l}e^{il\xi}, \tag{A.4}$$

where the $a_{N,l}$ are symmetric, $a_{N,l} = a_{N,-l}$, and can be calculated explicitly from (A.3). The product $\prod_{j=0}^{m} P_N(\sin^2(2^{j-1}\xi))$ is therefore a symmetric trigonometric polynomial of order $2^m(N-1)$,

$$\prod_{j=0}^{m} P_N\left(\sin^2(2^{j-1}\xi)\right) = \sum_{l=-(N-1)2^m}^{(N-1)2^m} J_{N,m;k}e^{ik\xi}. \tag{A.5}$$

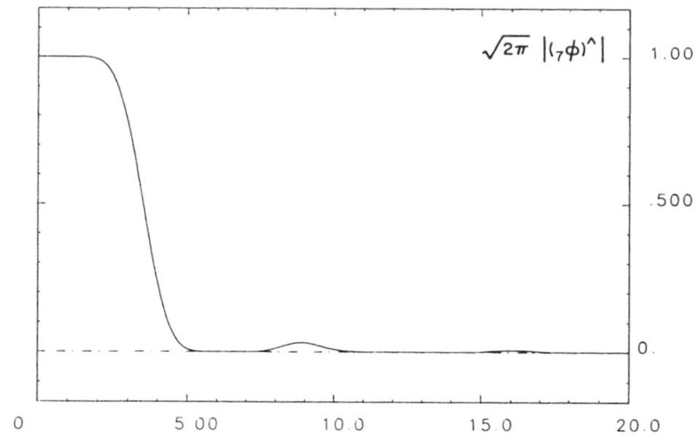

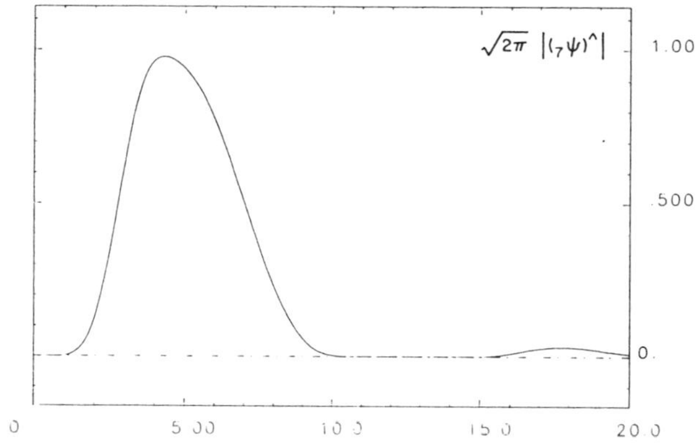

Figure 7. Continued

One easily checks that

$$J_{N,\,m;\,2k} = \sum_l a_{N,2l} J_{N,\,m-1;\,k-l},$$

(A.6)

$$J_{N,\,m;\,2k+1} = \sum_l a_{N,2l+1} J_{N,\,m-1;\,k-l},$$

with $J_{N,0;\,k} = a_{N,\,k}$, and where we implicitly make the assumption $a_{N,\,k} = 0$ for $|k| \geqq N$. The recursion (A.6) can be represented graphically, in a construction

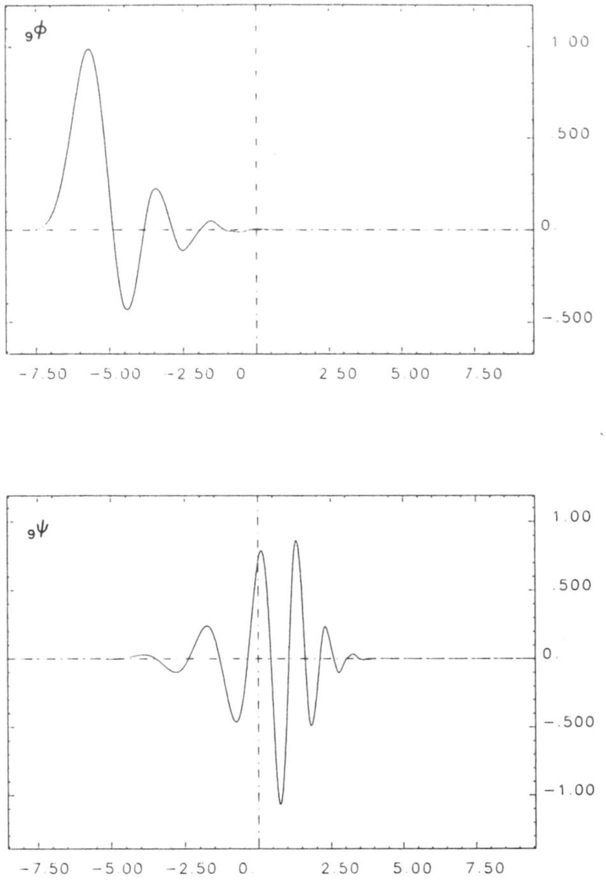

Figure 7. Continued

analogous to Figure 1. At lcvel 0 we start with $J_{N,0}$; each successive $J_{N,n}$ is calculated from $J_{N,n-1}$ by a tree algorithm (see Figure 8). To evaluate the left-hand side of (A.2) we need to compute

$$\int_0^{2\pi} d\xi \prod_{j=0}^{m} P_N\left(\sin^2(2^{j-1}\xi)\right) = J_{N,m;0}. \tag{A.7}$$

One can check directly from the recursion (A.6), or one can verify on the graphical representation (see Figure 8b) that only the $J_{N,m';l}$, $0 \leqq m' < m$, with $|l| \leqq N-2$ play a role in the computation of $J_{N,m;0}$. Define $d_N = 2N - 3$. Then

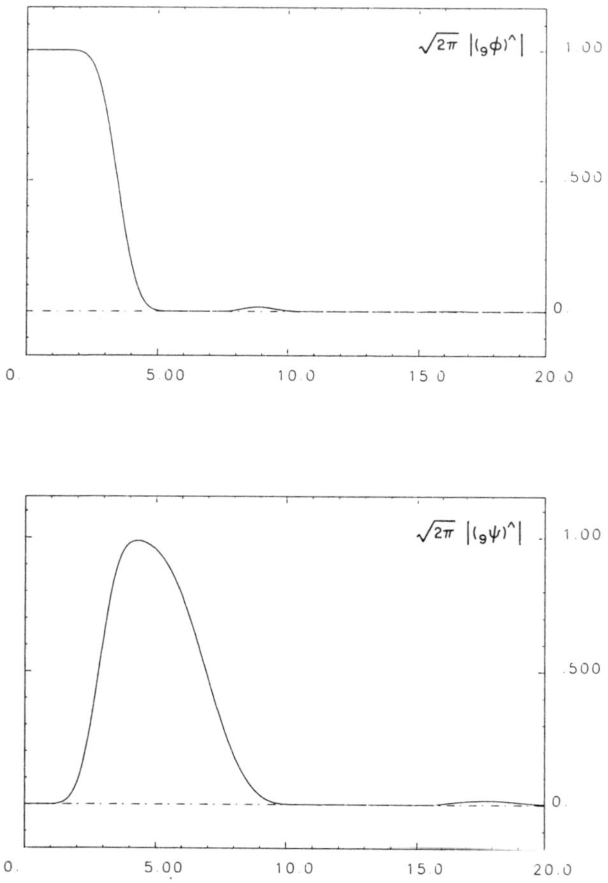

Figure 7. Continued

the set of relevant $J_{N,\,m';\,l}, \cdots, |l| \leqq N-2, \cdots,$ define a vector $j_{N,\,m'}$ in $\mathbb{R}^{d_N}$,

$$(j_{N,\,m'})_k = J_{N,\,m';\,k}. \tag{A.8}$$

Note that d_N is always odd, $d_N = 2m_N + 1$, and that we index vectors $v \in \mathbb{R}^{d_N}$ by $j = -m_N, -m_n + 1, \cdots, 0, \cdots, m_N$ (see (A.8).) The recursion (A.6) defines a matrix T_N such that, for all m,

$$j_{N,\,m+1} = T_N j_{N,\,m}. \tag{A.9}$$

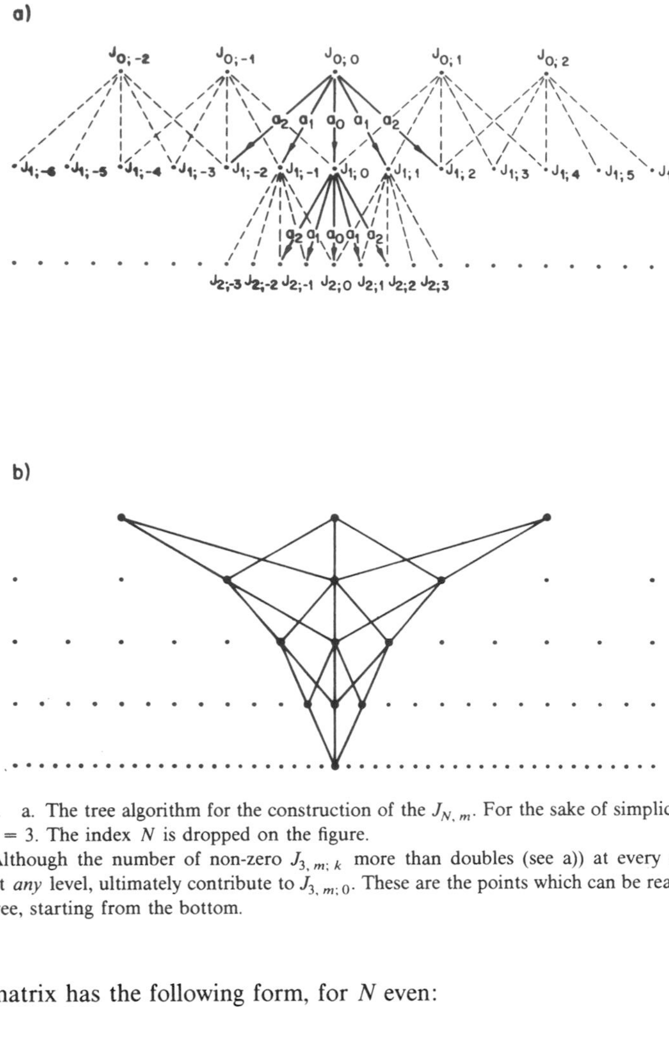

Figure 8. a. The tree algorithm for the construction of the $J_{N,m}$. For the sake of simplicity, we have taken $N = 3$. The index N is dropped on the figure.

b. Although the number of non-zero $J_{3,m;k}$ more than doubles (see a)) at every step, only 3 points, at *any* level, ultimately contribute to $J_{3,m;0}$. These are the points which can be reached from 0 by the tree, starting from the bottom.

This matrix has the following form, for N even:

(A.10)

$$T = \begin{pmatrix} a_{N-2} & a_{N-4} & a_{N-6} & \cdots & a_2 & a_0 & a_2 & \cdots & a_{N-2} & 0 & 0 & \cdots & 0 \\ a_{N-1} & a_{N-3} & a_{N-5} & \cdots & a_3 & a_1 & a_1 & \cdots & a_{N-3} & a_{N-1} & 0 & \cdots & 0 \\ 0 & a_{N-2} & a_{N-4} & \cdots & a_4 & a_2 & a_0 & \cdots & a_{N-4} & a_{N-2} & 0 & \cdots & 0 \\ 0 & a_{N-1} & a_{N-3} & \cdots & a_5 & a_3 & a_1 & \cdots & a_{N-5} & a_{N-3} & a_{N-1} & \cdots & 0 \\ \vdots & \vdots & \vdots & & \vdots & \vdots & \vdots & & \vdots & \vdots & \vdots & & \vdots \\ 0 & 0 & 0 & \cdots & 0 & 0 & 0 & \cdots & a_{N-3} & a_{N-5} & a_{N-7} & \cdots & a_{N-1} \\ 0 & 0 & 0 & \cdots & 0 & 0 & 0 & \cdots & a_{N-2} & a_{N-4} & a_{N-6} & \cdots & a_{N-2} \end{pmatrix}.$$

A completely analogous matrix is obtained for N odd. From (A.8)–(A.9) we have

$$J_{N,m;0} = \left(T_N^m j_{N,0}\right)_0.$$

Hence

$$\begin{aligned}\limsup_{m\to\infty} m^{-1}\log\left(J_{N,m;0}\right) &\leqq \limsup_{m\to\infty} m^{-1}\log\left[\|(T_N)^m\| \cdot \|j_{N,0}\|\right] \\ &\leqq \limsup_{m\to\infty} \log\left[\|(T_N)^m\|^{1/m}\right] = \log(\rho(T_N)),\end{aligned}$$

where $\rho(T_N)$ is the spectral radius of T_N. In view of (A.7) it then follows that (A.1) is convergent, i.e., ${}_N\phi \in C^\alpha$, if $\alpha < N - 1 - \frac{1}{2}\log_2[\rho(T_N)]$. It suffices therefore to compute $\rho(T_N)$, which can be done numerically, provided N is not too large. Note that the problem can be reduced considerably by using the fact that T_N commutes with the involution I,

$$I_{ij} = \delta_{i,-j}$$

(where, as before, $i, j = -m_N, \cdots, 0, \cdots, m_N$). This effectively reduces the problem of a $d_N \times d_N$ matrix to a $(m_N + 1) \times (m_N + 1)$ matrix.

If $N = 2$, then $d_N = 1$, and the matrix T_1 is given by a single number, $T_1 = a_{1;0} = 2$. Therefore one finds ${}_1\phi \in C^\alpha$ if $\alpha < \frac{1}{2}$. The cause of this simplification can be understood by looking at Figure 8b. For $N = 2$, the "tree" reduces to a single vertical line: only one possible path leads from $J_{N,0;0}$ to $J_{N,n;0}$ if $N = 2$. This is equivalent to saying that in the product $\prod_{j=0}^N P_N(\sin^2(2^{j-1}\xi))$ only one possible combination of terms has frequency zero. This is the idea which was borrowed from Riesz's lemma (see Remark 4 following Proposition 4.7).

Acknowledgment. Part of this work was done while I was visiting the Courant Institute (New York University) and the Mathematics Department of Yale University, in February and March of 1987, respectively. I would like to thank both these institutions for their hospitality and support. While working on this subject, I have had the pleasure of discussing these topics with many people, whom I would like to thank here. I am particularly grateful to Yves Meyer for explaining to me the multiresolution analysis concept when it was still very young and new, and for showing me how to use a technique of Riesz to obtain the estimates in the appendix, to Bob Hummel for introducing me to the Laplacian pyramid scheme, thus providing me with the background which led to the present construction, to Stéphane Mallat for letting me use his not yet published material on his discrete algorithm, and to Percy Deift, who listened to several of the early versions, and challenged me with many pertinent questions and made several useful suggestions. I also would like to thank R. Coifman, P. Jones and T. Steeger for their enthusiastic support and for suggesting various references.

The author is "Bevoegdverklaard Navorser" at the Belgium National Science Foundation (on leave). Part of the work for this paper was done at the Department of Theoretical Physics, Vrije Universiteit, Brussel from which she also is on leave. She would like to thank both institutions for their support.

Bibliography

[1] Grossman, A., and Morlet, J., *Decomposition of Hardy functions into square integrable wavelets of constant shape*, SIAM J. Math. Anal. 15, 1984, pp. 723–736.
Goupillaud, P., Grossmann, A., and Morlet, J., *Cyclo-octave and related transforms in seismic signal analysis*, Geoexploration 23, 1984, 85.

[2] The "continuous wavelet transform" is implicitly used in Calderón, A. P., *Intermediate spaces and interpolation, the complex method*, Studia Math. 24, 1964, pp. 113–190; see also Calderón, A. P., and Torchinsky, A., *Parabolic maximal functions associated to a distribution*, I, Adv. Math. 16, 1974, pp. 1–64.

[3] Wilson, K. G., and Kogut, J. B., Physics Reports 12C, 1974, p. 77.
Glimm, J., and Jaffe, A., *Quantum Physics: a Functional Integral Point of View*, Springer, New York, 1981.

[4] Meyer, Y., *Principe d'incertitude, bases hilbertiennes et algèbres d'opérateurs*, Séminaire Bourbaki, 1985–1986, nr. 662.

[5a] Tchamitchian, P., *Calcul symbolique sur les opérateurs de Calderón-Zygmund et bases inconditionnelles de* $L^2(\mathbb{R}^n)$, C. R. Acad. Sc. Paris. 303, série 1, 1986, pp. 215–218.

[5b] Tchamitchian, P., *Biorthogonalité et théorie des opérateurs*, to be published in Rev. Mat. Iberoamericana.

[6] Battle, G., and Federbush, P., *Ondelettes and phase cell cluster expansions; a vindication*, Comm. Math. Phys., 1987.
Federbush, P., *Quantum field theory in ninety minutes*, Bull. Am. Math. Soc. 17, 1987, pp. 93–103.

[7] Kronland-Martinet, R., Morlet, J., and Grossmann, A., *Analysis of sound patterns through wavelet transforms*, to be published in International Journal on Pattern Analysis and Artificial Intelligence.

[8] Mallat, S., *A theory for multiresolution signal decomposition: the wavelet transform*, Preprint GRASP Lab, Dept. of Computer and Information Science, Univ. of Pennsylvania, to be published.

[9] Grossmann, A., *Wavelet transforms and edge detection*, to be published in *Stochastic Processes in Physics and Engineering* (Ph. Blanchard, L. Streit and M. Hasewinkel, eds.),
Grossmann, A., Holschneider, M., Kronland-Martinet, R., and Morlet, J., *Detection of abrupt changes in sound signals with the help of wavelet transforms*, Preprint, Centre de Physique Théorique, CNRS, Marseille (France).

[10] Grossmann, A., Morlet, J., and Paul, T., *Transforms associated to square integrable group representations*, I. J. Math. Phys. 26, 1985, pp. 2473–2479; —II: *Examples*, Ann. Inst. H. Poincaré 45, 1986, pp. 293–309.

[11] Aslaksen, E. W., and Klauder, J. R., *Unitary representations of the affine group*, J. Math. Phys. 9, 1968, pp. 206–211; *Continuous representation theory using the affine group*, J. Math. Phys. 10, 1969, pp. 2267–2275.

[12] Paul, T., *Functions analytic on the half-plane as quantum mechanical states*, J. Math. Phys. 25, 1985, pp. 3252–3263.
Paul, T., *Affine coherent states and the radial Schrödinger equation*. I. *Radial harmonic oscillator and hydrogen atom*, to be published.

[13] Grossmann, A., and Kronland, R., Private demonstration.

[14] Daubechies, I., Grossmann, A., and Meyer, Y., *Painless non-orthogonal expansions*, J. Math. Phys. 27, 1986, pp. 1271–1283.

[15] Daubechies, I., *The wavelet transform, time-frequency localization and signal analysis*, Preprint AT & T Bell Laboratories; to be published.

[16] Meyer, Y., *Ondelettes et functions splines*, Séminaire EDP, Ecole Polytechnique, Paris, France, December, 1986.

[17] Lemarié, P. G., *Ondelettes à localisation exponentielle*, to be published, in Journ. de Math. Pures et Appl.

[18] Battle, G., *A block spin construction of ondelettes, Part I: Lemarié functions*, Comm. Math. Phys. 1987.

[19] Mallat, S., *Multiresolution approximation and wavelets*. Preprint GRASP Lab., Dept. of Computer and Information Science, Univ. of Pennsylvania, to be published.

[20a] Burt, P., and Adelson, E., *The Laplacian pyramid as a compact image code*, IEEE Trans. Comm. 31, 1983, pp. 482–540.

[20b] Burt, P., and Adelson, E., *A multiresolution spline with application to image mosaics*, ACM Trans. on Graphics, 2, 1983, pp. 217–236.

[21] Jaffard, S., Lemarié, P. G., Mallat, S., and Meyer, Y., *Multiscale analysis*, unpublished memorandum.

[22] Coifman, R., and Meyer, Y., *The discrete wavelet transform*, preprint. Dept. of Math., Yale University.

[23] Lemarié, P. G., and Meyer, Y., *Ondelettes et bases hilbertiennes*, Rev. Mat. Iberoamericana 2, 1986, pp. 1–18.

[24] Smith, M. J., and Barnwell, T. P., *Exact reconstruction techniques for tree-structured subband coders*, IEEE Trans. on ASSP 34, 1986, 434–441.

[25] Auscher, P., Thèse de Doctorat, Université de Paris-Dauphine 1988.

[26] Rioul, O., private communication.

[27] Polya, G., and Szegö, G., *Aufgaben and Lehrsätze aus der Analysis*, Vol. II, Springer, Berlin, 1971.

[28] Meyer, Y., *Wavelets with compact support*, Zygmund Lectures (University of Chicago) May 1987, and private communication.

[29] Deslauriers, G., and Dubuc, S., *Interpolation dyadique*, in *Fractals; Dimensions non Entières et Applications*, ed. G. Cherbit, Masson (Paris), 1987, pp. 44–45. See also other references listed there.

[30] Daubechies, I., and Lagarias, J., *Two-scale difference equations*. I. *Global regularity of solutions*, and —. II. *Infinite matrix products, local regularity and fractals*, Preprints AT & T Bell Laboratories; to be published.

Received December, 1987

SECTION VI

Multidimensional Wavelets

Introduction

Guido Weiss

We have considered affine systems $\{\psi_{j,k}\} = \{2^{-j/2}\psi(2^{-j}x - k)\}, j, k \in \mathbb{Z}$, obtained by first translating (by k) and then dilating (by 2^j) a function $\psi \in L^2(\mathbb{R})$. In the introduction to Section V, we have seen that there exists a characterization of those ψ for which $\{\psi_{j,k}\}$ is an orthonormal basis for $L^2(\mathbb{R})$ (or, more generally, a Parseval frame). We also described the important MRA method for constructing such systems. We shall now turn our attention to functions in $L^2(\mathbb{R}^d), d \geq 1$, and explore some of the natural extensions of these notions.

A natural extension of the notion of a wavelet to higher dimensions is to consider tensor products of one-dimensional wavelets. Let us be more specific in the two-dimensional case. Let ψ^1 and ψ^2 be two orthonormal wavelets in $L^2(\mathbb{R})$. For $x = (x_1, x_2) \in \mathbb{R}^2$ let $\psi(x_1, x_2) = \psi^1(x_1)\psi^2(x_2)$. Then the tensor product

$$\{2^{-j_1/2}2^{-j_2/2}\psi^1(2^{-j_1}x - k_1)\psi^2(2^{-j_2}x - k_2)\},$$

$j_1, j_2, k_1, k_2 \in \mathbb{Z}$, of the systems $\{\psi^1_{j_1,k_1}\}$ and $\{\psi^2_{j_2,k_2}\}$ can be written in the form

$$\{\psi_{a,k}(x)\} - \{|\det a|^{-1/2}\psi(a^{-1}x - k)\}, \tag{1}$$

where a ranges through the 2×2 matrices

$$\begin{pmatrix} 2^{j_1} & 0 \\ 0 & 2^{j_2} \end{pmatrix}$$

and $k = (k_1, k_2) \in \mathbb{Z}^2$.

An equally natural extension to the d-dimensional case is to form the system

$$\{\psi_{j,k}(x)\} = \{2^{-jd/2}\psi(2^{-j}x - k)\} \tag{2}$$

for $j \in \mathbb{Z}, k \in \mathbb{Z}^n$, when $\psi \in L^2(\mathbb{R}^d)$, and require that this system be an orthonormal basis (or Parseval frame, or simply a frame, ...) for $L^2(\mathbb{R}^d)$. In either case, we can ask whether such ψ can be characterized, whether there exists an appropriate associated MRA structure, or, more generally, what are the higher dimensional analogs of the properties of one-dimensional wavelets. When one begins doing this, one soon finds out that the d-dimensional situation presents several challenges. In order to describe some of these problems it is, perhaps, easier to consider the case of *continuous* wavelets.

In one dimension, a "continuous wavelet" is a function $\psi \in L^2(\mathbb{R})$ such that $\psi_{a,b}(x) = |a|^{-1/2}\psi(a^{-1}x - b), a, b \in \mathbb{R}, a \neq 0$, produces the "reproducing formula"

$$\|f\|_2^2 = \int_{\mathbb{R}} \int_{\mathbb{R}-\{0\}} |\langle f, \psi_{a,b}\rangle|^2 \frac{da}{|a|}\, db \tag{3}$$

for all $f \in L^2(\mathbb{R})$. The choice $da/|a|$ is made since this is the Haar measure of the multiplicative group of reals. Polarizing (2), we obtain the equivalent identity

$$\langle f, g\rangle = \int_{\mathbb{R}} \int_{\mathbb{R}-\{0\}} \langle f, \psi_{a,b}\rangle \langle \psi_{a,b}, g\rangle \frac{da}{|a|}\, db \tag{4}$$

for all $f, g \in L^2(\mathbb{R})$. This is a weak topology version of the equality

$$f = \int_{\mathbb{R}} \int_{\mathbb{R}-\{0\}} \langle f, \psi_{a,b}\rangle \psi_{a,b} \frac{da}{|a|}\, db, \tag{4'}$$

which justifies the term "reproduction formula" used above.

A natural extension of these notions to higher dimensions is to consider a subgroup D of $\mathrm{GL}_d(\mathbb{R})$, let $\psi \in L^2(\mathbb{R}^d)$, and form the family $\psi_{a,b}(x) = |\det a|^{-1/2}\psi(a^{-1}x - b), a \in D$, and $b \in \mathbb{R}^d$. We can then inquire what conditions on ψ imply the reproducing property

$$\|f\|_2^2 = \int_D \int_{\mathbb{R}^d} |\langle f, \psi_{a,b}\rangle|^2\, d\mu(a)\, db \tag{5}$$

for all $f \in L^2(\mathbb{R}^d)$, where μ is, say, the left Haar measure on D. It is relatively easy to show that this is the case if and only if

$$\Delta_\psi(\xi) = \int_D |\hat{\psi}(\xi a)|^2\, d\mu(a) = 1 \tag{6}$$

for a.e. $\xi \in \mathbb{R}^n$ (ξa is the matricial product of the row vector ξ with the matrix a). This condition is known as the *Calderón condition* associated with D (in [Cal64] Calderón introduced this equality when D is the d-dimensional affine group). If $d = 1$ and D is the dyadic group $\{2^j : j \in \mathbb{Z}\}$, this is precisely equation (I) in Theorem 1 in the introduction to Section V. If D is the multiplicative group of the positive reals, then (5) is the equality

$$1 = \int_0^\infty |\hat{\psi}(\xi a)|^2 \frac{da}{a}$$

for a.e. $\xi \in \mathbb{R}$. Since this measure is left (or right) invariant, this equality is verified for all nonzero ξ as long as $\int_0^\infty |\hat{\psi}(a)|^2 da/|a| = 1$. Thus, any ψ for which this integral is finite, appropriately normalized, provides us with a continuous wavelet in this case. This illustrates the importance of equation (II) in Theorem 1 in the introduction to Section V.

Though equation (6) characterizes the functions $\psi \in L^2(\mathbb{R}^d)$ for which (5) is true for all $f \in L^2(\mathbb{R}^d)$, it does not guarantee that such ψ exist. It can be easily shown, in fact, that such ψ cannot exist if D is a compact group. Groups D for which such continuous wavelets exist are called *admissible*. Conditions that are (almost) equivalent to admissibility are given in [LWWW02].

The collection $G = \{(a, b)\colon a \in D, b \in \mathbb{R}^d\}$ is a group relative to the operation $(c, d) \circ (a, b) = (ca, b + a^{-1}d)$ on G, arising from the action $x \to a(x + b)$ on $\mathbb{R}^d$, and the operator T defined by

$$(T_{(a,b)}\psi)(x) = \psi_{a,b}(x)$$

for $(a, b) \in G$ is a unitary representation of G acting on $L^2(\mathbb{R}^d)$. This representation was studied by many authors in Lie group theory and in quantum mechanics, where it is sometimes referred to as a *coherent state* (see [DM76] and [AAG00]). Consequently, notions involving "continuous wavelets" have been around for some time.

Our interest here, as mentioned above, is to use these continuous wavelets as a guide to find the related discrete wavelets in higher dimensions. We see, therefore, that there are many possibilities if we consider, for example, various possible discretizations of admissible dilation groups D as well as the translations by elements of $\mathbb{R}^d$. Results of this type are explained in [WW01]. It is fair to say, however, that there are still many open questions in this area. Higher-dimensional discrete wavelets obtained by dilations of the form $|\det a|^{-j/2}\psi(a^{-j}x)$, where a is an expanding $d \times d$ matrix and $j \in \mathbb{Z}$, applied to the translations of ψ by the elements $k \in \mathbb{Z}^d$ are fairly well understood. Their constructions and characterizations can be found in [Cal97], [DLS97], [SW95], [CCMW02] (this is a small representative list of many papers on the subject).

In these studies the emphasis is on singly generated wavelets and their construction and characterization. If one considers extensions to several dimensions of the MRA method, the situation is quite different. In the case where the dilations are powers of $2I$, I the $d \times d$ identity matrix (see equation (2)), an MRA wavelet system cannot be generated by a single $\psi \in L^2(\mathbb{R})$: one requires $2^d - 1$ generators. In the case when the dilations are powers of an expanding matrix having integer entries the number of generators required is $|\det a| - 1$. The *quincux* matrix $a = \begin{pmatrix} 1 & -1 \\ 1 & 1 \end{pmatrix}$, for example, is associated with singly generated MRA wavelets in two dimensions (observe that $|\det a| = 2$).

In the first two papers (the ones by Meyer and Gröchenig) several of these features can be encountered: the need for a finite number of generators of MRA wavelets, explicit use of tensor products of one-dimensional wavelets, and the introduction of more general dilation operators. The construction of "spline" wavelets in the Meyer paper is an important one since it is useful in applications.

The tensor product approach described above leads to filters (and, thus, scaling functions and wavelets) that are products of functions of the single variables $x_1, x_2, \ldots, x_d$, the components of $x = (x_1, x_2, \ldots, x_d) \in \mathbb{R}^d$. Such filters are referred to as *separable.* The paper of Kavacevic and Vetterli, among other things, shows how one can construct multidimensional filters that are not easily "decomposable" into functions of one variable and their finite linear combinations. Another paper that addresses this issue and explains these notions well is [Aya99] and the same is true for the fourth paper in this section.

References

[Aya99] A. Ayache, *Construction of non-separable dyadic compactly supported orthogonal wavelet bases for $L^2(\mathbb{R}^2)$ of arbitrary high regularity.* Rev. Mat. Iberoamericana **15** (1999), 37–58.

[AAG00] S. T. Ali, J.-P. Antoine, and J.-P. Gazeau, *Coherent States, Wavelets and Their Generalizations*, Graduate tests in contemprary physics, Springer-Verlag, New York, (2000).

[Cal64] A. P. Calderón, *Intermediate spaces and interpolation, the complex method,* Studia Math. **24** (1964), 113–190.

[Cal97] A. Calogero, *A Characterization of wavelets on general lattices*, J. Geom. Anal. **10** (1997), 597–621.

[CCMW02] C. K. Chui, W. Czaja, M. Maggioni, and G. L. Weiss, *General tight wavelet frames with matrix dilations and tightness preserving oversampling*, J. Fourier Anal. Appl. **8** (2002), 75–100.

[DLS97] X. Dai, D. R. Larson, and D. M. Speegle, *Wavelet sets in* $\mathbf{R}^{\mathrm{n}}$, J. Fourier Anal. Appl. **3** (1997), 451–456.

[DM76] M. Duflo and C. C. Moore, *On the regular representation of a non-unimodular locally compact group*, J. Funct. Anal. **21** (1976), 209–243.

[LWWW02] R. S. Laugesen, N. Weaver, G. L. Weiss, and E. N. Wilson, *A characterization of the higher dimensional groups associated with continuous wavelets*, J. Geom. Anal. **12** (2002), 89–102.

[SW95] M. Soardi, and D. Weiland, *Single wavelets in n-dimensions*, J. Fourier Anal. Appl. **2** (1995), 78–109.

[WW01] G. L. Weiss, and N. Wilson, *The Mathematical Theory of Wavelets*, Proceedings of the NATO-ASI Meeting, Harmonic Analysis 2000—A Celebration, Kluwer Academics, New York, 2001, 329–366.

Wavelets, Spline Functions, and Multiresolution Analysis

Yves Meyer
Translated by John Horváth

The first orthonormal basis of wavelets was obtained during the summer of 1985, and this accidental discovery remained mysterious. Nowadays the works of P. G. Lemarié and of S. Mallat have proved that the "historical basis" and its variants result simply from natural algorithms on nested spaces of spline functions.

We propose to describe this new approach to series of wavelets.

Let us recall the general characteristics we expect from what we will call an *orthonormal basis of wavelets.* In this account we shall deal with an orthonormal basis of the reference space $L^2(\mathbb{R}^n)$, which, however, remains efficient when one uses it for other purposes. That is to say that one starts with a classical function space (as $L^p(\mathbb{R}^n)$, $1 < p < \infty$, Sobolev spaces, Hölder spaces, and, more generally, Besov spaces, ...) and asks whether if the function f to be analyzed belongs to such a space B, does its decomposition with respect to the orthonormal basis still converge (and unconditionally) to f.

These classical function spaces B satisfy $\mathcal{S}(\mathbb{R}^n) \subset B \subset \mathcal{S}'(\mathbb{R}^n)$ and the two inclusions are continuous; $\mathcal{S}$ and $\mathcal{S}'$ denote the spaces of test functions and of tempered distributions of L. Schwartz. The orthonormal basis of wavelets is ψ_λ, $\lambda \in \Lambda$, and so we require the right to write $f = \sum_{\lambda\in\Lambda} \langle f|\psi_\lambda\rangle\, \psi_\lambda$ when f belongs to B, the series being convergent for the norm of B (this cannot happen for just any function space B, but only for a certain class B containing the examples mentioned above).

Because B could very well be a space of tempered distributions, it is appropriate to assume that our wavelets ψ_λ belong to $\mathcal{S}(\mathbb{R}^n)$.

One also wishes to be able to analyze the homogeneous Hölder spaces C^r, $r > 0$, by series of wavelets. More precisely, if f belongs to C^r, one requires that the series of wavelets $\sum_{\lambda\in\Lambda} \langle f|\psi_\lambda\rangle\, \psi_\lambda$ converge to f for the only reasonable topology on the homogeneous Hölder space, namely, the weak topology $\sigma(C^r, B_1^{-r,1})$. We recall that $C^r = B_\infty^{r,\infty}$ is the dual of

Translator's Note: The references have been updated, and new figures have been generated by Norbert Kaiblinger.

Author's Note: In contrast to the papers "Wavelets and Hilbert bases" and "Uncertainty Principle, Hilbert bases, and Algebras of Operators" reprinted earlier in this volume, mutiresolution analysis is present in this paper. Moreover Mallat's fundamental work is mentioned, and wavelets are related to quadrature mirror filters. At the end of the proof of Theorem 2, two open problems are discussed. The first one was solved by K. Gröchenig (and is reprinted in this volume), and the second one by P.-G. Lemarié (Rev. Mat. Iberoamericana, **7** (1991), 157–182). The construction of wavelets $\psi_{j,k}$ given by Theorem 4 has been much simplified by R. Coifman, P. Jones and S. Semmes (J. Amer. Math. Soc., **2** (1989), 553–564). Indeed they used a clever modification of the Haar system to give a "simple proof'" of the boundedness of the Cauchy integral on Lipschitz curves.

the homogeneous Besov space $\dot{B}_1^{-r,1}$. So we are led to assume that our wavelets ψ_λ, $\lambda \in \Lambda$, belong to $\dot{B}_1^{-r,1}$ for all $r > 0$. Since we already know that the ψ_λ, $\lambda \in \Lambda$, belong to $\mathcal{S}(\mathbb{R}^n)$, this requires that all the moments of ψ_λ be zero.

The last requirement is the numerical simplicity of the definition of the ψ_λ. The first choice which was made, and which we shall call the "historical wavelets," is the following. In one dimension one takes $\Lambda = \mathbb{Z}^2$ and one starts with a very particular function $\psi \in \mathcal{S}(\mathbb{R})$ whose oscillations, regularity, and decay are adjusted in such a way that the collection $\psi_{(j,k)}$ of functions $2^{j/2}\psi(2^j x - k)$, $(j,k) \in \mathbb{Z}^2$, is an orthonormal basis of $L^2(\mathbb{R})$. It is then automatic that the same wavelets $\psi_{(j,k)}$ provide a universal unconditional basis for most classical function spaces B.

This is the situation in one dimension, while in dimension n one must use a second function φ, which was discovered by P. G. Lemarié [11] and which was constructed starting from the function ψ.

The wavelets ψ_λ, $\lambda \in \Lambda$, of $L^2(\mathbb{R}^n)$ are indexed by $\Lambda = \mathbb{Z} \times \mathbb{Z}^n \times A$, where A is the set of the $2^n - 1$ sequences $\alpha = (\alpha_1, \ldots, \alpha_n)$ of zeros and ones with the exception of the sequence $(0, 0, \ldots, 0)$. One sets $\psi^{(1)} = \psi$, $\psi^{(0)} = \varphi$, and then

$$\psi^{(\alpha)}_{(j,k)}(x) = 2^{nj/2}\psi^{(\alpha_1)}(2^j x_1 - k_1)\,\psi^{(\alpha_2)}(2^j x_2 - k_2)\cdots\psi^{(\alpha_n)}(2^j x_n - k_n).$$

The search for new wavelet bases is motivated by a triple requirement.

(a) The numerical localization of the "historical wavelet" $\psi(x)$ is not very good (although ψ belongs to the Schwartz class $\mathcal{S}(\mathbb{R})$). One tries to find an analyzing wavelet with exponential decay (and at the same time with a limited regularity and with a finite number of moments equal to zero).

(b) The systematic use of the affine group to generate the wavelets $\psi_{(j,k)}$ starting with ψ is an inopportune limitation that prohibits the direct construction of wavelets on geometric objects different from $\mathbb{R}^n$.

(c) It is irritating that the orthonormal bases of wavelets were discovered by chance.

One tries to approximate complicated functions or distributions f (defined, for instance, on $\mathbb{R}^n$) by a sequence f_j, $j \in \mathbb{Z}$, of explicit functions. Each f_j is taken from a stock V_j of "simple functions." If one assigns (at the start) a privileged position to the reference space $L^2(\mathbb{R}^n)$, one asks that V_j be a closed linear subspace of $L^2(\mathbb{R}^n)$. The fact that the approximants f_j become more and more precise leads us to assume that $V_j \subset V_{j+1}$. Since the reference norm is at the start the norm in $L^2(\mathbb{R}^n)$, one defines f_j as the best quadratic approximation of f, that is, the orthogonal projection of f onto V_j.

The two, seemingly contradictory, requirements for a multiresolution analysis are

(a) the geometric and numerical simplicity of the spaces V_j considered;

(b) the quality of the approximation of f by the f_j for functional norms different from the reference $L^2(\mathbb{R}^n)$-norm.

The most often used examples of multiresolution analysis are the nested sequences of spaces of spline functions (of a given order) associated with increasingly finer meshes.

One denotes by W_j the orthogonal complement of V_j in V_{j+1}, and one then has a decomposition of $L^2(\mathbb{R}^n)$ into a Hilbert sum of the W_j (provided that one assumes, as we shall do, that $\bigcap V_j = \{0\}$ and that $\bigcup V_j$ is dense in $L^2(\mathbb{R}^n)$). The wavelets appear by looking, in each W_j, for a Hilbert basis which is the possibly best adapted to the geometric properties of this space W_j.

We shall make precise in the first section the definition of a multiresolution analysis, by introducing the geometric and arithmetic constraints on the nested spaces V_j we will use. The second section is dedicated to the description of examples of multiresolution analysis. The algorithm of the construction of the functions φ and ψ and of the wavelets will be given in section 3. Sections 4 and 5 are dedicated to explicit calculations on the most important examples. We shall find again the wavelets of [11] as well as those of G. Battle and of P. G. Lemarié [10], which are used by G. Battle and P. Federbush [2] in constructive field theory.

Section 6 is dedicated to the study of operators which are canonically associated with a multiresolution analysis. In the case of martingales, these operators are martingale transforms. In other cases of multiresolution analyses they are paraproduct operators.

The last section deals with wavelets adapted to the "$T(b)$ theorem" of David-Journé-Semmes. Then one can prove that the algebra A_b of singular integral operators satisfying $T(b) = 0$ (modulo the constant functions) and ${}^tT(b) = 0$ (modulo the constant functions), where $b \in L^\infty(\mathbb{R}^n)$, $\operatorname{Re} b \geq 1$, is in fact independent of b (up to a conjugacy isomorphism).

1. Definition of a Multiresolution Analysis

The geometric data of a multiresolution analysis are a lattice $\Gamma \subset \mathbb{R}^n$ as well as a linear endomorphism $M\colon \mathbb{R}^n \to \mathbb{R}^n$ having the following properties:

(1.1) all the eigenvalues λ_j, $1 \leq j \leq n$, of M satisfy $|\lambda_j| > 1$ $(1 \leq j \leq n)$;

(1.2) $M(\Gamma) \subset \Gamma$.

Let us recall that a lattice $\Gamma \subset \mathbb{R}^n$ is a discrete subgroup such that the quotient $\mathbb{R}^n/\Gamma$ is compact. The dual lattice Γ^* is then defined by the condition that $\exp(ix \cdot \gamma) = 1$ for every $\gamma \in \Gamma$. The compact quotient group $\mathbb{R}^n/\Gamma^*$ is denoted G. This compact group G is the dual of Γ. Finally we identify $L^2(G)$ with the subspace of functions in $L^2_{\text{loc}}(\mathbb{R}^n)$ which are Γ^*-periodic.

The reference space that we shall begin to use is the Hilbert space $L^2(\mathbb{R}^n)$.

Definition 1 *A multiresolution analysis* of $L^2(\mathbb{R}^n)$ associated with the couple (Γ, M) is by definition an increasing sequence V_j, $j \in \mathbb{Z}$, of closed subspaces of $L^2(\mathbb{R}^n)$ having the following properties:

(1.3) $\bigcap V_j = \{0\}$, $\bigcup V_j$ is dense in $L^2(\mathbb{R}^n)$;

(1.4) $f(x) \in V_j \Leftrightarrow f(Mx) \in V_{j+1}$;

(1.5) denoting by $R_\gamma\colon L^2(\mathbb{R}^n) \to L^2(\mathbb{R}^n)$ and by $\rho_\gamma\colon \ell^2(\Gamma) \to \ell^2(\Gamma)$ the operators of translation by $\gamma \in \Gamma$, one has $R_\gamma(V_0) = V_0$ (for every $\gamma \in \Gamma$), and there exists an isomorphism $T_0\colon V_0 \to \ell^2(\Gamma)$ such that for all $\gamma \in \Gamma$ one has $T_0 R_j = \rho_j T_0$.

This operator $T_0\colon V_0 \to \ell^2(\Gamma)$ is called the *sampling operator*. One obtains from it a sampling operator $T_j\colon V_j \to \ell^2(M^{-j}\Gamma)$ which is defined by $(T_j f)(M^{-j}\gamma) = (T_0 f_j)(\gamma)$, where $f_j(x) = f(M^{-j}x)$ belongs to V_0.

In many cases the functions f belonging to V_0 are uniformly continuous and the isomorphism $T_0\colon V_0 \to \ell^2(\Gamma)$ is given simply by the usual restriction to Γ of the function $f \in V_0$. Then the operator T_j is defined by restricting to $M^{-j}\Gamma$ the functions f belonging to V_j.

To describe explicitly the sampling operator T_0, one denotes by $\varepsilon_\gamma \in \ell^2(\Gamma)$ the sequence equal to 1 at γ and 0 elsewhere. One calls $g \in V_0$ the inverse image of ε_0 by T_0. One then has $T_0^{-1}(\varepsilon_\gamma) = g(x-\gamma)$, and property (1.5) means that the sequence $g(x-\gamma)$, $\gamma \in \Gamma$, is an unconditional basis of the Hilbert space V_0. Let us state precisely the concept of an unconditional basis of a Hilbert space H. We denote by e_k, $k \in \mathbb{N}$, a sequence of vectors of H such that for two constants $\delta_2 > \delta_1 > 0$ we have $\delta_1 \le \|e_k\| \le \delta_2$ (for all $k \in \mathbb{N}$). Then our sequence e_k, $k \in \mathbb{N}$, is an unconditional basis of H if the finite sums $\sum \alpha_k e_k$ are dense in H and if there exist two constants $C_2 \ge C_1 > 0$ such that

$$C_1\Big(\sum |\alpha_k|^2\Big)^{1/2} \le \Big\|\sum \alpha_k e_k\Big\| \le C_2\Big(\sum |\alpha_k|^2\Big)^{1/2}.$$

The flexibility of use of the sampling $T_0\colon V_0 \to \ell^2(\Gamma)$ comes from the possibility of its adaptation to other functional situations. One denotes by $E \subset V_0$ the space of the finite sums $\sum \alpha_\gamma g(x-\gamma)$, and one wishes that there should exist two constants $C_2' \ge C_1' > 0$ such that one still has

$$C_1'\Big(\sum |\alpha_\gamma|^p\Big)^{1/p} \le \Big\|\sum \alpha_\gamma g(x-\gamma)\Big\|_p \le C_2'\Big(\sum |\alpha_\gamma|^p\Big)^{1/p} \tag{1.6}$$

for every function in E.

A second wish is to have Bernstein-type inequalities for the functions $f \in V_0$, namely,

$$\Big\|\frac{\partial}{\partial x_j} f\Big\|_p \le C\,\|f\|_p \qquad (1 \le j \le n) \tag{1.7}$$

for $f \in E$ and $1 \le p \le \infty$.

Our first task will be the construction of a function $\varphi \in V_0$ such that the sequence $\varphi(x-\gamma)$, $\gamma \in \Gamma$ is an orthonormal basis of V_0.

Then we denote by $W_j \subset V_{j+1}$ the orthogonal complement of V_j in V_{j+1}. Thus one has

$$L^2(\mathbb{R}^n) = \bigoplus_{-\infty}^{\infty} W_j. \tag{1.8}$$

We then denote by R the finite commutative group Γ/M whose cardinality is $q = |\det M|$. We will show that there exist $q-1$ functions, denoted $\psi_1, \dots, \psi_{q-1}$, belonging to W_0 and such that the collection of functions $\psi_m(x-\gamma)$, $1 \le m < q$, $\gamma \in \Gamma$, is an orthonormal basis of W_0. We shall see that these functions ψ_m are not unique, which will allow us to choose them optimizing their localization and their decay at infinity. This leads us to the criterion given by the following definition.

Definition 2 A function $\psi \in L^2(\mathbb{R}^n)$ is called a *wavelet* if it satisfies the three conditions

(1.9) $|\psi(x)| \leq C(1+|x|)^{-n-1}$;

(1.10) $\int_{\mathbb{R}^n} \psi(x)\, dx = 0$;

(1.11) $|\partial/\partial x_j\, \psi(x)| \leq C(1+|x|)^{-n-2} \quad (1 \leq j \leq n)$.

Definition 3 An *orthonormal basis of wavelets associated with* (M, Γ) is, by definition, an orthonormal basis of $L^2(\mathbb{R}^n)$ of the form $q^{j/2}\psi_m(M^j x - \gamma)$, $j \in \mathbb{Z}$, $1 \leq m < q = |\det M|$, $\gamma \in \Gamma$, where the ψ_m satisfy (1.9), (1.10), and (1.11).

As we shall see in section 2, it is not true that starting from an arbitrary multiresolution analysis of $L^2(\mathbb{R}^n)$ one can always construct a wavelet basis associated with the decomposition (1.8).

In the opposite direction, we do not know whether, starting with a couple (M, Γ) satisfying $M\Gamma \subset \Gamma$ and with a sequence $\psi_1, \ldots, \psi_m$, $1 \leq m \leq q-1$ of wavelets such that the collection $q^{j/2}\psi_m(M^j x - \gamma)$, $j \in \mathbb{Z}$, $\gamma \in \Gamma$, $1 \leq m < q = |\det M|$ is an orthonormal basis of $L^2(\mathbb{R}^n)$, there exists necessarily a multiresolution analysis associated with (M, Γ), such that our orthonormal basis of wavelets is obtained by the procedure we will describe in section 3.

It often happens that one is led to require much more regularity, localization, and oscillation from the wavelets.

One starts with an integer $d \geq 1$ (which will be called the order of the analyzing wavelet ψ) and asks that ψ satisfy the following three conditions:

(1.12) $|\psi(x)| \leq C(1+|x|)^{-n-d}$;

(1.13) $\int x^d\, \psi(x)\, dx = 0$ for all the multi-indices $\alpha = (\alpha_1, \ldots, \alpha_n) \in \mathbb{N}^n$ such that $\alpha_1 + \alpha_2 + \cdots + \alpha_n = |\alpha| < d$;

(1.14) $|\partial^\beta \psi(x)| \leq C(1+|x|)^{-n-d-\beta}$ for all $\beta \in \mathbb{N}^n$ such that $|\beta| \leq d$.

The functions ψ of the Schwartz class $\mathcal{S}(\mathbb{R}^n)$ whose moments are all zero are therefore analyzing wavelets of infinite order.

The main interest of the wavelet bases is to supply universal unconditional bases for most classical function spaces. This is to say that the belonging of a function $f(x)$ to function spaces like $L^p(\mathbb{R}^n)$ for $1 < p < \infty$, like Besov spaces (and, in particular, Sobolev or Hölder spaces) or like the Hardy space (in its real version given by Stein and Weiss), are properties which can be read off immediately from the sequence of absolute values $|\alpha(m, j, \gamma)|$ of the coefficients of the decomposition of f with respect to the wavelet basis. The only limitations lie in the regularity of the analyzing wavelet, in its decay at infinity, and finally in the naive remark that certain Banach spaces, like $L^1(\mathbb{R}^n)$ or $C_0(\mathbb{R}^n)$, do not have unconditional bases.

It may be useful to recall the precise definition of an unconditional basis.

Definition 4 Let E be a Banach space and let e_k, $k \in \mathbb{N}$, be a sequence of (nonzero) elements of E. We say that this sequence is an *unconditional basis of* E if the following two conditions are satisfied:

(1.15) the finite linear combinations of the e_k, $k \in \mathbb{N}$, are dense in E;

(1.16) there exists a constant $C \geq 1$ such that for every integer $k \geq 1$, every sequence $\alpha_0, \dots, \alpha_k$ of scalars, and every sequence $\beta_0, \dots, \beta_k$ satisfying $|\beta_0| \leq |\alpha_0|, \dots, |\beta_k| \leq |\alpha_k|$ one has $\|\beta_0 e_0 + \cdots + \beta_k e_k\| \leq C \|\alpha_0 e_0 + \cdots + \alpha_k e_k\|$.

If $(e_k)_{k\in\mathbb{N}}$ is an unconditional basis of E, then every element $x \in E$ can be written uniquely as $x = \alpha_0 e_0 + \cdots + \alpha_k e_k + \cdots$. This series converges unconditionally to x and the coefficients are then unique.

The fact that a sequence $e_0, e_1, \dots, e_k, \dots$ is an unconditional basis always has an interpretation in the theory of operators. Assuming at the start that the sequence of the e_k, $k \in \mathbb{N}$, is independent (i.e., that the finite subsequences are independent) and total in E, one denotes by $\mathcal{A} \subset \mathcal{L}(E, E)$ the commutative algebra of continuous operators T from E into E such that $T(e_k) = \lambda_k e_k$ for all $k \in \mathbb{N}$. The problem of recognizing whether a bounded sequence is in fact the sequence λ_k, $k \in \mathbb{N}$, of eigenvalues of an operator T is in general extremely arduous. To say that the sequence e_k, $k \in \mathbb{N}$, is an unconditional basis of E amounts to saying that every sequence $(\lambda_k) \in \ell^\infty(\mathbb{N})$ satisfies the requirement. In other words, the theory of multipliers is trivial for an unconditional basis.

In the case of orthonormal bases of wavelets, the algebra $\mathcal{A}$ consists of generalized Calderón-Zygmund operators. Let us recall the definition of the latter operators. For this denote by $\mathcal{D} = \mathcal{D}(\mathbb{R}^n)$ the Schwartz class and by $\mathcal{D}'$ the dual space of distributions. To define precisely a Calderón-Zygmund operator we must start somewhere, for example, with a continuous linear operator $T\colon \mathcal{D} \to \mathcal{D}'$. According to the famous kernel theorem of L. Schwartz, T is then determined by a unique distribution $S \in \mathcal{D}'(\mathbb{R}^n \times \mathbb{R}^n)$ and the (bijective) correspondence between T and S is supplied by $\langle Tu, v\rangle = \langle S, u \otimes v\rangle$. The first bracket expresses the duality between $Tu \in \mathcal{D}'(\mathbb{R}^n)$ and $v \in \mathcal{D}(\mathbb{R}^n)$ when $u \in \mathcal{D}(\mathbb{R}^n)$. The second bracket expresses the duality between $\mathcal{D}(\mathbb{R}^n \times \mathbb{R}^n)$ and $\mathcal{D}'(\mathbb{R}^n \times \mathbb{R}^n)$.

We shall call $\Omega \subset \mathbb{R}^n \times \mathbb{R}^n$ the open subset of pairs $(x, y) \in \mathbb{R}^n \times \mathbb{R}^n$ such that $x \neq y$. Then an operator $T\colon \mathcal{D} \to \mathcal{D}'$ is a Calderón-Zygmund operator if the following four properties are satisfied:

(1.17) T can be extended to a continuous linear operator from $L^2(\mathbb{R}^n)$ into itself;

(1.18) the restriction $K(x, y)$ of $S(x, y)$ to the open set Ω is a continuous function satisfying, for a certain constant C, $|K(x, y)| \leq C|x - y|^{-n}$;

(1.19) there exists a constant C such that $|\partial/\partial x_j\, K(x, y)| \leq C|x-y|^{-n-1}$ for $1 \leq j \leq n$ and all $(x, y) \in \Omega$;

(1.20) there exists a constant C such that $|\partial/\partial y_j\, K(x, y)| \leq C|x-y|^{-n-1}$ for $1 \leq j \leq n$ and all $(x, y) \in \Omega$.

The continuity on $L^p(\mathbb{R}^n)$ of these operators ($1 < p < \infty$) is established by the "real variable methods" created by Calderón and Zygmund. The continuity on Besov spaces of these operators was studied by Lemarié [8] and M. Meyer [13]. Evidently one has to limit oneself to $|s| < 1$ unless one makes hypotheses analogous to (1.19) and (1.20) concerning the successive derivatives of $K(x, y)$. If $0 < s < 1$, the continuity of T on the Besov space $B_q^{s,p}$ ($1 \le p \le \infty$, $1 \le q \le \infty$) requires only a very weakened form of (1.17) called the "weak boundedness property" in the fundamental article of G. David and J. L. Journé [5]. This condition together with (1.19) and the condition $T(1) = 0$ (modulo the constants) means that for every function $\psi \in \mathcal{D}(\mathbb{R}^n)$ with integral zero, the distribution $\int S(u, x)\, \psi(u)\, du$ (which outside the support of ψ is a continuous function satisfying $O(|x|^{-n-1})$ at infinity) must have an integral on the whole space $\mathbb{R}^n$ equal to zero.

The condition $T(1) = 0$ is necessary and sufficient for the continuity of a Calderón-Zygmund operator on the space $\mathrm{BMO}(\mathbb{R}^n)$ of John and Nirenberg (condition (1.20) is not necessary).

Likewise the condition $T^*(1) = 0$ (modulo the constant functions) is necessary and sufficient for a Calderón-Zygmund operator to be continuous on the generalized Hardy space of Stein and Weiss.

Let us return to the Hilbert bases of wavelets. The set $\mathcal{J}$ will be the product $\mathbb{Z} \times \Gamma \times \{1, 2, \ldots, q-1\}$ that serves as the set of indices for the wavelets, which will be denoted simply ψ_α, $\alpha \in \mathcal{J}$. The operators T that satisfy $T(\psi_\alpha) = \lambda_\alpha \psi_\alpha$, $(\lambda_\alpha) \in \ell^\infty(\mathcal{J})$, are obviously continuous on $L^2(\mathbb{R}^n)$ because ψ_α, $\alpha \in \mathcal{J}$, is a Hilbert basis of $L^2(\mathbb{R}^n)$. The distributional kernel $S(x, y) \in \mathcal{D}'(\mathbb{R}^n \times \mathbb{R}^n)$ of such an operator is given by

$$S(x, y) = \sum_{\alpha \in \mathcal{J}} \lambda_\alpha\, \psi_\alpha(x)\, \psi_\alpha(y) \tag{1.21}$$

and this series converges in the sense of distributions thanks to (1.9) and (1.10). The properties (1.9), (1.10), and (1.11) imply then, through trivial estimates of geometric series, the properties (1.18), (1.19) and (1.20). That is to say that the operators which are diagonalized in a wavelet basis are Calderón-Zygmund operators (under the obviously necessary condition that the eigenvalues form a bounded sequence). Moreover, one has $T(1) = 0$ and $T^*(1) = 0$ because the integral of each wavelet is zero.

This is why the wavelet bases are unconditional bases for precisely the same function spaces as those for which one can prove the continuity of the generalized Calderón-Zygmund operators.

If we want to consider the case $s \in]1, 2[$ for the Besov spaces $B_q^{s,p}$, it is convenient to use wavelets of order 2 since the sufficient conditions for the continuity of Calderón-Zygmund operators will be completed by $T(x_j) = 0$ (modulo the affine functions) for $1 \le j \le n$.

But before going further, it is time to give examples of multiresolution analyses.

2. Examples of Multiresolution Analyses

The first example is didactic. Its aim is to show that a multiresolution analysis does not always lead to a wavelet basis. We place ourselves in one dimension, $\Gamma = \mathbb{Z}$ and $M(x) = 2x$. We denote by $V_0 \subset L^2(\mathbb{R})$ the closed subspace of $L^2(\mathbb{R})$ defined by the condition that the

Fourier transform $\hat{f}(\xi)$ of $f \in V_0$ is zero almost everywhere outside of $[-\pi, \pi]$. Let us recall that $\hat{f}(\xi) = \int_{-\infty}^{\infty} e^{-ix\xi} f(x)\, dx$.

The inclusion $V_j \subset V_{j+1}$ is checked with no difficulty. The sampling operator T_0 is the restriction operator of $f \in V_0$ to $\mathbb{Z}$. This operator T_0 is an isometric isomorphism between V_0 and $\ell^2(\mathbb{Z})$.

The space W_0 is then defined by the condition $\hat{f}(\xi) = 0$ if $|\xi| < \pi$ or $|\xi| > 2\pi$. It is then immediate to characterize the functions $\psi \in W_0$ such that the collection $\psi(x-k)$, $k \in \mathbb{N}$, is an orthonormal basis of W_0. This happens if and only if $|\hat{\psi}(\xi)| = 1$ for $\pi \leq |\xi| \leq 2\pi$. Such a function ψ cannot be a wavelet. Indeed, a wavelet belongs to $L^1(\mathbb{R})$ and its Fourier transform is continuous.

The second example is a correction of the first one. We still assume that the dimension is one, $\Gamma = \mathbb{Z}$, and $M(x) = 2x$. Let us start by defining V_0. We will then deduce the V_j, $j \in \mathbb{Z}$, thanks to the rule $f(x) \in V_0 \Leftrightarrow f(2^j x) \in V_j$, and we will have to verify that the V_j are indeed nested. To define V_0 one starts by constructing a function $\varphi \in \mathcal{S}(\mathbb{R})$ such that the different functions $\varphi(x-k)$, $k \in \mathbb{Z}$, are mutually orthogonal and such that $\|\varphi\|_2 = 1$. One then declares that these functions $\varphi(x-k)$, $k \in \mathbb{Z}$, form an orthonormal basis of V_0, which makes V_0 well defined. One shows without difficulty that the conditions concerning φ are equivalent to $\sum_{-\infty}^{\infty} |\hat{\varphi}(\xi + 2k\pi)|^2 = 1$. To satisfy this condition (with the ulterior aim of nesting the V_j), one imposes on $\hat{\varphi}$ the following properties

(2.1) $\hat{\varphi}(\xi) \geq 0$, $\hat{\varphi}(-\xi) = \hat{\varphi}(\xi)$, $\hat{\varphi}(\xi) = 1$ if $|\xi| \leq 2\pi/3$ and $\hat{\varphi}(\xi) = 0$ if $|\xi| \geq 4\pi/3$;

(2.2) $(\hat{\varphi}(\xi))^2 + (\hat{\varphi}(\xi - 2\pi))^2 = 1$ if $2\pi/3 \leq \xi \leq 4\pi/3$.

Then $\mathcal{F}V_0$ is the collection of functions $m(\xi)\, \hat{\varphi(\xi)}$, where $m(\xi)$ belongs to $L^2_{\text{loc}}(\mathbb{R})$ and is periodic with period 2π. The inclusion $\mathcal{F}V_{(-1)} \subset \mathcal{F}V_0$ follows from the identity $\hat{\varphi}(2\xi) = m_0(\xi)\, \hat{\varphi}(\xi)$, where $m_0(\xi)$ is the infinitely differentiable and 2π-periodic function that coincides with $\hat{\varphi}(2\xi)$ on $[-\pi, \pi]$. Note that $(m_0(\xi))^2 + (m_0(\xi + \pi))^2 = 1$ for all $\xi \in \mathbb{R}$.

It remains to describe the sampling operator $T_0 \colon V_0 \to \ell^2(\mathbb{Z})$. Here it is again the restriction to $\mathbb{Z}$ of the functions $f \in V_0$. Let us observe first that T_0 is well defined. This follows from the following more general lemma.

Lemma 1 *There exists a constant C such that for every $T > 0$ and every function $f \in L^p(\mathbb{R})$, $1 \leq p \leq \infty$, whose Fourier transform is supported in the interval $[-T, T]$, one has*

$$\left\{ \sum_{-\infty}^{\infty} T^{-1} \, |f(k\pi T^{-1})|^p \right\}^{1/p} \leq C\, \|f\|_p. \tag{2.3}$$

The structure of this statement implies at once the following seemingly stronger property: if $R \geq T$ we still have $\left\{ \sum_{-\infty}^{\infty} R^{-1}\, |f(k\pi R^{-1})|^p \right\}^{1/p} \leq C\, \|f\|_p$ when the Fourier transform of f is zero outside $[-T, T]$.

In our case, $T = 4\pi/3$ and the lemma implies the continuity of the operator T_0. To prove that T_0 is an isomorphism it is appropriate to construct the inverse operator, that is, the function g of the first section.

For this one forms first $\omega(\xi) = \sum_{-\infty}^{\infty} \hat{\varphi}(\xi + 2k\pi)$, which is 2π-periodic and does not vanish. Next one defines $g \in \mathcal{S}(\mathbb{R})$ by $\hat{g}(\xi) = \hat{\varphi}(\xi)/\omega(\xi)$. Then g belongs to V_0 and satisfies $g(0) = 1$ and $g(k) = 0$ for $k \in \mathbb{Z}$, $k \neq 0$. This last property is equivalent to $\sum_{-\infty}^{\infty} \hat{g}(\xi + 2k\pi) = 1$.

In this example the isomorphism between V_0 and $\ell^2(\mathbb{Z})$ defined by T_0 can be extended to an isomorphism between $V_0^{(p)}$ and $\ell^p(\mathbb{Z})$ for $1 \leq p \leq +\infty$. We denote by $V_0^{(p)}$ the subspace of $L^p(\mathbb{R})$, $1 \leq p \leq \infty$, consisting of the functions f whose Fourier transform (taken in the sense of distributions) is written $\hat{f}(\xi) = m(\xi)\,\hat{\varphi}(\xi)$, where $m(\xi)$ is a 2π-periodic distribution whose Fourier coefficients belong to $\ell^p(\mathbb{Z})$.

The continuity of $T_0\colon V_0^{(p)} \to \ell^p(\mathbb{Z})$ follows from (2.3). The fact that T_0 is an isomorphism results again from the existence of a lifting operator. The definition of this operator comes from the fact that the function g constructed above belongs to $\mathcal{S}(\mathbb{R})$. One can observe that the function $g(x)$ that would appear in the first example is $(\sin \pi x)/(\pi x)$.

This function does not belong to $L^1(\mathbb{R})$, and therefore T_0 is not an isomorphism between $V_0^{(1)}$ and $\ell^1(\mathbb{Z})$. Neither is this operator an isomorphism between $V_0^{(\infty)}$ and $\ell^\infty(\mathbb{Z})$. However, in this first example, T_0 remains an isomorphism between $V_0^{(p)}$ and $\ell^p(\mathbb{Z})$ if $1 < p < \infty$.

The third family of examples was discovered independently by G. Battle and P. G. Lemarié. G. Battle uses wavelet bases to perform more comfortably calculations needed in constructive field theory.

This third family of examples also has the advantage of making explicit the relations that can exist between multiresolution analyses, martingales, and algorithms of numerical analysis (spline functions and finite elements).

Here we still have $M(x) = 2x$, $\Gamma = \mathbb{Z}$, one fixes an integer $d \geq 1$, and $V_j^{(d)} \subset L^2(\mathbb{R})$ denotes the subspace of functions of class C^{d-1} on the whole real line, whose restriction to any interval $[k2^{-j}, (k+1)2^{-j}[$ coincides with a polynomial P_k of degree at most d. When d is odd, the isomorphism T_0 between $V_0^{(d)}$ and $\ell^2(\mathbb{Z})$ is simply the restriction to $\mathbb{Z} \subset \mathbb{R}$ of the functions $f \in V_0^{(d)}$. If $d = 0$, the functions of $V_j^{(0)}$ are the step functions whose (possible) jumps have $k2^{-j}$, $k \in \mathbb{Z}$, as abscissas. In this last case, the general algorithm, which we shall present in the next section, leads to the Haar system, while for $d \geq 1$ one obtains the wavelets of Battle and Lemarié.

Here is a last example. We situate ourselves in two dimensions and $\mathbb{R}^2$ is identified with the complex plane $\mathbb{C}$. The lattice Γ is the ring of Gaussian integers $m + in$, $m \in \mathbb{Z}$, $n \in \mathbb{Z}$, while $M(z) = (1+i)z$.

To define V_0 one considers a tiling of the plane with the help of isosceles right triangles obtained by the following procedure. One calls C the square whose vertices are 0, $1+i$, $2i$, and $-1+i$. One cuts C into four triangles (bounded by the four sides and the two diagonals of C), denoted by C_1, C_2, C_3, C_4. Finally one considers the partition of the plane obtained by translating the four triangles by $\gamma \in (1+i)\Gamma$. Then V_0 is the closed subspace of $L^2(\mathbb{R}^2)$ consisting of the continuous functions whose restriction to every triangle $C_j + \gamma$, $\gamma \in (1+i)\Gamma$, $1 \leq j \leq 4$, is an affine function.

One checks without difficulty that the operator of restriction to Γ of the functions $f \in V_0$ is an isomorphism between V_0 and $\ell^2(\Gamma)$. The function $g(z)$ is here the "pyramid function" equal to 1 at 0, to $1 - |x| - |y|$ if $|x| + |y| \leq 1$, and then to 0 if $|x| + |y| \geq 1$.

3. The Two Fundamental Algorithms

We begin by recalling the algorithm of construction of orthonormal bases of a Hilbert space starting with the symbolic calculus on positive definite matrices.

Let H be a Hilbert space and let $(e_j)_{j\in J}$ be a sequence of vectors in H having the following two properties:

(3.1) $0 < c \le \|e_j\| \le C$ for all $j \in J$;

(3.2) e_j, $j \in J$, is an unconditional basis of H.

We want to construct, starting with this sequence e_j, $j \in J$, an orthonormal basis f_j, $j \in J$, having the following functorial property: if one replaces the sequence e_j, $j \in J$, by the sequence $e'_j = e_{\sigma(j)}$, $j \in J$, where $\sigma\colon J \to J$ is an arbitrary bijection, then the sequence f_j has to be replaced by $f'_j = f_{\sigma(j)}$, $j \in J$.

The algorithm having this property is given by the following construction. For every $x \in H$ set $T(x) = \sum_{j\in J} \langle x|e_j\rangle\, e_j$. It is immediate to check that $T\colon H \to H$ is a self-adjoint, continuous linear operator, and that, more precisely, there exist two constants $C_2 \ge C_1 > 0$ such that

$$C_1\mathbf{1} \le T \le C_2\mathbf{1}. \tag{3.3}$$

Then $f_j = T^{-1/2}e_j$, $j \in J$, is the orthonormal basis we want to construct.

The matrix of T with respect to the unconditional basis of the e_j, $j \in J$, is the Gram matrix $G = (\langle e_j|e_k\rangle)_{(j,k)\in J\times J}$. This matrix itself is self-adjoint and positive. In fact G is the matrix of T with respect to the orthonormal basis f_j, $j \in J$. It follows that $G^{-1/2}$ is the matrix of $T^{-1/2}$ in this same orthonormal basis. Thus we have $T^{-1/2}(f_j) = \sum \gamma(j,k)\, f_k$, where the $\gamma(j,k)$ are the entries of $G^{-1/2}$. It follows that $f_j = \sum \gamma(j,k)\, T^{1/2}(f_k) = \sum \gamma(j,k)\, e_k$.

We can apply the algorithm of Gram to the particular case of multiresolution analysis. The set J is the lattice Γ and the elements e_j, $j \in J$, of H are the functions $g(x-\gamma)$, $\gamma \in \Gamma$. The Gram matrix will then be denoted A (G will denote the dual group of Γ), and its entries are $\alpha(\gamma,\gamma') = \int_{\mathbb{R}^n} g(x-\gamma)\, \bar{g}(x-\gamma')\, dx = a(\gamma-\gamma')$.

This matrix A is thus the matrix of a convolution operator acting on $\ell^2(\Gamma)$. The symbolic calculus on such matrices is done using the Fourier transformation $\mathcal{F}\colon \ell^2(\Gamma) \to L^2(G)$. One has

$$\begin{aligned}\sum_{\gamma\in\Gamma} a(\gamma)\, e^{i\gamma\cdot\xi} &= \sum_{\gamma\in\Gamma} \int_{\mathbb{R}^n} g(x-\gamma)\, \bar{g}(x)\, e^{i\gamma\cdot\xi}\, dx \\ &= \frac{1}{(2\pi)^n} \sum_{\gamma\in\Gamma} \int_{\mathbb{R}^n} |\hat{g}(u)|^2\, e^{-i\gamma\cdot(u-\xi)}\, du.\end{aligned}$$

The Poisson summation formula reads $\sum_{\gamma\in\Gamma} e^{i\gamma\cdot\xi} = (\mathrm{vol}\, G) \sum_{\gamma\in\Gamma^*} \delta(\xi-\gamma)$, where, by abuse of language, vol G denotes the volume in $\mathbb{R}^n$ of a fundamental domain of Γ^*. Thus we have

$$\sum_{\gamma\in\Gamma} a(\gamma)\, e^{i\gamma\cdot\xi} = \frac{\mathrm{vol}\, G}{(2\pi)^n} \sum_{\gamma\in\Gamma^*} |\hat{g}(\xi+\gamma)|^2 = \omega(\xi).$$

This means that after conjugation by the Fourier transformation the Gram matrix has become the operator of pointwise multiplication by $\omega(\xi)$ acting on $L^2(G)$. The symbolic calculus is then immediate and $A^{-1/2}$ corresponds to $1/\sqrt{\omega(\xi)}$ acting on $L^2(G)$ by pointwise multiplication. Finally one writes

$$\frac{1}{\sqrt{\omega(\xi)}} = \sum_{\gamma\in\Gamma} c(\gamma)\, e^{i\gamma\cdot\xi}$$

and one deduces from the functions $g(x-\gamma)$ the orthonormal basis $\varphi(x-\gamma)$ of V_0 by

$$\varphi(x-\gamma) = \sum_{\gamma'\in\Gamma} c(\gamma-\gamma')\, g(x-\gamma').$$

Passing again to the Fourier transform we obtain

$$\hat{\varphi}(\xi) = \frac{\hat{g}(\xi)}{\sqrt{\omega(\xi)}}. \tag{3.4}$$

The preceding calculations are summarized in the following theorem.

Theorem 1 *Let V_0 be a closed subspace of $L^2(\mathbb{R}^n)$ satisfying (1.5). One denotes by $g(x)$ the inverse image under the sampling map T_0 of the sequence in $\ell^2(\Gamma)$ equal to 1 at 0 and to 0 elsewhere. Then there exist two constants $C_2 \geq C_1 > 0$ such that one has*

$$C_1 \leq \sum_{\gamma\in\Gamma^*} |\hat{g}(\xi+\gamma)|^2 \leq C_2.$$

Set

$$\omega(\xi) = \frac{\text{vol } G}{(2\pi)^n} \sum_{\gamma\in\Gamma^*} |\hat{g}(\xi+\gamma)|^2$$

and define $\varphi \in V_0$ by

$$\hat{\varphi}(\xi) = \frac{\hat{g}(\xi)}{\sqrt{\omega(\xi)}}.$$

Then the collection $\varphi(x-\gamma)$, $\gamma \in \Gamma$, is an orthonormal basis of V_0.

The fact that $\omega(\xi)$ is bounded from above and from below is exactly the translation of (3.3).

An important particular case where Theorem 1 is applied occurs when the function $g(x)$ satisfies, for a certain constant $A > 0$ and a certain exponent $\alpha > 0$, the condition of exponential decay $|g(x)| \leq Ae^{-\alpha|x|}$. Then $\hat{g}(\xi)$ is real-analytic and so are $|\hat{g}(\xi)|^2$ and finally also $\omega(\xi)$ and $\hat{\varphi}(\xi)$. This implies that $|\varphi(x)| \leq A'e^{-\alpha'|x|}$ for a certain exponent $\alpha' > 0$. This situation will arise in the case of multiresolution analyses consisting of spline functions.

We intend to prove now the following theorem.

Theorem 2 *Let $(V_j)_{j\in\mathbb{Z}}$ be a multiresolution analysis associated with the couple (M,Γ). Set $q = |\det M|$. Then q is an integer ≥ 2. Denote by W_0 the orthogonal complement of V_0 in V_1. Then there exist $q-1$ functions $\psi_1, \ldots, \psi_{q-1}$ belonging to W_0 such that the collection of functions $\psi_m(x-\gamma)$, $\gamma \in \Gamma$, $1 \leq m \leq q-1$, is an orthonormal basis of W_0.*

This sequence ψ_m, $1 \leq m \leq q-1$, is constructed with the help of the function φ of Theorem 1 in the following way. We call $\mathcal{F}\colon L^2(\mathbb{R}^n) \to L^2(\mathbb{R}^n)$ the Fourier transformation.

Then $\mathcal{F}V_0$ is the collection of the functions $g(\xi)\,\hat{\varphi}(\xi)$; here and in what follows $g(\xi) \in L^2(G)$, that is, g belongs to $L^2_{\text{loc}}(\mathbb{R}^n)$ and is Γ^*-periodic. The norm of $f \in V_0$ is computed by

$$\|f\|_2^2 = (2\pi)^{-n} \int |g(\xi)|^2\, |\hat{\varphi}(\xi)|^2\, d\xi = \frac{1}{\text{vol } G} \int_G |g(\xi)|^2\, d\xi.$$

Similarly, $\mathcal{F}V_{(-1)}$ consists of functions $\sqrt{q}\hat{\varphi}(M^*\xi)\, g(M^*\xi)$, where $g(\xi)$ is still a Γ^*-periodic function and belongs to $L^2_{\text{loc}}(\mathbb{R}^n)$. Denote by R the finite commutative group $(M^*)^{-1}\Gamma^*/\Gamma^*$ included in $G = \mathbb{R}^n/\Gamma^*$. The cardinality of R is q and we shall index the functions ψ_m, using as indices the elements of R and agreeing that $\psi_0 = \varphi$.

Since ψ_m belongs to V_1, we necessarily have

$$\hat{\psi}_r(M^*\xi) = m_r(\xi)\, \hat{\varphi}(\xi), \text{ where } m_r(\xi) \in L^2(G),\ r \in R.$$

Thus our purpose is to write that $V_0 = V_{(-1)} \oplus W_{(-1)}$ and then that $W_{(-1)}$ has as an orthonormal basis the functions $q^{-1/2}\psi_r(M^{-1}x - \gamma)$, $r \in R^*$, $\gamma \in \Gamma$. We denote by R^* the group R without 0.

Finally we must write the orthogonal decomposition

$$g(\xi)\,\hat{\varphi}(\xi) = \sum_{r\in R} q^{1/2}\, m_r(\xi)\, g_r(M^*\xi)\, \hat{\varphi}(\xi), \tag{3.5}$$

where $g(\xi)$ is an arbitrary function in $L^2(G)$ extended to a Γ^*-periodic function on $\mathbb{R}^n$ and where the $g_r(\xi)$ belong equally to $L^2(G)$ and are uniquely determined by (3.5). For all the choices of the function $g_r(\xi)$ on the right-hand side of (3.5), the different terms of (3.5) are orthogonal in $L^2(\mathbb{R}^n)$ and, furthermore, we have for all $r \in R$

$$\|q^{1/2}\, m_r(\xi)\, g_r(M^*\xi)\, \hat{\varphi}(\xi)\|_{L^2(\mathbb{R}^n)} = \|g_r(\xi)\, \hat{\varphi}(\xi)\|_{L^2(\mathbb{R}^n)}. \tag{3.6}$$

This last property means that for all fixed r, the different functions $q^{1/2}\psi_r(M^{-1}x - \gamma)$, $\gamma \in \Gamma$, form an orthonormal sequence.

We write (3.6) using the canonical isometry between V_0 and $L^2(G; (\text{vol } G)^{-1}\, dx)$ defined by the orthonormal basis $\varphi(x - \gamma)$, $\gamma \in \Gamma$. Thus we obtain, for every $r \in R$,

$$\|q^{1/2}\, m_r(\xi)\, g_r(M^*\xi)\|_{L^2(G)} = \|g_r(\xi)\|_{L^2(G)}. \tag{3.7}$$

One squares (3.7) and uses the action of the finite group R. It follows that, for every $s \in R$,

$$q \int_G |m_r(\xi + s)|^2\, |g_r(M^*\xi)|^2\, d\xi = \int_G |g_r(\xi)|^2\, d\xi. \tag{3.8}$$

Adding all the relations (3.8) for all the $s \in R$ one obtains

$$\int_G \left\{ \sum_{s\in R} |m_r(\xi + s)|^2 \right\} |g_r(M^*\xi)|^2\, d\xi = \int_G |g_r(\xi)|^2\, d\xi = \int_G |g_r(M^*\xi)|^2\, d\xi. \tag{3.9}$$

The function $\sum_{s\in R} |m_r(\xi + s)|^2$ of ξ is itself $(M^*)^{-1}$-periodic. The equality (3.9), satisfied for every function $g_r(\xi) \in L^2(G)$, implies $\sum_{s\in R} |m_r(\xi + s)|^2 = 1$. In a completely analogous fashion the orthogonality between the different terms of (3.5) implies $\sum_{s\in R} m_r(\xi + s)\, \bar{m}_{r'}(\xi + s) = 0$ for $r' \neq r$.

Let us then consider the matrix

$$S(\xi) = \big(m_r(\xi + s)\big)_{(r,s)\in R\times R}. \tag{3.10}$$

The conditions we have written imply that this matrix is unitary.

Conversely, if $S(\xi)$ is unitary we can find, for any function $g(\xi) \in L^2(G)$, a unique sequence $g_r(\xi)$ belonging to $L^2(G)$ such that

$$g(\xi) = \sum_{r\in R} q^{1/2}\, m_r(\xi)\, g_r(M^*\xi).$$

It suffices to define $g_r(M^*\xi) = q^{-1/2} \sum_{r\in R} g(\xi + s)\, \bar{m}_r(\xi + s)$ and to observe that this last function is indeed $(M^*)^{-1}\Gamma$-periodic.

One can make explicit the properties of the matrix $S(\xi)$ by associating with it the unitary operator $U(\xi)\colon \ell^2(R) \to \ell^2(R)$ defined by $U(\xi)[f] = g$ when $g(r) = \sum_{s\in R} m_r(\xi+s)$ $f(s)$. Then one has $U(\xi+\tau) = U(\xi)\, R(\tau)$ for every $\tau \in R$, denoting by $R(\tau)\colon \ell^2(R) \to \ell^2(R)$ the operator of translation by τ.

The problem raised by Theorem 2 is the construction of this unitary matrix $S(\xi)$, $\xi \in G$, knowing the first column vector $m_0(\xi + s)$, $s \in R$. This first column vector is supplied by Theorem 1. If one seeks the functions $m_r(\xi)$ in $L^\infty(G)$, their construction poses no problem. Indeed, one calls $P \subset \mathbb{R}^n$ a parallelepiped which is a fundamental domain for the group $(M^*)^{-1}\Gamma$ and one constructs arbitrarily a unitary matrix $S(\xi)$, $\xi \in P$, whose first column vector is $m_0(\xi + s)$. The other column vectors are then $m_{r,s}(\xi)$, $\xi \in P$. The functions $m_r(\xi) \in L^\infty(G)$ are constructed setting $m_r(\xi+s) = m_{r,s}(\xi)$ for $\xi \in P$ and $s \in R$.

This brutal method will certainly not lead to a continuous choice of $S(\xi)$ that is Γ^*-periodic. But in the meantime we have proved Theorem 2.

We have left aside the irritating problem of knowing whether there exists an orthonormal basis of W_0 of the form $\psi_m(x - \gamma)$, $\gamma \in \Gamma$, $1 \le m < q$, for which the ψ_m are wavelets. This cannot be the case without a particular hypothesis concerning the multiresolution analysis. But this hypothesis could be the decay at infinity of the function $g(x)$, which was the starting point of the algorithms of Theorems 1 and 2.

An obvious corollary of Theorem 2 is the following assertion.

Corollary 1 *With the notation of Theorem 2, the functions*

$$q^{j/2}\, \psi_m(M^j x - \gamma), \quad j \in \mathbb{Z},\ 1 \le m \le q-1,\ \gamma \in \Gamma, \tag{3.11}$$

form an orthonormal basis of $L^2(\mathbb{R}^n)$.

Our second problem will be presented in a special case. Let, in dimension one, $\psi(x)$ be a function of the class $\mathcal{S}(\mathbb{R})$ such that the collection $2^{j/2}\psi(2^j x - k)$, $j \in \mathbb{Z}$, $k \in \mathbb{Z}$, is an orthonormal basis of $L^2(\mathbb{R})$. Does there exist a multiresolution analysis (where $\Gamma = \mathbb{Z}$ and $M(x) = 2x$) such that our collection can be constructed from this multiresolution analysis?

Since we cannot answer these questions, we will restrict ourselves to some examples. The question is to check that all known families of wavelets indeed arise from multiresolution analyses.

4. Application of the Preceding Algorithms

We will begin with examples in dimension one, and we will restrict ourselves to the case $\Gamma = \mathbb{Z}$ and $M(x) = 2x$. One then speaks of a dyadic multiresolution analysis.

At the start we have available the function $g \in V_0$ which figures in the definition of a multiresolution analysis. One constructs $\varphi \in V_0$ by $\hat{\varphi}(\xi) = \hat{g}(\xi)/\sqrt{\omega(\xi)}$ where $\omega(\xi) = \sum_{-\infty}^{\infty} |\hat{g}(\xi + 2k\pi)|^2$.

The finite group R consists of two elements 0 and π, and the addition is performed modulo 2π. The matrix $S(\xi)$ is here

$$\begin{pmatrix} m_0(\xi) & m_1(\xi) \\ m_0(\xi+\pi) & m_1(\xi+\pi) \end{pmatrix},$$

where $m_0(\xi)$ is defined by $\hat{\varphi}(2\xi) = m_0(\xi)\,\hat{\varphi}(\xi)$. The function $m_1(\xi)$ will serve to determine ψ by $\hat{\psi}(2\xi) = m_1(\xi)\,\hat{\varphi}(\xi)$.

We know that $|m_0(\xi)|^2 + |m_0(\xi+\pi)|^2 = 1$ and we want $m_1(\xi)$ to be 2π-periodic and $S(\xi)$ to be unitary. A natural choice of $m_1(\xi)$ is given by $m_1(\xi) = e^{-i\xi}\,\bar{m}_0(\xi+\pi)$. Let us suppose furthermore that $\hat{\varphi}(\xi) \geq 0$ for all $\xi \in \mathbb{R}$. Then we have $|\hat{\psi}(\xi)| = \sqrt{(\hat{\varphi}(\xi/2))^2 - (\hat{\varphi}(\xi))^2}$ and $\hat{\psi}(\xi) = e^{-i\xi/2}\,|\hat{\psi}(\xi)|$.

Let us return to the multiresolution analysis of the second section. The function we started with is precisely the function φ which Theorem 1 allows us to construct. It yields a wavelet ψ, belonging to the class $\mathcal{S}(\mathbb{R})$ of Schwartz, such that $\hat{\psi}(\xi)$ is supported by $2\pi/3 \leq |\xi| \leq 8\pi/3$; moreover, $\psi(x)$ is real and satisfies $\psi(1-x) = \psi(x)$. The graph of $\psi(x)$, for a particular choice of φ, is in figure 1.

We now consider the dyadic multiresolution analysis consisting of the nested sequence of spaces V_j, $j \in \mathbb{Z}$, composed of splines that are piecewise affine.

More precisely, $V_0 \subset L^2(\mathbb{R})$ is the space of square integrable continuous functions on the real line whose restriction to each interval $[k, k+1]$, $k \in \mathbb{Z}$, is an affine function. The group Γ is $\mathbb{Z}$, $M(x) = 2x$, and finally V_j is defined by $f(x) \in V_0 \Leftrightarrow f(2^j x) \in V_j$.

The function $g(x)$ required by condition (1.5) is the triangle function $\Delta(x) \in V_0$ defined by $\Delta(0) = 1$ and $\Delta(k) = 0$ if $k \in \mathbb{Z}$ and $k \neq 0$. Finally the sampling operator $T_0\colon V_0 \to \ell^2(\mathbb{Z})$ is the operator of restriction to $\mathbb{Z}$ of the functions of V_0. This means that every function $f \in V_0$ can be written uniquely as $f(x) = \sum_{-\infty}^{\infty} f(k)\,\Delta(x-k)$.

After these preliminary remarks, we can turn to the algorithms described by Theorems 1 and 2.

One has

$$\hat{\Delta}(\xi) = \left(\frac{\sin\xi/2}{\xi/2}\right)^2,$$

from which we get immediately

$$\hat{\varphi}(\xi) = \frac{\hat{\Delta}(\xi)}{\sqrt{\omega(\xi)}} = \hat{\Delta}(\xi)\left(1 - \frac{2}{3}\sin^2\frac{\xi}{2}\right)^{-1/2}.$$

Next, the algorithm of Theorem 2 yields

$$\hat{\psi}(\xi) = e^{-i\xi/2}\sin^2\xi/4\left(\frac{\sin\xi/4}{\xi/4}\right)^2\tilde{\omega}(\xi),$$

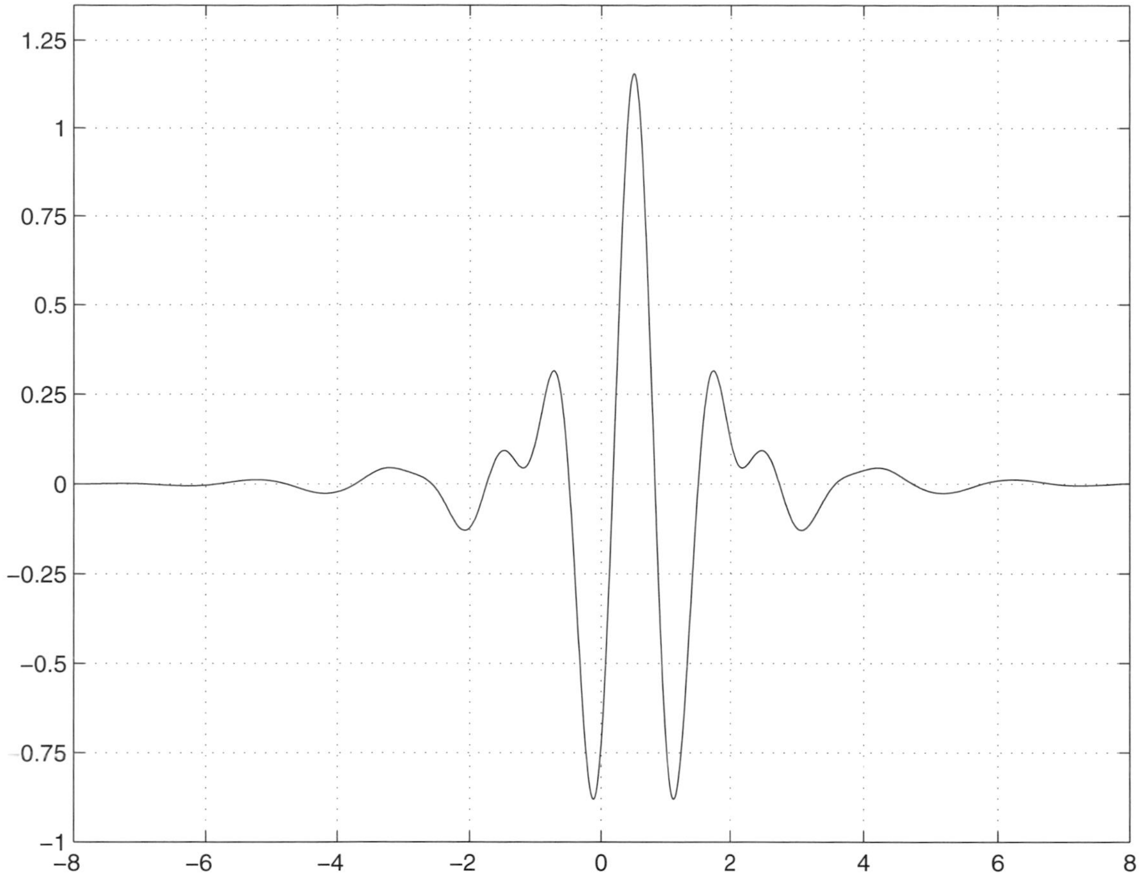

Figure 1. Wavelet.

where

$$\tilde{\omega}(\xi) = \frac{1}{\sqrt{6}} \left(\frac{7 - \cos \xi}{2 + \cos \xi} \right)^{1/2} \left(1 - \frac{2}{3} \sin^2 \frac{\xi}{4} \right)^{-1/2}.$$

The graph of $\psi(x)$ is given in figure 2. One has $|\psi(x)| \leq C\,(2 + \sqrt{3})^{-|x|}$ and this exponential decay is numerically effective.

The case of quadratic splines is completely different. One denotes by $V_0 \subset L^2(\mathbb{R})$ the space of functions of class $C^1(\mathbb{R})$ whose restriction to each interval $[k, k+1]$, $k \subset \mathbb{Z}$, is a polynomial of second degree. Then the (usual) restriction operator of the functions $f \in V_0$ is not surjective. For instance the sequence equal to 1 at 0 and to 0 at $k \neq 0$, $k \in \mathbb{Z}$, cannot be the restriction to $\mathbb{Z}$ of a function $f \in V_0$.

However, property (1.5) is satisfied. We denote by χ the characteristic function of the interval $[0, 1]$ and by g the convolutional product $\chi * \chi * \chi$. To check that every function

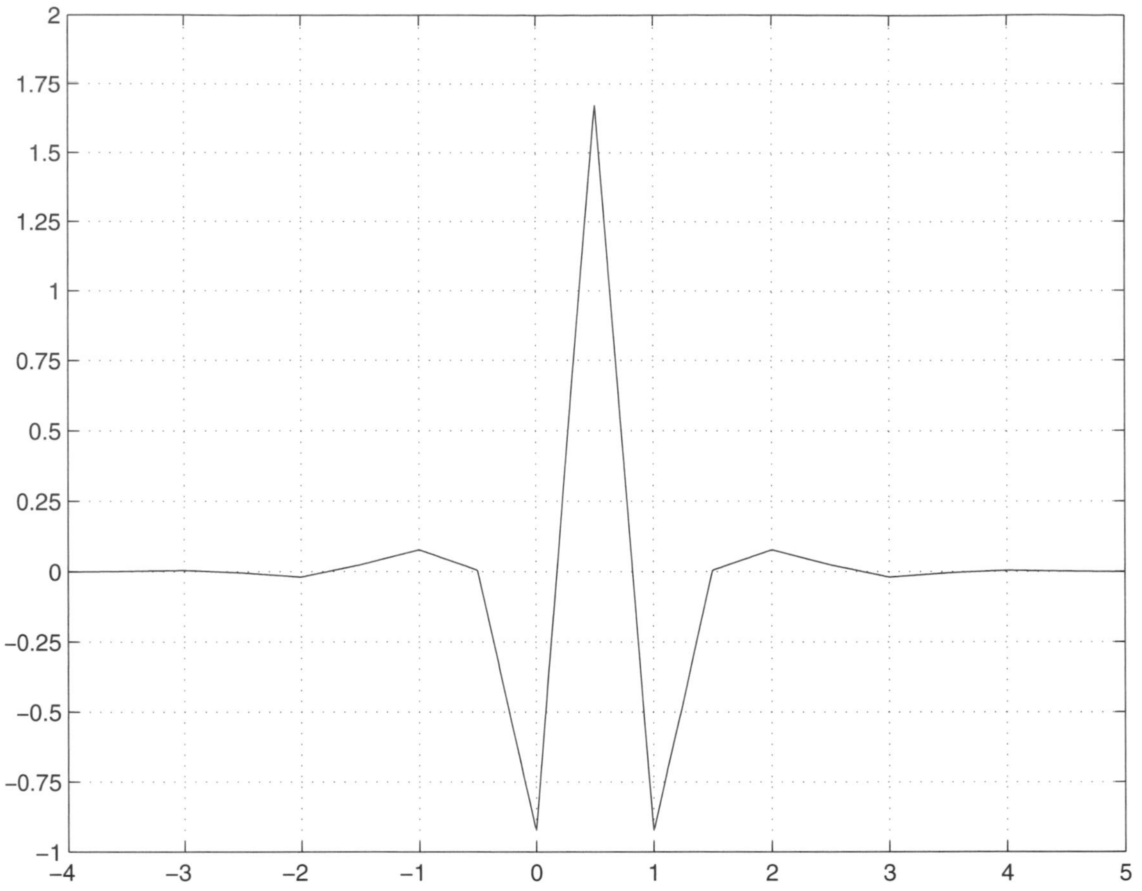

Figure 2. Wavelet.

$f \in V_0$ can be written uniquely as

$$f(x) = \sum_{-\infty}^{\infty} \alpha(k)\, g(x-k), \tag{4.1}$$

where $(\sum |\alpha_k|^2)^{1/2}$ and $\|f\|_2$ are two equivalent norms, we pass to the Fourier transforms and need to check that

$$\hat{f}(\xi) = m(\xi)\, \hat{g}(\xi), \tag{4.2}$$

where $m(\xi)$ is 2π-periodic and belongs to $L^2(0, 2\pi)$.

We begin by analyzing the Fourier transforms of the functions $f \in V_0$.

For this we observe that $f \in V_0$ is equivalent to the combination of the following conditions: $f \in L^2(\mathbb{R})$ and the third derivative in the sense of distributions of f is a sum $\sum_{-\infty}^{\infty} \gamma(k)\, \delta(x-k)$ where $\delta(x)$ is the Dirac mass at 0. Taking the Fourier transform, $\hat{f}(\xi)$ belongs to $L^2(\mathbb{R})$ and $\left(\frac{\xi}{2}\right)^3 \hat{f}(\xi) = \tilde{m}(\xi)$ is a 2π-periodic function belonging to $L^2(0, 2\pi)$.

It follows that $\tilde{m}(\xi) = (\sin \xi/2)^3 e^{-3i\xi/2} m(\xi)$, where $m(\xi)$ is 2π-periodic and belongs to $L^2(0, 2\pi)$. Finally

$$\hat{f}(\xi) = \left(\frac{\sin \xi/2}{\xi/2}\, e^{-i\xi/2}\right)^3 m(\xi) = m(\xi)\, \hat{g}(\xi)$$

as announced.

Let us now pass to the case of cubic splines. This time V_0 denotes the space of functions $f(x)$ of class $C^2(\mathbb{R})$ belonging to $L^2(\mathbb{R})$ whose restriction to each interval $[k, k+1]$, $k \in \mathbb{Z}$, is a polynomial of degree at most 3. Then the fourth derivative, in the sense of distributions, of $f(x)$ is a sum of Dirac masses at $k \in \mathbb{Z}$. It follows that $\xi^4 \hat{f}(\xi) = \tilde{m}(\xi)$, where $\tilde{m}(\xi)$ is a 2π-periodic function and belongs to $L^2(0, 2\pi)$. Since $\hat{f}(\xi)$ itself also belongs to $L^2(\mathbb{R})$, we have $\tilde{m}(\xi) = 2^4 (\sin \xi/2)^4 m(\xi)$, where $m(\xi)$ is also 2π-periodic and belongs to $L^2(0, 2\pi)$. Consequently the Fourier transforms of the functions in V_0 are characterized by

$$\hat{f}(\xi) = \left(\frac{\sin \xi/2}{\xi/2}\right)^4 m(\xi)$$

where $m(\xi)$ is 2π-periodic and belongs to $L^2(0, 2\pi)$.

This function $\left(\frac{\sin \xi/2}{\xi/2}\right)^4$ is the Fourier transform of $\Delta_2(x)$, where $\Delta_2 = \Delta * \Delta$ and $\Delta(x) = (1 - |x|)^+$. It follows that the functions f of V_0 can be written uniquely as $f(x) = \sum_{-\infty}^{\infty} \alpha_k \Delta_2(x - k)$ where α_k is a square summable sequence; in fact, the α_k are the Fourier coefficients of $m(\xi)$.

But in the case of cubic splines there exists a function $g \in V_0$ with exponential decay such that $g(0) = 1$ and $g(k) = 0$ if $k \in \mathbb{Z}$, $k \neq 0$. Indeed, one defines g by

$$\hat{g}(\xi) = \left(\frac{\sin \xi/2}{\xi/2}\right)^4 \left(1 - \frac{2}{3} \sin^2 \frac{\xi}{2}\right)^{-1}.$$

One can check that $|g(x)| \leq C\,(2 + \sqrt{3})^{-|x|}$. Then every function $f \in V_0$ can be written $f(x) = \sum_{-\infty}^{\infty} \beta_k\, g(x - k)$, where, this time, $\beta(k) = f(k)$. Thus, in the case of cubic splines, the sampling operator T_0 is simply the operator of restriction to $\mathbb{Z}$ of the functions $f \in V_0$.

The application of Theorems 1 and 2 is then straightforward. One observes that

$$\int_{-\infty}^{\infty} x^q\, \psi(x)\, dx = 0$$

for $q = 0$, 1, 2, or 3.

So the advantage of the wavelets constructed over cubic splines, compared to linear splines, is on the one hand a better regularity, and on the other a larger number of zero moments.

Before leaving the spline functions, let us recall the celebrated theorem of Holladay tying together the sampling with cubic splines and the quadratic approximation with linear splines. Let us denote by V_j (resp. $\mathcal{V}_j$) the multiresolution analysis consisting of linear splines (resp. cubic splines). Let $f(x)$ be a continuous function of the real variable x belonging to the Sobolev space $H^2(\mathbb{R})$. One defines $f_j \in \mathcal{V}_j$ by $f_j(k2^{-j}) = f(k2^{-j})$, $k \in \mathbb{Z}$.

This function f_j is unique, and the theorem of Holladay expresses the fact that f_j'' is the orthogonal projection of f'' onto V_j. In other words, f_j'' is the function of V_j yielding the best quadratic approximation of f''. The theorem of Holladay is one of the justifications for using the L^2-norm to define the multiresolution analysis.

5. Multiresolution Analyses in Several Dimensions

To lighten the notation, it will be enough to examine the case of dimension two. We begin with a new algorithm which gives a two-dimensional multiresolution analysis.

Thus let $V_j \subset L^2(\mathbb{R})$, $j \in \mathbb{Z}$, be a multiresolution analysis in one dimension. Let us denote by $V_j \hat{\otimes} V_j$ the closure in $L^2(\mathbb{R}^2)$ of the algebraic tensor product $V_j \otimes V_j$.

To simplify the discussion that follows, we will assume that $\Gamma = \mathbb{Z}$ and $M(x) = 2x$ in the multiresolution analysis V_j, $j \in \mathbb{Z}$. We denote by $\varphi \in V_0$ the function constructed thanks to Theorem 1. Then the functions $\varphi(x_1 - k_1)\varphi(x_2 - k_2)$, which will be denoted $\varphi(x - k)$, $k = (k_1, k_2)$, $k \in \mathbb{Z}^2$, form an orthonormal basis of $V_0 \hat{\otimes} V_0$.

It follows immediately that the sequence $\mathcal{V}_j = V_j \hat{\otimes} V_j$, $j \in \mathbb{Z}$, is a multiresolution analysis of $L^2(\mathbb{R}^2)$ associated with $\Gamma = \mathbb{Z}^2$ and $M(x) = 2x$.

Let us denote by $\mathcal{W}_j$ the orthogonal complement of $\mathcal{V}_j$ in $\mathcal{V}_{j+1}$. Then $\mathcal{W}_j = (V_j \hat{\otimes} W_j) \oplus (W_j \hat{\otimes} V_j) \oplus (W_j \hat{\otimes} W_j)$. Consequently the functions

$$\varphi(x_1 - k_1)\psi(x_2 - k_2), \quad \psi(x_1 - k_1)\varphi(x_2 - k_2), \quad \psi(x_1 - k_1)\psi(x_2 - k_2) \tag{5.1}$$

form an orthonormal basis of $\mathcal{W}_0$ when ψ and φ are given by the application of Theorems 1 and 2 to the multiresolution analysis V_j, $j \in \mathbb{Z}$.

The orthonormal basis of the $\mathcal{W}_j$ is obtained by a simple change of scale, and collecting the orthonormal bases of the $\mathcal{W}_j$ which are so obtained, one forms the orthonormal basis of $L^2(\mathbb{R}^2)$ associated with the multiresolution analysis of the $V_j \hat{\otimes} V_j$, $j \in \mathbb{Z}$. If the functions φ and ψ have sufficient regularity and decay at infinity, the algorithm (5.1) yields a wavelet basis of $L^2(\mathbb{R}^2)$. This wavelet basis is evidently compatible with Theorem 2 because $q = \det M = 4$, and $q - 1 = 3$ is in fact the number of sequences used in (5.1).

It is amusing to observe that essentially the same wavelets appear in a multiresolution analysis that seems to be different. We situate ourselves directly in $L^2(\mathbb{R}^2)$ and assume that $\Gamma = \mathbb{Z}^2$ and

$$M\begin{pmatrix} x_1 \\ x_2 \end{pmatrix} = \begin{pmatrix} 2x_2 \\ x_1 \end{pmatrix}.$$

We start with the same basic subspace $\mathcal{V}_0$ as above and define the $\mathcal{V}'_j$ successively by $f(x) \in \mathcal{V}'_j \Leftrightarrow f(Mx) \in \mathcal{V}'_{j+1}$ with $\mathcal{V}'_0 = \mathcal{V}_0$. We then have $\mathcal{V}'_{2j} = \mathcal{V}_j$ and the functions $2^j\varphi(2^j x_1 - k_1)\varphi(2^j x_2 - k_2)$ form an orthonormal basis of $\mathcal{V}_j$. Thus the functions

$$\sqrt{2}\, 2^j \varphi(2^j x_1 - k_2)\, \varphi(2^{j+1} x_2 - k_1)$$

form an orthonormal basis of $\mathcal{V}'_{2j+1}$, and with the change of variables $x \mapsto M(x)$ we get anew the functions

$$2^{j+1}\varphi(2^{j+1} x_1 - k_1)\, \varphi(2^{j+1} x_2 - k_2),$$

which form an orthonormal basis of $\mathcal{V}'_{2j+2} = \mathcal{V}_{j+1}$.

The algorithm of Theorem 2 yields a single bidimensional wavelet, $\varphi(x_1)\psi(x_2)$. The orthogonal complement of $\mathcal{V}'_{2j}$ in $\mathcal{V}'_{2j+1}$ consists of the functions $2^j\varphi(2^j x_1 - k_1)\, \psi(2^j x_2 - k_2)$, $k = (k_1, k_2) \in \mathbb{Z}^2$, while the orthogonal complement of $\mathcal{V}'_{2j+1}$ in $\mathcal{V}'_{2j+2}$ consists of the functions

$2^{j+1/2}\psi(2^j x_1 - k_1)\,\varphi(2^{j+1}x_2 - k_2)$. Setting $\psi(x_1, x_2) = \psi(x) = \varphi(x_1)\psi(x_2)$, the sequences we have just written are the same as $(\sqrt{2})^{2j}\psi(M^{2j}x - k)$ and $(\sqrt{2})^{2j+1}\psi(M^{2j+1}x - k)$. This result is in conformity with Theorem 2.

Here is a last example. Let e_0 and e_1 be two linearly independent vectors in $\mathbb{R}^2$ and Γ the lattice $\mathbb{Z}e_0 + \mathbb{Z}e_1$. We call T_0 the triangle whose vertices are 0, e_0, and e_1, and T_0' the symmetric image of T_0 with respect to 0. For every $\gamma \in \Gamma$ we set $T_\gamma = \gamma + T_0$, $T_\gamma' = \gamma + T_0'$, and these two families of triangles form a paving of the plane. We call V_0 the subspace of $L^2(\mathbb{R}^2)$ consisting of the functions f, continuous on $\mathbb{R}^2$, whose restriction to each triangle T_γ and to each triangle T_γ' is an affine function.

We still use the transformation $M(x) = 2x$, $x = (x_1, x_2)$, and define the V_j by $f(x) \in V_0 \Leftrightarrow f(2^j x) \in V_j$. It is then immediate to check that the sequence V_j so constructed is a multiresolution analysis of $L^2(\mathbb{R}^2)$.

The explicit calculations for this example were performed by S. Jaffard following the algorithms of Theorems 1 and 2.

6. Transformations of Martingales and Paraproducts

The use of multiresolution analyses composed of spline functions allows us to make explicit the relation between martingale transformations and paraproducts.

Let $m \in \mathbb{N}$ be an integer and $V_j^{(m)}$ the multiresolution analysis of $L^2(\mathbb{R})$ built with the help of spline functions of order m in each variable. More precisely, $f \in V_j^{(m)}(\mathbb{R})$ if f is a function of class C^{m-1} on the whole real line, if f belongs to $L^2(\mathbb{R})$, and if the restriction of f to each interval $[k2^{-j}, (k+1)2^{-j}[$ coincides with a polynomial of degree at most equal to m.

For $m = 0$, f is a step function belonging to $L^2(\mathbb{R})$ whose restriction to each interval $[k2^{-j}, (k+1)2^{-j}[$ is a constant.

Finally, $V_j^{(m)}(\mathbb{R}^n)$ is the completion in $L^2(\mathbb{R}^n)$ of the algebraic tensor product $V_j^{(m)} \otimes \cdots \otimes V_j^{(m)}$ of the spaces of spline functions in one variable.

One denotes by $E_j^{(m)}\colon L^2(\mathbb{R}^n) \to V_j^{(m)}(\mathbb{R}^n)$ the operator of orthogonal projection. Then $D_j^{(m)} = E_{j+1}^{(m)} - E_j^{(m)}$ is the operator of orthogonal projection onto $W_j^{(m)}(\mathbb{R}^n)$: this space is the orthogonal complement of $V_j^{(m)}(\mathbb{R}^n)$ in $V_{j+1}^{(m)}(\mathbb{R}^n)$.

Let $m_j(x) \in W_j^{(m)}(\mathbb{R}^n)$ be a sequence of functions satisfying

$$\|m_j(x)\|_\infty \le C. \tag{6.1}$$

We want to know under what supplementary condition concerning this bounded sequence $m_j(x)$ the operator defined by

$$M(f) = \sum_{-\infty}^{\infty} m_j(x)\,(E_j^{(m)} f)(x) \tag{6.2}$$

is bounded on $L^2(\mathbb{R}^n)$.

Observe that the orthogonality between W_j and V_j implies that $\int m_j(x)(E_j^{(m)}f)(x)\,dx = 0$ for each j.

The hypothesis $m_j \in W_j^{(m)}$ implies therefore that the oscillation of this function $m_j(x)$ is sufficiently strong to persist after multiplication by $E_j^{(m)}f$. This last function does not present any oscillating behavior.

We can observe that the various functions $m_j(x)$ are completely determined by $m(x) = \sum_{-\infty}^{\infty} m_j(x)$. In fact $m(x)$ belongs to the Besov space $B_\infty^{0,\infty}$, which is a space of distributions.

This distribution will be called the symbol of the operator M.

Theorem 3 *If $m = 0$, then the operator M is bounded on $L^2(\mathbb{R}^n)$ if and only if its symbol $m(x)$ belongs to the dyadic* BMO *space. If $m \geq 1$, the operator M is bounded on $L^2(\mathbb{R}^n)$ if and only if its symbol $m(x)$ belongs to the usual* BMO *space.*

If $m = 0$, the operator $E_j^{(0)}$ is no other than the conditional expectation operator with respect to the σ-algebra generated by the collection $\mathcal{Q}_j$ of the dyadic cubes given by

$$k_1 2^{-j} \leq x_1 < (k_1 + 1)2^{-j}, \ldots, k_n 2^{-j} \leq x_n < (k_n + 1)2^{-j}.$$

The series $\sum_{-\infty}^{\infty} m_j(x)$ is then a martingale and (6.2) represents a martingale transform. Finally, the dyadic version of BMO is defined by $\sup_j \sup_{Q \in \mathcal{Q}_j} \big((1/|Q|) \int_Q |m(x) - m_Q|^2\,dx\big)^{1/2} \leq C < \infty$. We have denoted by m_Q the mean of $m(x)$ on Q.

The usual BMO space is defined by $\big((1/|Q|) \int_Q |m(x) - m_Q|^2\,dx\big)^{1/2} \leq C$, where Q varies over the set of all cubes in $\mathbb{R}^n$.

When $m = 0$ the different terms of the series (6.2) represent again a martingale. In particular they are orthogonal to each other and $\|Mf\|_2 = \big(\sum_{-\infty}^{\infty} \|m_j E_j^{(0)} f\|_2^2\big)^{1/2}$.

One transforms this series by a lemma of L. Carleson [3], and one uses the obvious remark that

$$\sum_{j' \geq j} \int_Q |m_{j'}(x)|^2\,dx \leq C|Q|$$

for all $Q \in \mathcal{Q}_j$. The details can be found in [15].

Thus we will restrict ourselves to the case where $m \geq 1$ and we will not write the index m. Though the integral of each term of the series (6.2) is zero, it seems that one cannot adapt the reasoning used in the case $m = 0$.

One proceeds differently by calculating explicitly the distributional kernel $K(x, y)$ of the operator M. The restriction of $K(x, y)$ to the open set $\Omega \subset \mathbb{R}^n \times \mathbb{R}^n$ defined by $y \neq x$ is a function (also denoted $K(x, y)$) satisfying the inequalities

$$|K(x, y)| \leq C\,|x - y|^{-n}, \quad |\partial/\partial x_k K(x, y)| \leq C|x - y|^{-n-1},$$

and

$$\left|\frac{\partial}{\partial y_k} K(x, y)\right| \leq C|x - y|^{-n-1}.$$

This allows us to apply the condition of David and Journé concerning the continuity in L^2 of singular integral operators. The key point is that $M(1) = \sum_{-\infty}^{\infty} m_j(x) = m(x)$.

To write the details we shall restrict ourselves to the dimension $n = 1$ and we shall also assume that $m = 1$. The explicit calculation of $K(x, y)$ uses the construction of the function φ of Theorem 1. As we have said above, this function φ belonging to V_0 has an exponential decay at infinity. This is also true for its first derivative.

The distributional kernel of M is therefore

$$K(x,y) = \sum_{-\infty}^{\infty} \sum_{-\infty}^{\infty} m_j(x)\, 2^j\, \varphi(2^j x - k)\, \varphi(2^j y - k). \tag{6.3}$$

Since $m_j(x)$ belongs to W_j contained in V_{j+1}, one has a Bernstein-type inequality, namely,

$$\left\| \frac{d}{dx} m_j(x) \right\|_\infty \leq C\, 2^j\, \|m_j\|_\infty \leq C'\, 2^j.$$

It follows that the kernel defined by (6.3) satisfies the announced properties.

We infer from it immediately that the condition $M(1) \in \mathrm{BMO}$ is necessary for the continuity of the operator M on L^2 (theorem of J. Peetre).

Conversely, let us assume that $m(x)$ belongs to BMO. One begins by observing that $\|m_j\|_\infty \leq C\, \|m\|_{\mathrm{BMO}}$. Indeed, one has with the notation of Theorem 2

$$m_j(x) = \sum_{k \in \mathbb{Z}} \psi(2^j x - k) \int 2^j\, \psi(2^j y - k)\, m(y)\, dy.$$

Next one observes that the function $2^j \psi(2^j y - k)$ is a molecule in the sense of G. Weiss. This is due to the localization and oscillation of the wavelet ψ such as they are described by (1.9) and (1.10). Taking into account (1.11), ψ is even a "special atom" in the sense of G. de Souza [16]. The norm of $2^j \psi(2^j x - k)$ in the space $H^1(\mathbb{R})$ of Stein and Weiss is thus a finite constant independent of j and of k. The duality between $H^1(\mathbb{R})$ and $\mathrm{BMO}(\mathbb{R})$ implies

$$\left| \int 2^j \psi(2^j y - k) m(y)\, dy \right| \leq C \|m\|_{\mathrm{BMO}}.$$

One can even replace the BMO-norm by the $B_\infty^{0,\infty}$-norm. The space $B_\infty^{0,\infty}$ is the space (of Bloch) of the derivatives of the functions of the Zygmund class.

The criterion of David and Journé consists in checking the following three properties:

(6.4) $\|M(u)\|_2 \leq C\, \|u\|_2$ when $u = \psi(2^j x - k)$, $j \in \mathbb{Z}$, $k \in \mathbb{Z}$ (with a constant C independent of j and k);

(6.5) $M(1) \in \mathrm{BMO}$;

(6.6) $M^*(1) \in \mathrm{BMO}$.

The inequality (6.4) is one of the forms of the property of weak boundedness (called the "weak-boundedness property" in [5]).

Let us begin by checking (6.4). To simplify the writing of what follows, we write $u(x) = \psi(2^{\ell-1} x - k)$, reserving the index j to the expression of M by the series (6.2). If $j \leq \ell - 1$ one has $E_j(u) = 0$, while $E_j(u) = u$ if $j \geq \ell$. Thus $M(u) = \left(\sum_{j \geq \ell} m_j\right) u = (m - E_\ell m)\, u$.

To check (6.4) we have to show that

$$\|(m - E_\ell m)\, u\|_2 \leq C\, \|u\|_2. \tag{6.7}$$

For this one begins by observing that

$$\left(\int_I |m - E_\ell m|^2\, dx\right)^{1/2} \leq C\, |I|^{1/2}\, \|m\|_{\mathrm{BMO}} \tag{6.8}$$

for all intervals I of length $|I| = 2^{-\ell}$.

The implication (6.8) $\Rightarrow$ (6.7) is obtained by using a step function, with rapid decay at infinity, constructed on dyadic intervals I of length $2^{-\ell}$, which is a majorant of $|u|$. The checking of (6.8) is also simple. Since $m(x)$ belongs to BMO one has $\left(\int_I |m(x) - m_I|^2\, dx\right)^{1/2} \leq C|I|^{1/2}\|m\|_{\mathrm{BMO}}$ and it suffices to observe then that the mean m_I of $m(x)$ on I and the function $E_\ell m$, restricted to I, differ at most by $C\, \|m\|_{\mathrm{BMO}}$.

This finishes the proof of (6.4).

The checking of (6.5) is immediate since, as we have already observed,

$$M(1) = \sum_{-\infty}^{\infty} m_j(x) = m(x).$$

Finally, $M^*(1) = 0$ follows from the orthogonality between V_j and W_j.

7. Wavelets and the Algebra A_b of David-Journé-Semmes

Let $b(x) \in L^\infty(\mathbb{R})$ be a function with complex values such that $\mathrm{Re}\, b(x) \geq 1$.

We propose to prove the following theorem.

Theorem 4 *There exists an unconditional normalized basis $\psi_{(j,k)}$, $j \in \mathbb{Z}$, $k \in \mathbb{Z}$, of $L^2(\mathbb{R})$ having the following properties.*

(7.1) $\psi_{(j,k)}(x) = 2^{j/2}\, g_{(j,k)}(2^j x - k)$.

(7.2) *The functions $g_{(j,k)}$ satisfy, uniformly in j and k, the inequality*

$$|g_{(j,k)}(x)| \leq C\, \exp(-\alpha|x|),$$

and so do their first derivatives; C and α are two strictly positive constants.

(7.3) *One has*

$$\int_{-\infty}^{\infty} \psi_{(j,k)}(x)\, \psi_{(j',k')}(x)\, b(x)\, dx = \begin{cases} 0 & \textit{if } (j,k) \neq (j',k'), \\ 1 & \textit{if } (j,k) = (j',k'). \end{cases}$$

The normalization of the unconditional basis means that for two (nonzero) positive constants C and c one has

$$c \leq \|\psi_{(j,k)}\|_2 \leq C.$$

When $b(x) = 1$ we are back at the wavelets of the preceding sections.

The existence of the unconditional basis of the theorem is tied to that of an algebra of operators we shall now describe. In question is the subalgebra $A_b \subset \mathcal{L}(L^2(\mathbb{R}), L^2(\mathbb{R}))$

of operators T whose distributional kernel $K(x, y)$ satisfies the following properties. There exist a constant $C \geq 0$ and an exponent $\alpha \in]0, 1]$ such that the restriction to $x \neq y$ of $K(x, y)$ is a function which can be written $S(x, y)\, b(y)$ and where

(7.4) $|S(x, y)| \leq C|x - y|^{-1}$;

(7.5) $|S(x', y) - S(x, y)| \leq C|x' - x|^{\alpha}\, |x - y|^{-1-\alpha}$ if $|x' - x| \leq \frac{1}{2}\, |x - y|$;

(7.6) $|S(x, y') - S(x, y)| \leq C|y' - y|^{\alpha}\, |x - y|^{-1-\alpha}$ if $|y' - y| \leq \frac{1}{2}\, |x - y|$;

(7.7) $\int S(x, y)\, b(y)\, dy = 0$ (modulo the constants);

(7.8) $\int S(x, y)\, b(x)\, dx = 0$ (modulo the constants).

Since the integrals (7.7), (7.8) are not convergent in the usual sense, it is agreed to give them the following meaning. We call $u(x) \in \mathcal{D}_0$ a test function whose integral is zero. Then the left-hand side of (7.7) has to be integrated first against $u(x)$. The function of y given by $\int S(x, y) u(x)\, dx = v(y)$ belongs to $L^2(\mathbb{R})$ because T is bounded on $L^2(\mathbb{R})$ and is $\mathcal{O}(|y|^{-1-\alpha})$ at infinity due to (7.5). Then one can integrate it against $b(y) \in L^{\infty}(\mathbb{R})$. This is why the 0 on the right-hand side of (7.7) can be replaced by any constant (the integral of $u(x)$ being zero).

Then the operators $T \in A_b$ are operators that are "almost diagonal" with respect to the basis $\psi_{(j,k)}$ associated with b and given by Theorem 4.

To lighten the notation we will denote by $\mathcal{I}$ the collection of all the dyadic intervals $I = [k2^{-j}, (k + 1)2^{-j}[$ and write ψ_I instead of $\psi_{(j,k)}$.

To compute the matrix $\gamma(I, J)$ of the operator $T \in A_b$ with respect to the basis ψ_I given by Theorem 4, one is led to write

$$T\psi_I = \sum_{J \in \mathcal{I}} \gamma(I, J)\, \psi_J,$$

and one then has

$$\gamma(I, J) = \int (T\psi_I)\psi_J b(x)\, dx = \iint b(x)\psi_J(x) S(x, y)\psi_I(y) b(y)\, dx\, dy. \tag{7.9}$$

We shall see (as a corollary of the construction of the basis ψ_I) that one has $\int \psi_I(x) \times b(x)\, dx = 0$ (the integral is here absolutely convergent). This allows us to exploit the regularity in x and in y of the kernel as it is described by (7.5) and (7.6), and to obtain by "integration by parts" the following estimates for the $\gamma(I, J)$:

$$|\gamma(I, J)| \leq C(|I| \wedge |J|)^{\alpha} \frac{|I|^{1/2}\, |J|^{1/2}}{(|I| + |J| + \operatorname{dist}(I, J))^{1+\alpha}}. \tag{7.10}$$

Conversely, one shows without difficulty that these estimates imply $T \in A_b$ (the exponent that appears in (7.5) and (7.6) being positive and less than that of (7.10)).

The estimates (7.10) are independent of b. As a consequence we obtain the following result, which is new and was established by P. G. Lemarié.

Corollary 2 *The subalgebras $A_b \subset \mathcal{L}(L^2(\mathbb{R}^n), L^2(\mathbb{R}^n))$ are isomorphic by inner conjugation.*

We shall now prove Theorem 4. The proof is an imitation of the construction of the wavelets based on linear splines. But we shall have to replace systematically the scalar product of thc complex space $L^2(\mathbb{R}; dx)$ by

$$B(f,g) = \int_{-\infty}^{\infty} f(x)\, g(x)\, b(x)\, dx,$$

a bicontinuous and symmetric bilinear form. The favorable circumstance is the fundamental inequality

$$\text{Re } B(f, \bar{f}) \geq \int_{-\infty}^{\infty} |f(x)|^2\, dx.$$

Moreover, the proof we will describe uses the full force of the "$T(b)$ theorem" of David-Journé-Semmes. We present a simplified form of it which will suffice for our purposes.

Theorem 5 *For every function $b \in L^\infty(\mathbb{R})$ satisfying $\text{Re } b(x) \geq 1$ and every exponent $\alpha \in]0,1]$, there exists a constant $C(\alpha)$ such that the following property is true.*

If $S(x,y)$ is a continuous function on $\mathbb{R}^n \times \mathbb{R}^n$ satisfying $S(x,y) = \mathcal{O}(|x-y|^{-2})$ when $|y-x| \to +\infty$, and if the properties (7.4) to (7.8) are satisfied (with $C = 1$ in (7.4), (7.5), and (7.6)), then the operator $T\colon L^2(\mathbb{R}) \to L^2(\mathbb{R})$ whose kernel is $K(x,y) = S(x,y)\, b(y)$ satisfies

$$\|T\|_{\mathcal{L}(L^2,L^2)} \leq C(\alpha) \sup_{I \in \mathcal{I}} \|T(\Delta_I)\|_2, \tag{7.11}$$

where $\Delta_I(x) = 2^{j/2}\Delta(2^j x - k)$ and $\Delta(x) = (1-|x|)^+$.

In other words, the cancellation conditions (7.7) and (7.8) alone do not imply the continuity on L^2 of T. It is necessary to add the "weak continuity," which one tests by letting T act on the normalized test functions Δ_I.

The proof of Theorem 4 begins by some very simple remarks concerning symbolic calculus on accretive matrices. We have grouped these remarks in some lemmas whose proofs we recall for the convenience of the reader.

Definition 5 Let H be a Hilbert space over the field of complex numbers and $T\colon H \to H$ a continuous linear operator. We say that T is *accretive* if for every $x \in H$ we have $\text{Re } \langle T(x), x\rangle \geq 0$. We say that T is *δ-accretive* ($\delta > 0$) if we have $\text{Re } \langle T(x), x\rangle \geq \delta\, \|x\|^2$ for all $x \in H$.

Observe that

$$\text{Re } \langle T(x), x\rangle = \frac{1}{2}\langle T(x), x\rangle + \frac{1}{2}\langle x, T(x)\rangle = \left\langle \frac{T+T^*}{2} x, x \right\rangle.$$

Finally T is δ-accretive if and only if the self-adjoint operator $T + T^*$ satisfies, in the sense of the order relation between self-adjoint operators, $T + T^* \geq 2\delta\mathbf{1}$.

We will suppose $\delta > 0$ in all that follows.

Lemma 1 *Assume $\lambda > M^2/2\delta$, where $M \geq \|T\|$. Then we have*

$$\|T - \lambda\mathbf{1}\| < \lambda. \tag{7.12}$$

Indeed, one has

$$\|\lambda - T\| = \sqrt{\|(\lambda - T)(\lambda - T^*)\|} = \sqrt{\|TT^* - \lambda(T + T^*) + \lambda^2\|}.$$

But in the sense of positive self-adjoint operators we have

$$TT^* - \lambda(T + T^*) + \lambda^2 \le (M^2 - 2\lambda\delta + \lambda^2)\mathbf{1};$$

thus

$$\|TT^* - \lambda(T + T^*) + \lambda^2\| \le M^2 - 2\lambda\delta + \lambda^2 < \lambda^2$$

as announced.

Lemma 2 *Let H be a Hilbert space, $(e_j)_{j\in\mathbb{Z}}$ an orthonormal basis of H and $T\colon H \to H$ a continuous linear operator. Assume that T is δ-accretive (for a constant $\delta > 0$) and that $\|T\| \le M$. Assume furthermore, that the matrix of T with respect to this basis satisfies*

$$|m(j,k)| \le C\, e^{-\alpha|j-k|} \tag{7.13}$$

for a certain constant $C > 0$ and a certain exponent $\alpha > 0$. Then the matrix of T^{-1} with respect to this same basis satisfies

$$|m'(j,k)| \le C'\, e^{-\alpha'|j-k|},$$

where $C' > 0$, $\alpha' > 0$ depend only on δ, M, C, α.

Let us first observe that an accretive operator is always an isomorphism. One can see this by remarking that $T = T - \lambda + \lambda = \lambda(\mathbf{1} + \frac{T-\lambda}{\lambda}) = \lambda(\mathbf{1} + R)$, where $\|R\| < 1$ as soon as $\lambda > 0$ is sufficiently large. This allows us to calculate $T^{-1} = \lambda^{-1}(1 - R + R^2 - R^3 + \cdots)$.

We propose to estimate from above the coefficients $r_m(j,k)$ of R^m. On the one hand, we have $|r_m(j,k)| \le \|R\|^m$. On the other,

$$r_m(j,k) = \sum_{j_1} \cdots \sum_{j_m} r_m(j,j_1)\, r_m(j_1,j_2) \cdots r_m(j_m,k),$$

which implies

$$|r_m(j,k)| \le C^{m+1} \sum_{j_1} \cdots \sum_{j_m} e^{-\alpha|j-j_1|}\, e^{-\alpha|j_1-j_2|} \ldots e^{-\alpha|j_m-k|}. \tag{7.14}$$

To estimate the right-hand side we can, for every $\eta > 0$, denote by $H^\infty(\eta)$ the Banach algebra of holomorphic functions $f(z)$ that are 2π-periodic in $\operatorname{Re} z = x$ and bounded in $|\operatorname{Im} z| < \eta$. Then, if $\alpha > \eta$, $\sum_{-\infty}^{\infty} e^{-\alpha|j|}\, e^{ijz} = f(z)$ belongs to $H^\infty(\eta)$. Moreover, the generating series of the right-hand side of (7.14) is none other than f^{m+1}. Since $H^\infty(\eta)$ is a uniform algebra, one has

$$\|f^{m+1}\|_{H^\infty(\eta)} = \|f\|^{m+1}_{H^\infty(\eta)} = (C'(\alpha,\eta))^{m+1}.$$

But it is immediate to observe that, for every function $g \in H^\infty(\eta)$,

$$e^{\eta|j|}\, |\hat{g}(j)| \le \|g\|_{H^\infty(\eta)}.$$

Thus for all $\eta < \alpha$ and $C' = C'(\alpha, \eta, C)$

$$|r_m(j,k)| \le C'^{m+1}\, e^{-\eta|j-k|}.$$

We finish by writing

$$\sum_{m\ge 0} |r_m(j,k)| = \sum_{0\le m\le m_0} C'^{m+1}\, e^{-\eta|j-k|} + \sum_{m>m_0} \|R\|^m.$$

One defines m_0 by the condition that $C'^{m_0} = e^{\eta/2|j-k|}$ and the announced result follows.

Lemma 3 *Under the assumptions of the preceding lemma the entries of the matrix $T^{-1/2}$ satisfy*

$$|m''(i,j)| \le C''\, e^{-\alpha''|i-j|} \tag{7.15}$$

for a certain constant $C'' > 0$ and an exponent $\alpha'' > 0$.

Let us begin by recalling the definition of $T^{-1/2}$. We know that if T is accretive, then T is an isomorphism of H. But if $\zeta \in \mathbb{C}$ and if Re $\zeta > -\delta$, then $T + \zeta\mathbf{1}$ is still accretive. Therefore the spectrum of $T + \zeta\mathbf{1}$ does not contain 0, and it follows that the spectrum $\sigma(T)$ of T is contained in Re $z \ge \delta$. This spectrum is compact.

One argues in the subalgebra $\mathcal{A} \subset \mathcal{L}(H,H)$, which is the closure in the operator norm of the polynomials in T. This algebra $\mathcal{A}$ is a commutative Banach algebra and symbolic calculus yields the definition of $T^{1/2}$ and of $T^{-1/2}$.

We have

$$T^{-1/2} = \frac{1}{\pi}\int_0^\infty (T+\lambda)^{-1}\,\lambda^{-1/2}\,d\lambda = \frac{1}{\pi}\Big(\int_0^{2M} + \int_{2M}^\infty\Big).$$

If $0 \le \lambda \le 2M$, one applies the lemma to $(T+\lambda)^{-1}$ and the estimates are uniform in λ.

If $\lambda \ge 2M$, one write $(T+\lambda)^{-1} = \lambda^{-1}(1+\lambda^{-1}T)^{-1}$. One then applies the proof of Lemma 2 and obtains for the entries of $(1+\lambda^{-1}T)^{-1}$ an exponential decay, uniform in $\lambda \ge 2M$. It remains to integrate against $\lambda^{-3/2}\,d\lambda$, and this integral converges at infinity.

Lemma 4 *Let V be a complex Hilbert space and let $B\colon V \times V \to \mathbb{C}$ be a bicontinuous and symmetric bilinear form. Let $(v_j)_{j\in\mathbb{Z}}$ be an orthonormal basis of V (or more generally a normalized unconditional basis). Assume that for a certain $\delta > 0$ the matrix $((B(v_j, v_k)))_{(j,k)\in\mathbb{Z}\times\mathbb{Z}}$ is δ-accretive. Then there exists a normalized unconditional basis $\tilde v_j$ of V indexed by $\mathbb{Z}$ such that*

$$B(\tilde v_j, \tilde v_k) = 0 \text{ if } j \ne k, \qquad B(\tilde v_j, \tilde v_k) = 1 \text{ if } j = k.$$

If further, for a certain constant $C > 0$ and a certain exponent $\alpha > 0$ one has $|B(v_j, v_k)| \le C\, e^{-\alpha|j-k|}$, then

$$\tilde v_j = \sum_{-\infty}^{\infty} \xi(j,k)\, v_k, \quad \text{where } |\xi(j,k)| \le C'\, e^{-\alpha'|j-k|}$$

($C' > 0$ and $\alpha' > 0$ are two constants depending only on C, α, and the norm of the bilinear form B).

Indeed, let us consider the operator $T\colon V \to V$ defined formally by

$$T(f) = \sum_{-\infty}^{\infty} B(f, v_j)\, v_j.$$

If $(v_j)_{j \in \mathbb{Z}}$ is simply an unconditional basis of V, we change the Hilbert structure of V so that v_j becomes a Hilbert basis. Then T is continuous and δ-accretive. Then, due to the symmetry of B and to the definition of T one has $B(f, Tg) = B(Tf, g)$. It follows that, if $f \in V$, $g \in V$, for every $\lambda \geq 0$ one has

$$B(f, (T + \lambda)^{-1} g) = B((T + \lambda)^{-1} f, g).$$

Then we set $\tilde{v}_j = T^{-1/2}(v_j)$. We know (thanks to the properties of symbolic calculus on the operator T) that $T^{-1/2}\colon H \to H$ is an isomorphism. Then $\tilde{v}_j$, $j \in \mathbb{Z}$, is an unconditional basis of H satisfying $0 < c \leq \|\tilde{v}_j\| \leq C$ for two constants $C \geq c > 0$.

On the other hand,

$$B(\tilde{v}_j, \tilde{v}_k) = B(T^{-1/2} v_j, T^{-1/2} v_k) = B(T^{-1} v_j, v_k).$$

Set $w_j = T^{-1}(v_j)$. Then, due to the definition of T, we have $v_j = T(w_j) = \sum_k B(w_j, v_k)\, v_k$. But the uniqueness of the decomposition with respect to the basis v_k implies $B(w_j, v_k) = 0$ if $j \neq k$, $= 1$ if $j = k$.

In the intended application one obtains easily Re $\sum \sum B(v_j, v_k)\, \alpha_j\, \bar{\alpha}_k \geq \delta \sum |\alpha_j|^2$ from the following remark. We denote by $*$ the involution of V defined by $(\sum \alpha_j v_j)^* = \sum \bar{\alpha}_j v_j$. Then, setting $f = \sum \alpha_j v_j$ one has

$$\text{Re} \sum \sum B(v_j, v_k)\, \alpha_j\, \bar{\alpha}_k = \text{Re } B(f, f^*),$$

and Re $B(f, f^*) \geq \|f\|_2^2$ in the particular case where $V \subset L^2(\mathbb{R})$ and where $B(f,g) = \int_{-\infty}^{\infty} f(x) g(x) b(x)\, dx$.

The second part of the construction needed by Theorem 4 is described in the following lemma.

Lemma 5 *Let H be a complex Hilbert space and let $\{v_j, j \in \mathbb{Z}\} \cup \{w_k, k \in \mathbb{Z}\}$ be an orthonormal basis of H. This means that $H = V \oplus V'$, where V is the closed subspace of H generated by the v_j, $j \in \mathbb{Z}$, and where V' is generated in a similar fashion by the w_k. We denote by $*$ the involution of H defined by $v_j^* = v_j$ and $w_k^* = w_k$.*

Let $B\colon H \times H \to \mathbb{C}$ be a bicontinuous and symmetric bilinear form such that Re $B(f, f^*) \geq \|f\|^2$ *for all $f \in H$.*

One then defines W by

$$W = \{g \in H;\ B(f,g) = 0 \text{ for all } f \in V\}.$$

In these conditions H is the direct sum of V and of W (a sum which in general is no longer orthogonal).

Moreover, there exists a normalized unconditional basis $\tilde{w}_j$, $j \in \mathbb{Z}$, of W such that $B(\tilde{w}_j, \tilde{w}_k) = 0$ if $j \neq k$, $= 1$ if $j = k$. Finally, if the $B(v_j, v_{j'})$, $B(v_j, w_k)$, and $B(w_k, w_{k'})$ have exponential decay in $|j - j'|$, $|j - k|$, and $|k - k'|$ in the sense of Lemma 4, then the *matrices of passing to $\tilde{w}_j$ from the v_j and the w_k also have coefficients with exponential decay.*

The proof of Lemma 5 starts with the construction of a skew projection operator $P\colon H \to V$ whose image is V and whose kernel is W. One then defines P by applying Lemma 4 to V equipped with the orthonormal basis of the v_j, $j \in \mathbb{Z}$. One obtains the $\tilde{v}_j$ and sets

$$P(f) = \sum_{-\infty}^{\infty} B(f, \tilde{v}_j)\, \tilde{v}_j.$$

The properties of P are verified immediately.

One then defines

$$w'_k = w_k - P(w_k), \quad k \in \mathbb{Z}.$$

We shall verify successively that $B(w'_k, w'_\ell)$ is an accretive matrix and that the w'_k form an unconditional basis of W.

One considers $\sum\sum \alpha_k \bar{\alpha}_l\, B(w'_k, w'_\ell) = B(g,h)$, where $g = \sum \alpha_k\, w'_k$ and $h = \sum \bar{\alpha}_k\, w'_k$. But one has $(w'_k)^* = w^*_k - (P(w_k))^* = w_k + r_k$, where $r_k \in V$ (since $V^* = V$). Finally $(w'_k)^* = w'_k + r'_k$, where $r'_k \in V$. Since $g \in W$, one has $B(g, r'_k) = 0$. This implies $B(g,h) = B(g, g^*)$ and therefore Re $B(g,h) \geq \|g\|^2 = \sum |\alpha_k|^2$.

Let us verify now that the w'_k form a normalized unconditional basis of W. On the one hand, if $g \in W$, then $g = \sum \alpha_j v_j + \sum \beta_k w_k$, where $\|g\|^2 = \sum |\alpha_j|^2 + \sum |\beta_k|^2$. Furthermore, one has $g = \sum \beta_k w'_k + \sum \alpha_j v_j + \sum \beta_k P(w_k)$. But the last two terms belong to V, so we have $g = \sum \beta_k w'_k$ with $\sum |\beta_k|^2 \leq \|g\|^2$.

In the opposite direction, let us consider an arbitrary sum $\sum \beta_k w'_k$ and verify that $\|\sum \beta_k w'_k\|$ and $\left(\sum |\beta_k|^2\right)^{1/2}$ are equivalent norms.

This follows quite simply from $w'_k = w_k - P(w_k)$ and from the continuity of P. We denote by C the norm of P and have indeed

$$\left(\sum |\beta_k|^2\right)^{1/2} \leq \left\|\sum \beta_k\, w'_k\right\| \leq (1 + C^2)^{1/2} \left(\sum |\beta_k|^2\right)^{1/2}.$$

The end of the construction of the $\tilde{w}_k$ is done by applying Lemma 4 to the sequence w'_k (which takes over the role of the v_k) and to the Hilbert space W (which takes over the role of V).

Having concluded these preliminaries we can enter into the details of the construction of the wavelets defined by Theorem 4. The reader will observe that this construction can immediately be extended to any dimension.

One starts with the nested sequence V_j of spaces of spline functions that are piecewise affine on the dyadic intervals $[k2^{-j}, (k+1)2^{-j}]$, $k \in \mathbb{Z}$. These spaces V_j are the same as those we have used for the construction of piecewise affine wavelets.

One considers the symmetric and bicontinuous bilinear norm

$$B(f,g) = \int_{-\infty}^{\infty} f(x) g(x) b(x)\, dx, \qquad f \in L^2(\mathbb{R}), g \in L^2(\mathbb{R}).$$

One defines $f^* = \bar{f}$ and one has Re $B(f, f^*) \geq \|f\|^2$. Finally, $W_j \subset V_{j+1}$ is defined by

$$W_j = \{g \in V_{j+1}; B(g, f) = 0 \text{ for all } f \in V_j\}.$$

We denote by $\mathcal{W}_j$ the orthogonal complement of V_j in V_{j+1} for the canonical Hilbert structure of $L^2(\mathbb{R})$.

Then one constructs the wavelets $\psi_{j,k}^{(b)} \in W_j$ needed by Theorem 4 starting from the usual wavelets $\psi_{j,k} \in \mathcal{W}_j$ (relative to $b = 1$) and from the orthonormal basis $\varphi_{j,k}$ of V_j (defined by Theorem 1).

One then applies the procedure of Lemma 5. This way one shows that

$$\psi_{j,k}^{(b)}(x) = 2^{j/2} g_{j,k}(2^j x - k),$$

where $g_{j,k} \in V_1$ and satisfies, for $\alpha > 0$, $C > 0$,

$$|g_{j,k}(x)| \leq C\, e^{-\alpha|x|}.$$

The only thing that is not evident is the fact that this collection of the $\psi_{j,k}^{(b)}$ is an unconditional basis of $L^2(\mathbb{R})$.

In fact, it suffices to prove the existence of a constant C such that for every (double) sequence $\xi(j, k)$ of coefficients one has

$$\left(\sum_{-\infty}^{\infty}\sum_{-\infty}^{\infty} |\xi(j,k)|^2\right)^{1/2} \leq C \left\|\sum_{-\infty}^{\infty}\sum_{-\infty}^{\infty} \xi(j,k)\psi_{j,k}^{(b)}\right\|_2. \tag{7.16}$$

To prove this fundamental inequality one applies the $T(b)$ theorem of David-Journé-Semmes.

This is to say, one starts with a sequence $\lambda(j,k)$ satisfying $|\lambda(j,k)| \leq 1$ (to simplify the writing, we will only consider finite sequences). One then defines the operator T by $T(\psi_{j,k}^{(b)}) = \lambda(j,k)\,\psi_{j,k}^{(b)}$ and one wishes to prove the continuity of T.

For this purpose one writes the kernel $K(x,y) = S(x,y)\,b(y)$ of T, where

$$S(x,y) = \sum\sum \lambda(j,k)\psi_{j,k}^{(b)}(x)\psi_{j,k}^{(b)}(y).$$

It is then immediate to verify that $S(x,y)$ satisfies the conditions (1.18), (1.19), and (1.20) of the definition of Calderón-Zygmund operators.

One has

$$\int \psi_{j,k}^{(b)}(x) b(x)\, dx = 0.$$

Indeed, one has $B(\psi_{j,k}^{(b)}, f) = 0$ if $f \in V_j$. It suffices now to choose $f(x) = (1 - 2^{-m}|x|)^+$ and to let m tend to infinity. The dominated convergence theorem applies since $\psi_{j,k}^{(b)} \in L^1(\mathbb{R})$.

Then we use Theorem 5, which leads us to compute $T(\Delta_I)$ when $I = [\ell 2^{-q}, (\ell+1)2^{-q}[$ and $\Delta_I = 2^{q/2}\Delta(2^q x - \ell)$, $\Delta(x) = (1 - |x|)^+$. To perform this computation we observe that Δ_q belongs to V_q and that $B(\psi_{j,k}^{(b)}, \Delta_I) = 0$ if $j \geq q$. This implies $T(\Delta_I) = T_q(\Delta_I)$ where T_q is the "truncated operator" defined by the sequence $\lambda'(j,k) = \lambda(j,k)$ if $j < q$ and $\lambda'(j,k) = 0$ otherwise. The estimates concerning the kernel $K_q(x,y)$ of T_q are then immediate, and it follows that $|K_q(x,y)| \le C2^q(1+2^q|x-y|)^{-1}$. This estimate suffices to ensure $\|T_q(\Delta_I)\|_2 \le C$.

We have just proved that the operators T defined by $T(\psi^{(b)}_{j,k}) = \lambda(j,k)\psi^{(b)}_{j,k}$ are uniformly continuous on $L^2(\mathbb{R})$ under the lone condition that $|\lambda(j,k)| \le 1$. It follows that

$$\left\| \sum\sum \lambda(j,k)\,\xi(j,k)\,\psi^{(b)}_{j,k} \right\|_2 \le C \left\| \sum\sum \xi(j,k)\,\psi^{(b)}_{j,k} \right\|_2 . \tag{7.17}$$

Having arrived at this point, one takes the mean of the inequalities (7.17) with respect to all the (finite but arbitrarily long) sequences of ± 1 and one gets

$$\left(\sum\sum |\xi(j,k)|^2 \right)^{1/2} \le C' \left\| \sum\sum \xi(j,k)\,\psi^{(b)}_{j,k} \right\|_2 .$$

To prove the reverse inequality it suffices to take into account the orthogonality relations (7.3).

The last point to check is that the sequence of the $\psi^{(b)}_{j,k}$ is total in $L^2(\mathbb{R})$. But due to the construction of these functions all we have to check is that the (algebraic) direct sum of the subspaces W_j, $j \in \mathbb{Z}$, is dense in $L^2(\mathbb{R})$. Let us denote by E_j the operator defined by

$$E_j(f) = \sum_k B(f, \tilde{v}_{j,k})\,\tilde{v}_{j,k}$$

where $\tilde{v}_{j,k}$ is the unconditional basis of V_j given by Lemma 4. Then $E_j\colon L^2(\mathbb{R}) \to V_j$ is a (skew) projection operator whose kernel N_j is defined by $B(f,g) = 0$ for all $g \in V_j$. This implies that

$$E_{j_1}E_{j_2} = E_{j_2}E_{j_1} = E_{j_1} \quad \text{if } j_1 \le j_2.$$

Finally the image of $E_{j+1} - E_j$ is precisely W_j and, if $f \in V_N$ for a certain integer N, then one writes

$$f = E_N(f) = (E_N - E_{N-1})f + (E_{N-1} - E_{N-2})f + \cdots + (E_{-M+1} - E_{-M})f + E_{-M}(f).$$

But the estimates concerning the kernel of E_{-M} which follow from Lemma 4 yield immediately $\|E_{-M}(f)\|_2 \to 0$ when f is fixed and M tends to infinity.

Thus f can be approximated as closely as we want by functions that belong to the algebraic sum of the W_j.

Since the union of the V_n is dense in $L^2(\mathbb{R})$, the proof of Theorem 4 is now complete.

Ph. Tchamitchian discovered an elementary proof of Theorem 4 which does not use the $T(b)$ theorem but is based on the $T(1)$ theorem.

References

[1] G. Battle, *A block spin construction of ondelettes. I, Lemarié functions*, Comm. Math. Phys. **110** (1987), 601–615.

[2] G. A. Battle III and P. Federbush, *A phase cell cluster expansion for a hierarchical φ^4_3 model*, Comm. Math. Phys. **88** (1983), 263–293.

[3] L. Carleson, *An interpolation problem for bounded analytic functions*, Amer. J. Math. **80** (1958), 921–930.

[4] I. Daubechies, A. Grossmann, and Y. Meyer, *Painless nonorthogonal expansions*, J. Math. Phys. **27** (1986), 1271–1283.

[5] G. David and J. L. Journé, *A boundedness criterion for generalized Calderón-Zygmund operators*, Ann. Math. (2) **120** (1984), 371–397.

[6] G. David, J. L. Journé, and S. Semmes, *Opérateurs de Calderón-Zygmund, fonctions para-accrétives et interpolation*, Rev. Mat. Iberoamericana **1** (1985), 1–56.

[7] S. Jaffard, Centre de Mathématiques Appliquées, Ecole Polytechnique, oral communication.

[8] P. G. Lemarié, *Continuité sur les espaces de Besov des opérateurs définis par des intégrales singulières*, Ann. Inst. Fourier (Grenoble) **35** (1985), 175–187.

[9] P. G. Lemarié, *Thèse de troisième cycle* (Orsay Bât. 425, June 1984).

[10] P. G. Lemarié, *Ondelettes à localisation exponentielle*, J. Math. Pures Appl. **67** (1988), 227–236.

[11] P. G. Lemarié and Y. Meyer, *Ondelettes et bases hilbertiennes*, Rev. Mat. Iberoamericana **2** (1986), 1–18.

[12] S. Mallat, *A theory for multiresolution signal decomposition: the wavelet representation*, IEEE Trans. Pattern Anal. Machine Intell. **11** (1989) 674–693.

[13] M. Meyer, *Thèse de troisième cycle* (Orsay Bât. 425, June 1985).

[14] Y. Meyer, *Real analysis and operator theory*, Pseudodifferential Operators and Applications (Notre Dame, Ind., 1984), Proc. Sympos. Pure Math. **43**, Amer. Math. Soc., Providence, 1985, 219–235.

[15] Y. Meyer, *Wavelets and Operators*, Cambridge Studies in Advanced Mathematics **37**, Cambridge University Press, Cambridge, 1992.

[16] R. O'Neil, *Spaces formed with special atoms*, Proceedings of the Seminar on Harmonic Analysis (Pisa, 1980), Rend. Circ. Mat. Palermo (2), 1981, 139–144; G. S. De Souza, *Spaces formed by special atoms. I*, Rocky Mountain J. Math. **14** (1984), 423–431.

Multiscale Analyses and Wavelet Bases

Karlheinz Gröchenig
Translated by Robert D. Ryan

ABSTRACT. We extend to the n-dimensional case the algorithm for constructing wavelets that is described in [7] for one dimension. These wavelets will have the same smoothness and decay properties as the function that is used to construct the multiscale analysis.

1. Introduction

A wavelet basis is an orthogonal basis for $L^2(\mathbb{R}^n)$ of the form $2^{nj/2}\psi_\varepsilon(2^j x - k)$, where $j \in \mathbb{Z}$, $k \in \mathbb{Z}^n$, where ε ranges over a finite set E, and where ψ_ε enjoys good smoothness properties and good decay properties at infinity. We often require, for an integer $r \geq 2$, that

$$\sup_{x\in\mathbb{R}^n} (1 + |x|)^{n+r} |\partial^\alpha \psi_\varepsilon(x)| \leq C \tag{1}$$

for all multi-indices α of length $|\alpha| \leq r$.

Wavelet bases are used for signal and image processing [6] and are unconditional bases for many of the classic function spaces [1], [2].

At this time, we do not know how to construct all of the wavelet bases of $L^2(\mathbb{R}^n)$. The several examples that we do know are associated with certain regular multiscale analyses (see below). In this Note, we will show that every regular multiscale analysis of $L^2(\mathbb{R}^n)$ yields a wavelet basis for $L^2(\mathbb{R}^n)$. This result was already known for $n = 1$, but the proof that is given in [7] does not extend to the general case.

2. Definitions and Statements of Results

A multiscale analysis of $L^2(\mathbb{R}^n)$ is an increasing sequence V_j, $j \in \mathbb{Z}$, of closed subspaces of $L^2(\mathbb{R}^n)$ having the following properties.

(2.1) $\bigcap_{j\in\mathbb{Z}} V_j = \{0\}$, and $\bigcup_{j\in\mathbb{Z}} V_j$ is dense in $L^2(\mathbb{R}^n)$.

(2.2) If we denote by $D_t : L^2(\mathbb{R}^n) \to L^2(\mathbb{R}^n)$ the normalized dilation operator (that is, $D_t f(x) = t^{-n/2} f(x/t)$, $t > 0$), then $V_{j+1} = D_{1/2} V_j$.

(2.3) Denote by l the regular representation of $\mathbb{Z}^n$ (acting by translation on $l^2(\mathbb{Z}^n)$) and by L the regular representation of $\mathbb{R}^n$ (acting by translation on $L^2(\mathbb{R}^n)$). Then we require

Translator's note: The references have been updated.

that V_0 is invariant under the operators L_k, $k \in \mathbb{Z}^n$, and that the representation $k \to L_k$, $k \in \mathbb{Z}^n$, restricted to V_0 is equivalent to the regular representation l of $\mathbb{Z}^n$.

It is easy to verify that, if these two operators are equivalent, then they are unitarily equivalent.

We denote by $T : V_0 \to l^2(\mathbb{Z}^n)$ the intertwining operator between the two equivalent representations and by $\varphi \in V_0$ the inverse image of $\boldsymbol{e}_0$ under T, where $\boldsymbol{e}_0$ is the sequence in $l^2(\mathbb{Z}^n)$ whose value is 1 at 0 and 0 elsewhere.

The relation $TL_k = l_k T$, $k \in \mathbb{Z}^n$, implies that $L_k\varphi$, $k \in \mathbb{Z}^n$, is an orthonormal basis for V_0. A multiscale analysis is called r-regular ($r \geq 2$) if φ satisfies

$$\sup_{x\in\mathbb{R}^n} (1 + |x|)^{n+r} |\partial^\alpha \varphi(x)| \leq C \tag{2}$$

for all multi-indices α of length $|\alpha| \leq r$. The regularity of infinite order corresponds to functions of the Schwartz class.

An r-regular wavelet basis is an orthonormal basis for $L^2(\mathbb{R}^n)$ composed of functions $D_{2^j} L_k \psi_\varepsilon$, $j \in \mathbb{Z}$, $k \in \mathbb{Z}^n$, $\varepsilon \in E$, where E is a finite set with cardinality $2^n - 1$ and where the ψ_ε are r-regular.

The following theorem and its corollary relate these two concepts.

Theorem *Let V_j, $j \in \mathbb{Z}$, be an r-regular multiscale analysis. Then there exist functions ψ_ε, $\varepsilon \in E$, belonging to V_1 such that*

(a) *the $L_k\psi_\varepsilon$, $k \in \mathbb{Z}^n$, $\varepsilon \in E$, form an orthonormal basis for the orthogonal complement W_0 of V_0 in V_1;*

(b) *the ψ_ε, $\varepsilon \in E$, are r-regular.*

Denoting by W_j the orthogonal complement of V_j in V_{j+1}, we note that $L^2(\mathbb{R}^n)$ is the direct Hilbert sum of the W_j and that $W_j = D_{2^{-j}} W_0$. We thus have:

Corollary *$D_{2^j} L_k \psi_\varepsilon$, $j \in \mathbb{Z}$, $k \in \mathbb{Z}^n$, $\varepsilon \in E$, is an r-regular wavelet basis for $L^2(\mathbb{R}^n)$.*

3. Sketch of the Proof

We transform the problem using five steps, denoted by (a), (b), (c), (d), and (e).

(a) Since $D_{1/2} L_k \varphi$, $k \in \mathbb{Z}^n$, is an orthonormal basis for $V_1 \supset V_0$, we can expand φ and the ψ_ε in this basis and obtain the sequences of coefficients $\boldsymbol{a}$ and $\boldsymbol{b}_\varepsilon$ in $l^2(\mathbb{Z}^n)$, $\varepsilon \in E$, such that

$$\varphi = \sum_{k\in\mathbb{Z}^n} a(k) D_{1/2} L_k, \qquad \psi_\varepsilon = \sum_{k\in\mathbb{Z}^n} b_\varepsilon(k) D_{1/2} L_k. \tag{3}$$

The problem is to construct the sequences $\boldsymbol{b}_\varepsilon$ so that

$$\langle l_{2k} \boldsymbol{b}_\varepsilon, \boldsymbol{b}'_\varepsilon \rangle = \delta_{k,0}\delta_{\varepsilon,\varepsilon'}, \qquad k \in \mathbb{Z}^n, \quad \varepsilon, \varepsilon' \in E, \tag{4}$$

and

$$\langle l_{2k} \boldsymbol{a}, \boldsymbol{b}_\varepsilon \rangle = 0, \qquad k \in \mathbb{Z}^n, \quad \varepsilon \in E. \tag{5}$$

Furthermore, we know that

$$\langle l_{2k}\boldsymbol{a}, \boldsymbol{a}\rangle = \delta_{k,0}, \qquad k \in \mathbb{Z}^n. \tag{6}$$

In these three relations, $\delta_{u,v}$ is the Kronecker symbol.

(b) If $\boldsymbol{a}$ were at our disposal (instead of being imposed by the multiscale analysis), we could take as the set E the sequences $\varepsilon = (\varepsilon_1, \ldots, \varepsilon_n)$ such that $\varepsilon_j \in \{0, 1\}$ and $\varepsilon \neq (0, \ldots, 0)$. We write $\overline{E} = E \cup \{0\}$. We could take for $\boldsymbol{a}$ the vector $\boldsymbol{e}_0$ and for $\boldsymbol{b}_\varepsilon$ the vectors $\boldsymbol{e}_\varepsilon$, $\varepsilon \in E$.

We are going to resolve the real problem by constructing a *unitary* operator $\tilde{U}$ on $l^2(\mathbb{Z}^n)$ that commutes with all of the operators l_{2k}, $k \in \mathbb{Z}^n$, and that maps $\boldsymbol{e}_0$ to $\boldsymbol{a}$. We then define $\boldsymbol{b}_\varepsilon = \tilde{U}\boldsymbol{e}_\varepsilon$.

(c) We now consider the representation $k \to l_{2k}$ of $\mathbb{Z}^n$ on $l^2(\mathbb{Z}^n)$, and we decompose it in more detail. The subspace $\mathcal{H}_\varepsilon \subseteq l^2(\mathbb{Z}^n)$ generated by the $l_{2k}\boldsymbol{e}_\varepsilon$, $k \in \mathbb{Z}^n$, is invariant under the l_{2k}, and the restriction of $k \to l_{2k}$ to $\mathcal{H}_\varepsilon$ is equivalent to the regular representation of $\mathbb{Z}^n$ (on $l^2(\mathbb{Z}^n)$) for each $\varepsilon \in \overline{E}$.

By using another intertwining operator, we identify $l^2(\mathbb{Z}^n)$ with $\bigoplus_{\varepsilon\in\overline{E}} l^2(\mathbb{Z}^n)$, the direct sum of 2^n copies of $l^2(\mathbb{Z}^n)$, and $k \to l_{2k}$ is then identified with a diagonal matrix $\left(\begin{smallmatrix} l_k & 0 \\ 0 & l_k \end{smallmatrix}\right)$; the $\boldsymbol{e}_\varepsilon$, $\boldsymbol{a}$, $\boldsymbol{b}_\varepsilon$ finally become the column vectors $(\boldsymbol{e}_0\delta_{\varepsilon,\varepsilon'})_{\varepsilon'\in\overline{E}}$, $(\boldsymbol{a}_\varepsilon)_{\varepsilon\in\overline{E}}$, $(\boldsymbol{b}_{\varepsilon,\varepsilon'})_{\varepsilon'\in\overline{E}}$ whose components $\boldsymbol{a}_\varepsilon$, $\boldsymbol{b}_{\varepsilon,\varepsilon'}$ belong to $l^2(\mathbb{Z}^n)$.

The operator $\tilde{U}$ that is sought becomes $U = (_{\varepsilon,\varepsilon'})_{(\varepsilon,\varepsilon')\in\overline{E}\times\overline{E}}$, where

(3.1) the $U_{\varepsilon,\varepsilon'} : l^2(\mathbb{Z}^n) \to l^2(\mathbb{Z}^n)$ commute with all the l_k, $k \in \mathbb{Z}^n$;
(3.2) $U_{\varepsilon,0}\boldsymbol{e}_0 = \boldsymbol{a}_\varepsilon$ for all $\varepsilon \in \overline{E}$;
(3.3) $\sum_{\varepsilon\in\overline{E}} U^*_{\varepsilon,\varepsilon'} U_{\varepsilon,\varepsilon''} = \delta_{\varepsilon',\varepsilon''}1$.

Conditions (3.1) and (3.2) determine the first column of the matrix U, and we must expand this column into a unitary matrix of operators.

(d) Thanks to (3.1), the $U_{\varepsilon,\varepsilon'}$ translate (by the Fourier transform) into multiplication operators by functions $m_{\varepsilon,\varepsilon'}(\xi)$ acting on $L^2(\mathbb{T}^n)$, where $\mathbb{T}^n$ is the n-dimensional torus. Conditions (3.1), (3.2), and (3.3) become $\sum_{\varepsilon\in\overline{E}} |m_{\varepsilon,0}(\xi)|^2 = 1$, $M(\xi) = (m_{\varepsilon,\varepsilon'}(\xi))$ is unitary, and the Fourier coefficients of $m_{\varepsilon,\varepsilon'}$ are precisely the $b_{\varepsilon,\varepsilon'}(k)$, $k \in \mathbb{Z}^n$.

Consequently, it is necessary to complete a column vector, as in [7], into a unitary matrix. In [7], however, the coefficients of this unitary matrix were related by implicit relations, relations that have disappeared in our approach.

(e) The last step is made thanks to the following classic lemma.

Lemma *Let $q \geq 2$ be an integer and let K be a compact subset of the unit sphere $|z_1|^2 + \cdots + |z_q|^2 = 1$. Suppose that K is not equal to this unit sphere. Then there exists a matrix $M(z)$, $z \in K$, which is unitary and real analytic in a neighborhood of K and whose first column is exactly $z = (z_1, \ldots, z_q)$.*

In our case, $q = 2^n$, and $\xi \to (m_{\varepsilon,0}(\xi))_{\varepsilon\in\overline{E}}$ is a function that is at least of class C^1. The image of $\mathbb{T}^n$ under this mapping is thus a compact subset K with zero measure, since the dimension of the unit sphere is $q - 1 = 2^n - 1 > n$, if $n \geq 2$. If $n = 1$, there is nothing to prove [7]. The lemma allows us to construct the unitary matrix $m_{\varepsilon,\varepsilon'}(\xi)$, and then the theorem of P. Lévy is applicable to the Banach algebra A_r of continuous functions on $\mathbb{T}^n$ whose Fourier coefficients are $O(|k|^{-n-k})$. We conclude that the $b_{\varepsilon,\varepsilon'} \in l^2(\mathbb{Z}^n)$ have, in fact, the same decay properties as the a_ε and that, finally, the ψ_ε, $\varepsilon \in E$, have the same regularity and decay as φ.

References

[1] H. G. Feichtinger and K. Gröchenig, *A unified approach to atomic decompositions via integrable group representations*, Function spaces and applications (Lund, 1986), Lecture Notes in Math. **1302**, Springer, Berlin, 1988, 52–73.

[2] ———, *Banach spaces related to integrable group representations and their atomic decompositions I*, J. Funct. Anal. **86** (1989), 307–340.

[3] S. Jaffard, P.-G. Lemarié, S. Mallat, and Y. Meyer, *Multi-scale analysis* (unpublished technical report).

[4] P.-G. Lemarié and Y. Meyer, *Ondelettes et bases hilbertiennes*, Rev. Mat. Iberoamericana **2** (1986), 1–18.

[5] P. Lévy, *Sur la convergence absolue des séries de Fourier*, Compositio Math. **1** (1934), 1–14.

[6] S. G. Mallat, *A theory for multiresolution signal decomposition: the wavelet representation*, IEEE Trans. Pattern Anal. Machine Intell. **11** (1989), 674–693.

[7] Y. Meyer, *Ondelcttes, fonctions splines at analyses graduées*, Rend. Sem. Mat. Univ. Politec. Torino **45** (1987), 1–42.

[8] Y. Meyer, *Wavelets and Operators*, Cambridge University Press, Cambridge, 1992.

IEEE TRANSACTIONS ON INFORMATION THEORY, VOL. 38, NO. 2, MARCH 1992

Nonseparable Multidimensional Perfect Reconstruction Filter Banks and Wavelet Bases for $\mathcal{R}^n$

Jelena Kovačević, *Member, IEEE,* and Martin Vetterli, *Senior Member, IEEE*

Abstract—**Although filter banks have been in use for more than a decade, only recently have some results emerged, setting up the theory of general, nonseparable multidimensional filter banks. At the same time, wavelet theory emerged as a useful tool in many different fields of pure and applied mathematics as well as in signal analysis. Recently, it has been shown that the two theories are closely related. Not only does the filter bank perform a discrete wavelet transform, but also under certain conditions it can be used to construct continuous bases of compactly supported wavelets. For multidimensional filter banks, using arbitrary sampling lattices, conditions for perfect reconstruction are given. The orthogonal case is analyzed indicating orthogonality relations between the filters in the bank and their shifts on the sampling lattice. A linear phase condition follows, as a tool for testing or building banks containing linear phase (symmetric) filters. It is shown how, in some cases, nonseparable filters can be implemented in a separable fashion. The two-channel case in multiple dimensions is studied in detail: the form of a general orthogonal solution is given and possible linear phase solutions are presented, showing that orthogonality and symmetry are exclusive, independent of the number of dimensions (assuming real FIR filters). Attractive cascade structures with specific properties (orthogonality and linear phase) are proposed. For the four-channel two-dimensional case, filters being orthogonal and symmetric are obtained, a solution that is impossible using separable filters. We also discuss methods for obtaining multidimensional filters from their one-dimensional counterparts. Next, we make a connection to nonseparable wavelets through the construction of iterated filter banks. Assuming the L^2 convergence of the scaling function, we show that as in the one-dimensional case, the scaling function satisfies a two-scale equation, and the wavelets are orthogonal to each other and their scales and translates (as well as to the scaling function). Then, for the scaling function to exist, we show that it is necessary that the low-pass filter have a zero at aliasing frequencies. Following the discussion on the choice of the dilation matrix, an interesting "dragon" is constructed for the hexagonal case. For the two-channel case in multiple dimensions it is shown that the wavelets defined previously indeed constitute a basis for $L^2(\mathcal{R}^n)$ functions. Following the result on necessity of a zero, we conjecture that the low-pass filter can be made regular by putting a zero of sufficiently high order at aliasing frequencies. Based on this, a small orthonormal low-pass filter is designed for which we conjecture that it would lead to a continuous scaling function, and thus, wavelet basis. A biorthogonal example is also given.**

Manuscript received February 18, 1991; revised July 15, 1991. This work was supported by the National Science Foundation under Grants ECD-88-11111 and MIP-90-14189.

J. Kovačević is with AT&T Bell Laboratories, Murray Hill, NJ 07974.

M. Vetterli is with the Department of Electrical Engineering and the Center for Telecommunications Research, Columbia University, New York, NY 10027.

IEEE Log Number 9105035.

Index Terms—**Multidimensional, nonseparable filter banks, wavelets, multidimensional wavelets, filter banks.**

I. Introduction

SINCE the introduction of digital multirate filter banks for the compression of speech signals 15 years ago [8], they have been widely used mainly for subband coding of speech, still images, and video [2], [37], [43], [47]. The underlying theory progressed from cancellation of aliasing (or repeated spectra), to building systems achieving exact reconstruction of the signal, and from two-channel orthogonal banks [26], [30], to general multichannel systems [31], [33], [38], [39], [44]. For implementational reasons all of these efforts concentrated on filters having rational transfer functions.

Independent of this work, the theory of wavelets was developed in applied mathematics [9], [17], [23], [25]. With the work of Daubechies [9], Mallat [23], and Meyer [25], it became clear that filter banks and wavelets were closely related. Filter banks compute the equivalent of a discrete wavelet transform, and under certain conditions (*regularity* of the low-pass filter), they can be used to derive continuous bases of wavelets [9].

To explain briefly how a subband system works, refer to Fig. 1, where a general multidimensional N-channel system is shown. The input signal is fed through N branches, each one containing a bandpass filter, and subsequently subsampled by $N = \det \boldsymbol{D}$ to its new multidimensional Nyquist frequency ($\boldsymbol{D}$ is a sublattice of the input lattice and has a sampling density that is N times smaller). Then each channel signal (subband) is encoded, transmitted, and decoded. To resynthesize the original signal, one has to upsample all the subbands back to the original lattice and pass them through a set of bandpass synthesis filters. Note that the basic blocks in the system perform *filtering*, *sampling rate change*, and *coding/transmission/decoding*. In what follows, we will be concerned only with the former two parts and we will assume that coding/transmission/decoding is performed in a lossless fashion. Another assumption will be that the sampling density and the number of channels are the same so as to preserve the same number of samples through the various steps of the system. This will be referred to as a *critically sampled* filter bank.

Consider the sampling part of the system. While in one dimension sampling by N can be performed in only one way,

0018-9448/92$03.00 © 1992 IEEE

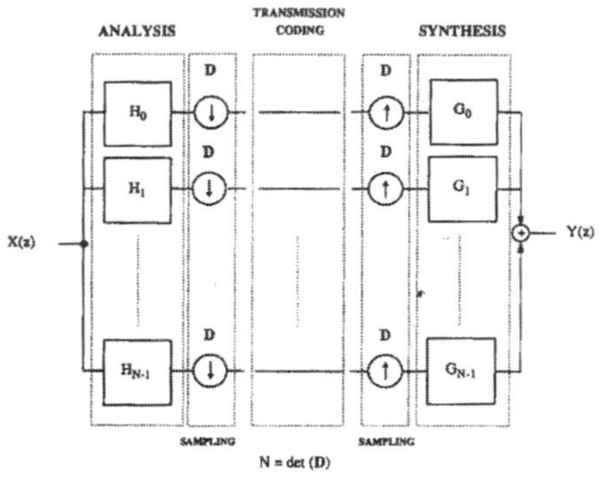

Fig. 1. An analysis/synthesis n-dimensional filter bank.

in two or more dimensions this is not true any more. Multidimensional sampling is represented by a lattice which can be separable or nonseparable. In most of the previous work on two- or three-dimensional multirate processing the sampling rate changes which are used are separable and can be performed along one dimension at a time. However, when dealing with multidimensional signals, true multidimensional processing is more appropriate. Recently, some results have emerged where nonseparable sampling is used, mostly for two-dimensional systems (see [2], [20], [21], [22], [37], [45], [46]). In [48], nonseparable filters were used on a hexagonally sampled input signal, which was then separably subsampled, leading to an interesting directional analysis of images. Aliasing was cancelled, and perfect reconstruction was well approximated.

As for the filtering part, there are a number of questions of interest that have to be addressed. They include design constraints, such as orthogonality, linear phase (symmetry), and regularity. At the same time, the filters themselves can be separable or nonseparable regardless of the sampling lattice. Obviously, while separable filters offer the advantage of low-complexity processing, their nonseparable counterparts have more degrees of freedom and hence allow better designs. In multirate filtering, a nonseparable filter can sometimes still be implemented in a separable fashion, and this will be explored as well. In what follows, unless stated otherwise, we will assume real finite impulse response (FIR), or compactly supported, filters.

If we consider the filter bank system as a whole, two issues of interest arise, namely alias-free reconstruction and perfect reconstruction. The former means that, from input to output, the system can be regarded as a shift-invariant filter, while the latter requires the reconstructed signal to be equal to the input signal (possibly within a delay and a scale factor).

Finally, we would like to use perfect reconstruction filter banks in order to derive wavelet bases. The key construction is the iteration of the filter bank along its low-pass branch. If the low-pass filter is regular (its iterated version converges to a well-defined, possibly smooth function), this construction leads to a wavelet basis, as was first shown in the one-dimensional case and subsampling by 2 by Daubechies [9]. This was also studied under the framework of multiresolution analysis in [23], [25]. In one dimension, subsampling by N in an N-branch filter bank leads, after iteration, to a scaling function $\phi(x)$ satisfying a two-scale equation of the form

$$\phi(x) = \sum_n c_n \phi(Nx - n), \tag{1}$$

as well as to $(N-1)$ wavelets which are also linear combinations of $\phi(Nx - n)$ [41]. The most studied case has been for $N = 2$.

In multiple dimensions the situation is more complicated. From a discrete filtering point of view, subsampling is defined by a sublattice of the original lattice (which we can assume, without loss of generality, to be $\mathscr{Z}^n$). The sublattice is represented by a dilation matrix $\boldsymbol{D}$ (the equivalent of the subsampling, or dilation factor, N in the one-dimensional case). The indexes of points belonging to the sublattice are given as weighted integer combinations of the columns of $\boldsymbol{D}$. For example, the following matrices are possible representations of the so-called two-dimensional quincunx sublattice [2], [7], [15], [37], [43]

$$\boldsymbol{D}_1 = \begin{pmatrix} 1 & 1 \\ 1 & -1 \end{pmatrix}, \quad \boldsymbol{D}_2 = \begin{pmatrix} 1 & -1 \\ 1 & 1 \end{pmatrix},$$
$$\boldsymbol{D}_3 = \begin{pmatrix} 2 & 1 \\ 0 & 1 \end{pmatrix}. \tag{2}$$

Its sampling lattice is given in Fig. 2 showing that one out of two points is retained (this is the only nonseparable sublattice with $|\det \boldsymbol{D}| = 2$). Thus, using this lattice would result in a two-dimensional nonseparable two-channel case.

Now, when iterating the filter bank, we iterate the subsampling by $\boldsymbol{D}$, that is, the overall subsampling corresponds to an integer power of $\boldsymbol{D}$. This can be very different for different matrices $\boldsymbol{D}$ (e.g., $\boldsymbol{D}_1^2 = 2\boldsymbol{I}$ while $\boldsymbol{D}_3^n$ will never be separable).

The scaling function derived from the iterated filter bank (if it exists) will also obey a two-scale equation (1) which now depends on $\boldsymbol{D}$. One necessary requirement on the matrix $\boldsymbol{D}$ is that it be a dilation in all dimensions (otherwise, an associated wavelet analysis would not increase resolution in all dimensions [7]). This is equivalent to requiring that all eigenvalues of $\boldsymbol{D}$ should have magnitude strictly greater than 1, and so, for example, $\boldsymbol{D}_3$ is not a valid dilation matrix (it dilates by 2 in the $[x, 0]$ direction, but does not dilate at all in the $[x, -x]$ direction).

The vastly different behavior of iterated filter banks depending on the matrix $\boldsymbol{D}$ was most strikingly demonstrated by Gröchenig and Madych [16], as well as Lawton and Resnikoff [49]. In [16], the authors showed iterations of very simple filters (essentially Haar filters) that produce fractal, self-similar tilings of the space for certain matrices, while giving simple geometric shapes for others.

This interplay of the lattice and the associated dilation matrix, added to the fact that multidimensional filters are hard to design because of the absence of factorization theorems, makes the construction of multidimensional nonseparable regular wavelets much more difficult than in one

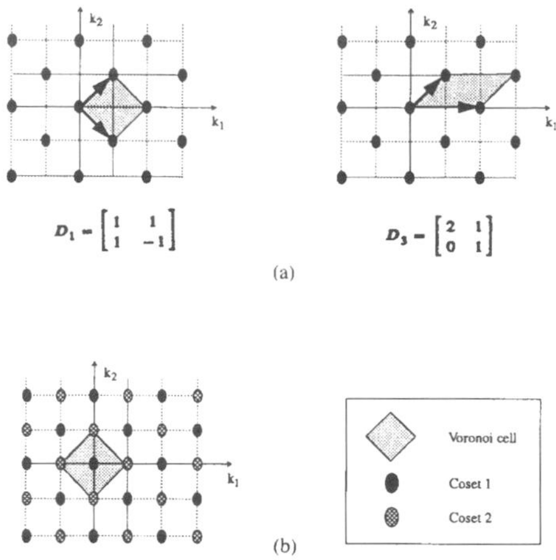

Fig. 2. (a) Quincunx lattice generated using two different matrices. Shaded regions represent fundamental parallelepipeds for each case. (b) Quincunx lattice with its Voronoi cell and cosets.

dimension. For example, a particular filter could be regular with respect to D_1, but not with respect to D_2 (see [7] for examples).

Thus, we see that while discrete signal processing is only concerned with the lattice, wavelet construction is concerned with the particular dilation matrix used to represent the lattice.

The present paper is concerned both with perfect reconstruction filter banks in multiple dimensions and with wavelets that can be obtained when iterating such filter banks. Therefore, we will have to develop the background for both subjects. While this leads to a few parts in the paper which are of a review or tutorial nature, it adds the benefit of a self-contained presentation. We will use both z-transform and Fourier domain notation, depending on which one is more appropriate. For example, synthesis of filter banks uses z-transforms, while discussion on wavelets is done in Fourier domain. While it is easy to go from one to the other, we sometimes indicate both next to each other.

Let us indicate below the principal results of the paper, along with an outline. Section II presents sampling on lattices, as well as sampling rate changes (the fundamental operations in multirate filter banks). Section III describes multidimensional multirate filter banks. There, we prove conditions for shift-invariance (so-called aliasing cancellation) and perfect reconstruction, in particular for the FIR case. We consider also orthogonal and linear phase (symmetric/antisymmetric filters) filter banks and their associated properties. Finally, separability of filters and of their polyphase components (the impulse responses on cosets of a lattice) is considered, showing possible alternatives. Section IV presents an extensive treatment of the two-channel case in multiple dimensions. This thorough treatment is done because it is the most natural generalization of the one-dimensional case. The form of a general unitary polyphase matrix is given, as well as an analysis of possible linear phase solutions. We show that, as in one dimension, orthogonality and linear phase (symmetry) are exclusive for the real compact support (FIR) case. Section V addresses the synthesis of filter banks and proposes attractive cascade structures for the two-channel multidimensional case satisfying particular design constraints (such as orthogonality or linear phase). A general construction for small size linear phase filters is given for an arbitrary number of dimensions. For the four-channel two-dimensional case we show how to obtain cascades leading to filter banks that are orthogonal and have linear phase [22], a solution which is impossible using separable filters. The transformation of one-dimensional filter banks into multidimensional ones is also described. In particular, the McClellan transformation, which can be used for biorthogonal filter banks, is discussed. Section VI develops the connection to nonseparable wavelet bases. First, the necessity of zeros at aliasing (or repeat) frequencies in the filters that are iterated is shown. Then, assuming that the iterated low-pass filter converges to a limit function in L^2, it is shown that all properties of one-dimensional wavelet bases generalize to higher dimensions, such as two-scale equations, orthogonality, and the fact that the wavelets form a basis for $L^2(\mathcal{R}^n)$ functions. Section VII shows some designs of orthogonal and biorthogonal wavelet bases, with numerical experiments indicating the conjectured regularity of the wavelet. Note that we do not address the regularity issue in any depth, and refer to the recent manuscript by Cohen and Daubechies [7] for a thorough treatment of the subject. For the sake of conciseness of the main text, most of the proofs are left to the Appendices, together with the definitions and notations.

Throughout the paper we will be dealing with a general multidimensional case. But at the end of each section the quincunx case will be used as an example to summarize all the important results of that section. Thus, the reader might go through this case study (Sections II-A, III-E, IV-C, V-C, VI-E, as well as Section VII on design) and get an overall idea of the basic concepts and results.

II. Sampling in Multiple Dimensions

In this section, some concepts and notions from the theory of lattices [5], [13] that will be used in the remainder of the paper are reviewed. Consider an analysis/synthesis filter bank as shown in Fig. 1. As pointed out in the Introduction, the two basic operations performed are *filtering* and *sampling*. The sampling process in n dimensions can be represented by a *lattice* defined as the set of all linear combinations of n basis vectors $a_1, a_2, \cdots, a_n$ with integer coefficients [5], [13], i.e., a lattice is the set of all vectors generated by Dk, $k \in \mathcal{Z}^n$, where D is the matrix characterizing the sampling process (its columns are the basis vectors $a_1, a_2, \cdots, a_n$). Since the elements of D belong to $\mathcal{Z}$ which is a principal ideal ring, then *unimodular matrices* would be all those with determinant equal to ± 1 [27]. Note that D is not unique for a given sampling pattern and that two matrices representing the same sampling process are related by a linear transformation represented by a unimodular matrix [5]. A *separable* lattice is a lattice that can be represented by a diagonal matrix; it will appear when one-dimensional systems are used in a separable fashion along each dimension. The number of input lattice samples contained in the *unit cell* (the set of points such that disjoint union of its copies shifted to all of the lattice points yields the input lattice) represents the reciprocal of the sampling density and is given by $N = \det \boldsymbol{D}$. An important unit cell is the *fundamental parallelepiped* $\mathscr{U}_c$ (the parallelepiped formed by n basis vectors). In what follows, $\mathscr{U}_c^t$ will denote the fundamental parallelepiped of the transposed lattice. Shifting the origin of the output lattice to any of the points of the input lattice yields a so-called *coset*. Clearly, there are exactly N distinct cosets obtained by shifting the origin of the output lattice to all of the points of the fundamental parallelepiped. The union of all cosets for a given lattice yields the input lattice.

Another important notion is that of the *reciprocal lattice* [5], [13]. This lattice is actually the Fourier transform of the original lattice and its points represent the points of replicated spectra in the frequency domain. If the matrix corresponding to the reciprocal lattice is denoted by $\boldsymbol{D}_r$, then $\boldsymbol{D}_r^t \cdot \boldsymbol{D} = A \cdot \boldsymbol{I}$ and $\det \boldsymbol{D} \cdot \det \boldsymbol{D}_r = A^n$. Now observe that the determinant of the matrix $\boldsymbol{D}$ represents the hypervolume of any unit cell of the corresponding lattice, as well as the reciprocal of the sampling density. One of the possible unit cells is the *Voronoi cell* which is actually the set of points closer to the origin than to any other lattice point. The meaning of the unit cell in the frequency domain is extremely important since if the signal to be sampled is bandlimited to that cell, no overlapping of spectra will occur and the signal can be reconstructed from its samples.

To conclude the discussion on multidimensional sampling let us examine some operations involving sampling that are going to be used later. First, *downsampling* will mean that the points on the sampling lattice are kept while all the others are discarded. The time, Fourier, and z-domain expressions for the output of a downsampler are given by [13], [46]

$$\begin{aligned} y(\boldsymbol{n}) &= x(\boldsymbol{D}\boldsymbol{n}), \\ \hat{Y}(\boldsymbol{\omega}) &= \frac{1}{N} \sum_{\boldsymbol{k} \in \mathscr{U}_c^t} \hat{X}\left((\boldsymbol{D}^t)^{-1} \cdot (\boldsymbol{\omega} - 2\pi \boldsymbol{k})\right), \\ Y(\boldsymbol{z}) &= \frac{1}{N} \sum_{\boldsymbol{k} \in \mathscr{U}_c^t} X(\boldsymbol{W}_{\boldsymbol{D}^{-1}}(2\pi \boldsymbol{k}) \circ \boldsymbol{z}^{\boldsymbol{D}^{-1}}) \end{aligned} \tag{3}$$

where $N = \det \boldsymbol{D}$, $\boldsymbol{\omega}$ is an n-dimensional real vector, $\boldsymbol{z}$ is an n-dimensional complex vector, and $\boldsymbol{n}$, $\boldsymbol{k}$ are n-dimensional integer vectors (the details of the notation are defined in Appendix A). Next, consider *upsampling*, i.e., the process that maps a signal on the input lattice to another one that is nonzero only at the points of the sampling lattice

$$\begin{aligned} y(\boldsymbol{n}) &= \begin{cases} x(\boldsymbol{D}^{-1}\boldsymbol{n}) & \text{if } \boldsymbol{n} = \boldsymbol{D}\boldsymbol{k} \\ 0 & \text{otherwise,} \end{cases} \\ \hat{Y}(\boldsymbol{\omega}) &= \hat{X}(\boldsymbol{D}^t \boldsymbol{\omega}), \\ Y(\boldsymbol{z}) &= X(\boldsymbol{z}^{\boldsymbol{D}}). \end{aligned} \tag{4}$$

Finally, combining (3) and (4), one obtains the expression for the output of a downsampler followed by an upsampler (that is, replacing by zeros all the samples that are not on the sublattice)

$$\begin{aligned} y(\boldsymbol{n}) &= \begin{cases} x(\boldsymbol{n}) & \text{if } \boldsymbol{n} = \boldsymbol{D}\boldsymbol{k} \\ 0 & \text{otherwise,} \end{cases} \\ \hat{Y}(\boldsymbol{\omega}) &= \frac{1}{N} \sum_{\boldsymbol{k} \in \mathscr{U}_c^t} \hat{X}\left(\boldsymbol{\omega} - 2\pi (\boldsymbol{D}^t)^{-1} \boldsymbol{k}\right), \\ Y(\boldsymbol{z}) &= \frac{1}{N} \sum_{\boldsymbol{k} \in \mathscr{U}_c^t} X(\boldsymbol{W}_{\boldsymbol{D}^{-1}}(2\pi \boldsymbol{k}) \circ \boldsymbol{z}). \end{aligned} \tag{5}$$

A. Quincunx Case

The reason the quincunx case is examined in detail is because it uses the simplest multidimensional sampling structure that is nonseparable. This is obvious from Fig. 2(a) where the same lattice is generated using two sets of basis vectors (corresponding to matrices $\boldsymbol{D}_1$ and $\boldsymbol{D}_2$ given in (2)). Since the determinant of either one of them equals 2, the corresponding critically sampled filter bank will have two channels. The same figure shows the fundamental parallelepipeds for both cases as well as the Voronoi cell. Since the reciprocal lattice for this case is again quincunx, its Voronoi cell will have the same diamond shape. This fact has been used in some image and video coding schemes [2], [43] since if restricted to this region; i) the spectra of the signal and its repeated occurrences that appear due to sampling will not overlap; and ii) due to the fact that the human eye is less sensitive to resolution along diagonals it is more appropriate for the low-pass filter to have diagonal cutoff. Note that the two vectors belonging to the unit cell are in both cases

$$\boldsymbol{n}_0 = \begin{pmatrix} 0 \\ 0 \end{pmatrix}, \qquad \boldsymbol{n}_1 = \begin{pmatrix} 1 \\ 0 \end{pmatrix}, \tag{6}$$

and are the same for the unit cell of the transposed lattice, a fact that is going to be used throughout the paper. Shifting the origin of the quincunx lattice to points determined by the unit cell vectors yields the two cosets for this lattice (see Fig. 2(b)). Obviously, their union gives back the original lattice.

Finally, let us state here some facts that are going to be used later. First, following the notation in Appendix A, one can express $\boldsymbol{z}^{\boldsymbol{D}_1}$ as

$$\boldsymbol{z}^{\boldsymbol{D}_1} = \boldsymbol{z}^{\begin{pmatrix} 1 & 1 \\ 1 & -1 \end{pmatrix}} = \left(\boldsymbol{z}^{\begin{pmatrix} 1 \\ 1 \end{pmatrix}}, \boldsymbol{z}^{\begin{pmatrix} 1 \\ -1 \end{pmatrix}} \right) = (z_1 z_2, z_1 z_2^{-1}). \tag{7}$$

Then, using (4) one can write the expressions for the output of an upsampler in Fourier and z-domains

$$\begin{aligned} \hat{Y}(\omega_1, \omega_2) &= \hat{X}(\omega_1 + \omega_2, \omega_1 - \omega_2), \\ Y(z_1, z_2) &= X(z_1 z_2, z_1 z_2^{-1}). \end{aligned} \tag{8}$$

Similarly, using (5), the output of a downsampler followed by an upsampler can be expressed as

$$\begin{aligned} \hat{Y}(\omega_1, \omega_2) &= \tfrac{1}{2}(\hat{X}(\omega_1, \omega_2) + \hat{X}(\omega_1 + \pi, \omega_2 + \pi)), \\ Y(z_1, z_2) &= \tfrac{1}{2}(X(z_1, z_2) + X(-z_1, -z_2)). \end{aligned} \tag{9}$$

III. Multidimensional Perfect Reconstruction Filter Banks

A. Polyphase and Modulation Domain Analysis

Consider a simple system consisting of a filter with the impulse response $h(0) = 1$, $h(1) = 2$, $h(2) = 1$, 0 otherwise, and of a downsampler by 2 (see Fig. 3). Then the impulse at time 0 ($x(n) = \delta(n)$) will produce the output $y(0) = 1$, $y(1) = 1$, 0 otherwise, while an impulse at time 1 ($x(n) = \delta(n-1)$) will produce the output $y(1) = 2$, 0 otherwise. It is then obvious that the system, due to downsampling, is periodically shift-variant and for it to be completely specified one needs two impulse responses. It is this shift-variance that leads to aliased versions (or overlapping repeated spectra) of the input signal in the output. A convenient way to take care of the shift-variance of such a multidimensional multirate system is to decompose both signals and filters into so-called *polyphase components*, each one corresponding to one of the cosets of the output lattice.

Then define the polyphase decomposition of the input signal as

$$X(z) = \sum_{k \in \mathscr{U}_c^t} z^k X_k(z^D) = p_i^t(z) \cdot x_p(z^D),$$

$$X_k(z) = \sum_{n \in \mathscr{Z}^n} x(Dn - k) \cdot z^{-n} \tag{10}$$

where $p_i(z) = \{z^k\}_{k \in \mathscr{U}_c^t}$ is a so-called vector of the *inverse polyphase transform* and $x_p(z)$ is the vector containing the polyphase components of the input signal $x_p(z) = \{X_k(z)\}_{k \in \mathscr{U}_c^t}$. Note that the vector $p_i(z)$ is noncausal and its causal version will be denoted by $p_{ic}(z)$. Similarly, define the polyphase components of the filter $H(z)$ as

$$H(z) = \sum_{k \in \mathscr{U}_c^t} z^{-k} H_k(z^D) = p_f^t(z) \cdot h_p(z^D),$$

$$H_k(z) = \sum_{n \in \mathscr{Z}^n} h(Dn + k) \cdot z^{-n} \tag{11}$$

where $p_f(z) = \{z^{-k}\}_{k \in \mathscr{U}_c^t}$ is the vector of the *forward polyphase transform* and $h_p(z)$ is the vector containing the polyphase components of the filter, $h_p(z) = \{H_k(z)\}_{k \in \mathscr{U}_c^t}$. Note that the polyphase components of signals and filters are defined in a reverse fashion so as to account for the action of convolution. Therefore, a single-input linear periodically shift-variant system can be expressed as a multi-input linear shift-invariant system (for an example, see Fig. 3). To summarize, signals at the output of the analysis bank can be represented in terms of the input signal, forward polyphase transform $p_f(z)$, and the *analysis polyphase matrix* $H_p(z)$ (that is, the matrix containing the polyphase components of the analysis filters), while the output signal can be represented in terms of the input channel signals, the *synthesis polyphase matrix* $G_p(z)$ (that is, the matrix containing the polyphase components of the synthesis filters, defined in a reverse order from the analysis polyphase matrix) and the inverse polyphase transform $p_i(z)$ (see Fig. 4). Then, the output of the synthesis bank is

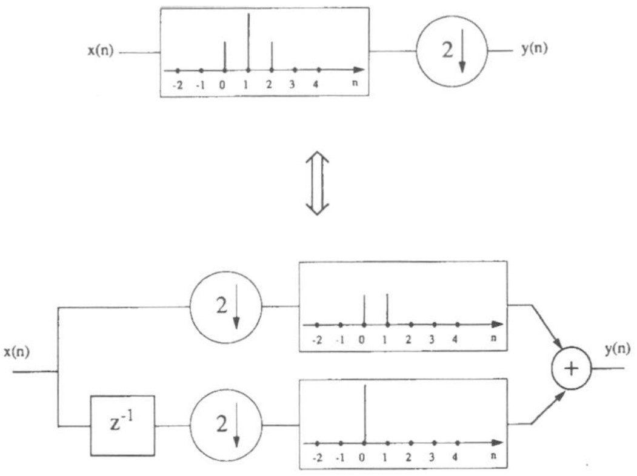

Fig. 3. Equivalent representations of a filter.

$$Y(z) = p_{ic}^t(z) \cdot G_p(z^D) \cdot H_p(z^D) \cdot x_p(z^D),$$

$$= p_{ic}^t(z) \cdot T_p(z^D) \cdot x_p(z^D) \tag{12}$$

where $T_p(z) = G_p(z) H_p(z)$ is a so-called *transfer polyphase matrix*. The last equation yields easily results on alias cancellation and perfect reconstruction.

Lemma 3.1: 1) Aliasing is cancelled if and only if the inverse polyphase transform vector p_i is the left eigenvector of the transfer polyphase matrix in the upsampled domain $T_p(z^D)$, i.e., if and only if $p_i^t \cdot T_p(z^D) = T(z) \cdot p_i^t$, where $T(z)$ is the corresponding eigenvalue.

2) Perfect reconstruction is achieved if and only if the eigenvalue $T(z)$ associated with the eigenvector p_i^t in 1) is a monomial, i.e., if and only if $T(z) = c \cdot z^{-k}$.

3) Perfect reconstruction with FIR filters is achieved if and only if the determinant of the analysis polyphase matrix is a monomial, i.e., if and only if $\det H_p(z) = z^{-k}$.

Proof: See Appendix B. Similar results have already appeared in [21], [45], [46].

The approach taken until now was to decompose signals and filters in such a way so that one can analyze the system as if it were shift-invariant. Here we proceed in a different manner, namely modulated versions of signals and filters are going to be used. Which approach is going to be applied to which problem depends on the nature of the problem itself. In the following sections both polyphase and modulation analysis will be employed intermittently.

The input signal and its modulated versions are given by

$$\hat{x}_m(\omega) = \left\{ \hat{X}\left(\omega - 2\pi (D^t)^{-1} k\right) \right\}_{k \in \mathscr{U}_c^t},$$

$$x_m(z) = \{X(W_{D^{-1}}(2\pi k) \circ z)\}_{k \in \mathscr{U}_c^t} \tag{13}$$

with notation as defined in Appendix A. Thus, the output of the system after upsampling and filtering in the synthesis

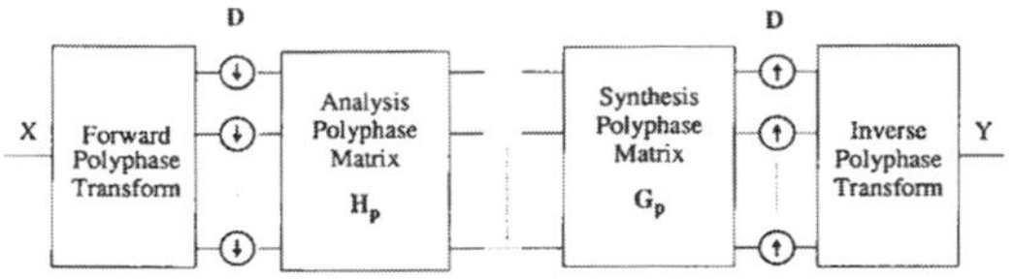

Fig. 4. Analysis/synthesis filter bank in the polyphase domain.

bank can be written as

$$\hat{Y}(\omega) = \frac{1}{N}(\hat{G}_0(\omega) \cdots \hat{G}_{N-1}(\omega)) \cdot \hat{H}_m(\omega) \cdot \hat{x}_m(\omega),$$

$$Y(z) = \frac{1}{N}(G_0(z) \cdots G_{N-1}(z)) \cdot H_m(z) \cdot x_m(z) \quad (14)$$

where $\hat{H}_m(\omega)$, $H_m(z)$ contain the modulated versions of all the filters, that is

$$\hat{H}_m(\omega) = \left\{\hat{H}_i\left(\omega - 2\pi(D^t)^{-1}k\right)\right\},$$

$$H_m(z) = \{H_i(W_{D^{-1}}(2\pi k) \circ z)\},$$

$$k \in \mathcal{U}_c^t, \quad i \in \{0, \cdots, N-1\}. \quad (15)$$

Let us note at this point that the relationship between polyphase and modulation domain analysis is in a way like "time-Fourier" domain relation; hence the two representations are related by Fourier transforms. For more details, see [39].

After having achieved the goal of obtaining perfect reconstruction one might impose some other requirements on the filter bank, some of the important ones being that the bank be orthogonal and/or linear phase. Note that when compared to the one-dimensional case the linear phase requirement is less constrained since linear phase in multiple dimensions means just centro-symmetry of the filter's impulse response [14].

B. Orthogonal Case

A filter with a rational transfer function is called *allpass* if it satisfies

$$\tilde{H}(z) \cdot H(z) = 1 \quad (16)$$

while a square matrix with rational entries is called *paraunitary* if it satisfies

$$\tilde{H}(z) \cdot H(z) = H(z) \cdot \tilde{H}(z) = c \cdot I. \quad (17)$$

On the unit hypercircles ($z_i = e^{j\omega_i}$, $i = 1, \cdots, n$) this matrix becomes orthogonal, and for the sake of simplicity it will be referred to as such throughout the paper. Now suppose that the analysis polyphase matrix is orthogonal. Then by choosing $G_p(z) = z^{-k}\tilde{H}_p(z)$, a perfect reconstruction system is obtained as can be seen from (12). This solution has important advantages both from the theoretical point of view (the filter bank calculates projections on orthogonal subspaces) and from the point of view of implementation (the synthesis filters are within shift-reversal the same as the analysis ones).

Suppose now that the modulation matrix defined in (15) is orthogonal. Then using (17) (with $c = N$)

$$\sum_{k \in \mathcal{U}_c^t} H_i(W_{D^{-1}}(2\pi k) \circ z)\tilde{H}_j(W_{D^{-1}}(2\pi k) \circ z) = N \cdot \delta_{ij}. \quad (18)$$

It is obvious that by choosing $(G_0(z) \cdots G_{N-1}(z))$ as the first row of $\tilde{H}_m(z)$, perfect reconstruction is achieved as can be seen from (14). Now observe that assuming real coefficients, $H_i(z)\tilde{H}_j(z)$ is the z-transform of the cross-correlation sequence $r_{ij}(n) = \langle h_i(k), h_j(k+n)\rangle$. Also note that $H_i(a_1 z_1, \cdots, a_n z_n)\tilde{H}_j(a_1 z_1, \cdots, a_n z_n)$ is the z-transform of $a_1^{-n_1} \cdot \cdots \cdot a_n^{-n_n} \cdot r_{ij}(n_1, \cdots, n_n)$. Using these facts one can see that (18) is the z-transform of

$$r_{ij}(n) \cdot \sum_{k \in \mathcal{U}_c^t} e^{j2\pi k^t D^{-1} n} = N \cdot \delta_{ij}\delta_n. \quad (19)$$

Analogously to the one-dimensional case, it can be shown that the sum in the previous equation is nonzero only at lattice points [34]

$$\sum_{k \in \mathcal{U}_c^t} e^{j2\pi k^t D^{-1} n} = \begin{cases} N & \text{if } n = Dn_0 \\ 0 & \text{otherwise.} \end{cases} \quad (20)$$

Using this fact, one can finally write (19) as

$$r_{ij}(Dn) = \langle h_i(k), h_j(k + Dn)\rangle = \delta_{ij}\delta_n \quad (21)$$

showing that each filter is orthogonal to its translates with respect to the lattice in question, and pairs of filters are orthogonal to each other and their shifts with respect to the lattice, that is, the set $S = \{h_i(k + Dn) \mid i = 0, \cdots, N-1, k, n \in \mathcal{Z}^n\}$ is an orthonormal set. This is the lattice extension of the well-known orthogonality relations with respect to shifts in the one-dimensional case [9], [41].

C. Linear Phase Case

Let us begin this section by introducing some notation. Suppose we circumscribe a parallelepiped around a polynomial represented in the space of its exponents. To facilitate the discussion (since we are interested in z-transform), a point (m, n) in this space will denote the polynomial term $z_1^{-m}z_2^{-n}$, i.e., the exponents will be taken with a minus sign. Then define $P = (p_1, p_2, \cdots, p_n)$ and $Q = (q_1, q_2, \cdots, q_n)$ as the corners on the main hyperdiagonal of the parallelepiped. The size of the polynomial in the ith direction is $l_i = q_i - p_i + 1$. When dealing with more than one polynomial at a time, their P's and Q's will be distinguished by a superscript. For example, $P^{(1)}$ is P of the first polynomial, while $p_2^{(1)}$ is the second coordinate of P of the first polynomial. Fig. 5 shows the above notation on an example in two dimensions.

The aim here is to derive a condition in terms of the analysis polyphase matrix that can be used to test linear phase of all the filters in the bank simultaneously. If a real filter is linear phase then it can be written as

$$H(z) = a \cdot D(z) \cdot \tilde{H}(z) \quad (22)$$

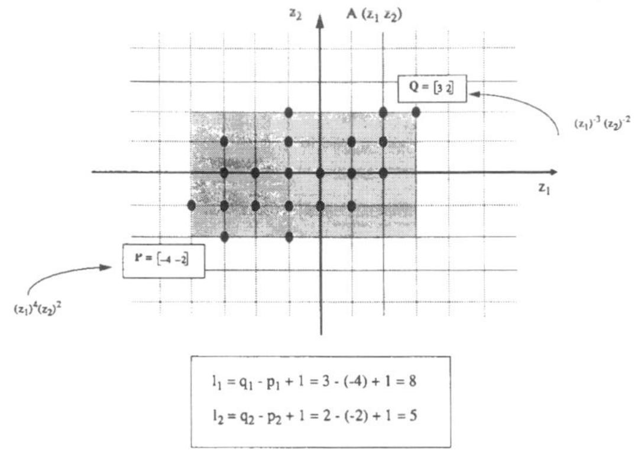

Fig. 5. Notation used for polynomials. P and Q denote the corners of the parallelogram circumscribed around the polynomial. It is represented in the space of its exponents. l_i denotes the length of the polynomial in the ith direction.

where a denotes the symmetry of the filter (1 for symmetric, -1 for antisymmetric) and

$$D(z) = z^{-(P+Q)} = \prod_{i=1}^{n} z_i^{-(p_i+q_i)}$$

with P and Q as defined above. A similar relation can be defined for the analysis polyphase matrix. Following the discussion in Section III-A one has that the vector containing the analysis filters can be written as

$$h(z) = (H_0(z) \cdots H_{N-1}(z))^t = H_p(z^D) \cdot p_f(z).$$

Suppose now that all of the filters are linear phase. Then each one can be expressed as in (22), and thus,

$$H_p(z^D) \cdot p_f(z)$$

$$= \begin{pmatrix} H_0(z) \\ \vdots \\ H_{N-1}(z) \end{pmatrix} = \begin{pmatrix} a_0 \cdot D_0(z) \cdot \tilde{H}_0(z) \\ \vdots \\ a_{N-1} \cdot D_{N-1}(z) \cdot \tilde{H}_{N-1}(z) \end{pmatrix},$$

$$= \begin{pmatrix} a_0 & & \\ & \ddots & \\ & & a_{N-1} \end{pmatrix} \begin{pmatrix} D_0(z) & & \\ & \ddots & \\ & & D_{N-1}(z) \end{pmatrix} \begin{pmatrix} \tilde{H}_0(z) \\ \vdots \\ \tilde{H}_{N-1}(z) \end{pmatrix},$$

$$= \begin{pmatrix} a_0 & & \\ & \ddots & \\ & & a_{N-1} \end{pmatrix} \begin{pmatrix} D_0(z) & & \\ & \ddots & \\ & & D_{N-1}(z) \end{pmatrix}$$

$$\cdot \tilde{H}_p^t(z^D) \cdot \tilde{p}_f^t(z), = a \cdot \Delta(z) \cdot \tilde{H}_p^t(z^D) \cdot \tilde{p}_f^t(z). \tag{23}$$

This can be used as a linear phase testing condition for the whole filter bank. Note that $\tilde{p}_f^t(z) = p_i(z)$. Basically (23) allows 1) to check linear phase of an already designed filter bank; or 2) if one wishes to design a filter bank containing linear phase filters, one can impose constraints on the polyphase matrix in such a way that (23) is satisfied. Another linear phase testing condition appeared in [21].

D. Separability Versus Nonseparability

As already noted in Section II, sampling lattices that are used can be both separable and nonseparable ones. In both cases an interesting question is whether it is possible to have all combinations separable/nonseparable sampling/filters/polyphase components.

To analyze separable sampling, let us first try to determine the number of free variables when the filters and polyphase components are separable/nonseparable. If a filter is separable it has $FV_1 = \sum_{i=1}^{n} (l_i - 1)$ free variables where l_i, $i = 1, \cdots, n$ is the size of the filter in the ith dimension. The reason that 1 is subtracted in each term is because of scaling. If, on the other hand, the filter is nonseparable, the number of free variables it possesses is $FV_2 = \prod_{i=1}^{n} l_i - 1$. Now, what happens if the polyphase components are separable? First note that if the sampling factor in the ith dimension is N_i, then the size of the polyphase component $H_{i_1 i_2 \cdots i_n}$, $i_j = 1, \cdots, N_j$ in the ith dimension can be expressed as $l_{i_1,\cdots,i_n j} = (l_j + N_j - i_j) \bmod N_j$. Thus, the number of free variables if the polyphase components are separable is $FV_3 = \sum_{i_1,\cdots,i_n}(\sum_j l_{i_1 \cdots i_n j} - n + 1) - 1$ which after some manipulations can be written as $FV_3 = \prod_{j=1}^{n} N_j(\sum_{i=1}^{n}(l_i/N_i) - (n-1)) - 1$. This number is in general different from FV_1 and FV_2 and satisfies the following:

$$FV_1 < FV_3 \leq FV_2. \tag{24}$$

In the above, equality holds just for very small filter sizes (basically when all polyphase components are either constants or one-dimensional polynomials since in that case there is no distinction between separable and nonseparable polyphase components). Thus, one may conclude that in general separable polyphase components would yield a nonseparable filter (in other words to get a separable filter assuming separable polyphase components one would have to reduce the number of free variables). Note that the assumed size of the filter in each dimension is at least two. Finally, if polyphase components are nonseparable the number of free variables is $FV_4 = \sum_{i_1,\cdots,i_n}\prod_{j=1}^{n} l_{i_1 \cdots i_n j} - 1 = \prod_{j=1}^{n}\sum_{i_j}((l_j + N_j - i_j) \bmod N_j) - 1 = \prod_{i=1}^{n} l_i - 1 = FV_2$, that is, if the polyphase components are nonseparable then the total number of free variables is the same as when the filter itself is nonseparable

$$FV_4 = FV_2. \tag{25}$$

Fact 3.1: For separable sampling, the following holds:

1) separable filter implies separable polyphase components;
2) nonseparable polyphase components will always yield a nonseparable filter.

Proof:

1) A separable filter can be expressed as follows:

$$H(z_1,\cdots,z_n) = H^{(1)}(z_1)\cdots H^{(n)}(z_n), = \sum_{i_1=0}^{N_1-1} z_1^{-i_1}H_{i_1}^{(1)}(z_1^{N_1})\cdots\sum_{i_n=0}^{N_n-1} z_n^{-i_n}H_{i_n}^{(n)}(z_n^{N_n}), = \sum_{i_1=0}^{N_1-1}\cdots\sum_{i_n=0}^{N_n-1} z_1^{-i_1}\cdots z_n^{-i_n}\cdot H_{i_1}^{(1)}(z_1^{N_1})\cdots H_{i_n}^{(n)}(z_n^{N_n})$$

where $H_{i_1}^{(1)}(z_1)\cdot\cdots\cdot H_{i_n}^{(n)}(z_n)$, corresponding to the polyphase components of the filter, are clearly separable.

2) Due to (25). □

For nonseparable sampling, things become more complicated. Consider, for example, what happens if the filter is separable. Writing it in terms of its polyphase components it is obvious that to get separable polyphase components one would have to reduce the number of degrees of freedom. It follows similarly for the nonseparable filter. Thus, in general both separable and nonseparable filters would have nonseparable polyphase components. On the other hand, if the polyphase components are separable, after upsampling they would in general become nonseparable, yielding in turn a nonseparable filter. For nonseparable polyphase components this is even more obvious. Therefore, starting from either separable or nonseparable polyphase components one would obtain a nonseparable filter. Since in this case, there are no implications as for separable sampling one can find examples for any combination separable/nonseparable filter/polyphase components. However, in general (that is, except for particular values of coefficients) one has that separable polyphase components do not yield separable filters and separable filters do not yield separable polyphase components.

E. Quincunx Case Revisited

Let us continue our analysis of the quincunx case. As we said in Section III-A, polyphase domain analysis is used to help overcome problems that arise when dealing with shift-variant systems. Thus, in this case polyphase decomposition would correspond to considering the system "living" on the two cosets of the quincunx lattice (see Fig. 2(b)). For example, if a filter's impulse response is denoted by $h(\boldsymbol{n}) = h(n_1, n_2)$ its two polyphase components can be written as (matrix $\boldsymbol{D}_1$ from (2) is used)

$$h_0(n_1, n_2) = h(\boldsymbol{Dn}) = h(n_1+n_2, n_1-n_2), \tag{26}$$

$$h_1(n_1, n_2) = h(\boldsymbol{Dn}+\boldsymbol{n}_1) = h(n_1+n_2+1, n_1-n_2), \tag{27}$$

where $\boldsymbol{n}_1$ (see (6)) belongs to the unit cell (in this case to the unit cell of the transposed lattice as well) and is the representative vector of one of the cosets. From (8), (26), and (27) it is obvious that the z-transform expression of a filter is then

$$H(z_1,z_2) = H_0(z^{D}) + z_1^{-1}H_1(z^{D}) = H_0(z_1z_2, z_1z_2^{-1}) + z_1^{-1}H_1(z_1z_2, z_1z_2^{-1}). \tag{28}$$

The vectors of forward and inverse polyphase transforms (causal versions) being $\boldsymbol{p}_f(z_1,z_2)^t = (1\; z_1^{-1})$ and $\boldsymbol{p}_{ic}(z_1,z_2)^t = (z_1^{-1}\; 1)$, respectively (see Section III-A), one gets (12) for the quincunx case

$$Y(z_1,z_2) = \begin{pmatrix} z_1^{-1} & 1\end{pmatrix}\boldsymbol{G}_p(z_1z_2, z_1z_2^{-1}) \cdot \boldsymbol{H}_p(z_1z_2, z_1z_2^{-1})\boldsymbol{x}_p(z_1z_2, z_1z_2^{-1}) \tag{29}$$

where the analysis polyphase matrix can be written as (H_{ij} denotes the jth polyphase component of the ith filter)

$$\boldsymbol{H}_p(z_1,z_2) = \begin{pmatrix} H_{00}(z_1,z_2) & H_{01}(z_1,z_2)\\ H_{10}(z_1,z_2) & H_{11}(z_1,z_2)\end{pmatrix}. \tag{30}$$

To give a gist of the modulation domain analysis let us first find the modulated versions of the input signal as given in (13)

$$\boldsymbol{x}_m(z_1,z_2) = \left\{X\left(\begin{pmatrix}e^0\\ e^0\end{pmatrix}\circ\begin{pmatrix}z_1\\ z_2\end{pmatrix}\right), X\left(\begin{pmatrix}z_1\\ z_2\end{pmatrix}\circ\begin{pmatrix}e^{-j\pi}\\ e^{-j\pi}\end{pmatrix}\right)\right\} = \{X(z_1,z_2), X(-z_1,-z_2)\}. \tag{31}$$

Then (14) for the output of the system is

$$\hat{Y}(\omega_1,\omega_2) = \frac{1}{2}(\hat{G}_0(\omega_1,\omega_2)\,\hat{G}(\omega_1,\omega_2)) \cdot\begin{pmatrix}\hat{H}_0(\omega_1,\omega_2) & \hat{H}_0(\omega_1+\pi,\omega_2+\pi)\\ \hat{H}_1(\omega_1,\omega_2) & \hat{H}_1(\omega_1+\pi,\omega_2+\pi)\end{pmatrix} \cdot\begin{pmatrix}\hat{X}(\omega_1,\omega_2)\\ \hat{X}(\omega_1+\pi,\omega_2+\pi)\end{pmatrix},$$

$$Y(z_1,z_2) = \frac{1}{2}(G_0(z_1,z_2)\,G_1(z_1,z_2)) \cdot\begin{pmatrix}H_0(z_1,z_2) & H_0(-z_1,-z_2)\\ H_1(z_1,z_2) & H_1(-z_1,-z_2)\end{pmatrix} \cdot\begin{pmatrix}X(z_1,z_2)\\ X(-z_1,-z_2)\end{pmatrix}. \tag{32}$$

Note the resemblance between this expression and its one-dimensional two-channel counterpart (see, for example, [9], [40]). This striking similarity that appears when analyzing the quincunx case time and again is basically due to the fact that a number of results depend heavily on the overall sampling density (number of channels for the critically sampled filter bank). Thus, as will be seen in Section IV, some of the results obtained for the classical one-dimensional two-channel filter bank will extend easily to the n-dimensional two-channel case.

To illustrate the statement that the polyphase and modulation domain representations are related through Fourier transform, consider the following relation between polyphase and modulation domain matrices:

$$\hat{H}_p(\omega_1+\omega_2,\omega_1-\omega_2)=\frac{1}{2}\hat{H}_m(\omega_1,\omega_2)\cdot\begin{pmatrix}1 & 1\\ 1 & -1\end{pmatrix}\begin{pmatrix}1 & 0\\ 0 & e^{j\omega_1}\end{pmatrix},$$

$$H_p(z_1z_2,z_1z_2^{-1})=\frac{1}{2}H_m(z_1,z_2)\cdot\begin{pmatrix}1 & 1\\ 1 & -1\end{pmatrix}\begin{pmatrix}1 & 0\\ 0 & z_1\end{pmatrix}. \quad (33)$$

Next, one can examine the relation between the two filters in the analysis bank if the system is orthogonal. Without going through the whole analysis from Section III-B just write (21) for the quincunx case ($\boldsymbol{D}$ represents quincunx sampling)

$$\langle h_0(\boldsymbol{k}), h_0(\boldsymbol{k}+\boldsymbol{D}\boldsymbol{n})\rangle=\delta_n, \quad (34)$$

$$\langle h_1(\boldsymbol{k}), h_1(\boldsymbol{k}+\boldsymbol{D}\boldsymbol{n})\rangle=\delta_n, \quad (35)$$

$$\langle h_0(\boldsymbol{k}), h_1(\boldsymbol{k}+\boldsymbol{D}\boldsymbol{n})\rangle=0. \quad (36)$$

Equations (34) and (35) show that each filter is orthogonal to its shifts on the quincunx lattice, while (36) states that filters h_0 and h_1 are orthogonal to each other as well as their shifts on the quincunx lattice. These facts are going to be used when constructing orthonormal bases of wavelets for the quincunx case in Section VI.

Finally, let us show the linear phase testing condition given by (23) for the quincunx case

$$H_p(z_1z_2,z_1z_2^{-1})\begin{pmatrix}1\\ z_1^{-1}\end{pmatrix}=\begin{pmatrix}a_0 & 0\\ 0 & a_1\end{pmatrix}\begin{pmatrix}D_0(z_1,z_2) & 0\\ 0 & D_1(z_1,z_2)\end{pmatrix}\cdot H_p(z_1^{-1}z_2^{-1},z_1^{-1}z_2)\begin{pmatrix}1\\ z_1\end{pmatrix}. \quad (37)$$

Here a_0 and a_1 demonstrate what kind of a symmetry filters h_0 and h_1 possess (1 for symmetric and -1 for antisymmetric) and D_0, D_1 are monomials related to the size of the filter defined in Section III-C.

IV. Two-Channel Case In Multiple Dimensions

Let us first point out that the two-channel case we are going to deal with in this section is the only true nonseparable one in terms of the sampling used. Among others, it can be characterized by one of the following sampling matrices:

$$\boldsymbol{D}_1=\begin{pmatrix}1 & 0 & \cdots & 0 & 1\\ 1 & 1 & \cdots & 0 & 0\\ 0 & 1 & \cdots & 0 & 0\\ \vdots & \vdots & \ddots & \vdots & \vdots\\ 0 & 0 & \cdots & 1 & 0\\ 0 & 0 & \cdots & 1 & (-1)^{n-1}\end{pmatrix},$$

$$\boldsymbol{D}_2=\begin{pmatrix}2 & 1 & 0 & \cdots & 0\\ 0 & 1 & 1 & \cdots & 0\\ 0 & 0 & 1 & \cdots & 0\\ \vdots & \vdots & \vdots & \ddots & \vdots\\ 0 & 0 & 0 & \cdots & 1\\ 0 & 0 & 0 & \cdots & 1\end{pmatrix}. \quad (38)$$

Since the overall sampling density is 2, the equivalent sampling factor per dimension would be $2^{1/n}$. In two dimensions the corresponding lattice is the quincunx lattice and in three dimensions the *FCO* or *face centered orthorhombic* lattice. Since they are natural extensions of the one-dimensional two-channel case they already found application in image and video processing [13], [19], [43].

Refer to the discussion on the modulation domain analysis in Section III-A. For the two-channel case the output of the system given by (14) can be written as

$$Y(z)=\frac{1}{2}(G_0(z)\,G_1(z))\cdot\begin{pmatrix}H_0(z) & H_0(-z)\\ H_1(z) & H_1(-z)\end{pmatrix}\cdot\begin{pmatrix}X(z)\\ X(-z)\end{pmatrix} \quad (39)$$

which, as already pointed out for the quincunx case in Section III-E, (32), bears striking resemblance to the one-dimensional case. Then it is easy to see that the result on alias cancellation (or no overlapping of spectra) obtained for the one-dimensional case [8] holds for an arbitrary number of dimensions:

Proposition 4.1: The classical QMF solution holds for the n-dimensional filter bank, i.e., the following choice of filters will yield alias cancellation:

$$H_0(z)=H(z) \quad \text{and} \quad H_1(z)=H(-z)$$
$$G_0(z)=H(z) \quad \text{and} \quad G_1(z)=-H(-z).$$

As already pointed out in Section III additional constraints can be imposed upon the filters in the bank beside requiring perfect reconstruction. In what follows two cases of interest are investigated, the first one yielding orthogonal filters and the second linear phase ones. Note that in the two-channel real FIR case these requirements are mutually exclusive as will be shown later.

A. Orthogonal Case

If the filter bank is orthogonal it can be shown that the filters involved have some important structural properties, namely that a single filter specifies completely the most general orthogonal system. Here we just state the theorem and its corollaries, for the proofs refer to Appendix C.

Theorem 4.1: The most general 2×2 real FIR orthogonal polyphase matrix H_p can be written in the following form:

$$H_p(z) = \begin{pmatrix} 1 & 0 \\ 0 & z^{-k} \end{pmatrix} \cdot \begin{pmatrix} H_{00}(z) & H_{01}(z) \\ c\tilde{H}_{01}(z) & -c\tilde{H}_{00}(z) \end{pmatrix} \quad (40)$$

where c is ± 1, $\boldsymbol{k}$ is large enough so as to make the entries in the second row causal, and H_{00} and H_{01} satisfy the PC property.

Corollary 4.1: The polyphase components of each filter are of the same size.

Corollary 4.2: The second filter is completely specified by modulating and reversing the first one, i.e.,

$$H_1(z) = -z^{-k'} \cdot \tilde{H}_0(-z)$$

where

$$z^{-k'} = z_1^{-(k_1+k_n+1)} z_2^{-(k_1+k_2)} \cdots z_{n-1}^{-(k_{n-2}+k_{n-1})} z_n^{-(k_{n-1}+(-1)^{n-1}k_n)}.$$

B. Linear Phase Solutions

Which solutions are possible if one wants a perfect reconstruction system where both filters are linear phase? The first polyphase component of each filter will be denoted by H_{i0} and the second one by H_{i1}, $i = 1, 2$. The whole analysis will be performed in the upsampled domain and thus the filters can be expressed as (see Section III-A and (28))

$$H_i(z) = H_{i0}(z^D) + z_1^{-1} H_{i1}(z^D) = a_i \cdot D_i(z) \cdot \tilde{H}_i(z)$$

where a_i and $D_i(z)$ characterize a linear phase filter as defined in (22). The determinant of the polyphase matrix in the upsampled domain is then

$$\begin{aligned} \det H_p(z^D) &= T(z^D) \\ &= H_{00}(z^D) H_{11}(z^D) - H_{10}(z^D) H_{01}(z^D) \end{aligned}$$

and the aim here is to force it to be a monomial in order to achieve perfect reconstruction.

The sizes of the filter H_i will be denoted by $P^{(i)}$, $Q^{(i)}$ (where P and Q are as introduced in Section III-C). Then for the filters' polyphase components in the upsampled domain the following holds: H_{i0} is of size $P^{(i)}$, $Q^{(i)}$ and H_{i1} is of size $P^{(i)} - (1, 0, \cdots, 0)$, $Q^{(i)} - (1, 0, \cdots, 0)$. The determinant $T(z^D)$ is of size $P^{(T)}$ and $Q^{(T)}$, where $P^{(T)} = P^{(1)} + P^{(2)} - (1, 0, \cdots, 0)$, $Q^{(T)} = Q^{(1)} + Q^{(2)} - (1, 0, \cdots, 0)$ and $l_k^{(T)} = l_k^{(1)} + l_k^{(2)} - 1$.

Here, we will just state some facts on the structure of linear phase solutions. For explicit proofs, refer to [19]. Note that $i(n)$ is ± 1 if n is even/odd, respectively.

Proposition 4.2: For a symmetric/antisymmetric filter the following holds:

1) $i(\sum_{k=1}^{n} l_k) = (-1)^n \Leftrightarrow H_{i0}(z^D) = a_i D_i(z) \tilde{H}_{i0}(z^D)$ and $H_{i1}(z^D) = a_i D_i(z) z_1^2 \tilde{H}_{i1}(z^D)$,
2) $i(\sum_{k=1}^{n} l_k) = (-1)^{n-1} \Leftrightarrow H_{i1}(z^D) = a_i D_i(z) z_1 \tilde{H}_{i0}(z^D)$

where a_i and $D_i(z)$ are associated with the filter as defined in (22).

This proposition basically states how the polyphase components of a symmetric/antisymmetric filter are related depending upon the size of the filter. For example, in one dimension it is easy to see that an odd length filter (case 1)) will have each polyphase component as a symmetric/antisymmetric filter, while if the length is even (case 2)) one can get the second polyphase component by shift-reversing the first one. To obtain perfect reconstruction with FIR filters the determinant T has to be a monomial (following Lemma 3.1). Now observe that taking all possible combinations of the two filters one can obtain T being symmetric/antisymmetric or T that does not have any specific symmetry. In what follows just the former case will be considered since all useful solutions found until now belong to that category.

Proposition 4.3: For perfect reconstruction with a symmetric/antisymmetric determinant T, T has to satisfy the following:

1) $i(l_k^{(T)}) = -1$, $\forall k$, i.e., T is of odd size in all directions;
2) $a_T = 1$, i.e., T has to be symmetric;
3) $i(\sum_{k=1}^{n}(p_k^{(T)} + q_k^{(T)})/2) = 1$, i.e., the degree of the center coefficient is even.

Proposition 4.4: For perfect reconstruction with a symmetric/antisymmetric T and linear phase filters, there are two possible cases:

1) both filters are as in Proposition 4.2, case 1) and they have the same symmetry, i.e., $a_1 a_2 = 1$; or
2) both filters are as in Proposition 4.2, case 2) and they have different symmetry, i.e., $a_1 a_2 = -1$.

Proposition 4.5: For a perfect reconstruction linear phase filter pair the following holds:

1) it is not possible to have the filters of the same symmetry and the same size;
2) It is possible to have the filters of opposite symmetry and the same size.

As an example consider the one-dimensional case. There, it has been shown that assuming a symmetric/antisymmetric T either both filters are of odd length and the same symmetry or they are of even length and opposite symmetry (the uninteresting case has not been considered) [42].

Proposition 4.6: In the two-channel real FIR case linear phase and orthogonality requirements are mutually exclusive (except for the trivial two-tap filters).

Proof: To prove this, it will be shown that it is not possible to have a linear phase filter whose polyphase components satisfy the PC property as required by Theorem 4.1. Since by Corollary 4.1 the polyphase components of each filter are of the same size, i.e., the filters are of the same size, then they have to have opposite symmetry and they belong to case 2) of Proposition 4.2 (by the previous discussion). Thus, substituting polyphase components into (90) one obtains

$$H_{i0}(z^D)\tilde{H}_{i0}(z^D) + H_{i1}(z^D)\tilde{H}_{i1}(z^D)$$
$$= H_{i0}(z^D)\tilde{H}_{i0}(z^D) + a_i D_i(z) z_1 \tilde{H}_{i0}(z^D) \cdot a_i \tilde{D}_i(z) z_1^{-1} H_{i0}(z^D) \quad (41)$$
$$= 2 \cdot H_{i0}(z^D)\tilde{H}_{i0}(z^D) = 1 \quad (42)$$

which is possible only if H_{i0} and H_{i1} are constants, i.e., if the filters are two-tap. □

C. Another Visit to the Quincunx Case

To summarize the important results of this section first note how in two dimensions matrices given in (38) reduce to those for the quincunx lattice introduced in (2). The input/output relation given in (39) for this case reduces to the expression already given in (32). The result from Theorem 4.1 for the quincunx case first appeared in [43]. Since it is going to be used later we state here how the two filters in the analysis bank are related (follows from Corollary 4.2)

$$\hat{H}_1(\omega_1, \omega_2) = -e^{-j((k_1+k_2+1)\omega_1 + (k_1-k_2)\omega_2)} \cdot \hat{H}_0(-\omega_1 + \pi, -\omega_2 + \pi),$$
$$H_1(z_1, z_2) = -z_1^{-(k_1+k_2+1)} z_2^{-(k_1-k_2)} \cdot H_0(-z_1^{-1}, -z_2^{-1}) \quad (43)$$

where the vector $\boldsymbol{k} = (k_1, k_2)$ is as stated in Section IV-A large enough to make the entries in the second row of the matrix in (40) causal.

The linear phase case was studied extensively in [43] where similar analysis on possible linear phase solutions was performed but in polyphase domain and for diamond shaped filters. Here we want to see how the Propositions 4.2–4.6 translate for this case. For example, case 1) of Proposition 4.2 would tell us that if $i(l_1 + l_2) = 1$, i.e., if the sizes of the filter in two dimensions are either both odd or both even, then both polyphase components will have the same kind of symmetry as the filter itself has. If, on the other hand, in one dimension filter length is even and in the other one odd, polyphase components do not necessarily have any kind of symmetry but instead one can obtain the second polyphase component by shift-reversing the first one. Next, Proposition 4.4 states that to have a perfect reconstruction pair both filters have either the same symmetry and their sizes are as in case 1) of Proposition 4.2 or they are of opposite symmetry and each is of odd length in one dimension and of even length in the other. The fact that in the two-channel case having linear phase and orthogonal filters is not possible (see Proposition 4.6) is as was said earlier due to the fact that the algebraic structure of the modulation/polyphase matrices is basically the same regardless of the number of dimensions. Similar reasoning can be used when constructing solutions for the four-channel two-dimensional case that are at the same time orthogonal and have linear phase, namely achieving both at the same time is feasible since it is feasible in the four-channel one-dimensional case as well (see Section V-A-2).

V. Synthesis of Multidimensional Filter Banks

A. Cascade Structures

When synthesizing filter banks one of the most obvious approaches is to try to find cascade structures that would generate filters of the desired form, the reason being that cascade structures: i) usually have very low complexity; ii) higher order filters are easily derived from the lower order ones; and iii) the coefficients can be quantized without affecting the desired form.

While in the orthogonal case forming a cascade that would achieve perfect reconstruction is trivial since one has just to combine orthogonal building blocks (i.e., orthogonal matrices and diagonal delay matrices), in the linear phase case this is not so simple. There one has to make use of the linear phase testing condition given in (23) or [21] to obtain possible cascades. As one of the possible approaches consider the generalization of the linear phase cascade structure proposed in [21], [22], [44]. Suppose that a linear phase system has been already designed and a higher order one is needed. Then choosing

$$\boldsymbol{H}_p''(z) = \boldsymbol{H}_p'(z) \cdot \boldsymbol{D}(z) \cdot \boldsymbol{R} \quad (44)$$

where $\boldsymbol{D}(z) = z^{-\boldsymbol{k}} \cdot \boldsymbol{J}\tilde{\boldsymbol{D}}(z)\boldsymbol{J}$ and $\boldsymbol{R}$ is *persymmetric* (i.e., $\boldsymbol{R} = \boldsymbol{JRJ}$), another linear phase system is obtained where the filters have the same symmetry as in $\boldsymbol{H}_p'$. This can be easily verified if substituted into (23). Although this cascade is in no way complete it can produce very useful filters as will be seen later (for examples, refer to Section V-A-1). Let us also point out that while building cascades in the polyphase domain one must bear in mind that using different sampling matrices for the same lattice will greatly affect the geometry of the filters obtained.

1) Cascade Structures for the Two-Channel Case in n Dimensions:

Lemma 5.1: The following cascade will produce a perfect reconstruction set containing two filters of the same size:

$$\boldsymbol{H}_p(z) = \boldsymbol{R}_0 \cdot \prod_{j=1}^{k} \prod_{i=1}^{n} \begin{pmatrix} 1 & 0 \\ 0 & z_i^{-1} \end{pmatrix} \boldsymbol{R}_{j_i}. \quad (45)$$

For the filters to be orthogonal the matrices $\boldsymbol{R}_{j_i}$ have to be unitary, while for them to be linear phase matrices have to be symmetric. In the latter case the filters obtained will have opposite symmetry.

Proof: For the orthogonal case it is obvious since all the blocks involved are orthogonal. For the linear phase case use condition (23). □

The one-dimensional orthogonal solution obtained by the above cascade is complete (all orthogonal solutions can be reached using it) [35]. The linear phase solution in one dimension generates filters of size $2(k+1)$ and was proposed in [28], [44]. The two-dimensional solution generates filters of sizes $2(k+1) \times (2k+1)$ and both orthogonal and linear phase filters were proposed in [43]. Note that unlike in the one-dimensional orthogonal case, for linear phase and higher dimensional cascades, completeness results are missing except for very small cases. In [43], it was shown that the smallest solutions both for orthogonal and linear phase cascades are general. The same will be shown in the next lemma, where the smallest size perfect reconstruction filter pairs are complete in *any* number of dimensions. Moreover, higher dimensional solutions will be generated from lower dimensional ones. In what follows, $z_u^{(i)}$ will denote z^D where $z = (z_1, \cdots, z_i)^t$.

Lemma 5.2: 1) The general solution for the perfect reconstruction linear phase set where one filter is of size 3 and the other of size 5 in dimensions $(1, \cdots, n)$ can be generated from a general solution for the perfect reconstruction linear phase set with the same sizes in dimensions $(1, \cdots, n-1)$ with

$$H_{00}(z_u^{(n)}) = H_{00}(z_u^{(n-1)}) + a \cdot z_1^{-1}(z_n^{-1} + z_n),$$

$$H_{01}(z_u^{(n)}) = H_{01}(z_u^{(n-1)}),$$

$$H_{10}(z_u^{(n)}) = H_{10}(z_u^{(n-1)}) + H_c(z_u^{(n-1)}) \cdot z_1^{-1}(z_n^{-1} + z_n) + c \cdot z_1^{-2}(z_n^{-2} + z_n^2),$$

$$H_{11}(z_u^{(n)}) = H_{11}(z_u^{(n-1)}) + d \cdot z_1^{-1}(z_n^{-1} + z_n),$$

where

$$H_{01}(z_u^{(n-1)}) = b, b \neq 0,$$

$$H_c(z_u^{(n-1)}) = \frac{d \cdot H_{00}(z_u^{(n-1)}) + a \cdot H_{11}(z_u^{(n-1)})}{b},$$

$$a \cdot d = b \cdot c,$$

$$t_{n-1} + 2ad \neq 0.$$

Here t_{n-1} is the only nonzero coefficient in the determinant of the polyphase matrix in dimensions $(1, \cdots, n-1)$, i.e., $\det \boldsymbol{H}_p(z_u^{(n-1)}) = t_{n-1} \cdot z_1^{-2}$. The determinant of the polyphase matrix in dimensions $(1, \cdots, n)$ is then $\det \boldsymbol{H}_p(z_u^{(n)}) = (t_{n-1} + 2ad) \cdot z_1^{-2} = t_n \cdot z_1^{-2}$.

2) The cascade of k polyphase matrices as above will generate a linear phase perfect reconstruction filter set of the same shape, where the first filter is of size $(2k+1)$ and the second one of size $(2k+3)$ in all dimensions, and where all the polyphase components are $\mathcal{A}$-polynomials.

Proof: For proof and properties of $\mathcal{A}$-polynomials, see Appendix D.

The one-dimensional solution as in Lemma 5.2 appeared in [44] while a two-dimensional diamond shaped filter pair was proposed in [43]. Basically, the previous lemma gives a possibility to generate n-dimensional filters of the above sizes from the $(n-1)$-dimensional ones of the same size. It should be noted that the initial 3/5 solution is completely general regardless of the number of dimensions. The way higher dimensional filters are constructed from the lower dimensional ones is that the lower dimensional solution is kept and then smaller size filters are stacked upon it. An example showing how to construct the two-dimensional (quincunx) solution from the one-dimensional one is given in Section V-C. Note that the cascade obtained in part 2) of the lemma produces a linear phase set where both filters are symmetric. Bearing in mind the fact that they are odd in all dimensions it becomes obvious that they belong to the class 1) solution of Proposition 4.4.

2) How to Generate Cascades Being Orthogonal and Linear Phase: This section will deal with the four-channel separable two-dimensional case; thus the matrix characterizing the sampling process is $\boldsymbol{D} = 2 \cdot \boldsymbol{I}$ and the corresponding sampling density is $N = \det \boldsymbol{D} = 4$.

Let us first present a cascade structure that will generate four linear phase/orthogonal filters of the same size, where two of them are symmetric and the other two antisymmetric [22]

$$\boldsymbol{H}_p(z_1, z_2) = \boldsymbol{W}_4 \prod_{i=1}^{k} \boldsymbol{D}(z_1, z_2) \boldsymbol{R}_i \qquad (46)$$

where $\boldsymbol{W}_4$ is the matrix representing the Walsh-Hadamard transform of size 4, $\boldsymbol{D}$ is the matrix of delays containing the vector of the forward polyphase transform along the diagonal, and $\boldsymbol{R}_i$ are scalar persymmetric matrices of the following form:

$$\boldsymbol{R}_i = \begin{pmatrix} a_{i0} & a_{i2} & a_{i1} & a_{i3} \\ a_{i2} & \pm a_{i0} & \pm a_{i3} & a_{i1} \\ a_{i1} & \pm a_{i3} & \pm a_{i0} & a_{i2} \\ a_{i3} & a_{i1} & a_{i2} & a_{i0} \end{pmatrix}. \qquad (47)$$

The "$-$" sign in (47) along with the requirement that the $\boldsymbol{R}_i$ be unitary allows one to design filters being both linear phase and orthogonal. In view of the fact that in the two-channel one-dimensional case these two requirements are mutually exclusive (see Proposition 4.6), it becomes obvious that one cannot design separable filters satisfying both properties in this four-channel two-dimensional case. This shows how using a true multidimensional solution offers greater freedom in design. For an example to the previous discussion, refer to [21]. At the same time, if a regular low-pass filter can be found (see Section VI), this cascade would allow one to generate orthogonal bases of linear phase (symmetric) wavelets in two dimensions, a construction which is not possible in current designs based on tensor products of one-dimensional systems.

The previous cascade generated filters of the same size. Now we show how to generate a cascade structure producing four linear phase filters of sizes $(2k+3) \times (2k+3)$, $(2k+3) \times (2k+1)$, $(2k+1) \times (2k+3)$ and $(2k+1) \times (2k+1)$. The basic building block is obtained for $k = 1$ and the corresponding filters' impulse responses are shown in Fig. 6. The cascade of such building blocks will produce again a linear phase set of sizes as above and of the same symmetry.

B. One to Multidimensional Transformations

Because of the difficulty of designing good filters in multiple dimensions, transformations mapping one-dimensional designs into multidimensional ones have been used for some time, the most popular being the McClellan transformation [14], [24].

In the context of filter banks and wavelets, one would like to transform a one-dimensional filter bank into a multidimensional one such that:

1) perfect reconstruction is preserved;
2) zeros at aliasing frequencies are preserved.

The first requirement is obvious, while the second one is necessary to achieve some degree of regularity (see Section VI-B).

Note that iteration of a one-dimensional filter with respect to a nonseparable lattice leads to a multidimensional filter (because upsampling transforms z into z^D; see (4)). This can be used in order to get multidimensional wavelets, and Cohen and Daubechies [7] have used this techniques to construct smooth wavelets from iterated one-dimensional filters (with respect to the dilation matrix D for the quincunx lattice). However, from a discrete filtering point of view, this is of little interest, since the filters are one-dimensional. In the following, we are going to consider transforms that lead to multidimensional filters.

1) Separable Polyphase Components: A first possible transform is obtained by designing a multidimensional filter having separable polyphase components, given as products of the polyphase components of a one-dimensional filter [1], [6]. To be specific, consider the quincunx subsampling case. Start with a one-dimensional filter having polyphase components $H_0(z)$ and $H_1(z)$, that is, a filter with a z-transform $H(z) = H_0(z^2) + z^{-1}H_1(z^2)$. Derive separable polyphase components

$$H_i(z_1, z_2) = H_i(z_1)H_i(z_2), \qquad i = 0, 1. \tag{48}$$

Then, the two-dimensional filter with respect to the quincunx lattice is given as (by upsampling the polyphase components with respect to D_1)

$$H(z_1, z_2) = H_0(z_1 z_2)H_0(z_1 z_2^{-1}) + z_1^{-1}H_1(z_1 z_2)H_1(z_1 z_2^{-1}). \tag{49}$$

It can be verified that an Nth-order zero at π in $H(e^{j\omega})$

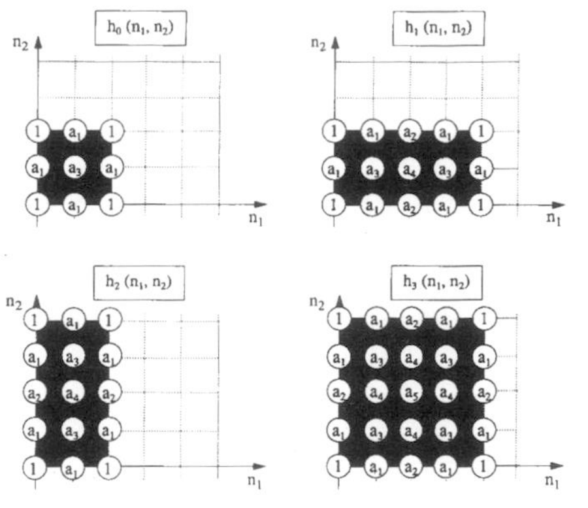

Fig. 6. The impulse responses of the four filters which form the first block in the cascade.

maps into an Nth-order zero at (π, π) for $H(e^{j\omega_1}, e^{j\omega_2})$, and thus property 2) is achieved. However, an orthogonal filter bank is mapped into an orthogonal two-dimensional bank, if and only if the polyphase components of the one-dimensional filter are all-pass functions (that is, $H_i(e^{j\omega})H_i(e^{-j\omega}) = c$, see (16)). Perfect reconstruction is thus not conserved in general. Note that the separable polyphase components lead to efficient implementations, reducing the number of operations from $O[l^2]$ to $O[l]$ per output where l is the filter size.

2) McClellan Transformation: The Fourier transform of a zero phase symmetric filter ($h(n) = h(-n)$) can be written as a function of $\cos(n\omega)$

$$\hat{H}(\omega) = \sum_{n=0}^{L} a(n)\cos(n\omega) \tag{50}$$

where $a(0) = h(0)$ and $a(n) = 2h(n)$, $n \neq 0$. Using Tchebycheff polynomials, one can replace $\cos(n\omega)$ by $T_n[\cos(\omega)]$ where $T_n[\cdot]$ is the nth Tchebycheff polynomial, and thus $\hat{H}(\omega)$ can be written as a polynomial of $\cos(\omega)$

$$\hat{H}(\omega) = \sum_{n=0}^{L} a(n)T_n[\cos(\omega)]. \tag{51}$$

The idea of the McClellan transformation is to replace $\cos(\omega)$ by a zero phase two-dimensional filter $\hat{F}(\omega_1, \omega_2)$, hence resulting in an overall zero phase two-dimensional filter [14], [24]

$$\hat{H}(\omega) = \sum_{n=0}^{L} a(n)T_n[\hat{F}(\omega_1, \omega_2)]. \tag{52}$$

In the context of filter banks, this transformation can only be applied to the biorthogonal case (because of the zero phase requirement). Typically, in the case of quincunx subsampling, $\hat{F}(\omega_1, \omega_2)$ is chosen as [3], [7]

$$\hat{F}(\omega_1, \omega_2) = \tfrac{1}{2}(\cos(\omega_1) + \cos(\omega_2)). \tag{53}$$

That the perfect reconstruction is preserved can be checked by considering the determinant of the matrix $\hat{H}_m(\omega)$ in (15) which is a monomial in the one-dimensional case since one starts with a perfect reconstruction filter bank. The transformation in (53) leads to a determinant of $\hat{H}_m(\omega_1, \omega_2)$ which is also a monomial, and thus, perfect reconstruction is conserved.

In addition to this, it is easy to see that pairs of zeros at π (that is, factors of the form $1 + \cos(\omega)$) map into zeros of order two at (π, π) in the transformed domain (or factors of the form $1 + 1/2\cos(\omega_1) + 1/2\cos(\omega_2)$).

Therefore, the McClellan transform is a powerful method to map one-dimensional biorthogonal solutions to multidimensional biorthogonal solutions, and this while conserving zeros at aliasing frequencies.

C. Cascade Structure for the Quincunx Case

As an illustration to Section V-A-1, the 3/5 set from Lemma 5.2 for the two-dimensional (quincunx) case will be constructed. That solution, in turn, is used in [19] for constructing filters for the FCO case (three-dimensional nonseparable two-channel case). We start from the one-dimensional solution for a general polyphase matrix in the upsampled domain [44] $H_{00}(z_1^2) = 1 + z_1^{-2}$, $H_{01}(z_1^2) = a_1$, $H_{10}(z_1^2) = 1 + a_2 z_1^{-2} + z_1^{-4}$, $H_{11}(z_1^2) = a_1(1 + z_1^{-2})$. To construct the two-dimensional solution one needs to evaluate the polynomial

$$H_c(z_1^2) = \frac{d \cdot (1 + z_1^{-2}) + a \cdot a_1(1 + z_1^{-2})}{a_1} = \left(a + \frac{d}{a_1}\right)(1 + z_1^{-2}) \qquad (54)$$

and also to express one of the variables a, c, and d using the two other ones. Thus, writing $c = ad/a_1$ the following is obtained (note that (z_1, z_2) in the upsampled domain is $(z_1 z_2, z_1 z_2^{-1})$ as given in (8))

$$H_{00}(z_1 z_2, z_1 z_2^{-1}) = 1 + z_1^{-2} + a \cdot z_1^{-1}(z_2^{-1} + z_2),$$

$$H_{01}(z_1 z_2, z_1 z_2^{-1}) = a_1,$$

$$H_{10}(z_1 z_2, z_1 z_2^{-1}) = 1 + a_2 z_1^{-2} + z_1^{-4} + \left(a + \frac{d}{a_1}\right) \cdot (1 + z_1^{-2}) z_1^{-1}(z_2^{-1} + z_2) + \frac{ad}{a_1} \cdot z_1^{-2}(z_2^{-2} + z_2^2),$$

$$H_{11}(z_1 z_2, z_1 z_2^{-1}) = a_1(1 + z_1^{-2}) + d \cdot z_1^{-1}(z_2^{-1} + z_2)$$

which yields the following impulse responses of the filters:

$$h_0(n_1, n_2) = \begin{pmatrix} & a & \\ 1 & a_1 & 1 \\ & a & \end{pmatrix}$$

$$h_1(n_1, n_2) = \begin{pmatrix} & & \frac{ad}{a_1} & & \\ & a + \frac{d}{a_1} & d & a + \frac{d}{a_1} & \\ 1 & a_1 & a_2 & a_1 & 1 \\ & a + \frac{d}{a_1} & d & a + \frac{d}{a_1} & \\ & & \frac{ad}{a_1} & & \end{pmatrix} \qquad (55)$$

which is the same solution as obtained in [43]. Using the result of Lemma 5.2 one can see that the cascades of the building blocks from (55) would generate diamond shaped filters of sizes $(2k + 1) \times (2k + 1)$ and $(2k + 3) \times (2k + 3)$ conserving the perfect reconstruction property. Fig. 7 shows how the smallest size filters are obtained. Note that if one would give up some freedom in design and impose additional circular symmetry, this solution would become the same as the one obtained by using McClellan transformation as explained in Section V-B-2. As will be seen in Section VII, this example will allow us to obtain a filter bank where the analysis low-pass can be made highly regular.

VI. Connection to Nonseparable Wavelet Bases

A. Constructing Wavelet Bases from Iterated Filter Banks

Let us first recall the one-dimensional case. There it is obvious that an orthogonal perfect reconstruction filter bank computes the discrete-time wavelet transform when the branch with the low-pass filter is iterated. Also, under certain conditions the same filter bank can be used to obtain a continuous-time wavelet transform [9], [23], [41], [42]. In multiple dimensions the basic ideas are the same, the only difference being that instead of dealing with dilation factors we deal with a dilation matrix.

Thus, consider Fig. 8. The equivalent low branch after i steps of filtering and sampling by $\boldsymbol{D}$ will contain the following filter and sampling by $\boldsymbol{D}^i$:

$$\hat{H}^{(i)}(\omega) = \prod_{k=0}^{i-1} \hat{H}_0\left((\boldsymbol{D}^t)^i \omega\right) \qquad i = 1, 2, \cdots \qquad (56)$$

where $\hat{H}^{(0)}(\omega) = 1$ and the fact that sampling by $\boldsymbol{D}$ followed by filtering by $\hat{H}(\omega)$ is equivalent to filtering by $\hat{H}(\boldsymbol{D}^t\omega)$ followed by sampling by $\boldsymbol{D}$ was used. The aim is to construct a continuous-time function corresponding to $h^{(i)}(\boldsymbol{n})$, the latter being the impulse response of the iterated filter $\hat{H}^{(i)}(\omega)$. Consequently, we define

$$f^{(i)}(\boldsymbol{x}) = N^{i/2} \cdot h^{(i)}(\boldsymbol{n}) \qquad \boldsymbol{D}^i \boldsymbol{x} \in \boldsymbol{n} + \left[-\tfrac{1}{2}, \tfrac{1}{2}\right)^n. \quad (57)$$

Obviously, $f^{(0)}(\boldsymbol{x})$ is just the indicator function over the

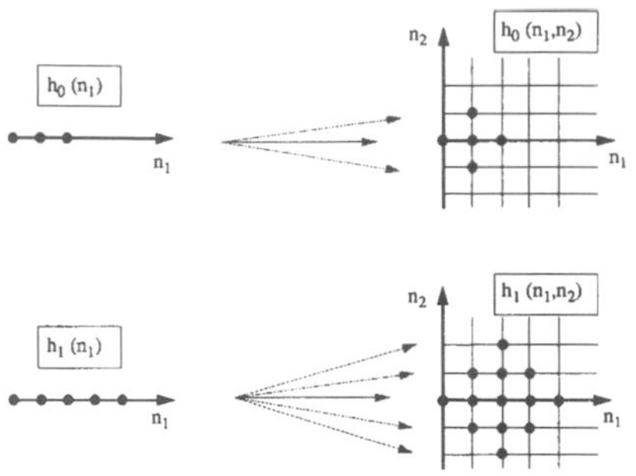

Fig. 7. The process of obtaining two-dimensional linear phase set from the one-dimensional one.

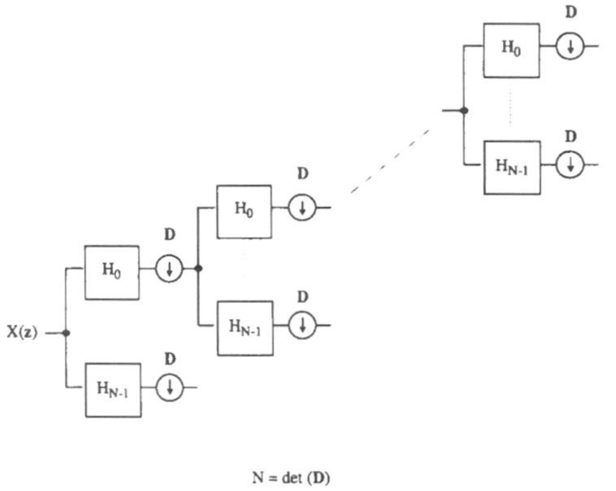

Fig. 8. Computation of the discrete wavelet transform using a filter bank.

hypercube $[-\frac{1}{2}, \frac{1}{2})^n$. The function in (57) is constant over regions of hypervolume $1/N^i$ and normalization by $N^{i/2}$ is chosen so that if $\|h^{(i)}(n)\|_2 = 1$, then $\|f^{(i)}(x)\|_2 = 1$ as well. Note that the shape of the hyperregions defined in (57) is not rectangular, rather it is determined by the shape of the unit cell belonging to the lattice D^{-i}. As in the one-dimensional case we are interested in the limiting behavior of the iterated function. Thus, we conjecture that the iterated filter can be made regular (i.e., the iterated function continuous) by putting a sufficient number of zeros at the points in frequency domain where the repeated spectra occur, i.e., at the points belonging to the reciprocal lattice of D (see the discussion in Section II).

In what follows, except for the necessity of having a zero at aliasing frequencies (see Section VI-B), we will not deal with the problem of achieving the continuous limit of $f^{(i)}(x)$. This is discussed in detail in [7] for the quincunx case. We will instead start by assuming that the limit of $f^{(i)}(x)$ exists and is in L^2 and will call it the *scaling function* associated with the discrete filter $h_0(n)$

$$\phi(x) = \lim_{i\to\infty} f^{(i)}(x), \qquad \phi(x) \in L^2. \tag{58}$$

Morever, we will assume for simplicity, that this limit function is continuous. We want to show now that the scaling function satisfies the so-called *two-scale equation* [9], [11], [32], [41]. Following (56) we can write the equivalent filter after i steps in terms of the equivalent filter after $(i-1)$ steps as

$$h^{(i)}(n) = \sum_k h_0(k) h^{(i-1)}(n - D^{i-1}k). \tag{59}$$

Using (57) we want to express the previous equation in terms of iterated functions and thus

$$h^{(i)}(n) = N^{-i/2} \cdot f^{(i)}(x), \tag{60}$$

$$h^{(i-1)}(n - D^{i-1}k) = N^{-(i-1)/2} \cdot f^{(i-1)}(Dx - k) \tag{61}$$

both for $D^i x \in n + [-\frac{1}{2}, \frac{1}{2})^n$. Substituting equations (60) and (61) into (59) yields

$$f^{(i)}(x) = \sqrt{N} \sum_k h_0(k) f^{(i-1)}(Dx - k). \tag{62}$$

Recall that the assumption is that the iterated function $f^{(i)}(x)$ converges to the scaling function. Hence, we can take the limits of both sides of (62) to obtain

$$\phi(x) = \sqrt{N} \sum_k h_0(k) \phi(Dx - k) \tag{63}$$

showing that indeed the scaling function satisfies the two-scale equation. Note that in the multidimensional case the change of scale involves the dilation matrix D. Until now we have been concerned only with the iterated low-pass filter. But what happens with other branches in the last stage of iteration? Following the same arguments as before, each one of them can be expressed as the iterated low-pass filter $h^{(j)}(n)$ followed by one filter $h_i(n)$. Since the former one tends to $\phi(x)$, $(N-1)$ *wavelets* can be defined, each one satisfying

$$\psi_i(x) = \sqrt{N} \sum_k h_i(k) \phi(Dx - k)$$
$$i = 1, 2, \cdots, N-1. \tag{64}$$

Refer to Section III-B. There the orthogonality relations between filters and their translates with respect to the sampling lattice were shown (see (21)). We want to use those results to obtain the same kind of relationships for the scaling function and the $(N-1)$ wavelets. Here, we just state the facts and outline the proof for the first one, the others would follow similarly:

1) $\langle \phi(x), \phi(x - l)\rangle = \delta_l$, that is, the scaling function is orthogonal to its integer translates;
2) $\langle \phi(D^i x - l), \phi(D^i x - k)\rangle, = N^{-i}\delta_{lk}$, i.e., the previous fact holds for all scales;
3) $\langle \psi_m(D^i x - l), \psi_n(D^i x - k)\rangle = N^{-i}\delta_{mn}\delta_{lk}$, wavelets are orthogonal to each other and their integer translates;
4) $\langle \phi(x), \psi_m(x - l)\rangle = 0$, the scaling function is orthogonal to each of the wavelets; and
5) $\langle \psi_m(D^i x - l), \psi_n(D^j x - k)\rangle = N^{-i}\delta_{mn}\delta_{ij}\delta_{lk}$, wavelets are orthogonal across scales.

To prove the first fact we use induction on the function $f^{(i)}$ and then take the limit (which exists by assumption). The first step $\langle f^{(0)}(x), f^{(0)}(x-k)\rangle = \delta_k$ is obvious since by definition $f^{(0)}(x)$ is just the indicator function on the hypercube $x \in [-\frac{1}{2}, \frac{1}{2})^n$. For the inductive step we can write

$$\begin{aligned}\langle f^{(i+1)}(x), f^{(i+1)}(x-l)\rangle &= \Big\langle \sqrt{N}\sum_k h_0(k) f^{(i)}(Dx-k), \\ &\qquad \sqrt{N}\sum_m h_0(m) f^{(i)}(Dx-Dl-m)\Big\rangle, \\ &= N\sum_k\sum_m h_0(k)h_0(m) \\ &\qquad \cdot \langle f^{(i)}(Dx-k), f^{(i)}(Dx-Dl-m)\rangle, \\ &= \sum_m h_0(m)h_0(Dl+m), \\ &= \langle h_0(m), h_0(Dl+m)\rangle = \delta_l\end{aligned}$$

where we used (21). Taking the limit of both sides of the previous equation we get exactly the first fact. □

We have thus verified that

$$S = \{N^{-m/2}\psi_i(D^{-m}x - n) \mid i = 1,\cdots, N-1, m \in \mathscr{Z}, \quad n \in \mathscr{Z}^n, x \in \mathscr{R}^n\} \tag{65}$$

is an orthonormal set. The only thing left to do is to show that the members of the set S constitute an orthonormal basis for $L^2(\mathscr{R}^n)$. The way to do it is to verify that S is a tight frame with framebound equal to one [9]. In Section VI-D we will show that this is indeed true for the two-channel case in any number of dimensions.

B. Necessity of Zeros at Aliasing Frequencies

In this section we want to extend Rioul's one-dimensional result on the necessity of a zero at π [29]. This result then makes it plausible why one would try to impose a zero of a sufficiently high order at aliasing frequencies (or points of repeated spectra).

Theorem 6.1: If the scaling function $\phi(x)$ exists for some $x \in \mathscr{R}^n$, then

$$\sum_{k\in\mathscr{Z}^n} h(Dk + k_i) = \frac{1}{\sqrt{N}}, \qquad k_i \in \mathscr{U}_c, \tag{66}$$

or in other words

$$\hat{H}(\omega = 0) = \sqrt{N}, \qquad \hat{H}\big(\omega = 2\pi(D^t)^{-1}n\big) = 0, \qquad n \in \mathscr{Z}^n \tag{67}$$

where $2\pi(D^t)^{-1}n$ are the aliasing frequencies or the points of repeated spectra.

Proof: Here we give just an outline of the proof; for more details refer to [18]. Note first that we can write (59) as

$$h^{(i)}(n) = \sum_k h^{(i-1)}(k) h_0(n - Dk) \tag{68}$$

and thus

$$h^{(i)}(Dn) = \sum_k h_0(Dk) h^{(i-1)}(n-k). \tag{69}$$

Using the same approach as in derivation (60)–(63), express $h^{(i-1)}$ and $h^{(i)}$ in terms of $f^{(i-1)}$ and $f^{(i)}$ and then take the limits (we are allowed to do so by assumption)

$$\phi(Dx) = \sqrt{N}\sum_k h_0(Dk)\phi(Dx). \tag{70}$$

Writing (69) for all the elements of the unit cell, i.e., $h^{(i)}(Dn + n_i)$, $n_i \in \mathscr{U}_c$, and following the same path as for $h^{(i)}(Dn)$ we finally obtain

$$\phi(Dx) = \sqrt{N}\sum_k h_0(Dk + k_i)\phi(Dx), \qquad \forall k_i \in \mathscr{U}_c. \tag{71}$$

Equating (71) for all the various coset representatives results in (66). □

C. Choice of the Dilation Matrix

In the Introduction we mentioned that the dilation matrix D must satisfy the following conditions:

1) $|\lambda_i| > 1$, $\forall i$, and
2) $D\mathscr{Z}^n \subseteq \mathscr{Z}^n$

where λ_i denote the eigenvalues of the matrix D. The first condition ensures that there is indeed a dilation in each dimension [7], [16]. Such a matrix will be called a *well-behaved* matrix. From the discrete filtering point of view, we would also like to use matrices that lead to separable sampling after a small number of iterations. That is the reason why we use the matrix D_1 rather than D_2 when dealing with the quincunx case (see Sections VI-E and VII). By the same token we would use the following matrices for the *hexagonal* and *FCO* lattices

$$D_{HEX} = \begin{pmatrix} 2 & 1 \\ 0 & -2 \end{pmatrix}, \qquad D_{HEX}^2 = \begin{pmatrix} 4 & 0 \\ 0 & 4 \end{pmatrix}, \tag{72}$$

$$D_{FCO} = \begin{pmatrix} 1 & 0 & 1 \\ -1 & -1 & 1 \\ 0 & -1 & 0 \end{pmatrix}, \qquad D_{FCO}^3 = \begin{pmatrix} 2 & 0 & 0 \\ 0 & 2 & 0 \\ 0 & 0 & 2 \end{pmatrix}. \tag{73}$$

At the same time, the examples discovered by Gröchenig and Madych in [16], as well as Lawton and Resnikoff [49], that use equivalents of Haar bases for filters and by construction their iterates are self-similar and tile the space, show that vastly different behavior is obtained when using different matrices for the same lattice. For example, the matrix D_2 ("twin dragon" in the Haar case) would lead to fractal support while D_1 would lead to parallelepiped support.

Fig. 9 gives an interesting example of a "dragon" for the hexagonal lattice. It has been generated using the lattice given in (72) with the following filter $H(z_1, z_2) = 1 + z_1^{-1} + z_1^{-1}z_2^{-1} + z_1^{-1}z_2$. The plot gives the sixth iteration (plotted on the rectangular grid since the sixth iteration corresponds to separable sampling). When the unit cell of this lattice is used as a basic filter, one obtains the patterns similar to those in [16] with the dilation matrix $\begin{pmatrix} 2 & 1 \\ 0 & 2 \end{pmatrix}$.

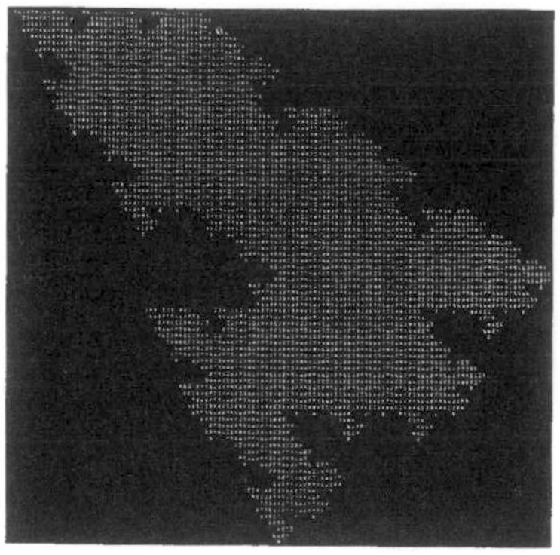

Fig. 9. A dragon obtained when iterating $H(z_1, z_2) = 1 + z_1^{-1} + z_1^{-1}z_2^{-1} + z_1^{-1}z_2$ with the dilation matrix as in (72). Sixth iteration is given plotted on the rectangular grid.

D. Two-Channel Orthonormal Bases

Assume that we are dealing with the n-dimensional two-channel case characterized by a matrix D, with $|\det D| = 2$ and D is a well-behaved matrix (see Section VI-C). Suppose now the following for $h_0(n)$, $h_1(n)$, $\phi(x)$, and $\psi(x)$ where n and x are n-dimensional integer and real vectors, respectively.

If:

1) filters h_0 and h_1 are orthogonal to each other and their translates (as given in (21));
2) the low-pass filter has a zero at aliasing frequencies (see (67));
3) the filters are FIR;
4) h_1 is specified from h_0 by the statement of Corollary 4.2;
5) the scaling function is the limit of the iterated functions as given by (58);
6) the wavelet is a linear combination of the scaling function and its shifts (see (64)); and
7) the scaling function and the wavelet are orthogonal to each other and their integer translates across scales (see the enumerated properties that appear after (64));

then (note that 4) follows from 1) and 7) follows from 5)).

Theorem 6.2: The orthonormal set of functions $S = \{\psi_m n \mid m \in \mathcal{Z}, n \in \mathcal{Z}^n, x \in \mathcal{R}^n\}$ where $\psi_{mn} = 2^{-m/2} \cdot \psi(D^{-m}x - n)$ is a basis for $L^2(\mathcal{R}^n)$, i.e., for $\forall f \in L^2(\mathcal{R}^n)$

$$\sum_{m \in \mathcal{Z}\, n \in \mathcal{Z}^n} |\langle \psi_{mn}, f \rangle|^2 = \|f\|_2^2. \tag{74}$$

Proof: The proof of the theorem is the n-dimensional version of the proof given in [9] with appropriate modifications pertaining to n-dimensional Fourier transform, and with the dilation matrix D instead of the dilation factor 2. For more details, we refer the reader to [18].

E. Final Visit to the Quincunx Case

For the purpose of the following analysis we will use the matrix D_1 as given by (2) in the Introduction, the reason being that when iterated this matrix would lead to separable sampling in every other step. Thus, this gives us an opportunity to check all of our results in the case that is very well understood, namely two dimensions with dilation factors of 2 in each one of them. As noted earlier, the quincunx case corresponds to a two-channel filter bank. Consequently, we will have a scaling function and one wavelet. To be consistent with our previous notation the low-pass filter is denoted by h_0 and the high-pass by h_1.

Let us first see how the "graphical" function defined in (57) looks like

$$f^{(i)}(x_1, x_2) = 2^{i/2} \cdot h_0^{(i)}(n_1, n_2)$$

$$\begin{pmatrix} 1 & 1 \\ 1 & -1 \end{pmatrix}^i \begin{pmatrix} x_1 \\ x_2 \end{pmatrix} \in \begin{pmatrix} n_1 \\ n_2 \end{pmatrix} + [-\tfrac{1}{2}, \tfrac{1}{2}) \times [-\tfrac{1}{2}, \tfrac{1}{2}). \tag{75}$$

As we said, the regions as defined above are not in general rectangular. To see that, consider what happens for $(n_1, n_2) = (0, 0)$ and the first few iterations. Fig. 10 shows these regions for $i = 1, 2, 3$. For $i = 0$ the basic support is just the square $[-\frac{1}{2}, \frac{1}{2}) \times [-\frac{1}{2}, \frac{1}{2})$ as can be seen from (75). In the first and the third iteration the regions are diamonds (tilted squares), while in the second one it is a square as we expected because that case (second iteration) corresponds to separable sampling. We conjecture that to make this graphical function continuous we have to place a zero of a sufficiently high order at the points of repeated spectra, i.e., at (π, π) (following the discussion in Section VI-B we know that at least one zero is necessary). For a design example using this criterion see the next section and Fig. 11 where the tenth iteration is given (the regions over which the function is plotted are square). The limit function as obtained in (63)

$$\phi(x) = \sqrt{2} \sum_k h_0(k)\phi(D_1 x - k) \tag{76}$$

thus satisfies a two-scale equation with respect to scale change given by D_1 and is orthogonal to its integer shifts. Similarly, the wavelet obtained in (64)

$$\psi(x) = \sqrt{2} \sum_k h_1(k)\phi(D_1 x - k) \tag{77}$$

together with its integer shifts and scales by D_1 will form an orthonormal set.

VII. Design of Compactly Supported Wavelets

The design of multidimensional filters is a difficult task from a signal processing point of view, but it becomes all the more involved by introducing the requirement that the low-pass filter be regular. To design a filter having a number of vanishing moments at a particular location in one dimension is made possible by the existence of the factorization theorem. Following our conjecture, we would like to do the same

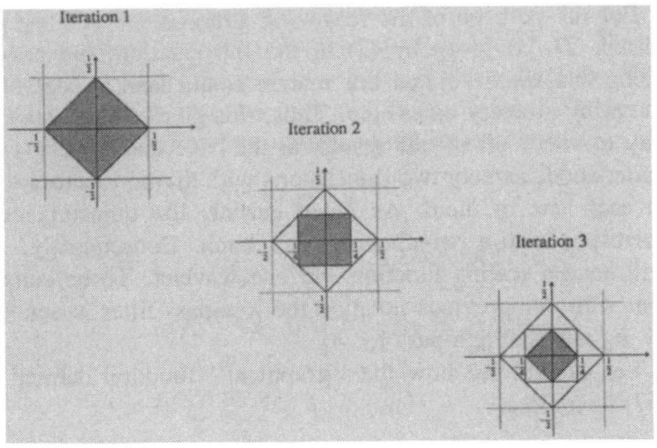

Fig. 10. Shaded regions show the basic supports obtained when iterating the continuous function as given in (75) for the quincunx case.

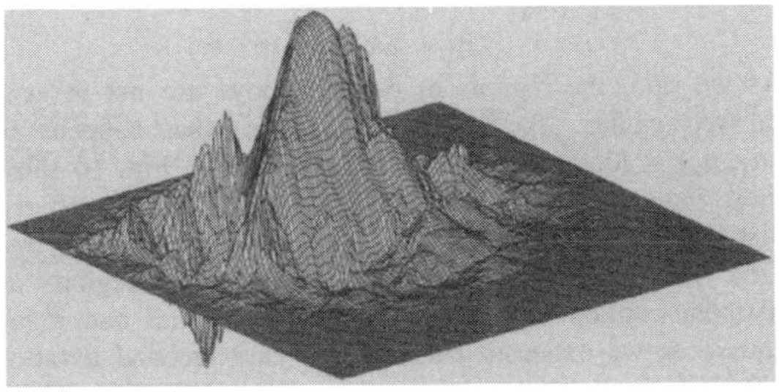

Fig. 11. Tenth iteration of a filter in (78) with coefficients as in (79). The dilation matrix used is D_1 from (2). The plot is given on the rectangular grid.

for the multidimensional case, but unfortunately factorization theorems are lacking. Thus solving the problem, except for very small cases, has to be done numerically. If one wants an orthogonal solution the equivalent system of equations becomes nonlinear. Therefore, here we show a very small orthogonal design example that was obtained algebraically solving a system of nonlinear equations. First note that the filter bank is obtained using the cascade structure from Lemma 5.1 where matrices R_{j_i} are unitary and the sampling matrix used is again D_1 from (2). Here we use the smallest (4 × 3) filer pair, since it was shown in [43], that it is general and thus we know that the search space is complete. The impulse response of the low-pass filter is

$$h_0(n_1, n_2) = \begin{pmatrix} & -a_1 & -a_0 a_1 & \\ -a_2 & -a_0 a_2 & -a_0 & 1 \\ & a_0 a_1 a_2 & -a_1 a_2 & \end{pmatrix}. \quad (78)$$

Following our conjecture (see Section VI) the approach would be to try to impose as high as possible an order of a zero at (π, π). An mth-order zero means that all the partial derivatives $(\partial^{k-1} H_\omega(\omega_1, \omega_2)/\partial^l \omega_1\, \partial^{k-l-1}\omega_2)|_{(\pi,\pi)} = 0$, $k = 1, \cdots, m-1$, $l = 0, \cdots, k-1$. Upon imposing a second-order zero, the following solutions are obtained:

$$a_0 = \mp\sqrt{3} \quad a_1 = \mp\sqrt{3} \quad a_2 = 2 \pm \sqrt{3}, \quad (79)$$

$$a_0 = \pm\sqrt{3} \quad a_1 = 0 \quad a_2 = 2 \pm \sqrt{3}. \quad (80)$$

The solution obtained in (80) is the one-dimensional Daubechies' D4 filter [9]. Hence, we conjecture that the first solution (actually there are two of them but they are related by reversal) would be the smallest "regular" two-dimensional filter. Fig. 11 shows the tenth iteration of the filter in (78) with coefficients as in (79) plotted using matrix D_1 from (2). The plot is given on the rectangular grid. As can be seen from the figure the obtained function looks continuous although not differentiable at some points. As one simple check of continuity we computed the largest first-order differences in the first seven iterations on the rectangular (or 14 on the quincunx) grid. They are given in Table I. As can be seen from the table the largest difference decreases with an almost constant rate which is a good indicator of a function being continuous. For the purpose of this calculation the filter was normalized so that its l^2 norm is 1, and the differences were computed for the iterated function as given in (75). Using the result of the Theorem 6.2 one can then conjecture that the above filter would lead to an orthonormal basis of a compactly supported wavelet. Let us note that Daubechies and Cohen discovered with the same example [7].

For larger size filters problems start to arise. First, the cascade structure given by Lemma 5.1 not being complete we are not even sure we are searching over the whole space of possible solutions. Next, even for the first larger size, namely a filter of size (6 × 5), the system of nonlinear equations could not be solved analytically. Thus, already for this case having five free variables one would have to resort to numerical solutions.

Turning to the linear phase case, we used the cascade structure given by Lemma 5.2, the reason being that the diamond-shaped structure allows us to have a highly regular low-pass obtained when convolving the filter $h_0(n_1, n_2)$ from (55) (with $a = 1$ and $a_1 = 4$)

$$h_0(n_1, n_2) = \begin{pmatrix} & 1 & \\ 1 & 4 & 1 \\ & 1 & \end{pmatrix} \quad (81)$$

a number of times with itself. Writing the Fourier-domain expression for (81) one obtains

$$\hat{H}_0(\omega_1, \omega_2) = 1 + e^{-j2\omega_1} + e^{-j(\omega_1+\omega_2)} + e^{-j(\omega_1-\omega_2)} + 4e^{-j\omega_1} \quad (82)$$

showing that it possesses a second-order zero at (π, π). Although not differentiable, it is continuous [12] (see Fig. 12). However, since the filter bank is not orthogonal, the synthesis low-pass is not equal to the analysis low-pass, but rather to the analysis high-pass modulated by $(-1)^{n_1+n_2}$. Thus, taking the synthesis low-pass from the cascade given in Lemma 5.2 we imposed a second-order zero at (π, π) along with some additional symmetry resulting in the following impulse response:

$$g_0(n_1, n_2) = \begin{pmatrix} & & 1 & & \\ & 2 & -4 & 2 & \\ 1 & -4 & -28 & -4 & 1 \\ & 2 & -4 & 2 & \\ & & 1 & & \end{pmatrix}. \quad (83)$$

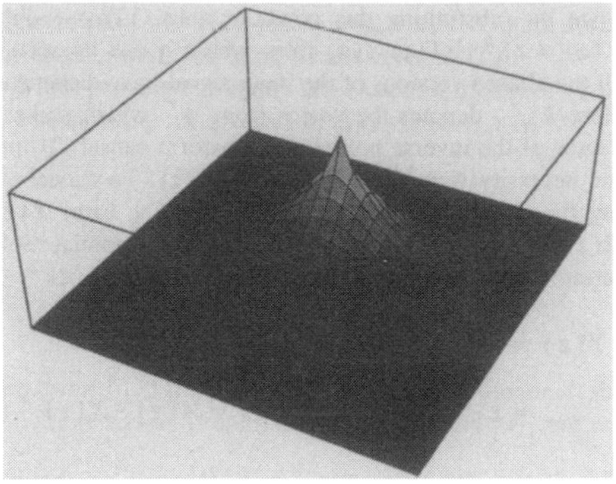

Fig. 12. Sixth iteration of the filter in (81). The dilation matrix used is D_1 from (2). The plot is given on the rectangular grid.

TABLE I
THE SUCCESSIVE LARGEST FIRST-ORDER DIFFERENCES FOR THE FILTER GIVEN IN (78) WITH COEFFICIENTS AS IN (79) COMPUTED ON THE RECTANGULAR GRID

Iteration Number	Largest first Order Difference	Rate of Convergence
2	1.25163960	
4	0.91034730	1.3749
6	0.62581208	1.4547
8	0.55111247	1.1355
10	0.51048814	1.0796
12	0.46069373	1.1081
14	0.40993778	1.1238

Unfortunately, it turns out that the above filter leads to a fractal iterated function (see Fig. 13(a) and (b)). Daubechies and Cohen in [7] show that a regular synthesis low-pass corresponding to the analysis low-pass as given in (81) would be a diamond of size 57, that is, impractical.

As we mentioned earlier, the filter given in (81) allows one to obtain filters with arbitrarily high regularity. When convolved just once ($\hat{H}_0^2(\omega_1, \omega_2)$) it already yields a continuous and differentiable iterated filter [7] as can be seen from Fig. 14 (sixth iteration is shown).

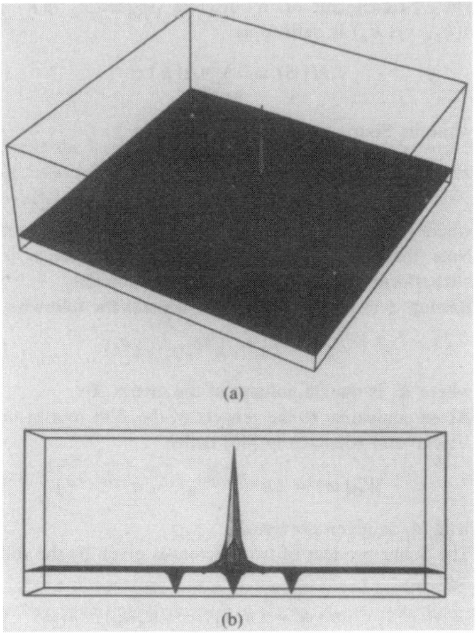

Fig. 13. (a) Sixth iteration of the filter in (83). (b) The same as (a) with a different viewpoint.

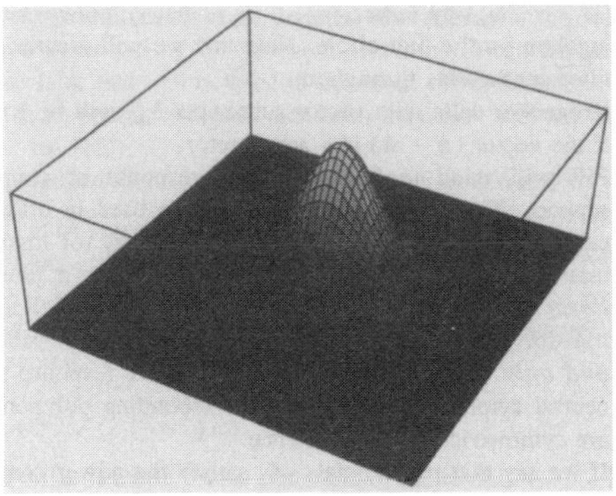

Fig. 14. Sixth iteration of a filter obtained when convolving (81) with itself. The plot is given on the rectangular grid.

VIII. CONCLUSION

This paper presented new results on multidimensional filter banks and their connection to multidimensional nonseparable wavelets.

Many results are similar to their one-dimensional counterparts, but with the dilation factor replaced by a dilation matrix $\boldsymbol{D}$. This matrix plays a central role: its nonuniqueness for a given lattice becomes important because iterated filter banks (which are the key to our construction of wavelets following [9]) lead to taking powers of $\boldsymbol{D}$. Thus, very unlike the one-dimensional case, a given filter can lead to vastly different scaling functions depending on $\boldsymbol{D}$.

Design techniques for filters leading to regular wavelets do not carry over to the multidimensional case easily, as expected. Therefore, the design of regular wavelets in multiple dimensions still poses a number of challenges. Some initial results and conjectures on regularity were given, indicating the directions for future work.

APPENDIX A

NOTATION AND DEFINITIONS

This appendix establishes the notation used in the paper. Most of the definitions involving multidimensional z-transform and sampling are taken from [46].

1) Boldface lower/upper case letters will denote vectors and matrices, respectively. Raising an n-dimensional complex vector $\boldsymbol{z} = (z_1, \cdots, z_n)$ to an n-dimensional integer vector $\boldsymbol{k} = (k_1, \cdots, k_n)$ yields

$$z^k = z_1^{k_1} z_2^{k_2} \cdot \cdots \cdot z_n^{k_n}. \quad (84)$$

2) The z-transform of a discrete sequence $h(k) = h(k_1, \cdots, k_n)$ is defined as

$$H(z) = \sum_{k \in \mathcal{Z}^n} h(k) z^{-k} \tag{85}$$

while its Fourier transform is given by

$$\hat{H}(\omega) = \sum_{k \in \mathcal{Z}^n} h(k)\, e^{-j\langle \omega, k\rangle} \tag{86}$$

where $\langle \omega, k \rangle$ denotes the inner product of the two vectors. Note that the Fourier transform of a sequence is its z-transform evaluated on the unit hypercircles.

3) Raising z to a matrix power D denotes the following:

$$z^D = (z^{d_1}, z^{d_2}, \cdots, z^{d_n}) \tag{87}$$

where d_i is the ith column of the matrix D.

4) As an equivalent to the powers of the Nth root of unity, Viscito and Allebach in [46] define

$$W_D(\omega) = (e^{-j\langle \omega, d_1\rangle}, \cdots, e^{-j\langle \omega, d_n\rangle}) \tag{88}$$

with d_i as given previously.

5) The Schur product of two vectors is given by the following:

$$n \circ k = (n_1 k_1, \cdots, n_n k_n). \tag{89}$$

6) $\tilde{H}(z)$ will mean transposition of the matrix, conjugation of coefficients and substitution of $z = (z_1, \cdots, z_n)$ by $(z_1^{-1}l, \cdots, z_n^{-1})$, or equivalently substitution of $\omega = (\omega_1, \cdots, \omega_n)$ by $(-\omega_1, \cdots, -\omega_n)$, that is, complex conjugation on the unit circle. Note that we will assume real filter coefficients throughout.

7) Kronecker delta with vector subscripts δ_{nk} will be 1 only if the vector $(n - k)$ is a zero vector.

8) FIR will stand for finite impulse response or compact support. The term *orthogonal* will be used in different contexts, among others to denote paraunitary (or lossless) matrices (see Section III-B). If we say that a filter is *linear phase* it will mean that the angle in its Fourier transform is a linear function of ω, and this is possible *if and only if* the coefficients of its impulse response have central symmetry, i.e., if the corresponding polynomials are symmetric or antisymmetric.

9) If we say that polynomials A_i satisfy the power complementary property it will mean

$$\sum_i A_i(z) \tilde{A}_i(z) = 1 \tag{90}$$

and the abbreviation PC will be used. In particular on the unit hypercircles ($z_i = e^{j\omega_i}$, $i = 1, \cdots, n$) (90) means that the magnitudes squared of the Fourier transforms of A and B sum up to one, i.e.,

$$\sum_i |\hat{A}_i(\omega)|^2 = 1. \tag{91}$$

10) The identity matrix will be denoted by I, and the matrix with 1's along the antidiagonal by J.

Appendix B

Proof of Lemma 3.1

We prove the lemma point by point.

1) Noting that $p_i^t(z) x_p(z^D) = X(z)$ sufficiency is obvious since by substituting this condition into (12) one obtains $Y(z) = z^{-k_{N-1}} T(z) X(z)$ from where it can be seen that all the aliased versions of the input signal have disappeared. Here k_{N-1} denotes the vector from $\mathcal{U}_c^t$ which makes the vector of the inverse polyphase transform causal. To prove the necessity write $p_i^t \cdot T_p(z^D) = v^t(z)$. To cancel aliasing the output of the bank has to be of the form $Y(z) = A(z)X(z)$ where $A(z)$ is just a scalar polynomial. Substituting $v(z)$ into (12) and equating it to $A(z)X(z)$

$$\begin{aligned} Y(z) &= z^{-k_{N-1}} p_i^t(z) \cdot T_p(z^D) \cdot x_p(z^D) \\ &= z^{-k_{N-1}} v^t(z) \cdot x_p(z^D) = A(z) \cdot X(z). \end{aligned} \tag{92}$$

Substituting the expression for $X(z)$ as given above into the right-hand side of (92) yields

$$\begin{aligned} A(z) \cdot X(z) &= A(z) \cdot \sum_{k \in \mathcal{U}_c^t} z^k X_k(z^D) \\ &= \sum_{k \in \mathcal{U}_c^t} b_k(z) \cdot X_k(z^D), \qquad (93) \\ &= b^t(z) \cdot x_p(z^D) \\ &= z^{-k_{N-1}} v^t(z) x_p(z^D) \qquad (94) \end{aligned}$$

from where it is obvious that since $b_k(z) = A(z) \cdot z^k = z^{-k_{N-1}} v_k(z)$ the vector $v(z)$ can be written as

$$\begin{aligned} v^t(z) &= \left(A(z) z^{k_{N-1}} z^{k_0} \cdots A(z) z^{k_{N-1}} z^{k_{N-1}}\right) \\ &= T(z) \cdot p_i^t \end{aligned} \tag{95}$$

which completes the proof of the first part.

2) If the system is perfect reconstruction then $Y(z) = cz^{-n}X(z)$. When compared to $Y(z) = z^{-k_{N-1}}T(z)X(z)$ as obtained in the proof of the first part it is obvious that $T(z) = c \cdot z^{-n} z^{k_{N-1}} = c \cdot z^{-k}$. On the other hand, if $T(z) = c \cdot z^{-k}$ then substituting it into $Y(z) = z^{-k_{N-1}} T(z) X(z)$ results in $Y(z) = z^{-k_{N-1}} \cdot c \cdot z^{-k} \cdot X(z) = c \cdot z^{-n} \cdot X(z)$, i.e., perfect reconstruction is achieved.

3) Sufficiency is obvious since by choosing $G_p(z) = \mathrm{Adj}(H_p(z))$ the transfer matrix T_p becomes a diagonal matrix of delays and

$$\begin{aligned} Y(z) &= z^{-k_{N-1}} p_i^t(z) \cdot z^{-k} \cdot I \cdot x_p(z^D) \\ &= z^{-n} \cdot X(z). \end{aligned}$$

To prove the necessity, use the result of the second part of the lemma and write $T(z) = c \cdot z^{-k}$. Then it can be seen that $T_p(z^D) = c \cdot z^{-k} I$. Thus, $\det G_p \det H_p = (c \cdot z^{-k})^n$ and since filters in both banks are FIR all the polynomial factors in both $\det G_p$ and $\det H_p$ have to be monomials. □

Appendix C

Proof of Theorem 4.1 and its Corollaries

This result for a one-dimensional system appeared in filter bank literature in [36] and in wavelet literature in [9], for a two-dimensional one in [43] and since the approach here is quite similar, we just outline the proof, for more details refer to [19].

Let us first prove a fact that will be used in the proof of the main theorem. Expanding the product in (17) yields

$$\tilde{H}_{00}(z)H_{00}(z) + \tilde{H}_{10}(z)H_{10}(z) = 1, \tag{96}$$

$$\tilde{H}_{00}(z)H_{01}(z) + \tilde{H}_{10}(z)H_{11}(z) = 0, \tag{97}$$

$$\tilde{H}_{01}(z)H_{00}(z) + \tilde{H}_{11}(z)H_{10}(z) = 0, \tag{98}$$

$$\tilde{H}_{01}(z)H_{01}(z) + \tilde{H}_{11}(z)H_{11}(z) = 1. \tag{99}$$

Proposition C.1: If the polyphase matrix is orthogonal, then H_{00} and H_{10} are relatively prime. Similarly, H_{01} and H_{11} are relatively prime.

Proof: Take the first statement. The second one is proved similarly. Suppose that H_{00} and H_{10} are not coprime. Then using the fact that a polynomial in n variables which is not identically zero can be resolved into the product of irreducible factors in only one way [4], one can take the common factors out and call their product $P(z_1, z_2, \cdots, z_n)$. Substituting this into (96) results in

$$P(z)\tilde{P}(z) \cdot \left(H'_{00}(z)\tilde{H}'_{00}(z) + H'_{10}(z)\tilde{H}'_{10}(z)\right) = 1 \tag{100}$$

which for all the vanishing points of $P(z)$ goes to zero, contradicting the fact that the right side of (100) identically equals 1. □

Proof of Theorem 4.1: Consider (97). Since by Proposition C.1 H_{00} and H_{10} are relatively prime, one can conclude that $H_{11}(z)$ must contain all the polynomial factors of $\tilde{H}_{00}(z)$ except for the monomial z^k which makes it causal. Thus, $H_{11}(z) = c_1 \cdot z^k \cdot \tilde{H}_{00}(z)$. Similarly $H_{10}(z) = c_2 \cdot z^{k'} \cdot \tilde{H}_{01}(z)$. Substituting this into (98) and cancelling $H_{00}(z)\tilde{H}_{01}(z)$ (it is allowed to do so since they are not identically equal to zero) yields $c_2 = -(1/c_1)$ and $k' = k$. Thus, $H_{11}(z) = c \cdot z^k \cdot \tilde{H}_{00}(z)$, and $H_{10}(z) = -(1/c) \cdot z^k \cdot \tilde{H}_{01}(z)$. Substituting this into (99) and using (96) it can be seen that c has to be either 1 or -1. This finally gives us the complete specification of the system as given in the statement of the theorem. □

Proof of Corollary 4.1: Take the polyphase components of the first filter (the proof for the second one is analogous). Since they are causal their corner coefficients are $P^{(0)} = P^{(1)} = (0, \cdots, 0)$ and $Q^{(0)}, Q^{(1)}$. Consider now the PC condition as given in the statement of the theorem. The polynomial $A(z) = H_{00}(z)\tilde{H}_{00}(z)$ will have the corner coefficients $P^{(A)} = -Q^{(0)}$ and $Q^{(A)} = Q^{(0)}$, while the corner coefficients of $B(z) = H_{01}(z)\tilde{H}_{01}(z)$ will be $P^{(B)} = -Q^{(1)}$ and $Q^{(B)} = Q^{(1)}$. For the PC property to hold the polynomials $A(z)$ and $B(z)$ must be of the same size for all coefficients (except the center one) to cancel, and thus both polyphase components have to be of the same size. □

Proof of Corollary 4.2: Using the matrix D_1 from (38) write

$$\tilde{H}_0(-z) = \tilde{H}_{00}(z^{D_1}) - z_1\tilde{H}_{01}(z^{D_1})$$

where $z^{D_1} = (z_1z_2, z_2z_3, \cdots, z_1z_n^{(-1)^{n-1}})$. Using the expressions obtained in the proof of the theorem for the polyphase components of the second filter write

$$\begin{aligned} H_1(z) &= H_{10}(z^{D_1}) + z_1^{-1}H_{11}(z^{D_1}), \\ &= z_1^{-(k_1+k_n)}z_2^{-(k_1+k_2)} \cdots z_n^{-(k_{n-1}+(-1)^{n-1}k_n)} \\ &\quad \cdot \tilde{H}_{01}(z^{D_1}), \\ &\quad - z_1^{-1}z_1^{-(k_1+k_n)}z_2^{-(k_1+k_2)} \cdots z_n^{-(k_{n-1}+(-1)^{n-1}k_n)} \\ &\quad \cdot \tilde{H}_{00}(z^{D_1}), \\ &= -z^{-k'}\tilde{H}_0(-z). \end{aligned}$$

□

Appendix D

Proof of Lemma 5.2

To prove the lemma a polynomial having a specific structure that will henceforth be denoted as an A-polynomial is first defined. A polynomial is an A-polynomial if it satisfies the following recursion:

$$AP^{(k)}(z_n) = AP^{(k)}(z_{n-1}) + \sum_{i=1}^{k} AP^{(k-i)}(z_{n-1})z_1^{-i}(z_n^{-i} + z_n^i) \tag{101}$$

where

$$AP^{(k)}(z_1) = z_1^{-2k}AP^{(k)}(z_1^{-1}) = \sum_{i=0}^{k} p_i z_1^{-2i}. \tag{102}$$

Here $z_i = (z_1, \cdots, z_i)$. Note that in one dimension an A-polynomial is just a symmetric polynomial having only even degree terms, in two dimensions it is a diamond shaped symmetric polynomial having again just even degree terms, a.s.o. Also note that the sum of two A-polynomials of the same degree is again an A-polynomial. A prefix "A" associated with the name of the polynomial will indicate that it is an A-polynomial, a superscript indicates its degree (superscript (0) would mean a constant). By d_c the degree of the center coefficient will be denoted (the degree is the sum of all the exponents) and by a_c the coefficient itself. Here the statements of some facts that will be used in the proof of the lemma are given. Since their proofs are just technical they are omitted here (for details, see [18]).

Proposition D.1: For $AP^{(k)}$ the following is true:

1) it is of size $(2k + 1)$ in all dimensions;
2) it has just even degree terms;
3) $i(d_c) = i(k)$;
4) $AP^{(k)}(z_n) = z_1^{-2k}AP^{(k)}(z_n^{-1})$, i.e., all A-polynomials are symmetric polynomials;
5) $z_1^{-2m}AP^{(k)} = AQ^{(k+2m)}$;
6) $AP^{(k)}AQ^{(m)} = AR^{(k+m)}$.

For the sake of clarity in the statement of the lemma the prefix "A" was omitted as well as the superscripts associated with the polynomials. For the proof they are reintroduced. Also note that the whole analysis will be performed in the upsampled domain.

Proof:

1) To prove the lemma let us write the general expression for the polyphase matrix containing polyphase components of sizes 3 and 5 in n dimensions (recall that $z_u^{(i)}$ denotes z^D where $z = (z_1, \cdots, z_i)'$)

$$H_p(z_u^{(n)}) = \begin{pmatrix} AH_{00}^{(1)}(z_u^{(n)}) & AH_{01}^{(0)}(z_u^{(n)}) \\ AH_{10}^{(2)}(z_u^{(n)}) & AH_{11}^{(1)}(z_u^{(n)}) \end{pmatrix}. \tag{103}$$

Due to their required size and the fact that they have to be linear phase it is obvious that the above polyphase components indeed are A-polynomials. Therefore, we can expand each one using (101). $AH_{01}^{(0)}(z_u^{(n)})$ having a superscript (0) is obviously a constant that will henceforth be denoted by b. Note that (101) implies that the first term can be taken as the corresponding polyphase component in $(n-1)$ dimensions. Then the determinant of (103) can be written as

$$\det H_p(z_u^{(n)}) = AH_{00}^{(1)}(z_u^{(n)})AH_{11}^{(1)}(z_u^{(n)}) - AH_{10}^{(2)}(z_u^{(n)})b,$$

$$= \overbrace{\underbrace{AH_{00}^{(1)}(z_u^{(n-1)})AH_{11}^{(1)}(z_u^{(n-1)}) - AH_{10}^{(2)}(z_u^{(n-1)})b + 2a\,dz_1^{-2}}_{T_1}}^{\det H_p(z_u^{(n-1)})}$$
$$+ \underbrace{(a \cdot AH_{11}^{(1)}(z_u^{(n-1)}) + d \cdot AH_{00}^{(1)}(z_u^{(n-1)}) - b \cdot AH_c^{(1)}(z_u^{(n-1)}))}_{T_2} \cdot z_1^{-1}(z_n^{-1} + z_n)$$
$$+ \underbrace{(ad - bc)}_{T_3} \cdot z_1^{-2}(z_n^{-2} + z_n^2). \tag{104}$$

a) Let us first consider the case when $b \neq 0$. Observe that T_2 and T_3 have to equal zero since they appear in pairs, and thus

$$a \cdot d = b \cdot c, \tag{105}$$

$$AH_c^{(1)}(z_u^{(n-1)}) = \frac{d \cdot AH_{00}^{(1)}(z_u^{(n-1)}) + a \cdot AH_{11}^{(1)}(z_u^{(n-1)})}{b}. \tag{106}$$

Here we still do not know what b, $AH_{00}^{(1)}(z_u^{(n-1)})$, and $AH_{11}^{(1)}(z_u^{(n-1)})$ are. Substituting (105) and (106) into (104) we are left with

$$\det H_p(z_u^{(n)}) = \det H_p(z_u^{(n-1)}) + 2a\,dz_1^{-2}$$

which obviously has to have a single nonzero coefficient. Using Proposition D.1 it can be shown that $\det H_p(z_u^{(n-1)})$ is again an A-polynomial with index (2) and it possesses a center term z_1^{-2}. Hence, we just have to find a solution which would make this A-polynomial become a single term. Note that achieving this is equivalent to finding a general solution in $(n-1)$ dimensions, which in turn yields b, $AH_{00}^{(1)}(z_u^{(n-1)})$, $AH_{11}^{(1)}(z_u^{(n-1)})$ and finally $AH_c^{(1)}(z_u^{(n-1)})$. This proves the lemma for the case when b is nonzero.

b) We want to show now by contradiction that b cannot be zero. Thus suppose that $b = 0$. This means that $ad = 0$ as well, i.e., $a = 0$ or $d = 0$ or both. Without loss of generality suppose that $a = 0$. Plugging all this into T_2 we see that $d \cdot AH_{00}^{(1)}(z_u^{(n-1)})$ has to equal zero as well since it appears in pairs. Now $AH_{00}^{(1)}(z_u^{(n-1)})$ cannot be zero since otherwise the whole determinant would be zero as well. Thus, we conclude that both a and d have to be zero. What we want to show now is that the only term left, i.e., $AH_{00}^{(1)}(z_u^{(n-1)})\,AH_{11}^{(1)}(z_u^{(n-1)})$ cannot have a single nonzero coefficient. To do this we can express each of these A-polynomials in terms of A-polynomials in $(n-2)$ dimensions and upon multiplying them observe that forcing the coefficient that appear in pairs to be zero, we are left with $AH_{00}^{(1)}(z_u^{(n-2)})\,AH_{11}^{(1)}(z_u^{(n-2)})$. Continuing this procedure we finally arrive to the same product in one dimension for which it is trivial to show that it cannot be forced to have a single nonzero coefficient.

2) To prove the second part of the lemma we have to use double induction, on the number of dimensions and on k. Both in the first and in the inductive step on n, the first step for $k = 1$ is trivial since it is just the 3/5 solution from the first part of the lemma. Thus, in each case, we will consider only inductive step on k. Also bear in mind that each polyphase component in the 3/5 solution from the first part of the lemma is an A-polynomial.

- $n = 1$:

$$\begin{pmatrix} AH_{00}^{(k-1)}(z_1^2) & AH_{01}^{(k-2)}(z_1^2) \\ AH_{10}^{(k)}(z_1^2) & AH_{11}^{(k-1)}(z_1^2) \end{pmatrix} \cdot \begin{pmatrix} AH_{00}^{(1)}(z_1^2) & AH_{01}^{(0)}(z_1^2) \\ AH_{10}^{(2)}(z_1^2) & AH_{11}^{(1)}(z_1^2) \end{pmatrix} = \begin{pmatrix} AH_{00}^{(k)}(z_1^2) & AH_{01}^{(k-1)}(z_1^2) \\ AH_{10}^{(k+1)}(z_1^2) & AH_{11}^{(k)}(z_1^2) \end{pmatrix}$$

where we used the results on A-polynomials stated previously.

- *Inductive step:* The proof is completely analogous to the first step. □

ACKNOWLEDGMENT

The authors would like to thank Dr. I. Daubechies and Dr. A. Cohen of AT&T Bell Labs for useful discussions.

REFERENCES

[1] R. Ansari, "Two-dimensional IIR filters for exact reconstruction in tree-structured subband decomposition," *Electron. Lett.*, vol. 23, no. 12, pp. 633–634, June 1987.
[2] R. Ansari, H. Gaggioni, and D. J. LeGall, "HDTV coding using a non-rectangular subband decomposition," in *Proc. SPIE Conf. Vis. Commun. Image Processing*, Cambridge, MA, Nov. 1988, pp. 821–824.
[3] M. Antonini, M. Barlaud, and P. Mathieu," Image coding using lattice vector quantization of wavelet coefficients," in *Proc. IEEE Int. Conf. ASSP*, Toronto, Canada, May 1991, pp. 2273–2276.
[4] M. Bocher, *Introduction to Higher Algebra*. New York: Macmillan, 1907.
[5] J. W. Cassels, *An Introduction to the Geometry of Numbers*. Berlin; Springer-Verlag, 1971.
[6] T. Chen and P. P. Vaidyanathan, "Multidimensional multirate filters derived from one-dimensional filters," *Electron. Lett.*, Jan. 1991.
[7] A. Cohen and I. Daubechies, "Nonseparable bidimensional wavelet bases," submitted to *Rev. Math. Iberoamericana*, 1991.
[8] A. Croisier, D. Esteban, and C. Galand, "Perfect channel splitting by use of interpolation/decimation/tree decomposition techniques," in *Int. Conf. Inform. Sci. Syst.*, Patras, Greece, Aug. 1976, pp. 443–446.
[9] I. Daubechies, "Orthonormal bases of compactly supported wavelets," *Commun. Pure Appl. Math.*, vol. 41, pp. 909–996, Nov. 1988.
[10] I. Daubechies and A. Cohen, Private communication.
[11] I. Daubechies and J. Lagarias, "Two-scale difference equations: II. Local regularity, infinite products of matrices and fractals," *SIAM J. Math. Anal.*, 1992, to appear.
[12] G. Deslauriers, J. Dubois, and S. Dubuc, "Multidimensional iterative interpolation," *Can. J. Math.*, vol. 43, no. 2, pp. 297–312, 1991.

[13] E. Dubois, "The sampling and reconstruction of time-varying imagery with application in video systems," *Proc. IEEE*, vol. 73, no. 4, pp. 502–522, Apr. 1985.

[14] D. E. Dudgeon and R. M. Mersereau, *Multidimensional Digital Signal Processing*. Englewood Cliffs, NJ; Prentice-Hall 1984.

[15] J.-C. Feauveau, "Analyse multirésolution par ondelettes non orthogonales et bancs de filtres numériques," Ph.D. dissertation, Univ. Paris Sud, 1990.

[16] K. Gröchenig and W. R. Madych, "Multiresolution analysis, Haar bases and self-similar tilings of R^n," *IEEE Trans. Inform. Theory,* vol. 38, pt. II, pp. 556–568, Mar. 1992.

[17] A. Grossmann and J. Morlet, "Decomposition of Hardy functions into square integrable wavelets of constant shape," *SIAM J. Math. Anal.*, vol. 15, no. 4, pp. 723–736, 1984.

[18] J. Kovačević, "Filter Banks and Wavelets: Extensions and Applications," Ph.D. dissertation, Columbia Univ., New York, 1991.

[19] J. Kovačević and M. Vetterli, "A theory of multidimensional multirate FIR filter banks," Center for Telecommun. Res. Elect. Eng. Dep., Columbia Univ., Tech. Rep., 1991.

[20] J. Kovačević, M. Vetterli, and G. Karlsson, "Design of multidimensional filter banks for non-separable subsampling," in *Proc. IEEE Int. Symp. Circuits Syst.*, New Orleans, LA, May 1990, pp. 2004–2008.

[21] G. Karlsson and M. Vetterli, "Theory of two-dimensional multirate filter banks," *IEEE Trans. Acoust., Speech, Signal Processing*, vol. 38, no. 6, pp. 925–937, June 1990.

[22] G. Karlsson, M. Vetterli, and J. Kovačević, "Non-separable two-dimensional perfect reconstruction filter banks," in *Proc. SPIE Conf. Vis. Commun. Image Processing*, Cambridge, MA, Nov. 1988, pp. 187–199.

[23] S. Mallat, "A theory of multiresolution signal decomposition: The wavelet representation," *IEEE Trans. Pattern Recognition Machine Intell.*, vol. 11, no. 7, pp. 674–693, July 1989.

[24] J. McClellan, "The design of two-dimensional filters by transformations, in *Proc. 7th Ann. Princeton Conf. ISS*, Princeton, NJ, 1973, pp. 274–251.

[25] Y. Meyer, *Ondelettes*. Paris; Hermann, 1990.

[26] F. Mintzer, "Filters for distortion-free two-band multirate filter banks," *IEEE Trans. Acoust., Speech, Signal Processing*, vol. ASSP-33, no. 3, pp. 626–630, June 1985.

[27] M. Newman, *Integral Matrices*. New York: Academic, 1972.

[28] T. Q. Nguyen and P. P. Vaidyanathan, "Two-channel perfect-reconstruction FIR QMF structures which yield linear-phase analysis and synthesis filters," *IEEE Trans. Acoust., Speech, Signal Processing*, vol. 37, no. 5, pp. 676–690, May 1989.

[29] O. Rioul, "Dyadic up-scaling schemes: Simple criteria for regularity," *SIAM J. Math. Anal.,* Feb. 1991, submitted for publication.

[30] M. J. T. Smith and T. P. Barnwell III, "Exact reconstruction for tree-structured subband coders," *IEEE Trans. Acoust., Speech, Signal Processing*, vol. ASSP-34, no. 3, pp. 431–441, June 1986.

[31] —, "A new filter bank theory for time-frequency representation," *IEEE Trans. Acoust., Speech, Signal Processing*, vol. ASSP-35, no. 3, pp. 314–327, Mar. 1987.

[32] G. Strang, "Wavelets and dilation equations: A brief introduction," *SIAM Rev.*, vol. 31, no. 4, pp. 614–627, Dec. 1990.

[33] P. P. Vaidyanathan, "Theory and design of M-channel maximally decimated quadrature mirror filters with arbitrary M, having the perfect reconstruction property," *IEEE Trans. Acoust., Speech, Signal Processing*, vol. ASSP-35, no. 4, pp. 476–492, Apr. 1987.

[34] —, *Robust Digital Filters and Multirate Filter Banks*. Englewood Cliffs, NJ: Prentice Hall, 1992, to appear.

[35] P. P. Vaidyanathan and P. Q. Hoang, "Lattice structures for optimal design and robust implementation of two-channel perfect reconstruction filter banks," *IEEE Trans. Acoust., Speech, Signal Processing*, vol. 36, no. 1, pp. 81–94, 1988.

[36] P. P. Vaidyanathan and Z. Doğanata, "The role of lossless systems in modern digital signal processing: A tutorial," *IEEE Trans. Education*, vol. 32, no. 3, pp. 181–197, Aug. 1989.

[37] M. Vetterli, "Multi-dimensional sub-band coding: Some theory and algorithms," *Signal Processing*, vol. 6, no. 2, pp. 97–112, Feb. 1984.

[38] —, "Filter banks allowing perfect reconstruction," *Signal Processing*, vol. 10, no. 3, pp. 219–244, Apr. 1986.

[39] —, "A theory of multirate filter banks," *IEEE Trans. Acoust., Speech, Signal Processing*, vol. ASSP-35, pp. 356–372, Mar. 1987.

[40] —, "Multirate filter banks for subband coding," in *Subband Image Coding*, J. W. Woods, Ed. New York: Kluwer, 1990.

[41] —, "Wavelets and filter banks for discrete-time signal processing," in *Wavelets and Their Applications*, R. Coifman *et al.*, Eds. New York: Jones and Bartlett, 1991.

[42] M. Vetterli and C. Herley, "Wavelets and filter banks: Theory and design," *IEEE Trans. Acoust., Speech, Signal Processing*, to appear Sept. 1992.

[43] M. Vetterli, J. Kovačević, and D. J. LeGall, "Perfect reconstruction filter banks for HDTV representation and coding," *Image Commun.*, vol. 2, no. 3, pp. 349–364, Oct. 1990.

[44] M. Vetterli and D. J. LeGall, "Perfect reconstruction FIR filter banks: some properties and factorizations," *IEEE Trans. Acoust., Speech, Signal Processing*, vol. 37, no. 7, pp. 1057–1071, July 1989.

[45] E. Viscito and J. P. Allebach, "Design of perfect reconstruction multi-dimensional filter banks using cascaded Smith form matrices," in *Proc. IEEE Int. Symp. Circuits Syst.*, Espoo, Finland, June 1988, pp. 381–384.

[46] E. Viscito and J. P. Allebach, "The analysis and design of multidimensional FIR perfect reconstruction filter banks for arbitrary sampling lattices," *IEEE Trans. Circuits Syst.*, vol. 38, no. 1, pp. 29–42, Jan. 1991.

[47] J. W. Woods and S. D. O'Neil, "Sub-band coding of images," *IEEE Trans. Acoust., Speech, Signal Processing*, vol. ASSP-34, pp. 1278–1288, Oct. 1986.

[48] E. P. Simoncelli and E. H. Adelson, "Non-separable extensions of quadrature mirror filters to multiple dimensions," *Proc. IEEE,* vol. 78, pp. 652–664, Apr. 1990.

[49] W. M. Lawton and H. L. Resnikoff, "Multidimensional wavelet bases," submitted to *SIAM J. Anal.*, 1991.

Multiresolution Analysis, Haar Bases, and Self-Similar Tilings of R^n

K. Gröchenig and W. R. Madych

Abstract—**Orthonormal bases for $L^2(R^n)$ are constructed that have properties that are similar to those enjoyed by the classical Haar basis for $L^2(R)$. For example, each basis consists of appropriate dilates and translates of a finite collection of "piecewise constant" functions. The construction is based on the notion of multiresolution analysis and reveals an interesting connection between the theory of compactly supported wavelet bases and the theory of self-similar tilings.**

Index Terms—**Multiresolution analysis, multivariate Haar basis, self-similar tilings, wavelets, fractals.**

I. Introduction

RECALL that the Haar system on $L^2(R)$ is the collection of functions

$$2^{k/2}\psi(2^k x - j), \qquad j, k \in Z, \tag{1}$$

where

$$\psi(x) = \begin{cases} 1, & \text{if } 0 \le x < 1/2, \\ -1, & \text{if } 1/2 \le x < 1, \\ 0, & \text{otherwise}, \end{cases}$$

where Z denotes the set of integers. Note the role played by the dilation $x \to 2x$ and the translations $x \to x - j$. It is well known that this collection is a complete orthonormal system for $L^2(R)$.

The point of this paper is to construct analogous systems for $L^2(R^n)$, $n \ge 2$, where the dilation noted above is replaced by appropriate linear transformations of R^n and the integers Z are replaced by an appropriate lattice in R^n. The motivation and framework for our construction is outlined in Section II.

We remind the reader that there are obvious generalizations of the Haar basis to higher dimensions. However, the general case is by no means obvious and offers some interesting surprises.

The plan of this paper is as follows. In Section II, we briefly review the concepts of multiresolution analysis and wavelet basis, introduced by Mallat [7] and Meyer [9], for $L^2(R^n)$ and explain how the classical Haar system and our construction fit into this scheme. In short, these bases are simply wavelets whose corresponding scaling functions are characteristic functions of appropriate sets. In Section III, we show how such scaling functions are related to certain self-similar tilings of R^n and indicate how to construct such tilings. Essentially the celebrated two scale functional equations reduce to simple iterated function systems in this case. Thurston [10, Sections 8–10] considers self-similar tilings generated by similarities, that is matrices which are constant multiples of rotations, which do not necessarily preserve some lattice. The construction presented here is different because it requires matrices which leave a lattice invariant but includes many cases that are not similarities. In Section IV, we construct the promised bases from appropriate scaling functions or, equivalently, certain self-similar tilings of R^n. Representative examples in R and R^2 together with several general observations are presented in Section V. We conclude the paper with miscellaneous remarks and citations to the literature.

II. Multiresolution Analysis and Wavelet Bases

In what follows Γ is a lattice in R^n, that is, Γ is the image of the integer lattice Z^n under some nonsingular linear transformation. We say that a linear transformation A on R^n is an *acceptable dilation for* Γ if it satisfies the following properties:

- A leaves Γ invariant. In other words, $A\Gamma \subset \Gamma$. Here
$$A\Gamma = \{y : y = Ax \quad \text{and} \quad x \in \Gamma\};$$
- all the eigenvalues, λ_i, of A satisfy $|\lambda_i| > 1$.

These properties imply that $|\det A|$ is an integer q which is ≥ 2.

Such an A induces a unitary dilation operator $U_A : f \to U_A f$ on $L^2(R^n)$, defined by

$$U_A f(x) = |\det A|^{-1/2} f(A^{-1}x). \tag{2}$$

If V is a subspace of $L^2(R^n)$ we use the customary notation $U_A V$ to denote the image of V under U_A, that is, $U_A V = \{f : f = U_A g,\ g \in V\}$. The translation operator τ_y is defined by $\tau_y f(x) = f(x - y)$.

A *wavelet basis* associated to (Γ, A) is a complete orthonormal basis of $L^2(R^n)$ whose members are A dilates of Γ translates of a finite collection $\psi_1, \cdots, \psi_m$ of orthonormal functions. More specifically, the members of the basis are the functions

$$U_A^{-j}\tau_\gamma \psi_i(x) = |\det A|^{j/2}\psi_i(A^j x - \gamma), \qquad j \in Z,\ \gamma \in \Gamma,\quad i = 1, \cdots, m. \tag{3}$$

The ψ_i's are called the basic wavelets.

Manuscript received February 1, 1991; revised July 10, 1991. This work was supported in part by Grant AFOSR-90-0311.

The authors are with the Department of Mathematics U-9, University of Connecticut, Storrs, CT 06269-3009.

IEEE Log Number 9105033.

0018-9448/92$03.00 © 1992 IEEE

The classical Haar system defined by (1) is the simplest example of such a basis for $L^2(R)$. In this case $m = 1$, $\Gamma = Z$, and A is the dyadic dilation $Ax = 2x$.

There is a generic recipe, due to Y. Meyer, for the construction of wavelet bases. The main ingredient is the notion of multiresolution analysis.

A *multiresolution analysis* $\mathscr{V}$ associated with (Γ, A) is an increasing family $\cdots \subset V_{j-1} \subset V_j \subset V_{j+1} \subset \cdots$, $j \in Z$, of closed subspaces of $L^2(R^n)$ with the following properties:

1) $\cup_{j\in Z} V_j$ is dense in $L^2(R^n)$, and $\cap_j V_j = \{0\}$,
2) $f(x) \in V_j$, if and only if $f(Ax) \in V_{j+1}$. In other words
$$V_j = U_A^{-j} V_0, \qquad j \in Z. \tag{4}$$
3) V_0 is invariant under τ_γ. More specifically, if $f(x)$ is in V_0 then so is $f(x-\gamma)$ for all γ in Γ.
4) There is a function $\phi \in V_0$, called the scaling function, such that $\{\tau_\gamma \phi, \gamma \in \Gamma\}$ is a complete orthonormal basis for V_0.

From the definition it should be clear that a multiresolution analysis is determined by the scaling function ϕ. Since $V_0 \subset V_1$ there is a sequence $\{a_\gamma\}$ in $l^2(\Gamma)$ such that
$$\phi(x) = \sum_{\gamma\in\Gamma} a_\gamma U_A^{-1} \tau_\gamma \phi(x)$$
$$= \sum_{\gamma\in\Gamma} a_\gamma |\det A|^{1/2} \phi(Ax - \gamma). \tag{5}$$

It is known that, under certain conditions, these coefficients determine the scaling function ϕ uniquely. On the other hand, in spite of the fact that the orthogonality relations
$$\langle \tau_\gamma \phi, \phi \rangle = \delta_{\gamma 0} \tag{6}$$
impose certain restrictions on the sequence $\{a_\gamma\}$, these conditions are not sufficient to guarantee that the a_γ's are acceptable scaling coefficients; in short, the nature of the scaling coefficients is not completely understood in the general case.

The simplest example of a multiresolution analysis in one dimension with $(\Gamma, A) = (Z, 2I)$, where I is the identity, is given by the scaling function $\phi(x) = \chi_{[0,1)}(x)$. Then V_0 is the closed subspace of $L^2(R)$ consisting of all functions that are constant on the intervals $[j, j+1)$, $j \in Z$, and the subspaces V_k consist of those functions that are constant on subintervals $[j2^{-k}, (j+1)2^{-k})$, $j \in Z$. The scaling relation is given by
$$\phi(x) = \phi(2x) + \phi(2x-1). \tag{7}$$

Given a multiresolution analysis $\mathscr{V}$, define W_j as the orthogonal complement of V_j in V_{j+1}, thus $V_j \oplus W_j = V_{j+1}$ for all j. It follows that
$$W_j = U_A^{-j} W_0 \quad \text{and} \quad L^2(R^n) = \bigoplus_{j\in Z} W_j. \tag{8}$$

If $q = |\det A|$ then a result which can be found in Meyer's paper [8] says the following: There exist $q-1$ functions $\psi_1, \cdots, \psi_{q-1}$ such that $\{\tau_\gamma \psi_i;\ \gamma \in \Gamma,\ i = 1, \cdots, q-1\}$ *is a complete orthonormal basis of* W_0.

In view of (8) this theorem implies that the collection $\{U_A^j \tau_\gamma \psi_i : j \in Z,\ \gamma \in \Gamma,\ i = 1, \cdots, q-1\}$ is a wavelet basis of $L^2(R^n)$. Furthermore, since $W_0 \subset V_1$ there are sequences $\{a_{i\gamma}\}$ in $l^2(\Gamma)$ such that
$$\psi_i(x) = \sum_{\gamma\in\Gamma} a_{i\gamma} |\det A|^{1/2} \phi(Ax - \gamma),$$
$$i = 1, \cdots, q-1. \tag{9}$$

For example, in the specific case of multiresolution analysis mentioned above where the scaling relation is given by (7), $q = 2$ and the corresponding basic wavelet is given by
$$\psi(x) = \phi(2x) - \phi(2x-1)$$
$$= \chi_{[0,1)}(2x) - \chi_{[0,1)}(2x-1).$$

This is the basic Haar wavelet.

Thus, the generic recipe to construct a wavelet basis can be briefly summarized as follows: Start with a multiresolution analysis with scaling relation (5) and look for basic wavelets which are of form (9).

Due to the work of Cohen [2], Daubechies [3], Mallat [7], Meyer [9] and others, the algorithm outlined above is well understood in the case when $n = 1$ and $Ax = 2x$. In this case, multiresolution analyses can be constructed that have desired continuity and support properties. The coefficients for the basic wavelet can always be expressed in terms of the original scaling coefficients via a simple formula.

The construction of the basic wavelets in the general case is not so clear. For example, the structure of scaling sequences which will produce multiresolution analyses with desired properties is not well understood. Also, although it is clear that the coefficients of (9) should have some relationship to the coefficients in the basic scaling relation (5), except for certain examples, there are no known formulas for the coefficients in (9) in the general case.

We are now ready to state the questions addressed in this paper precisely: Given (Γ, A), what are the multiresolution analyses whose scaling functions are characteristic functions of measurable sets Q? What are the corresponding basic wavelets whose support is in Q and how can they be constructed explicitly? In what follows we will refer to such multiresolution analyses as *simple* and to such wavelets as *elementary*.

Before we start, let us make the following simplification: Since every lattice $\Gamma \subset R^n$ is of the form $\Gamma = EZ^n$ for some invertible real-valued $n \times n$ matrix E, without loss of generality, we may and do restrict our attention to the case $\Gamma = Z^n$. In this case the matrices A must have integer entries.

III. Self-Similar Tilings and Scaling Functions

In this section, we establish a connection between self-similar tilings and multiresolution analyses that have a characteristic function for a scaling function.

Given a measurable set S, χ_S denotes its characteristic or indicator function and $|S|$ denotes its Lebesgue measure. The notation $S \simeq T$ means that the sets S and T are equal up to a set of measure zero, in other words, $|S \setminus T| = |T$

$\setminus S| = 0$. If $S \cap T \simeq \emptyset$ we say that S and T are essentially disjoint. Also recall that $q = |\det A|$.

We begin with two technical lemmas which are elementary and are probably folklore.

Lemma 1: Suppose Q is a measurable subset of R^n such that

$$\bigcup_{k \in Z^n} (Q + k) \simeq R^n.$$

Then the following are equivalent:

1) $Q \cap (Q + k) \simeq \emptyset$ whenever k is a nonzero element in Z^n,
2) $|Q| = 1$.

Proof: Let $f(x) = \Sigma_{j \in Z^n} \chi_Q(x - j)$. Then if Q satisfies property 1) it follows that $f \equiv 1$ and we may write

$$|Q| = \int_{R^n} \chi_Q(x)\,dx = \int_{Q_0} f(x)\,dx = |Q_0| = 1,$$

where $Q_0 = [0, 1]^n$.

To see the converse let f and Q_0 be as previously defined. Note that assumption implies $f(x) \geq 1$. Also observe that

$$\int_{Q_0} f(x)\,dx = \int_{R^n} \chi_Q(x)\,dx = |Q| = 1$$

implies that $f \equiv 1$, which in turn implies the desired result. □

Lemma 2: The number of disjoint cosets in Z^n / AZ^n is q.

Proof: Let $Q_0 = [0, 1]^n$ and let $k_1 + AZ^n, \cdots, k_m + AZ_n$ be an enumeration of the cosets in Z^n/AZ^n. Express R^n as a union of essentially disjoint subsets as follows

$$\bigcup_{k \in Z^n} \left\{ Ak + \bigcup_{i=1}^{m} (k_i + Q_0) \right\}$$

$$= \bigcup_{k \in Z^n} \left\{ \bigcup_{i=1}^{m} (k_i + Ak + Q_0) \right\} = \bigcup_{k \in Z^n} (k + Q_0) = R^n.$$

Since $A^{-1}R^n = R^n$, applying A^{-1} to both sides of the last equality results in

$$\bigcup_{k \in Z^n} \left\{ k + \bigcup_{i=1}^{m} A^{-1}(k_i + Q_0) \right\} = R^n.$$

It should now be clear that

$$Q = \bigcup_{i=1}^{m} A^{-1}(k_i + Q_0)$$

satisfies the hypothesis and 1) in Lemma 1. Hence, $|Q| = 1$, and, since Q is the union of disjoint subsets $A^{-1}(k_1 + Q_0), \cdots, A^{-1}(k_m + Q_0)$ each of which has measure $1/q$, it follows that $m = q$. □

Theorem 1: Suppose $\phi = c\chi_Q$ is a scaling function for a multiresolution analysis associated with (Z^n, A). Here χ_Q is the characteristic function of a measurable set Q and $c = |Q|^{-1/2}$. Then Q satisfies the following properties.

1)
$$Q \cup (Q + k) \simeq \emptyset, \qquad \text{for } k \neq 0,\ k \in Z^n. \quad (10)$$

2) There is a collection of q lattice points $k_1, \cdots, k_q$ that are representatives of distinct cosets in Z^n/AZ^n such that
$$AQ \simeq \bigcup_{i=1}^{q} (k_i + Q). \quad (11)$$

3)
$$\bigcup_{k \in Z^n} (Q + k) \simeq R^n. \quad (12)$$

4) There is a compact set K such that $Q \simeq \dot{K}$.

Conversely, the characteristic function of a bounded measurable set Q that satisfies properties 1), 2), and 3) is the scaling function of a multiresolution analysis associated with (Z^n, A).

Remarks:

- Properties (10) and (12) mean that translates of Q by the integer lattice form a tiling of R^n. Sets Q that enjoy property (11) are sometimes said to be self-similar in the affine sense.
- In view of Lemma 1 properties (10) and (12) imply that $|Q| = 1$. Thus $c = 1$, $\phi = \chi_Q$ and satisfies the functional equation
$$\phi(x) = \sum_{i=1}^{q} \phi(Ax - k_i), \quad (13)$$
where the k_i's are those lattice points whose existence is implied by property 2).
- In view of Lemma 2 the k_i's in (11) are a full set of coset representatives, namely,
$$\bigcup_{i=1}^{q} (k_i + AZ^n) = Z^n. \quad (14)$$

Proof: Suppose $\phi = c\chi_Q$ is a scaling function for a multiresolution analysis associated with (Z^n, A).

The disjointness of the translates of Q (10) follows from (6): $\langle L_k\phi, \phi\rangle = c^2 |Q \cap (Q + k)| = \delta_{k0}$.

The second property follows from the scaling relation for χ_Q,

$$\chi_Q(x) = \sum_k a_k q^{1/2} \chi_Q(Ax - k)$$
$$= \sum_k a_k q^{1/2} \chi_{A^{-1}(Q+k)}(x),$$

where $a_{k_i} q^{1/2} = 1$ for exactly q lattice points k_i and $a_k = 0$ for the remaining coefficients. The fact concerning the coefficients a_k, which of course immediately implies (11), follows from (10) and the formula $|AQ| = q|Q|$.

That the k_i's are representatives of distinct cosets of Z^n/AZ^n follows from the orthogonality of $\phi(A^{-1}x - k)$, $k \in Z^n$. To wit, suppose k_1 and k_2 are not in distinct cosets. Then there is a lattice point k so that $k_1 = k_2 + Ak$ and this in turn implies that

$$AQ = \bigcup_{i=1}^{q} (k_i + Q) \quad \text{and}$$

$$A(Q + k) = \bigcup_{i=1}^{q} (k_i + Ak + Q)$$

are not disjoint which contradicts the orthogonality of $\phi(A^{-1}x - k)$ and $\phi(A^{-1}x)$.

The covering property (12) is a consequence of the density of $\bigcup_j V_j$ in $L^2(R^n)$ and 2). To see this, let $P_j f$ be the projection of $f \in L^2(R^n)$ onto V_j, in other words, $P_j f = \sum_k |Q|^{-j} \langle U_A^{-j} \tau_k \chi_Q, f \rangle U_A^{-j} \tau_k \chi_Q$. Then $P_j f \to f$ as $j \to \infty$ and

$$\| P_j f \|_2^2 = \sum_k |Q|^{-j^0} |\langle U_A^{-j} \tau_k \chi_Q, f \rangle|^2 \to \| f \|_2^2 \quad (15)$$

as $j \to \infty$. Hence, if B is any measurable set of finite measure, if follows from (15) that

$$\sum_k |Q|^{-1} q^j | A^{-j}(k + Q) \cap B |^2$$
$$= \sum_k |Q|^{-1} q^{-j} |(k + Q) \cap A^j B|^2 \to |B| \quad (16)$$

as $j \to \infty$. Now, in view of 2),

$$\cup_k (Q + k) = \cup_k (A^{-j}(Q + k)),$$

for all j in Z^n. Hence, we may write

$$\begin{aligned} |B| &\geq |\cup_k (Q + k) \cap B| \\ &= |\cup_k (A^{-j}(Q + k)) \cap B| \\ &= \sum_k q^{-j} |(k + Q) \cap A^j B| \\ &\geq \sum_k q^{-j} |Q| \left(\frac{|(k + Q) \cap A^j B|}{|Q|} \right)^2 \\ &= \sum_k |Q|^{-1} q^{-j} |(k + Q) \cap A^j B|^2. \end{aligned}$$

By virtue of (16) the last expression converges to $|B|$ as $j \to \infty$ and, as a consequence, we may conclude that $|B| = |\cup_k (Q + k) \cap B|$ and, since B was arbitrary, the desired result follows.

Property 4) is a consequence of Lemmas 3 and 5.

To see the converse let

$$V_0 = \left\{ f \in L^2(R^n) : f(x) = \sum_{k \in Z^n} c_k \chi_Q(x - k) \right\}$$

and let $V_j = U_A^{-j} V_0$ for each integer j. Then $\mathscr{V} = \{V_j\}_{j \in Z}$ is a family of closed subplaces of $L^2(R^n)$ and using both the above definition and the properties of the set Q it is easy to see that this family satisfies the following properties:

- $\mathscr{V}$ is an increasing family, namely, $V_j \subset V_{j+1}$ for all integers j,
- $\cap_{j \in Z} V_j = \{0\}$,
- $V_j = U_A^{-j} V_0$ for each integer j,
- V_0 is invariant under τ_k for each k in Z^n,
- $\{\tau_k \chi_Q : k \in Z^n\}$ is a complete orthonormal basis for V_0.

In view of this, to see that $\mathscr{V}$ is a multiresolution analysis associated with (Z^n, A) with scaling function χ_Q, it suffices to show that $\cup_{j \in Z} V_j$ is dense in $L^2(R^n)$.

To this end, let $P_j f$ be the orthogonal projection of f onto V_j, let $\phi_j(x) = |\det A|^j \chi_Q(-A^j x)$, and observe that

$$P_j f(x) = \phi_j * f(A^{-j}k), \quad \text{whenever } x - A^{-j}k \in A^{-j}Q, \quad (17)$$

where

$$\phi_j * f(x) = \int_{R^n} \phi_j(x - y) f(y)\, dy$$

is the convolution of ϕ_j and f. Now, it is easy to see that

$$\phi_j * f - f \to 0 \quad \text{in} \quad L^2(R^n) \text{ as } j \to \infty, \quad (18)$$

for all f in $L^2(R^n)$. In view of (17), the difference between $P_j f$ and $\phi_j * f$ may be expressed as

$$\begin{aligned} E_j(x) &= \phi_j * f(x) - P_j f(x) \\ &= \int_{R_n} \Big\{ f(x - y) - \sum_{k \in Z^n} \chi_{A^{-j}Q}(x - A^{-j}k) f \\ &\quad \cdot (A^{-j}k - y) \Big\} \phi_j(y)\, dy. \end{aligned} \quad (19)$$

If we call the expression in braces $F_j(x, y)$ and take f to be continuous with compact support and $|y| \leq 1$ then for any positive ϵ we may write

$$\int_{R^n} |F_j(x, y)|^2\, dx < \epsilon^2, \quad (20)$$

for sufficiently large j. Hence, in this case, if we take the $L^2(R^n)$ norm of E^j and apply the integral variant of Minkowski's inequality to the right-hand side of (19), inequality (20) implies that

$$\| E_j \|_{L^2(R^n)} < \epsilon,$$

for sufficiently large j. In other words,

$$\phi_j * f - P_j f \to 0 \quad \text{in} \quad L^2(R^n) \qquad \text{as } j \to \infty, \quad (21)$$

whenever f is continuous with compact support. Since such f are dense in $L^2(R^n)$, (21) together with (18) imply that $\cup_{j \in Z} V_j$ is dense in $L^2(R^n)$. □

Suppose

$$\phi = c\chi_Q \quad (22)$$

is a scaling function for a multiresolution analysis associated with (Z^n, A). In view of this, we know that $|Q| = c = 1$ and

$$\phi(x) = \sum_{k \in \mathscr{K}} \phi(Ax - k), \quad (23)$$

where $\phi = \chi_Q$ and $\mathscr{K} = \{k_1, \cdots, k_q\}$ is a collection of representatives of distinct cosets in Z^n / AZ^n. Thus, if we are interested in constructing a multiresolution analysis whose scaling function ϕ is of the form (22), a reasonable approach seems to be the following: find an appropriate collection of lattice points $\mathcal{K}$ and solve the functional equation (23).

Since the solution of (23) is a fixed point of the mapping

$$\psi(x) \rightarrow \sum_{k \in \mathcal{K}} \psi(Ax - k),$$

it is quite natural to apply fixed point iteration to solve for ϕ. Namely, start with an initial function ϕ_0 and define the sequence $\phi_1, \phi_2, \phi_3, \cdots$ via

$$\phi_{N+1}(x) = \sum_{k \in \mathcal{K}} \phi_N(x) \tag{24}$$

and hope that the sequence converges to ϕ. Since the desired solution is the characteristic function of a set Q whose Z^n translates tile, it is reasonable to begin the iteration with $\phi_0 = \chi_{Q_0}$ where Q_0 has the same properties.

Suppose Q_0 satisfies (10) and (12) and $\mathcal{K} = \{k_1, \cdots, k_q\}$ is a collection of representatives of distinct cosets of Z^n / AZ^n. If $\phi_0 = \chi_{Q_0}$ and ϕ_1 is related to ϕ_0 via (24) then $\phi_1 = \chi_{Q_1}$ where

$$Q_1 = \bigcup_{k \in \mathcal{K}} A^{-1}(Q_0 + k)$$

also satisfies (10) and (12); that Q_1, satisfies (12) follows from the fact that $\mathcal{K}$ is a full collection of distinct representatives of Z^n / AZ^n and that it satisfies (10) follows from $|Q_1| = 1$. By induction we may conclude that ϕ_{N+1}, $N = 0, 1, 2, \cdots$, is the characteristic function of the set Q_{N+1} defined by

$$Q_{N+1} = \bigcup_{k \in \mathcal{K}} A^{-1}(Q_N + k). \tag{25}$$

Observe that (25) looks like an iterated function system in the sense of Barnsley [1]. Convergence of schemes like (25) is usually considered in terms of the following metric defined on the space of subsets of R^n:

$$\rho(P, Q) = \max\{r(P, Q), r(Q, P)\} \tag{26}$$

where

$$r(P, Q) = \sup_{x \in P} \inf_{y \in Q} |x - y|.$$

It is well known that when equipped with the metric ρ the class of compact subsets of R^n is a complete metric space. If the mapping $x \rightarrow A^{-1}x$ is a contraction and Q_0 is compact then the iteration (25) converges to a compact set Q; for example, see [1].

Unfortunately, in our considerations the mapping $x \rightarrow A^{-1}x$ is not necessarily a contraction, see Example 1) in Section V-C. Nevertheless since all the eigenvalues of A^{-1} are less than one in absolute value it follows that

$$\| A^{-j}x \| \leq C\lambda^j j^s \| x \|, \tag{27}$$

for all $x \in R^n$, where C, s, λ are positive constants and $\lambda < 1$. This is easily seen by writing A in its Jordan form. Inequality (27) allows us to state the following.

Lemma 3: Suppose $\mathcal{K} = \{k_1, \cdots, k_m\}$ is a finite collection of lattice points in Z^n and Q is the compact set defined by

$$Q = \left\{ x \in R^n : x = \sum_{j=1}^{\infty} A^{-j}\epsilon_j,\ \epsilon_j \in \mathcal{K} \right\}. \tag{28}$$

If Q_0 is any compact set then the sequence of sets $Q_1, Q_2, \cdots$, defined by

$$Q_{N+1} = \bigcup_{i=1}^{m} A^{-1}(k_i + Q_N) \tag{29}$$

converges in the metric ρ to the set Q.

Remarks:

- Note that Q is well defined and bounded by virtue of (27).
- It follows immediately from the definition that the set Q satisfies the self-similarity relation

$$Q = \bigcup_{i=1}^{m} A^{-1}(k_i + Q).$$

Proof: Using (25) repeatedly, we see that

$$\begin{aligned} Q_N &= \bigcup_{i=1}^{m} (A^{-1}k_i + A^{-1}Q_{N-1}) \\ &= \bigcup_{i=1}^{m} \bigcup_{j=1}^{m} (A^{-1}k_i + A^{-2}k_j + A^{-2}Q_{N-2}) \\ &= \cdots = \bigcup_{(\epsilon_1, \epsilon_2, \cdots, \epsilon_N) \in \mathcal{K}^N} \left(\sum_{j=1}^{N} A^{-j}\epsilon_j + A^{-N}Q_0 \right), \end{aligned} \tag{30}$$

where $\mathcal{K}^N$ is the collection of all N tuples $(\epsilon_1, \epsilon_2, \cdots, \epsilon_N)$ whose components are in $\mathcal{K}$. Hence, if x is any element in Q_N then there is a point x_0 in Q_0 such that $x = \sum_{j=1}^{N} A^{-j}\epsilon_j + A^{-N}x_0$. So by choosing an element y in Q of the form $y = \sum_{j=1}^{N} A^{-j}\epsilon_j + A^{-N}y_0$ where $y_0 \in Q$ we may write

$$\begin{aligned} \inf_{y \in Q} |x - y| &\leq \inf_{y_0 \in Q} |A^{-N}(x_0 - y_0)| \\ &\leq C\lambda^N N^s \inf_{y_0 \in Q} |x_0 - y_0| \\ &\leq C\lambda^N N^s r(Q_0, Q), \end{aligned} \tag{31}$$

where the inequalities follow by virtue of (27) and the definition of $r(Q_0, Q)$. Taking supremum over $x \in Q_N$ shows that $r(Q_N, Q)$ can be made arbitrarily small by choosing N sufficiently large. Similar reasoning shows that $r(Q, Q_N)$ can also be made arbitrarily small by choosing N sufficiently large. It now follows that the sequence of compact sets $\{Q_N\}$ converges to Q in the sense of the metric ρ and, as a consequence, Q is compact. □

In what follows, we will often be considering collections $\mathcal{K} = \{k_1, \cdots, k_m\}$ of lattice points in conjunction with the scaling relation (13) or the self-similarity (11). In this context it is a minor inconvenience if 0 is not in $\mathcal{K}$. For example, the set defined by (28) does not contain finite sums of the specified form. In this particular case this inconvenience can be remedied by re-expressing the elements x as follows:

$$x = (A - I)^{-1}k_1 + \sum_{j=1}^{\infty} A^{-j}(\epsilon_j + k_1),$$

where the ϵ_j's are in $\mathscr{K}$. In the arguments used below there is no loss of generality in assuming that $\mathscr{K}$ contains 0 in view of the following easily verifiable lemma.

Lemma 4: Suppose $\mathscr{K}_1$ is a collection of lattice points and $\mathscr{K}_2 = x_0 + \mathscr{K}_1$ for some point x_0 in R^n. If $\phi(x)$ satisfies

$$\phi(x) = \sum_{k \in \mathscr{K}_1} \phi(Ax - k),$$

then $\psi(x) = \phi(x - (A - I)^{-1}x_0)$ satisfies

$$\psi(x) = \sum_{k \in \mathscr{K}_2} \psi(Ax - k).$$

Similarly, if Q satisfies (11) where the union is taken over k_i in $\mathscr{K}_1$ then $Q + (A - I)^{-1}x_0$ satisfies (11) where the union is taken over k_i in $\mathscr{K}_2$.

Next we define two sequences of useful measures. If $\mathscr{K} = \{k_1, \cdots, k_q\}$ is a collection of lattice points in Z^n that contains 0 then the sequences of measures $\{\mu_N\}$ and $\{\nu_N\}$, $N = 1, 2, \cdots$, are defined as follows:

$$\mu_N(x) = \frac{1}{q} \sum_{k \in \mathscr{K}} \delta(x - A^{-N}k), \tag{32}$$

where $\delta(x)$ is the unit Dirac measure at the origin and

$$\nu_{N+1} = \mu_{N+1} * \nu_N, \tag{33}$$

where $\nu_1 = \mu_1$. Note that the support of ν_N is the finite collection of points

$$\mathscr{Q}_N = \left\{ x \in R^N : x = \sum_{j=1}^{N} A^{-j}\epsilon_j,\ \epsilon_j \in \mathscr{K} \right\}. \tag{34}$$

Note that $\{\mathscr{Q}_N\}$, $N = 0, 1, 2, \cdots$, is the sequence of sets generated via (25) with $Q_0 = \{0\}$.

Suppose $\mathscr{K} = \{k_1, \cdots, k_q\}$ is a full collection of representatives of distinct cosets of Z^n/AZ^n. To avoid needless repetition of these phrases we say that such a collection is a *full collection of digits*. Of course we refer to elements of such a set $\mathscr{K}$ as *digits*.

Lemma 5: Suppose $\mathscr{K} = \{k_1, \cdots, k_q\}$ is a collection of q distinct lattice points in Z^n. Then any integrable solution of

$$\phi(x) = \sum_{k \in \mathscr{K}} \phi(Ax - k) \tag{35}$$

is unique up to multiplication by a constant and has support in the compact set

$$Q = \left\{ x \in R^n : x = \sum_{j=1}^{\infty} A^{-j}\epsilon_j,\ \epsilon_j \in \mathscr{K} \right\}.$$

Proof: Without loss of generality, assume $\mathscr{K}$ contains zero. Suppose ϕ is an integrable solution of (35). Note that (35) may be re-expressed as

$$\phi(x) = \mu_1 * \phi_1(x) = \int \phi_1(x - y)\, d\mu_1(y), \tag{36}$$

where $\phi_N(x) = q^N \phi(A^N x)$ and μ_1 is defined by (32). By induction it follows that for $N = 1, 2, \cdots$,

$$\phi(x) = \nu_N * \phi_N(x), \tag{37}$$

where ν_N is defined by (33). Now, if ψ is in $L^p(R^n)$, $1 \le p < \infty$, $\phi_N * \psi$ converges to $c\psi$ in $L^p(R^n)$ as $N \to \infty$ where $c = \int_{R^n} \phi(x)dx$. Since for each N ν_N has total variation one it follows that $\nu_N * \phi_N * \psi - c\nu_N * \psi$ converges to zero in $L^p(R^n)$ as $N \to \infty$. Since for any $N = 1, 2, \cdots, \phi = \nu_N * \phi_N = c\nu_N + (\nu_N * \phi_N + c\nu_N)$ and $\nu_N * \phi_N - c\nu_N$ converges weakly to zero we may conclude that $c\nu_N$ converges weakly to ϕ as $N \to \infty$. Thus the support of ϕ must be in Q. Since the sequence μ_N is uniquely defined by the functional equation (35) and it converges weakly to a constant multiple of ϕ it follows that ϕ is unique up to multiplication by constants. □

Theorem 2: Suppose $\mathscr{K} = \{k_1, \cdots, k_q\}$ is a full collection of digits and suppose the compact set Q is defined by

$$Q = \left\{ x \in R^N : x = \sum_{j=1}^{\infty} A^{-j}\epsilon_j,\ \epsilon_j \in \mathscr{K} \right\}. \tag{38}$$

Then the set Q has the following properties.

1) If Q_0 is any compact set then the sequence of sets Q_1, $Q_2, \cdots$, defined by (25) converges in the metric ρ to Q.
2) $AQ \simeq \cup_{i=1}^{q}(k_i + Q)$.
3) $\cup_{k \in Z^n}(Q + k) \simeq R^n$.
4) $(Q + k_i) \cap (Q + k_j) \simeq \emptyset$ wherever both k_i and k_j are in $\mathscr{K}$ and $k_i \ne k_j$.
5) $\phi = |Q|^{-1/2}\chi_Q$ is the unique solution, in the $L^1(R^n)$ sense, of

$$\phi(x) = \sum_{k \in \mathscr{K}} \phi(Ax - k),$$

which has $L^2(R^n)$ norm 1.
6) The sequence of measures $\{\nu_N\}$, $N = 0, 1, 2, \cdots$, defined by (33) converges weakly to $|Q|^{-1}\chi_Q$. In other words,

$$\lim_{N \to \infty} \int_{R^n} \psi(x)\, d\nu_N(x) = \frac{1}{|Q|} \int_{R^n} \psi(x)\chi_Q(x)\, dx,$$

for all functions ψ that are continuous and bounded on R^n.
7) If $\{Q_N\}$, $N = 0,1,2,\cdots$, is the sequence of sets generated via (25) with $Q_0 = [-1/2, 1/2]^n$, then the corresponding sequence of characteristic functions $\{\chi_{Q_N}\}$ converges weakly to $|Q|^{-1}\chi_Q$. In other words,

$$\lim_{N \to \infty} \int_{R^n} \psi(x)\chi_{Q_n}(x)d(x) = \frac{1}{|Q|} \int_{R^n} \psi(x)\chi_Q(x)\, dx,$$

for all functions ψ that are continuous and bounded on R^n.

Remarks:

- note that Property 3) implies that $|Q| \geq 1$;
- as will be clear from the proof, Property 7) holds if $Q_0 = [-1/2, 1/2]^n$ is replaced by any compact set Q_0 that satisfies (10) and (12).

Proof: The first two assertions are simply implied by Lemma 3.

To see item 3) let $Q_0 = [-1/2, 1/2]^n$ and observe that for each N, $N = 0, 1, 2, \cdots, Q_N$ satisfies (12) by virtue of the fact that $\mathcal{K}$ is a full collection of representatives of distinct cosets of Z^n/AZ^n, namely, $Z^n = \cup_{k\in\mathcal{K}}(k + AZ^n)$. This implies that for every x in R^n there is a sequence of lattice points $\{m_N\}$ such that $x - m_N$ is in Q_N. Since the Q_N's are all contained in a fixed ball, $\{m_N\}$ is a bounded sequence as well, and thus, contains a constant subsequence $m_{N_j} = m$, $j = 1, 2, \cdots$. Finally, since $x - m \in Q_{N_i}$ for infinitely many indexes, it follows that $x - m$ is in Q by virtue of the fact that $\{Q_N\}$ converges to Q in the ρ metric.

Since 3) implies that $|Q| \geq 1$, item 4) follows from the identity

$$\left| \bigcup_{k\in\mathcal{K}} (k + Q) \right| = q|Q|,$$

which is implied by 2).

That χ_Q satisfies the identity in item 5) follows immediately from 2) and 4). The fact concerning uniqueness is the assertion of Lemma 5.

To see item 6) let ψ be any bounded continuous function on R^n. Write

$$\int_{R^n} \psi(x)\, d\nu_N(x) = \frac{1}{q^N} \sum_{x\in\mathcal{Q}_N} \psi(x)$$
$$= \frac{1}{|Q|} \left\{ |A^{-N}Q| \sum_{x\in\mathcal{Q}_N} \psi(x) \right\}$$
$$= \frac{1}{|Q|} \int_{R^n} \psi_N(x)\, dx,$$

where

$$\psi_N(x) = \sum_{y\in\mathcal{Q}_N} \psi(y)\chi_Q(A^N(x - y))$$

is a simple function which converges to $\psi(x)\chi_Q(x)$ almost everywhere and is dominated by a constant multiple of χ_Q. The dominated convergence theorem now implies the desired result.

Item 7) follows from an argument analogous to the one used to show 6). □

It should now be clear how to construct scaling functions ϕ for multiresolution analyses associated with (Z^n, A), which have characteristic functions as scaling functions.

- Start with compact set Q_0 and a full collection $\mathcal{K} = (k_1, \cdots, k_q\}$ of representatives of distinct cosets of Z^n/AZ^n.
- Find Q as the limit of the iteration (25).
- If $|Q| = 1$ the algorithm is successful and $\phi = \chi_Q$ is the scaling function of a multiresolution analysis associated with (Z^n, A). Otherwise the algorithm fails.

The reason one must check the condition $|Q| = 1$ is that the requirement that $\mathcal{K}$ be a full collection of representatives of distinct cosets of Z^n is a necessary but not sufficient condition on this set of indices. Indeed, examples show that Q need not satisfy this condition and the algorithm may fail.

The following theorem gives various equivalent conditions that guarantee that this algorithm be successful.

Theorem 3: Suppose $\mathcal{K} = (k_1, \cdots, k_q\}$ is a collection of representatives of distinct cosets of Z^n/AZ^n and the compact set Q is defined by (38). Then the following statements are equivalent:

1) χ_Q is a scaling function for a multiresolution analysis associated with (Z^n, A).
2) $|Q| = 1$.
3) $Q \cap (k + Q) \simeq \emptyset$ for all k in Z^n which are different from 0.
4) The sequence of measures $\{\nu_N\}$, $N = 0, 1, 2, \cdots$, defined by (33) converges weakly to χ_Q. In other words,

$$\lim_{N\to\infty} \int_{R^n} \psi(x)\, d\nu_N(x) = \int_{R^n} \psi(x)\chi_Q(x)\, dx,$$

for all functions ψ that are continuous and bounded on R^n.

5) If $\{Q_N\}$, $N = 0, 1, 2, \cdots$, is the sequence of sets generated via (25) with $Q_0 = [0, 1]^n$, then the corresponding sequence of characteristic functions $\{\chi_{Q_N}\}$ converges to χ_Q in measure.
6) (Cohen's condition) Let $\hat{\mu}$ be the Fourier transform of the measure $\mu = \mu_1$ where μ_1 is defined by (32), namely

$$\hat{\mu}(\xi) = \frac{1}{q} \sum_{k\in\mathcal{K}} e^{-i\langle k, \xi\rangle}. \tag{39}$$

There exists a compact set K that contains a neighborhood of the origin and which satisfies

- $\cup_{k\in Z^n}(2\pi k + K) = R^n$,
- $K \cap (2\pi k + K) \simeq \emptyset$, whenever $k \neq 0$,

such that if $B = A^*$ then

$$|\hat{\mu}(B^{-j}\xi)| > 0 \tag{40}$$

holds for all $\xi \in K$ and $j \geq 1$.

Proof: That items 1), 2), and 3) are equivalent is essentially the content of Lemma 1 and Theorem 1.

That items 2) and 4) are equivalent follows immediately from item 6) of Theorem 2.

To see that 2) and 5) are equivalent observe that $|Q_n| = 1$, $n = 1, 2, \cdots$, and by virtue of dominated convergence

$$\int_{R^n} \chi_{Q_N}(x)\,dx = \int_{R^n} \chi_Q(x)\,dx,$$

so it is quite clear that 5) implies 2).

To see that 2) implies 5), let ϵ be any positive number and observe that the regularity of Lebesgue measure and the compactness of Q imply that there is a positive δ such that $|Q + B_\delta| < \epsilon$, where $B_\delta = \{x \in R^n : |x| < \delta\}$. Since $\{Q_N\}$ converges to Q in the ρ metric it follows that for N sufficiently large $Q_N \subseteq Q + B_\delta$. In this case $Q_N \cup Q \subseteq Q + B_\delta$ so

$$|Q_N \cup Q| \le |Q + B_\delta| < 1 + \epsilon.$$

This and the fact that $|Q_N| = |Q| = 1$ gives

$$|Q_N \cup Q| = |Q_N| + |Q| - |Q_N \cap Q| = 2 - |Q_N \cap Q| < 1 + \epsilon$$

so $|Q_N \cap Q| > 1 - \epsilon$. Thus, we may conclude that

$$|Q_N \Delta Q| = |Q_N \cup Q| - |Q_N \cap Q| < 2\epsilon,$$

whenever N is sufficiently large. Since $|\chi_Q - \chi_{Q_N}| = \chi_{Q_N \Delta Q}$, the last inequality implies the desired result.

To see how $\hat{\mu}$ fits in let $\{Q_N\}$, $N = 0, 1, 2, \cdots$, be the sequence of sets generated by (25) with $Q_0 = [1/2, 1/2]^n$ and let $\phi_N = \chi_{Q_n}$, $N = 0, 1, 2, \cdots$, and $\phi = |Q|^{-1}\chi_Q$. Recall that the Fourier transform $\phi \to \hat{\phi}$ is defined by

$$\hat{\phi}(\xi) = \int_{R^n} \phi(x) e^{-i\langle \xi, x\rangle}\,dx.$$

Let ν_N be the measure defined by (33) and observe that

$$\hat{\nu}_N(\xi) = \prod_{j=1}^{N} \hat{\mu}(B^{-j}\xi) \quad \text{and}$$

$$\hat{\phi}_N(\xi) = \hat{\nu}_N(\xi)\hat{\phi}_0(B^{-N}\xi).$$

In view of items 6) and 7) of Theorem 2 we see that

$$\lim_{N\to\infty} \hat{\nu}_N(\xi) = \hat{\phi}(\xi) \quad \text{and} \quad \lim_{N\to\infty} \hat{\phi}_N(\xi) = \hat{\phi}(\xi),$$

for all ξ in R^n and

$$\hat{\nu}_N(\xi)\hat{\phi}(B^{-N}\xi) = \hat{\phi}(\xi), \tag{41}$$

for $N = 1, 2, \cdots$.

Now, to see that item 1) implies 6) let $\phi = \chi_Q$ and recall that the fact that the family $\{\phi(x-k)\}_{k \in Z^n}$ is orthonormal implies that

$$\hat{\phi}(0) = 1 \quad \text{and} \quad \sum_{k \in Z^n} |\hat{\phi}(\xi - 2\pi k)|^2 = 1,$$

for almost all ξ in R^n. From this and the smoothness properties of $\hat{\phi}$ it follows that there is a finite subset $\mathscr{Z}$ of Z^n such that

$$\sum_{k \in \mathscr{Z}} |\hat{\phi}(\xi - 2\pi k)|^2 > 1/2,$$

for all ξ in the cube $\Omega = [-\pi, \pi]^n$. The last inequality implies that if N is the number of elements in $\mathscr{Z}$ then for each ξ in Ω there is an open cube Ω_ξ and a lattice point k_ξ in $\mathscr{Z}$ such that

$$|\hat{\phi}(\eta - 2\pi k_\xi)|^2 > \frac{1}{2N},$$

for all η in Ω_ξ. Let $\Omega_{\xi_0}, \Omega_{\xi_1}, \cdots, \Omega_{\epsilon_m}$ be any finite subcollection of these cubes which covers Ω and such that $\Omega_{\xi_0} = \Omega_0$ is a cube centered at 0. Finally, let K_0 be the closure of $\Omega_{\xi_0} \cap \Omega$ and let K_j be the closure of $(\Omega_{\xi_j} \cap \Omega) \setminus (\cup_{l=0}^{j-1} K_l)$. It is clear that $K = \cup_{j=0}^{m}(k_{\xi_j} + K_j)$ does the desired job by virtue of (41).

To complete the proof we show that item 6) implies 2).

Let $\phi = |Q|^{-1}\chi_Q$. Since $\|\phi\|_{L^2(R^n)} = |Q|^{-1/2}$, to see the desired result it suffices to show that $\|\phi\|_{L^2(R^n)} = 1$. Since $\|\phi_N\|_{L^2(R^n)} = 1$, $N = 1, 2, \cdots$, Plancherel's formula will imply the desired result if

$$\lim_{N\to\infty} \int_{R^n} |\hat{\phi}_N(\xi)|^2\,d\xi = \int_{R^n} |\hat{\phi}(\xi)|^2\,d\xi. \tag{42}$$

This is where the set K comes in. First observe that

$$\sum_{k \in Z^n} |\hat{\phi}_0(\xi - 2\pi k)|^2 = 1,$$

so, by virtue of the fact that $\hat{\nu}_N$ is $2\pi B^N Z^n$ periodic, setting $\Omega = [-\pi, \pi]^n$ we may write

$$\begin{aligned}
\int_{R^n} |\hat{\phi}_N(\xi)|^2\,d\xi
&= \int_{R^n} |\hat{\nu}_N(\xi)\hat{\phi}_0(B^{-N}\xi)|^2\,d\xi \\
&= \sum_{k \in Z^n} \int_{B^N\Omega} |\hat{\nu}_N(\xi)|^2 |\hat{\phi}_0(B^{-N}\xi - 2\pi k)|^2\,d\xi \\
&= \int_{B^N\Omega} |\hat{\nu}_N(\xi)|^2\,d\xi = \int_{B^N K} |\hat{\nu}_N(\xi)|^2\,d\xi \\
&= \int_{R^n} |\hat{\nu}_N(\xi)\chi_{B^N K}(\xi)|^2\,d\xi
\end{aligned}$$

or, more briefly,

$$\int_{R^n} |\hat{\phi}_N(\xi)|^2\,d\xi = \int_{R^n} |\hat{\nu}_N(\xi)\chi_{B^N K}(\xi)|^2\,d\xi. \tag{43}$$

Next observe that there is a positive constant C such that

$$|\hat{\phi}(\xi)| > C, \qquad \text{for all } \xi \text{ in } K. \tag{44}$$

(If this were not the case then $\hat{\phi}(\xi) = 0$ for some ξ in K. The hypothesis and (41) then imply that $\hat{\phi}(B^{-N}\xi) = 0$ for $N = 1, 2, \cdots$, contradicting the fact that $\hat{\phi}(0) = 1$.) Inequality (44) implies that $|\hat{\phi}(B^{-N}\xi)| > C$ for all ξ in $B^N K$ so by virtue of (41), we may write

$$|\hat{\nu}_N(\xi)\chi_{B^N k}(\xi)| \le C^{-1}|\phi(\xi)|. \tag{45}$$

Finally, since K contains a neighborhood of 0, it follows that

$$\lim_{N\to\infty} \hat{\nu}_N(\xi)\chi_{B^N k}(\xi) = \hat{\phi}(\xi)$$

for all ξ in R^n so that (45) implies that

$$\lim_{N\to\infty}\int_{R^n}|\hat{\nu}_N(\xi)\chi_{B^N K}(\xi)|^2\,d\xi=\int_{R^n}|\hat{\phi}(\xi)|^2\,d\xi$$

by virtue of dominated convergence. In view of (43), this gives the desired result. □

IV. Wavelet Bases of Haar Type

To construct a piecewise constant wavelet basis associated with (Z^n, A) we use the results of Section III and follow Meyer's recipe outlined in Section II.

First let Q be any set satisfying items 1)–4) of Theorem 1. Then χ_Q is a scaling function for a multiresolution analysis $\mathscr{V}=\{\cdots, V_0, V_1, \cdots\}$ associated with (Z^n, A). Next we need to identify the subspace W_0, the orthogonal complement of V_0 in V_1. Since V_0 is the collection of all functions of the form

$$f(x)=\sum_{k\in Z^n} a_k\chi_Q(x-k),$$

where $\{a_k\}$ is in $l^2(Z^n)$, it is not difficult to see the following.

Lemma 6: W_0 is the collection of all functions of the form

$$f(x)=\sum_{k\in Z^n} c_k q^{1/2}\chi_Q(Ax-k), \tag{46}$$

where the sequence of coefficients $\{c_k\}$ is in $l^2(R^n)$ and satisfies

$$\sum_{k\in\mathscr{K}} c_{k+al}=0, \qquad \text{for all } l \text{ in } Z^n. \tag{47}$$

Here $\mathscr{K}=\{k_1,\cdots,k_q\}$ is the collection of coset representatives appearing in the self-similarity relation (11) for Q.

From (46) and (47) it is not difficult to construct a collection of $q-1$ basic wavelets whose existence is guaranteed by Meyer's result. Indeed one can easily verify the following.

Lemma 7: Suppose $U=(u_{ij})$ is a unitary matrix $q\times q$ matrix whose first row is constant, namely $u_{1j}=q^{-1/2}$, $j=1,\cdots,q$. Let $\mathscr{K}=\{k_1,\cdots,k_q\}$ be the collection of coset representatives as in Lemma 6. Then the collection of functions $\{\psi_1,\cdots,\psi_{q-1}\}$ defined by

$$\psi_{i-1}(x)=\sum_{j=1}^{q} u_{ij}q^{1/2}\chi_Q(Ax-k_j) \qquad i=2,\cdots,q \tag{48}$$

is a collection of elementary basic wavelets corresponding to the simple multiresolution analysis associated with (Z^n, A) whose scaling function is χ_Q. In other words, the support of ψ_i is contained in Q, $i=1,\cdots,q-1$, and the collection $\{\tau_k\psi_i,\ k\subset Z^n,\ i=1,\cdots,q-1\}$ is a complete orthonormal system for W_0.

Conversely, every collection of elementary basic wavelets that arises from the multiscale analysis associated with (Z^n, A) whose scaling function is χ_Q is of form (48).

In the case $q=2$, there is essentially only one matrix $U=(u_{ij})$ that satisfies the property described in the lemma. Namely, $u_{11}=u_{12}=1/\sqrt{2}$ and $u_{21}=-u_{22}=\pm 1/\sqrt{2}$. When $q\ge 3$ there are many such examples. Specifically, we mention the case

$$u_{ij}=\frac{1}{q^{1/2}}\cos\frac{(i-1)(j-1)2\pi}{q},$$

$i=1,\cdots,q$ and $j=1,\cdots,q$.

As a corollary we state the following.

Theorem 4: Suppose $\psi_1,\cdots,\psi_{q-1}$ is the collection of functions defined in Lemma 7. Then the system

$$U_A^{-j}\tau_k\psi_i \qquad j\in Z,\ k\in Z^n,\ i=1,\cdots,q-1$$

is a complete orthonormal basis for $L^2(R^n)$.

V. Examples

A. Generalities

Numerical experiments lead to various observations. These include the following.

- Fixing the dilation matrix A but varying the choice of digits can result in wildly varying Q's. Some cases appear totally unrelated while others appear to be some sort of dilates of each other.
- Certain choices of dilation matrix A can give rise to Q's that are simple parallelepipeds when an appropriate choice of digits $\mathscr{K}$ is used. Other choices of A never give rise to such simple Q's; the corresponding Q's are always "fractals."

The following proposition sheds some light on the first item.

Lemma 8: Suppose $\mathscr{K}_1$ and $\mathscr{K}_2$ are two full sets of digits for A. Let Q_1 and Q_2 be the self-similar sets satisfying (11), with $\mathscr{K}=\mathscr{K}_1$ and $\mathscr{K}_2$, respectively. If there is a linear transformation B that commutes with A and such that $\mathscr{K}_2=B\mathscr{K}_1$ then $Q_2=BQ_1$.

Remark: Examples show that the hypothesis that B commutes with A is essential for the conclusion.

Proof: Write

$$ABQ_1=BAQ_1=B\Big\{\bigcup_{k\in\mathscr{K}_1}(k+Q_1)\Big\}$$
$$=\bigcup_{k\in\mathscr{K}_1}(Bk+BQ_1)=\bigcup_{k\in\mathscr{K}_2}(k+BQ_1).$$

Since the solution of (11) is unique the last string of equalities implies the desired result. □

Concerning the second item, here is a characterization of dilation matrices A that can give rise to self-similar sets Q, which are simple parallelepipeds with the appropriate choice of digits $\mathscr{K}$.

Lemma 9: The self-similar tile Q resulting as the limit of the iteration (25) can be a parallelepiped, if and only if the dilation factorizes as $A=CDPC^{-1}$. Here C is an integer-valued, invertible matrix with determinant ± 1, P a permutation matrix, $(Px)_i=x_{\pi(i)}$ for some permutation $\pi\in S_n$, and D is a diagonal matrix with entries $d_i\in Z$ along the diagonal such that on each cycle of π $|d_i|>1$ for at least one i; in other words, such that $|d_i d_{\pi(i)} d_{\pi^2(i)} \cdots d_{\pi^{n-1}(i)}| > 1$ for each $i = 1, 2, \cdots, n$.

Proof: Assume that A is of the described form, then A is indeed a dilation. Since C, P, D are integer-valued, A leaves Z^n invariant. The conditions on D, P guarantee that DP, hence, $A = CDPC^{-1}$ have all eigenvalues $|\lambda_i| > 1$. To see this, let $D^{(k)}$ be the diagonal matrix with the permuted entries $d_i^{(k)} = d_{\pi^k(i)}$, and check that $P^k D = D^{(k)} P^k$. After rearranging, $(DP)^n = DD'D'' \cdots D^{(n-1)}P^n = \Pi_{k=1}^n D^{(k)}$, one obtains a diagonal matrix $\tilde{D}$ with elements $d_i = d_i d_{\pi(i)} d_{\pi^2(i)} \cdots d_{\pi^{n-1}(i)}$. The eigenvalues of DP are nth roots of D_0, and its eigenvalues $|\lambda_i| > 1$, if and only if in the cycle determined by $i\,|\,d_{\pi^k(i)}| > 1$ for at least one k.

P just interchanges the coordinate axes and thus $PQ_0 = Q_0$. Therefore, $DPQ_0 = DQ_0$ is a parallelepiped with edges parallel to the coordinate axes and side-lengths $|d_i|$. Obviously $DPQ_0 = \cup_{i=1}^q (k_i + Q_0)$ for an appropriate set $\{k_i, i = 1, \cdots, q\} \subseteq Z^n$ of the form $\{1 \in Z^n: 0 \le l_i < d_i \text{ or } d_i \le l_i < 0\}$. In the general case, $C \ne Id$, set $Q = CQ_0$ and $k_i' = Ck_i$, then $CDPC^{-1}Q = \cup_{i=1}^q (Ck_i + CQ_0) = \cup_{i=1}^q (k_i' + Q)$.

Conversely, assume that Q is a parallelepiped $Q = \{x: x = \sum_{i=1}^n x_i e_i,\ a_i \le x_i \le b_i,\ i = 1, 2, \cdots, n\}$ for some numbers a_i, b_i and n linearly independents vectors $e_i \in R^n$. Upon writing $a = \sum a_i e_i$ and C for the invertible matrix with columns $(b_i - a_i)e_i$, one obtains $Q = CQ_0 + a$.

The self-similarity of Q (11) becomes $C^{-1}A(CQ_0 + a) = \cup_{i=1}^q C^{-1}(k_i + a + CQ_0)$ or

$$C^{-1}ACQ_0 = \bigcup_{i=1}^{q} \left(C^{-1}(k_i + a - Aa) + Q_0\right). \quad (49)$$

In other words, the parallelepiped $C^{-1}ACQ_0$ is a union of unit cubes. This is only possible if the edges of $C^{-1}ACQ_0$ are parallel to the coordinate axes and $C^{-1}(k_i + a - Aa) \in \{l \in Z^n: 0 \le l_i < d_i \text{ or } -d_i \le l_i < 0\}$ for some $d_i \in Z$. Therefore $A' = C^{-1}AC$ maps the coordinate axes onto themselves, $A'\delta_i = d_i \delta_{\pi(i)}$, where $d_i \in Z$, $\pi \in S_n$ is a permutation, and δ_i is the ith unit vector. Thus $A' = DP$ and $A = C^{-1}DPC$. Finally, $R^n = C^{-1}R^n = C^{-1}(\cup_{k \in Z^n}(k + Q)) = C^{-1}(\cup_{k \in Z^n}(k + a + CQ_0)) = \cup_{k \in Z^n}(C^{-1}a + C^{-1}k + Q_0)) = \cup_{k \in Z^n}(C^{-1}k + Q_0))$ implies that $C^{-1}Z^n = Z^n$. A symmetric argument yields $CZ^n = Z^n$, from which $C \in SL(n, Z)$ follows. □

Remarks:

- As we have just seen, the choice $\mathcal{K} = \{l \in Z^n: 0 \le l_i < d_i \text{ or } -d_i \le l_i < 0\}$ produces a parallelepiped as the self-similar tile when A is of the appropriate form. It is, however, not the only possible choice of digits in this case. As we will see (Example 3 in Section V-C), other sets of digits can be used and yield more interesting tilings.
- Since the set Q depends on the choice of digits, it should be clear that there may be many different multiresolution analyses associated with (Z^n, A), which consists of simple functions. In certain cases, Lemmas 4 and 8 are useful in relating different multiresolution analyses that arise in this way.

The following examples represent a very small sampling of a very large smorgasbord.

B. Univariate Examples

The case $A = 2$ is relatively uninteresting in this context. Since all sets of possible digits are related via shifts and multiplications by integers, the fact that $(A - I)^{-1} = I$ together with Lemmas 4 and 8 implies that the only Q's that lead to scaling functions are integer translates of the interval [0, 1]. Thus, we may conclude that *there is only one multiresolution analysis associated with (Z, 2) that consists of simple functions; its scaling function is the characteristic function of the interval [0, 1] and the corresponding elementary wavelet basis is the classical Haar system.* For a similar result obtained from a different point of view see [3].

The case $A = 3$ is already more interesting. The choice of digits {0, 1, 2} leads to $Q = [0, 1]$. Since $(A - I)^{-1} = 1/2$, the choice of digits {1, 2, 3} leads to $Q = [1/2, 3/2]$. These Q's lead to two different multiresolution analyses associated with $(Z, 3)$. The choice of digits {0, 1, 5} leads to a disconnected set Q that can be shown to have measure one by using item 6) of Theorem 3; the corresponding characteristic function is the scaling function of yet another multiresolution analysis associated with $(Z, 3)$. Using Lemma 7, one can construct many different elementary wavelet bases corresponding to each of these examples.

These examples should give the flavor of what happens in the general univariate case.

C. Two-Dimensional Examples

1) The dilation

$$A = \begin{pmatrix} 0 & 2 \\ -1 & 0 \end{pmatrix}$$

is of the type described in Lemma 9. Choosing $k_1 = (0, -1)$, $k_2 = (1, -1)$ one obtains $Q = [0, 1]^2$ as the basic tile, since $AQ = [0, 2] \times [-1, 0] = (0, -1) + Q \cup (1, -1) + Q$. The basic wavelet ψ is defined by $\psi(x) = 1$ for $x \in [0, 1] \times [0, 1/2)$, $\psi(x) = -1$ for $x \in [0, 1] \times [1/2, 1]$ and $\psi(x) = 0$, elsewhere. The corresponding elementary wavelet basis for $L^2(R^2)$ is $2^{j/2}\psi(A^j x - k)$, $j \in Z$, $k \in Z^2$; it is generated by one function only.

In this case one can easily determine all the simple multiresolution analyses associated with (Z^2, A). Indeed we have the following:

There are exactly three different simple multiresolution analyses associated with (Z^2, A). Their scaling functions are the characteristic functions of the squares Q, $Q + x_0$ and $Q - x_0$ where $Q = [0, 1]^2$ and $x_0 = (-1/3, 1/3)$.

To see this observe that any acceptable pair of digits must be Z^2 translates of one of the pairs $\{(0, 0), (l, m)\}$ where $l = 2j + 1$, and $j, m \in Z$. Now, the pair $\{(0, 0), (l, m)\}$ is the image of the pair $\{(0, 0), (1, 0)]$ under the linear transformation whose matrix is given by

$$B = \begin{pmatrix} l & -2m \\ m & l \end{pmatrix}$$

and which commutes with A. In view of Lemma 8 and the fact that the determinant of B is $l^2 + 2m^2$, it follows that the only such pairs that can give rise to a Q of measure one are $\{(0,0), (1,0)\}$ and $\{(0,0), (-1,0)\}$. Since these pairs are Z^2 translates of each other we may conclude by virtue of Lemma 4 that all the acceptable sets Q are of the form $Q_0 + (A - I)^{-1}k$, $k \in Z^2$. Since

$$(A - I)^{-1} = -\frac{1}{3}\begin{pmatrix} 1 & 2 \\ -1 & 1 \end{pmatrix},$$

it is easy to check that each such square is a Z^2 translate of one of the three squares previously given.

2) The matrix

$$A = \begin{pmatrix} 1 & -1 \\ 1 & 1 \end{pmatrix}$$

rotates a vector by $\pi/4$ and stretches it by a factor of $\sqrt{2}$. Choosing $k_1 = (0,0)$, $k_2 = (1,0)$, the algorithm (25) produces the set Q shown in Fig. 1, which is known as the twin dragon set in the flowery language of fractals [1].

In order to check that the characteristic function of Q is a scaling function for a multiresolution analysis we verify Cohen's condition. In this case $\hat{\mu}(\xi_1, \xi_2) = (1 + e^{i\xi_1})/2$ has zeroes at $\xi_1 = (2n+1)\pi$, ξ_2 arbitrary. The natural guess $K = \Omega = [-\pi, \pi]^2$ does not satisfy (40), because $\hat{\mu}(\pm A^{-1}(\pi, \pi)) = \hat{\mu}(\pm\pi, 0) = 0$. This can be avoided by moving a neighborhood of the critical points $\pm(\pi, \pi)$ by $\mp 2\pi$ as follows: Let $U_{\pm 1} = \Omega \cap B_{\pm 1}$ where $B_{\pm 1} = \{x: |x \mp (\pi, \pi)| < \delta\}$. Then $K = (\Omega \setminus (U_1 \cup U_{-1})) \cup (U_1 + (-2\pi, 0)) \cup (U_{-1} + (2\pi, 0))$ does the job.

The function ψ defined by $\psi(x) = 1$ for $x \in A^{-1}Q$, $\psi(x) = -1$ for $x \in A^{-1}Q + (1/2, -1/2)$, $\psi(x) = 0$ elsewhere is the basic elementary wavelet in this case, see Fig. 2. The collection $2^{j/2}\psi(A^j x - k)$, $j \in Z$, $k \in Z^2$, which is generated by the one basic wavelet is a complete orthonormal basis for $L^2(R^2)$.

By using reasoning similar to that used in the previous example we can easily determine all the simple multiresolution analyses associated with (Z^2, A). Indeed, we have the following.

There are exactly two different simple multiresolution analyses associated with (Z^2, A). Their scaling functions are the characteristic functions of the twin dragon set Q described above and BQ where B is a rotation about the origin by the angle $\pi/2$.

3) At first glance the usual homogeneous dilation $A = 2I$, as in the univariate case, does not seem very interesting. If $k_1 = (0,0)$, $k_2 = (1,0)$, $k_3 = (0,1)$, $k_4 = (1,1)$, then clearly $Q = [0,1]^2$ and χQ is the scaling function for multiscale analysis $\mathscr{V}_1$ associated with (Z^2, A), which is the obvious generalization of the univariate dyadic multiscale analysis considered in Section V-B. The corresponding elementary wavelet basis is generated by three basic wavelets that are easily constructed using the recipe given in Section IV.

Choosing as the digits $k_1 = (0,0)$, $k_2 = (1,1)$, $k_3 = (0,1)$, $k_4 = (1,2)$, one obtains the parallelogram with corners k_i, $i = 1, 2, 3, 4$ as the self-similar set Q. The characteristic function of Q is the scaling function for a multiscale analysis $\mathscr{V}_2$ associated with (Z^2, A), which is different from $\mathscr{V}_1$ previously mentioned.

However, if the digits are chosen $k_1 = (0,0)$, $k_2 = (1,0)$, $k_3 = (0,1)$, $k_4 = (-1,-1)$, then the algorithm (25) converges to a Cantor-like set Q, which is shown in Fig. 3. To see that χ_Q is a scaling function for a multiresolution analysis associated with (Z^n, A), we check Cohen's condition: The zeros of $\hat{\mu}(\xi_1, \xi_2) = (1 + e^{i\xi_1} + e^{i\xi_2} + e^{-i(\xi_1+\xi_2)})/4$ are at the points $((2k+1)\pi, l\pi)$ or $(l\pi, (2k+1)\pi)$, $l, k \in Z$. Then $K = [-\pi, \pi]^2$ does the job since $\hat{\mu}$ does not vanish $2^{-j}K \subseteq \frac{1}{2}K = [-\pi/2, \pi/2]^2$. It should be clear that the corresponding multiresolution analysis is different from $\mathscr{V}_1$

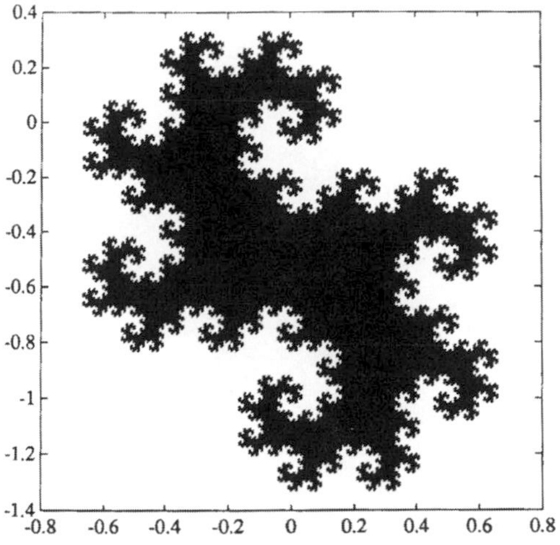

Fig. 1. Tile described in Example 2) in Section V-C.

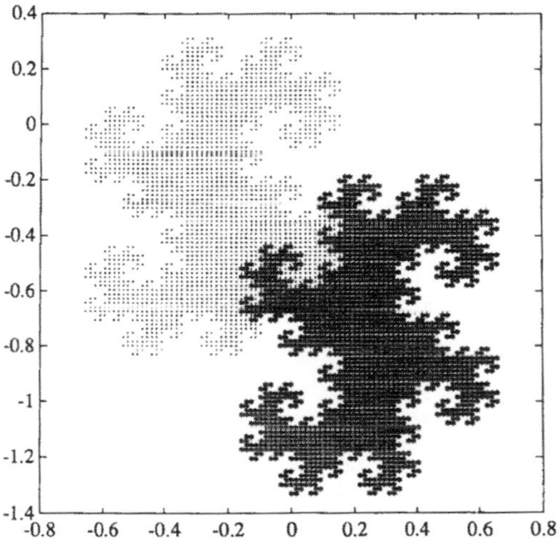

Fig. 2. Wavelet corresponding to tile in Fig. 1.

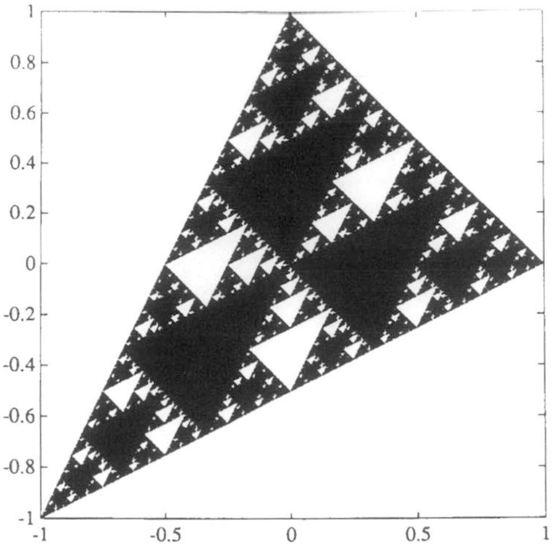

Fig. 3. Tile described in last paragraph of Example 3) in Section V-C.

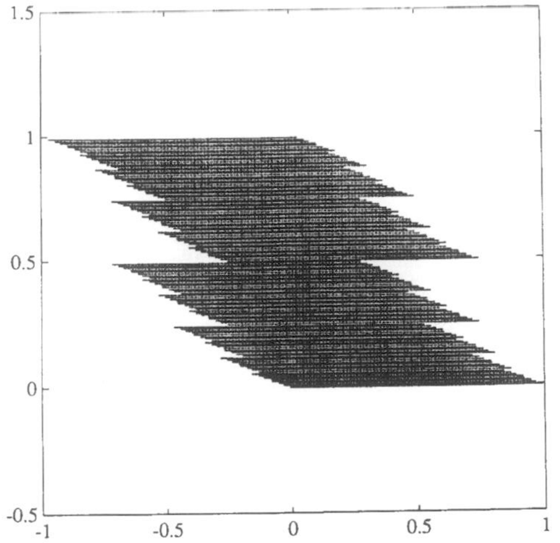

Fig. 5. One of the tiles described in Example 4) in Section V-C.

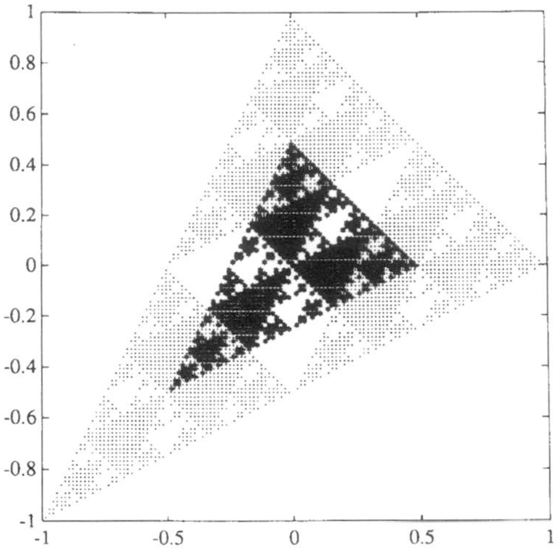

Fig. 4. Wavelet corresponding to tile in Fig. 2.

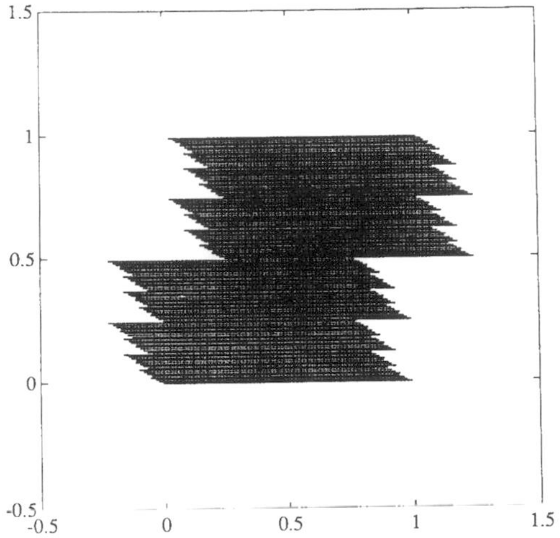

Fig. 6. Second tile described in Example 4) in Section V-C.

and $\mathscr{V}_2$ previously mentioned. The corresponding elementary wavelet bases are generated by three basic wavelets that are easily constructed using the recipe given in Section IV, see Fig. 4.

4) Figs. 5 and 6 show the tiles obtained from the matrix

$$A = \begin{pmatrix} 2 & 1 \\ 0 & 2 \end{pmatrix},$$

with the digits $k_1 = (0,0)$, $k_2 = (1,0)$, $k_3 = (0,1)$, $k_4 = (1,1)$ and the digits $k_1 = (0,0)$, $k_2 = (1,0)$, $k_3 = (1,1)$, $k_4 = (2,1)$, respectively. It should be clear from the picture that the first set has measure one; this can also be checked by verifying Cohen's condition with $K = [-\pi, \pi]^2$. The set in Fig. 6 is the image of the first under the linear map that sends $(1,0)$ to $(1,0)$ and $(0,1)$ to $(1,1)$; this transformation maps the first set of digits into the second, commutes with A, and has determinant one.

5) Figs. 7 and 8 show the tiles obtained from the matrices

$$\begin{pmatrix} 3 & 0 \\ 0 & 3 \end{pmatrix} \text{ and } \begin{pmatrix} 2 & -1 \\ 1 & -2 \end{pmatrix}$$

with the digits $\{(0,0), (0,1), (0,2), (1,0), (1,2), (2,0), (2,1), (2,2), (4,4)\}$ and $\{(0,0), (1,0), (0,-1)\}$, respectively.

VI. Miscellaneous Remarks

For more details and background concerning dyadic multiresolution analysis and wavelet bases, including the classi-

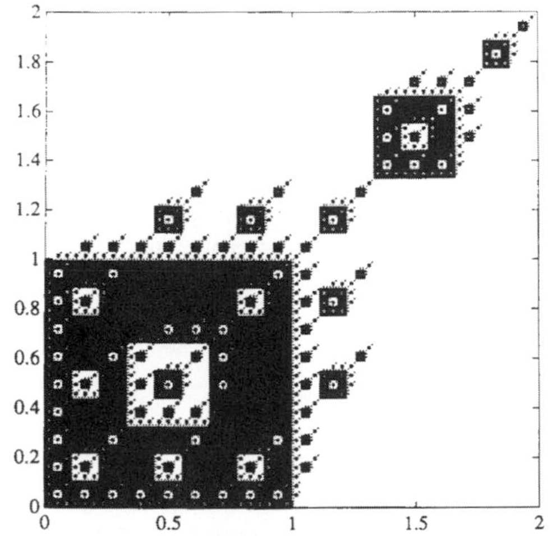

Fig. 7. Tile corresponding to $A = 3I$ described in Example 5) in Section V-C.

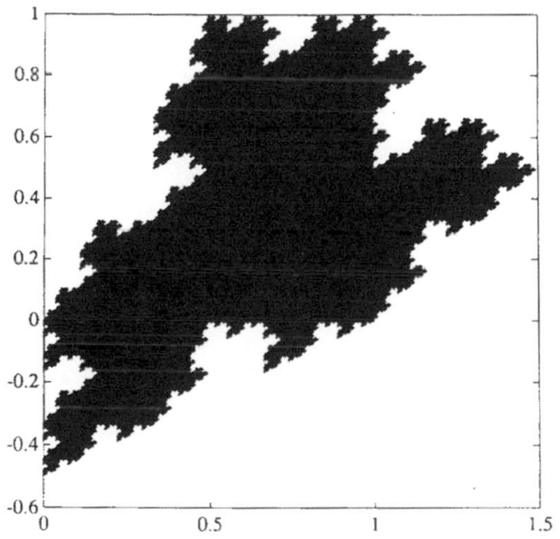

Fig. 8. Second tile described in Example 5) in Section V-C.

cal Haar system, see [3], [6], [7], [9] and the references cited there; for the more general case see [8].

The fixed point iteration (24) is called the *cascade algorithm* in [3] in the case considered there. Tilings of R^n that are not necessarily self-similar arise naturally in other contexts also; for example, see [10] and [4]. The proof of the equivalence of item 6) to the other items in Theorem 3 is an easy modification of the argument found in [2], we include it for the sake of completeness.

In the case $|\det A| = 2$, the martingale version of a classical theorem of Littlewood and Paley, see [5, Theorem 5.3.8], implies that the elementary wavelet bases constructed here are also unconditional bases for $L^p(R^n)$, $1 < p < \infty$.

We emphasize that the specific examples considered here represent a small selection of a wealth of interesting examples. The numerical experiments were done with Matlab software on a Sun 3 workstation.

Acknowledgment

The authors wish to acknowledge Cohen who showed us his condition and made [2] available to us. After this paper was submitted for publication, W. Lawton kindly brought to our attention the related work [11]. These authors propose an interesting framework for multiresolution analyses associated with (Γ, A) whose scaling sequences are finite and, in that spirit, obtain some of the results found here.

References

[1] M. Barnsley, *Fractals Everywhere*. New York: Academic Press, 1988.

[2] A. Cohen, "Ondelettes, analyses multirésolutions et filtres miroirs en quadrature," *Ann. Inst. Poincaré*, vol. 7, pp. 439–459, 1990.

[3] I. Daubechies, "Orthonormal bases of compactly supported wavelets," *Comm. in Pure and Applied Math.*, vol. 41, pp. 909–996, 1988.

[4] C. deBoor, C. K. Hollig, and S. Riemenschneider, "Convergence of cardinal series," *Proc. Amer. Math. Soc.*, vol. 98, pp. 457–460, 1986.

[5] R. E. Edwards and G. I. Gaudry, *Littlewood-Paley and Multiplier Theory*. New York: Springer, 1977.

[6] P. G. Lemarie and Y. Meyer, "Ondelette et bases hilbertiennes," *Revista Matematica Iberoamericana*, vol. 2, pp. 1–18, 1986.

[7] S. Mallat, "Multiresolution approximations and wavelet orthonormal bases of $L^2(R)$," *Trans. Amer. Math. Soc.*, vol. 315, pp. 69–88, 1989.

[8] Y. Meyer, "Ondelettes, fonctions splines et analyses graduées," Rapport CEREMADE n.8703, 1987.

[9] —, *Ondelettes et Opérateurs*. Paris: Hermann, 1990.

[10] W. P. Thurston, "Groups, Tilings, and Finite State Automata," in *Lecture Notes, Summer Meetings Amer. Math. Soc.*, Bolder, CO 1989.

[11] W. M. Lawton and H. L. Resnikoff, "Multidimensional wavelet bases," Aware Inc. tech. rep., submitted to *SIAM J. Math. Anal.*

SECTION VII
Selected Applications

Introduction

Mladen Victor Wickerhauser

1. The Selections

Over the past decade, wavelet transforms have been widely applied. Good implementations of the discrete wavelet transform (DWT) were built into software systems such as Matlab and S-Plus, and the DWT became a frequently used tool for data analysis and signal processing. There are certain problems, though, on which this tool works particularly well. The most common ingredient in those problems is some complicated object that can be closely approximated by a few superposed wavelets. This compilation includes four seminal articles that introduced some of these stand out DWT applications. I have taken a random and sparse sampling of relevant articles and books published around the same time, in order to place the results in context and illustrate their influence.

2. Fast Evaluation of Singular Integral Operators

Discrete wavelet transforms are fast algorithms, costing $O(1)$ arithmetic operations per output regardless of the number of inputs. This is even better than the fast Fourier transform, which costs $O(\log N)$ operations per output given N inputs, as much as a complete wavelet packet or multiscale local cosine analysis. A general linear transformation of N inputs, by contrast, costs $O(N)$ operations per output. It was seen right away that DWTs could be advantageous in high-dimensional problems.

An example is the evaluation of an integral operator $f \mapsto Tf$ with

$$Tf(x) = \int t(x,y)f(y)\,dy,$$

where t is a smooth function except on some thin subset of its domain. The gravitational potential, with $t(x,y) = 1/|x-y|$, is one such operator. To simulate the time evolution of a many-particle system interacting by gravitation requires repeated recomputation and then evaluation of T. A great deal of work in the 1980s (see [App85], [BH86], [Zha87]) culminated in V. Rokhlin's fast multipole algorithm [GR87], [CGR88].

The seminal 1991 article in this compilation, "Fast Wavelet Transforms and Numerical Algorithms, I" by Beylkin, Coifman, and Rokhlin, shows that the fast multipole hierarchical decomposition is in essence a multiresolution analysis. It may be performed fast by an orthogonal pair of conjugate quadrature (mirror) filters. Sparsity of the resulting matrix is guaranteed for Calderon-Zygmund singular integral operators [Mey92] such as the gravitational potential, if the underlying wavelets representing the operator have many vanishing moments.

Subsequently, more complex wavelet-like transforms were brought to bear on ever nastier linear operators to get sparse matrix approximations [Alp93], [Bey92], [ABCR93], [Wic94], [MBH99]. Sparse matrix multiplication makes linear algebra feasible even in very high dimensions. The wide class of operators that reduce to sparse matrices in wavelet bases made possible fast algorithms for such difficult problems as numerical homogenization [BB95], electromagnetic scattering [Rok93], general trigonometric approximation [Bey95], and Hilbert transforms [BT96]. The special properties of wavelets also permit linear superpositions to be used in nonlinear functions [DM93], [BK97], [BBG98].

Multiresolution decomposition into wavelets with many desirable analytic properties has provided an elegant path into operator theory. The existence of fast discrete wavelet transforms has made this a smooth path to efficient numerical methods as well.

3. Improved Transform Coding Image Compression

Digital images also have the potential to be enormously complicated, but when they are pictures of interest to humans they must actually be relatively simple. Among many techniques for efficient storage or transmission of such pictures is *transform coding image compression* [Wic92]. The Joint Photographic Experts Group (JPEG) algorithm [JPEG1], [JPEG2], [Wal91] is perhaps the most common, since it is used in the JPG files found throughout the World Wide Web. But JPEG is an approximation algorithm. The errors it introduces, while nearly invisible to the eye, interfere with edge detection and similar image analysis.

The advantages of wavelets are nicely explained in Devore, Jawerth, and Lucier's foundation article, "Image Compression Through Wavelet Transform Coding" [DJL92]. The absence of JPEG's block artifacts allows compressed images to be used for automatic fingerprint identification systems, and so the United States Federal Bureau of Investigation (FBI) and Great Britain's Scotland Yard collaborated to design a custom wavelet and scalar quantization (WSQ) image compression standard [FBI93], [BBH93]. This relied on symmetric biorthogonal wavelets that were verified as suitable for high resolution images [CDF92], [MBA90], and for which a convenient boundary treatment existed [Bri96].

Subsequently, a more efficient implementation of the biorthogonal wavelet transform used in WSQ was found by Sweldens [Swe96]. Also, redundancy removers that partitioned wavelet coefficients into hierarchical subsets called zero-trees [Sha93], [SP96] were matched to this family of transforms, producing a remarkably simple and efficient coder. The result became a new standard, called JPEG-2000 [JPEG2000].

There are other boundary treatments using wavelets on intervals [CDV93], more general transforms such as wavelet packets [CW92], lapped orthogonal transforms [Mal90], and multiwavelets [SS94], plus various methods for progressive transmission and error correction coding of wavelet-compressed images that are making their way into proprietary, state-of-the-art coders for pictures and video. It is safe to say that every advantage of wavelets will be exploited in the fierce competition for better image quality and coding efficiency.

4. Easy Generic "De-Noising"

Digitally sampled signals that vary smoothly with time appear rough and may be hard to detect when measurement errors are present in each sample. The model of identically distributed independent normal errors, or "additive Gaussian white noise," is an extreme case of rough noise that is frequently used in practice. There are classical digital signal processing algorithms (DSP) to compute Gaussian white noise power, based on the discrete Fourier transform (DFT). With knowledge of the signal, we may design matched filters in the frequency domain and obtain minimax linear estimators for signal detection [Cou93], [ZT95].

But there are examples where the signal to be detected contains added noise that is correlated in time from sample to sample. Such noise may be smoother than Gaussian white noise, though still rougher than the signal. In addition, the signal itself or even its smoothness may be unknown. In these cases, very complicated estimators have been devised [Sto82].

A much simpler way to build estimators was described in Donoho and Johnstone's "Adapting to Unknown Smoothness by Wavelet Shrinkage" [DJ95], which is reprinted in this compilation. It followed a number of papers [Don93a], [Don93b], [DJ94] on the remarkable properties of wavelet coefficient thresholding, the bounded nonlinear operation of reducing or removing small-amplitude wavelet components of a noisy signal.

In practice, there still remains the problem of setting a threshold for wavelet shrinkage. There is a universal value that depends on the noise power, and there are techniques to adjust for correlated noise [JS97]. When the signal to be detected is known, there is the oracle method which selects a threshold to minimize the estimator variance [CRM99], [Cai99], [DWP02].

Other wavelet transforms, principally the continuous wavelet transform [Tor95], have also found use in signal estimation and detection. Examples include speech and music [KMG87], [GK96], NMR spectra [GKM92], and even gravitational waves [IT97].

5. Roughness, Volatility, and Turbulence

How do we estimate the roughness of a continuous function? One way is to calculate the Hölder exponent at each point. Jaffard's seminal 1989 paper "Exposants de Hölder en des points donnés et coéfficients d'ondelettes", included in this compilation, describes an elegant way to estimate the exponents from the asymptotic decay of wavelet coefficient amplitudes as scale tends to zero. The slower the decay, the smaller the exponent and the rougher the function. This fact leads to an elegant proof [HT91], [JMR01] of Gerver's famous result on the almost nowhere differentiability of Riemann's function [Ger70].

For a more detailed analysis of roughness, we may inquire about the distribution of Hölder exponents over the domain of a function. The *singularity spectrum* is one way to describe this distribution; it gives the fractal dimension [Man77] of domain subsets where the function has a particular Hölder exponent. This spectrum is useful in distinguishing physical phenomena [SM88], and it can be computed efficiently from time series using DWT [MBA94]. The asymptotic behavior of wavelet coefficients can also be used to detect *fractional Brownian motion*, or to synthesize examples [Fla92], [AS96] that are used in mathematical finance [MCF97].

When a theory predicts a certain degree of roughness, wavelet coefficient asymptotics may be used to test it. Kolmogorov's famous $-5/3$ power law for the velocity power spectrum in fully developed turbulence [Kol41], [Kol61] may be tested this way. We may also compute the singularity spectrum of portions of simulated or measured flows to determine if they are turbulent or laminar [AAGGHF89], [FGMPW92, ZSW91], for example.

References

[AAGGHF89] F. Argoul, A. Arneodo, G. Grasseau, Y. Gagne, E. Hopfinger, and U. Frisch, *Wavelet analysis of turbulence reveals the multifractal nature of the Richardson cascade*, Nature **338** (1989), 51–53.

[ABCR93] B. Alpert, G. Beylkin, R. R. Coifman, and V. Rokhlin, *Wavelet-like bases for the fast solution of second-kind integral equations*, SIAM J. Sci. Comput. **14** (1993), 159–184.

[Alp93] B. K. Alpert, *A class of bases for the sparse representation of integral operators*, SIAM J. Math. Anal. **24** (1993), 246–262.

[App85] A. W. Appel, *An efficient program for many-body simulation*, SIAM J. Sci. Comput. **6** (1985), 85–103.

[AS96] P. Abry and F. Sellan, *The wavelet-based synthesis for the fractional Brownian motion proposed by Y. Meyer and F. Sellan: Remarks and fast implementations*, Appl. Comput. Harmon. Anal. **3** (1996), 377–383.

[BB95] M. E. Brewster and G. Beylkin, *A multiresolution strategy for numerical homogenization*, Appl. Comput. Harmon. Anal. **2** (1995), 327–349.

[BBG98] G. Beylkin, M. E. Brewster, and A. C. Gilbert, *A multiresolution strategy for numerical reduction and homogenization of nonlinear odes*, Appl. Comput. Harmon. Anal. **5** (1998), 450–486.

[BBH93] J. N. Bradley, C. M. Brislawn, and T. E. Hopper, *FBI wavelet/scalar quantization standard for gray-scale fingerprint image compression*, Visual Information Processing II (F. O. Huck and R. D. Juday, eds.), Proc. SPIE Vol. 1961, SPIE, Orlando, Fl., 1993, 293–304.

[Bey92] G. Beylkin, *On the representation of operators in bases of compactly supported wavelets*, SIAM J. Numer. Anal. **629** (1992), 1716–1740.

[Bey95] G. Beylkin, *On the fast Fourier transform of of functions with singularities*, Appl. Comput. Harmon. Anal. **2** (1995), 363–381.

[BH86] J. Barnes and P. Hut, *A hierarchical $O(N \log N)$ force-calculation algorithm*, Nature **324** (1986), 446–449.

[BK97] G. Beylkin and J. M. Keiser, *on the adaptive numerical solution of nonlinear partial differential equations in wavelet bases*, J. Comput. Physics **132** (1997), 233–259.

[BT96] G. Beylkin and B. Torrésani, *Implementation of operators via filter banks, Hardy wavelets and autocorrelation shell*, Appl. Comput. Harmon. Anal. **3** (1996), 164–185.

[Bri96] C. Brislawn, *Classification of nonexpansive symmetric extension transforms for multirate filter banks*, Appl. Comput. Harmon. Anal. **3** (1996), 337–357.

[Cai99] T. T. Cai, *Adaptive wavelet estimation: a block thresholding and oracle inequality approach*, Ann. Statist. **27** (1999), 898–924.

[CDF92] A. Cohen, I. Daubechies, and J.-C. Feauveau, *Biorthogonal bases of compactly supported wavelets*, Comm. Pure Appl. Math. **45** (1992), 485–500.

[CDV93] A. Cohen, I. Daubechies, and P. Vial, *Wavelets on the interval and fast wavelet transforms*, Appl. Comput. Harmon. Anal. **1** (1993), 54–81.

[CGR88] J. Carrier, L. Greengard, and V. Rokhlin, *A fast adaptive multipole algorithm for particle simulations*, SIAM J. Sci. Comput. **93** (1988), 669–686.

[CRM99] I. Cohen, S. Raz, and D. Malah, *Translation-invariant denoising using the minimum description length criterion*, Signal Proc. **75** (1999), 201–223.

[CW92] R. R. Coifman and M. V. Wickerhauser, *Entropy based algorithms for best basis selection*, IEEE Trans. Inform. Th. **32** (1992), 712–718.

[DJ94] D. L. Donoho and I. M. Johnstone, *Ideal spatial adaptation via wavelet shrinkage*, Biometrika **81** (1994), 425–455.

[DJ95] D. L. Donoho and I. M. Johnstone, *Adapting to unknown smoothness via wavelet shrinkage*, J. Amer. Statist. Assoc. **90** (1995), 1200–1224.

[DJL92] R. Devore, B. Jawerth, and B. J. Lucier, *Image compression through wavelet transform coding*, IEEE Trans. Inform. Th. **38** (1992), 719–746.

[DM93] W. Dahmen and C. A. Micchelli, *Using the refinement equation for evaluating integrals of wavelets*, SIAM J. Numer. Anal. **30** (1993), 507–537.

[Don93a] D. L. Donoho, *Wavelet shrinkage and WVD: A 10-minute tour*, Progress in Wavelet Analysis and Applications (Toulouse, 1992) (Y. Meyer and Sylvie Roques, eds.), Editions Frontières, Gif-sur-Yvette, France, 1993, 109–128.

[Don93b] D. L. Donoho, *Unconditional bases are optimal bases for data compression and for statistical estimation*, Appl. Comput. Harmon. Anal. **1** (1993), 100–115.

[DWP02] J. O. Deasy, M. V. Wickerhauser, and M. Picard, *Accelerating Monte Carlo simulations of radiation therapy dose distributions using wavelet threshold de-noising*, Medical Physics **29** (2002), 2366–2373.

[FGMPW92] M. Farge, E. Goirand, Y. Meyer, F. Pascal, and M. V. Wickerhauser, *Improved predictability of two-dimensional turbulent flows using wavelet packet compression*, Fluid Dynamics Research **10** (1992), 229–250.

[Fla92] P. Flandrin, *Wavelet analysis and synthesis of fractional Brownian motion*, IEEE Trans. Inform. Th. **32** (1992), 910–917.

[Cou93] L. W. Couch II, *Digital and Analog Communications Systems*, 4th ed., Macmillan, New York, 1993.

[FBI93] IAFIS-IC-0110v2, *WSQ gray-scale fingerprint image compression specification*, Version 2, US Department of Justice, Federal Bureau of Investigation, 16 February 1993.

[Ger70] J. L. Gerver, *The differentiability of the Riemann function at certain rational multiples of π*, Amer. J. Math. **92** (1970), 33–55.

[GK96] P. Guillemain and R. Kronland-Martinet, *Characterization of acoustic signals through continuous linear time-frequency representations*, Proc. IEEE **84** (1996), 561–585.

[GKM92] P. Guillemain, R. Kronland-Martinet, and B. Martens, *Estimation of spectral lines with the help of the wavelet transform—applications in N.M.R. spectroscopy*, Wavelets and Applications (Marseille, 1989) (Y. Meyer, ed.), RMA Res. Notes Appl. Math. **20**, Masson, Paris, 1992, 38–60.

[GR87] L. Greengard and V. Rokhlin, *A fast algorithm for particle simulations*, J. Comp. Phys. **73** (1987), 325–348.

[HT91] M. Holschneider and P. Tchamitchian, *Pointwise analysis of Riemann's "nondifferentiable" function*, Invent. Math. **105** (1991), 157–176.

[IT97] J. M. Innocent and B. Torrésani, *Wavelets and binary coalescence detection*, Appl. Comput. Harmon. Anal. **4** (1997), 113–116.

[JMR01] S. Jaffard, Y. Meyer, and R. D. Ryan, *Wavelets: Tools for Science and Technology*, rev. ed., SIAM, Philadelphia, 2001.

[JPEG1] ISO/IEC, *JTC1 draft international standard 10918-1: Digital compression and coding of continuous-tone still images, part 1: Requirements and guidelines*, ISO/IEC CD 10918-1, available from ANSI Sales (212)642-4900, November 1991, Alternate number SC2 N2215.

[JPEG2] ISO/IEC, *JTC1 draft international standard 10918-2: Digital compression and coding of continuous-tone still images, part 2: Compliance testing*, ISO/IEC CD 10918-2, available from ANSI Sales (212)642-4900, December 1991.

[JPEG2000] ISO/IEC, *JPEG 2000 final committee draft 15444-1*, Available from www.jpeg.org, March 2000.

[JS97] I. M. Johnstone and B. W. Silverman, *Wavelet threshold estimators for data with correlated noise*, J. Roy. Statist. Soc. Ser. B **59** (1997), 319–351.

[Kol41] A. N. Kolmogoroff, *The local structure of turbulence in incompressible viscous fluids for very large Reynolds numbers*, C. R. (Doklady) Acad. Sci. URSS (N.S.) **30** (1941), 301–305.

[KMG87] R. Kronland-Martinet, J. Morlet, and A. Grossmann, *Analysis of sound patterns through wavelet transforms*, Internat. J. Pattern Recog. Artif. Int. **1** (1987), 273–302.

[Kol61] A. N. Kolmogorov, *A refinement of previous hypotheses concerning the local structure of turbulence in viscous incompressible fluids at high Reynolds number*, J. Fluid Mech. **13** (1961), 82–85.

[Mal90] H. Malvar, *Lapped transforms for efficient transform/subband coding*, IEEE Trans. Acoustics, Speech, Signal Proc. **38** (1990), 969–978.

[Man77] B. Mandelbrot, *The Fractal Geometry of Nature*, W. H. Freeman, New York, 1977.

[MBA90] P. Mathieu, M. Barlaud, and M. Antonini, *Compression d'images par transformée en ondelette et quantification vectorielle*, Traitment du Signal **7** (1990), 101–115.

[MBA94] J. F. Muzy, E. Bacry, and A. Arneodo, *The multifractal formalism revisited with wavelets*, Internat. J. Bifur. Chaos Appl. Sci. Engrg. **4** (1994), 245–302.

[MBH99] L. Monzón, G. Beylkin, and W. Hereman, *Compactly supported wavelets based on almost interpolating and nearly linear phase filters (Coiflets)*, Appl. Comput. Harmon. Anal. **7** (1999), 184–210.

[MCF97] B. B. Mandelbrot, L. Calvet, and A. Fisher, *A multifractal model of asset returns*, Technical report, Yale University, New Haven, 1997. Cowles Foundation Discussion Paper #1164.

[Mey92] Y. Meyer, Wavelets and Operators, Cambridge University Press, Cambridge, 1992.

[Rok93] V. Rokhlin, *Diagonal forms of translation operators for the Helmholtz equation in three dimensions*, Appl. Comput. Harmon. Anal. **1** (1993), 82–93.

[Sha93] J. M. Shapiro, *Embedded image coding using zerotrees of wavelet coefficients*, IEEE Trans. Signal Proc. **41** (1993), 3445–3462.

[SM88] H. E. Stanley and P. Meakin, *Multifractal phenomena in physics and chemistry*, Nature **335** (1988), 405–409.

[SP96] A. Said and W. A. Pearlman, *A new fast and efficient image codec based on set partitioning in hierarchical trees*, IEEE Trans. Circuits Systems Video Tech. **6** (1996), 243–250.

[SS94] G. Strang and V. Strela, *Orthogonal multiwavelets with vanishing moments*, Opt. Engrg., **33** (1994), 2104–2107.

[Sto82] C. Stone, *Optimal global rates of convergence for nonparametric estimators*, Ann. Statist. **10** (1982), 1040–1053.

[Swe96] W. Sweldens, *The lifting scheme: A custom-design construction of biorthogonal wavelets*, Appl. Comput. Harmon. Anal. **3** (1996), 186–200.

[Tor95] B. Torrésani, *Analyse continue par ondelette*, InterÉditions/CNRS Éditions, Paris, 1995.

[Wal91] G. K. Wallace, *The JPEG still picture compression standard*, Comm. ACM **34** (1991), 30–44.

[Wic92] M. V. Wickerhauser, *High-resolution still picture compression*, Digital Signal Processing: a Review Journal **2** (1992), 204–226.

[Wic94] M. V. Wickerhauser, *An adapted waveform functional calculus*, Proc. of the Cornelius Lanczos Centenary (Raleigh, NC, December 1993) (M. Chu, R. Plemmons, D. Brown, and D. Ellison, eds.), SIAM, Philadelphia, 1994, 418–421.

[Zha87] F. Zhao, *An $O(N)$ algorithm for three-dimensional N-body simulations* preprint, MIT Artificial Intelligence Laboratory, Cambridge, October 1987.

[ZSW91] L. Zubair, K. R. Sreenivasan, and M. V. Wickerhauser, *Characterization and compression of turbulent signals and images using wavelet packets*, Studies in Turbulence (T. Gadsky, S. Sirkar, and C. Speziale, eds.), Springer, New York, 1991, 489–513.

[ZT95] R. E. Ziemer and W. E. Tranter, Fourth edition, *Principles of Communications Systems, Modulation and Noise*, 4th ed., John Wiley, New York, 1995.

Fast Wavelet Transforms and Numerical Algorithms I

G. BEYLKIN, R. COIFMAN, AND V. ROKHLIN

Yale University

Abstract

A class of algorithms is introduced for the rapid numerical application of a class of linear operators to arbitrary vectors. Previously published schemes of this type utilize detailed analytical information about the operators being applied and are specific to extremely narrow classes of matrices. In contrast, the methods presented here are based on the recently developed theory of wavelets and are applicable to all Calderon-Zygmund and pseudo-differential operators. The algorithms of this paper require order $O(N)$ or $O(N \log N)$ operations to apply an $N \times N$ matrix to a vector (depending on the particular operator and the version of the algorithm being used), and our numerical experiments indicate that many previously intractable problems become manageable with the techniques presented here.

1. Introduction

The purpose of this paper is to introduce a class of numerical algorithms designed for rapid application of dense matrices (or integral operators) to vectors. As is well known, applying directly a dense $N \times N$-matrix to a vector requires roughly N^2 operations, and this simple fact is a cause of serious difficulties encountered in large-scale computations. For example, the main reason for the limited use of integral equations as a numerical tool in large-scale computations is that they normally lead to dense systems of linear algebraic equations, and the latter have to be solved, either directly or iteratively. Most iterative methods for the solution of systems of linear equations involve the application of the matrix of the system to a sequence of recursively generated vectors, which tends to be prohibitively expensive for large-scale problems. The situation is even worse if a direct solver for the linear system is used, since such solvers normally require $O(N^3)$ operations. As a result, in most areas of computational mathematics dense matrices are simply avoided whenever possible. For example, finite difference and finite element methods can be viewed as devices for reducing a partial differential equation to a sparse linear system. In this case, the cost of sparsity is the inherently high condition number of the resulting matrices.

For translation invariant operators, the problem of excessive cost of applying (or inverting) the dense matrices has been met by the Fast Fourier Transform (FFT) and related algorithms (fast convolution schemes, etc.). These methods use algebraic properties of a matrix to apply it to a vector in order $N \log(N)$ operations. Such schemes are exact in exact arithmetic, and are fragile in the sense that they depend on the exact algebraic properties of the operator for their applicability. A more recent group of fast algorithms (see [1], [2], [5], [9]) uses explicit analytical properties of specific operators to rapidly apply them to arbitrary vectors. The

Communications on Pure and Applied Mathematics, Vol. XLIV, 141–183 (1991)
© 1991 John Wiley & Sons, Inc.
CCC 0010-3640/91/020141-43$04.00

algorithms in this group are approximate in exact arithmetic (though they are capable of producing any prescribed accuracy), do not require that the operators in question be translation invariant, and are considerably more adaptable than the algorithms based on the FFT and its variants.

In this paper, we introduce a radical generalization of the algorithms of [1], [2], [5], [9]. We describe a method for the fast numerical application to arbitrary vectors of a wide variety of operators. The method normally requires order $O(N)$ operations, and is directly applicable to all Calderon-Zygmund and pseudo-differential operators. While each of the algorithms of [1], [2], [5], [9] addresses a particular operator and uses an analytical technique specifically tailored to it, we introduce several numerical tools applicable in all of these (and many other) situations. The algorithms presented here are meant to be a general tool similar to FFT. However, they do not require that the operator be translation invariant, and are approximate in exact arithmetic, though they achieve any prescribed finite accuracy. In addition, the techniques of this paper generalize to certain classes of multi-linear transformations (see Section 4.6 below).

We use a class of orthonormal "wavelet" bases generalizing the Haar functions and originally introduced by Stromberg [10] and Meyer [7]. The specific wavelet basis functions used in this paper were constructed by I. Daubechies [4] and are remarkably well adapted to numerical calculations. In these bases (for a given accuracy) integral operators satisfying certain analytical estimates have a band-diagonal form and can be applied to arbitrary functions in a "fast" manner. In particular, Dirichlet and Neumann boundary value problems for certain elliptic partial differential equations can be solved in order N calculations, where N is the number of nodes in the discretization of the boundary of the region. Other applications include an $O(N \log(N))$ algorithm for the evaluation of Legendre series and similar schemes (comparable in speed to FFT in the same dimensions) for other special function expansions. In general, the scheme of this paper can be viewed as a method for the conversion (whenever regularity permits) of dense matrices to a sparse form.

Once the sparse form of the matrix is obtained, applying it to an arbitrary vector is an order $O(N)$ procedure, while the construction of the sparse form in general requires $O(N^2)$ operations. On the other hand, if the structure of the singularities of the matrix is known *a priori* (as for Green's functions of elliptic operators or for Calderon-Zygmund operators) the compression of the operator to a banded form is an order $O(N)$ procedure. The non-zero entries of the resulting compressed matrix mimic the structure of the singularities of the original kernel.

Effectively, this paper provides two schemes for the numerical evaluation of integral operators. The first is a straightforward realization ("standard form") of the matrix of the operator in the wavelet basis. This scheme is an order $N \log(N)$ procedure (even for such simple operators as multiplication by a function). While this straightforward realization of the matrix is a useful numerical tool in itself, its range of applicability is significantly extended by the second scheme, which we describe in this paper in more detail. This realization ("non-standard form") leads to an order N scheme. The estimates for the latter follow from the more subtle analysis of the proof of the "$T(1)$ theorem" of David and Journé (see [3]). We also present two numerical examples showing that our algorithms can be useful for certain operators which are outside the class for which we provide proofs. The paper is organized as follows. In Section 2 we use the well-known Haar basis to describe a simplified version of the algorithm. In Section 3 we summarize the relevant facts from the theory of wavelets. Section 4 contains an analysis of a class of integral operators for which we obtain an order N algorithm and a description of a version of the algorithm for bilinear operators. Section 5 contains a detailed description and a complexity analysis of the scheme. Finally, in Section 6 we present several numerical applications.

Generalizations to higher dimensions and numerical operator calculus containing $O(N \log(N))$ implementations of pseudodifferential operators and their inverses will appear in a sequel to this paper.

2. The Algorithm in the Haar System

The Haar functions $h_{j,k}$ with integer indices j and k are defined by[1]

$$(2.1)\qquad h_{j,k}(x) = \begin{cases} 2^{-j/2} & \text{for } 2^j(k-1) < x < 2^j(k-1/2) \\ -2^{-j/2} & \text{for } 2^j(k-1/2) \leqq x < 2^j k \\ 0 & \text{elsewhere.} \end{cases}$$

Clearly, the Haar function $h_{j,k}(x)$ is supported in the dyadic interval $I_{j,k}$

$$(2.2)\qquad I_{j,k} = [2^j(k-1), 2^j k].$$

We will use the notation $h_{j,k}(x) = h_{I_{j,k}}(x) = h_I(x) = 2^{-j/2}h(2^{-j}x - k + 1)$, where $h(x) = h_{0,1}(x)$. We index the Haar functions by dyadic intervals $I_{j,k}$ and observe that the system $h_{I_{j,k}}(x)$ forms an orthonormal basis of $L^2(\mathbf{R})$ (see, for example, [8]).

We also introduce the normalized characteristic function $\chi_{I_{j,k}}(x)$

$$(2.3)\qquad \chi_{I_{j,k}}(x) = \begin{cases} |I_{j,k}|^{-1/2} & \text{for } x \in I_{j,k} \\ 0 & \text{elsewhere,} \end{cases}$$

where $|I_{j,k}|$ denotes the length of $I_{j,k}$, and will use the notation $\chi_{j,k} = \chi_{I_{j,k}}$.

Given a function $f \in L^2(\mathbf{R})$ and an interval $I \subset \mathbf{R}$, we define its Haar coefficient d_I of f

[1] We define the basis so that the dyadic scale with the index j is finer than the scale with index $j+1$. This choice of indexing is convenient for numerical applications.

$$d_I = \int_{-\infty}^{+\infty} f(x) h_I(x)\, dx, \tag{2.4}$$

and "average" s_I of f on I as

$$s_I = \int_{-\infty}^{+\infty} f(x) \chi_I(x)\, dx, \tag{2.5}$$

and observe that

$$d_I = (s_{I'} - s_{I''}) \frac{1}{\sqrt{2}}, \tag{2.6}$$

where I' and I'' are the left and the right halves of I.

To obtain a numerical method for calculating the Haar coefficients of a function we proceed as follows. Suppose we are given $N = 2^n$ "samples" of a function, which can for simplicity be thought of as values of scaled averages

$$s_k^0 = 2^{n/2} \int_{2^{-n}(k-1)}^{2^{-n}k} f(x)\, dx, \tag{2.7}$$

of f on intervals of length 2^{-n}. We then get the Haar coefficients for the intervals of length 2^{-n+1} via (2.6), and obtain the coefficients

$$d_k^1 = \frac{1}{\sqrt{2}} (s_{2k-1}^0 - s_{2k}^0). \tag{2.8}$$

We also compute the "averages"

$$s_k^1 = \frac{1}{\sqrt{2}} (s_{2k-1}^0 + s_{2k}^0) \tag{2.9}$$

on the intervals of length 2^{-n+1}. Repeating this procedure, we obtain the Haar coefficients

$$d_k^{j+1} = \frac{1}{\sqrt{2}} (s_{2k-1}^j - s_{2k}^j) \tag{2.10}$$

and averages

$$s_k^{j+1} = \frac{1}{\sqrt{2}} (s_{2k-1}^j + s_{2k}^j) \tag{2.11}$$

for $j = 0, \cdots, n-1$ and $k = 1, \cdots, 2^{n-j-1}$. This is illustrated by the pyramid scheme

$$\begin{array}{ccccccc} \{s_k^0\} & \rightarrow & \{s_k^1\} & \rightarrow & \{s_k^2\} & \rightarrow & \{s_k^3\} \cdots \\ & \searrow & & \searrow & & \searrow & \\ & & \{d_k^1\} & & \{d_k^2\} & & \{d_k^3\} \cdots . \end{array} \tag{2.12}$$

It is easy to see that evaluating the whole set of coefficients d_I, s_I in (2.12) requires $2(N-1)$ additions and $2N$ multiplications.

In two dimensions, there are two natural ways to construct Haar systems. The first is simply the tensor product $h_{I \times J} = h_I \otimes h_J$, so that each basis function $h_{I \times J}$ is supported on the rectangle $I \times J$. The second basis is defined by associating three basis functions: $h_I(x)h_{I'}(y)$, $h_I(x)\chi_{I'}(y)$, and $\chi_I(x)h_{I'}(y)$ to each square $I \times I'$, where I and I' are two dyadic intervals of the same length.

We consider an integral operator,

$$T(f)(x) = \int K(x, y) f(y)\, dy, \tag{2.13}$$

and expand its kernel (formally) as a function of two variables in the two-dimensional Haar series

$$K(x, y) = \sum_{I,I'} \alpha_{II'} h_I(x) h_{I'}(y) + \sum_{I,I'} \beta_{II'} h_I(x) \chi_{I'}(y) + \sum_{I,I'} \gamma_{II'} \chi_I(x) h_{I'}(y), \tag{2.14}$$

where the sum extends over all dyadic squares $I \times I'$ with $|I| = |I'|$, and where

$$\alpha_{II'} = \iint K(x, y) h_I(x) h_{I'}(y)\, dx\, dy, \tag{2.15}$$

$$\beta_{II'} = \iint K(x, y) h_I(x) \chi_{I'}(y)\, dx\, dy, \tag{2.16}$$

and

$$\gamma_{II'} = \iint K(x, y) \chi_I(x) h_{I'}(y)\, dx\, dy. \tag{2.17}$$

When $I = I_{j,k}$, $I' = I_{j,k}$ (see (2.2)), we will also use the notation

$$\alpha^j_{k,k'} = \alpha_{I_{j,k} I_{j,k'}}, \tag{2.18}$$

$$\beta^j_{k,k'} = \beta_{I_{j,k} I_{j,k'}}, \tag{2.19}$$

$$\gamma^j_{k,k'} = \gamma_{I_{j,k} I_{j,k'}}, \tag{2.20}$$

defining the matrices $\alpha^j = \{\alpha^j_{i,l}\}$, $\beta^j = \{\beta^j_{i,l}\}$, $\gamma^j = \{\gamma^j_{i,l}\}$, with $i, l = 1, 2, \cdots, 2^{n-j}$. Substituting (2.14) into (2.13), we obtain

$$\begin{aligned} T(f)(x) = \sum_I h_I(x) \sum_{I'} \alpha_{II'} d_{I'} + \sum_I h_I(x) \sum_{I'} \beta_{II'} s_{I'} \\ + \sum_I \chi_I(x) \sum_{I'} \gamma_{II'} d_{I'} \end{aligned} \tag{2.21}$$

(recall that in each of the sums in (2.21) I and I' always have the same length).

To discretize (2.21), we define projection operators

$$P_j f = \sum_{|I| = 2^{j-n}} \langle f, \chi_I \rangle \chi_I, \qquad j = 0, \cdots, n \tag{2.22}$$

and approximate T by

$$T \sim T_0 = P_0 T P_0, \tag{2.23}$$

where P_0 is the projection operator on the finest scale. An alternative derivation of (2.21) consists of expanding T_0 in a "telescopic" series

$$\begin{aligned} T_0 = P_0 T P_0 &= \sum_{j=1}^{n} (P_{j-1} T P_{j-1} - P_j T P_j) + P_n T P_n \\ &= \sum_{j=1}^{n} [(P_{j-1} - P_j) T (P_{j-1} - P_j) + (P_{j-1} - P_j) T P_j \\ &\qquad + P_j T (P_{j-1} - P_j)] + P_n T P_n. \end{aligned} \tag{2.24}$$

Defining the operators Q_j with $j = 1, 2, \cdots, n$, by the formula

$$Q_j = P_{j-1} - P_j, \tag{2.25}$$

we can rewrite (2.24) in the form

$$T_0 = \sum_{j=1}^{n} (Q_j T Q_j + Q_j T P_j + P_j T Q_j) + P_n T P_n. \tag{2.26}$$

The latter can be viewed as a decomposition of the operator T into a sum of contributions from different scales. Comparing (2.14) and (2.26), we observe that while the term $P_n T P_n$ (or its equivalent) is absent in (2.14), it appears in (2.26) to compensate for the finite number of scales.

Observation 2.1. Clearly, expression (2.21) can be viewed as a scheme for the numerical application of the operator T to arbitrary functions. To be more specific, given a function f, we start with discretizing it into samples s_k^0, $k = 1, 2, \cdots, N$, (where $N = 2^n$), which are then converted into a vector $\tilde{f} \in R^{2N-2}$ consisting of all coefficients s_k^j, d_k^j and ordered as follows

$$\tilde{f} = (d_1^1, d_2^1, \cdots, d_{N/2}^1, s_1^1, s_2^1, \cdots, s_{N/2}^1, d_1^2, d_2^2, \cdots, d_{N/4}^2, s_1^2, s_2^2, \cdots, s_{N/4}^2, \cdots, d_1^n, s_1^n). \tag{2.27}$$

Then, we construct the matrices $\alpha^j, \beta^j, \gamma^j$ for $j = 1, 2, \cdots, n$ (see (2.15)–(2.20) and Observation 3.2) corresponding to the operator T, and evaluate the vectors $\hat{s}^j = \{\hat{s}_k^j\}$, $\hat{d}^j = \{\hat{d}_k^j\}$ via the formulae

$$\hat{d}^j = \alpha^j(d^j) + \beta^j(s^j) \tag{2.28}$$

$$\hat{s}^j = \gamma^j(d^j), \tag{2.29}$$

where $d^j = \{d_k^j\}$, $s^j = \{s_k^j\}$, $k = 1, 2, \cdots, 2^{n-j}$, with $j = 1, \cdots, n$. Finally, we define an approximation T_0^N to T_0 by the formula

$$T_0^N(f)(x) = \sum_{j=1}^{n} \sum_{k=1}^{2^{n-j}} (\hat{d}_k^j h_{j,k}(x) + \hat{s}_k^j \chi_{j,k}(x)). \tag{2.30}$$

Clearly, $T_0^N(f)$ is a restriction of the operator T_0 in (2.23) on a finite-dimensional subspace of $L^2(\mathbf{R})$. A rapid procedure for the numerical evaluation of the operator T_0^N is described (in a more general situation) in Section 3 below.

It is convenient to organize the matrices $\alpha^j, \beta^j, \gamma^j$ with $j = 1, 2, \cdots, n$ into a single matrix, depicted in Figure 1, and for reasons that will become clear in Section 4, the matrix in Figure 1 will be referred to as the non-standard form of the operator T, while (2.14) will be referred to as the "non-standard" representation of T (note that the (2.14) is *not* the matrix realization of the operator T_0 in the Haar basis).

3. Wavelets with Vanishing Moments and Associated Quadratures

3.1. Wavelets with Vanishing Moments

Though the Haar system leads to simple algorithms, it is not very useful in actual calculations, since the decay of $\alpha_{II'}$, $\beta_{II'}$, $\gamma_{II'}$ away from diagonal is not

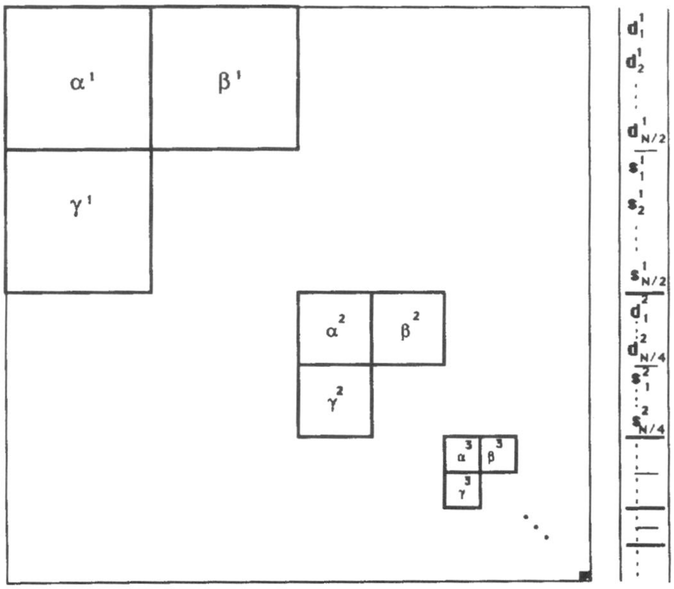

Figure 1. Representation of the decomposed matrix. Submatrices α, β, and γ on different scales are the only nonzero submatrices. In fact, most of the entries of these submatrices can be set to zero given the desired accuracy (see examples in Figures 2–8).

sufficiently fast (see below). To have a faster decay, it is necessary to use a basis in which the elements have several vanishing moments. In our algorithms, we use the orthonormal bases of compactly supported wavelets constructed by I. Daubechies [4] following the work of Y. Meyer [7] and S. Mallat [6]. We now describe these orthonormal bases.

Consider functions ψ and φ (corresponding to h and χ in Section 2), which satisfy the following relations:

$$\varphi(x) = \sqrt{2} \sum_{k=0}^{2M-1} h_{k+1}\varphi(2x-k), \tag{3.1}$$

$$\psi(x) = \sqrt{2} \sum_{k=0}^{2M-1} g_{k+1}\varphi(2x-k), \tag{3.2}$$

where

$$g_k = (-1)^{k-1} h_{2M-k+1}, \qquad k = 1, \cdots, 2M \tag{3.3}$$

and

$$\int \varphi(x)\, dx = 1. \tag{3.4}$$

The coefficients $\{h_k\}_{k=1}^{k=2M}$ are chosen so that the functions

$$\psi^j_k(x) = 2^{-j/2}\psi(2^{-j}x - k + 1), \tag{3.5}$$

where j and k are integers, form an orthonormal basis and, in addition, the function ψ has M vanishing moments

$$\int \psi(x)x^m\, dx = 0, \qquad m = 0, \cdots, M-1. \tag{3.6}$$

We will also need the dilations and translations of the scaling function φ,

$$\varphi^j_k(x) = 2^{-j/2}\varphi(2^{-j}x - k + 1). \tag{3.7}$$

Note that the Haar system is a particular case of (3.1)–(3.6) with $M = 1$ and $h_1 = h_2 = 1/\sqrt{2}$, $\varphi = \chi$ and $\psi = h$, and that the expansion (2.14)–(2.17) and the non-standard form in (2.26) in Section 2 can be rewritten in any wavelet basis by simply replacing functions χ and h by φ and ψ respectively.

Remark 3.1. Several classes of functions φ, ψ have been constructed in recent years, and we refer the reader to [4] for a detailed description of some of them.

Remark 3.2. Unlike the Haar basis, the functions φ_I, φ_J can have overlapping supports for $J \neq I$. As a result, the pyramid structure (2.12) "spills out" of the interval $[1, N]$ on which the structure is originally defined. Therefore, it is technically convenient to replace the original structure with a periodic one with period N. This is equivalent to replacing the original wavelet basis with its periodized version (see [8]).

3.2. Wavelet-Based Quadratures

In the preceding subsection, we introduce a procedure for calculating the coefficients s^j_k, d^j_k for all $j \geqq 1$, $k = 1, 2, \cdots, N$, given the coefficients s^0_k for $k = 1, 2, \cdots, N$. In this subsection, we introduce a set of quadrature formulae for the efficient evaluation of the coefficients s^0_k corresponding to smooth functions f. The simplest class of procedures of this kind is obtained under the assumption that there exists an integer constant τ_M such that the function φ satisfies the condition

(3.8) $$\int \varphi(x+\tau_M)x^m\,dx = 0, \quad \text{for} \quad m = 1, 2, \cdots, M-1,$$

(3.9) $$\int \varphi(x)\,dx = 1,$$

i.e., that the first $M-1$ "shifted" moments of φ are equal to zero, while its integral is equal to 1. Recalling the definition of s_k^0,

(3.10) $$\begin{aligned} s_k^0 &= 2^{n/2}\int f(x)\varphi(2^n x - k + 1)\,dx \\ &= 2^{n/2}\int f(x + 2^{-n}(k-1))\varphi(2^n x)\,dx, \end{aligned}$$

expanding f into a Taylor series around $2^{-n}(k-1+\tau_M)$, and using (3.8), we obtain

(3.11) $$s_k^0 = 2^{-n/2} f(2^{-n}(k-1+\tau_M)) + O(2^{-n(M+1/2)}).$$

In effect, (3.11), is a one-point quadrature formula for the evaluation of s_k^0. Applying the same calculation to s_k^j with $j \geqq 1$, we easily obtain

(3.12) $$s_k^j = 2^{(-n+j)/2} f(2^{-n+j}(k-1+\tau_M)) + O(2^{-(n-j)(M+1/2)}),$$

which turns out to be extremely useful for the rapid evaluation of the coefficients of compressed forms of matrices (see Section 4 below).

Though the compactly supported wavelets found in [4] do not satisfy the condition (3.8), a slight variation of the procedure described there produces a basis satisfying (3.8), in addition to (3.1)–(3.6). Coefficients of the filters $\{h_k\}$ corresponding to $M = 2, 4, 6$ and appropriate choices of τ_M can be found in Appendix A, and we would like to thank I. Daubechies for providing them to us.

It turns out that the filters in Table 1 are 50% longer than those in the original wavelets found in [4], given the same order M. Therefore, it might be desirable to adapt the numerical scheme so that the "shorter" wavelets could be used. Such an adaptation (by means of appropriately designed quadrature formulae for the evaluation of the integrals (3.10)) is presented in Appendix B.

Remark 3.3. We do not discuss in this paper wavelet-based quadrature formulae for the evaluation of singular integrals, since such schemes tend to be problem-specific. Note, however, that for all integrable kernels quadrature formulae of the type developed in this paper are adequate with minor modifications.

3.3. Fast Wavelet Transform

For the rest of this section, we treat the procedures being discussed as linear transformations in R^N, viewed as the Euclidean space of all periodic sequences with the period N.

Replacing the Haar basis with a basis of wavelets with vanishing moments, and assuming that the coefficients s_k^0, $k = 1, 2, \cdots, N$ are given, we replace the expressions (2.8)–(2.11) with the formulae and

$$s_k^j = \sum_{n=1}^{n=2M} h_n s_{n+2k-2}^{j-1}, \tag{3.13}$$

$$d_k^j = \sum_{n=1}^{n=2M} g_n s_{n+2k-2}^{j-1}, \tag{3.14}$$

where s_k^j and d_k^j are viewed as periodic sequences with the period 2^{n-j} (see also Remark 3.2 above). As is shown in [4], the formulae (3.13) and (3.14) define an orthogonal mapping $O_j: R^{2^{n-j+1}} \to R^{2^{n-j+1}}$, converting the coefficients s_k^{j-1} with $k = 1, 2, \cdots, 2^{n-j+1}$ into the coefficients s_k^j, d_k^j with $k = 1, 2, \cdots, 2^{n-j}$, and the inverse of O_j is given by the formulae

$$\begin{aligned} s_{2n}^{j-1} &= \sum_{k=1}^{k=M} h_{2k} s_{n-k+1}^j + \sum_{k=1}^{k=M} g_{2k} d_{n-k+1}^j, \\ s_{2n-1}^{j-1} &= \sum_{k=1}^{k=M} h_{2k-1} s_{n-k+1}^j + \sum_{k=1}^{k=M} g_{2k-1} d_{n-k+1}^j. \end{aligned} \tag{3.15}$$

Obviously, given a function f of the form

$$\begin{aligned} f(x) = \sum_{k=1}^{2^{n-j}} s_k^j 2^{(n-j)/2} \varphi(2^{n-j}x - (k-1)) \\ + \sum_{k=1}^{2^{n-j}} d_k^j 2^{(n-j)/2} \psi(2^{n-j}x - (k-1)), \end{aligned} \tag{3.16}$$

it can be expressed in the form

$$f(x) = \sum_{l=1}^{2^{n-j+1}} s_l^{j-1} 2^{(n-j+1)/2} \varphi(2^{n-j+1}x - (l-1)), \tag{3.17}$$

with s_l^{j-1}, $l = 1, 2, \cdots, 2^{n-j+1}$ given by (3.15).

Observation 3.1. Given the coefficients s_k^0, $k = 1, 2, \cdots, N$, recursive application of the formulae (3.13), (3.14) yields a numerical procedure for evaluating the coefficients s_k^j, d_k^j for all $j = 1, 2, \cdots, n$, $k = 1, 2, \cdots, 2^{n-j}$, with a cost proportional to N. Similarly, given the values d_k^j for all $j = 1, 2, \cdots, n$, $k = 1, 2, \cdots, 2^{n-j}$, and s_1^n (note that the vector s^n contains only one element) we can reconstruct the coefficients s_k^0 for all $k = 1, 2, \cdots, N$ by using (3.15) recursively for $j = n, n-1, \cdots, 0$. The cost of the latter procedure is also $O(N)$. Finally, given an expansion of the form

$$f(x) = \sum_{j=0}^{n} \sum_{k=1}^{2^{n-j}} s_k^j 2^{(n-j)/2} \varphi(2^{n-j}x - (k-1)) + \sum_{j=0}^{n} \sum_{k=1}^{2^{n-j}} d_k^j 2^{(n-j)/2} \psi(2^{n-j}x - (k-1)), \tag{3.18}$$

it costs $O(N)$ to evaluate all coefficients s_k^0, $k = 1, 2, \cdots, N$ by the recursive application of the formula (3.17) with $j = n, n-1, \cdots, 0$.

Observation 3.2. It is easy to see that the entries of the matrices α^j, β^j, γ^j with $j = 1, 2, \cdots, n$, are the coefficients of the two-dimensional wavelet expansion of the function $K(x, y)$, and can be obtained by a two-dimensional version of the pyramid scheme (2.12), (3.13), (3.14). Indeed, the definitions (2.15)–(2.17) of these coefficients can be rewritten in the form

$$\alpha_{i,l}^j = 2^{-j} \int_{-\infty}^{\infty} \int_{-\infty}^{\infty} K(x, y) \psi(2^{-j}x - (i-1)) \psi(2^{-j}y - (l-1))\, dx\, dy, \tag{3.19}$$

$$\beta_{i,l}^j = 2^{-j} \int_{-\infty}^{\infty} \int_{-\infty}^{\infty} K(x, y) \psi(2^{-j}x - (i-1)) \varphi(2^{-j}y - (l-1))\, dx\, dy, \tag{3.20}$$

$$\gamma_{i,l}^j = 2^{-j} \int_{-\infty}^{\infty} \int_{-\infty}^{\infty} K(x, y) \varphi(2^{-j}x - (i-1)) \psi(2^{-j}y - (l-1))\, dx\, dy, \tag{3.21}$$

and we will define an additional set of coefficients $s_{i,l}^j$ by the formula

$$s_{i,l}^j = 2^{-j} \int_{-\infty}^{\infty} \int_{-\infty}^{\infty} K(x, y) \varphi(2^{-j}x - (i-1)) \varphi(2^{-j}y - (l-1))\, dx\, dy. \tag{3.22}$$

Now, given a set of coefficients $s_{i,l}^0$ with $i, l = 1, 2, \cdots, N$, repeated application of the formulae (3.13), (3.14) produces

$$\alpha_{i,l}^j = \sum_{k,m=1}^{2M} g_k g_m s_{k+2i-2,m+2l-2}^{j-1}, \tag{3.23}$$

(3.24) $$\beta^j_{i,l} = \sum_{k,m=1}^{2M} g_k h_m s^{j-1}_{k+2i-2,m+2l-2},$$

(3.25) $$\gamma^j_{i,l} = \sum_{k,m=1}^{2M} h_k g_m s^{j-1}_{k+2i-2,m+2l-2},$$

(3.26) $$s^j_{i,l} = \sum_{k,m=1}^{2M} h_k h_m s^{j-1}_{k+2i-2,m+2l-2},$$

with $i, l = 1, 2, \cdots, 2^{n-j}, j = 1, 2, \cdots, n$. Clearly, formulae (3.23)–(3.26) are a two-dimensional version of the pyramid scheme (2.12), and provide an order N^2 scheme for the evaluation of the elements of all matrices $\alpha^j, \beta^j, \gamma^j$ with $j = 1, 2, \cdots, n$.

4. Integral Operators and Accuracy Estimates

4.1. Non-Standard Form of Integral Operators

In order to describe methods for "compression" of integral operators, we restrict our attention to several specific classes of operators frequently encountered in analysis. In particular, we give exact estimates for pseudo-differential and Calderon-Zygmund operators.

We start with several simple observations. The non-standard form of a kernel $K(x, y)$ is obtained by evaluating the expressions

(4.1) $$\alpha_{II'} = \iint K(x, y)\psi_I(x)\psi_{I'}(y)\, dx\, dy,$$

(4.2) $$\beta_{II'} = \iint K(x, y)\psi_I(x)\varphi_{I'}(y)\, dx\, dy,$$

and

(4.3) $$\gamma_{II'} = \iint K(x, y)\varphi_I(x)\psi_{I'}(y)\, dx\, dy.$$

(See Figure 1.) Suppose now that K is smooth on the square $I \times I'$. Expanding K into a Taylor series around the center of $I \times I'$, combining (3.6) with (4.1)–(4.3), and remembering that the functions ψ_I, $\psi_{I'}$ are supported on the intervals I, I' respectively, we obtain the estimate

(4.4) $$|\alpha_{II'}| + |\beta_{II'}| + |\gamma_{II'}| \leqq C|I|^{M+1} \sup_{(x,y)\in I\times I'} (|\partial_x^M K(x, y)| + |\partial_y^M K(x, y)|).$$

Obviously, the right-hand side of (4.4) is small whenever either $|I|$ or the derivatives involved are small, and we use this fact to "compress" matrices of integral operators by converting them to the non-standard form and discarding the coefficients that are smaller than a chosen threshold.

To be more specific, consider pseudo-differential operators and Calderon-Zygmund operators. These classes of operators are given by integral or distributional kernels that are smooth away from the diagonal, and the case of Calderon-Zygmund operators is particularly simple. These operators have kernels $K(x, y)$ which satisfy the estimates

$$|K(x, y)| \leqq \frac{1}{|x-y|}, \tag{4.5}$$

$$|\partial_x^M K(x, y)| + |\partial_y^M K(x, y)| \leqq \frac{C_M}{|x-y|^{1+M}}, \tag{4.6}$$

for some $M \geqq 1$. To illustrate the use of the estimates (4.5) and (4.6) for the compression of operators, we let $M = 1$ in (4.6) and consider

$$\beta_{II'} = \iint K(x, y) h_I(x) \chi_{I'}(y)\, dx\, dy, \tag{4.7}$$

where we assume that the distance between I and I' is greater than $|I|$. Since

$$\int h_I(x)\, dx = 0, \tag{4.8}$$

we have

$$\begin{aligned} |\beta_{II'}| &\leqq \left| \iint [K(x, y) - K(x_I, y)] h_I(x) \chi_{I'}(y)\, dx\, dy \right| \\ &\leqq 2C_M \frac{|I|^2}{|x_I - y_{I'}|^2}, \end{aligned} \tag{4.9}$$

where x_I denotes the center of the interval I. In other words, the coefficient $\beta_{II'}$ decays quadratically as a function of the distance between the intervals I, I', and for sufficiently large N and finite precision of calculations, most of the matrix can be discarded, leaving only a band around the diagonal. However, algorithms using the above estimates (with $M = 1$) tend to be quite inefficient, due to the slow decay of the matrix elements with their distance from the diagonal. The following simple proposition generalizes the estimate (4.9) for the case of larger M, and provides an analytical tool for efficient numerical compression of a wide class of operators.

PROPOSITION 4.1. *Suppose that in the expansion* (2.14), *the wavelet basis has* M *vanishing moments, i.e., the functions* φ *and* ψ (*replacing* χ *and* h) *satisfy the conditions* (3.1)–(3.6). *Then for any kernel* K *satisfying the conditions* (4.5) *and* (4.6), *the coefficients* $\alpha^j_{i,l}$, $\beta^j_{i,l}$, $\gamma^j_{i,l}$ *in the non-standard form* (*see* (2.18)–(2.20) *and Figure* 1) *satisfy the estimate*

$$|\alpha^j_{i,l}| + |\beta^j_{i,l}| + |\gamma^j_{i,l}| \leqq \frac{C_M}{1 + |i - l|^{M+1}}, \tag{4.10}$$

for all

$$|i - l| \geqq 2M. \tag{4.11}$$

Remark 4.1. For most numerical applications, the estimate (4.10) is quite adequate, as long as the singularity of K is integrable across each row and each column (see the following section). To obtain a more subtle analysis of the operator T_0 (see (2.23) above) and correspondingly tighter estimates, some of the ideas arising in the proof of the "$T(1)$" theorem of David and Journé are required. We discuss these issues in more detail in Section 4.5 below.

Similar considerations apply in the case of pseudo-differential operators. Let T be a pseudo differential operator with symbol $\sigma(x, \xi)$ defined by the formula

$$T(f)(x) = \sigma(x, D)f = \int e^{ix\xi}\sigma(x, \xi)\hat{f}(\xi)\, d\xi = \int K(x, y)f(y)\, dy, \tag{4.12}$$

where K is the distributional kernel of T. Assuming that the symbols σ of T and σ^* of T^* satisfy the standard conditions

$$|\partial^\alpha_\xi \partial^\beta_x \sigma(x, \xi)| \leqq C_{\alpha,\beta}(1 + |\xi|)^{\lambda-\alpha+\beta}, \tag{4.13}$$

$$|\partial^\alpha_\xi \partial^\beta_x \sigma^*(x, \xi)| \leqq C_{\alpha,\beta}(1 + |\xi|)^{\lambda-\alpha+\beta}, \tag{4.14}$$

we easily obtain the inequality

$$|\alpha^j_{i,l}| + |\beta^j_{i,l}| + |\gamma^j_{i,l}| \leqq \frac{2^{\lambda j}C_M}{(1 + |i - l|)^{M+1}}, \tag{4.15}$$

for all integer i, l.

Remark 4.2. A simple case of the estimate (4.15) is provided by the operator $T = d/dx$, in which case it is obvious that

$$(4.16)\quad \beta^j_{il} = \left\langle \psi^j_i, \frac{d}{dx}\varphi^j_l \right\rangle = 2^{-j} \int \psi(2^{-j}x - i + 1) \times \varphi'(2^{-j}x - l + 1) 2^{-j}\, dx = 2^{-j}\beta_{i-l},$$

where the sequence $\{\beta_i\}$ is defined by the formula

$$(4.17)\qquad \beta_i = \int \psi(x - i)\varphi'(x)\, dx,$$

provided a sufficiently smooth wavelet $\varphi(x)$ is used.

4.2. Numerical Calculations and Compression of Operators

Suppose now that we approximate the operator T_0^N by the operator $T_0^{N,B}$ obtained from T_0^N by setting to zero all coefficients of matrices $\alpha = \{\alpha_{II'}\}$, $\beta = \{\beta_{II'}\}$, $\gamma = \{\gamma_{II'}\}$ outside of bands of width $B \geqq 2M$ around their diagonals. It is easy to see that

$$(4.18)\qquad \|T_0^{N,B} - T_0^N\| \leqq \frac{C}{B^M} \log_2 N,$$

where C is a constant determined by the kernel K. In other words, the matrices α, β, γ can be approximated by banded matrices α^B, β^B, γ^B respectively, and the accuracy of the approximation is

$$(4.19)\qquad \frac{C}{B^M} \log_2 N.$$

In most numerical applications, the accuracy ϵ of calculations is fixed, and the parameters of the algorithm (in our case, the band width B and order M) have to be chosen in such a manner that the desired precision of calculations is achieved. If M is fixed, then B has to be such that

$$(4.20)\qquad \|T_0^{N,B} - T_0^N\| \leqq \frac{C}{B^M} \log_2 N \leqq \epsilon,$$

or, equivalently,

$$(4.21)\qquad B \geqq \left(\frac{C}{\epsilon} \log_2 N\right)^{1/M}.$$

In other words, T_0^N has been approximated to precision ϵ with its truncated version, which can be applied to arbitrary vectors for a cost proportional to $N((C/\epsilon)\log_2 N)^{1/M}$, which for all practical purposes does not differ from N. A considerably more detailed investigation (see Remark 4.1 above and Section 4.5 below) permits the estimate (4.21) to be replaced with the estimate

$$B \geqq \left(\frac{C}{\epsilon}\right)^{1/M} \tag{4.22}$$

making the application of the operator T_0^N to an arbitrary vector with arbitrary fixed accuracy into a procedure of order exactly $O(N)$.

Whenever sufficient analytical information about the operator T is available, the evaluation of those entries in the matrices α, β, γ that are smaller than a given threshold can be avoided altogether, resulting in an $O(N)$ algorithm (see Section 4.3 below for a more detailed description of this procedure).

Remark 4.3. Both Proposition 4.1 and the subsequent discussion assume that the kernel K is non-singular everywhere outside the diagonal, on which it is permitted to have integrable singularities. Clearly, it can be generalized to the case when the singularities of K are distributed along a finite number of bands, columns, rows, etc. While the analysis is not considerably complicated by this generalization, the implementation of such a procedure on the computer is significantly more involved (see Section 5 below).

4.3. Rapid Evaluation of the Non-Standard Form of an Operator

In this subsection, we construct an efficient procedure for the evaluation of the elements of the non-standard form of an operator T lying within a band of width B around the diagonal. The procedure assumes that T satisfies conditions (4.5) and (4.6) of Section 4, and has an operation count proportional to $N \cdot B$ (as opposed to the $O(N^2)$ estimate for the general procedure described in Observation 3.2).

To be specific, consider the evaluation of the coefficients $\beta^j_{i,l}$ for all $j = 1, 2, \cdots, n$, and $|i - l| \leqq B$. According to (3.24),

$$\beta^j_{i,l} = \sum_{k,m=1}^{2M} g_k h_m s^{j-1}_{k+2i-2,m+2l-2}, \tag{4.23}$$

which involves the coefficients $s^{j-1}_{i',l'}$ in a band of size $3B$ defined by the condition $|i' - l'| \leqq 3B$. Clearly, (3.26) could be used recursively to obtain the required coefficients $s^{j-1}_{i',l'}$, and the resulting procedure would require order N^2 operations. We therefore compute the coefficients $s^{j-1}_{i',l'}$ directly by using appropriate quadratures. In particular, the application of the one-point quadrature (3.12) to $K(x, y)$, combined with the estimate (4.6), gives

$$\begin{aligned} s^j_{i',l'} = {} & 2^{n-j+1} K(2^{-n+j-1}(i' - 1 + \tau_M), 2^{-n+j-1}(l' - 1 + \tau_M)) \\ & + O\left(\frac{1}{|i' - l'|^{M+1}}\right). \end{aligned} \tag{4.24}$$

If the wavelets used do not satisfy the moment condition (3.8), more complicated quadratures have to be used (see Appendix B to this paper).

4.4. The Standard Matrix Realization in the Wavelet Basis

While the evaluation of the operator T via the non-standard form (i.e., via the matrices α^j, β^j, γ^j) is an efficient tool for applying it to arbitrary functions, it is not a representation of T in any basis. There are obvious advantages to obtaining a mechanism for the compression of operators that is simply a representation of the operator in a suitably chosen basis, even at the cost of certain sacrifices in the speed of calculations (provided that the cost stays $O(N)$ or $O(N \log N)$). It turns out that simply representing the operator T in the basis of wavelets satisfying the conditions (3.6) results (to any fixed accuracy) in a matrix containing no more than $O(N \log N)$ non-zero elements. Indeed, the elements of the matrix representing T in this basis are of the form,

$$T_{IJ} = \langle T\psi_I, \psi_J \rangle, \tag{4.25}$$

with I, J all possible pairs of diadic intervals in $\mathbf{R}$, not necessarily such that $|I| = |J|$. Combining estimates (4.5), (4.6) with (3.6), we see that

$$|T_{IJ}| \leqq C_M \left(\frac{|I|}{|J|} \right)^{1/2} \left(\frac{|I|}{d(I, J)} \right)^{M+1}, \tag{4.26}$$

where C_M is a constant depending on M, K, and the choice of the wavelets ($d(I, J)$ denotes the distance between I, J) and it is assumed that $|I| \leqq |J|$. It is easy to see that for large N and fixed $\epsilon \geqq 0$, only $O(N \log N)$ elements of the matrix (4.25) will be greater than ϵ, and by discarding all elements that are smaller than a predetermined threshold, we compress it to $O(N \log N)$ elements.

Remark 4.4. A considerably more detailed investigation (see [8]) shows that in fact the number of elements in the compressed matrix is asymptotically proportional to N, as long as the images of the constant function under the mappings T and T^* are smooth. Fortunately, the latter is always the case for pseudo-differential and many other operators.

Numerically, evaluation of the compressed form of the matrix $\{T_{IJ}\}$ starts with the calculation of the coefficients s^0 (see (2.7)) via an appropriately chosen quadrature formula. For example, if the wavelets used satisfy the conditions (3.8), (3.9), the one-point formula (3.10) is quite adequate. Other quadrature formulae for this purpose can be found in Appendix B to this paper. Once the coefficients s^0 have been obtained, the subsequent calculations can be carried out in one of three ways.

1. The naive approach is to construct the full matrix of the operator T in the basis associated with wavelets by following the pyramid (2.12). After that, the elements of the resulting matrix that are smaller than a predetermined

threshold, are discarded. Clearly, this scheme requires $O(N^2)$ operations, and does not require any prior knowledge of the structure of T.

2. When the structure of singularities of the kernel K is known, the locations of the coefficients of the matrix $\{T_{IJ}\}$ exceeding the threshold ϵ can be determined *a priori*. After that, these can be evaluated by simply using appropriate quadrature formulae on each of the supports of the corresponding basis functions. The resulting procedure requires order $O(N\log(N))$ operations when the operator in question is either Calderon-Zygmund or pseudo-differential, and is easily adaptable to other distributions of singularities of the kernel.
3. The third approach is to start with the non-standard form of the operator T, compress it, and then convert the compressed version into the standard form. The conversion procedure starts with the formula

$$\beta^j_{k,l} = 2^{-j}\int_{-\infty}^{\infty}\left(\int_{-\infty}^{\infty} K(x,y)\psi(2^{-j}x-(k-1))\,dx\right) \times \varphi(2^{-j}y-(l-1))\,dy, \tag{4.27}$$

which is an immediate consequence of (2.16), (2.19). Combining (4.27) with (3.14), we immediately obtain

$$T_{IJ} = 2^{-(2j+1)/2}\int_{-\infty}^{\infty}\int_{-\infty}^{\infty} K(x,y) \times \psi(2^{-j}x-(k-1))\psi(2^{-(j+1)}y-(i-1))\,dx\,dy = \sum_{l=1}^{2M} g_l\beta^j_{k,l+2i-2}, \tag{4.28}$$

where $I = I_{j,k}$ and $J = I_{j+1,i}$. Similarly, we define the set of coefficients $\{S_{I,J}\}$ via the formula

$$S_{IJ} = \sum_{l=1}^{2M} h_l\beta^j_{k,l+2i-2}, \tag{4.29}$$

and observe that these are the coefficients s^{j+1}_i in the pyramid scheme (2.12). In general, given the coefficients S_{IJ} on step m (that is, for all pairs (I, J) such that $|J| = 2^m|I|$), we move to the next step by applying the formula (4.28) recursively.

Remark 4.5. Clearly, the above procedure amounts to simply applying the pyramid scheme (2.12) to each row of the matrix β^j.

4.5. Uniform Estimates for Discretizations of Calderon-Zygmund Operators

As has been observed in Remark 4.1, the estimates (4.10) are adequate for most numerical purposes. However, they can be strengthened in two important respects.

1. The condition (4.11) can be eliminated under a weak cancellation condition (4.30).
2. The condition (4.10) does not by itself guarantee either the boundedness of the operator T, or the uniform (in N) boundedness of its discretizations T_0. In this section, we provide the necessary and sufficient conditions for the boundedness of T, or, equivalently, for the uniform boundedness of its discretizations T_0. This condition is, in fact, a reformulation of the "$T(1)$" theorem of David and Journé.

UNIFORM BOUNDEDNESS OF THE MATRICES α, β, γ. We start by observing that estimates (4.5), (4.6) are not sufficient to conclude that $\alpha^j_{i,l}$, $\beta^j_{i,l}$, $\gamma^j_{i,l}$ are bounded for $|i-l| \leqq 2M$ (for example, consider $K(x,y) = 1/(|x-y|)$). We therefore need to assume that T defines a bounded operator on L^2 or a substantially weaker condition

$$\left|\int_{I\times I} K(x,y)\,dx\,dy\right| \leqq C|I| \tag{4.30}$$

for all dyadic intervals I (this is the "weak cancellation condition"; see [8]). Under this condition and the conditions (4.5), (4.6) Proposition 4.1 can be extended to

$$|\alpha^j_{i,l}| + |\beta^j_{i,l}| + |\gamma^j_{i,l}| \leqq \frac{C_M}{1+|i-l|^{M+1}} \tag{4.31}$$

for all i, l (see [8]).

UNIFORM BOUNDEDNESS OF THE OPERATORS T_0. We have seen in (2.26) a decomposition of the operator T_0 into a sum of contributions from the different scales j. More precisely, the matrices α^j, β^j, γ^j act on the vector $\{s^j_k\}$, $\{d^j_k\}$, where d^j are coordinates of the function with respect to the orthogonal set of functions $2^{-j/2}\psi(2^{-j}x-k)$, and the s^j are auxiliary quantities needed to calculate the d^j_k. The remarkable feature of the non-standard form is the decoupling achieved among the scales j followed by a simple coupling performed in the reconstruction formulas (3.17). (The standard form, by contrast, contains matrix entries reflecting "interactions" between all pairs of scales.) In this subsection, we analyze this coupling mechanism in the simple case of the Haar basis, in effect reproducing the proof of the "$T(1)$" theorem (see [3]).

For simplicity, we will restrict our attention to the case where $\alpha = \gamma = 0$, and β satisfies conditions (4.31) (which are essentially equivalent to (4.5), (4.6), and (4.30)). In this case, for the Haar basis we have

$$T(f)(x) = \sum_{I} h_I(x) \sum_{I'} \beta_{II'} s_{I'} \tag{4.32}$$

which can be rewritten in the form

$$T(f) = \sum_{I} h_I(x) \sum_{I'} \beta_{II'}(s_{I'} - s_I) + \sum_{I} \beta_I s_I h_I(x), \tag{4.33}$$

where

$$\beta_I = \sum_{I'} \beta_{II'} = \frac{1}{|I|^{1/2}} \int\int h_I(x) K(x, y)\, dx\, dy = \langle h_I, \beta(x) \rangle \frac{1}{|I|^{1/2}} \tag{4.34}$$

and

$$\beta(x) = \int K(x, y)\, dy = T(1)(x). \tag{4.35}$$

It is easy to see (by expressing s_I in terms of d_I) that the operator

$$B_1(f)(x) = \sum_{I} h_I(x) \sum_{I'} \beta_{II'}(s_{I'} - s_I) \tag{4.36}$$

is bounded on L^2 whenever (4.31) is satisfied with $M = 1$. We are left with the "diagonal" operator

$$B_2(f)(x) = \sum_{I} \beta_I s_I h_I(x), \tag{4.37}$$

$$\beta_I = \frac{1}{|I|^{1/2}} \langle \beta(x), h_I \rangle, \tag{4.38}$$

with

$$s_I = \langle f, \chi_I \rangle. \tag{4.39}$$

Clearly

$$\| B_2(f) \|_2^2 = \sum \beta_I^2 s_I^2. \tag{4.40}$$

If we choose $f = \chi_J$ where J is a dyadic interval we find $s_I \doteq |I|^{1/2}$ for $I \subseteq J$ from which we deduce that a necessary condition for B_2 to define a bounded operator on $L^2(\mathbf{R})$ is given as

$$\sum_{I \subseteq J} |I| \beta_I^2 = \sum_{I \subseteq J} \langle \beta, h_I \rangle^2 \leqq c|J| \tag{4.41}$$

but since the h_I for $I \subseteq J$ are orthogonal in $L^2(J)$,

$$\sum_{I \subseteq J} \langle \beta, h_I \rangle^2 = \int_J |\beta(x) - m_J(\beta)|^2, \tag{4.42}$$

with

$$m_J(\beta) = \frac{1}{|J|} \int_J \beta(x)\, dx. \tag{4.43}$$

Combining (4.41) with (4.42), we obtain

$$\frac{1}{|J|} \int_J |\beta(x) - m_J(\beta)|^2 \, dx \leqq C. \tag{4.44}$$

Expression (4.44) is usually called bounded dyadic mean oscillation condition (BMO) on β, and is necessary for the boundedness of B_2 on L^2. It has been proved by Carleson (see, for example, [8]) that the condition (4.41) is necessary and sufficient for the following inequality to hold

$$\sum \beta_I^2 s_I^2 \leqq C \int |f|^2 \, dx, \qquad s_I = \langle f, \chi_I \rangle. \tag{4.45}$$

Combining these remarks we obtain:

THEOREM 4.1 (G. David, J. L. Journé). *Suppose that the operator*

$$T(f) = \int K(x, y) f(y)\, dy \tag{4.46}$$

satisfies the conditions (4.5), (4.6), (4.30). *Then the necessary and sufficient condition for T to be bounded on L^2 is that*

$$\beta(x) = T(1)(x), \tag{4.47}$$

$$\gamma(y) = T^*(1)(y) \tag{4.48}$$

belong to dyadic BMO, i.e., satisfy condition (4.44).

We have shown that the operator T in Theorem 4.1 can be decomposed as a sum of three terms

(4.49) $$T = B_1 + B_2 + B_3,$$

where

(4.50) $$B_2(f) = \sum_I h_I \beta_I s_I,$$

(4.51) $$B_3(f) = \sum_{I'} \chi_{I'} \gamma_{I'} d_{I'},$$

and

(4.52) $$B_1(f) = T(f) - B_2(f) - B_3(f),$$

with $|I|^{1/2}\beta_I = \langle h_I, \beta \rangle$, $|I'|^{1/2}\gamma_{I'} = \langle h_{I'}, \gamma \rangle$, $\beta = T(1)$, and $\gamma = T^*(1)$.

The principal term B_1, when converted to the standard form, has a band structure with decay rate independent of N. The terms B_2, B_3 are bounded only when β, γ are in BMO (see [8]).

4.6. Algorithms for Bilinear Functionals

The terms B_2 and B_3 in (4.49) are bilinear transformations in (β, f), (γ, f), respectively. Such "pseudo products" occur frequently as differentials (in the direction β) of non-linear functionals of f (see [3]). In this section, we show that pseudo-products can be implemented in order N operation (or for the same cost as ordinary multiplication). To be specific, we have the following proposition.

PROPOSITION 4.2. *Let $K(x, y, z)$ satisfy the conditions*

(4.53) $$|K(x, y, z)| \leqq \frac{1}{(x-y)^2 + (x-z)^2},$$

(4.54) $$|\partial_x^M K| + |\partial_y^M K| + |\partial_z^M K| \leqq \frac{C_M}{(|x-y| + |x-z|)^{M+2}},$$

for some $M \geqq 1$, and the bilinear functional $B(f, g)$ be defined by the formula

(4.55) $$B(f, g)(x) = \int K(x, y, z) f(y) g(z)\, dy\, dz.$$

Then the bilinear functional $B(f, g)$ can be applied to a pair of arbitrary functions f, g for a cost proportional to N, (where N is the number of samples in the discretization of the functions f and g), with the proportionality coefficient depending on M and the desired accuracy, and independent of the kernel K.

The following is an outline of an algorithm implementing such a procedure. As in the linear case, we use the wavelet basis with M vanishing moments and write

$$K(x, y, z) = \sum_{I,Q} \alpha_{I,Q}\psi_I(x)\psi_Q(y, z) + \beta_{I,Q}\psi_I(x)\varphi_Q(y, z) + \gamma_{I,Q}\varphi_I(x)\psi_Q(y, z) \tag{4.56}$$

where $Q = J \times J'$, $|I| = |J| = |J'|$ and $\psi_Q(y, z)$ is a wavelet basis in two variables (i.e., $\varphi_J(y)\psi_{J'}(z)$, $\psi_J(y)\varphi_{J'}(z)$, $\psi_J(y)\psi_{J'}(z)$) and

$$\varphi_Q(y, z) = \varphi_J(y)\varphi_{J'}(z). \tag{4.57}$$

Substituting in (4.56) into (4.55) we obtain

$$\begin{aligned} B(f, g)(x) = &\sum_I \psi_I(x) \\ &\times \left\{ \sum_{J,J'} \alpha^{(1)}_{I,J,J'} s_J(f) d_{J'}(g) + \alpha^{(2)}_{I,J,J'} d_J(f) s_{J'}(g) + \alpha^{(3)}_{I,J,J'} d_J(f) d_{J'}(g) \right\} \\ &+ \sum_I \psi_I(x) \sum_{J,J'} \beta_{I,J,J'} s_J(f) s_{J'}(g) + \sum_I \varphi_I(x) \\ &\times \left\{ \sum_{J,J'} \gamma^{(1)}_{I,J,J'} s_J(f) d_{J'}(g) + \gamma^{(2)}_{I,J,J'} d_J(f) s_{J'}(g) + \gamma^{(3)}_{I,J,J'} d_J(f) d_{J'}(g) \right\} \end{aligned} \tag{4.58}$$

where $\alpha^{(1)}_{I,J,J'}$, $\alpha^{(2)}_{I,J,J'}$, $\alpha^{(3)}_{I,J,J'}$, $\beta_{I,J,J'}$, and $\gamma^{(1)}_{I,J,J'}$, $\gamma^{(2)}_{I,J,J'}$, $\gamma^{(3)}_{I,J,J'}$, denote the coefficients of the function $K(x, y, z)$ in the three-dimensional wavelet basis. Therefore, combining (4.58) with Observation 3.1, we obtain an order $O(N)$ algorithm for the evaluation of (4.55) on an arbitrary pair of vectors.

It easily follows from the estimates (4.53) and (4.54) that

$$|\alpha_{I,J,J'}| + |\beta_{I,J,J'}| + |\gamma_{I,J,J'}| \leqq \left(\frac{C|I|}{\text{dist}(I, J) + \text{dist}(I, J')} \right)^{2+M} \tag{4.59}$$

resulting in banded matrices and a "compressed" version having $O(N)$ entries (also, compare (4.59) with (4.10)).

Similar results can be obtained for many classes of non-linear functionals whose differentials satisfy the conditions analogous to (4.53) and (4.54).

5. Description of the Algorithm

In this section, we describe an algorithm for rapid application of a matrix T_0 discretizing an integral operator T to an arbitrary vector. It is assumed that T

satisfies the estimates (4.5), (4.6), or the more general conditions described in Remark 4.3. The scheme consists of four steps.

Step 1. Evaluate the coefficients of the matrices $\alpha^j, \beta^j, \gamma^j, j = 1, 2, \cdots, n$ corresponding to T_0 (see (2.18)–(2.20) above), and discard all elements of these matrices whose absolute values are smaller than ϵ. The number of elements remaining in all matrices α^j, β^j, γ^j is proportional to N (see estimates (4.21), (4.22)).

Depending on the *a priori* information available about the operator T, one of two procedures is used, as follows.

1. If the *a priori* information is limited to that specified in Remark 4.3 (i.e., the singularities of K are distributed along a finite number of bands, rows, and columns, but their exact locations are not known), then the extremely simple procedure described in Observation 3.2 is utilized. The resulting cost of this step is $O(N^2)$, and it should only be used when the second scheme (see below) can not be applied.
2. If the operator T satisfies the estimates (4.5), (4.6) for some $M \geqq 1$, and the wavelets employed satisfy the condition (3.8), then the more efficient procedure described in Section 4.3 is used. While the implementation of this scheme is somewhat involved, it results in an order $O(N)$ algorithm, and should be used whenever possible.

Step 2. Evaluate the coefficients s_k^j, d_k^j for all $j = 1, 2, \cdots, n, k = 1, 2, \cdots, 2^{n-j}$ (see formulae (3.13), (3.14) and Observation 3.1).

Step 3. Apply the matrices α^j, β^j, γ^j to the vectors s^j, d^j, obtaining the vectors $\hat{s}^j$, $\hat{d}^j$ for $j = 1, 2, \cdots, n$ (see formulae (2.28), (2.30)).

Step 4. Use the vectors $\hat{s}^j$, $\hat{d}^j$ to evaluate $T_0(f)$ via the formula (3.15) (see Observation 3.1).

Remark 5.1. It is clear that Steps 2–4 in the above scheme require order $O(N)$ operations, and that Step 1 requires either order $O(N)$ or $O(N^2)$ operations, depending on the *a priori* information available about the operator T. It turns out, however, that even when Step 1 requires order N operations, it is still the dominant part of the algorithm in terms of the actual operation count. In most applications, a single operator has to be applied to a relatively large number of vectors, and in such cases it makes sense to produce the nonstandard form of the operator T and store it. After that, it can be retrieved and used whenever necessary, for a very small cost (see also Section 6 below).

Remark 5.2. In the above procedure, Step 1 requires $O(N^2)$ operations whenever the structure of the operator T is not described by the estimates (4.5), (4.6). Clearly, it is not the only structure of T for which an order $O(N)$ procedure can be constructed. In fact, this can be done for any structure of T described in Remark 4.3, *provided that the location of singularities of T is known a priori*. The data structures required for the construction of such an algorithm are fairly involved, but conceptually the scheme is not substantially different from that described in Section 4.3.

6. Numerical Results

A FORTRAN program has been written implementing the algorithm of the preceding section, and numerical experiments have been performed on the SUN-3/50 computer equipped with the MC68881 floating-point accelerator. All calculations were performed in three ways: in single precision using the standard (direct) method, in double precision using the algorithm of this paper with the matrices α, β, γ truncated at various thresholds (see Section 4.2 above), and in double precision using the standard method. The latter was used as the standard against which the accuracy of the other two calculations was measured.

We applied the algorithm to a number of operators; the results of six such experiments are presented in this section and summarized in Tables 1–6, and illustrated in Figures 2–9. Column 1 of each of the tables contains the number N of nodes in the discretization of the operator, columns 2 and 3 contain CPU times T_s, T_w required by the standard (order $O(N^2)$) and the "fast" ($O(N)$) schemes to multiply a vector by the resulting discretized matrix respectively, and column 4 contains the CPU T_d time used by our scheme to produce the non-standard form of the operator. Columns 5 and 6 contain the L_2 and L_∞ errors of the direct calculation respectively, and columns 7 and 8 contain the same information for the result obtained via the algorithm of this paper. Finally, column 9 contains the compression coefficients C_{comp} obtained by our scheme, defined by the ratio of N^2 to the number of non-zero elements in the non-standard form of T. In all cases, the experiments were performed for $N = 64$, 128, 256, 512, and 1024, and in all Figures 2–9, the matrices are depicted for $N = 256$.

Example 1. In this example, we compress matrices of the form

$$A_{ij} = \begin{cases} \dfrac{1}{i-j} & i \neq j, \\ 0 & i = j, \end{cases}$$

and convert them to a system of coordinates spanned by wavelets with six first moments equal to zero. Setting to zero all entries in the resulting matrix whose absolute values are smaller than 10^{-7}, we obtain the matrix whose non-zero elements are shown in black in Figure 2. The results of this set of experiments are tabulated in Table 1. The standard form of the operator A with $N = 256$ is depicted in Figure 9.

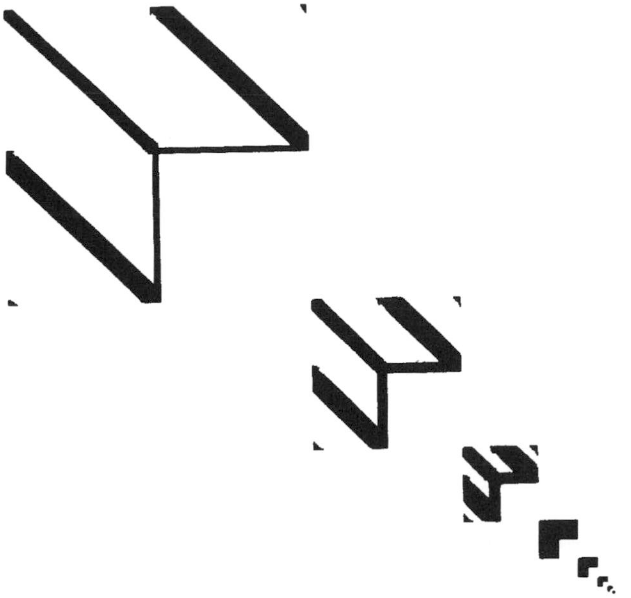

Figure 2. Entries above the threshold of 10^{-7} of the decomposed matrix of Example 1 are shown black. Note that the width of the bands does not grow with the size of the matrix.

Table 1. Numerical results for Example 1.

Input size	Time			Error of single precision multiplication		Error of FWT multiplication		Compression coefficient
(N)	T_s	T_w	T_d	L_2-norm	L_∞-norm	L_2-norm	L_∞-norm	C_{comp}
64	0.12	0.16	7.76	$1.26\cdot10^{-7}$	$3.65\cdot10^{-7}$	$8.89\cdot10^{-8}$	$1.72\cdot10^{-7}$	1.39
128	0.48	0.38	32.62	$2.17\cdot10^{-7}$	$8.64\cdot10^{-7}$	$1.12\cdot10^{-7}$	$9.94\cdot10^{-7}$	2.22
256	1.92	0.80	96.44	$2.81\cdot10^{-7}$	$1.12\cdot10^{-6}$	$1.25\cdot10^{-7}$	$5.30\cdot10^{-7}$	3.93
512	7.68	1.80	252.72	$4.21\cdot10^{-7}$	$1.75\cdot10^{-6}$	$1.23\cdot10^{-7}$	$5.16\cdot10^{-7}$	7.33
1024	30.72	3.72	605.74	$6.64\cdot10^{-7}$	$3.90\cdot10^{-6}$	$1.36\cdot10^{-7}$	$5.04\cdot10^{-7}$	14.09

Example 2. Here, we compress matrices of the form

$$A_{ij} = \begin{cases} \dfrac{\log|i - 2^{n-1}| - \log|j - 2^{n-1}|}{i-j} & i \neq j;\, i \neq 2^{n-1};\, j \neq 2^{n-1} \\ 0 & \text{otherwise} \end{cases}$$

where $i, j = 1, \cdots, N$ and $N = 2^n$.

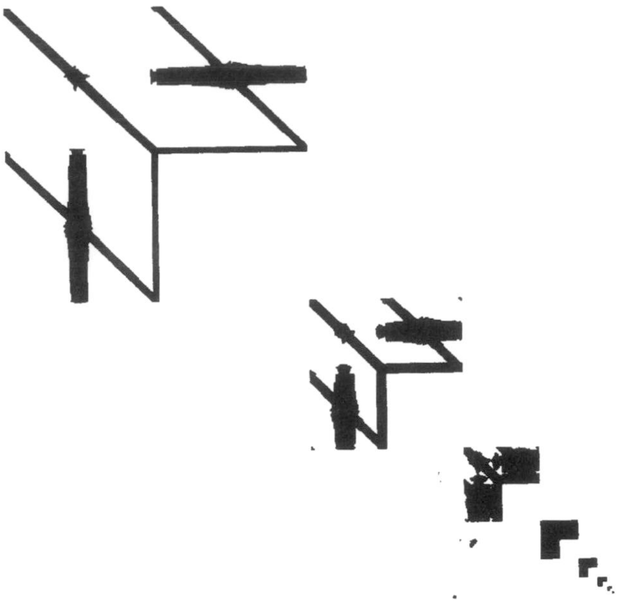

Figure 3. Entries above the threshold of 10^{-7} of the decomposed matrix of Example 2. Vertical and horizontal bands in the middle of submatrices as well as the diagonal bands are due to the singularities of the kernel (matrix). Note that in this case the kernel is not a convolution.

This matrix is not a convolution and its singularities are more complicated. The decomposition of this matrix using wavelets with six vanishing moments displaying entries above the threshold of 10^{-7} is shown in Figure 3, and the numerical results of these experiments are tabulated in Table 2. In this case, the structure of the singularities of the matrix is not known *a priori*, and its non-standard form was obtained by converting the whole matrix to the wavelet system of coordinates,

Table 2. Numerical results for Example 2.

Input size	Time			Error of single precision multiplication		Error of FWT multiplication		Compression coefficient
(N)	T_s	T_w	T_d	L_2-norm	L_∞-norm	L_2-norm	L_∞-norm	C_{comp}
64	0.12	0.16	8.62	$1.87 \cdot 10^{-7}$	$7.53 \cdot 10^{-7}$	$8.24 \cdot 10^{-8}$	$2.87 \cdot 10^{-7}$	1.23
128	0.48	0.34	35.06	$3.18 \cdot 10^{-7}$	$8.62 \cdot 10^{-7}$	$1.14 \cdot 10^{-7}$	$3.79 \cdot 10^{-7}$	2.02
256	1.92	0.84	142.82	$4.30 \cdot 10^{-7}$	$2.03 \cdot 10^{-6}$	$1.33 \cdot 10^{-7}$	$4.72 \cdot 10^{-7}$	3.76
512	7.68	1.72	574.86	$6.63 \cdot 10^{-7}$	$4.42 \cdot 10^{-6}$	$1.44 \cdot 10^{-7}$	$4.80 \cdot 10^{-7}$	7.50
1024	30.72	3.30	2,298.7	$9.25 \cdot 10^{-7}$	$6.06 \cdot 10^{-6}$	$1.71 \cdot 10^{-7}$	$6.77 \cdot 10^{-7}$	15.68

and discarding the elements that are smaller than the threshold (see Section 4.2). Correspondingly, the cost of constructing the non-standard form of the operator is proportional to N^2 (see column 4 of Table 2). The standard form of the operator A with $N = 256$ is depicted in Figure 10.

Example 3. In this example, we compress and rapidly apply to arbitrary vectors the matrix converting the coefficients of a finite Chebyshev expansion into the coefficients of a finite Legendre expansion representing the same polynomial (see [1]). The matrix is given by the formulae

$$A_{ij} = M^N_{2i2j}$$

where $i, j = 1, \cdots, N$ and $N = 2^n$ and M^N_{ij} is defined as

$$M^N_{ij} = \begin{cases} \frac{1}{\pi}\Lambda^2(j/2) & \text{if } 0 = i \leqq j < N \text{ and } j \text{ is even} \\ \frac{2}{\pi}\Lambda((j-i)/2)\Lambda((j+i)/2) & \text{if } 0 < i \leqq j < N \text{ and } i + j \text{ is even} \\ 0 & \text{otherwise,} \end{cases}$$

where $\Lambda(z) = \Gamma(z + 1/2)/\Gamma(z + 1)$ and $\Gamma(z)$ is the gamma function. Alternatively,

$$A_{ij} = \begin{cases} \frac{1}{\pi}\Lambda^2(j) & \text{if } 0 = i \leqq j < N \\ \frac{2}{\pi}\Lambda(j-i)\Lambda(j+i) & \text{if } 0 < i \leqq j < N. \\ 0 & \text{otherwise} \end{cases}$$

We used the threshold of 10^{-6} and wavelets with five vanishing moments to obtain the numerical results depicted in Table 3 and Figure 4. As a corollary, we obtain an algorithm for the rapid evaluation of Legendre expansions of the same complexity (and roughly the same actual efficiency) as that described in [1].

Example 4. Here,

$$A_{ij} = \begin{cases} \log(i-j)^2 & i \neq j \\ 0 & i = j \end{cases}.$$

We use wavelets with six vanishing moments and set to zero everything below 10^{-6}. Table 4 and Figure 5 and Figure 6 describe the results of these experiments.

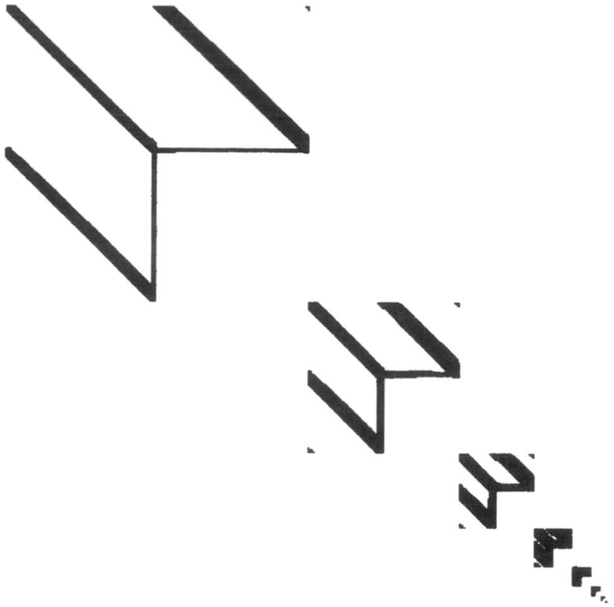

Figure 4. Entries above the threshold of 10^{-6} of the decomposed matrix of Example 3. This matrix is one of two transition matrices to compute Legendre expansion from Chebyshev expansion.

Example 5. In this example,

$$A_{ij} = \begin{cases} \dfrac{1}{i - j + \frac{1}{2}(\cos ij)} & i \neq j \\ 0 & i = j \end{cases}.$$

Table 3. Numerical results for Example 3.

Input size (N)	Time			Error of single precision multiplication		Error of FWT multiplication		Compression coefficient
	T_s	T_w	T_d	L_2-norm	L_∞-norm	L_2-norm	L_∞-norm	C_{comp}
64	0.12	0.12	10.28	$2.64 \cdot 10^{-7}$	$7.19 \cdot 10^{-7}$	$8.09 \cdot 10^{-7}$	$2.34 \cdot 10^{-6}$	1.73
128	0.48	0.30	42.70	$6.19 \cdot 10^{-7}$	$3.94 \cdot 10^{-6}$	$1.66 \cdot 10^{-6}$	$8.02 \cdot 10^{-6}$	2.89
256	1.92	0.66	133.66	$1.28 \cdot 10^{-6}$	$5.23 \cdot 10^{-6}$	$2.51 \cdot 10^{-6}$	$1.21 \cdot 10^{-5}$	5.18
512	7.68	1.40	344.60	$2.24 \cdot 10^{-6}$	$1.35 \cdot 10^{-5}$	$3.75 \cdot 10^{-6}$	$3.31 \cdot 10^{-5}$	9.70
1024	30.72	2.78	805.90	$4.45 \cdot 10^{-6}$	$2.42 \cdot 10^{-5}$	$6.40 \cdot 10^{-6}$	$9.00 \cdot 10^{-5}$	18.60

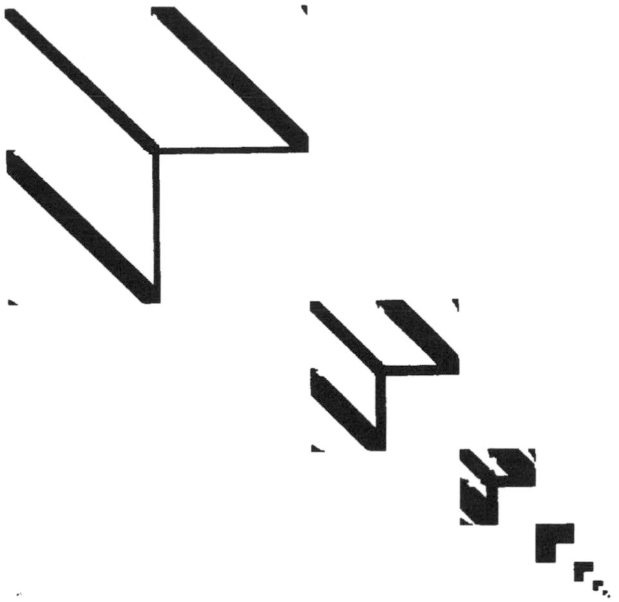

Figure 5. Entries above the threshold of 10^{-6} of the decomposed matrix of Example 4.

and it is easy to see that this operator does not satisfy the condition (4.10). Nonetheless, when a low order version of our scheme is applied to it, the results are quite satisfactory, albeit with an expectedly low accuracy (we used wavelets with two vanishing moments, and set the threshold to 10^{-3}). The results of these numerical experiments can be seen in Figure 7 and Table 5.

Table 4. Numerical results for Example 4.

Input size (N)	Time			Error of single precision multiplication		Error of FWT multiplication		Compression coefficient
	T_s	T_w	T_d	L_2-norm	L_∞-norm	L_2-norm	L_∞-norm	C_{comp}
64	0.12	0.14	8.84	$2.22 \cdot 10^{-5}$	$6.31 \cdot 10^{-5}$	$1.13 \cdot 10^{-6}$	$2.33 \cdot 10^{-6}$	1.37
128	0.48	0.34	38.42	$6.23 \cdot 10^{-5}$	$1.62 \cdot 10^{-4}$	$2.07 \cdot 10^{-6}$	$5.19 \cdot 10^{-6}$	2.19
256	1.92	0.84	120.22	$2.11 \cdot 10^{-4}$	$6.99 \cdot 10^{-4}$	$2.99 \cdot 10^{-6}$	$8.46 \cdot 10^{-6}$	3.82
512	7.68	1.76	310.86	$7.90 \cdot 10^{-4}$	$2.47 \cdot 10^{-3}$	$4.08 \cdot 10^{-6}$	$1.23 \cdot 10^{-5}$	7.04
1024	30.72	3.70	736.8	$2.65 \cdot 10^{-3}$	$9.44 \cdot 10^{-3}$	$6.53 \cdot 10^{-6}$	$2.19 \cdot 10^{-5}$	13.43

Scale $j = 1$, first column of the matrix α^1_{ij}.

0.42592E+00	0.31311E+00	-.10042E+00	0.32510E-01	-.75046E-02	0.95151E-03
-.30115E-04	-.16471E-05	0.00000E+00	0.00000E+00	0.00000E+00	0.00000E+00
0.00000E+00	0.00000E+00	0.00000E+00	0.00000E+00	0.00000E+00	0.00000E+00
0.00000E+00	0.00000E+00	0.00000E+00	0.00000E+00	0.00000E+00	0.00000E+00
0.00000E+00	0.00000E+00	0.00000E+00	0.00000E+00	0.00000E+00	0.00000E+00
0.00000E+00	0.00000E+00	0.00000E+00	0.00000E+00	0.00000E+00	0.00000E+00
0.00000E+00	0.00000E+00	0.00000E+00	0.00000E+00	0.00000E+00	0.00000E+00
0.00000E+00	0.00000E+00	0.00000E+00	0.00000E+00	0.00000E+00	0.00000E+00
0.00000E+00	0.00000E+00	0.00000E+00	0.00000E+00	0.00000E+00	0.00000E+00
0.00000E+00	0.00000E+00	0.00000E+00	0.00000E+00	0.00000E+00	0.00000E+00
0.00000E+00	0.00000E+00	0.00000E+00	0.00000E+00	0.00000E+00	0.00000E+00
0.00000E+00	0.00000E+00	0.00000E+00	0.00000E+00	0.00000E+00	0.00000E+00
0.00000E+00	0.00000E+00	0.00000E+00	0.00000E+00	0.00000E+00	0.00000E+00
0.00000E+00	0.00000E+00	0.00000E+00	0.00000E+00	0.00000E+00	0.00000E+00
0.00000E+00	0.00000E+00	0.00000E+00	0.00000E+00	0.00000E+00	0.00000E+00
0.00000E+00	0.00000E+00	0.00000E+00	0.00000E+00	0.00000E+00	0.00000E+00
0.00000E+00	0.00000E+00	0.00000E+00	0.00000E+00	0.00000E+00	0.00000E+00
0.00000E+00	0.00000E+00	0.00000E+00	0.00000E+00	0.00000E+00	0.00000E+00
0.00000E+00	0.00000E+00	0.00000E+00	0.00000E+00	0.00000E+00	0.00000E+00
0.00000E+00	0.00000E+00	0.00000E+00	0.00000E+00	0.00000E+00	0.00000E+00
0.00000E+00	0.00000E+00	0.00000E+00	0.17856E-02	-.77315E-02	0.32623E-01
-.10046E+00	0.31312E+00				

Scale j=1, first column of the matrix β^1_{ij}.

-.10075E+00	0.27192E-01	-.53507E-02	0.11903E-02	-.11031E-03	0.25509E-04
0.65200E-05	0.34067E-05	0.18585E-05	0.10718E-05	0.00000E+00	0.00000E+00
0.00000E+00	0.00000E+00	0.00000E+00	0.00000E+00	0.00000E+00	0.00000E+00
0.00000E+00	0.00000E+00	0.00000E+00	0.00000E+00	0.00000E+00	0.00000E+00
0.00000E+00	0.00000E+00	0.00000E+00	0.00000E+00	0.00000E+00	0.00000E+00
0.00000E+00	0.00000E+00	0.00000E+00	0.00000E+00	0.00000E+00	0.00000E+00
0.00000E+00	0.00000E+00	0.00000E+00	0.00000E+00	0.00000E+00	0.00000E+00
0.00000E+00	0.00000E+00	0.00000E+00	0.00000E+00	0.00000E+00	0.00000E+00
0.00000E+00	0.00000E+00	0.00000E+00	0.00000E+00	0.00000E+00	0.00000E+00
0.00000E+00	0.00000E+00	0.00000E+00	0.00000E+00	0.00000E+00	0.00000E+00
0.00000E+00	0.00000E+00	0.00000E+00	0.00000E+00	0.00000E+00	0.00000E+00
0.00000E+00	0.00000E+00	0.00000E+00	0.00000E+00	0.00000E+00	0.00000E+00
0.00000E+00	0.00000E+00	0.00000E+00	0.00000E+00	0.00000E+00	0.00000E+00
0.00000E+00	0.00000E+00	0.00000E+00	0.00000E+00	0.00000E+00	0.00000E+00
0.00000E+00	0.00000E+00	0.00000E+00	0.00000E+00	0.00000E+00	0.00000E+00
0.00000E+00	0.00000E+00	0.00000E+00	0.00000E+00	0.00000E+00	0.00000E+00
0.00000E+00	0.00000E+00	0.00000E+00	0.00000E+00	0.00000E+00	0.00000E+00
0.00000E+00	0.00000E+00	0.00000E+00	0.00000E+00	0.00000E+00	0.00000E+00
0.00000E+00	0.00000E+00	0.00000E+00	0.00000E+00	0.00000E+00	0.00000E+00
0.00000E+00	0.00000E+00	0.00000E+00	0.00000E+00	0.00000E+00	0.00000E+00
0.00000E+00	0.00000E+00	0.00000E+00	-.54776E+01	0.10421E+01	-.77699E+00
0.29165E+00	0.90294E-01				

Figure 6. Entries of the first column of matrices α and β (on the fine scale) of Example 4. We observe fast decay away from the diagonal. The threshold is 10^{-6}. Note the large numbers at the end of the columns due to periodization (see Remark 3.2).

Example 6. Here,

$$A_{ij} = \begin{cases} \dfrac{i\cos(\log i^2) - j\cos(\log j^2)}{(i-j)^2} & i \neq j \\ 0 & i = j \end{cases}.$$

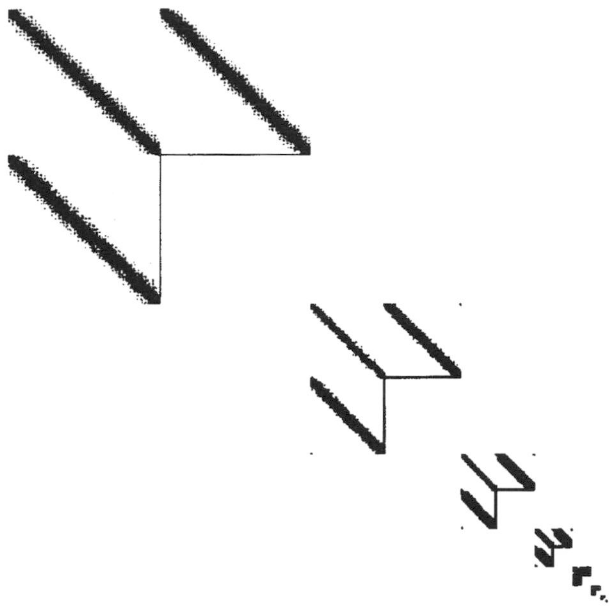

Figure 7. Entries above the threshold of 10^{-3} of the decomposed matrix of Example 5.

Like in the preceding example, the operator being compressed satisfies the condition (4.10) with $M = 1$, and fails to do so for any larger M. Using wavelets with two vanishing moments, and setting the threshold to 10^{-3}, we obtain the results depicted in Figure 8 and Table 6. Again, the compression rate for this reasonably large threshold is quite satisfactory.

Table 5. Numerical results for Example 5.

Input size (N)	Time			Error of single precision multiplication		Error of FWT multiplication		Compression coefficient
	T_s	T_w	T_d	L_2-norm	L_∞-norm	L_2-norm	L_∞-norm	C_{comp}
64	0.12	0.10	2.84	$1.93 \cdot 10^{-7}$	$5.04 \cdot 10^{-7}$	$1.18 \cdot 10^{-3}$	$3.11 \cdot 10^{-3}$	1.99
128	0.48	0.18	9.00	$2.65 \cdot 10^{-7}$	$9.27 \cdot 10^{-7}$	$1.54 \cdot 10^{-3}$	$4.36 \cdot 10^{-3}$	3.51
256	1.92	0.42	23.62	$3.76 \cdot 10^{-7}$	$1.83 \cdot 10^{-6}$	$2.02 \cdot 10^{-3}$	$8.33 \cdot 10^{-3}$	6.58
512	7.68	0.88	55.62	$4.93 \cdot 10^{-7}$	$2.46 \cdot 10^{-6}$	$3.19 \cdot 10^{-3}$	$3.91 \cdot 10^{-2}$	12.81
1024	30.72	1.74	123.84	$7.53 \cdot 10^{-7}$	$4.78 \cdot 10^{-6}$	$3.99 \cdot 10^{-3}$	$7.57 \cdot 10^{-2}$	25.19

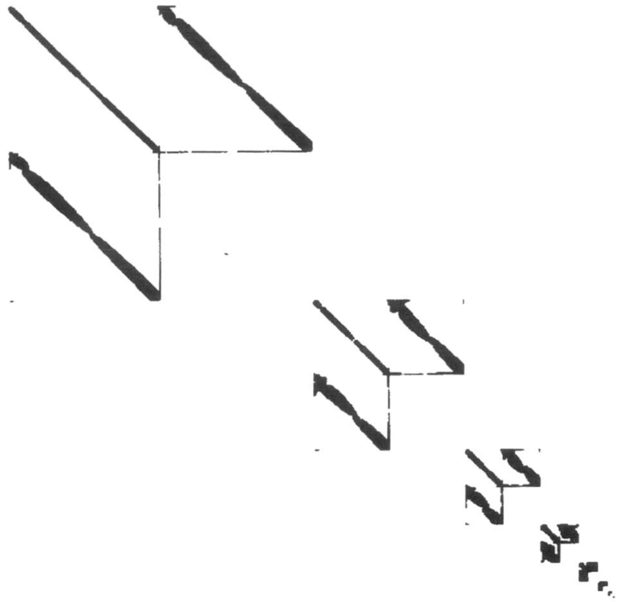

Figure 8. Entries above the threshold of 10^{-3} of the decomposed matrix of Example 6.

The following observations can be made from Tables 1–6 and Figures 2–7.

1. The CPU times required by the algorithm of this paper to apply the matrix to a vector grow linearly with N, while those for the direct algorithm grow quadratically (as expected).
2. The accuracy of the method is in agreement with the estimates of Section 4, and when the threshold is set to 10^{-6}, the actual accuracies obtained tend

Table 6. Numerical results for Example 6.

Input size (N)	Time			Error of single precision multiplication		Error of FWT multiplication		Compression coefficient
	T_s	T_w	T_d	L_2-norm	L_∞-norm	L_2-norm	L_∞-norm	C_{comp}
64	0.12	0.10	4.22	$2.59 \cdot 10^{-7}$	$8.76 \cdot 10^{-7}$	$2.42 \cdot 10^{-3}$	$4.58 \cdot 10^{-3}$	2.37
128	0.48	0.20	16.60	$3.71 \cdot 10^{-7}$	$1.07 \cdot 10^{-6}$	$2.81 \cdot 10^{-3}$	$8.61 \cdot 10^{-3}$	4.13
256	1.92	0.38	66.70	$5.03 \cdot 10^{-7}$	$2.12 \cdot 10^{-6}$	$3.62 \cdot 10^{-3}$	$1.38 \cdot 10^{-2}$	8.25
512	7.68	0.82	263.72	$8.71 \cdot 10^{-7}$	$3.10 \cdot 10^{-6}$	$3.68 \cdot 10^{-3}$	$1.60 \cdot 10^{-2}$	14.80
1024	30.72	1.50	1,107.6	$1.12 \cdot 10^{-6}$	$5.52 \cdot 10^{-6}$	$4.56 \cdot 10^{-3}$	$4.12 \cdot 10^{-2}$	33.07

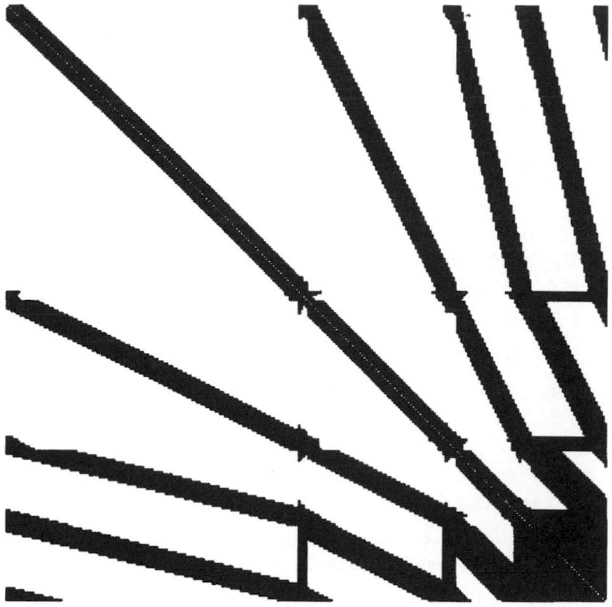

Figure 9. Entries above the threshold of 10^{-7} of the standard form for Example 1. Different bands represent "interactions" between scales.

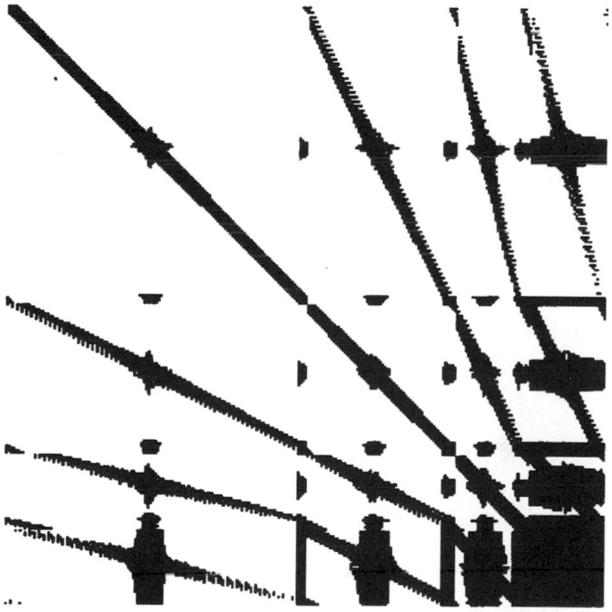

Figure 10. Entries above the threshold of 10^{-7} of the standard form for Example 2.

to be slightly better than those obtained by the direct calculation in single precision.

3. In many cases, the algorithm becomes more efficient than the direct computation for $N = 100$, and for $N = 1000$ the gain is roughly of the factor of 10.
4. Even when the operator fails to satisfy the condition (4.10), the application of the algorithm with a reasonably large threshold and small M leads to satisfactory compression factors.
5. Combining the linear asymptotic CPU time estimate of the algorithm of this paper with the actual timings in Tables 1–6, we observe that whenever the algorithm of this paper is applicable, large-scale problems become tractable, even with relatively modest computing resources.

7. Extensions and Generalizations

7.1. Numerical Operator Calculus

In this paper, we construct a mechanism for the rapid application to arbitrary vectors of a wide variety of dense matrices. It turns out that in addition to the application of matrices to vectors, our techniques lead to algorithms for the rapid multiplication of operators (or, rather, their standard forms). The asymptotic complexity of the resulting procedure is also proportional to N. When applied recursively, it permits a whole range of matrix functions (polynomials, exponentials, inverses, square roots, etc.) to be evaluated for a cost proportional to N, converting the operator calculus into a competitive numerical tool (as opposed to the purely analytical apparatus it has been). These (and several related) algorithms have been implemented, and are described in a paper currently in preparation.

7.2. Generalizations to Higher Dimensions

The construction of the present paper is limited to the one-dimensional case, i.e., the integral operators being compressed are assumed to act on $L^2(\mathbf{R})$. Its generalization to problems in higher dimensions is fairly straightforward, and is being implemented. When combined with the Lippman-Schwinger equation, or with the classical pseudo-differential calculus, these techniques should lead to algorithms for the rapid solution of a wide variety of elliptic partial differential equations in regions of complicated shapes, of second kind integral equations in higher-dimensional domains, and of several related problems.

7.3. Non-Linear Operators

While the present paper discusses the "compression" of linear and bilinear operators, extensions to multilinear functionals (defined on the functions in one, as well as higher dimensions) is not difficult to obtain. These methods (together with some of their applications) will be described in a forthcoming paper. The underlying theory can be found in [3].

Appendix A

The following table contains filter coefficients $\{h_k\}_{k=1}^{k=3M}$ for $M = 2, 4, 6$ for one particular choice of the shift τ. These coefficients have $M - 1$ vanishing moments,

$$M_l = \sum_{k=1}^{k=3M} h_k(k - \tau_M)^l = 0, \qquad l = 1, \cdots, M-1$$

where τ_M is the shift, and have been provided to the authors by I. Daubechies (see also Section 3.2 above). For $M = 2$ there are explicit expressions for $\{h_k\}_{k=1}^{k=3M}$, and with $\tau_2 = 5$, they are

$$h_1 = \frac{\sqrt{15} - 3}{16\sqrt{2}}, \quad h_2 = \frac{1 - \sqrt{15}}{16\sqrt{2}}, \quad h_3 = \frac{3 - \sqrt{15}}{8\sqrt{2}},$$

$$h_4 = \frac{\sqrt{15} + 3}{8\sqrt{2}}, \quad h_5 = \frac{\sqrt{15} + 13}{16\sqrt{2}}, \quad h_6 = \frac{9 - \sqrt{15}}{16\sqrt{2}},$$

and for $M = 4, 6$, the coefficients $\{h_k\}$ are presented in the table below.

	k	Coefficients h_k		k	Coefficients h_k
$M = 2$	1	0.038580777747887	$M = 6$	1	−0.0016918510194918
$\tau_2 = 5$	2	−0.12696912539621	$\tau_6 = 8$	2	−0.00348787621998426
	3	−0.077161555495774		3	0.019191160680044
	4	0.60749164138568		4	0.021671094636352
	5	0.74568755893443		5	−0.098507213321468
	6	0.22658426519707		6	−0.056997424478478
				7	0.45678712217269
				8	0.78931940900416
$M = 4$	1	0.0011945726958388		9	0.38055713085151
$\tau_8 = 8$	2	−0.01284557955324		10	−0.070438748794943
	3	0.024804330519353		11	−0.056514193868065
	4	0.050023519962135		12	0.036409962612716
	5	−0.15535722285996		13	0.0087601307091635
	6	−0.071638282295294		14	−0.011194759273835
	7	0.57046500145033		15	−0.0019213354141368
	8	0.75033630585287		16	0.0020413809772660
	9	0.28061165190244		17	0.00044583039753204
	10	−0.0074103835186718		18	−0.00021625727664696
	11	−0.014611552521451			
	12	−0.0013587990591632			

Appendix B

In this appendix we construct quadrature formulae using the compactly supported wavelets of [4] which do not satisfy condition (3.8). These quadrature formulae are similar to the quadrature formula (3.12) in that they do not require explicit evaluation of the function $\varphi(x)$ and are completely determined by the filter coefficients $\{h_k\}_{k=1}^{k=2M}$. Our interest in these quadrature formulae stems from the fact that for a given number M of vanishing moments of the basis functions, the wavelets of [4] have the support of length $2M$ compared with $3M$ for the wavelets satisfying condition (3.8). Since our algorithms depend linearly on the size of the support, using wavelets of [4] and quadrature formulae of this appendix makes these algorithms $\approx$50% faster.

We use these quadrature formulae to evaluate the coefficients s_k^j of smooth functions without the pyramid scheme (2.12), where s_k^j are computed via (3.13) for $j = 1, \cdots, n$.

First, we explain how to compute $\{s_k^0\}_{k=1}^{k=N}$. Recalling the definition of s_k^0,

$$\begin{aligned} s_k^0 &= 2^{n/2} \int f(x)\varphi(2^n x - k + 1)\,dx \\ &= 2^{n/2} \int f(x + 2^{-n}(k-1))\varphi(2^n x)\,dx, \end{aligned} \tag{B.1}$$

we look for the coefficients $\{c_l\}_{l=0}^{l=M-1}$ such that

$$2^{n/2} \int f(x + 2^{-n}(k-1))\varphi(2^n x)\,dx = 2^{-n/2} \sum_{l=0}^{l=M-1} c_l f(l + 2^{-n}(k-1)), \tag{B.2}$$

for polynomials of degree less than M. Using (B.2), we arrive at the linear algebraic system for the coefficients c_l,

$$\sum_{l=0}^{l=M-1} l^m c_l = \int x^m \varphi(x)\,dx, \qquad m = 0, 1, \cdots, M-1, \tag{B.3}$$

where the moments of the function $\varphi(x)$ are computed in terms of the filter coefficients $\{h_k\}_{k=1}^{k=2M}$.

Given the coefficients c_l, we obtain the quadrature formula for computing s_k^0,

$$s_k^0 = 2^{-n/2} \sum_{l=0}^{l=M-1} c_l f(l + 2^{-n}(k-1)) + O(2^{-n(M+1/2)}). \tag{B.4}$$

The moments of the function φ are obtained by differentiating (an appropriate number of times) its Fourier transform $\hat{\varphi}$,

$$\hat{\varphi}(\xi) = (2\pi)^{-1/2} \int dx\, e^{ix\xi} \varphi(x), \tag{B.5}$$

and setting $\xi = 0$. The expressions for the moments $\int x^m \varphi(x)\, dx$ in terms of the filter coefficients $\{h_k\}_{k=1}^{k=2M}$ are found using a formula for $\hat{\varphi}$ [4],

$$(2\pi)^{1/2} \hat{\varphi}(\xi) = \prod_{j=1}^{\infty} m_0(2^{-j}\xi), \tag{B.6}$$

where

$$m_0(\xi) = 2^{-1/2} \sum_{k=1}^{k=2M} h_k e^{i(k-1)\xi}. \tag{B.7}$$

The moments $\int x^m \varphi(x)\, dx$ are obtained numerically (within the desired accuracy) by recursively generating a sequence of vectors, $\{\mathcal{M}^r_m\}_{m=0}^{m=M-1}$ for $r = 1, 2, \cdots$,

$$\mathcal{M}^{r+1}_m = \sum_{j=0}^{j=m} \binom{m}{j} 2^{m-j(r+1)} \mathcal{M}^r_{m-j} \mathcal{M}^1_j, \tag{B.8}$$

starting with

$$\mathcal{M}^1_m = 2^{-m-1/2} \sum_{k=1}^{k=2M} h_k (k-1)^m, \qquad m = 0, \cdots, M-1. \tag{B.9}$$

Each vector $\{\mathcal{M}^r_m\}_{m=0}^{m=M-1}$ represents M moments of the product in (B.6) with r terms.

We now derive formulae to compute the coefficients s^j_k of smooth functions without the pyramid scheme (2.12). Let us formulate the following

PROPOSITION B1. *Let s^j_m be the coefficients of a smooth function at some scale j. Then*

$$s^{j+1}_m = 2^{1/2} \sum_{l=1}^{l=M} q_l s^j_{2m+2l-3} + O(2^{-(n-j)M}), \tag{B.10}$$

is a formula to compute the coefficients s^{j+1}_m at the scale $j+1$ from those at the *scale j. The coefficients $\{q_l\}_{l=1}^{l=M}$ in (B.10) are solutions of the linear algebraic system*

$$\sum_{l=1}^{l=M} q_l(2l-1)^m = M_m, \qquad m = 0, \cdots, M-1, \tag{B.11}$$

and where M_m are the moments of the coefficients h_k scaled for convenience by $1/H(0) = 2^{-1/2}$,

$$M_m = 2^{-1/2} \sum_{k=1}^{k=2M} h_k k^m, \qquad m = 0, \cdots, M-1. \tag{B.12}$$

Using Proposition B1 we prove the following

LEMMA B1. *Let s^j_m be the coefficients of a smooth function at some scale j. Then*

$$s^{j+r}_m = 2^{r/2} \sum_{l=1}^{l=M} q^r_l s^j_{2^r(m+l-2)+1} + O(2^{-(n-j-r)M}), \tag{B.13}$$

is a formula to compute the coefficients s^{j+r}_m at the scale $j+r$ from those at the scale j, with $r \geqq 1$. The coefficients $\{q^r_l\}_{l=1}^{l=M}$ in (B.13) are obtained by recursively generating the sequence of vectors $\{q^1_l\}_{l=1}^{l=M}, \cdots, \{q^r_l\}_{l=1}^{l=M}$ as solutions of the linear algebraic system

$$\sum_{l=1}^{l=M} q^r_l(2l-1)^m = M^r_m, \qquad m = 0, \cdots, M-1, \tag{B.14}$$

where the sequence of the moments $\{M^1_m = M_m\}, \{M^2_m\}, \cdots, \{M^r_m\}$ is computed via

$$M^{r+1}_m = \sum_{j=0}^{j=m} \binom{m}{j} M_{m-j} L^r_j, \tag{B.15}$$

where

$$L^r_j = \sum_{l=1}^{l=M} q^r_l (l-1)^j. \tag{B.16}$$

We note that for $r = 1$ (B.13) reduces to (B.10).

Proof of Proposition B1: Let $H(\xi)$ denote the Fourier transform of the filter with the coefficients $\{h_k\}_{k=1}^{k=2M}$,

$$H(\xi) = \sum_{k=1}^{k=2M} h_k e^{ik\xi}. \tag{B.17}$$

Clearly, the moments M_m in (B.12) can be written as

$$M_m = 2^{-1/2}\left(-i\frac{d}{d\xi}\right)^m H(\xi)|_{\xi=0}, \qquad m = 0, \cdots, M-1. \tag{B.18}$$

Also, the trigonometric polynomial $H(\xi)$ can always be written as the product,

$$H(\xi) = \tilde{H}(\xi)Q(\xi), \tag{B.19}$$

where we choose Q to be of the form

$$Q(\xi) = \sum_{l=1}^{l=M} q_l e^{i(2l-1)\xi}, \tag{B.20}$$

and $\tilde{H}$ to have zero moments

$$\left(-i\frac{d}{d\xi}\right)^m \tilde{H}(\xi)|_{\xi-0}, \qquad m = 1, \cdots, M-1. \tag{B.21}$$

By differentiating (B.19) appropriate number of times, setting $\xi = 0$ and using (B.21) we arrive at (B.11). Solving (B.11), we find the coefficients $\{q_l\}_{l=1}^{l=M}$.

Since moments of $\tilde{H}$ vanish, the convolution with the coefficients of the filter $\tilde{H}$ reduces to the one-point quadrature formula of the type in (3.12). Thus applying H reduces to applying Q and scaling the result by $1/H(0) = 2^{-1/2}$. Clearly, there are only M coefficients of Q compared to $2M$ of H, and the particular form of the filter Q (B.20) was chosen so that only every second entry of s_k^j, starting with $k = 1$, is multiplied by a coefficient of the filter Q.

Proof of Lemma B1: Lemma B1 is proved by induction. Since for $r = 1$ (B.13) reduces to (B.10), we have to show that given (B.13), it also holds if r is increased by one.

Let $\tilde{s}_k^j$ be the subsequence consisting of every 2^r entry of s_k^j starting with $k = 1$. Applying filter $\{q_l^r\}_{l=1}^{l=M}$ to s_k^j in (B.13) is equivalent to applying filter P^r to $\tilde{s}_k^j$, where

$$P^r(\xi) = \sum_{l=1}^{l=M} q_l^r e^{i(l-1)\xi}. \tag{B.22}$$

To obtain (B.13), where r is increased by one, we use the quadrature formula (B.10) of Proposition B1. Therefore, the result is obtained by convolving $\tilde{s}_k^j$ with the coefficients of the filter $Q(\xi)P^r(\xi)$, where $Q(\xi)$ is defined in (B.20).

Let us construct a new filter Q^{r+1} by factoring $Q(\xi)P^r(\xi)$ similar to (B.19),

$$Q(\xi)P^r(\xi) = \tilde{\tilde{H}}(\xi)Q^{r+1}(\xi), \tag{B.23}$$

where we chose Q^{r+1} to be of the form

$$Q^{r+1}(\xi) = \sum_{l=1}^{l=M} q_l^{r+1} e^{i(2l-1)\xi}, \tag{B.24}$$

and $\tilde{\tilde{H}}$ to have zero moments

$$\left(-i\frac{d}{d\xi}\right)^m \tilde{\tilde{H}}(\xi)|_{\xi=0} = 0, \qquad m = 1, \cdots, M-1. \tag{B.25}$$

Again, since moments of $\tilde{\tilde{H}}$ vanish, the convolution with the coefficients of the filter $\tilde{\tilde{H}}$ reduces to scaling the result by $2^{-1/2}$.

To compute moments M_m^{r+1} of $Q(\xi)P^r(\xi)$ we differentiate $Q(\xi)P^r(\xi)$ appropriate number of times, set $\xi = 0$, and arrive at (B.15) and (B.16). To obtain the linear algebraic system (B.14) for the coefficients q_l^{r+1}, we differentiate (B.23) appropriate number of times, set $\xi = 0$, and use (B.25).

Recalling that the filter P^r is applied to the subsequence $\tilde{s}_k^j$, we arrive at (B.13), where r is increased by one.

Acknowledgments. The research of the first and second authors was partially supported by ONR grant N00014-88-K0020. The research of the third author was partially supported by ONR grants N00014-89-J1527, N00014-86-K0310, and IBM grant P00038436.

The permanent address of the first author is: Schlumberger-Doll Research, Ridgefield, CT 06897.

Bibliography

[1] Alpert, B., and Rokhlin, V., *A Fast Algorithm for the Evaluation of Legendre Expansions*, Yale University Technical Report, YALEU/DCS/RR-671, 1989.

[2] Carrier, J., Greengard, L., and Rokhlin, V., *A Fast Adaptive Multipole Algorithm for Particle Simulations*, Yale University Technical Report, YALEU/DCS/RR-496, 1986, SIAM J. Sci. Stat. Comp. 9, 1988, pp. 669–686.

[3] Coifman, R., and Meyer, Y., *Nonlinear Harmonic Analysis, Operator Theory, and P.D.E.*, Ann. Math. Studies, E. Stein, ed., Princeton, 1986.

[4] Daubechies, I., *Orthonormal bases of compactly supported wavelets*, Comm. Pure Appl. Math. 41, 1988, pp. 909–996.

[5] Greengard, L., and Rokhlin, V., *A fast algorithm for particle simulations*, J. Comp. Phys. 73 (1) 325, 1987, pp. 325–348.
[6] Mallat, S., *Review of Multifrequency Channel Decomposition of Images and Wavelet Models*, Technical Report 412, Robotics Report 178, New York Univ., 1988.
[7] Meyer, Y., *Principe d'incertitude, bases Hilbertiennes et algèbres d'opérateurs*, Séminaire Bourbaki 662, 1985–86, Astérisque (Société Mathématique de France).
[8] Meyer, Y., *Wavelets and operators*, in *Analysis at Urbana*, Vol. 1, E. Berkson, N. T. Peck, and J. Uhl, eds., London Math. Soc., Lecture Notes Series 137, 1989, pp. 256–365.
[9] O'Donnel, S. T., and Rokhlin, V., *A Fast Algorithm for the Numerical Evaluation of Conformation Mappings*, Yale University Technical Report, YALEU/DSC/RR-554, 1987, SIAM J. Sci. Stat. Comp., 1989, pp. 475–487.
[10] Stromberg, J. O., *A modified Haar system and higher order spline systems*, Conf. in Harmonic Analysis in honor of Antoni Zygmund, Wadworth Math. Series II, W. Beckner et al., eds., pp. 475–493.

Received January 1990.

COMPRESSION OF WAVELET DECOMPOSITIONS

By Ronald A. DeVore, Björn Jawerth, and Vasil Popov[1]

We characterize functions with a given degree of nonlinear approximation by linear combinations with n terms of a function φ, its dilates and their translates. This gives a unified viewpoint of recent results on nonlinear approximation by spline functions and give their extension to functions of several variables. Our approach is formulated in terms of wavelet decompositions.

1. Introduction. There has recently been great interest in the numerical applications of wavelet decompositions. Such applications call for the efficient recovery of a function f from the coefficients in such a decomposition. We shall discuss in this paper nonlinear methods for accomplishing the recovery when the error is measured in the metric of L_p, $0 < p < \infty$.

Let φ be a bounded function defined on $\mathbb{R}^d$. Together with φ we have its dyadic dilates $\varphi(2^k x)$, $k \in \mathbb{Z}$, and their translates $\varphi_I(x) := \varphi(2^k x - j)$, $j \in \mathbb{Z}^d$. Here, we index these functions by the dyadic cube $I = j2^{-k} + 2^{-k}\Omega$ with $\Omega := [0, 1]^d$. We shall say that $j2^{-k}$ **corresponds** to I. We shall also use the notation $\mathcal{D}_k$ to denote the set of dyadic cubes I whose sidelength $\ell(I)$ is 2^{-k} and $\mathcal{D}$ to denote the union of the $\mathcal{D}_k$, $k \in \mathbb{Z}$.

It is now well understood that under certain conditions on φ (described below), a general function f has a decomposition in terms of the φ_I, $I \in \mathcal{D}$:

$$f = \sum_{I \in \mathcal{D}} a_I(f)\varphi_I \tag{1.1}$$

where the a_I are appropriate coefficient functionals. Moreover, for many spaces of functions X, the norm of $f \in X$ is equivalent to an appropriate

Manuscript received 16 November 1989.

[1]The first author was supported by NSF Grant DMS 8620108. The second author was supported by NSF Grant DMS 8803585. The first two authors were supported by AFoSR Contract 89-0455.

American Journal of Mathematics 114 (1992), 737–785.

norm applied to the sequence (a_I). We call the decompositions (1.1) **wavelet decompositions** and the functions φ **wavelets**.

While the history of wavelet decompositions is too extensive to be described in complete detail here, we shall mention some aspects of their development that relates to the present work. Many of the fundamental aspects are part of the development of Littlewood-Paley theory during the past fifty–sixty years. Continuous analogues of (1.1) are often referred to as "Calderón's reproducing formula" and appear in the seminal work of Calderón [5]. Similar continuous representations are also an important tool in the paper by Hörmander [22] and the book by Peetre [31].

Discrete wavelet decompositions can be constructed from discrete analogues of Calderón's reproducing formula. This is the starting point for Frazier and Jawerth [19–21] in their development of the "φ transform." (For some earlier antecedents see also Coifman-Rochberg [9] and [25], [31].) They have given very general criteria for a function φ to be a wavelet and, in particular, they show that any function φ with sufficient smoothness, vanishing moments, and decay at ∞ is a wavelet for L^p, $1 < p < \infty$ (for H^p, $0 < p \leq 1$) in the sense of (1.1). In these constructions, only one function is needed in the multivariate case and one can often give a simple description of the coefficient functionals as translated dilates of one coefficient functional. We note, however, that in general the collection of functions $\{\varphi_I\}_{I\in\mathfrak{D}}$ need not be even linearly independent. Related results can also be found in Daubechies, Grossman, and Mcycr [13].

Another example of a discrete decomposition (1.1) is the classical Haar decomposition which uses the univariate function $\varphi(x) := -1$ on $[0, \frac{1}{2})$, $\varphi(x) := 1$ on $[\frac{1}{2}, 1]$. In this case, the functions $|I|^{-1/2}\varphi_I$, $I \in \mathfrak{D}$, form a complete orthonormal system for $L_2(\mathbb{R})$. When a function φ has this orthogonality property, we shall call it an **orthogonal wavelet**. Strömberg [36] used a modification of the classical Franklin functions of piecewise polynomials to construct orthogonal wavelets which are an unconditional basis for the Hardy space H^1. The Strömberg wavelets can be prescribed with an arbitrary order of differentiability but do not have compact support. Another construction of an orthogonal wavelet, and a starting point for the systematic development of a theory for orthogonal wavelets, was given in the paper by Lemarié and Meyer [26]. More recent constructions of orthogonal wavelets are based on multiresolution analysis as introduced by Mallat [27]. An important step in the construction of orthogonal wavelets was made by Daubechies [12] who used multi-resolution analysis to construct compactly supported orthogonal wavelets for any prescribed order of differentiability. The previous constructions of smooth orthogonal wavelets were not of compact support. More recently, several authors (see for example Battle [1], Chui and Wang [7], and Jia and Micchelli [24]) have constructed wavelets in which orthogonality holds between different dyadic levels but not at a fixed dyadic level. That is $|I|^{-1/2}\varphi_I$ is orthogonal to $|J|^{-1/2}\varphi_J$ if $|I| \neq |J|$ but not necessarily when $|I| = |J|$. These are sometimes called prewavelets. The construction of orthogonal wavelets and prewavelets in $\mathbb{R}^d$ is usually accomplished by using certain tensor products of their one dimensional constructions. This results in $2^d - 1$ functions φ (not one) in the wavelet decomposition (1.1).

In the early 1970's, it was already noticed by Ciesielski (cf. [8]) in his construction of unconditional bases for certain Banach spaces, that B-splines are wavelets. It is rather easy to show that other spline functions, such as the multivariate box splines [2], are wavelets in several variables (see Section 4). These constructions are very similar to multiresolution analysis since they are based on the refinement property (4.1) for these splines. However, rather than using the L_2 projector as is done in multiresolution analysis, one uses quasi-interpolant projectors which are bounded on L_p spaces.

The importance of wavelets is not so much that they provide the decomposition (1.1), but rather that they provide a simultaneous (approximate) time and frequency localization which yields a description of many spaces, such as L_p, H_p, and the Sobolev, Besov, and Triebel-Lizorkin spaces, in terms of the coefficients in wavelet decompositions, and a description of important classes of operators, such as the Calderón-Zygmund operators, in terms of almost diagonal matrices. The predecessor to such results is the Littlewood-Paley theory for Fourier series and the subsequent development of the Littlewood-Paley g function (see the book by Stein [37]) in Harmonic Analysis. In this way, the wavelet φ gives atomic decompositions of spaces. For the wavelet description of the above mentioned spaces, and corresponding aspects of operator theory, we refer to the book by Meyer [29], for orthogonal wavelets, and Frazier-Jawerth [21], for the general φ-transform, and the references in there. Wavelet descriptions of smoothness spaces can also be obtained by using B-splines and box splines (see for example Oswald [30] and DeVore-Popov [16] and Section 4).

The present paper is not concerned with wavelet decompositions, per se. For the most part, we assume the existence of the wavelet decomposition and the description of function spaces by wavelet coefficients. Our interest is rather in the compression of the wavelet decomposition. That is, we wish to replaee the infinite sum (1.1) by a finite sum in an efficient manner. (We also study the case when the coefficients in such a finite sum are arbitrary, i.e. not necessarily the wavelet coefficients). Since the terms to be chosen may come from any dyadic level and their position generally depends on f, we are dealing with a problem of nonlinear approximation. This is to be contrasted with the better studied question of linear approximation where one takes all terms of the sum (1.1) below a given dyadic level. That is the approximation is done by taking the partial sum of (1.1) corresponding to all dyadic cubes whose side length is $\leq 2^k$ for some fixed $k \in \mathbb{Z}$.

While the present paper is the first general treatment of nonlinear approximation based on wavelets, there are several other works that deal with similar questions and motivated our investigation. The study of univariate approximation by splines with n free knots or rational functions of degree n has been extensively studied in the last 20 years, culminating with the work of Peller, Pekaarski, Petrushev, and DeVore-Popov (see for example [15] or [32]). The work of Peller, who considers rational approximation of functions in BMO and the subsequent investigations of Coifman and Rochberg (cf. [33]) is based on discrete representations for functions in terms of the Bergman kernel. DeVore and Popov [17] prove theorems of the type obtained in this paper for the special case when $\varphi := \chi_\Omega$, $\Omega := [0, 1]^d$.

We can describe the problem of interest to us as follows. We let Σ_n denote the nonlinear manifold consisting of all functions $S = \Sigma_{I\in\Lambda} a_I\varphi_I$ where the set Λ consists of at most n dyadic cubes (from arbitrary levels). We approximate functions f by elements from Σ_n; the error of approximating f in a quasinormed space X is

$$\sigma_n(f)_X := \inf_{S\in\Sigma_n} \|f - S\|_X. \tag{1.2}$$

The main result of this paper is that with certain conditions on φ we have for each $0 < p < \infty$ and for a certain range of α:

$$\sum_{n=1}^{\infty} [n^{\alpha/d}\sigma_n(f)_{L_p(\mathbb{R}^d)}]^r \frac{1}{n} < \infty \leftrightarrow f \in B^\alpha \tag{1.3}$$

where $B^\alpha := B^\alpha_\tau(L_\tau)$, $\tau := (\alpha/d + 1/p)^{-1}$ are the Besov spaces (see Section 4 for their definition).

Our approach to this problem is based on some general results from approximation theory. It is well known (see for example [15]) that to prove (1.3), it is sufficient to show that the following holds for some $\beta > \alpha$:

$$(1.4) \qquad \begin{array}{ll} \text{(i) } \sigma_n(f)_p \le Cn^{-\beta/d}|f|_{B^\beta}, & f \in B^\beta \\ \text{(ii) } |S|_{B^\beta} \le Cn^{\beta/d}\|S\|_{L_p(\mathbb{R}^d)}, & S \in \Sigma_n. \end{array}$$

The estimate (i) is called a direct theorem or Jackson estimate, while (ii) is the inverse theorem or Bernstein estimate. In Section 2, we shall show that the direct estimate (i) holds for very general functions φ and for $0 < p < \infty$. Our proof is quite simple since it constructs the approximation by merely taking the terms in the wavelet decomposition of f that are biggest in a certain L_p sense. However, it has the disadvantage that the constants in inequality (i) depend on p and grow as $p \to \infty$.

The direct theorem of Section 2 applies to a wide variety of functions φ and thereby gives a simple proof of the direct theorems for univariate nonlinear approximation by splines and rational functions mentioned above. Moreover, our theorem includes natural extensions of these results to higher dimensions.

In Section 3, we shall discuss general strategies for proving the inverse estimate (1.4)(ii). Among other things, we show that (ii) is valid in the case of orthogonal wavelets. This inverse theorem together with the direct theorem of Section 2 proves the characterization (1.3) for orthogonal wavelets. We remark that while for convenience we state our result for a single wavelet function φ, the results we obtain in Section 3 apply equally well to the case of multivariate orthogonal wavelets where $2^d - 1$ functions φ are used for the wavelet basis.

The results in Sections 2, 3 do not give a satisfactory treatment of the case where φ is a B-spline or box spline since in this case the lack of linear independence between dyadic levels plays a damaging role. However, we shall give another approach to inverse theorems in Section 5 which applies to a fairly broad class of wavelets φ and will include the case of spline functions. This treatment applies to functions φ that satisfy the refinement equation (4.1).

In Section 4, we develop the properties of functions φ needed in our second approach. In particular, we show that such functions φ give the wavelet decomposition (1.1). Moreover, we show that the norms of functions $f \in B^\alpha$ can be described simply in terms of a sequence norm applied to the coefficients of the wavelet decomposition. The results of this section are obtained by linear approximation and are analogues of the well known results of Meyer [29] for orthogonal wavelets and Frazier-Jawerth [21] for the φ-transform. We have included detailed proofs since there is no ready reference to the particular case treated in Sections 4, 5, 6. This also gives us an opportunity to indicate how approximation theoretic techniques can be used to prove wavelet decompositions and characterize smoothness spaces by the coefficients in such decompositions.

In Section 6, we shall give a second proof of the direct theorem (1.4)(i) which applies to the wavelets φ of the type discussed in Sections 4, 5. While this proof is much more substantial than the simple proof of Section 2, it has the advantage that the constants appearing in (1.4)(i) do not deteriorate as $p \to \infty$. This is accomplished by taking a more delicate approximation than simply using the biggest coefficients in the wavelet decomposition. The technique put forward in Section 6 also has potential numerical implications.

We give in Section 7, some applications of the results of Sections 4, 5, and 6. As a special case, our results include the characterization (1.3) in the case that $\varphi(x) := N(x; 0, 1, \ldots, r)$ is the univariate B-spline of degree $r - 1$ with n equally spaced knots. In this way, we recover the univariate results on nonlinear approximation by splines already mentioned above. More interestingly, our results apply equally well when φ is a multivariate box spline or multivariate B-spline (see Section 7). In this way, we give a natural extension of the univariate free knot spline approximation to higher space dimension.

2. A direct theorem for general φ. The direct theorem of this section will apply to wavelets φ of all types described in the introduction. The essential property possessed by each of these wavelets is that they allow the simple characterization of many classical function spaces in terms of sequence norms applied to the coefficients. We shall need this characterization only for the Besov spaces B^α. We postpone the definition of the Besov spaces to Section 4 since at this point we only need the description of their seminorm in terms of the coefficients of the wavelet decomposition (1.1). The characterization which we now describe can be found in Meyer [29] for orthogonal wavelets, in Frazier and Jawerth [21] for the φ transform, and will be proved in Section 4 for wavelets based on the refinement equation, provided (in all three of these cases) that φ has suitable smoothness and decay.

We fix a value of p with $0 < p < \infty$ and let $B^\beta := B^\beta_\tau(L_\tau)$, $\tau := (\beta/d + 1/p)^{-1}$ be the Besov spaces. In this section, we shall assume that for the fixed value $0 < p < \infty$, φ has the property that each f in the Besov space B^β, has a representation

$$f = \sum_{I \in \mathcal{D}} a_I(f)\varphi_I \tag{2.1}$$

where the coefficients satisfy

$$|f|_\beta := \left(\sum_{I \in \mathcal{D}} |a_I(f)|^\tau |I|^{\tau/p}\right)^{1/\tau} \tag{2.2}$$

$$= \left(\sum_{I \in \mathcal{D}} |a_I(f)|^\tau |I|^{(-\beta\tau/d)+1}\right)^{1/\tau} \leq C_0 |f|_{B^\beta}.$$

This assumption is valid for the wavelets described in the introduction provided $p > 1$. If φ has suitable size and smoothness then a fixed dilate of φ satisfies (2.1) and (2.2) (see [21, Section 4]). In fact, the stronger result (which we do not need in this section) that the left side of (2.2) is equivalent to the right is valid. In the case of the wavelet of Section 4, the results of this section apply for all $0 < p < \infty$.

We shall also need an assumption about the size of φ. For $0 < r < \infty$, let

$$M_r(g)(x) := \sup_{I \ni x} \left(\frac{1}{|I|} \int_I |g|^r du\right)^{1/r}, \qquad x \in \mathbb{R}^d \tag{2.3}$$

be the modified Hardy-Littlewood maximal function. It is well known that M_r boundedly maps L_q into itself for all $q > r$. There is a stronger result due to Fefferman and Stein [18] for sequences of functions. If $(g_\nu)_{\nu \in \Lambda}$ is a sequence of functions indexed on some set Λ, and if $r < \min(q, s)$, then

$$\left\| \left(\sum_{\nu \in \Lambda} [M_r(g_\nu)(x)]^s\right)^{1/s} \right\|_{L_q(\mathbb{R}^d)} \leq C \left\| \left(\sum_{\nu \in \Lambda} |g_\nu(x)|^s\right)^{1/s} \right\|_{L_q(\mathbb{R}^d)} \tag{2.4}$$

where C depends only on q, s, r, and d.

We shall assume that for some $0 < r < \min(1, p)$ and $\Omega := [0, 1]^d$, we have

$$|\varphi(x)| \le CM_r(\chi_\Omega)(x), \qquad x \in \mathbb{R}^d \tag{2.5}$$

with C a constant. It follows from (2.5) that

$$|\varphi_I(x)| \le CM_r(\chi_I, x) \tag{2.6}$$

for all $I \in \mathcal{D}$.

Since φ is bounded, the condition (2.5) is equivalent to

$$|\varphi_I(x)| \le C|x|^{-d/r}. \tag{2.7}$$

In the case φ has compact support, (2.5) is automatically satisfied.

Before stating our main result of this section, we make some simple observations about $|f|_\beta$ of (2.2). We fix β and let

$$F(x) := \left\{\sum_{I\in\mathcal{D}} (|I|^{-\beta/d}|a_I(f)|\chi_I(x))^\tau\right\}^{1/\tau}. \tag{2.8}$$

Then, we have

$$\|F\|_{L_\tau(\mathbb{R}^d)} \le C|f|_\beta \le C_0|f|_{B^\beta} \tag{2.9}$$

for some constants $C_0 \ge C \ge 1$.

THEOREM 2.1. *If for* $\beta > 0$, φ *satisfies* (2.1), (2.2), *and* (2.5), *then*

$$\sigma_n(f)_p \le C(p, \varphi)n^{-\beta/d}|f|_{B^\beta}. \tag{2.10}$$

Proof. We first note that we can assume that the constant of (2.5) is 1 since we can always multiply φ by a constant. We can also assume that $|f|_{B^\alpha} = 1/C_0$ where C_0 is the constant in (2.9). With these assumptions, $|f|_\beta \le 1$ and hence there are at most n coefficients for which $|a_I(f)|^\tau|I|^{\tau/p} \ge 1/n$. We denote the set of such I by Λ and let $S = \sum_{I\in\Lambda} a_I(f)\varphi_I$. Then, $S \in \Sigma_n$ and the error $E := f - S = \sum_{I\in\mathcal{D}} a_I^+\varphi_I$ where $a_I^+ := a_I(f)$, $I \notin \Lambda$, and $a_I^+ := 0$, $I \in \Lambda$. We also have $|a_I^+| \le \epsilon|I|^{-1/p}$ with $\epsilon := n^{-1/\tau}$.

From (2.5) and the Fefferman-Stein inequality, we have for $\tilde{E}(x) := \sum_{I\in\mathcal{D}} |a_I^+|\chi_I(x)$,

$$\|E\|_{L_p(\mathbb{R}^d)} \le C\|\tilde{E}\|_{L_p(\mathbb{R}^d)}. \tag{2.11}$$

Therefore, it will be enough to bound the right side of (2.11). For this purpose, we fix $t > 0$ and estimate the distribution function of $\tilde{E}$:

$$\mu(\tilde{E}, t) := |\{x : |\tilde{E}(x)| > t\}|. \quad \text{Clearly} \tag{2.12}$$

$$|\tilde{E}(x)| \le \sum_{\ell(I)\ge 2^N} |a_I^+|\chi_I(x) + \sum_{\ell(I)<2^N} |a_I^+|\chi_I(x) =: \Sigma_1(x) + \Sigma_2(x) \tag{2.13}$$

where N will be chosen in a moment. (We do not indicate the dependence of Σ_1 and Σ_2 on N.) We have

$$\Sigma_1(x) \le \epsilon \sum_{\ell(I)\ge 2^N} |I|^{-1/p}\chi_I(x) \le C_1\epsilon 2^{-Nd/p}, \tag{2.14}$$

where we have used the fact that $\sum_{\ell(I)\ge 2^N} |I|^{-1/p}\chi_I(x) \le C_1 2^{-Nd/p}$ because a point x is in exactly one dyadic cube of any given sidelength.

Now, we choose N as the smallest integer such that

$$C_1\epsilon 2^{-Nd/p} \le t/2. \quad \text{Then,} \tag{2.15}$$

$$\mu(\tilde{E}, t) \le \mu(\Sigma_2, t/2). \tag{2.16}$$

Moreover, from the definition of F, we have for each x

$$\begin{aligned} \Sigma_2(x) &\le \sum_{\ell(I)<2^N} |I|^{\beta/d}\chi_I(x)|I|^{-\beta/d}|a_I^+|\chi_I(x) \\ &\le \left\{\sup_{\ell(I)<2^N} |a_I^+||I|^{-\beta/d}\chi_I(x)\right\}\left\{\sum_{\ell(I)<2^N} |I|^{\beta/d}\chi_I(x)\right\} \\ &\le C\left\{\sum_{I\in\mathcal{D}} (|I|^{-\beta/d}|a_I(f)|\chi_I(x))^\tau\right\}^{1/\tau} 2^{N\beta} = C2^{N\beta}F(x). \end{aligned} \tag{2.17}$$

Using this with the definition N (see (2.15)) gives

$$\Sigma_2(x) \le C\epsilon^{p\beta/d}t^{-p\beta/d}F(x),$$

from which it follows that

$$\mu(\tilde{E}, t) \le \mu(F, C_2\epsilon^{-p\beta/d} t^{(p\beta/d)+1}). \tag{2.18}$$

We use this with the change of variables $u := C_2\epsilon^{-p\beta/d} t^{(p\beta/d)+1}$ and obtain

$$\begin{aligned} \|\tilde{E}\|_p^p &= \int_0^\infty \mu(\tilde{E}, t) t^p \frac{dt}{t} \le \int_0^\infty \mu(F, C_2\epsilon^{-p\beta/d} t^{(p\beta/d)+1}) t^p \frac{dt}{t} \\ &= C_2^{-\tau} \int_0^\infty \mu(F, u)\epsilon^{p\beta\tau/d} u^\tau \frac{du}{u} = C_2^{-\tau} \|F\|_{L_\tau(\mathbb{R}^d)}^\tau \epsilon^{p\beta\tau/d} \\ &\le C_2^{-\tau}\epsilon^{p\beta\tau/d} \end{aligned} \tag{2.19}$$

because $\|F\|_{L_\tau(\mathbb{R}^d)} \le C_0|f|_{B^\beta} \le 1$ by (2.9). Taking a p-th root in (2.19) proves the theorem. ■

If we take for φ a B-spline or box spline (see Section 7 for definitions) of degree r, the elements in Σ_n are piecewise polynomials of degree $\le r$ with at most Cn pieces and Theorem 2.1 applies. Results of this type are summarized in Section 7.

We can also apply Theorem 2.1 to rational functions. In Frazier and Jawerth [21] it is shown that if ψ satisfies (2.7) (and some minimal smoothness condition) then some fixed dilate φ of ψ satisfies (2.1) and (2.2) for $p > 1$. When $p \le 1$, (2.1) and (2.2) hold for approximation in H_p and the Besov spaces defined via H_τ smoothness provided that in addition φ has mean value zero. The same proof as above would give the direct estimate (2.10) for these Besov spaces. For example, we can take φ as a suitable fixed dilate of $\psi(x) := (1 + |x|^2)^{-k}$. From Theorem 2.1, we obtain

Corollary 2.2. *Let φ be a rational function as described above. Then for the space Σ_n defined by this φ, we have*

$$\sigma_n(f)_p \le C(p, \varphi) n^{-\beta/d} |f|_{B^\beta}. \tag{2.20}$$

If φ has degree m, then the space Σ_n in the corollary is contained in $\mathcal{R}_{nm}$, the space of rational functions of degree $\le mn$, however it is actually a much thinner space than $\mathcal{R}_{nm}$ because the latter space is a manifold of dimension $\approx n^d$ while the former only has dimension $\approx n$.

3. An inverse theorem for orthogonal wavelets. We shall prove in this section a simple inverse theorem for the case of orthogonal wavelets. This result combined with Theorem 2.1 then gives the characterization (1.3) for orthogonal wavelets (with some restriction on α depending on the smoothness of φ). When φ is an orthogonal wavelet, then each f has a unique representation

$$f = \sum_{I \in \mathcal{D}} a_I(f)\varphi_I \tag{3.1}$$

where $a_I(f) = \lambda \int_{\mathbb{R}^d} |I|^{-1}\varphi_I(x)f(x)dx$ and $\lambda^{-1} = \int_{\mathbb{R}^d} |\varphi(x)|^2 dx$. It is well known (see for example [21]) that under suitable smoothness and size conditions on φ, we have a discrete Littlewood-Paley theory for the expansion (3.1) and, in particular,

$$\|f\|_{L_p(\mathbb{R}^d)} \approx \left\| \left(\sum_{I \in \mathcal{D}} |a_I(f)|^2 \chi_I(x) \right)^{1/2} \right\|_{L_p(\mathbb{R}^d)} \tag{3.2}$$

provided $1 < p < \infty$ with constants of equivalency depending on p. If $p \le 1$ the right side of (3.2) is equivalent to the H_p norm of f.

In this section, we shall also assume that

$$|f|_{B^\beta} \le C\left(\sum_{I \in \mathcal{D}} |a_I(f)|^\tau |I|^{\tau/p} \right)^{1/\tau} \tag{3.3}$$

$$= C\left\| \left(\sum_{I \in \mathcal{D}} (|I|^{-\beta/d}|a_I(f)|\chi_I)^\tau \right)^{1/\tau} \right\|_{L_\tau(\mathbb{R}^d)}.$$

We have discussed in the previous section the connection between $|f|_{B^\beta}$ and the sum in (3.3). Here, we only need further remark that (3.3) is valid in the case of orthogonal wavelets provided φ has suitable size and smoothness; roughly speaking, φ should have slightly better smoothness and decay than membership in B^β. We refer the reader to [21] for a discussion of sufficient conditions on φ so that both (3.2) and (3.3) hold.

THEOREM 3.1. *If φ satisfies (3.2) and (3.3), then for any $S \in \Sigma_n$ we have the inverse estimate*

$$|S|_{B^\beta} \le Cn^{\beta/d}\|S\|_{L_p(\mathbb{R}^d)}, \qquad 1 < p < \infty,\ n = 1, 2, \ldots. \tag{3.4}$$

Remark. The proof below also shows that (3.4) is valid for $p \leq 1$ with L_p replaced by H_p and the Besov spaces B^β replaced by $B^\beta_\tau(H_\tau)$.

Proof. We can write $S = \Sigma_{I\in\Lambda}\, a_I(S)\varphi_I$ where Λ is a set of at most n dyadic cubes. We order the cubes of Λ in order of nondecreasing size: $I_1, \ldots, I_m$, where $m \leq n$. We let $E_j := I_j \backslash \cup\{I_\nu : \nu < j\}$. Then the sets E_j, $j = 1, \ldots, m$ are disjoint. From (3.3), we find

$$(3.5)\qquad |S|^\tau_{B^\beta} \leq C \int_{\mathbb{R}^d} \Big(\sum_{I\in\Lambda} (|I|^{-\beta/d} |a_I(S)| \chi_I(x))^\tau \Big) dx$$

$$= C \sum_{i=1}^m \int_{E_i} \Big(\sum_{I\in\Lambda} (|I|^{-\beta/d} |a_I(S)| \chi_I(x))^\tau \Big) dx$$

$$\leq C \sum_{i=1}^m \int_{E_i} \Big(\sum_{I\in\Lambda} (|I|^{-\beta/d} \chi_I(x))^\tau \Big) \sup_{i\in\Lambda} (|a_I(S)| \chi_I(x))^\tau dx$$

$$\leq C \sum_{i=1}^m |I_i|^{-\beta\tau/d} \int_{E_i} \sup_{I\in\Lambda} (|a_I(S)| \chi_I(x))^\tau dx$$

where we have used the fact that $\Sigma_{I\in\Lambda} |I|^{-\beta\tau/d} \chi_I(x) \leq C |I_i|^{-\beta\tau/d}$ for $x \in E_i$ because $\chi_I(x) = 0$ for $\ell(I) < \ell(I_i)$ and each x appears in at most one I at a given dyadic level.

Now,

$$(3.6)\qquad \sup_{I\in\Lambda} |a_I(S)| \chi_I(x) \leq \Big(\sum_{I\in\Lambda} |a_I(S)|^2 \chi_I(x) \Big)^{1/2} =: \tilde{S}(x).$$

Using this in (3.5) gives

$$(3.7)\qquad |S|^\tau_{B^\beta} \leq C \sum_{i=1}^m |I_i|^{-\beta\tau/d} \int_{E_i} |(\tilde{S}(x)|^\tau dx$$

$$\leq C \sum_{i=1}^m |I_i|^{-\beta\tau/d} |E_i|^{1-\tau/p} \Big(\int_{E_i} |\tilde{S}(x)|^p dx \Big)^{\tau/p}$$

$$= C \sum_{i=1}^m \Big(\int_{E_i} |\tilde{S}(x)|^p dx \Big)^{\tau/p}$$

$$\le Cm^{1-\tau/p}\left(\sum_{i=1}^{m} \int_{E_i} |\tilde{S}(x)|^p dx\right)^{\tau/p} \le Cn^{\beta\tau/d}\|\tilde{S}\|^{\tau}_{L_p(\mathbb{R}^d)}$$

where we have used the fact that $|E_i| \le |I_i|$ and $1 - \tau/p = \beta\tau/d$. Since by (3.2), $\|\tilde{S}\|_{L_p(\mathbb{R}^d)} \le C\|S\|_{L_p(\mathbb{R}^d)}$, we have proven (3.4). ■

For example, we can take for φ the Haar functions defined earlier, in which case the characterization (1.3) holds for all $1 < p < \infty$ and all $0 < \alpha < 1/p$.

4. Besov spaces and wavelet decompositions. The remainder of this paper will be concerned with a class of wavelets which contain as a special case various spline functions. In this section, we shall introduce the basic properties of these wavelets including the decomposition (1.1). We shall then prove in the next two sections the direct and inverse theorems for these wavelets.

We shall assume from here on out that φ is supported on a cube $\Omega' := [-\rho, \rho]^d$ with ρ a positive integer. In what follows, for a dyadic cube I which corresponds to $j2^{-k}$, we let $I' := j2^{-k} + 2^{-k}\Omega'$. Then φ_I is supported on I'. Our first main assumption about the function φ is that each given φ_I can be written as a finite linear combination of the functions φ_J at the next dyadic level, that is, $\ell(J) = \ell(I)/2$, where $\ell(Q)$ denotes the sidelength of a cube Q. What is the same thing, we should have

$$\varphi(x) = \sum_j c_j \varphi(2x - j) \tag{4.1}$$

where the sum is taken over a finite number of $j \in \mathbb{Z}^d$. Functions φ which satisfy (4.1) for some constants c_j are the basis for the study of subdivision algorithms of Computer Aided Design (see for example the recent article of Cavaretta, Dahmen, and Micchelli [6]) and they also form the basic assumption of Daubechies [12] in her development of orthogonal wavelets. However, we should emphasize that φ itself cannot be an orthogonal wavelet since its translated dilates are not linearly independent.

Now, for $k \in \mathbb{Z}$, let $\mathbb{S}_k$ be the span of the functions φ_I, $I \in \mathfrak{D}_k$. From the support of the φ_I, it follows that any sum $\sum_{I\in\mathfrak{D}_k} a_I\varphi_I$ will converge uniformly on compact sets. We are interested in when the spaces $\mathbb{S}_k$ can be used for approximation. This will be the case provided φ has some additional properties. Our second main assumption about φ is that for some positive integer r, we have

$$(4.2)\qquad \begin{aligned}&\text{(i) } \hat{\varphi}(0) = 1, \quad \hat{\varphi}(2\pi j) = 0, \quad j \in \mathbb{Z}^d, \quad j \neq 0,\\ &\text{(ii) } D^\nu \hat{\varphi}(2\pi j) = 0, \quad j \in \mathbb{Z}^d, \quad j \neq 0, \quad |\nu| < r.\end{aligned}$$

Conditions of this type were first introduced by Schoenberg [34] and later studied by Strang and Fix [35] and de Boor and Jia [4] among others. They imply that the spaces $\mathbb{S}_k$ contain the space $\mathbb{P}_r$ of polynomials of total degree $< r$. In particular from the Poisson summation formula and (4.2)(i), we see that the φ_I are a partition of unity:

$$\sum_{I \in \mathcal{D}_k} \varphi_I \equiv 1, \qquad k \in \mathbb{Z}.$$

Our last main assumption about φ concerns linear independence. Let Q be a cube in $\mathbb{R}^d$ and let Λ_Q denote the set of all $j \in \mathbb{Z}^d$ for which $\varphi(\cdot - j)$ is not identically zero on Q. We shall assume that

$$(4.3)\qquad \text{for all } Q \in \mathcal{D}, \text{ the functions } \varphi(\cdot - j), \; j \in \Lambda_Q \text{ are linearly independent over } Q.$$

It follows from this that the functions $\varphi(\cdot - j), j \in \mathbb{Z}^d$, are globally linearly independent and hence are a basis for $\mathbb{S}_0$.

In what follows in this section, we shall mention some well known results concerning the space $\mathbb{S}_k$ and we extend these in certain directions. Let c_j be the dual functionals to the basis $\varphi(\cdot - j)$ of $\mathbb{S}_0$: $c_j(\varphi(\cdot - i)) = \delta_{i,j}$ for the Kronecker delta $\delta := (\delta_{i,j})$. Then each $S \in \mathbb{S}_0$ has the representation

$$(4.4)\qquad S = \sum_{j \in \mathbb{Z}^d} c_j(S)\varphi(\cdot - j).$$

If we consider together with S the functions $S(\cdot + j)$, we find that $c_j(S) = c_0(S(\cdot + j))$, $j \in \mathbb{Z}^d$. Then by translation and dilation we obtain for any $S \in \mathbb{S}_k$, $k \in \mathbb{Z}$,

$$(4.5)\qquad S = \sum_{J \in \mathcal{D}_k} c_J(S)\varphi_J$$

where c_0 is a fixed linear functional on $\mathbb{S}_0$ and $c_J(S) := c_0(S(2^{-k}(\cdot + j))$ when J corresponds to $j2^{-k}$.

We shall need some estimates for the linear functionals c_I. If Q is a dyadic subcube of Ω' on which φ does not vanish identically, we shall call Q a **support cube** of φ; similarly, Q is a support cube of φ_I if $Q \subset I'$ and φ_I does not vanish identically on Q. Given such a cube, any two quasinorms are equivalent on the finite dimensional space

$$\mathbb{S}_0(Q) := \mathbb{S}_0 \upharpoonright_Q := \operatorname{span}\{\varphi(\cdot - j) : j \in \Lambda_Q\} \upharpoonright_Q.$$

In particular, let $\|S\|_0 := \max_{j \in \Lambda_Q} |c_j(S)|$ and $\|S\|_1 := (\frac{1}{|Q|} \int_Q |S|^p dx)^{1/p}$ where $0 < p \le \infty$ is fixed. Then,

$$\frac{1}{C} \|\cdot\|_1 \le \|\cdot\|_0 \le C\|\cdot\|_1. \tag{4.6}$$

It follows that if Q is a support cube of φ, then

$$|c_0(S)| \le C\left(\frac{1}{|Q|} \int_Q |S(x)|^p dx\right)^{1/p}, \qquad S \in \mathbb{S}_0,$$

for some constant C depending only on Q, φ, and p. Since $(\frac{1}{|Q|} \int_Q |S(x)|^p dx)^{1/p} \le (\frac{1}{|Q|} \int_Q |S(x)|^r dx)^{1/r}$ for $p \le r$, the constant in (4.6) depends only on p when p is small.

By translation and dilation we obtain that for any $I \in \mathcal{D}_k$ and Q which is a dyadic subcube of I on which φ_I does not vanish identically, we have

$$|c_I(S)| \le C\left(\frac{1}{|Q|} \int_Q |S(x)|^p dx\right)^{1/p}, \qquad S \in \mathbb{S}_k, \tag{4.7}$$

where the constant C of (4.7) depends only on $|Q|/|I|$ (because there are only a finite number of subcubes of I of sidelength greater than any fixed number), φ, and p if p is small.

The inequality (4.7) gives the following important information about the representations

$$\varphi_I = \sum_{J \in \mathcal{D}_k} c_J(\varphi_I)\varphi_J, \tag{4.8}$$

which as we know from (4.1) holds for any k and any I provided $\ell(I) \geq 2^{-k}$:

$$\text{(4.9)} \qquad \text{for any } J \text{ appearing in (4.8) supp } \varphi_J \subset \text{supp } \varphi_I.$$

(Here and later supp g is the smallest closed set outside of which g vanishes identically.) Indeed, if φ_I vanishes identically on an open set containing the point x_0, then there exists a closed dyadic cube Q containing x_0 on which φ vanishes identically. Now if J appears in (4.8) and φ_J does not vanish identically on Q, then by (4.7), $|c_J(\varphi_I)| \leq C \frac{1}{|Q|} \int_Q |\varphi_I| dx = 0$.

We need an improvement of (4.7) in the special case that Q is a cube of the same sidelength as I. We return to the case of φ and Q a dyadic cube of sidelength one which is a support cube for φ. Let C_p be a constant such that (4.6) is valid with $\|\cdot\|_1$ the L_p norm. If $E \subset Q$ has measure $\leq \frac{1}{2C_p^p C_\infty^p}$ then from (4.6) for any $S \in \mathbb{S}_0$, we have

$$\int_E |S(x)|^p dx \leq \|S\|^p_{L_\infty(Q)} |E| \leq C_\infty^p \|S\|_0^p |E|$$

$$\leq C_\infty^p C_p^p |E| \int_Q |S(x)|^p dx \leq \frac{1}{2} \int_Q |S(x)|^p dx.$$

We therefore obtain by dilation and translation from this and (4.7), the following inequality: there is a constant $0 < \delta < 1$ which depends only on φ, d, and p such that for any I and any support cube Q of φ_I with $\ell(Q) = \ell(I)$ and any set $E \subset Q$ of measure $\leq \delta |Q|$, we have

$$\text{(4.10)} \qquad |c_I(S)| \leq C\left(\frac{1}{|I|} \int_{Q \setminus E} |S(x)|^p dx\right)^{1/p}, \qquad S \in \mathbb{S}_k$$

where the constant C depends only on φ and p, if p is small.

Here is another simple application of (4.7).

LEMMA 4.1. *For $S \in \mathbb{S}_k$, we have*

$$\text{(4.11)} \qquad \|S\|_{L_p(\mathbb{R}^d)} \approx \begin{cases} \left(\sum_{J \in \mathfrak{D}_k} |c_J(S)|^p |J|\right)^{1/p}, & 0 < p \leq \infty \\ \sup_{J \in \mathfrak{D}_k} |c_J(S)|, & p - \infty \end{cases}$$

with constants of equivalency depending at most on φ *and* p *when* p *is small.*

Proof. We assume $0 < p < \infty$ (the case $p = \infty$ is handled similarly). That the left side of (4.11) does not exceed a multiple of the right follows by integrating the inequality $(\Sigma_{J\in\mathcal{D}_k} |c_J(S)||\varphi_J|)^p \le C\, \Sigma_{J\in\mathcal{D}_k} |c_J(S)|^p|\varphi_J|^p$ which holds because at most C terms of the left sum are nonzero at any point $x \in \mathbb{R}^d$. Here C depends only on φ. That the right side of (4.11) does not exceed the left follows from (4.7) with $Q = I$. ■

We next develop quasi-interpolant operators for approximation by the elements of $\mathbb{S}_k$. We take $I = Q = \Omega$ in (4.7); then we can extend the functional c_Ω to a functional γ_Ω defined on $L_1(\Omega)$ and retain this inequality. Thus, by dilation and translation we obtain functionals γ_J which extend the c_J (i.e. $\gamma_J(S) = c_J(S)$, $J \in \mathcal{D}_k$, $S \in \mathbb{S}_k$) and we have

$$(4.12) \qquad |\gamma_J(f)| \le C \frac{1}{|J|}\int_J |f(x)|dx, \qquad J \in \mathcal{D}_k$$

for each $f \in L_1(\text{loc})$. For such f, we define

$$(4.13) \qquad Q_k(f) := \sum_{J\in\mathcal{D}_k} \gamma_J(f)\varphi_J.$$

The operators Q_k are called quasi-interpolants and are known to have good approximation properties. As mentioned earlier, the condition (4.2) implies that $\mathbb{S}_k$ contains all polynomials of degree $< r$ (see e.g. [4]). Using this, it is easy to prove (see e.g. [35]) that for any f in the Sobolev space $W_p^r(\mathbb{R}^d)$,

$$(4.14) \qquad \|f - Q_k(f)\|_{L_p(\mathbb{R}^d)} \le C2^{-kr}|f|_{W_p^r(\mathbb{R}^d)}, \qquad k \in \mathbb{Z}.$$

We want to mention some refinements of (4.14) which can be obtained in the same way one obtains results for spline functions from their quasi-interpolants (see [16]). We first discuss estimates in terms of the modulus of smoothness of f. Let $\mathcal{I}$ denote the identity operator, $\tau(h)$ the translation operator $(\tau(h)(f, x) := f(x + h))$ and $\Delta_h^r := (\tau(h) - \mathcal{I})^r$, $r = 1, 2, \ldots$ be the difference operators. If D is a domain in $\mathbb{R}^d$, the modulus of smoothness of order r of a function $f \in L_p(D)$, $0 <$

$p \leq \infty$, is

$$\omega_r(f, t)_p := \sup_{|h| \leq t} \|\Delta_h^r(f, \cdot)\|_p(D(rh)) \tag{4.15}$$

where $D(h)$ is the set of x such that the line segment $[x, x + h]$ is contained in D. In most of what follows $D = \mathbb{R}^d$.

For $f \in L_p(\mathbb{R}^d)$, $0 < p \leq \infty$, we define the error of approximation of f by the elements of $\mathbb{S}_k$:

$$s_k(f)_p := \inf_{S \in \mathbb{S}_k} \|f - S\|_{L_p(\mathbb{R}^d)}, \qquad k \in \mathbb{Z}. \tag{4.16}$$

It is well known that for $1 \leq p \leq \infty$ the K-functional for the spaces (L_p, W_p^r) on the domain $\mathbb{R}^d$ is equivalent to $\omega_r(f, t)_p$. It follows therefore from (4.14) (see e.g. [16]) that

$$s_k(f)_p \leq \|f - Q_k(f)\|_p \leq C\omega_r(f, 2^{-k})_p, \qquad k \in \mathbb{Z}, \qquad 1 \leq p \leq \infty \tag{4.17}$$

where C in this and the next two inequalities depend on φ, r, and p if p is small.

It is important for us to extend this last estimate to the case $p < 1$. This can be done in exactly the same way that this was accomplished for B-splines in [16]. Namely, we use the fact that for a fixed value of $\gamma > 0$ (which in its application is chosen smaller than all p under consideration), for each $f \in L_\gamma(\text{loc})$ there is a piecewise polynomial R_k of degree $< r$ on the partition $\mathfrak{D}_k$ satisfying

$$\|f - R_k(f)\|_p \leq C\omega_r(f, 2^{-k})_p, \qquad k \in \mathbb{Z}, \qquad \gamma \leq p \leq \infty. \tag{4.18}$$

One way to construct such R_k is to define it on I to be a best $L_\gamma(I)$ approximation to f (see [16]). However, in what follows we only require property (4.18) for R_k.

It follows that $S_k := Q_k(R_k(f))$ is well defined for each $f \in L_\gamma$ and is in $\mathbb{S}_k$ and provides the approximation error

$$s_k(f)_p \leq \|f - S_k\|_p \leq C\omega_r(f, 2^{-k})_p, \qquad k \in \mathbb{Z}, \qquad \gamma \leq p \leq \infty. \tag{4.19}$$

We refer the reader to [16, Theorem 4.5] for the method of proof for (4.19), but mention that it follows from the fact that Q_k is a projector onto $\mathbb{S}_k$ and that $\mathbb{S}_k$ locally contains all polynomials of degree $< r$.

We also have an inverse inequality to (4.19). To formulate this inequality, we shall assume that

$$\varphi \in W^s_\infty(\mathbb{R}^d). \tag{4.20}$$

Later, in Section 7, we shall note that it is also possible to improve these inverse estimates slightly in the case φ is a spline function. It follows that $|\varphi_I|_{W^s_\infty(\mathbb{R}^d)} \leq C2^{ks}$, if $I \in \mathcal{D}_k$. Therefore

$$|\Delta^s_h(\varphi_I, x)| \leq C \min(1, 2^{ks}|h|^s), \qquad I \in \mathcal{D}_k \tag{4.21}$$

with C depending only on φ.

Now let $S \in \mathbb{S}_k$, $S = \sum_{J \in \mathcal{D}_k} c_J(S)\varphi_J$. Since $\Delta^s_h(S, x) = \sum_{J \in \mathcal{D}_k} c_J(S)\Delta^s_h(\varphi_J, x)$, we obtain for any $0 < p < \infty$

$$|\Delta^s_h(S, x)|^p \leq C^p \sum_{J \in \mathcal{D}_k} |c_J(S)|^p |\Delta^s_h(\varphi_J, x)|^p$$

where we used the fact that for $x \in \mathbb{R}^d$, $\Delta^s_h(\varphi_J, x) \neq 0$ for at most C of the cubes J with C depending only on φ. Since $\Delta^s_h(\varphi_J, \cdot)$ is supported on a set of measure $\leq C|J|$, we obtain from (4.21) and Lemma 4.1 that for each $S \in \mathbb{S}_k$ and $0 < p < \infty$,

$$\begin{aligned} \omega_s(S, t)_p &\leq C \min(1, 2^{ks}t^s)\left(\sum_{J \in \mathcal{D}_k} |c_J(S)|^p |J|\right)^{1/p} \\ &\leq C \min(1, 2^{ks}t^s)\|S\|_{L_p(\mathbb{R}^d)} \end{aligned} \tag{4.22}$$

with C depending only on φ and p if p is small. The estimate (4.22) also holds for $p = \infty$ with a similar proof.

It will be useful to observe that $s_k(f)_p \to \|f\|_{L_p(\mathbb{R}^d)}$ as $k \to -\infty$ provided $p < \infty$. The same is true for $p = \infty$ if f has compact support. To see this, we note first that $s_k(f)_p \leq \|f\|_{L_p(\mathbb{R}^d)}$. Now, if $S_k \in \mathbb{S}_k$ satisfy $\|S_k\|_{L_p(\mathbb{R}^d)} \leq C\|f\|_{L_p(\mathbb{R}^d)}$, then by (4.7) (with $Q = I \in \mathcal{D}_k$), $|c_I(S_k)| \leq C2^{kd/p}\|S_k\|_{L_p(\mathbb{R}^d)}$ and therefore $S_k \to 0$, $k \to -\infty$, uniformly on every compact set K. Now, choose K so that $\|f\|_{L_p(\mathbb{R}^d \setminus K)} < \epsilon$. Then, if $p \geq 1$,

$\|f - S_k\|_{L_p(K)} \geq \|f\|_{L_p(K)} - \|S_k\|_{L_p(K)} \geq \|f\|_{L_p(K)} - \epsilon$, provided k is large enough and negative. A similar argument applies when $p < 1$. This proves our assertion.

Now suppose that for each $k \in \mathbb{Z}$, $S_k \in \mathbb{S}_k$ satisfies $\|f - S_k\|_p \leq 2s_k(f)_p$. By the remarks of the above paragraph, we can assume $S_k \equiv 0$ provided k is large enough negative. For any k_0, we can write $f = f - S_k + S_{k_0} + \sum_{j=k_0}^{k-1} (S_{j+1} - S_j)$. Then, $S_{j+1} - S_j \in \mathbb{S}_{j+1}$ and $\|S_{j+1} - S_j\|_{L_p(\mathbb{R}^d)} \leq C[s_j(f)_p + s_{j+1}(f)_p]$ and $\|S_{k_0}\|_{L_p(\mathbb{R}^d)} \leq C[s_{k_0}(f)_p + \|f\|_{L_p(\mathbb{R}^d)}]$. Therefore, from (4.22), we obtain

$$\omega_s(f, 2^{-k})_p \leq C2^{-ks}\left(2^{k_0 s\mu}\|f\|^{\mu}_{L_p(\mathbb{R}^d)} + \sum_{j=k_0}^{k} [2^{js}s_j(f)_p]^{\mu}\right)^{1/\mu}$$

where $\mu := \min(1, p)$ and C in this and the next inequality depends only on s, φ and p if p is small. Here, we have used the subadditivity of $\|\cdot\|_p$ for $1 \leq p \leq \infty$ and we used the subadditivity of $\|\cdot\|_p^p$ when $0 < p < 1$. We can allow $k_0 \to -\infty$ and obtain

$$\omega_s(f, 2^{-k})_p \leq C2^{-ks}\left(\sum_{j=-\infty}^{k} [2^{jd}s_j(f)_p]^{\mu}\right)^{1/\mu}. \tag{4.23}$$

Inequalities (4.19) and (4.23) are companion inequalities which serve to characterize the class of functions f which have a given order of approximation by the elements of $\mathbb{S}_k$ in terms of the smoothness of f. To formulate these results and also for our later use, we introduce the Besov spaces.

A Besov space is a collection of functions f with common smoothness. If $0 < \alpha < r$ and $0 < q, p \leq \infty$, the Besov space $B_q^{\alpha}(L_p(D))$, with D a domain in $\mathbb{R}^d$, consists of all functions f such that

$$|f|_{B_q^{\alpha}(L_p(D))} := \left(\int_0^{\infty} [t^{-\alpha}\omega_r(f, t)_p]^q dt/t\right)^{1/q} < \infty \tag{4.24}$$

with the usual change to sup when $q = \infty$. It follows that (4.24) is a semi-(quasi)norm for $B_q^{\alpha}(L_p(D))$. If we add $\|f\|_{L_p(D)}$ to (4.24), we obtain the (quasi)norm for $B_q^{\alpha}(L_p(D))$. It is well known that different values of $r > \alpha$ give equivalent norms (see e.g. [16]).

Let us next mention some properties of the Besov spaces and the real method of interpolation. These were proved in [16] for the case D a cube in $\mathbb{R}^d$. An almost identical proof applies for the case $D = \mathbb{R}^d$. For any $0 < p, q, s \leq \infty$ and $0 < \alpha < \beta$, we have

$$(L_p, B_q^\beta(L_p))_{\alpha/\beta,s} = B_s^\alpha(L_p). \tag{4.25}$$

Another important result on interpolation applies to the following special Besov spaces. Let $0 < p \leq \infty$ be fixed for the moment and define for each $\alpha > 0$ the Besov space $B^\alpha := B_\tau^\alpha(L_\tau)$ where $\tau := \tau(\alpha, p) := (\alpha/d + 1/p)^{-1}$. Of course the spaces B^α also depend on p but in our applications p will be fixed and clearly understood. We note for later use that if $\tau(\alpha) \leq 1$, then $|\cdot|_{B^\alpha}^\tau$ is subadditive. This follows from the subadditivity of $\omega_r(\cdot, t)_\tau^\tau$.

The spaces B^α are an interpolation family (see [16]); namely, if $0 < \alpha < \beta$ and $0 < p < \infty$, then

$$(L_p, B^\beta)_{\alpha/\beta,\tau} = B^\alpha, \qquad \text{if} \quad \tau = \tau(\alpha) := (\alpha/d + 1/p)^{-1}. \tag{4.26}$$

We return to the approximation by the elements of $\mathbb{S}_k$. It follows from (4.19) and (4.23) that the following are equivalent (with constants of equivalency independent of f) for any $0 < p, q \leq \infty$ and $0 < \alpha < \min(s, r)$:

$$\text{(4.27)}\qquad \begin{aligned} &\text{(i)} \ |f|_{B_q^\alpha(L_p(\mathbb{R}^d))} < \infty \\ &\text{(ii)} \ \left(\sum_{k\in\mathbb{Z}} [2^{k\alpha}\omega_r(f, 2^{-k})_p]^q\right)^{1/q} < \infty \\ &\text{(iii)} \ \left(\sum_{k\in\mathbb{Z}} [2^{k\alpha}s_k(f)_p]^q\right)^{1/q} < \infty \\ &\text{(iv)} \ \left(\sum_{k\in\mathbb{Z}} [2^{k\alpha}\|S_{k+1} - S_k\|_p]^q\right)^{1/q} < \infty \end{aligned}$$

with the usual change to the ℓ_∞ norm when $q = \infty$. In (iv), $S_k := Q_k(R_k(f))$ where R_k is any piecewise polynomial satisfying (4.18) and the constants of equivalency in (4.27) depends on the constant of (4.18). For what follows this constant is considered to be fixed. The sum (ii) is simply a discretization of $|f|_{B_q^\alpha(\mathbb{R}^d)}$, and the equivalence of (ii) and (iii) follows from (4.19) and (4.23) by applying discrete versions of the Hardy inequalities. See [16] for a proof of this as well as the equivalence of (iii) and (iv).

Now if $f \in L_p(\mathbb{R}^d)$, $0 < p < \infty$, then with convergence in L_p, we have for any k_0:

$$f = S_{k_0} + \sum_{k=k_0}^{\infty} (S_{k+1} - S_k) = \sum_{k=k_0}^{\infty} \sum_{I \in \mathfrak{D}_k} a_I \varphi_I \tag{4.28}$$

where we have used (4.1) to write $S_k - S_{k-1}$ as a linear combination of the φ_I with $I \in \mathfrak{D}_k$. We can also let $k_0 \to -\infty$ and obtain

$$f = \sum_{k \in \mathbb{Z}} (S_{k+1} - S_k) = \sum_{k \in \mathbb{Z}} \sum_{I \in \mathfrak{D}_k} a_I \varphi_I = \sum_{I \in \mathfrak{D}} a_I \varphi_I, \tag{4.29}$$

where we have used our previous remark that we can assume $\|S_j\|_{L_p(\mathbb{R}^d)} = 0$ if j is large enough negative. Thus, (4.29) is the wavelet decomposition (1.1) for the function φ. We emphasize that this decomposition is not unique. In fact, in our construction we had freedom in the choice of R_k but of course there are also many other possible ways to create such a decomposition.

We wish to relate the size of the coefficients a_I in (4.29) to the norm of f in Besov spaces and in L_p spaces. As noted earlier, such results have been established (under certain assumptions on φ) by Frazier and Jawerth [21]. In the case of present interest to us they follow easily from properties of the S_k. Again let $0 < p \leq \infty$ be fixed and let B^α be the scale of Besov spaces described above and let $\tau := \tau(\alpha) := (\alpha/d + 1/p)^{-1}$ as before.

Lemma 4.2. *Let φ be a compactly supported function which satisfies (4.1), (4.2) and (4.3) and let $0 < p < \infty$. If $f \in B^\alpha$ has the representation $f = \sum_{I \in \mathfrak{D}} b_I \varphi_I$, for some constants b_I, and if $0 < \alpha < \min(r, s)$, then*

$$|f|_{B^\alpha} \leq C\left(\sum_{I \in \mathfrak{D}} |b_I|^\tau |I|^{\tau/p}\right)^{1/\tau} \tag{4.30}$$

with C depending only on φ, and p if p is small. In addition, if a_I are the coefficients of f given in the representation (4.29) and if $0 < \alpha < r$

and $\tau(\alpha) \geq \gamma$, *then*,

$$|f|_{B^\alpha} \approx \left(\sum_{I\in\mathcal{D}} |a_I|^\tau |I|^{\tau/p}\right)^{1/\tau} \tag{4.31}$$

where the constants of equivalency depend only on φ *and* τ *if* τ *is small.*

Remark. We recall for the reader that the construction of the S_k was made for all $f \in L_\gamma$ and γ can be chosen arbitrarily small.

Proof. To prove (4.30), we let $T_k := \sum_{I\in\mathcal{D}_k} b_I\varphi_I$, $k \in \mathbb{Z}$. Then for the error of approximation by elements of $\mathbb{S}_k$, we have

$$s_k(f)_\tau \leq \left(\sum_{j=k+1}^{\infty} \|T_j\|_\tau^\mu\right)^{1/\mu} \tag{4.32}$$

where $\mu := \min(1, \tau)$. We apply a discrete Hardy inequality (see (5.2)–(5.3) of [16]) to (4.32) and obtain

$$\sum_{k=-\infty}^{\infty} [2^{k\alpha} s_k(f)_\tau]^\tau \leq C \sum_{k=-\infty}^{\infty} 2^{k\alpha\tau}\|T_k\|_\tau^\tau. \tag{4.33}$$

From (4.27), the left side of (4.33) is equivalent to $|f|_{B^\alpha}^\tau$, while from (4.11) the right side of (4.33) is equivalent to the right side of (4.30) raised to the power τ. Therefore, (4.30) follows.

From (4.29), (4.27)(iv) and (4.11), we have

$$|f|_{B^\alpha}^\tau \approx \sum_{k\in\mathbb{Z}} 2^{k\alpha\tau}\|S_{k+1} - S_k\|_{L_\tau(\mathbb{R}^d)}^\tau \approx \sum_{k\in\mathbb{Z}} 2^{k\alpha\tau} \sum_{I\in\mathcal{D}_k} |a_I|^\tau |I|. \tag{4.34}$$

Since $2^{k\alpha\tau} = |I|^{-\alpha\tau/d}$ for each $I \in \mathcal{D}_k$ and $-\alpha\tau/d + 1 = \tau/p$, (4.31) follows from (4.34). ∎

Again, we fix $0 < p < \infty$. Another important property of the Besov spaces is that they imbed continuously into $L_p(\mathbb{R}^d)$. Namely, if $f \in B^\alpha$, then

$$\|f\|_{L_p(\mathbb{R}^d)} \leq C|f|_{B^\alpha}, \tag{4.35}$$

where C is a constant depending only on φ, α, and p if p is small. While these embeddings are known (see [16]), they can also be derived easily from what we have developed in this section. For this one needs the following inequality for $S \in \mathbb{S}_k$:

$$(4.36) \qquad \|S\|_{L_p(R^d)} \leq C\left(\sum_{I\in\mathcal{D}_k} |c_I(S)|^p|I|\right)^{1/p}$$

$$\leq C2^{k\alpha/d}\left(\sum_{I\in\mathcal{D}_k} |c_I(S)|^\tau|I|\right)^{1/\tau} \leq C2^{k\alpha/d}\|S\|_{L_\tau(\mathbb{R}^d)}$$

which follows from (4.11) and the fact that $\tau < p$. With (4.36), the inequality (4.35) can be proved in a similar way to the embedding for a cube proved in [16]. We note that the right side of (4.35) involves the seminorm for B^α not the norm.

Finally, we mention a simple property of dyadic cubes which will be useful in what follows. If I is a dyadic cube and $\nu \in \mathbb{Z}^d$, we let $I(\nu) := \nu\ell(I) + I$ be the dyadic translates of I.

LEMMA 4.3. *If $J \subset I$ are dyadic cubes and Q is an arbitrary cube with sides parallel to the axes which intersects J and $I(\nu)$ for some $\nu \in \mathbb{Z}^d$, then $Q \cap J(\nu) \neq \emptyset$.*

Proof. Let $\pi_i(x) := x_i$ be the i-th coordinate projection of $x = (x_1, \ldots, x_d)$. For each $i = 1, \ldots, d$, the interval $\pi_i(Q)$ intersects each of the intervals $\pi_i(J)$ and $\pi_i(I(\nu)) = \nu_i\ell(I) + \pi_i(I)$. Since $J \subset I$, the interval $\pi_i(Q)$ will therefore intersect $\pi_i(J(\nu))$. Let ξ_i be in $\pi_i(Q) \cap \pi_i(J(\nu))$, $i = 1, \ldots, d$. Then $\xi := (\xi_1, \ldots, \xi_d)$ is in $Q \cap J(\nu)$. ∎

5. Inverse theorems. In this section, we shall prove the inverse estimate (1.4)(ii) for functions φ which satisfy the conditions of Section 4. We shall assume in addition that the coefficients c_j which appear in the refinement equation (4.1) are all nonnegative. This implies that the function φ is nonnegative (see [6]). Since the φ_I, $I \in \mathcal{D}_k$, are a partition of unity, we have $0 \leq \varphi_I \leq 1$, for all $I \in \mathcal{D}$.

Before proving the inverse theorem, we derive a representation for the functions $S \in \Sigma_n$ which will be important in the proof of this inverse estimate. From the definition of this space, we know that

$$(5.1) \qquad S = \sum_{I\in\Lambda} a_I\varphi_I$$

where $|\Lambda| \le n$; we can assume that all coefficients in (5.1) are nonzero. It will be enough to consider functions S for which the only cubes appearing in (5.1) have sidelength ≤ 1. The inverse inequality will then follow for other S by dilation. We let Λ_j be the set of cubes $I \in \Lambda$ for which $\ell(I) = 2^{-j}$, i.e. $I \in \mathcal{D}_j$, and let $S_j := \sum_{I\in\Lambda_j} a_I\varphi_I$. In other words,

$$(5.2)\qquad \begin{aligned} S &= S_0 + \cdots + S_m, \qquad \text{and} \\ S_j &= \sum_{J\in\Lambda_j} c_J(S_j)\varphi_J, \qquad j = 0, 1, \ldots, m, \end{aligned}$$

where the c_J are the dual functionals which appear in (4.5).

We recall the support cubes Q of φ_J are the dyadic cubes on which φ_J does not vanish identically. We will be particularly interested in the support cubes J_0 of J for which $\ell(J_0) = \ell(J)$. These can each be identified with a $\nu \in \mathbb{Z}^d$ where ν satisfies, $J_0 = J + \nu\ell(J)$. We note that the set of ν which correspond to the support cubes of φ_J are independent of J, that is, for $\tilde{J} \neq J$, the support cubes of $\varphi_{\tilde{J}}$ will determine exactly the same values ν.

If $I \in \mathcal{D}_j$ and $k > j$, then we can rewrite φ_I at the dyadic level k as in (4.8). We will frequently use, without further elaboration, the fact that for any $K \in \mathcal{D}_k$ such that φ_K appears in this sum with a nonzero coefficient, each support cube K_0 of φ_K is a support cube of φ_I. Indeed, we can bound the coefficient of φ_K by an integral over K_0 (see (4.7)) and if φ_I were to vanish identically on K_0 then this integral and hence the coefficient would also vanish.

If $I \in \mathcal{D}$, we let

$$(5.3)\qquad \Sigma(I) := \{Q \subset I : Q \text{ is a support cube of } \varphi_J, \quad \ell(Q) = \ell(J) < \ell(I), J \in \Lambda\}$$

and

$$(5.4)\qquad E(I) := \cup\{Q : Q \in \Sigma(I)\}.$$

The set $E(I)$ is determined by the maximal cubes in $\Sigma(I)$, i.e. those cubes in this set which are not contained in any larger cubes of this set.

We shall modify the representation (5.2) of S by deleting cubes from Λ_j and adding to Λ_{j+1} the cubes that are necessary for rewriting the deleted φ_I at the next dyadic level. We do this iteratively first for $j = 0$ then $j = 1$ and so on. Our criteria for deciding when to delete a cube I will be based on $|E(I)|$. We let $\delta > 0$ be a constant which is sufficiently small that (4.10) is valid. We shall require that $\delta \leq 2^{-d}$ and pose another restriction on δ later. We will use δ to decide whether to remove a cube or not. We say that a cube I is of **low density** if $|E(I)| \leq \delta|I|$, otherwise it is a cube of **high density**.

We begin with the case $j = 0$. We let $\mathcal{B}_0$ denote the collection of "high density" cubes $I \in \mathcal{D}_0$. We shall say a dyadic cube Q has distance $\leq k$ from the dyadic cube I with $\ell(I) = \ell(Q)$ if $Q = \nu\ell(I) + I$ with $|\nu_j| \leq k$, $j = 1, \ldots, d$. We recall that φ is supported on $[-\rho, \rho]^d$. We let $\tilde{\mathcal{B}}_0$ denote the collection of cubes $I \in \mathcal{D}_0$ whose distance from one of the cubes in $\mathcal{B}_0$ is $\leq 100\rho$ (we are exaggerating the constant 100 in order to make later arguments more obvious).

We now delete from Λ_0 any cube I such that all the support cubes $I_0 \in \mathcal{D}_0$ of φ_I are in $\tilde{\mathcal{B}}_0$. We denote the resulting set by Λ_0'. Meanwhile, we add to the set Λ_1 any cube $J \in \mathcal{D}_1$ which appears with a nonzero coefficient in the representation

$$\varphi_I = \sum_{J \in \mathcal{D}_1} c_J(\varphi_I)\varphi_J \tag{5.5}$$

for one of the cubes I which we have deleted from Λ_0. We denote this new collection of cubes, i.e. the collection of all cubes in Λ_1 together with the new cubes, by Λ_1''. This gives us a new sequence Λ_0', Λ_1'', Λ_2, $\ldots$, Λ_m.

The above construction gives a new decomposition for S:

$$S = S_0' + S_1'' + S_2 + \cdots + S_m \tag{5.6}$$

where S_0' is the part of the sum S_0 which remains after the cubes have been deleted, i.e.

$$S_0' = \sum_{J \in \Lambda_0'} c_J(S_0)\varphi_J = \sum_{J \in \Lambda_0'} c_J(S_0')\varphi_J \tag{5.7}$$

and S_1'' is the sum S_1 together with the rewrites of the terms deleted from S_0.

We now describe the general inductive step in our construction. Suppose that for some $1 \leq k < m$, the sets $\Lambda_0', \ldots, \Lambda_{k-1}', \Lambda_k''$ have been constructed. We let $\mathcal{B}_k$ denote the set of all cubes in $\mathcal{D}_k$ which have "high density." We further let $\tilde{\mathcal{B}}_k$ denote the set of all cubes in $\mathcal{D}_k$ which have distance $\leq 100\rho$ from one of the cubes in $\mathcal{B}_k$. We delete from Λ_k'' any cube I such that all the support cubes I_0, $\ell(I_0) = \ell(I)$, of φ_I are contained in $\tilde{\mathcal{B}}_k$. We denote the resulting collection by Λ_k'. We add to Λ_{k+1} any cube $J \in \mathcal{D}_{k+1}$ which appears with a nonzero coefficient in the representation

$$\varphi_I = \sum_{J \in \mathcal{D}_{k+1}} c_J(\varphi_I)\varphi_J \tag{5.8}$$

for one of the cubes I which we have deleted from Λ_k''. We denote this new collection of cubes by Λ_{k+1}''.

The inductive construction stops when Λ_{m-1}' and Λ_m'' have been constructed. In this case, we let $\Lambda_m' := \Lambda_m''$.

For $j = 1, \ldots, m$, the above construction gives a decomposition for S:

$$S = S_0' + \cdots + S_{j-1}' + S_j'' + S_{j+1} + \cdots + S_m. \tag{5.9}$$

We let $S_m' := S_m''$. Then, we have

$$S = S_0' + \cdots + S_m', \quad S_j' = \sum_{J \in \Lambda_j'} \gamma_J \varphi_J, \tag{5.10}$$

$$\gamma_J := c_J(S_j') = c_J(S_j''), \qquad J \in \Lambda_j'.$$

This is our desired decomposition of S.

The following lemmas record the important properties of our construction. We let $\Lambda := \cup_{j=0}^m \Lambda_j$, $\Lambda' := \cup_{j=0}^m \Lambda_j'$ and $\Lambda'' := \cup_{j=0}^m \Lambda_j''$. We also want to single out certain cubes which have favorable properties. Given a dyadic cube $I \in \mathcal{D}_j$, there is a largest integer $k \leq j$ for which there is a cube $I^* \in \Lambda_k$ such that I is a support cube of φ_{I^*}. We denote by I_*^* the support cube of φ_{I^*} which contains I. While I^* is not necessarily unique, the cube I_*^* is. In general I^* does not need to exist but as we will use later without further mention, if I_0 is a support cube of some φ_I, $I \in \Lambda'$, $\ell(I_0) = \ell(I)$, then there always is an I^*. Indeed, if I itself is not in Λ then it must have occurred by rewriting some φ_J, $J \in \Lambda$. As noted earlier, I is a support cube of φ_J. Thus there are $J \in \Lambda$, $\ell(J) \geq \ell(I)$ such that φ_J has I as a support cube. The smallest such J will be an I^*.

We shall also use the generic notation I^{**} to denote cubes in Λ' such that $\ell(I^{**}) > \ell(I)$ and $\varphi_{I^{**}}$ contains I as a support cube. We call the cubes I^{**} **blocking** cubes. We also use the notation I^{**}_* for the support cube of $\varphi_{I^{**}}$ which contains I. We say that a cube $I \in \mathcal{D}$ is **good** if

$$\text{(5.11)} \qquad \ell(I^*) < \ell(I^{**}) \text{ for all "blocking" cubes.}$$

An equivalent formulation of condition (5.11) is that

$$I^*_* \text{ is strictly contained in } I^{**}_*.$$

It will be useful to say a few remarks about how new cubes $J \in \Lambda''_j$ (i.e. cubes not in Λ_j) appear. Clearly, there must be φ_I, $I \in \Lambda''_{j-1}$ which have been rewritten as in (5.8) and the φ_J term appears in the sum with a nonzero coefficient. As noted earlier, for any φ_I which contributes to φ_J in this manner, each support cube J_0 of φ_J is a support cube of φ_I. Now this cube I is either in Λ_{j-1} or φ_I itself is obtained by rewriting one or more φ_K, $K \in \Lambda''_{j-2}$, and so on. Again, for any φ_K which contributes to φ_J in this manner, each support cube J_0 of φ_J must also be a support cube of φ_K.

Looking at this from another viewpoint, given any $T = \Sigma_J c_J(S''_k)\varphi_J$ which is part of the sum for S''_k, we can follow through our entire construction on T, that is, we rewrite and delete according to our construction. This then gives a new representation of T. We call this new representation the rewrite of T. This rewrite represents T in terms of the φ_I, $I \in \Lambda'$, $\ell(I) \leq \ell(J)$.

LEMMA 5.1. *If $I \in \Lambda'_j$ and I_0 is one of the support cubes of φ_I with $\ell(I_0) = \ell(I)$, then there is a $k \leq j$ and a set of cubes $\Lambda^*(I, I_0) \subset \Lambda''_k$ with the following properties*:

(5.12) (i) *the cube $J_0 \in \mathcal{D}_k$ which contains I_0 is "good,"*
(ii) *J_0 is a support cube of each φ_J, $J \in \Lambda^*(I, I_0)$,*
(iii) *each support cube of $J \in \Lambda^*(I, I_0)$ has "low density,"*

(iv) $\gamma_I\varphi_I = c_I(S'_j)\varphi_I$
is a term in the rewrite of $T := \sum_{J\in\Lambda^*(I,I_0)} c_J(S''_k)\varphi_J$

(v) $\Lambda^*(I, I_0)$ *is explicitly determined by the construction below.*

Proof. If there is a cube $K \in \Lambda$, $\ell(K) = \ell(I)$, and φ_K has I_0 as a support cube then we take $\Lambda^*(I, I_0) = \{I\}$. Since cubes in Λ' have all their support cubes of "low density," the properties of the lemma are trivially verified with $J_0 = I_0$. In particular, this settles the case $I \in \Lambda \cap \Lambda'$. Assume then that I does not have a K as described above. Let $I_0^* \in \Lambda_{k_0}$ be one of the cubes mentioned above for I_0. Recall that while I_0^* is not unique, the integer k_0 is. Now, since I is not in Λ, it is new and therefore $c_I(S'_j)\varphi_I$ is obtained by rewriting certain terms $c_J(S''_{j-1})\varphi_J$. If J contributes to $c_I(S'_j)\varphi_I$ by such a rewriting, then as noted earlier each support cube of φ_I is also a support cube of φ_J. If $\ell(J) < 2^{-k_0}$, then each of the J are new and hence each of these can be obtained by rewriting. In this way for each $k_0 \leq k < j$ we obtain a set of cubes $\Lambda_k(I, I_0) \subset \Lambda''_k$ such that

(a) each φ_J, $J \in \Lambda_k(I, I_0)$ contains I_0 as a support cube
(b) each φ_J, $J \in \Lambda_k(I, I_0)$, contributes to φ_I when it is rewritten according to our construction.

In particular (ii) holds for $\Lambda_k(I, I_0)$ (instead of $\Lambda^*(I, I_0)$). Also, if T is defined as in (iv) for $\Lambda_k(I, I_0)$ (instead of $\Lambda^*(I, I_0)$), then our construction shows that $c_I\varphi_I$ is a term in the rewrite of T. Therefore, for each of these sets $\Lambda_k(I, I_0)$, we have (iv).

Now, let $k \geq k_0$ be the smallest integer such that there is a $\varphi_{\tilde{J}}$, $\tilde{J} \in \Lambda'_k$, that has I_0 as a support cube. We can take $\Lambda^*(I, I_0) := \Lambda_k(I, I_0)$. To see this, we first claim that all the support cubes K, $\ell(K) = \ell(J)$, of all of the φ_J, $J \in \Lambda_k(I, I_0)$, have "low density." To prove the claim, let $J_0 \in \mathcal{D}_k$ be the dyadic cube at this level which contains I_0. Then each φ_J, $J \in \Lambda_k(I, I_0)$, has J_0 as a support cube. Now, since $\tilde{J} \in \Lambda'_k$ its distance from the cubes in $\mathcal{B}_k$ is $\geq 98\rho$. Since all φ_J, $J \in \Lambda_k(I, I_0)$ share J_0 as a support cube, each of these J have distance $\geq 94\rho$ from the cubes in $\mathcal{B}_k$. It follows from the definition of $\mathcal{B}_k$ that all of their support cubes K are of "low density" which is our claim. Next, we claim that J_0 is "good." Suppose that some φ_K, $K \in \Lambda'$ with $\ell(K) > 2^{-k}$ has J_0 as a support cube. Then the support cube K_0 of φ_K which contains J_0 also contains I_0. Therefore, by the definition of k, we must have $\ell(K) > 2^{-k_0}$. But we also have $2^{-k_0} \geq \ell(J_0^*)$ because $\varphi_{I_0^*}$ has J_0 as a support cube. This verifies property (5.11) and shows that J_0 is "good." ■

Next, we make some further comments about the construction in Lemma 5.1. Let $\mathcal{G} := \cup\Lambda^*(I, I_0)$ where the union is taken over all I and I_0 and let $\mathcal{G}_j := \mathcal{G} \cap \mathcal{D}_j$. A given $J \in \mathcal{G}$ may occur for several pairs I, I_0. We choose one of those pairs (in an arbitrary manner) and fix the corresponding cube J_0, i.e. the cube at the same dyadic level as J which contains I_0. We say that J_0 is the **corresponding support cube** of J. The next lemma gives some properties of the cubes J_0.

LEMMA 5.2. (i) *If $J \neq \tilde{J}$, $\ell(J) \leq \ell(\tilde{J})$, are in $\mathcal{G}$ and if their corresponding support cubes J_0 and $\tilde{J}_0$ have the same position number ν, then either $J_0 \cap \tilde{J}_0 = \emptyset$, or $J = \tilde{J}$, or there is a φ_K, $K \in \Lambda$, with support cube K_0 satisfying $J_0 \subset K_0 \subset \tilde{J}_0$.* (ii) *A point $x \in \mathbb{R}^d$ is in at most $(2\rho)^d$ of the sets $J_0 \backslash E(J_0)$ where the J_0 are the cubes which correspond to $J \in \mathcal{G}$.*

Proof. Suppose that neither $J = \tilde{J}$ nor $J \cap \tilde{J} \neq \emptyset$ is valid. Let I, I_0 be the pair which determines J_0 and $\tilde{I}$, $\tilde{I}_0$ be the pair which determines $\tilde{J}_0$. If J_0 and $\tilde{J}_0$ are not disjoint then $J_0 \subset \tilde{J}_0$. Now recalling the construction of J_0, we let $K_0 := I_0^*$ be the cube in that construction which is a support cube of some φ_K with $K \in \Lambda$ and K_0 contains J_0. Suppose that $\tilde{J}_0 \subset K_0$. We know by the construction of $\tilde{J}_0$ that there is a cube $\tilde{K} \in \Lambda'$ such that $\varphi_{\tilde{K}}$ has a support cube which contains $\tilde{J}_0$. But then, unless $J_0 = \tilde{J}_0$, we have a contradiction to the minimal level property in the construction of J_0. Therefore, $J_0 = \tilde{J}_0$ and $J = \tilde{J}$ so that (i) has been verified.

To prove (ii), it is enough to prove that x can be in at most one of the sets $J_0 \backslash E(J_0)$ for J_0 which are identified with the same ν. Now, suppose that $x \in J_0 \backslash E(J_0)$, $\tilde{J}_0 \backslash E(\tilde{J}_0)$, with J_0, $\tilde{J}_0$, both having the same ν with respect to J and $\tilde{J}$ and suppose that $J \neq \tilde{J}$. We can assume that $\ell(J) \leq \ell(\tilde{J})$. Then, by (i) there is a cube K_0 which is a support cube of some φ_K, $K \in \Lambda$, and $J_0 \subset K_0 \subset \tilde{J}_0$. Since $K_0 \subset E(\tilde{J}_0)$, we have a contradiction. ■

If $J \in \Lambda_k''$ and J is rewritten in our construction, then the φ_K, $K \in \Lambda_{k+1}''$ which appear in expression (4.5) for J may either remain, and therefore $K \in \Lambda_{k+1}'$, or they themselves are rewritten. Continuing in this manner, we obtain a decomposition

$$\varphi_J = \sum_{I \in \Phi(J)} a_I \varphi_I \tag{5.13}$$

where the $I \in \Phi(J)$ are cubes in Λ' which are obtained when carrying out our entire construction.

LEMMA 5.3. *If $1 \le p < \infty$ and φ_J, $J \in \Lambda''$, has the decomposition* (5.13), *then*

$$\left(\sum_{I \in \Phi(J)} |a_I|^p |I|\right)^{1/p} \le C|J|^{1/p} \tag{5.14}$$

with C depending only on φ.

Proof. Since the mask coefficients c_j in (4.1) are nonnegative, the same applies to the a_I. We first note that the $a_I \le C_0$ with C_0 depending only on φ. Indeed, $\sup_x \varphi(x) =: C_1 > 0$ and if x_I is a point where $\varphi_I(x_1) > C_1/2$, then $a_I\varphi_I(x_i) \le \varphi_J(x_i) \le 1$ (recall that $0 \le \varphi \le 1$), so that $a_I \le 2C_1^{-1}$. We also have

$$|J| \int_{\mathbb{R}^d} \varphi = \int_{\mathbb{R}^d} \varphi_J = \sum_{I \in \Phi(J)} a_I \int_{\mathbb{R}^d} \varphi_I = \left(\sum_{I \in \Phi(J)} a_I |I|\right) \int_{\mathbb{R}^d} \varphi, \tag{5.15}$$

which gives (5.14) in the case $p = 1$. For $1 < p < \infty$, we have $a_I^p \le C_0^{p-1} a_I$ and therefore (5.14) follows from (5.15). ∎

We shall use the generic notation Λ^* for the sets $\Lambda^*(I, I_0)$. Each such set Λ^* contains at most $C(\varphi)$ cubes J because each φ_J has J_0 as a support cube (where J_0 is the cube at the dyadic level of J which contains I_0). Now a given set Λ^* may occur for several pairs I, I_0. We let $\Phi(\Lambda^*)$ be the set of all I which are associated to Λ^* in this way (for any I_0). Each such I is in $\Phi(J)$, $J \in \Lambda^*$.

LEMMA 5.4. *For a constant C depending only on φ, we have*

$$\left(\sum_{I \in \Lambda'} |\gamma_I|^p |I|\right)^{1/p} \le C\left(\sum_{j=0}^{m} \sum_{J \in \mathcal{G}_j} |c_J(S_j'')|^p |J|\right)^{1/p}. \tag{5.16}$$

Proof. If $\Lambda^* \subset \mathcal{D}_k$ then from (5.12) for each $I \in \Phi(\Lambda^*)$, we have

$$\gamma_I = \sum_{J\in\Lambda^*} c_J(S_k'')a_I(\varphi_J)$$

with the coefficients $a_I(\varphi_J)$ those that appear in the representation (5.13). We have estimated the a_I in Lemma 5.3. With $M := \max_{J\in\Lambda^*} |c_J(S_k'')|$, we have

$$|\gamma_I|^p \le \left(M \sum_{J\in\Lambda'} |a_I(\varphi_J)|\right)^p \le C^p M^p \sum_{J\in\Lambda'} |a_I(\varphi_J)|^p.$$

Using Lemma 5.3 and the fact that $\Phi(\Lambda^*) \subset \Phi(J)$, $J \in \Lambda^*$, we obtain

$$(5.17)\qquad \left(\sum_{I\in\Phi(\Lambda^*)} |\gamma_I|^p|I|\right)^{1/p} \le CM\left(\sum_{J\in\Lambda'} |J|\right)^{1/p}$$

$$\le CM|J| \le C\left(\sum_{J\in\Lambda'} |c_J(S_k'')|^p|J|\right)^{1/p}.$$

If we raise inequality (5.17) to the power p and sum over all Λ^*, we obtain inequality (5.16) because each $J \in \Lambda^*$ is in $\mathcal{G}_k$ for some k and each J can appear in at most $C(\varphi)$ of the Λ^*. ■

As our final preparation for the proof of the inverse theorem, we need to count how many cubes there are in Λ'.

LEMMA 5.5. *We have* $|\Lambda'| \le C(\varphi, \delta)n$.

Proof. First of all, at level j, the number of cubes $|\tilde{\mathcal{B}}_j|$ in $\tilde{\mathcal{B}}_j$ does not exceed a constant multiple (with constant depending only on d and ρ) of $\mathcal{B}_j$. Therefore, the number of new cubes in Λ'_{j+1} (they are all obtained by rewriting φ_I with $I \in \tilde{\mathcal{B}}_j$) does not exceed a constant multiple of $|\mathcal{B}_j|$. In other words, if N is the total number of cubes added at all levels, then

$$(5.18)\qquad N \le C(\varphi) \sum_{j=0}^{m-1} |\mathcal{B}_j|.$$

Now for each $I \in \mathcal{B}_j$, we have from the definition of this set that

$|E(I)| \geq \delta|I|$. Hence,

$$N \leq C(\varphi, \delta) \sum_{j=0}^{m-1} \sum_{I \in \mathcal{D}_j} \frac{|E(I)|}{|I|}. \tag{5.19}$$

Now, every cube that appears in computing $E(I)$ is from our original set Λ. Hence, from the definition of $E(I)$, we find

$$|E(I)| \leq \sum_{\substack{J \in \Lambda \\ J' \cap I \neq \emptyset \\ \ell(J) < \ell(I)}} |J'| \leq C(\varphi) \sum_{\substack{J \in \Lambda \\ J' \cap I \neq \emptyset \\ \ell(J) < \ell(I)}} |J|, \tag{5.20}$$

where, as before, J' denotes the cube of sidelength $2\rho\ell(J)$ outside of which φ_J vanishes. Moreover, for a fixed value of j, any cube J which appears in the right most sum of (5.20) does so for at most C different cubes $I \in \mathcal{D}_j$ with C depending only on φ. Therefore, the inner sum in (5.19) does not exceed a constant multiple of $\Sigma_{\nu=j+1}^{m} 2^{jd}2^{-\nu d}N_\nu$ where N_ν is the number of cubes in Λ_ν. We use this in (5.19) to obtain

$$N \leq C(\varphi, \delta) \sum_{j=0}^{m-1} \sum_{\nu=j+1}^{m} 2^{jd}2^{-\nu d}N_\nu.$$

Interchanging the orders of summation in this last sum and realizing that $N_1 + \cdots + N_m \leq n$ shows that $N \leq C(\varphi, \delta)n$. It follows that the number of cubes in Λ' does not exceed $C(\varphi, \delta)n$. ∎

THEOREM 5.6. *If* $1 \leq p < \infty$, $\varphi \in W_\infty^s$ *satisfies* (4.1–3), *and the mask coefficients* c_j *of* (4.1) *are nonnegative, then for each* $0 < \beta < \min(r, s)$, (1.4)(ii) *holds with a constant C depending only on* φ *and p.*

Proof. It is enough to prove (1.4)(ii) in the case $S \in \Sigma_n$ can be written in terms of φ_I with $\ell(I) \leq 1$. The general case follows from this by a change of variable since both the L_p norm and the Besov space norm scale the same under a dilation. We use the above procedure to write

$$S = S_0' + \cdots + S_m', \qquad S_j' \in \mathbb{S}_j, \tag{5.21}$$

$$S_j' = \sum_{I \in \Lambda_j'} c_I(S_j')\varphi_I, \qquad j = 0, 1, \ldots, m.$$

We continue to use the notation $\gamma_I := c_I(S_j') = c_I(S_j'')$, $I \in \Lambda_j'$. Since the number of cubes in Λ' does not exceed $C(\varphi, \delta)n$, according to (4.30) and Hölder's inequality, we have

$$(5.22)\quad |S|_{B^\beta} \le C(\varphi)\Bigl(\sum_{I\in\Lambda'} |\gamma_I|^\tau |I|^{\tau/p}\Bigr)^{1/\tau} \le C(\varphi)|\Lambda'|^{1/\tau - 1/p}\Bigl(\sum_{I\in\Lambda'} |\gamma_I|^p |I|\Bigr)^{1/p}$$
$$\le C(\varphi, \delta)n^{\beta/d}\Bigl(\sum_{I\in\Lambda'} |\gamma_I|^p |I|\Bigr)^{1/p}$$

where we have used the fact that $1/\tau - 1/p = \beta/d$.

In view of Lemma 5.4, in order to complete the proof of the theorem, it is enough to show that

$$(5.23)\quad \Bigl(\sum_{j=0}^{m} \sum_{J\in\mathcal{G}_j} |c_J(S_j'')|^p |J|\Bigr)^{1/p} \le C\|S\|_{L_p(\mathbb{R}^d)}.$$

Now let us fix for the moment a value of $j = 0, \ldots, m$. For each $J \subset \mathcal{G}_j$, let J_0, $\ell(J) = \ell(J_0)$, be the support cube which corresponds to J. Then, J_0 is "good" and has "low density." Therefore

$$(5.24)\quad |E(J_0)| \le \delta |J_0|,$$

where $E(J_0)$ is defined by (5.4). We let $F_j := \cup\{J_0 \backslash E(J_0) : J \in \mathcal{G}_j\}$. According to (4.10), we have

$$(5.25)\quad \Bigl(\sum_{J\in\mathcal{G}_j} |c_J(S_j'')|^p |J|\Bigr)^{1/p} \le C(\varphi)\Bigl\{\sum_{J\in\mathcal{G}_j} \int_{J_0\backslash E(J_0)} |S_j''|^p dx\Bigr\}^{1/p}$$
$$\le C(\varphi)\|S_j''\|_{L_p(F_j)},$$

where we have used the fact that a given J_0 can correspond to at most $C(\varphi)$ different $J \in \mathcal{G}_j$. Therefore,

$$(5.26)\quad \Bigl(\sum_{j=0}^{m} \sum_{J\in\mathcal{G}_j} |c_J(S_j'')|^p |J|\Bigr)^{1/p} \le C(\varphi)\Bigl(\sum_{j=0}^{m} \|S_j''\|^p_{L_p(F_j)}\Bigr)^{1/p}.$$

We first consider the case $p = 1$ since it is simpler than the general case and still indicates the main idea of the proof. Since all the S_ν with $\nu > j$ vanish on F_j, we have $S = S_0' + \cdots + S_{j-1}' + S_j''$ on this set. Therefore,

$$\|S_j''\|_{L_1(F_j)} \le \|S\|_{L_1(F_j)} + \sum_{\nu=0}^{j-1} \|S_\nu'\|_{L_1(F_j)}.$$

According to Lemma 5.2, part (ii), a point $x \in \mathbb{R}^d$ appears in at most $(2\rho)^d$ of the sets F_j. Hence, we can sum over all $j = 0, \ldots, m$ and obtain

$$(5.27) \quad \sum_{j=0}^{m} \sum_{J \in \mathcal{G}_j} |c_J(S_j'')||J| \le C(\varphi) \sum_{j=0}^{m} \|S_j''\|_{L_1(F_j)}$$

$$\le C(\varphi)\|S\|_{L_1(\mathbb{R}^d)} + C(\varphi) \sum_{j=0}^{m} \sum_{\nu=0}^{j-1} \|S_\nu'\|_{L_1(F_j)}$$

$$\le C(\varphi)\|S\|_{L_1(\mathbb{R}^d)} + C(\varphi) \sum_{j=0}^{m} \|S_j'\|_{L_1(E_j)},$$

where $E_j := \cup_{\nu=j+1}^{m} F_\nu$. We fix j for the moment and estimate

$$(5.28) \qquad \|S_j'\|_{L_1(E_j)} = \sum_{I \in \mathcal{D}_j} \int_{I \cap E_j} |S_j'|\,dx = \Sigma.$$

To estimate Σ, we note that if I is not a support cube of some φ_J, $J \in \Lambda_j'$, then $S_j' \equiv 0$ on I. On the other hand if I is a support cube of some φ_J, $J \in \Lambda_j'$, then we claim that $I \cap E_j \subset E(I)$. Indeed, each of the cubes K_0 such that $K_0 \backslash E(K_0)$ contributes to the makeup of E_j is "good." If $K_0 \subset I$, then according to (5.11), $K_0 \subset (K_0)_*^* \subset I$ and the last containment is strict. Since $K_0^* \in \Lambda$ and $(K_0)_*^*$ is one of the support cubes of $\varphi_{K_0^*}$, we have $K_0 \subset E(I)$ which gives our claim. It follows that $|I \cap E_j| \le |E(I)| \le \delta|I|$. Now, on the cube I, we have $S_j' = \sum_{J' \cap I \ne \emptyset} c_J(S_j')\varphi_J$. Since there are only $C(\varphi)$ terms in this sum and $|I| = |J|$ and $0 \le \varphi_J \le 1$ for the J in the sum, we have

$$(5.29) \qquad \int_{I \cap E_j} |S_j'|\,dx \le C(\varphi)\,\delta \sum_{\substack{J \in \Lambda_j' \\ J' \cap I = \emptyset}} |c_J(S_j')||J|.$$

If we sum these last estimates over $I \in \mathcal{D}_j$ and we realize that each cube J appears in at most $C(\varphi)$ of these sums, we obtain

$$\Sigma \le C(\varphi)\delta \sum_{J \in \Lambda_j'} |c_J(S_j')||J|. \tag{5.30}$$

We use this in (5.28) and (5.27) to find

$$\begin{aligned} \sum_{j=0}^{m} \sum_{J \in \mathcal{G}_j} |c_J(S_j'')||J| &\le C(\varphi)\|S\|_{L_1(\mathbb{R}^d)} + C(\varphi)\delta \sum_{j=0}^{m} \sum_{J \in \Lambda_j'} |c_J(S_j')||J| \\ &\le C(\varphi)\|S\|_{L_1(\mathbb{R}^d)} + C(\varphi)\delta \sum_{j=0}^{m} \sum_{J \in \mathcal{G}_j} |c_J(S_j'')||J| \end{aligned} \tag{5.31}$$

where the last inequality follows from Lemma 5.4. We now choose δ so small that the constant in front of the sum on the right side of (5.31) is $\le \frac{1}{2}$. If we then move that sum to the left side, we obtain (5.23) and complete the proof in the case $p = 1$.

We now consider the case $1 < p < \infty$. The proof is similar to the case $p = 1$ just given, but because we no longer have subadditivity of $\|\cdot\|_p^p$, it is slightly more involved. We again need to prove (5.23). Arguing as in the case $p = 1$ (see (5.26) and (5.27)), we arrive at

$$\begin{aligned} \sum_{j=0}^{m} \sum_{I \in \mathcal{G}_j} |c_I(S_j'')|^p|I| &\le 2^p C(\varphi)\|S\|_{L_p(\mathbb{R}^d)}^p \\ &\quad + 2^p C(\varphi) \sum_{j=0}^{m} \|S_0' + \cdots + S_{j-1}'\|_{L_p(F_j)}^p \end{aligned} \tag{5.32}$$

where $F_j := \cup\{K_0 \backslash E(K_0) : K \in \mathcal{G}_j\}$.

We fix a cube $K \in \mathcal{G}_j$ and estimate $\|S_0' + \cdots + S_{j-1}'\|_{L_p(K_0 \backslash E(K_0))}^p$. Let K_1 be the smallest cube which is the support cube of some φ_J, $J \in \Lambda$, $\ell(K_1) = \ell(J)$, such that K_1 contains K_0. If $K_1, \ldots, K_s$ have been defined, we let K_{s+1} be the smallest cube which is the support cube of some φ_J, $J \in \Lambda$, $\ell(K_{s+1}) = \ell(J)$, such that K_{s+1} *strictly* contains K_s, and for which there are $I \in \Lambda'$, $\ell(K_s) < \ell(I) \le \ell(K_{s+1})$, such that φ_I has K_0 as a support cube. Let $A(K_0, s)$ denote the set of $I \in \mathcal{D}$ such that φ_I has K_0 as a support cube and $\ell(K_s) < \ell(I) \le \ell(K_{s+1})$. We note first that because K_0 is good, we have $A(K_0, 0) \cap \Lambda' = \emptyset$.

Now if $s \geq 1$, and $I \in A(K_0, s) \cap \Lambda'$, we let I_0 be the support cube of φ_I, $\ell(I_0) = \ell(I)$, which contains K_0 and let $\Lambda^*(I, I_0)$ be the set of cubes constructed in Lemma 5.1. The construction in Lemma 5.1 shows that the sets $\Lambda^*(I, I_0)$, $I \in A(K_0, s) \cap \Lambda'$ consist of cubes at the same dyadic level k_s. Let J_s be the cube in $\mathfrak{D}_{k_s}$ which contains K_0. Then J_s contains all of the I_0 and J_s is "good." Going further, we define

$$T_s := \sum_J c_J(S''_{k_s})\varphi_J, \tag{5.33}$$

where the sum is taken over all cubes $J \in \mathcal{G}_{k_s}$ such that J_s is a support cube of φ_J. Then, there are at most $(2\rho)^d$ terms in the sum (5.33). If we rewrite T_s, then each $\gamma_I\varphi_I$, $I \in A(K_0, s) \cap \Lambda'$, appears in the rewrite. Therefore on K_0, we have

$$S'_0 + \cdots + S'_{j-1} = T_1 + \cdots + T_s + \cdots. \tag{5.34}$$

It follows from (5.34) that

$$\begin{aligned} \|S'_0 + \cdots S'_{j-1}\|^p_{L_p(K_0\setminus E(K_0))} &= \|T_1 + \cdots + T_s + \cdots\|^p_{L_p(K_0\setminus E(K_0))} \\ &\leq 2^p\{\|T_1\|^p_{L_p(K_0\setminus E(K_0))} + \|T_2 + \cdots + T_s + \cdots\|^p_{L_p(K_0\setminus E(K_0))}\} \\ &\leq \sum_{s=1}^{\infty} 2^{sp}\|T_s\|^p_{L_p(K_0\setminus E(K_0))}. \end{aligned} \tag{5.35}$$

Since there are at most $C(\varphi)$ terms in the sum for T_s, $s = 1, 2, \ldots,$ and the functions φ_I are all bounded by 1, we have

$$\int_{K_0\setminus E(K_0)} |T_s|^p dx \leq C^p \sum_{J\in A(K_0,s)\cap\mathcal{G}_{k_s}} |c_J(S''_{k_s})|^p|(K_0\setminus E(K_0)) \cap J'|. \tag{5.36}$$

We use (5.36) in (5.35) and sum over all $K_0 \in \mathcal{G}_j$ and then over all $j = 0, 1, \ldots, m$, and obtain that

$$\begin{aligned} &\sum_{j=0}^{m} \|S'_0 + \cdots + S'_{j-1}\|^p_{L_p(F_j)} \\ &\qquad \leq C^p \sum_{J\in\mathcal{G}_j} |c_J(S''_j)|^p\left(\sum_{s=1}^{\infty} 2^{sp}|E(J, s) \cap J'|\right) \end{aligned} \tag{5.37}$$

where

$$E(J, s) := \cup\{K_0 : K_0 \text{ a support cube of some } K \in \mathcal{G},$$
$$\ell(K_0) = \ell(K), \text{ and } J \in A(K_0, s)\}.$$

Here, we have used again the fact that a point $x \in \mathbb{R}^d$ appears in at most $(2\rho)^d$ of the cubes $K_0 \backslash E(K_0)$ by Lemma 5.2, part (ii).

We prove next by induction that for $J \in \mathcal{G}$, and any support cube J_* of φ_J, we have

$$|E(J, s) \cap J_*| \le \delta^s |J|. \tag{5.38}$$

When $s = 1$ this follows because for any K_0 appearing in $E(J, s)$ with $K_0 \subset J_*$, there is a cube $K_1 \in \Lambda$ such that $K_0 \subset K_1 \subset J_*$. Hence $K_0 \subset E(J_*)$. Since J_* has "low density," we have

$$|E(J, 1) \cap J_*| \le |E(J_*)| \le \delta |J|$$

because all support cubes of φ_J have "low density." More generally, we claim that

$$|E(J, s) \cap J_*| \le \delta |E(J, s - 1) \cap J_*|. \tag{5.39}$$

Indeed, if $K_0 \subset J_*$ contributes to $E(J, s)$, then with our previous notation, there are cubes K_i, $i = 1, \ldots, s$, which are support cubes of some cubes in Λ at the same dyadic level as K_i and there are cubes J_i, $i = 1, \ldots, s$, which are support cubes of some $\tilde{J}_i \in \mathcal{G}$, $\ell(J_i) = \ell(\tilde{J}_i)$, such that

$$K_0 \subset K_1 \subset J_1 \subset \cdots \subset K_s \subset J_s$$

and $J_s = J_*$. Also each of the containments $K_i \subset J_i$ is strict. It follows that K_0 contributes to $E(\tilde{J}_{s-1}, s - 1)$. Therefore, if we sum over all distinct $K_0 \subset J_{s-1}$, we find by our induction hypothesis that

$$\sum_{K_0 \subset E(J,s) \cap J_{s-1}} |K_0| \le |E(\tilde{J}_{s-1}, s - 1) \cap J_{s-1}| \le \delta^{s-1} |J_{s-1}|. \tag{5.40}$$

Moreover, each J_{s-1} is contained in $E(J, 1) \cap J_*$ and therefore if we sum in (5.40) over distinct $J_{s-1} \subset J_*$ and use our induction hypothesis again, we obtain (5.38).

It follows from (5.38) that $|E(J, s) \cap J'| \leq (2\rho)^d \delta^s$ because there are at most $(2\rho)^d$ support cubes of J. We use this in (5.37) and find

$$(5.41) \quad \sum_{j=0}^{m} \|S_0' + \cdots + S_{j-1}'\|_{L_p(F_j)}^p \leq C^p (2\rho)^d \left(\sum_{s=1}^{\infty} 2^{sp}\delta^s\right) \sum_{J \in \mathcal{G}_j} |c_J(S_j'')|^p |J|.$$

Given any $\epsilon > 0$, we can choose δ sufficiently small so that

$$(5.42) \quad \sum_{j=0}^{m} \|S_0' + \cdots + S_{j-1}'\|_{L_p(F_j)}^p \leq \epsilon \sum_{j=0}^{m} \sum_{J \in \mathcal{G}_j} |c_J(S_j'')|^p |J|.$$

Using this in (5.32) and taking ϵ sufficiently small, we obtain the desired results (5.23). ■

In Section 2, Theorem 2.1, we have proved the direct theorem for functions φ which includes the φ of Section 4. Therefore, this together with Theorem 5.1 gives the following.

THEOREM 5.7. *If $\varphi \in W_\infty^s(\mathbb{R}^d)$ is a compactly supported function which satisfies* (4.1–3), *and the mask coefficients of* (4.1) *are nonnegative, then for each* $1 \leq p < \infty$ *and each* $0 < \alpha < \min(r, s)$, *we have*

$$(5.43) \quad \sum_{n=1}^{\infty} [n^{\alpha/d}\sigma_n(f)_{L_p(\mathbb{R}^d)}]^\tau \frac{1}{n} < \infty \leftrightarrow f \in B^\alpha.$$

6. Another direct theorem. We shall give another proof of the direct theorem for functions φ which satisfy the conditions of Section 4. While it is more involved than the proof of Section 2, it has the advantage of yielding constants which are independent of p, $1 \leq p < \infty$. We also feel that the construction of good approximants given in this section may have numerical significance.

We wish to prove

$$(6.1) \quad \sigma_n(f)_p \leq C|f|_{B^\beta} n^{-\beta/d}, \qquad 0 < p < \infty$$

with C a constant depending only on φ and β. We shall need the following *combinatorial* lemma which was proved in [17]. For a collection Γ of dyadic cubes and $I \in \Gamma$, we let $B_I = B_I(\Gamma)$ denote the set of cubes $J \in \Gamma$ such that $J \subset I$ and J is maximal, that is, J is not contained in another cube $\tilde{J}$ with these properties.

LEMMA 6.1. *If $\Gamma = \{I\}$ is an arbitrary finite collection of dyadic cubes, then there is a second collection $\tilde{\Gamma}$ of dyadic cubes with the following properties*:

$$\text{(6.2)} \qquad \begin{aligned} &\text{(i) } \Gamma \subset \tilde{\Gamma}, \\ &\text{(ii) } |B_I(\tilde{\Gamma})| \leq 2^d, \qquad \textit{for all } I \in \tilde{\Gamma}, \\ &\text{(iii) } |\tilde{\Gamma}| \leq 2^d |\Gamma|. \end{aligned}$$

Moreover, for any cube $I \in \tilde{\Gamma}$, each child of I contains at most one cube from $B_I(\tilde{\Gamma})$.

The last property of $\tilde{\Gamma}$, although not explicitly stated in [17], is part of the construction of $\tilde{\Gamma}$. We note also for further use that (ii) holds for all dyadic cubes I (not necessarily from $\tilde{\Gamma}$). Indeed, we have two possibilities. Either I is contained in a minimal cube I^* from $\tilde{\Gamma}$, in which case, any cube of $\tilde{\Gamma}$ which is maximal in I will be maximal in I^*. The other possibility is that I is not contained in any cube of $\tilde{\Gamma}$. In this case, our claim follows because $|B_{\mathbb{R}^d}(\tilde{\Gamma})| \leq 2^d$ as was shown in [17].

We fix $0 < p < \infty$ and β and as before let $\tau := \tau(\beta) := (\beta/d + 1/p)^{-1}$. If $I \in \mathcal{D}$ or $I = \mathbb{R}^d$ and if g is a function with a representation $g = \Sigma_J d_J\varphi_J$, we let

$$\text{(6.3)} \qquad \begin{aligned} \Lambda(g, I) &:= \{J \in \mathcal{D} : J' \cap I \neq \emptyset,\ d_J \neq 0,\ \ell(J) \leq \ell(I)\} \\ \lambda(g, I) &:= \sum_{J \in \Lambda(g,I)} |d_J|^\tau |J|^{\tau/p} \end{aligned}$$

where J' is defined as before (φ_J vanishes outside of J' and J' has side length $2\rho\ell(J)$). We note that Λ and λ depend on the particular representation of g, but the representation meant will always be clear in our usage.

THEOREM 6.2. *If φ satisfies* (4.1–3), *then for each* $0 < \beta < \min(r, s)$, (6.1) *holds.*

Proof. Let $f \in B^\beta$. We begin with the representation (4.29). By

(4.31), $|f|_{B^\beta}$ is equivalent to $\lambda(f, \mathbb{R}^d)^{1/\tau}$ with the coefficients $d_J := a_J$ for all J. It is therefore enough to prove that $\sigma_n(f)_p \leq Cn^{-\beta/d}$ whenever $\lambda(f, \mathbb{R}^d) = 1$. We let $\epsilon := n^{-1}$. Then there are at most n cubes J for which the coefficients of (4.29) for f satisfy $|a_I|^\tau |I|^{\tau/p} \geq \epsilon$. We subtract these terms from f. Then, it is enough to show that the resulting function f_0 satisfies $\sigma_{Cn}(f_0)_p \leq Cn^{-\beta/d}$ with the constant C depending only on d and ρ.

We can represent $f_0 = \Sigma_{I \in \mathcal{D}} a_I \varphi_I$ where the coefficients are the same as those of f except that some of them have been made 0. From here on in the proof, the a_I refer to the coefficients of f_0. We choose a dyadic cube I_1 as small as possible so that $\lambda(f_0, I_1) > \epsilon$. We further define

$$(6.4) \qquad \begin{aligned} g_1 &:= \sum_{J \in \Lambda(f_0, I_1)} a_J \varphi_J \\ f_1 &:= f_0 - g_1. \end{aligned}$$

Then, f_1 has a representation $f_1 = \Sigma_{J \in \mathcal{D}} a_J^{(1)} \varphi_J$ with the coefficients $a_J^{(1)} = a_J$ if $J \notin \Lambda(f, I_1)$ and $= 0$ otherwise. We now repeat the above procedure with f_0 replaced by f_1. Namely, we let I_2 be the smallest dyadic cube for which $\lambda(f_1, I_2) > \epsilon$ and let $f_2 := f_1 - g_2$ where g_2 is defined as in (6.4) except that the sum is now taken over all $J \in \Lambda(f_1, I_2)$. We continue in this way to obtain dyadic cubes I_j and functions g_j, f_j. Since the sets $\Lambda(f_{j-1}, I_j)$ are disjoint, and $\Sigma_j \lambda(f_{j-1}, I_j) \leq \lambda(f_0, \mathbb{R}^d) \leq 1$, we generate in this way $m - 1 \leq n$ intervals I_j. We let $I_m := \mathbb{R}^d$ and $g_m := f_{m-1}$.

Let $\Gamma := \{I_1, \ldots, I_m\}$. We apply Lemma 6.1 to the sequence Γ and obtain a new sequence which for simplicity we shall continue to denote by Γ and refer to it as the **new** Γ when we wish to distinguish it from the **old** Γ. We cycle through the above definitions (6.4) for the **new** Γ and thereby obtain **new** functions f_j, g_j. Let us now point out some properties of the functions and cubes which we have constructed.

First, we note that

$$(6.5) \qquad \lambda(f_{j-1}, I_j) \leq C\epsilon$$

for each old f_{j-1}, I_j. Indeed, if I is one of the children of I_j then $\lambda(f_{j-1}, I) \leq \epsilon$ because of the definition of I_j. There are 2^d such children and hence these sums contribute $\leq 2^d\epsilon$ to $\lambda(f_{j-1}, I_j)$. The only terms that we have not accounted for in $\lambda(f_{j-1}, I_j)$ correspond to the cubes J of length

$\ell(J) = \ell(I_j)$. There are at most $(2\rho)^d$ such cubes in $\Lambda(f_{j-1}, I_j)$ and, by our initial assumption on the size of the coefficients a_I, each of these terms contributes an amount $\leq \epsilon$ to the sum $\lambda(f_{j-1}, I_j)$. Therefore, we have (6.5) for the **old** f_j, I_j. We also have (6.5) for the **new** f_{j-1}, I_j because each such **new** sum is part of one of the **old** sums.

From this point on we shall work only with the **new** sequence Γ and the **new** functions f_j, g_j. We use the abbreviated notation $\Lambda(I_j) := \Lambda(f_{j-1}, I_j)$. Since $g_j = \Sigma_{I\in\Lambda(I_j)} a_I^{(j)}\varphi_I$, where $a_I^{(j)} = a_J$ or $= 0$, it follows from (6.5) and (4.30) that

$$|g_j|_{B^\beta} \leq C\epsilon^{1/\tau}. \tag{6.6}$$

Another important property of our construction is

$$f_0 = g_1 + g_2 + \cdots + g_m. \tag{6.7}$$

We can now complete the proof of the theorem in the case $p \leq 1$. Indeed, from (4.35) and (6.6), we have $\|g_j\|_p \leq C|g_j|_{B^\beta} \leq C\epsilon^{1/\tau}$, $j = 1, \ldots, m$. Hence,

$$\|f_0\|_p^p \leq \sum_{j=1}^{m} \|g_j\|_p^p \leq Cm\epsilon^{p/\tau} = Cn^{-\beta p/d}$$

as desired.

In the case $p > 1$, the situation is more involved and we need to further modify the functions g_j to obtain functions with more restricted supports. We recall that φ_J is supported on J'. We introduce the notation $I'' := \cup\{I + \nu\ell(I) : \nu \in [-3\rho - 1, 3\rho + 1]^d\}$ for a cube I. Clearly g_j is supported on I_j''. We shall also use without further mention the fact that when $J \subset I$, then $J' \subset I'$.

We fix j and we consider the $I_k \in \Gamma$ with $k < j$, $I_k'' \cap I_j'' \neq \emptyset$ and I_k maximal with respect to these properties, i.e. I_k is not contained in any other cube $I_{k'}$ with these same properties. The number of these cubes I_k is $\leq C$ with C depending only on d and ρ. Indeed any such I_k is contained in a dyadic translate $Q = \nu\ell(I_j) + I_j$ with $\nu \in [-6\rho - 2, 6\rho + 2]^d$. There are at most C such Q and within any Q there are at most 2^d maximal I_k because of property (6.2)(ii) (recall that (6.2)(ii) holds for any dyadic cube I and note that if $I_k'' \cap I_j'' \neq \emptyset$ and $I_k \subset I_{k'}$, then $I_{k'}'' \cap I_j'' \neq \emptyset$ as well). Let the largest of these I_k have sidelength 2^{s_1} and let $D_{1,j}$ be the union of all the cubes I''_k, $k < j$, for which $\ell(I_k) = 2^{s_1}$. Because of (4.1), we can rewrite the terms $c_J \varphi_J$, which appear in g_j and whose J have length $\ell(J) > 2^{s_1}$, in terms of φ_K with $\ell(K) = 2^{s_1}$. If we do this we obtain the representation

$$(6.8) \quad g_j = \sum_{\substack{\ell(K)=2^{s_1} \\ K' \cap \overline{D}_{1,j} \neq \emptyset}} b_K \varphi_K + \sum_{\substack{\ell(K)=2^{s_1} \\ K' \cap \overline{D}_{1,j} = \emptyset}} b_K \varphi_K + \sum_{\substack{J \in \Lambda(I_j) \\ \ell(J) < 2^{s_1}}} a_J^{(j)} \varphi_J = S_{1,j} + g_{1,j},$$

where $S_{1,j}$ is by definition the first sum in (6.8) and $\overline{D}_{1,j}$ is the closure of the set $D_{1,j}$.

We note the important properties of $g_{1,j}$ and $S_{1,j}$. First note that $S_{1,j}$ has $\le C$ terms, where C depends only on ρ and d. Also, if $x \in K'$ with φ_K appearing in the representation (6.8) of g_j, then $x \in J'$ for some J such that φ_J appears in the original representation of g_j. Indeed, when we rewrite a φ_J in terms of lower level φ_K then by (4.9) only K with $K' \subset J'$ appear. The final property we wish to point out for $g_{1,j}$ is that

$$(6.9) \qquad |g_{1,j}|_{B^\beta} \le C\epsilon^{1/\tau}$$

with C depending only on ρ and d. Indeed, $\|g_j\|_{L_p(\mathbb{R}^d)} \le C\epsilon^{1/\tau}$ because of (6.6) and the embedding inequality (4.35). When we write g_j as in (6.8), then the last sum Σ satisfies $\|\Sigma\|_{L_p(\mathbb{R}^d)} \le C\epsilon^{1/\tau}$ for the same reasons. It follows that S, the sum of the first two sums of (6.8), also satisfies $\|S\|_{L_p(\mathbb{R}^d)} \le C\epsilon^{1/\tau}$. Hence, by (4.11), the coefficients b_K of $S_{1,j}$ all satisfy $|b_K|^p|K| \le C^p\epsilon^{p/\tau}$. Since $S_{1,j}$ has at most C terms, we have by (4.30) that $|S_{1,j}|_{B^\beta} \le C\epsilon^{1/\tau}$. Then (6.9) follows because $g_{1,j} = g_j - S_{1,j}$.

We now repeat the above procedure by replacing g_j by $g_{1,j}$ and by replacing $D_{1,j}$ by $D_{2,j}$, where $D_{2,j}$ is the union of the I''_k for the maximal cubes which remain (i.e. after deleting the cubes I_k which determined $D_{1,j}$) of the largest remaining length, say 2^{s_2}. We continuc this process until we have exhausted all maximal cubes; this occurs after say m_j steps with $m_j \le C$ and C depending only on d and ρ. At each stage of our construction, the function $g_{i-1,j}$ has the representation

$$(6.10) \quad \begin{aligned} g_{i-1,j} &= \sum_{\substack{\ell(K)=2^{s_i} \\ K' \cap \overline{D}_{i,j} \neq \emptyset}} b_K \varphi_K + \sum_{\substack{\ell(K)=2^{s_i} \\ K' \cap \overline{D}_{i,j} = \emptyset}} b_K \varphi_K \\ &\quad + \sum_{\substack{J \in \Lambda(I_j) \\ \ell(J) < 2^{s_i}}} a_J^{(j)} \varphi_J = S_{i,j} + g_{i,j}. \end{aligned}$$

We denote by $\tilde{g}_J := = g_{m_j,j}$ the final function that results when we have completed all these steps. Then $\tilde{g}_j$ is supported on I_j'' and satisfies

$$|\tilde{g}_j|_{B^\beta} \leq C\epsilon^{1/\tau}. \tag{6.11}$$

We define also the function $S_j := S_{1,j} + \cdots + S_{m_j,j}$ which is a linear combination of at most C of the φ_J with C depending only on ρ and d. Then,

$$g_J = \tilde{g}_J + S_j. \tag{6.12}$$

We now define $S := \Sigma_{j=1}^{m} S_j$. Then, $S \in \Sigma_{Cn}$. Furthermore,

$$f_0 - S = \sum_{j=1}^{m} \tilde{g}_J. \tag{6.13}$$

We now show that

(6.14) if $x \in \mathbb{R}^d$ then $\tilde{g}_j(x) \neq 0$ for at most C values of j.

For the proof of (6.14), we fix $x \in \mathbb{R}$ and let Λ be the set of cubes $I_j \in \Gamma$ such that $\tilde{g}_j(x) \neq 0$. We shall use the following remarks about the cubes I''. If I is any dyadic cube, then $I'' = \cup_\nu I(\nu)$ where we recall our notation $I(\nu) := I + \nu\ell(I)$. If a point $y \in I(\nu)$ then we say y corresponds to ν for I. A point on the boundary of one of these cubes corresponds to more than one ν. Now, let I and J be two dyadic cubes for which a point $y \in J'' \cap I''$ has exactly the same corresponding ν for both I and J. Then the smaller of the cubes I'', J'' is contained in the larger. Now let Λ_ν denote the collection of cubes $I_j \in \Lambda$ such that $x \in I_j''$ and x corresponds to ν for I_j. Clearly there at most C sets Λ_ν. To complete the proof of (6.14), we shall show that Λ_ν contains at most one cube. Indeed, suppose I_j, I_k were both in Λ_ν and $k < j$. By our earlier comments, this implies that $I_k(\nu) \subset I_j(\nu)$ and $I_k'' \subset I_j''$. Now let $k' \geq k$ be such that $I_k \subset I_{k'}$ with $I_{k'}$ maximal (under containment) among the $I_p, p < j$. Then, $I_{k'}'' \cap I_j'' \neq \emptyset$ and therefore $I_{k'}''$ is contained in one of the sets $D_{i,j}$ in the definition of $\tilde{g}_j$. We consider the representation $g_{i-1,j}$ as in (6.10). If $\tilde{g}_j(x) \neq 0$, then by our earlier comments preceding (6.9), $x \in K'$ for one of the terms φ_K appearing in the second or third sum. This is certainly not the case for the terms of the second sum since they all vanish on $\overline{D}_{i,j}$

and hence on the set I_k'' which contains x. On the other hand, if $x \in J'$ for one of the terms φ_J (with nonzero coefficient $a_J^{(j)}$) appearing in the third sum, then $J' \cap I_j \neq \emptyset$. Therefore the cube J' intersects both I_j and $I_j(\nu)$ and $I_k(\nu)$. We apply Lemma 4.3 with $Q = J'$ and find that J' intersects I_k. Hence $J' \cap I_{k'} \neq \emptyset$ as well. But since $\ell(J) < \ell(I_{k'})$, this means that J is in $\Lambda(I_s)$ for one of the cubes $I_s \in \Gamma$ with $I_s \subset I_{k'}$. But then $a_J^{(j)} = 0$ and we have the desired contradiction and have therefore proven (6.14).

In view of (6.14) and the embedding inequality (4.35), we can estimate for $0 < p < \infty$,

$$(6.15)\quad \|f_0 - S\|_p^p \le C \sum_{j=1}^{m} \|\tilde{g}_j\|_p^p \le C \sum_{j=1}^{m} |\tilde{g}_j|_{B^\alpha}^p \le Cn\epsilon^{p/\tau} = Cn^{-p\beta/d}.$$

■

7. Some examples. We consider first the univariate case. Let $N_r(x) := N_r(x; -r/2, -r/2 + 1, \ldots, r/2) := r[-r/2, \ldots, r/2](\cdot - x)_+^r$, $r = 1, 2, \ldots$, be the univariate B-spline of order r with knots at the points $-r/2, \ldots, r/2$. Here, $[t_0, \ldots, t_r]$ denotes the divided difference operator for the points $t_0, \ldots, t_r$. Another way to define N_r is inductively on r: $N_1 := \chi_{[-1/2, 1/2]}$ is the characteristic function of $[-1/2, 1/2]$ and $N_r := N_{r-1} * N_1$, $r = 2, 3, \ldots$. It follows that $N_r \in W_\infty^s(\mathbb{R})$ with $s = r - 1$.

The translates $N(x - j)$, $j \in \mathbb{Z}$ are a basis for the space $\mathcal{S}_0$ of univariate spline functions with knots at the integers when r is even and at the half integers when r is odd. The elements in $\mathcal{S}_0$ are piecewise polynomials of degree $< r$ with continuity C^{r-2} at each break point. In particular, it follows that $N_r(x/2)$ is in $\mathcal{S}_0$ and therefore can be expressed as a linear combination of the $N_r(x - j)$. Also, it is well known that the mask coefficient c_j in (4.1) are nonnegative. Therefore, condition (4.1) is satisficd. It is easy to see and well known that the only $N_r(x - j)$ needed in the representation (4.1) are those whose support is contained in the support of $N_r(x/2)$.

The Fourier transform of N_r is $\hat{N}_r(x) = (\sin x/2)^r/(x/2)^r$. Therefore condition (4.2) is satisfied. The local linear independence of the B-splines is well known and therefore (4.3) is satisfied as well. We therefore have that Theorem 6.4 holds for $\varphi = N_r$ and $s = r - 1$. That is, we have the characterization (1.3) for all $\alpha < r - 1$.

Actually, the range of α for which (1.3) holds can be extended by giving a more careful analysis of the smoothness of $\varphi = N_r$. Namely, the restriction that $\alpha < s$ comes about because of (4.23). It is easy however to improve (4.23) to

$$\omega_r(f, 2^{-k})_p \leq C2^{-k\lambda}\left(\sum_{j=-\infty}^{k} [2^{j\lambda}s_j(f)_p]^\mu\right)^{1/\mu}, \tag{7.1}$$

with $\lambda := r - 1 + 1/p$. For this, it is enough to note that $\Delta_h^r(\varphi_J, x) = 0$ except for points x for which $[x, x + rh]$ intersects more than one of the intervals which define the polynomial pieces of φ_J. The set Γ_J of such x has measure $\leq C \min(|h|, |J|)$. On the other hand, $|\Delta_h^r(\varphi_J, x)| \leq C \min(1, 2^{k(r-1)}|h|^{r-1})$ if $|J| = 2^{-k}$. Therefore, for any $S = \sum_{J\in\mathcal{D}_k} c_J\varphi_J$, $\varphi = N_r$, we have as in the derivation of (4.22)

$$\omega_r(S, t)_p \leq C \min(1, 2^{k\lambda}t^\lambda)\|S\|_{L_p(\mathbb{R}^d)}. \tag{7.2}$$

When this is used in place of (4.22) in the derivation of (4.23), we obtain (7.1).

We see therefore that the assumptions of Theorem 5.7 are satisfied. It follows that (5.43) holds for approximation in L_p for any $1 \leq p < \infty$. Actually, this theorem holds also for $0 < p < 1$. Indeed, the direct theorem follows from Theorem 6.2 which is valid for all $0 < p < \infty$. Moreover, the inverse inequality (1.4)(ii) has been shown by Petrushev [32] to be valid for all spline functions with $\leq n$ knots and therefore in particular for the splines in $\Sigma_{n,r}$. Therefore, we obtain the characterization (5.43) for all $0 < p < \infty$. It is interesting to note that this characterization (for φ a univariate B-spline) is exactly the same as the characterization of the nonlinear approximation by splines with n free knots given recently by Petrushev [32] and DeVore and Popov [15].

There are two extensions of these results to the multivariate setting which we wish to mention. The first is for the tensor product B-spline $N_r(x) := N_r(x_1) \cdots N_r(x_d)$. The function $\varphi = N_r$ satisfies (4.1)–(4.3) (with nonnegative mask coefficients) and (4.20) with $s = r - 1$. Therefore, Theorem 5.7 is valid for this φ provided $0 < \alpha < r - 1$. A more delicate argument, similar to that given above for the univariate B-spline can extend the range of α to $0 < \alpha < r - 1 + 1/p$. We refer the reader to [16] for details.

A second class of multivariate functions φ to which our results apply are the box splines introduced in [2]. Let $T := \{t_i\}_{i=1}^m$ be a set of vectors in $\mathbb{R}^d$ which span $\mathbb{R}^d$. We assume that the components of T are integers. The box spline $M := M_T$ is the function defined by the distributional equation

$$(7.3) \qquad \int_{\mathbb{R}^d} M(x)f(x)dx = \int_{Q_m} f\left(\sum_{i=1}^{m} y_i t_i\right)dy, \qquad f \in C_0^\infty(\mathbb{R}^d),$$

where $Q_m := [-1/2, 1/2]^m$. Then (see [3]), M is a piecewise polynomial which is supported on the set

$$\left\{x : x = \sum_{i=1}^{m} y_i t_i : -\frac{1}{2} \leq y_i \leq \frac{1}{2}, \, t_i \in T\right\}.$$

The box spline M also satisfies the refinement identity (4.1) with non-negative mask coefficients (see [10]).

The box spline M is in C^s, $s := s_0 - 2$ where s_0 is the smallest integer for which there are s_0 vectors in T whose removal results in a set of vectors which do not span R^d (see [3]). Also the Fourier transform of M is $\hat{M}(x) = \Pi_{i=1}^{m} \sin(t_i \cdot x/2/t_i) \cdot (x/2)$ where $x \cdot y$ is the scalar product of the two vectors x and y. It follows that (4.2) is satisfied for $r := s_0 - 1$.

Dahmen and Micchelli [10] and Jia [23] have characterized when the translates of M are locally linearly independent. Namely, the following condition introduced by de Boor and Höllig [3] should be satisfied:

$$(7.4) \qquad |\det(Y_d)| = 1$$

for each $d \times d$ matrix Y_d whose columns are d vectors from T which span $\mathbb{R}^d$. Therefore, under this condition, (4.3) will be fulfilled. In summary, Theorem 5.7 gives the following results for the nonlinear approximation by translates and dilates of the box spline $\varphi := M$.

THEOREM 7.1. *If the box spline* $\varphi := M := M_T$ satisfies the linear independence condition (7.4), *then for* $1 \leq p < \infty$ *and* $0 < \alpha < s$, *we have*

$$(7.5) \qquad \sum_{n=1}^{m} [n^{\alpha/d}\sigma_n(f)_{L_p(\mathbb{R}^d)}]^r \frac{1}{n} < \infty \leftrightarrow f \in B^\alpha.$$

In a manner completely analogous to that mentioned above for the univariate B-splines, it is possible to extend the restriction of α in Theorem 7.1 to $0 < \alpha < \lambda$ with $\lambda := s + 1/p$.

UNIVERSITY OF SOUTH CAROLINA
UNIVERSITY OF SOUTH CAROLINA
BULGARIAN ACADEMY OF SCIENCE, SOFIA, BULGARIA

REFERENCES

[1] G. Battle, A block spin construction of ondelettes. Part I: Lemarié functions, *Communications Math. Phys.*, **110** (1987), 601–615.

[2] C. de Boor and R. DeVore, Approximation by smooth multivariate splines, *Transactions AMS*, **276** (1983), 775–788.

[3] ——— and K. Höllig, *B*-splines from parallelpipeds, *J. Analyse Math.*, **42** (1982), 99–115.

[4] ——— and R. Q. Jia, Controlled approximation and a characterization of the local approximation order, *Proceedings AMS*, **95** (1985), 547–553.

[5] A. P. Calderón, Intermediate space and interpolation, the complex method, *Studia Math.*, **24** (1964), 113–190.

[6] A. Cavaretta, W. Dahmen and C. Micchelli, Subdivision for Computer Aided Geometric Design, Memoirs, *Amer. Math. Soc.*, **93** (1991).

[7] C. K. Chui and J. Z. Wang, A general framework of compactly supported splines and wavelets, *J. Approx. Th.*, (to appear).

[8] Z. Ciesielski, Constructive function theory and spline systems, *Studia Math.*, **52** (1973), 277–302.

[9] R. Coifman and R. Rochberg, Representation theorems for holomorphic and harmonic functions in L^p, *Astérisque*, **77** (1980), 11–66.

[10] W. Dahmen and C. Micchelli, Subdivision algorithms for the generation of box-spline surfaces, *Computer Aided Geometric Design*, **1** (1984), 191–215.

[11] ——— and ———, Local linear independence of translates of a box spline, *Studia Math.*, **82** (1985), 243–262.

[12] I. Daubechies, Orthonormal basis of compactly supported wavelets, *Communications on Pure & Applied Math.*, **41** (1988), 909–996.

[13] ———, A. Grossman, and Y. Meyer, Painless nonorthogonal expansions, *J. Math. Physics*, **27** (1986), 1271–1283.

[14] R. DeVore, Degree of approximation, in *Approximation II*, G. G. Lorentz, C. K. Chui, and L. L. Schumaker, eds., Academic Press, New York, 1976, 117–162.

[15] ——— and V. Popov, Interpolation spaces and nonlinear approximation, in *Functions Spaces and Approximation*, M. Cwikel, J. Peetre, Y. Sagher, H. Wallin, eds., Springer Lecture Notes in Math., **1302** (1988), 191–207.

[16] ——— and ———, Interpolation of Besov spaces, *Trans. AMS*, **305** (1988), 397–414.

[17] ——— and ———, Free multivariate splines, *Const. Approx.*, **3** (1987), 239–248.

[18] C. Fefferman and E. Stein, Some maximal inequalities, *Amer. J. Math.*, **93** (1971), 107–115.

[19] M. Frazier and B. Jawerth, Decomposition of Besov spaces, *Indiana Math. J.*, **34** (1985), 777–799.

[20] ______ and ______, The φ-transform and applications to distribution spaces, in *Functions Spaces and Approximation*, M. Cwikel, J. Peetre, Y. Sagher, H. Wallin, eds., Springer Lecture Notes in Math., **1302** (1988), 223–246.

[21] ______ and ______, A discrete transform and decompositions of distribution spaces, *J. of Functional Analysis*, **93** (1990), 34–170.

[22] L. Hörmander, Estimates for translation invariant operators in L^p-spaces, *Acta Math.*, **104** (1960), 93–140.

[23] R-Q. Jia, Local linear independence of the translates of a box spline, *Constructive Approximation*, **1** (1985), 175–182.

[24] ______ and C. A. Micchelli, Using the refinement equation for the construction of pre-wavelets II: power of two, In: Curves and Surfaces, (P. J. Laurent, A. LeMéhanté, and L. Schumaker, eds.) Academic Press, New York, (1991), pp. 209–246.

[25] B. Jawerth, On Besov spaces, Lund Technical Report, **1** (1977).

[26] P. G. Lemarié and Y. Meyer, Ondelettes et bases Hilbertiennes, *Revista Matematica Ibero Americana*, **2** (1986), 1–18.

[27] S. Mallat, Multiresolution approximations and wavelet orthonormal bases of $L^2(\mathbb{R})$, *Transactions AMS*, **315** (1989), 69–87.

[28] Y. Meyer, Wavelets and operators, in *Analysis at Urbana*, I. E. Berkson et al., editors, LMS Lecture Notes 137, Cambridge Univ. Press, Cambridge, 1989.

[29] ______, *Ondelettes et Opérateurs*, I, II, Hermann Ed., Paris, 1990.

[30] P. Oswald, On the degreee of nonlinear spline approximation in Besov-Sobolev spaces, *J. Approx. Th.*, **61** (1990), 131–157.

[31] J. Peetre, *New Thoughts on Besov Spaces*, Duke Univ. Math. Series, Duke Univ. Press, Durham, N.C., 1976.

[32] P. Petrushev, Direct and converse theorems for spline and rational approximation and Besov spaces, in *Functions Spaces and Approximation*, M. Cwikel, J. Peetre, Y. Sagher, H. Wallin, eds., Springer Lecture Notes in Math., **1302** (1988), 363–377.

[33] R. Rochberg, Toeplitz and Hankel operators, wavelets, NWO sequences, and almost diagonalization of operators, *Symposia in Pure Mathematics*, to appear.

[34] I. J. Schoenberg, *Cardinal Spline Interpolation*, SIAM CBMS, 1973.

[35] G. Strang and G. F. Fix, A Fourier analysis of the finite element method, in *Constructive Aspects of Functional Analysis*, G. Geymonant, ed., C.I.M.E. II Cilo, 1971, 793–840.

[36] J. O. Strömberg, A modified Franklin system and higher-order spline systems on $\mathbb{R}^n$ as unconditional bases for Hardy spaces, in Conference on Harmonic Analysis in Honor of Antoni Zygmund, vol. II, W. Beckner et al., editors, Univ. of Chicago Press, Chicago, 1981, 475–484.

[37] E. Stein, *Singular Integrals and the Differentiability Properties of Functions*, Princeton Univ. Press, Princeton, N.J., 1970.

Adapting to Unknown Smoothness via Wavelet Shrinkage

David L. DONOHO and Iain M. JOHNSTONE

We attempt to recover a function of unknown smoothness from noisy sampled data. We introduce a procedure, *SureShrink*, that suppresses noise by thresholding the empirical wavelet coefficients. The thresholding is adaptive: A threshold level is assigned to each dyadic resolution level by the principle of minimizing the Stein unbiased estimate of risk (*Sure*) for threshold estimates. The computational effort of the overall procedure is order $N \cdot \log(N)$ as a function of the sample size N. *SureShrink* is smoothness adaptive: If the unknown function contains jumps, then the reconstruction (essentially) does also; if the unknown function has a smooth piece, then the reconstruction is (essentially) as smooth as the mother wavelet will allow. The procedure is in a sense optimally smoothness adaptive: It is near minimax simultaneously over a whole interval of the Besov scale; the size of this interval depends on the choice of mother wavelet. We know from a previous paper by the authors that traditional smoothing methods—kernels, splines, and orthogonal series estimates—even with optimal choices of the smoothing parameter, would be unable to perform in a near-minimax way over many spaces in the Besov scale. Examples of *SureShrink* are given. The advantages of the method are particularly evident when the underlying function has jump discontinuities on a smooth background.

KEY WORDS: Besov, Hölder, Sobolev, Triebel spaces; Compactly supported wavelets; Denoising; James–Stein estimator; Minimax decision theory; Nonparametric regression; Nonlinear estimation; Orthonormal bases; Stein unbiased risk estimate; Thresholding; White noise model.

1. INTRODUCTION

Suppose that we are given N noisy samples of a function f,

$$y_i = f(t_i) + z_i, \qquad i = 1, \ldots, N, \tag{1}$$

with $t_i = (i-1)/N$, z_i iid $N(0, \sigma^2)$. Our goal is to estimate the vector $\mathbf{f} = (f(t_i))_{i=1}^N$ with small mean squared error (MSE); that is, to find an estimate $\hat{\mathbf{f}}$ depending on $y_1, \ldots, y_N$ with small *risk* $R(\hat{\mathbf{f}}, \mathbf{f}) = N^{-1} \cdot E\|\hat{\mathbf{f}} - \mathbf{f}\|_2^2 = E \text{ Ave}_i(\hat{f}(t_i) - f(t_i))^2$.

To develop a nontrivial theory, one usually specifies some fixed class $\mathcal{F}$ of functions to which f is supposed to belong. Then one may seek an estimator $\hat{f}$ attaining the *minimax risk* $R(N, \mathcal{F}) = \inf_{\hat{\mathbf{f}}} \sup_f R(\hat{\mathbf{f}}, \mathbf{f})$.

This approach has led to many theoretical developments of considerable interest (see, for example, Stone 1982, Nemirovskii, Polyak, and Tsybakov 1985; Nussbaum 1985). But from a practical point of view, it has the difficulty that it rarely corresponds with the usual situation where one is given data, but no knowledge of an a priori class $\mathcal{F}$.

To repair this difficulty, one may suppose that $\mathcal{F}$ is an unknown member of a *scale* of function classes and may attempt to behave in a way that is simultaneously near minimax across the entire scale. An example is the L^2 Sobolev scale, a set of function classes indexed by parameters m (degree of differentiability) and C (quantitative limit on the mth derivative):

$$W_2^m(C) = \left\{ f : \|f\|_2^2 + \left\| \frac{d^m}{dt^m} f \right\|_2^2 \le C^2 \right\}. \tag{2}$$

Here and later, $\|f\|_p^p = \int_0^1 |f(t)|^p \, dt$. Work of Efroimovich and Pinsker (1984) and Golubev and Nussbaum (1990), for example, shows how to construct estimates that are simultaneously minimax over a whole range of m and C. Those methods perform asymptotically as well when m and C are unknown as they would if these quantities were known.

Such results are limited to the case of L^2 smoothness measures. There are many other scales of function spaces, such as the Sobolev spaces,

$$W_p^m(C) = \left\{ f : \|f\|_p^p + \left\| \frac{d^m}{dt^m} f \right\|_p^p \le C^p \right\}. \tag{3}$$

If $p < 2$, then linear methods cannot attain the optimal rate of convergence over such a class when m and C are known (Nemirovskii 1985; Donoho and Johnstone 1994b). Thus adaptive linear methods cannot attain the optimal rate of convergence either. If one admits that not only the degree but also the type of smoothness are unknown, then it is not known how to estimate smooth functions adaptively.

In Section 2 we introduce a method, *SureShrink*, which is very simple to implement and attains much broader adaptivity properties than previously proposed methods. The properties apply over function classes measuring smoothness in traditional ways, such as (2), and also in less common but practically relevant ways, such as (3) and the Besov norms (see Sec. 3). The method is based on new results in multivariate normal decision theory that are interesting in their own right.

David L. Donoho and Iain M. Johnstone are Professors, Department of Statistics, Stanford University, CA 94305. The first author was supported at University of California Berkeley by National Science Foundation Grant DMS 88-10192, by NASA Contract NCA2-488, and by a grant from the ATT Foundation. The second author was supported in part by National Science Foundation Grants DMS 84-51750, 86-00235, and 92-09130 by National Institutes of Health, Public Health Service Grant CA 59039, and by a grant from the ATT Foundation. An early version of this article was presented as "Wavelets + Decision Theory = Optimal Smoothing" at the Wavelets and Applications Workshop, Luminy, France, March 10, 1991, and at the Workshop on Trends in the Analysis of Curve Data, University of Heidelberg, March 22, 1991. The authors thank P. Diaconis, T. Gasser, and R. Tibshirani for helpful comments and discussion.

© 1995 American Statistical Association
Journal of the American Statistical Association
December 1995, Vol. 90, No. 432, Theory and Methods

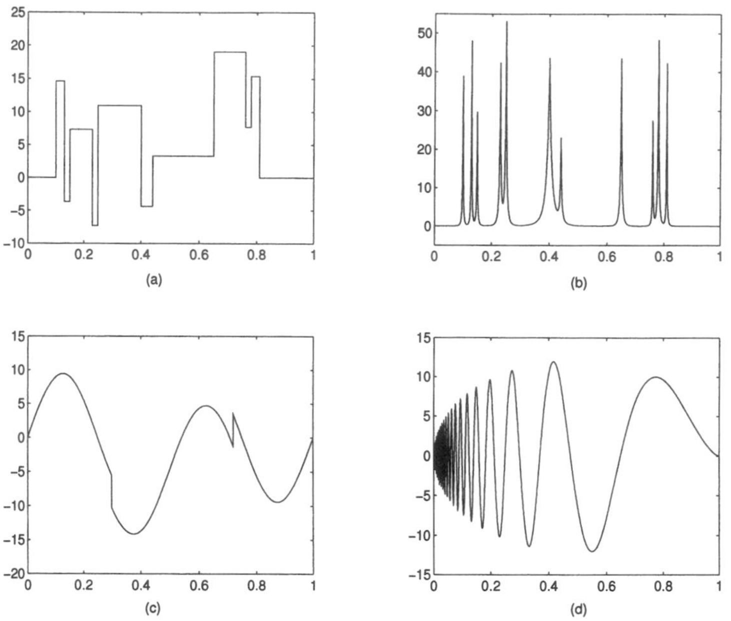

Figure 1. Four Spatially Variable Functions. (a) Blocks; (b) Bumps; (c) HeaviSine; (d) Doppler, N = 2,048. Formulas in Table 1.

SureShrink has the following ingredients (described in greater detail in Sec. 2):

1. *Discrete wavelet transform of noisy data.* The N noisy data are transformed via the discrete wavelet transform, to obtain N noisy wavelet coefficients $(y_{j,k})$.
2. *Thresholding of noisy wavelet coefficients.* Let

$$\eta_t(y) = \operatorname{sgn}(y)(|y| - t)_+ \qquad (4)$$

denote the *soft threshold,* which sets to zero data y below t in absolute value and pulls other data toward the origin by an amount t. The wavelet coefficients $y_{j,k}$ are subjected to soft thresholding with a level-dependent threshold level t_j^*.

Table 1. Formulas for Test Functions

Blocks	
	$f(t) = \sum h_j K(t - t_j) \quad K(t) = (1 + \operatorname{sgn}(t))/2.$
	$(t_j) = (.1, .13, .15, .23, .25, .40, .44, .65, .76, .78, .81)$
	$(h_j) = (4, -5, 3, -4, 5, -4.2, 2.1, 4.3, -3.1, 2.1, -4.2)$
Bumps	
	$f(t) = \sum h_j K((t - t_j)/w_j) \quad K(t) = (1 + \lvert t \rvert)^{-4}.$
	$(t_j) = t_{\text{Blocks}}$
	$(h_j) = (4, 5, 3, 4, 5, 4.2, 2.1, 4.3, 3.1, 5.1, 4.2)$
	$(w_j) = (.005, .005, .006, .01, .01, .03, .01, .01, .005, .008, .005)$
HeaviSine	
	$f(t) = 4 \sin 4\pi t - \operatorname{sgn}(t - .3) - \operatorname{sgn}(.72 - t).$
Doppler	
	$f(t) = \sqrt{t(1 - t)} \sin(2\pi(1 + \varepsilon)/(t + \varepsilon)), \ \varepsilon = .05.$

3. *Stein's unbiased estimate of risk (Sure) for threshold choice.* The level-dependent thresholds are found by regarding the different resolution levels (different j) of the wavelet transform as independent multivariate normal estimation problems. Within one level (fixed j), one has data $y_{j,k} = w_{j,k} + \varepsilon z_{j,k}, k = 0, \ldots, 2^j - 1$ and one wishes to estimate $(w_{j,k})_{k=0}^{2^j-1}$. Stein's unbiased estimate of risk for $\hat{\theta}_k^{(t)} = \eta_t(y_{j,k})$ gives an estimate of the risk for a particular threshold value t; minimizing this in t gives a selection of the threshold level for that level j. (A fixed threshold modification of this recipe is used in case the data vector has a very small l_2 norm.)

We briefly describe some examples of the method in action. Figure 1 depicts four specific functions f that we wavelet analyze repeatedly in this article:

- *Blocks.* A piecewise constant function, with jumps at $\{.1, .13, .15, .23, .25, .40, .44, .65, .76, .78, .81\}$.
- *Bumps.* A sum of bumps $\sum_{j=1}^{11} h_j K((t - t_j)/w_j)$ with locations t_j at the same places as jumps in *Blocks;* the heights h_j and widths s_j vary, and the individual bumps are of the form $K(t) = 1/(1 + |t|)^4$.
- *HeaviSine.* A sinusoid of period 1 with two jumps, at $t_1 = .3$ and $t_2 = .72$.
- *Doppler.* The variable-frequency signal $f(t) = \sqrt{t(1-t)}$ $\sin(2\pi \cdot 1.05/t + .05)$.

Precise formulas appear in Table 1. These examples have been chosen to represent various *spatially nonhomogeneous* phenomena. We regard *Blocks* as a caricature of the acoustic

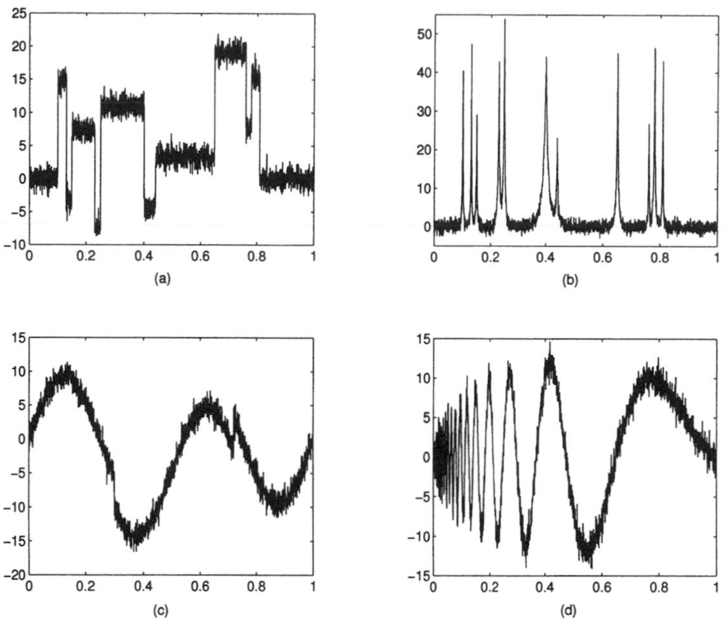

Figure 2. Four Functions With Gaussian White Noise, $\sigma = 1$, Rescaled to Have Signal-to-Noise Ratio $SD(f)/\sigma = 7$. (a) Noisy Blocks; (b) noisy Bumps; (c) noisy HeaviSine; (d) noisy Doppler.

impedance of a layered medium in geophysics and also of a 1-d profile along certain images arising in image-processing problems. We regard *Bumps* as a caricature of spectra arising, for example, in NMR, infrared, and absorption spectroscopy.

Figure 2 displays noisy versions of the same functions. The noise is independent $N(0, 1)$. Figure 3 displays the outcome of applying *SureShrink* (as described in Definition 1) in this case. The results are qualitatively appealing; the reconstructions jump where the true object jumps and are smooth where the true object is smooth. We emphasize that the same computer program, with the same parameters, produced all four reconstructions; no user intervention was permitted or required. *SureShrink* is automatically smoothness adaptive.

Section 3 gives a theoretical result showing that this smoothness adaptation is near optimal. *SureShrink* is asymptotically near minimax over large intervals of the Besov, Sobolev, and Triebel scales. Its speed of convergence is always the optimum one for the best smoothness condition obeyed by the true function, as long as the optimal rate is less than some "speed limit" set by the regularity of the wavelet basis. (By using increasingly higher-order wavelets, that is, wavelets with more vanishing moments and more smoothness, the "speed limit" may be expanded arbitrarily. The cost of such an expansion is a computational effort linearly proportional to the smoothness of the wavelet used.)

Linear methods like kernel, spline, and orthogonal series estimates, even with ideal choice of bandwidth, are unable to converge at the minimax speed over the members of the Besov, Sobolev, and Triebel scales involving L^p smoothness measures with $p < 2$. Thus *SureShrink* can achieve advantages over classical methods even at the level of rates. In fact such advantages are plainly visible in concrete problems where the object to be recovered exhibits significant spatial homogeneity. To illustrate this, Figure 4 shows an example of what can be accomplished by a representative adaptive linear method. This method applies the James–Stein shrinker (which may be interpreted as an adaptive linear shrinker; see Sec. 4.1) to the set of wavelet coefficients at each resolution level. It has a number of pleasant theoretical properties; it automatically achieves the minimax rate for linear estimates over large intervals of the Besov, Triebel, Sobolev, and Hölder scales. Nevertheless, Figure 4 shows that this adaptive linear method performs significantly worse than *SureShrink* in cases of significant spatial variability. A small simulation study described in Section 5 shows that for N in the range 10^3–10^4, *SureShrink* achieves the same level of performance with N samples that adaptive linear methods achieve for $2 \cdot N$ or $4 \cdot N$ samples.

To avoid possible confusion, we emphasize that the method *SureShrink* described in this article differs from variants *RiskShrink* and *VisuShrink* discussed by Donoho and Johnstone (1994a) and Donoho, Johnstone, Kerkyacharian, and Picard (1995) only in the choice of thresholds. Through use of a data based choice of threshold, *SureShrink* is more explicitly adaptive to unknown smoothness and has

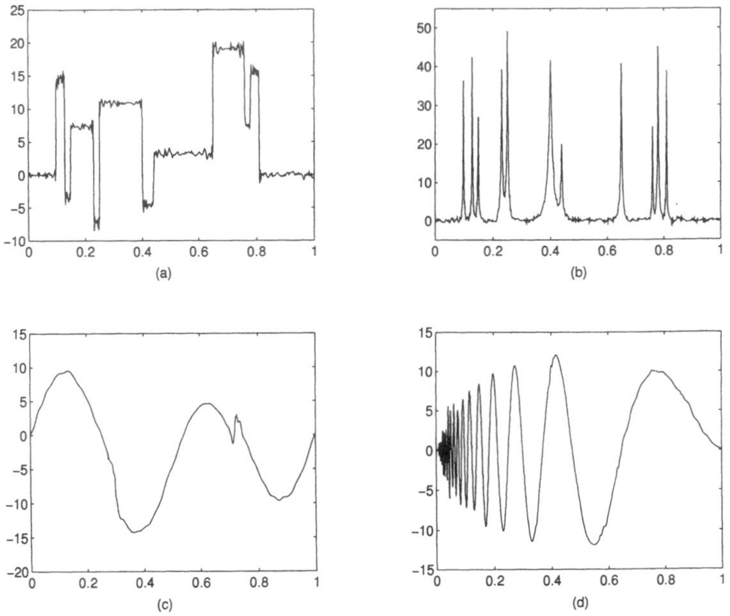

Figure 3. SureShrink Reconstruction Using Soft Thresholding, Most Nearly Symmetric Daubechies Wavelet with N = 8, and Cutoff L = 5. (a) SureShrink [Blocks]; (b) SureShrink [Bumps]; (c) SureShrink [HeaviSine]; (d) SureShrink [Doppler].

better large-sample MSE properties. For further comparative discussion, see Section 5.

2. SURESHRINK

We now describe in detail the ingredients of our procedure.

2.1 Discrete Wavelet Transform

Suppose we have data $y = (y_i)_{i=0}^{N-1}$, with $N = 2^J$. We consider here a family of discrete wavelet transforms, indexed by two integer parameters L and M, and one additional adjective "periodic" or "boundary adjusted." The parameter M represents the number of vanishing moments of the wavelet and L the coarsest resolution level considered; see (7). The construction relies heavily on concepts of Cohen, Daubechies, Jawerth, and Vial (1993), Daubechies (1992), and Meyer (1990, 1991). For a fixed value of M and L, we get a matrix $\mathcal{W}$; this matrix yields a vector $\mathbf{w}$ of the *wavelet coefficients* of $\mathbf{y}$ via

$$\mathbf{w} = \mathcal{W}\mathbf{y}.$$

For simplicity in exposition, we use the periodic version; in this case the transform is exactly orthogonal, so we have the inversion formula $\mathbf{y} = \mathcal{W}^T\mathbf{w}$. Brief comments on the minor changes needed for the boundary corrected version have been given by Donoho and Johnstone (1994a, sec. 4.6).

A crucial detail: The transform is implemented not by matrix multiplication, but by a sequence of special finite-length filtering steps that result in an order $O(N)$ transform. The choice of wavelet transform is essentially a choice of filter. (See Strang 1989 and especially Daubechies 1992 for the full story, and the appendix of Donoho, Johnstone, Kerkyacharian, and Picard 1995 for a brief summary relevant to our implementation.)

The vector $\mathbf{w}$ has $N = 2^J$ elements; it is convenient to index dyadically $N - 1 = 2^J - 1$ of the elements following the scheme

$$w_{j,k}: j = 0, \ldots, J-1; \qquad k = 0, \ldots, 2^j - 1;$$

we label the remaining element $w_{-1,0}$. To interpret these coefficients, let $\mathbf{W}_{j,k}$ denote the (j,k)th row of $\mathcal{W}$. The inversion formula $\mathbf{y} = \mathcal{W}^T\mathbf{w}$ becomes

$$y_i = \sum_{j,k} w_{j,k}\mathbf{W}_{j,k}(i),$$

expressing $\mathbf{y}$ as a sum of basis elements $\mathbf{W}_{j,k}$ with coefficients $w_{j,k}$.

In the special case $L = 0$ and $M = 0$, the transform reduces to the *discrete Haar transform.* Then, if $j \geq 0, \mathbf{W}_{j,k}(i)$ is proportional to 1 for $2^{-j}k \leq i/N < 2^{-j}(k + 1/2)$ and -1 for $2^{-j}(k + 1/2) \leq i/N < 2^{-j}(k+1)$. $\mathbf{W}_{-1,0}$ is proportional to the constant function 1. Thus the wavelet coefficients measure the differences of the function across various scales, and the function is reconstructed from building blocks of zero-mean localized square waves. Figure 5 shows a schematic of W for the (artificially small) sample size $N = 16$.

In the case $M > 0$, the building blocks of the transform are smoother than square waves. In that case, the vector

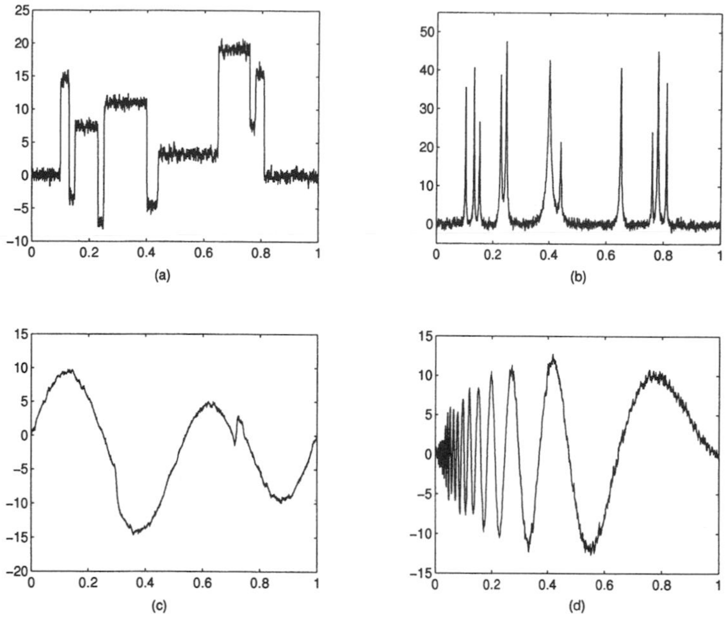

Figure 4. Reconstructions From Noisy Data Using WaveJS, Defined in Section 4.1. L = 5, S8 wavelet. (a) WaveJS [Blocks]; (b) WaveJS [Bumps]; (c) WaveJS [HeaviSine]; (d) WaveJS [Doppler].

$\mathbf{W}_{j,k}$, plotted as a function of i, has a continuous, wiggly, localized appearance that motivates the label "wavelet." For j and k bounded away from extreme cases by the condition

$$L < j \ll J, \qquad 0 \ll k \ll 2^j, \tag{5}$$

we have the approximation

$$\sqrt{N} \cdot \mathbf{W}_{j,k}(i) \approx 2^{j/2}\psi(2^j t) \qquad t = i/N - k2^{-j}, \tag{6}$$

where ψ is the mother wavelet arising in a wavelet transform on $\mathbb{R}$, as described by Daubechies (1988, 1992). This approximation improves with increasing N. ψ is an oscillating function of compact support. We thus speak of $\mathbf{W}_{j,k}$ as being localized to a spatial interval of size 2^{-j} and to have a frequency near 2^j. The basis element $\mathbf{W}_{j,k}$ has an increasingly smooth visual appearance, the larger the parameter M in the construction of the matrix $\mathcal{W}$. Daubechies (1988, 1992) has shown how the parameter M controls the smoothness (i.e., number of derivatives) of ψ; the smoothness is proportional to M.

The vectors $\mathbf{W}_{j,k}$ outside the range of (5) come in two types. First, there are those at $j < L$. These no longer resemble dilations of a mother wavelet ψ, and may no longer be localized. In fact, they may have support including all of $(0,1)$. They are, qualitatively, low-frequency terms. Second, there are those terms at $j \geq L$ that have k near the boundaries 0 and 2^j. These cases fail to satisfy (6). If the transform is periodized, this is because $\mathbf{W}_{j,k}$ is actually approximated by dilation of a circularly wrapped version of ψ. If the transform is boundary-adjusted, this is because the boundary element $\mathbf{W}_{j,k}$ is actually approximated by a boundary wavelet as defined by Cohen et al. (1993).

Figure 6 displays $\mathbf{W}_{j,k}$ for $j = 6, k = 32$ (and $N = 2{,}048$), in four cases corresponding to specific wavelet filter sequences. The smoother wavelets have broader support.

The usual displays of wavelet transforms use S. Mallat's idea of multiresolution decomposition (Mallat 1989a,b). This adapts in the present situation as follows. Let $\mathbf{x} = (x_i)_{i=0}^{N-1}$ be the data, let

$$V_L\mathbf{x} = \sum_{j<L} w_{j,k}\mathbf{W}_{j,k} \tag{7}$$

denote the partial reconstruction from "gross structure" terms, and for $j \geq L$, let

$$W_j\mathbf{x} = \sum_{0\leq k<2^j} w_{j,k}\mathbf{W}_{j,k} \tag{8}$$

denote the partial reconstruction from terms at resolution level j, or scale 2^{-j}. Then $\mathbf{x}$ can be recovered from these components via $\mathbf{x} = V_L\mathbf{x} + \sum_{L\leq j<J} W_j\mathbf{x}$, and it is usual to examine the behavior of the components by displaying the graphs of $V_L\mathbf{x}$ and of $W_j\mathbf{x}$ for $j = L, L+1, \ldots, J-1$. In Figure 7 we do this for our four functions and the S8 wavelet. In Figure 8 we look just at the *Blocks* and *HeaviSine* functions to contrast the Haar and Daubechies D4 transforms.

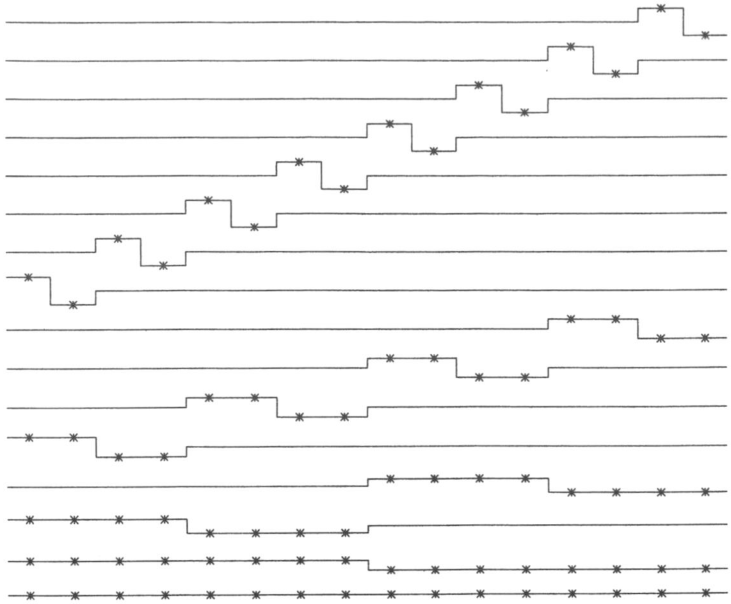

Figure 5. Schematic of Nonzero Entries of the Haar Wavelet Transform Matrix, M = L = 0, N = 16. Lines connect entries in rows $i \rightarrow W_{j,k}(i)$, with stars representing nonzero values (magnitudes equal $\sqrt{2^j/N}$.)

A less usual way to display wavelet transforms is to look at the wavelet coefficients directly. We do this in Figure 9. The display at level j depicts $w_{j,k}$ by a vertical line of height proportional to $w_{j,k}$ at horizontal position $k/2^j$. The low-resolution coefficients at $j < L$ are not displayed. The coefficients displayed are those of the S8 wavelet analysis of the four functions under consideration.

Note the considerable sparsity of the wavelet coefficient plots. In all of these plots more than 2,000 coefficients are displayed, but only a small fraction are nonzero at the resolution of the 300 dot-per-inch laser printer. It is also of interest to note the position of the nonzero coefficients, which at high-resolution number j cluster around the discontinuities and spatial nonhomogeneities of the function f. This is an instance of the data-compression properties of the wavelet transform. Indeed, the transform preserves the sum of squares, but in the wavelet coefficients this sum of squares is concentrated in a much smaller fraction of the components than in the raw data.

For comparison, we display in Figure 10 the Haar coefficients of the object; the compression is very pronounced for object *Blocks,* and in fact better than in the S8 case, but the compression is less effective for object *HeaviSine*—much less so than for the S8-based transform.

2.2 Thresholding of Noisy Wavelet Coefficients

The orthogonality of the (periodized) discrete wavelet transform has a fundamental statistical consequence: $\mathcal{W}$ transforms white noise into white noise. Hence if $(y_{j,k})$ are the wavelet coefficients of $(y_i)_{i=0}^{N-1}$ collected according to model (1) and $w_{j,k}$ are the wavelet coefficient of $(f(t_i))$, then

$$y_{j,k} = w_{j,k} + \sigma z_{j,k}, \tag{9}$$

where $z_{j,k}$ is an iid $N(0,1)$ noise sequence. Hence the wavelet coefficients of a noisy sample are themselves just noisy versions of the noiseless wavelet coefficients.

Moreover, $\mathcal{W}$ transforms estimators in one domain into estimators in the other domain, with isometry of risks. If $\hat{w}_{j,k}$ are estimates of the wavelet coefficients, then there is an estimate $\hat{\mathbf{f}}$ of $\mathbf{f} = (f(t_i))$ in the other domain obtained by

$$\hat{\mathbf{f}} = \mathcal{W}^T \hat{\mathbf{w}},$$

and the losses obey the Parseval relation

$$\|\hat{\mathbf{w}} - \mathbf{w}\|_2 = \|\hat{\mathbf{f}} - \mathbf{f}\|_2.$$

The connection also goes in the other direction: If $\hat{\mathbf{f}}$ is any estimator of $\mathbf{f}$, then $\hat{\mathbf{w}} = \mathcal{W}\hat{\mathbf{f}}$ defines an estimator with isometric risk.

The foregoing data-compression remarks were meant to suggest that most of the coefficients in a noiseless wavelet transform are effectively zero. Accepting this slogan, one reformulates the problem of recovering f as one of recovering those few coefficients of f that are significantly nonzero against a Gaussian white noise background.

This motivates the use of a thresholding scheme that "kills" small $y_{j,k}$ and "keeps" large $y_{j,k}$. The particular soft

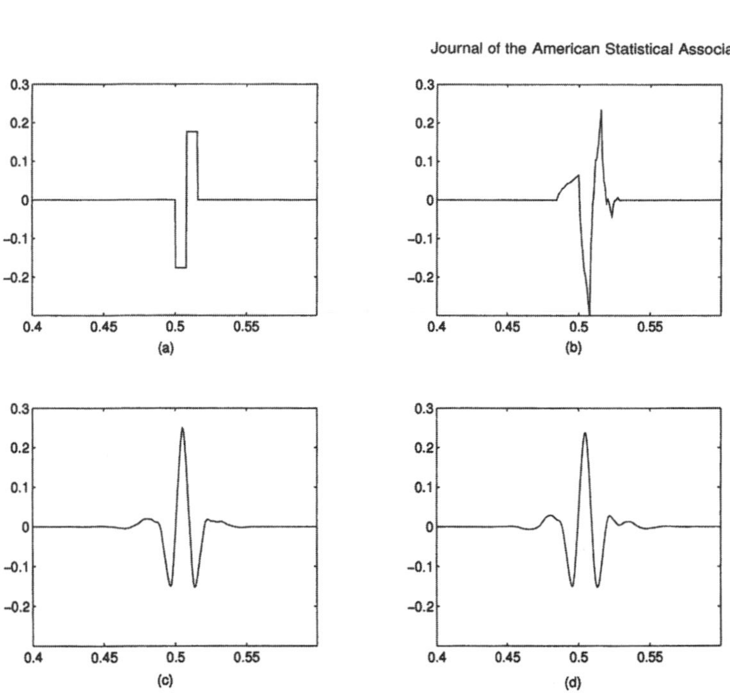

Figure 6. "Typical" Rows of the Wavelet Transform Matrix $\mathcal{W}$ Corresponding to $j = 6$, $k = 32$ in Four Cases. (a) Haar wavelet $L = 0$, $M = 0$; (b) Daubechies D4 wavelet $L = 2$, $M = 2$; (c) Coiflet C3 $M = 9$; (d) Daubechies "nearly linear phase" S8 wavelet $M = 9$.

thresholding scheme (4) that we introduced earlier is an instance of this.

Figure 3 has already shown the results such a scheme can provide in the domain of the original data (using S8). Figure 11 illustrates how this works in the wavelet domain using the Haar transform of a noisy version of *Blocks*.

The reconstruction obtained here is via the device of selecting from the noisy wavelet coefficients at level j a threshold t_j^* and applying this threshold to all the empirical wavelet coefficients at level j; the reconstruction is then $\hat{\mathbf{f}} = \mathcal{W}^T\hat{\mathbf{w}}$. Obviously, the choice of threshold t_j^* is crucial.

2.3 Threshold Selection by SURE

Let $\boldsymbol{\mu} = (\mu_i : i = 1, \ldots, d)$ be a d-dimensional vector, and let $x_i \sim N(\mu_i, 1)$ be multivariate normal observations with that mean vector. Let $\hat{\boldsymbol{\mu}} = \hat{\boldsymbol{\mu}}(\mathbf{x})$ be a particular fixed estimator of $\boldsymbol{\mu}$. Stein (1981) introduced a method for estimating the loss $\|\hat{\boldsymbol{\mu}} - \boldsymbol{\mu}\|^2$ in an unbiased fashion. Stein showed that for a nearly arbitrary, nonlinear biased estimator, one can nevertheless estimate its loss unbiasedly.

Write $\hat{\boldsymbol{\mu}}(\mathbf{x}) = \mathbf{x} + \mathbf{g}(\mathbf{x})$, where $\mathbf{g} = (g_i)_{i=1}^d$ is a function from R^d into R^d. Stein (1981) showed that when $\mathbf{g}(\mathbf{x})$ is weakly differentiable, then

$$E_{\boldsymbol{\mu}}\|\hat{\boldsymbol{\mu}}(\mathbf{x}) - \boldsymbol{\mu}\|^2 = d + E_{\boldsymbol{\mu}}\{\|\mathbf{g}(\mathbf{x})\|^2 + 2\nabla \cdot \mathbf{g}(\mathbf{x})\}, \quad (10)$$

where

$$\nabla \cdot \mathbf{g} \equiv \sum_i \frac{\partial}{\partial x_i} g_i.$$

Now consider the soft threshold estimator $\hat{\mu}_i^{(t)}(\mathbf{x}) = \eta_t(x_i)$ and apply Stein's result. $\hat{\boldsymbol{\mu}}^{(t)}$ is weakly differentiable in Stein's sense, and so we get from (10) that the quantity

$$\mathrm{SURE}(t;\mathbf{x}) = d - 2 \cdot \#\{i : |x_i| \le t\} + \sum_{i=1}^{d} (|x_i| \wedge t)^2 \quad (11)$$

is an unbiased estimate of risk: $E_{\boldsymbol{\mu}}\|\hat{\boldsymbol{\mu}}^{(t)}(\mathbf{x}) - \boldsymbol{\mu}\|^2 = E_{\boldsymbol{\mu}}\mathrm{SURE}(t;\mathbf{x})$. Here $a \wedge b = \min(a, b)$.

Consider using this estimator of risk to *select* a threshold,

$$t^S = \operatorname{argmin}_{0 \le t \le \sqrt{2\log d}} \mathrm{SURE}(t;\mathbf{x}). \quad (12)$$

Arguing heuristically, one expects that for large dimension d, a sort of statistical regularity will set in, the law of large numbers will ensure that SURE is close to the true risk, and that t^S will be almost the optimal threshold for the case at hand. Theory developed later will show that this hope is justified (and explains the choice of upper bound at $\sqrt{2\log d}$.)

Computational evidence that t^S is a reasonable threshold selector is given in Figure 12. A vector $\boldsymbol{\mu}$ of dimension $d = 128$ consists of 16 consecutive 4's, followed by all zeros. White Gaussian noise of variance 1 was added (Fig. 12a). The profile of SURE(t) is displayed in Fig. 12c; it quite closely resembles the actual loss (Fig. 12d), which of course we know in this (artificial) example. The SURE principle was used to select a threshold that is applied to the data, resulting in an estimate of the mean vector (Fig. 12b). This

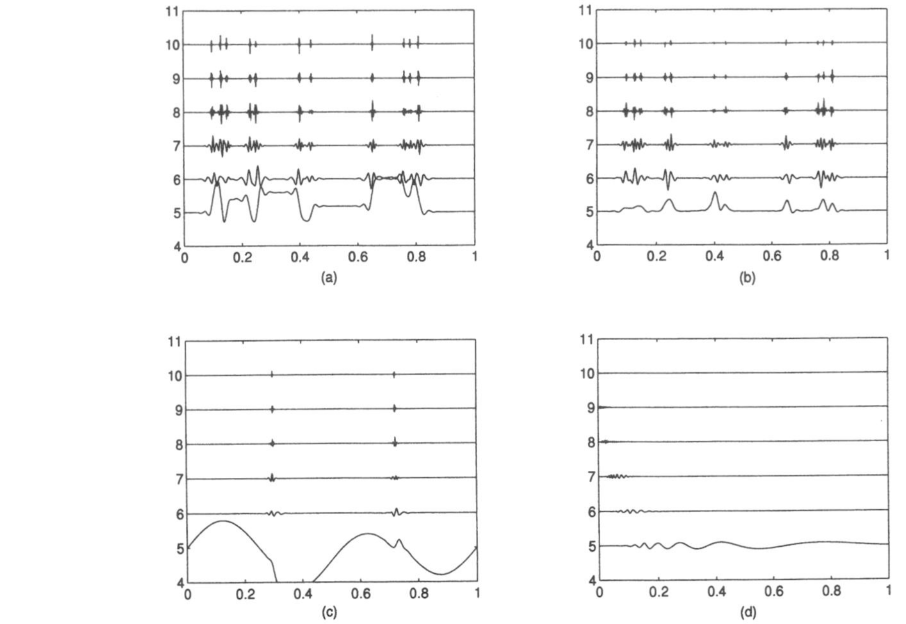

Figure 7. Mallat's Multiresolution Decomposition of the Four Basic Functions—(a) Blocks, (b) Bumps, (c) HeaviSine, (d) Doppler—S8 Wavelet. The line with baseline height j ($5 \leq j \leq 10$) plots the partial reconstruction $(W_j x)_i$ (cf. (8)) against i/N, i = 0, ... N − 1, and the bottom line similarly shows the "gross structure" decomposition $V_L x$ (cf. (7); here L = 4.)

estimate is sparse and much less noisy than the raw data (Fig. 12a). Note also the shrinkage of the nonzero part of the signal.

The optimization problem (12) is computationally straightforward. Suppose, without any loss of generality, that the x_i have been reordered in order of increasing $|x_i|$. Then on intervals of t that lie between two values of $|x_i|$, SURE(t) is strictly increasing, indeed quadratic. Therefore, the minimum value t^S is one of the data values $|x_i|$. There are only d such values; when they have been already arranged in increasing order, the collection of all values SURE($|x_i|$) may be computed in order $O(d)$ additions and multiplications, with appropriate arrangement of the calculations. It may cost as much as order $O(d\log(d))$ calculations to arrange the $|x_i|$ in order; so the whole effort to calculate t^S is order $O(d\log(d))$. This is scarcely worse than the order $O(d)$ calculations required simply to apply either form of thresholding.

2.4 Threshold Selection in Sparse Cases

The SURE principle just described has a serious drawback in situations of extreme sparsity of the wavelet coefficients. In such cases the noise contributed to the SURE profile by the many coordinates at which the signal is zero swamps the information contributed to the SURE profile by the few coordinates where the signal is nonzero. Consequently, *SureShrink* uses a Hybrid scheme.

Figure 13a depicts the results of a small-scale simulation study. A vector μ of dimension d = 1,024 contained $\lfloor \varepsilon \cdot d \rfloor$ nonzero elements, all of size C. Independent $N(0, 1)$ noise was added. The SURE estimator t^S was applied. Amplitudes $C = 3, 5$, and 7 were tried, and sparsities $\varepsilon = \{.005, .01, .02(.02).20, .25\}$ were studied. 25 replications were tried at each parameter combination, and the root MSE's are displayed in Figure 13. Evidently, the root MSE's do not tend to zero linearly as the sparsity tends to zero. For the theoretical results of Section 3, such behavior would be unacceptable.

In contrast, Figure 13b portrays the results of the same experiment, with a "fixed thresholding" estimator $\hat{\mu}^F$, where the threshold is set to $t_d^F = \sqrt{2\log(d)}$ independent of the data. The losses tend to be larger than SURE for "dense" situations $\varepsilon \gg 0$, but much smaller for ε near zero. We have developed the rationale for the choice $\sqrt{2\log(d)}$ in earlier work (Donoho and Johnstone 1994a). To summarize, first, the maximum of N iid standard Gaussian variates is smaller than $\sqrt{2\log N}$, with probability increasing to 1 as N increases. Thus with high probability, a pure noise signal is correctly estimated as being identically zero. Second, the threshold t_d^F is also asymptotically optimal in a MSE sense for mimicking the MSE of an "oracle" that knows which coordinates of the mean vector are larger than the standard deviation of the noise.

Figure 13c displays the results of applying a hybrid method that we label $\hat{\mu}^*$, which is designed to behave like $\hat{\mu}^S$ in dense situations and like $\hat{\mu}^F$ in sparse ones. Its performance is roughly as desired.

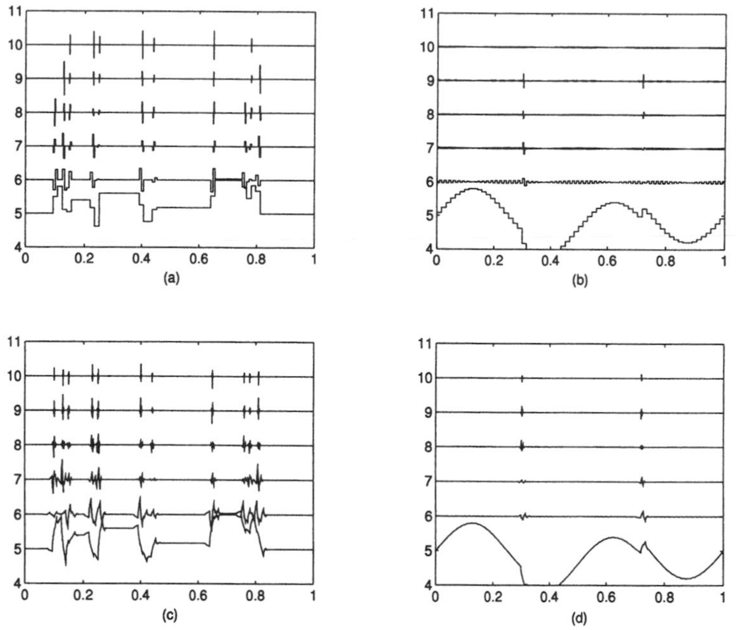

Figure 8. Mallat's Multiresolution Decomposition of Blocks and HeaviSine Using the Haar (a and b) and D4 (c and d) wavelets. Here $L = 4$.

In detail, the Hybrid method works as follows: Define $s_d^2 = d^{-1}\sum_i (x_i^2 - 1)$ and let γ_d be a critical value, which for the present we take as $\log_2^{3/2} d/\sqrt{d}$. Let I denote a random subset of half the indices in $\{1, \ldots, d\}$ and let I' denote its complement. Let t_I^S and $t_{I'}^S$ denote the minimizers of SURE for the given subsets of indices, only with an additional restriction on the search range,

$$t_I^S = \mathrm{argmin}_{0 \le t \le t_d^F} \mathrm{SURE}(t, (x_i)_{i \in I}),$$

and similarly for $t_{I'}^S$. Define the estimate

$$\begin{aligned} \hat{\boldsymbol{\mu}}^*(\mathbf{x})_i &= \eta_{t_d^F}(x_i) && s_d^2 \le \gamma_d, \\ &= \eta_{t_I^S}(x_i) && i \in I' \text{ and } s_d^2 > \gamma_d, \\ &= \eta_{t_{I'}^S}(x_i) && i \in I \text{ and } s_d^2 > \gamma_d. \end{aligned} \quad (13)$$

In other words, we use one half-sample to estimate the threshold for use with the other half-sample; but unless there is convincing evidence that the signal is nonnegligible, we set the threshold to $\sqrt{2\log(d)}$.

This half-sample scheme was developed for the proof of Theorems 3 and 4 in Sections 3.2 and 3.3. In practice, the half-sample aspect of the estimate seems unnecessary; the simpler estimator $\hat{\mu}^+$ derived from

$$\begin{aligned} \hat{\mu}^+(\mathbf{x})_i &= \eta_{t_d^F}(x_i) && s_d^2 \le \gamma_d \\ &= \eta_{t^S}(x_i) && s_d^2 > \gamma_d, \end{aligned} \quad (14)$$

offers the same performance benefits in simulations and in fact is used in all the examples in this article; see Figure 13d.

Note Added in Proof. We have since developed a proof of Theorems 3 and 4 that applies to the more natural estimator $\hat{\mu}^+$, making the introduction of random half-sampling unnecessary. It is hoped to include details in the written version of our March 1995 lectures in Oberwolfach.

We now apply this multivariate normal theory in our wavelet setting.

Definition 1. The term *SureShrink* refers to the following estimator $\hat{\mathbf{f}}^*$ of $\mathbf{f}$. Assuming that $N = 2^J$ and that the noise is normalized so that it has standard deviation $\sigma = 1$, we set $\mathbf{x}_j = (y_{j,k})_{0 \le k < 2^j}$ and

$$\begin{aligned} \hat{w}^*_{j,k} &= y_{j,k}, && j < L, \\ &= (\boldsymbol{\mu}^*(\mathbf{x}_j))_k, && L \le j < L; \end{aligned}$$

the estimator $\hat{\mathbf{f}}^*$ derives from this via inverse discrete wavelet transform. We use $\hat{\mathbf{f}}^+$ to denote the variant using (14) used in practice.

Note that $\hat{\mathbf{f}}^*$ is fully automatic, modulo the choice of specific wavelet transform. Moreover, with appropriate arrangement of the work, the whole computational effort involved is order $O(N\log(N))$, scarcely worse than linear in the sample size N. Extensive experience with computations on a Macintosh show that performance is quite reasonable even on personal computers. The Matlab command *SureShrink* takes a few seconds to complete on an array of size $N = 4{,}096$.

3. MAIN ADAPTIVITY RESULT

In this section we investigate the adaptivity of *SureShrink*

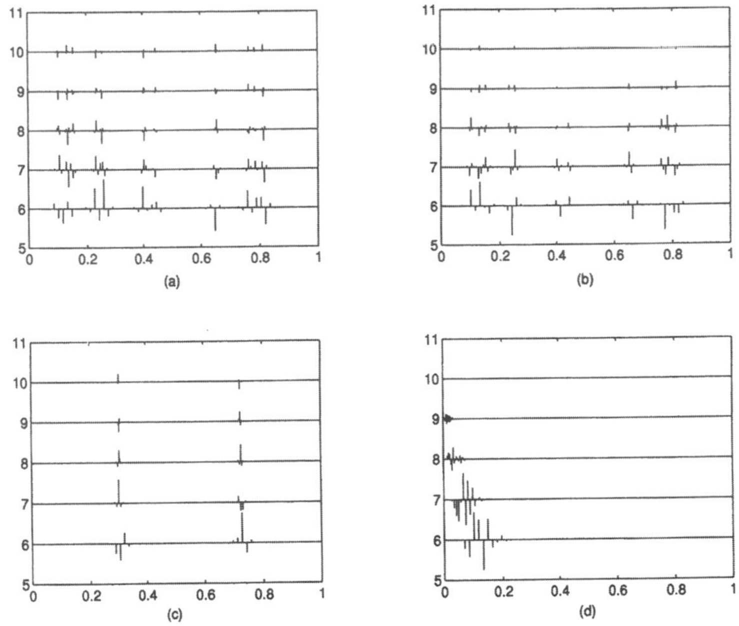

Figure 9. Plot of Wavelet Coefficients Using S8. (a) Blocks; (b) Bumps; (c) HeaviSine; (d) Doppler. The display at level j depicts w_{jk} by a vertical line of height proportional to $w_{j,k}$ at horizontal position $k2^{-j}$.

to unknown degree of smoothness. To state our result, we must define Besov spaces. We follow DeVore and Popov (1988). Let $\Delta_h^{(r)} f$ denote the rth difference $\sum_{k=0}^{r} \binom{r}{k} (-1)^k f(t+kh)$. The rth modulus of smoothness of f in $L^p[0,1]$ is

$$w_{r,p}(f;h) = \|\Delta_h^{(r)} f\|_{L^p[0,1-rh]}.$$

The *Besov* seminorm of index (σ, p, q) is derived for $r > \sigma$ by

$$|f|_{B^\sigma_{p,q}} = \left(\int_0^1 \left(\frac{w_{r,p}(f;h)}{h^\sigma} \right)^q \frac{dh}{h} \right)^{1/q}$$

if $q < \infty$, and by

$$|f|_{B^\sigma_{p,\infty}} = \sup_{0<h<1} \frac{w_{r,p}(f;h)}{h^\sigma}$$

if $q = \infty$. The *Besov ball* $B^\sigma_{p,q}(C)$ (resp. *space* $B^\sigma_{p,q}$) is then the class of functions $f : [0,1] \to \mathbb{R}$ satisfying $f \in L^p[0,1]$ and $|f|_{B^\sigma_{p,q}} \le C$ (resp. $|f|_{B^\sigma_{p,q}} < \infty$). Standard references on Besov spaces are works of Peetre (1975) and Triebel (1983, 1990).

This measure of smoothness includes, for various settings (σ, p, q), other commonly used measures. For example, let C^δ denote the *Hölder class* of functions with $|f(s) - f(t)| \le c|s-t|^\delta$ for some $c > 0$. Then f has for a given $m = 0, 1, \ldots$ a distributional derivative $f^{(m)}$ satisfying $f^{(m)} \in C^\delta$, $0 < \delta < 1$, if and only if $|f|_{B^{m+\delta}_{\infty,\infty}} < \infty$. Similarly, with W_2^m the L^2 Sobolev space as in Section 1, $f \in W_2^m$ iff $|f|_{B^m_{2,2}} < \infty$.

The Besov scale essentially includes other less traditional spaces as well. For example, recall the definition of total variation of a function,

$$TV(f) = \sup \Big\{ \sum |f(t_i) - f(t_{i-1})| : 0 = t_0 < t_1 < \cdots < t_{k-1} \le t_k = 1, k \in \mathcal{N} \Big\}$$

and note that the space of functions of bounded variation is a superset of $B^1_{1,1}$ and a subset of $B^1_{1,\infty}$. Similarly, all the L^p Sobolev spaces W_p^m contain $B^m_{p,1}$ and are contained in $B^m_{p\infty}$.

For the theoretical results, we permit alternate choices of cutoff γ_d for the pretest in *SureShrink;* for example, $\gamma_d = d^\gamma, 0 < \gamma < \frac{1}{2}$. We recall that $a_N \asymp b_N$ means $\liminf_N |a_N/b_N| > 0$ and $\limsup_N |a_N/b_N| < \infty$. Let the minimax risk be denoted by

$$R(N; B^\sigma_{p,q}(C)) = \inf_{\hat{\mathbf{f}}} \sup_{B^\sigma_{p,q}(C)} R(\hat{\mathbf{f}}, \mathbf{f}).$$

We note that for the ranges of (σ, p, q) considered later, $R(N; B^\sigma_{p,q}(C)) \asymp N^{-r}$, with $r = \sigma/(\sigma + \frac{1}{2})$.

Theorem 1. Let the discrete wavelet analysis correspond to a wavelet ψ having r null moments and r continuous derivatives, $r > \max(1, \sigma)$. Then, *SureShrink* is si-

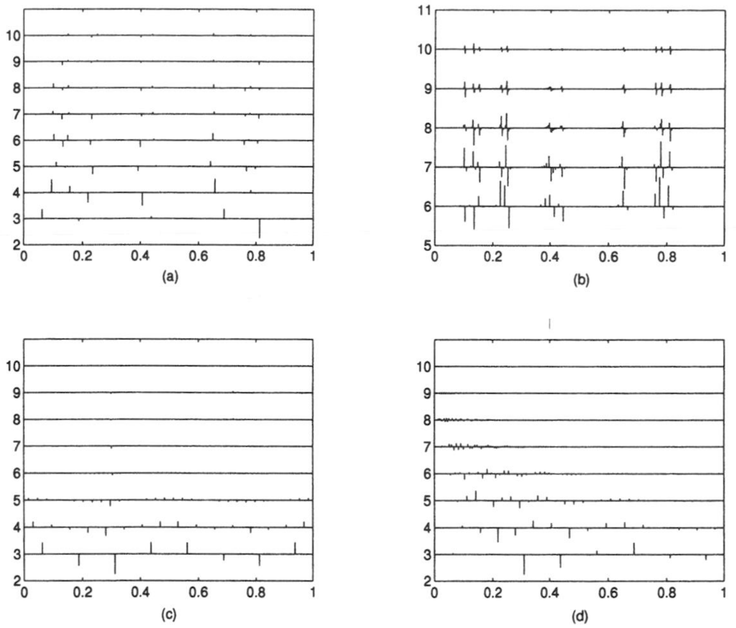

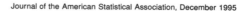

Figure 10. Wavelet Coefficients Using the Haar Wavelet. (a) Blocks; (b) Bumps; (c) HeaviSine; (d) Doppler. Compare amounts of compression with Figure 9.

multaneously nearly minimax,

$$\sup_{B^{\sigma}_{p,q}(C)} R(\hat{\mathbf{f}}^*, \mathbf{f}) \asymp R(N; B^{\sigma}_{p,q}(C)) \qquad N \to \infty,$$

for all $p, q \in [1, \infty]$, for all $C \in (0, \infty)$, and for all $\sigma_0 < \sigma < r$. In particular,

$$\gamma_d = \log^{3/2} d/\sqrt{d}, \Rightarrow$$

$$\sigma_0 = \max\left(\frac{1}{p}, 2\left(\frac{1}{p} - \frac{1}{2}\right)_+\right)$$

and

$$\gamma_d = d^{\gamma}, \qquad 0 < \gamma < \frac{1}{2}, \Rightarrow$$

$$\sigma_0 = \max\left(\frac{1}{p}, 2\left(\frac{1}{p} - \frac{1}{2}\right)_+ + \gamma - \frac{1}{2}\right).$$

In words, this estimator, which "knows nothing" about the a priori degree, type, or amount of regularity of the object, nevertheless achieves the optimal rate of convergence that one could attain by knowing such regularity. Over a Hölder class, it attains the optimal rate; over an L^2 Sobolev class, it achieves the optimal rate; and over Sobolev classes with $p < 2$, it also achieves the optimal rate.

We mentioned in Section 1 that no linear estimator achieves the optimal rate over all L^p Sobolev classes; as a result, the modification of *SureShrink* achieves something that usual estimates could not, even if the optimal bandwidth were known a priori.

Many other results along these lines could be proved, for other (σ, p, q). One particularly interesting result, because it refers to the Haar basis, is the following.

Theorem 2. Let $\mathcal{V}(C)$ denote the class of all functions on the unit interval of total variation $\leq C$. Let $\hat{f}^*$ denote the application of *SureShrink* in the Haar basis, with $\gamma_d = d^{\gamma}, 0 < \gamma < \frac{1}{2}$. This "*HaarShrink*" estimator is simultaneously nearly minimax,

$$\sup_{\mathcal{V}(C)} R(\hat{\mathbf{f}}^*, \mathbf{f}) \asymp R(N; \mathcal{V}(C)) \qquad N \to \infty$$

for all $C \in (0, \infty)$.

Again, without knowing any a priori limit on the total variation, the estimator behaves essentially as well as one could by knowing this limit. Figure 11 shows the plausibility of this result.

3.1 Estimation in Sequence Space

Our proof of Theorem 1 uses a method of sequence spaces described in earlier work (Donoho and Johnstone 1995). The key idea is to approximate the problem of estimating a function from finite noisy data by the problem of estimating an infinite sequence of wavelet coefficients contaminated with white noise.

The heuristic for this replacement is as follows. From (6) and (9), the empirical wavelet coefficient is $y_{j,k} = w_{j,k}$

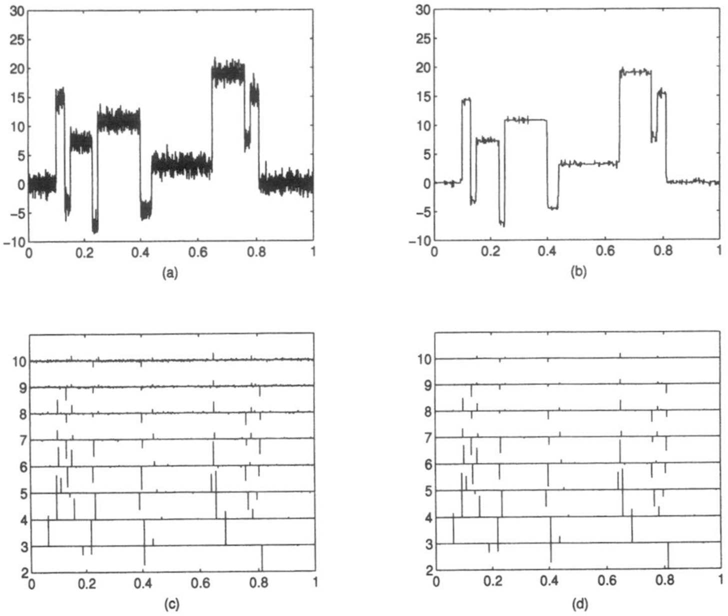

Figure 11. Component Steps of SureShrink Reconstruction. (a) Raw data: a noisy version of Blocks; (b) SureShrink reconstruction using Haar wavelet, and L = 2; (c) raw wavelet coefficients of the data; (d) the same coefficients after thresholding.

$+ \sigma z_{j,k}$, where the discrete $w_{j,k}$ obeys

$$w_{j,k} \approx \sqrt{N} \int f(t)\psi_{j,k}(t)\, dt$$

for a certain wavelet $\psi_{j,k}(t)$. In terms of the continuous wavelet coefficients $\theta_{j,k} = \int f(t)\psi_{j,k}(t)\, dt$, it is thus tempting to act as though our observations were actually

$$\sqrt{N} \cdot \theta_{j,k} + \sigma z_{j,k}$$

or, equivalently,

$$\theta_{j,k} + \varepsilon z_{j,k},$$

where $\varepsilon = \sigma/\sqrt{N}$ and $z_{j,k}$ is still a standard iid, $N(0, 1)$ sequence. Moreover, due to the Parseval relation, $\|\hat{\mathbf{f}} - \mathbf{f}\|_2 = \|\hat{\mathbf{w}} - \mathbf{w}\|_2$, and the foregoing approximation, we are also tempted to act as if the loss $N^{-1}\|\hat{\mathbf{f}} - \mathbf{f}\|_2^2$ were the same as $\|\hat{\theta} - \theta\|_2^2$.

These (admittedly vague) approximation heuristics lead to the study of the following sequence space problem. We observe an infinite sequence of data,

$$y_{j,k} = \theta_{j,k} + z_{j,k} \qquad j \geq 0, \qquad k = 0, \ldots, 2^j - 1, \quad (15)$$

where $z_{j,k}$ are iid. $N(0, \varepsilon^2)$ and $\theta = (\theta_{j,k})$ is unknown. We wish to estimate θ with small squared error loss $\|\hat{\theta} - \theta\|_2^2 = \sum(\hat{\theta}_{j,k} - \theta_{j,k})^2$. We let $\Theta(\sigma, p, q, C)$ denote the set of all wavelet coefficient sequences $\theta = (\theta_{j,k})$ arising from an $f \in B^\sigma_{p,q}(C)$. Finally, we search for a method $\hat{\theta}$ that is simultaneously nearly minimax over a range of $\Theta(\sigma, p, q, C)$.

Suppose that we can solve this sequence problem. Under certain conditions on σ, p, and q, this will imply Theorem 1. Specifically, if σ_0 is big enough and the wavelet is of regularity $r > \sigma_0$, an estimator that is simultaneously near minimax in the sequence space problem $\sigma_0 < \sigma < r$ may be applied to the empirical wavelet coefficients in the original problem under study and will also be simultaneously near minimax in the original function space problem. We have already discussed the approximation arguments necessary to establish this correspondence (Donoho 1992; Donoho and Johnstone 1994c), and for reasons of space we omit them. (See also Brown and Low 1990.)

3.2 Adaptive Estimation over Besov Bodies

The collections $\Theta(\sigma, p, q, C)$ of wavelet expansions $\theta = \theta(f)$ arising from functions $f \in B^\sigma_{p,q}(C)$ are related to certain simpler sets that we have called (Donoho and Johnstone 1994c) *Besov bodies*. These are sets $\|\theta\|_{\mathbf{b}^s_{p,q}} \leq C$, where

$$\|\theta\|^q_{\mathbf{b}^s_{p,q}} = \sum_{j \geq 0} \left(2^{js} \left(\sum_{0 \leq k < 2^j} |\theta_{j,k}|^p \right)^{1/p} \right)^q \qquad (16)$$

and $s = \sigma + \frac{1}{2} - 1/p$.

Consider the problem of estimating θ when it is observed in a Gaussian white noise and is known a priori to lie in a

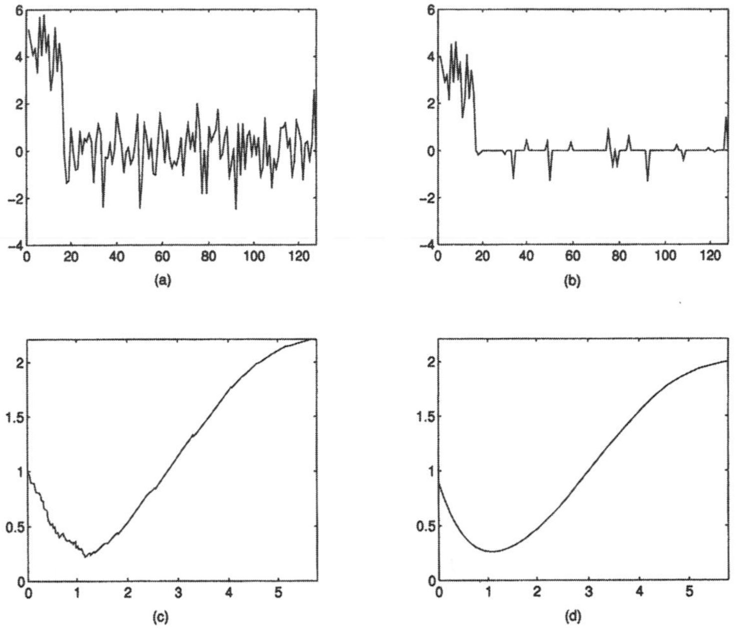

Figure 12. Illustration of Choice of Threshold Using SURE(t). See Section 2.3. (a) raw data; (b) estimate; (c) estimated risk versus lambda; (d) loss versus lambda.

certain convex set $\Theta^s_{p,q}(C) \equiv \{\theta : \|\theta\|_{\mathbf{b}^s_{p,q}} \le C\}$. We often put for short $\Theta^s_{p,q} = \Theta^s_{p,q}(C)$. The difficulty of estimation in this setting is measured by the *minimax risk*

$$R^*(\varepsilon;\Theta^s_{p,q}) = \inf_{\hat{\theta}} \sup_{\Theta^s_{p,q}} E\|\hat{\theta}-\theta\|_2^2, \tag{17}$$

and the minimax risk among threshold estimates is

$$R^*_T(\varepsilon;\Theta^s_{p,q}) = \inf_{(t_j)} \sup_{\Theta^s_{p,q}} E\|\hat{\theta}_{(t_j)}-\theta\|_2^2, \tag{18}$$

where $\hat{\theta}_{(t_j)}$ stands for the estimator $(\eta_t, (y_{j,k}))_{j,k}$. We have shown (Donoho and Johnstone 1995) that $R^*_T \le \Lambda(p)\cdot R^* \cdot (1+o(1))$ with, for example, $\Lambda(1) \approx 1.6$. Hence threshold estimators are nearly minimax. Furthermore, they show that the minimax risk and minimax threshold risk over sets $\Theta^s_{p,q}(C)$ are equivalent, to within constants, to that over sets $\Theta(\sigma,p,q,C)$, provided that σ is large enough and we make the calibration $s = \sigma + \frac{1}{2} - 1/p$.

We may construct a *SureShrink*-style estimator in this problem by applying $\boldsymbol{\mu}^*$ level by level. Let $\mathbf{x}_j = (y_{j,k}/\varepsilon)_{k=0}^{2^j-1}$. Then set

$$\hat{\theta}^*_{j,k}(\mathbf{y}) = y_{j,k}, \qquad j < L \tag{19}$$

and

$$\hat{\theta}^*_{j,k}(\mathbf{y}) = \varepsilon \cdot \hat{\mu}^*(\mathbf{x}_j)_k \qquad j \ge L. \tag{20}$$

This is a particular adaptive threshold estimator.

Theorem 3. If either (a) $\gamma_d = d^{-1/2}\log^{3/2} d$ and $s > |\frac{1}{p} - \frac{1}{2}|$, or (b) $\gamma_d = d^{-\gamma}, 0 < \gamma < \frac{1}{2}$, and $s > |\frac{1}{p} - \frac{1}{2}| + \gamma - \frac{1}{2}$, then

$$\sup_{\Theta^*_{p,q}(C)} E_\theta\|\hat{\theta}^* - \theta\|_2^2 \le R^*_T(\varepsilon;\Theta^s_{p,q}(C))(1+o(1))$$

as $\varepsilon \to 0$.

In short, without knowing s, p, q, or C, one obtains results as good asymptotically as if one did know those parameters. The result is effective across an infinite range of all the parameters in question. Because the minimax risk is close to the minimax threshold risk, this solves the problem of adapting across a scale of Besov bodies.

This theorem, together with the approximation arguments alluded to in Section 3.1, proves Theorems 1 and 2.

3.3 Adaptive Estimation at a Single Resolution Level

Theorem 3 depends on an analysis of adaptive threshold selection by the SURE principle. Return to the setup of Section 2.3.

Let $\tilde{R}(\boldsymbol{\mu})$ denote the ideal threshold risk, which we could achieve with information about the optimal threshold to use,

$$\tilde{R}(\boldsymbol{\mu}) = \inf_t d^{-1} \cdot \sum_i r(t,\mu_i),$$

where $r(t,\mu)$ is the risk $E(\eta_t(x)-\mu)^2$ in the scalar setting $x = \mu + z, z \sim N(0,1)$. Of course, we can never hope to actually know the ideal threshold t attaining this expression; however, the following result says that the adaptive

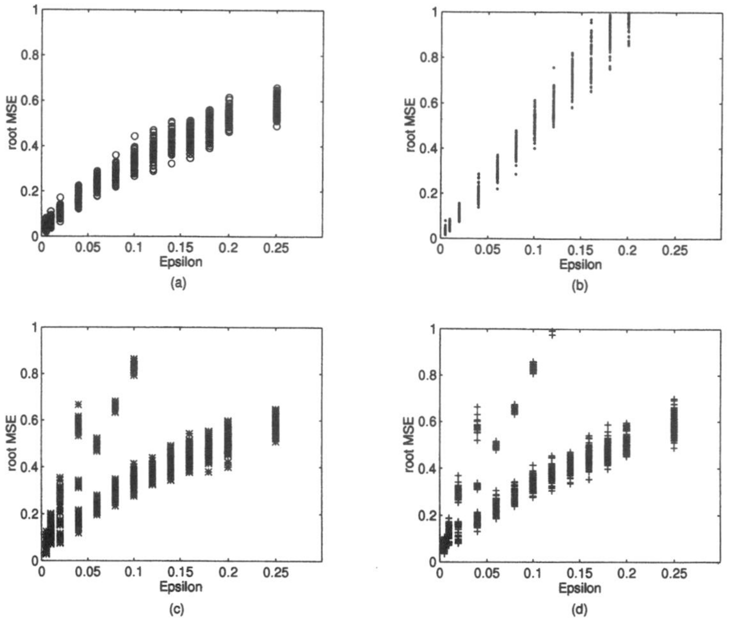

Figure 13. Root MSE's for Simulated Data at Varying Levels of Sparsity When Threshold is Chosen by (a) SURE, (b) $\sqrt{2 \log d}$, and (c) and (d) Two Variants of the Hybrid Method Described, Along With the Design, in Section 2.4.

estimator $\hat{\mu}^*$ performs almost as if we did know this ideal threshold. Let $\tau^2 = d^{-1} \sum \mu_i^2$; when τ^2 is small, the hybrid method switches to the fixed threshold t_d^F and leads to terms involving the fixed threshold risk

$$R_F(\boldsymbol{\mu}) = d^{-1} \sum_i r(t_d^F, \mu_i).$$

Theorem 4. Suppose that $\gamma_d \leq 1$ and $\gamma_d^2 d / \log d \to \infty$. Then (a) uniformly in $\boldsymbol{\mu} \in \mathbb{R}^d$, we have

$$d^{-1} E_{\boldsymbol{\mu}} \|\hat{\boldsymbol{\mu}}^* - \boldsymbol{\mu}\|_2^2 \leq \tilde{R}(\boldsymbol{\mu}) + R_F(\boldsymbol{\mu}) I\{\tau^2 \leq 3\gamma_d\} + c(\log d)^{3/2} d^{-1/2}, \text{ and}$$

(b) uniformly in $\tau^2 \leq \frac{1}{3}\gamma_d$, we have

$$d^{-1} E_{\boldsymbol{\mu}} \|\hat{\boldsymbol{\mu}}^* - \boldsymbol{\mu}\|_2^2 \leq R_F(\boldsymbol{\mu}) + O(d^{-1}(\log d)^{-3/2}).$$

These results can be thought of as asymptotic oracle inequalities; they describe the ability of the SURE-based estimator $\hat{\mu}^*$ to mimic the risk of an "ideal" estimator constructed with special knowledge of the optimal threshold $t = t(\boldsymbol{\mu})$.

4. COMPARISON WITH ADAPTIVE LINEAR SHRINKAGE

We now briefly explain in an informal fashion why *SureShrink* may be expected to compare favorably to adaptive linear shrinkage.

4.1 Adaptive Linear Estimation via James-Stein

In the multivariate normal setting of Section 2.3, the simplest linear shrinkage estimate $\hat{\boldsymbol{\mu}}^c = c\mathbf{x}$ has MSE

$$E\|\hat{\boldsymbol{\mu}}^c - \boldsymbol{\mu}\|^2 = c^2 d + (1-c)^2 \|\boldsymbol{\mu}\|^2.$$

If $\boldsymbol{\mu}$ were known, we could choose

$$\tilde{c}(\boldsymbol{\mu}) = 1 - d/(d + \|\boldsymbol{\mu}\|^2)$$

to minimize the MSE at $\boldsymbol{\mu}$. Because $\boldsymbol{\mu}$ is unknown (it is, after all, the quantity we are trying to estimate), this linear shrinker represents an unattainable ideal. The James–Stein (positive part) estimate is $\hat{\mu}_i^{\mathrm{JS}} = c^{\mathrm{JS}}(\mathbf{x}) \cdot x_i, i = 1, \ldots, d$, where the shrinkage coefficient is

$$c^{\mathrm{JS}}(\mathbf{x}) = (1 - (d-2)/\|\mathbf{x}\|_2^2)_+.$$

From $E\|\mathbf{x}\|^2 = d + \|\boldsymbol{\mu}\|^2$, we see that the James–Stein shrinkage coefficient $c^{\mathrm{JS}}(\mathbf{x})$ is essentially an estimate of the ideal shrinkage coefficient $\tilde{c}$. This ideal estimator (not a statistic!) $\tilde{\boldsymbol{\mu}}^{\mathrm{IS}}(\mathbf{x}) = \tilde{c}(\boldsymbol{\mu})\mathbf{x}$ has MSE

$$E_{\boldsymbol{\mu}} \|\tilde{\boldsymbol{\mu}}^{\mathrm{IS}} - \boldsymbol{\mu}\|_2^2 = \frac{d\|\boldsymbol{\mu}\|^2}{d + \|\boldsymbol{\mu}\|^2}.$$

In fact, the James–Stein estimate does an extremely good job of approaching this ideal.

Theorem 5. For all $d > 2$, and for all $\boldsymbol{\mu} \in R^d$,

$$E_{\boldsymbol{\mu}} \|\hat{\boldsymbol{\mu}}^{\mathrm{JS}} - \boldsymbol{\mu}\|_2^2 \leq 2 + E_{\boldsymbol{\mu}} \|\tilde{\boldsymbol{\mu}}^{\mathrm{IS}} - \boldsymbol{\mu}\|_2^2. \qquad (21)$$

We pay a price of at most 2 for using the James–Stein shrinker rather than the ideal shrinker. In high dimensions

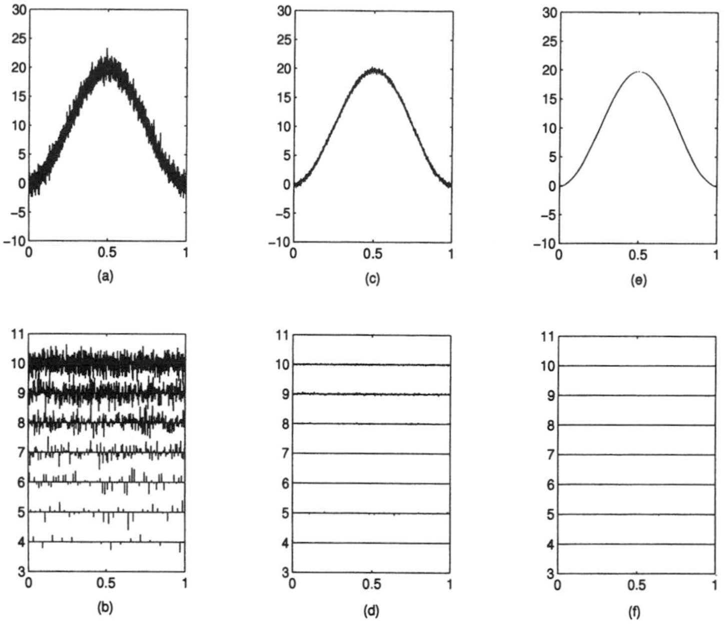

Figure 14. Comparison of WaveJS (c) and SureShrink (e) Reconstructions on Noisy Version (a) of f(t) = c sin $^2\pi$ t, and the Concomitant Action on Noisy Wavelet Coefficients (b) of WaveJS (d) and SureShrink (f).

d, this price is negligible. Because (21) bounds the risk of a genuine estimator in terms of an ideal estimator, it is an example of an "oracle inequality" (cf. Donoho and Johnstone 1994).

Now return to the function estimation setting setting of (1) and Definition 1 with observed wavelet coefficients $\mathbf{x}_j = (y_{j,k})_{0\le k<2^j}, 0 \le j < n$. We apply James–Stein shrinkage separately on each resolution level in the wavelet domain:

$$\hat{w}_j^{\mathrm{JS}} = \begin{cases} \mathbf{x}_j & j < L \\ \hat{\boldsymbol{\mu}}^{\mathrm{JS}}(\mathbf{x}_j) = \left(1 - \frac{2^j - 2}{\|\mathbf{x}_j\|^2}\right)_+ \mathbf{x}_j & j \ge L. \end{cases}$$

Inverting the wavelet transform via

$$\hat{f}^{\mathrm{WJS}}(t_i) = \sum_{j,k} \hat{w}_{j,k}^{\mathrm{JS}} \mathcal{W}_{jk}(t_i)$$

gives an estimate that we call *WaveJS.*

Figures 14 and 15 show *WaveJS* in action on two spatially homogeneous functions: a low-frequency sinusoid $f(t) = c\sin^2 \pi t$ (with $L = 4$) and a high-frequency sinusoid $f(t) = c\sin 50\pi t$ (with $L = 6$), scaled in each case to have signal-to-noise ratio 7. In the first case, there is essentially no signal in levels $j \ge 4$, and so all coefficients are shrunk heavily. (Whether they are shrunk exactly to zero depends on whether $|\mathbf{x}_j|^2 < 2^j - 2$, a threshold that lies in the central part of the $\chi^2_{2^j}$ distribution.) For the high-frequency sinusoid, in levels $j \ge 6$ the signal is concentrated at level 6, so little shrinkage occurs there. The low-frequency oscillation in the reconstruction reflects the fact that no shrinkage is applied at levels $j < 6$. (Note also that *SureShrink* performs essentially as well as *WaveJS*, both visually and in MSE terms, on these homogeneous examples. At the higher levels, the pretest leads to the use of a $\sqrt{2\log 2^j}$ threshold, which in turn shrinks *almost* all coefficients to zero.)

Corresponding to this pleasant visual performance, a number of nice adaptivity properties of *WaveJS* follow immediately from Theorem 5. We state one such in the sequence-space setting of Sections 3.1 and 3.2. Consider model (15) and, as before, set $\mathbf{x}_j = (y_{j,k}/\varepsilon)$. Make the calibration $J = \log_2 \varepsilon^{-2}$ (so that $N = 2^J = \varepsilon^{-2}$). The corresponding WaveJS estimator is defined by

$$\begin{aligned} \hat{\theta}_{j,k}^{\mathrm{WJS}}(y) &= y_{jk} & j < L, \\ &= \varepsilon\hat{\boldsymbol{\mu}}^{\mathrm{JS}}(\mathbf{x}_j) & L \le j \le J, \\ &= 0 & j > J. \end{aligned}$$

Let $\hat{\theta}_L$ denote an estimator that is linear in the data y; the minimax risk among linear estimators is

$$R_L^*(\varepsilon; \Theta) = \inf_{\hat{\theta}_L} \sup_{\Theta} E\|\hat{\theta} - \theta\|^2.$$

We now state a linear adaptivity result for *WaveJS* that is analogous in form to that given for *SureShrink* in Theorem 3.

Theorem 6. If $\sigma > 1/p$, then

$$\sup_{\Theta^s_{p,q}(C)} E_\theta \|\hat{\theta}^{\mathrm{WJS}} - \theta\|_2^2 \le R_L^*(\varepsilon; \Theta^s_{p,q}(C))(1 + o(1))$$

as $\varepsilon \to 0$.

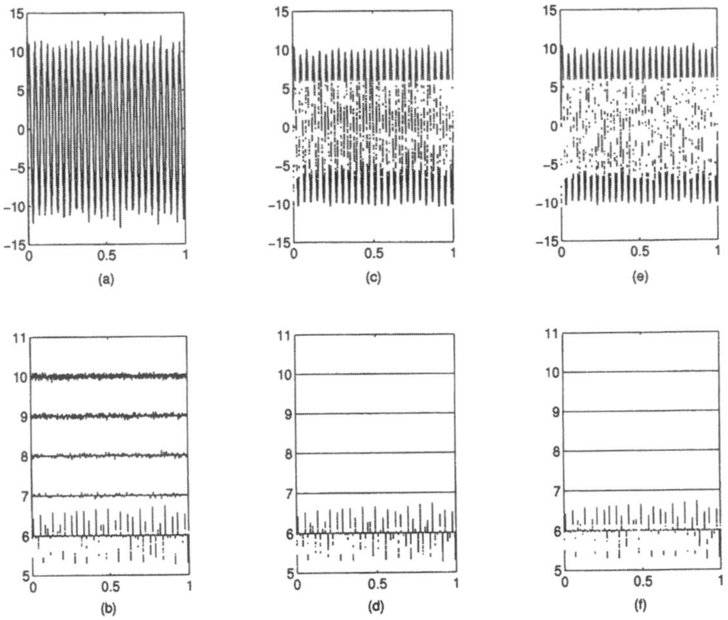

Figure 15. Comparison of WaveJS (c) and SureShrink (e) Reconstructions on Noisy Version (a) of f(t) = c sin 50πt, and the Concomitant Action on Noisy Wavelet Coefficients (b) of WaveJS (d) and SureShrink (f).

In the same way that sequence-space Theorem 3 leads to function-space Theorem 1, one may deduce a function-space version of Theorem 6 for the *WaveJS* estimator $\hat{\mathbf{f}}$. We indicate some of the features of this without going into full detail. Consider the ideal linear shrinkage estimator (again not a statistic)

$$(\tilde{w}_{j,k})_{0 \le k < 2^j} = \tilde{\boldsymbol{\mu}}^{\mathrm{IS}}(\mathbf{x}_j),$$

with inverse wavelet transform $\tilde{\mathbf{f}}^{\mathrm{ID}}$. Then, as an immediate corollary of Theorem 5, for all $N = 2^J$, and for all f,

$$R(\hat{\mathbf{f}}^{\mathrm{WJS}}, \mathbf{f}) \le R(\tilde{\mathbf{f}}^{\mathrm{ID}}, \mathbf{f}) + \frac{2^L + 2\log_2(N)}{N}.$$

Now the ideal estimator achieves within a constant factor of the minimax risk for *linear* estimators. Moreover, the minimax linear risk measured as earlier behaves like $N^{-r'}$, where $r' = \sigma/(\sigma + \frac{1}{2})$ if $p \ge 2$ and $s/(s + \frac{1}{2})$ if $p \le 2$. The ideal estimator is not, however, a statistic, whereas the James–Stein estimate *is* a statistic; and because $(2^L + 2\log_2(N))/N = o(N^{-r})$, it follows that $\hat{\mathbf{f}}^{\mathrm{WJS}}$ achieves the optimal rate of convergence for *linear* estimates over the whole Besov scale. This is in fact a better adaptivity result than previously established for adaptive linear schemes, because it holds over a very broad scale of spaces. Note, however, that the linear rate is slower than the nonlinear rate $N^{-\sigma/(\sigma+1/2)}$ if $p < 2$. This is one of the theoretical reasons for preferring the nonlinear *SureShrink* to *WaveJS*.

Moreover, such an estimate is not very good in practice, as we have seen in Figure 4. The *WaveJS* reconstruction is much noisier than *SureShrink*. This can be seen in Figure 16, which compares the action of *WaveJS* and *SureShrink* on wavelet coefficients of the Doppler signal: if in one resolution level there are significant coefficients that need to be kept, then the James–Stein estimate keeps all the coefficients, incurring a large variance penalty.

To obtain estimators with acceptable performance on spatially variable functions, one must, like *SureShrink*, adaptively keep large coordinates and kill small ones. An adaptive linear estimator does not do this, because it operates on coordinates at each level by the same multiplicative factor.

4.2 Other Adaptive Linear Estimation Methods

To a considerable degree, *WaveJS* serves as a representative for other, more familiar adaptive linear smoothing regimens. Examples include (Priestly–Chao) kernel smoothers, smoothing splines, and truncated Fourier inversion. Each case can be thought of as applying a linear shrinkage of Fourier coefficients. In each case there is a smoothing parameter (i.e., bandwidth, regularization weight, and truncation point) that controls the degree of shrinkage. This smoothing parameter might be chosen to minimize an unbiased estimate of MSE (just as the James–Stein shrinkage factor does approximately). This was done with smoothing splines and Fourier inversion (Donoho et al. 1995, figs. 5 and 6); these two figures are qualitatively quite similar to Figure 4 for *WaveJS* in this article. This and the structural similarities of all these adaptive linear methods underly our

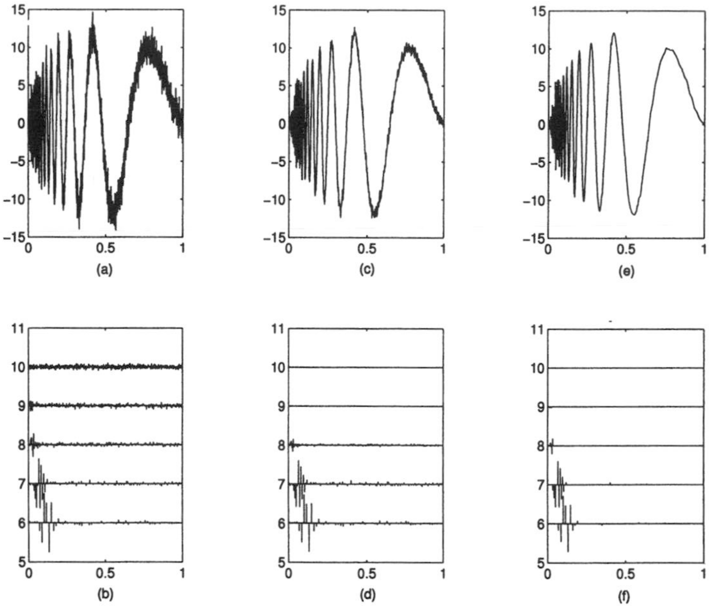

Figure 16. Comparison of WaveJS (c) and SureShrink (e) Reconstructions on Noisy Version (a) of the Doppler Signal, and the Concomitant Action on Noisy Wavelet Coefficients (b) of WaveJS (d) and SureShrink (f).

belief that the advantages of *SureShrink* over *WaveJS* apply quite generically to other adaptive linear methods. In the following subsection we briefly discuss a refined form of Fourier inversion for which impressive theoretical results exist.

4.3 Linear Adaptation Using Fourier Coronae

Suppose that we identify zero with 1, so that $[0, 1]$ has a circular interpretation. Work by Efroimovich and Pinsker (1984), and other recent Russian literature, would consider the use of adaptive linear estimators based on empirical Fourier coefficients $(\hat{v}_l)$. One divides the frequency domain into coronas, $l_i \leq l < l_{i+1}$, and within each corona one uses a linear shrinker,

$$\hat{f}_l = c_i \cdot \hat{v}_l \qquad l_i \leq l < l_{i+1}.$$

The weights are chosen adaptively by an analysis of the Fourier coefficients in the corresponding coronas. Letting $\mathbf{v}_i$ denote the vector of coefficients belonging to the ith corona, the choice used by Efroimovich and Pinsker (1984) is essentially

$$c_i = c^{\text{EP}}(\mathbf{v}_i) = (\|\mathbf{v}_i\|_2^2 - d)/\|\mathbf{v}_i\|_2^2.$$

We propose an adaptive linear scheme that differs from the Efroimovich–Pinsker choice in two ways. First, we propose to use dyadic coronas $l_i = 2^{i+L}$. Such dyadic Fourier coronas occur frequently in Littlewood–Paley theory: (Frazier, Jawerth, Weiss 1991; Peetre 1975; Triebel 1983). Second, within each corona, we shrink via the James–Stein estimate $c_i = c^{\text{JS}}(\mathbf{v}_i)$, which has nicer theoretical properties than the Efroimovich–Pinsker choice. The estimator that we get in this way we shall label LPJS.

LPJS is an adaptive linear estimator. Indeed, from Theorem 5, its risk is at most a term $[4\log_2(N)]/N$ worse than an ideal linear estimator $\tilde{\mathbf{f}}^{\text{LP}IS}$ defined in the obvious way. This ideal linear estimator, based on constant shrinkage in dyadic coronas, has performance not worse than a constant factor times the performance of so-called Pinsker weights, and hence we conclude that, except for constants, LPJS replicates the adaptive-rate advantages of the Efroimovich–Pinsker choice of coronas. LPJS offers advantages the Efroimovich–Pinsker choice does not. It achieves the optimal rate of *linear* estimators over a whole range of L^2 Sobolev, Hölder, and Besov spaces. Theoretically, LPJS is a very good adaptive linear estimator.

But in practice LPJS is disappointing; Figure 17 shows the LPJS reconstructions of our basic examples. The answers are significantly noisier than what can be obtained by *SureShrink*. Instead, the result is comparable to the (disappointing) performance of WaveJS. There is a deeper reason for the similarity between the LPJS and WaveJS, which derives from the Littlewood–Paley theory (Frazier et al. 1991).

5. DISCUSSION

5.1 Simulation Results

A small-scale simulation experiment was conducted to investigate the performance of the methods we have discussed. For each of the four objects under study, we applied nine different methods to noisy versions of the data:

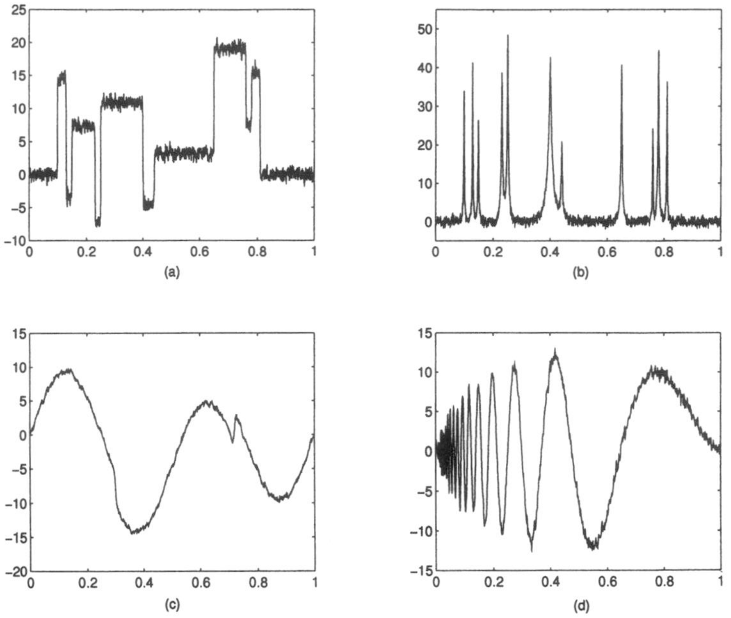

Figure 17. LPJS Reconstruction With Cutoff L = 5. (a) Blocks; (b) Bumps; (c) Heavisine; (d) Doppler.

SureShrink in the Haar, Db4, C3, and S8 wavelet bases, *VisuShrink* (Donoho et al. 1995 and Sec. 5.2 herein) in the S8 basis, *WaveJS* in the S8 wavelet basis, LPJS; and finally, the procedure *RiskShrink* (Donoho and Johnstone 1994a) using the C3 and S8 wavelet bases (denoted as ThrC3 and ThrS8). *RiskShrink* uses a fixed threshold chosen to yield minimax performance for MSE against an "oracle." These threshold values have been tabulated (Donoho and Johnstone 1994a). Dyadic sample sizes $N = 2^J$, from $N = 128$ to $N = 16{,}384$, were studied, and in all cases $L = 6$ was used ($L = 5$ for LPJS).

Sample results are given in Table 2, which reports the root mean square over 20 replications (10 if $N = 8{,}192, 1$ if $N = 16{,}384$) of the root loss $N^{-1/2}\|\hat{\mathbf{f}} - \mathbf{f}\|_2$ (not its square). The data are plotted in Figure 18.

In all examples, the *SureShrink* variants are generally competitive or close to best, with the partial exceptions of *Blocks* (where Haar *SureShrink* is clearly more appropriate) and *Doppler* (where the short and relatively rough filter D4 is noticeably inferior to C3 and S8). The adaptive linear methods *WaveJS* and LPJS are (predictably) less successful on the very nonhomogeneous examples, sometimes needing two to four times the sample size to obtain similar risks. The most extreme case is object *Blocks,* where the performance of shrinkage in the Haar basis at sample size 1,024 is comparable to the performance of LPJS at sample size 8,192 and 16,384. The results for *SureShrink* and *RiskShrink* are remarkably similar for *Doppler* and for *HeaviSine. VisuShrink* quite generally pays a high MSE price for its visually smooth appearance, with the high threshold incurring a larger bias.

5.2 Visual Quality of Reconstruction

The reader may have noticed that *SureShrink* reconstructions contain structure at all scales. This is inherent in the method, which has no a priori knowledge about the smoothness (or lack of smoothness) of the object. Occasional spurious fine-scale structure must sneak into the reconstruction; otherwise, the method would not be able to adapt spatially to true fine-scale structure.

Some readers may be actually annoyed at the tendency of *SureShrink* to show a small amount of spurious fine-scale structure and will demand a more thorough explanation. The presence of this fine-scale structure is demanded by the task of minimizing the l^2 norm loss, which always involves a trade-off between noise and bias. The l^2 norm balances these in equilibrium, which ensures that some noise artifacts will be visible.

Enhanced visual quality can be obtained by keeping the noise term in the trade-off to a minimum. This may be obtained by uniformly applying the threshold $\sqrt{2\log(N)}$ without any adaptive selection. This ensures that essentially all "pure noise" wavelet coefficients (i.e., coefficients where $w_{j,k} = 0$) are set to zero by the thresholding. The resulting curve shows most of the structure and very little noise. Further discussion of threshold selection by the $\sqrt{2\log(N)}$

Table 2. Root Mean Square Errors of Estimation Using Various Threshold Methods

	Haar	*DB4*	*Coif3*	*Symm8*	*WvJS*	*LPJS*	*ThrC3*	*ThrS8*	*VisS8*	*SD's*
Blocks										
128	.81	.88	.87	.89	.94	.93	.89	.91	1.18	.062
256	.68	.78	.80	.80	.92	.91	.87	.88	1.20	.039
512	.59	.76	.77	.78	.85	.85	.78	.80	1.12	.034
1024	.47	.64	.63	.64	.77	.76	.69	.72	1.03	.023
2048	.41	.50	.54	.56	.67	.68	.60	.60	.85	.016
4096	.29	.41	.43	.44	.58	.58	.50	.50	.72	.011
8192	.24	.33	.34	.37	.50	.50	.41	.42	.59	.008
16384	.22	.27	.29	.30	.43	.42	.34	.35	.49	*NA*
Bumps										
128	.93	.94	.94	.94	.99	.99	1.00	.99	1.35	.062
256	.85	.85	.85	.85	.99	.98	.96	.97	1.41	.039
512	.76	.74	.73	.74	.94	.94	.90	.92	1.38	.034
1024	.70	.63	.70	.70	.84	.84	.78	.79	1.16	.023
2048	.67	.55	.51	.52	.70	.69	.65	.66	.96	.016
4096	.58	.43	.42	.41	.54	.53	.53	.54	.77	.011
8192	.51	.32	.31	.32	.42	.42	.42	.43	.61	.008
16384	.38	.25	.22	.22	.33	.33	.33	.34	.48	*NA*
Heavi										
128	.77	.73	.74	.73	.74	.61	.75	.75	.73	.062
256	.62	.55	.55	.56	.56	.49	.56	.55	.56	.039
512	.55	.42	.43	.44	.45	.40	.42	.43	.45	.034
1024	.46	.32	.33	.33	.34	.34	.33	.33	.34	.023
2048	.35	.26	.26	.27	.28	.28	.25	.26	.28	.016
4096	.28	.20	.20	.20	.23	.23	.20	.20	.23	.011
8192	.22	.15	.17	.17	.20	.20	.17	.17	.20	.008
16384	.18	.12	.11	.12	.16	.16	.12	.13	.16	*NA*
Doppler										
128	.93	.88	.83	.82	.93	.94	.82	.80	.92	.062
256	.88	.83	.73	.74	.92	.90	.74	.75	.98	.039
512	.89	.73	.65	.63	.77	.75	.61	.58	.80	.034
1024	.72	.64	.54	.50	.59	.61	.50	.49	.65	.023
2048	.61	.45	.35	.38	.47	.47	.39	.38	.51	.016
4096	.50	.36	.28	.28	.38	.36	.31	.30	.42	.011
8192	.38	.28	.20	.19	.27	.28	.24	.23	.31	.008
16384	.32	.18	.15	.15	.21	.21	.18	.18	.25	*NA*

NOTE: Haar, DB4, Coif3, Symm8 = *SureShrink* using the indicated wavelet filter; WvJS = *WaveJS* using S8; LPJS = LPJS (Sec. 4.3); ThrC3 and ThrS8 = *RiskShrink* using C3 and S8 filters; and VisS8 = *VisuShrink* using S8. Sample sizes $N = 2^j$ for $j = 7(1)14$, with 20 replications for each N, except for 10 replications at $N = 8{,}192$ and 1 replication only at $N = 16{,}384$. Standard deviations shown in column SD are average over estimators and target function of the 36 individual SD's obtained for each sample size N.

rule (called *VisuShrink*) and the connection with optimum "visual quality" may be found in other work (Donoho et al. 1995).

5.3 Hard Thresholding

Could one use "hard thresholding,"

$$\xi_t(y) = y \quad |y| \geq t,$$
$$= 0 \quad |y| < t,$$

in place of soft thresholding η_t? Indeed, hard thresholding seems more natural to nonstatisticians. We prefer soft thresholding because of various statistical advantages (e.g., continuity of the rule; simplicity of the SURE formula). But in principle, the foregoing results could have equally well been derived for hard thresholding. Because hard thresholding is not continuous, let alone weakly differentiable, a more complicated SURE formula would be required to implement the idea on data. The proofs would also be more complicated.

5.4 Estimated Noise Level

For practical use, it is important to estimate the noise level σ from the data rather than to assume that the noise level is known. In practice we derive an estimate from the finest scale empirical wavelet coefficients: $\hat{\sigma} = \text{median}(|y_{J-1,k}| : 0 \leq k < 2^{J-1})/.6745$. We believe that it is important to use a robust estimator like the median, in case the fine-scale wavelet coefficients contain a small proportion of strong "signals" mixed in with "noise."

5.5 Other Literature

There are a number of interesting comparisons or extensions that we have not discussed for lack of space:

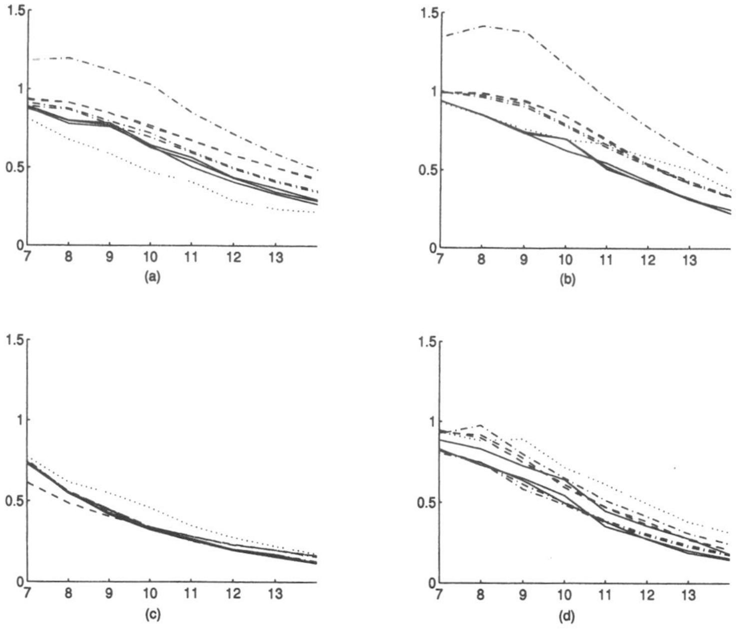

Figure 18. Root MSE's (y Axis) Plotted Against $\log_2 N$ for SureShrink in the Haar basis (dotted line) and Db4, C3, and S8 Bases (Solid Line); WaveJS and LPJS in the S8 Basis (Dashed Line); RiskShrink in C3 and S8 (Dash-Dot line); and VisuShrink in S8 (Dash-Dot Line). (a) Blocks; (b) Bumps; (c) HeaviSine; (d) Doppler.

- We have not compared our results here with the considerable literature on the use of cross-validation to select bandwidth of fixed-kernel smoothers (compare Johnstone and Hall 1992 and the many references therein).
- Neither have we considered how some of the recent more stable bandwidth selection methods discussed and cited there might be applied in the present context.
- Nor have we discussed in detail earlier applications of SURE with linear estimates (compare Li 1985 and references therein).
- Other spatially adaptive, nonlinear regression methods based on variable bandwidth kernels (Brockmann, Gasser, and Herrmann 1993), local polynomials (Cleveland 1979; Fan and Gijbels 1995) or adaptive splines (Friedman 1991).
- Finally we have not discussed applications of wavelet thresholding in density estimation (Johnstone, Kerkyacharian, and Picard 1992).

5.6 Software

Software for the methods described in this article forms part of the WaveLab package of MATLAB M-files, data sets, demonstrations, and documentation. It and further technical reports are available either by anonymous ftp to `playfair.stanford.edu` in directories `pub/donoho` and `pub/johnstone`, or via WWW and URL `http://stat.stanford.edu/`. In particular, the figures in this paper may be reproduced (or modified) by using the M-files in the directory `~/Papers/Adapt` in the WaveLab release. The *SureShrink* method is also implemented in the S+WAVELETS toolkit, by A. G. Bruce and H.-Y. Gao (1995), which runs within S-PLUS.

6. CONCLUSION

A curious divergence has recently developed between minimax adaptation theory and the heuristics guiding algorithm development. As we have seen, the former established adaptivity properties for essentially linear methods (up to the data-based choice of linear shrinkage or bandwidth parameter). The latter was concerned in part with spatial adaptivity, using local polynomials or splines and strongly nonlinear methods. The results in this article can be thought of as a step toward reversing this divergence. First, wavelet bases have spatial adaptation properties that are essentially equivalent with those of local polynomials and splines— these are certainly reflected in *SureShrink,* which has a fast and at least potentially practical algorithm. Second, we establish rate-optimal adaptation results for *SureShrink. SureShrink* is (necessarily) strongly nonlinear, but fortunately in the wavelet domain, the nonlinearity can take the relatively simple form of coordinatewise thresholding.

APPENDIX: PROOFS OF THEOREMS 3–6

We proceed in reverse: first collecting tools, then establishing Theorem 4, and finally returning to Theorem 3.

Exponential Inequalities

We first recall two basic exponential inequalities for bounded variates from Hoeffding (1962) and note a corresponding inequality for chi-squared variates:

(A) Let $Y_1, \ldots, Y_n$ be independent, $a_i \le Y_i \le b_i$, and $\bar{Y}_n = n^{-1} \sum_1^n Y_i$ and $\mu = E\bar{Y}_n$. For $t > 0$,

$$P\{|\bar{Y}_n - \mu| \ge t\} \le 2\exp\left\{-2n^2t^2 \Big/ \sum_1^n (b_i - a_i)^2\right\}. \quad \text{(A.1)}$$

(B) Let $X_1, \ldots, X_m$ be sampled *without replacement* from $\{c_1, \ldots, c_n\}$. Suppose that $a \le c_i \le b$ for all i. Set $\bar{X}_m = m^{-1} \sum_1^m X_i$ and $\mu = n^{-1} \sum_1^n c_i$. For $t > 0$,

$$P\{|\bar{X}_m - \mu| \ge t\} \le 2\exp\{-2nt^2/(b-a)^2\}. \quad \text{(A.2)}$$

(C1) Let $Z_1, \ldots, Z_n$ be iid $N(0,1)$. Then by elementary arguments,

$$P\{|\Sigma\alpha_j(z_j^2 - 1)| > t\} \le 2e^{2s^2\Sigma\alpha_j^2 - |s|t}$$

for

$$|s| \le 1/(4\max(|\alpha_j|)).$$

(C2) If all $\alpha_j = n^{-1}$, then, by optimizing over s,

$$P\left\{\left|n^{-1}\sum (z_j^2 - 1)\right| > t\right\} \le 2e^{-nt(t\wedge 1)/8}. \quad \text{(A.3)}$$

Preparatory Propositions

We use (A) to bound the deviation of the unbiased risk estimate (11) from its expectation. To recapitulate the setting of Section 2.3, suppose that $x_i \sim N(\mu_i, 1), i = 1, \ldots, d$ are independent. Let F_d denote the empirical distribution function of $\{\mu_i\}$. As earlier, let $r(t, \mu_i) = E[\eta_t(x_i) - \mu_i]^2$ denote the MSE of the soft threshold estimate of a single coordinate and define

$$r(t, F) = \int r(t, \mu) F(d\mu).$$

In particular,

$$r(t, F_d) = d^{-1}\Sigma r(t, \mu_i) = d^{-1}E_{\boldsymbol{\mu}}\|\hat{\boldsymbol{\mu}}^{(t)} - \boldsymbol{\mu}\|^2. \quad \text{(A.4)}$$

Stein's unbiased risk estimate (11),

$$\begin{aligned} U_d(t) &= d^{-1}\mathrm{SURE}(t, \mathbf{x}), \\ &= 1 - 2d^{-1}\sum_i I\{x_i^2 \le t^2\} + d^{-1}\sum_i x_i^2 \wedge t^2, \\ &= d^{-1}\sum_i 1 - 2I\{x_i^2 \le t^2\} + x_i^2 \wedge t^2, \end{aligned} \quad \text{(A.5)}$$

has expectation $r(t, F_d)$. We study the deviation

$$Z_d(t) = U_d(t) - r(t, F_d)$$

uniformly for $0 \le t \le t_d = \sqrt{2\log d}$.

Proposition 1. Uniformly in $\boldsymbol{\mu} \in \mathbb{R}^d$,

$$E_{\boldsymbol{\mu}} \sup_{0 \le t \le t_d} |U_d(t) - r(t, F_d)| = O\left(\frac{\log^{3/2} d}{d^{1/2}}\right).$$

Proof. Combining (A.4) and (A.5) with the bound $r(t, \mu_i) \le 1 + t^2$, we can write $Z_d(t) = d^{-1}\sum_1^d Y_i(t)$ with zero mean summands that are uniformly bounded: $|Y_i(t)| \le 2 + t^2$. Hoeffding's inequality (A.1) gives, for a fixed t and (for now) arbitrary $r_d > 1$,

$$P\{|Z_d(t)| > r_d d^{-1/2}\} \le 2\exp\{-r_d^2/2(t^2 + 2)^2\}. \quad \text{(A.6)}$$

For distinct $t < t'$, let $N_d(t, t') = \#\{i : t < |x_i| \le t'\}$ and

$$\begin{aligned} U_d(t) - U_d(t') &= 2d^{-1}\Sigma I\{t^2 < x_i^2 \le t'^2\} \\ &\quad + d^{-1}\sum_i x_i^2 \wedge t^2 - x_i^2 \wedge t'^2 \\ &\le d^{-1}(2 + t'^2 - t^2)N_d(t, t'). \end{aligned}$$

We may bound $r(t, F_d) - r(t', F_d)$ by recalling that for $t \le t_d$, $(\partial/\partial t)r(t, F_d) \le 5t_d$. Then, so long as $|t - t'| < \delta_d$,

$$|Z_d(t) - Z_d(t')| \le 2d^{-1}(1 + \delta_d t_d)N_d(t, t') + 5\delta_d t_d.$$

Now choose $t_j = j\delta_d \in [0, t_d]$; clearly,

$$A_d = \{\sup_{[0,t_d]} |Z_d(t)| \ge 3r_d d^{-1/2}\} \subset D_d \cup E_d,$$

where $D_d = \{\sup_j |Z_d(t_j)| \ge r_d d^{-1/2}\}$ and

$$E_d = \{\sup_j \sup_{|t - t_j| \le \delta_d} |Z(t) - Z(t_j)| \ge 2r_d d^{-1/2}\}.$$

Choose δ_d so that $\delta_d t_d = o(d^{-1/2})$; then E_d is contained in

$$\begin{aligned} E_d' &= \{\sup_j 2d^{-1}N_d(t_j, t_j - \delta_d) \ge r_d d^{-1/2}\} \\ &\subset \{\sup_j d^{-1}|N_d(t_j, t_j + \delta_d) - EN_d| \ge r_d d^{-1/2}/3\} \\ &= E_d'', \end{aligned}$$

say, for large d where we used $EN_d(t_j, t_j + \delta_d) \le c_0 d\delta_d = O(r_d d^{1/2})$. Again, from Hoeffding's inequality (A.1),

$$P\{d^{-1}|N_d(t_j, t_j + \delta_d) - EN_d| \ge r_d d^{-1/2}/3\} \le e^{-2r_d^2/9}. \quad \text{(A.7)}$$

Finally, using (A.6), (A.7), and the cardinality of $\{t_j\}$,

$$\begin{aligned} P(A_d) &\le P(D_d) + P(E_d'') \\ &\le 2t_d\delta_d^{-1}(\exp\{-r_d^2/2(t_d^2 + 2)^2\} + \exp\{-2r_d^2/9\}). \end{aligned}$$

Set $r_d = (2b\log d)^{1/2}(t_d^2 + 2) = O(\log^{3/2} d)$. Then

$$P(A_d) \le \frac{3t_d}{\delta_d d^b}. \quad \text{(A.8)}$$

Let $\|Z_d\| = \sup\{|Z_d(t)| : 0 \le t \le t_d\}$ and $r_d^0 = (2\log d)^{1/2}(t_d^2 + 2)$. We may rephrase (A.8) as

$$P_{\boldsymbol{\mu}}\{\sqrt{d}\|Z_d\|/r_d^0 > s\} \le 3t_d\delta_d^{-1}e^{-s^2\log d},$$

which suffices for the L_1 convergence claimed.

Proposition 2. Uniformly in $\boldsymbol{\mu} \in \mathbb{R}^d$,

$$E_I\|r(\cdot, F_I) - r(\cdot, F)\|_\infty = O\left(\frac{\log^{3/2} d}{d^{1/2}}\right).$$

Proof. This is similar to that of Proposition 1, but is simpler and uses Hoeffding's inequality (A.2). In the notation of (A.2), set $n = d, c_i = r(t, \mu_i) \le 1 + t_d^2$, and $m = d/2$, so that $\mu = r(t, F_d), \bar{X}_m = r(t, F_I)$, and

$$Z(t) := r(t, F_I) - r(t, F_d) = \bar{X}_m - \mu,$$

$$P\{|Z(t)| > r_d d^{-1/2}\} \le 2\exp\{-2r_d^2/(1+t_d^2)^2\}.$$

Because $|(\partial/\partial t)r(t,F)| \le 5t_d$ for any F, it follows that for $|t'-t| < \delta_d$,

$$|Z(t') - Z(t)| \le 10\delta_d t_d.$$

Thus if δ_d is small enough so that $10\delta_d t_d \le r_d d^{-1/2}$ and $r_d = (2b\log d)^{1/2}(t_d^2+1)$, then

$$P\{\|Z_d\| \ge 2r_d d^{-1/2}\} \le P\{\sup_{j:j\delta_d \le t_d} |Z_d(j\delta_d)| \ge r_d d^{-1/2}\} \le \frac{2t_d}{\delta_d}\frac{1}{d^{4b}}.$$

As for Proposition 1, this yields the result.

Lemma 1. Let $x_i \sim N(\mu_i, 1), i = 1, \ldots, d$, be independent, $s_d^2 = d^{-1}\Sigma(x_i^2 - 1)$, and $\tau^2 = d^{-1}\Sigma\mu_i^2$. Then, if $\gamma_d^2 d/\log d \to \infty$,

$$\sup_{\tau^2 \ge 3\gamma_d} (1+\tau^2)P\{s_d^2 \le \gamma_d\} = o(d^{-1/2}). \tag{A.9}$$

Proof. This is a simple statement about the tails of the noncentral chi-squared distribution. Write $x_i = z_i + \mu_i$, where $z_i \sim N(0,1)$. The event in (A.9) may be rewritten as

$$\begin{aligned} A_d &= \{d^{-1}\Sigma(z_i^2-1) + d^{-1}\Sigma 2\mu_i z_i \le -(\tau^2 - \gamma_d)\} \\ &\subset \{d^{-1}\Sigma(z_i^2-1) \le -\tau^2/3\} \cup \{d^{-1}\Sigma 2\mu_i z_i \le -\tau^2/3\} \\ &= B_d \cup C_d, \end{aligned} \tag{A.10}$$

from the lower bound on τ^2 in (A.9).

By elementary inequalities, for $\tilde\Phi(z) = \int_z^\infty e^{-s^2/2}\,ds/\sqrt{2\pi}$,

$$P(C_d) = \tilde\Phi\left(\frac{\tau^2}{3}\frac{d^{1/2}}{2\tau}\right) \le c_1 e^{-c_2 d\tau^2}, \tag{A.11}$$

and it is easily verified that $(1+\tau^2)e^{-c_2 d\tau^2} \le 2e^{-3c_2 d\gamma_d} = o(d^{-1/2})$ for $\tau^2 \ge 3\gamma_d$ and d large.

For B_d, apply the exponential inequality (A.3) to obtain

$$\begin{aligned} (1+\tau^2)P(B_d) &\le 2(1+\tau^2)\exp\{-d\tau^2(\tau^2 \wedge 3)/72\} \\ &\le c_3\exp\{-\gamma_d^2 d/8\} = o(d^{-1/2}), \end{aligned}$$

because $\gamma_d^2 d/\log d \to \infty$.

Proof of Theorem 4(a)

Decompose the risk of $\hat{\boldsymbol\mu}^*$ according to the outcome of the pretest event $A_d = \{s_d^2 \le \gamma_d\}$, with the goal of showing that

$$\begin{aligned} R_{1d}(\boldsymbol\mu) &= d^{-1}E[\|\boldsymbol\mu^* - \boldsymbol\mu\|^2, A_d] \\ &\le R_F(\boldsymbol\mu)I\{\tau^2 \le 3\gamma_d\} + o(d^{-1/2}) \end{aligned} \tag{A.12}$$

and

$$R_{2d}(\boldsymbol\mu) = d^{-1}E[\|\boldsymbol\mu^* - \boldsymbol\mu\|^2, A_d^c] \le \tilde R(\boldsymbol\mu) + c(\log d)^{3/2}d^{-1/2}. \tag{A.13}$$

On event A_d, fixed thresholding is used:

$$R_{1d} = d^{-1}E\left[\sum_i (\eta(x_i, t_d^F) - \mu_i)^2, A_d\right] \le R_F(\boldsymbol\mu).$$

If $\tau^2 \ge 3\gamma_d$, then we first note that on event A_d,

$$d^{-1}\sum_i \eta(x_i, t_d^F)^2 \le d^{-1}\Sigma x_i^2 \le 1 + \gamma_d.$$

Using Lemma 1, (A.12) follows from

$$R_{1d} \le 2(1+\gamma_d+\tau^2)P(A_d) = o(d^{-1/2}).$$

On event A_d^c, the adaptive, half-sample–based thresholding applies. Let $E_{\boldsymbol\mu}$ denote expectation over the distribution of (x_i) and let E_I denote expectation over the random choice of half-sample I. Then

$$dR_{2d} \le E_I\left\{\sum_{i\in I} E_{\boldsymbol\mu}[\eta(X_i, \hat t_{I'}) - \mu_i]^2 + \sum_{i\in I'} E_{\boldsymbol\mu}[\eta(X_i, \hat t_I) - \mu_i]^2.\right\}$$

Let F_I, (resp. $F_{I'}, F_d$) denote the empirical distribution functions of $\{\mu_i : i \in I\}$ (resp. of $\mu_i, i \in I', \{1, \ldots, d\}$), and set $r(t,F) = \int r(t,\boldsymbol\mu)F(d\boldsymbol\mu)$. Then, using the symmetry between I and I', we have

$$\begin{aligned} R_{2d} &= \frac{1}{2}E_I E_{\boldsymbol\mu}\{r(\hat t_{I'}, F_I) + r(\hat t_I, F_{I'})\} \\ &= E_I E_{\boldsymbol\mu} r(\hat t_I, F_{I'}). \end{aligned}$$

Thus to complete the proof of (A.13), it suffices to show that

$$R_{3d}(\boldsymbol\mu) := E_I E_{\boldsymbol\mu} r(\hat t_I, F_{I'}) - \tilde R(\boldsymbol\mu) \le c(\log d)^{3/2}d^{-1/2}. \tag{A.14}$$

There is a natural decomposition,

$$\begin{aligned} R_{3d} &= E_{\boldsymbol\mu}E_I[r(\hat t_I, F_{I'}) - r(\hat t_I, F_I)] \\ &\quad + E_{\boldsymbol\mu}E_I[r(\hat t_I, F_I) - r_{\min}(F_I)] \\ &\quad + E_I[r_{\min}(F_I) - r_{\min}(F_d)], \\ &= S_{1d} + S_{2d} + S_{3d}, \end{aligned}$$

where we have set $r_{\min}(F) = \inf\{r(t,F), 0 \le t \le t_d^F\}$ and note that $r_{\min}(F_d) = \tilde R(\boldsymbol\mu)$. We use

$$r(t, F_I) - r(t, F_d) = \frac{1}{2}[r(t,F_I) - r(t,F_{I'})] \tag{A.15}$$

together with the simple observation that $|r_{\min}(F) - r_{\min}(G)| \le \|r(\cdot,F) - r(\cdot,G)\|_\infty$ to conclude that

$$S_{1d} + S_{3d} \le 3E_I\|r(\cdot,F_I) - r(\cdot,F_d)\|_\infty = O\left(\frac{\log^{3/2} d}{d^{1/2}}\right)$$

using Proposition 2.

Finally, let $U_{d/2}(t,I)$ denote the unbiased risk estimate derived from subset I. Then, using Proposition 1,

$$\begin{aligned} S_{2d} &\le E_{\boldsymbol\mu}E_I|r(\hat t_I, F_I) - U_{d/2}(\hat t_I, I)| + |U_{d/2}(\hat t_I, I) - r_{\min}(F_I)| \\ &\le 2E_I E_{\boldsymbol\mu}\|r(\cdot, F_I) - U_{d/2}(\cdot, I)\|_\infty = O\left(\frac{\log^{3/2} d}{d^{1/2}}\right). \end{aligned}$$

Putting all together, we obtain (A.14).

Proof of Theorem 4(b)

When $\|\boldsymbol\mu\|$ is small, the pretest of $s_d^2 \le \gamma_d$ will with high probability lead to use of the fixed threshold t_d^F. The $O(d^{-1/2}\log^{3/2} d)$ error term in Theorem 4, which arises from empirical process fluctuations connected with minimization of SURE, is then not germane and can be improved to $O(d^{-1})$.

We decompose the risk of $\boldsymbol\mu^*$ as in (A.12) and (A.13), but now $P(A_d) \nearrow 1$ as $d \nearrow \infty$. On A_d, fixed thresholding is used and so

$$dR_{1d} \le \sum_1^d r(t_d^F, \mu_i).$$

We use large derivation inequalities to bound the remaining term R_{2d}. Using symmetry between I and I',

$$dR_{2d} \le 2E_I \sum_{i\in I'} E_{\boldsymbol\mu}\{(\eta(X_i, \hat t_I) - \mu_i)^2, A_d^c\}.$$

Using the Cauchy–Schwartz inequality and noting from the limited translation structure of $\eta(\cdot,t)$ that $E_\mu(\eta(X,t)-\mu)^4 \le c(t_d^F)^4$ for $t \le t_d^F$, we get

$$dR_{2d} \le cd(t_d^F)^2 P_{\boldsymbol{\mu}}(A_d^c)^{1/2}.$$

Arguing similarly to (A.10), we note that the hypothesized bound on $\boldsymbol{\mu}$ implies that

$$A_d^c \subset \left\{d^{-1}\sum(z_i^2-1) > \gamma_d/3\right\} \cup \left\{d^{-1}\sum 2\mu_i z_i > \gamma_d/3\right\} = B_d \cup C_d.$$

The chi-squared exponential inequality (A.3) and standard Gaussian inequalities give

$$P(B_d) \le \exp\{-d\gamma_d^2/72\}$$

and

$$P(C_d) \le \exp\{-\gamma_d d/24\},$$

which imply that $d\log d \cdot P_d(A_d^c)^{1/2} = o(\log^{-3/2} d)$, which shows negligibility of R_{2d} and completes the proof.

Proof of Theorem 3

The idea behind the proof is as follows. For a given $\Theta_{p,q}^s(C)$ and noise level ε, there is a certain level at which the least favorable cases are found; that is, j_* at (A.26). For j near j_*, the unbiased risk estimate picks the threshold in an asymptotically efficient manner. At levels where total signal is negligible, the pretest picks t_j^F with high probability, so that risk and modulus of continuity bounds for $\sqrt{2\log d}$ thresholding can be used to show that the remaining terms are of lower order.

We make use of the definitions (19) and (20) to write

$$E_\theta\|\hat{\theta}^* - \theta\|^2 = 2^L\varepsilon^2 + \varepsilon^2\sum_{j\ge L} E\|\hat{\boldsymbol{\mu}}^*(x_j) - \boldsymbol{\mu}_j\|^2,$$

where $\boldsymbol{\mu}_j = (\mu_{jk}) = (\theta_{jk}/\varepsilon)$ and $\theta_j = (\theta_{jk})$. We use the abbreviations $t_j^F = t_{2^j}^F = \sqrt{2\log 2^j}$ and $\gamma_j = \gamma_{2^j}$. For a $L = j_0(\varepsilon,\sigma,p,q) \nearrow \infty$ to be specified later, we use Theorem 4(a) for levels $j \le j_0$ and Theorem 4(b) for $j > j_0$:

$$E_\theta\|\hat{\theta}^* - \theta\|^2 \le O(\varepsilon^2) + S_{1\varepsilon} + S_{2\varepsilon}, \tag{A.16}$$

$$S_{1\varepsilon} \le \varepsilon^2 \sum_{j\le j_0}\left\{\inf_{t_j}\sum_k r(t_j,\mu_{jk}) + I\{\tau_j^2 \le 3\gamma_j\} \times \sum_k r(t_j^F,\mu_{jk}) + cj^{3/2}2^{j/2}\right\} \tag{A.17}$$

$$= S_{11\varepsilon} + S_{12\varepsilon} + S_{13\varepsilon}, \tag{A.18}$$

$$S_{2\varepsilon} \le \varepsilon^2 \sum_{j>j_0}\left\{\sum_k r(t_j^F,\mu_{jk}) + cj^{-3/2}\right\} \tag{A.19}$$

$$= S_{21\varepsilon} + S_{22\varepsilon}, \tag{A.20}$$

say. Maximizing now over $\Theta_{p,q}^s$,

$$\sup_\Theta S_{11\varepsilon} \le \inf_{(t_j)}\sup_\Theta E\|\hat{\theta}_{(t_j)} - \theta\|^2 \tag{A.21}$$

$$= R_T^*(\varepsilon, \Theta_{p,q}^s) \asymp \varepsilon^{2r}, \tag{A.22}$$

where $r = 2\sigma/(2\sigma+1)$. It remains to verify that j_0 can be chosen so that all other terms are $o(\varepsilon^{2r})$. Define $j_\varepsilon = (1-r)\log C^2\varepsilon^{-2}$. The term $S_{13\varepsilon}$ is negligible if

$$j_0 + 3\log_2 j_0 - 2j_\varepsilon \to -\infty. \tag{A.23}$$

Because $S_{22\varepsilon} = O(\varepsilon^2)$, it remains to consider $S_{12\varepsilon}$ and $S_{21\varepsilon}$.

We begin by borrowing some notation and results from Donoho et al. (1994). Suppose that $y \sim N_d(\xi, \varepsilon^2 I)$ and $\xi \in \Xi \subset \mathbb{R}^d$. Define a modulus of continuity,

$$\Omega(\delta;\Xi) = \sup\{\|\xi\|_2 : \xi \in \Xi, \|\xi\|_\infty \le \delta\}.$$

Let $\hat{\xi}_{F,i}(y) = \eta(y_i, t_d^F\varepsilon)$ denote the fixed threshold estimator; using arguments similar to those of section 5 of Donoho et al. (1995), we have

$$\sup_\Xi E\|\hat{\xi}_F - \xi\|^2 \le c\Omega^2(2t_d^F\varepsilon;\Xi). \tag{A.24}$$

We shall denote an l_p-ball in $\mathcal{R}^n$ by $B_{p,n}(r) = \{\xi \in \mathbb{R}^n : \|\xi\|_p \le r\}$. Return to the sequence-space setting of Section 3. If $\theta = (\theta_{jk})$, then define $\theta^{(j)}$ to be the same vector with all coefficients set to zero that are not at resolution level j. With a slight abuse of notation, we define

$$\Theta^{(j)} \equiv \{\theta^{(j)} : \theta \in \Theta_{p,q}^s(C)\} \equiv B_{p,2^j}(C2^{-sj}). \tag{A.25}$$

The constraint $\tau^2 \le 3\gamma_j$ in (A.17) corresponds to a set

$$\bar{\Theta}^{(j)} \equiv B_{2,2^j}(c_{j\varepsilon}), \qquad c_{j\varepsilon}^2 = 3\varepsilon^2 2^j\gamma_j.$$

From Donoho et al. (1995), we borrow the inequality

$$\Omega(\delta;\Theta^{(j)}) \le \delta^r C^{1-r}2^{-\eta|j-j_*|}, \tag{A.26}$$

where $\eta = \eta(\sigma,p) > 0, r = 2\sigma/(2\sigma+1)$, and $j_* = 2(1-r)\log_2 C/\delta$. Again, note from Donoho et al. (1995) that because $2^j(2\varepsilon t_j^F)^2 \ge c_{j\varepsilon}^2$,

$$\Omega(2\varepsilon t_j^F, \bar{\Theta}^{(j)}) = c_{j\varepsilon}. \tag{A.27}$$

The bound (A.24) shows that

$$S_{12\varepsilon} \le c\sum_{j\le j_0}\Omega^2(2t_j^F\varepsilon;\Theta^{(j)} \cap \bar{\Theta}^{(j)}),$$

and we plan to use (A.26) and (A.27) in conjunction with $\Omega(\delta,\cap\Theta_i) \le \min_i \Omega(\delta,\Theta_i)$.

For small j, we note that $\Omega(2\varepsilon t_j^F, \bar{\Theta}^{(j)}) = o(\varepsilon^{2r})$ if

$$j + \log_2\gamma_j - j_\varepsilon \to -\infty. \tag{A.28}$$

Choose $\alpha > 0$ small. If we choose γ_j so that $\gamma_j \le 2^{-\alpha j}$ for all j, then (A.28) will certainly hold for all $j \le j_{1\varepsilon} := j_\varepsilon/(1-\alpha/2)$. For large j (i.e., $j \ge j_{1\varepsilon}$), we use (A.26) after noting that $j_* \le j_\varepsilon$,

$$\Omega^2(2\varepsilon t_j^F;\Theta^{(j)}) \le c\varepsilon^{2r}j^r 2^{-2\eta|j-j_\varepsilon|} = o(\varepsilon^{2r}).$$

Because the bounds for both $\Theta^{(j)}$ and $\bar{\Theta}^{(j)}$ decay geometrically with j, it follows that the sum $S_{12\varepsilon} = o(\varepsilon^{2r})$ also.

We now turn to $S_{21\varepsilon}$. So that we may apply Theorem 4(b) to obtain the bound (A.20), it is necessary that j_0 be chosen so that $\Theta^{(j)} \subset \frac{1}{3}\bar{\Theta}^{(j)}$ for all $j \ge j_0$. Because $B_{p,n}(r_1) \subset B_{2,n}(r_2)$ iff $n^{1-2/p\vee 2}r_1^2 \le r_2^2$, this requires that $2^{j(1-2/p\vee 2)}C^2 2^{-2sj} \le \frac{1}{3}\varepsilon^2 2^j\gamma_j$ for $j \ge j_0$, which in turn amounts to the requirement that

$$(2s + 2/2\vee p)j_0 + \log_2\gamma_{j0}/3 \ge (2\sigma+1)j_\varepsilon. \tag{A.29}$$

Using (A.24), (A.26), and $j_* \le j_\varepsilon$,

$$S_{21\varepsilon} \le c\sum_{j>j_0}\Omega^2(2\varepsilon t_j^F;\Theta^{(j)}) \tag{A.30}$$

$$\le c\varepsilon^{2r}C^{2-2r}\sum_{j>j_0} j^r 2^{-2\eta|j-j_\varepsilon|} = o(\varepsilon^{2r}), \tag{A.31}$$

as long as $j_0 - j_\varepsilon \to \infty$.

Suppose that γ_j equals either $3.2^{-\gamma j}$ (for some fixed $\gamma \in (0, \frac{1}{2})$) or $j^{3/2}2^{-j/2}$. In either case, set $j_0 = aj_\varepsilon$. Condition (A.23) is satisfied if $a < 2$. Condition (A.29) holds if $a \geq \underline{a} = (2\sigma + 1)/(2s - \gamma + 2/2 \vee p)$. Simple algebra shows that it is possible to choose a to satisfy the two conditions simultaneously as long as $s > |1/p - \frac{1}{2}| + \gamma - \frac{1}{2}$.

Proof of Theorem 5

We first recall that the risk of the positive-part James–Stein estimator is no larger than that of the original James–Stein estimator, $\bar{\mu}^{\mathrm{JS}}$, in which the restriction that the shrinkage coefficient be positive is dropped. Then using Stein's (1981) unbiased estimate of risk (or, alternatively, Lehmann 1983, p. 300) and Jensen's inequality, we have for $d > 2$,

$$E_{\mu}\|\hat{\mu}^{\mathrm{JS}} - \mu\|_2^2 \leq E_{\mu}\|\bar{\mu}^{\mathrm{JS}} - \mu\|_2^2 \;=\; d - (d-2)^2 E_{\mu}\|\mathbf{x}\|_2^{-2}$$
$$\leq\; d - (d-2)^2/(\|\mu\|_2^2 + d).$$

By direct calculation,

$$E_{\mu}\|\bar{\mu}^{\mathrm{IS}} - \mu\|_2^2 = \|\mu\|_2^2/(\|\mu\|_2^2 + d). \qquad \text{(A.32)}$$

The difference of the two expressions is thus bounded by $(2d - 4)/(\|\mu\|_2^2 + d) \leq 2$.

Proof of Theorem 6

We use a notation and decomposition similar to that of the proof of Theorem 3. Thus

$$E_\theta\|\hat{\theta}^{\mathrm{WJS}} - \theta\|^2 \;=\; 2^L\varepsilon^2 + \varepsilon^2\sum_{j=L}^{J} E\|\hat{\mu}^{\mathrm{JS}}(x_j) - \mu_j\|^2 + \sum_{j>J}\|\theta_j\|^2$$
$$\leq\; O(\varepsilon^2) + T_{1\varepsilon} + T_{2\varepsilon}.$$

The maximum over $\Theta^s_{p,q}$ of the tail term $T_{2\varepsilon}$ is given via the tail n-width,

$$\Delta(\varepsilon, \|\cdot\|; \Theta) = \sup\{\|\theta\| : \theta \in \Theta, \theta_{jk} = 0 \text{ if } j \leq J(\varepsilon)\}$$

(recall that $J(\varepsilon) = \log_2 \varepsilon^{-2}$). From theorem 5 of Donoho et al. (1994), we have

$$T_{2\varepsilon} = \Delta^2(\varepsilon, \|\cdot\|_2; \Theta^s_{p,q}) \leq c(\varepsilon^2)^{(\sigma - (1/p - 1/2)_+)}.$$

Applying (21) to $T_{1\varepsilon}$ yields

$$T_{1\varepsilon} \leq 2J\varepsilon^2 + \varepsilon^2 \sum_{j=L}^{J} E\|\hat{\mu}^{\mathrm{IS}}(\mathbf{x}_j) - \mu_j\|^2. \qquad \text{(A.33)}$$

Write $\hat{\theta}_{\mathrm{DL}}$ for a diagonal linear estimate of the form $\hat{\theta}_j(y) = c_j y_j$. By the definition of the ideal estimator, the second term on the right of (A.33) is bounded for $\theta \in \Theta^s_{p,q}$ by

$$\inf_{\hat{\theta}_{\mathrm{DL}}} E\|\hat{\theta}_{\mathrm{DL}} - \theta\|^2 \leq R^*_L(\varepsilon, \Theta^s_{p,q}).$$

From other work (Donoho and Johnstone 1995, sec. 6), it is known that $R^*_L(\varepsilon, \Theta^s_{p,q}) \sim \varepsilon^{2r'}$, where $r' = \sigma/(\sigma + 1/2)$ if $p \geq 2$, and $s/(s + 1/2)$ if $p \leq 2$. Thus in particular $r' < 1$, and so all other remainder terms are $o(\varepsilon^{2r'})$, and the proof is complete.

[Received June 1993. Revised January 1995.]

REFERENCES

Brockmann, M., Gasser, T., and Herrmann, E. (1993), "Locally Adaptive Bandwidth Choice for Kernel Regression Estimators," *Journal of the American Statistical Association,* 88, 1302–1309.

Brown, L. D., and Low, M. G. (1995), "Asymptotic Equivalence of Nonparametric Regression and White Noise," submitted to *The Annals of Statistics.*

Bruce, A. G., and Gao, H.-Y. (1995), S + WAVELETS Toolkit Statistics, a division of MathSoft Inc., Seattle WA.

Cleveland, W. S. (1979), "The Robust Locally Weighted Regression and Smoothing Scatterplots," *Journal of the American Statistical Association,* 74, 829–836.

Cohen, A., Daubechies, I., Jawerth, B., and Vial, P. (1993), "Multiresolution Analysis, Wavelets, and Fast Algorithms on an Interval," *Comptes Rendus Academie des Sciences, Paris (A),* 316, 417–421.

Daubechies, I. (1988), "Orthonormal Bases of Compactly Supported Wavelets," *Communications in Pure and Applied Mathematics,* 41, 909–996.

——— (1992), *Ten Lectures on Wavelets,* CBMS-NSF Series in Applied Mathematics, No. 61, Philadelphia: SIAM.

DeVore, R. A., and Popov, V. A. (1988), "Interpolation of Besov Spaces," *Transactions of the American Mathematical Society,* 305, 397–414.

Donoho, D. (1992), "Interpolating Wavelet Transforms," Technical Report 408, Stanford University, Dept. of Statistics.

Donoho, D. L., and Johnstone, I. M. (1994a), "Ideal Spatial Adaptation via Wavelet Shrinkage," *Biometrika,* 81, 425–455.

——— (1994b), "Minimax Risk Over l_p-Balls for l_q-Error," *Probability Theory and Related Fields,* 99, 277–303.

——— (in press), "Neo-Classical Minimax Problems, Thresholding, and Adaptation," *Bernoulli.*

——— (1995), "Minimax Estimation via Wavelet Shrinkage," submitted to *The Annals of Statistics.*

Donoho, D. L., Johnstone, I. M., Kerkyacharian, G., and Picard, D. (1994), "Universal Near Minimaxity of Wavelet Shrinkage," technical report, Stanford University, Dept. of Statistics.

——— (1995), "Wavelet Shrinkage: Asymptopia?" (with discussion), *Journal of the Royal Statistical Society,* Ser. B, 57, 301–369.

Efroimovich, S., and Pinsker, M. (1984), "A Learning Algorithm for Nonparametric Filtering," *Automat. i Telemeh.,* 11, 58–65 (in Russian).

Fan, J., and Gijbels, I. (1995), "Data-Driven Bandwidth Selection in Local Polynomial Fitting, Variable Bandwidth, and Spatial Adaptation," *Journal of the Royal Statistical Society,* Ser. B, 57, 371–394.

Frazier, M., Jawerth, B., and Weiss, G. (1991), *Littlewood–Paley Theory and the Study of Function Spaces,* NSF-CBMS Regional Conference Series in Mathematics, 79, Providence, RI: American Mathematical Society.

Friedman, J. (1991), "Multivariate Adaptive Regression Splines" (with discussion), *The Annals of Statistics,* 19, 1–67.

Golubev, G. K., and Nussbaum, M. (1990), "A Risk Bound in Sobolev Class Regression," *The Annals of Statistics,* 18, 758–778.

Johnstone, I. M., and Hall, P. G. (1992), "Empirical Functionals and Efficient Smoothing Parameter Selection" (with discussion), *Journal of the Royal Statistical Society,* Ser. B, 54, 475–530.

Johnstone, I., Kerkyacharian, G., and Picard, D. (1992), "Estimation d'une Densité de Probabilité par Méthode d'Ondelettes," *Comptes Rendus Academie des Sciences Paris (A),* 315, 211–216.

Lehmann, E. L. (1983), *Theory of Point Estimation,* New York: John Wiley.

Li, K. C. (1985), "From Stein's Unbiased Risk Estimates to the Method of Generalized Cross-Validation," *The Annals of Statistics,* 13, 1352–1377.

Mallatt, S. G. (1989a), "The Multifrequency Channel Decompositions of Images and Wavelet Models," *IEEE Transactions on Acoustic Signal Speech Processes,* 37, 2091–2110.

——— (1989b), "A Theory for Multiresolution Signal Decomposition: The Wavelet Representation," *IEEE Transactions on Pattern Analysis and Machine Intelligence,* 11, 674–693.

Meyer, Y. (1990), *Ondelettes et Opérateurs; I: Ondelettes, II: Opérateurs de Calderón-Zygmund, III:* (with R. Coifman) *Opérateurs multilinéaires,* Paris: Hermann. (English translation of Vol. I published by Cambridge University Press).

——— (1991), "Ondelettes sur l'Intervalle," *Revista Matematica Iberoamericana,* 7, 115–133.

Nemirovskii, A. (1985), "Nonparametric Estimation of Smooth Regression Function," *Izv. Akad. Nauk. SSR Teckhn. Kibernet,* 3, 50–60 (in Russian), *Journal of Computer and System Sciences,* 23, 1–11 (in English).

Nemirovskii, A., Polyak, B., and Tsybakov, A. (1985), "Rate of Convergence of Nonparametric Estimates of Maximum-Likelihood Type," *Problems of Information Transmission,* 21, 258–272.

Nussbaum, M. (1985), "Spline Smoothing and Asymptotic Efficiency in l_2," *The Annals of Statistics,* 13, 984–997.
Peetre, J. (1975), *New Thoughts on Besov Spaces,* Duke University, Durham, NC: Dept. of Mathematics.
Stein, C. (1981), "Estimation of the Mean of a Multivariate Normal Distribution," *The Annals of Statistics,* 9, 1135–1151.
Stone, C. (1982), "Optimal Global Rates of Convergence for Nonparametric Estimators," *The Annals of Statistics,* 10, 1040–1053.
Strang, G. (1989), "Wavelets and Dilation Equations: A Brief Introduction," *SIAM Review,* 31, 614–627.
Tribel, H. (1983), *Theory of Function Spaces,* Basel: Birkhäuser Verlag.
——— (1990), *Theory of Function Spaces II,* Basel: Birkhäuser Verlag.

Hölder Exponents at Given Points and Wavelet Coefficients

Stéphane Jaffard
Translated by Robert D. Ryan

ABSTRACT. We prove that, for any $\varepsilon > 0$ and any $f \in C^\varepsilon(\mathbb{R}^n)$, the Hölder exponent of f at a given point x_0 can be explicitly computed, up to a logarithmic factor, by size conditions on the wavelet coefficients of f.

1. The Wavelets

For ease of notation, we will limit the discussion to the one-dimensional case; however, the statements and proofs extend immediately to the general case.

Let $\psi(x)$ be a Lipschitz function of the real variable x that also satisfies the following conditions:

$$|\psi(x)| \le C(1+|x|)^{-3}, \quad |\psi'(x)| \le C(1+|x|)^{-3}; \tag{1}$$

$$\text{the set } \psi_{(j,k)} = 2^{j/2}\psi(2^j x - k),\ j,k \in \mathbb{Z}, \text{ is an orthonormal basis for } L^2(\mathbb{R}). \tag{2}$$

If

$$\int_{-\infty}^{\infty} |f(x)|(1+|x|)^{-3}\,dx < \infty,$$

then the wavelet coefficients

$$c(j,k) = \int_{-\infty}^{\infty} f(x)\overline{\psi}_{(j,k)}(x)\,dx$$

make sense. If, in addition, $f(x)$ satisfies $|f(x) - f(x_0)| \le C|x - x_0|^\alpha$ for some exponent $\alpha \in (0,1]$, then this implies immediately that the condition

$$|c(j,k)| \le C' 2^{-j(\alpha+1/2)}(1 + |2^j x_0 - k|^\alpha) \tag{3}$$

is satisfied for all $j \in \mathbb{R}$ and all $k \in \mathbb{Z}$.

We propose to examine, conversely, if (3) implies that

$$|f(x) - f(x_0)| \le C|x - x_0|^\alpha.$$

Translator's Note: A few typos were corrected with the author's concurrence.

2. Statement of the Theorem

Theorem 1 *For each* $\varepsilon > 0$ *and each function* $f(x)$ *in the class* C^ε, *condition* (3) *implies that*

$$|f(x) - f(x_0)| \leq C|x - x_0|^\alpha \log \frac{2}{|x - x_0|} \tag{4}$$

for $|x - x_0| \leq 1$ *and that this estimate is optimal.*

Before sketching the proof of (4), several remarks can be useful.

First, the assumption that $f \in C^\varepsilon(\mathbb{R})$ cannot be replaced by the weaker hypothesis that f is uniformly continuous on $\mathbb{R}$. Under this latter assumption, (4) would be replaced with $|f(x) - f(x_0)| \leq \omega(|x - x_0|)$, where $\omega(h)$ is the modulus of continuity of f, and this means that (3) serves no purpose.

Second, the estimate (4) is optimal in the following sense: If $0 < \varepsilon < \alpha$, then there exists a function $f \in C^\varepsilon(\mathbb{R})$ such that (3) is true and such that

$$\overline{\lim_{x \to x_0}} \left(|f(x) - f(x_0)/|x - x_0|^\alpha \log \frac{2}{|x - x_0|} \right) > 0.$$

Finally, the problem that we are considering is naturally related to the following question. Suppose that $f(x)$ is a function of a real variable such that

$$|f(x) - f(x_0)| \leq C|x - x_0|^\alpha$$

for some $\alpha \in (0, 1]$ and some constant $C > 0$. What can we then say about the Hilbert transform of f, $\tilde{f} = \mathrm{H}(f)$? Without knowing more about f, there is nothing to conclude. However, if we assume that $f(x)$ belongs to the Hölder space C^ε, then $|f(x) - f(x_0)| \leq C|x - x_0|^\alpha$ implies that

$$|\tilde{f}(x) - \tilde{f}(x_0)| \leq C|x - x_0|^\alpha \log \frac{2}{|x - x_0|}, \qquad \text{for } |x - x_0| \leq 1. \tag{5}$$

Furthermore, this estimate is optimal.

What is the relation between this statement and that of Theorem 1? If $\psi_{(j,k)}$ is the orthonormal wavelet basis constructed in [1] from a function ψ belonging to the Schwartz class $\mathcal{S}(\mathbb{R})$ and all of whose moments vanish, then $\tilde{\psi}_{(j,k)} = \mathrm{H}(\psi_{(j,k)})$ is an orthonormal wavelet basis of the same kind. Let's imagine that, for all functions $f \in C^\alpha(\mathbb{R})$, the two conditions $|f(x) - f(x_0)| \leq C|x - x_0|^\alpha$ and (3) are equivalent. Then condition (3) would also be equivalent to $|\tilde{f}(x) - \tilde{f}(x_0)| \leq C|x - x_0|^\alpha$, since $\tilde{f}(x) = \sum\sum c(j,k)\tilde{\psi}_{(j,k)}(x)$. The implication is that the Hilbert transform preserves the condition $|f(x) - f(x_0)| \leq C|x - x_0|^\alpha$, which is not the case. This means that the logarithm that one finds in the right-hand side of (4) is the same as that in (5).

3. An Outline of the Proof of Theorem 1

If we write

$$\Delta_j(x) = \sum_k c(j,k)\psi_{(j,k)}(x),$$

then (3) implies immediately that

$$|\Delta_j(x)| \le C2^{-\alpha j}(1+2^j|x-x_0|)^\alpha \quad \text{and} \quad |\Delta_j'(x)| \le C2^{(1-\alpha)j}(1+2^j|x-x_0|)^\alpha.$$

Furthermore, since $|c(j,k)| \le C'2^{-j(1/2+\varepsilon)}$, $|\Delta_j(x)| \le C''2^{-j\varepsilon}$. We have

$$f(x) = \sum_{-\infty}^{\infty} \Delta_j(x),$$

and

$$|f(x)-f(x_0)| \le \sum_{j\le j_0} |\Delta(x)-\Delta_j(x_0)| + \sum_{j_0\le j<j_1} (|\Delta_j(x)|+|\Delta_j(x_0)|) + 2\sum_{j\ge j_1} \|\Delta_j\|_\infty.$$

We define j_0 by $2^{-j_0} \le |x-x_0| < 2\cdot 2^{-j_0}$, and we let $j_1 = \alpha j_0/\varepsilon$. The inequality $|\Delta_j(x) - \Delta_j(x_0) \le C2^{j(1-\alpha)}|x-x_0|$ holds for $j \le j_0$, and this lets us estimate $\sum_{j\le j_0} |\Delta(x)-\Delta_j(x_0)|$ by $C|x-x_0|^\alpha$. If $j_0 \le j < j_1$, then $|\Delta_j(x)|+|\Delta_j(x_0)|$ is dominated by $C2^{-\alpha j}(1+2^j|x-x_0|)^\alpha \le C'|x-x_0|^\alpha$, and it finally come down to counting the number of terms, namely, $j_1 - j_0 = ((\alpha/\varepsilon)-1)j_0 \le C|\log|x-x_0||$. The last series is estimated by $C\sum_{j\ge j_1} 2^{-\varepsilon j} \le C'|x-x_0|^\alpha$.

References

[1] P.-G. Lemarié and Y. Meyer, *Ondelettes et bases hilbertiennes*, Revista Matematica Iberoamericana **2** (1986), 1–18.

[2] J. O. Stromberg, *A modified Franklin system and higher-order spline systems on $\mathbb{R}^n$ as unconditional bases for Hardy spaces*, in Conference in Harmonic Analysis in Honor of Antoni Zygmund, II, W. Beckner et al., eds., Wadsworth, Belmont, CA, (1983), 475–493,

C.E.R.M.A., Ecole national des Ponts-et-Chaussées, B.P. n° 105, 93194, Noisy-le-Grand and C.M.A.P., Ecole Polytechnique, 91128 Palaiseau Cedex.

The author wishes to thank Yves Meyer for allowing the use of his unpublished proof that the estimate (4) is optimal.

Note submitted 16 December 1988, accepted 20 December 1988.

IEEE TRANSACTIONS ON SIGNAL PROCESSING, VOL. 41, NO. 12, DECEMBER 1993 3445

Embedded Image Coding Using Zerotrees of Wavelet Coefficients

Jerome M. Shapiro

Abstract—**The embedded zerotree wavelet algorithm (EZW) is a simple, yet remarkably effective, image compression algorithm, having the property that the bits in the bit stream are generated in order of importance, yielding a fully embedded code. The embedded code represents a sequence of binary decisions that distinguish an image from the "null" image. Using an embedded coding algorithm, an encoder can terminate the encoding at any point thereby allowing a target rate or target distortion metric to be met exactly. Also, given a bit stream, the decoder can cease decoding at any point in the bit stream and still produce exactly the same image that would have been encoded at the bit rate corresponding to the truncated bit stream. In addition to producing a fully embedded bit stream, EZW consistently produces compression results that are competitive with virtually all known compression algorithms on standard test images. Yet this performance is achieved with a technique that requires absolutely no training, no pre-stored tables or codebooks, and requires no prior knowledge of the image source.**

The EZW algorithm is based on four key concepts: 1) a discrete wavelet transform or hierarchical subband decomposition, 2) prediction of the absence of significant information across scales by exploiting the self-similarity inherent in images, 3) entropy-coded successive-approximation quantization, and 4) universal lossless data compression which is achieved via adaptive arithmetic coding.

I. Introduction and Problem Statement

THIS paper addresses the two-fold problem of 1) obtaining the best image quality for a given bit rate, and 2) accomplishing this task in an embedded fashion, i.e., in such a way that all encodings of the same image at lower bit rates are embedded in the beginning of the bit stream for the target bit rate.

The problem is important in many applications, particularly for progressive transmission, image browsing [25], multimedia applications, and compatible transcoding in a digital hierarchy of multiple bit rates. It is also applicable to transmission over a noisy channel in the sense that the ordering of the bits in order of importance leads naturally to prioritization for the purpose of layered protection schemes.

Manuscript received April 28, 1992; revised June 13, 1993. The guest editor coordinating the review of this paper and approving it for publication was Prof. Martin Vetterli.

The author is with the David Sarnoff Research Center, Princeton, NJ 08543.

IEEE Log Number 9212175.

A. Embedded Coding

An embedded code represents a sequence of binary decisions that distinguish an image from the "null," or all gray, image. Since, the embedded code contains all lower rate codes "embedded" at the beginning of the bit stream, effectively, the bits are "ordered in importance." Using an embedded code, an encoder can terminate the encoding at any point thereby allowing a target rate or distortion metric to be met exactly. Typically, some target parameter, such as bit count, is monitored in the encoding process. When the target is met, the encoding simply stops. Similarly, given a bit stream, the decoder can cease decoding at any point and can produce reconstructions corresponding to all lower-rate encodings.

Embedded coding is similar in spirit to binary finite-precision representations of real numbers. All real numbers can be represented by a string of binary digits. For each digit added to the right, more precision is added. Yet, the "encoding" can cease at any time and provide the "best" representation of the real number achievable within the framework of the binary digit representation. Similarly, the embedded coder can cease at any time and provide the "best" representation of an image achievable within its framework.

The embedded coding scheme presented here was motivated in part by universal coding schemes that have been used for lossless data compression in which the coder attempts to optimally encode a source using no prior knowledge of the source. An excellent review of universal coding can be found in [3]. In universal coders, the encoder must *learn* the source statistics as it progresses. In other words, the source model is incorporated into the actual bit stream. For lossy compression, there has been little work in universal coding. Typical image coders require extensive training for both quantization (both scalar and vector) and generation of nonadaptive entropy codes, such as Huffman codes. The embedded coder described in this paper attempts to be universal by incorporating all learning into the bit stream itself. This is accomplished by the exclusive use of adaptive arithmetic coding.

Intuitively, for a given rate or distortion, a nonembedded code should be more efficient than an embedded code, since it is free from the constraints imposed by embedding. In their theoretical work [9], Equitz and Cover proved that a successively refinable description can only be optimal if the source possesses certain Markovian

1053-587X/93$03.00 © 1993 IEEE

properties. Although optimality is never claimed, a method of generating an embedded bit stream with no apparent sacrifice in image quality has been developed.

B. Features of the Embedded Coder

The EZW algorithm contains the following features

- A discrete wavelet transform which provides a compact multiresolution representation of the image.
- Zerotree coding which provides a compact multiresolution representation of *significance maps*, which are binary maps indicating the positions of the significant coefficients. Zerotrees allow the successful prediction of insignificant coefficients across scales to be efficiently represented as part of exponentially growing trees.
- Successive Approximation which provides a compact *multiprecision* representation of the significant coefficients and facilitates the embedding algorithm.
- A prioritization protocol whereby the ordering of importance is determined, in order, by the precision, magnitude, scale, and spatial location of the wavelet coefficients. Note in particular, that larger coefficients are deemed more important than smaller coefficients regardless of their scale.
- Adaptive multilevel arithmetic coding which provides a fast and efficient method for entropy coding strings of symbols, and requires no training or prestored tables. The arithmetic coder used in the experiments is a customized version of that in [31].
- The algorithm runs sequentially and stops whenever a target bit rate or a target distortion is met. A target bit rate can be met *exactly*, and an operational rate-vs.-distortion function (RDF) can be computed point-by-point.

C. Paper Organization

Section II discusses how wavelet theory and multiresolution analysis provide an elegant methodology for representing "trends" and "anomalies" on a statistically equal footing. This is important in image processing because edges, which can be thought of as anomalies in the spatial domain, represent extremely important information despite that fact that they are represented in only a tiny fraction of the image samples. Section III introduces the concept of a *zerotree* and shows how zerotree coding can efficiently encode a significance map of wavelet coefficients by predicting the absence of significant information across scales. Section IV discusses how successive approximation quantization is used in conjunction with zerotree coding, and arithmetic coding to achieve efficient embedded coding. A discussion follows on the protocol by which EZW attempts to order the bits in order of importance. A key point there is that the definition of importance for the purpose of ordering information is based on the magnitudes of the uncertainty intervals as seen from the viewpoint of what the decoder can figure out. Thus, there is no additional overhead to transmit this ordering information. Section V consists of a simple 8 × 8 example illustrating the various points of the EZW algorithm. Section VI discusses experimental results for various rates and for various standard test images. A surprising result is that using the EZW algorithm, terminating the encoding at an arbitrary point in the encoding process does not produce any artifacts that would indicate where in the picture the termination occurs. The paper concludes with Section VII.

II. Wavelet Theory and Multiresolution Analysis

A. Trends and Anomalies

One of the oldest problems in statistics and signal processing is how to choose the size of an analysis window, block size, or record length of data so that statistics computed within that window provide good models of the signal behavior within that window. The choice of an analysis window involves trading the ability to analyze "anomalies," or signal behavior that is more localized in the time or space domain and tends to be wide band in the frequency domain, from "trends," or signal behavior that is more localized in frequency but persists over a large number of lags in the time domain. To model data as being generated by random processes so that computed statistics become meaningful, stationary and ergodic assumptions are usually required which tend to obscure the contribution of anomalies.

The main contribution of wavelet theory and multiresolution analysis is that it provides an elegant framework in which both anomalies and trends can be analyzed on an equal footing. Wavelets provide a signal representation in which some of the coefficients represent long data lags corresponding to a narrow band, low frequency range, and some of the coefficients represent short data lags corresponding to a wide band, high frequency range. Using the concept of *scale*, data representing a continuous tradeoff between time (or space in the case of images) and frequency is available.

For an introduction to the theory behind wavelets and multiresolution analysis, the reader is referred to several excellent tutorials on the subject [6], [7], [17], [18], [20], [26], [27].

B. Relevance to Image Coding

In image processing, most of the image area typically represents spatial "trends," or areas of high statistical spatial correlation. However "anomalies," such as edges or object boundaries, take on a perceptual significance that is far greater than their numerical energy contribution to an image. Traditional transform coders, such as those using the DCT, decompose images into a representation in which each coefficient corresponds to a fixed size spatial area and a fixed frequency bandwidth, where the bandwidth and spatial area are effectively the same for all coefficients in the representation. Edge information tends to disperse so that many non-zero coefficients are required to represent edges with good fidelity. However, since the edges represent relatively insignificant energy with respect to the entire image, traditional transform coders, such as those using the DCT, have been fairly successful at medium and high bit rates. At extremely low bit rates, however, traditional transform coding techniques, such as JPEG [30], tend to allocate too many bits to the "trends," and have few bits left over to represent "anomalies." As a result, blocking artifacts often result.

Wavelet techniques show promise at extremely low bit rates because trends, anomalies, and information at all "scales" in between are available. A major difficulty is that fine detail coefficients representing possible anomalies constitute the largest number of coefficients, and therefore, to make effective use of the multiresolution representation, much of the information is contained in representing the *position* of those few coefficients corresponding to significant anomalies.

The techniques of this paper allow coders to effectively use the power of multiresolution representations by efficiently representing the positions of the wavelet coefficients representing significant anomalies.

C. A Discrete Wavelet Transform

The discrete wavelet transform used in this paper is identical to a hierarchical subband system, where the subbands are logarithmically spaced in frequency and represent an octave-band decomposition. To begin the decomposition, the image is divided into four subbands and critically subsampled as shown in Fig. 1. Each coefficient represents a spatial area corresponding to approximately a 2×2 area of the original image. The low frequencies represent a bandwidth approximately corresponding to $0 < |\omega| < \pi/2$, whereas the high frequencies represent the band from $\pi/2 < |\omega| < \pi$. The four subbands arise from separable application of vertical and horizontal filters. The subbands labeled LH_1, HL_1, and HH_1 represent the finest scale wavelet coefficients. To obtain the next coarser scale of wavelet coefficients, the subband LL_1 is further decomposed and critically sampled as shown in Fig. 2. The process continues until some final scale is reached. Note that for each coarser scale, the coefficients represent a larger spatial area of the image but a narrower band of frequencies. At each scale, there are three subbands; the remaining lowest frequency subband is a representation of the information at all coarser scales. The issues involved in the design of the filters for the type of subband decomposition described above have been discussed by many authors and are not treated in this paper. Interested readers should consult [1], [6], [32], [35], in addition to references found in the bibliographies of the tutorial papers cited above.

It is a matter of terminology to distinguish between a transform and a subband system as they are two ways of describing the same set of numerical operations from differing points of view. Let x be a column vector whose elements represent a scanning of the image pixels, let X be a column vector whose elements are the array of coefficients resulting from the wavelet transform or subband decomposition applied to x. From the transform point of view, X represents a linear transformation of x represented by the matrix W, i.e.,

$$X = Wx. \tag{1}$$

Although not actually computed this way, the effective filters that generate the subband signals from the original signal form basis functions for the transformation, i.e., the rows of W. Different coefficients in the same subband represent the projection of the entire image onto translates of a prototype subband filter, since from the subband point of view, they are simply regularly spaced different outputs of a convolution between the image and a subband filter. Thus, the basis functions for each coefficient in a given subband are simply translates of one another.

In subband coding systems [32], the coefficients from a given subband are usually grouped together for the purposes of designing quantizers and coders. Such a grouping suggests that statistics computed from a subband are in some sense representative of the samples in that sub-

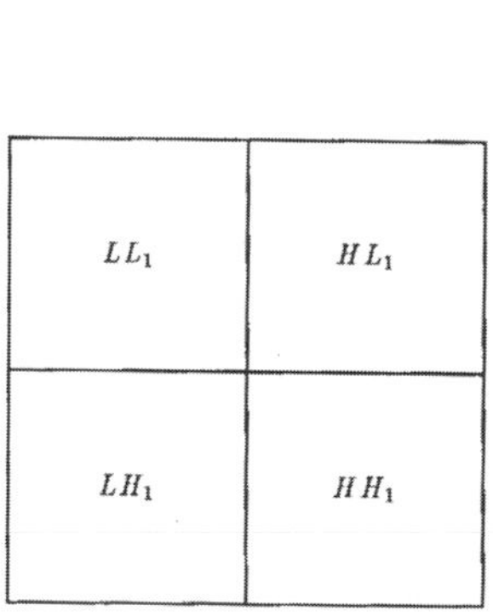

Fig. 1. First stage of a discrete wavelet transform: The image is divided into four subbands using separable filters. Each coefficient represents a spatial area corresponding to approximately a 2×2 area of the original picture. The low frequencies represent a bandwidth approximately corresponding to $0 < |\omega| < \pi/2$, whereas the high frequencies represent the band from $\pi/2 < |\omega| < \pi$. The four subbands arise from separable application of vertical and horizontal filters.

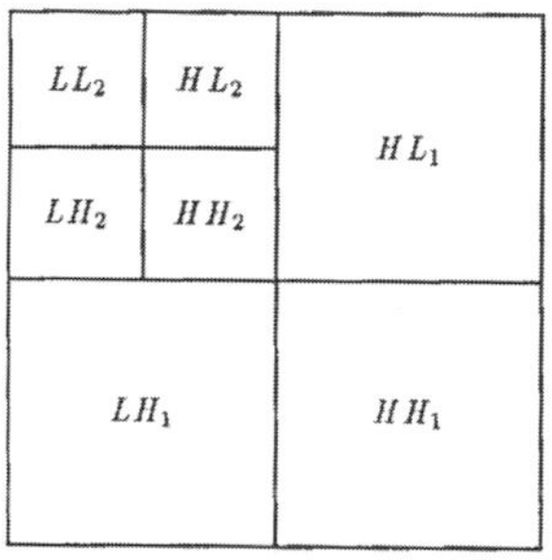

Fig. 2. A two-scale wavelet decomposition: The image is divided into four subbands using separable filters. Each coefficient in the subbands LL_2, LH_2, HL_2 and HH_2 represents a spatial area corresponding to approximately a 4×4 area of the original picture. The low frequencies at this scale represent a bandwidth approximately corresponding to $0 < |\omega| < \pi/4$, whereas the high frequencies represent the band from $\pi/4 < |\omega| < \pi/2$.

band. However this statistical grouping once again implicitly de-emphasizes the outliers, which tend to represent the most significant anomalies or edges. In this paper, the term "wavelet transform" is used because each wavelet coefficient is individually and deterministically compared to the same set of thresholds for the purpose of measuring significance. Thus, each coefficient is treated as a distinct, potentially important piece of data regardless of its scale, and no statistics for a whole subband are used in any form. The result is that the small number of "deterministically" significant fine scale coefficients are not obscured because of their "statistical" insignificance.

The filters used to compute the discrete wavelet transform in the coding experiments described in this paper are based on the 9-tap symmetric quadrature mirror filters (QMF) whose coefficients are given in [1]. This transformation has also been called a QMF-pyramid. These filters were chosen because in addition to their good localization properties, their symmetry allows for simple edge treatments, and they produce good results empirically. Additionally, using properly scaled coefficients, the transformation matrix for a discrete wavelet transform obtained using these filters is so close to unitary that it can be treated as unitary for the purpose of lossy compression. Since unitary transforms preserve L_2 norms, it makes sense from a numerical standpoint to compare all of the resulting transform coefficients to the same thresholds to assess significance.

III. Zerotrees of Wavelet Coefficients

In this section, an important aspect of low bit rate image coding is discussed: the coding of the positions of those coefficients that will be transmitted as nonzero values. Using scalar quantization followed by entropy coding, in order to achieve very low bit rates, i.e., less than 1 bit/pel, the probability of the most likely symbol after quantization—the zero symbol—must be extremely high. Typically, a large fraction of the bit budget must be spent on encoding the *significance map*, or the binary decision as to whether a sample, in this case a coefficient of a 2-D discrete wavelet transform, has a zero or nonzero quantized value. It follows that a significant improvement in encoding the significance map translates into a corresponding gain in compression efficiency.

A. Significance Map Encoding

To appreciate the importance of significance map encoding, consider a typical transform coding system where a decorrelating transformation is followed by an entropy-coded scalar quantizer. The following discussion is not intended to be a rigorous justification for significance map encoding, but merely to provide the reader with a sense of the relative coding costs of the position information contained in the significance map relative to amplitude and sign information.

A typical low-bit rate image coder has three basic components: a transformation, a quantizer and data compression, as shown in Fig. 3.

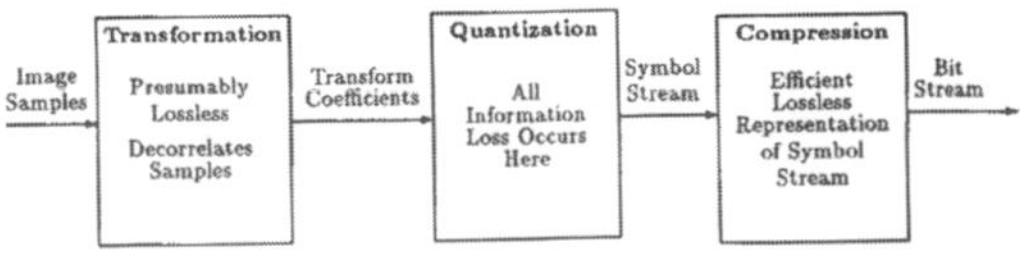

Fig. 3. A generic transform coder.

The original image is passed through some transformation to produce transform coefficients. This transformation is considered to be lossless, although in practice this may not be the case exactly. The transform coefficients are then quantized to produce a stream of symbols, each of which corresponds to an index of a particular quantization bin. Note that virtually all of the information loss occurs in the quantization stage. The data compression stage takes the stream of symbols and attempts to losslessly represent the data stream as efficiently as possible.

The goal of the transformation is to produce coefficients that are decorrelated. If we could, we would ideally like a transformation to remove all dependencies between samples. Assume for the moment that the transformation is doing its job so well that the resulting transform coefficients are not merely uncorrelated, but statistically independent. Also, assume that we have removed the mean and coded it separately so that the transform coefficients can be modeled as zero-mean, independent, although perhaps not identically distributed random variables. Furthermore, we might additionally constrain the model so that the probability density functions (PDF) for the coefficients are symmetric.

The goal is to quantize the transform coefficients so that the entropy of the resulting distribution of bin indexes is small enough so that the symbols can be entropy-coded at some target low bit rate, say for example 0.5 bits per pixel (bpp.). Assume that the quantizers will be symmetric midtread, perhaps nonuniform, quantizers, although different symmetric midtread quantizers may be used for different groups of transform coefficients. Letting the central bin be index 0, note that because of the symmetry, for a bin with a nonzero index magnitude, a positive or negative index is equally likely. In other words, for each nonzero index encoded, the entropy code is going to require at least one-bit for the sign. An entropy code can be designed based on modeling probabilities of bin indices as the fraction of coefficients in which the absolute value of a particular bin index occurs. Using this simple model, and assuming that the resulting symbols are independent, the entropy of the symbols H can be expressed as

$$H = -p \log_2 p - (1 - p) \log_2 (1 - p) + (1 - p)[1 + H_{NZ}], \quad (2)$$

where p is the probability that a transform coefficient is quantized to zero, and H_{NZ} represents the conditional entropy of the absolute values of the quantized coefficients conditioned on them being nonzero. The first two terms in the sum represent the first-order binary entropy of the significance map, whereas the third term represents the conditional entropy of the distribution of nonzero values multiplied by the probability of them being nonzero. Thus, we can express the true cost of encoding the actual symbols as follows:

$$\text{Total Cost} = \text{Cost of Significance Map} + \text{Cost of Nonzero Values}. \quad (3)$$

Returning to the model, suppose that the target is $H = 0.5$. What is the minimum probability of zero achievable? Consider the case where we only use a 3-level quantizer, i.e. $H_{NZ} = 0$. Solving for p provides a lower bound on the probability of zero given the independence assumption

$$p_{\min}(H_{NZ} = 0, H = 0.5) = 0.916. \quad (4)$$

In this case, under the most ideal conditions, 91.6% of the coefficients must be quantized to zero. Furthermore, 83% of the bit budget is used in encoding the significance map. Consider a more typical example where $H_{NZ} = 4$, the minimum probability of zero is

$$p_{\min}(H_{NZ} = 4, H = 0.5) = 0.954. \quad (5)$$

In this case, the probability of zero must increase, while the cost of encoding the significance map is still 54% of the cost.

As the target rate decreases, the probability of zero increases, and the fraction of the encoding cost attributed to the significance map increases. Of course, the independence assumption is unrealistic and in practice, there are often additional dependencies between coefficients that can be exploited to further reduce the cost of encoding the significance map. Nevertheless, the conclusion is that no matter how optimal the transform, quantizer or entropy coder, under very typical conditions, the cost of determining the positions of the few significant coefficients represents a significant portion of the bit budget at low rates, and is likely to become an increasing fraction of the total cost as the rate decreases. As will be seen, by employing an image model based on an extremely simple and easy to satisfy hypothesis, we can efficiently encode significance maps of wavelet coefficients.

B. Compression of Significance Maps using Zerotrees of Wavelet Coefficients

To improve the compression of significance maps of wavelet coefficients, a new data structure called a *zerotree* is defined. A wavelet coefficient x is said to be *insignificant* with respect to a given threshold T if $|x| < T$. The zerotree is based on the hypothesis that if a wavelet coefficient at a coarse scale is insignificant with respect to a given threshold T, then *all* wavelet coefficients of the same orientation in the same spatial location at finer scales are likely to be insignificant with respect to T. Empirical evidence suggests that this hypothesis is often true.

More specifically, in a hierarchical subband system, with the exception of the highest frequency subbands, every coefficient at a given scale can be related to a set of coefficients at the next finer scale of similar orientation. The coefficient at the coarse scale is called the *parent*, and all coefficients corresponding to the same spatial location at the next finer scale of similar orientation are called *children*. For a given parent, the set of all coefficients at all finer scales of similar orientation corresponding to the same location are called *descendants*. Similarly, for a given child, the set of coefficients at all coarser scales of similar orientation corresponding to the same location are called *ancestors*. For a QMF-pyramid subband decomposition, the parent–child dependencies are shown in Fig. 4. A wavelet tree descending from a coefficient in subband $HH3$ is also seen in Fig. 4. With the exception of the lowest frequency subband, all parents have four children. For the lowest frequency subband, the parent–child relationship is defined such that each parent node has three children.

A scanning of the coefficients is performed in such a way that no child node is scanned before its parent. For an N-scale transform, the scan begins at the lowest frequency subband, denoted as LL_N, and scans subbands HL_N, LH_N, and HH_N, at which point it moves on to scale $N - 1$, etc. The scanning pattern for a 3-scale QMF-pyramid can be seen in Fig. 5. Note that each coefficient within a given subband is scanned before any coefficient in the next subband.

Given a threshold level T to determine whether or not a coefficient is significant, a coefficient x is said to be an element of a *zerotree* for threshold T if itself and *all* of its descendents are insignificant with respect to T. An element of a zerotree for threshold T is a *zerotree root* if it is not the descendant of a previously found zerotree root for threshold T, i.e., it is not *predictably insignificant* from the discovery of a zerotree root at a coarser scale at the same threshold. A zerotree root is encoded with a special symbol indicating that the insignificance of the coefficients at finer scales is completely predictable. The significance map can be efficiently represented as a string of symbols from a 3-symbol alphabet which is then entropy-coded. The three symbols used are 1) zerotree root, 2) isolated zero, which means that the coefficient is insignificant but has some significant descendant, and 3) significant. When encoding the finest scale coefficients, since coefficients have no children, the symbols in the string come from a 2-symbol alphabet, whereby the zerotree symbol is not used.

As will be seen in Section IV, in addition to encoding the significance map, it is useful to encode the sign of significant coefficients along with the significance map. Thus, in practice, four symbols are used: 1) zerotree root, 2) isolated zero, 3) positive significant, and 4) negative significant. This minor addition will be useful for embedding. The flow chart for the decisions made at each coefficient are shown in Fig. 6.

Note that it is also possible to include two additional symbols such as "positive/negative significant, but descendants are zerotrees" etc. In practice, it was found that

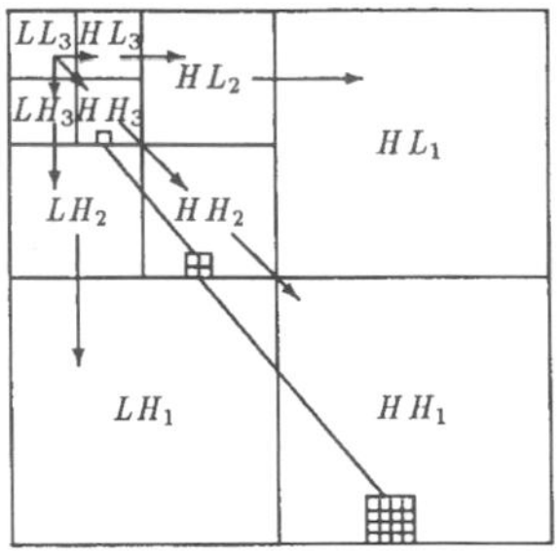

Fig. 4. Parent–child dependencies of subbands: Note that the arrow points from the subband of the parents to the subband of the children. The lowest frequency subband is the top left, and the highest frequency subband is at the bottom right. Also shown is a wavelet tree consisting of all of the descendents of a single coefficient in subband $HH3$. The coefficient in $HH3$ is a zerotree root if it is insignificant and *all* of its descendants are insignificant.

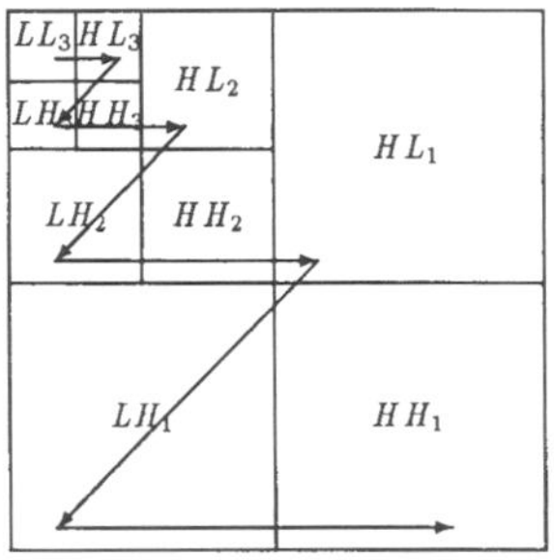

Fig. 5. Scanning order of the subbands for encoding a significance map: Note that parents must be scanned before children. Also note that all positions in a given subband are scanned before the scan moves to the next subband.

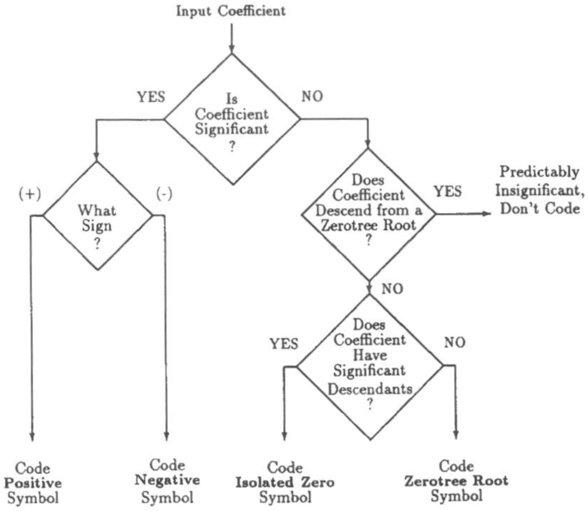

Fig. 6. Flow chart for encoding a coefficient of the significance map.

at low bit rates, this addition often increases the cost of coding the significance map. To see why this may occur, consider that there is a cost associated with partitioning the set of positive (or negative) significant samples into those whose descendents are zerotrees and those with significant descendants. If the cost of this decision is C bits, but the cost of encoding a zerotree is less than $C/4$ bits, then it is more efficient to code four zerotree symbols separately than to use additional symbols.

Zerotree coding reduces the cost of encoding the significance map using self-similarity. Even though the image has been transformed using a decorrelating transform the *occurrences* of insignificant coefficients are not independent events. More traditional techniques employing transform coding typically encode the binary map via some form of run-length encoding [30]. Unlike the zerotree symbol, which is a single "terminating" symbol and applies to all tree-depths, run-length encoding requires a symbol for each run-length which much be encoded. A technique that is closer in spirit to the zerotrees is the end-of-block (EOB) symbol used in JPEG [30], which is also a "terminating" symbol indicating that all remaining DCT coefficients in the block are quantized to zero. To see why zerotrees may provide an advantage over EOB symbols, consider that a zerotree represents the insignificance information in a given orientation over an approximately square spatial area at all finer scales up to and including the scale of the zerotree root. Because the wavelet transform is a hierarchical representation, varying the scale in which a zerotree root occurs automatically adapts the spatial area over which insignificance is represented. The EOB symbol, however, always represents insignificance over the same spatial area, although the number of frequency bands within that spatial area varies. Given a fixed block size, such as 8×8, there is exactly one scale in the wavelet transform in which if a zerotree root is found at that scale, it corresponds to the same spatial area as a block of the DCT. If a zerotree root can be identified at a coarser scale, then the insignificance pertaining to that orientation can be predicted over a larger area. Similarly, if the zerotree root does not occur at this scale, then looking for zerotrees at finer scales represents a hierarchical divide and conquer approach to searching for one or more smaller areas of insignificance over the same spatial regions as the DCT block size. Thus, many more coefficients can be predicted in smooth areas where a root typically occurs at a coarse scale. Furthermore, the zerotree approach can isolate interesting non-zero details by immediately eliminating large insignificant regions from consideration.

Note that this technique is quite different from previous attempts to exploit self-similarity in image coding [19] in that it is far easier to predict insignificance than to predict significant detail across scales. The zerotree approach was developed in recognition of the difficulty in achieving meaningful bit rate reductions for significant coefficients via additional prediction. Instead, the focus here is on reducing the cost of encoding the significance map so that, for a given bit budget, more bits are available to encode expensive significant coefficients. In practice, a large fraction of the insignificant coefficients are efficiently encoded as part of a zerotree.

A similar technique has been used by Lewis and Knowles (LK) [15], [16]. In that work, a "tree" is said to be zero if its energy is less than a perceptually based threshold. Also, the "zero flag" used to encode the tree is not entropy-coded. The present work represents an improvement that allows for embedded coding for two reasons. Applying a deterministic threshold to determine significance results in a zerotree symbol which guarantees that no descendant of the root has a magnitude larger than the threshold. As a result, there is no possibility of a significant coefficient being obscured by a statistical energy measure. Furthermore, the zerotree symbol developed in this paper is part of an alphabet for entropy coding the significance map which further improves its compression. As will be discussed subsequently, it is the first property that makes this method of encoding a significance map useful in conjunction with successive-approximation. Recently, a promising technique representing a compromise between the EZW algorithm and the LK coder has been presented in [34].

C. Interpretation as a Simple Image Model

The basic hypothesis—if a coefficient at a coarse scale is insignificant with respect to a threshold then all of its descendants, as defined above, are also insignificant—can be interpreted as an extremely general image model. One of the aspects that seems to be common to most models used to describe images is that of a "decaying spectrum." For example, this property exists for both stationary autoregressive models, and non-stationary fractal, or "nearly-$1/f$" models, as implied by the name which refers to a generalized spectrum [33]. The model for the zerotree hypothesis is even more general than "decaying spectrum" in that it allows for some deviations to "decaying spectrum" because it is linked to a specific threshold. Consider an example where the threshold is 50, and we are considering a coefficient of magnitude 30, and whose largest descendant has a magnitude of 40. Although a higher frequency descendant has a larger magnitude (40) than the coefficient under consideration (30), i.e., the "decaying spectrum" hypothesis is violated, the coefficient under consideration can still be represented using a zerotree root since the whole tree is still insignificant (magnitude less than 50). Thus, assuming the more common image models have some validity, the zerotree hypothesis should be satisfied easily and extremely often. For those instances where the hypothesis is violated, it is safe to say that an *informative*, i.e., unexpected, event has occurred, and we should expect the cost of representing this event to be commensurate with its self-information.

It should also be pointed out that the improvement in encoding significance maps provided by zerotrees is specifically *not* the result of exploiting any linear dependencies between coefficients of different scales that were not removed in the transform stage. In practice, the linear correlation between the values of parent and child wavelet coefficients has been found to be extremely small, implying that the wavelet transform is doing an excellent job of producing nearly uncorrelated coefficients. However, there is likely additional dependency between the *squares* (or magnitudes) of parents and children. Experiments run on about 30 images of all different types, show that the correlation coefficient between the square of a child and the square of its parent tends to be between 0.2 and 0.6 with a string concentration around 0.35. Although this dependency is difficult to characterize in general for most images, even without access to specific statistics, it is reasonable to *expect* the magnitude of a child to be smaller than the magnitude of its parent. In other words, it can be reasonably conjectured based on experience with real-world images, that had we known the details of the statistical dependencies, and computed an "optimal" estimate, such as the conditional expectation of the child's magnitude given the parent's magnitude, that the "optimal" estimator would, with very high probability, predict that the child's magnitude would be the smaller of the two. Using only this mild assumption, based on an inexact statistical characterization, given a fixed threshold, and conditioned on the knowledge that a parent is insignificant with respect to the threshold, the "optimal" estimate of the significance of the rest of the descending wavelet tree is that it is entirely insignificant with respect to the same threshold, i.e., a zerotree. On the other hand, if the parent is significant, the "optimal" estimate of the significance of descendants is highly dependent on the details of the estimator whose knowledge would require more detailed information about the statistical nature of the image. Thus, under this mild assumption, using zerotrees to predict the insignificance of wavelet coefficients at fine scales given the insignificance of a root at a coarse scale is more likely to be successful in the absence of additional information than attempting to predict significant detail across scales.

This argument can be made more concrete. Let x be a child of y, where x and y are zero-mean random variables, whose probability density functions (PDF) are related as

$$p_x(x) = ap_y(ax), \qquad a > 1. \tag{6}$$

This states that random variables x and y have the same PDF shape, and that

$$\sigma_y^2 = a^2\sigma_x^2. \tag{7}$$

Assume further that x and y are uncorrelated, i.e.,

$$E[xy] = 0. \tag{8}$$

Note that nothing has been said about treating the subbands as a group, or as stationary random processes, only that there is a similarity relationship between random variables of parents and children. It is also reasonable because for intermediate subbands a coefficient that is a child with respect to one coefficient is a parent with respect to others; the PDF of that coefficient should be the same in either case. Let $u = x^2$ and $v = y^2$. Suppose that u and

v are correlated with correlation coefficient ρ. We have the following relationships:

$$E[u] = \sigma_x^2 \tag{9}$$

$$E[v] = \sigma_y^2 \tag{10}$$

$$\sigma_u^2 = E[x^4] - \sigma_x^4 \tag{11}$$

$$\sigma_v^2 = E[y^4] - \sigma_y^4. \tag{12}$$

Notice in particular that

$$\sigma_v^2 = a^4 \sigma_u^2. \tag{13}$$

Using a well known result, the expression for the best linear unbiased estimator (BLUE) of u given v to minimize error variance is given by

$$\hat{u}_{\text{BLUE}}(v) = E[u] - \rho \frac{\sigma_u}{\sigma_v}(E[v] - v) \tag{14}$$

$$= \frac{1-\rho}{a^2}\sigma_y^2 + \rho \frac{v}{a^2}. \tag{15}$$

If it is observed that the magnitude of the parent is below the threshold T, i.e., $v = y^2 < T^2$, then the BLUE can be upper bounded by

$$\hat{u}_{\text{BLUE}}(v | v < T^2) < \frac{1-\rho}{a^2}\sigma_y^2 + \rho \frac{T^2}{a^2}. \tag{16}$$

Consider two cases a) $T \geq \sigma_y$ and b) $T < \sigma_y$. In case (a), we have

$$\hat{u}_{\text{BLUE}}(v | v < T^2) \leq \frac{T^2}{a^2} < T^2, \tag{17}$$

which implies that the BLUE of x^2 given $|y| < T$ is less than T^2, for *any* ρ, including $\rho = 0$. In case (b), we can only upper bound the right hand side of (16) by T^2 if ρ exceeds the lower bound

$$\rho \geq \frac{1 - \frac{a^2T^2}{\sigma_y^2}}{1 - \frac{T^2}{\sigma_y^2}} \triangleq \rho_0. \tag{18}$$

Of course, a better nonlinear estimate might yield different results, but the above analysis suggests that for threshold exceeding the standard deviation of the parent, which by (6) exceeds the standard deviation of all descendants, if it is observed that a parent is insignificant with respect to the threshold, then, using the above BLUE, the estimates for the magnitudes of all descendants is that they are less than the threshold, and a zerotree is expected regardless of the correlation between squares of parents and squares of children. As the threshold decreases, more correlation is required to justify *expecting* a zerotree to occur. Finally, since the lower bound $\rho_0 \to 1$ as $T \to 0$, as the threshold is reduced, it becomes increasingly difficult to expect zerotrees to occur, and more knowledge of the particular statistics are required to make inferences. The implication of this analysis is that at very low bit rates, where the probability of an insignificant sample must be high and thus, the significance threshold T must also be large, expecting the occurrence of zerotrees and encoding significance maps using zerotree coding is reasonable without even knowing the statistics. However, letting T decrease, there is some point below which the advantage of zerotree coding diminishes, and this point is dependent on the specific nature of higher order dependencies between parents and children. In particular, the stronger this dependence, the more T can be decreased while still retaining an advantage using zerotree coding. Once again, this argument is not intended to "prove" the optimality of zerotree coding, only to suggest a rationale for its demonstrable success.

D. *Zerotree-Like Structures in Other Subband Configurations*

The concept of predicting the insignificance of coefficients from low frequency to high frequency information corresponding to the same spatial localization is a fairly general concept and not specific to the wavelet transform configuration shown in Fig. 4. Zerotrees are equally applicable to quincunx wavelets [2], [13], [23], [29], in which case each parent would have two children instead of four, except for the lowest frequency, where parents have a single child.

Also, a similar approach can be applied to linearly spaced subband decompositions, such as the DCT, and to other more general subband decompositions, such as wavelet packets [5] and Laplacian pyramids [4]. For example, one of many possible parent–child relationship for linearly spaced subbands can be seen in Fig. 7. Of course, with the use of linearly spaced subbands, zerotree-like coding loses its ability to adapt the spatial extent of the insignificance prediction. Nevertheless, it is possible for zerotree-like coding to outperform EOB-coding since more coefficients can be predicted from the subbands along the diagonal. For the case of wavelet packets, the situation is a bit more complicated, because a wider range of tilings of the "space-frequency" domain are possible. In that case, it may not always be possible to define similar parent-child relationships because a high-frequency coefficient may in fact correspond to a larger spatial area than a co-located lower frequency coefficient. On the other hand, in a coding scheme such as the "best-basis" approach of Coifman *et al.* [5], had the image-dependent best basis resulted in such a situation, one wonders if the underlying hypothesis—that magnitudes of coefficients tend to decay with frequency—would be reasonable anyway. These zerotree-like extensions represent interesting areas for further research.

IV. Successive-Approximation

The previous section describes a method of encoding significance maps of wavelet coefficients that, at least empirically, seems to consistently produce a code with a lower bit rate than either the empirical first-order entropy,

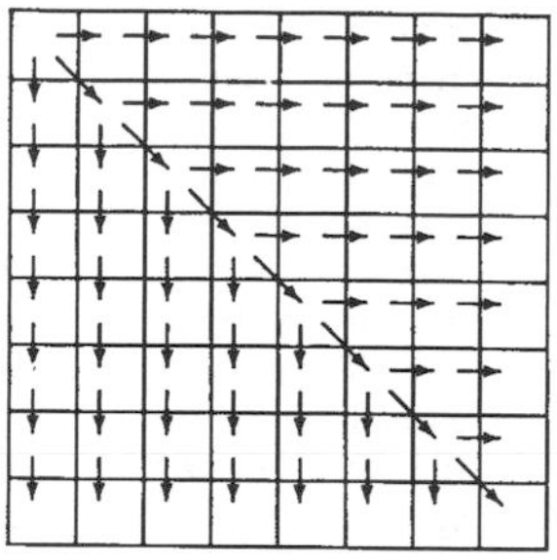

Fig. 7. Parent–child dependencies for linearly spaced subbands systems, such as the DCT. Note that the arrow points from the subband of the parents to the subband of the children. The lowest frequency subband is the top left, and the highest frequency subband is at the bottom right.

or a run-length code of the significance map. The original motivation for employing successive-approximation in conjunction with zerotree coding was that since zerotree coding was performing so well encoding the significance map of the wavelet coefficients, it was hoped that more efficient coding could be achieved by zerotree coding more significance maps.

Another motivation for successive-approximation derives directly from the goal of developing an embedded code analogous to the binary-representation of an approximation to a real number. Consider the wavelet transform of an image as a mapping whereby an amplitude exists for each coordinate in scale-space. The scale-space coordinate system represents a coarse-to-fine "logarithmic" representation of the domain of the function. Taking the coarse-to-fine philosophy one-step further, successive-approximation provides a coarse-to-fine, multiprecision "logarithmic" representation of amplitude information, which can be thought of as the range of the image function when viewed in the scale-space coordinate system defined by the wavelet transform. Thus, in a very real sense, the EZW coder generates a representation of the image that is coarse-to-fine in both the domain and range simultaneously.

A. Successive-Approximation Entropy-Coded Quantization

To perform the embedded coding, successive-approximation quantization (SAQ) is applied. As will be seen, SAQ is related to bit-plane encoding of the magnitudes. The SAQ sequentially applies a sequence of thresholds $T_0, \cdots, T_{N-1}$ to determine significance, where the thresholds are chosen so that $T_i = T_{i-1}/2$. The initial threshold T_0 is chosen so that $|X_j| < 2T_0$ for all transform coefficients x_j.

During the encoding (and decoding), two separate lists of wavelet coefficients are maintained. At any point in the process, the *dominant* list contains the *coordinates* of those coefficients that have not yet been found to be significant in the same relative order as the initial scan. This scan is such that the subbands are ordered, and within each subband, the set of coefficients are ordered. Thus, using the ordering of the subbands shown in Fig. 5, all coefficients in a given subband appear on the initial dominant list prior to coefficients in the next subband. The *subordinate* list contains the *magnitudes* of those coefficients that have been found to be significant. For each threshold, each list is scanned once.

During a *dominant pass*, coefficients with coordinates on the dominant list, i.e., those that have not yet been found to be significant, are compared to the threshold T_i to determine their significance, and if significant, their sign. This significance map is then zerotree coded using the method outlined in Section III. Each time a coefficient is encoded as significant, (positive or negative), its magnitude is appended to the subordinate list, and the coefficient in the wavelet transform array is set to zero so that the significant coefficient does not prevent the occurrence of a zerotree on future dominant passes at smaller thresholds.

A dominant pass is followed by a *subordinate pass* in which all coefficients on the subordinate list are scanned and the specifications of the magnitudes available to the decoder are refined to an additional bit of precision. More specifically, during a subordinate pass, the *width* of the effective quantizer step size, which defines an uncertainty interval for the true magnitude of the coefficient, is cut in half. For each magnitude on the subordinate list, this refinement can be encoded using a binary alphabet with a "1" symbol indicating that the true value falls in the upper half of the old uncertainty interval and a "0" symbol indicating the lower half. The string of symbols from this binary alphabet that is generated during a subordinate pass is then entropy coded. Note that prior to this refinement, the width of the uncertainty region is exactly equal to the current threshold. After the completion of a subordinate pass the magnitudes on the subordinate list are sorted in decreasing magnitude, to the extent that the decoder has the information to perform the same sort.

The process continues to alternate between dominant passes and subordinate passes where the threshold is halved before each dominant pass. (In principle one could divide by other factors than 2. This factor of 2 was chosen here because it has nice interpretations in terms of bit plane encoding and numerical precision in a familiar base 2, and good coding results were obtained).

In the decoding operation, each decoded symbol, both during a dominant and a subordinate pass, refines and reduces the width of the uncertainty interval in which the true value of the coefficient (or coefficients, in the case of a zerotree root) may occur. The reconstruction value used can be anywhere in that uncertainty interval. For minimum mean-square error distortion, one could use the centroid of the uncertainty region using some model for the PDF of the coefficients. However, a practical approach, which is used in the experiments, and is also MINMAX optimal, is to simply use the center of the uncertainty interval as the reconstruction value.

The encoding stops when some target stopping condition is met, such as when the bit budget is exhausted. The encoding can cease at any time and the resulting bit stream contains all lower rate encodings. Note, that if the bit stream is truncated at an arbitrary point, there may be bits at the end of the code that do not decode to a valid symbol since a codeword has bene truncated. In that case, these bits do not reduce the width of an uncertainty interval or any distortion function. In fact, it is very likely that the first L bits of the bit stream will produce exactly the same image as the first $L + 1$ bits which occurs if the additional bit is insufficient to complete the decoding of another symbol. Nevertheless, terminating the decoding of an embedded bit stream at a specific point in the bit stream produces exactly the same image that would have resulted had that point been the initial target rate. This ability to cease encoding or decoding anywhere is extremely useful in systems that are either rate-constrained or distortion-constrained. A side benefit of the technique is that an operational rate vs. distortion plot for the algorithm can be computed on-line.

B. Relationship to Bit Plane Encoding

Although the embedded coding system described here is considerably more general and more sophisticated than simple bit-plane encoding, consideration of the relationship with bit-plane encoding provides insight into the success of embedded coding.

Consider the successive-approximation quantizer for the case when all thresholds are powers of two, and all wavelet coefficients are integers. In this case, for each coefficient that eventually gets coded as significant, the sign and *bit position* of the most-significant binary digit (MSBD) are measured and encoded during a dominant pass. For example, consider the 10-bit representation of the number 41 as 0000101001. Also, consider the binary digits as a sequence of binary decisions in a binary tree. Proceeding from left to right, if we have not yet encountered a "1," we expect the probability distribution for the next digit to be strongly biased toward "0." The digits to the left and including the MSBD are called the *dominant bits*, and are measured during dominant passes. After the MSBD has been encountered, we expect a more random and much less biased distribution between a "0" and a "1," although we might still expect $P(0) > P(1)$ because most PDF models for transform coefficients decay with amplitude. Those binary digits to the right of the MSBD are called the *subordinate bits* and are measured and encoded during the subordinate pass. A zeroth-order approximation suggests that we should expect to pay close to one bit per "binary digit" for subordinate bits, while dominant bits should be far less expensive.

By using successive-approximation beginning with the largest possible threshold, where the probability of zero is extremely close to one, and by using zerotree coding, whose efficiency increases as the probability of zero increases, we should be able to code dominant bits with very few bits, since they are most often part of a zerotree.

In general, the thresholds need not be powers of two. However, by factoring out a constant mantissa, M, the starting threshold T_0 can be expressed in terms of a threshold that is a power of two

$$T_0 = M2^E, \tag{19}$$

where the exponent E is an integer, in which case, the dominant and subordinate bits of appropriately scaled wavelet coefficients are coded during dominant and subordinate passes, respectively.

C. Advantage of Small Alphabets for Adaptive Arithmetic Coding

Note that the particular encoder alphabet used by the arithmetic coder at any given time contains either 2, 3, or 4 symbols depending whether the encoding is for a subordinate pass, a dominant pass with no zerotree root symbol, or a dominant pass with the zerotree root symbol. This is a real advantage for adapting the arithmetic coder. Since there are never more than four symbols, all of the possibilities typically occur with a reasonably measurable frequency. This allows an adaptation algorithm with a short memory to learn quickly and constantly track changing symbol probabilities. This adaptivity accounts for some of the effectiveness of the overall algorithm. Contrast this with the case of a large alphabet, as is the case in algorithms that do not use successive approximation. In that case, it takes many events before an adaptive entropy coder can reliably estimate the probabilities of unlikely symbols (see the discussion of the zero-frequency problem in [3]). Furthermore, these estimates are fairly unreliable because images are typically statistically nonstationary and local symbol probabilities change from region to region.

In the practical coder used in the experiments, the arithmetic coder is based on [31]. In arithmetic coding, the encoder is separate from the model, which in [31], is basically a histogram. During the dominant passes, simple Markov conditioning is used whereby one of four histograms is chosen depending on 1) whether the previous coefficient in the scan is known to be significant, and 2) whether the parent is known to be significant. During the subordinate passes, a single histogram is used. Each histogram entry is initialized to a count of one. After encoding each symbol, the corresponding histogram entry is incremented. When the sum of all the counts in a histogram reaches the maximum count, each entry is incremented and integer divided by two, as described in [31]. It should be mentioned, that for practical purposes, the coding gains provided by using this simple Markov conditioning may not justify the added complexity and using a single histogram strategy for the dominant pass performs almost as well (0.12 dB worse for Lena at 0.25 bpp.). The choice of maximum histogram count is probably more critical, since that controls the learning rate for the adaptation. For the experimental results presented, a maximum count of 256 was used, which provides an intermediate tradeoff between the smallest possible probability, which is the reciprocal of the maximum count, and the learning rate, which is faster with a smaller maximum histogram count.

D. Order of Importance of the Bits

Although importance is a subjective term, the order of processing used in EZW implicitly defines a precise ordering of importance that is tied to, in order, *precision*, *magnitude*, *scale*, and *spatial location* as determined by the initial dominant list.

The primary determination of ordering importance is the numerical *precision* of the coefficients. This can be seen in the fact that the uncertainty intervals for the magnitude of all coefficients are refined to the same precision before the uncertainty interval for any coefficient is refined further.

The second factor in the determination of importance is *magnitude*. Importance by magnitude manifests itself during a dominant pass because prior to the pass, all coefficients are insignificant and presumed to be zero. When they are found to be significant, they are all assumed to have the same magnitude, which is greater than the magnitudes of those coefficients that remain insignificant. Importance by magnitude manifests itself during a subordinate pass by the fact that magnitudes are refined in descending order of the center of the uncertainty intervals, i.e., the decoder's interpretation of the magnitude.

The third factor, *scale*, manifests itself in the *a priori* ordering of the subbands on the initial dominant list. Until the significance of the magnitude of a coefficient is discovered during a dominant pass, coefficients in coarse scales are tested for significance before coefficients in fine scales. This is consistent with prioritization by the decoder's version of magnitude since for all coefficients not yet found to be significant, the magnitude is presumed to be zero.

The final factor, *spatial location*, merely implies that two coefficients that cannot yet be distinguished by the decoder in terms of either precision, magnitude, or scale, have their relative importance determined arbitrarily by the initial scanning order of the subband containing the two coefficients.

In one sense, this embedding strategy has a strictly non-increasing operational distortion-rate function for the distortion metric defined to be the sum of the widths of the uncertainty intervals of all of the wavelet coefficients. Since a discrete wavelet transform is an invertible representation of an image, a distortion function defined in the wavelet transform domain is also a distortion function defined on the image. This distortion function is also not without a rational foundation for low-bit rate coding, where noticeable artifacts must be tolerated, and perceptual metrics based on just-noticeable differences (JND's) do not always predict which artifacts human viewers will prefer. Since minimizing the widths of uncertainty intervals minimizes the largest possible errors, artifacts, which result from numerical errors large enough to exceed perceptible thresholds, are minimized. Even using this distortion function, the proposed embedding strategy is not optimal, because truncation of the bit stream in the middle of a pass causes some uncertainty intervals to be twice as large as others.

Actually, as it has been described thus far, EZW is unlikely to be optimal for *any* distortion function. Notice that in (19), dividing the thresholds by two simply decrements E leaving M unchanged. While there must exist an optimal starting M which minimizes a given distortion function, how to find this optimum is still an open question and seems highly image dependent. Without knowledge of the optimal M and being forced to choose it based on some other consideration, with probability one, either increasing or decreasing M would have produced an embedded code which has a lower distortion for the same rate. Despite the fact that without trial and error optimization for M, EZW is probably suboptimal, it is nevertheless quite effective in practice.

Note also that using the width of the uncertainty interval as a distance metric is exactly the same metric used in finite-precision fixed-point approximations of real numbers. Thus, the embedded code can be seen as an "image" generalization of finite-precision fixed-point approximations of real numbers.

E. Relationship to Priority-Position Coding

In a technique based on a very similar philosophy, Huang *et al.* discusses a related approach to embedding, or ordering the information in importance, called priority-position coding (PPC) [10]. They prove very elegantly that the entropy of a source is equal to the average entropy of a particular ordering of that source plus the average entropy of the position information necessary to reconstruct the source. Applying a sequence of decreasing thresholds, they attempt to sort by amplitude all of the DCT coefficients for the entire image based on a partition of the range of amplitudes. For each coding pass, they transmit the significance map which is arithmetically encoded. Additionally, when a significant coefficient is found they transmit its value to its full precision. Like the EZW algorithm, PPC implicitly defines importance with respect to the magnitudes of the transform coefficients. In one sense, PPC is a generalization of the successive-approximation method presented in this paper, because PPC allows more general partitions of the amplitude range of the transform coefficients. On the other hand, since PPC sends the value of a significant coefficient to full precision, its protocol assigns a greater importance to the least significant bit of a significant coefficient than to the identification of new significant coefficients on next PPC pass. In contrast, as a top priority, EZW tries to reduce the width of the largest uncertainty interval in all coefficients before increasing the precision further. Additionally, PPC makes no attempt to predict insignificance from low frequency to high frequency, relying solely on the arithmetic coding to encode the significance map. Also unlike EZW, the probability estimates needed for the arithmetic coder were derived via training on an image database instead of adapting to the image itself. It would be interesting to experiment with variations which combine advantages of EZW (wavelet transforms, zerotree coding, importance defined by a decreasing sequence of uncertainty intervals, and adaptive arithmetic coding using small alphabets) with the more general approach to partitioning the range of amplitudes found in PPC. In practice, however, it is unclear whether the finest grain partitioning of the amplitude range provides any coding gain, and there is certainly a much higher computational cost associated with more passes. Additionally, with the exception of the last few low-amplitude passes, the coding results reported in [10] did use power-of-two amplitudes to define the partition suggesting that, in practice, using finer partitioning buys little coding gain.

V. A Simple Example

In this section, a simple example will be used to highlight the order of operations used in the EZW algorithm. Only the string of symbols will be shown. The reader interested in the details of adaptive arithmetic coding is referred to [31]. Consider the simple 3-scale wavelet transform of an 8 × 8 image. The array of values is shown in Fig. 8. Since the largest coefficient magnitude is 63, we can choose our initial threshold to be anywhere in (31.5, 63]. Let $T_0 = 32$. Table I shows the processing on the first dominant pass. The following comments refer to Table I:

1) The coefficient has magnitude 63 which is greater than the threshold 32, and is positive so a positive symbol is generated. After decoding this symbol, the decoder knows the coefficient in the interval [32, 64) whose center is 48.

2) Even though the coefficient 31 is insignificant with respect to the threshold 32, it has a significant descendant two generations down in subband $LH1$ with magnitude 47. Thus, the symbol for an isolated zero is generated.

3) The magnitude 23 is less than 32 and all descendants which include (3, −12, -14, 8) in subband $HH2$ and all coefficients in subband $HH1$ are insignificant. A zerotree symbol is generated, and no symbol will be generated for any coefficient in subbands $HH2$ and $HH1$ during the current dominant pass.

4) The magnitude 10 is less than 32 and all descendants (-12, 7, 6, -1) also have magnitudes less than 32. Thus a zerotree symbol is generated. Notice that this tree has a violation of the "decaying spectrum" hypothesis since a coefficient (-12) in subband $HL1$ has a magnitude greater than its parent (10). Nevertheless, the entire tree has magnitude less than the threshold 32 so it is still a zerotree.

5) The magnitude 14 is insignificant with respect to 32. Its children are (-1, 47, -3, 2). Since its child with magnitude 47 is significant, an isolated zero symbol is generated.

6) Note that no symbols were generated from subband $HH2$ which would ordinarily precede subband $HL1$ in the scan. Also note that since subband $HL1$ has no descendants, the entropy coding can resume using a 3-symbol alphabet where the IZ and ZTR symbols are merged into the Z (zero) symbol.

7) The magnitude 47 is significant with respect to 32. Note that for the future dominant passes, this position will be replaced with the value 0, so that for the next dominant pass at threshold 16, the parent of this coefficient, which has magnitude 14, can be coded using a zerotree root symbol.

During the first dominant pass, which used a threshold of 32, four significant coefficients were identified. These coefficients will be refined during the first subordinate pass. Prior to the first subordinate pass, the uncertainty interval for the magnitudes of all of the significant coefficients is the interval [32, 64). The first subordinate pass will refine these magnitudes and identify them as being either in interval [32, 48), which will be encoded with the symbol "0," or in the interval [48, 64), which will be encoded with the symbol "1." Thus, the decision boundary is the magnitude 48. It is no coincidence that these symbols are exactly the first bit to the right of the MSBD in the binary representation of the magnitudes. The order

63	−34	49	10	7	13	−12	7
−31	23	14	−13	3	4	6	−1
15	14	3	−12	5	−7	3	9
−9	−7	−14	8	4	−2	3	2
−5	9	−1	47	4	6	−2	2
3	0	−3	2	3	−2	0	4
2	−3	6	−4	3	6	3	6
5	11	5	6	0	3	−4	4

Fig. 8. Example of 3-scale wavelet transform of an 8 × 8 image.

TABLE I
PROCESSING OF FIRST DOMINANT PASS AT THRESHOLD $T = 32$. SYMBOLS ARE POS FOR POSITIVE SIGNIFICANT, NEG FOR NEGATIVE SIGNIFICANT, IZ FOR ISOLATED ZERO, ZTR FOR ZEROTREE ROOT, AND Z FOR A ZERO WHEN THERE ARE NO CHILDREN. THE RECONSTRUCTION MAGNITUDES ARE TAKEN AS THE CENTER OF THE UNCERTAINTY INTERVAL

Comment	Subband	Coefficient Value	Symbol	Reconstruction Value
(1)	LL3	63	POS	48
	HL3	-34	NEG	-48
(2)	LH3	-31	IZ	0
(3)	HH3	23	ZTR	0
	HL2	49	POS	48
(4)	HL2	10	ZTR	0
	HL2	14	ZTR	0
	HL2	-13	ZTR	0
	LH2	15	ZTR	0
(5)	LH2	14	IZ	0
	LH2	-9	ZTR	0
	LH2	-7	ZTR	0
(6)	HL1	7	Z	0
	HL1	13	Z	0
	HL1	3	Z	0
	HL1	4	Z	0
	LH1	-1	Z	0
(7)	LH1	47	POS	48
	LH1	-3	Z	0
	LH1	-2	Z	0

TABLE II
PROCESSING OF THE FIRST SUBORDINATE PASS. MAGNITUDES ARE PARTITIONED INTO THE UNCERTAINTY INTERVALS [32,48) AND [48,64), WITH SYMBOLS "0" AND "1" RESPECTIVELY

Coefficient Magnitude	Symbol	Reconstruction Magnitude
63	1	56
34	0	40
49	1	56
47	0	40

of operations in the first subordinate pass is illustrated in Table II.

The first entry has magnitude 63 and is placed in the upper interval whose center is 56. The next entry has magnitude 34, which places it in the lower interval. The third entry 49 is in the upper interval, and the fourth entry 47 is in the lower interval. Note that in the case of 47, using the center of the uncertainty interval as the reconstruction value, when the reconstruction value is changed from 48 to 40, the reconstruction error actually increases from 1 to 7. Nevertheless, the uncertainty interval for this coefficient decreases from width 32 to width 16. At the conclusion of the processing of the entries on the subordinate list corresponding to the uncertainty interval [32, 64), these magnitudes are reordered for future subordinate passes in the order (63, 49, 34, 47). Note that 49 is moved ahead of 34 because from the decoder's point of view, the reconstruction values 56 and 40 are distinguishable. However, the magnitude 34 remains ahead of magnitude 47 because as far as the decoder can tell, both have magnitude 40, and the initial order, which is based first on importance by scale, has 34 prior to 47.

The process continues on to the second dominant pass at the new threshold of 16. During this pass, only those coefficients not yet found to be significant are scanned. Additionally, those coefficients previously found to be significant are treated as zero for the purpose of determining if a zerotree exists. Thus, the second dominant pass consists of encoding the coefficient -31 in subband $LH3$ as negative significant, the coefficient 23 in subband $HH3$ as positive significant, the three coefficients in subband $HL2$ that have not been previously found to be significant (10, 14, -13) are each encoded as zerotree roots, as are all four coefficients in subband $LH2$ and all four coefficients in subband $HH2$. The second dominant pass terminates at this point since all other coefficients are predictably insignificant.

The subordinate list now contains, in order, the magnitudes (63, 49, 34, 47, 31, 23) which, prior to this subordinate pass, represent the three uncertainty intervals [48, 64), [32, 48) and [16, 31), each having equal width 16. The processing will refine each magnitude by creating two new uncertainty intervals for each of the three current uncertainty intervals. At the end of the second subordinate pass, the order of the magnitudes is (63, 49, 47, 34, 31, 23), since at this point, the decoder could have identified 34 and 47 as being in different intervals. Using the center of the uncertainty interval as the reconstruction value, the decoder lists the magnitudes as (60, 52, 44, 36, 28, 20).

The processing continues alternating between dominant and subordinate passes and can stop at any time.

VI. EXPERIMENTAL RESULTS

All experiments were performed by encoding and decoding an actual bit stream to verify the correctness of the algorithm. After a 12-byte header, the entire bit stream is arithmetically encoded using a single arithmetic coder with an adaptive model [31]. The model is initialized at each new threshold for each of the dominant and subordinate passes. From that point, the encoder is fully adaptive. Note in particular that there is no training of any kind, and no ensemble statistics of images are used in any way (unless one calls the zerotree hypothesis an ensemble statistic). The 12-byte header contains 1) the number of wavelet scales, 2) the dimensions of the image, 3) the maximum histogram count for the models in the arithmetic coder, 4) the image mean and 5) the initial threshold. Note that after the header, there is no overhead except for an extra symbol for end-of-bit-stream, which is always maintained at minimum probability. This extra symbol is not needed for storage on computer medium if the end of a file can be detected.

The EZW coder was applied to the standard black and white 8 bpp. test images, 512 × 512 "Lena" and the 512 × 512 "Barbara," which are shown in Figs. 9(a) and 11(a). Coding results for "Lena" are summarized in Table III and Fig. 9. Six scales of the QMF-pyramid were used. Similar results are shown for "Barbara" in Table IV and Fig. 10. Additional results for the 256 × 256 "Lena" are given in [22].

Quotes of PSNR for the 512 × 512 "Lena" image are so abundant throughout the image coding literature that it is difficult to definitively compare these results with other coding results.[1] However, a literature search has only found two published results where authors generate an actual bit stream that claims higher PSNR performance at rates between 0.25 and 1 bit/pixel [12] and [21], the latter of which is a variation of the EZW algorithm. For the "Barbara" image, which is far more difficult than "Lena," the performance using EZW is substantially better, at least numerically, than the 27.82 dB for 0.534 bpp. reported in [28].

The performance of the EZW coder was also compared to a widely available version of JPEG [14]. JPEG does not allow the user to select a target bit rate but instead allows the user to choose a "Quality Factor." In the experiments shown in Fig. 11, "Barbara" is encoded first using JPEG to a file size of 12 866 bytes, or a bit rate of 0.39 bpp. The PSNR in this case is 26.99 dB. The EZW encoder was then applied to "Barbara" with a target file

[1] Actually there are multiple versions of the luminance only "Lena" floating around, and the one used in [22] is darker and slightly more difficult than the "official" one obtained by this author from RPI after [22] was published. Also note that this should not be confused with results using only the green component of an RGB version which are also commonly cited.

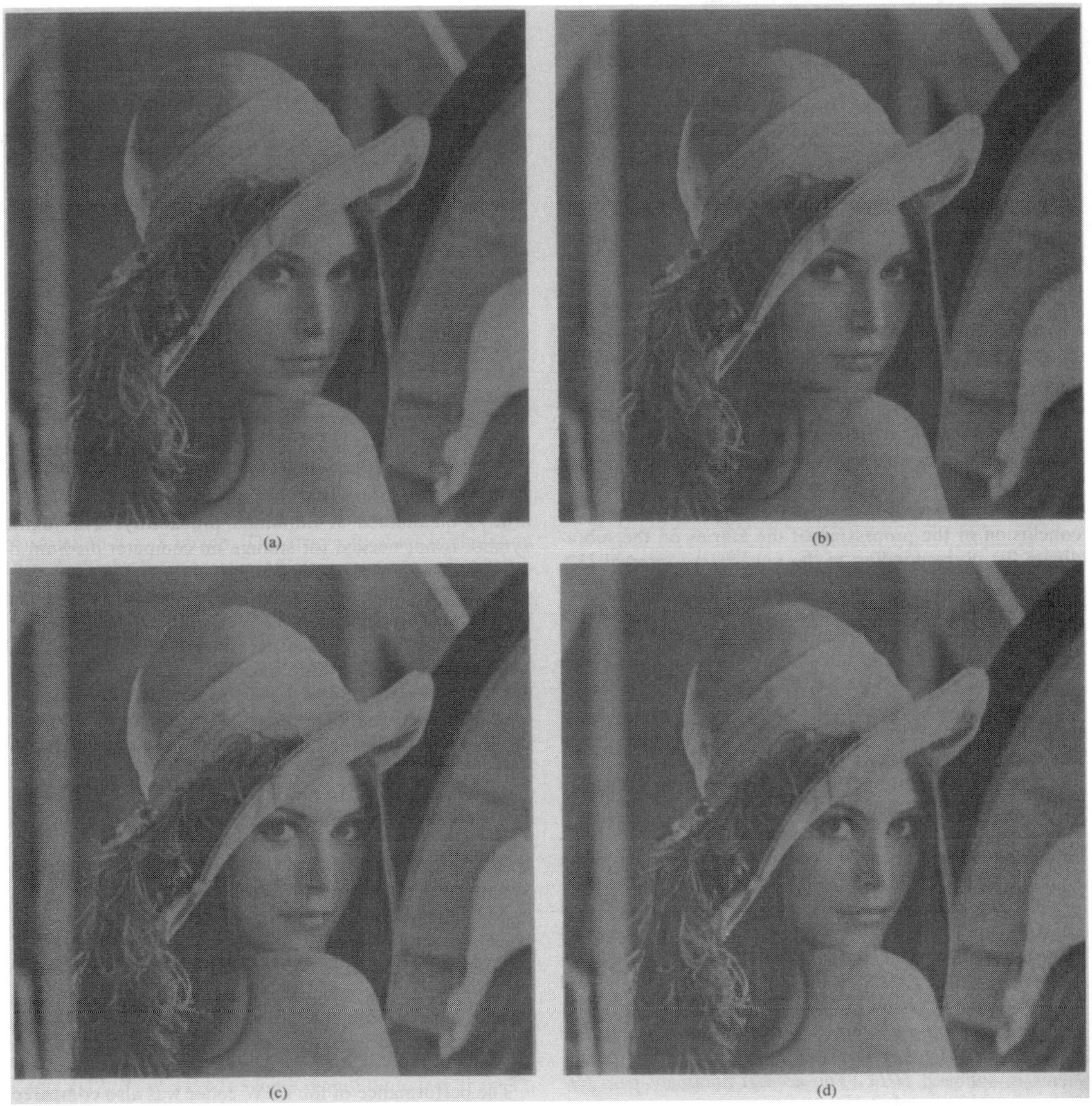

Fig. 9. Performance of EZW Coder operating on "Lena." (a) Original 512 × 512 "Lena" image at 8 bits/pixel (b) 1.0 bits/pixel, 8 : 1 Compression, PSNR = 39.55 dB, (c) 0.5 bits/pixel 16 : 1 Compression, PSNR = 36.28, (d) 0.25 bits/pixel, 32 : 1 Compression, PSNR = 33.17 dB, (e) 0.0625 bits/pixel, 128 : 1 Compression, PSNR = 27.54 dB, (f) 0.015625 bits/pixel, 512 : 1 Compression, PSNR = 23.63 dB.

size of exactly 12 866 bytes. The resulting PSNR is 29.39 dB, significantly higher than for JPEG. The EZW encoder was then applied to "Barbara" using a target PSNR to obtain exactly the same PSNR of 26.99. The resulting file size is 8820 bytes, or 0.27 bpp. Visually, the 0.39 bpp. EZW version looks better than the 0.39 bpp. JPEG version. While there is some loss of resolution in both, there are noticeable blocking artifacts in the JPEG version. For the comparison at the same PSNR, one could probably argue in favor of the JPEG.

Another interesting figure of merit is the number of significant coefficients retained. DeVore *et al.* used wavelet transform coding to progressively encode the same image [8]. Using 68 272 bits, (8534 bytes, 0.26 bpp.), they re-

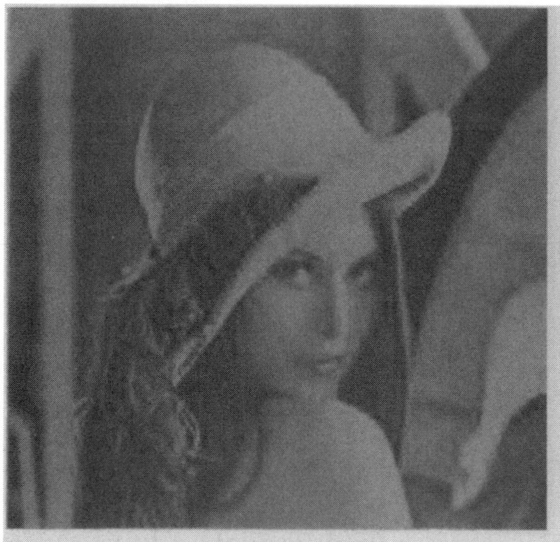

(e)

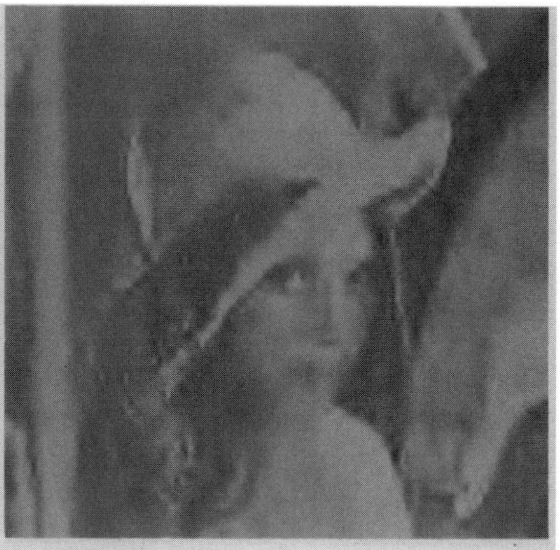

(f)

Fig. 9. (*Continued.*)

TABLE III
CODING RESULTS FOR 512 × 512 LENA SHOWING PEAK-SIGNAL-TO-NOISE (PSNR) AND THE NUMBER OF WAVELET COEFFICIENTS THAT WERE CODED AS NONZERO

# bytes	R	Compression	MSE	PSNR (dB)	signif. coef
32768	1.0	8:1	7.21	39.55	39446
16384	0.5	16:1	15.32	36.28	19385
8192	0.25	32:1	31.33	33.17	9774
4096	0.125	64:1	61.67	30.23	4950
2048	0.0625	128:1	114.5	27.54	2433
1024	0.03125	256:1	188.2.7	25.38	1253
512	0.015625	512:1	281.7	23.63	616
256	0.0078125	1024:1	440.2	21.69	265

TABLE IV
CODING RESULTS FOR 512 × 512 BARBARA SHOWING PEAK-SIGNAL-TO-NOISE (PSNR) AND THE NUMBER OF WAVELET COEFFICIENTS THAT WERE CODED AS NONZERO

# bytes	R	Compression	MSE	PSNR (dB)	signif. coef
32768	1.0	8:1	19.92	35.14	40766
16384	0.5	16:1	57.57	30.53	20554
8192	0.25	32:1	136.8	26.77	10167
4096	0.125	64:1	257.1	24.03	4522
2048	0.0625	128:1	318.5	23.10	2353
1024	0.03125	256:1	416.2	21.94	1259
512	0.015625	512:1	546.8	20.75	630
256	0.0078125	1024:1	772.5	19.54	291

tained 2019 coefficients and achieved a RMS error of 15.30 (MSE = 234, 24.42 dB), whereas using the embedded coding scheme, 9774 coefficients are retained, using only 8192 bytes. The PSNR for these two examples differs by over 8 dB. Part of the difference can be attributed to fact that the Haar basis was used in [8]. However, closer examination shows that the zerotree coding provides a much better way of encoding the positions of the significant coefficients than was used in [8].

An interesting and perhaps surprising property of embedded coding is that when the encoding or decoding is terminated during the middle of a pass, or in the middle of the scanning of a subband, there are no artifacts produced that would indicate where the termination occurs. In other words, some coefficients in the same subband are represented with twice the precision of the others. A possible explanation of this phenomena is that at low rates, there are so few significant coefficients that any one does not make a perceptible difference. Thus, if the last pass is a dominant pass, setting some coefficient that might be significant to zero may be imperceptible. Similarly, the fact that some have more precision than others is also imperceptible. By the time the number of significant coefficients becomes large, the picture quality is usually so good that adjacent coefficients with different precisions are imperceptible.

Another interesting property of the embedded coding is that because of the implicit global bit allocation, even at extremely high compression ratios, the performance scales. At a compression ratio of 512 : 1, the image quality of "Lena" is poor, but still recognizable. This is not the case with conventional block coding schemes, where at such high compression ratios, there would be insufficient bits to even encode the DC coefficients of each block.

The unavoidable artifacts produced at low bit rates using this method are typical of wavelet coding schemes coded to the same PSNR's. However, subjectively, they are not nearly as objectionable as the blocking effects typical of block transform coding schemes.

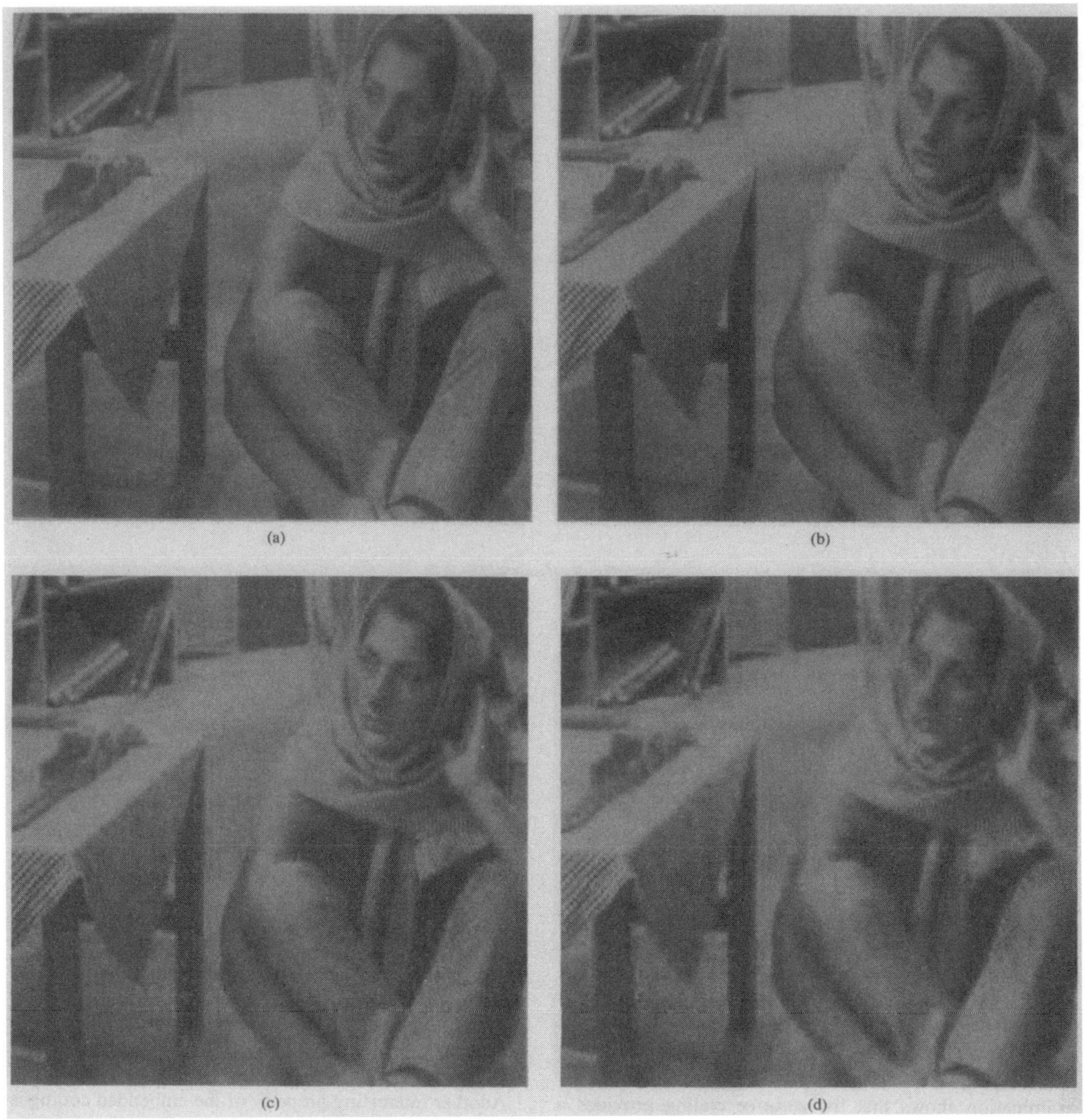

Fig. 10. Performance of EZW Coder operating on "Barbara" at (a) 1.0 bits/pixel, 8 : 1 Compression, PSNR = 35.14 dB (b) 0.5 bits/pixel, 16 : 1 Compression, PSNR = 30.53 dB, (c) 0.125 bits/pixel, 64 : 1 Compression, PSNR = 24.03 dB, (d) 0.0625 bits/pixel, 128 : 1 Compression, PSNR = 23.10 dB.

VII. Conclusion

A new technique for image coding has been presented that produces a fully embedded bit stream. Furthermore, the compression performance of this algorithm is competitive with virtually all known techniques. The remarkable performance can be attributed to the use of the following four features:

- a discrete wavelet transform, which decorrelates most sources fairly well, and allows the more significant bits of precision of most coefficients to be efficiently encoded as part of exponentially growing zerotrees,
- zerotree coding, which by predicting insignificance across scales using an image model that is easy for

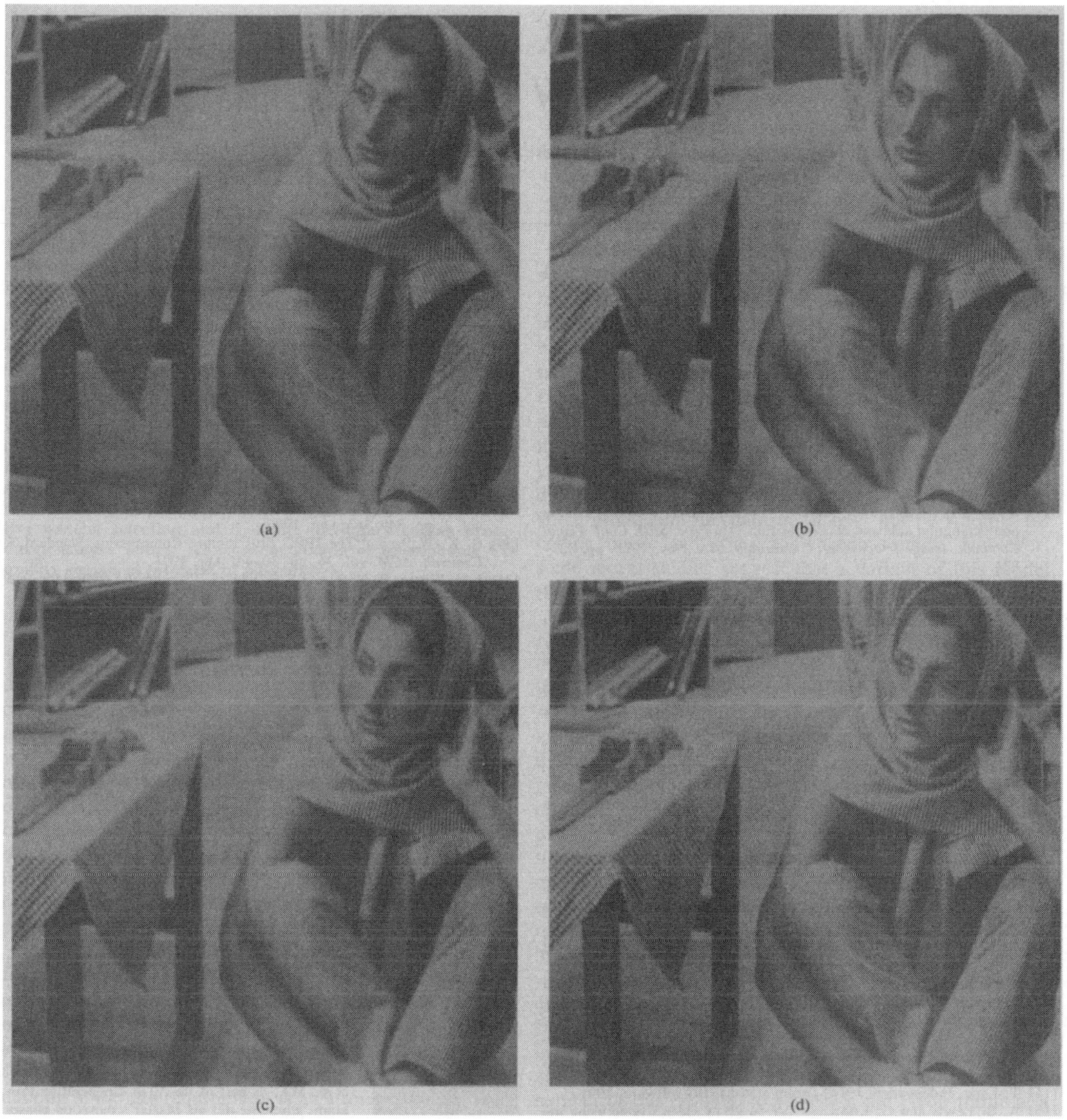

Fig. 11. Comparison of EZW and JPEG operating on "Barbara" (a) Original 512 × 512 (b) EZW at 12 866 bytes, 0.39 bits/pixel, 29.39 dB, (c) EZW at 8820 bytes, 0.27 bits/pixel, 26.99 dB, (d) JPEG at 12 866 bytes, 0.39 bits/pixel, 26.99 dB.

most images to satisfy, provides substantial coding gains over the first-order entropy for significance maps,

- successive-approximation, which allows the coding of multiple significance maps using zerotrees, and allows the encoding or decoding to stop at any point,
- adaptive arithmetic coding, which allows the entropy coder to incorporate learning into the bit stream itself.

The precise rate control that is achieved with this algorithm is a distinct advantage. The user can choose a bit rate and encode the image to *exactly* the desired bit rate. Furthermore, since no training of any kind is required, the algorithm is fairly general and performs remarkably well with most types of images.

ACKNOWLEDGMENT

The author would like to thank Joel Zdepski who suggested incorporating the sign of the significant values into the significance map to aid embedding, Rajesh Hingorani who wrote much of the original C code for the QMF-pyramids, Allen Gersho who provided the original "Barbara" image, and Gregory Wornell whose fruitful discussions convinced me to develop a more mathematical analysis of zerotrees in terms of bounding an optimal estimator. I would also like to thank the editor and the anonymous reviewers whose comments led to a greatly improved manuscript.

REFERENCES

[1] E. H. Adelson, E. Simoncelli, and R. Hingorani, "Orthogonal pyramid transforms for image coding," *Proc. SPIE*, vol. 845, Cambridge, MA, Oct. 1987, pp. 50-58.
[2] R. Ansari, H. Gaggioni, and D. J. LeGall, "HDTV coding using a nonrectangular subband decomposition," in *Proc. SPIE Conf. Visual Commun. Image Processing*, Cambridge, MA, Nov. 1988, pp. 821-824.
[3] T. C. Bell, J. G. Cleary, and I. H. Witten, *Text Compression*. Englewood Cliffs, NJ: Prentice-Hall, 1990.
[4] P. J. Burt and E. H. Adelson, "The Laplacian pyramid as a compact image code," *IEEE Trans. Commun.*, vol. 31, pp. 532-540, 1983.
[5] R.R. Coifman and M. V. Wickerhauser, "Entropy-based algorithms for best basis selection," *IEEE Trans. Informat. Theory*, vol. 38, pp. 713-718, Mar. 1992.
[6] I. Daubechies, "Orthonormal bases of compactly supported wavelets," *Commun. Pure Appl. Math.*, vol. 41, pp. 909-996, 1988.
[7] —, "The wavelet transform, time-frequency localization and signal analysis," *IEEE Trans. Informat. Theory*, vol. 36, pp. 961-1005, Sept. 1990.
[8] R. A. DeVore, B. Jawerth, and B. J. Lucier, "Image compression through wavelet transform coding," *IEEE Trans. Informat. Theory*, vol. 38, pp. 719-746, Mar. 1992.
[9] W. Equitz and T. Cover, "Successive refinement of information," *IEEE Trans. Informat. Theory*, vol. 37, pp. 269-275, Mar. 1991.
[10] Y. Huang, H. M. Driezen, and N. P. Galatsanos, "Prioritized DCT for Compression and Progressive Transmission of Images," *IEEE Trans. Image Processing*, vol. 1, pp. 477-487, Oct. 1992.
[11] N. S. Jayant and P. Noll, *Digital Coding of Waveforms*. Englewood Cliffs, NJ: Prentice-Hall, 1984.
[12] Y. H. Kim and J. W. Modestino, "Adaptive entropy coded subband coding of images," *IEEE Trans. Image Processing*, vol. 1, pp. 31-48, Jan. 1992.
[13] J. Kovačević and M. Vetterli, "Nonseparable multidimensional perfect reconstruction filter banks and wavelet bases for $\mathcal{R}$," *IEEE Trans. Informat. Theory*, vol. 38, pp. 533-555, Mar. 1992.
[14] T. Lane, Independent JPEG Group's free JPEG software, 1991.
[15] A. S. Lewis and G. Knowles, "A 64 kB/s video Codec using the 2-D wavelet transform," in *Proc. Data Compression Conf.*, Snowbird, Utah, IEEE Computer Society Press, 1991.
[16] —, "Image compression using the 2-D wavelet transform," *IEEE Trans. Image Processing*, vol. 1, pp. 244-250, Apr. 1992.
[17] S. Mallat, "A theory for multiresolution signal decomposition: The wavelet representation," *IEEE Trans. Pattern Anal. Mach. Intell.*, vol. 11, pp. 674-693, July 1989.
[18] —, "Multifrequency channel decompositions of images and wavelet models," *IEEE Trans. Acoust. Speech and Signal Processing*, vol. 37, pp. 2091-2110, Dec. 1990.
[19] A Pentland and B. Horowitz, "A practical approach to fractal-based image compression," in *Proc. Data Compression Conf.*, Snowbird, Utah, IEEE Computer Society Press, 1991.
[20] O. Rioul and M. Vetterli, "Wavelets and signal processing," *IEEE Signal Processing Mag.*, vol. 8, pp. 14-38, Oct. 1991.
[21] A. Said and W. A. Pearlman, "Image Compression using the Spatial-Orientation Tree," in *Proc. IEEE Int. Symp. Circuits and Syst.*, Chicago, IL, May 1993, pp. 279-282.
[22] J. M. Shapiro, "An Embedded Wavelet Hierarchical Image Coder," *Proc. IEEE Int. Conf. Acoust., Speech, Signal Processing*, San Francisco, CA, Mar. 1992.
[23] —, "Adaptive multidimensional perfect reconstruction filter banks using McClellan transformations," *Proc. IEEE Int. Symp. Circuits Syst.*, San Diego, CA, May 1992.
[24] —, "An embedded hierarchical image coder using zerotrees of wavelet coefficients," in *Proc. Data Compression Conf.*, Snowbird, Utah, IEEE Computer Society Press, 1993.
[25] —, "Application of the embedded wavelet hierarchical image coder to very low bit rate image coding," *Proc. IEEE Int. Conf. Acoust., Speech, Signal Processing*, Minneapolis, MN, Apr. 1993.
[26] Special issue of *IEEE Trans. Informat. Theory*, Mar. 1992.
[27] G. Strang, "Wavelets and dilation equations: A brief introduction," *SIAM Rev.*, vol. 4, pp. 614-627, Dec. 1989.
[28] J. Vaisey and A. Gersho, "Image compression with variable block size segmentation," *IEEE Trans. Signal Processing.*, vol. 40, pp. 2040-2060, Aug. 1992.
[29] M. Vetterli, J. Kovačević, and D. J. LeGall, "Perfect reconstruction filter banks for HDTV representation and coding," *Image Commun.*, vol. 2, pp. 349-364, Oct. 1990.
[30] G. K. Wallace, "The JPEG Still Picture Compression Standard," *Commun. ACM*, vol. 34, pp. 30-44, Apr. 1991.
[31] I. H. Witten, R. Neal, and J. G. Cleary, "Arithmetic coding for data compression," *Comm. ACM*, vol. 30, pp. 520-540, June 1987.
[32] J. W. Woods, Ed., *Subband Image Coding*. Boston, MA: Kluwer, 1991.
[33] G. W. Wornell, "A Karhunen-Loéve expansion for $1/f$ processes via wavelets," *IEEE Trans. Informat. Theory*, vol. 36, pp. 859-861, July 1990.
[34] Z. Xiong, N. Galatsanos, and M. Orchard, "Marginal analysis prioritization for image compression based on a hierarchical wavelet decomposition," in *Proc. IEEE Int. Conf. Acoust., Speech, Signal Processing*, Minneapolis, MN, Apr. 1993.
[35] W. Zettler, J. Huffman, and D. C. P. Linden, "Applications of compactly supported wavelets to image compression," *SPIE Image Processing Algorithms*, Santa Clara, CA 1990.

Jerome M. Shapiro (S'85–M'90) was born April 29, 1962 in New York City. He received the B.S., M.S., and Ph.D. degrees in electrical engineering from the Massachusetts Institute of Technology, Cambridge, MA, in 1985, 1985, and 1990, respectively.

From 1982 to 1984, he was at GenRad, Concord, MA, as part of the VI-A Cooperative Program, where he did his Master's thesis on phase-locked loop frequency synthesis. From 1985 to 1987, he was a Research Assistant in the Video Image Processing Group of the MIT Research Laboratory of Electronics. From 1988 to 1990, while pursuing his doctoral studies, he was a Research Assistant in the Sensor Processor Technology Group of MIT Lincoln Laboratory, Lexington, MA. In 1990, he joined the Digital HDTV Research Group of the David Sarnoff Research Center, a Subsidiary of SRI International, Princeton, NJ. His research interests are in the areas of video and image data compression, digital signal processing, adaptive filtering and systolic array algorithms.